Five Millennium Catalog of Lunar Eclipses –1999 to +3000

2nd Edition

Fred Espenak and Jean Meeus

August, 2021

Five Millennium Catalog of Lunar Eclipses: -1999 to +3000

Second Edition

Five Millennium Canon of Lunar Eclipses:: –1999 to +3000 (2nd Edition) is published through Astropixels Publishing. For more information, contact the publisher:

Astropixels Publishing
P.O. Box 16197
Portal, AZ 85632

Astropixels Publishing Website: *astropixels.com/pubs*

This book may be ordered at: *astropixels.com/pubs/5MKLE2.html*

Printed in the United States of America

ISBN 978-1-941983-40-9

Astropixels Publication: AP041 (Version 1.0a)

Front Cover: This almanac from *Calendarium Romanum magnum* (Johann Stoeffler, 1518) shows predictions for 3 lunar eclipses in 1570, 1572 and 1573.

Back Cover: Photo of Fred Espenak is Copyright © 1999 by Patricia Espenak

Table of Contents

Preface – 1st Edition

Theodor von Oppolzer's 1887 *Canon der Finsternisse* (Canon of Eclipses) stands as one of the greatest accomplishments in computational astronomy of the nineteenth century. It contains the elements of all 8000 solar eclipses and 5200 lunar eclipses occurring between the years −1207 and +2161, together with maps showing the approximate positions of the central lines of total, annular and hybrid solar eclipses.

To make this remarkable achievement possible, a number of approximations were used in the calculations and maps. Furthermore, the 19th century ephemerides for the Sun and Moon, which are critical to eclipse predictions, cannot compare to the accuracy and precision of modern ephemerides. Finally, the 1887 canon did not include the shifts in longitude and timing of ancient eclipses due to Earth's variable rotation rate and the Moon's secular acceleration.

Nevertheless, von Oppolzer's canon remained the seminal reference on predictions of eclipses until well into the 20th century. With the arrival of the electronic computer, the time was ripe to produce updated solar[1] and lunar eclipse canons. In 1979, Meeus and Mucke published *Canon of Lunar Eclipses: −2002 to +2526* containing the Besselian elements of 10,936 lunar eclipses. It was intended mainly for historical research and served as the modern day successor of von Oppolzer's great canon.

The Meeus-Mucke work also contains data on penumbral eclipses that are not included in von Oppolzer's canon. Neither of these publications offers diagrams or maps to illustrate the geometry or visibility of each eclipse. Espenak's *Fifty Year Canon of Lunar Eclipses* (1989) includes individual Moon-shadow geometry diagrams and eclipse visibility maps of all lunar eclipses, but it covers a relatively short period from +1986 to +2035.

Both of these recent lunar eclipse canons are based on *Newcomb's Tables of the Sun* (1895) and Brown's lunar theory (1905), subject to later modifications in the *Improved Lunar Ephemeris* (1954). These were the best ephemerides of their day but they have since been superseded.

The *Canon of Lunar Eclipses 1500 B.C.–A.D. 3000* (Liu and Fiala, 1992) uses modern theories of the Sun and the Moon prepared by the *Bureau des Longitudes* of Paris. However, it does not contain individual eclipse geometry diagrams or maps. Instead, it offers a series of map templates to approximate the geographic regions of eclipse visibility, and an optional computer program do generate these figures for any eclipse in the document.

The *Five Millennium Canon of Lunar Eclipses: −1999 to +3000* (NASA TP–2009–214172, Espenak and Meeus, 2009) contains individual eclipse geometry diagrams and visibility maps for every lunar eclipse (12,064 eclipses) over a period covering 5000 years from −1999 to +3000. It is based on the modern *Bureau des Longitudes* theories of the Sun and the Moon and uses the most current determination of historical values of ΔT in the eclipse visibility maps.

The present work is a catalog to supplement the 12,064 eclipse diagrams and visibility maps in the *Five Millennium Canon of Lunar Eclipses*. It includes additional information for each eclipse that could not be included in the lunar eclipse canon because of size limits. The data tabulated for each eclipse include the catalog number; canon plate number; calendar date; Terrestrial Dynamical Time of greatest eclipse; ΔT; lunation number; Saros number; eclipse type; Quincena Solar Eclipse parameter; gamma; penumbral and umbral eclipse magnitudes; durations of penumbral, partial, and total eclipse phases; and the geographic coordinates of greatest eclipse (latitude and longitude).

A primary goal of this work is to assist historians and archeologists in the identification and dating of eclipses found in references and records from antiquity. For example, an ancient mechanical calculator known as the Antikythera mechanism was apparently designed to calculate eclipses and other astronomical phenomena (Freeth, et. al., 2008).

---

[1] Several new solar eclipse canons were published in the second half of the 20th century. Meeus, Grosjean, and Vanderleen published *Canon of Solar Eclipses* (1966) containing the Besselian elements of all solar eclipses from +1898 to +2510, together with central line tables and maps. The aim of this work was principally to provide data on future eclipses. The Mucke and Meeus work *Canon of Solar Eclipses: −2003 to +2526* (1983) was intended mainly for historical research and served as a modern day successor of von Oppolzer's great canon. Espenak's *Fifty Year Canon of Solar Eclipses* (1987) includes individual detailed maps and central path data for all solar eclipses from +1986 to +2035. Finally, the Espenak and Meeus work *Five Millennium Canon of Solar Eclipses: −1999 to +3000* (2006) contains individual maps of every solar eclipse and uses modern ephemerides of the Sun and Moon.

The decoding of this device was possible in part by comparing its combination of wheel positions with the dates of lunar eclipses.

The *Canon* and the *Catalog* both use the same solar and lunar ephemerides, and values of ΔT. This 1-to-1 correspondence between them will enhance the value of each. The researcher may now search, evaluate, and compare eclipses graphically (*Canon*) or textually (*Catalog*).

— Fred Espenak and Jean Meeus, 2009 January

Preface – 2nd Edition

The *Five Millennium Catalog of Lunar Eclipses: –1999 to +3000* was first published in October 2009 (NASA TP-2009-214173). As a NASA Technical Publication it had a limited publication run and distribution. The available supply of hard copies was depleted within 12 months of publication although the PDF version continues to be available.

More than a decade later, the *Five Millennium Catalog of Lunar Eclipses* is available again in hard copy in this Second Edition through Astropixels Publishing. The *Catalog* serves to complement the 2-volume *Five Millennium Canon of Lunar Eclipses* which contains a figure and map of every eclipse. The *Catalog* and the *Canon* both use the same solar and lunar ephemerides as well as the same value of ΔT. This 1-to-1 correspondence between them enhances the value of each..

A new section (Section 4: Eclipses and the Moon's Orbit) has also been added to this second edition.

— Fred Espenak and Jean Meeus, 2021 August

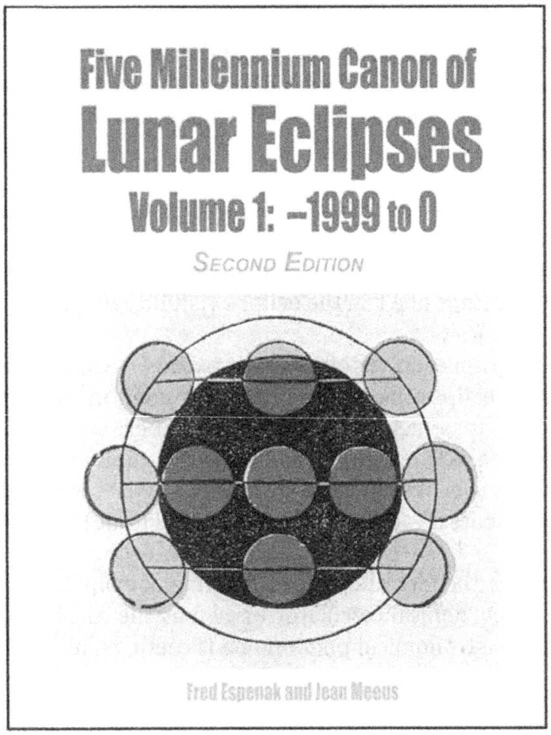

See: *http://astropixels.com/pubs/5MCLE2.html*

Section 1: Eclipse Tables and Predictions

1.1 Introduction

Earth will experience 12,064 eclipses of the Moon during the 5000-year period from –1999 to +3000 (2000 BCE to 3000 CE[2]). The Appendix consists of a series of tables that summarize the principal characteristics of each lunar eclipse over this time interval. As such, the *Catalog* serves to complement the 2-volume *Five Millennium Canon of Lunar Eclipses* which contains a figure and map of every eclipse.

1.2 Explanation of the Lunar Eclipse Tables

Each line in the tables corresponds to a single lunar eclipse and provides concise parameters to characterize the eclipse. The calendar date and Dynamical Time of the instant of greatest eclipse are given along with the adopted value of delta T (ΔT). The lunation number (since 2000 Jan 06) and the Saros series are listed along with the eclipse type (N=Penumbral, P=Partial, or T=Total). Gamma is the distance from the axis of Earth's shadow cone to the center of the Moon at greatest eclipse, while the penumbral and umbral eclipse magnitudes are defined as the fraction of the Moon's diameter immersed in each shadow at that instant. The duration of the penumbral, partial, and total eclipse phases are given in minutes. Finally, the geographic latitude and longitude are given for the location where the Moon lies in the zenith at greatest eclipse. A more detailed description of each field in the tables follows.

1.2.1 Catalog Number

The catalog number is the sequential number assigned to each eclipse in the tables from 1 to 12,064.

1.2.2 Canon Plate

The canon plate identifies the plate number on which each eclipse appears in the Five Millennium Canon of Lunar Eclipses. The Canon consists of 604 plates with 20 lunar eclipse diagrams per plate.

1.2.3 Calendar Date

All eclipse dates from 1582 Oct 15 onwards use the modern Gregorian calendar currently found throughout most of the world. The older Julian calendar is used for eclipse dates prior to 1582 Oct 04. Because of the Gregorian Calendar Reform, the day following 1582 Oct 04 (Julian calendar) is 1582 Oct 15 (Gregorian calendar).

Pope Gregory XIII decreed the use of the Gregorian calendar in 1582 in order to correct a problem in a drift of the seasons. It adopts the convention of a year containing 365 days. Every fourth year is a leap year of 366 days if it is divisible by 4 (e.g., 2004, 2008, etc.). However, whole century years (e.g., 1700, 1800, 1900) are excluded from the leap year rule unless they are also divisible by 400 (e.g., 2000). This dating scheme was designed to keep the vernal equinox on or within a day of March 21.

Prior to the Gregorian Calendar Reform of 1582, the Julian calendar was in wide use. It was less complicated than the Gregorian calendar in that all years divisible by 4 were counted as 366-day leap years, but this simplicity came at a

[2] The terms BCE and CE are abbreviations for "Before the Common Era" and "Common Era," respectively. They are the secular equivalents to the BC and AD dating conventions. A major advantage of the BCE/CE convention is that both terms are suffixes, whereas BC and AD are used as a suffix and prefix, respectively.

cost. After more than 16 centuries of use, the Julian calendar date of the vernal equinox had drifted 11 days from March 21. It was this failure in the Julian calendar that prompted the Gregorian Calendar Reform.

The Julian calendar does not include the year 0, so the year 1 BCE is followed by the year 1 CE. This is awkward for arithmetic dating calculations. In this publication, dates are counted using the astronomical numbering system, which recognizes the year 0. Historians should note the numerical difference of one year between astronomical dates and BCE dates. Thus, the astronomical year 0 corresponds to 1 BCE, and the astronomical year –100 corresponds to 101 BCE, etc.

There are a number of ways to write the calendar date through variations in the order of day, month, and year. The International Organization for Standardization's (ISO) 8601 advises a numeric date representation, which organizes the elements from the largest to the smallest. The exact format is YYYY-MM-DD where YYYY is the calendar year, MM is the month of the year between 01 (January) and 12 (December), and DD is the day of the month between 01 and 31. For example, the 27th day of April in the year 1943 would then be expressed as 1943-04-27. The authors of this report support the ISO convention, but have replaced the month number with the three-letter English abbreviation of the month name for additional clarity. From the previous example, the date used will be expressed as 1943 Apr 27.

1.2.4 Terrestrial Dynamical Time (TD) of Greatest Eclipse

The instant of greatest eclipse occurs when the distance between the axis of Earth's umbral shadow cone and the center of the Moon reaches a minimum. Because of Earth's flattening, the instant of greatest eclipse differs slightly from the instant of greatest magnitude. In practice, Earth's atmosphere diffuses the edges of the penumbral and umbral shadows such that the difference between greatest eclipse and greatest magnitude cannot be distinguished observationally.

Lunar eclipses occur when the Moon is near one of the nodes of its orbit, and therefore, moving at an angle of about 5° to the ecliptic. Hence, unless the eclipse is perfectly central, the instant of greatest eclipse does not coincide with that of ap- parent ecliptic conjunction with Earth's shadow (i.e., Full Moon), nor with the time of conjunction in Right Ascension.

Greatest eclipse is given in Terrestrial Dynamical Time, or TD (Sect. 2.3), which is a time system based on International Atomic Time. As such, TD is the atomic time equivalent to its predecessor Ephemeris Time (Sect. 2.2) and is used in the theories of motion for bodies in the Solar System. To determine the geographic visibility of an eclipse, TD is converted to Universal Time (Sect. 2.4) using the parameter ΔT (Sects. 2.6 and 2.7).

1.2.5 Delta T (ΔT)

Delta T (ΔT) is the arithmetic difference, in seconds, between Terrestrial Dynamical Time (Sect. 2.3) and Universal Time (Sect. 2.4). For more information on ΔT, see Section 2.6.

1.2.6 Lunation Number

The lunation number is the number of synodic months, or lunations, since New Moon on 2000 Jan 06. It can be converted into the Brown Lunation Number (counted from New Moon on 1923 Jan 17) by adding 953.

1.2.7 Saros Series Number

Each lunar eclipse belongs to a Saros series (Sect. 5.3) using a numbering system first introduced by van den Bergh (1955). This system has been expanded to include negative values from the past, as well as additional series in the future. The eclipses with an odd Saros number take place at the descending node of the Moon's orbit, while those with an even Saros number take place at the ascending node. This relationship is reversed for solar eclipses.

The Saros is a period of 223 synodic months (~18 years, 11 days, and 8 hours). Eclipses separated by this interval belong to the same Saros series and share similar geometry and characteristics.

1.2.8 Lunar Eclipse Type

The first character in this 2-character parameter gives the lunar eclipse type. The three basic types of lunar eclipses are:

1) **N = Penumbral Lunar Eclipse** (Moon traverses through Earth's penumbral shadow; Moon is partly or completely within the penumbra)
2) **P = Partial Lunar Eclipse** (Part of the Moon traverses through Earth's umbral shadow)
3) **T = Total Lunar Eclipse** (Entire Moon traverses through Earth's umbral shadow)

The second character of the lunar eclipse type is a qualifier defined as follows.

1) **m** = Middle eclipse of Saros series.
2) **+** = Central total eclipse[3] (Moon's center passes north of shadow axis).
3) **–** = Central total eclipse (Moon's center passes south of shadow axis).
4) ***** = Total penumbral eclipse[4].
5) **b** = Saros series begins (first penumbral eclipse in a Saros series).
6) **e** = Saros series ends (last penumbral eclipse in a Saros series).

Qualifiers 1 through 3 are used exclusively with total lunar eclipses while qualifiers 4 through 6 are only used with penumbral eclipses.

1.2.9 Quincena Solar Eclipse Parameter (QSE)

A solar eclipse always occurs within ~15 days of a lunar eclipse. The Quincena Solar Eclipse parameter (QSE) identifies the type of the solar eclipse and whether it precedes or succeeds a particular lunar eclipse. There are four basic types of solar eclipses:

1) **p = partial solar eclipse** (Moon's penumbral shadow traverses Earth; umbral/antumbral shadow completely misses Earth)
2) **a = annular solar eclipse** (Moon's antumbral shadow traverses Earth; Moon is too far from Earth to completely cover the Sun)
3) **t = total solar eclipse** (Moon's umbral shadow traverses Earth; Moon is close enough to Earth to completely cover the Sun)
4) **h = hybrid solar eclipse** (Moon's umbral and antumbral shadows traverse different parts of Earth; eclipse appears either total or annular along different sections of its path—hybrid eclipses are also known as annular-total eclipses)

The QSE is a two character string consisting of one or more of the above solar eclipse types. The first character in the QSE identifies a solar eclipse preceding a lunar eclipse, while the second character identifies a solar eclipse succeeding a lunar eclipse. In most instances, one of the two characters is "–" indicating a single solar eclipse either precedes or

[3] A central total eclipse is one in which some portion of the Moon's disk passes through the axis of the umbral shadow. Most total eclipses are non-central in that the Moon's disk misses the umbral axis.

[4] It is possible for a penumbral eclipse to occur in which the Moon's entire disk is enveloped within the penumbra. Such events are called total penumbral eclipses.

succeeds the lunar eclipse. On rare occasions, a double quincena occurs in which a lunar eclipse is both preceded and succeeded by solar eclipses.

1.2.10 Gamma

The quantity "gamma" is the distance from the axis of Earth's umbral shadow cone to the center of the Moon at the instant of greatest eclipse, in units of Earth's equatorial radius. This distance is positive or negative, depending on whether the Moon passes north or south of the shadow cone axis.

The change in the value of gamma after one Saros period is larger when Earth is near aphelion (June–July) than when it is near perihelion (December–January). Table 1-1 illustrates this point using eclipses from Saros series 129 (near aphelion— table on left) and Saros series 134 (near perihelion—table on right).

Table 1-1. Variation in Gamma at Aphelion vs. Perihelion

Date	Gamma	Δ Gamma		Date	Gamma	Δ Gamma
1928 Jun 03	-0.3175	-		1928 Nov 27	0.3952	-
1946 Jun 14	-0.2324	0.0851		1946 Dec 08	0.3864	-0.0088
1964 Jun 25	-0.1461	0.0863		1964 Dec 19	0.3801	-0.0063
1982 Jul 06	-0.0579	0.0882		1982 Dec 30	0.3758	-0.0043
2000 Jul 16	0.0302	0.0881		2001 Jan 09	0.3720	-0.0038
2018 Jul 27	0.1168	0.0866		2019 Jan 21	0.3684	-0.0036
2036 Aug 07	0.2004	0.0836		2037 Jan 31	0.3619	-0.0065
2054 Aug 18	0.2806	0.0802		2055 Feb 11	0.3526	-0.0093

A similar situation exists in the case of solar eclipses. The explanation can be found in van den Bergh (1955).

1.2.11 Eclipse Magnitude

The eclipse magnitude is defined as the fraction of the Moon's diameter immersed in Earth's shadows. It is equal to the distance from the edge of the shadow to the edge of the Moon deepest in the shadow, in units of the Moon's diameter. Because there are two shadows—penumbral and umbral—there are two corresponding eclipse magnitudes. The eclipse magnitudes appearing in the figures are given at the instant of greatest eclipse when the Moon passes closest to the axis of the two shadow cones.

The penumbral eclipse magnitude of penumbral eclipses in this Catalog of Lunar Eclipses (hereafter referred to as the Catalog) ranges from 0.0004 to 1.0858 (Sect. 3.5). For most penumbral eclipses the penumbral magnitude is less than 1, meaning only a fraction of the Moon's disk enters the penumbra. When the penumbral magnitude is greater than or equal to 1.0, the Moon's entire disk is immersed in the penumbra and the event is termed a total penumbral eclipse (Sect. 3.11). It should be noted that penumbral eclipses are subtle events (Sect. 1.7.1). In fact, they cannot be detected visually—with or without optical aid—unless the eclipse magnitude is greater than ~0.6. The umbral eclipse magnitude of a penumbral eclipse is always negative. It is a measure of the distance of the Moon's limb to the edge of the umbral shadow in units of the Moon's diameter.

During a partial lunar eclipse, some fraction of the Moon's disk enters the umbral shadow. The umbral magnitude for partial eclipses in the Catalog ranges from 0.0001 to 0.9998 (Sect. 3.6). Of course, the Moon also passes through the penumbra during a partial eclipse, so the penumbral magnitude is usually greater than 1.

In the case of a total lunar eclipse, the Moon's entire disk passes through Earth's umbral shadow. During totality, the Moon can take on a range of colors from bright orange, to deep red, dark brown, or even very dark gray (Sect. 1.7.2).

The only light reaching the Moon at this time is heavily filtered and attenuated by Earth's atmosphere. The umbral magnitude for total eclipses in the Catalog ranges from 1.0001 to 1.8821 (Sect. 3.7).

1.2.12 Eclipse Phase Durations

Lunar eclipses are punctuated by one or more phases that begin and end with the contact of the Moon with the edge of the penumbral and umbral shadows (Sect. 1.5). The penumbral phase duration is the time interval between the first and last external contacts of the Moon with the penumbra. The partial phase duration is the time interval between the first and last external contacts of the Moon with the umbra. Finally, the total phase duration is the time interval between the first and last internal contacts of the Moon with the umbra. This last duration is called "totality" and is the period when the Moon's entire disk lies within Earth's umbral shadow. All eclipse phase durations are listed in decimal minutes.

1.2.13 Greatest in Zenith: Latitude and Longitude

The geographic latitude and longitude listed corresponds to the location where the Moon appears in the zenith at greatest eclipse.

1.3 Solar and Lunar Coordinates

The coordinates of the Moon used in the eclipse predictions have been calculated on the basis of the VSOP87 theory constructed by Bretagnon and Francou (1988) at the *Bureau des Longitudes*, Paris. This theory gives the ecliptic longitude and latitude of each planet, and its radius vector, as sums of periodic terms. In the calculations presented here, the complete set of periodic terms of version D[5] of VSOP87 were used.

For the Moon, use has been made of the theory ELP-2000/82 of Chapront-Touzé and Chapront (1983), again of the *Bureau des Longitudes*. This theory contains a total of 37,862 periodic terms, namely 20,560 for the Moon's longitude, 7,684 for the latitude, and 9,618 for the distance to Earth. However, many of these terms are very small: some have an amplitude of only 0.00001 arcsec for the longitude or the latitude, and of just 2 cm for the distance. In the computer program used for the Canon, all periodic terms were neglected with coefficients smaller than 0.0005 arcsec in longitude and latitude, and smaller than 1 m in distance. Because of neglecting the very small periodic terms, the Moon's positions calculated in the program have a mean error (as compared to the full ELP theory) of about 0.0006 seconds of time in right ascension, and about 0.006 arcsec in declination. The corresponding error in the calculated times of the contacts of a lunar eclipse is of the order of 1/40 second, which is much smaller than the uncertainties in predicted values of ΔT, and several orders of magnitude smaller than the uncertainty due to the diffuse edge of Earth's shadows caused by the terrestrial atmosphere.

Improved expressions for the mean arguments L', D, M, M' and F have been taken from Chapront, Chapront-Touzé, and Francou [2002]. A major consequence of this work is to bring the secular acceleration of the Moon's longitude (–25.858 "/cy^2, where arcsec/cy^2 is arc seconds per Julian century squared[6]) into good agreement with Lunar Laser Ranging (LLR) observations from 1972 to 2001 (Sect. 1.4).

A fundamental parameter used in eclipse predictions is the Moon's radius k expressed in units of Earth's equatorial radius. The Moon's actual radius varies as a function of position angle and libration because of irregularities in the surface topography of the Moon and the projection of these features along the line of sight onto the lunar limb profile. The current work uses a value for the lunar radius of k=0.2724880 representing a mean over the Moon's topographic

---

5 Version D of VSOP87 gives solar (and planetary) positions referred to the mean equinox of the date.

6 This unit, arcsec/cy^2, is used in discussing secular changes in the Moon's longitude over long time intervals.

features. This is the same value used for partial solar eclipses in the *Five Millennium Canon of Solar Eclipses: –1999 to +3000* (Espenak and Meeus, 2006).

The center of figure of the Moon does not coincide exactly with its center of mass. The magnitude of this difference is ~0.5 arcsec and is much smaller than the uncertainty in the edges of the penumbral and umbral shadows because of diffusion by Earth's atmosphere (Sect. 1.5). Thus, the difference between center-of-figure and center-of-mass positions has no practical impact on the present work.

1.4 Secular Acceleration of the Moon

Ocean tides are caused by the gravitational pull of the Moon (and, to a lesser extent, the Sun). The resulting tidal bulge in Earth's oceans is dragged ahead of the Moon in its orbit due to the daily rotation of Earth. As a consequence, the ocean mass offset from the Earth-Moon line exerts a pull on the Moon and accelerates it in its orbit. Conversely, the Moon's gravitational tug on this mass exerts a torque that decelerates the rotation of Earth. The length of the day gradually increases as energy is transferred from Earth to the Moon, causing the lunar orbit and period of revolution about Earth to increase.

This secular acceleration of the Moon is small but it has a cumulative effect on the Moon's position when extrapolated over many centuries. Direct measurements of the acceleration have only been possible since 1969 using the Apollo retro-reflectors left on the Moon. The results from Lunar Laser Ranging show that the Moon's mean distance from Earth is increasing by 3.8 cm per year (Dickey, et al., 1994). The corresponding acceleration in the Moon's ecliptic longitude is –25.858 arcsec/cy^2 (Chapront, Chapront-Touzé, and Francou, 2002). This is the value we have adopted in our lunar ephemeris calculations.

There is a close correlation between the Moon's secular acceleration and changes in the length of the day. However, the relationship is not exact because the lunar orbit is inclined anywhere from about 18.5° to 28.5° to Earth's equator. The parameter ΔT (sections 2.6 and 2.7) is a measure of the accumulated difference in time between an ideal clock and one based on Earth's rotation as it gradually slows down. Published determinations of ΔT from historical eclipse records have assumed a secular acceleration of –26 "/cy^2 (Morrison and Stephenson, 2004). Since we have adopted a slightly different value for the secular acceleration, we must make a small correction "c" to the published values of ΔT as follows:

$$c = -\,0.91072\ (\,-25.858 + 26.0\,)\ u^2, \qquad (1\text{-}1)$$

where: u = (year–1955)/100

Then:

$$\Delta T\ (\text{corrected}) = \Delta T + c \qquad (1\text{-}2)$$

Evaluation of the correction at 1,000 year intervals over the period spanned by the *Canon* is found in Table 1-2.

Table 1-2. Corrections to ΔT Due to Secular Acceleration

Year	Correction (seconds)
-2000	-202
-1000	-113
0	-49
+1000	-12
+2000	0
+3000	-14

Clearly the correction is only important for negative years although it is significantly smaller than the actual uncertainty in ΔT itself (Sect. 2.8).

The secular acceleration of the Moon is very poorly known and may not be constant. Careful records for its derivation only go back about a century. Before then, spurious and often incomplete eclipse and occultation observations from medieval and ancient manuscripts comprise the data base. In any case, the current value implies an increase in the length of day [LOD] of about 2.3 milliseconds per century. Such a small amount may seem insignificant, but it has very measurable cumulative effects. At this rate, time as measured through Earth's rotation is losing about 84 seconds per century^2 when compared to atomic time.

1.5 Lunar Eclipse Contacts

The major stages of a lunar eclipse are defined by the instants when the Moon's limb is externally or internally tangent to the penumbral or umbral shadows.

Penumbral lunar eclipses have two primary contacts. Neither of these events is observable.

P_1 — Instant of first exterior tangency of the Moon with the Penumbra. (Penumbral Eclipse Begins)
P_4 — Instant of last exterior tangency of the Moon with the Penumbra. (Penumbral Eclipse Ends)

Partial lunar eclipses have two additional contacts as the Moon's limb enters and exits the umbral shadow (U_1 and U_4, respectively). At these two instants, the partial phase of the eclipse begins and ends.

P_1 — Instant of first exterior tangency of the Moon with the Penumbra. (Penumbral Eclipse Begins)
U_1 — Instant of first exterior tangency of the Moon with the Umbra. (Partial Umbral Eclipse Begins)
U_4 — Instant of last exterior tangency of the Moon with the Umbra. (Partial Umbral Eclipse Ends)
P_4 — Instant of last exterior tangency of the Moon with the Penumbra. (Penumbral Eclipse Ends)

Total lunar eclipses have two additional umbral contacts at the instants when the Moon's entire disk is first and last internally tangent to the umbra (U_2 and U_3, respectively). These are the times when the total phase of the eclipse begins and ends, i.e., totality.

P_1 — Instant of first exterior tangency of the Moon with the Penumbra. (Penumbral Eclipse Begins)
U_1 — Instant of first exterior tangency of the Moon with the Umbra. (Partial Umbral Eclipse Begins)
U_2 — Instant of first interior tangency of the Moon with the Umbra. (Total Umbral Eclipse Begins)
U_3 — Instant of last interior tangency of the Moon with the Umbra. (Total Umbral Eclipse Ends)
U_4 — Instant of last exterior tangency of the Moon with the Umbra. (Partial Umbral Eclipse Ends)
P_4 — Instant of last exterior tangency of the Moon with the Penumbra. (Penumbral Eclipse Ends)

The instant when the Moon passes closest to the shadow axis is known as the instant of greatest eclipse. This corresponds to the maximum phase of the eclipse.

1.6 Enlargement of Earth's Shadows

In the early 1700s, Philippe de La Hire made a curious observation about Earth's umbra. The predicted radius of the shadow needed to be enlarged by about 1/41 in order to fit timings made during a lunar eclipse (La Hire 1707). Additional observations over the next two centuries revealed that the shadow enlargement was somewhat variable from one eclipse to the next. According to Chauvenet (1891):

"This fractional increase of the breath of the shadow was given by Lambert as 1/40, and by Mayer as 1/60. Beer and Maedler found 1/50 from a number of observations of eclipses of lunar spots in the very favorable eclipse of December 26, 1833."

Chauvenet adopted a value of 1/50, which has become the standard enlargement factor for lunar eclipse predictions published by many national institutes worldwide. The enlargement enters into the definitions of the penumbral and umbral shadow radii as follows.

penumbral radius: $R_p = 1.02 * (0.998340 * P_m + S_s + P_s)$ (1-3)
umbral radius: $R_u = 1.02 * (0.998340 * P_m - S_s + P_s)$ (1-4)

where: P_m = Equatorial horizontal parallax of the Moon,
S_s = Geocentric semi-diameter of the Sun, and
P_s = Equatorial horizontal parallax of the Sun.

The factor 1.02 is the enlargement of the shadows by 1/50. Earth's true figure approximates that of an oblate ellipsoid with a flattening of ~1/300. Furthermore, the axial tilt of the planet towards or away from the Sun throughout the year means the shape of the penumbral and umbral shadows vary although the effect is small. It is sufficient to use a mean radius of Earth at latitude 45° to approximate the departure from perfectly circular shadows. The *Astronomical Almanac*[7] uses a factor of 0.998340 to scale the Moon's equatorial horizontal parallax to account for this (i.e., 0.998340 ≈ 1 – 0.5 * 1/300).

In an analysis of 57 eclipses covering a period of 150 years, Link (1969) found a mean shadow enlargement of 2.3%. Furthermore, timings of crater entrances and exits through the umbra during four lunar eclipses from 1972 to 1982 (Table 1-3) closely support the Chauvenet value of 2%. From a physical point of view, there is no abrupt boundary between the umbra and penumbra. The shadow density actually varies continuously as a function of radial distance from the central axis out to the extreme edge of the penumbra. However, the density variation is most rapid near the theoretical edge of the umbra. Kuhl's (1928) contrast theory demonstrates that the edge of the umbra is perceived at the point of inflexion in the shadow density. This point appears to be equivalent to a layer in Earth's atmosphere at an altitude of about 120 to 150 km. The net enlargement of Earth's radius of 1.9% to 2.4% corresponds to an umbral shadow enlargement of 1.5% to 1.9%, in reasonably good agreement with the conventional value.

Table 1-3. Umbral Shadow Enlargement from Craters Timings

Lunar Eclipse Date	Crater Entrances Percent Enlargement	Crater Exits Percent Enlargement	Sky & Telescope Reference
1972 Jan 30	1.69 [420]	1.68 [295]	Oct 1972, p.264
1975 May 24	1.79 [332]	1.61 [232]	Oct 1975, p.219
1982 Jul 05	2.02 [538]	2.24 [159]	Dec 1982, p.618
1982 Dec 30	1.74 [298]	1.74 [90]	Apr 1983, p.387

Note: Values in "[ ]" are the number of observations included in each shadow enlargement measurement.

Some authorities dispute Chauvenet's shadow enlargement convention. Danjon (1951) notes that the only reasonable way of accounting for a layer of opaque air surrounding Earth is to increase the planet's radius by the altitude of the layer. This can be accomplished by proportionally increasing the parallax of the Moon. The radii of the umbral and penumbral shadows are then subject to the same absolute correction and not the same relative correction employed in the traditional Chauvenet 1/50 convention. Danjon estimates the thickness of the occulting layer to be 75 km and this results in an enlargement of Earth's radius and the Moon's parallax of about 1/85.

Since 1951, the French almanac *Connaissance des Temps* has adopted Danjon's method for the enlargement Earth's shadows in their eclipse predictions as shown below.

penumbral radius: $R_p = 1.01 * Pm + S_s + P_s$ (1-5)
umbral radius: $R_u = 1.01 * Pm - S_s + P_s$ (1-6)

[7] The *Astronomical Almanac* is published annually by the Almanac Office of the U.S. Naval Observatory.

where: P_m = Equatorial horizontal parallax of the Moon,
S_s = Geocentric semi-diameter of the Sun,
P_s = Equatorial horizontal parallax of the Sun, and
$1.01 \approx 1 + 1/85 - 1/594$.

The factor 1.01 combines the 1/85 shadow enlargement term with a 1/594 term[8] to correct for Earth's oblateness at a latitude of 45°.

Danjon's method correctly models the geometric relationship between an enlargement of Earth's radius and the corresponding increase in the size of its shadows. Meeus and Mucke (1979) and Espenak (2006) both use Danjon's method. However, the resulting umbral and penumbral eclipse magnitudes are smaller by approximately 0.006 and 0.026, respectively, as compared to predictions using the traditional Chauvenet convention of 1/50.

For instance, the umbral magnitude of the partial lunar eclipse of 2008 Aug 16 was 0.813 according to the *Astronomical Almanac* (2008) using Chauvenet's method, but only 0.806 according to Meeus and Mucke (1979) using Danjon's method.

Of course, the small magnitude difference between the two methods is difficult to observe because the edge of the umbral shadow is poorly defined. The choice of shadow enlargement method has the largest effect in certain limiting cases where a small change in magnitude can shift an eclipse from one type to another. For example, an eclipse that is barely total according to Chauvenet's method will be a large magnitude partial eclipse if calculated using Danjon's method. Table 1–4 shows five such instances where a shallow total eclipse calculated with Chauvenet's method becomes a deep partial eclipse with Danjon's.

Table 1-4. Total (Chauvenet) vs. Partial (Danjon) Lunar Eclipses: 1501–3000

Calendar Date	Umbral Magnitude (Chauvenet)	Umbral Magnitude (Danjon)	Magnitude Difference
1540 Sep 16	1.0007	0.9947	0.0060
1856 Oct 13	1.0017	0.9960	0.0057
2196 Jul 10	1.0007	0.9960	0.0047
2413 Nov 08	1.0042	0.9993	0.0049
2669 Feb 08	1.0016	0.9951	0.0065

Similarly, small umbral magnitude partial eclipses using Chauvenet's method must be reclassified as penumbral eclipses of large penumbral magnitude when calculated with Danjon's method. A recent example was the eclipse of 1988 Mar 03, which was partial with an umbral magnitude of 0.0028 according to Chauvenet's method, but was penumbral with an umbral magnitude –0.0017[9] by Danjon's method. A similar case will occur on 2042 Sep 29. For a list of all such cases from 1501 through 3000, see Table 1-5.

[8] *Connaissance des Temps* uses a value of 1/297 for Earth's flattening. At latitude 45°: 1/594 = 0.5 * 1/297.

[9] A negative umbral magnitude means that the Moon lies fully outside the umbral shadow and is, therefore, a penumbral eclipse.

Table 1-5. Partial (Chauvenet) vs. Penumbral (Danjon) Lunar Eclipses: 1501–3000

Calendar Date	Umbral Magnitude (Chauvenet)	Umbral Magnitude (Danjon)	Magnitude Difference
1513 Sep 15	0.0036	-0.0003	0.0039
1900 Jun 13	0.0012	-0.0040	0.0052
1988 Mar 03	0.0028	-0.0017	0.0045
2042 Sep 29	0.0027	-0.0031	0.0058
2429 Dec 11	0.0020	-0.0033	0.0053
2581 Oct 13	0.0017	-0.0054	0.0071
2678 Aug 24	0.0007	-0.0036	0.0043
2733 Aug 17	0.0037	-0.0040	0.0077

Finally, in some cases, the shadow enlargement convention can make the difference between a shallow penumbral eclipse (Chauvenet) or no eclipse at all (Danjon). Table 1–6 lists nine small magnitude penumbral eclipses over a 500-year interval as determined using Chauvenet's method (Liu and Fiala, 1992). When the eclipse predictions are repeated using Danjon's method, no lunar eclipses are found on these dates.

Table 1-6. Penumbral Lunar Eclipses (Chauvenet): 1801–2300

Calendar Date	Penumbral Magnitude (Chauvenet)
1864 Apr 22	0.0237
1872 Jun 21	0.0008
1882 Oct 26	0.0059
1951 Feb 21	0.0068
2016 Aug 18	0.0165
2042 Oct 28	0.0077
2194 Mar 07	0.0085
2219 Apr 30	0.0008
2288 Feb 18	0.0204

Practically speaking, the faint and indistinct edge of the penumbral shadow makes the penumbral eclipse contacts (P_1 and P_4) completely unobservable. So the small magnitude differences discussed here are only of academic interest. Still, it is important to note which shadow enlargement convention is used because it is critical in comparing predictions from different sources.

In the *Catalog*, Earth's penumbral and umbral shadow sizes have been calculated by using Danjon's enlargement method. Other sources using Danjon's method include Meeus and Mucke (1979), Espenak (2006) and *Connaissance des Temps*. Several sources using Chauvenet's method are Espenak (1989), Liu and Fiala (1992), and *Astronomical Almanac.*

1.7 Visual Appearances of Lunar Eclipses

The visual appearance of penumbral, partial, and total lunar eclipses differs significantly from each other. While penumbral eclipses are pale and difficult to see, partial eclipses are easy naked-eye events, while total eclipses are colorful and dramatic.

1.7.1 Appearance of Penumbral and Partial Lunar Eclipses

Earth's penumbral shadow forms a diverging cone that expands into space in the anti-solar direction. From within this zone, Earth blocks part but not the entire disk of the Sun. Thus, some fraction of the Sun's direct rays continues to reach the most deeply eclipsed parts of the Moon during a penumbral eclipse.

The primary penumbral contacts (P_1 and P_4), as well as the early and late stages of a penumbral eclipse, are completely invisible to the eye with or without optical aid. A penumbral magnitude greater than ~0.6 is needed before skilled observers can detect faint shading across the Moon's disk.

Even when one edge of the Moon is 9/10 of the way into the penumbral shadow, approximately 10% of the Sun's rays still reach the Moon's deepest limb. Under such conditions, the Moon remains relatively bright with only a subtle gradient across its disk. The penumbral eclipse only becomes readily apparent when it is within ~0.05 magnitudes of becoming a partial eclipse.

In comparison, partial eclipses are easy to see with the naked eye. The lunar limb extending into the umbral shadow usually appears very dark or black. This is primarily due to a contrast effect because the remaining portion of the Moon in the penumbra may be brighter by a factor of ~500x. Because the umbral shadow's diameter is typically ~2.7x the Moon's diameter, it appears as though a semi-circular bite has been taken out of the Moon.

Aristotle (384–322 BCE) first proved that Earth was round using the curved umbral shadow seen at partial eclipses. In comparing observations of several eclipses, he noted that Earth's shadow was round no matter where the eclipse took place, whether the Moon was high in the sky or low near the horizon. Aristotle correctly reasoned that only a sphere casts a round shadow from every angle.

1.7.2 Appearance of Total Lunar Eclipses

The total lunar eclipse is the most dramatic and visually compelling type of lunar eclipse. The Moon's appearance can vary enormously throughout the period of totality and from one eclipse to the next. Obviously, the geometry of the Moon's path through the umbra plays an important role. Not as apparent is the effect that Earth's atmosphere has on a total eclipse. Although the physical mass of Earth blocks all direct sunlight from the umbra, the planet's atmosphere filters, attenuates and bends some of the Sun's rays into the shadow.

The molecules in Earth's atmosphere scatter short wavelength light (e.g., yellow, green, blue) more than long wavelength light (e.g., orange, red). This process, which is responsible for making sunsets red, also gives total eclipses their characteristic red-orange color. However, the exact color can vary considerably in both hue and brightness.

Because the lowest layers of the atmosphere are thicker than the upper layers, they absorb more sunlight and refract it through larger angles. About 75% of the atmosphere's mass is concentrated in the bottom 10 km (troposphere) as well as most of the water vapor, which can form massive clouds that block even more light. Just above the troposphere lies the stratosphere (~10 km to 50 km), a rarified zone above most of the planet's weather systems. The stratosphere is subject to important photochemical reactions due to the high levels of solar ultraviolet radiation that penetrates the region. The troposphere and stratosphere act together as a ring-shaped lens that refracts heavily reddened sunlight into Earth's umbral shadow. Because the higher atmospheric layers in the stratosphere contain less gas, they refract sunlight through progressively smaller angles into the outer parts of the umbra. In contrast, lower atmospheric layers containing more gas refract sunlight through larger angles to reach the inner parts of the umbra.

As a consequence of this lensing effect, the amount of light refracted into the umbra tends to increase radially from center to edge. However, inhomogeneities in the form of asymmetric amounts of cloud and dust at differing latitudes can cause significant variations in brightness throughout the umbra.

Besides water (clouds, mist, precipitation), Earth's atmosphere also contains aerosols or tiny particles of organic debris, meteoric dust, volcanic ash, and photochemical droplets. This material can significantly attenuate sunlight before it is refracted into the umbra. For instance, major volcanic eruptions in 1963 (Agung) and 1982 (El Chichón) each dumped huge quantities of gas and ash into the stratosphere and were followed by several years of very dark eclipses (Keen, 1983).

The same thing occurred after the eruption of the Philippine volcano Pinatubo in 1991. While most of the solid ash fell to Earth several days after circulating through the troposphere, a sizable volume of sulphur dioxide (SO_2) and water vapor reached the stratosphere where it produced sulfuric acid (H2SO4). This high-altitude volcanic haze layer can severely dim sunlight that must travel several hundred kilometers horizontally through the layer before being refracted into the umbral shadow. Consequently, the total eclipses following large volcanic eruptions are unusually dark. For instance, the total lunar eclipse of 1992 Dec 09 (1.5 years after Pinatubo) was so dark that the Moon's dull gray disk was difficult to see with the naked eye (Espenak, 2008, personal observation).

1.7.3 Danjon Scale of Lunar Eclipse Brightness

The French astronomer A. Danjon proposed a useful five-point scale for evaluating the visual appearance and brightness of the Moon during total lunar eclipses. The L values for various luminosities are defined as follows:

L=0	Very dark eclipse (Moon is almost invisible, especially at mid-totality)
L=1	Dark eclipse, gray or brownish in coloration (Details are distinguishable only with difficulty)
L=2	Deep red or rust-colored eclipse (Very dark central shadow, while outer umbra is relatively bright)
L=3	Brick-red eclipse (Umbral shadow usually has a bright or yellow rim)
L=4	Very bright copper-red or orange eclipse (Umbral shadow has a bluish, very bright rim)

The Danjon scale illustrates the range of colors and brightness the Moon can take on during a total lunar eclipse. It is also a useful tool to visual observers in characterizing the appearance of an eclipse. The evaluation of an L value is best done with the naked eye, binoculars, or a small telescope near the time of mid-totality. It is also helpful to examine the Moon's appearance just after the beginning and just before the end of totality. The Moon is then near the edge of the shadow, providing an opportunity to assign an L value to the outer umbra. In making such evaluations, the instrumentation and the time must be also recorded.

1.8 Five Millennium Catalog of Lunar Eclipses on EclipseWise.com

The tables in the *Catalog* are available via the website *EclipseWise.com* at:

http://eclipsewise.com/lunar/LEcatalog/LEcatalog.html

Organized into 100-year intervals, the tables have individual links to eclipse figures, maps, and tables.

A ASCII text file listing basic details about each eclipse in the Catalog is also available on-line at:

http://eclipsewise.com/lunar/LEpubs/5MKLE.html

All the information found on the Figures and maps is included (calendar date, Dynamical Time, eclipse type, Saros number, gamma, duration of each phase). In addition, a number of other useful data are listed including ΔT, and the geographic coordinates of greatest. The eclipse type is augmented to indicate the first/last eclipse in a Saros series..

Section 2: Time

2.1 Greenwich Mean Time

For thousands of years, time has been measured using the length of the solar day. This is the interval between two successive returns of the Sun to an observer's local meridian. Unfortunately, the length of the apparent solar day can vary by tens of seconds over the course of a year. Earth's elliptical orbit around the Sun and the 23.5° inclination of Earth's axis of rotation are responsible for these variations. Apparent solar time was eventually replaced by mean solar time because it provides for a uniform time scale. The key to mean solar time is the mean solar day which has a constant length of 24 hours throughout the year.

Mean solar time on the 0° longitude meridian in Greenwich, England is known as Greenwich Mean Time (GMT). At the International Meridian Conference of 1884, GMT[10] was adopted as the reference time for all clocks around the world. It was also agreed that all longitudes would be measured east or west with respect to the Greenwich meridian. In 1972, Greenwich Mean Time was replaced by Coordinated Universal Time (UTC) as the international time reference. Nevertheless, UTC is colloquially referred to as Greenwich Mean Time although this is technically not correct.

2.2 Ephemeris Time

During the twentieth century, it was found that the rotational period of Earth (length of the day) was gradually slowing down. For the purposes of orbital calculations, time using Earth's rotation was abandoned for a more uniform time scale based on Earth's orbit about the Sun. In 1952, the International Astronomical Union (IAU) introduced Ephemeris Time to address this problem. The ephemeris second was defined as a fraction of the tropical year for 1900 Jan 01 as calculated from Newcomb's Tables of the Sun [1895]. Ephemeris Time was used for Solar System ephemeris calculations until it was replaced by Terrestrial Dynamical Time in 1979.

2.3 Terrestrial Dynamical Time

Terrestrial Dynamical Time (TD) was introduced by the International Astronomical Union (IAU) in 1979 as the coordinate time scale for an observer on the surface of Earth. It takes into account relativistic effects and is based on International Atomic Time (TAI) which is a high-precision standard using several hundred atomic clocks worldwide. As such Terrestrial Dynamical Time is the atomic time equivalent to its predecessor Ephemeris Time (ET) and is used in the theories of motion for bodies in the Solar System. To ensure continuity with Ephemeris Time, TD was defined to match ET for the date 1977 Jan 01. In 1991, the IAU refined the definition of TD to make it more precise. It was also renamed Terrestrial Time (TT) although we prefer and use the older name Terrestrial Dynamical Time.

2.4 Universal Time

For many centuries, the fundamental unit of time was the rotational period of Earth with respect to the Sun. Greenwich Mean Time (GMT) was the standard time reference based on the mean solar time on the 0° longitude meridian in Greenwich, England. Universal Time (UT) is the modern counterpart to GMT and is determined from Very Long Baseline Interferometry (VLBI) observations of the diurnal motion of quasars. Unfortunately, Universal Time is not a uniform time scale because Earth's rotational period is (on average) gradually increasing.

---

[10] GMT was originally reckoned from noon to noon. In 1925, some countries shifted GMT by 12 hours so that it would begin at Greenwich midnight. This new definition is the one in common usage for world time and in the navigational publications of English-speaking countries. The designation Greenwich Mean Astronomical Time (GMAT) is reserved for the reckoning of time from noon (and previously called GMT).

The change is primarily due to tidal friction between Earth's oceans and its rocky mantle through the gravitational attraction of the Moon and, to a lesser extent, the Sun. This secular acceleration (Sect. 1.4) gradually transfers angular momentum from Earth to the Moon. As Earth loses energy and slows down, the Moon gains this energy and its orbital period and distance from Earth increase. Shorter period fluctuations in terrestrial rotation also exist which can produce an accumulated clock error of ±20 seconds in one or more decades. These decade variations are attributed to several geophysical mechanisms including fluid interactions between the core and mantle of Earth. Climatological changes and variations in sea-level may also play significant roles since they alter Earth's moment of inertia.

The secular acceleration of the Moon implies an increase in the length of day [LOD] of about 2.3 milliseconds per century. Such a small amount may seem insignificant, but it has very measurable cumulative effects. At this rate, time as measured through Earth's rotation is losing about 84 seconds per century^2 when compared to atomic time.

2.5 Coordinated Universal Time

Coordinated Universal Time (UTC) is the present day basis of all civilian time throughout the world. Derived from International Atomic Time (TAI), the length of the UTC second is defined in terms an atomic transition of the element cesium and is accurate to approximately one nanosecond (billionth of a second) per day. Since most daily life is still organized around the solar day, UTC was defined to closely parallel Universal Time. However, the two time systems are intrinsically incompatible since UTC is uniform while UT is based on Earth's rotation which is gradually slowing. In order to keep the two times within 0.9 seconds of each other, a leap second is added to UTC about once every 12 to 18 months.

2.6 Delta T (ΔT)

The orbital positions of the Sun and Moon required by eclipse predictions, are calculated using Terrestrial Dynamical Time (TD) because it is a uniform time scale. World time zones and daily life, however, are based on Universal Time (UT)[11]. In order to convert eclipse predictions from TD to UT, the difference between these two time scales must be known. The parameter delta-T (ΔT) is the arithmetic difference, in seconds, between the two as:

$$\Delta T = TD - UT \qquad (2\text{-}1)$$

Past values of ΔT can be deduced from the historical records. In particular, hundreds of eclipse observations (both solar and lunar) were recorded in early European, Middle Eastern and Chinese annals, manuscripts, and canons. In spite of their relatively low precision, these data represent the only evidence for the value of ΔT prior to 1600 CE. In the centuries following the introduction of the telescope (circa 1609 CE), thousands of high quality observations have been made of lunar occultations of stars. The number and accuracy of these timings increase from the seventeenth through the twentieth century, affording valuable data in the determination of ΔT. A detailed analysis of these measurements fitted with cubic splines for ΔT from –500 to +1950 is presented in Table 2-1 and includes the standard error for each value (Morrison and Stephenson, 2004).

[11] World time zones are actually based on Coordinated Universal Time (UTC). It is an atomic time synchronized and adjusted to stay within 0.9 seconds of astronomically determined Universal Time (UT). Occasionally, a "leap second" is added to UTC to keep it in sync with UT (which changes due to variations in Earth's rotation rate).

Table 2-1. Values of ΔT Derived from Historical Records

Year	ΔT (seconds)	Standard Error (seconds)
-500	17190	430
-400	15530	390
-300	14080	360
-200	12790	330
-100	11640	290
0	10580	260
100	9600	240
200	8640	210
300	7680	180
400	6700	160
500	5710	140
600	4740	120
700	3810	100
800	2960	80
900	2200	70
1000	1570	55
1100	1090	40
1200	740	30
1300	490	20
1400	320	20
1500	200	20
1600	120	20
1700	9	5
1750	13	2
1800	14	1
1850	7	<1
1900	-3	<1
1950	29	<0.1

In modern times, the determination of ΔT is made using atomic clocks and radio observations of quasars, so it is completely independent of the lunar ephemeris. Table 2-2 gives the value of ΔT every five years from 1955 to 2005 (*Astronomical Almanac* for 2006, page K9).

Table 2-2. Recent Values of ΔT from Direct Observations

Year	ΔT (seconds)	5-Year Change (seconds)	Average 1-Year Change (seconds)
1955.0	+31.1	—	—
1960.0	+33.2	2.1	0.42
1965.0	+35.7	2.5	0.50
1970.0	+40.2	4.5	0.90
1975.0	+45.5	5.3	1.06
1980.0	+50.5	5.0	1.00
1985.0	+54.3	3.8	0.76
1990.0	+56.9	2.6	0.52
1995.0	+60.8	3.9	0.78
2000.0	+63.8	3.0	0.60
2005.0	+64.7	0.9	0.18

The average annual change of ΔT was 0.99 second from 1965 to 1980. However, the average annual increase was just 0.63 second from 1985 to 2000, and only 0.18 second from 2000 to 2005. Future changes and trends in ΔT cannot be predicted with certainty since theoretical models of the physical causes are not of high enough precision. Extrapolations from the table weighted by the long period trend from tidal braking of the Moon offer reasonable estimates of +67 seconds in 2010, +93 seconds in 2050, +203 seconds in 2100, and +442 seconds in the year 2200.

Outside the period of observations (500 BCE to 2005 CE), the value of ΔT can be extrapolated from measured values using the long-term mean parabolic trend:

$$\Delta T = -20 + 32\,u^2 \text{ seconds,} \qquad (2\text{-}2)$$

where: u = (year–1820)/100, and is defined as time measured in centuries.

2.7 Polynomial Expressions for ΔT

Using the ΔT values derived from the historical record and from direct observations (Tables 2-1 and 2-2, respectively), a series of polynomial expressions have been created to simplify the evaluation of ΔT for any time during the interval -1999 to +3000. We define the decimal year "y" as follows:

$$y = \text{year} + (\text{month} - 0.5)/12 \qquad (2\text{-}3)$$

This gives "y" for the middle of the month, which is accurate enough given the precision in the known values of ΔT. The following polynomial expressions can be used to calculate the value of ΔT (in seconds) over the interval of the *Canon*.

Before the year -500, calculate:

$$\Delta T = -20 + 32\,u^2, \qquad (2\text{-}4)$$

where: u = (y–1820)/100.

Between years -500 and +500, we use the data from Table 2-1, except that for the year -500 we changed the value 17190 to 17203.7 in order to avoid a discontinuity with the previous formula at that epoch. The value for ΔT is given by a polynomial of the 6th degree, which reproduces the values in Table 2-1 with an error not larger than 4 seconds:

$$\Delta T = 10583.6 - 1014.41\,u + 33.78311\,u^2 - 5.952053\,u^3 - 0.1798452\,u^4 + 0.022174192\,u^5 + 0.0090316521\,u^6, \qquad (2\text{-}5)$$

where: u = y/100.

Between years +500 and +1600, we again use the data from Table 2-1. Calculate u = (y–1000)/100. The value for ΔT is given by the following polynomial of the 6th degree with a divergence from Table 2-1 not larger than 4 seconds:

$$\Delta T = 1574.2 - 556.01\,u + 71.23472\,u^2 + 0.319781\,u^3 - 0.8503463\,u^4 - 0.005050998\,u^5 + 0.0083572073\,u^6, \qquad (2\text{-}6)$$

where: u = (y–1000)/100.

Between years +1600 and +1700, calculate:

$$\Delta T = 120 - 0.9808\,t - 0.01532\,t^2 + t^3 / 7129, \qquad (2\text{-}7)$$

where: t = y - 1600.

Between years +1700 and +1800, calculate:

$$\Delta T = 8.83 + 0.1603\,t - 0.0059285\,t^2 + 0.00013336\,t^3 - t^4 / 1174000, \qquad (2\text{-}8)$$

where: t = y – 1700.

Between years +1800 and +1860, calculate:

$$\Delta T = 13.72 - 0.332447\ t + 0.0068612\ t^2 + 0.0041116\ t^3 - 0.00037436\ t^4 + 0.0000121272\ t^5 - 0.0000001699\ t^6 + 0.000000000875\ t^7, \qquad (2\text{-}9)$$

where: t = y – 1800.

Between years 1860 and 1900, calculate:

$$\Delta T = 7.62 + 0.5737\ t - 0.251754\ t^2 + 0.01680668\ t^3 - 0.0004473624\ t^4 + t^5 / 233174, \qquad (2\text{-}10)$$

where: t = y – 1860.

Between years 1900 and 1920, calculate:

$$\Delta T = -2.79 + 1.494119\ t - 0.0598939\ t^2 + 0.0061966\ t^3 - 0.000197\ t^4, \qquad (2\text{-}11)$$

where: t = y – 1900.

Between years 1920 and 1941, calculate:

$$\Delta T = 21.20 + 0.84493\ t - 0.076100\ t^2 + 0.0020936\ t^3, \qquad (2\text{-}12)$$

where: t = y – 1920.

Between years 1941 and 1961, calculate:

$$\Delta T = 29.07 + 0.407\ t - t^2 / 233 + t^3 / 2547, \qquad (2\text{-}13)$$

where: t = y – 1950.

Between years 1961 and 1986, calculate:

$$\Delta T = 45.45 + 1.067\ t - t^2 /260 - t^3 / 718, \qquad (2\text{-}14)$$

where: t = y – 1975.

Between years 1986 and 2005, calculate:

$$\Delta T = 63.86 + 0.3345\ t - 0.060374\ t^2 + 0.0017275\ t^3 + 0.000651814\ t^4 + 0.00002373599\ t^5, \qquad (2\text{-}15)$$

where: t = y – 2000.

Between years 2005 and 2050, calculate:

$$\Delta T = 62.92 + 0.32217\ t + 0.005589\ t^2, \qquad (2\text{-}16)$$

where: t = y – 2000.

This expression is derived from estimated values of ΔT in the years 2010 and 2050. The value for 2010 (66.9 seconds) is based on a linearly extrapolation from 2005 using 0.39 seconds/year (average from 1995 to 2005). The value for 2050 (93 seconds) is linearly extrapolated from 2010 using 0.66 seconds/year (average rate from 1901 to 2000).

Between years 2050 and 2150, calculate:

$$\Delta T = -20 + 32\ ((y-1820)/100)^2 - 0.5628\ (2150 - y), \qquad (2\text{-}17)$$

The last term is introduced to eliminate the discontinuity at 2050.

After 2150, calculate:

$$\Delta T = -20 + 32\ u^2, \qquad (2\text{-}18)$$

where: u = (y–1820)/100.

All values of ΔT based on Morrison and Stephenson [2004] assume a value for the Moon's secular acceleration of –26 arcsec/cy². However, the ELP-2000/82 lunar ephemeris employed in the *Canon* uses a slightly different value of –25.858 arcsec/cy². Thus, a small correction "c" must be added to the values derived from the polynomial expressions for ΔT before they can be used in the *Canon*:

$$c = -0.000012932\ (y - 1955)^2. \qquad (2\text{-}19)$$

Since the values of ΔT for the interval 1955 to 2005 were derived independent of any lunar ephemeris, no correction is needed for this period.

2.8 Uncertainty in ΔT

The uncertainty in the value of ΔT is of particular interest in the calculation of eclipse paths in the distant past and future. Unfortunately, estimating the standard error in ΔT prior to 1600 CE is a difficult problem. It depends on a number of factors which include the accuracy of determining ΔT from historical eclipse records and modeling the physical processes producing changes in Earth's rotation. Morrison and Stephenson [2004] propose a simple parabolic relation to estimate the standard error (σ) which is valid over the period 1000 BCE to 1200 CE:

$$\sigma = 0.8\ t^2 \text{ seconds}, \qquad (2\text{-}20)$$

where: t = (year–1820)/100.

Table 2-3 gives the errors in ΔT along with the corresponding uncertainties in the longitude of an eclipse path.

Table 2-3. Uncertainty of ΔT, Part I

Year	σ (seconds)	Longitude
-1000	636	2.65°
-500	431	1.79°
0	265	1.10°
+500	139	0.58°
+1000	54	0.22°
+1200	31	0.13°

The decade fluctuations in ΔT result in an uncertainty of ~20 seconds (0.08°) for the period 1300 CE to 1600 CE.

During the telescopic era (1600 CE to present), records of astronomical observations pin down the decade fluctuations with increasing reliability. The uncertainties in ΔT are presented in Table 2-4 [Stephenson and Houlden, 1986].

Table 2-4. Uncertainty of ΔT, Part II

Year	σ (seconds)	Longitude
+1700	5	0.021°
+1800	1	0.004°
+1900	0.1	0.0004°

The estimation in the uncertainty of ΔT prior to 1000 BCE must rely on a certain amount of modeling and theoretical arguments since no measurements of ΔT are available for this period. Huber (2000) has proposed a Brownian motion model including drift to estimate the standard error in ΔT for periods outside the epoch of measured values. The intrinsic variability in the LOD (length of day) during the 2500 years of observations (500 BCE to 2000 CE) is 1.780 ms/cy with a standard error of 0.56 ms/cy. This rate is not due entirely to tidal friction but includes a drift in LOD from imperfectly understood effects such as changes in sea level due to variations in polar ice caps. Presumably, the same mechanisms operating during the present era also operated prior to 1000 BCE as well as one millennium into the future.

Huber's derived estimate for the total standard error (fluctuations plus drift) in ΔT is as follows.

$$\sigma = 365.25\ N\ SQRT\,[\,(\,N\ Q\,/\,3)\ (\,1 + N\,/\,M\,)\,]\,/\,1000, \qquad (2\text{-}21)$$

where:

N = Difference between target year and calibration year
M = 2500 years (–500 to +2000);
this covers period of observed ΔT measurements
Q = 0.058 ms^2 /yr

The calibration year is taken as –500 for target years before 500 BCE, while the calibration year is 2005 CE for target years in the future. Evaluation of this expression at 1,000 year intervals is found in Table 2-5. It shows estimates in the standard error of ΔT along with the equivalent shift in longitude

Table 2-5. Uncertainty of ΔT, Part III

Year	σ (seconds)	Longitude
-4500	20,717	86.3°
-4000	16,291	67.9°
-3500	12,378	51.6°
-3000	8,978	37.4°
-2500	6,094	25.4°
-2000	3,732	15.6°
-1500	1,900	7.9°
-1000	622	2.6°
—	—	—
+2500	612	2.6°
+3000	1,885	7.9°
+3500	3,711	15.6°
+4000	6,068	25.3°
+4500	8,946	37.3°
+5000	12,341	51.4°

Section 3: Lunar Eclipse Statistics

3.1 Statistical Distribution of Lunar Eclipse Types

Eclipses of the Moon can only occur during the Full Moon phase. It is then possible for the Moon to pass through Earth's penumbral and umbral shadows thereby producing an eclipse. There are three types of lunar eclipses:

1) **Penumbral** — Moon passes partially or completely into Earth's penumbral shadow.
2) **Partial** — Moon passes partially into Earth's umbral shadow.
3) **Total** — Moon passes completely into Earth's umbral shadow

During the 5,000-year period from –1999 to +3000 (2000 BCE to 3000 CE), Earth will experience 12,064 eclipses of the Moon. The statistical distribution of the three eclipse types over this interval is shown in Table 3-1.

Table 3-1. Distribution of Basic Lunar Eclipse Types

Eclipse Type	Abbreviation	Number	Percent
All Eclipses	—	12064	100.0%
Penumbral	N	4378	36.3%
Partial	P	4207	34.9%
Total	T	3479	28.8%

During most penumbral eclipses, only part of the Moon passes through Earth's penumbral shadow. Examples of such partial penumbral eclipses include: 2009 Feb 09, 2009 Jul 07, 2009 Aug 06, and 2012 Nov 28. However, it is also possible to have a penumbral eclipse in which the Moon passes completely within Earth's penumbral shadow without entering the inner umbral shadow. Such total penumbral eclipses are quite rare compared to normal (or partial) penumbral eclipses. During the 21st century, there are 87 partial penumbral eclipses, but only 5 total penumbral eclipses: 2006 Mar 14, 2053 Aug 29, 2070 Apr 25, 2082 Aug 08, and 2099 Sep 29. Table 3-2 shows the distribution of the two penumbral eclipse types during the period covered by the Canon. For more on total penumbral eclipses, see Section 3.11.

Table 3-2. Statistics of Penumbral Lunar Eclipses

Eclipse Type	Number	Percent
All Penumbral Eclipses	4378	100.0%
Partial Penumbral	4237	96.8%
Total Penumbral	141	3.2%

Total lunar eclipses through Earth's umbral shadow can be categorized as:

1) **Central** — The Moon passes through the central axis of Earth's umbral shadow.
2) **Non-Central** — The Moon misses the central axis of Earth's umbral shadow.

Using the above categories, the distribution of the 3,479 total eclipses is shown in Table 3-3.

Table 3-3. Statistics of Total Lunar Eclipses

Eclipse Type	Number	Percent
All Total Eclipses	3479	100.0%
Central Total	2074	59.6%
Non-Central Total	1405	40.4%

Examples of central total eclipses include: 2000 Jul 16, 2011 Jun 15, 2018 Jul 27, and 2022 May 16. Several examples of non-central total eclipses are: 2010 Dec 21, 2011 Dec 10, 2014 Apr 15, and 2014 Oct 08.

3.2 Distribution of Lunar Eclipse Types by Century

Table 3-4 summarizes 5,000 years of eclipses by eclipse type over 100-year intervals. The average century contains 241 lunar eclipses of which 88 are penumbral, 84 are partial and 70 are total. Individual centuries deviate from these mean values in interesting ways. For instance, the total number of eclipses in a century varies from a minimum of 225 (century beginning –0899) to a maximum of 259 (centuries beginning 1101 and 2801). The number of penumbral eclipses varies from 76 (century 0301) to 100 (century 2801), while the number of partial eclipses ranges from 54 (century –1599) to 102 (centuries –0699, 1101, and 2801). Finally, the number of total lunar eclipses per century varies from 57 (centuries 0001 and 2801) to 89 (century 0801). Note that century 2801 not only has the most number of eclipses (259), it also has the maximum number of penumbral (100) and partial (102) eclipses, but the minimum number of total eclipses (57).

The last column of Table 3-4 lists the number of total lunar eclipse tetrads per century. A tetrad is a grouping of four consecutive total lunar eclipses each separated by six lunations. Their frequency is quite variable with some centuries having 0 tetrads, while others may have as many as 8. For more on tetrads, see Section 3.16.

When the data in Table 3-4 is displayed graphically (Figure 3-1), other important relationships are revealed. The most apparent feature is a periodic oscillation in the number of all lunar eclipses (Figure 3-1a), as well as in the individual types. By inspection, the period is about 600 years. Using tables from von Oppolzer's Canon der Finsternisse (1887) the Dutch astronomer G. van den Bergh (1954) calculated a period of 586 years. By studying the distributions of tetrad eclipse groups (Section 3.16), Meeus derived an empirical expression showing that this period is slowly decreasing (Meeus 2004b). While its current value is actually 565 years, the period was 618 years in –1999 and it will decrease to 552 years by +3000. A theoretical study by T. Hughes (Hughes 2004) demonstrates that the period is caused by the eccentricity of Earth's orbit, which is gradually decreasing.

There are other interesting features in Figure 3-1. For instance, the number of total lunar eclipses as well as deep central total eclipses is highest when the number of all lunar eclipses is at a minimum. In contrast, numbers of penumbral eclipses, partial eclipses and non-central total eclipses all appear to be synchronized with the overall number of lunar eclipses. Finally, the number of total penumbral eclipses, and total lunar eclipse tetrads are most frequent during epochs when the overall number of lunar eclipses is at a minimum.

Table 3-4a. Number of Lunar Eclipse Types by Century: –1999 to -1000 (2000 BCE to 1001 BCE)

Century Interval	All Lunar Eclipses	All Penum. Eclipses	Partial Penum. Eclipses	Total Penum. Eclipses	All Partial Eclipses	All Total Eclipses	Central Total Eclipses	Non-Central Total Eclipses	Total Eclipse Tetrads
-1999 to -1900	242	88	84	4	92	62	13	49	1
-1899 to -1800	255	98	97	1	97	60	11	49	0
-1799 to -1700	254	94	94	0	99	61	12	49	0
-1699 to -1600	244	90	87	3	87	67	21	46	3
-1599 to -1500	226	85	81	4	54	87	62	25	8
-1499 to -1400	228	86	78	8	60	82	55	27	6
-1399 to -1300	239	83	80	3	90	66	23	43	2
-1299 to -1200	251	92	91	1	99	60	9	51	0
-1199 to -1100	252	96	95	1	95	61	13	48	0
-1099 to -1000	240	88	81	7	86	66	18	48	4

Table 3-4b. Number of Lunar Eclipse Types by Century: -0999 to +3000 (1000 BCE to 3000 CE)

Century Interval	All Lunar Eclipses	All Penum. Eclipses	Partial Penumb. Eclipses	Total Penum. Eclipses	All Partial Eclipses	All Total Eclipses	Central Total Eclipses	Non-Central Total Eclipses	Total Eclipse Tetrads
-0999 to -0900	228	80	72	8	61	87	59	28	7
-0899 to -0800	225	79	76	3	62	84	57	27	7
-0799 to -0700	239	87	85	2	88	64	21	43	1
-0699 to -0600	253	91	91	0	102	60	10	50	0
-0599 to -0500	255	95	95	0	100	60	11	49	0
-0499 to -0400	242	88	86	2	90	64	19	45	1
-0399 to -0300	229	80	79	1	61	88	54	34	7
-0299 to -0200	228	79	76	3	64	85	55	30	6
-0199 to -0100	240	85	84	1	86	69	21	48	3
-0099 to 0000	253	94	93	1	98	61	11	50	0
0001 to 0100	251	93	92	1	101	57	7	50	0
0101 to 0200	239	84	82	2	88	67	23	44	3
0201 to 0300	228	83	75	8	63	82	53	29	6
0301 to 0400	227	76	72	4	69	82	58	24	4
0401 to 0500	244	83	82	1	95	66	18	48	3
0501 to 0600	254	97	96	1	95	62	12	50	0
0601 to 0700	255	97	94	3	100	58	10	48	0
0701 to 0800	240	84	82	2	88	68	23	45	3
0801 to 0900	228	79	74	5	60	89	65	24	8
0901 to 1000	230	82	79	3	70	78	45	33	6
1001 to 1100	246	87	84	3	97	62	14	48	0
1101 to 1200	259	98	98	0	102	59	10	49	0
1201 to 1300	251	95	94	1	96	60	12	48	0
1301 to 1400	231	81	73	8	73	77	41	36	6
1401 to 1500	228	80	77	3	65	83	60	23	4
1501 to 1600	233	82	79	3	74	77	39	38	6
1601 to 1700	249	91	88	3	97	61	11	50	0
1701 to 1800	256	98	98	0	98	60	13	47	0
1801 to 1900	249	90	88	2	97	62	11	51	0
1901 to 2000	229	83	74	9	65	81	48	33	5
2001 to 2100	228	86	81	5	57	85	61	24	8
2101 to 2200	238	81	77	4	88	69	24	45	4
2201 to 2300	252	94	93	1	97	61	11	50	0
2301 to 2400	253	95	95	0	98	60	12	48	0
2401 to 2500	237	83	77	6	85	69	23	46	4
2501 to 2600	227	82	77	5	58	87	64	23	8
2601 to 2700	230	81	80	1	70	79	44	35	7
2701 to 2800	241	86	83	3	91	64	16	48	0
2801 to 2900	259	100	99	1	102	57	9	48	0
2901 to 3000	249	89	89	0	97	63	13	50	1

Figure 3–1. Number of Lunar Eclipses Per Century: -1999 to +3000

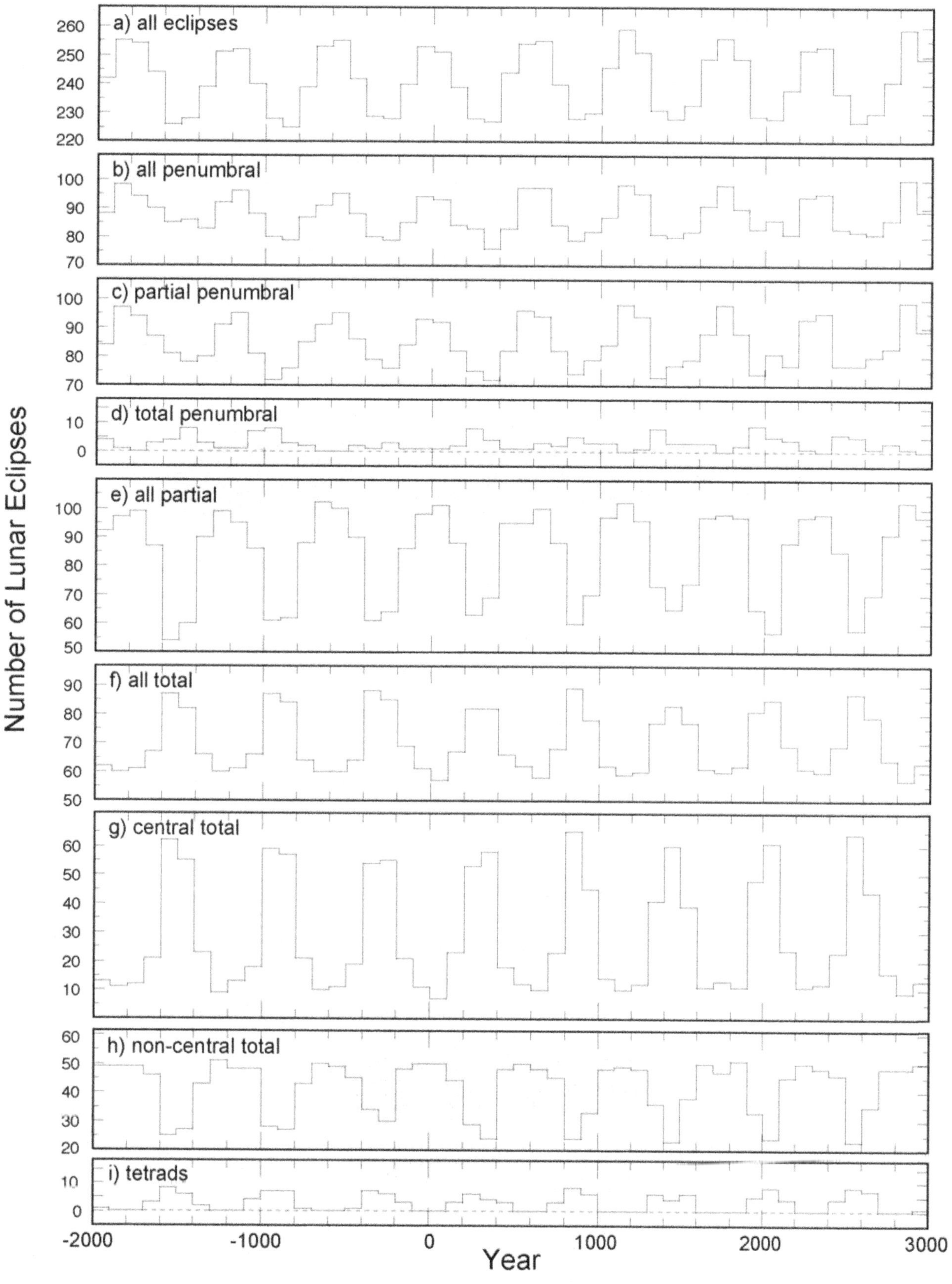

3.3 Distribution of Lunar Eclipse Types by Month

Table 3-5 summarizes 5,000 years of eclipses by eclipse type in each month of the year. The first value in each column is the number of eclipses of a given type for the corresponding month. The second value, in square brackets [ ], is the number of eclipses divided by the number of days in that month. This normalization allows direct comparison of eclipse frequencies in different months.

A brief examination of the normalized values in the column "Number of All Lunar Eclipses" shows that eclipses are equally distributed around the year. The same holds true for partial eclipses; however, the columns for penumbral and total eclipses reveal something interesting. Penumbral eclipses are about 6% more likely during the period of December– January–February compared to the months June–July–August. This effect is attributed to Earth's elliptical orbit. Earth currently reaches perihelion in early January and aphelion in early July. Consequently, the Sun's apparent diameter varies from 1,952 to 1,887 arcsec between perihelion and aphelion. The Sun's larger apparent diameter at perihelion makes Earth's penumbral shadow larger so penumbral eclipses are more frequent at that time.

The opposite argument holds true for total eclipses that are about 4% more likely during the period April to August, compared to the months November to February. In this case, the Sun's smaller apparent size around aphelion increases the diameter of Earth's umbral shadow so the frequency of total eclipses is slightly higher at that time.

Table 3-5. Lunar Eclipse Types by Month: –1999 to +3000 (2000 BCE to 3000 CE)

Month	Number of All Lunar Eclipses	Number of Penumbral Eclipses	Number of Partial Eclipses	Number of Total Eclipses
January	1027 [33.1]	379 [12.2]	352 [11.4]	296 [9.5]
February	936 [33.4]	343 [12.2]	333 [11.9]	260 [9.3]
March	1028 [33.2]	376 [12.1]	361 [11.6]	291 [9.4]
April	986 [32.9]	354 [11.8]	342 [11.4]	290 [9.7]
May	1025 [33.1]	370 [11.9]	358 [11.5]	297 [9.6]
June	992 [33.1]	351 [11.7]	347 [11.6]	294 [9.8]
July	1025 [33.1]	368 [11.9]	356 [11.5]	301 [9.7]
August	1015 [32.7]	361 [11.6]	355 [11.5]	299 [9.6]
September	990 [33.0]	364 [12.1]	341 [11.4]	285 [9.5]
October	1023 [33.0]	377 [12.2]	356 [11.5]	290 [9.4]
November	993 [33.1]	350 [11.7]	357 [11.9]	286 [9.5]
December	1024 [33.0]	385 [12.4]	349 [11.3]	290 [9.4]

3.4 Lunar Eclipse Frequency and the Calendar Year

There are 2 to 5 lunar eclipses in every calendar year. Table 3-6 shows the distribution in the number of eclipses per year for the 5,000 years covered in the Canon.

Table 3-6. Number of Lunar Eclipses per Year

Number of Eclipses per Year	Number of Years	Percent
2	3,541	70.8%
3	887	17.7%
4	539	10.8%
5	33	0.7%

When two eclipses occur in one calendar year, they can be in any combination of N, P, or T (penumbral, partial, or total, respectively). Table 3-7 lists the frequency of each eclipse combination, along with five recent years when the combination occurs. The table makes no distinction in the order of any two eclipses. For example, the eclipse combination PT includes all years where the order is either PT or TP.

Over 71% of all years containing two eclipses are composed of the combinations PP (34.2%) or TT (37.2%). The NT pair is the rarest combination with only six instances, all within the first two millennia of the Canon, and with five of the six involving a total penumbral eclipse.

Table 3-7. Two Lunar Eclipses in One Year

Eclipse Combinations[a]	Number of Years	Percent	Examples (Years)[b]
NN	116	3.3%	..., 1987, 2016, 2042, 2045, 2053, ...
NP	373	10.5%	..., 1999, 2005, 2006, 2012, 2017, ...
NT	6	0.2%	[-1989, -1971, -1423, -1403, -1385, -0186]
PP	1211	34.2%	..., 1980, 2039, 2041, 2046, 2057, ...
PT	516	14.6%	..., 1997, 2008, 2010, 2019, 2021, ...
TT	1319	37.2%	..., 2003, 2007, 2011, 2014, 2015, ...

a. N = Penumbral, P = Partial, and T = Total.
b. When years are surrounded by square brackets [ ], there are no other examples outside this range.

When three eclipses occur in one calendar year, there are nine possible combinations of N, P, or T. Table 3-8 lists the frequency of each eclipse combination along with five recent years when each combination occurs. The table makes no distinction in the order of eclipses in any combination. For example, the eclipse combination NPT includes all years where the order is NPT, NTP, PNT, PTN, TPN, and TNP. The rarest combinations—NNT and NTT—each occurs six times or less in the five-millennium span of this work. Interestingly, no PPP combinations occur.

Table 3-8. Three Lunar Eclipses in One Year

Eclipse Combinations[a]	Number of Years	Percent	Examples (Years)[b]
NNN	382	43.1%	..., 2002, 2027, 2031, 2049, 2060, ...
NNP	159	17.9%	..., 1958, 2013, 2147, 2168, 2186, ...
NNT	4	0.5%	[-1991, -1405, -0168, 0418]
NPP	158	17.8%	..., 1871, 2075, 2140, 2151, 2169, ...
NPT	27	3.0%	..., 1963, 2001, 2048, 2066, 2504, ...
NTT	6	0.7%	[-1953, -1432, -1367, -0846, 0977, 2466]
PPT	51	5.7%	..., 1898, 2028, 2113, 2178, 2531, ...
PTT	75	8.5%	..., 1852, 2094, 2159, 2224, 2243, ...
TTT	25	2.8%	..., 1917, 1982, 2485, 2550, 2615]

a. N = Penumbral, P = Partial, and T = Total.
b. When a year is bounded by a square bracket "[" or "]", there are no other examples beyond that year.

When four eclipses occur in one calendar year, there are three possible combinations of eclipse types N, P, and T. Table 3-9 lists the frequency of each eclipse combination along with five recent years when each combination occurs. The table makes no distinction in the order of eclipses in the seven combinations. The rarest combination NNPP occurs just 32 times.

Table 3-9. Four Lunar Eclipses in One Year

Eclipse Combinationsa	Number of Years	Percent	Examples (Years)
NNNN	390	72.4%	..., 1944, 2020, 2038, 2056, 2085, ...
NNNP	117	21.7%	..., 1991, 2009, 2150, 2197, 2215, ...
NNPP	32	5.9%	..., 1684, 2205, 2411, 2726, 2819, ...

The maximum number of five lunar eclipses in one calendar year is quite rare. Over the 5,000-year span of the Canon, there are only 33 years containing five lunar eclipses. They occur in two possible combinations of eclipse types where four out of the five eclipses are always of type N. The first eclipse of such a quintet always occurs in the first half of January, while the last eclipse falls in the latter half of December. Table 3-10 lists the frequency of the two eclipse combinations, along with recent years when each combination occurs. The rarest combination NNNPP (actually either PNPNN or NNPNP) occurred just three times. Once again, the table makes no distinction in the order of eclipses in any combination.

Table 3-10. Five Lunar Eclipses in One Year

Eclipse Combinations[a]	Number of Years	Percent	Examples (Years)[b]
NNNNP	30	90.9%	..., 1879, 2132, 2262, 2400, 2653, ...
NNNPP	3	9.1%	[0475, 1694, 1749]

a. N = Penumbral, P = Partial, and T = Total.
b. When years are surrounded by square brackets [ ], there are no other examples outside this range.

3.5 Extremes in Eclipse Magnitude—Penumbral Lunar Eclipses

The penumbral eclipse magnitude is defined as the fraction of the Moon's diameter immersed in Earth's penumbral shadow. It is a unitless parameter that can be expressed numerically by

$$Mp = xp/dm, \qquad (3\text{-}1)$$

where Mp is the penumbral eclipse magnitude, dm is the apparent diameter of the Moon, and xp is the distance measured from the edge of the penumbral shadow to the edge of the Moon deepest in the penumbra.

The penumbral eclipse magnitude reaches its maximum value at the instant of greatest eclipse. This is the value shown in the figures in the Appendix. A search through the 12,064 eclipses in the Canon reveals some interesting cases involving extreme values of the penumbral eclipse magnitude.

Seven penumbral eclipses have a maximum magnitude less than 0.0020 (Table 3-11). These events are all the first or last members in a Saros series. The smallest magnitude was the penumbral eclipse of –0780 Dec 13 with a magnitude of just 0.0004.

Table 3-11. Penumbral Lunar Eclipses with Magnitude ≤ 0.0020

Date (Dynamical Time)	Saros	Gamma	Penumbral Eclipse Magnitude	Penumbral Eclipse Duration
-0780 Dec 13	61	-1.5529	0.0004	5.2m
-0411 Jan 23	27	1.5791	0.0013	11.0m
-0331 May 07	74	1.5502	0.0008	7.8m
-0103 Feb 07	80	1.5540	0.0019	12.0m
0859 May 20	112	1.5700	0.0007	8.1m
2027 Jul 18	110	-1.5758	0.0014	11.8m
2791 Feb 11	175	-1.5670	0.0006	7.4m

Table 3-12 lists the nine penumbral eclipses having a maximum magnitude greater than 1.0800. The greatest penumbral eclipse occurred on 1322 Nov 24 with a maximum magnitude of 1.0951. Because the penumbral magnitudes of these eclipses are all greater than 1, they are classified as total penumbral eclipses, i.e., the Moon is completely immersed within the penumbral shadow.

Table 3-12. Total Penumbral Lunar Eclipses with Magnitude ≥ 1.0800

Date (Dynamical Time)	Saros	Gamma	Penumbral Eclipse Magnitude	Penumbral Eclipse Duration
-1517 Jan 12	0	-0.9901	1.0858	291.2m
-1348 Jan 03	32	0.9948	1.0831	294.4m
-1058 Oct 29	19	0.9917	1.0860	291.6m
0348 Oct 24	90	0.9890	1.0946	296.1m
1322 Nov 24	95	0.9897	1.0951	296.5m
1340 Dec 04	95	0.9970	1.0820	295.1m
1988 Mar 03	113	0.9885	1.0907	293.8m
2429 Dec 11	132	-0.9904	1.0853	290.1m
2732 Apr 02	163	-0.9928	1.0814	293.2m

3.6 Extremes in Eclipse Magnitude—Partial Lunar Eclipses

The umbral eclipse magnitude is defined as the fraction of the Moon's diameter immersed in Earth's umbral shadow. It is a unitless parameter that can be expressed numerically by

$$Mu = xu/dm \qquad (3\text{-}2)$$

where Mu is the umbral eclipse magnitude, dm is the apparent diameter of the Moon, and xu is the distance measured from the edge of the umbral shadow to the edge of the Moon deepest in the umbra.

The umbral eclipse magnitude reaches its maximum value at the instant of greatest eclipse. This is the value shown in the figures in the Appendix. A search through the 12,064 eclipses in the Canon reveals some interesting cases involving extreme values of the umbral eclipse magnitude in partial and total lunar eclipses.

Eleven partial eclipses have a maximum magnitude (at greatest eclipse) less than or equal to 0.0020 (Table 3-13). The partial eclipse with the smallest magnitude (at greatest eclipse) occurred on 1553 Jul 25 and had a magnitude of just 0.0001.

Table 3-13. Partial Lunar Eclipses with Magnitude ≤ 0.0020

Date (Dynamical Time)	Saros	Gamma	Umbral Eclipse Magnitude	Partial Eclipse Duration
-0602 May 03	31	0.9940	0.0015	9.7m
-0519 May 15	32	-1.0261	0.0004	4.3m
-0292 Mar 27	66	1.0016	0.0010	7.5m
1050 Feb 09	108	0.9861	0.0013	8.9m
1416 Nov 05	97	1.0107	0.0019	9.9m
1430 Sep 02	128	1.0229	0.0008	6.3m
1553 Jul 25	131	-1.0253	0.0001	2.5m
1890 Nov 26	114	-0.9994	0.0017	9.8m
2157 Feb 24	145	-0.9868	0.0005	5.6m
2421 Jun 16	156	1.0225	0.0011	7.3m
2627 May 22	160	0.9914	0.0011	8.0m

Nine partial eclipses have a maximum magnitude (at greatest eclipse) greater than or equal to 0.9980 (Table 3-14). The partial eclipse with the largest magnitude (at greatest eclipse) occurred on –1972 Jun 27 with a magnitude of 0.9998.

Table 3-14. Partial Lunar Eclipses with Magnitude ≥ 0.9980

Date (Dynamical Time)	Saros	Gamma	Umbral Eclipse Magnitude	Partial Eclipse Duration
-1972 Jun 27	-2	-0.4505	0.9998	209.0m
-1021 May 16	24	-0.4758	0.9986	192.6m
-0247 Oct 03	52	-0.4532	0.9988	202.7m
0145 Sep 18	78	0.4511	0.9994	205.0m
0274 Apr 08	65	0.4523	0.9993	207.1m
0778 Sep 11	98	0.4670	0.9993	195.0m
1001 Mar 12	87	0.4728	0.9995	191.1m
1165 May 27	107	-0.4727	0.9986	194.5m
2413 Nov 08	152	0.4733	0.9993	189.5m

3.7 Extremes in Eclipse Magnitude—Total Lunar Eclipses

Twelve total eclipses have a maximum magnitude less than or equal to 1.0020 (Table 3-15). The smallest magnitude was the total eclipse of 1529 Oct 17 with a value of just 1.0001 (note the upcoming total eclipse of 2015 Apr 04 and its magnitude of 1.0008). Because the Moon's passage through the umbral shadow is so shallow, these events all have short total phases less than 7 min.

Table 3-15. Total Lunar Eclipses with Magnitude ≤ 1.0020

Date (Dynamical Time)	Saros	Gamma	Umbral Eclipse Magnitude	Total Eclipse Duration
-1573 Jan 31	18	0.4776	1.0008	4.2m
-1338 Jun 09	28	0.4622	1.0012	5.5m
-0318 Aug 10	62	0.4474	1.0009	4.9m
-0200 Mar 30	48	-0.4790	1.0019	6.6m
0767 Apr 19	92	0.4581	1.0002	2.4m
0792 Dec 03	96	0.4413	1.0004	3.1m
1529 Oct 17	109	0.4775	1.0001	1.7m
2015 Apr 04	132	0.4460	1.0008	4.7m
2155 Sep 11	130	-0.4752	1.0003	2.6m
2366 May 25	146	0.4817	1.0007	3.9m
2565 Sep 11	156	0.4700	1.0009	4.6m
2905 Jun 08	164	0.4748	1.0018	6.5m

Eight total eclipses have a maximum magnitude greater than or equal to 1.8700. Their total phase durations all last 98 to 99 min and gamma values are all close to 0.0 indicating that the Moon passes centrally through the middle of the umbral shadow. These eclipses all take place when the Moon is near perigee—the time when the ratio of the Moon's to umbra's diameters is at maximum. The total eclipse with the largest magnitude (1.8821) occurs on 2756 Jun 05.

Table 3-16. Total Lunar Eclipses with Magnitude ≥ 1.8700

Date (Dynamical Time)	Saros	Gamma	Umbral Eclipse Magnitude	Total Eclipse Duration
-0731 Apr 19	39	-0.0054	1.8766	99.4m
-0630 Nov 06	45	-0.0002	1.8703	98.3m
-0051 Apr 01	60	-0.0009	1.8774	99.7m
0564 Sep 06	85	0.0017	1.8764	98.9m
1092 Apr 24	97	0.0080	1.8702	99.3m
1226 Aug 09	106	-0.0021	1.8770	99.6m
1631 May 15	115	0.0052	1.8721	99.8m
2756 Jun 05	152	0.0018	1.8821	99.3m

3.8 Greatest Duration—Penumbral Lunar Eclipses

Nine penumbral eclipses each have durations of 292 min or more. Because the penumbral eclipse magnitude of each of these events is greater than 1, they are all classified as total penumbral eclipses. Each eclipse in Table 3-17 occurs near lunar apogee and within several weeks of aphelion. Earth's penumbra is then largest, while the Moon exhibits a small angular diameter and travels relatively slowly, thereby extending the length of the eclipse.

Table 3-17. Penumbral Lunar Eclipses with a Duration ≥ 292 min

Date (Dynamical Time)	Saros	Gamma	Penumbral Eclipse Magnitude	Penumbral Eclipse Duration
-1367 Dec 23	32	1.0089	1.0588	292.2m
-1348 Jan 03	32	0.9948	1.0831	294.4m
0348 Oct 24	90	0.9890	1.0946	296.1m
1322 Nov 24	95	0.9897	1.0951	296.5m
1340 Dec 04	95	0.9970	1.0820	295.1m
1358 Dec 16	95	1.0039	1.0693	293.6m
1376 Dec 26	95	1.0104	1.0569	292.1m
1988 Mar 03	113	0.9885	1.0907	293.8m
2732 Apr 02	163	-0.9928	1.0814	293.2m

3.9 Greatest Duration—Partial Lunar Eclipses

Seven partial eclipses have durations of 209 min or more. All of these events occur with the Moon close to apogee so its slow trajectory through the umbra tends to prolong the partial phase.

Table 3-18. Partial Lunar Eclipses with a Duration ≥ 209 min

Date (Dynamical Time)	Saros	Gamma	Umbral Eclipse Magnitude	Partial Eclipse Duration
-1274 Jun 21	19	0.4611	0.9768	209.3m
-1133 May 24	22	-0.4592	0.9814	209.6m
-0764 Jan 26	31	0.4540	0.9771	209.1m
-0387 Oct 20	59	-0.4458	0.9908	209.3m
0303 Sep 12	71	0.4510	0.9864	209.6m
1382 Sep 23	117	-0.4458	0.9977	209.6m
2669 Feb 08	144	-0.4424	0.9951	210.0m

3.10 Greatest Duration—Total Lunar Eclipses

Eighteen total eclipses have durations greater than or equal to 106 min. The most recent one was 2000 Jul 16, while the next is 2123 Jun 09. All of these eclipses occur near aphelion and with the Moon close to apogee. Consequently, the Moon has a small angular diameter coupled with a relatively slow orbital motion with respect to the umbra. Furthermore, the umbral shadow is largest when Earth is at aphelion. Such conditions prolong the Moon's passage through the umbra to produce long total lunar eclipses (Meeus, 2002).

Earth's tropical longitude of perihelion is currently increasing by 1.72° per century. This means that the date of aphelion occurred ~17 days earlier for each 1000 years backwards from the present. This too is reflected in the dates seen in Table 3-19.

Table 3-19. Total Lunar Eclipses with a Duration ≥ 106 min

Date (Dynamical Time)	Saros	Gamma	Umbral Eclipse Magnitude	Total Eclipse Duration
-1921 Apr 07	1	0.0302	1.7681	106.2m
-1646 Jun 24	13	0.0164	1.7946	106.3m
-1505 May 27	16	-0.0171	1.7948	106.5m
-1364 Apr 28	19	0.0187	1.7910	106.4m
-0380 Jun 06	53	-0.0308	1.7703	106.2m
-0239 May 09	56	0.0301	1.7713	106.1m
-0098 Apr 11	59	-0.0016	1.8212	106.1m
0054 Aug 07	68	-0.0147	1.7924	106.2m
0177 Jun 28	71	-0.0198	1.7874	106.5m
0318 May 31	74	0.0160	1.7951	106.6m
0459 May 03	77	0.0033	1.8167	106.5m
1443 Jun 12	111	-0.0098	1.8097	106.2m
1584 May 24	114	-0.0065	1.8145	106.1m
1859 Aug 13	126	0.0038	1.8148	106.5m
2000 Jul 16	129	0.0302	1.7684	106.4m
2123 Jun 09	132	0.0406	1.7487	106.1m
2141 Jun 19	132	-0.0446	1.7415	106.1m
2264 May 12	135	0.0121	1.7979	106.2m

3.11 Total Penumbral Lunar Eclipses

During most penumbral eclipses, only part of the Moon passes through Earth's penumbral shadow. However, it is also possible to have a penumbral eclipse in which the entire Moon passes completely within Earth's penumbral shadow without entering the inner umbral shadow. The geocentric apparent diameter of the Moon ranges from an extreme minimum of 1763.0 arcsec (apogee) to an extreme maximum of 2011.8 arcsec (perigee). The penumbral annulus formed by the zone between the outer edges of the penumbra and umbra also undergoes extremes ranging from 1887.7 arcsec (aphelion) to 1951.9 arcsec (perihelion). From these values, it is apparent that the Moon cannot fit entirely within the penumbral annulus when it is near perigee (Meeus, 1997). As a consequence of the restrictive geometry and conditions, total penumbral eclipses are quite rare and account for just 3.2 %, or 141 out of 4378 penumbral eclipses in the Canon. Table 3-20 lists the dates of all 141 total penumbral eclipses.

The frequency of total penumbral eclipse varies considerably with time. If the 5,000-year period covered by the Canon is divided into 100-year intervals, the number of total penumbral eclipses per century varies from 0 to a maximum of 9 (Table 3.4). For instance, the number of total penumbral eclipses in centuries beginning in 1701, 1801, and 1901 are 0, 2, and 9, respectively. The number of total penumbral eclipses appears to be correlated with the number of total eclipses, as well as the number of tetrads per century, and inversely correlated with the overall number of all lunar eclipses per century. When a century has a relatively large number of lunar eclipses, it has fewer total lunar eclipses and few or no total penumbral lunar eclipses or tetrads (Figure 3.1)

Table 3-20. Total Penumbral Lunar Eclipses: –1999 to +3000 (2000 BCE to 3000 CE)

-1989 Dec 01	-1142 Jun 02	-0274 Oct 02	0682 Apr 27	1455 Oct 25	2103 Jan 23
-1981 Jul 07	-1058 Oct 29	-0263 Aug 31	0711 Oct 01	1496 Jul 25	2121 Feb 02
-1971 Dec 12	-1040 Nov 08	-0205 Jan 26	0729 Oct 11	1502 Apr 22	2128 Mar 16
-1953 Dec 23	-1022 Nov 19	-0186 Jun 22	0805 Mar 19	1513 Sep 15	2139 Feb 13
-1840 Jun 08	-1018 Mar 14	-0045 May 25	0816 Aug 11	1542 Aug 25	2222 Aug 23
-1699 May 11	-1007 Aug 07	0096 Apr 26	0822 May 09	1607 Mar 13	2429 Dec 11
-1669 Sep 05	-1004 Nov 30	0154 Sep 09	0823 Sep 24	1637 Jul 07	2447 Dec 22
-1604 Mar 13	-1001 May 05	0197 Nov 12	0862 Sep 12	1665 Jul 27	2458 May 28
-1575 Aug 17	-0986 Dec 11	0215 Nov 24	0945 Mar 31	1806 Jun 30	2466 Jan 01
-1564 Jul 16	-0968 Dec 21	0219 Mar 18	0974 Sep 04	1900 Jun 13	2484 Jan 13
-1558 Apr 13	-0964 Oct 10	0233 Dec 04	0996 Jan 08	1901 May 03	2498 Sep 30
-1517 Jan 12	-0949 Jan 02	0248 Aug 21	1014 Jan 19	1908 Dec 07	2502 Jan 24
-1499 Jan 22	-0946 Oct 21	0251 Dec 15	1032 Jan 30	1926 Dec 19	2520 Feb 04
-1457 Oct 30	-0943 Sep 17	0265 Apr 18	1097 Jul 27	1944 Dec 29	2523 Nov 24
-1452 Jul 08	-0931 Jan 12	0269 Dec 26	1221 May 08	1948 Oct 18	2538 Feb 15
-1439 Nov 10	-0925 Sep 29	0288 Jan 06	1322 Nov 24	1963 Jan 09	2562 Nov 12
-1435 Mar 05	-0860 Apr 06	0312 Oct 02	1340 Dec 04	1981 Jan 20	2656 Oct 25
-1421 Nov 21	-0838 Jul 29	0330 Oct 13	1344 Mar 29	1988 Mar 03	2714 Mar 23
-1406 Aug 09	-0831 Sep 10	0348 Oct 24	1358 Dec 16	1999 Jan 31	2732 Apr 02
-1403 Dec 01	-0755 Feb 15	0388 Mar 09	1373 Sep 02	2006 Mar 14	2761 Sep 06
-1385 Dec 13	-0726 Jul 22	0417 Aug 13	1376 Dec 26	2053 Aug 29	2885 Jun 18
-1367 Dec 23	-0461 Apr 05	0512 Jun 15	1390 Apr 29	2070 Apr 25	
-1348 Jan 03	-0432 Sep 08	0653 May 18	1395 Jan 06	2082 Aug 08	
-1283 Jun 30	-0356 Feb 14	0681 Jun 07	1413 Jan 17	2099 Sep 29	

3.12 Lunar Eclipse Duos

A duo is a pair of eclipses separated by one lunation (synodic month). Of the 12,064 eclipses in the Canon, 3,054 eclipses (25.3%) belong to a duo. One eclipse of a duo always passes north of Earth's shadow axis, while the other eclipse passes to the south. In most cases, both eclipses in a duo are penumbral eclipses; however, there are 51 instances (3.3% of duos) in the Canon where one eclipse is penumbral and the other is partial. In each of these pairs, the penumbral magnitude of the penumbral eclipse is quite small, as is the umbral magnitude of the partial eclipse. The dates and eclipse combinations of all lunar eclipse duos are listed in Table 3-21.

Table 3-21. Lunar Eclipse Duos of Two Types

–1952 May-Jun - NP	–0668 May-Jun - NP	0475 Jun-Jul - NP	1296 Apr-May - NP	2288 Jul-Aug - PN
–1811 Apr-May - NP	–0527 Apr-May - NP	0483 Jul-Aug - PN	1427 Apr-May - PN	2411 Jun-Jul - PN
–1774 May-May - PN	–0519 May-Jun - PN	0606 May-Jun - PN	1430 Aug-Sep - NP	2429 Jun-Jul - PN
–1670 Mar-Apr - NP	–0252 Jul-Jul - NP	0624 Jun-Jul - PN	1553 Jun-Jul - NP	2552 May-Jun - PN
–1492 Mar-Apr - PN	–0111 Jun-Jul - NP	0747 Apr-May - PN	1608 Jul-Aug - PN	2819 Jun-Jul - NP
–1358 Jun-Jul - PN	–0074 Jun-Jul - PN	0888 Mar-Apr - PN	1694 Jun-Jul - NP	2827 Jul-Aug - PN
–1217 May-Jun - PN	0030 May-Jun - NP	0891 Jul-Aug - NP	1749 Jun-Jul - PN	2960 Jun-Jun - NP
–1076 Apr-May - PN	0067 May-Jun - PN	1014 Jun-Jul - NP	1835 May-Jun - NP	
–0935 Mar Apr - PN	0153 Mar-Apr - NP	1022 Jul-Aug - PN	1958 Apr-May - NP	
–0932 Jul-Aug - NP	0208 Apr-May - PN	1155 May-Jun - NP	2013 Apr-May - PN	
–0809 Jun-Jul - NP	0428 Jun-Jul - NP	1163 Jun-Jul - PN	2147 Aug-Sep - PN	

3.13 Lunar Eclipses Duos in One Calendar Month

There are 57 instances where both members of an eclipse duo occur in one calendar month. In most cases, both eclipses in the duos are penumbral. In three instances, the duo consists of the NP combination (–1774 May, –0252 Jul, and 2960 Jun). The year and month of each duo appears in Table 3-22.

Table 3-22. Two Lunar Eclipses in One Calendar Month

-1883 Sep	-1033 Jan	-0078 Mar	0768 Sep	1694 Dec	2566 Aug
-1861 Jul	-0765 Aug	0019 Dec	0790 Jul	1705 Nov	2577 Jul
-1807 Aug	-0686 May	0074 Jan	0877 May	1716 Oct	2838 Jan
-1796 Jul	-0675 Apr	0269 Jul	1116 Jan	1817 May	2848 Dec
-1785 Jun	-0599 Mar	0291 May	1126 Dec	1904 Mar	2957 Aug
-1785 Jun	-0512 Jan	0356 May	1213 Oct	2172 Oct	2960 Jun
-1706 Mar	-0502 Dec	0367 Apr	1289 Sep	2208 May	2968 Jul
-1286 Aug	-0447 Jan	0443 Mar	1300 Aug	2284 Apr	
-1275 Jul	-0252 Jul	0595 Jan	1311 Jul	2295 Mar	
-1185 Mar	-0165 May	0605 Dec	1376 Jul	2382 Jan	

3.14 January–March Lunar Eclipse Duos

The mean length of one synodic month is 29.5306 days (in year 2000). Because this is longer than the month of February, it is possible to have one member of an eclipse duo in January followed by the second in March. There are six instances of such a rare January/March duo in the Canon: –1109, –0523, –0002, 1915, 2306, and 2371. In all cases, both eclipses in the duos are penumbral.

3.15 Total Lunar Eclipse Multiplets

A total lunar eclipse is usually preceded or succeeded by at least one other total lunar eclipse. Of the 3479 total eclipses in the Canon, 1798 of them (51.7%) are part of a doublet. Another 1023 eclipses (29.4%) belong to a triplet. Finally, 568 total eclipses (16.3%) are part of a quadruplet known as a tetrad. In comparison, only 90 total eclipses (2.6%) occur as solitary singlets.

A key feature of total lunar eclipse multiplets is that the individual members are always separated by six lunations. A summary of the muliplet statistics appears in Table 3-23.

Table 3-23. Total Lunar Eclipse Multiplets

Total Eclipse Multiplet	Number of Eclipses Per Multiplet	Number of Multiplets	Number of Total Eclipses	Percent of Total Eclipses	Recent Examples
All Total Eclipses	-	-	3479	100.0 %	
Singlet	1	90	90	2.6 %	1997 Sep 16; 2021 May 26
Doublet	2	899	1798	51.7 %	1978; 1996; 2022; 2040
Triplet	3	341	1023	29.4 %	2007 to 2008; 2010 to 2011
Tetrad	4	142	568	16.3 %	2003 to 2004; 2014 to 2015

3.16 Lunar Eclipse Tetrads

When four consecutive lunar eclipses are all total, the group is termed a tetrad. As discussed previously, 16.3% (568 out of 3479) of all total eclipses are members of a tetrad. These 142 groupings of total eclipses occur because of the eccentricity of Earth's orbit in conjunction with the timing of eclipse seasons (Section 3.18). During the first 1,000 years of the Canon, the first eclipse of every tetrad occurs sometime from December to May. In later millennia, the first eclipse date occurs later in the year because of precession—during the 3rd Millennium, the period for the first date extends from February to July. For a detailed description of tetrad geometry and an explanation of why tetrads happen, see Meeus (2004b).

Italian astronomer Giovanni Schiaparelli first pointed out that the frequency of tetrads is variable over time. He noticed that tetrads were relatively plentiful during one 300-year interval, while none occurred during the next 300 years. For example, there are no tetrads from 1582 to 1908, but 17 tetrads occur during the following 2 1/2 centuries from 1909 to 2156. This can be seen graphically in Figure 3-1 (i). The 565-year period of the tetrad "seasons" is tied to the slowly decreasing eccentricity of Earth's orbit. Consequently, the tetrad period is gradually decreasing (Hughes, 2004). In the distant future, when Earth's eccentricity is 0, tetrads will no longer be possible.

The umbral magnitudes of the total eclipses making up a tetrad are all relatively small. For the 300-year period 1901 to 2200, the largest umbral magnitude of a tetrad eclipse is 1.4251 on 1949 Apr 13. For comparison, some other total eclipses during this period are much deeper. Two examples are the total eclipses of 2000 Jul 16 and 2029 Jun 26 with umbral magnitudes of 1.7684 and 1.8436, respectively.

Table 3-24 lists the date of the first total lunar eclipse in each of the 142 tetrads in the Canon.

Table 3-24. Date of First Eclipse in Lunar Eclipse Tetrads

-1991 Dec 22	-0999 Mar 14	-0283 Mar 18	0766 Apr 29	1457 Mar 11	2137 Mar 07
-1661 Apr 12	-0981 Mar 25	-0272 Feb 16	0784 May 09	1475 Mar 22	2155 Mar 19
-1643 Apr 22	-0963 Apr 04	-0243 Jan 26	0795 Apr 09	1493 Apr 02	2448 Jun 17
-1625 May 04	-0945 Apr 16	-0225 Feb 06	0802 May 21	1504 Mar 01	2466 Jun 28
-1596 Apr 13	-0923 Feb 12	-0207 Feb 16	0813 Apr 19	1515 Jan 30	2477 May 28
-1585 Mar 13	-0916 Mar 26	-0189 Feb 28	0824 Mar 18	1522 Mar 12	2495 Jun 08
-1578 Apr 24	-0905 Feb 24	-0168 Dec 27	0842 Mar 30	1533 Feb 09	2506 May 08
-1574 Feb 10	-0887 Mar 06	-0149 Jan 07	0860 Apr 09	1562 Jan 20	2524 May 19
-1567 Mar 24	-0876 Feb 03	0162 Apr 17	0878 Apr 20	1580 Jan 31	2542 May 30
-1556 Feb 21	-0869 Mar 17	0180 Apr 27	0889 Mar 21	1909 Jun 04	2564 Mar 29
-1538 Mar 04	-0858 Feb 14	0198 May 08	0900 Feb 18	1927 Jun 15	2571 May 11
-1520 Mar 14	-0847 Jan 14	0227 Apr 19	0918 Feb 28	1949 Apr 13	2582 Apr 09
-1473 Mar 06	-0840 Feb 25	0238 Mar 18	0936 Mar 11	1967 Apr 24	2589 May 21
-1462 Feb 02	-0829 Jan 25	0245 Apr 29	0947 Feb 08	1985 May 04	2600 Apr 20
-1444 Feb 13	-0782 Jan 15	0256 Mar 28	0965 Feb 18	2003 May 16	2611 Mar 20
-1433 Jan 13	-0453 May 06	0267 Feb 26	0976 Jan 19	2014 Apr 15	2618 May 01
-1415 Jan 23	-0377 Apr 06	0285 Mar 08	0994 Jan 30	2032 Apr 25	2629 Mar 31
-1405 Dec 24	-0359 Apr 17	0332 Feb 28	1305 May 09	2043 Mar 25	2640 Feb 29
-1386 Jan 04	-0341 Apr 28	0350 Mar 10	1323 May 21	2050 May 06	2647 Apr 11
-1368 Jan 15	-0330 Mar 28	0361 Feb 06	1341 May 31	2061 Apr 04	2658 Mar 11
-1057 Apr 24	-0319 Feb 24	0390 Jan 17	1352 Apr 30	2072 Mar 04	2676 Mar 22
-1039 May 05	-0312 Apr 07	0408 Jan 29	1370 May 11	2090 Mar 15	2987 Jul 02
-1028 Apr 03	-0301 Mar 07	0437 Jan 08	1399 Apr 20	2101 Feb 14	
-1010 Apr 14	-0290 Feb 04	0455 Jan 19	1428 Mar 31	2119 Feb 25	

3.17 Lunar Eclipses on February 29

There are six instances of a lunar eclipse occurring on February 29. Two eclipses are penumbral, two are partial, and two are total. A list of eclipses on February 29 with physical parameters appears in Table 3-25.

Table 3-25. Lunar Eclipses on February 29

Date (Dynamical Time)	Eclipse Type	Saros	Gamma	Umbral Eclipse Magnitude
-0664 Feb 29	N	22	-1.3596	0.3819[a]
-0588 Feb 29	N	62	1.4929	0.1363[a]
-0124 Feb 29	P	50	-0.5634	0.8114
1420 Feb 29	P	94	-0.9205	0.1614
2268 Feb 29	T-	137	-0.1142	1.6602
2640 Feb 29	T	143	0.3353	1.2478

a. Penumbral Eclipse Magnitude

3.18 Eclipse Seasons

The 5.1° inclination of the lunar orbit around Earth means that the Moon's orbit crosses the ecliptic at two points or nodes. If Full Moon takes place within about 16° of a node[12], then a lunar eclipse will be visible from a portion of Earth.

The Sun makes one complete circuit of the ecliptic in 365.24 days, so its average angular velocity is 0.99° per day. At this rate, it takes 34.5 days for the Sun—and at the opposite node, Earth's umbral and penumbral shadows—to cross the 34° wide eclipse zone centered on each node. Because the Moon's orbit with respect to the Sun has a mean duration of 29.53 days, there will always be one, and possibly two, lunar eclipses during each 34.5-day interval when the Sun (and Earth's shadows) pass through the nodal eclipse zones. These time periods are called eclipse seasons.

The mid-point of each eclipse season is separated by 173.3 days because this is the mean time for the Sun to travel from one node to the next. The period is a little less than half a calendar year because the lunar nodes slowly regress westward by 19.3° per year.

3.19 Quincena

The mean time interval between New Moon and Full Moon is 14.77 days. This is less than half the duration of an eclipse season. As a consequence, the same Sun-node alignment geometry responsible for producing a lunar eclipse always results in a complementary solar eclipse within a fortnight. The solar eclipse may either precede or succeed the lunar eclipse. In either case, the pair of eclipses is referred to here as a quincena[13]. The Quincena Solar Eclipse (QSE) parameter identifies the type of the solar eclipse and whether it precedes or succeeds a particular lunar eclipse. There are four basic types of solar eclipses:

---

[12] The actual value ranges from 15.2° to 16.8° because of the eccentricity of the Moon's and Earth's orbits. If a lunar eclipse occurs within 9.4° to 11.0° of a node, the eclipse will be partial. A lunar eclipse within 4.2° to 5.0° central will be total.

[13] Quincena is a Spanish word for a period of about 15 days.

1) p = partial solar eclipse
(Moon's penumbral shadow traverses Earth; umbral/antumbral shadow completely misses Earth)
2) a = annular solar eclipse
(Moon's antumbral[14] shadow traverses Earth; Moon is too far from Earth to completely cover the Sun)
3) t = total solar eclipse
(Moon'sumbral shadow traverses Earth; Moon is close enough to Earth to completely cover the Sun)
4) h = hybrid solar eclipse
(Moon's umbral and antumbral shadows traverse different parts of Earth; eclipse appears either total or annular along different sections of its path—hybrid eclipses are also known as annular-total eclipses)

The QSE is a two character string consisting of one or more of the above solar eclipse types. The first character in the QSE identifies a solar eclipse preceding a lunar eclipse while the second character identifies a solar eclipse succeeding a lunar eclipse. In most instances, one of the two characters is "–" indicating a single solar eclipse either precedes or succeeds the lunar eclipse. On rare occasions, a double quincena occurs in which a lunar eclipse is both preceded and succeeded by solar eclipses.

3.20 Quincena Combinations with Total Lunar Eclipses

A total lunar eclipse can be preceded or succeeded by a total solar eclipse (7.7%), an annular solar eclipse (10.2%), a hybrid solar eclipse (0.4%), or a partial solar eclipse (42.5%). Double quincenas (a lunar eclipse is both preceded and succeeded by a solar eclipse) occur with a frequency of 39.1% and usually consist of two partial solar eclipses (38.7%). In rare instances, a double quincena consists of a total and a partial solar eclipse (0.4%). A complete list of all QSE combinations with total lunar eclipses appears in Table 3-26. The most recent years when each quincena combination occurs are given in the last column.

Table 3-26 Quincena Combinations with Total Lunar Eclipses

Quincena Solar Eclipse	QSE	Number	Percent	Examples (Years) [a]
– total	–t	127	3.7%	..., 1985, 2003, 2043, 2061, 2072,...
total –	t–	139	4.0%	..., 1957, 1968, 2015, 2033, 2044,...
– annular	–a	178	5.1%	..., 1891, 2003, 2014, 2021, 2032,...
annular –	a–	178	5.1%	..., 1990, 2008, 2026, 2044, 2102,...
– hybrid	–h	12	0.3%	..., 1627, 1645, 1768, 1909, 2050,...
hybrid –	h–	5	0.1%	[–1989, –1848, –1642, 0163, 1986]
– partial	–p	730	21.0%	..., 2000, 2007, 2010, 2014, 2018,...
partial –	p–	749	21.5%	..., 2001, 2004, 2011, 2015, 2019,...
total – partial	tp	8	0.2%	..., –0434, –0159, 1248, 1928, 2912]
partial – total	pt	6	0.2%	[–1310, –1169, –1028, –0026, 2195, 2459]
partial – partial	pp	1347	38.7%	..., 1982, 2000, 2011, 2018, 2029,...

a. When a year is bounded by a square bracket "[" or "]", there are no other examples beyond that year.

[14] The cone-shaped umbra gradually narrows to a point. Extending the sides of the umbra beyond this vertex forms an inverted cone. This zone is known as the antumbra. It corresponds to the region in the Moon's shadow where the Moon appears smaller than the Sun. The Moon is then seen in complete silhouette against the Sun's photospheric disk.

3.21 Quincena Combinations with Partial Lunar Eclipses

A partial lunar eclipse can be preceded or succeeded by a total solar eclipse (37.6%), an annular solar eclipse (55.1%), or a hybrid solar eclipse (6.9%). In rare instances, a partial lunar eclipse can be followed by a partial solar eclipse (0.3%). Double quincenas do not occur with partial lunar eclipses. A list of quincena solar eclipse combinations with partial lunar eclipses appears in Table 3-27. The most recent years when each quincena combination occurs are given in the last column.

Table 3-27. Quincena Combinations with Partial Lunar Eclipses

Quincena Solar Eclipse	QSE	Number	Percent	Examples (Years)
– total	–t	782	18.6%	..., 1992, 1999, 2010, 2017, 2021,...
total –	t–	801	19.0%	..., 2001, 2008, 2019, 2026, 2037,...
– annular	–a	1149	27.3%	..., 1995, 2006, 2009, 2013, 2024,...
annular –	a–	1171	27.8%	..., 1994, 2005, 2012, 2023, 2030,...
– hybrid	–h	167	4.0%	..., 1912, 1930, 2209, 2350, 2368,...
hybrid –	h–	124	2.9%	..., 1827, 1845, 2164, 2182, 2323,...
– partial	–p	13	0.3%	..., –0754, –0196, 2086, 2607, 2625,...

3.22 Quincena Combinations with Penumbral Lunar Eclipses

A penumbral lunar eclipse can be preceded or succeeded by a total solar eclipse (39.5%), an annular solar eclipse (51.9%), or a hybrid solar eclipse (8.7%). There are no instances of a quincena involving a partial solar and a penumbral lunar eclipse, nor are there any double quincenas in the Canon. A list of quincena solar eclipse combinations with penumbral lunar eclipses appears in Table 3-28. The most recent years when each quincena combination occurs are given in the last column.

Table 3-28. Quincena Combinations with Penumbral Lunar Eclipses

Quincena Solar Eclipse	QSE	Number	Percent	Examples (Years)
– total	–t	878	20.1%	..., 2002, 2006, 2009, 2020, 2024,...
total –	t–	848	19.4%	..., 1998, 2009, 2012, 2016, 2027,...
– annular	–a	1144	26.1%	..., 1999, 2002, 2017, 2020, 2031,...
annular –	a–	1128	25.8%	..., 2001, 2002, 2009, 2013, 2016,...
– hybrid	–h	176	4.0%	..., 1846, 1908, 2013, 2031, 2049,...
hybrid –	h–	204	4.7%	..., 1987, 2005, 2023, 2172, 2190,...

Section 4: Eclipses and the Moon's Orbit

4.1 Introduction

The Moon revolves around Earth in an elliptical orbit with a mean eccentricity of 0.0549. Thus, the Moon's center-to-center distance from Earth varies with mean values of 363,396 km at perigee to 405,504 km at apogee. The lunar orbital period with respect to the stars (sidereal month) is 27.32166 days (27d 07h 43m 12s). However, there are three other orbital periods or months that are crucial to the understanding and prediction of eclipses. These three cycles and the harmonics between them determine when, where, and how lunar (and solar) eclipses occur.

The mutual gravitational force between the Sun and Moon is over twice as large as between the Moon and Earth. For this reason, the Sun plays a dominant role in perturbing the Moon's motion. The ever changing distances and relative positions between the Sun, Moon, and Earth, the inclination of the Moon's orbit, the oblateness of Earth, and (to a lesser extent) the gravitational attraction of the other planets all act to throw the Moon's orbital parameters into a constant state of change. Although the Moon's position and velocity can be described by the classic Keplerian orbital elements, such osculating elements are only valid for a single instant in time (Chapront-Touze' and Chapront, 1991). Nevertheless, these instantaneous parameters are of value in understanding the Moon's complex motions particularly with respect to the three major orbital cycles that govern eclipses.

4.2 Synodic Month

The most familiar lunar cycle is the synodic month because it governs the well-known cycle of the Moon's phases. The Moon has no light of its own but shines by reflected sunlight. As a consequence, the geometry of its orbital position relative to the Sun and Earth determines the Moon's apparent phase.

The mean length of the synodic month is 29.53059 days (29d 12h 44m 03s). This is nearly 2.21 days longer than the sidereal month. As the Moon revolves around Earth, both objects also progress in orbit around the Sun. After completing one revolution with respect to the stars, the Moon must continue a little farther along its orbit to catch up to the same position it started from relative to the Sun and Earth. This explains why the mean synodic month is longer than the sidereal month.

According to astronomical convention, New Moon is defined as the instant when the geocentric ecliptic longitudes of the Sun and Moon are equal. When the synodic month is measured from New Moon to New Moon, it is sometimes referred to as a lunation, and we will follow that usage here. Historically, the phases of the Moon have been used as the basis of lunar calendars by many cultures around the world. The major problem with such calendars is that the year, based on the solar calendar, is not evenly divisible by a whole number of lunations. Consequently, most lunar calendars are actually lunisolar calendars (e.g., Chinese, Hebrew, and Hindu) that include intercalary months to keep the seasons in step with the year.

The duration of the lunation actually varies from its mean value by up to seven hours. For instance, Table 4-1 contains details for all lunations in 2008. The first column lists the decimal date of every New Moon throughout the year (Terrestrial Dynamical Time), while the second column gives the duration of each lunation. The third column is the difference between the actual and mean lunation. The first lunation of the year (Jan 08) was 03h 23m longer than the mean. Continuing through 2008, the length of each lunation drops and reaches a minimum of 05h 48m shorter than the mean value (Jun 03). The duration now increases with each succeeding lunation until the maximum value of the year is reached of 06h 49m longer than the mean (Dec 27).

Table 4-1 New Moon and Lunation Length in 2008

Date of New Moon (Dynamical Time)	Length of Lunation	Difference From Mean Lunation	Moon's True Anomaly
2008 Jan 08.4849	29d 16h 07m	+03h 23m	242.4°
2008 Feb 07.1567	29d 13h 30m	+00h 46m	280.0°
2008 Mar 07.7190	29d 10h 41m	-02h 03m	310.8°
2008 Apr 06.1642	29d 08h 23m	-04h 21m	332.7°
2008 May 05.5134	29d 07h 04m	-05h 40m	349.4°
2008 Jun 03.8081	29d 06h 56m	-05h 48m	4.4°
2008 Jul 03.0970	29d 07h 54m	-04h 50m	20.1°
2008 Aug 01.4261	29d 09h 45m	-02h 59m	39.2°
2008 Aug 30.8327	29d 12h 14m	-00h 30m	64.9°
2008 Sep 29.3426	29d 15h 02m	+02h 18m	98.7°
2008 Oct 28.9687	29d 17h 41m	+04h 57m	133.4°
2008 Nov 27.7053	29d 19h 28m	+06h 44m	161.9°
2008 Dec 27.5163	29d 19h 33m	+06h 49m	186.6°

What is the cause of this odd behavior? The last column in Table 4-1 gives a clue; it contains the Moon's true anomaly at the instant of New Moon. The true anomaly is the angle between the Moon's position and the point of perigee along its orbit. In other words, it is the orbital longitude of the Moon with respect to perigee. Table 4-1 shows that when New Moon occurs near perigee (true anomaly = 0°), the length of the lunation is at a minimum (e.g., Jun 03). Similarly, when New Moon occurs near apogee (true anomaly = 180°), the length of the lunation reaches a maximum (e.g., Dec 27).

Figure 4–1: Length of Lunation for 2008 – 2010

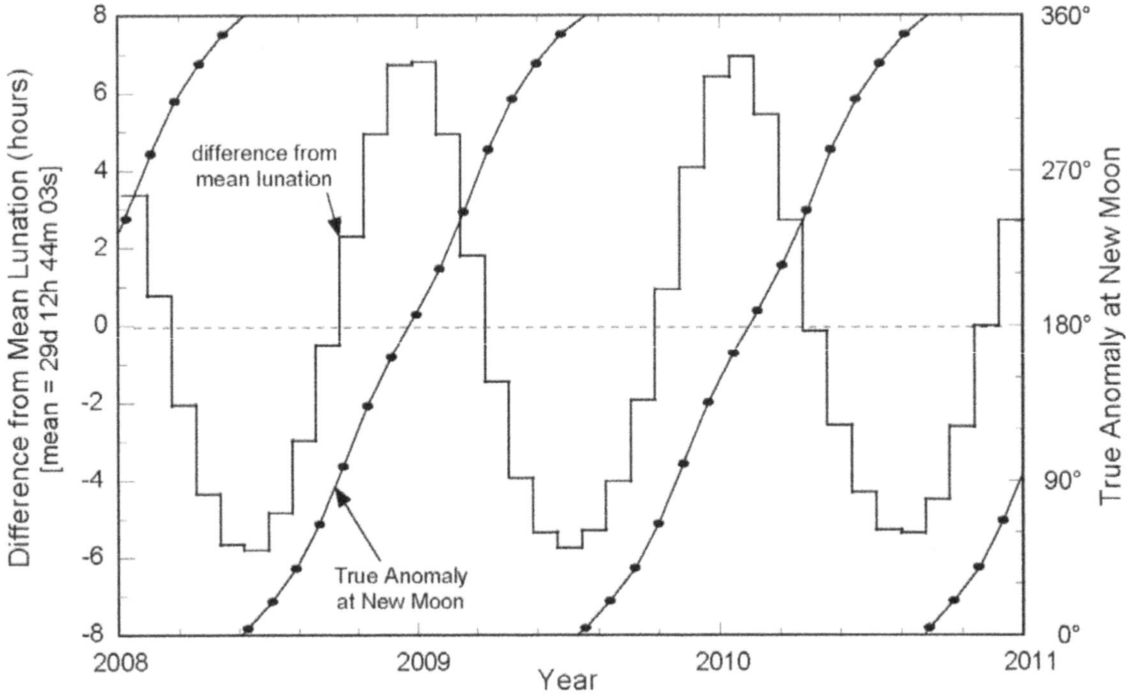

This relationship is quite apparent when viewed graphically. Figure 4-1 plots the difference from mean lunation (histogram) and the Moon's true anomaly (diagonal curves) for every New Moon from 2008 through 2010. The left-hand scale is for the difference from mean lunation, while the right-hand scale is for the true anomaly. The shortest lunations are clearly correlated with New Moon at perigee, while the longest lunations occur at apogee. From the figure, the length of this cycle appears to be about 412 days. The reason why must wait until the next section.

The Moon's orbital period with respect to perigee is the anomalistic month and has a duration of approximately 27.55 days. The lock-step rhythm between the lunation length and true anomaly can be explained with the help of the anomalistic month and Figure 4-2. It illustrates the Moon's orbit around Earth and Earth's orbit around the Sun. The relative sizes and distances of the Sun, Moon, and Earth as well as the eccentricity of the Moon's orbit are all exaggerated for clarity. The major axis of the Moon's orbit marks the positions of perigee and apogee.

Figure 4–2: Moon's Orbit and the Synodic Month

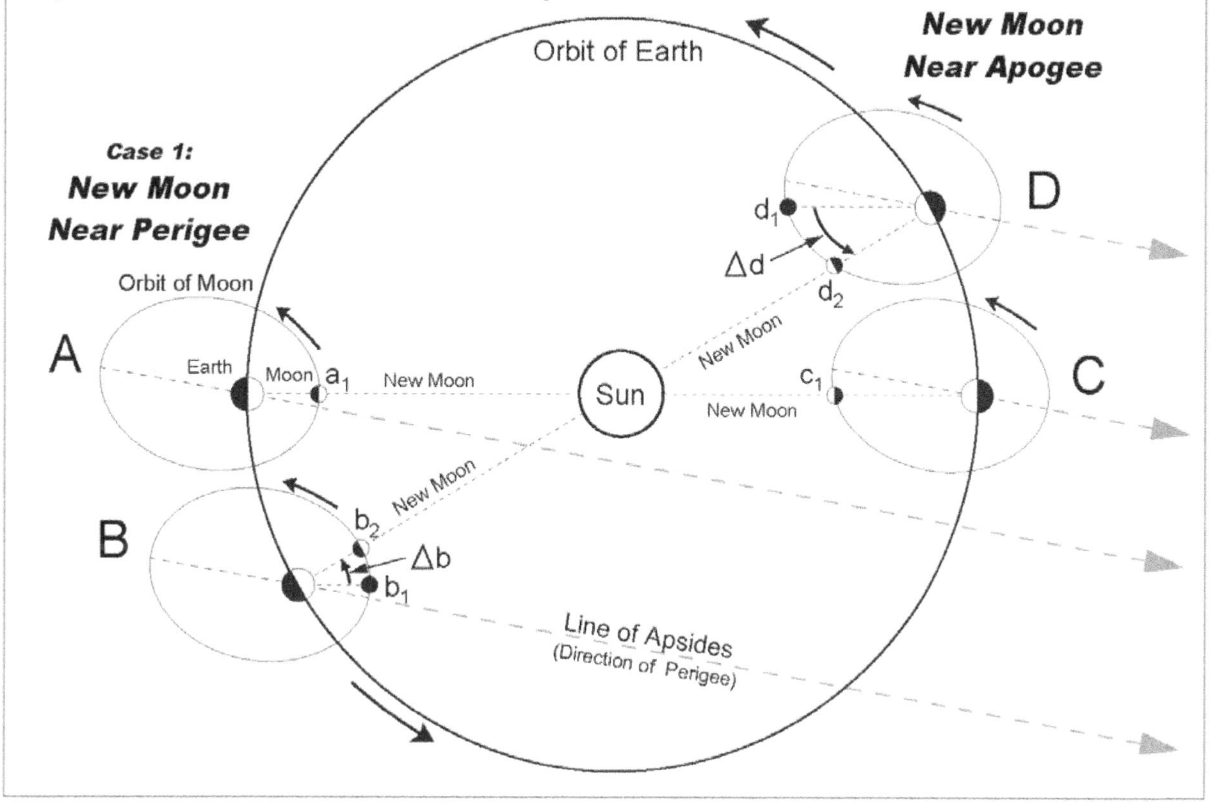

Two distinct cases—each consisting of two revolutions of the Moon around Earth—are depicted in Figure 4-2. The first case covers the New Moon geometry around perigee. The orbit marked A shows New Moon taking place near perigee at position a_1. One anomalistic month later (orbit B), the Moon has returned to the same position relative to perigee (marked b_1). However, Earth has traveled about 30° around its orbit so the Sun's direction relative to the Moon's major axis has shifted. The Moon must travel an additional distance of Δb in its orbit before reaching the New Moon phase at b_2. This graphically demonstrates why the synodic month is longer (~1.98 days) than the anomalistic month.

The second case takes place about half a year later. New Moon then occurs near apogee (orbit C, position c_1). After one anomalistic month, the Moon has returned to the same location with respect to apogee (orbit D, position d_1). Once again, Earth has traveled about 30° around its orbit so the Moon must revolve an additional distance of Δd before reaching the New Moon phase at position d_2.

An inspection of orbits B and D reveals that the orbital arc Δd is longer that Δb. This means that the Moon must cover a greater orbital distance to reach New Moon near apogee as compared to perigee. Furthermore, the Moon's orbital velocity is slower at apogee so it takes longer to travel a given distance. Thus, the length of the lunation is shorter than average when New Moon occurs near perigee and longer than average when New Moon occurs near apogee.

Earth's elliptical orbit around the Sun also factors into the length of the lunation. With an eccentricity of 0.0167, Earth's orbit is about one third as elliptical as the Moon's orbit. Nevertheless, it affects the length of the lunation by producing shorter lunations near aphelion and longer lunations near perihelion.

During the 5000-year period covered in this catalog, there are 61841 complete lunations. The shortest lunation began on –1602 Jun 03 and lasted 29.26574 days (29d 06h 22m 40s; 6h 21m 23s shorter than the mean). The longest lunation began on –1868 Nov 27 and lasted 29.84089 days (29d 20h 10m 53s; 7h 26m 50s longer than the mean). Thus, the duration of the lunation varies over a range of 13h 48m 13s during this time interval.

Figure 4–3: Length of Lunation Over 5000 Years

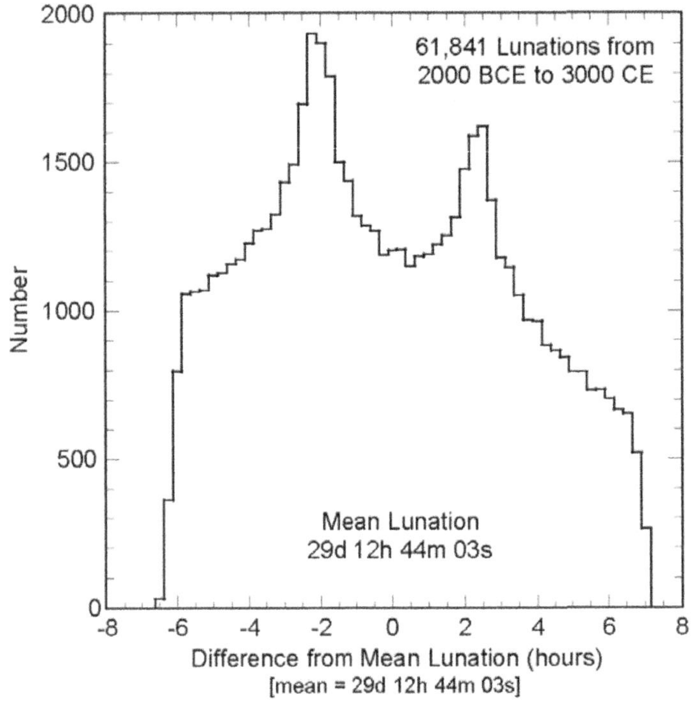

The histogram presented in Figure 4-3 shows the distribution in the length of the lunation over 5000 years. To create the histogram, the durations of individual lunations were binned into 30-minute groups. It might seem reasonable to expect a simple bell-shaped Gaussian curve. However, the results are surprising because the distribution in lunation length has two distinct peaks. This bifurcation can be understood if the lunation length, which depends primarily on the Moon's distance, is considered as a series of sine functions. The extremes of a sine function always occur more frequently than the mean, which is just what is seen in Figure 4-3. For a more detailed discussion, see Meeus (1997).

4.3 Anomalistic Month

The anomalistic month is defined as the revolution of the Moon around its elliptical orbit as measured from perigee to perigee. The length of this period can vary by several days from its mean value of 27.55455 days (27d 13h 18m 33s). Figure 4-4 plots the difference of the anomalistic month from the mean value for the 3-year interval 2008 through 2010. Also plotted is the difference between the mean longitudes of the Sun and perigee. This is just the angle between the Sun and the Moon's major axis in the direction of perigee. The left-hand scale is the length of the anomalistic month minus the mean value, while the right-hand scale is for the difference in longitude (Sun–perigee). For comparison, the lunation length minus its mean value is also plotted (light gray).

Figure 4–4: Length of Anomalistic Month for 2008 – 2010

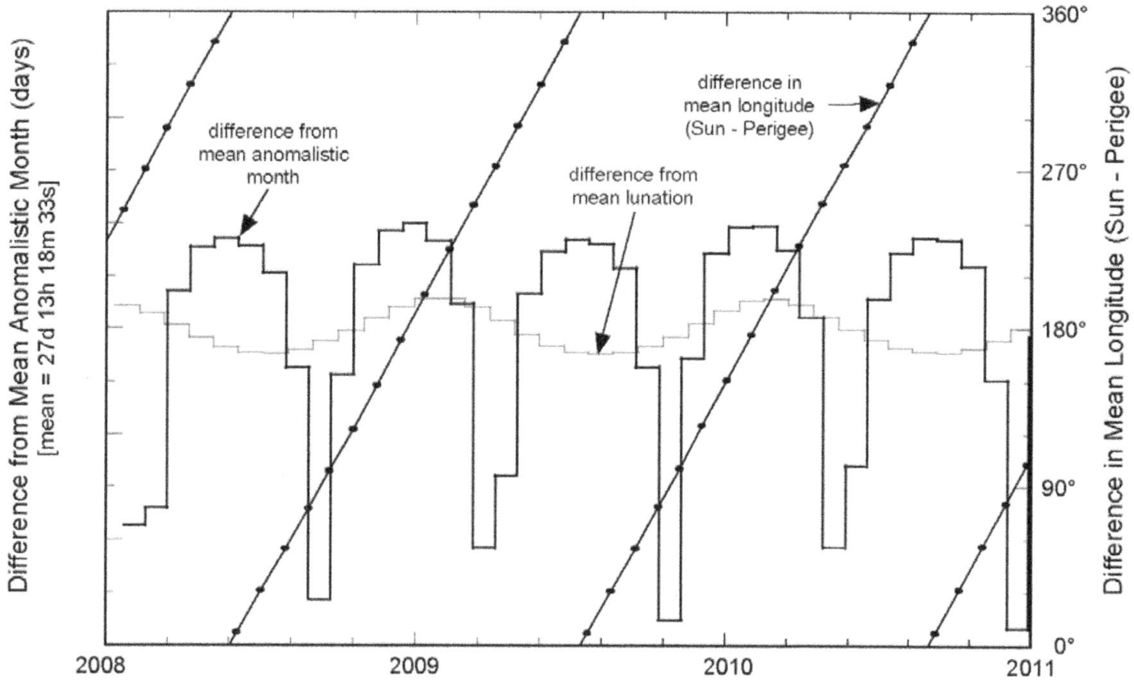

The variation in the length of the anomalistic month is much larger than that of the lunation. Figure 4-4 shows the anomalistic month is typically within 1 day of its mean value. But once or twice every 7 to 8 months, the anomalistic month is significantly shorter than the mean by 2 to nearly 3 days. The difference in longitude of the Sun and perigee show that the shortest anomalistic months are correlated with values of 90° and 270°, when the line of apsides is perpendicular to the Sun's direction.

In comparison, the longest anomalistic months take place when the difference in longitude passes through 0° or 180°. The line of apsides is then directed towards or away from the Sun. The maximum duration of the anomalistic month is then about 28.5 days (1.0 day longer than the mean). The Earth–Sun distance also influences the anomalistic month by causing greater extremes near perihelion. This currently occurs in early January each year.

In an earlier discussion on the synodic month, it was assumed that the lunar orbit's line of apsides has a fixed and permanent direction in space. In fact, the length of the mean anomalistic month (27.55 days) exceeds the mean sidereal month (27.32 days) by 0.23 days. Thus, the Moon's major axis slowly shifts with a mean rate of 0.11140° per day in the direct sense, that is, in the same direction as the Moon's orbital motion. This corresponds to an average of 40.7° per year, so it takes 8.85 years (3231.6 days) for the line of apsides to make one complete revolution with respect to the stars.

What impact do the varying length of the anomalistic month and the direct (eastward) rotation of the Moon's elliptical orbit have on the length of the lunation? To answer this, one must first consider Earth's elliptical orbit around the Sun, which has a mean eccentricity of 0.0167. The center-to-center distance between Earth and the Sun varies with mean values of 147,098,074 km at perihelion to 152,097,701 km at aphelion. The direction of Earth's orbital line of apsides also changes but at a rate far slower than the Moon's. Having a direct (eastward) shift with a mean value of 0.0172° per year, it takes about 20,500 years for Earth's major axis to make one complete revolution. This is only 0.0004 of the lunar rate, so it can be treated as fixed for the purpose of the following discussion.

At certain times, the perigee of the lunar orbit and the perihelion of Earth's orbit can have the same ecliptic longitude. Ignoring the 5.1° tilt of the Moon's obit, the major axes are then essentially parallel to each other and point in the same direction. As time passes, the major axis of the lunar orbit slowly rotates east with respect to Earth's major axis until it becomes perpendicular to it 2.21 years later. In another 2.21 years (4.42 years from the start), the major axes of the orbits are again parallel to each other, but the perigee and the perihelion are 180° apart as they point in opposite directions. After an additional period of 2.21 years, the axes are once more perpendicular. Finally, the Moon's perigee and Earth's perihelion again share the same ecliptic longitude after a total interval of 8.85 years.

Figure 4–5: Length of Lunation for 2008 – 2027

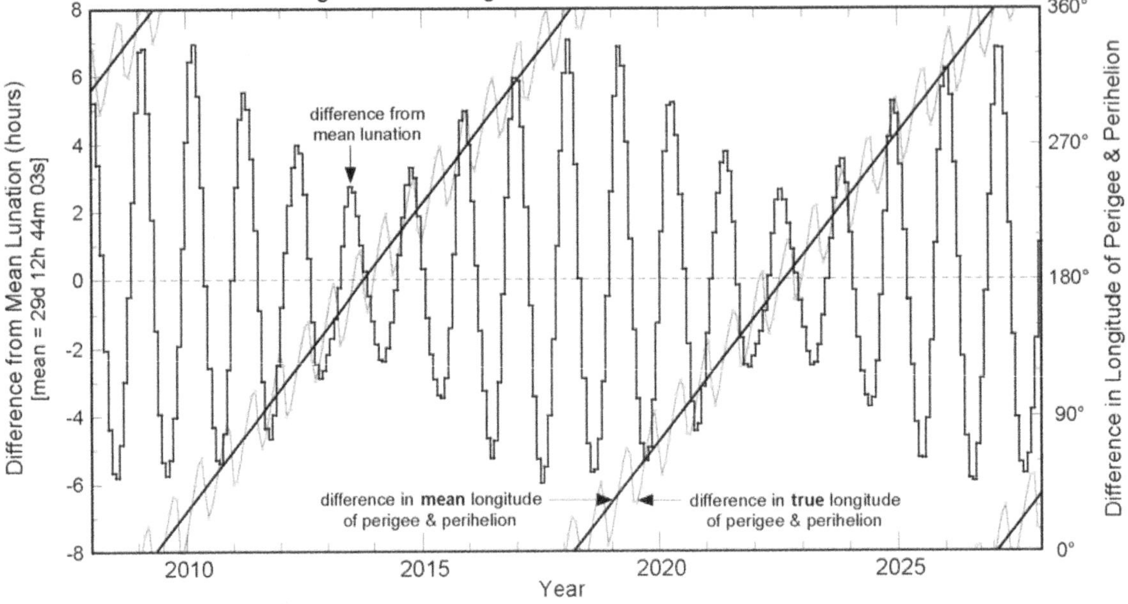

The length of each lunation minus the mean lunation is plotted in Figure 4-5 for the 20-year period from 2008 through 2027. The periodic rhythm between the lunation length and the true anomaly, as described earlier (via Figure 4-1), can now be seen over the course of two decades. The 412-day mean period of this cycle corresponds to the time between two consecutive alignments of the major axis in the direction of the Sun. It is slightly longer than a year because of the slow eastward shift of the Moon's major axis.

An interesting feature revealed in Figure 4-5 is how the extremes in the lunation length slowly vary over a period of nearly 9 years. The envelope defined by the minima and maxima appears to oscillate over a range of values from ±2 h to ±6 h. This behavior is evidence revealing the influence of the 8.85-year cycle in the alignment of the major axes of the orbits of the Moon and Earth.

The amplitude of the envelope is due to the eccentricity of Earth's orbit. When Earth is at perihelion, its orbital velocity is at its maximum value so Earth travels a larger distance around its orbit in a given time as compared to aphelion. Thus, the Moon must travel a greater distance to align with the Sun, which results in a longer lunation. Near aphelion, the opposite conditions produce a shorter lunation.

Using the axis scale on the right, the diagonal lines in Figure 4-5 plot the angle between the Moon's perigee and Earth's perihelion. This is the difference between the Moon's mean longitude of perigee and Earth's true longitude of perihelion. When the angle between the perigee and perihelion is 0°, the length of the lunation varies from a minimum of 29.273 days (–6.17 hours from mean) to a maximum of 29.820 days (+6.93 hours from mean). Similarly, when the angle between the perigee and perihelion is 180°, the length of the lunation varies from a minimum of 29.452 days (–1.88 hours from mean) to a maximum of 29.628 days (+2.33 hours from mean). To summarize, the greatest extremes in the length of the lunation occur when the longitudes of the Moon's perigee and Earth's perihelion are equal. The smallest extremes in the lunation length occur when their longitudes differ by 180°.

Although the Moon's major axis rotates eastward at a mean rate of 0.1114° per day, the true rate varies considerably. Figure 4-5 illustrates the variation by plotting the difference between the true longitudes of the Moon's perigee and

Earth's perihelion. This quasi-sinusoidal oscillation about the difference in the mean longitudes shows peak departures of ±30° from average. Indeed, the Moon's major axis can swing both east and west of its mean value, taking on an actual retrograde shift west during some anomalistic months.

This dynamic behavior is due to the gravitational pull of the Sun on the Moon as it orbits Earth. Consequently, a continuous torque is applied to the lunar orbit in an unsuccessful effort to permanently align the major axis towards the Sun. The annual orbit of the Earth–Moon system around the Sun coupled with the Moon's synodic orbit around Earth mean that the conditions for such a permanent alignment are always changing. The overall effect is to twist and distort the shape and orientation of the Moon's elliptical orbit.

Figure 4–6: Moon's Eccentricity for 2008 – 2010

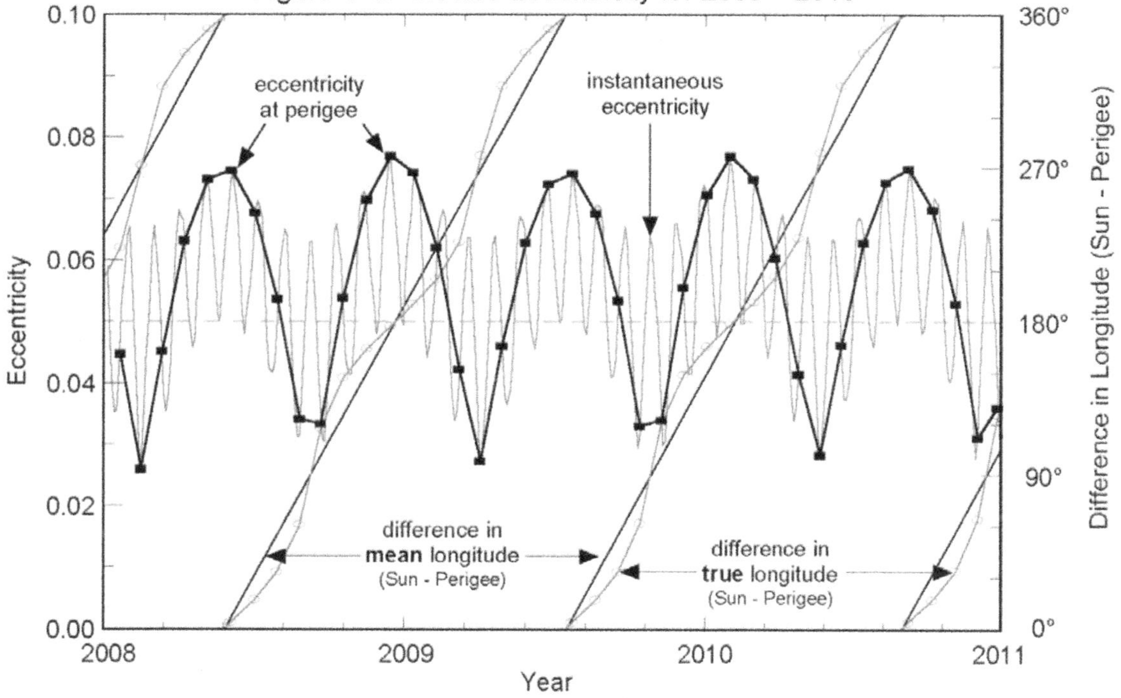

It was stated earlier that the Moon's mean orbital eccentricity is 0.0549, but this too is subject to large changes because of solar perturbations. Figure 4-6 plots the variation in the Moon's orbital eccentricity from 2008 through 2010. The instantaneous eccentricity (light gray curve) oscillates with a period tied to the synodic month and ranges from 0.0266 to 0.0762 over this 3-year interval. Superimposed on the instantaneous eccentricity is the eccentricity at the instant of perigee, which occurs at the beginning of each anomalistic month (heavy black curve). The straight diagonal lines represent the difference between the mean longitudes of the Sun and perigee. In other words, it is the angle between the Moon's perigee-directed major axis and the Sun. Oscillating about this line is the difference between the true longitudes of Sun and perigee. The scale for these angles appears along the right side of Figure 4-6. The extreme range of the Moon's orbital eccentricity at perigee during the 5000 years of the catalog is 0.0255 to 0.0775.

Figure 4-6 shows that the eccentricity reaches a maximum when the major axis of the lunar orbit is pointed directly towards or directly away from the Sun (angles of 0° and 180°, respectively). This occurs at a mean interval of 205.9 days, which is somewhat longer than half a year because of the eastward shift of the major axis. The eccentricity reaches a minimum when the major axis of the lunar orbit is perpendicular to the Sun (angles of 90° and 270°).

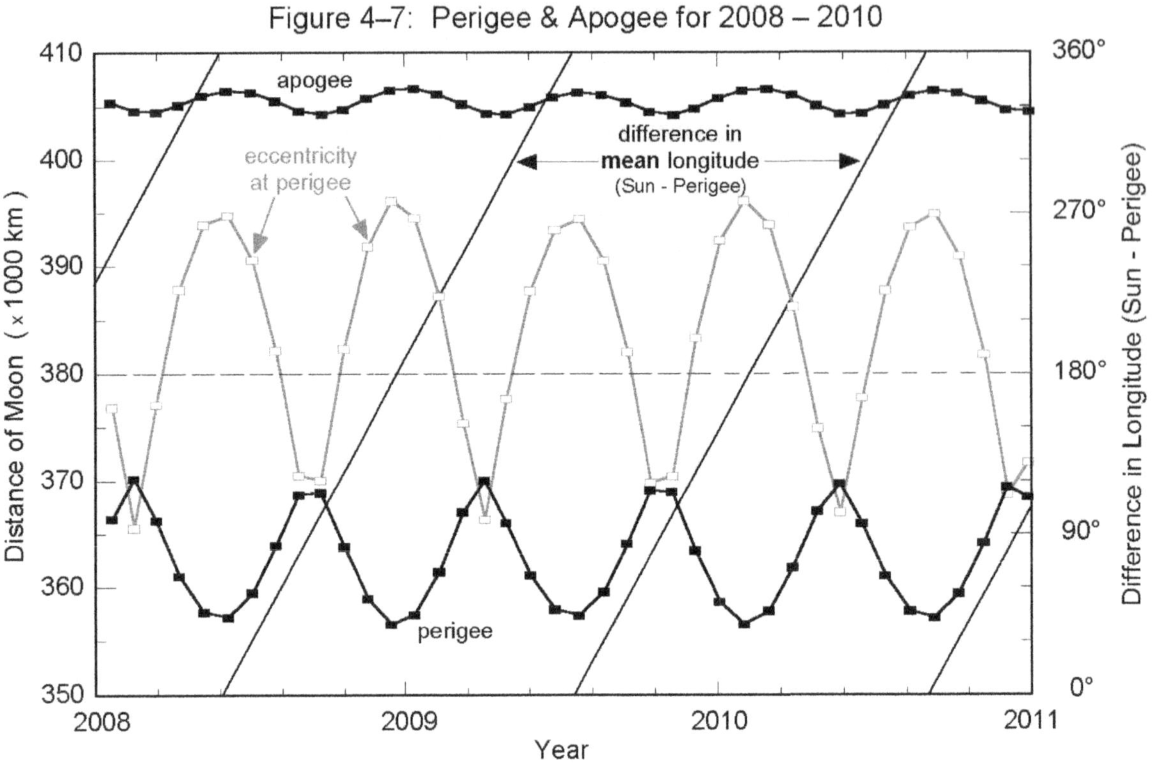

Such changes in orbital eccentricity produce significant variations in the Moon's distance at perigee and apogee. Figure 4-7 plots the Moon's distance for all perigees and apogees from 2008 through 2010. Also shown is the orbital eccentricity at perigee as well as the angle between the perigee directed major axis and the Sun. The closest perigee (minimum perigee distance) and farthest apogee (maximum apogee distance) occur when the eccentricity is at maximum. This corresponds to times when the Moon's major axis points directly towards or directly away from the Sun (angles of 0° and 180°, respectively). The farthest perigee (maximum perigee distance) and closest apogee (minimum apogee distance) occur when the eccentricity is at minimum. At such times, the major axis is oriented perpendicular to the Sun. During the 3-year interval covered in Figure 4-7, the Moon's perigee distance ranges from 356,568 to 370,216 km while the apogee distance ranges from 404,168 to 406,602 km.

Over the 5000-year period of the catalog, there are 66,276 perigees and apogees. During this epoch, the distance of the Moon's perigee varies from 356,355 to 370,399 km while the apogee varies from 404,042 to 406,725 km. The minimum and maximum extremes in orbital eccentricity are 0.0255 to 0.0775 and the extremes in the length of the anomalistic month are 24.629 days (2.925 days shorter than the mean) to 28.565 days (1.011 days longer than the mean). A histogram showing the distribution in the length of the anomalistic month is presented in Figure 4-8 where the durations of individual anomalistic months have been binned into 2-hour groups. The sharply asymmetric distribution shows that anomalistic months longer than the mean cluster over a much shorter range of values compared to anomalistic months shorter than the mean.

Figure 4–8: Length of Anomalistic Month Over 5000 Years

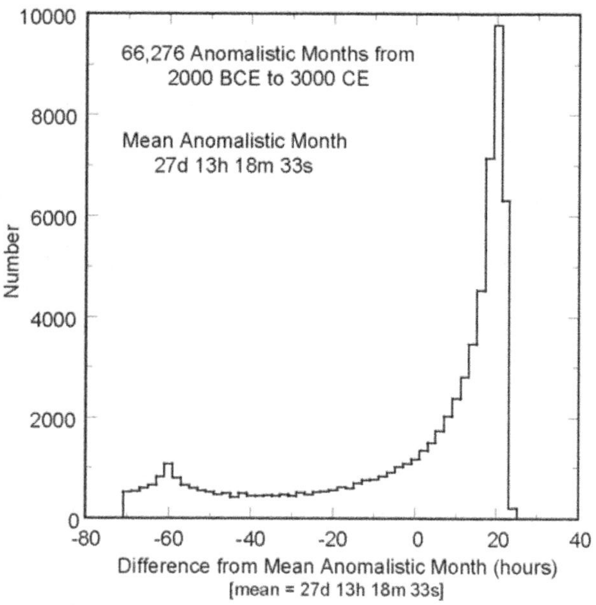

4.4 Draconic Month

The plane of the Moon's orbit is inclined at a mean angle of 5.145° to the plane of Earth's orbit around the Sun. The intersection of these planes defines two points or nodes on the celestial sphere. The node where the Moon's path crosses the ecliptic from south to north is the ascending node, while the node where the Moon's path crosses the ecliptic from north to south is the descending node.

Figure 4–9: Length of Draconic Month for 2008 – 2010

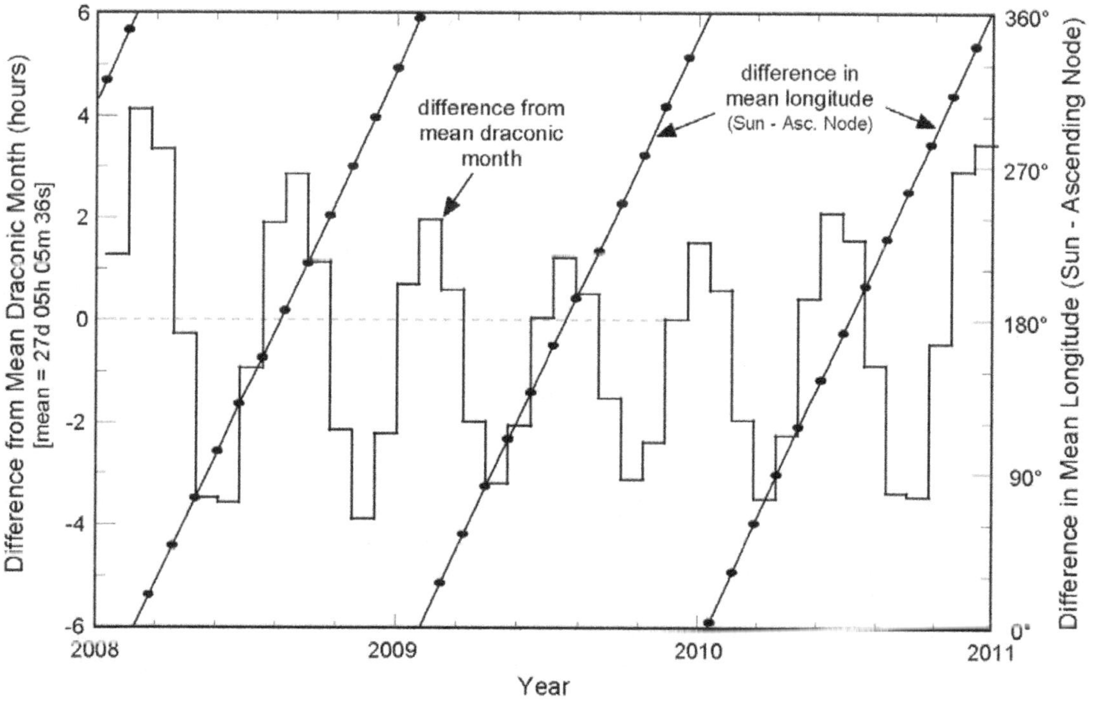

The draconic month is defined as one revolution of the Moon about its orbit with respect to the ascending node. The mean length of this nodical period is 27.21222 days (27d 05h 05m 36s). However, the actual duration can vary by over 6 h from the mean. Figure 4-9 plots the duration of the draconic month minus its mean value for 2008 through 2010. The shortest month over this 3-year period is 27.05115 days (27d 01h 14m), while the longest month is 27.38409 days (27d 09h 13m).

The most significant characteristic of this variation is that it is synchronized with the ascending node relative to the Sun's position along the ecliptic. The mean angle between the Sun and the ascending node (i.e., difference in mean longitude) is also plotted in Figure 4-9 (diagonal lines) to illustrate this relationship. The longitude difference at the start of each draconic month is plotted as a black dot. Longitude values can be read using the scale along the right side of the figure. The longest draconic months occur when the difference in the mean longitudes of the Sun and the ascending node is either 0° or 180°. In contrast, the shortest months occur when the angle between the Sun and the ascending node is either 90° or 270°.

The mean draconic month is 0.10944 day (2h 36m 36s) shorter than the sidereal month. Consequently, the lunar nodes slowly rotate west or retrograde (opposite the Moon's orbital motion) along the ecliptic at a rate of 0.05295° per day. One complete rotation of the ascending node about the ecliptic requires 18.6 years (6793.48 days) with respect to the fixed stars.

Figure 4–10: Lunar Orbit Inclination for 2008 – 2010

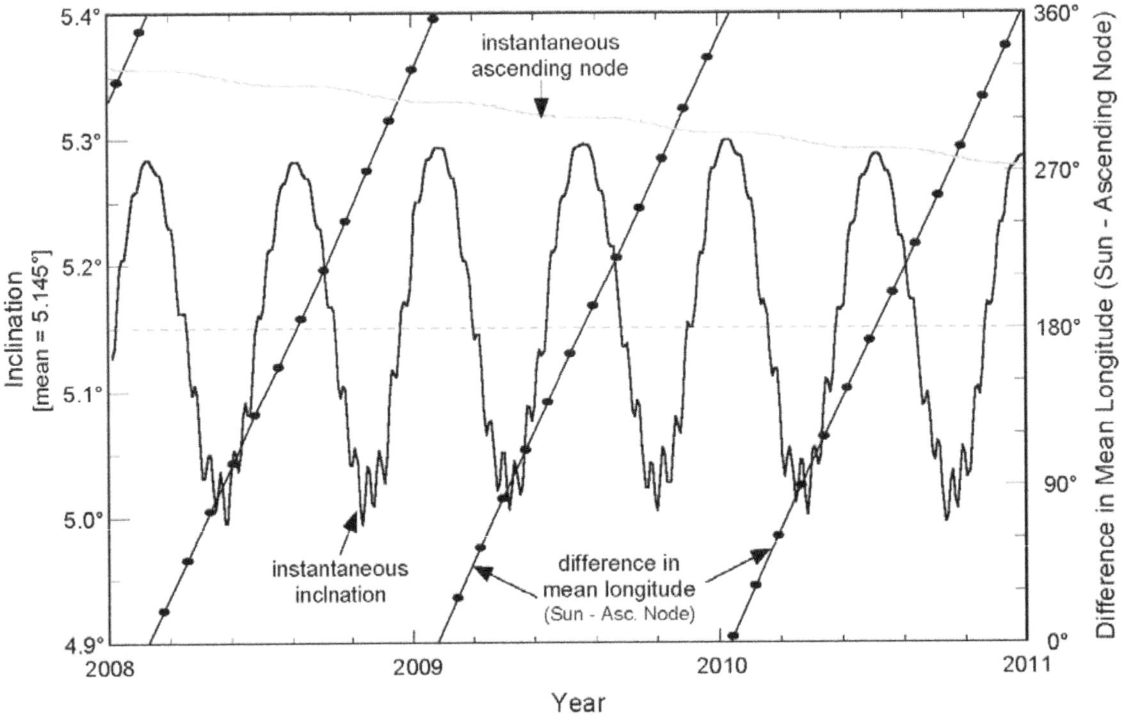

Figure 4-10 plots the instantaneous inclination of the lunar orbit over the 3-year period 2008–2010. The mean angle between the Sun and the ascending node (i.e., difference in mean longitude) is also plotted. The largest inclination of 5.30° occurs when the difference in longitude is either 0° or 180°. In other words, the inclination is always near its maximum value for both solar and lunar eclipses. The smallest inclination of 5.00° occurs when the difference in longitude is either 90° or 270°. Note the small monthly oscillations in the inclination when near its minimum. The figure also plots the longitude of the instantaneous ascending node. Its westward motion draws to a near standstill whenever the Sun aligns with either of the nodes. This corresponds to a difference in longitude of either 0° or 180°.

The mean interval in the periodic variation of both the draconic month and the orbital inclination is 173.3 days. This is the average time it takes for the Sun to travel from one node to the other. It is also equivalent to the interval between the midpoints of two eclipse seasons. The period is slightly less than half a year because of the retrograde motion of the nodes.

Figure 4–11: Draconic Month and Perigee for 2008 – 2017

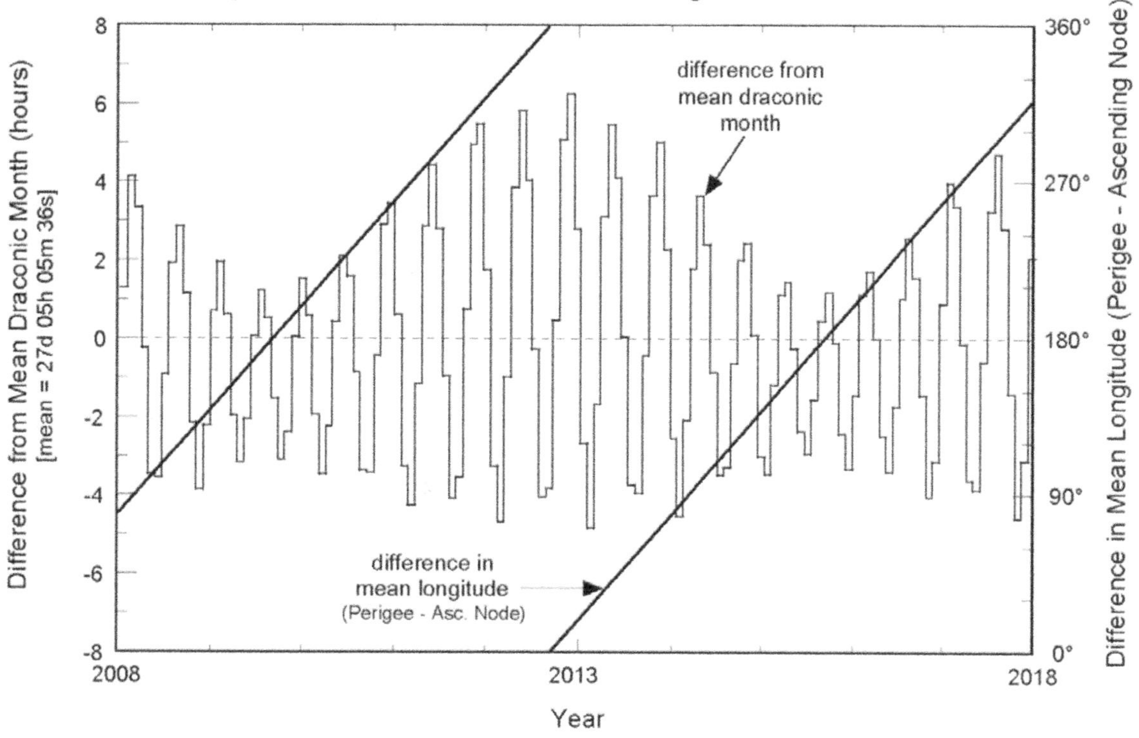

The length of the draconic month is strongly modulated by the position of the nodes with respect to the major axis of the Moon's orbit. The histogram in Figure 4-11 shows how the draconic month changes from 2008 through 2017. The 173-day alignment of the Sun with a node appears as the rapid oscillation in the month length. The quasi-sinusoidal envelopes surrounding the minima and maxima form two longer period oscillations. Over the 10-year period covered in this figure, the minimum month duration varies from 27.089 to 27.011 days (3.0 to 4.8 hours shorter than the mean). The maximum month duration ranges from 27.261 to 27.472 days (1.2 to 6.2 hours longer than the mean).

The difference in the mean longitudes of perigee and the ascending node appear as diagonal lines in Figure 4-11. This is the angle between these orbital parameters measured along the ecliptic. The greatest extremes in the draconic month occur when the angle between perigee and the ascending node is 0°. Likewise, the smallest extremes of the month take place when the difference in longitude is 180°. The mean rates of the major axis and the ascending node are 0.11140° east and 0.05295° west per day, respectively. Therefore, the mean period between alignments of the axis and node is 2190.4 days or 6.0 years. This period is clearly seen in Figure 4-11.

There are 67,111 draconic months during the 5000 years covered in this catalog. The shortest and longest months are 27.004 days (0.208 days or 5.0 hours shorter than the mean) and 27.487 days (0.275 days or 6.6 hours longer than the mean), respectively. A histogram of the distribution in the length of the draconic month over the five millennia appears in Figure 4-12 where the duration of individual draconic months have been binned into 30-min groups. The width and bifurcated symmetry of the distribution resemble the distribution for the lunation (synodic month) in Figure 4-4.

Figure 4–12: Length of Draconic Month Over 5000 Years

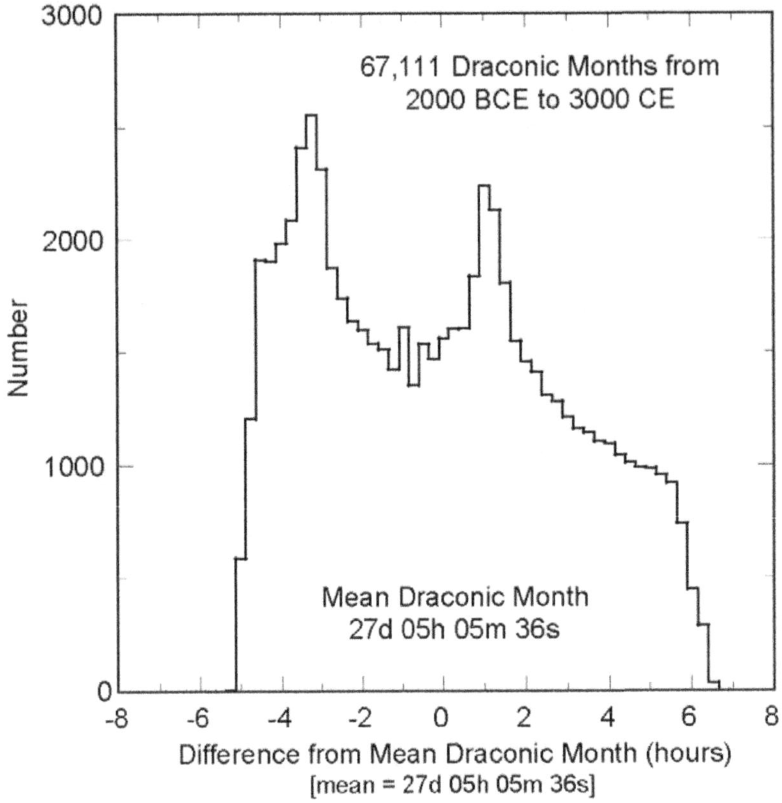

4.5 Eclipse Cycles

The interaction and harmonics of the synodic, anomalistic, and draconic months not only determine how frequently eclipses occur, but they also control the geometric characteristics and classification of each eclipse. The commensurability of these periods over long time scales results in several important eclipse cycles, which will be the subject of the next section.

Section 5: Lunar Eclipse Periodicity

5.1 Interval Between Two Successive Eclipses

The time interval between any two successive lunar eclipses can be either 1, 5 or 6 lunations (synodic months). The distribution of these intervals among the 12,063 in the *Canon* is found in Table 5-1.

Table 5-1. Interval Between Successive Eclipses

Number of Lunations	Number of Eclipses	Percent
1	1527	12.7%
5	2909	24.1%
6	7627	63.2%

5.2 Lunar Eclipse Repetition

Lunar eclipses separated by 1, 5, or 6 lunations are usually quite dissimilar. They are frequently of unlike types (i.e., penumbral, partial, or total) with diverse Sun–Moon–Earth alignment geometries, and with different lunar orbital characteristics (i.e., longitude of perigee and longitude of ascending node). More importantly, these short periods are of no value as predictors of future eclipses because they do not repeat in a recognizable pattern.

A simple lunar eclipse repetition cycle can be found by requiring that certain orbital parameters be repeated. The Moon must be in the full phase with the same longitude of perigee and same longitude of the ascending node. These conditions are met by searching for an integral multiple in the Moon's three major periods—the synodic, anomalistic, and draconic months. A fourth condition might require that an eclipse occur at approximately the same time of year to preserve the axial tilt of Earth and thus, the same season, as well as the distance from the Sun. This last factor controls the apparent diameter of Earth's umbral and penumbral shadows.

5.3 Saros Series

The Saros arises from a harmonic between three of the Moon's orbital cycles. All three periods are subject to slow variations over long time scales, but there current values (2000 CE) are:

Synodic Month (New Moon to New Moon)	= 29.530589 days	= 29d 12h 44m 03s
Anomalistic Month (perigee to perigee)	= 27.554550 days	= 27d 13h 19m 33s
Draconic Month (node to node)	= 27.212221 days	= 27d 05h 06m 36s

One Saros is equal to 223 synodic months, however, 239 anomalistic months and 242 draconic months and are also equal (within a few hours) to this same period:

223 Synodic Months	= 6585.3223 days	= 6585d 07h 43m
239 Anomalistic Months	= 6585.5375 days	= 6585d 12h 54m
242 Draconic Months	= 6585.3575 days	= 6585d 08h 35m

With a period of approximately 6,585.32 days (~18 years 11 days 8 hours), the Saros is a valuable tool in investigating the periodicity and recurrence of eclipses. It was first known to the Chaldeans as an interval when lunar eclipses repeat, but the Saros is applicable to solar eclipses as well.

Any two eclipses separated by one Saros cycle share similar characteristics. They occur at the same node with the Moon at nearly the same distance from Earth and at the same time of year. Because the Saros period is not equal to a whole number of days, its biggest drawback as an eclipse predictor is that subsequent eclipses are visible from different parts of the globe. The extra 1/3 day displacement means that Earth must rotate an additional ~8 hours or ~120° with each cycle. For solar eclipses, this results in a shift of each succeeding eclipse path by ~120° west. Thus, a Saros series returns to approximately the same geographic region every three Saros periods (~54 years and 34 days). This triple Saros cycle is known as the Exeligmos.

Figure 5–1. Lunar Eclipses from Saros 136: 1932 to 2022

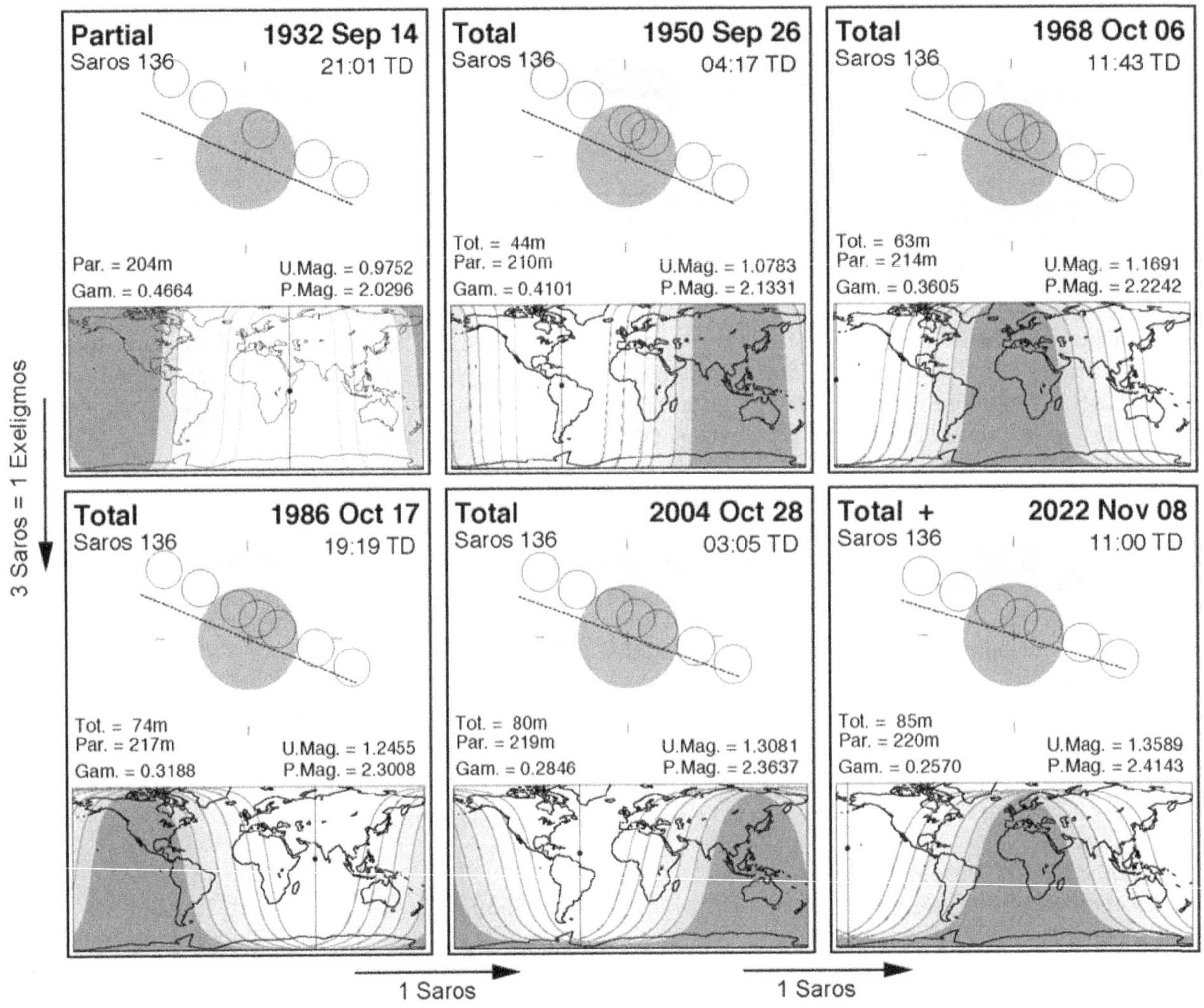

Figure Figure 5-1 shows the path of the Moon through Earth's shadows and the geographic regions of visibility for six lunar eclipses belonging to Saros 136 from 1932 through 2022. The 1932 Sep 14 eclipse was a large magnitude partial eclipse with a visibility zone centered on eastern Europe and Africa. One Saros period later the eclipse was total (1950 Sep 26) with visibility centered on the Americas. After another Saros interval, the eclipse was a larger magnitude event centered on the Pacific Ocean (1968 Oct 06). Finally, after the lapse of one more Saros period (3 Saros = Exeligmos = ~54.1 years), the zone of visibility returned to eastern Europe and Africa with a deeper non-central total lunar eclipse (1986 Oct 17). The same 54.1-year time interval in geographic visibility repeats for the Americas (1950 and 2004) and for the Pacific (1968 and 2022). The westward migration in the zone of eclipse visibility illustrates the effect of the extra 1/3-day in the Saros period. The southward shift of the Moon's path with respect to the shadow axis is due to the progressive decrease in gamma from 0.4664 (1932) to 0.2570 (2022). During this interval, lunar eclipses in the series change from partial (1932) to non-central total (1950) to central total (2022). Although the Moon's path through Earth's shadows is similar from one eclipse to the next, it is not exact. The Moon-shadow geometry changes slowly as do the characteristics of each lunar eclipse in a Saros series.

Saros series do not last indefinitely because the synodic, draconic, and anomalistic months are not perfectly commensurate with one another. In particular, the Moon's node shifts eastward by about 0.48° with each eclipse in a series. The following narrative describes the life cycle of a typical Saros series at the Moon's ascending node. The series begins when the Full Moon occurs approximately 17° east of the node.

A small fraction of the Moon's disk passes through the northern edge of the penumbra and a small penumbral eclipse occurs. One Saros period later, the lunar trajectory shifts a little further south, the Moon passes deeper into the penumbra (gamma decreases) and a penumbral eclipse of slightly larger magnitude results. After ~10 penumbral eclipses, the first partial lunar eclipse occurs as the Moon's southern limb passes through the umbra. Around 20 more partial eclipses occur as the Moon swings increasingly deeper into the umbra after each Saros and the magnitude of each event grows larger. Finally, the Moon passes completely into the umbra and a shallow total eclipse occurs. Over the course of the next 2 centuries, a total lunar eclipse occurs every 18.031 years (= Saros), as the Moon is displaced progressively southward through the umbral shadow with each eclipse. Halfway through this period, the Moon passes through the center of the umbra producing long total eclipses. The last of ~13 total eclipses in the series takes place just inside the southern edge of the umbra. The Saros series winds down now with another string of ~20 partial eclipses following by a final set of ~10 penumbral eclipses. The last eclipse of the series is a small magnitude penumbral event just inside the southern edge of the penumbral shadow. In all, this typical Saros series produces 73 eclipses spanning nearly 13 centuries.

The scenario for a Saros series occurring at the Moon's descending node is similar except that gamma increases as each successive eclipse shifts the Moon's trajectory farther north of the previous one. The magnitude of the shift in gamma is again tied to aphelion and perihelion.

Because of the ellipticity of the orbits of Earth and the Moon, the exact duration and number of eclipses in a complete Saros series is not constant. A series may last 1,226 to 1,587 years and is composed of 69 to 89 eclipses. A series can begin with 6 to 25 penumbral eclipses, followed by 6 to 24 partial eclipses. During the mid-life of a Saros series, there are 11 to 29 total eclipses. Finally, a series ends with 6 to 24 partial eclipses followed by 7 to 25 penumbral eclipses. At present (2008), there are 40 active Saros series numbered 110 to 149. The number of eclipses in each of these series ranges from 70 to 83, however, the majority of them (80%) are composed of 70 to 73 eclipses.

5.4 Gamma And Saros Series

Gamma changes monotonically throughout any single Saros series. As mentioned previously (Sect. 1.2.6), the change in gamma is larger when Earth is near its aphelion (June to July) than when it is near perihelion (December to January). For odd numbered series (descending node), gamma increases, while for even numbered series (ascending node), gamma decreases. This simple rule describes the current behavior of gamma, but this has not always been the case. The eccentricity of Earth's orbit is presently 0.0167, and is slowly decreasing. It was 0.0181 in the year -2000 and will be 0.0163 in +3000. In the past, when the eccentricity was larger, there were Saros series in which the trend in gamma reversed for one or more Saros cycles before resuming its original direction. These instances occur near perihelion when the Sun's apparent motion is highest and may, in fact, overtake the eastward shift of the node. The resulting effect is a relative shift west of the node after one Saros cycle instead of the usual eastward shift. Consequently, gamma reverses direction.

The most unusual case of this occurs in Saros series 13. It began in -2313 with 11 penumbral eclipses, followed by 7 partial eclipses. The series then reverted back to 3 more penumbral eclipse followed by another 11 partial eclipses. The series then produced 13 total, 20 partial, and 8 penumbral eclipses. To understand this odd behavior we must look to gamma. The value of gamma increased positively for the first 15 eclipses of the series. It then reversed direction after the fifth partial and decreased for 5 sequential eclipses while changing from partial back to penumbral. Finally resuming its northward motion, gamma began to increase again and continued to do so for the remainder of the series. Saros 13 produced 73 ending in the year -1015.

Among 200 Saros series examined (-20 to 183), there are many other examples of temporary shifts in the monotonic nature of gamma, although none as unusual as Saros 13. In fact, the first 38 Saros series (series -20 to 17) with members represented in the Canon, experience short reversals in gamma. The reversals are short and rarely last for more than four or five eclipses in a series. Some series even have two separate reversals in gamma (e.g., series -19, -15, -14, -12, -11, -5, -1, 4, 7, and 25). The most recent eclipse with a gamma reversal was on 1648 Jan 10 (Saros 138). The next and last in the Canon will occur on 2353 Jan 20 (Saros 120). In past millennia, the gamma reversals were more frequent because Earth's orbital eccentricity was larger.

5.5 Saros Series Statistics

Lunar eclipses belonging to 204 different Saros series fall within the five millennium span of the Canon. One series (183) has only two eclipses represented, while two series (-20 and 182) have three eclipses each. Another 81 have a larger, but incomplete, subset of their members included (-18 to 19, 12, 25, 139, 142, and 144 to 181). Finally, 120 complete Saros series are contained within the Canon (20 to 23, 26 to 138, 140, 141, and 143).

The number of lunar eclipses in each of these series ranges from 69 to 89. Almost a quarter (22.6%) of the series contain 72 eclipses, another quarter (24.0%) has 73 eclipses, and a sixth (15.7%) consists of 71 eclipses. In other words, nearly 2/3 (62.3%) of all Saros series are composed of 71 to 73 eclipses. If all Saros series with 70 to 74 eclipses are considered, then the percentage jumps to 75.0%. The remaining quarter have either 69 eclipses (Saros 181) or 75 to 89 eclipses.

Table 5-2 presents the statistical distribution of the number of eclipses in each Saros series. The approximate duration (years) as a function of the number of eclipses, is listed along with the first five Saros series containing the corresponding number of eclipses.

All Saros series begin and end with a number of penumbral eclipses. Among the 204 Saros series with members falling within the scope of this Canon, the number of penumbral eclipses in the initial phase ranges from 6 to 25. Similarly, the number of penumbral eclipses in the final phase varies from 7 to 25. The initial penumbral eclipse sequence is followed by a series of 6 to 24 partial eclipses, while the final penumbral eclipses are preceded by a sequence of 6 to 24 partial eclipses. The middle life of a Saros series is composed of total (umbral) eclipses, which range in number from 11 to 29.

Saros 13 is an exception to the normal progression of eclipse types through a series. After beginning with 11 penumbral eclipses followed by 7 partial eclipses, the series reverts back to produce 3 more penumbral eclipses. It then resumes the normal pattern to produce 11 partial eclipses followed by 13 total, 20 partial, and 8 penumbral eclipses. This odd behavior is caused by a reversal in gamma (Sect.5.4).

Figure 5-2 presents data on the number of penumbral, partial, and total eclipses in Saros series 1 through 150. The information is plotted in four separate histograms showing the numbers of all eclipses (Fig. 5-2a), penumbral eclipses (Fig. 5-2b), partial eclipses (Fig. 5-2c), and total eclipses (Fig. 5-2d). Several interesting relationships are revealed in these diagrams. Figure 5-2a shows that Saros series with large numbers of lunar eclipses tend to cluster into tight groups or 3 or 4 series separated by 18 to 19 series. However, most series are composed of 72 to 73 eclipses as already shown in Table 5-2.

Figure 5-2b displays the numbers of both leading (solid) and following (dashed) penumbral eclipses in a series. Note how the two histograms are displaced from each other, but their overlapping sections coincide with Saros series with large numbers of eclipses. Figure 5-2c shows the numbers of leading (solid) and following (dashed) partial eclipses in a series, which are nearly in phase with each other and peak during periods when the number of all lunar eclipses is near the minimum. Finally, Figure 5-2d presents the number of total eclipses in each series, which is inversely correlated with the number of partial lunar eclipses.

Table 5-2. Number of Lunar Eclipses in Saros Series

Number of Eclipses	Duration (Years)	Number of Series	Saros Series
69	1226.0	1	181
70	1244.0	10	126, 147, 148, 163, 166, ...
71	1262.1	32	59, 92, 93, 94, 95, ...
72	1280.1	46	-4, -1, 17, 20, 34, ...
73	1298.1	49	-20, -19, -17, -9, -6, ...
74	1316.2	16	-8, -7, 0, 10, 11, ...
75	1334.2	4	-18, -16, 9, 48
76	1352.2	4	3, 44, 46, 85
77	1370.3	2	5, 140
78	1388.3	2	61, 137
79	1406.3	4	100, 139, 174, 176
81	1442.4	2	156, 158
82	1460.4	5	63, 103, 119, 121, 138
83	1478.4	4	-10, 29, 101, 120
84	1496.5	7	-12, 64, 66, 82, 83, ...
85	1514.5	6	4, 24, 26, 27, 43, ...
86	1532.5	4	6, 8, 47, 65
87	1550.5	3	-15, -13, 25
88	1568.6	2	-14, -11
89	1586.6	1	7

To generalize these relationships, it appears that Saros series that are rich in the number of total eclipses are also rich in the number of penumbral eclipses, but poor in the number of partial eclipses. Conversely, Saros series poor in total eclipses are also poor in penumbral eclipses, but rich in partial eclipses. Finally, Saros series with large numbers of eclipses are centered on series with large numbers of total eclipses. Figure 5-2a also shows that the extremes in the minimum and maximum numbers of lunar eclipses in a Saros series is gradually decreasing.

A concise summary of all 204 Saros series (–20 to 183) is presented in Tables 5-3 to 5-8. The number of lunar eclipses in each series is listed followed by the calendar dates of the first and last eclipses in the Saros. Finally, the chronological sequence of lunar eclipse types in the series is tabulated. The number and type of eclipses varies from one Saros series to the next as reflected in the sequence diversity. Note that the tables make no distinction between central and non-central total lunar eclipses. The following abbreviations are used in the eclipse sequences:

T = Total Lunar Eclipse
P = Partial Lunar Eclipse
N = Partial Penumbral Lunar Eclipse
N* = Total Penumbral Lunar Eclipse

Figure 5–2. Number of Lunar Eclipses vs. Saros Series

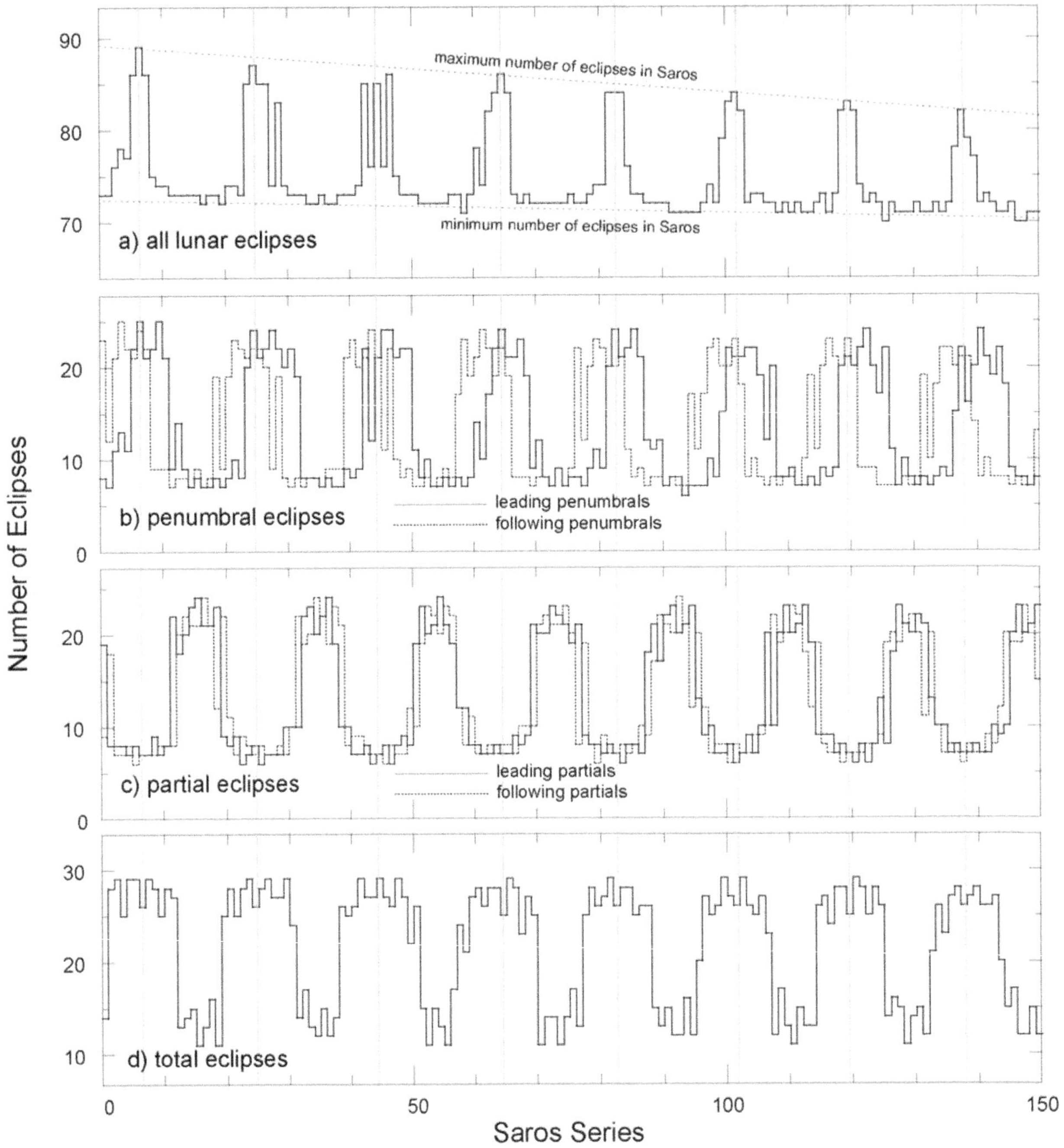

Table 5-3. Summary of Saros Series –20 to 25

Saros Series	Number of Eclipses	First Eclipse	Last Eclipse	Lunar Eclipse Sequence
-20	73	-3250 Apr 01	-1952 May 18	7N 1N* 20P 13T 23P 9N
-19	73	-3221 Mar 11	-1923 Apr 28	7N 9P 27T 22P 8N
-18	75	-3228 Jan 29	-1894 Apr 08	9N 10P 27T 8P 2N* 19N
-17	73	-3109 Mar 03	-1811 Apr 20	7N 8P 26T 8P 24N
-16	75	-3116 Jan 20	-1782 Mar 31	9N 7P 29T 9P 21N
-15	87	-3322 Aug 12	-1771 Feb 27	22N 8P 29T 6P 1N* 21N
-14	88	-3239 Aug 24	-1670 Mar 23	22N 1N* 6P 27T 7P 25N
-13	87	-3192 Aug 14	-1641 Mar 03	21N 7P 29T 8P 22N
-12	84	-3199 Jul 03	-1703 Dec 18	23N 8P 29T 6P 18N
-11	88	-3134 Jul 05	-1565 Feb 01	24N 1N* 6P 26T 8P 23N
-10	83	-3069 Jul 06	-1591 Dec 10	21N 8P 29T 7P 18N
-9	73	-3058 Jun 05	-1760 Jul 23	21N 10P 28T 7P 7N
-8	74	-2993 Jun 07	-1677 Aug 05	23N 1N* 7P 24T 10P 9N
-7	74	-2946 May 28	-1630 Jul 27	9N 2N* 19P 27T 9P 8N
-6	73	-2917 May 08	-1619 Jun 25	8N 23P 18T 17P 7N
-5	73	-2852 May 09	-1554 Jun 26	9N 1N* 21P 11T 23P 8N
-4	72	-2787 May 10	-1507 Jun 17	6N 1N* 22P 14T 22P 7N
-3	73	-2776 Apr 09	-1478 May 28	7N 24P 13T 21P 1N* 7N
-2	73	-2711 Apr 11	-1413 May 29	8N 22P 11T 23P 9N
-1	72	-2646 Apr 12	-1366 May 20	6N 21P 15T 23P 7N
0	74	-2653 Mar 01	-1337 Apr 30	8N 12P 25T 18P 2N* 9N
1	73	-2570 Mar 14	-1272 Apr 30	8N 19P 14T 9P 1N* 22N
2	73	-2523 Mar 03	-1225 Apr 22	7N 8P 28T 18P 12N
3	76	-2567 Dec 30	-1214 Mar 21	11N 8P 29T 7P 1N* 20N
4	78	-2646 Oct 06	-1131 Apr 02	13N 8P 25T 7P 1N* 24N
5	77	-2455 Dec 22	-1084 Mar 24	11N 7P 29T 8P 22N
6	86	-2624 Aug 04	-1091 Feb 10	22N 8P 29T 6P 1N* 20N
7	89	-2595 Jul 16	-1008 Feb 22	25N 7P 26T 7P 24N
8	86	-2494 Aug 08	-0961 Feb 13	21N 7P 29T 7P 22N
9	75	-2501 Jun 26	-1167 Sep 05	22N 9P 28T 7P 9N
10	74	-2454 Jun 17	-1138 Aug 15	24N 1N* 7P 25T 8P 9N
11	74	-2371 Jun 29	-1055 Aug 27	20N 1N* 8P 28T 8P 9N
12	73	-2360 May 28	-1062 Jul 17	9N 22P 27T 8P 7N
13	73	-2313 May 20	-1015 Jul 06	11N 7P 3N* 11P 13T 20P 8N
14	73	-2230 Jun 01	-0932 Jul 19	8N 1N* 20P 14T 22P 8N
15	73	-2219 Apr 30	-0921 Jun 19	7N 23P 15T 21P 7N
16	73	-2172 Apr 21	-0874 Jun 08	8N 24P 11T 21P 1N* 8N
17	72	-2089 May 04	-0809 Jun 11	6N 1N* 21P 13T 24P 7N
18	73	-2078 Apr 02	-0780 May 21	7N 21P 16T 21P 8N
19	73	-2031 Mar 24	-0733 May 11	8N 23P 11T 12P 8N* 11N
20	72	-1948 Apr 05	-0668 May 12	6N 1N* 9P 25T 22P 9N
21	74	-1955 Feb 22	-0639 Apr 23	8N 8P 28T 11P 2N* 17N
22	74	-1926 Feb 02	-0610 Apr 02	10N 9P 25T 7P 1N* 22N
23	73	-1825 Feb 25	-0527 Apr 14	7N 1N* 6P 28T 9P 22N
24	85	-2031 Sep 16	-0516 Mar 14	20N 7P 29T 8P 21N
25	87	-2038 Aug 06	-0487 Feb 21	24N 8P 26T 7P 22N

Table 5-4. Summary of Saros Series 26 to 70

Saros Series	Number of Eclipses	First Eclipse	Last Eclipse	Lunar Eclipse Sequence
26	85	-1919 Sep 09	-0404 Mar 06	20N 1N* 6P 28T 8P 22N
27	85	-1926 Jul 28	-0411 Jan 23	22N 7P 29T 7P 20N
28	74	-1897 Jul 09	-0581 Sep 06	24N 7P 27T 7P 9N
29	83	-1814 Jul 21	-0336 Dec 24	21N 1N* 7P 27T 8P 19N
30	74	-1803 Jun 19	-0487 Aug 18	20N 10P 29T 7P 8N
31	73	-1774 May 30	-0476 Jul 17	22N 10P 24T 10P 7N
32	73	-1673 Jun 23	-0375 Aug 09	12N 7N* 10P 14T 22P 8N
33	73	-1662 May 22	-0364 Jul 10	8N 22P 17T 19P 7N
34	72	-1615 May 13	-0335 Jun 19	8N 23P 13T 20P 1N* 7N
35	72	-1532 May 25	-0252 Jul 01	7N 1N* 20P 12T 24P 8N
36	73	-1521 Apr 24	-0223 Jun 11	7N 22P 15T 22P 7N
37	72	-1492 Apr 03	-0212 May 10	8N 24P 12T 19P 1N* 8N
38	72	-1391 Apr 27	-0111 Jun 03	6N 1N* 19P 14T 23P 9N
39	73	-1380 Mar 26	-0082 May 14	7N 10P 26T 21P 9N
40	73	-1369 Feb 24	-0071 Apr 12	9N 10P 25T 8P 1N* 20N
41	73	-1268 Mar 18	0030 May 06	7N 1N* 7P 26T 9P 23N
42	74	-1275 Feb 04	0041 Apr 05	9N 7P 29T 9P 20N
43	85	-1463 Sep 07	0052 Mar 04	22N 8P 27T 7P 21N
44	76	-1199 Jan 06	0153 Mar 27	11N 1N* 6P 27T 7P 24N
45	85	-1351 Aug 29	0164 Feb 25	21N 7P 29T 7P 21N
46	76	-1358 Jul 19	-0006 Oct 08	24N 8P 27T 6P 1N* 10N
47	86	-1275 Jul 31	0258 Feb 05	23N 1N* 6P 26T 8P 22N
48	75	-1228 Jul 21	0106 Sep 30	21N 8P 29T 7P 10N
49	73	-1217 Jun 21	0081 Aug 08	22N 9P 27T 7P 1N* 7N
50	73	-1134 Jul 03	0164 Aug 20	21N 1N* 8P 22T 12P 9N
51	73	-1105 Jun 13	0193 Jul 31	9N 2N* 19P 26T 10P 7N
52	72	-1076 May 23	0204 Jun 29	8N 23P 15T 19P 1N* 6N
53	72	-0993 Jun 05	0287 Jul 12	9N 1N* 20P 11T 23P 8N
54	72	-0946 May 26	0334 Jul 03	6N 1N* 21P 15T 22P 7N
55	72	-0935 Apr 25	0345 Jun 01	7N 24P 13T 20P 1N* 7N
56	72	-0852 May 07	0428 Jun 13	7N 1N* 21P 11T 23P 9N
57	73	-0823 Apr 16	0475 Jun 05	7N 19P 17T 22P 8N
58	73	-0812 Mar 16	0486 May 04	8N 12P 24T 12P 6N* 11N
59	71	-0711 Apr 09	0551 May 06	7N 12P 21T 8P 1N* 22N
60	73	-0700 Mar 08	0598 Apr 27	8N 8P 27T 11P 19N
61	78	-0780 Dec 13	0609 Mar 26	14N 8P 28T 7P 1N* 20N
62	74	-0624 Feb 08	0692 Apr 06	10N 7P 26T 7P 24N
63	82	-0722 Nov 03	0739 Mar 29	17N 7P 28T 8P 22N
64	84	-0783 Aug 20	0714 Feb 04	22N 8P 28T 7P 19N
65	86	-0736 Aug 11	0797 Feb 16	24N 7P 25T 8P 22N
66	84	-0671 Aug 12	0826 Jan 27	21N 7P 29T 8P 19N
67	73	-0660 Jul 11	0638 Aug 30	21N 9P 28T 7P 1N* 7N
68	72	-0595 Jul 14	0685 Aug 20	23N 8P 23T 10P 8N
69	73	-0530 Jul 15	0768 Sep 01	18N 1N* 9P 27T 10P 8N
70	72	-0519 Jun 13	0761 Jul 21	9N 21P 25T 10P 1N* 6N

Table 5-5. Summary of Saros Series 71 to 115

Saros Series	Number of Eclipses	First Eclipse	Last Eclipse	Lunar Eclipse Sequence
71	72	-0472 Jun 04	0808 Jul 11	11N 1N* 20P 11T 21P 1N* 7N
72	72	-0389 Jun 17	0891 Jul 25	7N 1N* 20P 14T 22P 8N
73	72	-0378 May 16	0902 Jun 23	7N 23P 14T 21P 7N
74	72	-0331 May 07	0949 Jun 13	9N 22P 11T 21P 1N* 8N
75	72	-0266 May 08	1014 Jun 15	7N 21P 14T 23P 7N
76	73	-0255 Apr 07	1043 May 26	8N 19P 17T 20P 9N
77	72	-0190 Apr 09	1090 May 16	8N 21P 13T 8P 2N* 20N
78	72	-0125 Apr 10	1155 May 18	7N 9P 25T 19P 12N
79	73	-0132 Feb 27	1166 Apr 17	9N 8P 28T 8P 1N* 19N
80	74	-0103 Feb 07	1213 Apr 06	11N 8P 26T 6P 1N* 22N
81	74	-0020 Feb 19	1296 Apr 19	9N 7P 27T 9P 22N
82	84	-0208 Sep 21	1289 Mar 08	20N 8P 29T 7P 20N
83	84	-0197 Aug 22	1300 Feb 05	24N 7P 26T 7P 20N
84	84	-0096 Sep 13	1401 Feb 28	20N 1N* 6P 28T 8P 21N
85	76	-0103 Aug 02	1249 Oct 23	22N 8P 28T 7P 11N
86	73	-0074 Jul 13	1224 Aug 30	24N 8P 25T 8P 8N
87	73	0027 Aug 06	1325 Sep 23	20N 1N* 7P 26T 10P 9N
88	72	0038 Jul 05	1318 Aug 12	12N 18P 26T 9P 7N
89	72	0067 Jun 15	1347 Jul 23	11N 21P 15T 17P 1N* 7N
90	72	0150 Jun 27	1430 Aug 04	9N 3N* 17P 13T 22P 8N
91	72	0179 Jun 07	1459 Jul 15	7N 22P 15T 21P 7N
92	71	0208 May 17	1470 Jun 13	8N 23P 12T 20P 1N* 7N
93	71	0291 May 30	1553 Jun 25	7N 1N* 20P 12T 24P 7N
94	71	0320 May 09	1582 Jun 06	6N 21P 16T 21P 7N
95	71	0349 Apr 19	1611 May 26	7N 23P 12T 12P 6N* 11N
96	71	0432 May 01	1694 Jun 07	7N 13P 20T 20P 11N
97	72	0443 Mar 31	1723 May 20	7N 9P 27T 12P 17N
98	74	0436 Feb 18	1752 Apr 28	10N 10P 25T 7P 1N* 21N
99	72	0555 Mar 24	1835 May 12	7N 1N* 7P 26T 8P 23N
100	79	0439 Dec 06	1846 Apr 11	15N 7P 29T 8P 20N
101	83	0360 Sep 11	1839 Feb 28	22N 8P 27T 6P 1N* 19N
102	84	0461 Oct 05	1958 Apr 04	20N 1N* 6P 26T 8P 23N
103	82	0472 Sep 03	1933 Feb 10	21N 7P 29T 7P 18N
104	72	0483 Aug 04	1763 Sep 22	22N 9P 26T 7P 1N* 7N
105	73	0566 Aug 16	1864 Oct 15	21N 1N* 7P 25T 9P 10N
106	73	0595 Jul 27	1893 Sep 25	19N 10P 27T 9P 8N
107	72	0606 Jun 26	1886 Aug 14	12N 20P 23T 10P 7N
108	72	0689 Jul 08	1969 Aug 27	17N 3N* 10P 12T 22P 8N
109	71	0736 Jun 27	1998 Aug 08	7N 1N* 20P 17T 19P 7N
110	72	0747 May 28	2027 Jul 18	8N 23P 13T 20P 1N* 7N
111	71	0830 Jun 10	2092 Jul 19	8N 1N* 20P 11T 23P 8N
112	72	0859 May 20	2139 Jul 12	7N 21P 15T 22P 7N
113	71	0888 Apr 29	2150 Jun 10	7N 23P 13T 18P 2N* 8N
114	71	0971 May 13	2233 Jun 22	7N 1N* 19P 13T 12P 6N* 13N
115	72	1000 Apr 21	2280 Jun 13	7N 9P 26T 19P 11N

Table 5-6. Summary of Saros Series 116 to 160

Saros Series	Number of Eclipses	First Eclipse	Last Eclipse	Lunar Eclipse Sequence
116	73	0993 Mar 11	2291 May 14	9N 9P 27T 8P 1N* 19N
117	71	1094 Apr 03	2356 May 15	8N 9P 24T 7P 23N
118	73	1105 Mar 02	2403 May 07	9N 7P 28T 8P 21N
119	82	0935 Oct 14	2396 Mar 25	20N 8P 28T 6P 1N* 19N
120	83	1000 Oct 16	2479 Apr 07	21N 7P 25T 7P 1N* 22N
121	82	1047 Oct 06	2508 Mar 18	19N 1N* 6P 29T 7P 20N
122	74	1022 Aug 14	2338 Oct 29	22N 8P 28T 7P 9N
123	72	1087 Aug 16	2367 Oct 08	23N 1N* 6P 25T 8P 9N
124	73	1152 Aug 17	2450 Oct 21	20N 8P 28T 8P 9N
125	72	1163 Jul 17	2443 Sep 09	17N 13P 26T 9P 7N
126	70	1228 Jul 18	2472 Aug 19	21N 1N* 8P 14T 19P 7N
127	72	1275 Jul 09	2555 Sep 02	10N 1N* 18P 16T 20P 7N
128	71	1304 Jun 18	2566 Aug 02	7N 23P 15T 19P 1N* 6N
129	71	1351 Jun 10	2613 Jul 24	9N 1N* 21P 11T 21P 8N
130	71	1416 Jun 10	2678 Jul 26	7N 1N* 20P 14T 22P 7N
131	72	1427 May 10	2707 Jul 07	7N 22P 15T 20P 8N
132	71	1492 May 12	2754 Jun 26	8N 21P 12T 11P 7N* 12N
133	71	1557 May 13	2819 Jun 29	6N 1N* 13P 21T 20P 10N
134	72	1550 Apr 01	2830 May 28	8N 10P 26T 10P 1N* 17N
135	71	1615 Apr 13	2877 May 18	9N 10P 23T 7P 1N* 21N
136	72	1680 Apr 13	2960 Jun 01	7N 1N* 7P 27T 8P 22N
137	78	1564 Dec 17	2953 Apr 20	15N 8P 28T 7P 20N
138	82	1521 Oct 15	2982 Mar 30	21N 1N* 7P 26T 6P 21N
139	79	1658 Dec 09	3065 Apr 13	16N 7P 27T 8P 21N
140	77	1597 Sep 25	2968 Jan 06	20N 8P 28T 7P 14N
141	72	1608 Aug 25	2888 Oct 11	24N 7P 26T 7P 8N
142	73	1709 Sep 19	3007 Nov 17	20N 1N* 7P 26T 9P 10N
143	72	1720 Aug 18	3000 Oct 05	19N 10P 27T 8P 8N
144	71	1749 Jul 29	3011 Sep 04	21N 1N* 9P 20T 12P 1N* 7N
145	71	1832 Aug 11	3094 Sep 16	15N 3N* 10P 15T 20P 8N
146	72	1843 Jul 11	3123 Aug 29	9N 20P 17T 19P 7N
147	70	1890 Jul 02	3134 Jul 28	8N 23P 12T 19P 1N* 7N
148	70	1973 Jul 15	3217 Aug 09	7N 1N* 20P 12T 23P 7N
149	71	1984 Jun 13	3246 Jul 20	7N 21P 15T 21P 7N
150	71	2013 May 25	3275 Jun 30	8N 23P 12T 15P 3N* 10N
151	71	2096 Jun 06	3358 Jul 13	7N 1N* 18P 14T 21P 10N
152	72	2107 May 07	3387 Jun 23	8N 10P 25T 15P 1N* 13N
153	71	2136 Apr 16	3398 May 22	9N 10P 24T 8P 1N* 19N
154	71	2237 May 10	3499 Jun 16	7N 8P 25T 8P 23N
155	73	2212 Mar 18	3510 May 17	9N 8P 28T 8P 20N
156	81	2060 Nov 08	3503 Apr 05	20N 8P 27T 6P 1N* 19N
157	73	2306 Mar 01	3604 Apr 27	11N 6P 26T 8P 22N
158	81	2154 Oct 21	3597 Mar 17	20N 7P 28T 8P 18N
159	73	2147 Sep 09	3445 Nov 07	23N 8P 26T 7P 9N
160	72	2248 Oct 03	3528 Nov 19	21N 7P 25T 8P 11N

Table 5-7. Summary of Saros Series 161 to 183

Saros Series	Number of Eclipses	First Eclipse	Last Eclipse	Lunar Eclipse Sequence
161	73	2259 Sep 02	3557 Oct 31	20N 9P 27T 8P 9N
162	71	2288 Aug 12	3550 Sep 19	19N 12P 24T 9P 7N
163	70	2371 Aug 27	3615 Sep 20	19N 2N* 8P 13T 20P 8N
164	71	2400 Aug 05	3662 Sep 11	9N 1N* 18P 18T 18P 7N
165	71	2411 Jul 06	3673 Aug 11	9N 22P 14T 19P 7N
166	70	2494 Jul 18	3738 Aug 13	9N 1N* 19P 11T 22P 8N
167	71	2541 Jul 09	3803 Aug 16	7N 20P 15T 21P 8N
168	71	2552 Jun 08	3814 Jul 15	8N 22P 13T 18P 2N* 8N
169	70	2635 Jun 22	3879 Jul 17	7N 1N* 19P 13T 13P 4N* 13N
170	71	2664 Jun 01	3926 Jul 09	7N 11P 24T 19P 10N
171	71	2675 May 01	3937 Jun 07	8N 10P 26T 8P 1N* 18N
172	70	2758 May 15	4002 Jun 08	8N 9P 23T 8P 22N
173	72	2787 Apr 24	4067 Jun 11	8N 7P 27T 9P 21N
174	79	2635 Dec 16	4042 Apr 18	18N 8P 27T 7P 19N
175	74	2791 Feb 11	4107 Apr 20	14N 7P 25T 7P 21N
176	79	2747 Dec 09	4154 Apr 11	17N 1N* 6P 28T 8P 19N
177	73	2704 Oct 05	4002 Dec 03	21N 8P 28T 6P 1N* 9N
178	70	2769 Oct 07	4013 Nov 01	22N 7P 24T 8P 1N* 8N
179	73	2816 Sep 27	4114 Nov 26	20N 8P 27T 8P 10N
180	71	2827 Aug 28	4089 Oct 03	19N 11P 26T 8P 7N
181	69	2892 Aug 29	4118 Sep 13	21N 1N* 8P 15T 17P 7N
182	70	2957 Aug 31	4201 Sep 26	10N 4N* 14P 15T 20P 7N
183	70	2968 Jul 30	4212 Aug 26	8N 21P 16T 18P 1N* 6N

5.6 Saros and Other Periods

The numbering system used for the Saros series was introduced by van den Bergh in his book *Periodicity and Variation of Solar (and Lunar) Eclipses* (1955). He assigned the number 1 to a pair of solar and lunar eclipse series that were in progress during the second millennium BCE based on an extrapolation from Oppolzer's *Canon der Finsternisse* [1887].

There is an interval of 1, 5 or 6 synodic months between any sequential pair of lunar eclipses. Interestingly, the number of lunations between two eclipses permits the determination of the Saros series number of the second eclipse when the Saros series number of the first eclipse is known. Let the Saros series number of the first eclipse in a pair be "s". The Saros series number of the second eclipse can be found from the relationships in Table 5-8 [Meeus, Grosjean, and Vanderleen, 1966].

Table 5-8. Some Eclipse Periods and their Relationships to the Saros Number

Number of Synodic Months	Length of Time	Saros Series Number	Period Name
1	~1 month	s + 38	Lunation
5	~5 months	s – 33	Short Semester
6	~6 months	s + 5	Semester
135	~11 years - 1 month	s + 1	Tritos
223	~18 years + 11 days	s	Saros
235	~19 years	s + 10	Metonic Cycle
358	~29 years - 20 days	s + 1	Inex
669	~54 years + 33 days	s	Exeligmos (Triple Saros)

5.7 Saros and Inex

A number of different eclipse cycles were investigated by van den Bergh, but the most useful were the Saros and the Inex. The Inex is equal to 358 synodic months (~29 years less 20 days), which is very nearly 388.5 draconic months.

358 Synodic Months = 10,571.9509 days = 10,571d 22h 49m
388.5 Draconic Months = 10,571.9479 days = 10,571d 22h 55m

The extra 0.5 in the number of draconic months means that eclipses separated by one Inex period occur at opposite nodes. Consequently, an eclipse occurring in the northern half of Earth's shadows will be followed one Inex later by an eclipse occurring in the southern half of Earth's shadows, and vice versa.

The mean time difference between 358 synodic months and 388.5 draconic months making up an Inex is only 6 min. In comparison, the mean difference between these two cycles in the Saros is 52 min. This means that after one Inex, the shift of the Moon with respect to the node (+0.04°) is much smaller than for the Saros (–0.48°). While a Saros series lasts 12 to 15 centuries, an Inex series typically lasts 225 centuries and contains about 780 eclipses.

5.8 Saros–Inex Panorama

Van den Bergh placed all 5,200 lunar eclipses in von Oppolzer's Canon der Finsternisse (1887) into a large two-dimensional matrix. (Von Oppolzer did not include penumbral lunar eclipses in his Canon der Finsternisse.) Each Saros series was arranged as a separate column containing every eclipse in chronological order. The individual Saros columns were then staggered so that the horizontal rows each corresponded to a different Inex series. This "Saros-Inex Panorama" proved useful in organizing eclipses. For instance, one step down in the panorama is a change of one Saros period (6585.32 days) later, while one step to the right is a change of one Inex period (10571.95 days) later. The rows and columns were then numbered with the Saros and Inex numbers.

The panorama also made it possible to predict the approximate circumstances of lunar (and solar) eclipses occurring before or after the period spanned by von Oppolzer's Canon. The time interval "t" between any two lunar eclipses can be found through an integer combination of Saros and Inex periods via the following relationship:

$$t = a\,i + b\,s, \qquad (5\text{–}1)$$

where: t = interval in days,
i = Inex period of 10571.95 days (358 synodic months),
s = Saros period of 6585.32 days (223 synodic months), and
a, b = integers (negative, zero, or positive).

From this equation, a number of useful combinations of Inex and Saros periods can be employed to extend von Oppolzer's *Canon* from –1207 back to –1600 using nothing more than simple arithmetic (van den Bergh, 1954). The ultimate goal of the effort was to a produce an eclipse canon for dating historical events prior to –1207. Periods formed by various combinations of Inex and Saros were evaluated in order to satisfy one or more of the following conditions:

1) The deviation from a multiple of 0.5 draconic months should be small
   (i.e., Moon should be nearly the same distance from the node).
2) The deviation from an integral multiple of anomalistic months should be small
   (i.e., Moon should be nearly the same distance from Earth).
3) The deviation from an integral multiple of anomalistic years should be small
   (i.e., eclipse should occur on nearly the same calendar date).

No single Inex–Saros combination meets all three criteria, but there are periods that do a reasonably good job for any one of them. Note that secular changes in the Moon's elements cause a particular period to be of high accuracy for a limited number of centuries. The direct application of the Saros–Inex panorama allows for the determination of eclipse dates in the past (or future); however, the application of the longer Saros–Inex combinations permit the rapid estimation of a number of eclipse characteristics without lengthy calculations. Table 5-9 lists several of the most useful periods.

Table 5-9. Some Useful Eclipse Periods

Period Name	Period (Inex + Saros)	Period (years)	Use
Heliotrope	58i + 6s	1,787	Geographic longitude of eclipse
Accuratissima	58i + 9s	1,841	Geographic latitude of eclipse
Horologia	110i + 7s	3,310	Time of ecliptic conjunction

Modern digital computers using high precision solar and lunar ephemerides can directly predict the dates and circumstances of eclipses. Nevertheless, the Saros and Inex cycles remain useful tools in understanding the periodicity and frequency of eclipses.

5.9 Secular Variations in the Saros and Inex

Because of long secular variations in the average ellipticity of the Moon's and Earth's orbits, the mean lengths of the synodic, draconic, and anomalistic months are slowly changing. The mean synodic and draconic months are increasing by approximately 0.2 and 0.4 s per millennium, respectively. Meanwhile, the anomalistic month is decreasing by about 0.8 s per millennium.

Although small, the cumulative effects of such changes has an impact on both the Saros and Inex. Table 5-10 shows how the number of draconic and anomalistic months change with respect to 223 synodic months (Saros period) over an interval of 7000 years. Of particular interest is the last column, which shows the mean shift of the Moon's node after a period of 1 Saros. It is gradually increasing, which means that the average number of eclipses in a typical Saros series is decreasing. This explains the trend in the number of lunar eclipses seen in Figure 5-2a.

Table 5-10: Number of Anomalistic and Draconic Months in 1 Saros

Year	Anomalistic Months (223 Lunations)	Draconic Months (223 Lunations)	Node Shift (after 1 Saros)
-3000	238.991679	241.998742	0.4529
-2000	238.991763	241.998730	0.4571
-1000	238.991854	241.998717	0.4618
0	238.991950	241.998703	0.4668
1000	238.992051	241.998688	0.4722
2000	238.992157	241.998673	0.4779
3000	238.992267	241.998656	0.4838
4000	238.992379	241.998639	0.4899

Table 5-11 shows how the number of draconic months is changing with respect to 358 synodic months (Inex period) over a 7000-year interval. The mean shift in the lunar node after 1 Inex is much smaller than the Saros and is gradually decreasing. This explains why the lifetime of the Inex is so much longer than the Saros and is still increasing.

Table 5-11: Number of Draconic Months in 1 Inex

Year	Draconic Months (358 Lunations)	Node Shift (after 1 Inex)
-3000	388.500223	-0.0801
-2000	388.500204	-0.0734
-1000	388.500183	-0.0659
0	388.500160	-0.0578
1000	388.500136	-0.0491
2000	388.500111	-0.0400
3000	388.500085	-0.0305
4000	388.500057	-0.0207

Although the Inex possesses a long lifespan, its mean duration is not easily characterized because of the decreasing nodal shift seen in Table 5-11. If the instantaneous mean durations of the synodic and draconic months for the years –2000, +2000, and +4000 are used to calculate the mean duration of the Inex, the resulting lengths are about 14,500, 26,600, and 51,000 years, respectively (Meeus, 2004a).

Abbreviations

arcsec	Arc second
AT	Hybrid eclipse that begins as annular, then changes to total.
ATA	Hybrid eclipse that begins as annular, changes to total, and then reverts back to annular.
BCE	Before the Common Era
CE	Common Era
cm	Centimeter
ET	Ephemeris Time
GMAT	Greenwich Mean Astronomical Time
GMT	Greenwich Mean Time
IAU	International Astronomical Union
ISO	International Standards Organization
LLR	Lunar Laser Ranging
LOD	Length of Day
m	Meter (or minutes in tables)
min	Minutes
ms	Millisecond
s	Second
arcsec/cy2	Arc seconds per Julian century squared
TA	Hybrid eclipse that begins as total and ends as annular.
TAI	International Atomic Time
TD	Terrestrial Dynamical Time
TT	Terrestrial Time
UT	Universal Time
UTC	Coordinated Universal Time
VLBI	Very Long Baseline Interferometry
yr	Year

References

Astronomical Almanac for 2006, Almanac Office, U.S. Naval Observatory, U.S. Government Printing Office, Washington; London: HM Stationery Office (2004).

Astronomical Almanac for 2008, Almanac Office, U.S. Naval Observatory, U.S. Government Printing Office, Washington; London: HM Stationery Office (2006).

Bretagnon P., G. Francou., "Planetary Theories in rectangular and spherical variables: VSOP87 solution", *Astron. Astrophys.*, **202** (309), (1988).

Brown, E.W., "Theory of the Motion of the Moon", *Mem. Royal Astron. Soc.*, Vol. LVII, Part II, pp. 136-141, London (1905).

Chapront-Touzé, M., and J. Chapront, "The Lunar Ephemeris ELP 2000," *Astron. Astrophys.*, **124**(1), pp. 50–62 (1983).

Chapront-Touzé, M., and J. Chapront, *Lunar Tables and Programs from 4000 B.C. to A.D. 8000*, Willmann-Bell, Richmond, Virginia (1991).

Chapront, J., M. Chapront-Touzé, and G. Francou, "A new determination of lunar orbital parameters, precession constant and tidal acceleration from LLR measurements," *Astron. Astrophys.*, **387**, pp. 700–709 (2002).

Chauvenet, W.A., *Manual of Spherical and Practical Astronomy, Vol. 1*, edition of 1891, (Dover reprint, New York, 1960).

Danjon, A., "Les éclipses de Lune par la pénombre en 1951," *L'Astronomie*, **65**, 51–53 (1951).

Dickey, J.O., P.L. Bender, J.E. Faller, X.X. Newhall, R.L. Ricklefs, J.G. Ries,, P.J. Shelus, C. Veillet, A.L. Whipple, J.R. Wiant, J.G. Williams, and C.E. Yoder, "Lunar Laser Ranging: a Continuing Legacy of the Apollo Program," *Science*, **265**, pp. 482–490 (1994).

Espenak, F., *Fifty Year Canon of Solar Eclipses: 1986–2035*, Sky Publishing Corp., Cambridge, Massachusetts (1987).

——, *Fifty Year Canon of Lunar Eclipses: 1986–2035*, Sky Publishing Corp., Cambridge, Massachusetts (1989).

——, Eclipses During 2007, *Observer's Handbook - 2007*, Royal Astronomical Society of Canada (2006).

——, and J. Meeus, *Five Millennium Canon of Solar Eclipses: –1999 to +3000 (2000 BCE to 3000 CE)*, NASA Tech. Pub. 2006–214141, NASA Goddard Space Flight Center, Greenbelt, Maryland (2006).

——, and ——, *Five Millennium Catalog of Lunar Eclipses: –1999 to +3000 (2000 BCE to 3000 CE)*, NASA Tech. Pub. 2009–214173, NASA Goddard Space Flight Center, Greenbelt, Maryland (2009).

Freeth, T., A. Jones, J.M. Steele, and Y. Bitsakis, "Calendars with Olympiad display and eclipse prediction on the Antikythera Mechanism," *Nature*, 07/2008, **454**, pp. 614–617, (2008).

Gingerich, O., (Translator) Canon of Eclipses, Dover Publications, New York (1962) (from the original T.R. von Oppolzer, book, *Canon der Finsternisse*, Wien, [1887]).

Hughes, T., "Addendum," In: *Mathematical Astronomy Morsels III*, Willmann-Bell, Richmond, Virginia, pp. 137–140, (2004b).

Huber, P.J., "Modeling the Length of Day and Extrapolating the Rotation of the Earth," In: *Papers in Honor of Jean Meeus, Astronomical Amusements*, F. Bònoli, S. De Meis, and A. Panaino, Eds., IsIAO, Rome (2000).

Improved Lunar Ephemeris 1952–1959, Nautical Almanac Office, U.S. Naval Observatory, Washington, DC (1954).

Keen, R.A., "Volcanic Aerosols and Lunar Eclipses," *Science*, **222**, pp. 1011–1013 (1983).

Kühl, A., 1928, "Ueber den Einfluss des Grenzkontrastes auf Präzisionsmessungen," *Physikalische Zeitschrift*, **29**, 1– 34.

La Hire, P., *Tabulae Astronomicae* (Paris 1707).

Link, F., *Eclipse Phenomena in Astronomy*, Springer-Verlag, New York (1969).

Liu, B.-L., and A.D. Fiala, *Canon of Lunar Eclipses 1500 B.C.–A.D. 3000*, Willmann-Bell, Richmond, Virginia, p. 215 (1992).

Meeus, J., and H. Mucke, *Canon of Lunar Eclipse: -2002 to +2526*, Astronomisches Büro, Vienna, Austria (1979).

Meeus, J., *Mathematical Astronomy Morsels*, Willmann-Bell, Richmond, Virginia, pp. 108–110, (1997).

——, *More Mathematical Astronomy Morsels*, Willmann-Bell, Richmond, Virginia, pp. 120–126 (2002).

——, *Mathematical Astronomy Morsels III*, Willmann-Bell, Richmond, Virginia, pp. 109–111, (2004a).

——, C.C. Grosjean, and W. Vanderleen, *Canon of Solar Eclipses*, Pergamon Press, Oxford, United Kingdom (1966).

Morrison, L., and Stephenson, F.R., "Historical Values of the Earth's Clock Error DT and the Calculation of Eclipses," J. *Hist. Astron.*, Vol. 35 Part 3, August 2004, No. 120, pp. 327–336 (2004).

Mucke, H., and J. Meeus, *Canon of Solar Eclipses: -2003 to +2526*, Astronomisches Büro, Vienna, Austria (1983).

Newcomb, S., "Tables of the Motion of the Earth on its Axis Around the Sun," *Astron. Papers Amer. Eph.*, Vol. 6, Part I (1895).

Stephenson, F.R., *Historical Eclipses and Earth's Rotation*, Cambridge University Press, Cambridge (1997).

——, and M.A. Houlden, *Atlas of Historical Eclipse Maps, East Asia 1500 BC–AD 1900*, Cambridge University Press, Cambridge/New York (1986).

van den Bergh, *Eclipses in the Second Millennium B.C. –1600 to –1207*, Tjeenk Willink, and Haarlem, Netherlands (1954).

——, *Periodicity and Variation of Solar (and Lunar) Eclipses*, Tjeenk Willink, and Haarlem, Netherlands (1955).

von Oppolzer, T.R., *Canon der Finsternisse*, Vienna, Austria (1887).

Photo 1–1 captures the beginning, middle and end of totality during the total lunar eclipse of 2004 October 27. ©2004 F. Espenak

Photo 1–2 Various phases of the total lunar eclipse of 2000 January 20–21 appear in this composite image. ©2000 F. Espenak

Five Millennium Catalog of Lunar Eclipses –1999 to +3000

Appendix

Key to Lunar Eclipse Catalog

Cat Num — sequential Catalog Number assigned to each eclipse from 1 to 12,064

Calendar Date — Julian or Gregorian date of Greatest Eclipse (Julian date used prior to 1582 Oct 04)

TD of Greatest Eclipse — Terrestrial Dynamical Time of Greatest Eclipse

ΔT — arithmetic difference between Terrestrial Dynamical Time Universal Time (seconds)

Luna Num — number of synodic months, or lunations, since New Moon on 2000 Jan 06

Saros Num — Saros Series Number of eclipse

Ecl Type — Lunar Eclipse Type

N = Penumbral Lunar Eclipse
P = Partial Lunar Eclipse
T = Total Lunar Eclipse
m = Middle eclipse of Saros series
+ = Central total eclipse (Moon's center passes north of shadow axis)
– = Central total eclipse (Moon's center passes south of shadow axis)
* = Total penumbral eclipse
b = Saros series begins (first penumbral eclipse in a Saros series)
e = Saros series ends (last penumbral eclipse in a Saros series)

QSE — Quincena Solar Eclipse Parameter

p = Partial Solar Eclipse
a = Annular Solar Eclipse
t = Total Solar Eclipse
h = Hybrid Solar Eclipse

Gamma — minimum distance from center of Moon to axis of Earth's umbral shadow cone

Pen Mag — Penumbral Magnitude; fraction of the Moon's diameter immersed in penumbra

Um Mag — Umbral Magnitude; fraction of the Moon's diameter immersed in umbra

Phase Durations
Pen — duration from contact P1 to P4 (minutes)
Par — duration from contact U1 to U4 (minutes)
Total — duration from contact U2 to U3 (minutes)

Greatest in Zenith
Lat & Long — latitude and longitude where Moon appears in zenith at greatest eclipse

Cat Num	Canon Plate	Calendar Date	TD of Greatest Eclipse	ΔT s	Luna Num	Saros Num	Ecl. Type	QSE	Gamma	Pen. Mag.	Um. Mag.	Eclipse Phase Durations Pen. m	Eclipse Phase Durations Par. m	Eclipse Phase Durations Total m	Greatest in Zenith Lat.	Greatest in Zenith Lng.
1	001	-1999 Jun 26	14:13:28	46437	-49456	017	N	t-	-1.0981	0.8791	-0.1922	268.8	-	-	24S	22W
2	001	-1999 Nov 21	20:23:49	46427	-49451	-016	N	-a	-1.1155	0.8143	-0.1921	233.4	-	-	15N	114W
3	001	-1998 May 17	05:47:36	46416	-49445	-011	P	-t	0.8988	1.2105	0.2069	281.7	102.7	-	13S	105E
4	001	-1998 Nov 11	05:15:58	46404	-49439	-006	P	-a	-0.4644	2.0382	0.9740	343.4	200.8	-	12N	113E
5	001	-1997 May 06	18:57:01	46392	-49433	-001	T+	pp	0.1003	2.6513	1.6963	322.8	213.5	98.2	11S	92W
6	001	-1997 Oct 31	06:54:09	46380	-49427	004	T+	pp	0.2207	2.5057	1.4014	374.4	228.7	90.6	9N	88E
7	001	-1996 Apr 25	11:57:19	46368	-49421	009	P	a-	-0.6385	1.6562	0.7159	288.6	166.0	-	7S	13E
8	001	-1996 Oct 19	06:00:26	46357	-49415	014	P	t-	0.8915	1.2694	0.1756	308.8	100.1	-	5N	101E
9	001	-1995 Mar 16	18:59:06	46347	-49410	-019	N	-a	1.2015	0.6538	-0.3471	217.0	-	-	10N	87W
10	001	-1995 Apr 15	04:12:15	46345	-49409	019	N	a-	-1.4029	0.2681	-0.7010	139.5	-	-	4S	130E
11	001	-1995 Sep 08	21:00:34	46335	-49404	-014	N	-a	-1.2651	0.5382	-0.4652	196.3	-	-	13S	121W
12	001	-1995 Oct 08	10:05:04	46333	-49403	024	N	a-	1.5232	0.0836	-0.9571	83.8	-	-	1N	40E
13	001	-1994 Mar 06	02:07:38	46323	-49398	-009	P	-t	0.4871	1.9933	0.9355	347.6	202.1	-	13N	166E
14	001	-1994 Aug 29	11:25:40	46311	-49392	-004	P	-a	-0.5326	1.8619	0.8992	300.4	180.5	-	16S	23E
15	001	-1993 Feb 23	02:59:48	46300	-49386	001	T-	pp	-0.2503	2.4408	1.3571	371.8	228.2	87.7	16N	153E
16	001	-1993 Aug 19	03:45:11	46288	-49380	006	T+	p-	0.1639	2.5396	1.5744	320.2	209.8	93.2	19S	137E
17	001	-1992 Feb 12	04:09:50	46276	-49374	011	P	t-	-0.9653	1.1167	0.0571	287.5	57.2	-	18N	135E
18	001	-1992 Aug 07	17:18:34	46264	-49368	016	P	h-	0.9154	1.1818	0.1748	276.1	93.7	-	20S	67W
19	001	-1991 Jan 02	00:22:41	46254	-49363	-017	N	-a	1.0940	0.8451	-0.1440	233.6	-	-	25N	171W
20	001	-1991 Jun 28	09:30:53	46243	-49357	-012	N	-t	-1.1239	0.8298	-0.2379	261.3	-	-	24S	48E
21	002	-1991 Dec 22	15:38:22	46231	-49351	-007	T	-p	0.4019	2.1044	1.1362	308.3	193.5	52.6	23N	42W
22	002	-1990 Jun 17	09:59:52	46219	-49345	-002	T	-t	-0.3722	2.2071	1.1436	360.9	217.8	59.8	22S	40E
23	002	-1990 Dec 12	06:49:16	46207	-49339	003	T-	p-	-0.2724	2.3547	1.3616	323.9	206.7	80.9	21N	90E
24	002	-1989 Jun 06	14:31:12	46196	-49333	008	T	h-	0.4062	2.1212	1.1041	338.4	207.0	49.8	19S	28W
25	002	-1989 Dec 01	17:21:02	46184	-49327	013	N*	h-	-1.0024	1.0436	-0.0058	271.9	-	-	18N	69W
26	002	-1988 Apr 26	19:06:30	46174	-49322	-020	N	-a	-1.3775	0.3035	-0.6432	144.1	-	-	9S	95W
27	002	-1988 May 26	02:18:15	46172	-49321	018	N	a-	1.1372	0.7530	-0.2107	222.0	-	-	16S	155E
28	002	-1988 Oct 21	01:10:42	46162	-49316	-015	N	-t	1.3197	0.4900	-0.6163	210.7	-	-	6N	173E
29	002	-1987 Apr 16	12:20:37	46150	-49310	-010	P	-a	-0.6515	1.6347	0.6897	288.9	164.2	-	4S	8E
30	002	-1987 Oct 10	01:04:52	46139	-49304	-005	P	-t	0.6260	1.7498	0.6697	337.7	178.9	-	1N	175E
31	002	-1986 Apr 06	03:30:53	46127	-49298	000	T+	pp	0.0913	2.6827	1.6980	334.6	218.5	99.9	1N	141E
32	002	-1986 Sep 29	07:21:46	46115	-49292	005	T-	p-	-0.0990	2.6872	1.6656	341.6	219.4	98.8	5S	82E
33	002	-1985 Mar 26	12:28:30	46103	-49286	010	P	t-	0.8902	1.2455	0.2037	296.5	105.2	-	6N	8E
34	002	-1985 Sep 18	20:27:52	46092	-49280	015	P	a-	-0.7787	1.4156	0.4422	279.0	137.0	-	9S	114W
35	002	-1984 Feb 13	22:07:18	46082	-49275	-018	N	-t	-1.2531	0.6009	-0.4830	228.9	-	-	18N	135W
36	002	-1984 Aug 09	03:11:22	46070	-49269	-013	N	-a	1.1725	0.6924	-0.2798	213.1	-	-	20S	144E
37	002	-1984 Sep 07	12:31:19	46068	-49268	025	N	a-	-1.4394	0.1994	-0.7660	117.9	-	-	14S	5E
38	002	-1983 Feb 02	00:35:27	46058	-49263	-008	P	-t	-0.5000	1.9646	0.9169	338.8	197.0	-	21N	172W
39	002	-1983 Jul 29	15:17:47	46047	-49257	-003	P	-a	0.4792	1.9890	0.9685	331.2	197.2	-	22S	38W
40	002	-1982 Jan 22	10:03:40	46035	-49251	002	T+	p-	0.2411	2.4129	1.4182	328.9	210.7	86.1	23N	45E
41	003	-1982 Jul 18	20:21:53	46023	-49245	007	T	pp	-0.2744	2.3880	1.3214	364.6	224.6	83.7	24S	115W
42	003	-1981 Jan 12	00:54:10	46011	-49239	012	P	a-	0.9159	1.1598	0.1947	257.5	93.6	-	25N	179W
43	003	-1981 Jul 07	20:46:01	46000	-49233	017	N*	t-	-1.0219	1.0195	-0.0534	285.2	-	-	25S	122W
44	003	-1981 Dec 03	05:05:53	45990	-49228	-016	N	-a	-1.1187	0.8094	-0.1989	233.6	-	-	18N	114E
45	003	-1980 May 27	12:41:51	45978	-49222	-011	P	-t	0.9745	1.0700	0.0696	267.8	60.5	-	17S	1W
46	003	-1980 Nov 21	13:37:31	45966	-49216	-006	P	-a	-0.4654	2.0378	0.9708	344.6	201.2	-	16N	15W
47	003	-1979 May 17	02:13:09	45954	-49210	-001	T+	pp	0.1770	2.5092	1.5567	320.4	210.9	93.3	14S	156E
48	003	-1979 Nov 10	14:56:40	45943	-49204	004	T+	pp	0.2178	2.5117	1.4059	375.0	229.0	91.0	13N	35W
49	003	-1978 May 06	19:21:07	45931	-49198	009	P	a-	-0.5647	1.7917	0.8515	295.2	177.0	-	11S	101W
50	003	-1978 Oct 30	14:05:11	45919	-49192	014	P	t-	0.8853	1.2804	0.1873	309.1	103.0	-	9N	22W
51	003	-1977 Mar 28	02:15:29	45909	-49187	-019	N	-a	1.2691	0.5303	-0.4717	198.5	-	-	5N	161E
52	003	-1977 Apr 26	11:26:40	45907	-49186	019	N	a-	-1.3358	0.3921	-0.5788	167.3	-	-	8S	19E
53	003	-1977 Sep 20	05:17:52	45898	-49181	-014	N	-a	-1.2801	0.5107	-0.4925	191.2	-	-	9S	113E
54	003	-1977 Oct 19	10:20:40	45896	-49180	024	N	a-	1.5134	0.1009	-0.9384	91.6	-	-	5N	88W
55	003	-1976 Mar 16	09:04:23	45886	-49175	-009	P	-t	0.5519	1.8742	0.8169	343.2	193.5	-	9N	59E
56	003	-1976 Sep 08	19:47:02	45874	-49169	-004	P	-a	-0.5580	1.8162	0.8516	298.3	176.9	-	13S	105W
57	003	-1975 Mar 05	09:52:15	45863	-49163	001	T-	pp	-0.1908	2.5483	1.4683	373.3	231.5	96.0	12N	48E
58	003	-1975 Aug 29	11:55:30	45851	-49157	006	T+	p-	0.1314	2.6018	1.6318	321.8	211.0	95.3	15S	13E
59	003	-1974 Feb 22	11:25:58	45839	-49151	011	P	t-	-0.9146	1.2066	0.1534	294.6	92.1	-	15N	24E
60	003	-1974 Aug 19	01:01:49	45827	-49145	016	P	h-	0.8775	1.2546	0.2410	283.5	109.3	-	18S	176E

Cat Num	Canon Plate	Calendar Date	TD of Greatest Eclipse	ΔT s	Luna Num	Saros Num	Ecl. Type	QSE	Gamma	Pen. Mag.	Um. Mag.	Eclipse Phase Durations Pen. m	Par. m	Total m	Greatest in Zenith Lat.	Lng.
61	004	-1973 Jan 13	08:47:38	45818	-49140	-017	N	-a	1.1079	0.8169	-0.1671	229.5	-	-	25N	62E
62	004	-1973 Jul 09	16:06:48	45806	-49134	-012	N	-t	-1.1976	0.6966	-0.3752	243.2	-	-	25S	52W
63	004	-1972 Jan 03	00:18:02	45794	-49128	-007	T	-p	0.4132	2.0826	1.1165	307.5	192.6	49.0	24N	172W
64	004	-1972 Jun 27	16:31:27	45782	-49122	-002	P	-t	-0.4505	2.0636	0.9998	355.3	209.0	-	24S	59W
65	004	-1972 Dec 22	15:28:24	45771	-49116	003	T-	p-	-0.2679	2.3631	1.3695	324.9	207.4	81.7	23N	41W
66	004	-1971 Jun 16	21:23:00	45759	-49110	008	T	a-	0.3272	2.2651	1.2501	341.3	213.2	73.4	22S	133W
67	004	-1971 Dec 12	01:44:10	45747	-49104	013	N*	h-	-1.0035	1.0424	-0.0087	272.7	-	-	21N	164E
68	004	-1970 May 08	02:26:38	45737	-49099	-020	N	-a	-1.4473	0.1743	-0.7704	110.3	-	-	13S	152E
69	004	-1970 Jun 06	09:32:50	45736	-49098	018	N	a-	1.0603	0.8931	-0.0688	237.5	-	-	19S	45E
70	004	-1970 Nov 01	09:10:02	45726	-49093	-015	N	-t	1.3233	0.4841	-0.6234	209.6	-	-	10N	51E
71	004	-1969 Apr 27	19:44:29	45714	-49087	-010	P	-a	-0.7248	1.5004	0.5551	281.8	150.7	-	8S	106W
72	004	-1969 Oct 21	09:12:41	45702	-49081	-005	P	-t	0.6298	1.7423	0.6630	336.4	177.8	-	5N	51E
73	004	-1968 Apr 16	10:42:21	45691	-49075	000	T+	pp	0.0198	2.8146	1.8283	336.4	220.1	102.0	3S	30E
74	004	-1968 Oct 09	15:47:22	45679	-49069	005	T-	p-	-0.0907	2.7019	1.6812	340.6	218.9	98.9	0S	47W
75	004	-1967 Apr 05	19:16:42	45667	-49063	010	P	t-	0.8205	1.3738	0.3313	307.9	132.0	-	1N	97W
76	004	-1967 Sep 29	05:02:47	45655	-49057	015	P	a-	-0.7646	1.4420	0.4676	280.3	140.1	-	5S	115E
77	004	-1966 Feb 24	05:09:10	45646	-49052	-018	N	-t	-1.3026	0.5076	-0.5715	211.8	-	-	14N	118E
78	004	-1966 Aug 20	11:10:02	45634	-49046	-013	N	-a	1.2130	0.6210	-0.3567	204.0	-	-	17S	23E
79	004	-1966 Sep 18	20:59:01	45632	-49045	025	N	a-	-1.4224	0.2325	-0.7368	127.1	-	-	10S	124W
80	004	-1965 Feb 13	08:08:41	45622	-49040	-008	P	-t	-0.5373	1.8928	0.8518	334.0	191.6	-	18N	73E
81	005	-1965 Aug 09	22:45:19	45611	-49034	-003	P	-a	0.5294	1.9003	0.8731	329.2	191.0	-	20S	151W
82	005	-1964 Feb 02	18:10:18	45599	-49028	002	T+	p-	0.2119	2.4635	1.4746	328.3	211.8	89.6	21N	78W
83	005	-1964 Jul 29	03:15:13	45587	-49022	007	T-	pp	-0.2132	2.5030	1.4312	368.2	228.5	92.9	23S	141E
84	005	-1963 Jan 22	09:20:50	45576	-49016	012	P	a-	0.8943	1.1977	0.2360	260.3	102.4	-	24N	53E
85	005	-1963 Jul 18	03:25:58	45564	-49010	017	P	t-	-0.9513	1.1501	0.0754	298.5	67.0	-	25S	137E
86	005	-1963 Dec 13	13:43:39	45554	-49005	-016	N	-a	-1.1247	0.7992	-0.2105	233.0	-	-	21N	16W
87	005	-1962 Jun 07	19:40:04	45542	-48999	-011	N	-t	1.0474	0.9348	-0.0629	252.8	-	-	19S	108W
88	005	-1962 Dec 02	21:54:58	45531	-48993	-006	P	-a	-0.4687	2.0328	0.9638	345.5	201.2	-	19N	140W
89	005	-1961 May 28	09:31:42	45519	-48987	-001	T+	-p	0.2521	2.3705	1.4197	317.3	207.2	85.3	17S	44E
90	005	-1961 Nov 21	22:58:46	45507	-48981	004	T+	pp	0.2154	2.5164	1.4101	375.4	229.3	91.3	16N	157W
91	005	-1960 May 17	02:43:37	45496	-48975	009	P	a-	-0.4899	1.9292	0.9884	301.1	186.4	-	15S	146E
92	005	-1960 Nov 09	22:13:21	45484	-48969	014	P	t-	0.8815	1.2866	0.1950	309.0	104.8	-	13N	146W
93	005	-1959 Apr 07	09:23:16	45474	-48964	-019	N	-a	1.3423	0.3967	-0.6068	174.6	-	-	1N	51E
94	005	-1959 May 06	18:36:13	45472	-48963	019	N	a-	-1.2657	0.5220	-0.4514	191.4	-	-	12S	92W
95	005	-1959 Sep 30	13:44:26	45463	-48958	-014	N	-a	-1.2885	0.4952	-0.5077	188.0	-	-	5S	16W
96	005	-1959 Oct 30	02:58:31	45461	-48957	024	N	a-	1.5077	0.1103	-0.9270	95.4	-	-	10N	142E
97	005	-1958 Mar 27	15:49:23	45451	-48952	-009	P	-t	0.6252	1.7395	0.6823	337.1	181.4	-	5N	45W
98	005	-1958 Sep 20	04:17:44	45439	-48946	-004	P	-a	-0.5760	1.7842	0.8176	296.7	174.2	-	9S	125E
99	005	-1957 Mar 16	16:35:22	45427	-48940	001	T-	pp	-0.1234	2.6701	1.5937	374.2	234.1	102.2	8N	56W
100	005	-1957 Sep 09	20:13:35	45416	-48934	006	T+	pp	0.1059	2.6510	1.6762	323.3	211.9	96.6	12S	114W
101	006	-1956 Mar 04	18:36:11	45404	-48928	011	P	t-	-0.8581	1.3071	0.2602	301.8	117.8	-	12N	86W
102	006	-1956 Aug 29	08:51:57	45392	-48922	016	P	h-	0.8463	1.3153	0.2951	289.7	120.4	-	15S	57E
103	006	-1955 Jan 23	17:07:04	45383	-48917	-017	N	-a	1.1268	0.7796	-0.1992	224.3	-	-	24N	63W
104	006	-1955 Feb 22	03:20:00	45381	-48916	021	N	a-	-1.5328	0.0425	-0.9519	57.7	-	-	15N	143E
105	006	-1955 Jul 19	22:47:07	45371	-48911	-012	N	-t	-1.2670	0.5715	-0.5045	223.4	-	-	25S	153W
106	006	-1954 Jan 13	08:52:18	45359	-48905	-007	T	-a	0.4290	2.0525	1.0888	306.5	191.3	43.2	24N	59E
107	006	-1954 Jul 08	23:09:54	45348	-48899	-002	P	-t	-0.5236	1.9297	0.8654	349.1	198.9	-	24S	160W
108	006	-1953 Jan 03	00:00:33	45336	-48893	003	T-	p-	-0.2583	2.3808	1.3870	326.1	208.5	83.3	24N	170W
109	006	-1953 Jun 28	04:23:00	45324	-48887	008	T+	p-	0.2538	2.3990	1.3856	343.0	217.4	86.5	23S	121E
110	006	-1953 Dec 23	10:01:00	45313	-48881	013	N*	h-	-1.0001	1.0493	-0.0030	274.3	-	-	22N	39E
111	006	-1952 May 18	09:48:07	45303	-48876	-020	Ne	-a	-1.5162	0.0472	-0.8960	58.0	-	-	16S	39E
112	006	-1952 Jun 16	16:52:31	45301	-48875	018	P	a-	0.9869	1.0272	0.0666	250.4	56.8	-	21S	67W
113	006	-1952 Nov 11	17:10:18	45291	-48870	-015	N	-t	1.3259	0.4794	-0.6285	208.7	-	-	14N	71W
114	006	-1951 May 08	03:06:11	45280	-48864	-010	P	-a	-0.7995	1.3638	0.4176	273.5	133.7	-	12S	140E
115	006	-1951 Oct 31	17:24:42	45268	-48858	-005	P	-t	0.6306	1.7402	0.6623	335.3	177.3	-	9N	75W
116	006	-1950 Apr 27	17:47:12	45256	-48852	000	T-	pp	-0.0560	2.7494	1.7608	337.7	220.5	101.6	8S	79W
117	006	-1950 Oct 21	00:19:31	45245	-48846	005	T-	p-	-0.0873	2.7076	1.6883	339.5	218.4	98.8	4N	178W
118	006	-1949 Apr 17	01:57:10	45233	-48840	010	P	t-	0.7453	1.5125	0.4687	319.0	154.2	-	3S	160E
119	006	-1949 Oct 10	13:44:03	45221	-48834	015	P	a-	-0.7556	1.4589	0.4838	281.0	142.0	-	1S	18W
120	006	-1948 Mar 06	12:02:07	45212	-48829	-018	N	-t	-1.3590	0.4017	-0.6726	189.8	-	-	11N	12E

Cat Num	Canon Plate	Calendar Date	TD of Greatest Eclipse	ΔT s	Luna Num	Saros Num	Ecl. Type	QSE	Gamma	Pen. Mag.	Um. Mag.	Eclipse Phase Durations Pen. m	Par. m	Total m	Greatest in Zenith Lat.	Lng.
121	007	-1948 Apr 05	03:37:38	45210	-48828	020	N	t-	1.5198	0.1056	-0.9667	100.5	-	-	2N	136E
122	007	-1948 Aug 30	19:16:55	45200	-48823	-013	N	-a	1.2460	0.5632	-0.4201	196.1	-	-	14S	100W
123	007	-1948 Sep 29	05:34:07	45198	-48822	025	N	a-	-1.4118	0.2538	-0.7192	132.8	-	-	6S	105E
124	007	-1947 Feb 23	15:33:11	45188	-48817	-008	P	-t	-0.5826	1.8062	0.7720	328.3	184.4	-	15N	40W
125	007	-1947 Aug 20	06:19:46	45177	-48811	-003	P	-a	0.5727	1.8242	0.7902	327.4	185.0	-	18S	94E
126	007	-1946 Feb 13	02:08:34	45165	-48805	002	T+	p-	0.1751	2.5281	1.5452	327.8	212.9	93.2	19N	161E
127	007	-1946 Aug 09	10:16:29	45154	-48799	007	T-	pp	-0.1590	2.6051	1.5279	371.0	231.2	98.6	21S	35E
128	007	-1945 Feb 02	17:38:57	45142	-48793	012	P	a-	0.8659	1.2481	0.2898	263.9	112.6	-	22N	72W
129	007	-1945 Jul 29	10:15:45	45130	-48787	017	P	t-	-0.8880	1.2673	0.1906	309.2	104.7	-	24S	34E
130	007	-1945 Dec 24	22:16:54	45121	-48782	-016	N	-a	-1.1334	0.7836	-0.2271	231.8	-	-	22N	145W
131	007	-1944 Jun 18	02:42:46	45109	-48776	-011	N	-h	1.1174	0.8054	-0.1902	236.9	-	-	21S	145E
132	007	-1944 Dec 13	06:09:27	45097	-48770	-006	P	-a	-0.4736	2.0244	0.9541	346.3	201.1	-	21N	95E
133	007	-1943 Jun 07	16:52:12	45086	-48764	-001	T	-p	0.3259	2.2347	1.2849	313.7	202.3	73.6	20S	68W
134	007	-1943 Dec 02	06:59:23	45074	-48758	004	T+	pp	0.2115	2.5233	1.4173	375.7	229.6	91.9	20N	81E
135	007	-1942 May 28	10:05:45	45062	-48752	009	T	a-	-0.4149	2.0674	1.1253	306.2	194.1	51.1	18S	33E
136	007	-1942 Nov 21	06:22:58	45051	-48746	014	P	h-	0.8782	1.2917	0.2023	308.6	106.4	-	17N	90E
137	007	-1941 Apr 18	16:25:15	45041	-48741	-019	N	-h	1.4191	0.2567	-0.7488	142.9	-	-	3S	58W
138	007	-1941 May 18	01:44:16	45039	-48740	019	N	h-	-1.1948	0.6538	-0.3227	212.3	-	-	16S	159E
139	007	-1941 Oct 11	22:16:40	45029	-48735	-014	N	-a	-1.2932	0.4862	-0.5160	185.9	-	-	0S	147W
140	007	-1941 Nov 10	11:29:55	45027	-48734	024	N	a-	1.5025	0.1185	-0.9162	98.4	-	-	14N	12E
141	008	-1940 Apr 06	22:28:13	45018	-48729	-009	P	-t	0.7028	1.5973	0.5400	329.5	165.6	-	1N	148W
142	008	-1940 Sep 30	12:54:52	45006	-48723	-004	P	-a	-0.5890	1.7613	0.7929	295.6	172.2	-	4S	6W
143	008	-1939 Mar 26	23:10:33	44995	-48717	001	T-	pp	-0.0493	2.8043	1.7314	374.5	235.6	105.8	4N	157W
144	008	-1939 Sep 20	04:38:42	44983	-48711	006	T+	pp	0.0866	2.6887	1.7092	324.6	212.5	97.4	8S	118E
145	008	-1938 Mar 16	01:37:58	44971	-48705	011	P	t-	-0.7937	1.4221	0.3815	309.3	139.7	-	8N	166E
146	008	-1938 Sep 09	16:49:58	44960	-48699	016	P	h-	0.8221	1.3630	0.3362	294.8	128.1	-	11S	65W
147	008	-1937 Feb 04	01:16:41	44950	-48694	-017	N	-a	1.1544	0.7263	-0.2473	217.0	-	-	22N	173E
148	008	-1937 Mar 05	10:56:43	44948	-48693	021	N	a-	-1.4792	0.1377	-0.8505	102.6	-	-	11N	27E
149	008	-1937 Jul 31	05:37:10	44938	-48688	-012	N	-t	-1.3280	0.4616	-0.6187	203.3	-	-	24S	103E
150	008	-1936 Jan 24	17:18:03	44927	-48682	-007	T	-a	0.4521	2.0089	1.0477	304.9	189.2	32.1	23N	68W
151	008	-1936 Jul 19	05:56:25	44915	-48676	-002	P	-t	-0.5907	1.8070	0.7418	342.6	187.8	-	24S	97E
152	008	-1935 Jan 13	08:25:06	44904	-48670	003	T-	p-	-0.2430	2.4090	1.4153	327.5	209.9	85.6	23N	64E
153	008	-1935 Jul 08	11:29:27	44892	-48664	008	T+	p-	0.1844	2.5257	1.5135	343.8	220.1	94.5	24S	13E
154	008	-1934 Jan 02	18:10:11	44880	-48658	013	P	h-	-0.9913	1.0656	0.0128	276.8	26.7	-	23N	84W
155	008	-1934 Jun 28	00:16:33	44869	-48652	018	P	a-	0.9163	1.1565	0.1964	261.2	95.6	-	22S	179W
156	008	-1934 Nov 23	01:12:12	44859	-48647	-015	N	-t	1.3273	0.4765	-0.6308	208.0	-	-	18N	167E
157	008	-1933 May 19	10:25:50	44847	-48641	-010	P	-a	-0.8753	1.2255	0.2777	264.0	111.5	-	16S	28E
158	008	-1933 Nov 12	01:40:00	44836	-48635	-005	P	-t	0.6294	1.7413	0.6656	334.4	177.2	-	14N	160E
159	008	-1932 May 08	00:48:41	44824	-48629	000	T-	pp	-0.1337	2.6083	1.6167	338.1	219.6	98.3	12S	173E
160	008	-1932 Oct 31	08:56:42	44813	-48623	005	T-	p-	-0.0872	2.7067	1.6892	338.3	217.8	98.6	9N	51E
161	009	-1931 Apr 27	08:31:07	44801	-48617	010	P	t-	0.6654	1.6599	0.6144	329.4	173.0	-	7S	59E
162	009	-1931 Oct 20	22:30:56	44790	-48611	015	P	a-	-0.7511	1.4673	0.4918	281.2	142.8	-	4N	152W
163	009	-1930 Mar 17	18:47:44	44780	-48606	-018	N	-t	-1.4213	0.2850	-0.7846	161.2	-	-	7N	92W
164	009	-1930 Apr 16	10:00:50	44778	-48605	020	N	t-	1.4404	0.2505	-0.8199	152.6	-	-	2S	37E
165	009	-1930 Sep 11	03:30:46	44768	-48600	-013	N	-a	1.2728	0.5168	-0.4721	189.4	-	-	10S	134E
166	009	-1930 Oct 10	14:14:24	44766	-48599	025	N	a-	-1.4059	0.2663	-0.7100	136.2	-	-	1S	28W
167	009	-1929 Mar 06	22:51:19	44757	-48594	-008	P	-t	-0.6339	1.7088	0.6812	321.6	175.3	-	11N	152W
168	009	-1929 Aug 31	14:02:16	44745	-48588	-003	P	-a	0.6084	1.7621	0.7213	325.9	179.3	-	14S	23W
169	009	-1928 Feb 24	10:00:35	44733	-48582	002	T+	p-	0.1325	2.6032	1.6262	327.2	213.9	96.3	16N	41E
170	009	-1928 Aug 19	17:25:21	44722	-48576	007	T-	pp	-0.1117	2.6946	1.6122	373.1	233.0	102.1	18S	74W
171	009	-1927 Feb 13	01:49:25	44710	-48570	012	P	a-	0.8312	1.3100	0.3550	268.2	123.4	-	19N	164E
172	009	-1927 Aug 08	17:15:18	44699	-48564	017	P	t-	-0.8322	1.3706	0.2920	317.5	127.5	-	22S	72W
173	009	-1926 Jan 04	06:41:30	44689	-48559	-016	N	-a	-1.1486	0.7560	-0.2553	229.0	-	-	23N	88E
174	009	-1926 Feb 02	17:47:59	44687	-48558	022	N	a-	1.5407	0.0197	-0.9580	38.4	-	-	23N	76W
175	009	-1926 Jun 29	09:52:43	44677	-48553	-011	N	-h	1.1822	0.6858	-0.3083	220.4	-	-	22S	36E
176	009	-1926 Jul 28	21:01:47	44676	-48552	027	N	h-	-1.5162	0.0912	-0.9392	88.1	-	-	24S	129W
177	009	-1926 Dec 24	14:16:18	44666	-48547	-006	P	-h	-0.4837	2.0062	0.9351	346.5	200.3	-	23N	27W
178	009	-1925 Jun 19	00:17:31	44654	-48541	-001	T	-p	0.3961	2.1056	1.1562	309.6	196.5	56.7	22S	179E
179	009	-1925 Dec 13	14:55:19	44643	-48535	004	T+	pp	0.2042	2.5363	1.4313	376.0	230.1	93.0	22N	38W
180	009	-1924 Jun 07	17:29:26	44631	-48529	009	T	p-	-0.3417	2.2028	1.2587	310.6	200.2	70.4	21S	79W

Cat Num	Canon Plate	Calendar Date	TD of Greatest Eclipse	ΔT s	Luna Num	Saros Num	Ecl. Type	QSE	Gamma	Pen. Mag.	Um. Mag.	Eclipse Phase Durations Pen. m	Par. m	Total m	Greatest in Zenith Lat.	Lng.
181	010	-1924 Dec 01	14:32:33	44620	-48523	014	P	h-	0.8742	1.2973	0.2111	308.1	108.3	-	20N	34W
182	010	-1923 Apr 28	23:19:12	44610	-48518	-019	Ne	-h	1.5009	0.1078	-0.9001	94.2	-	-	7S	164W
183	010	-1923 May 28	08:48:40	44608	-48517	019	N	h-	-1.1217	0.7897	-0.1904	231.1	-	-	19S	51E
184	010	-1923 Oct 22	06:56:11	44598	-48512	-014	N	-a	-1.2928	0.4864	-0.5148	185.4	-	-	4N	81E
185	010	-1923 Nov 20	20:04:34	44597	-48511	024	N	a-	1.4993	0.1227	-0.9088	99.7	-	-	17N	118W
186	010	-1922 Apr 18	04:58:14	44587	-48506	-009	P	-t	0.7862	1.4445	0.3866	319.7	143.8	-	3S	112E
187	010	-1922 Oct 11	21:38:54	44575	-48500	-004	P	-a	-0.5964	1.7484	0.7786	295.0	170.9	-	0N	140W
188	010	-1921 Apr 07	05:39:20	44564	-48494	001	T+	pp	0.0302	2.8377	1.7681	373.9	235.6	106.2	0N	103E
189	010	-1921 Oct 01	13:10:59	44552	-48488	006	T+	pp	0.0741	2.7140	1.7300	325.8	213.0	97.9	3S	13W
190	010	-1920 Mar 26	08:35:29	44541	-48482	011	P	t-	-0.7246	1.5457	0.5112	316.3	158.0	-	4N	59E
191	010	-1920 Sep 20	00:53:59	44529	-48476	016	P	h-	0.8037	1.3999	0.3668	298.9	133.6	-	7S	172E
192	010	-1919 Feb 14	09:20:46	44519	-48471	-017	N	-a	1.1871	0.6637	-0.3046	208.0	-	-	19N	51E
193	010	-1919 Mar 15	18:28:43	44517	-48470	021	N	a-	-1.4210	0.2415	-0.7407	134.0	-	-	7N	89W
194	010	-1919 Aug 10	12:34:17	44508	-48465	-012	N	-t	-1.3829	0.3631	-0.7216	182.3	-	-	22S	2W
195	010	-1919 Sep 09	05:36:35	44506	-48464	026	N	t-	1.5633	0.0306	-1.0512	53.8	-	-	11S	101E
196	010	-1918 Feb 04	01:35:40	44496	-48459	-007	P	-a	0.4818	1.9530	0.9942	303.0	186.3	-	21N	166E
197	010	-1918 Jul 30	12:52:49	44485	-48453	-002	P	-t	-0.6502	1.6982	0.6320	336.0	176.2	-	23S	8W
198	010	-1917 Jan 24	16:41:03	44473	-48447	003	T-	p-	-0.2209	2.4493	1.4557	329.1	211.6	88.5	22N	61W
199	010	-1917 Jul 19	18:45:27	44462	-48441	008	T+	pp	0.1221	2.6396	1.6283	343.9	221.5	99.1	24S	97W
200	010	-1916 Jan 14	02:10:19	44450	-48435	013	P	t-	-0.9760	1.0938	0.0407	280.4	47.6	-	23N	155E
201	011	-1916 Jul 08	07:48:07	44438	-48429	018	P	a-	0.8512	1.2758	0.3159	270.1	118.8	-	23S	67E
202	011	-1916 Dec 03	09:11:29	44429	-48424	-015	N	-t	1.3306	0.4698	-0.6361	206.5	-	-	21N	46E
203	011	-1915 May 29	17:44:39	44417	-48418	-010	P	-a	-0.9511	1.0875	0.1374	253.1	80.1	-	19S	84W
204	011	-1915 Nov 22	09:56:28	44406	-48412	-005	P	-t	0.6279	1.7425	0.6698	333.3	177.2	-	17N	34E
205	011	-1914 May 19	07:46:48	44394	-48406	000	T-	-p	-0.2133	2.4641	1.4690	337.8	217.2	91.5	15S	66E
206	011	-1914 Nov 11	17:35:59	44383	-48400	005	T-	p-	-0.0882	2.7036	1.6887	337.1	217.2	98.4	13N	81W
207	011	-1913 May 08	15:00:57	44371	-48394	010	P	t-	0.5828	1.8126	0.7649	338.9	188.7	-	11S	42W
208	011	-1913 Nov 01	07:21:56	44360	-48388	015	P	a-	-0.7498	1.4698	0.4944	281.1	143.0	-	8N	73E
209	011	-1912 Mar 28	01:26:18	44350	-48383	-018	N	-t	-1.4890	0.1585	-0.9065	121.2	-	-	3N	165E
210	011	-1912 Apr 26	16:19:27	44348	-48382	020	N	t-	1.3567	0.4031	-0.6657	190.8	-	-	6S	60W
211	011	-1912 Sep 21	11:51:57	44338	-48377	-013	N	-a	1.2931	0.4824	-0.5119	184.2	-	-	6S	6E
212	011	-1912 Oct 20	22:59:59	44337	-48376	025	N	a-	-1.4048	0.2699	-0.7095	137.3	-	-	3N	161W
213	011	-1911 Mar 17	06:02:25	44327	-48371	-008	P	-t	-0.6917	1.5994	0.5783	313.7	163.7	-	7N	98E
214	011	-1911 Sep 10	21:52:33	44315	-48365	-003	P	-a	0.6365	1.7139	0.6666	324.9	174.5	-	11S	143W
215	011	-1910 Mar 06	17:43:30	44304	-48359	002	T+	pp	0.0818	2.6934	1.7220	326.6	214.7	98.8	12N	77W
216	011	-1910 Aug 31	00:44:07	44292	-48353	007	T-	pp	-0.0731	2.7679	1.6804	374.8	234.0	103.9	15S	174E
217	011	-1909 Feb 24	09:51:39	44281	-48347	012	P	a-	0.7901	1.3839	0.4319	273.1	134.6	-	16N	42E
218	011	-1909 Aug 20	00:24:50	44269	-48341	017	P	t-	-0.7839	1.4601	0.3796	324.1	143.2	-	19S	179E
219	011	-1908 Jan 15	14:58:54	44260	-48336	-016	N	-a	-1.1689	0.7189	-0.2926	224.9	-	-	22N	37W
220	011	-1908 Feb 14	01:51:59	44258	-48335	022	N	a-	1.5115	0.0728	-0.9042	73.6	-	-	20N	162E
221	012	-1908 Jul 09	17:09:26	44248	-48330	-011	N	-h	1.2420	0.5753	-0.4176	203.3	-	-	23S	74W
222	012	-1908 Aug 08	04:26:05	44246	-48329	027	N	h-	-1.4624	0.1897	-0.8405	125.6	-	-	22S	118E
223	012	-1907 Jan 03	22:16:52	44237	-48324	-006	P	-h	-0.4984	1.9793	0.9082	346.3	198.9	-	23N	148W
224	012	-1907 Jun 29	07:46:09	44225	-48318	-001	T	-a	0.4639	1.9814	1.0318	305.0	189.6	26.5	23S	66E
225	012	-1907 Dec 23	22:47:02	44214	-48312	004	T+	pp	0.1935	2.5549	1.4519	376.1	230.8	94.4	23N	157W
226	012	-1906 Jun 19	00:55:24	44202	-48306	009	T-	p-	-0.2707	2.3343	1.3876	314.2	204.9	82.5	23S	168E
227	012	-1906 Dec 12	22:38:44	44191	-48300	014	P	h-	0.8670	1.3086	0.2263	308.1	111.6	-	22N	156W
228	012	-1905 Jun 08	15:54:50	44179	-48294	019	N	h-	-1.0508	0.9220	-0.0625	247.4	-	-	21S	58W
229	012	-1905 Nov 02	15:39:00	44169	-48289	-014	N	-a	-1.2901	0.4904	-0.5092	185.6	-	-	8N	52W
230	012	-1905 Dec 02	04:38:02	44167	-48288	024	N	a-	1.4949	0.1288	-0.8989	101.7	-	-	21N	113E
231	012	-1904 Apr 28	11:24:14	44158	-48283	-009	P	-t	0.8720	1.2874	0.2287	308.0	113.5	-	7S	13E
232	012	-1904 Oct 22	06:26:51	44146	-48277	-004	P	-a	-0.6006	1.7412	0.7702	294.6	170.1	-	5N	86E
233	012	-1903 Apr 17	12:03:29	44135	-48271	001	T+	pp	0.1136	2.6832	1.6165	372.2	234.0	103.0	4S	4E
234	012	-1903 Oct 11	21:48:28	44123	-48265	006	T+	pp	0.0661	2.7307	1.7426	326.9	213.4	98.1	1N	144W
235	012	-1902 Apr 06	15:26:31	44112	-48259	011	P	t-	-0.6490	1.6817	0.6528	323.0	173.9	-	0S	46W
236	012	-1902 Oct 01	09:05:16	44100	-48253	016	P	h-	0.7919	1.4245	0.3856	302.0	137.0	-	3S	47E
237	012	-1901 Feb 25	17:15:41	44091	-48248	-017	N	-a	1.2280	0.5861	-0.3772	196.4	-	-	16N	70W
238	012	-1901 Mar 27	01:54:06	44089	-48247	021	N	a-	-1.3563	0.3575	-0.6192	160.6	-	-	3N	157E
239	012	-1901 Aug 21	19:41:17	44079	-48242	-012	N	-t	-1.4295	0.2798	-0.8091	161.5	-	-	19S	110W
240	012	-1901 Sep 20	13:15:40	44077	-48241	026	N	t-	1.5417	0.0728	-1.0141	83.0	-	-	7S	16W

Cat Num	Canon Plate	Calendar Date	TD of Greatest Eclipse	ΔT s	Luna Num	Saros Num	Ecl. Type	QSE	Gamma	Pen. Mag.	Um. Mag.	Eclipse Phase Durations Pen. m	Eclipse Phase Durations Par. m	Eclipse Phase Durations Total m	Greatest in Zenith Lat.	Greatest in Zenith Lng.
241	013	-1900 Feb 15	09:44:19	44068	-48236	-007	P	-a	0.5192	1.8834	0.9270	300.3	182.2	-	19N	43E
242	013	-1900 Aug 09	19:59:34	44056	-48230	-002	P	-t	-0.7020	1.6037	0.5365	329.6	164.5	-	21S	116W
243	013	-1899 Feb 04	00:46:57	44045	-48224	003	T-	p-	-0.1913	2.5035	1.5102	330.9	213.5	91.9	20N	176E
244	013	-1899 Jul 30	02:09:52	44033	-48218	008	T+	pp	0.0657	2.7428	1.7319	343.4	221.9	101.3	22S	151E
245	013	-1898 Jan 24	10:00:54	44022	-48212	013	P	t-	-0.9537	1.1347	0.0817	285.2	67.3	-	22N	37E
246	013	-1898 Jul 19	15:26:32	44010	-48206	018	P	a-	0.7913	1.3860	0.4256	277.4	135.4	-	23S	49W
247	013	-1898 Dec 14	17:07:48	44001	-48201	-015	N	-t	1.3364	0.4581	-0.6456	203.9	-	-	23N	74W
248	013	-1897 Jun 10	01:04:51	43989	-48195	-010	N	-a	-1.0253	0.9530	-0.0001	241.1	-	-	22S	164E
249	013	-1897 Jul 09	08:15:12	43987	-48194	028	N	a-	1.4840	0.1081	-0.8387	87.0	-	-	22S	58E
250	013	-1897 Dec 03	18:13:37	43978	-48189	-005	P	-t	0.6268	1.7427	0.6737	332.1	177.1	-	20N	92W
251	013	-1896 May 29	14:43:12	43966	-48183	000	T	-a	-0.2931	2.3196	1.3205	336.5	213.3	80.2	18S	40W
252	013	-1896 Nov 22	02:17:20	43955	-48177	005	T-	p-	-0.0902	2.6984	1.6864	335.8	216.7	98.2	17N	147E
253	013	-1895 May 18	21:28:01	43943	-48171	010	P	t-	0.4983	1.9690	0.9188	347.3	201.8	-	15S	141W
254	013	-1895 Nov 11	16:15:10	43932	-48165	015	P	a-	-0.7504	1.4684	0.4935	280.7	142.7	-	12N	63W
255	013	-1894 Apr 08	07:59:26	43922	-48160	-018	Ne	-t	-1.5608	0.0246	-1.0362	48.2	-	-	2S	64E
256	013	-1894 May 07	22:35:43	43920	-48159	020	N	t-	1.2706	0.5607	-0.5070	221.6	-	-	10S	157W
257	013	-1894 Oct 02	20:19:24	43911	-48154	-013	N	-a	1.3076	0.4582	-0.5412	180.6	-	-	2S	123W
258	013	-1894 Nov 01	07:48:24	43909	-48153	025	N	a-	-1.4064	0.2682	-0.7138	137.2	-	-	8N	64E
259	013	-1893 Mar 28	13:09:19	43899	-48148	-008	P	-t	-0.7536	1.4827	0.4677	304.8	149.3	-	3N	12W
260	013	-1893 Sep 22	05:48:49	43888	-48142	-003	P	-a	0.6587	1.6763	0.6228	324.4	170.4	-	6S	96E
261	014	-1892 Mar 17	01:21:33	43876	-48136	002	T+	pp	0.0264	2.7925	1.8262	325.7	214.9	100.2	8N	166E
262	014	-1892 Sep 10	08:11:00	43865	-48130	007	T-	pp	-0.0416	2.8281	1.7359	376.1	234.6	104.8	11S	61E
263	014	-1891 Mar 06	17:44:59	43853	-48124	012	P	a-	0.7416	1.4715	0.5224	278.6	146.0	-	13N	79W
264	014	-1891 Aug 30	07:44:56	43842	-48118	017	P	t-	-0.7436	1.5350	0.4526	329.0	154.4	-	16S	67E
265	014	-1890 Jan 25	23:05:46	43832	-48113	-016	N	-a	-1.1970	0.6673	-0.3443	218.5	-	-	21N	160W
266	014	-1890 Feb 24	09:45:03	43830	-48112	022	N	a-	1.4743	0.1409	-0.8356	102.0	-	-	17N	42E
267	014	-1890 Jul 21	00:35:44	43821	-48107	-011	N	-a	1.2950	0.4779	-0.5145	186.4	-	-	22S	173E
268	014	-1890 Aug 19	12:02:13	43819	-48106	027	N	a-	-1.4170	0.2730	-0.7572	149.0	-	-	20S	3E
269	014	-1889 Jan 15	06:06:31	43809	-48101	-006	P	-t	-0.5212	1.9374	0.8665	345.4	196.3	-	23N	94E
270	014	-1889 Jul 10	15:22:29	43798	-48095	-001	P	-a	0.5255	1.8687	0.9183	300.3	182.3	-	23S	49W
271	014	-1888 Jan 04	06:31:35	43786	-48089	004	T+	pp	0.1771	2.5837	1.4832	376.3	231.6	96.4	24N	86E
272	014	-1888 Jun 29	08:24:20	43775	-48083	009	T-	p-	-0.2027	2.4607	1.5108	317.1	208.5	90.4	24S	54E
273	014	-1888 Dec 23	06:42:08	43763	-48077	014	P	h-	0.8568	1.3250	0.2472	308.3	116.0	-	24N	82E
274	014	-1887 Jun 18	23:01:01	43752	-48071	019	P	h-	-0.9809	1.0527	0.0632	261.9	57.0	-	23S	166W
275	014	-1887 Nov 13	00:25:20	43743	-48066	-014	N	-a	-1.2857	0.4974	-0.5001	186.2	-	-	12N	174E
276	014	-1887 Dec 12	13:10:04	43741	-48065	024	N	a-	1.4886	0.1382	-0.8851	104.8	-	-	23N	16W
277	014	-1886 May 09	17:44:44	43731	-48060	-009	P	-t	0.9614	1.1241	0.0641	293.6	61.6	-	11S	85W
278	014	-1886 Nov 02	15:19:24	43720	-48054	-004	P	-a	-0.6014	1.7401	0.7684	294.5	169.9	-	9N	49W
279	014	-1885 Apr 28	18:24:44	43708	-48048	001	T+	pp	0.1999	2.5233	1.4594	369.4	230.5	95.4	8S	95W
280	014	-1885 Oct 23	06:30:03	43697	-48042	006	T+	pp	0.0620	2.7402	1.7483	328.0	213.8	98.3	6N	83E
281	015	-1884 Apr 16	22:16:03	43685	-48036	011	P	h-	-0.5708	1.8224	0.7988	328.8	187.1	-	4S	152W
282	015	-1884 Oct 11	17:21:36	43674	-48030	016	P	h-	0.7849	1.4400	0.3956	304.5	139.0	-	2N	80W
283	015	-1883 Mar 08	01:03:56	43664	-48025	-017	N	-a	1.2748	0.4978	-0.4607	182.1	-	-	12N	171E
284	015	-1883 Apr 06	09:14:50	43662	-48024	021	N	a-	-1.2868	0.4823	-0.4892	183.6	-	-	1S	44E
285	015	-1883 Sep 01	02:56:47	43653	-48019	-012	N	-t	-1.4689	0.2095	-0.8835	140.8	-	-	16S	139E
286	015	-1883 Sep 30	21:01:40	43651	-48018	026	N	t-	1.5258	0.1044	-0.9873	99.4	-	-	2S	135W
287	015	-1882 Feb 25	17:44:55	43641	-48013	-007	P	-a	0.5631	1.8017	0.8474	297.1	176.8	-	15N	79W
288	015	-1882 Aug 21	03:17:11	43630	-48007	-002	P	-t	-0.7455	1.5243	0.4561	323.7	153.3	-	19S	134E
289	015	-1881 Feb 15	08:43:08	43618	-48001	003	T-	pp	-0.1543	2.5713	1.5782	332.8	215.6	95.3	18N	56E
290	015	-1881 Aug 10	09:45:33	43607	-47995	008	T+	pp	0.0177	2.8309	1.8202	342.5	221.6	101.9	21S	36E
291	015	-1880 Feb 04	17:41:24	43596	-47989	013	P	t-	-0.9239	1.1892	0.1365	291.1	86.6	-	20N	79W
292	015	-1880 Jul 29	23:12:52	43584	-47983	018	P	a-	0.7374	1.4853	0.5241	283.3	147.7	-	22S	167W
293	015	-1880 Dec 25	00:58:50	43575	-47978	-015	N	-t	1.3465	0.4380	-0.6627	199.5	-	-	25N	168E
294	015	-1879 Jun 20	08:27:21	43563	-47972	-010	N	-a	-1.0972	0.8229	-0.1339	227.8	-	-	24S	52E
295	015	-1879 Jul 19	15:57:40	43561	-47971	028	N	a-	1.4295	0.2098	-0.7403	120.3	-	-	22S	59W
296	015	-1879 Dec 14	02:26:55	43552	-47966	-005	P	-t	0.6295	1.7353	0.6708	330.4	176.3	-	23N	144E
297	015	-1878 Jun 09	21:39:41	43540	-47960	000	T	-a	-0.3719	2.1775	1.1736	334.4	207.9	62.2	21S	146W
298	015	-1878 Dec 03	10:57:29	43529	-47954	005	T-	p-	-0.0903	2.6964	1.6881	334.5	216.2	98.1	20N	16E
299	015	-1877 May 30	03:54:06	43517	-47948	010	T	t-	0.4136	2.1259	1.0727	354.5	212.3	43.3	18S	120E
300	015	-1877 Nov 23	01:08:26	43506	-47942	015	P	a-	-0.7508	1.4671	0.4931	280.3	142.6	-	16N	162E

Cat Num	Canon Plate	Calendar Date	TD of Greatest Eclipse	ΔT s	Luna Num	Saros Num	Ecl. Type	QSE	Gamma	Pen. Mag.	Um. Mag.	Eclipse Phase Durations Pen. m	Eclipse Phase Durations Par. m	Eclipse Phase Durations Total m	Greatest in Zenith Lat.	Greatest in Zenith Lng.
301	016	-1876 May 18	04:50:52	43495	-47936	020	N	t-	1.1826	0.7217	-0.3451	247.3	-	-	14S	107E
302	016	-1876 Oct 13	04:52:46	43485	-47931	-013	N	-a	1.3169	0.4435	-0.5606	178.6	-	-	3N	106E
303	016	-1876 Nov 11	16:39:15	43483	-47930	025	N	a-	-1.4107	0.2615	-0.7227	135.8	-	-	12N	70W
304	016	-1875 Apr 07	20:09:37	43474	-47925	-008	P	-t	-0.8219	1.3545	0.3453	294.2	130.3	-	1S	120W
305	016	-1875 Oct 02	13:52:31	43462	-47919	-003	P	-a	0.6738	1.6514	0.5921	324.4	167.5	-	2S	28W
306	016	-1874 Mar 28	08:52:26	43451	-47913	002	T-	pp	-0.0358	2.7728	1.8114	324.4	214.5	100.0	4N	51E
307	016	-1874 Sep 21	15:46:41	43439	-47907	007	T-	pp	-0.0177	2.8742	1.7777	377.1	234.9	105.1	7S	56W
308	016	-1873 Mar 18	01:31:20	43428	-47901	012	P	a-	0.6873	1.5699	0.6231	284.3	157.1	-	8N	162E
309	016	-1873 Sep 10	15:15:52	43417	-47895	017	P	t-	-0.7113	1.5950	0.5111	332.6	162.2	-	12S	47W
310	016	-1872 Feb 06	07:04:40	43407	-47890	-016	N	-a	-1.2308	0.6053	-0.4063	210.0	-	-	19N	79E
311	016	-1872 Mar 06	17:30:33	43405	-47889	022	N	a-	1.4319	0.2186	-0.7576	126.5	-	-	13N	77W
312	016	-1872 Jul 31	08:09:25	43396	-47884	-011	N	-a	1.3427	0.3902	-0.6018	169.3	-	-	21S	58E
313	016	-1872 Aug 29	19:47:20	43394	-47883	027	N	a-	-1.3779	0.3447	-0.6854	165.8	-	-	16S	115W
314	016	-1871 Jan 25	13:48:29	43384	-47878	-006	P	-t	-0.5493	1.8853	0.8152	343.8	192.6	-	22N	22W
315	016	-1871 Jul 20	23:04:45	43373	-47872	-001	P	-a	0.5822	1.7653	0.8135	295.5	174.5	-	23S	166W
316	016	-1870 Jan 14	14:08:08	43361	-47866	004	T+	pp	0.1543	2.6239	1.5265	376.5	232.6	98.7	24N	28W
317	016	-1870 Jul 10	15:58:39	43350	-47860	009	T-	pp	-0.1398	2.5782	1.6244	319.5	210.9	95.4	24S	60W
318	016	-1869 Jan 03	14:39:46	43339	-47854	014	P	h-	0.8416	1.3503	0.2777	309.2	122.1	-	25N	38W
319	016	-1869 Jun 30	06:11:04	43327	-47848	019	P	h-	-0.9149	1.1765	0.1817	274.4	95.3	-	24S	86E
320	016	-1869 Nov 24	09:10:50	43318	-47843	-014	N	-a	-1.2825	0.5020	-0.4929	186.5	-	-	16N	41E
321	017	-1869 Dec 23	21:37:30	43316	-47842	024	N	a-	1.4782	0.1548	-0.8635	110.4	-	-	25N	144W
322	017	-1868 May 20	00:04:59	43306	-47837	-009	N	-t	1.0499	0.9625	-0.0992	277.0	-	-	15S	177E
323	017	-1868 Nov 13	00:13:00	43295	-47831	-004	P	-a	-0.6014	1.7403	0.7682	294.6	169.9	-	13N	175E
324	017	-1867 May 09	00:43:26	43283	-47825	001	T	pp	0.2887	2.3594	1.2977	365.3	225.0	81.8	12S	168E
325	017	-1867 Nov 02	15:14:41	43272	-47819	006	T+	pp	0.0610	2.7436	1.7485	329.1	214.2	98.4	10N	50W
326	017	-1866 Apr 28	05:02:45	43261	-47813	011	P	h-	-0.4893	1.9696	0.9507	333.7	198.0	-	8S	104E
327	017	-1866 Oct 23	01:41:56	43249	-47807	016	P	h-	0.7822	1.4475	0.3983	306.3	139.8	-	6N	153E
328	017	-1865 Mar 19	08:44:44	43240	-47802	-017	N	-a	1.3280	0.3980	-0.5562	163.9	-	-	9N	53E
329	017	-1865 Apr 17	16:32:08	43238	-47801	021	N	a-	-1.2138	0.6141	-0.3529	203.7	-	-	5S	68W
330	017	-1865 Sep 12	10:23:19	43228	-47796	-012	N	-t	-1.4991	0.1560	-0.9406	122.2	-	-	12S	25E
331	017	-1865 Oct 12	04:55:09	43226	-47795	026	N	t-	1.5164	0.1237	-0.9721	108.3	-	-	2N	105E
332	017	-1864 Mar 08	01:36:56	43217	-47790	-007	P	-a	0.6140	1.7074	0.7549	292.9	169.7	-	12N	160E
333	017	-1864 Aug 31	10:45:24	43206	-47784	-002	P	-t	-0.7811	1.4596	0.3905	318.4	142.9	-	15S	20E
334	017	-1863 Feb 25	16:29:24	43194	-47778	003	T-	pp	-0.1098	2.6529	1.6599	334.6	217.5	98.4	15N	63W
335	017	-1863 Aug 20	17:31:26	43183	-47772	008	T-	pp	-0.0229	2.8212	1.8106	341.3	220.8	101.5	18S	82W
336	017	-1862 Feb 15	01:10:16	43171	-47766	013	P	t-	-0.8853	1.2598	0.2075	298.2	106.0	-	17N	167E
337	017	-1862 Aug 10	07:08:20	43160	-47760	018	P	a-	0.6905	1.5719	0.6095	287.9	156.9	-	20S	73E
338	017	-1861 Jan 05	08:44:19	43151	-47755	-015	N	-t	1.3613	0.4090	-0.6881	193.0	-	-	25N	51E
339	017	-1861 Jul 01	15:52:55	43139	-47749	-010	N	-a	-1.1657	0.6992	-0.2617	213.3	-	-	25S	60W
340	017	-1861 Jul 30	23:46:14	43137	-47748	028	N	a-	1.3801	0.3022	-0.6516	143.4	-	-	21S	177W
341	018	-1861 Dec 25	10:37:35	43128	-47743	-005	P	-t	0.6352	1.7224	0.6629	328.3	175.1	-	24N	21E
342	018	-1860 Jun 20	04:37:27	43116	-47737	000	T	-a	-0.4486	2.0394	1.0304	331.5	200.8	27.3	23S	108E
343	018	-1860 Dec 13	19:36:06	43105	-47731	005	T-	p-	-0.0887	2.6973	1.6930	333.3	215.8	98.1	22N	115W
344	018	-1859 Jun 09	10:19:36	43094	-47725	010	T	tp	0.3290	2.2828	1.2262	360.5	220.6	72.7	20S	22E
345	018	-1859 Dec 03	10:00:51	43082	-47719	015	P	a-	-0.7507	1.4667	0.4942	280.0	142.6	-	19N	28E
346	018	-1858 May 29	11:07:47	43071	-47713	020	N	t-	1.0952	0.8818	-0.1847	268.7	-	-	17S	10E
347	018	-1858 Oct 24	13:29:07	43062	-47708	-013	N	-a	1.3233	0.4340	-0.5744	177.5	-	-	7N	25W
348	018	-1858 Nov 23	01:28:40	43060	-47707	025	N	a-	-1.4143	0.2556	-0.7302	134.7	-	-	15N	156E
349	018	-1857 Apr 19	03:08:59	43050	-47702	-008	P	-t	-0.8916	1.2238	0.2199	282.4	105.5	-	5S	132E
350	018	-1857 Oct 13	22:01:10	43039	-47696	-003	P	-a	0.6843	1.6348	0.5704	324.7	165.4	-	2N	152W
351	018	-1856 Apr 07	16:19:37	43027	-47690	002	T-	pp	-0.1016	2.6498	1.6927	322.6	213.2	98.0	1S	64W
352	018	-1856 Oct 01	23:29:24	43016	-47684	007	T+	pp	0.0003	2.9081	1.8078	377.9	235.0	105.0	3S	174W
353	018	-1855 Mar 28	09:10:38	43005	-47678	012	P	a-	0.6272	1.6793	0.7344	290.1	167.6	-	4N	44E
354	018	-1855 Sep 20	22:57:00	42993	-47672	017	P	t-	-0.6870	1.6402	0.5551	334.9	167.5	-	8S	165W
355	018	-1854 Feb 16	14:50:56	42984	-47667	-016	N	-a	-1.2738	0.5265	-0.4852	198.1	-	-	16N	39W
356	018	-1854 Mar 18	01:05:07	42982	-47666	022	N	a-	1.3815	0.3111	-0.6653	150.1	-	-	9N	167E
357	018	-1854 Aug 11	15:54:40	42973	-47661	-011	N	-a	1.3817	0.3185	-0.6735	153.5	-	-	19S	59W
358	018	-1854 Sep 10	03:44:40	42971	-47660	027	N	a-	-1.3479	0.3999	-0.6302	177.1	-	-	13S	124E
359	018	-1853 Feb 05	21:18:28	42961	-47655	-006	P	-t	-0.5862	1.8173	0.7480	341.2	187.1	-	20N	136W
360	018	-1853 Aug 01	06:55:16	42950	-47649	-001	P	-a	0.6323	1.6743	0.7208	291.0	166.6	-	22S	75E

Cat Num	Canon Plate	Calendar Date	TD of Greatest Eclipse	ΔT s	Luna Num	Saros Num	Ecl. Type	QSE	Gamma	Pen. Mag.	Um. Mag.	Eclipse Phase Durations Pen. m	Eclipse Phase Durations Par. m	Eclipse Phase Durations Total m	Greatest in Zenith Lat.	Greatest in Zenith Lng.
361	019	-1852 Jan 25	21:35:43	42938	-47643	004	T+	pp	0.1243	2.6772	1.5835	376.6	233.7	101.3	22N	141W
362	019	-1852 Jul 20	23:38:26	42927	-47637	009	T-	pp	-0.0820	2.6863	1.7282	321.3	212.4	98.2	24S	176W
363	019	-1851 Jan 13	22:31:01	42916	-47631	014	P	h-	0.8202	1.3866	0.3199	310.8	129.9	-	24N	156W
364	019	-1851 Jul 10	13:23:39	42904	-47625	019	P	h-	-0.8517	1.2955	0.2948	285.6	119.8	-	25S	24W
365	019	-1851 Dec 04	17:56:56	42895	-47620	-014	N	-a	-1.2796	0.5058	-0.4860	186.5	-	-	19N	91W
366	019	-1850 Jan 03	06:00:58	42893	-47619	024	N	a-	1.4638	0.1784	-0.8344	117.7	-	-	25N	90E
367	019	-1850 May 31	06:23:46	42884	-47614	-009	N	-t	1.1386	0.8008	-0.2630	257.4	-	-	18S	80E
368	019	-1850 Nov 24	09:07:10	42872	-47608	-004	P	-a	-0.6011	1.7407	0.7688	294.6	170.0	-	17N	40E
369	019	-1849 May 20	07:03:18	42861	-47602	001	T	-t	0.3769	2.1966	1.1368	360.0	217.3	58.5	15S	71E
370	019	-1849 Nov 14	00:01:20	42850	-47596	006	T+	pp	0.0623	2.7426	1.7447	330.1	214.6	98.5	14N	176E
371	019	-1848 May 08	11:49:34	42838	-47590	011	T	h-	-0.4066	2.1192	1.1047	337.6	206.7	49.9	12S	1W
372	019	-1848 Nov 02	10:05:23	42827	-47584	016	P	h-	0.7827	1.4488	0.3953	307.6	139.7	-	11N	25E
373	019	-1847 Mar 29	16:20:41	42817	-47579	-017	N	-a	1.3856	0.2903	-0.6600	141.1	-	-	4N	64W
374	019	-1847 Apr 27	23:47:34	42816	-47578	021	N	a-	-1.1383	0.7506	-0.2124	221.4	-	-	9S	180W
375	019	-1847 Sep 22	17:58:01	42806	-47573	-012	N	-t	-1.5230	0.1137	-0.9861	104.9	-	-	8S	90W
376	019	-1847 Oct 22	12:52:58	42804	-47572	026	N	t-	1.5102	0.1368	-0.9623	114.0	-	-	7N	17W
377	019	-1846 Mar 19	09:21:02	42795	-47567	-007	P	-a	0.6716	1.6010	0.6499	287.7	160.4	-	8N	42E
378	019	-1846 Sep 11	18:24:42	42783	-47561	-002	P	-t	-0.8086	1.4094	0.3396	314.0	134.0	-	12S	97W
379	019	-1845 Mar 09	00:06:45	42772	-47555	003	T-	pp	-0.0584	2.7475	1.7542	336.4	219.1	100.8	11N	179W
380	019	-1845 Sep 01	01:27:01	42761	-47549	008	T-	pp	-0.0567	2.7593	1.7488	340.0	219.8	100.6	15S	157E
381	020	-1844 Feb 26	08:28:57	42749	-47543	013	P	t-	-0.8390	1.3447	0.2927	306.0	124.7	-	14N	55E
382	020	-1844 Aug 20	15:12:24	42738	-47537	018	P	a-	0.6502	1.6467	0.6828	291.6	163.8	-	17S	49W
383	020	-1843 Jan 15	16:20:59	42729	-47532	-015	N	-t	1.3833	0.3666	-0.7262	183.0	-	-	25N	64W
384	020	-1843 Jul 11	23:22:55	42717	-47526	-010	N	-a	-1.2301	0.5835	-0.3822	197.8	-	-	25S	174W
385	020	-1843 Aug 10	07:42:32	42715	-47525	028	N	a-	1.3372	0.3831	-0.5748	160.5	-	-	19S	62E
386	020	-1842 Jan 04	18:41:51	42706	-47520	-005	P	-t	0.6471	1.6977	0.6439	325.5	172.6	-	25N	101W
387	020	-1842 Jul 01	11:38:36	42695	-47514	000	P	-a	-0.5214	1.9086	0.8939	328.0	192.4	-	24S	1E
388	020	-1842 Dec 25	04:09:24	42683	-47508	005	T-	p-	-0.0818	2.7076	1.7080	332.0	215.5	98.3	23N	116E
389	020	-1841 Jun 20	16:48:26	42672	-47502	010	T+	pp	0.2475	2.4344	1.3739	365.3	226.8	88.7	22S	77W
390	020	-1841 Dec 14	18:50:51	42661	-47496	015	P	a-	-0.7482	1.4702	0.4996	280.0	143.2	-	21N	105W
391	020	-1840 Jun 08	17:26:00	42649	-47490	020	N*	t-	1.0079	1.0420	-0.0245	287.0	-	-	20S	86W
392	020	-1840 Nov 03	22:09:20	42640	-47485	-013	N	-a	1.3261	0.4307	-0.5814	177.5	-	-	12N	157W
393	020	-1840 Dec 03	10:17:51	42638	-47484	025	N	a-	-1.4183	0.2489	-0.7380	133.3	-	-	18N	22E
394	020	-1839 Apr 29	10:05:09	42629	-47479	-008	P	-t	-0.9651	1.0866	0.0874	268.9	67.5	-	10S	25E
395	020	-1839 Oct 24	06:14:49	42617	-47473	-003	P	-a	0.6897	1.6272	0.5581	325.6	164.4	-	7N	82E
396	020	-1838 Apr 18	23:42:07	42606	-47467	002	T-	pp	-0.1721	2.5186	1.5653	320.3	210.9	93.5	5S	178W
397	020	-1838 Oct 13	07:19:54	42595	-47461	007	T+	pp	0.0117	2.8886	1.7853	378.6	235.1	104.9	2N	66E
398	020	-1837 Apr 08	16:45:03	42583	-47455	012	P	a-	0.5632	1.7961	0.8525	295.7	177.1	-	0S	72W
399	020	-1837 Oct 02	06:46:28	42572	-47449	017	P	t-	-0.6687	1.6742	0.5884	336.4	171.2	-	3S	75E
400	020	-1836 Feb 27	22:29:12	42563	-47444	-016	N	-a	-1.3223	0.4376	-0.5743	182.7	-	-	13N	155W
401	021	-1836 Mar 28	08:33:32	42561	-47443	022	N	a-	1.3268	0.4117	-0.5652	171.6	-	-	5N	52E
402	021	-1836 Aug 21	23:48:31	42551	-47438	-011	N	-a	1.4147	0.2582	-0.7341	138.5	-	-	16S	179W
403	021	-1836 Sep 20	11:50:44	42550	-47437	027	N	a-	-1.3239	0.4437	-0.5861	185.1	-	-	9S	0E
404	021	-1835 Feb 16	04:38:33	42540	-47432	-006	P	-t	-0.6306	1.7353	0.6671	337.6	179.6	-	17N	112E
405	021	-1835 Aug 11	14:53:28	42529	-47426	-001	P	-a	0.6763	1.5945	0.6389	286.6	158.8	-	19S	46W
406	021	-1834 Feb 05	04:54:28	42518	-47420	004	T+	pp	0.0867	2.7440	1.6543	376.5	234.7	103.6	20N	108E
407	021	-1834 Aug 01	07:25:29	42506	-47414	009	T-	pp	-0.0307	2.7829	1.8201	322.7	213.2	99.4	22S	66E
408	021	-1833 Jan 25	06:14:39	42495	-47408	014	P	h-	0.7917	1.4359	0.3752	313.1	139.2	-	23N	87E
409	021	-1833 Jul 21	20:43:12	42484	-47402	019	P	t-	-0.7949	1.4028	0.3959	295.0	137.2	-	24S	134W
410	021	-1833 Dec 16	02:39:48	42474	-47397	-014	N	-a	-1.2799	0.5035	-0.4847	185.6	-	-	21N	137E
411	021	-1832 Jan 14	14:18:16	42472	-47396	024	N	a-	1.4439	0.2120	-0.7950	127.4	-	-	25N	35W
412	021	-1832 Jun 10	12:44:28	42463	-47391	-009	N	-t	1.2250	0.6437	-0.4227	234.8	-	-	20S	17W
413	021	-1832 Dec 04	17:59:34	42452	-47385	-004	P	-a	-0.6024	1.7382	0.7669	294.5	169.9	-	20N	94W
414	021	-1831 May 30	13:24:50	42440	-47379	001	P	-t	0.4638	2.0363	0.9778	353.4	207.4	-	18S	27W
415	021	-1831 Nov 24	08:46:54	42429	-47373	006	T+	pp	0.0631	2.7421	1.7421	331.1	215.0	98.6	17N	43E
416	021	-1830 May 19	18:37:15	42418	-47367	011	T	a-	-0.3234	2.2700	1.2592	340.5	213.3	74.4	16S	105W
417	021	-1830 Nov 13	18:29:55	42407	-47361	016	P	h-	0.7849	1.4464	0.3895	308.6	139.2	-	15N	103W
418	021	-1829 Apr 09	23:51:09	42397	-47356	-017	N	-a	1.4482	0.1738	-0.7732	110.1	-	-	0N	179W
419	021	-1829 May 09	07:01:55	42395	-47355	021	N	a-	-1.0609	0.8910	-0.0687	236.9	-	-	13S	69E
420	021	-1829 Oct 04	01:41:02	42386	-47350	-012	N	-t	-1.5403	0.0834	-1.0191	90.1	-	-	3S	151E

Cat Num	Canon Plate	Calendar Date	TD of Greatest Eclipse	ΔT s	Luna Num	Saros Num	Ecl. Type	QSE	Gamma	Pen. Mag.	Um. Mag.	Eclipse Phase Durations Pen. m	Par. m	Total m	Greatest in Zenith Lat.	Lng.
421	022	-1829 Nov 02	20:54:20	42384	-47349	026	N	t-	1.5071	0.1439	-0.9578	117.1	-	-	11N	140W
422	022	-1828 Mar 29	16:57:59	42375	-47344	-007	P	-a	0.7348	1.4845	0.5343	281.3	148.4	-	4N	75W
423	022	-1828 Sep 22	02:14:17	42363	-47338	-002	P	-t	-0.8286	1.3730	0.3028	310.4	126.9	-	7S	143E
424	022	-1827 Mar 19	07:33:04	42352	-47332	003	T+	pp	0.0014	2.8523	1.8585	337.8	220.2	101.9	7N	67E
425	022	-1827 Sep 11	09:33:07	42341	-47326	008	T-	pp	-0.0825	2.7118	1.7013	338.7	218.8	99.4	11S	34E
426	022	-1826 Mar 08	15:36:35	42329	-47320	013	P	t-	-0.7844	1.4448	0.3931	314.5	142.7	-	10N	54W
427	022	-1826 Aug 31	23:25:39	42318	-47314	018	P	a-	0.6173	1.7079	0.7424	294.4	168.9	-	14S	175W
428	022	-1825 Jan 26	23:50:13	42309	-47309	-015	N	-t	1.4112	0.3130	-0.7752	169.5	-	-	23N	177W
429	022	-1825 Feb 25	16:58:50	42307	-47308	023	N	t-	-1.5387	0.0766	-1.0068	86.0	-	-	14N	74W
430	022	-1825 Jul 23	06:58:47	42298	-47303	-010	N	-a	-1.2890	0.4781	-0.4929	181.6	-	-	25S	71E
431	022	-1825 Aug 21	15:46:39	42296	-47302	028	N	a-	1.3010	0.4516	-0.5106	173.5	-	-	17S	60W
432	022	-1824 Jan 16	02:41:44	42286	-47297	-005	P	-t	0.6634	1.6647	0.6171	322.1	169.2	-	24N	139E
433	022	-1824 Jul 11	18:42:29	42275	-47291	000	P	-a	-0.5909	1.7842	0.7634	323.9	182.5	-	24S	106W
434	022	-1823 Jan 04	12:39:16	42264	-47285	005	T-	p-	-0.0714	2.7242	1.7296	330.8	215.3	98.7	24N	12W
435	022	-1823 Jun 30	23:20:30	42253	-47279	010	T+	pp	0.1688	2.5808	1.5161	368.9	231.0	98.3	23S	176W
436	022	-1823 Dec 25	03:35:55	42241	-47273	015	P	a-	-0.7416	1.4812	0.5128	280.4	144.8	-	23N	123E
437	022	-1822 Jun 19	23:49:58	42230	-47267	020	P	t-	0.9244	1.1955	0.1286	302.1	86.8	-	22S	176E
438	022	-1822 Nov 15	06:49:40	42221	-47262	-013	N	-a	1.3286	0.4277	-0.5876	177.6	-	-	16N	71E
439	022	-1822 Dec 14	19:02:39	42219	-47261	025	N	a-	-1.4190	0.2479	-0.7397	133.3	-	-	21N	110W
440	022	-1821 May 10	17:02:58	42209	-47256	-008	N	-t	-1.0379	0.9509	-0.0440	254.0	-	-	13S	82W
441	023	-1821 Nov 04	14:29:35	42198	-47250	-003	P	-a	0.6937	1.6219	0.5490	326.4	163.7	-	11N	44W
442	023	-1820 Apr 29	07:03:25	42187	-47244	002	T-	-p	-0.2443	2.3847	1.4344	317.3	207.4	86.3	9S	69E
443	023	-1820 Oct 23	15:15:07	42176	-47238	007	T+	pp	0.0191	2.8762	1.7706	379.0	235.1	104.8	6N	55W
444	023	-1819 Apr 19	00:12:57	42164	-47232	012	P	a-	0.4938	1.9230	0.9802	301.1	185.9	-	4S	173E
445	023	-1819 Oct 12	14:44:42	42153	-47226	017	P	t-	-0.6569	1.6959	0.6100	337.0	173.3	-	1N	46W
446	023	-1818 Mar 10	05:55:35	42144	-47221	-016	N	-a	-1.3795	0.3329	-0.6795	161.5	-	-	9N	91E
447	023	-1818 Apr 08	15:53:11	42142	-47220	022	N	a-	1.2656	0.5245	-0.4535	192.3	-	-	0N	61W
448	023	-1818 Sep 02	07:53:38	42132	-47215	-011	N	-a	1.4396	0.2125	-0.7799	125.9	-	-	13S	58E
449	023	-1818 Oct 01	20:06:36	42131	-47214	027	N	a-	-1.3070	0.4744	-0.5547	190.3	-	-	4S	126W
450	023	-1817 Feb 27	11:46:51	42121	-47209	-006	P	-t	-0.6838	1.6371	0.5700	332.5	169.1	-	14N	3E
451	023	-1817 Aug 22	23:01:24	42110	-47203	-001	P	-a	0.7127	1.5290	0.5711	282.9	151.6	-	17S	169W
452	023	-1816 Feb 16	12:03:43	42099	-47197	004	T+	pp	0.0413	2.8253	1.7399	376.2	235.4	105.4	18N	1W
453	023	-1816 Aug 11	15:19:00	42087	-47191	009	T+	pp	0.0147	2.8147	1.8469	323.9	213.6	99.6	20S	54W
454	023	-1815 Feb 04	13:51:25	42076	-47185	014	P	h-	0.7567	1.4968	0.4426	316.0	149.2	-	21N	28W
455	023	-1815 Aug 01	04:08:03	42065	-47179	019	P	t-	-0.7435	1.5005	0.4869	303.1	150.6	-	23S	113E
456	023	-1815 Dec 26	11:18:57	42056	-47174	-014	N	-a	-1.2838	0.4943	-0.4898	183.5	-	-	22N	7E
457	023	-1814 Jan 24	22:28:30	42054	-47173	024	N	a-	1.4178	0.2569	-0.7441	139.1	-	-	24N	158W
458	023	-1814 Jun 21	19:07:40	42044	-47168	-009	N	-t	1.3083	0.4923	-0.5769	208.7	-	-	22S	114W
459	023	-1814 Jul 21	10:04:42	42043	-47167	029	Nb	t-	-1.5402	0.0632	-0.9991	76.6	-	-	25S	23E
460	023	-1814 Dec 16	02:49:50	42033	-47162	-004	P	-a	-0.6052	1.7325	0.7622	294.2	169.6	-	22N	133E
461	024	-1813 Jun 10	19:49:53	42022	-47156	001	P	-t	0.5485	1.8804	0.8230	345.8	195.3	-	20S	125W
462	024	-1813 Dec 05	17:30:53	42011	-47150	006	T+	pp	0.0633	2.7428	1.7410	332.1	215.5	98.7	20N	89W
463	024	-1812 May 30	01:27:55	41999	-47144	011	T-	p-	-0.2416	2.4185	1.4109	342.2	218.0	88.4	19S	150E
464	024	-1812 Nov 24	02:53:47	41988	-47138	016	P	h-	0.7871	1.4438	0.3841	309.5	138.8	-	18N	129E
465	024	-1811 Apr 20	07:17:34	41979	-47133	-017	Ne	-a	1.5145	0.0508	-0.8934	60.1	-	-	4S	66E
466	024	-1811 May 19	14:16:11	41977	-47132	021	P	a-	-0.9823	1.0338	0.0767	250.5	60.8	-	16S	42W
467	024	-1811 Oct 14	09:30:50	41968	-47127	-012	N	-t	-1.5522	0.0625	-1.0420	78.1	-	-	1N	32E
468	024	-1811 Nov 13	04:57:17	41966	-47126	026	N	t-	1.5051	0.1482	-0.9551	119.0	-	-	15N	98E
469	024	-1810 Apr 10	00:29:23	41956	-47121	-007	P	-a	0.8025	1.3601	0.4102	273.7	132.7	-	1S	169E
470	024	-1810 Oct 03	10:12:12	41945	-47115	-002	P	-t	-0.8425	1.3473	0.2772	307.6	121.6	-	3S	21E
471	024	-1809 Mar 30	14:52:26	41934	-47109	003	T+	pp	0.0662	2.7340	1.7391	338.8	220.5	101.2	3N	46W
472	024	-1809 Sep 22	17:48:39	41923	-47103	008	T-	pp	-0.1013	2.6772	1.6669	337.4	217.7	98.3	7S	92W
473	024	-1808 Mar 18	22:34:42	41912	-47097	013	P	t-	-0.7225	1.5582	0.5066	323.1	159.6	-	6N	161W
474	024	-1808 Sep 11	07:47:16	41900	-47091	018	P	a-	0.5909	1.7571	0.7900	296.5	172.6	-	10S	58E
475	024	-1807 Feb 06	07:10:16	41891	-47086	-015	N	-t	1.4468	0.2451	-0.8380	150.6	-	-	21N	72E
476	024	-1807 Mar 07	23:49:11	41889	-47085	023	N	t-	-1.4800	0.1827	-0.8975	131.5	-	-	10N	179W
477	024	-1807 Aug 02	14:41:37	41880	-47080	-010	N	-a	-1.3418	0.3840	-0.5926	164.8	-	-	23S	45W
478	024	-1807 Aug 31	23:59:26	41878	-47079	028	N	a-	1.2719	0.5072	-0.4594	183.4	-	-	13S	175E
479	024	-1806 Jan 26	10:32:26	41869	-47074	-005	P	-t	0.6883	1.6158	0.5746	317.7	163.9	-	23N	21E
480	024	-1806 Jul 23	01:53:11	41857	-47068	000	P	-a	-0.6539	1.6719	0.6447	319.6	171.6	-	24S	146E

Cat Num	Canon Plate	Calendar Date	TD of Greatest Eclipse	ΔT s	Luna Num	Saros Num	Ecl. Type	QSE	Gamma	Pen. Mag.	Um. Mag.	Eclipse Phase Durations Pen. m	Par. m	Total m	Greatest in Zenith Lat.	Lng.
481	025	-1805 Jan 15	21:02:10	41846	-47062	005	T-	pp	-0.0540	2.7535	1.7642	329.6	215.1	99.2	23N	138W
482	025	-1805 Jul 12	05:57:54	41835	-47056	010	T+	pp	0.0949	2.7187	1.6497	371.6	233.6	103.6	24S	84E
483	025	-1804 Jan 05	12:15:52	41824	-47050	015	P	a-	-0.7307	1.5000	0.5341	281.3	147.2	-	23N	8W
484	025	-1804 Jun 30	06:19:06	41813	-47044	020	P	t-	0.8438	1.3437	0.2761	314.6	124.3	-	23S	78E
485	025	-1804 Nov 25	15:30:19	41803	-47039	-013	N	-a	1.3302	0.4261	-0.5918	177.9	-	-	19N	60W
486	025	-1804 Dec 25	03:43:16	41801	-47038	025	N	a-	-1.4172	0.2514	-0.7365	134.5	-	-	22N	119E
487	025	-1803 May 20	23:59:59	41792	-47033	-008	N	-t	-1.1122	0.8128	-0.1785	237.2	-	-	17S	172E
488	025	-1803 Jun 19	10:21:59	41790	-47032	030	Nb	t-	1.5508	0.0236	-0.9987	45.1	-	-	21S	16E
489	025	-1803 Nov 14	22:46:49	41781	-47027	-003	P	-a	0.6948	1.6213	0.5453	327.5	163.6	-	15N	170W
490	025	-1802 May 10	14:23:15	41770	-47021	002	T	-p	-0.3182	2.2477	1.2998	313.6	202.7	75.1	13S	44W
491	025	-1802 Nov 03	23:14:11	41758	-47015	007	T+	pp	0.0233	2.8692	1.7621	379.4	235.1	104.7	10N	177W
492	025	-1801 Apr 30	07:38:52	41747	-47009	012	T	a-	0.4228	2.0532	1.1104	306.0	193.4	48.2	8S	59E
493	025	-1801 Oct 23	22:49:56	41736	-47003	017	P	h-	-0.6499	1.7085	0.6232	336.9	174.5	-	5N	170W
494	025	-1800 Mar 20	13:14:07	41727	-46998	-016	N	-a	-1.4420	0.2186	-0.7944	132.7	-	-	5N	22W
495	025	-1800 Apr 18	23:07:28	41725	-46997	022	N	a-	1.2007	0.6444	-0.3352	211.4	-	-	4S	172W
496	025	-1800 Sep 12	16:07:11	41716	-46992	-011	N	-a	1.4587	0.1777	-0.8150	115.1	-	-	9S	68W
497	025	-1800 Oct 12	04:29:49	41714	-46991	027	N	a-	-1.2952	0.4956	-0.5326	193.4	-	-	0N	106E
498	025	-1799 Mar 09	18:46:06	41704	-46986	-006	P	-t	-0.7436	1.5267	0.4607	325.9	155.0	-	10N	104W
499	025	-1799 Sep 02	07:17:36	41693	-46980	-001	P	-a	0.7422	1.4761	0.5157	279.7	145.2	-	13S	65E
500	025	-1798 Feb 26	19:03:40	41682	-46974	004	T-	pp	-0.0120	2.8769	1.7958	375.5	235.5	105.9	14N	107W
501	026	-1798 Aug 22	23:21:06	41671	-46968	009	T+	pp	0.0523	2.7484	1.7754	324.9	213.6	99.0	17S	176W
502	026	-1797 Feb 15	21:20:35	41660	-46962	014	P	h-	0.7146	1.5707	0.5230	319.4	159.6	-	18N	141W
503	026	-1797 Aug 12	11:39:35	41648	-46956	019	P	t-	-0.6985	1.5865	0.5662	310.1	160.9	-	21S	1W
504	026	-1796 Jan 06	19:52:12	41639	-46951	-014	N	-a	-1.2929	0.4754	-0.5044	179.8	-	-	23N	122W
505	026	-1796 Feb 05	06:31:39	41637	-46950	024	N	a-	1.3853	0.3134	-0.6814	152.2	-	-	22N	80E
506	026	-1796 Jul 02	01:36:07	41628	-46945	-009	N	-t	1.3864	0.3506	-0.7218	178.8	-	-	22S	148E
507	026	-1796 Jul 31	17:01:15	41626	-46944	029	N	t-	-1.4851	0.1672	-0.9007	123.9	-	-	23S	82W
508	026	-1796 Dec 26	11:34:15	41617	-46939	-004	P	-a	-0.6131	1.7172	0.7484	293.6	168.5	-	23N	1E
509	026	-1795 Jun 21	02:19:43	41606	-46933	001	P	-t	0.6299	1.7308	0.6740	337.3	180.9	-	22S	136E
510	026	-1795 Dec 16	02:10:11	41594	-46927	006	T+	pp	0.0600	2.7493	1.7464	333.1	216.1	99.0	22N	140E
511	026	-1794 Jun 10	08:23:06	41583	-46921	011	T-	pp	-0.1621	2.5630	1.5580	343.0	220.8	96.7	21S	45E
512	026	-1794 Dec 05	11:14:06	41572	-46915	016	P	h-	0.7871	1.4449	0.3831	310.7	139.0	-	21N	3E
513	026	-1793 May 30	21:32:54	41561	-46909	021	P	a-	-0.9050	1.1746	0.2195	262.2	100.6	-	19S	153W
514	026	-1793 Oct 25	17:27:27	41552	-46904	-012	N	-t	-1.5586	0.0513	-1.0544	70.8	-	-	6N	90W
515	026	-1793 Nov 24	13:01:07	41550	-46903	026	N	t-	1.5041	0.1506	-0.9537	120.0	-	-	19N	25W
516	026	-1792 Apr 20	07:54:23	41540	-46898	-007	P	-a	0.8751	1.2271	0.2769	264.5	111.4	-	5S	55E
517	026	-1792 Oct 13	18:19:09	41529	-46892	-002	P	-t	-0.8500	1.3333	0.2638	305.6	118.5	-	2N	103W
518	026	-1791 Apr 09	22:02:55	41518	-46886	003	T+	pp	0.1373	2.6041	1.6078	339.1	219.6	98.0	1S	156W
519	026	-1791 Oct 03	02:13:16	41507	-46880	008	T-	pp	-0.1137	2.6543	1.6445	336.2	216.8	97.4	2S	139E
520	026	-1790 Mar 30	05:23:07	41496	-46874	013	P	t-	-0.6532	1.6855	0.6338	331.9	175.4	-	2N	94E
521	027	-1790 Sep 22	16:17:48	41485	-46868	018	P	a-	0.5716	1.7934	0.8246	297.9	175.1	-	6S	72W
522	027	-1789 Feb 17	14:23:07	41475	-46863	-015	N	-t	1.4885	0.1659	-0.9120	124.4	-	-	18N	37W
523	027	-1789 Mar 19	06:32:35	41473	-46862	023	N	t-	-1.4150	0.3002	-0.7767	166.6	-	-	6N	78E
524	027	-1789 Aug 13	22:30:16	41464	-46857	-010	N	-a	-1.3894	0.2995	-0.6828	147.3	-	-	21S	164W
525	027	-1789 Sep 12	08:18:41	41462	-46856	028	N	a-	1.2484	0.5526	-0.4185	191.0	-	-	9S	48E
526	027	-1788 Feb 06	18:18:04	41453	-46851	-005	P	-t	0.7184	1.5572	0.5227	312.6	157.2	-	21N	97W
527	027	-1788 Aug 02	09:09:00	41442	-46845	000	P	-a	-0.7119	1.5687	0.5350	315.1	159.7	-	22S	36E
528	027	-1787 Jan 26	05:18:27	41431	-46839	005	T-	pp	-0.0306	2.7937	1.8099	328.4	215.0	99.6	22N	97E
529	027	-1787 Jul 22	12:41:32	41420	-46833	010	T+	pp	0.0264	2.8467	1.7731	373.3	234.8	105.8	23S	18W
530	027	-1786 Jan 15	20:48:49	41408	-46827	015	P	a-	-0.7139	1.5295	0.5663	282.9	150.9	-	23N	137W
531	027	-1786 Jul 11	12:57:00	41397	-46821	020	P	t-	0.7696	1.4804	0.4119	324.7	148.5	-	23S	23W
532	027	-1786 Dec 07	00:06:45	41388	-46816	-013	N	-a	1.3349	0.4185	-0.6014	177.0	-	-	22N	169E
533	027	-1785 Jan 05	12:16:15	41386	-46815	025	N	a-	-1.4098	0.2650	-0.7229	138.2	-	-	22N	10W
534	027	-1785 Jun 01	07:02:00	41377	-46810	-008	N	-h	-1.1832	0.6809	-0.3073	219.1	-	-	20S	64E
535	027	-1785 Jun 30	17:19:23	41375	-46809	030	N	h-	1.4759	0.1602	-0.8605	115.6	-	-	22S	89W
536	027	-1785 Nov 26	07:02:14	41366	-46804	-003	P	-h	0.6971	1.6183	0.5400	328.4	163.4	-	19N	65E
537	027	-1784 May 20	21:43:09	41355	-46798	002	T	-a	-0.3928	2.1101	1.1640	309.2	196.7	58.0	16S	156W
538	027	-1784 Nov 14	07:15:24	41343	-46792	007	T+	pp	0.0259	2.8649	1.7571	379.5	235.1	104.7	14N	61E
539	027	-1783 May 10	15:01:12	41332	-46786	012	T	p-	0.3488	2.1895	1.2460	310.4	199.8	69.0	12S	55W
540	027	-1783 Nov 03	06:59:54	41321	-46780	017	P	h-	-0.6460	1.7149	0.6309	336.4	174.9	-	10N	65E

Cat Num	Canon Plate	Calendar Date	TD of Greatest Eclipse	ΔT s	Luna Num	Saros Num	Ecl. Type	QSE	Gamma	Pen. Mag.	Um. Mag.	Eclipse Phase Durations Pen. m	Eclipse Phase Durations Par. m	Eclipse Phase Durations Total m	Greatest in Zenith Lat.	Greatest in Zenith Lng.
541	028	-1782 Mar 31	20:21:55	41312	-46775	-016	Ne	-h	-1.5119	0.0908	-0.9234	86.9	-	-	1N	132W
542	028	-1782 Apr 30	06:15:29	41310	-46774	022	N	h-	1.1312	0.7731	-0.2088	229.4	-	-	8S	78E
543	028	-1782 Sep 24	00:31:09	41301	-46769	-011	N	-a	1.4703	0.1564	-0.8365	107.9	-	-	5S	164E
544	028	-1782 Oct 23	13:00:22	41299	-46768	027	N	a-	-1.2883	0.5075	-0.5192	194.8	-	-	5N	24W
545	028	-1781 Mar 21	01:35:12	41290	-46763	-006	P	-t	-0.8109	1.4029	0.3377	317.5	135.5	-	6N	151E
546	028	-1781 Sep 13	15:41:55	41278	-46757	-001	P	-a	0.7653	1.4350	0.4720	277.2	139.8	-	9S	63W
547	028	-1780 Mar 09	01:55:13	41267	-46751	004	T-	pp	-0.0724	2.7637	1.6871	374.2	234.8	104.7	11N	147E
548	028	-1780 Sep 02	07:30:28	41256	-46745	009	T+	-p	0.0832	2.6943	1.7159	325.8	213.5	98.0	14S	60E
549	028	-1779 Feb 26	04:41:16	41245	-46739	014	P	h-	0.6646	1.6591	0.6182	323.2	170.4	-	15N	106E
550	028	-1779 Aug 22	19:18:32	41234	-46733	019	P	t-	-0.6603	1.6600	0.6330	316.0	168.8	-	18S	117W
551	028	-1778 Jan 17	04:19:46	41225	-46728	-014	N	-a	-1.3073	0.4466	-0.5286	174.3	-	-	22N	110E
552	028	-1778 Feb 15	14:27:35	41223	-46727	024	N	a-	1.3462	0.3820	-0.6066	166.2	-	-	19N	40W
553	028	-1778 Jul 13	08:10:29	41214	-46722	-009	N	-t	1.4594	0.2183	-0.8575	143.0	-	-	23S	48E
554	028	-1778 Aug 12	00:05:28	41212	-46721	029	N	t-	-1.4359	0.2602	-0.8133	153.7	-	-	21S	171E
555	028	-1777 Jan 06	20:13:25	41202	-46716	-004	P	-a	-0.6255	1.6936	0.7264	292.5	166.7	-	23N	129W
556	028	-1777 Jul 02	08:56:57	41191	-46710	001	P	-t	0.7058	1.5914	0.5349	328.2	164.5	-	23S	36E
557	028	-1777 Dec 27	10:45:26	41180	-46704	006	T+	pp	0.0540	2.7608	1.7571	334.1	216.7	99.4	24N	11E
558	028	-1776 Jun 20	15:22:56	41169	-46698	011	T-	pp	-0.0855	2.7026	1.6997	342.9	222.2	101.2	23S	62W
559	028	-1776 Dec 15	19:31:04	41158	-46692	016	P	t-	0.7851	1.4493	0.3860	312.0	139.9	-	23N	122W
560	028	-1775 Jun 10	04:52:44	41147	-46686	021	P	a-	-0.8297	1.3122	0.3584	272.1	125.7	-	22S	95E
561	029	-1775 Nov 05	01:26:57	41138	-46681	-012	N	-t	-1.5629	0.0436	-1.0625	65.3	-	-	10N	148E
562	029	-1775 Dec 04	21:01:40	41136	-46680	026	N	t-	1.5007	0.1569	-0.9473	122.6	-	-	22N	146W
563	029	-1774 May 01	15:16:31	41126	-46675	-007	P	-a	0.9497	1.0906	0.1395	253.7	80.8	-	9S	58W
564	029	-1774 May 30	21:59:04	41125	-46674	031	Nb	a-	-1.5341	0.0130	-0.9275	30.4	-	-	20S	162W
565	029	-1774 Oct 25	02:32:02	41115	-46669	-002	P	-t	-0.8533	1.3267	0.2584	304.2	117.1	-	6N	132E
566	029	-1773 Apr 21	05:08:26	41104	-46663	003	T+	-p	0.2119	2.4683	1.4700	338.7	217.5	91.6	5S	94E
567	029	-1773 Oct 14	10:44:03	41093	-46657	008	T-	pp	-0.1217	2.6391	1.6302	335.0	216.0	96.7	2N	9E
568	029	-1772 Apr 09	12:04:16	41082	-46651	013	P	t-	-0.5785	1.8227	0.7705	340.2	189.4	-	2S	9W
569	029	-1772 Oct 03	00:55:31	41071	-46645	018	P	a-	0.5580	1.8191	0.8489	298.8	176.7	-	2S	156E
570	029	-1771 Feb 27	21:25:14	41062	-46640	-015	Ne	-t	1.5389	0.0708	-1.0017	81.7	-	-	15N	145W
571	029	-1771 Mar 29	13:06:10	41060	-46639	023	N	t-	-1.3417	0.4333	-0.6406	197.5	-	-	2N	24W
572	029	-1771 Aug 24	06:27:42	41051	-46634	-010	N	-a	-1.4294	0.2292	-0.7591	130.2	-	-	18S	76E
573	029	-1771 Sep 22	16:46:43	41049	-46633	028	N	a-	1.2324	0.5842	-0.3913	196.3	-	-	5S	81W
574	029	-1770 Feb 17	01:54:18	41039	-46628	-005	P	-t	0.7570	1.4828	0.4551	306.2	147.8	-	18N	148E
575	029	-1770 Aug 13	16:33:08	41028	-46622	000	P	-a	-0.7619	1.4803	0.4398	310.8	147.4	-	20S	76W
576	029	-1769 Feb 06	13:26:34	41017	-46616	005	T+	pp	0.0004	2.8463	1.8680	327.1	214.7	99.8	20N	26W
577	029	-1769 Aug 02	19:33:51	41006	-46610	010	T-	pp	-0.0346	2.8339	1.7555	374.3	235.0	105.5	22S	122W
578	029	-1768 Jan 27	05:14:46	40995	-46604	015	P	a-	-0.6914	1.5693	0.6090	285.0	155.4	-	21N	96E
579	029	-1768 Jul 21	19:42:06	40984	-46598	020	P	t-	0.7003	1.6081	0.5384	332.9	166.2	-	23S	125W
580	029	-1768 Dec 17	08:40:32	40975	-46593	-013	N	-a	1.3411	0.4078	-0.6136	175.5	-	-	24N	40E
581	030	-1767 Jan 15	20:42:47	40973	-46592	025	N	a-	-1.3980	0.2865	-0.7011	143.8	-	-	22N	137W
582	030	-1767 Jun 11	14:06:43	40964	-46587	-008	N	-h	-1.2532	0.5512	-0.4346	198.8	-	-	22S	44W
583	030	-1767 Jul 11	00:23:39	40962	-46586	030	N	h-	1.4050	0.2898	-0.7299	153.1	-	-	23S	163E
584	030	-1767 Dec 06	15:15:25	40953	-46581	-003	P	-h	0.7004	1.6130	0.5331	329.0	162.9	-	21N	59W
585	030	-1766 Jun 01	05:04:33	40941	-46575	002	T	-a	-0.4665	1.9742	1.0292	304.2	189.4	25.4	19S	92E
586	030	-1766 Nov 25	15:17:28	40930	-46569	007	T+	pp	0.0280	2.8609	1.7533	379.5	235.1	104.7	18N	61W
587	030	-1765 May 21	22:22:53	40919	-46563	012	T	p-	0.2745	2.3263	1.3816	314.1	204.9	82.1	16S	168W
588	030	-1765 Nov 14	15:12:57	40908	-46557	017	P	h-	-0.6439	1.7178	0.6358	335.6	175.0	-	14N	60W
589	030	-1764 May 10	13:20:30	40897	-46551	022	N	h-	1.0599	0.9053	-0.0794	245.9	-	-	12S	31W
590	030	-1764 Oct 04	09:02:08	40888	-46546	-011	N	-a	1.4771	0.1439	-0.8488	103.4	-	-	0S	34E
591	030	-1764 Nov 02	21:35:07	40886	-46545	027	N	a-	-1.2844	0.5136	-0.5110	195.2	-	-	9N	155W
592	030	-1763 Mar 31	08:15:18	40877	-46540	-006	P	-t	-0.8845	1.2675	0.2030	307.0	107.4	-	1N	49E
593	030	-1763 Sep 24	00:14:04	40866	-46534	-001	P	-a	0.7820	1.4055	0.4401	275.3	135.6	-	5S	166E
594	030	-1762 Mar 20	08:39:14	40855	-46528	004	T-	pp	-0.1392	2.6391	1.5666	372.2	232.9	100.9	6N	44E
595	030	-1762 Sep 13	15:47:21	40844	-46522	009	T+	-p	0.1073	2.6527	1.6691	326.7	213.3	96.9	10S	66W
596	030	-1761 Mar 09	11:55:41	40833	-46516	014	P	h-	0.6083	1.7591	0.7246	327.1	180.7	-	11N	4W
597	030	-1761 Sep 03	03:05:25	40821	-46510	019	P	t-	-0.6295	1.7199	0.6862	320.8	174.8	-	14S	125E
598	030	-1760 Jan 28	12:39:40	40812	-46505	-014	N	-a	-1.3284	0.4056	-0.5649	166.3	-	-	20N	16W
599	030	-1760 Feb 26	22:15:48	40810	-46504	024	N	a-	1.3002	0.4634	-0.5192	180.9	-	-	16N	159W
600	030	-1760 Jul 23	14:51:58	40801	-46499	-009	N	-t	1.5259	0.0980	-0.9814	97.0	-	-	22S	53W

Cat Num	Canon Plate	Calendar Date	TD of Greatest Eclipse	ΔT s	Luna Num	Saros Num	Ecl. Type	QSE	Gamma	Pen. Mag.	Um. Mag.	Eclipse Phase Durations Pen. m	Par. m	Total m	Greatest in Zenith Lat.	Lng.
601	031	-1760 Aug 22	07:17:57	40799	-46498	029	N	t-	-1.3937	0.3405	-0.7385	175.2	-	-	18S	62E
602	031	-1759 Jan 17	04:44:54	40790	-46493	-004	P	-a	-0.6441	1.6587	0.6933	290.9	164.0	-	23N	102E
603	031	-1759 Jul 12	15:42:26	40779	-46487	001	P	-t	0.7753	1.4639	0.4073	318.8	146.1	-	23S	67W
604	031	-1758 Jan 06	19:11:58	40768	-46481	006	T+	pp	0.0416	2.7837	1.7796	335.2	217.4	100.0	24N	116W
605	031	-1758 Jul 01	22:30:36	40757	-46475	011	T-	pp	-0.0144	2.8322	1.8309	342.0	222.2	102.5	24S	170W
606	031	-1758 Dec 27	03:41:34	40746	-46469	016	P	t-	0.7791	1.4608	0.3967	313.7	142.0	-	24N	115E
607	031	-1757 Jun 21	12:16:23	40735	-46463	021	P	a-	-0.7570	1.4453	0.4921	280.5	144.1	-	23S	17W
608	031	-1757 Nov 16	09:29:33	40726	-46458	-012	N	-t	-1.5647	0.0401	-1.0656	62.6	-	-	14N	26E
609	031	-1757 Dec 16	04:59:34	40724	-46457	026	N	t-	1.4953	0.1663	-0.9372	126.1	-	-	24N	94E
610	031	-1756 May 11	22:34:52	40715	-46452	-007	N	-a	1.0266	0.9504	-0.0023	241.0	-	-	12S	170W
611	031	-1756 Jun 10	05:23:17	40713	-46451	031	N	a-	-1.4647	0.1411	-0.8010	99.1	-	-	22S	86E
612	031	-1756 Nov 04	10:50:47	40704	-46446	-002	P	-t	-0.8528	1.3266	0.2602	303.2	117.1	-	10N	5E
613	031	-1755 May 01	12:06:30	40692	-46440	003	T	-a	0.2916	2.3232	1.3224	337.3	213.7	80.5	9S	13W
614	031	-1755 Oct 24	19:21:55	40681	-46434	008	T-	pp	-0.1247	2.6330	1.6254	333.9	215.4	96.3	7N	123W
615	031	-1754 Apr 20	18:38:47	40670	-46428	013	P	t-	-0.4989	1.9693	0.9161	347.9	201.8	-	6S	110W
616	031	-1754 Oct 14	09:39:21	40659	-46422	018	P	a-	0.5493	1.8356	0.8642	299.4	177.7	-	3N	23E
617	031	-1753 Apr 09	19:35:43	40648	-46416	023	N	t-	-1.2645	0.5736	-0.4975	224.0	-	-	2S	124W
618	031	-1753 Sep 04	14:32:10	40639	-46411	-010	N	-a	-1.4630	0.1706	-0.8237	113.4	-	-	15S	47W
619	031	-1753 Oct 04	01:20:51	40637	-46410	028	N	a-	1.2219	0.6056	-0.3741	199.9	-	-	1S	148E
620	031	-1752 Feb 28	09:24:50	40628	-46405	-005	P	-t	0.8012	1.3984	0.3774	298.7	135.7	-	15N	33E
621	032	-1752 Aug 24	00:03:32	40617	-46399	000	P	-a	-0.8059	1.4029	0.3558	306.8	134.7	-	17S	169E
622	032	-1751 Feb 16	21:28:04	40606	-46393	005	T+	pp	0.0376	2.7753	1.8025	325.7	214.3	99.5	17N	148W
623	032	-1751 Aug 13	02:34:35	40595	-46387	010	T-	pp	-0.0887	2.7371	1.6540	374.8	234.3	103.6	20S	131E
624	032	-1750 Feb 06	13:31:39	40584	-46381	015	P	a-	-0.6613	1.6232	0.6656	287.8	161.1	-	19N	29W
625	032	-1750 Aug 02	02:38:15	40573	-46375	020	P	t-	0.6390	1.7213	0.6503	339.2	179.0	-	21S	130E
626	032	-1750 Dec 28	17:07:22	40564	-46370	-013	N	-a	1.3528	0.3870	-0.6355	171.8	-	-	25N	87W
627	032	-1749 Jan 27	05:00:17	40562	-46369	025	N	a-	-1.3793	0.3206	-0.6666	152.0	-	-	21N	98E
628	032	-1749 Jun 22	21:17:33	40553	-46364	-008	N	-h	-1.3189	0.4297	-0.5543	176.9	-	-	24S	153W
629	032	-1749 Jul 22	07:37:06	40551	-46363	030	N	h-	1.3400	0.4086	-0.6103	179.2	-	-	22S	54E
630	032	-1749 Dec 17	23:23:10	40542	-46358	-003	P	-h	0.7078	1.6000	0.5192	329.0	161.5	-	23N	178E
631	032	-1748 Jun 11	12:28:41	40531	-46352	002	P	-a	-0.5386	1.8419	0.8971	298.6	180.8	-	22S	21W
632	032	-1748 Dec 05	23:16:41	40520	-46346	007	T+	pp	0.0323	2.8524	1.7459	379.3	235.1	104.7	21N	178E
633	032	-1747 Jun 01	05:43:31	40508	-46340	012	T+	p-	0.1997	2.4646	1.5178	317.2	208.9	91.0	19S	80E
634	032	-1747 Nov 24	23:27:10	40498	-46334	017	P	h-	-0.6419	1.7200	0.6409	334.7	175.2	-	17N	175E
635	032	-1746 May 21	20:22:41	40486	-46328	022	P	h-	0.9870	1.0409	0.0527	261.1	52.1	-	15S	139W
636	032	-1746 Oct 15	17:40:14	40477	-46323	-011	N	-a	1.4790	0.1401	-0.8522	101.8	-	-	4N	98W
637	032	-1746 Nov 14	06:12:38	40475	-46322	027	N	a-	-1.2819	0.5169	-0.5051	195.1	-	-	13N	74E
638	032	-1745 Apr 11	14:47:59	40466	-46317	-006	P	-t	-0.9633	1.1227	0.0586	294.0	59.0	-	3S	53W
639	032	-1745 Oct 05	08:53:51	40455	-46311	-001	P	-a	0.7928	1.3869	0.4193	274.2	132.7	-	1S	34E
640	032	-1744 Mar 30	15:15:32	40444	-46305	004	T-	pp	-0.2128	2.5020	1.4337	369.3	229.6	93.5	2N	58W
641	033	-1744 Sep 24	00:11:36	40433	-46299	009	T+	-p	0.1248	2.6233	1.6346	327.7	213.2	96.0	6S	165E
642	033	-1743 Mar 19	19:03:16	40422	-46293	014	P	h-	0.5453	1.8716	0.8435	330.9	190.4	-	7N	114W
643	033	-1743 Sep 13	11:00:00	40411	-46287	019	P	t-	-0.6058	1.7667	0.7264	324.8	179.1	-	10S	4E
644	033	-1742 Feb 07	20:52:02	40402	-46282	-014	N	-a	-1.3560	0.3525	-0.6133	155.4	-	-	18N	140W
645	033	-1742 Mar 09	05:57:07	40400	-46281	024	N	a-	1.2478	0.5567	-0.4201	195.7	-	-	12N	84E
646	033	-1742 Sep 02	14:39:33	40389	-46275	029	N	t-	-1.3588	0.4073	-0.6771	191.0	-	-	15S	51W
647	033	-1741 Jan 28	13:09:56	40380	-46270	-004	P	-a	-0.6678	1.6142	0.6508	288.7	160.1	-	21N	25W
648	033	-1741 Jul 23	22:35:57	40369	-46264	001	P	-t	0.8388	1.3475	0.2907	309.2	125.3	-	22S	171W
649	033	-1740 Jan 18	03:32:07	40358	-46258	006	T+	pp	0.0246	2.8149	1.8106	336.3	218.2	100.5	23N	118E
650	033	-1740 Jul 12	05:45:25	40347	-46252	011	T+	pp	0.0515	2.7636	1.7634	340.5	221.2	101.7	24S	81E
651	033	-1739 Jan 06	11:44:12	40336	-46246	016	P	t-	0.7671	1.4828	0.4185	316.2	145.7	-	25N	6W
652	033	-1739 Jul 01	19:45:44	40325	-46240	021	P	a-	-0.6884	1.5712	0.6181	287.5	158.0	-	24S	130W
653	033	-1739 Nov 26	17:31:36	40316	-46235	-012	N	-t	-1.5671	0.0350	-1.0693	58.5	-	-	17N	96W
654	033	-1739 Dec 26	12:50:54	40314	-46234	026	N	t-	1.4848	0.1848	-0.9172	132.8	-	-	25N	24W
655	033	-1738 May 23	05:53:10	40305	-46229	-007	N	-a	1.1032	0.8110	-0.1440	226.6	-	-	16S	78E
656	033	-1738 Jun 21	12:51:55	40303	-46228	031	N	a-	-1.3987	0.2635	-0.6810	134.1	-	-	24S	28W
657	033	-1738 Nov 15	19:10:36	40294	-46223	-002	P	-t	-0.8522	1.3265	0.2627	302.1	117.3	-	14N	122W
658	033	-1737 May 12	19:02:34	40283	-46217	003	T	-a	0.3720	2.1773	1.1734	334.9	208.2	62.3	13S	120W
659	033	-1737 Nov 05	04:03:35	40272	-46211	008	T-	pp	-0.1252	2.6312	1.6254	332.8	214.8	96.1	11N	105E
660	033	-1736 May 01	01:06:59	40261	-46205	013	T	t-	-0.4147	2.1245	1.0701	354.9	212.4	42.6	10S	150E

Cat Num	Canon Plate	Calendar Date	TD of Greatest Eclipse	ΔT s	Luna Num	Saros Num	Ecl. Type	QSE	Gamma	Pen. Mag.	Um. Mag.	Eclipse Phase Durations Pen. m	Par. m	Total m	Greatest in Zenith Lat.	Lng.
661	034	-1736 Oct 24	18:28:29	40250	-46199	018	P	a-	0.5449	1.8441	0.8720	299.6	178.1	-	7N	112W
662	034	-1735 Apr 20	01:58:14	40239	-46193	023	N	t-	-1.1811	0.7253	-0.3433	247.8	-	-	6S	138E
663	034	-1735 Sep 14	22:45:05	40230	-46188	-010	N	-a	-1.4895	0.1250	-0.8752	97.8	-	-	11S	173W
664	034	-1735 Oct 14	10:01:17	40228	-46187	028	N	a-	1.2169	0.6167	-0.3667	202.1	-	-	4N	15E
665	034	-1734 Mar 10	16:46:27	40219	-46182	-005	P	-t	0.8536	1.2989	0.2845	289.8	119.1	-	11N	80W
666	034	-1734 Sep 04	07:43:13	40208	-46176	000	P	-a	-0.8415	1.3410	0.2873	303.5	122.6	-	14S	53E
667	034	-1733 Feb 28	05:21:35	40197	-46170	005	T+	pp	0.0822	2.6907	1.7232	324.1	213.4	98.4	14N	92E
668	034	-1733 Aug 24	09:43:25	40186	-46164	010	T-	pp	-0.1359	2.6529	1.5652	375.0	233.1	100.6	17S	22E
669	034	-1732 Feb 17	21:40:58	40175	-46158	015	P	a-	-0.6249	1.6885	0.7336	291.1	167.3	-	16N	153W
670	034	-1732 Aug 12	09:43:34	40164	-46152	020	P	t-	0.5841	1.8227	0.7502	344.1	188.6	-	19S	22E
671	034	-1731 Jan 08	01:27:15	40155	-46147	-013	N	-a	1.3692	0.3571	-0.6660	166.0	-	-	25N	148E
672	034	-1731 Feb 06	13:08:18	40153	-46146	025	N	a-	-1.3539	0.3670	-0.6198	162.3	-	-	19N	25W
673	034	-1731 Jul 03	04:33:36	40144	-46141	-008	N	-a	-1.3812	0.3149	-0.6678	152.6	-	-	25S	97E
674	034	-1731 Aug 01	14:59:46	40142	-46140	030	N	a-	1.2813	0.5161	-0.5023	198.6	-	-	21S	58W
675	034	-1731 Dec 28	07:25:21	40133	-46135	-003	P	-t	0.7190	1.5795	0.4985	328.5	159.2	-	24N	57E
676	034	-1730 Jun 22	19:56:47	40122	-46129	002	P	-a	-0.6074	1.7156	0.7706	292.6	171.0	-	24S	134W
677	034	-1730 Dec 17	07:12:40	40111	-46123	007	T+	pp	0.0394	2.8384	1.7338	379.0	235.1	104.6	23N	59E
678	034	-1729 Jun 12	13:06:26	40100	-46117	012	T+	pp	0.1272	2.5991	1.6496	319.6	211.6	96.4	21S	32W
679	034	-1729 Dec 06	07:41:29	40089	-46111	017	P	a-	-0.6393	1.7231	0.6475	333.7	175.5	-	20N	51E
680	034	-1728 Jun 01	03:23:00	40078	-46105	022	P	t-	0.9133	1.1781	0.1859	274.8	96.5	-	18S	114E
681	035	-1728 Oct 26	02:23:24	40069	-46100	-011	N	-a	1.4777	0.1421	-0.8493	102.3	-	-	9N	129E
682	035	-1728 Nov 24	14:51:35	40067	-46099	027	N	a-	-1.2799	0.5188	-0.5000	194.7	-	-	16N	58W
683	035	-1727 Apr 21	21:14:40	40058	-46094	-006	N	-t	-1.0462	0.9706	-0.0935	278.2	-	-	7S	152W
684	035	-1727 Oct 15	17:38:49	40047	-46088	-001	P	-a	0.7994	1.3758	0.4062	273.6	130.9	-	4N	100W
685	035	-1726 Apr 10	21:47:07	40036	-46082	004	T	pp	-0.2905	2.3573	1.2929	365.3	224.6	81.1	2S	159W
686	035	-1726 Oct 05	08:42:15	40025	-46076	009	T+	-p	0.1365	2.6043	1.6107	328.7	213.2	95.3	1S	35E
687	035	-1725 Mar 31	02:06:25	40014	-46070	014	P	a-	0.4775	1.9929	0.9708	334.4	199.0	-	3N	138E
688	035	-1725 Sep 24	19:00:40	40003	-46064	019	P	t-	-0.5880	1.8025	0.7560	328.2	182.3	-	6S	119W
689	035	-1724 Feb 19	04:56:39	39994	-46059	-014	N	-a	-1.3902	0.2874	-0.6737	140.8	-	-	15N	97E
690	035	-1724 Mar 19	13:32:13	39992	-46058	024	N	a-	1.1895	0.6607	-0.3106	210.2	-	-	8N	33W
691	035	-1724 Sep 12	22:09:51	39981	-46052	029	N	t-	-1.3312	0.4604	-0.6290	202.6	-	-	11S	165W
692	035	-1723 Feb 07	21:24:26	39972	-46047	-004	P	-a	-0.6999	1.5542	0.5927	285.6	154.4	-	19N	150W
693	035	-1723 Aug 03	05:40:26	39961	-46041	001	P	-t	0.8939	1.2467	0.1893	300.0	102.4	-	21S	82E
694	035	-1722 Jan 28	11:41:42	39950	-46035	006	T-	pp	-0.0002	2.8598	1.8555	337.3	218.9	101.0	22N	5W
695	035	-1722 Jul 23	13:09:47	39939	-46029	011	T+	pp	0.1105	2.6550	1.6554	338.6	219.4	99.1	23S	31W
696	035	-1721 Jan 17	19:37:38	39928	-46023	016	P	t-	0.7485	1.5169	0.4526	319.4	151.2	-	24N	125W
697	035	-1721 Jul 13	03:21:41	39917	-46017	021	P	a-	-0.6249	1.6879	0.7343	293.3	168.7	-	24S	115E
698	035	-1721 Dec 08	01:33:28	39908	-46012	-012	N	-t	-1.5697	0.0292	-1.0731	53.3	-	-	20N	142E
699	035	-1720 Jan 06	20:36:26	39906	-46011	026	N	t-	1.4697	0.2113	-0.8883	141.6	-	-	25N	141W
700	035	-1720 Jun 02	13:08:58	39897	-46006	-007	N	-a	1.1809	0.6699	-0.2880	209.5	-	-	18S	33W
701	036	-1720 Jul 01	20:22:29	39895	-46005	031	N	a-	-1.3346	0.3825	-0.5649	160.0	-	-	25S	141W
702	036	-1720 Nov 26	03:33:39	39886	-46000	-002	P	-h	-0.8497	1.3293	0.2689	301.2	118.2	-	17N	111E
703	036	-1719 May 23	01:54:21	39875	-45994	003	T	-a	0.4546	2.0277	1.0201	331.5	200.6	22.3	16S	135E
704	036	-1719 Nov 15	12:48:09	39864	-45988	008	T-	pp	-0.1238	2.6326	1.6291	331.7	214.4	96.1	15N	28W
705	036	-1718 May 12	07:31:52	39853	-45982	013	T	tp	-0.3284	2.2837	1.2276	360.8	220.9	73.0	14S	51E
706	036	-1718 Nov 05	03:21:21	39842	-45976	018	P	a-	0.5439	1.8461	0.8736	299.5	178.1	-	12N	113E
707	036	-1717 May 01	08:19:01	39831	-45970	023	N	t-	-1.0955	0.8813	-0.1851	268.6	-	-	10S	40E
708	036	-1717 Sep 26	07:03:15	39822	-45965	-010	N	-a	-1.5110	0.0882	-0.9176	82.8	-	-	6S	61E
709	036	-1717 Oct 25	18:45:04	39820	-45964	028	N	a-	1.2152	0.6215	-0.3654	203.3	-	-	8N	118W
710	036	-1716 Mar 21	00:04:00	39811	-45959	-005	P	-t	0.9101	1.1919	0.1839	279.5	96.8	-	7N	169E
711	036	-1716 Sep 14	15:29:15	39800	-45953	000	P	-h	-0.8710	1.2900	0.2299	300.7	111.0	-	10S	66W
712	036	-1715 Mar 10	13:07:40	39789	-45947	005	T+	pp	0.1337	2.5937	1.6313	322.2	212.0	95.9	10N	27W
713	036	-1715 Sep 03	17:01:59	39779	-45941	010	T-	pp	-0.1749	2.5835	1.4915	374.9	231.7	96.9	13S	89W
714	036	-1714 Feb 28	05:41:40	39768	-45935	015	P	a-	-0.5815	1.7670	0.8145	294.9	174.1	-	13N	85E
715	036	-1714 Aug 23	16:59:56	39757	-45929	020	P	t-	0.5377	1.9086	0.8349	347.7	195.6	-	16S	89W
716	036	-1713 Jan 19	09:37:27	39748	-45924	-013	N	-a	1.3929	0.3139	-0.7095	156.8	-	-	24N	25E
717	036	-1713 Feb 17	21:06:34	39746	-45923	025	N	a-	-1.3213	0.4266	-0.5599	174.5	-	-	16N	147W
718	036	-1713 Jul 14	11:58:34	39737	-45918	-008	N	-a	-1.4370	0.2120	-0.7699	126.0	-	-	25S	15W
719	036	-1713 Aug 12	22:33:22	39735	-45917	030	N	a-	1.2304	0.6095	-0.4087	213.2	-	-	19S	173W
720	036	-1712 Jan 08	15:18:57	39726	-45912	-003	P	-t	0.7366	1.5470	0.4664	327.0	155.1	-	25N	62W

Cat Num	Canon Plate	Calendar Date	TD of Greatest Eclipse	ΔT s	Luna Num	Saros Num	Ecl. Type	QSE	Gamma	Pen. Mag.	Um. Mag.	Eclipse Phase Durations Pen. m	Eclipse Phase Durations Par. m	Eclipse Phase Durations Total m	Greatest in Zenith Lat.	Greatest in Zenith Lng.
721	037	-1712 Jul 03	03:29:16	39715	-45906	002	P	-a	-0.6729	1.5960	0.6501	286.2	160.0	-	24S	112E
722	037	-1712 Dec 27	15:02:57	39704	-45900	007	T+	pp	0.0514	2.8151	1.7132	378.4	234.9	104.4	24N	59W
723	037	-1711 Jun 22	20:31:26	39693	-45894	012	T+	pp	0.0568	2.7299	1.7771	321.4	213.2	99.3	23S	145W
724	037	-1711 Dec 16	15:52:03	39682	-45888	017	P	a-	-0.6327	1.7329	0.6618	333.0	176.5	-	22N	73W
725	037	-1710 Jun 12	10:23:58	39671	-45882	022	P	t-	0.8409	1.3135	0.3166	287.2	124.0	-	21S	7E
726	037	-1710 Nov 06	11:11:13	39662	-45877	-011	N	-a	1.4737	0.1488	-0.8413	104.3	-	-	13N	5W
727	037	-1710 Dec 05	23:30:18	39660	-45876	027	N	a-	-1.2769	0.5225	-0.4925	194.6	-	-	19N	172E
728	037	-1709 May 03	03:36:10	39651	-45871	-006	N	-t	-1.1322	0.8129	-0.2516	259.1	-	-	11S	110E
729	037	-1709 Oct 27	02:28:46	39640	-45865	-001	P	-a	0.8020	1.3719	0.4007	273.5	130.1	-	8N	126E
730	037	-1708 Apr 21	04:13:56	39629	-45859	004	T	-t	-0.3727	2.2048	1.1439	360.0	217.5	59.8	6S	101E
731	037	-1708 Oct 15	17:18:35	39618	-45853	009	T+	-p	0.1426	2.5954	1.5973	329.9	213.4	94.9	3N	96W
732	037	-1707 Apr 10	09:03:58	39608	-45847	014	T	a-	0.4042	2.1245	1.1083	337.4	206.6	50.6	1S	30E
733	037	-1707 Oct 05	03:08:40	39597	-45841	019	P	t-	-0.5772	1.8252	0.7730	330.8	184.4	-	2S	117E
734	037	-1706 Mar 01	12:54:06	39588	-45836	-014	N	-a	-1.4305	0.2112	-0.7454	121.3	-	-	12N	24W
735	037	-1706 Mar 30	21:02:22	39586	-45835	024	N	a-	1.1266	0.7737	-0.1926	224.1	-	-	3N	148W
736	037	-1706 Sep 24	05:47:32	39575	-45829	029	N	t-	-1.3098	0.5021	-0.5920	211.3	-	-	7S	78E
737	037	-1705 Feb 19	05:31:52	39566	-45824	-004	P	-a	-0.7378	1.4839	0.5241	281.7	147.0	-	16N	87E
738	037	-1705 Aug 14	12:54:55	39555	-45818	001	P	-t	0.9414	1.1599	0.1019	291.4	75.8	-	18S	28W
739	037	-1704 Feb 08	19:42:44	39544	-45812	006	T-	pp	-0.0313	2.8026	1.7983	338.3	219.5	101.1	19N	126W
740	037	-1704 Aug 02	20:42:26	39533	-45806	011	T+	pp	0.1634	2.5577	1.5585	336.4	217.2	95.2	21S	146W
741	038	-1703 Jan 28	03:21:17	39522	-45800	016	P	t-	0.7227	1.5640	0.5002	323.4	158.1	-	22N	119E
742	038	-1703 Jul 23	11:05:03	39511	-45794	021	P	a-	-0.5673	1.7941	0.8394	297.9	176.9	-	24S	2W
743	038	-1703 Dec 18	09:29:57	39502	-45789	-012	N	-t	-1.5765	0.0153	-1.0840	38.5	-	-	21N	22E
744	038	-1702 Jan 17	04:11:58	39500	-45788	026	N	t-	1.4468	0.2520	-0.8449	154.0	-	-	25N	105E
745	038	-1702 Jun 13	20:27:31	39491	-45783	-007	N	-a	1.2554	0.5350	-0.4266	190.4	-	-	20S	145W
746	038	-1702 Jul 13	03:59:44	39489	-45782	031	N	a-	-1.2763	0.4912	-0.4597	179.8	-	-	25S	104E
747	038	-1702 Dec 07	11:54:39	39480	-45777	-002	P	-h	-0.8491	1.3284	0.2720	300.0	118.5	-	20N	15W
748	038	-1701 Jun 03	08:45:28	39469	-45771	003	P	-h	0.5364	1.8796	0.8678	327.0	190.8	-	19S	31E
749	038	-1701 Nov 26	21:32:42	39459	-45765	008	T-	-p	-0.1229	2.6328	1.6321	330.6	214.0	96.0	18N	161W
750	038	-1700 May 22	13:54:01	39448	-45759	013	T-	pp	-0.2406	2.4458	1.3875	365.6	227.4	89.9	17S	47W
751	038	-1700 Nov 15	12:15:37	39437	-45753	018	P	a-	0.5438	1.8462	0.8738	299.4	178.0	-	15N	22W
752	038	-1699 May 11	14:36:06	39426	-45747	023	N*	t-	-1.0060	1.0446	-0.0201	287.0	-	-	14S	57W
753	038	-1699 Oct 06	15:28:58	39417	-45742	-010	N	-a	-1.5261	0.0633	-0.9479	70.6	-	-	2S	68W
754	038	-1699 Nov 05	03:32:28	39415	-45741	028	N	a-	1.2170	0.6197	-0.3701	203.6	-	-	12N	108E
755	038	-1698 Apr 01	07:15:02	39406	-45736	-005	P	-t	0.9729	1.0738	0.0717	267.4	61.2	-	3N	58E
756	038	-1698 Sep 25	23:22:45	39395	-45730	000	P	-h	-0.8936	1.2515	0.1855	298.8	100.6	-	6S	174E
757	038	-1697 Mar 21	20:47:23	39384	-45724	005	T+	-p	0.1911	2.4861	1.5283	319.9	209.8	91.6	6N	144W
758	038	-1697 Sep 15	00:29:36	39373	-45718	010	T-	pp	-0.2062	2.5280	1.4319	374.7	230.2	93.2	9S	157E
759	038	-1696 Mar 10	13:33:53	39363	-45712	015	P	a-	-0.5311	1.8583	0.9080	298.9	181.0	-	9N	36W
760	038	-1696 Sep 03	00:26:30	39352	-45706	020	P	t-	0.4987	1.9807	0.9057	350.3	200.7	-	13S	158E
761	039	-1695 Jan 29	17:39:01	39343	-45701	-013	N	-a	1.4227	0.2593	-0.7642	143.7	-	-	23N	96W
762	039	-1695 Feb 28	04:55:17	39341	-45700	025	N	a-	-1.2819	0.4988	-0.4874	188.0	-	-	12N	94E
763	039	-1695 Jul 24	19:31:03	39332	-45695	-008	N	-a	-1.4876	0.1190	-0.8626	94.9	-	-	24S	129W
764	039	-1695 Aug 23	06:17:06	39330	-45694	030	N	a-	1.1865	0.6899	-0.3282	224.2	-	-	16S	70E
765	039	-1694 Jan 18	23:03:45	39321	-45689	-003	P	-t	0.7607	1.5023	0.4225	324.6	148.9	-	24N	178W
766	039	-1694 Jul 14	11:08:16	39310	-45683	002	P	-a	-0.7332	1.4861	0.5386	279.8	148.1	-	25S	4W
767	039	-1693 Jan 07	22:47:52	39299	-45677	007	T+	pp	0.0681	2.7828	1.6841	377.7	234.7	103.9	24N	176W
768	039	-1693 Jul 04	03:59:35	39288	-45671	012	T-	pp	-0.0105	2.8169	1.8602	322.6	213.8	100.0	24S	102E
769	039	-1693 Dec 27	23:59:54	39278	-45665	017	P	a-	-0.6230	1.7483	0.6821	332.5	178.1	-	23N	165E
770	039	-1692 Jun 22	17:26:19	39267	-45659	022	P	t-	0.7701	1.4460	0.4439	298.2	144.6	-	22S	100W
771	039	-1692 Nov 16	20:00:11	39258	-45654	-011	N	-a	1.4692	0.1563	-0.8321	106.6	-	-	17N	139W
772	039	-1692 Dec 16	08:05:51	39256	-45653	027	N	a-	-1.2709	0.5314	-0.4793	195.4	-	-	21N	42E
773	039	-1691 May 13	09:54:47	39247	-45648	-006	N	-t	-1.2196	0.6529	-0.4123	236.3	-	-	15S	13E
774	039	-1691 Nov 06	11:21:08	39236	-45642	-001	P	-a	0.8027	1.3712	0.3987	273.6	129.9	-	13N	9W
775	039	-1690 May 02	10:39:50	39225	-45636	004	P	-t	-0.4559	2.0505	0.9928	353.6	208.2	-	10S	2E
776	039	-1690 Oct 27	01:57:43	39214	-45630	009	T+	-p	0.1458	2.5914	1.5894	331.0	213.7	94.7	8N	132E
777	039	-1689 Apr 21	16:00:07	39203	-45624	014	T	a-	0.3284	2.2609	1.2498	339.7	212.6	73.2	6S	77W
778	039	-1689 Oct 16	11:21:25	39192	-45618	019	P	t-	-0.5709	1.8393	0.7819	332.9	185.6	-	3N	8W
779	039	-1688 Mar 11	20:42:52	39183	-45613	-014	N	-a	-1.4782	0.1216	-0.8309	92.6	-	-	8N	144W
780	039	-1688 Apr 10	04:26:27	39182	-45612	024	N	a-	1.0578	0.8975	-0.0642	237.4	-	-	1S	98E

Cat Num	Canon Plate	Calendar Date	TD of Greatest Eclipse	ΔT s	Luna Num	Saros Num	Ecl. Type	QSE	Gamma	Pen. Mag.	Um. Mag.	Eclipse Phase Durations Pen. m	Par. m	Total m	Greatest in Zenith Lat.	Lng.
781	040	-1688 Oct 04	13:33:20	39171	-45606	029	N	t-	-1.2949	0.5313	-0.5668	217.2	-	-	2S	41W
782	040	-1687 Mar 01	13:28:51	39162	-45601	-004	P	-a	-0.7841	1.3981	0.4399	276.5	136.6	-	12N	35W
783	040	-1687 Aug 24	20:21:14	39151	-45595	001	P	-t	0.9800	1.0893	0.0308	283.8	42.0	-	15S	142W
784	040	-1686 Feb 19	03:31:51	39140	-45589	006	T-	pp	-0.0716	2.7287	1.7245	339.2	219.8	100.4	16N	115E
785	040	-1686 Aug 14	04:26:03	39129	-45583	011	T+	pp	0.2081	2.4757	1.4766	334.2	214.6	90.7	19S	97E
786	040	-1685 Feb 08	10:54:45	39118	-45577	016	P	t-	0.6894	1.6247	0.5616	328.0	166.4	-	20N	4E
787	040	-1685 Aug 03	18:55:11	39108	-45571	021	P	a-	-0.5152	1.8905	0.9344	301.6	183.3	-	22S	120W
788	040	-1684 Jan 28	11:40:15	39097	-45565	026	N	t-	1.4181	0.3030	-0.7906	168.0	-	-	23N	8W
789	040	-1684 Jun 24	03:46:41	39088	-45560	-007	N	-a	1.3283	0.4033	-0.5626	168.0	-	-	22S	104E
790	040	-1684 Jul 23	11:41:29	39086	-45559	031	N	a-	-1.2223	0.5924	-0.3626	196.0	-	-	24S	13W
791	040	-1684 Dec 17	20:13:54	39077	-45554	-002	P	-h	-0.8510	1.3225	0.2709	298.3	118.0	-	22N	141W
792	040	-1683 Jun 13	15:35:47	39066	-45548	003	P	-h	0.6178	1.7327	0.7160	321.5	178.5	-	21S	74W
793	040	-1683 Dec 07	06:16:55	39055	-45542	008	T-	-p	-0.1228	2.6312	1.6339	329.5	213.6	96.0	21N	67E
794	040	-1682 Jun 02	20:16:30	39044	-45536	013	T-	pp	-0.1534	2.6072	1.5464	369.1	231.9	100.0	20S	144W
795	040	-1682 Nov 26	21:09:36	39034	-45530	018	P	a-	0.5436	1.8463	0.8744	299.3	178.0	-	19N	157W
796	040	-1681 May 22	20:55:07	39023	-45524	023	P	t-	-0.9173	1.2066	0.1435	302.6	91.4	-	17S	154W
797	040	-1681 Oct 17	23:58:43	39014	-45519	-010	N	-a	-1.5374	0.0451	-0.9710	59.9	-	-	3N	162E
798	040	-1681 Nov 16	12:20:37	39012	-45518	028	N	a-	1.2201	0.6153	-0.3769	203.5	-	-	16N	25W
799	040	-1680 Apr 11	14:22:24	39003	-45513	-005	N	-t	1.0394	0.9489	-0.0475	253.5	-	-	2S	52W
800	040	-1680 Oct 06	07:22:03	38992	-45507	000	P	-h	-0.9107	1.2229	0.1515	297.5	91.6	-	1S	51E
801	041	-1679 Apr 01	04:21:26	38981	-45501	005	T+	-p	0.2536	2.3692	1.4156	317.0	206.6	84.8	2N	99E
802	041	-1679 Sep 25	08:05:39	38970	-45495	010	T-	pp	-0.2305	2.4852	1.3856	374.5	228.9	89.7	5S	41E
803	041	-1678 Mar 21	21:18:27	38960	-45489	015	T	a-	-0.4746	1.9611	1.0125	303.1	187.8	16.7	5N	154W
804	041	-1678 Sep 14	08:04:12	38949	-45483	020	P	t-	0.4683	2.0370	0.9611	351.9	204.2	-	9S	41E
805	041	-1677 Feb 10	01:29:58	38940	-45478	-013	N	-a	1.4603	0.1904	-0.8333	124.3	-	-	20N	145E
806	041	-1677 Mar 11	12:34:35	38938	-45477	025	N	a-	-1.2355	0.5839	-0.4024	202.4	-	-	9N	23W
807	041	-1677 Aug 05	03:12:29	38929	-45472	-008	Ne	-a	-1.5319	0.0378	-0.9438	53.7	-	-	23S	115E
808	041	-1677 Sep 03	14:10:47	38927	-45471	030	N	a-	1.1499	0.7571	-0.2610	232.6	-	-	12S	50W
809	041	-1676 Jan 30	06:38:13	38918	-45466	-003	P	-t	0.7927	1.4431	0.3645	320.8	139.9	-	22N	67E
810	041	-1676 Jul 24	18:54:02	38907	-45460	002	P	-a	-0.7882	1.3861	0.4366	273.3	135.3	-	24S	121W
811	041	-1675 Jan 18	06:22:49	38897	-45454	007	T+	pp	0.0931	2.7349	1.6402	376.6	234.1	103.0	23N	70E
812	041	-1675 Jul 14	11:32:56	38886	-45448	012	T-	pp	-0.0728	2.7048	1.7436	323.4	213.6	98.9	24S	12W
813	041	-1674 Jan 07	08:01:19	38875	-45442	017	P	a-	-0.6071	1.7746	0.7140	332.4	180.8	-	23N	44E
814	041	-1674 Jul 04	00:31:48	38864	-45436	022	P	t-	0.7028	1.5722	0.5645	307.8	160.5	-	23S	152E
815	041	-1674 Nov 28	04:49:30	38855	-45431	-011	N	-a	1.4651	0.1626	-0.8234	108.4	-	-	20N	87E
816	041	-1674 Dec 27	16:37:29	38853	-45430	027	N	a-	-1.2611	0.5468	-0.4590	197.1	-	-	22N	87W
817	041	-1673 May 24	16:11:52	38844	-45425	-006	N	-t	-1.3074	0.4924	-0.5738	208.7	-	-	18S	84W
818	041	-1673 Jun 23	06:33:50	38843	-45424	032	N	t-	1.5143	0.1076	-0.9486	99.5	-	-	22S	61E
819	041	-1673 Nov 17	20:16:01	38834	-45419	-001	P	-a	0.8013	1.3742	0.4010	274.0	130.3	-	16N	145W
820	041	-1672 May 12	17:02:20	38823	-45413	004	P	-t	-0.5422	1.8907	0.8358	345.8	196.2	-	14S	96W
821	042	-1672 Nov 06	10:40:38	38812	-45407	009	T+	-p	0.1452	2.5944	1.5889	332.2	214.2	94.8	12N	1W
822	042	-1671 May 01	22:53:22	38801	-45401	014	T+	p-	0.2491	2.4040	1.3977	341.3	217.3	87.3	10S	177E
823	042	-1671 Oct 26	19:38:10	38790	-45395	019	P	t-	-0.5689	1.8454	0.7834	334.6	186.3	-	7N	135W
824	042	-1670 Mar 23	04:25:38	38781	-45390	-014	Ne	-a	-1.5309	0.0230	-0.9259	40.5	-	-	4N	98E
825	042	-1670 Apr 21	11:48:03	38780	-45389	024	P	a-	0.9863	1.0268	0.0690	249.6	57.7	-	5S	15W
826	042	-1670 Oct 15	21:25:15	38769	-45383	029	N	t-	-1.2850	0.5513	-0.5502	221.2	-	-	2N	161W
827	042	-1669 Mar 12	21:19:09	38760	-45378	-004	P	-a	-0.8354	1.3033	0.3462	270.3	123.1	-	8N	155W
828	042	-1669 Sep 05	03:56:22	38749	-45372	001	N*	-t	1.0119	1.0309	-0.0281	277.0	-	-	12S	103E
829	042	-1668 Mar 01	11:12:32	38738	-45366	006	T-	-p	-0.1178	2.6441	1.6397	339.7	219.6	98.6	13N	2W
830	042	-1668 Aug 24	12:18:49	38727	-45360	011	T+	-p	0.2456	2.4068	1.4077	331.9	212.1	85.9	16S	23W
831	042	-1667 Feb 18	18:16:25	38717	-45354	016	P	t-	0.6476	1.7011	0.6387	333.3	175.7	-	17N	108W
832	042	-1667 Aug 14	02:54:28	38706	-45348	021	T	a-	-0.4707	1.9731	1.0151	304.5	188.0	18.3	20S	118E
833	042	-1666 Feb 07	18:57:50	38695	-45342	026	N	t-	1.3812	0.3689	-0.7212	184.1	-	-	21N	118W
834	042	-1666 Jul 05	11:10:39	38686	-45337	-007	N	-a	1.3965	0.2806	-0.6900	142.3	-	-	22S	8W
835	042	-1666 Aug 03	19:30:41	38684	-45336	031	N	a-	-1.1749	0.6816	-0.2777	208.9	-	-	23S	131W
836	042	-1666 Dec 29	04:27:39	38675	-45331	-002	P	-h	-0.8576	1.3076	0.2613	295.8	115.7	-	23N	96E
837	042	-1665 Jun 24	22:28:52	38665	-45325	003	P	-h	0.6954	1.5929	0.5710	315.1	163.8	-	23S	178W
838	042	-1665 Dec 18	14:57:40	38654	-45319	008	T-	-p	-0.1261	2.6233	1.6299	328.3	213.1	95.7	23N	63W
839	042	-1664 Jun 13	02:38:41	38643	-45313	013	T-	pp	-0.0666	2.7680	1.7042	371.5	234.5	105.2	22S	119E
840	042	-1664 Dec 07	06:02:00	38632	-45307	018	P	a-	0.5423	1.8481	0.8772	299.3	178.2	-	21N	69E

Cat Num	Canon Plate	Calendar Date	TD of Greatest Eclipse	ΔT s	Luna Num	Saros Num	Ecl. Type	QSE	Gamma	Pen. Mag.	Um. Mag.	Eclipse Phase Durations Pen. m	Par. m	Total m	Greatest in Zenith Lat.	Lng.
841	043	-1663 Jun 02	03:14:41	38622	-45301	023	P	t-	-0.8284	1.3692	0.3070	316.0	130.4	-	20S	109E
842	043	-1663 Oct 28	08:32:43	38613	-45296	-010	N	-a	-1.5445	0.0343	-0.9862	52.5	-	-	7N	31E
843	043	-1663 Nov 26	21:08:00	38611	-45295	028	N	a-	1.2234	0.6102	-0.3840	203.3	-	-	20N	158W
844	043	-1662 Apr 22	21:26:27	38602	-45290	-005	N	-h	1.1091	0.8185	-0.1729	237.5	-	-	6S	160W
845	043	-1662 May 22	07:37:52	38600	-45289	033	Nb	h-	-1.5444	0.0332	-0.9848	53.3	-	-	18S	43E
846	043	-1662 Oct 17	15:27:46	38591	-45284	000	P	-h	-0.9213	1.2059	0.1297	297.1	85.2	-	3N	72W
847	043	-1661 Apr 12	11:50:29	38580	-45278	005	T	-p	0.3207	2.2442	1.2942	313.4	202.3	74.4	2S	16W
848	043	-1661 Oct 06	15:49:22	38570	-45272	010	T-	-p	-0.2483	2.4541	1.3514	374.3	227.8	86.8	1S	77W
849	043	-1660 Apr 01	04:56:16	38559	-45266	015	T	a-	-0.4128	2.0739	1.1265	307.1	194.2	51.3	1N	88E
850	043	-1660 Sep 24	15:51:52	38548	-45260	020	T	t-	0.4451	2.0798	1.0033	352.8	206.5	9.3	5S	78W
851	043	-1659 Feb 20	09:10:11	38539	-45255	-013	N	-a	1.5055	0.1074	-0.9163	94.4	-	-	17N	29E
852	043	-1659 Mar 21	20:04:12	38537	-45254	025	N	a-	-1.1821	0.6822	-0.3046	217.3	-	-	5N	138W
853	043	-1659 Sep 13	22:15:03	38527	-45248	030	N	a-	1.1209	0.8102	-0.2077	238.5	-	-	8S	173W
854	043	-1658 Feb 09	14:03:05	38518	-45243	-003	P	-t	0.8318	1.3705	0.2934	315.8	127.1	-	20N	45W
855	043	-1658 Aug 05	02:46:59	38507	-45237	002	P	-a	-0.8374	1.2971	0.3453	267.2	121.9	-	22S	119E
856	043	-1657 Jan 29	13:50:58	38496	-45231	007	T+	pp	0.1239	2.6763	1.5858	375.3	233.2	101.2	22N	43W
857	043	-1657 Jul 25	19:11:47	38485	-45225	012	T-	-p	-0.1299	2.6024	1.6363	323.8	212.7	96.3	23S	128W
858	043	-1656 Jan 18	15:56:49	38475	-45219	017	P	a-	-0.5856	1.8109	0.7563	332.8	184.2	-	23N	76W
859	043	-1656 Jul 14	07:40:33	38464	-45213	022	P	t-	0.6391	1.6922	0.6784	316.3	173.3	-	23S	44E
860	043	-1656 Dec 08	13:37:05	38455	-45208	-011	N	-a	1.4629	0.1652	-0.8181	109.0	-	-	23N	45W
861	044	-1655 Jan 07	01:03:37	38453	-45207	027	N	a-	-1.2463	0.5714	-0.4293	200.2	-	-	23N	146E
862	044	-1655 Jun 03	22:29:51	38444	-45202	-006	N	-t	-1.3935	0.3352	-0.7326	174.9	-	-	21S	180E
863	044	-1655 Jul 03	13:10:04	38442	-45201	032	N	t-	1.4430	0.2408	-0.8199	147.6	-	-	22S	39W
864	044	-1655 Nov 28	05:08:48	38433	-45196	-001	P	-a	0.8017	1.3735	0.4001	274.1	130.2	-	20N	81E
865	044	-1654 May 23	23:27:40	38423	-45190	004	P	-t	-0.6268	1.7344	0.6818	336.8	181.6	-	18S	166E
866	044	-1654 Nov 17	19:23:28	38412	-45184	009	T+	-p	0.1443	2.5975	1.5890	333.4	214.7	95.0	16N	133W
867	044	-1653 May 13	05:47:00	38401	-45178	014	T+	pp	0.1691	2.5488	1.5467	342.0	220.4	96.1	14S	71E
868	044	-1653 Nov 07	03:56:46	38390	-45172	019	P	t-	-0.5690	1.8471	0.7811	336.0	186.6	-	11N	99E
869	044	-1652 May 01	19:06:32	38380	-45166	024	P	a-	0.9114	1.1626	0.2083	260.9	98.0	-	9S	128W
870	044	-1652 Oct 26	05:22:20	38369	-45160	029	N	t-	-1.2796	0.5626	-0.5415	223.6	-	-	7N	78E
871	044	-1651 Mar 23	04:59:02	38360	-45155	-004	P	-a	-0.8949	1.1939	0.2376	262.4	103.8	-	4N	88E
872	044	-1651 Sep 15	11:43:49	38349	-45149	001	N	-t	1.0345	0.9895	-0.0697	271.8	-	-	8S	16W
873	044	-1650 Mar 12	18:42:09	38339	-45143	006	T-	-p	-0.1720	2.5447	1.5399	339.8	218.6	95.1	9N	117W
874	044	-1650 Sep 04	20:21:35	38328	-45137	011	T	-p	0.2756	2.3519	1.3526	329.8	209.7	81.3	13S	145W
875	044	-1649 Mar 02	01:27:51	38317	-45131	016	P	t-	0.5981	1.7915	0.7299	339.1	185.4	-	13N	142E
876	044	-1649 Aug 25	11:01:55	38306	-45125	021	T	a-	-0.4330	2.0433	1.0833	306.7	191.6	42.0	17S	5W
877	044	-1648 Feb 19	02:06:00	38296	-45119	026	N	t-	1.3364	0.4492	-0.6369	201.5	-	-	18N	133E
878	044	-1648 Jul 15	18:37:22	38287	-45114	-007	N	-a	1.4615	0.1641	-0.8119	110.4	-	-	22S	121W
879	044	-1648 Aug 14	03:26:08	38285	-45113	031	N	a-	-1.1330	0.7609	-0.2032	219.5	-	-	20S	109E
880	044	-1647 Jan 08	12:37:16	38276	-45108	-002	P	-h	-0.8686	1.2846	0.2441	292.6	111.7	-	23N	27W
881	045	-1647 Jul 05	05:24:22	38265	-45102	003	P	-h	0.7699	1.4592	0.4316	307.9	146.0	-	23S	77E
882	045	-1647 Dec 28	23:34:13	38255	-45096	008	T-	-p	-0.1333	2.6078	1.6187	327.0	212.6	95.3	24N	167E
883	045	-1646 Jun 24	09:04:57	38244	-45090	013	T+	pp	0.0164	2.8616	1.7946	372.8	235.2	106.3	23S	21E
884	045	-1646 Dec 18	14:51:43	38233	-45084	018	P	a-	0.5392	1.8532	0.8837	299.4	178.7	-	23N	64W
885	045	-1645 Jun 13	09:37:51	38222	-45078	023	P	t-	-0.7416	1.5282	0.4666	327.2	156.6	-	22S	12E
886	045	-1645 Nov 08	17:08:13	38213	-45073	-010	N	-a	-1.5498	0.0266	-0.9978	46.4	-	-	11N	99W
887	045	-1645 Dec 08	05:53:16	38212	-45072	028	N	a-	1.2257	0.6067	-0.3890	203.3	-	-	22N	69E
888	045	-1644 May 03	04:29:39	38203	-45067	-005	N	-h	1.1801	0.6859	-0.3009	219.4	-	-	10S	91E
889	045	-1644 Jun 01	14:25:07	38201	-45066	033	N	h-	-1.4610	0.1846	-0.8302	123.5	-	-	21S	60W
890	045	-1644 Oct 27	23:36:20	38192	-45061	000	P	-h	-0.9292	1.1935	0.1131	297.0	80.0	-	8N	163E
891	045	-1643 Apr 22	19:15:48	38181	-45055	005	T	-p	0.3916	2.1127	1.1658	309.2	196.7	58.2	6S	130W
892	045	-1643 Oct 16	23:39:54	38171	-45049	010	T-	-p	-0.2606	2.4326	1.3277	374.1	226.9	84.5	4N	163E
893	045	-1642 Apr 12	12:28:33	38160	-45043	015	T	p-	-0.3465	2.1953	1.2485	311.0	200.0	69.3	3S	27W
894	045	-1642 Oct 05	23:47:30	38149	-45037	020	T	h-	0.4277	2.1118	1.0352	353.1	208.0	30.1	0S	161E
895	045	-1641 Mar 03	16:39:41	38140	-45032	-013	Ne	-h	1.5584	0.0104	-1.0135	29.8	-	-	14N	86W
896	045	-1641 Apr 02	03:25:44	38138	-45031	025	N	h-	-1.1229	0.7914	-0.1964	232.4	-	-	0N	109E
897	045	-1641 Sep 25	06:28:36	38128	-45025	030	N	a-	1.0987	0.8508	-0.1668	242.6	-	-	4S	61E
898	045	-1640 Feb 20	21:15:30	38119	-45020	-003	P	-t	0.8803	1.2807	0.2052	308.8	107.9	-	17N	155W
899	045	-1640 Aug 15	10:48:01	38108	-45014	002	P	-a	-0.8800	1.2204	0.2658	261.5	108.2	-	20S	2W
900	045	-1639 Feb 08	21:08:10	38098	-45008	007	T+	pp	0.1639	2.6005	1.5148	373.5	231.7	98.1	19N	153W

Cat Num	Canon Plate	Calendar Date	TD of Greatest Eclipse	ΔT s	Luna Num	Saros Num	Ecl. Type	QSE	Gamma	Pen. Mag.	Um. Mag.	Eclipse Phase Durations Pen. m	Eclipse Phase Durations Par. m	Eclipse Phase Durations Total m	Greatest in Zenith Lat.	Greatest in Zenith Lng.
901	046	-1639 Aug 05	02:57:27	38087	-45002	012	T-	-p	-0.1808	2.5119	1.5404	324.0	211.4	92.5	21S	115E
902	046	-1638 Jan 28	23:43:50	38076	-44996	017	P	a-	-0.5563	1.8615	0.8133	333.7	188.6	-	21N	167E
903	046	-1638 Jul 25	14:55:15	38065	-44990	022	P	t-	0.5811	1.8019	0.7817	323.6	183.4	-	22S	65W
904	046	-1638 Dec 19	22:22:31	38057	-44985	-011	N	-a	1.4632	0.1630	-0.8171	108.0	-	-	24N	177W
905	046	-1637 Jan 18	09:24:17	38055	-44984	027	N	a-	-1.2265	0.6049	-0.3901	204.6	-	-	22N	21E
906	046	-1637 Jun 15	04:48:10	38046	-44979	-006	N	-t	-1.4786	0.1800	-0.8897	130.1	-	-	23S	84E
907	046	-1637 Jul 14	19:49:31	38044	-44978	032	N	t-	1.3742	0.3695	-0.6961	181.3	-	-	23S	140W
908	046	-1637 Dec 09	14:01:08	38035	-44973	-001	P	-a	0.8024	1.3721	0.3988	274.2	130.1	-	22N	53W
909	046	-1636 Jun 03	05:53:38	38024	-44967	004	P	-t	-0.7114	1.5781	0.5274	326.5	163.4	-	20S	67E
910	046	-1636 Nov 28	04:05:58	38014	-44961	009	T+	-p	0.1428	2.6014	1.5905	334.6	215.2	95.2	19N	95E
911	046	-1635 May 23	12:41:18	38003	-44955	014	T+	pp	0.0886	2.6946	1.6961	341.9	221.9	101.1	17S	34W
912	046	-1635 Nov 17	12:16:33	37992	-44949	019	P	t-	-0.5708	1.8455	0.7763	337.1	186.7	-	15N	28W
913	046	-1634 May 13	02:25:16	37982	-44943	024	P	a-	0.8361	1.2993	0.3477	270.8	123.9	-	13S	120E
914	046	-1634 Nov 06	13:21:38	37971	-44937	029	N	t-	-1.2759	0.5701	-0.5358	225.2	-	-	11N	44W
915	046	-1633 Apr 03	12:34:00	37962	-44932	-004	P	-a	-0.9578	1.0782	0.1222	253.1	75.8	-	0S	29W
916	046	-1633 Sep 26	19:39:37	37952	-44926	001	N	-t	1.0510	0.9593	-0.1000	267.7	-	-	4S	137W
917	046	-1632 Mar 23	02:02:37	37941	-44920	006	T-	-p	-0.2329	2.4335	1.4279	339.3	216.6	88.9	5N	130E
918	046	-1632 Sep 15	04:33:38	37930	-44914	011	T	-p	0.2985	2.3099	1.3105	328.0	207.7	77.3	8S	90E
919	046	-1631 Mar 12	08:28:31	37920	-44908	016	P	t-	0.5405	1.8970	0.8361	345.0	195.3	-	10N	35E
920	046	-1631 Sep 04	19:18:21	37909	-44902	021	T	p-	-0.4025	2.1002	1.1381	308.4	194.1	53.2	13S	131W
921	047	-1630 Mar 01	09:04:05	37898	-44896	026	N	t-	1.2832	0.5449	-0.5374	219.6	-	-	14N	27E
922	047	-1630 Jul 27	02:11:34	37889	-44891	-007	Ne	-a	1.5195	0.0604	-0.9212	67.9	-	-	21S	125E
923	047	-1630 Aug 25	11:30:27	37888	-44890	031	N	a-	-1.0988	0.8261	-0.1428	227.8	-	-	17S	14W
924	047	-1629 Jan 19	20:39:54	37879	-44885	-002	P	-h	-0.8857	1.2500	0.2157	288.4	105.1	-	22N	149W
925	047	-1629 Jul 16	12:24:08	37868	-44879	003	P	-h	0.8394	1.3346	0.3010	300.2	124.8	-	23S	29W
926	047	-1628 Jan 09	08:05:14	37857	-44873	008	T-	-p	-0.1455	2.5831	1.5987	325.6	211.9	94.5	24N	39E
927	047	-1628 Jul 04	15:35:10	37847	-44867	013	T+	pp	0.0954	2.7186	1.6479	373.0	234.4	103.9	24S	78W
928	047	-1628 Dec 28	23:35:40	37836	-44861	018	P	a-	0.5314	1.8666	0.8989	299.9	179.8	-	24N	165E
929	047	-1627 Jun 23	16:06:05	37825	-44855	023	P	t-	-0.6582	1.6812	0.6198	336.3	175.8	-	24S	86W
930	047	-1627 Nov 19	01:45:01	37816	-44850	-010	N	-a	-1.5532	0.0218	-1.0058	42.2	-	-	15N	130E
931	047	-1627 Dec 18	14:34:38	37815	-44849	028	N	a-	1.2260	0.6067	-0.3899	203.8	-	-	24N	61W
932	047	-1626 May 14	11:32:41	37806	-44844	-005	N	-h	1.2520	0.5521	-0.4309	198.6	-	-	13S	17W
933	047	-1626 Jun 12	21:16:44	37804	-44843	033	N	h-	-1.3797	0.3326	-0.6797	162.9	-	-	23S	165W
934	047	-1626 Nov 08	07:47:43	37795	-44838	000	P	-h	-0.9340	1.1864	0.1026	297.3	76.5	-	12N	39E
935	047	-1625 May 04	02:38:48	37784	-44832	005	T	-a	0.4646	1.9774	1.0329	304.1	189.6	27.0	10S	117E
936	047	-1625 Oct 28	07:36:18	37774	-44826	010	T-	-p	-0.2681	2.4198	1.3132	374.0	226.4	83.1	8N	41E
937	047	-1624 Apr 22	19:54:17	37763	-44820	015	T	p-	-0.2750	2.3265	1.3797	314.5	205.0	82.0	7S	142W
938	047	-1624 Oct 16	07:51:51	37753	-44814	020	T	a-	0.4167	2.1317	1.0555	352.9	208.7	37.5	4N	38E
939	047	-1623 Apr 12	10:39:36	37742	-44808	025	N	h-	-1.0583	0.9107	-0.0785	247.1	-	-	4S	2W
940	047	-1623 Oct 05	14:50:24	37731	-44802	030	N	a-	1.0826	0.8801	-0.1369	245.2	-	-	0N	67W
941	048	-1622 Mar 03	04:18:33	37722	-44797	-003	P	-t	0.9356	1.1784	0.1045	299.9	78.2	-	13N	97E
942	048	-1622 Aug 26	18:57:15	37712	-44791	002	P	-a	-0.9158	1.1560	0.1985	256.5	94.4	-	17S	126W
943	048	-1621 Feb 20	04:18:27	37701	-44785	007	T+	pp	0.2099	2.5136	1.4327	371.1	229.5	93.2	16N	97E
944	048	-1621 Aug 16	10:49:00	37690	-44779	012	T-	-p	-0.2260	2.4316	1.4546	324.0	209.8	87.9	19S	4W
945	048	-1620 Feb 09	07:24:41	37680	-44773	017	P	a-	-0.5209	1.9232	0.8817	334.8	193.5	-	19N	50E
946	048	-1620 Aug 04	22:15:48	37669	-44767	022	P	t-	0.5286	1.9016	0.8747	329.8	191.5	-	21S	177W
947	048	-1620 Dec 30	07:01:57	37660	-44762	-011	N	-a	1.4689	0.1507	-0.8257	103.7	-	-	25N	53E
948	048	-1619 Jan 28	17:36:29	37659	-44761	027	N	a-	-1.1991	0.6524	-0.3369	210.6	-	-	20N	103W
949	048	-1619 Jun 25	11:11:37	37650	-44756	-006	Ne	-t	-1.5588	0.0339	-1.0380	57.3	-	-	25S	13W
950	048	-1619 Jul 25	02:36:46	37648	-44755	032	N	t-	1.3116	0.4869	-0.5836	206.4	-	-	22S	117E
951	048	-1619 Dec 19	22:48:31	37639	-44750	-001	P	-a	0.8076	1.3623	0.3896	273.8	128.9	-	24N	175E
952	048	-1618 Jun 14	12:25:23	37629	-44744	004	P	-t	-0.7918	1.4299	0.3807	315.3	141.6	-	23S	32W
953	048	-1618 Dec 09	12:44:28	37618	-44738	009	T+	-p	0.1442	2.5997	1.5870	335.7	215.7	95.3	22N	35W
954	048	-1617 Jun 03	19:39:11	37607	-44732	014	T+	pp	0.0099	2.8376	1.8421	340.9	222.1	102.7	20S	141W
955	048	-1617 Nov 28	20:34:34	37597	-44726	019	P	t-	-0.5719	1.8446	0.7730	338.2	186.9	-	18N	154W
956	048	-1616 May 23	09:42:31	37586	-44720	024	P	a-	0.7590	1.4399	0.4902	279.8	143.7	-	16S	9E
957	048	-1616 Nov 16	21:23:25	37575	-44714	029	N	t-	-1.2743	0.5736	-0.5333	226.0	-	-	14N	166W
958	048	-1615 Apr 13	20:00:41	37567	-44709	-004	N	-a	-1.0272	0.9511	-0.0053	241.5	-	-	4S	143W
959	048	-1615 May 13	02:56:56	37565	-44708	034	N	a-	1.4720	0.1273	-0.8140	94.3	-	-	12S	111E
960	048	-1615 Oct 07	03:45:18	37556	-44703	001	N	-t	1.0604	0.9418	-0.1170	264.9	-	-	1N	99E

Cat Num	Canon Plate	Calendar Date	TD of Greatest Eclipse	ΔT s	Luna Num	Saros Num	Ecl. Type	QSE	Gamma	Pen. Mag.	Um. Mag.	Eclipse Phase Durations Pen. m	Par. m	Total m	Greatest in Zenith Lat.	Lng.
961	049	-1614 Apr 03	09:13:48	37545	-44697	006	T	-a	-0.3005	2.3100	1.3032	338.0	213.2	78.6	0N	20E
962	049	-1614 Sep 26	12:55:27	37535	-44691	011	T	-p	0.3143	2.2809	1.2816	326.3	206.0	74.1	4S	38W
963	049	-1613 Mar 23	15:19:42	37524	-44685	016	P	t-	0.4757	2.0155	0.9552	350.9	204.7	-	5N	70W
964	049	-1613 Sep 16	03:42:35	37513	-44679	021	T	p-	-0.3786	2.1452	1.1809	309.6	195.9	59.9	9S	101E
965	049	-1612 Mar 11	15:54:06	37503	-44673	026	N	t-	1.2231	0.6532	-0.4251	237.5	-	-	10N	78W
966	049	-1612 Sep 04	19:41:58	37492	-44667	031	N	a-	-1.0713	0.8790	-0.0949	234.4	-	-	14S	138W
967	049	-1611 Jan 30	04:35:35	37483	-44662	-002	P	-h	-0.9092	1.2036	0.1758	283.1	95.2	-	20N	92E
968	049	-1611 Jul 26	19:29:34	37473	-44656	003	P	-h	0.9030	1.2211	0.1812	292.2	98.8	-	22S	136W
969	049	-1610 Jan 19	16:30:35	37462	-44650	008	T-	-p	-0.1626	2.5492	1.5697	324.0	211.0	93.3	23N	88W
970	049	-1610 Jul 15	22:11:14	37452	-44644	013	T+	pp	0.1691	2.5851	1.5108	372.4	232.2	98.4	24S	178W
971	049	-1609 Jan 09	08:14:15	37441	-44638	018	P	a-	0.5196	1.8873	0.9216	300.7	181.4	-	24N	35E
972	049	-1609 Jul 04	22:40:59	37430	-44632	023	P	t-	-0.5795	1.8256	0.7641	343.5	190.1	-	24S	174E
973	049	-1609 Nov 30	10:19:54	37421	-44627	-010	N	-a	-1.5577	0.0148	-1.0154	35.0	-	-	18N	0W
974	049	-1609 Dec 29	23:10:14	37420	-44626	028	N	a-	1.2221	0.6140	-0.3832	205.4	-	-	25N	169E
975	049	-1608 May 24	18:36:25	37411	-44621	-005	N	-h	1.3239	0.4183	-0.5613	174.4	-	-	16S	125W
976	049	-1608 Jun 23	04:13:17	37409	-44620	033	N	h-	-1.3009	0.4761	-0.5342	191.5	-	-	24S	90E
977	049	-1608 Nov 18	15:59:05	37400	-44615	000	P	-h	-0.9381	1.1802	0.0938	297.5	73.4	-	15N	86W
978	049	-1607 May 14	10:00:50	37390	-44609	005	P	-a	0.5388	1.8404	0.8976	298.3	180.8	-	14S	4E
979	049	-1607 Nov 07	15:34:49	37379	-44603	010	T	-p	-0.2737	2.4098	1.3025	373.8	225.9	81.9	12N	80W
980	049	-1606 May 04	03:17:23	37369	-44597	015	T-	p-	-0.2018	2.4612	1.5137	317.4	208.9	90.8	11S	105E
981	050	-1606 Oct 27	16:02:01	37358	-44591	020	T	a-	0.4101	2.1433	1.0682	352.3	208.9	41.3	8N	87W
982	050	-1605 Apr 23	17:46:41	37347	-44585	025	P	h-	-0.9888	1.0391	0.0479	261.4	49.8	-	8S	112W
983	050	-1605 Oct 16	23:19:22	37337	-44579	030	N	a-	1.0716	0.8997	-0.1163	246.7	-	-	5N	164E
984	050	-1604 Mar 13	11:10:32	37328	-44574	-003	N*	-t	0.9992	1.0610	-0.0113	288.5	-	-	9N	8W
985	050	-1604 Sep 06	03:15:18	37317	-44568	002	P	-a	-0.9448	1.1045	0.1439	252.3	81.0	-	13S	108E
986	050	-1603 Mar 02	11:17:33	37307	-44562	007	T+	pp	0.2656	2.4089	1.3330	368.0	226.1	85.1	13N	9W
987	050	-1603 Aug 26	18:49:14	37296	-44556	012	T-	-p	-0.2636	2.3656	1.3829	324.1	208.1	83.0	16S	126W
988	050	-1602 Feb 19	14:57:23	37286	-44550	017	P	a-	-0.4775	1.9993	0.9645	336.3	198.8	-	16N	64W
989	050	-1602 Aug 16	05:42:53	37275	-44544	022	P	t-	0.4822	1.9901	0.9565	335.2	197.9	-	18S	70E
990	050	-1601 Jan 10	15:36:46	37266	-44539	-011	N	-a	1.4789	0.1304	-0.8421	96.4	-	-	25N	76W
991	050	-1601 Feb 09	01:42:32	37264	-44538	027	N	a-	-1.1660	0.7102	-0.2733	217.6	-	-	18N	134E
992	050	-1601 Aug 05	09:30:22	37254	-44532	032	N	t-	1.2539	0.5954	-0.4803	226.6	-	-	20S	13E
993	050	-1601 Dec 31	07:31:53	37245	-44527	-001	P	-a	0.8158	1.3469	0.3751	273.0	126.9	-	25N	44E
994	050	-1600 Jun 24	19:00:47	37235	-44521	004	P	-t	-0.8697	1.2865	0.2383	303.1	114.2	-	24S	132W
995	050	-1600 Dec 19	21:19:42	37224	-44515	009	T+	-p	0.1477	2.5941	1.5800	336.7	216.2	95.3	23N	165W
996	050	-1599 Jun 14	02:41:15	37213	-44509	014	T-	pp	-0.0664	2.7326	1.7397	339.1	220.9	101.4	22S	112E
997	050	-1599 Dec 09	04:49:23	37203	-44503	019	P	t-	-0.5711	1.8471	0.7737	339.4	187.4	-	21N	82E
998	050	-1598 Jun 03	17:03:20	37192	-44497	024	P	a-	0.6841	1.5765	0.6282	287.3	159.0	-	19S	103W
999	050	-1598 Nov 28	05:24:52	37182	-44491	029	N	t-	-1.2722	0.5775	-0.5295	226.8	-	-	18N	72E
1000	050	-1597 Apr 25	03:23:27	37173	-44486	-004	N	-a	-1.0991	0.8196	-0.1376	227.9	-	-	9S	103E
1001	051	-1597 May 24	10:19:10	37171	-44485	034	N	a-	1.4020	0.2562	-0.6861	132.3	-	-	16S	2W
1002	051	-1597 Oct 18	11:57:26	37162	-44480	001	N	-h	1.0653	0.9323	-0.1254	263.0	-	-	5N	26W
1003	051	-1596 Apr 13	16:18:14	37152	-44474	006	T	-a	-0.3728	2.1781	1.1697	335.9	208.2	61.8	4S	89W
1004	051	-1596 Oct 06	21:24:52	37141	-44468	011	T	-p	0.3242	2.2625	1.2636	325.0	204.8	72.0	0N	168W
1005	051	-1595 Apr 02	22:01:42	37131	-44462	016	T	t-	0.4040	2.1470	1.0868	356.6	213.5	47.2	1N	174W
1006	051	-1595 Sep 26	12:15:20	37120	-44456	021	T	p-	-0.3618	2.1771	1.2108	310.5	197.1	64.0	5S	29W
1007	051	-1594 Mar 22	22:36:05	37110	-44450	026	N	t-	1.1560	0.7743	-0.3002	255.1	-	-	6N	179E
1008	051	-1594 Sep 16	04:00:56	37099	-44444	031	N	a-	-1.0504	0.9198	-0.0590	239.5	-	-	10S	95E
1009	051	-1593 Feb 10	12:23:39	37090	-44439	-002	P	-h	-0.9396	1.1445	0.1234	276.5	80.1	-	18N	27W
1010	051	-1593 Aug 07	02:41:53	37080	-44433	003	P	-h	0.9597	1.1204	0.0740	284.3	64.3	-	20S	114E
1011	051	-1592 Jan 31	00:47:15	37069	-44427	008	T-	-p	-0.1877	2.5007	1.5262	322.3	209.9	91.3	21N	147E
1012	051	-1592 Jul 26	04:54:09	37058	-44421	013	T+	pp	0.2370	2.4626	1.3842	371.1	228.9	89.9	22S	81E
1013	051	-1591 Jan 19	16:44:55	37048	-44415	018	P	a-	0.5011	1.9200	0.9565	302.0	183.8	-	23N	93W
1014	051	-1591 Jul 15	05:24:02	37037	-44409	023	P	t-	-0.5065	1.9598	0.8979	349.1	200.8	-	24S	73E
1015	051	-1591 Dec 10	18:51:36	37029	-44404	-010	N	-a	-1.5642	0.0040	-1.0282	18.3	-	-	20N	129W
1016	051	-1590 Jan 09	07:38:32	37027	-44403	028	N	a-	1.2132	0.6305	-0.3668	208.4	-	-	25N	42E
1017	051	-1590 Jun 05	01:43:22	37018	-44398	-005	N	-a	1.3940	0.2885	-0.6885	146.0	-	-	19S	126E
1018	051	-1590 Jul 04	11:17:24	37016	-44397	033	N	a-	-1.2269	0.6111	-0.3977	213.3	-	-	25S	17W
1019	051	-1590 Nov 30	00:10:23	37008	-44392	000	P	-t	-0.9415	1.1748	0.0866	297.8	70.7	-	18N	150E
1020	051	-1589 May 25	17:21:52	36997	-44386	005	P	-a	0.6141	1.7016	0.7599	291.6	170.1	-	17S	109W

Cat Num	Canon Plate	Calendar Date	TD of Greatest Eclipse	ΔT s	Luna Num	Saros Num	Ecl. Type	QSE	Gamma	Pen. Mag.	Um. Mag.	Eclipse Phase Durations Pen. m	Par. m	Total m	Greatest in Zenith Lat.	Lng.
1021	052	-1589 Nov 18	23:36:43	36987	-44380	010	T	-p	-0.2768	2.4041	1.2969	373.5	225.6	81.3	16N	158E
1022	052	-1588 May 14	10:37:01	36976	-44374	015	T-	pp	-0.1261	2.6008	1.6520	319.8	211.8	96.6	15S	7W
1023	052	-1588 Nov 07	00:17:20	36965	-44368	020	T	a-	0.4066	2.1488	1.0755	351.4	208.8	43.2	13N	147E
1024	052	-1587 May 04	00:48:42	36955	-44362	025	P	t-	-0.9158	1.1743	0.1806	275.0	95.3	-	12S	140E
1025	052	-1587 Oct 27	07:54:32	36944	-44356	030	N	a-	1.0650	0.9110	-0.1035	247.1	-	-	9N	33E
1026	052	-1586 Mar 24	17:55:08	36936	-44351	-003	N	-t	1.0682	0.9336	-0.1373	274.6	-	-	5N	112W
1027	052	-1586 Sep 17	11:40:02	36925	-44345	002	P	-a	-0.9685	1.0625	0.0989	248.8	67.5	-	9S	21W
1028	052	-1585 Mar 13	18:11:00	36915	-44339	007	T	-p	0.3265	2.2946	1.2237	364.1	221.4	72.5	9N	115W
1029	052	-1585 Sep 07	02:56:13	36904	-44333	012	T	-p	-0.2951	2.3107	1.3223	324.3	206.6	78.0	12S	110E
1030	052	-1584 Mar 01	22:22:28	36894	-44327	017	T	a-	-0.4268	2.0890	1.0609	337.9	204.1	38.5	12N	178W
1031	052	-1584 Aug 26	13:17:34	36883	-44321	022	T	t-	0.4428	2.0657	1.0254	339.8	202.9	25.2	15S	45W
1032	052	-1583 Jan 21	00:03:49	36874	-44316	-011	N	-a	1.4958	0.0973	-0.8710	83.4	-	-	24N	157E
1033	052	-1583 Feb 19	09:40:01	36872	-44315	027	N	a-	-1.1252	0.7822	-0.1956	225.9	-	-	15N	13E
1034	052	-1583 Aug 15	16:33:16	36862	-44309	032	N	t-	1.2037	0.6899	-0.3908	242.4	-	-	18S	94W
1035	052	-1582 Jan 10	16:07:11	36853	-44304	-001	P	-a	0.8308	1.3187	0.3481	271.2	122.9	-	25N	85W
1036	052	-1582 Jul 06	01:45:30	36843	-44298	004	P	-t	-0.9408	1.1556	0.1081	290.5	78.1	-	25S	126E
1037	052	-1582 Dec 31	05:48:22	36832	-44292	009	T+	-p	0.1561	2.5791	1.5641	337.6	216.5	94.9	24N	68E
1038	052	-1581 Jun 25	09:48:25	36822	-44286	014	T-	pp	-0.1395	2.5974	1.6063	336.7	218.6	97.6	23S	4E
1039	052	-1581 Dec 20	12:59:37	36811	-44280	019	P	t-	-0.5670	1.8551	0.7807	340.7	188.5	-	23N	41W
1040	052	-1580 Jun 14	00:26:05	36801	-44274	024	P	a-	0.6100	1.7122	0.7645	293.8	171.2	-	21S	144E
1041	053	-1580 Dec 08	13:23:50	36790	-44268	029	N	t-	-1.2680	0.5848	-0.5214	228.1	-	-	20N	49W
1042	053	-1579 May 05	10:40:47	36781	-44263	-004	N	-a	-1.1749	0.6813	-0.2776	211.3	-	-	13S	9W
1043	053	-1579 Jun 03	17:41:12	36780	-44262	034	N	a-	1.3319	0.3858	-0.5583	160.7	-	-	19S	114W
1044	053	-1579 Oct 28	20:17:20	36771	-44257	001	N	-h	1.0650	0.9321	-0.1242	262.1	-	-	10N	154W
1045	053	-1578 Apr 24	23:15:47	36760	-44251	006	T	-h	-0.4497	2.0381	1.0276	332.6	201.3	26.1	8S	164E
1046	053	-1578 Oct 18	06:01:03	36750	-44245	011	T	-p	0.3293	2.2528	1.2546	323.8	203.9	70.8	5N	61E
1047	053	-1577 Apr 14	04:36:26	36739	-44239	016	T	pp	0.3268	2.2886	1.2285	361.7	221.1	73.1	3S	85E
1048	053	-1577 Oct 07	20:55:00	36729	-44233	021	T	p-	-0.3506	2.1985	1.2303	311.1	197.8	66.4	0S	162W
1049	053	-1576 Apr 02	05:10:15	36718	-44227	026	N	t-	1.0821	0.9081	-0.1626	272.0	-	-	2N	78E
1050	053	-1576 Sep 26	12:27:13	36708	-44221	031	N	a-	-1.0362	0.9484	-0.0352	243.3	-	-	5S	34W
1051	053	-1575 Feb 20	20:04:45	36699	-44216	-002	P	-h	-0.9765	1.0733	0.0589	268.5	55.6	-	15N	144W
1052	053	-1575 Aug 17	10:01:01	36689	-44210	003	N*	-h	1.0099	1.0315	-0.0215	276.7	-	-	17S	3E
1053	053	-1574 Feb 10	08:57:26	36678	-44204	008	T-	-p	-0.2187	2.4412	1.4718	320.4	208.3	88.3	18N	23E
1054	053	-1574 Aug 06	11:45:42	36668	-44198	013	T	-p	0.2976	2.3535	1.2710	369.4	224.8	78.9	21S	23W
1055	053	-1573 Jan 31	01:09:05	36657	-44192	018	T	a-	0.4776	1.9620	1.0008	303.6	186.7	4.2	22N	140E
1056	053	-1573 Jul 26	12:15:10	36647	-44186	023	T	t-	-0.4392	2.0835	1.0210	353.4	208.8	23.6	23S	31W
1057	053	-1572 Jan 20	15:59:14	36636	-44180	028	N	a-	1.1990	0.6565	-0.3409	212.7	-	-	24N	83W
1058	053	-1572 Jun 15	08:54:34	36627	-44175	-005	N	-a	1.4610	0.1644	-0.8105	111.1	-	-	21S	17E
1059	053	-1572 Jul 14	18:29:38	36626	-44174	033	N	a-	-1.1584	0.7361	-0.2715	230.3	-	-	25S	126W
1060	053	-1572 Dec 10	08:16:54	36617	-44169	000	P	-t	-0.9481	1.1633	0.0740	297.4	65.6	-	21N	28E
1061	054	-1571 Jun 05	00:45:31	36606	-44163	005	P	-a	0.6874	1.5670	0.6255	284.3	157.6	-	20S	139E
1062	054	-1571 Nov 29	07:37:33	36596	-44157	010	T	-p	-0.2804	2.3970	1.2908	373.0	225.3	80.7	19N	36E
1063	054	-1570 May 25	17:56:21	36585	-44151	015	T-	pp	-0.0505	2.7404	1.7896	321.5	213.5	99.7	18S	119W
1064	054	-1570 Nov 18	08:34:21	36575	-44145	020	T	a-	0.4040	2.1524	1.0816	350.3	208.6	44.8	16N	21E
1065	054	-1569 May 15	07:46:57	36564	-44139	025	P	t-	-0.8407	1.3137	0.3169	287.6	124.2	-	15S	33E
1066	054	-1569 Nov 07	16:33:55	36554	-44133	030	N	a-	1.0613	0.9170	-0.0956	247.0	-	-	13N	99W
1067	054	-1568 Apr 04	00:28:58	36545	-44128	-003	N	-t	1.1451	0.7918	-0.2778	256.8	-	-	1N	147E
1068	054	-1568 Sep 27	20:13:36	36535	-44122	002	P	-a	-0.9850	1.0336	0.0670	246.4	55.9	-	4S	151W
1069	054	-1567 Mar 24	00:55:18	36524	-44116	007	T	-t	0.3951	2.1662	1.1001	359.0	214.9	50.5	5N	141E
1070	054	-1567 Sep 17	11:11:15	36514	-44110	012	T	-p	-0.3191	2.2694	1.2754	324.7	205.3	73.4	8S	15W
1071	054	-1566 Mar 13	05:40:48	36503	-44104	017	T	a-	-0.3696	2.1906	1.1691	339.4	209.2	61.9	8N	71E
1072	054	-1566 Sep 06	21:00:26	36493	-44098	022	T	p-	0.4112	2.1272	1.0802	343.7	206.7	44.1	11S	163W
1073	054	-1565 Feb 01	08:24:51	36484	-44093	-011	N	-a	1.5179	0.0547	-0.9095	62.6	-	-	22N	31E
1074	054	-1565 Mar 02	17:31:06	36482	-44092	027	N	a-	-1.0785	0.8649	-0.1073	234.6	-	-	11N	107W
1075	054	-1565 Aug 26	23:43:49	36472	-44086	032	N	t-	1.1597	0.7732	-0.3125	255.1	-	-	15S	156E
1076	054	-1564 Jan 22	00:36:47	36463	-44081	-001	P	-a	0.8503	1.2823	0.3130	268.9	117.3	-	23N	147E
1077	054	-1564 Jul 16	08:37:25	36453	-44075	004	N*	-t	-1.0068	1.0343	-0.0129	277.6	-	-	25S	22E
1078	054	-1563 Jan 10	14:09:53	36442	-44069	009	T+	-p	0.1698	2.5542	1.5387	338.4	216.7	94.0	24N	58W
1079	054	-1563 Jul 05	17:03:01	36432	-44063	014	T-	pp	-0.2075	2.4720	1.4823	333.7	215.3	91.4	24S	105W
1080	054	-1563 Dec 30	21:03:50	36421	-44057	019	P	t-	-0.5583	1.8713	0.7965	342.4	190.3	-	23N	163W

Cat Num	Canon Plate	Calendar Date	TD of Greatest Eclipse	ΔT s	Luna Num	Saros Num	Ecl. Type	QSE	Gamma	Pen. Mag.	Um. Mag.	Eclipse Phase Durations Pen. m	Par. m	Total m	Greatest in Zenith Lat.	Lng.
1081	055	-1562 Jun 25	07:53:42	36411	-44051	024	P	a-	0.5396	1.8414	0.8938	299.2	180.8	-	23S	31E
1082	055	-1562 Dec 19	21:19:03	36401	-44045	029	N	t-	-1.2606	0.5976	-0.5071	230.2	-	-	22N	168W
1083	055	-1561 May 16	17:56:32	36392	-44040	-004	N	-a	-1.2513	0.5422	-0.4188	191.7	-	-	16S	120W
1084	055	-1561 Jun 15	01:05:21	36390	-44039	034	N	a-	1.2636	0.5123	-0.4340	183.2	-	-	21S	133E
1085	055	-1561 Nov 09	04:40:29	36381	-44034	001	N	-h	1.0627	0.9353	-0.1187	261.6	-	-	14N	79E
1086	055	-1560 May 05	06:07:29	36371	-44028	006	P	-h	-0.5302	1.8916	0.8785	328.2	191.9	-	12S	58E
1087	055	-1560 Oct 28	14:42:27	36360	-44022	011	T	-p	0.3307	2.2497	1.2526	322.9	203.3	70.4	9N	71W
1088	055	-1559 Apr 24	11:05:02	36350	-44016	016	T+	pp	0.2451	2.4388	1.3782	366.0	227.3	89.2	7S	15W
1089	055	-1559 Oct 18	05:39:56	36340	-44010	021	T	p-	-0.3439	2.2117	1.2420	311.4	198.2	67.8	4N	65E
1090	055	-1558 Apr 13	11:39:29	36329	-44004	026	N*	t-	1.0033	1.0509	-0.0165	287.8	-	-	3S	22W
1091	055	-1558 Oct 07	21:00:06	36319	-43998	031	N	a-	-1.0277	0.9662	-0.0219	245.8	-	-	1S	165W
1092	055	-1557 Mar 04	03:38:01	36310	-43993	-002	N	-h	-1.0205	0.9892	-0.0185	258.8	-	-	11N	101E
1093	055	-1557 Aug 28	17:28:02	36300	-43987	003	N	-h	1.0527	0.9563	-0.1032	269.7	-	-	14S	110W
1094	055	-1556 Feb 21	16:59:01	36289	-43981	008	T-	-p	-0.2573	2.3678	1.4034	318.0	206.2	83.8	15N	99W
1095	055	-1556 Aug 16	18:46:42	36279	-43975	013	T	-t	0.3500	2.2593	1.1727	367.5	220.5	65.2	18S	130W
1096	055	-1555 Feb 10	09:23:13	36268	-43969	018	T	a-	0.4460	2.0188	1.0598	305.7	190.2	35.8	19N	15E
1097	055	-1555 Aug 05	19:17:23	36258	-43963	023	T	t-	-0.3804	2.1920	1.1286	356.3	214.5	56.3	21S	138W
1098	055	-1554 Jan 31	00:11:21	36247	-43957	028	N	a-	1.1789	0.6933	-0.3039	218.4	-	-	22N	153E
1099	055	-1554 Jun 26	16:10:36	36239	-43952	-005	Ne	-a	1.5248	0.0466	-0.9268	59.5	-	-	22S	93W
1100	055	-1554 Jul 26	01:50:08	36237	-43951	033	N	a-	-1.0957	0.8510	-0.1560	243.9	-	-	24S	123E
1101	056	-1554 Dec 21	16:20:04	36228	-43946	000	P	-t	-0.9567	1.1476	0.0580	296.5	58.4	-	22N	94W
1102	056	-1553 Jun 16	08:10:54	36218	-43940	005	P	-a	0.7593	1.4353	0.4935	276.2	142.8	-	21S	26E
1103	056	-1553 Dec 10	15:37:41	36207	-43934	010	T	-p	-0.2851	2.3874	1.2831	372.3	224.8	79.8	21N	84W
1104	056	-1552 Jun 05	01:14:16	36197	-43928	015	T+	pp	0.0258	2.7871	1.8337	322.6	214.1	100.2	20S	130E
1105	056	-1552 Nov 28	16:53:31	36187	-43922	020	T	a-	0.4022	2.1541	1.0866	349.1	208.3	45.9	20N	105W
1106	056	-1551 May 25	14:43:20	36176	-43916	025	P	t-	-0.7646	1.4551	0.4546	299.1	146.3	-	18S	73W
1107	056	-1551 Nov 18	01:15:14	36166	-43910	030	N	a-	1.0584	0.9210	-0.0891	246.6	-	-	17N	129E
1108	056	-1550 Apr 15	06:58:20	36157	-43905	-003	N	-t	1.2249	0.6450	-0.4236	235.4	-	-	3S	47E
1109	056	-1550 Oct 09	04:53:04	36147	-43899	002	P	-a	-0.9969	1.0132	0.0439	244.8	45.4	-	0S	77E
1110	056	-1549 Apr 04	07:35:04	36136	-43893	007	P	-t	0.4681	2.0299	0.9685	352.9	206.4	-	1N	39E
1111	056	-1549 Sep 28	19:32:38	36126	-43887	012	T	-p	-0.3374	2.2385	1.2390	325.3	204.3	69.3	4S	143W
1112	056	-1548 Mar 23	12:53:35	36115	-43881	017	T	a-	-0.3066	2.3030	1.2879	340.6	213.7	77.3	4N	40W
1113	056	-1548 Sep 17	04:50:57	36105	-43875	022	T	p-	0.3862	2.1762	1.1230	346.9	209.5	54.0	7S	78E
1114	056	-1547 Mar 13	01:14:08	36095	-43869	027	N	a-	-1.0244	0.9616	-0.0053	244.0	-	-	8N	135E
1115	056	-1547 Sep 06	07:04:56	36084	-43863	032	N	t-	1.1239	0.8413	-0.2491	264.9	-	-	11S	44E
1116	056	-1546 Feb 01	08:57:23	36075	-43858	-001	P	-a	0.8775	1.2315	0.2637	265.4	108.6	-	22N	21E
1117	056	-1546 Jul 27	15:39:19	36065	-43852	004	N	-t	-1.0657	0.9262	-0.1209	264.9	-	-	23S	84W
1118	056	-1545 Jan 21	22:22:29	36055	-43846	009	T+	-p	0.1904	2.5164	1.5007	339.0	216.5	92.4	23N	178E
1119	056	-1545 Jul 17	00:25:07	36044	-43840	014	T-	-p	-0.2706	2.3559	1.3670	330.4	211.3	83.1	24S	143E
1120	056	-1544 Jan 11	04:59:08	36034	-43834	019	P	t-	-0.5430	1.8993	0.8247	344.6	193.2	-	23N	78E
1121	057	-1544 Jul 05	15:25:43	36023	-43828	024	T	a-	0.4724	1.9651	1.0168	303.7	188.4	19.3	23S	83W
1122	057	-1544 Dec 30	05:08:40	36013	-43822	029	N	t-	-1.2484	0.6188	-0.4835	233.7	-	-	23N	74E
1123	057	-1543 May 27	01:09:54	36004	-43817	-004	N	-a	-1.3288	0.4015	-0.5625	167.8	-	-	19S	130E
1124	057	-1543 Jun 25	08:32:05	36003	-43816	034	N	a-	1.1976	0.6349	-0.3145	202.0	-	-	22S	20E
1125	057	-1543 Nov 19	13:06:38	35994	-43811	001	N	-h	1.0589	0.9407	-0.1104	261.2	-	-	17N	49W
1126	057	-1542 May 16	12:55:45	35984	-43805	006	P	-h	-0.6125	1.7422	0.7259	322.4	179.7	-	16S	46W
1127	057	-1542 Nov 08	23:28:34	35973	-43799	011	T	-p	0.3290	2.2520	1.2564	322.0	203.0	70.7	13N	155E
1128	057	-1541 May 05	17:28:05	35963	-43793	016	T+	pp	0.1592	2.5967	1.5354	369.3	231.8	99.5	11S	114W
1129	057	-1541 Oct 29	14:30:16	35952	-43787	021	T	p-	-0.3415	2.2167	1.2457	311.6	198.3	68.2	9N	70W
1130	057	-1540 Apr 23	18:04:21	35942	-43781	026	P	t-	0.9200	1.2023	0.1380	302.1	89.7	-	7S	121W
1131	057	-1540 Oct 18	05:38:24	35932	-43775	031	N	a-	-1.0242	0.9747	-0.0176	247.5	-	-	4N	64E
1132	057	-1539 Mar 14	11:05:12	35923	-43770	-002	N	-h	-1.0704	0.8946	-0.1068	247.4	-	-	7N	13W
1133	057	-1539 Sep 08	01:03:10	35913	-43764	003	N	-h	1.0880	0.8947	-0.1710	263.7	-	-	10S	134E
1134	057	-1538 Mar 04	00:54:55	35902	-43758	008	T	-p	-0.3012	2.2850	1.3253	315.4	203.4	77.3	12N	140E
1135	057	-1538 Aug 28	01:55:56	35892	-43752	013	T	-t	0.3956	2.1777	1.0873	365.4	216.0	47.6	15S	121E
1136	057	-1537 Feb 21	17:30:36	35881	-43746	018	T	p-	0.4091	2.0855	1.1286	308.0	194.0	51.6	16N	108W
1137	057	-1537 Aug 17	02:29:32	35871	-43740	023	T	t-	-0.3289	2.2868	1.2226	358.3	218.5	71.5	19S	113E
1138	057	-1536 Feb 11	08:13:02	35861	-43734	028	N	a-	1.1511	0.7442	-0.2528	225.8	-	-	20N	31E
1139	057	-1536 Aug 05	09:20:52	35850	-43728	033	N	a-	-1.0401	0.9528	-0.0538	254.4	-	-	22S	10E
1140	057	-1535 Jan 01	00:15:14	35842	-43723	000	P	-t	-0.9712	1.1208	0.0316	294.4	43.3	-	23N	147E

Cat Num	Canon Plate	Calendar Date	TD of Greatest Eclipse	ΔT s	Luna Num	Saros Num	Ecl. Type	QSE	Gamma	Pen. Mag.	Um. Mag.	Eclipse Phase Durations Pen. m	Par. m	Total m	Greatest in Zenith Lat.	Lng.
1141	058	-1535 Jun 26	15:41:38	35831	-43717	005	P	-a	0.8271	1.3115	0.3686	267.8	125.6	-	23S	88W
1142	058	-1535 Dec 20	23:32:21	35821	-43711	010	T	-p	-0.2945	2.3689	1.2672	371.3	224.1	78.0	23N	156E
1143	058	-1534 Jun 16	08:35:13	35811	-43705	015	T+	pp	0.0990	2.6545	1.6977	323.0	213.5	98.2	22S	18E
1144	058	-1534 Dec 10	01:11:25	35800	-43699	020	T	a-	0.3985	2.1587	1.0952	347.9	208.3	47.9	22N	130E
1145	058	-1533 Jun 05	21:37:35	35790	-43693	025	P	t-	-0.6877	1.5986	0.5938	309.7	164.1	-	21S	178W
1146	058	-1533 Nov 29	09:57:55	35780	-43687	030	N	a-	1.0561	0.9236	-0.0834	246.0	-	-	20N	3W
1147	058	-1532 Apr 25	13:20:20	35771	-43682	-003	N	-t	1.3095	0.4895	-0.5786	208.3	-	-	7S	51W
1148	058	-1532 May 25	03:39:12	35769	-43681	035	N	t-	-1.5074	0.1193	-0.9347	104.7	-	-	19S	91E
1149	058	-1532 Oct 19	13:39:15	35761	-43676	002	P	-a	-1.0035	1.0023	0.0306	244.0	38.0	-	4N	57W
1150	058	-1531 Apr 14	14:08:31	35750	-43670	007	P	-t	0.5468	1.8834	0.8263	345.3	195.2	-	4S	62W
1151	058	-1531 Oct 09	04:01:25	35740	-43664	012	T	-p	-0.3490	2.2199	1.2154	326.1	203.8	66.4	1N	87E
1152	058	-1530 Apr 03	20:01:21	35729	-43658	017	T-	p-	-0.2382	2.4254	1.4163	341.5	217.4	88.5	0N	150W
1153	058	-1530 Sep 28	12:47:59	35719	-43652	022	T	p-	0.3672	2.2140	1.1547	349.7	211.7	60.0	3S	44W
1154	058	-1529 Mar 24	08:51:59	35709	-43646	027	P	a-	-0.9653	1.0675	0.1056	253.4	70.9	-	4N	18E
1155	058	-1529 Sep 17	14:34:14	35698	-43640	032	N	t-	1.0946	0.8973	-0.1975	272.6	-	-	7S	70W
1156	058	-1528 Feb 12	17:09:24	35690	-43635	-001	P	-a	0.9115	1.1684	0.2020	260.7	96.1	-	19N	103W
1157	058	-1528 Aug 06	22:50:26	35679	-43629	004	N	-t	-1.1178	0.8306	-0.2166	252.6	-	-	22S	167E
1158	058	-1527 Feb 01	06:26:14	35669	-43623	009	T+	-p	0.2177	2.4665	1.4507	339.3	216.0	89.7	21N	57E
1159	058	-1527 Jul 27	07:55:57	35659	-43617	014	T	-p	-0.3270	2.2520	1.2637	326.9	206.9	72.9	23S	30E
1160	058	-1526 Jan 21	12:45:57	35648	-43611	019	P	t-	-0.5213	1.9389	0.8648	347.1	196.9	-	22N	39W
1161	059	-1526 Jul 16	23:04:56	35638	-43605	024	T	a-	0.4108	2.0787	1.1292	307.2	194.2	51.8	23S	162E
1162	059	-1525 Jan 10	12:51:57	35628	-43599	029	N	t-	-1.2309	0.6494	-0.4500	238.5	-	-	23N	42W
1163	059	-1525 Jun 07	08:22:57	35619	-43594	-004	N	-a	-1.4057	0.2622	-0.7053	137.9	-	-	22S	20E
1164	059	-1525 Jul 06	16:02:15	35617	-43593	034	N	a-	1.1347	0.7521	-0.2009	217.8	-	-	23S	94W
1165	059	-1525 Nov 30	21:33:18	35609	-43588	001	N	-h	1.0556	0.9450	-0.1025	260.7	-	-	21N	177W
1166	059	-1524 May 26	19:41:44	35598	-43582	006	P	-t	-0.6958	1.5913	0.5712	315.3	164.1	-	19S	150W
1167	059	-1524 Nov 19	08:15:38	35588	-43576	011	T	-p	0.3273	2.2542	1.2607	321.2	202.8	71.1	17N	22E
1168	059	-1523 May 15	23:48:50	35578	-43570	016	T+	pp	0.0718	2.7578	1.6953	371.6	234.6	105.1	15S	149E
1169	059	-1523 Nov 08	23:23:18	35567	-43564	021	T	p-	-0.3411	2.2179	1.2461	311.7	198.3	68.2	13N	155E
1170	059	-1522 May 05	00:27:10	35557	-43558	026	P	t-	0.8341	1.3585	0.2970	314.9	128.4	-	11S	140E
1171	059	-1522 Oct 29	14:20:14	35547	-43552	031	N	a-	-1.0241	0.9767	-0.0193	248.5	-	-	8N	69W
1172	059	-1521 Mar 25	18:26:18	35538	-43547	-002	N	-h	-1.1259	0.7895	-0.2057	233.8	-	-	3N	126W
1173	059	-1521 Apr 24	04:57:22	35536	-43546	036	Nb	h-	1.5463	0.0298	-0.9884	50.4	-	-	6S	74E
1174	059	-1521 Sep 19	08:46:32	35528	-43541	003	N	-h	1.1156	0.8470	-0.2246	258.9	-	-	6S	16E
1175	059	-1520 Mar 14	08:41:24	35517	-43535	008	T	-p	-0.3534	2.1870	1.2317	312.1	199.6	67.3	8N	21E
1176	059	-1520 Sep 07	09:16:04	35507	-43529	013	T	-t	0.4321	2.1126	1.0184	363.6	211.8	22.4	11S	9E
1177	059	-1519 Mar 04	01:28:24	35497	-43523	018	T	p-	0.3644	2.1666	1.2115	310.6	198.1	64.5	12N	130E
1178	059	-1519 Aug 27	09:53:02	35487	-43517	023	T	p-	-0.2859	2.3663	1.3012	359.5	221.2	80.7	15S	0E
1179	059	-1518 Feb 21	16:05:37	35476	-43511	028	N	a-	1.1167	0.8073	-0.1897	234.4	-	-	16N	88W
1180	059	-1518 Aug 16	17:01:46	35466	-43505	033	P	a-	-0.9918	1.0413	0.0348	262.6	42.6	-	19S	107W
1181	060	-1517 Jan 12	08:04:28	35457	-43500	000	N*	-t	-0.9901	1.0858	-0.0025	291.2	-	-	23N	29E
1182	060	-1517 Jul 07	23:15:38	35447	-43494	005	P	-a	0.8922	1.1928	0.2484	258.8	104.9	-	23S	158E
1183	060	-1516 Jan 01	07:24:05	35437	-43488	010	T	-p	-0.3065	2.3450	1.2468	370.0	223.0	75.5	24N	38E
1184	060	-1516 Jun 26	15:57:57	35426	-43482	015	T+	-p	0.1700	2.5263	1.5655	322.9	211.9	93.9	23S	94W
1185	060	-1516 Dec 20	09:26:54	35416	-43476	020	T	a-	0.3922	2.1679	1.1091	346.7	208.5	51.0	24N	6E
1186	060	-1515 Jun 16	04:33:31	35406	-43470	025	P	t-	-0.6129	1.7381	0.7285	318.9	178.4	-	23S	77E
1187	060	-1515 Dec 09	18:39:54	35395	-43464	030	N	a-	1.0530	0.9274	-0.0759	245.6	-	-	23N	134W
1188	060	-1514 May 06	19:40:24	35387	-43459	-003	N	-t	1.3950	0.3324	-0.7354	174.3	-	-	11S	149W
1189	060	-1514 Jun 05	10:04:48	35385	-43458	035	N	t-	-1.4269	0.2683	-0.7886	155.4	-	-	22S	7W
1190	060	-1514 Oct 30	22:28:20	35377	-43453	002	P	-a	-1.0078	0.9954	0.0216	243.6	32.0	-	9N	169E
1191	060	-1513 Apr 25	20:40:13	35366	-43447	007	P	-t	0.6274	1.7335	0.6805	336.5	181.3	-	8S	163W
1192	060	-1513 Oct 20	12:34:23	35356	-43441	012	T	-p	-0.3567	2.2081	1.1989	327.1	203.5	64.3	5N	43W
1193	060	-1512 Apr 14	03:04:38	35346	-43435	017	T-	pp	-0.1649	2.5572	1.5536	341.8	220.1	96.3	4S	102E
1194	060	-1512 Oct 08	20:51:39	35335	-43429	022	T	p-	0.3544	2.2403	1.1755	352.0	213.2	63.6	1N	167W
1195	060	-1511 Apr 03	16:24:01	35325	-43423	027	P	a-	-0.9005	1.1841	0.2266	262.6	101.9	-	1S	98W
1196	060	-1511 Sep 27	22:11:46	35315	-43417	032	N	t-	1.0719	0.9409	-0.1578	278.4	-	-	3S	173E
1197	060	-1510 Feb 23	01:12:03	35306	-43412	-001	P	-a	0.9532	1.0914	0.1262	254.5	76.9	-	16N	134E
1198	060	-1510 Aug 18	06:13:29	35296	-43406	004	N	-t	-1.1611	0.7513	-0.2962	241.4	-	-	19S	55E
1199	060	-1509 Feb 12	14:19:27	35285	-43400	009	T+	-p	0.2530	2.4018	1.3859	339.2	214.8	85.4	18N	63W
1200	060	-1509 Aug 07	15:35:44	35275	-43394	014	T	-p	-0.3770	2.1602	1.1719	323.4	202.3	60.5	21S	86W

Cat Num	Canon Plate	Calendar Date	TD of Greatest Eclipse	ΔT s	Luna Num	Saros Num	Ecl. Type	QSE	Gamma	Pen. Mag.	Um. Mag.	Eclipse Phase Durations Pen. m	Par. m	Total m	Greatest in Zenith Lat.	Lng.
1201	061	-1508 Feb 01	20:22:22	35265	-43388	019	P	t-	-0.4915	1.9930	0.9198	350.2	201.5	-	20N	154W
1202	061	-1508 Jul 27	06:51:14	35255	-43382	024	T	p-	0.3547	2.1825	1.2313	310.0	198.6	67.0	22S	44E
1203	061	-1507 Jan 20	20:26:21	35244	-43376	029	N	t-	-1.2058	0.6937	-0.4021	245.1	-	-	22N	156W
1204	061	-1507 Jun 17	15:37:17	35236	-43371	-004	N	-a	-1.4808	0.1266	-0.8451	97.4	-	-	24S	90W
1205	061	-1507 Jul 16	23:38:10	35234	-43370	034	N	a-	1.0767	0.8607	-0.0965	231.0	-	-	22S	151E
1206	061	-1507 Dec 11	06:00:07	35226	-43365	001	N	-h	1.0534	0.9468	-0.0964	259.9	-	-	23N	56E
1207	061	-1506 Jun 07	02:26:22	35215	-43359	006	P	-t	-0.7790	1.4408	0.4165	306.8	144.0	-	21S	107E
1208	061	-1506 Nov 30	17:04:02	35205	-43353	011	T	-p	0.3251	2.2569	1.2660	320.3	202.6	71.5	20N	111W
1209	061	-1505 May 27	06:07:39	35195	-43347	016	T-	pp	-0.0171	2.8588	1.7948	372.6	235.4	106.5	18S	52E
1210	061	-1505 Nov 20	08:18:21	35184	-43341	021	T	p-	-0.3426	2.2152	1.2430	311.7	198.2	67.9	16N	20E
1211	061	-1504 May 15	06:48:18	35174	-43335	026	P	t-	0.7460	1.5190	0.4599	326.1	155.5	-	14S	43E
1212	061	-1504 Nov 08	23:05:03	35164	-43329	031	N	a-	-1.0272	0.9728	-0.0265	248.8	-	-	12N	158E
1213	061	-1503 Apr 05	01:43:30	35155	-43324	-002	N	-h	-1.1854	0.6775	-0.3120	218.0	-	-	2S	122E
1214	061	-1503 May 04	11:48:23	35154	-43323	036	N	h-	1.4663	0.1742	-0.8392	119.9	-	-	10S	32W
1215	061	-1503 Sep 29	16:36:00	35145	-43318	003	N	-h	1.1375	0.8097	-0.2675	255.0	-	-	2S	104W
1216	061	-1502 Mar 25	16:23:45	35135	-43312	008	T	-a	-0.4095	2.0821	1.1308	308.3	194.8	52.2	3N	97W
1217	061	-1502 Sep 18	16:45:04	35125	-43306	013	P	-t	0.4613	2.0606	0.9632	361.9	208.2	-	7S	105W
1218	061	-1501 Mar 15	09:18:51	35114	-43300	018	T	p-	0.3139	2.2587	1.3050	313.2	202.1	75.2	8N	10E
1219	061	-1501 Sep 07	17:26:56	35104	-43294	023	T-	p-	-0.2505	2.4315	1.3657	360.0	222.9	86.5	12S	115W
1220	061	-1500 Mar 03	23:47:48	35094	-43288	028	N	a-	1.0746	0.8846	-0.1125	244.2	-	-	13N	154E
1221	062	-1500 Aug 27	00:53:41	35083	-43282	033	P	a-	-0.9518	1.1148	0.1084	268.7	74.1	-	16S	134E
1222	062	-1499 Jan 22	15:42:24	35075	-43277	000	N*	-t	-1.0172	1.0352	-0.0517	286.3	-	-	21N	86W
1223	062	-1499 Jul 18	06:57:33	35065	-43271	005	P	-a	0.9507	1.0866	0.1400	250.0	79.9	-	22S	41E
1224	062	-1498 Jan 11	15:07:57	35054	-43265	010	T	-p	-0.3249	2.3093	1.2150	368.2	221.5	71.2	23N	78W
1225	062	-1498 Jul 07	23:24:30	35044	-43259	015	T+	-p	0.2370	2.4055	1.4402	322.3	209.4	87.2	24S	154E
1226	062	-1498 Dec 31	17:38:01	35034	-43253	020	T	a-	0.3816	2.1849	1.1313	345.7	209.1	55.4	24N	117W
1227	062	-1497 Jun 27	11:30:20	35024	-43247	025	P	t-	-0.5400	1.8747	0.8597	327.1	190.2	-	24S	29W
1228	062	-1497 Dec 21	03:18:40	35013	-43241	030	N	a-	1.0466	0.9371	-0.0620	245.8	-	-	24N	96E
1229	062	-1496 May 17	01:55:43	35005	-43236	-003	N	-t	1.4836	0.1698	-0.8981	126.5	-	-	14S	115E
1230	062	-1496 Jun 15	16:29:54	35003	-43235	035	N	t-	-1.3460	0.4186	-0.6417	192.0	-	-	23S	104W
1231	062	-1496 Nov 10	07:21:16	34994	-43230	002	P	-a	-1.0091	0.9938	0.0185	243.8	29.6	-	13N	34E
1232	062	-1495 May 06	03:09:16	34984	-43224	007	P	-t	0.7108	1.5786	0.5292	326.1	163.4	-	11S	97E
1233	062	-1495 Oct 30	21:11:11	34974	-43218	012	T	-p	-0.3606	2.2030	1.1896	328.2	203.6	63.1	9N	174W
1234	062	-1494 Apr 25	10:06:02	34964	-43212	017	T-	pp	-0.0888	2.6942	1.6958	341.4	221.6	100.9	8S	6W
1235	062	-1494 Oct 20	05:00:46	34954	-43206	022	T	p-	0.3467	2.2569	1.1871	353.9	214.3	65.5	6N	68E
1236	062	-1493 Apr 14	23:51:16	34943	-43200	027	P	a-	-0.8311	1.3094	0.3559	271.4	125.1	-	5S	148E
1237	062	-1493 Oct 09	05:57:06	34933	-43194	032	N	t-	1.0553	0.9730	-0.1291	282.6	-	-	2N	55E
1238	062	-1492 Mar 05	09:06:31	34924	-43189	-001	P	-a	1.0012	1.0028	0.0386	246.8	43.1	-	12N	14E
1239	062	-1492 Apr 03	16:52:15	34923	-43188	037	Nb	a-	-1.5316	0.0199	-0.9253	37.6	-	-	1S	107W
1240	062	-1492 Aug 28	13:46:42	34914	-43183	004	N	-t	-1.1973	0.6851	-0.3626	231.4	-	-	16S	60W
1241	063	-1491 Feb 22	22:02:45	34904	-43177	009	T	-p	0.2959	2.3231	1.3071	338.7	212.9	78.7	15N	180E
1242	063	-1491 Aug 17	23:25:38	34894	-43171	014	T	-p	-0.4197	2.0819	1.0935	320.0	197.8	45.6	18S	155E
1243	063	-1490 Feb 12	03:49:47	34884	-43165	019	P	t-	-0.4546	2.0601	0.9880	353.6	206.7	-	18N	93E
1244	063	-1490 Aug 07	14:44:36	34873	-43159	024	T	p-	0.3041	2.2762	1.3231	312.2	201.9	76.6	20S	75W
1245	063	-1489 Feb 01	03:53:08	34863	-43153	029	N	t-	-1.1741	0.7500	-0.3419	253.1	-	-	20N	91E
1246	063	-1489 Jul 28	07:19:43	34853	-43147	034	N	a-	1.0234	0.9608	-0.0010	242.3	-	-	21S	35E
1247	063	-1489 Dec 22	14:22:43	34844	-43142	001	N	-h	1.0553	0.9409	-0.0975	258.1	-	-	24N	70W
1248	063	-1488 Jun 17	09:11:54	34834	-43136	006	P	-t	-0.8603	1.2940	0.2650	297.1	117.9	-	23S	5E
1249	063	-1488 Dec 11	01:49:55	34824	-43130	011	T	-p	0.3257	2.2541	1.2664	319.3	202.2	71.5	22N	117E
1250	063	-1487 Jun 06	12:28:06	34814	-43124	016	T-	pp	-0.1043	2.7000	1.6339	372.5	234.4	103.7	21S	45W
1251	063	-1487 Nov 30	17:11:44	34804	-43118	021	T	p-	-0.3427	2.2151	1.2430	311.8	198.2	67.9	19N	115W
1252	063	-1486 May 26	13:11:30	34793	-43112	026	P	t-	0.6584	1.6788	0.6216	335.5	175.8	-	17S	55W
1253	063	-1486 Nov 20	07:50:38	34783	-43106	031	N	a-	-1.0312	0.9667	-0.0353	249.0	-	-	16N	25E
1254	063	-1485 Apr 16	08:55:28	34775	-43101	-002	N	-h	-1.2499	0.5565	-0.4277	199.0	-	-	6S	11E
1255	063	-1485 May 15	18:37:55	34773	-43100	036	N	h-	1.3834	0.3241	-0.6852	160.6	-	-	14S	136W
1256	063	-1485 Oct 11	00:33:19	34764	-43095	003	N	-h	1.1523	0.7849	-0.2973	252.6	-	-	3N	135E
1257	063	-1484 Apr 04	23:58:43	34754	-43089	008	T	-a	-0.4724	1.9648	1.0170	303.7	188.5	19.4	1S	147E
1258	063	-1484 Sep 29	00:24:04	34744	-43083	013	P	-t	0.4825	2.0232	0.9230	360.6	205.2	-	2S	138E
1259	063	-1483 Mar 25	17:00:38	34734	-43077	018	T+	p-	0.2561	2.3640	1.4114	315.9	205.9	84.3	4N	108W
1260	063	-1483 Sep 18	01:12:25	34723	-43071	023	T-	p-	-0.2237	2.4809	1.4146	360.1	223.8	90.1	8S	127E

Cat Num	Canon Plate	Calendar Date	TD of Greatest Eclipse	ΔT s	Luna Num	Saros Num	Ecl. Type	QSE	Gamma	Pen. Mag.	Um. Mag.	Eclipse Phase Durations Pen. m	Par. m	Total m	Greatest in Zenith Lat.	Lng.
1261	064	-1482 Mar 15	07:21:28	34713	-43065	028	N	a-	1.0264	0.9734	-0.0241	254.7	-	-	9N	38E
1262	064	-1482 Sep 07	08:54:39	34703	-43059	033	P	a-	-0.9182	1.1763	0.1699	273.3	91.7	-	12S	12E
1263	064	-1481 Feb 02	23:12:49	34695	-43054	000	N	-t	-1.0500	0.9742	-0.1108	279.8	-	-	19N	161E
1264	064	-1481 Jul 29	14:44:44	34684	-43048	005	P	-a	1.0048	0.9886	0.0393	241.2	42.9	-	21S	77W
1265	064	-1480 Jan 22	22:44:56	34674	-43042	010	T	-p	-0.3495	2.2619	1.1722	365.9	219.2	64.7	22N	167E
1266	064	-1480 Jul 18	06:55:31	34664	-43036	015	T	-p	0.2998	2.2930	1.3226	321.3	206.2	78.0	23S	40E
1267	064	-1479 Jan 11	01:44:18	34654	-43030	020	T	a-	0.3661	2.2102	1.1624	344.9	210.1	60.9	24N	121E
1268	064	-1479 Jul 07	18:31:17	34644	-43024	025	P	t-	-0.4710	2.0041	0.9835	334.1	199.5	-	24S	135W
1269	064	-1479 Dec 31	11:53:05	34633	-43018	030	N	a-	1.0361	0.9541	-0.0406	246.7	-	-	25N	33W
1270	064	-1478 May 28	08:13:03	34625	-43013	-003	Ne	-t	1.5699	0.0118	-1.0566	33.8	-	-	17S	19E
1271	064	-1478 Jun 26	22:59:32	34623	-43012	035	N	t-	-1.2686	0.5624	-0.5016	220.1	-	-	25S	157E
1272	064	-1478 Nov 21	16:14:29	34615	-43007	002	P	-a	-1.0103	0.9923	0.0158	244.0	27.4	-	16N	101W
1273	064	-1477 May 17	09:38:22	34604	-43001	007	P	-t	0.7948	1.4230	0.3767	314.2	140.8	-	15S	2W
1274	064	-1477 Nov 11	05:49:58	34594	-42995	012	T	-p	-0.3623	2.2017	1.1846	329.4	203.9	62.5	13N	54E
1275	064	-1476 May 05	17:06:09	34584	-42989	017	T-	pp	-0.0108	2.8350	1.8411	340.3	221.8	102.6	12S	114W
1276	064	-1476 Oct 30	13:13:03	34574	-42983	022	T	p-	0.3424	2.2670	1.1930	355.5	215.1	66.5	10N	57W
1277	064	-1475 Apr 25	07:15:31	34564	-42977	027	P	a-	-0.7586	1.4408	0.4906	279.6	143.7	-	9S	34E
1278	064	-1475 Oct 19	13:49:08	34554	-42971	032	N	t-	1.0444	0.9942	-0.1104	285.3	-	-	6N	66W
1279	064	-1474 Mar 16	16:52:15	34545	-42966	-001	N	-a	1.0562	0.9016	-0.0620	237.0	-	-	8N	105W
1280	064	-1474 Apr 15	00:24:55	34543	-42965	037	N	a-	-1.4689	0.1345	-0.8099	96.9	-	-	6S	138E
1281	065	-1474 Sep 08	21:30:45	34535	-42960	004	N	-t	-1.2256	0.6332	-0.4147	222.9	-	-	12S	178W
1282	065	-1473 Mar 06	05:35:26	34525	-42954	009	T	-a	0.3466	2.2301	1.2138	337.5	209.8	68.1	11N	64E
1283	065	-1473 Aug 29	07:25:31	34514	-42948	014	T	-p	-0.4550	2.0173	1.0286	316.9	193.6	25.6	15S	33E
1284	065	-1472 Feb 23	11:04:22	34504	-42942	019	T	t-	-0.4079	2.1452	1.0744	357.3	212.5	43.7	14N	18W
1285	065	-1472 Aug 17	22:46:58	34494	-42936	024	T+	p-	0.2610	2.3567	1.4011	313.9	204.2	82.9	18S	163E
1286	065	-1471 Feb 11	11:10:08	34484	-42930	029	N	t-	-1.1340	0.8213	-0.2663	262.5	-	-	17N	19W
1287	065	-1471 Aug 07	15:08:06	34474	-42924	034	P	a-	0.9763	1.0497	0.0829	251.6	62.9	-	19S	83W
1288	065	-1470 Jan 01	22:41:55	34465	-42919	001	N	-h	1.0608	0.9281	-0.1049	255.6	-	-	25N	165E
1289	065	-1470 Jun 28	15:59:33	34455	-42913	006	P	-t	-0.9386	1.1529	0.1187	286.1	80.8	-	24S	98W
1290	065	-1470 Dec 22	10:34:13	34445	-42907	011	T	-p	0.3280	2.2481	1.2639	318.3	201.8	71.1	24N	15W
1291	065	-1469 Jun 17	18:48:43	34435	-42901	016	T-	pp	-0.1909	2.5422	1.4738	371.2	231.5	96.5	23S	141W
1292	065	-1469 Dec 12	02:04:27	34425	-42895	021	T	p-	-0.3424	2.2153	1.2438	311.9	198.4	68.0	22N	112E
1293	065	-1468 Jun 05	19:36:46	34414	-42889	026	P	t-	0.5714	1.8375	0.7818	343.3	191.6	-	20S	153W
1294	065	-1468 Nov 30	16:35:04	34404	-42883	031	N	a-	-1.0351	0.9606	-0.0435	249.1	-	-	19N	107W
1295	065	-1467 Apr 26	16:06:11	34396	-42878	-002	N	-h	-1.3162	0.4325	-0.5470	176.8	-	-	10S	99W
1296	065	-1467 May 26	01:30:38	34394	-42877	036	N	h-	1.3017	0.4724	-0.5334	190.3	-	-	17S	118E
1297	065	-1467 Oct 21	08:35:15	34386	-42872	003	N	-h	1.1629	0.7677	-0.3188	251.1	-	-	7N	12E
1298	065	-1466 Apr 16	07:30:32	34376	-42866	008	P	-a	-0.5382	1.8426	0.8978	298.4	180.7	-	5S	31E
1299	065	-1466 Oct 10	08:09:24	34365	-42860	013	P	-t	0.4984	1.9950	0.8926	359.5	202.9	-	2N	19E
1300	065	-1465 Apr 06	00:36:24	34355	-42854	018	T+	p-	0.1939	2.4780	1.5260	318.3	209.2	91.4	1S	136E
1301	066	-1465 Sep 29	09:07:03	34345	-42848	023	T-	p-	-0.2039	2.5173	1.4509	359.8	224.2	92.3	3S	6E
1302	066	-1464 Mar 25	14:44:12	34335	-42842	028	P	h-	0.9703	1.0767	0.0784	265.9	63.6	-	4N	75W
1303	066	-1464 Sep 17	17:06:04	34325	-42836	033	P	a-	-0.8927	1.2231	0.2167	276.4	102.5	-	8S	113W
1304	066	-1463 Feb 13	06:31:05	34316	-42831	000	N	-t	-1.0915	0.8970	-0.1860	270.9	-	-	17N	50E
1305	066	-1463 Aug 08	22:40:18	34306	-42825	005	N	-a	1.0519	0.9037	-0.0487	232.8	-	-	19S	163E
1306	066	-1463 Sep 07	07:21:03	34304	-42824	043	N	a-	-1.5290	0.0344	-0.9300	49.7	-	-	13S	34E
1307	066	-1462 Feb 02	06:12:40	34296	-42819	010	T	-t	-0.3817	2.2003	1.1156	362.9	216.0	54.0	20N	54E
1308	066	-1462 Jul 29	14:32:50	34286	-42813	015	T	-a	0.3562	2.1921	1.2163	320.1	202.5	66.3	22S	75W
1309	066	-1461 Jan 22	09:43:29	34276	-42807	020	T	a-	0.3440	2.2476	1.2061	344.3	211.5	67.5	23N	1E
1310	066	-1461 Jul 19	01:35:19	34266	-42801	025	T	t-	-0.4056	2.1272	1.1004	340.3	207.1	49.0	24S	118E
1311	066	-1460 Jan 11	20:22:09	34255	-42795	030	N	a-	1.0210	0.9793	-0.0103	248.4	-	-	25N	160W
1312	066	-1460 Jul 07	05:32:38	34245	-42789	035	N	t-	-1.1942	0.7010	-0.3670	243.1	-	-	25S	58E
1313	066	-1460 Dec 02	01:07:16	34237	-42784	002	P	-a	-1.0115	0.9904	0.0132	244.1	25.1	-	19N	125E
1314	066	-1459 May 27	16:08:46	34227	-42778	007	P	-t	0.8779	1.2690	0.2254	300.9	111.1	-	18S	102W
1315	066	-1459 Nov 21	14:30:00	34217	-42772	012	T	-p	-0.3620	2.2039	1.1838	330.7	204.4	62.5	17N	77W
1316	066	-1458 May 17	00:06:08	34206	-42766	017	T+	pp	0.0683	2.7275	1.7376	338.3	220.7	101.4	15S	139E
1317	066	-1458 Nov 10	21:27:54	34196	-42760	022	T	p-	0.3409	2.2716	1.1940	356.9	215.6	66.8	14N	178E
1318	066	-1457 May 06	14:37:31	34186	-42754	027	P	a-	-0.6839	1.5766	0.6291	287.1	158.9	-	13S	79W
1319	066	-1457 Oct 30	21:46:14	34176	-42748	032	N*	t-	1.0372	1.0085	-0.0981	287.2	-	-	11N	173E
1320	066	-1456 Mar 27	00:29:32	34168	-42743	-001	N	-a	1.1176	0.7888	-0.1746	224.8	-	-	4N	138E

Cat Num	Canon Plate	Calendar Date	TD of Greatest Eclipse	ΔT s	Luna Num	Saros Num	Ecl. Type	QSE	Gamma	Pen. Mag.	Um. Mag.	Eclipse Phase Durations Pen. m	Par. m	Total m	Greatest in Zenith Lat.	Lng.
1321	067	-1456 Apr 25	07:52:24	34166	-42742	037	N	a-	-1.4022	0.2568	-0.6872	132.6	-	-	10S	23E
1322	067	-1456 Sep 19	05:24:57	34157	-42737	004	N	-h	-1.2467	0.5944	-0.4534	216.1	-	-	8S	61E
1323	067	-1455 Mar 16	12:59:13	34147	-42731	009	T	-a	0.4040	2.1250	1.1083	335.6	205.4	50.3	7N	49W
1324	067	-1455 Sep 08	15:34:12	34137	-42725	014	P	-a	-0.4838	1.9647	0.9757	314.1	189.9	-	11S	91W
1325	067	-1454 Mar 05	18:10:22	34127	-42719	019	T	p-	-0.3546	2.2425	1.1729	361.0	218.2	64.7	11N	126W
1326	067	-1454 Aug 29	06:57:22	34117	-42713	024	T+	p-	0.2244	2.4250	1.4669	315.2	205.8	87.2	14S	38E
1327	067	-1453 Feb 22	18:17:59	34107	-42707	029	N	t-	-1.0861	0.9069	-0.1762	272.9	-	-	14N	128W
1328	067	-1453 Aug 18	23:03:28	34097	-42701	034	P	a-	0.9355	1.1273	0.1552	259.4	85.4	-	17S	157E
1329	067	-1452 Jan 13	06:54:46	34088	-42696	001	N	-h	1.0723	0.9040	-0.1231	251.8	-	-	24N	41E
1330	067	-1452 Jul 08	22:51:31	34078	-42690	006	N*	-t	-1.0123	1.0205	-0.0194	274.3	-	-	25S	158E
1331	067	-1451 Jan 01	19:11:48	34068	-42684	011	T	-p	0.3366	2.2303	1.2500	316.9	200.9	69.5	24N	144W
1332	067	-1451 Jun 28	01:14:49	34058	-42678	016	T	pp	-0.2729	2.3931	1.3219	369.0	226.9	84.5	24S	121E
1333	067	-1451 Dec 22	10:52:47	34048	-42672	021	T	p-	-0.3382	2.2225	1.2519	312.2	198.8	69.0	23N	21W
1334	067	-1450 Jun 17	02:05:46	34038	-42666	026	P	t-	0.4865	1.9929	0.9383	349.4	203.6	-	22S	108E
1335	067	-1450 Dec 12	01:16:53	34028	-42660	031	N	a-	-1.0374	0.9573	-0.0484	249.5	-	-	21N	122E
1336	067	-1449 May 07	23:14:14	34019	-42655	-002	N	-h	-1.3853	0.3035	-0.6718	149.3	-	-	14S	151E
1337	067	-1449 Jun 06	08:24:54	34018	-42654	036	N	h-	1.2195	0.6216	-0.3812	214.2	-	-	19S	13E
1338	067	-1449 Nov 01	16:41:48	34009	-42649	003	N	-t	1.1690	0.7584	-0.3317	250.5	-	-	11N	111W
1339	067	-1448 Apr 26	14:56:35	33999	-42643	008	P	-a	-0.6090	1.7115	0.7690	292.0	170.8	-	9S	83W
1340	067	-1448 Oct 20	16:02:59	33989	-42637	013	P	-t	0.5078	1.9787	0.8747	358.8	201.4	-	6N	101W
1341	068	-1447 Apr 16	08:05:43	33979	-42631	018	T+	pp	0.1266	2.6014	1.6493	320.3	211.8	96.5	5S	21E
1342	068	-1447 Oct 09	17:10:08	33969	-42625	023	T-	p-	-0.1903	2.5421	1.4761	359.1	224.2	93.6	1N	117W
1343	068	-1446 Apr 05	21:59:59	33959	-42619	028	P	h-	0.9092	1.1892	0.1899	277.1	97.7	-	0N	173E
1344	068	-1446 Sep 29	01:26:11	33949	-42613	033	P	a-	-0.8736	1.2580	0.2518	278.4	109.6	-	4S	119E
1345	068	-1445 Feb 24	13:40:09	33940	-42608	000	N	-t	-1.1399	0.8069	-0.2738	259.4	-	-	13N	59W
1346	068	-1445 Aug 20	06:42:31	33930	-42602	005	N	-a	1.0937	0.8288	-0.1269	225.0	-	-	16S	41E
1347	068	-1445 Sep 18	15:45:52	33928	-42601	043	N	a-	-1.5050	0.0793	-0.8869	75.1	-	-	9S	95W
1348	068	-1444 Feb 13	13:32:47	33920	-42596	010	T	-t	-0.4206	2.1262	1.0467	359.3	211.8	35.1	17N	57W
1349	068	-1444 Aug 08	22:16:31	33910	-42590	015	T	-a	0.4067	2.1024	1.1208	318.8	198.5	51.2	20S	168E
1350	068	-1443 Feb 01	17:35:40	33900	-42584	020	T	p-	0.3153	2.2970	1.2620	343.9	213.3	74.5	21N	118W
1351	068	-1443 Jul 29	08:46:13	33890	-42578	025	T	p-	-0.3468	2.2384	1.2052	345.4	212.9	67.9	22S	10E
1352	068	-1442 Jan 22	04:45:20	33880	-42572	030	P	a-	1.0006	1.0141	0.0296	250.9	38.1	-	23N	74E
1353	068	-1442 Jul 18	12:11:18	33870	-42566	035	N	t-	-1.1243	0.8315	-0.2408	262.0	-	-	24S	42W
1354	068	-1442 Dec 13	09:57:08	33861	-42561	002	P	-a	-1.0151	0.9840	0.0065	243.8	17.7	-	21N	9W
1355	068	-1441 Jun 07	22:42:36	33851	-42555	007	P	-t	0.9586	1.1197	0.0785	286.3	66.8	-	20S	158E
1356	068	-1441 Dec 02	23:07:29	33841	-42549	012	T	-p	-0.3633	2.2027	1.1801	331.8	204.8	62.1	20N	153E
1357	068	-1440 May 27	07:07:44	33831	-42543	017	T+	pp	0.1471	2.5812	1.5947	335.6	218.2	97.1	18S	32E
1358	068	-1440 Nov 21	05:42:31	33821	-42537	022	T	p-	0.3399	2.2748	1.1945	358.2	216.1	67.0	18N	53E
1359	068	-1439 May 16	21:59:12	33811	-42531	027	P	a-	-0.6081	1.7146	0.7691	293.6	171.5	-	16S	169E
1360	068	-1439 Nov 10	05:45:27	33801	-42525	032	N*	t-	1.0317	1.0191	-0.0885	288.5	-	-	15N	52E
1361	069	-1438 Apr 07	07:59:45	33792	-42520	-001	N	-a	1.1842	0.6667	-0.2970	209.8	-	-	0S	23E
1362	069	-1438 May 06	15:16:54	33791	-42519	037	N	a-	-1.3332	0.3835	-0.5608	160.3	-	-	13S	91W
1363	069	-1438 Sep 30	13:29:26	33782	-42514	004	N	-h	-1.2604	0.5690	-0.4784	211.3	-	-	3S	62W
1364	069	-1437 Mar 27	20:11:40	33772	-42508	009	P	-a	0.4697	2.0050	0.9875	332.6	199.0	-	3N	160W
1365	069	-1437 Sep 19	23:53:00	33762	-42502	014	P	-a	-0.5048	1.9263	0.9369	311.9	186.9	-	7S	143E
1366	069	-1436 Mar 16	01:04:18	33752	-42496	019	T	pp	-0.2921	2.3565	1.2881	364.6	223.7	80.3	7N	128E
1367	069	-1436 Sep 08	15:16:52	33742	-42490	024	T+	p-	0.1954	2.4796	1.5188	316.1	206.9	89.9	10S	88W
1368	069	-1435 Mar 05	01:16:22	33732	-42484	029	N*	t-	-1.0301	1.0075	-0.0711	284.0	-	-	10N	125E
1369	069	-1435 Aug 29	07:06:57	33722	-42478	034	P	a-	0.9019	1.1916	0.2142	265.8	99.7	-	14S	34E
1370	069	-1434 Jan 23	15:02:40	33713	-42473	001	N	-h	1.0887	0.8708	-0.1502	246.8	-	-	23N	81W
1371	069	-1434 Jul 20	05:46:19	33703	-42467	006	N	-t	-1.0826	0.8945	-0.1512	261.4	-	-	24S	54E
1372	069	-1433 Jan 13	03:45:44	33693	-42461	011	T	-p	0.3488	2.2059	1.2299	315.4	199.8	67.0	24N	87E
1373	069	-1433 Jul 09	07:44:17	33683	-42455	016	T	-t	-0.3520	2.2495	1.1754	365.8	220.7	65.7	24S	23E
1374	069	-1432 Jan 02	19:36:18	33673	-42449	021	T	p-	-0.3304	2.2361	1.2669	312.7	199.6	70.8	24N	152W
1375	069	-1432 Jun 27	08:40:12	33663	-42443	026	T	t-	0.4051	2.1418	1.0879	354.1	212.6	47.3	23S	8E
1376	069	-1432 Dec 22	09:54:11	33653	-42437	031	N	a-	-1.0367	0.9592	-0.0477	250.5	-	-	22N	8W
1377	069	-1431 May 18	06:23:55	33645	-42432	-002	N	-a	-1.4536	0.1765	-0.7953	114.7	-	-	17S	42E
1378	069	-1431 Jun 16	15:24:54	33643	-42431	036	N	a-	1.1409	0.7646	-0.2357	233.3	-	-	21S	93W
1379	069	-1431 Nov 12	00:48:37	33635	-42426	003	N	-t	1.1741	0.7505	-0.3425	250.0	-	-	15N	125E
1380	069	-1430 May 07	22:22:18	33625	-42420	008	P	-a	-0.6803	1.5798	0.6390	284.8	158.9	-	13S	163E

Cat Num	Canon Plate	Calendar Date	TD of Greatest Eclipse	ΔT s	Luna Num	Saros Num	Ecl. Type	QSE	Gamma	Pen. Mag.	Um. Mag.	Eclipse Phase Durations Pen. m	Par. m	Total m	Greatest in Zenith Lat.	Lng.
1381	070	-1430 Nov 01	00:00:34	33615	-42414	013	P	-t	0.5139	1.9679	0.8631	358.1	200.3	-	11N	137E
1382	070	-1429 Apr 27	15:29:33	33604	-42408	018	T+	pp	0.0554	2.7325	1.7798	321.9	213.6	99.5	9S	93W
1383	070	-1429 Oct 21	01:20:42	33594	-42402	023	T-	p-	-0.1820	2.5568	1.4916	358.3	224.0	94.2	6N	118E
1384	070	-1428 Apr 16	05:07:17	33584	-42396	028	P	t-	0.8420	1.3134	0.3124	288.3	123.6	-	4S	64E
1385	070	-1428 Oct 09	09:54:17	33574	-42390	033	P	a-	-0.8605	1.2818	0.2762	279.4	114.1	-	1N	10W
1386	070	-1427 Mar 06	20:37:34	33566	-42385	000	N	-t	-1.1970	0.7012	-0.3773	244.5	-	-	9N	166W
1387	070	-1427 Aug 30	14:54:15	33556	-42379	005	N	-a	1.1275	0.7684	-0.1908	218.2	-	-	13S	83W
1388	070	-1427 Sep 29	00:19:20	33554	-42378	043	N	a-	-1.4879	0.1116	-0.8563	88.8	-	-	4S	135E
1389	070	-1426 Feb 23	20:43:51	33546	-42373	010	P	-t	-0.4672	2.0380	0.9640	354.8	206.2	-	14N	167W
1390	070	-1426 Aug 20	06:06:48	33536	-42367	015	T	-a	0.4507	2.0246	1.0370	317.6	194.4	29.1	17S	49E
1391	070	-1425 Feb 13	01:20:22	33526	-42361	020	T	p-	0.2797	2.3590	1.3307	343.6	215.4	81.5	18N	125E
1392	070	-1425 Aug 09	16:02:54	33516	-42355	025	T	p-	-0.2939	2.3387	1.2989	349.8	217.4	79.4	20S	100W
1393	070	-1424 Feb 02	13:00:08	33506	-42349	030	P	a-	0.9731	1.0618	0.0827	254.5	63.1	-	21N	51W
1394	070	-1424 Jul 28	18:57:00	33496	-42343	035	N	t-	-1.0602	0.9513	-0.1254	277.5	-	-	23S	144W
1395	070	-1424 Dec 23	18:43:33	33487	-42338	002	N	-a	-1.0210	0.9730	-0.0043	243.1	-	-	23N	141W
1396	070	-1423 Jun 18	05:20:59	33477	-42332	007	N	-t	1.0362	0.9764	-0.0630	270.5	-	-	22S	57E
1397	070	-1423 Dec 13	07:42:22	33467	-42326	012	T	-a	-0.3659	2.1989	1.1743	332.9	205.1	61.4	22N	23E
1398	070	-1422 Jun 07	14:12:30	33457	-42320	017	T+	pp	0.2241	2.4386	1.4549	332.2	214.4	89.8	21S	76W
1399	070	-1422 Dec 02	13:56:30	33447	-42314	022	T	p-	0.3390	2.2774	1.1951	359.3	216.7	67.2	20N	71W
1400	070	-1421 May 28	05:20:02	33437	-42308	027	P	a-	-0.5314	1.8547	0.9105	299.4	181.9	-	19S	57E
1401	071	-1421 Nov 21	13:47:24	33427	-42302	032	N*	t-	1.0284	1.0253	-0.0824	289.2	-	-	18N	70W
1402	071	-1420 Apr 17	15:23:36	33419	-42297	-001	N	-a	1.2551	0.5370	-0.4275	191.3	-	-	5S	91W
1403	071	-1420 May 16	22:39:01	33417	-42296	037	N	a-	-1.2628	0.5133	-0.4321	183.5	-	-	17S	157E
1404	071	-1420 Oct 10	21:41:35	33409	-42291	004	N	-h	-1.2687	0.5535	-0.4932	208.0	-	-	1N	173E
1405	071	-1419 Apr 07	03:16:46	33399	-42285	009	P	-h	0.5402	1.8762	0.8574	328.5	190.5	-	1S	92E
1406	071	-1419 Sep 30	08:19:42	33389	-42279	014	P	-a	-0.5198	1.8989	0.9094	310.0	184.6	-	3S	14E
1407	071	-1418 Mar 27	07:50:41	33379	-42273	019	T-	pp	-0.2238	2.4813	1.4140	367.8	228.5	91.9	3N	24E
1408	071	-1418 Sep 19	23:43:05	33369	-42267	024	T+	p-	0.1721	2.5238	1.5603	316.9	207.7	91.8	6S	143E
1409	071	-1417 Mar 16	08:06:53	33359	-42261	029	P	t-	-0.9670	1.1208	0.0468	295.3	53.0	-	6N	20E
1410	071	-1417 Sep 09	15:17:38	33349	-42255	034	P	a-	0.8747	1.2442	0.2614	270.9	109.6	-	10S	91W
1411	071	-1416 Feb 03	23:01:10	33341	-42250	001	N	-h	1.1136	0.8220	-0.1925	239.9	-	-	21N	159E
1412	071	-1416 Jul 30	12:48:27	33330	-42244	006	N	-t	-1.1458	0.7817	-0.2702	248.3	-	-	23S	53W
1413	071	-1415 Jan 23	12:11:11	33321	-42238	011	T	-p	0.3686	2.1673	1.1957	313.5	198.1	62.5	22N	40W
1414	071	-1415 Jul 19	14:21:25	33311	-42232	016	T	-t	-0.4245	2.1180	1.0407	361.9	213.3	33.1	24S	77W
1415	071	-1414 Jan 13	04:12:24	33301	-42226	021	T	p-	-0.3166	2.2607	1.2930	313.5	200.8	73.6	23N	79E
1416	071	-1414 Jul 08	15:21:54	33291	-42220	026	T	t-	0.3287	2.2818	1.2283	357.5	219.1	72.5	24S	93W
1417	071	-1413 Jan 02	18:25:14	33281	-42214	031	N	a-	-1.0317	0.9686	-0.0390	252.2	-	-	23N	136W
1418	071	-1413 May 29	13:32:41	33272	-42209	-002	Ne	-a	-1.5230	0.0476	-0.9213	60.1	-	-	20S	67W
1419	071	-1413 Jun 27	22:28:45	33271	-42208	036	N	a-	1.0642	0.9044	-0.0940	249.1	-	-	23S	160E
1420	071	-1413 Nov 23	08:57:24	33262	-42203	003	N	-t	1.1770	0.7461	-0.3489	250.0	-	-	19N	2E
1421	072	-1412 May 18	05:45:10	33252	-42197	008	P	-a	-0.7542	1.4437	0.5038	276.6	144.0	-	17S	50E
1422	072	-1412 Nov 11	08:02:12	33242	-42191	013	P	-t	0.5169	1.9623	0.8576	357.6	199.7	-	15N	15E
1423	072	-1411 May 07	22:49:37	33232	-42185	018	T-	pp	-0.0186	2.8007	1.8466	323.0	214.3	100.3	13S	154E
1424	072	-1411 Oct 31	09:37:50	33222	-42179	023	T-	p-	-0.1783	2.5629	1.4992	357.2	223.5	94.4	10N	8W
1425	072	-1410 Apr 27	12:08:38	33212	-42173	028	P	t-	0.7708	1.4451	0.4420	299.1	144.7	-	8S	44W
1426	072	-1410 Oct 20	18:28:57	33202	-42167	033	P	a-	-0.8523	1.2964	0.2916	279.7	116.7	-	5N	141W
1427	072	-1409 Mar 18	03:26:50	33194	-42162	000	N	-t	-1.2598	0.5847	-0.4915	225.9	-	-	5N	90E
1428	072	-1409 Sep 10	23:12:59	33184	-42156	005	N	-a	1.1554	0.7191	-0.2437	212.5	-	-	9S	150E
1429	072	-1409 Oct 10	08:59:12	33182	-42155	043	N	a-	-1.4762	0.1337	-0.8357	97.0	-	-	0N	3E
1430	072	-1408 Mar 06	03:46:32	33174	-42150	010	P	-t	-0.5208	1.9368	0.8683	349.3	198.8	-	10N	86E
1431	072	-1408 Aug 30	14:04:44	33164	-42144	015	P	-a	0.4874	1.9603	0.9667	316.6	190.6	-	13S	72W
1432	072	-1407 Feb 23	08:58:18	33154	-42138	020	T+	p-	0.2378	2.4325	1.4109	343.3	217.4	87.9	15N	8E
1433	072	-1407 Aug 19	23:26:30	33144	-42132	025	T-	pp	-0.2477	2.4268	1.3804	353.4	220.9	87.0	17S	147E
1434	072	-1406 Feb 12	21:08:15	33134	-42126	030	P	a-	0.9398	1.1203	0.1464	258.9	83.0	-	19N	174W
1435	072	-1406 Aug 09	01:50:47	33124	-42120	035	N*	t-	-1.0028	1.0588	-0.0223	290.1	-	-	21S	111E
1436	072	-1405 Jan 04	03:23:38	33116	-42115	002	N	-a	-1.0321	0.9525	-0.0243	241.4	-	-	23N	89E
1437	072	-1405 Jun 29	12:05:01	33106	-42109	007	N	-t	1.1098	0.8408	-0.1973	253.7	-	-	23S	45W
1438	072	-1405 Dec 24	16:11:25	33096	-42103	012	T	-a	-0.3726	2.1874	1.1615	333.7	205.1	59.5	23N	104W
1439	072	-1404 Jun 17	21:21:55	33086	-42097	017	T	-p	0.2980	2.3019	1.3204	328.1	209.5	79.0	22S	175E
1440	072	-1404 Dec 12	22:05:23	33076	-42091	022	T	p-	0.3348	2.2858	1.2023	360.5	217.5	68.4	23N	166E

Cat Num	Canon Plate	Calendar Date	TD of Greatest Eclipse	ΔT s	Luna Num	Saros Num	Ecl. Type	QSE	Gamma	Pen. Mag.	Um. Mag.	Eclipse Phase ---- Durations ---- Pen. m	Par. m	Total m	Greatest in Zenith Lat.	Lng.
1441	073	-1403 Jun 07	12:43:32	33066	-42085	027	T	a-	-0.4566	1.9917	1.0481	304.1	190.2	32.4	21S	56W
1442	073	-1403 Dec 01	21:48:27	33056	-42079	032	N*	t-	1.0246	1.0320	-0.0751	289.8	-	-	21N	169E
1443	073	-1402 Apr 28	22:41:16	33048	-42074	-001	N	-a	1.3303	0.3998	-0.5662	167.8	-	-	8S	157E
1444	073	-1402 May 28	05:59:06	33046	-42073	037	N	a-	-1.1910	0.6460	-0.3012	203.6	-	-	20S	45E
1445	073	-1402 Oct 22	06:01:18	33038	-42068	004	N	-h	-1.2717	0.5474	-0.4980	206.3	-	-	6N	46E
1446	073	-1401 Apr 18	10:12:50	33028	-42062	009	P	-h	0.6170	1.7361	0.7157	322.9	179.0	-	5S	15W
1447	073	-1401 Oct 11	16:54:38	33018	-42056	014	P	-a	-0.5288	1.8823	0.8928	308.6	183.0	-	2N	117W
1448	073	-1400 Apr 06	14:26:26	33008	-42050	019	T-	pp	-0.1471	2.6215	1.5550	370.5	232.3	100.3	2S	78W
1449	073	-1400 Sep 30	08:17:46	32998	-42044	024	T+	p-	0.1557	2.5552	1.5891	317.5	208.1	92.9	2S	12E
1450	073	-1399 Mar 26	14:50:10	32988	-42038	029	P	t-	-0.8973	1.2466	0.1770	306.4	101.0	-	2N	83W
1451	073	-1399 Sep 19	23:35:02	32978	-42032	034	P	a-	0.8536	1.2856	0.2974	275.1	116.6	-	6S	143E
1452	073	-1398 Feb 14	06:54:14	32970	-42027	001	N	-h	1.1436	0.7635	-0.2445	231.5	-	-	18N	39E
1453	073	-1398 Aug 10	19:56:11	32960	-42021	006	N	-t	-1.2034	0.6790	-0.3790	234.8	-	-	21S	161W
1454	073	-1397 Feb 03	20:30:24	32950	-42015	011	T	-p	0.3939	2.1185	1.1515	311.2	195.9	55.7	20N	166W
1455	073	-1397 Jul 30	21:04:42	32940	-42009	016	P	-t	-0.4917	1.9964	0.9157	357.5	204.9	-	22S	179W
1456	073	-1396 Jan 24	12:41:54	32930	-42003	021	T	p-	-0.2976	2.2948	1.3287	314.5	202.3	77.1	22N	49W
1457	073	-1396 Jul 18	22:12:07	32920	-41997	026	T+	p-	0.2584	2.4109	1.3574	359.6	223.6	86.3	23S	164E
1458	073	-1395 Jan 13	02:48:25	32910	-41991	031	N	a-	-1.0211	0.9883	-0.0198	255.0	-	-	22N	98E
1459	073	-1395 Jul 08	05:41:21	32900	-41985	036	P	a-	0.9934	1.0337	0.0367	261.7	43.8	-	23S	51E
1460	073	-1395 Dec 03	17:03:38	32892	-41980	003	N	-t	1.1817	0.7382	-0.3580	249.4	-	-	22N	120W
1461	074	-1394 May 29	13:08:59	32882	-41974	008	P	-a	-0.8274	1.3093	0.3697	267.4	125.8	-	20S	63W
1462	074	-1394 Nov 22	16:04:46	32872	-41968	013	P	-t	0.5194	1.9573	0.8535	356.9	199.2	-	18N	106W
1463	074	-1393 May 19	06:07:12	32862	-41962	018	T-	pp	-0.0942	2.6629	1.7068	323.3	213.8	98.6	16S	43E
1464	074	-1393 Nov 11	17:58:21	32852	-41956	023	T-	p-	-0.1767	2.5647	1.5032	356.0	223.0	94.4	14N	135W
1465	074	-1392 May 07	19:04:14	32842	-41950	028	P	t-	0.6957	1.5843	0.5784	309.3	162.6	-	12S	151W
1466	074	-1392 Oct 31	03:08:55	32832	-41944	033	P	a-	-0.8478	1.3040	0.3006	279.5	118.0	-	9N	87E
1467	074	-1391 Mar 28	10:06:47	32824	-41939	000	N	-t	-1.3292	0.4563	-0.6179	202.1	-	-	1N	13W
1468	074	-1391 Apr 27	00:54:47	32822	-41938	038	N	t-	1.5130	0.1101	-0.9462	100.8	-	-	8S	123E
1469	074	-1391 Sep 21	07:39:42	32814	-41933	005	N	-a	1.1767	0.6818	-0.2846	208.0	-	-	5S	21E
1470	074	-1391 Oct 20	17:44:35	32812	-41932	043	N	a-	-1.4691	0.1474	-0.8232	101.6	-	-	5N	131W
1471	074	-1390 Mar 17	10:41:40	32804	-41927	010	P	-t	-0.5811	1.8235	0.7605	342.6	189.2	-	6N	20W
1472	074	-1390 Sep 10	22:10:26	32794	-41921	015	P	-a	0.5169	1.9094	0.9096	316.0	187.3	-	9S	164E
1473	074	-1389 Mar 06	16:27:21	32784	-41915	020	T+	p-	0.1875	2.5214	1.5065	343.0	219.2	93.8	11N	106W
1474	074	-1389 Aug 31	06:57:59	32774	-41909	025	T-	pp	-0.2089	2.5014	1.4484	356.6	223.4	92.1	14S	33E
1475	074	-1388 Feb 24	05:07:51	32764	-41903	030	P	a-	0.8989	1.1926	0.2241	264.1	101.4	-	15N	65E
1476	074	-1388 Aug 19	08:53:39	32755	-41897	035	P	t-	-0.9526	1.1532	0.0676	300.3	63.7	-	18S	4E
1477	074	-1387 Jan 14	11:56:56	32746	-41892	002	N	-a	-1.0484	0.9221	-0.0539	238.6	-	-	22N	40W
1478	074	-1387 Jul 09	18:56:49	32736	-41886	007	N	-t	1.1776	0.7158	-0.3213	236.2	-	-	23S	149W
1479	074	-1386 Jan 04	00:35:26	32726	-41880	012	T	-a	-0.3826	2.1695	1.1427	334.2	204.8	56.4	23N	129E
1480	074	-1386 Jun 29	04:35:26	32717	-41874	017	T	-p	0.3692	2.1703	1.1904	323.5	203.6	63.4	23S	66E
1481	075	-1386 Dec 24	06:10:50	32707	-41868	022	T	p-	0.3284	2.2978	1.2138	361.7	218.5	70.2	24N	44E
1482	075	-1385 Jun 18	20:08:59	32697	-41862	027	T	a-	-0.3833	2.1261	1.1825	308.1	196.7	60.6	23S	168W
1483	075	-1385 Dec 13	05:47:07	32687	-41856	032	N*	t-	1.0185	1.0423	-0.0633	290.7	-	-	23N	49E
1484	075	-1384 May 09	05:55:00	32679	-41851	-001	N	-a	1.4078	0.2586	-0.7095	137.2	-	-	12S	46E
1485	075	-1384 Jun 07	13:19:54	32677	-41850	037	N	a-	-1.1200	0.7773	-0.1723	220.9	-	-	22S	67W
1486	075	-1384 Nov 01	14:26:06	32669	-41845	004	N	-h	-1.2714	0.5469	-0.4967	205.5	-	-	10N	82W
1487	075	-1383 Apr 28	17:04:15	32659	-41839	009	P	-t	0.6966	1.5910	0.5685	316.0	164.0	-	9S	121W
1488	075	-1383 Oct 22	01:34:03	32649	-41833	014	P	-a	-0.5347	1.8712	0.8822	307.4	181.8	-	6N	111E
1489	075	-1382 Apr 17	20:57:12	32639	-41827	019	T-	pp	-0.0668	2.7686	1.7026	372.3	234.7	105.2	6S	178W
1490	075	-1382 Oct 11	16:58:10	32629	-41821	024	T+	p-	0.1442	2.5775	1.6091	318.0	208.4	93.6	3N	120W
1491	075	-1381 Apr 06	21:26:00	32619	-41815	029	P	t-	-0.8208	1.3848	0.3195	317.1	132.6	-	2S	175E
1492	075	-1381 Oct 01	07:59:35	32609	-41809	034	P	a-	0.8391	1.3150	0.3216	278.3	121.1	-	1S	15E
1493	075	-1380 Feb 25	14:37:57	32601	-41804	001	N	-h	1.1819	0.6900	-0.3115	220.6	-	-	15N	78W
1494	075	-1380 Mar 26	01:54:14	32599	-41803	039	Nb	h-	-1.5235	0.0736	-0.9487	78.9	-	-	2N	109E
1495	075	-1380 Aug 21	03:12:46	32591	-41798	006	N	-t	-1.2529	0.5914	-0.4728	221.9	-	-	18S	89E
1496	075	-1379 Feb 14	04:40:05	32581	-41792	011	T	-p	0.4276	2.0545	1.0919	308.4	192.7	44.2	17N	71E
1497	075	-1379 Aug 10	03:58:27	32571	-41786	016	P	-t	-0.5503	1.8906	0.8065	353.1	196.1	-	20S	77E
1498	075	-1378 Feb 03	21:02:42	32561	-41780	021	T-	p-	-0.2716	2.3416	1.3773	315.7	204.2	81.3	20N	175W
1499	075	-1378 Jul 30	05:10:39	32552	-41774	026	T+	pp	0.1939	2.5292	1.4755	360.8	226.4	94.7	22S	58E
1500	075	-1377 Jan 24	11:03:25	32542	-41768	031	P	a-	-1.0047	1.0185	0.0104	258.8	23.1	-	21N	27W

Cat Num	Canon Plate	Calendar Date	TD of Greatest Eclipse	ΔT s	Luna Num	Saros Num	Ecl. Type	QSE	Gamma	Pen. Mag.	Um. Mag.	Eclipse Phase Durations Pen. m	Par. m	Total m	Greatest in Zenith Lat.	Lng.
1501	076	-1377 Jul 19	13:00:38	32532	-41762	036	P	a-	0.9268	1.1554	0.1594	272.1	89.3	-	22S	60W
1502	076	-1377 Dec 15	01:06:57	32524	-41757	003	N	-t	1.1880	0.7268	-0.3698	248.1	-	-	24N	118E
1503	076	-1376 Jun 08	20:33:09	32514	-41751	008	P	-a	-0.9003	1.1757	0.2357	257.2	102.4	-	22S	175W
1504	076	-1376 Dec 03	00:08:15	32504	-41745	013	P	-t	0.5216	1.9524	0.8503	356.2	198.8	-	21N	132E
1505	076	-1375 May 29	13:23:23	32494	-41739	018	T-	-p	-0.1702	2.5247	1.5661	323.0	212.2	94.1	19S	68W
1506	076	-1375 Nov 22	02:21:21	32484	-41733	023	T-	p-	-0.1764	2.5638	1.5052	354.6	222.5	94.3	17N	98E
1507	076	-1374 May 19	01:56:43	32474	-41727	028	P	t-	0.6189	1.7270	0.7179	318.7	177.7	-	16S	104E
1508	076	-1374 Nov 11	11:52:57	32464	-41721	033	P	a-	-0.8463	1.3058	0.3042	278.8	118.3	-	13N	45W
1509	076	-1373 Apr 08	16:38:42	32456	-41716	000	N	-t	-1.4043	0.3177	-0.7546	170.8	-	-	3S	114W
1510	076	-1373 May 08	07:20:23	32454	-41715	038	N	t-	1.4334	0.2567	-0.8008	152.3	-	-	12S	24E
1511	076	-1373 Oct 02	16:13:15	32446	-41710	005	N	-a	1.1924	0.6548	-0.3151	204.7	-	-	0S	109W
1512	076	-1373 Nov 01	02:34:39	32445	-41709	043	N	a-	-1.4660	0.1536	-0.8179	103.7	-	-	9N	95E
1513	076	-1372 Mar 27	17:30:10	32436	-41704	010	P	-t	-0.6472	1.6995	0.6418	334.6	177.0	-	2N	125W
1514	076	-1372 Sep 21	06:22:44	32426	-41698	015	P	-a	0.5401	1.8697	0.8641	315.7	184.5	-	5S	39E
1515	076	-1371 Mar 16	23:50:58	32417	-41692	020	T+	pp	0.1319	2.6202	1.6118	342.4	220.7	98.3	7N	141E
1516	076	-1371 Sep 10	14:37:09	32407	-41686	025	T-	pp	-0.1772	2.5628	1.5034	359.2	225.3	95.4	10S	84W
1517	076	-1370 Mar 06	13:00:47	32397	-41680	030	P	a-	0.8522	1.2757	0.3122	269.8	118.0	-	11N	56W
1518	076	-1370 Aug 30	16:04:30	32387	-41674	035	P	t-	-0.9090	1.2354	0.1455	308.6	92.4	-	15S	105W
1519	076	-1369 Jan 25	20:22:14	32379	-41669	002	N	-a	-1.0708	0.8805	-0.0946	234.6	-	-	21N	167W
1520	076	-1369 Feb 24	05:30:25	32377	-41668	040	N	a-	1.5278	0.0316	-0.9231	47.5	-	-	16N	57E
1521	077	-1369 Jul 21	01:57:06	32369	-41663	007	N	-t	1.2389	0.6030	-0.4335	218.5	-	-	22S	105E
1522	077	-1368 Jan 15	08:49:30	32359	-41657	012	T	-a	-0.3997	2.1384	1.1110	334.4	203.8	50.4	23N	6E
1523	077	-1368 Jul 09	11:56:46	32349	-41651	017	T	-a	0.4346	2.0499	1.0710	318.6	197.0	40.1	23S	45W
1524	077	-1367 Jan 03	14:08:05	32339	-41645	022	T	p-	0.3164	2.3198	1.2358	363.0	219.9	73.3	24N	75W
1525	077	-1367 Jun 29	03:39:02	32330	-41639	027	T	p-	-0.3138	2.2541	1.3098	311.2	201.7	75.6	24S	78E
1526	077	-1367 Dec 23	13:41:14	32320	-41633	032	N*	t-	1.0089	1.0588	-0.0444	292.2	-	-	24N	70W
1527	077	-1366 May 20	13:05:37	32311	-41628	-001	Ne	-a	1.4866	0.1154	-0.8555	93.2	-	-	15S	63W
1528	077	-1366 Jun 18	20:41:50	32310	-41627	037	N	a-	-1.0508	0.9060	-0.0466	235.9	-	-	24S	178W
1529	077	-1366 Nov 12	22:55:36	32302	-41622	004	N	-a	-1.2682	0.5517	-0.4895	205.6	-	-	14N	149E
1530	077	-1365 May 09	23:48:01	32292	-41616	009	P	-t	0.7812	1.4372	0.4121	307.1	143.5	-	13S	136E
1531	077	-1365 Nov 02	10:19:51	32282	-41610	014	P	-a	-0.5359	1.8687	0.8805	306.6	181.3	-	10N	23W
1532	077	-1364 Apr 28	03:20:47	32272	-41604	019	T+	pp	0.0187	2.8569	1.7910	373.0	235.5	106.4	10S	83E
1533	077	-1364 Oct 22	01:44:20	32262	-41598	024	T+	p-	0.1376	2.5905	1.6200	318.4	208.6	93.9	7N	106E
1534	077	-1363 Apr 17	03:57:45	32252	-41592	029	P	t-	-0.7403	1.5305	0.4692	326.9	156.8	-	6S	75E
1535	077	-1363 Oct 11	16:30:26	32243	-41586	034	P	a-	0.8306	1.3329	0.3347	280.7	123.6	-	3N	115W
1536	077	-1362 Mar 07	22:16:20	32234	-41581	001	N	-h	1.2252	0.6073	-0.3878	207.7	-	-	11N	165E
1537	077	-1362 Apr 06	08:58:11	32233	-41580	039	N	h-	-1.4530	0.1999	-0.8163	128.0	-	-	2S	0E
1538	077	-1362 Sep 01	10:35:34	32225	-41575	006	N	-t	-1.2963	0.5148	-0.5556	209.4	-	-	14S	24W
1539	077	-1361 Feb 25	12:43:42	32215	-41569	011	T	-a	0.4667	1.9805	1.0223	305.2	188.7	22.2	14N	52W
1540	077	-1361 Aug 21	11:00:37	32205	-41563	016	P	-t	-0.6021	1.7973	0.7099	348.6	187.0	-	17S	30W
1541	078	-1360 Feb 15	05:14:50	32195	-41557	021	T-	p-	-0.2385	2.4014	1.4388	317.1	206.3	85.9	17N	60E
1542	078	-1360 Aug 09	12:20:10	32185	-41551	026	T+	pp	0.1375	2.6329	1.5788	361.2	228.0	99.5	20S	50W
1543	078	-1359 Feb 03	19:09:19	32175	-41545	031	P	a-	-0.9814	1.0613	0.0530	263.8	52.2	-	19N	149W
1544	078	-1359 Jul 29	20:29:17	32166	-41539	036	P	a-	0.8668	1.2651	0.2696	280.2	114.0	-	21S	173W
1545	078	-1359 Dec 25	09:04:12	32157	-41534	003	N	-t	1.1988	0.7069	-0.3894	245.5	-	-	25N	1W
1546	078	-1358 Jun 20	04:01:01	32148	-41528	008	P	-a	-0.9702	1.0479	0.1068	246.3	70.2	-	24S	72E
1547	078	-1358 Jul 19	11:06:21	32146	-41527	046	Nb	a-	1.5374	0.0127	-0.9395	30.3	-	-	22S	33W
1548	078	-1358 Dec 14	08:08:27	32138	-41522	013	P	-t	0.5266	1.9420	0.8424	355.0	197.9	-	23N	11E
1549	078	-1357 Jun 09	20:38:58	32128	-41516	018	T-	-p	-0.2460	2.3873	1.4253	322.0	209.3	86.4	22S	178W
1550	078	-1357 Dec 03	10:44:35	32118	-41510	023	T-	p-	-0.1755	2.5637	1.5087	353.2	222.0	94.3	20N	29W
1551	078	-1356 May 29	08:46:45	32108	-41504	028	P	t-	0.5409	1.8721	0.8591	327.3	190.4	-	19S	0W
1552	078	-1356 Nov 21	20:37:47	32098	-41498	033	P	a-	-0.8449	1.3072	0.3080	278.0	118.7	-	17N	178W
1553	078	-1355 Apr 18	23:04:07	32090	-41493	000	N	-t	-1.4837	0.1712	-0.8996	127.1	-	-	7S	147E
1554	078	-1355 May 18	13:43:17	32089	-41492	038	N	t-	1.3517	0.4076	-0.6516	189.7	-	-	15S	74W
1555	078	-1355 Oct 13	00:53:43	32080	-41487	005	N	-a	1.2025	0.6378	-0.3353	202.7	-	-	4N	118E
1556	078	-1355 Nov 11	11:27:53	32079	-41486	043	N	a-	-1.4654	0.1549	-0.8170	104.1	-	-	13N	40W
1557	078	-1354 Apr 08	00:12:17	32071	-41481	010	P	-t	-0.7190	1.5653	0.5126	325.1	161.1	-	3S	132E
1558	078	-1354 Oct 02	14:42:01	32061	-41475	015	P	-a	0.5567	1.8421	0.8308	315.9	182.5	-	1S	88W
1559	078	-1353 Mar 28	07:07:31	32051	-41469	020	T+	pp	0.0693	2.7319	1.7297	341.4	221.5	101.3	2N	29E
1560	078	-1353 Sep 21	22:24:14	32041	-41463	025	T-	pp	-0.1531	2.6102	1.5446	361.5	226.6	97.5	6S	157E

Cat Num	Canon Plate	Calendar Date	TD of Greatest Eclipse	ΔT s	Luna Num	Saros Num	Ecl. Type	QSE	Gamma	Pen. Mag.	Um. Mag.	Eclipse Phase Durations Pen. m	Par. m	Total m	Greatest in Zenith Lat.	Lng.
1561	079	-1352 Mar 16	20:45:21	32031	-41457	030	P	a-	0.7983	1.3724	0.4136	275.9	133.5	-	7N	174W
1562	079	-1352 Sep 09	23:25:44	32022	-41451	035	P	t-	-0.8739	1.3019	0.2079	315.0	109.6	-	11S	142E
1563	079	-1351 Feb 05	04:39:43	32013	-41446	002	N	-a	-1.0991	0.8282	-0.1459	229.2	-	-	18N	68E
1564	079	-1351 Mar 06	13:28:52	32012	-41445	040	N	a-	1.4866	0.1063	-0.8464	86.5	-	-	12N	64W
1565	079	-1351 Jul 31	09:05:59	32004	-41440	007	N	-t	1.2939	0.5019	-0.5343	200.6	-	-	20S	3W
1566	079	-1351 Aug 29	23:35:55	32002	-41439	045	N	t-	-1.5496	0.0504	-1.0208	68.4	-	-	15S	140E
1567	079	-1350 Jan 25	16:56:22	31994	-41434	012	T	-a	-0.4219	2.0978	1.0701	334.1	202.3	40.7	21N	117W
1568	079	-1350 Jul 20	19:24:37	31984	-41428	017	P	-a	0.4950	1.9387	0.9603	313.5	189.8	-	23S	158W
1569	079	-1349 Jan 14	21:58:10	31974	-41422	022	T	p-	0.2988	2.3516	1.2684	364.5	221.7	77.5	23N	167E
1570	079	-1349 Jul 10	11:13:14	31964	-41416	027	T-	p-	-0.2475	2.3764	1.4308	313.6	205.4	85.3	24S	36W
1571	079	-1348 Jan 03	21:30:17	31955	-41410	032	N*	t-	0.9948	1.0831	-0.0171	294.4	-	-	25N	172E
1572	079	-1348 Jun 29	04:07:11	31945	-41404	037	P	a-	-0.9846	1.0293	0.0730	248.9	59.1	-	25S	70E
1573	079	-1348 Nov 23	07:25:41	31937	-41399	004	N	-a	-1.2652	0.5556	-0.4826	205.5	-	-	17N	20E
1574	079	-1347 May 20	06:30:27	31927	-41393	009	P	-t	0.8656	1.2840	0.2557	296.6	116.1	-	16S	33E
1575	079	-1347 Nov 12	19:07:53	31917	-41387	014	P	-a	-0.5355	1.8688	0.8819	305.9	181.1	-	14N	156W
1576	079	-1346 May 09	09:41:00	31907	-41381	019	T+	pp	0.1067	2.6956	1.6295	372.6	234.4	103.6	13S	14W
1577	079	-1346 Nov 02	10:34:07	31898	-41375	024	T+	p-	0.1342	2.5976	1.6254	318.7	208.8	94.1	11N	28W
1578	079	-1345 Apr 28	10:25:28	31888	-41369	029	P	t-	-0.6558	1.6838	0.6260	335.6	176.2	-	10S	25W
1579	079	-1345 Oct 23	01:05:56	31878	-41363	034	P	a-	0.8263	1.3430	0.3404	282.4	124.8	-	8N	114E
1580	079	-1344 Mar 18	05:46:10	31870	-41358	001	N	-h	1.2763	0.5105	-0.4785	191.4	-	-	7N	50E
1581	080	-1344 Apr 16	15:57:02	31868	-41357	039	N	h-	-1.3770	0.3367	-0.6740	163.3	-	-	7S	107W
1582	080	-1344 Sep 11	18:08:19	31860	-41352	006	N	-t	-1.3312	0.4538	-0.6224	198.4	-	-	10S	139W
1583	080	-1343 Mar 07	20:38:48	31850	-41346	011	P	-a	0.5135	1.8925	0.9385	301.2	183.4	-	10N	173W
1584	080	-1343 Aug 31	18:12:52	31841	-41340	016	P	-t	-0.6458	1.7187	0.6281	344.4	178.3	-	14S	140W
1585	080	-1342 Feb 25	13:18:04	31831	-41334	021	T-	p-	-0.1983	2.4744	1.5133	318.7	208.5	90.4	13N	62W
1586	080	-1342 Aug 20	19:40:01	31821	-41328	026	T+	pp	0.0886	2.7229	1.6683	361.1	228.6	102.1	17S	162W
1587	080	-1341 Feb 15	03:04:22	31811	-41322	031	P	a-	-0.9502	1.1186	0.1104	270.1	75.0	-	16N	91E
1588	080	-1341 Aug 10	04:06:48	31801	-41316	036	P	a-	0.8133	1.3632	0.3679	286.7	130.8	-	19S	71E
1589	080	-1340 Jan 05	16:55:33	31793	-41311	003	N	-t	1.2138	0.6788	-0.4165	241.5	-	-	25N	119W
1590	080	-1340 Jun 30	11:31:56	31783	-41305	008	N	-a	-1.0375	0.9253	-0.0175	234.6	-	-	25S	42W
1591	080	-1340 Jul 29	18:52:09	31782	-41304	046	N	a-	1.4822	0.1149	-0.8387	90.0	-	-	20S	151W
1592	080	-1340 Dec 24	16:05:25	31774	-41299	013	P	-t	0.5346	1.9256	0.8294	353.4	196.6	-	24N	108W
1593	080	-1339 Jun 20	03:56:14	31764	-41293	018	T	-a	-0.3197	2.2542	1.2883	320.4	205.3	74.8	23S	71E
1594	080	-1339 Dec 13	19:07:44	31754	-41287	023	T-	p-	-0.1736	2.5650	1.5142	351.8	221.6	94.5	22N	155W
1595	080	-1338 Jun 09	15:34:58	31744	-41281	028	T	t-	0.4622	2.0185	1.0012	334.9	201.0	5.5	21S	104W
1596	080	-1338 Dec 03	05:23:53	31735	-41275	033	P	a-	-0.8442	1.3071	0.3108	277.2	118.9	-	20N	50E
1597	080	-1337 Apr 30	05:25:04	31727	-41270	000	Ne	-t	-1.5661	0.0193	-1.0502	43.3	-	-	12S	50E
1598	080	-1337 May 29	20:04:21	31725	-41269	038	N	t-	1.2682	0.5618	-0.4994	220.0	-	-	18S	171W
1599	080	-1337 Oct 24	09:38:41	31717	-41264	005	N	-a	1.2089	0.6276	-0.3483	201.6	-	-	9N	15W
1600	080	-1337 Nov 22	20:21:56	31715	-41263	043	N	a-	-1.4658	0.1542	-0.8177	103.8	-	-	16N	175W
1601	081	-1336 Apr 18	06:50:50	31707	-41258	010	P	-t	-0.7941	1.4251	0.3772	314.1	140.8	-	7S	29E
1602	081	-1336 Oct 12	23:06:50	31697	-41252	015	P	-a	0.5682	1.8237	0.8072	316.4	181.1	-	4N	144E
1603	081	-1335 Apr 07	14:21:05	31688	-41246	020	T+	pp	0.0034	2.8499	1.8534	340.0	221.4	102.4	2S	82W
1604	081	-1335 Oct 02	06:16:58	31678	-41240	025	T-	pp	-0.1343	2.6475	1.5763	363.5	227.7	99.0	1S	37E
1605	081	-1334 Mar 28	04:25:02	31668	-41234	030	P	a-	0.7397	1.4776	0.5232	281.9	147.5	-	3N	68E
1606	081	-1334 Sep 21	06:55:23	31658	-41228	035	P	t-	-0.8456	1.3556	0.2580	319.9	121.2	-	6S	28E
1607	081	-1333 Feb 16	12:46:45	31650	-41223	002	N	-a	-1.1356	0.7607	-0.2125	221.6	-	-	15N	55W
1608	081	-1333 Mar 17	21:17:51	31649	-41222	040	N	a-	1.4380	0.1945	-0.7564	116.1	-	-	8N	176E
1609	081	-1333 Aug 11	16:25:39	31640	-41217	007	N	-t	1.3410	0.4154	-0.6206	183.4	-	-	18S	114W
1610	081	-1333 Sep 10	07:07:49	31639	-41216	045	N	t-	-1.5132	0.1177	-0.9544	103.8	-	-	11S	26E
1611	081	-1332 Feb 06	00:51:45	31631	-41211	012	T	-a	-0.4527	2.0412	1.0135	333.2	199.6	18.2	19N	123E
1612	081	-1332 Jul 31	03:02:15	31621	-41205	017	P	-a	0.5483	1.8410	0.8627	308.5	182.5	-	21S	86E
1613	081	-1331 Jan 25	05:37:28	31611	-41199	022	T	pp	0.2731	2.3984	1.3162	366.2	224.0	82.8	22N	51E
1614	081	-1331 Jul 20	18:54:52	31601	-41193	027	T-	p-	-0.1873	2.4877	1.5403	315.3	207.8	91.4	23S	152W
1615	081	-1330 Jan 14	05:12:31	31592	-41187	032	P	t-	0.9750	1.1176	0.0211	297.4	35.8	-	24N	57E
1616	081	-1330 Jul 10	11:35:12	31582	-41181	037	P	a-	-0.9212	1.1477	0.1871	260.3	93.3	-	25S	43W
1617	081	-1330 Dec 04	15:57:40	31574	-41176	004	N	-a	-1.2615	0.5606	-0.4738	205.5	-	-	20N	109W
1618	081	-1329 May 31	13:09:02	31564	-41170	009	P	-t	0.9515	1.1282	0.0962	284.0	73.1	-	19S	68W
1619	081	-1329 Nov 24	03:58:19	31554	-41164	014	P	-a	-0.5337	1.8710	0.8860	305.3	181.1	-	18N	70E
1620	081	-1328 May 19	15:58:04	31545	-41158	019	T+	pp	0.1968	2.5306	1.4639	371.0	231.3	95.9	17S	110W

Cat Num	Canon Plate	Calendar Date	TD of Greatest Eclipse	ΔT s	Luna Num	Saros Num	Ecl. Type	QSE	Gamma	Pen. Mag.	Um. Mag.	Eclipse Phase Durations Pen. m	Eclipse Phase Durations Par. m	Eclipse Phase Durations Total m	Greatest in Zenith Lat.	Greatest in Zenith Lng.
1621	082	-1328 Nov 12	19:27:16	31535	-41152	024	T+	p-	0.1340	2.5987	1.6253	319.0	208.8	94.1	15N	163W
1622	082	-1327 May 08	16:52:14	31525	-41146	029	P	t-	-0.5694	1.8406	0.7861	343.0	191.8	-	14S	124W
1623	082	-1327 Nov 02	09:44:53	31515	-41140	034	P	a-	0.8257	1.3463	0.3396	283.7	125.0	-	12N	18W
1624	082	-1326 Mar 29	13:12:17	31507	-41135	001	N	-a	1.3311	0.4070	-0.5762	171.8	-	-	3N	64W
1625	082	-1326 Apr 27	22:54:49	31506	-41134	039	N	a-	-1.2987	0.4776	-0.5279	191.0	-	-	11S	146E
1626	082	-1326 Sep 23	01:47:49	31498	-41129	006	N	-t	-1.3598	0.4040	-0.6777	188.8	-	-	6S	104E
1627	082	-1325 Mar 19	04:26:40	31488	-41123	011	P	-a	0.5667	1.7931	0.8428	296.4	176.6	-	6N	68E
1628	082	-1325 Sep 12	01:34:37	31478	-41117	016	P	-t	-0.6817	1.6542	0.5606	340.6	170.2	-	10S	108E
1629	082	-1324 Mar 07	21:12:55	31468	-41111	021	T-	p-	-0.1515	2.5598	1.5999	320.3	210.6	94.4	10N	177E
1630	082	-1324 Aug 31	03:10:43	31459	-41105	026	T+	pp	0.0479	2.7978	1.7429	360.5	228.6	103.3	13S	84E
1631	082	-1323 Feb 25	10:49:39	31449	-41099	031	P	a-	-0.9118	1.1891	0.1807	277.2	95.3	-	13N	27W
1632	082	-1323 Aug 20	11:54:43	31439	-41093	036	P	a-	0.7675	1.4472	0.4519	291.6	142.7	-	16S	47W
1633	082	-1322 Jan 16	00:38:28	31431	-41088	003	N	-t	1.2351	0.6388	-0.4548	235.4	-	-	24N	125E
1634	082	-1322 Jul 11	19:07:26	31422	-41082	008	N	-a	-1.1010	0.8101	-0.1352	222.3	-	-	25S	157W
1635	082	-1322 Aug 10	02:45:24	31420	-41081	046	N	a-	1.4326	0.2066	-0.7486	119.7	-	-	18S	90E
1636	082	-1321 Jan 04	23:56:45	31412	-41076	013	P	-t	0.5477	1.8995	0.8074	351.3	194.4	-	24N	134E
1637	082	-1321 Jul 01	11:15:59	31402	-41070	018	T	-a	-0.3907	2.1262	1.1557	318.2	200.1	57.6	24S	40W
1638	082	-1321 Dec 25	03:26:11	31392	-41064	023	T-	p-	-0.1670	2.5747	1.5289	350.3	221.3	95.0	23N	80E
1639	082	-1320 Jun 19	22:24:36	31383	-41058	028	T	t-	0.3855	2.1618	1.1395	341.5	209.5	57.2	23S	153E
1640	082	-1320 Dec 13	14:07:49	31373	-41052	033	P	a-	-0.8408	1.3116	0.3187	276.7	120.0	-	22N	82W
1641	083	-1319 Jun 09	02:25:36	31363	-41046	038	N	t-	1.1848	0.7161	-0.3475	245.2	-	-	20S	92E
1642	083	-1319 Nov 03	18:27:26	31355	-41041	005	N	-a	1.2122	0.6228	-0.3556	201.3	-	-	13N	149W
1643	083	-1319 Dec 03	05:15:37	31354	-41040	043	N	a-	-1.4660	0.1535	-0.8179	103.6	-	-	19N	50E
1644	083	-1318 Apr 29	13:26:08	31345	-41035	010	P	-t	-0.8725	1.2789	0.2354	301.4	113.3	-	11S	72W
1645	083	-1318 Oct 24	07:37:04	31336	-41029	015	P	-a	0.5743	1.8149	0.7935	317.3	180.4	-	8N	14E
1646	083	-1317 Apr 18	21:28:33	31326	-41023	020	T-	pp	-0.0685	2.7277	1.7367	337.9	220.3	101.2	6S	168E
1647	083	-1317 Oct 13	14:17:10	31316	-41017	025	T-	pp	-0.1225	2.6718	1.5954	365.3	228.5	99.8	3N	85W
1648	083	-1316 Apr 07	11:58:29	31307	-41011	030	P	a-	0.6755	1.5936	0.6430	288.0	160.2	-	1S	48W
1649	083	-1316 Oct 01	14:33:45	31297	-41005	035	P	t-	-0.8242	1.3963	0.2956	323.5	129.0	-	2S	89W
1650	083	-1315 Feb 26	20:45:40	31289	-41000	002	N	-a	-1.1782	0.6820	-0.2904	212.0	-	-	12N	177W
1651	083	-1315 Mar 28	05:00:41	31287	-40999	040	N	a-	1.3848	0.2914	-0.6581	141.0	-	-	3N	58E
1652	083	-1315 Aug 21	23:55:41	31279	-40994	007	N	-t	1.3807	0.3427	-0.6933	167.2	-	-	15S	132E
1653	083	-1315 Sep 20	14:49:53	31278	-40993	045	N	t-	-1.4839	0.1719	-0.9009	124.5	-	-	7S	92W
1654	083	-1314 Feb 16	08:38:47	31269	-40988	012	P	-a	-0.4896	1.9736	0.9459	331.7	195.9	-	16N	5E
1655	083	-1314 Aug 11	10:47:10	31260	-40982	017	P	-a	0.5960	1.7534	0.7750	303.7	175.1	-	19S	31W
1656	083	-1313 Feb 05	13:08:38	31250	-40976	022	T+	pp	0.2413	2.4559	1.3752	367.9	226.5	88.4	20N	62W
1657	083	-1313 Aug 01	02:42:53	31240	-40970	027	T-	p-	-0.1327	2.5891	1.6394	316.6	209.5	95.2	22S	90E
1658	083	-1312 Jan 25	12:46:34	31231	-40964	032	P	t-	0.9484	1.1643	0.0719	301.4	65.5	-	23N	57W
1659	083	-1312 Jul 20	19:09:34	31221	-40958	037	P	a-	-0.8638	1.2555	0.2901	269.9	114.7	-	24S	157W
1660	083	-1312 Dec 15	00:27:24	31213	-40953	004	N	-a	-1.2599	0.5614	-0.4688	204.8	-	-	21N	123E
1661	084	-1311 Jun 10	19:48:44	31203	-40947	009	N	-t	1.0353	0.9765	-0.0595	269.7	-	-	21S	169W
1662	084	-1311 Dec 04	12:47:19	31194	-40941	014	P	-a	-0.5335	1.8704	0.8876	304.5	180.9	-	20N	63W
1663	084	-1310 May 30	22:15:34	31184	-40935	019	T	pt	0.2860	2.3673	1.2997	368.1	226.2	82.3	19S	154E
1664	084	-1310 Nov 24	04:20:42	31174	-40929	024	T+	p-	0.1340	2.5991	1.6250	319.3	209.0	94.1	18N	63E
1665	084	-1309 May 19	23:17:34	31165	-40923	029	P	t-	-0.4809	2.0016	0.9499	349.2	204.3	-	17S	138E
1666	084	-1309 Nov 13	18:25:56	31155	-40917	034	P	a-	0.8273	1.3449	0.3349	284.6	124.6	-	16N	150W
1667	084	-1308 Apr 08	20:32:09	31147	-40912	001	N	-a	1.3916	0.2933	-0.6846	146.8	-	-	1S	177W
1668	084	-1308 May 08	05:50:48	31145	-40911	039	N	a-	-1.2177	0.6242	-0.3769	214.3	-	-	14S	40E
1669	084	-1308 Oct 03	09:35:09	31137	-40906	006	N	-t	-1.3814	0.3669	-0.7199	181.1	-	-	2S	15W
1670	084	-1307 Mar 29	12:07:49	31128	-40900	011	P	-a	0.6257	1.6831	0.7362	290.8	167.9	-	2N	50W
1671	084	-1307 Sep 22	09:06:57	31118	-40894	016	P	-t	-0.7091	1.6053	0.5092	337.6	163.5	-	6S	7W
1672	084	-1306 Mar 19	04:58:44	31108	-40888	021	T-	pp	-0.0974	2.6585	1.6995	321.7	212.4	97.7	6N	58E
1673	084	-1306 Sep 11	10:51:51	31099	-40882	026	T+	pp	0.0148	2.8587	1.8034	359.6	228.1	103.5	9S	33W
1674	084	-1305 Mar 08	18:24:32	31089	-40876	031	P	h-	-0.8657	1.2737	0.2651	285.2	114.4	-	9N	143W
1675	084	-1305 Aug 31	19:52:34	31079	-40870	036	P	a-	0.7290	1.5180	0.5225	295.3	151.3	-	13S	168W
1676	084	-1304 Jan 27	08:12:26	31071	-40865	003	N	-t	1.2633	0.5860	-0.5056	226.7	-	-	22N	11E
1677	084	-1304 Jul 22	02:49:06	31062	-40859	008	N	-a	-1.1594	0.7043	-0.2438	209.8	-	-	24S	87E
1678	084	-1304 Aug 20	10:47:39	31060	-40858	046	N	a-	1.3900	0.2858	-0.6715	139.6	-	-	15S	32W
1679	084	-1303 Jan 15	07:42:58	31052	-40853	013	P	-t	0.5657	1.8641	0.7766	348.5	191.4	-	24N	17E
1680	084	-1303 Jul 11	18:38:55	31042	-40847	018	T	-a	-0.4582	2.0047	1.0292	315.6	193.9	26.0	24S	151W

Cat Num	Canon Plate	Calendar Date	TD of Greatest Eclipse	ΔT s	Luna Num	Saros Num	Ecl. Type	QSE	Gamma	Pen. Mag.	Um. Mag.	Eclipse Phase Durations Pen. m	Par. m	Total m	Greatest in Zenith Lat.	Lng.
1681	085	-1302 Jan 04	11:42:00	31033	-40841	023	T-	p-	-0.1572	2.5900	1.5495	348.9	221.2	95.9	24N	44W
1682	085	-1302 Jul 01	05:15:56	31023	-40835	028	T	p-	0.3108	2.3016	1.2740	347.1	216.3	76.8	23S	49E
1683	085	-1302 Dec 24	22:49:06	31013	-40829	033	P	a-	-0.8350	1.3203	0.3312	276.4	121.9	-	23N	148E
1684	085	-1301 Jun 20	08:48:24	31004	-40823	038	N	t-	1.1024	0.8685	-0.1978	266.5	-	-	22S	5W
1685	085	-1301 Nov 15	03:18:13	30996	-40818	005	N	-a	1.2137	0.6209	-0.3595	201.4	-	-	17N	77E
1686	085	-1301 Dec 14	14:07:05	30994	-40817	043	N	a-	-1.4647	0.1555	-0.8150	104.2	-	-	21N	83W
1687	085	-1300 May 09	20:01:35	30986	-40812	010	P	-t	-0.9513	1.1323	0.0927	287.0	72.3	-	15S	173W
1688	085	-1300 Nov 03	16:09:07	30976	-40806	015	P	-a	0.5783	1.8097	0.7840	318.3	180.2	-	12N	116W
1689	085	-1299 Apr 29	04:36:07	30967	-40800	020	T-	pp	-0.1414	2.5915	1.6053	335.2	218.1	97.4	10S	59E
1690	085	-1299 Oct 23	22:21:41	30957	-40794	025	T-	pp	-0.1146	2.6884	1.6075	366.8	229.1	100.3	8N	151E
1691	085	-1298 Apr 18	19:27:45	30947	-40788	030	P	a-	0.6072	1.7172	0.7699	293.7	171.4	-	6S	163W
1692	085	-1298 Oct 12	22:19:31	30938	-40782	035	P	t-	-0.8087	1.4261	0.3227	326.1	134.3	-	2N	153E
1693	085	-1297 Mar 10	04:34:42	30930	-40777	002	N	-a	-1.2286	0.5894	-0.3825	199.4	-	-	8N	64E
1694	085	-1297 Apr 08	12:36:03	30928	-40776	040	N	a-	1.3256	0.3997	-0.5490	163.6	-	-	1S	59W
1695	085	-1297 Sep 02	07:37:26	30920	-40771	007	N	-h	1.4116	0.2860	-0.7501	153.0	-	-	12S	15E
1696	085	-1297 Oct 01	22:42:18	30919	-40770	045	N	h-	-1.4620	0.2121	-0.8610	137.5	-	-	3S	148E
1697	085	-1296 Feb 27	16:13:13	30910	-40765	012	P	-a	-0.5358	1.8888	0.8609	329.3	190.4	-	12N	110W
1698	085	-1296 Aug 21	18:43:33	30901	-40759	017	P	-a	0.6350	1.6822	0.7033	299.4	168.3	-	16S	152W
1699	085	-1295 Feb 15	20:28:30	30891	-40753	022	T+	pp	0.2008	2.5293	1.4503	369.7	229.1	94.1	17N	174W
1700	085	-1295 Aug 11	10:38:41	30882	-40747	027	T-	p-	-0.0846	2.6788	1.7264	317.4	210.4	97.4	19S	30W
1701	086	-1294 Feb 04	20:12:33	30872	-40741	032	P	t-	0.9148	1.2236	0.1359	306.3	89.2	-	20N	170W
1702	086	-1294 Aug 01	02:49:12	30862	-40735	037	P	a-	-0.8113	1.3543	0.3838	278.2	130.5	-	22S	87E
1703	086	-1294 Dec 26	08:54:36	30854	-40730	004	N	-a	-1.2614	0.5561	-0.4692	203.1	-	-	22N	4W
1704	086	-1293 Jun 22	02:27:33	30845	-40724	009	N	-t	1.1186	0.8259	-0.2148	253.0	-	-	22S	90E
1705	086	-1293 Dec 15	21:35:39	30835	-40718	014	P	-a	-0.5345	1.8672	0.8872	303.7	180.6	-	22N	164E
1706	086	-1292 Jun 10	04:33:40	30825	-40712	019	T	-t	0.3747	2.2052	1.1361	363.9	218.8	58.8	21S	58E
1707	086	-1292 Dec 04	13:13:31	30816	-40706	024	T+	p-	0.1334	2.6003	1.6258	319.6	209.1	94.2	21N	71W
1708	086	-1291 May 30	05:45:27	30806	-40700	029	T	t-	-0.3933	2.1610	1.1118	353.9	213.8	52.9	20S	39E
1709	086	-1291 Nov 24	03:07:55	30797	-40694	034	P	a-	0.8304	1.3407	0.3278	285.2	123.7	-	19N	79E
1710	086	-1290 Apr 20	03:49:00	30789	-40689	001	N	-a	1.4553	0.1739	-0.7990	113.8	-	-	5S	72E
1711	086	-1290 May 19	12:46:57	30787	-40688	039	N	a-	-1.1352	0.7734	-0.2237	233.9	-	-	18S	66W
1712	086	-1290 Oct 14	17:28:23	30779	-40683	006	N	-t	-1.3978	0.3391	-0.7522	175.0	-	-	3N	135W
1713	086	-1289 Apr 09	19:43:51	30769	-40677	011	P	-a	0.6893	1.5649	0.6208	284.1	157.0	-	2S	166W
1714	086	-1289 Oct 03	16:47:30	30760	-40671	016	P	-t	-0.7302	1.5675	0.4694	335.0	157.8	-	1S	125W
1715	086	-1288 Mar 29	12:37:16	30750	-40665	021	T-	pp	-0.0375	2.7683	1.8097	322.9	213.7	99.7	1N	59W
1716	086	-1288 Sep 21	18:43:39	30741	-40659	026	T-	pp	-0.0105	2.8665	1.8112	358.6	227.5	103.2	5S	153W
1717	086	-1287 Mar 19	01:50:47	30731	-40653	031	P	h-	-0.8132	1.3705	0.3613	293.6	132.2	-	5N	103E
1718	086	-1287 Sep 11	03:59:23	30721	-40647	036	P	a-	0.6969	1.5769	0.5813	298.0	157.7	-	9S	68E
1719	086	-1286 Feb 06	15:36:47	30713	-40642	003	N	-t	1.2990	0.5194	-0.5698	214.9	-	-	20N	101W
1720	086	-1286 Aug 02	10:37:32	30704	-40636	008	N	-a	-1.2126	0.6083	-0.3431	197.1	-	-	22S	31W
1721	087	-1286 Aug 31	18:58:19	30702	-40635	046	N	a-	1.3538	0.3533	-0.6062	154.2	-	-	12S	157W
1722	087	-1285 Jan 26	15:19:54	30694	-40630	013	P	-t	0.5918	1.8136	0.7313	344.9	186.9	-	22N	98W
1723	087	-1285 Jul 23	02:07:14	30684	-40624	018	P	-a	-0.5207	1.8929	0.9119	312.6	186.9	-	23S	96E
1724	087	-1284 Jan 15	19:50:45	30675	-40618	023	T-	p-	-0.1404	2.6176	1.5832	347.6	221.2	97.2	23N	167W
1725	087	-1284 Jul 11	12:11:43	30665	-40612	028	T+	pp	0.2406	2.4332	1.4000	351.7	221.4	88.7	23S	56W
1726	087	-1283 Jan 04	07:24:46	30656	-40606	033	P	a-	-0.8239	1.3384	0.3536	276.9	125.3	-	23N	18E
1727	087	-1283 Jun 30	15:15:26	30646	-40600	038	N*	t-	1.0232	1.0155	-0.0540	284.4	-	-	23S	103W
1728	087	-1283 Nov 25	12:10:19	30638	-40595	005	N	-a	1.2144	0.6205	-0.3615	201.7	-	-	20N	57W
1729	087	-1283 Dec 24	22:55:33	30637	-40594	043	N	a-	-1.4612	0.1612	-0.8079	106.1	-	-	22N	144E
1730	087	-1282 May 21	02:35:37	30629	-40589	010	N	-t	-1.0320	0.9824	-0.0537	270.6	-	-	18S	86E
1731	087	-1282 Nov 15	00:44:33	30619	-40583	015	P	-a	0.5790	1.8104	0.7810	319.6	180.4	-	16N	114E
1732	087	-1281 May 10	11:40:59	30609	-40577	020	T-	pp	-0.2179	2.4491	1.4671	331.7	214.5	90.5	14S	50W
1733	087	-1281 Nov 04	06:30:30	30600	-40571	025	T-	pp	-0.1110	2.6970	1.6124	368.1	229.6	100.6	12N	27E
1734	087	-1280 Apr 29	02:53:40	30590	-40565	030	P	a-	0.5355	1.8474	0.9027	298.9	181.3	-	10S	83E
1735	087	-1280 Oct 23	06:12:46	30581	-40559	035	P	t-	-0.7989	1.4452	0.3400	327.7	137.4	-	7N	32E
1736	087	-1279 Mar 20	12:16:37	30573	-40554	002	N	-a	-1.2840	0.4876	-0.4842	183.6	-	-	4N	54W
1737	087	-1279 Apr 18	20:07:03	30571	-40553	040	N	a-	1.2632	0.5140	-0.4344	183.8	-	-	5S	174W
1738	087	-1279 Sep 12	15:28:24	30563	-40548	007	N	-h	1.4361	0.2409	-0.7951	140.6	-	-	8S	105W
1739	087	-1279 Oct 12	06:42:34	30562	-40547	045	N	h-	-1.4454	0.2425	-0.8305	146.2	-	-	2N	25E
1740	087	-1278 Mar 09	23:39:27	30553	-40542	012	P	-h	-0.5879	1.7933	0.7652	325.9	183.1	-	8N	136E

Cat Num	Canon Plate	Calendar Date	TD of Greatest Eclipse	ΔT s	Luna Num	Saros Num	Ecl. Type	QSE	Gamma	Pen. Mag.	Um. Mag.	Eclipse Phase Durations Pen. m	Par. m	Total m	Greatest in Zenith Lat.	Lng.
1741	088	-1278 Sep 02	02:48:11	30544	-40536	017	P	-a	0.6677	1.6224	0.6428	295.5	162.1	-	12S	86E
1742	088	-1277 Feb 27	03:38:12	30534	-40530	022	T+	pp	0.1525	2.6171	1.5399	371.4	231.6	99.3	13N	77E
1743	088	-1277 Aug 22	18:42:30	30525	-40524	027	T-	pp	-0.0432	2.7563	1.8009	318.0	210.8	98.4	16S	153W
1744	088	-1276 Feb 16	03:29:47	30515	-40518	032	P	t-	0.8737	1.2966	0.2138	312.0	110.4	-	17N	80E
1745	088	-1276 Aug 11	10:36:01	30506	-40512	037	P	a-	-0.7653	1.4415	0.4655	285.2	142.3	-	20S	31W
1746	088	-1275 Jan 05	17:16:13	30498	-40507	004	N	-a	-1.2682	0.5409	-0.4790	199.8	-	-	23N	130W
1747	088	-1275 Feb 04	06:43:19	30496	-40506	042	N	a-	1.5391	0.0551	-0.9871	69.0	-	-	21N	31E
1748	088	-1275 Jul 02	09:10:54	30488	-40501	009	N	-t	1.1970	0.6845	-0.3612	234.5	-	-	23S	12W
1749	088	-1275 Jul 31	21:55:22	30487	-40500	047	Nb	t-	-1.5592	0.0102	-1.0160	29.5	-	-	23S	159E
1750	088	-1275 Dec 26	06:19:52	30479	-40495	014	P	-a	-0.5392	1.8570	0.8801	302.6	179.9	-	23N	33E
1751	088	-1274 Jun 21	10:54:18	30469	-40489	019	P	-t	0.4611	2.0476	0.9768	358.6	209.3	-	23S	39W
1752	088	-1274 Dec 15	22:03:38	30459	-40483	024	T+	p-	0.1310	2.6046	1.6303	320.0	209.4	94.5	23N	156E
1753	088	-1273 Jun 10	12:16:10	30450	-40477	029	T	t-	-0.3074	2.3178	1.2704	357.4	220.8	77.8	22S	60W
1754	088	-1273 Dec 05	11:47:27	30440	-40471	034	P	a-	0.8324	1.3383	0.3231	286.0	123.2	-	22N	52W
1755	088	-1272 Apr 30	11:02:54	30432	-40466	001	Ne	-a	1.5218	0.0497	-0.9190	61.3	-	-	9S	39W
1756	088	-1272 May 29	19:45:21	30431	-40465	039	N	a-	-1.0535	0.9217	-0.0720	250.4	-	-	20S	172W
1757	088	-1272 Oct 25	01:27:54	30423	-40460	006	N	-t	-1.4083	0.3217	-0.7733	171.2	-	-	7N	103E
1758	088	-1271 Apr 20	03:15:02	30413	-40454	011	P	-a	0.7572	1.4392	0.4973	276.3	143.2	-	6S	78E
1759	088	-1271 Oct 14	00:36:23	30404	-40448	016	P	-t	-0.7448	1.5415	0.4418	333.0	153.7	-	3N	116E
1760	088	-1270 Apr 09	20:08:29	30394	-40442	021	T+	pp	0.0280	2.7857	1.8271	323.7	214.3	100.1	3S	174W
1761	089	-1270 Oct 03	02:45:26	30385	-40436	026	T-	pp	-0.0286	2.8333	1.7781	357.5	226.7	102.7	1S	84E
1762	089	-1269 Mar 30	09:06:09	30375	-40430	031	P	t-	-0.7525	1.4823	0.4723	302.6	149.1	-	1N	8W
1763	089	-1269 Sep 22	12:16:30	30365	-40424	036	P	a-	0.6726	1.6216	0.6258	299.7	162.0	-	5S	58W
1764	089	-1268 Feb 17	22:51:28	30357	-40419	003	N	-t	1.3419	0.4393	-0.6471	199.1	-	-	17N	149E
1765	089	-1268 Mar 18	14:53:58	30356	-40418	041	Nb	t-	-1.5566	0.0338	-1.0298	56.3	-	-	4N	95W
1766	089	-1268 Aug 12	18:33:13	30348	-40413	008	N	-a	-1.2597	0.5239	-0.4312	184.7	-	-	20S	151W
1767	089	-1268 Sep 11	03:17:21	30346	-40412	046	N	a-	1.3245	0.4083	-0.5536	164.9	-	-	8S	77E
1768	089	-1267 Feb 05	22:50:25	30338	-40407	013	P	-t	0.6236	1.7525	0.6756	340.4	181.0	-	20N	149E
1769	089	-1267 Aug 02	09:41:15	30329	-40401	018	P	-a	-0.5775	1.7917	0.8047	309.6	179.4	-	22S	18W
1770	089	-1266 Jan 26	03:54:43	30319	-40395	023	T-	p-	-0.1190	2.6538	1.6256	346.2	221.4	98.6	21N	72E
1771	089	-1266 Jul 22	19:10:59	30310	-40389	028	T+	pp	0.1741	2.5583	1.5190	355.6	225.1	96.4	23S	162W
1772	089	-1265 Jan 15	15:55:55	30300	-40383	033	P	a-	-0.8087	1.3640	0.3837	277.9	129.7	-	22N	110W
1773	089	-1265 Jul 11	21:47:47	30291	-40377	038	P	t-	0.9479	1.1553	0.0824	299.4	70.1	-	23S	158E
1774	089	-1265 Dec 06	21:00:15	30283	-40372	005	N	-a	1.2168	0.6167	-0.3663	201.6	-	-	22N	170E
1775	089	-1264 Jan 05	07:37:57	30281	-40371	043	N	a-	-1.4531	0.1753	-0.7922	110.5	-	-	22N	14E
1776	089	-1264 May 31	09:13:39	30273	-40366	010	N	-t	-1.1101	0.8376	-0.1955	252.7	-	-	21S	15W
1777	089	-1264 Nov 25	09:19:01	30264	-40360	015	P	-a	0.5801	1.8098	0.7774	320.8	180.6	-	19N	16W
1778	089	-1263 May 20	18:47:46	30254	-40354	020	T	-p	-0.2935	2.3084	1.3300	327.5	209.7	79.9	17S	158W
1779	089	-1263 Nov 14	14:40:25	30245	-40348	025	T-	pp	-0.1086	2.7029	1.6152	369.3	230.1	100.8	15N	97W
1780	089	-1262 May 10	10:17:48	30235	-40342	030	T	a-	0.4618	1.9817	1.0389	303.6	189.6	29.2	13S	30W
1781	090	-1262 Nov 03	14:10:31	30226	-40336	035	P	t-	-0.7926	1.4572	0.3509	328.6	139.3	-	11N	89W
1782	090	-1261 Mar 31	19:48:00	30218	-40331	002	N	-a	-1.3474	0.3715	-0.6006	162.5	-	-	1S	170W
1783	090	-1261 Apr 30	03:31:26	30216	-40330	040	N	a-	1.1957	0.6380	-0.3108	202.7	-	-	9S	72E
1784	090	-1261 Sep 23	23:30:44	30208	-40325	007	N	-h	1.4524	0.2109	-0.8248	131.5	-	-	3S	132E
1785	090	-1261 Oct 23	14:51:04	30207	-40324	045	N	h-	-1.4347	0.2618	-0.8104	151.3	-	-	6N	99W
1786	090	-1260 Mar 20	06:54:02	30199	-40319	012	P	-h	-0.6485	1.6823	0.6537	321.2	173.0	-	4N	25E
1787	090	-1260 Sep 12	11:02:54	30189	-40313	017	P	-a	0.6929	1.5766	0.5964	292.3	156.9	-	8S	40W
1788	090	-1259 Mar 09	10:37:07	30180	-40307	022	T+	pp	0.0957	2.7203	1.6449	372.8	233.7	103.4	9N	30W
1789	090	-1259 Sep 02	02:54:58	30170	-40301	027	T-	pp	-0.0090	2.8206	1.8621	318.5	210.9	98.6	13S	82E
1790	090	-1258 Feb 26	10:37:54	30161	-40295	032	P	t-	0.8245	1.3843	0.3066	318.3	130.3	-	14N	29W
1791	090	-1258 Aug 22	18:29:17	30151	-40289	037	P	a-	-0.7252	1.5180	0.5362	291.1	151.4	-	17S	151W
1792	090	-1257 Jan 17	01:33:30	30143	-40284	004	N	-a	-1.2792	0.5178	-0.4963	195.1	-	-	22N	106E
1793	090	-1257 Feb 15	14:24:00	30141	-40283	042	N	a-	1.5001	0.1232	-0.9123	102.1	-	-	18N	85W
1794	090	-1257 Jul 13	15:57:05	30134	-40278	009	N	-t	1.2718	0.5500	-0.5011	213.9	-	-	22S	114W
1795	090	-1257 Aug 12	05:16:34	30132	-40277	047	N	t-	-1.5136	0.0972	-0.9356	91.0	-	-	20S	47E
1796	090	-1256 Jan 06	14:59:34	30124	-40272	014	P	-a	-0.5480	1.8389	0.8656	301.1	178.6	-	23N	97W
1797	090	-1256 Jul 01	17:19:40	30115	-40266	019	P	-t	0.5433	1.8978	0.8250	352.3	197.7	-	23S	136W
1798	090	-1256 Dec 26	06:50:51	30105	-40260	024	T+	p-	0.1266	2.6124	1.6386	320.4	209.8	94.9	24N	24E
1799	090	-1255 Jun 20	18:51:18	30096	-40254	029	T-	pp	-0.2241	2.4699	1.4241	359.5	225.5	91.7	23S	159W
1800	090	-1255 Dec 15	20:24:55	30086	-40248	034	P	a-	0.8333	1.3375	0.3205	286.9	123.1	-	24N	179E

Cat Num	Canon Plate	Calendar Date	TD of Greatest Eclipse	ΔT s	Luna Num	Saros Num	Ecl. Type	QSE	Gamma	Pen. Mag.	Um. Mag.	Eclipse Phase Durations Pen. m	Par. m	Total m	Greatest in Zenith Lat.	Lng.
1801	091	-1254 Jun 10	02:46:55	30077	-40242	039	P	a-	-0.9732	1.0675	0.0767	264.2	62.8	-	23S	81E
1802	091	-1254 Nov 05	09:30:33	30069	-40237	006	N	-t	-1.4162	0.3087	-0.7893	168.3	-	-	11N	19W
1803	091	-1253 May 01	10:42:31	30059	-40231	011	P	-a	0.8285	1.3075	0.3672	267.3	125.4	-	10S	36W
1804	091	-1253 Oct 25	08:31:39	30050	-40225	016	P	-t	-0.7547	1.5237	0.4233	331.5	150.7	-	8N	5W
1805	091	-1252 Apr 20	03:34:47	30040	-40219	021	T+	pp	0.0974	2.6588	1.6995	323.9	213.9	98.4	7S	71E
1806	091	-1252 Oct 13	10:54:31	30031	-40213	026	T-	pp	-0.0417	2.8089	1.7546	356.3	226.0	102.2	4N	40W
1807	091	-1251 Apr 09	16:14:53	30021	-40207	031	P	t-	-0.6870	1.6031	0.5918	311.3	164.4	-	4S	118W
1808	091	-1251 Oct 02	20:41:49	30012	-40201	036	P	a-	0.6545	1.6549	0.6592	300.7	164.9	-	0S	173E
1809	091	-1250 Feb 28	05:55:40	30004	-40196	003	N	-t	1.3927	0.3445	-0.7390	177.9	-	-	13N	41E
1810	091	-1250 Mar 29	21:41:13	30002	-40195	041	N	t-	-1.4926	0.1509	-0.9121	117.9	-	-	0N	161E
1811	091	-1250 Aug 24	02:36:43	29994	-40190	008	N	-a	-1.3005	0.4510	-0.5081	172.9	-	-	17S	87E
1812	091	-1250 Sep 22	11:44:45	29993	-40189	046	N	a-	1.3018	0.4511	-0.5131	172.6	-	-	4S	52W
1813	091	-1249 Feb 17	06:11:28	29985	-40184	013	P	-t	0.6638	1.6757	0.6047	334.8	173.0	-	17N	37E
1814	091	-1249 Aug 13	17:22:52	29975	-40178	018	P	-a	-0.6275	1.7031	0.7100	306.6	171.7	-	19S	135W
1815	091	-1248 Feb 06	11:49:34	29966	-40172	023	T-	pp	-0.0888	2.7059	1.6842	344.8	221.5	100.1	19N	48W
1816	091	-1248 Aug 02	02:18:00	29956	-40166	028	T+	pp	0.1146	2.6707	1.6251	358.7	227.6	100.9	21S	91E
1817	091	-1247 Jan 26	00:20:11	29947	-40160	033	P	a-	-0.7871	1.4013	0.4259	279.6	135.5	-	21N	124E
1818	091	-1247 Jul 22	04:25:54	29937	-40154	038	P	t-	0.8770	1.2873	0.2108	312.1	110.2	-	22S	58E
1819	091	-1247 Dec 17	05:48:19	29929	-40149	005	N	-a	1.2207	0.6097	-0.3739	200.9	-	-	24N	37E
1820	091	-1246 Jan 15	16:14:59	29928	-40148	043	N	a-	-1.4408	0.1968	-0.7688	116.9	-	-	22N	116W
1821	092	-1246 Jun 11	15:53:46	29920	-40143	010	N	-t	-1.1875	0.6944	-0.3362	232.5	-	-	23S	116W
1822	092	-1246 Dec 06	17:52:44	29910	-40137	015	P	-a	0.5811	1.8093	0.7744	321.9	180.9	-	22N	145W
1823	092	-1245 Jun 01	01:54:31	29901	-40131	020	T	-p	-0.3702	2.1662	1.1909	322.6	203.4	63.5	20S	93E
1824	092	-1245 Nov 25	22:51:36	29892	-40125	025	T-	pp	-0.1078	2.7055	1.6156	370.4	230.5	100.9	19N	140E
1825	092	-1244 May 20	17:41:12	29882	-40119	030	T	a-	0.3872	2.1180	1.1765	307.6	196.4	59.7	17S	143W
1826	092	-1244 Nov 13	22:11:15	29873	-40113	035	P	t-	-0.7884	1.4651	0.3585	329.0	140.6	-	15N	149E
1827	092	-1243 Apr 11	03:14:07	29865	-40108	002	N	-a	-1.4142	0.2492	-0.7236	135.1	-	-	5S	76E
1828	092	-1243 May 10	10:54:04	29863	-40107	040	N	a-	1.1271	0.7645	-0.1853	219.5	-	-	13S	41W
1829	092	-1243 Oct 04	07:41:35	29855	-40102	007	N	-h	1.4629	0.1913	-0.8439	125.0	-	-	1N	7E
1830	092	-1243 Nov 02	23:05:17	29854	-40101	045	N	h-	-1.4275	0.2744	-0.7965	154.2	-	-	10N	136E
1831	092	-1242 Mar 31	14:00:14	29846	-40096	012	P	-t	-0.7151	1.5606	0.5313	315.1	159.7	-	0S	84W
1832	092	-1242 Sep 23	19:25:35	29836	-40090	017	P	-a	0.7121	1.5417	0.5608	289.6	152.7	-	4S	168W
1833	092	-1241 Mar 20	17:27:12	29827	-40084	022	T+	pp	0.0320	2.8364	1.7628	373.7	235.1	105.8	5N	134W
1834	092	-1241 Sep 13	11:15:28	29817	-40078	027	T+	pp	0.0184	2.8049	1.8431	318.8	210.8	98.4	9S	45W
1835	092	-1240 Mar 08	17:37:53	29808	-40072	032	P	t-	0.7679	1.4855	0.4129	325.0	148.5	-	10N	136W
1836	092	-1240 Sep 02	02:31:01	29798	-40066	037	P	a-	-0.6928	1.5804	0.5929	296.0	158.1	-	13S	87E
1837	092	-1239 Jan 27	09:43:56	29791	-40061	004	N	-a	-1.2963	0.4833	-0.5248	188.2	-	-	20N	18W
1838	092	-1239 Feb 25	21:57:46	29789	-40060	042	N	a-	1.4547	0.2033	-0.8255	129.6	-	-	14N	160E
1839	092	-1239 Jul 23	22:48:53	29781	-40055	009	N	-t	1.3409	0.4261	-0.6305	191.2	-	-	21S	142E
1840	092	-1239 Aug 22	12:44:33	29780	-40054	047	N	t-	-1.4739	0.1733	-0.8662	121.3	-	-	17S	66W
1841	093	-1238 Jan 16	23:33:02	29772	-40049	014	P	-a	-0.5623	1.8107	0.8412	299.3	176.6	-	22N	134E
1842	093	-1238 Jul 12	23:51:12	29762	-40043	019	P	-t	0.6200	1.7582	0.6830	345.3	184.3	-	23S	126E
1843	093	-1237 Jan 06	15:31:05	29753	-40037	024	T+	p-	0.1166	2.6304	1.6575	320.9	210.3	95.6	24N	107W
1844	093	-1237 Jul 02	01:32:52	29743	-40031	029	T-	pp	-0.1450	2.6144	1.5697	360.5	228.3	99.6	24S	99E
1845	093	-1237 Dec 27	04:56:29	29734	-40025	034	P	a-	0.8302	1.3437	0.3255	288.3	124.4	-	25N	51E
1846	093	-1236 Jun 20	09:53:36	29724	-40019	039	P	a-	-0.8957	1.2086	0.2200	275.7	103.9	-	24S	27W
1847	093	-1236 Nov 15	17:35:03	29716	-40014	006	N	-t	-1.4221	0.2990	-0.8011	166.1	-	-	15N	142W
1848	093	-1235 May 11	18:07:55	29707	-40008	011	P	-a	0.9016	1.1729	0.2335	256.9	101.9	-	14S	150W
1849	093	-1235 Nov 04	16:33:11	29698	-40002	016	P	-t	-0.7598	1.5144	0.4139	330.4	149.0	-	12N	127W
1850	093	-1234 May 01	10:54:14	29688	-39996	021	T+	-p	0.1718	2.5229	1.5624	323.5	212.3	94.0	11S	41W
1851	093	-1234 Oct 24	19:12:05	29679	-39990	026	T-	pp	-0.0486	2.7957	1.7424	355.0	225.3	101.8	8N	167W
1852	093	-1233 Apr 20	23:15:03	29669	-39984	031	P	t-	-0.6155	1.7351	0.7221	320.0	178.3	-	8S	134E
1853	093	-1233 Oct 14	05:15:25	29660	-39978	036	P	a-	0.6425	1.6766	0.6813	301.1	166.6	-	4N	43E
1854	093	-1232 Mar 10	12:50:32	29652	-39973	003	N	-t	1.4505	0.2370	-0.8436	149.0	-	-	10N	65W
1855	093	-1232 Apr 09	04:20:37	29650	-39972	041	N	t-	-1.4225	0.2793	-0.7833	158.9	-	-	4S	59E
1856	093	-1232 Sep 03	10:48:40	29643	-39967	008	N	-a	-1.3344	0.3909	-0.5723	162.2	-	-	13S	38W
1857	093	-1232 Oct 02	20:20:10	29641	-39966	046	N	a-	1.2859	0.4814	-0.4849	177.8	-	-	1N	177E
1858	093	-1231 Feb 27	13:26:20	29633	-39961	013	P	-t	0.7097	1.5885	0.5234	328.1	162.7	-	13N	73W
1859	093	-1231 Aug 24	01:10:06	29624	-39955	018	P	-a	-0.6720	1.6245	0.6251	303.8	163.8	-	16S	107E
1860	093	-1230 Feb 16	19:39:20	29614	-39949	023	T-	pp	-0.0534	2.7675	1.7524	343.4	221.6	101.3	16N	167W

Cat Num	Canon Plate	Calendar Date	TD of Greatest Eclipse	ΔT s	Luna Num	Saros Num	Ecl. Type	QSE	Gamma	Pen. Mag.	Um. Mag.	Eclipse Phase Durations Pen. m	Par. m	Total m	Greatest in Zenith Lat.	Lng.
1861	094	-1230 Aug 13	09:30:39	29605	-39943	028	T+	pp	0.0603	2.7735	1.7215	361.2	229.2	103.4	19S	19W
1862	094	-1229 Feb 06	08:36:48	29595	-39937	033	P	a-	-0.7587	1.4509	0.4803	282.0	142.5	-	18N	1W
1863	094	-1229 Aug 02	11:11:57	29586	-39931	038	P	t-	0.8123	1.4080	0.3276	322.5	135.1	-	20S	45W
1864	094	-1229 Dec 28	14:31:03	29578	-39926	005	N	-a	1.2290	0.5946	-0.3892	199.1	-	-	25N	93W
1865	094	-1228 Jan 27	00:44:05	29577	-39925	043	N	a-	-1.4224	0.2295	-0.7341	125.9	-	-	20N	116E
1866	094	-1228 Jun 21	22:40:26	29569	-39920	010	N	-t	-1.2600	0.5602	-0.4683	210.8	-	-	24S	141E
1867	094	-1228 Jul 21	11:53:49	29567	-39919	048	N	t-	1.5223	0.0946	-0.9647	93.3	-	-	21S	56W
1868	094	-1228 Dec 17	02:21:11	29559	-39914	015	P	-a	0.5858	1.8015	0.7647	322.7	180.6	-	24N	88E
1869	094	-1227 Jun 11	09:06:01	29550	-39908	020	T	-a	-0.4439	2.0298	1.0569	317.0	196.0	36.0	22S	16W
1870	094	-1227 Dec 06	07:00:14	29540	-39902	025	T-	pp	-0.1053	2.7108	1.6195	371.3	231.0	101.2	21N	17E
1871	094	-1226 Jun 01	01:04:07	29531	-39896	030	T	p-	0.3118	2.2560	1.3152	310.9	201.9	76.1	20S	105E
1872	094	-1226 Nov 25	06:13:34	29522	-39890	035	P	t-	-0.7851	1.4707	0.3648	329.2	141.6	-	18N	28E
1873	094	-1225 Apr 22	10:32:15	29514	-39885	002	Ne	-a	-1.4869	0.1166	-0.8577	93.8	-	-	9S	36W
1874	094	-1225 May 21	18:13:18	29512	-39884	040	N	a-	1.0558	0.8963	-0.0553	235.0	-	-	16S	152W
1875	094	-1225 Oct 15	16:01:09	29504	-39879	007	N	-a	1.4674	0.1826	-0.8517	121.8	-	-	6N	120W
1876	094	-1225 Nov 14	07:23:21	29503	-39878	045	N	a-	-1.4225	0.2824	-0.7863	155.8	-	-	14N	10E
1877	094	-1224 Apr 10	20:56:55	29495	-39873	012	P	-t	-0.7884	1.4266	0.3962	307.1	141.3	-	4S	169E
1878	094	-1224 Oct 04	03:57:42	29486	-39867	017	P	-a	0.7244	1.5194	0.5380	287.7	149.8	-	1N	62E
1879	094	-1223 Mar 31	00:08:15	29476	-39861	022	T-	pp	-0.0388	2.8232	1.7512	373.9	235.3	105.9	1N	123E
1880	094	-1223 Sep 23	19:43:32	29467	-39855	027	T+	pp	0.0394	2.7681	1.8030	319.2	210.7	98.1	4S	174W
1881	095	-1222 Mar 20	00:30:19	29457	-39849	032	P	t-	0.7045	1.5992	0.5319	331.6	165.1	-	6N	118E
1882	095	-1222 Sep 13	10:40:01	29448	-39843	037	P	a-	-0.6672	1.6304	0.6370	300.1	163.2	-	9S	37W
1883	095	-1221 Feb 07	17:47:04	29440	-39838	004	N	-a	-1.3201	0.4366	-0.5651	178.9	-	-	17N	140W
1884	095	-1221 Mar 09	05:23:26	29439	-39837	042	N	a-	1.4017	0.2970	-0.7251	154.6	-	-	10N	46E
1885	095	-1221 Aug 04	05:46:48	29431	-39832	009	N	-t	1.4038	0.3135	-0.7489	166.4	-	-	19S	36E
1886	095	-1221 Sep 02	20:20:31	29429	-39831	047	N	t-	-1.4414	0.2364	-0.8099	141.5	-	-	14S	178E
1887	095	-1220 Jan 28	08:00:20	29421	-39826	014	P	-a	-0.5820	1.7726	0.8070	296.9	173.9	-	20N	7E
1888	095	-1220 Jul 23	06:29:44	29412	-39820	019	P	-t	0.6911	1.6291	0.5514	337.8	169.2	-	22S	25E
1889	095	-1219 Jan 17	00:05:44	29402	-39814	024	T+	p-	0.1022	2.6563	1.6844	321.4	211.0	96.5	23N	125E
1890	095	-1219 Jul 12	08:22:00	29393	-39808	029	T-	pp	-0.0713	2.7493	1.7053	360.5	229.5	103.5	24S	4W
1891	095	-1218 Jan 06	13:22:52	29384	-39802	034	P	a-	0.8237	1.3562	0.3371	290.1	126.7	-	25N	76W
1892	095	-1218 Jul 01	17:05:14	29374	-39796	039	P	a-	-0.8213	1.3444	0.3575	285.3	129.4	-	25S	135W
1893	095	-1218 Nov 27	01:40:07	29366	-39791	006	N	-t	-1.4270	0.2907	-0.8108	164.2	-	-	18N	96E
1894	095	-1217 May 23	01:32:39	29357	-39785	011	P	-a	0.9752	1.0378	0.0986	245.3	67.5	-	17S	97E
1895	095	-1217 Jun 21	08:08:20	29356	-39784	049	Nb	a-	-1.5092	0.0615	-0.8846	66.2	-	-	25S	2W
1896	095	-1217 Nov 16	00:36:42	29348	-39779	016	P	-t	-0.7635	1.5073	0.4076	329.4	147.8	-	15N	111E
1897	095	-1216 May 11	18:11:44	29338	-39773	021	T+	-p	0.2471	2.3855	1.4231	322.4	209.5	86.3	14S	152W
1898	095	-1216 Nov 04	03:34:41	29329	-39767	026	T-	pp	-0.0517	2.7890	1.7376	353.8	224.6	101.5	12N	66E
1899	095	-1215 May 01	06:10:12	29320	-39761	031	P	t-	-0.5406	1.8736	0.8584	328.0	190.6	-	12S	28E
1900	095	-1215 Oct 24	13:54:29	29310	-39755	036	P	a-	0.6346	1.6907	0.6961	301.1	167.6	-	9N	89W
1901	096	-1214 Mar 21	19:36:06	29302	-39750	003	N	-t	1.5151	0.1170	-0.9608	105.7	-	-	6N	168W
1902	096	-1214 Apr 20	10:52:33	29301	-39749	041	N	t-	-1.3467	0.4184	-0.6441	192.3	-	-	8S	42W
1903	096	-1214 Sep 14	19:08:36	29293	-39744	008	N	-a	-1.3619	0.3424	-0.6249	152.8	-	-	9S	165W
1904	096	-1214 Oct 14	05:02:08	29291	-39743	046	N	a-	1.2751	0.5022	-0.4661	181.2	-	-	5N	44E
1905	096	-1213 Mar 10	20:30:51	29284	-39738	013	P	-t	0.7647	1.4845	0.4254	319.8	148.6	-	9N	178E
1906	096	-1213 Sep 04	09:06:12	29274	-39732	018	P	-a	-0.7085	1.5607	0.5550	301.5	156.6	-	12S	14W
1907	096	-1212 Feb 28	03:20:05	29265	-39726	023	T-	pp	-0.0095	2.8449	1.8364	341.8	221.4	101.9	12N	76E
1908	096	-1212 Aug 23	16:51:15	29255	-39720	028	T+	pp	0.0135	2.8627	1.8043	363.2	230.1	104.2	15S	130W
1909	096	-1211 Feb 16	16:45:44	29246	-39714	033	P	a-	-0.7235	1.5131	0.5473	285.0	150.3	-	15N	125W
1910	096	-1211 Aug 12	18:06:30	29237	-39708	038	P	t-	0.7542	1.5165	0.4323	331.0	152.7	-	18S	149W
1911	096	-1210 Jan 07	23:08:43	29229	-39703	005	N	-a	1.2411	0.5724	-0.4113	196.2	-	-	25N	137E
1912	096	-1210 Feb 06	09:05:31	29227	-39702	043	N	a-	-1.3983	0.2727	-0.6889	136.7	-	-	18N	10W
1913	096	-1210 Jul 03	05:31:26	29220	-39697	010	N	-t	-1.3298	0.4315	-0.5955	186.6	-	-	25S	38E
1914	096	-1210 Aug 01	18:50:11	29218	-39696	048	N	t-	1.4570	0.2148	-0.8452	138.7	-	-	20S	161W
1915	096	-1210 Dec 28	10:45:49	29210	-39691	015	P	-a	0.5930	1.7890	0.7510	323.1	179.9	-	24N	39W
1916	096	-1209 Jun 22	16:20:50	29201	-39685	020	P	-a	-0.5156	1.8972	0.9261	310.9	187.3	-	24S	125W
1917	096	-1209 Dec 17	15:05:33	29191	-39679	025	T-	pp	-0.1006	2.7198	1.6279	372.2	231.5	101.6	23N	105W
1918	096	-1208 Jun 11	08:29:33	29182	-39673	030	T+	p-	0.2382	2.3911	1.4502	313.4	205.9	86.6	22S	8W
1919	096	-1208 Dec 05	14:16:12	29173	-39667	035	P	t-	-0.7815	1.4767	0.3721	329.3	142.8	-	21N	94W
1920	096	-1207 Jun 01	01:32:01	29163	-39661	040	P	a-	0.9844	1.0283	0.0744	248.9	59.7	-	19S	96E

Cat Num	Canon Plate	Calendar Date	TD of Greatest Eclipse	ΔT s	Luna Num	Saros Num	Ecl. Type	QSE	Gamma	Pen. Mag.	Um. Mag.	Eclipse Phase ---- Durations ---- Pen. m	Par. m	Total m	Greatest in Zenith Lat.	Lng.
1921	097	-1207 Oct 26	00:26:41	29156	-39656	007	N	-a	1.4681	0.1808	-0.8522	120.8	-	-	10N	112E
1922	097	-1207 Nov 24	15:43:55	29154	-39655	045	N	a-	-1.4187	0.2880	-0.7780	156.8	-	-	17N	116W
1923	097	-1206 Apr 22	03:47:23	29146	-39650	012	P	-t	-0.8657	1.2855	0.2536	297.4	115.8	-	9S	64E
1924	097	-1206 Oct 15	12:35:58	29137	-39644	017	P	-a	0.7319	1.5058	0.5242	286.4	147.9	-	5N	70W
1925	097	-1205 Apr 11	06:41:26	29128	-39638	022	T-	pp	-0.1154	2.6817	1.6111	373.2	234.2	102.9	4S	22E
1926	097	-1205 Oct 05	04:18:52	29118	-39632	027	T+	-p	0.0540	2.7427	1.7747	319.6	210.6	97.7	0N	55E
1927	097	-1204 Mar 30	07:16:42	29109	-39626	032	P	t-	0.6354	1.7235	0.6611	338.0	179.8	-	1N	14E
1928	097	-1204 Sep 23	18:55:38	29100	-39620	037	P	a-	-0.6478	1.6688	0.6698	303.4	166.8	-	5S	163W
1929	097	-1203 Feb 18	01:42:49	29092	-39615	004	N	-a	-1.3503	0.3779	-0.6176	166.7	-	-	14N	100E
1930	097	-1203 Mar 19	12:43:30	29090	-39614	042	N	a-	1.3431	0.4013	-0.6143	177.1	-	-	6N	66W
1931	097	-1203 Aug 14	12:52:41	29082	-39609	009	N	-t	1.4593	0.2146	-0.8535	139.5	-	-	17S	72W
1932	097	-1203 Sep 13	04:04:06	29081	-39608	047	N	t-	-1.4155	0.2871	-0.7656	155.9	-	-	10S	61E
1933	097	-1202 Feb 07	16:19:03	29073	-39603	014	P	-a	-0.6093	1.7206	0.7590	293.9	169.9	-	18N	119W
1934	097	-1202 Aug 03	13:16:41	29064	-39597	019	P	-t	0.7552	1.5128	0.4324	330.1	152.6	-	20S	78W
1935	097	-1201 Jan 28	08:31:26	29054	-39591	024	T+	pp	0.0807	2.6952	1.7245	322.1	211.8	97.6	21N	3W
1936	097	-1201 Jul 23	15:20:35	29045	-39585	029	T-	pp	-0.0045	2.8718	1.8281	359.8	229.4	104.5	23S	109W
1937	097	-1200 Jan 17	21:39:45	29036	-39579	034	P	a-	0.8102	1.3811	0.3616	292.9	131.2	-	24N	160E
1938	097	-1200 Jul 12	00:25:01	29026	-39573	039	P	a-	-0.7525	1.4700	0.4842	292.9	147.2	-	24S	114E
1939	097	-1200 Dec 07	09:44:03	29019	-39568	006	N	-t	-1.4320	0.2818	-0.8202	162.0	-	-	20N	26W
1940	097	-1199 Jan 06	03:55:50	29017	-39567	044	N	t-	1.5724	0.0121	-1.0659	33.7	-	-	25N	64E
1941	098	-1199 Jun 02	08:56:29	29009	-39562	011	N	-a	1.0494	0.9020	-0.0377	232.1	-	-	19S	15W
1942	098	-1199 Jul 01	15:37:46	29008	-39561	049	N	a-	-1.4404	0.1879	-0.7586	114.2	-	-	25S	115W
1943	098	-1199 Nov 26	08:43:21	29000	-39556	016	P	-t	-0.7650	1.5038	0.4057	328.5	147.3	-	18N	12W
1944	098	-1198 May 23	01:25:00	28991	-39550	021	T	-a	0.3251	2.2437	1.2788	320.5	205.2	73.9	17S	97E
1945	098	-1198 Nov 15	12:01:58	28981	-39544	026	T-	pp	-0.0523	2.7868	1.7378	352.4	224.0	101.3	16N	62W
1946	098	-1197 May 12	12:58:45	28972	-39538	031	T	t-	-0.4610	2.0212	1.0031	335.5	201.4	8.9	15S	76W
1947	098	-1197 Nov 04	22:39:11	28963	-39532	036	P	a-	0.6308	1.6973	0.7037	300.6	167.9	-	13N	138E
1948	098	-1196 Apr 30	17:19:40	28953	-39526	041	N	t-	-1.2666	0.5654	-0.4974	220.9	-	-	12S	141W
1949	098	-1196 Sep 25	03:35:03	28945	-39521	008	N	-a	-1.3842	0.3036	-0.6678	144.7	-	-	5S	67E
1950	098	-1196 Oct 24	13:48:42	28944	-39520	046	N	a-	1.2682	0.5157	-0.4543	183.5	-	-	9N	90W
1951	098	-1195 Mar 21	03:30:44	28936	-39515	013	P	-t	0.8242	1.3724	0.3191	310.2	130.4	-	5N	71E
1952	098	-1195 Sep 14	17:08:26	28927	-39509	018	P	-a	-0.7393	1.5074	0.4954	299.6	149.7	-	8S	136W
1953	098	-1194 Mar 10	10:54:44	28918	-39503	023	T+	pp	0.0405	2.7846	1.7827	339.9	220.7	101.5	9N	39W
1954	098	-1194 Sep 04	00:18:48	28908	-39497	028	T-	pp	-0.0270	2.8409	1.7763	364.8	230.4	104.1	12S	116E
1955	098	-1193 Feb 28	00:47:14	28899	-39491	033	P	a-	-0.6817	1.5874	0.6263	288.5	158.5	-	12N	113E
1956	098	-1193 Aug 24	01:10:38	28890	-39485	038	P	t-	0.7036	1.6112	0.5233	337.8	165.7	-	15S	103E
1957	098	-1192 Jan 19	07:37:48	28882	-39480	005	N	-a	1.2601	0.5374	-0.4459	191.2	-	-	24N	10E
1958	098	-1192 Feb 17	17:17:45	28880	-39479	043	N	a-	-1.3671	0.3290	-0.6306	149.5	-	-	15N	135W
1959	098	-1192 Jul 13	12:32:11	28873	-39474	010	N	-t	-1.3926	0.3156	-0.7102	160.7	-	-	25S	68W
1960	098	-1192 Aug 12	01:58:12	28871	-39473	048	N	t-	1.3996	0.3203	-0.7403	167.3	-	-	18S	91E
1961	099	-1191 Jan 07	19:02:47	28863	-39468	015	P	-a	0.6061	1.7654	0.7265	323.0	178.2	-	24N	163W
1962	099	-1191 Jul 02	23:41:28	28854	-39462	020	P	-a	-0.5834	1.7722	0.8022	304.4	177.6	-	24S	124E
1963	099	-1191 Dec 27	23:05:11	28845	-39456	025	T-	pp	-0.0914	2.7366	1.6447	373.0	232.2	102.3	24N	135E
1964	099	-1190 Jun 22	15:57:31	28835	-39450	030	T+	p-	0.1662	2.5236	1.5819	315.3	208.8	93.4	23S	121W
1965	099	-1190 Dec 16	22:15:17	28826	-39444	035	P	t-	-0.7747	1.4880	0.3857	329.6	144.9	-	22N	146E
1966	099	-1189 Jun 12	08:50:27	28817	-39438	040	P	a-	0.9132	1.1606	0.2035	261.4	97.2	-	21S	15W
1967	099	-1189 Nov 06	08:58:09	28809	-39433	007	N	-a	1.4650	0.1855	-0.8456	121.9	-	-	14N	17W
1968	099	-1189 Dec 06	00:04:43	28807	-39432	045	N	a-	-1.4141	0.2947	-0.7677	157.9	-	-	20N	118E
1969	099	-1188 May 02	10:31:21	28800	-39427	012	P	-t	-0.9471	1.1372	0.1032	285.4	75.7	-	12S	39W
1970	099	-1188 Oct 25	21:20:11	28790	-39421	017	P	-a	0.7351	1.4997	0.5182	285.5	146.9	-	10N	157E
1971	099	-1187 Apr 21	13:08:29	28781	-39415	022	T-	pp	-0.1970	2.5316	1.4621	371.4	231.3	95.8	8S	78W
1972	099	-1187 Oct 15	13:00:51	28772	-39409	027	T+	-p	0.0631	2.7275	1.7566	320.0	210.5	97.4	5N	77W
1973	099	-1186 Apr 10	13:56:04	28762	-39403	032	P	t-	0.5596	1.8601	0.8025	344.0	193.0	-	3S	88W
1974	099	-1186 Oct 05	03:18:40	28753	-39397	037	P	a-	-0.6351	1.6948	0.6904	306.0	169.2		0S	60E
1975	099	-1185 Mar 01	09:31:26	28745	-39392	004	N	-a	-1.3872	0.3071	-0.6822	150.6	-	-	11N	19W
1976	099	-1185 Mar 30	19:57:23	28744	-39391	042	N	a-	1.2781	0.5174	-0.4920	197.9	-	-	2N	177W
1977	099	-1185 Aug 25	20:06:37	28736	-39386	009	N	-t	1.5075	0.1291	-0.9448	109.4	-	-	14S	178E
1978	099	-1185 Sep 24	11:55:25	28735	-39385	047	N	t-	-1.3963	0.3256	-0.7333	166.1	-	-	5S	59W
1979	099	-1184 Feb 19	00:30:29	28727	-39380	014	P	-a	-0.6429	1.6569	0.6992	290.2	164.5	-	15N	117E
1980	099	-1184 Aug 13	20:13:26	28718	-39374	019	P	-t	0.8114	1.4110	0.3279	322.6	135.0	-	18S	177E

Cat Num	Canon Plate	Calendar Date	TD of Greatest Eclipse	ΔT s	Luna Num	Saros Num	Ecl. Type	QSE	Gamma	Pen. Mag.	Um. Mag.	Eclipse Phase Durations Pen. m	Par. m	Total m	Greatest in Zenith Lat.	Lng.
1981	100	-1183 Feb 07	16:50:35	28708	-39368	024	T+	pp	0.0541	2.7432	1.7738	322.8	212.6	98.6	18N	128W
1982	100	-1183 Aug 02	22:27:19	28699	-39362	029	T+	pp	0.0565	2.7761	1.7327	358.4	228.4	103.4	21S	143E
1983	100	-1182 Jan 28	05:49:26	28690	-39356	034	P	a-	0.7918	1.4151	0.3953	296.3	136.9	-	22N	37E
1984	100	-1182 Jul 23	07:51:56	28680	-39350	039	P	a-	-0.6890	1.5862	0.6012	299.0	160.6	-	23S	2E
1985	100	-1182 Dec 18	17:43:32	28673	-39345	006	N	-t	-1.4403	0.2665	-0.8352	157.9	-	-	22N	146W
1986	100	-1181 Jan 17	11:47:28	28671	-39344	044	N	t-	1.5576	0.0389	-1.0387	60.3	-	-	24N	54W
1987	100	-1181 Jun 13	16:22:17	28663	-39339	011	N	-a	1.1219	0.7695	-0.1714	217.5	-	-	21S	128W
1988	100	-1181 Jul 12	23:13:29	28662	-39338	049	N	a-	-1.3764	0.3058	-0.6417	143.9	-	-	25S	130E
1989	100	-1181 Dec 07	16:48:20	28654	-39333	016	P	-t	-0.7680	1.4970	0.4013	327.3	146.4	-	21N	134W
1990	100	-1180 Jun 02	08:38:34	28645	-39327	021	T	-a	0.4023	2.1037	1.1356	317.7	199.4	54.1	20S	13W
1991	100	-1180 Nov 25	20:29:26	28636	-39321	026	T-	pp	-0.0533	2.7834	1.7375	351.0	223.4	101.1	19N	170E
1992	100	-1179 May 22	19:45:19	28626	-39315	031	T	t-	-0.3807	2.1702	1.1489	342.1	210.3	59.0	18S	180W
1993	100	-1179 Nov 15	07:26:24	28617	-39309	036	P	a-	0.6286	1.7004	0.7084	300.0	168.0	-	17N	5E
1994	100	-1178 May 11	23:40:52	28608	-39303	041	N	t-	-1.1820	0.7211	-0.3424	246.1	-	-	16S	122E
1995	100	-1178 Oct 06	12:09:06	28600	-39298	008	N	-a	-1.4003	0.2760	-0.6993	138.6	-	-	0S	64W
1996	100	-1178 Nov 04	22:39:51	28599	-39297	046	N	a-	1.2653	0.5218	-0.4495	184.5	-	-	14N	136E
1997	100	-1177 Apr 01	10:22:20	28591	-39292	013	P	-t	0.8908	1.2475	0.1998	298.7	104.8	-	1N	34W
1998	100	-1177 Sep 26	01:18:55	28582	-39286	018	P	-a	-0.7625	1.4680	0.4499	298.4	144.1	-	4S	99E
1999	100	-1176 Mar 20	18:21:43	28572	-39280	023	T+	pp	0.0977	2.6766	1.6808	337.6	219.3	99.6	4N	153W
2000	100	-1176 Sep 14	07:55:34	28563	-39274	028	T-	pp	-0.0590	2.7854	1.7146	366.3	230.5	103.3	8S	0E
2001	101	-1175 Mar 10	08:41:12	28554	-39268	033	P	a-	-0.6333	1.6740	0.7174	292.4	167.0	-	8N	8W
2002	101	-1175 Sep 03	08:23:28	28545	-39262	038	P	t-	0.6599	1.6932	0.6016	343.2	175.4	-	11S	7W
2003	101	-1174 Jan 29	16:00:00	28537	-39257	005	N	-a	1.2843	0.4927	-0.4900	184.2	-	-	22N	116W
2004	101	-1174 Feb 28	01:22:02	28535	-39256	043	N	a-	-1.3299	0.3962	-0.5615	163.2	-	-	11N	102E
2005	101	-1174 Jul 24	19:40:05	28528	-39251	010	N	-t	-1.4506	0.2089	-0.8162	131.6	-	-	24S	176W
2006	101	-1174 Aug 23	09:15:59	28526	-39250	048	N	t-	1.3489	0.4137	-0.6476	188.1	-	-	15S	20W
2007	101	-1173 Jan 19	03:11:40	28518	-39245	015	P	-a	0.6249	1.7311	0.6917	322.3	175.5	-	23N	75E
2008	101	-1173 Jul 14	07:08:15	28509	-39239	020	P	-a	-0.6469	1.6556	0.6862	297.7	167.0	-	24S	11E
2009	101	-1172 Jan 08	06:58:38	28500	-39233	025	T-	pp	-0.0776	2.7617	1.6704	373.7	232.8	103.1	24N	17E
2010	101	-1172 Jul 02	23:29:34	28491	-39227	030	T+	pp	0.0978	2.6498	1.7068	316.5	210.5	97.3	24S	125E
2011	101	-1172 Dec 27	06:11:00	28481	-39221	035	P	t-	-0.7648	1.5047	0.4054	330.2	148.0	-	23N	27E
2012	101	-1171 Jun 22	16:10:57	28472	-39215	040	P	a-	0.8442	1.2890	0.3283	272.5	121.6	-	22S	126W
2013	101	-1171 Nov 16	17:32:45	28464	-39210	007	N	-a	1.4601	0.1932	-0.8354	123.9	-	-	18N	147W
2014	101	-1171 Dec 16	08:24:29	28463	-39209	045	N	a-	-1.4080	0.3037	-0.7545	159.6	-	-	22N	8W
2015	101	-1170 May 13	17:10:03	28455	-39204	012	N	-t	-1.0314	0.9838	-0.0526	270.8	-	-	16S	141W
2016	101	-1170 Nov 06	06:08:24	28446	-39198	017	P	-a	0.7356	1.4987	0.5177	284.9	146.6	-	14N	24E
2017	101	-1169 May 02	19:30:56	28437	-39192	022	T	pt	-0.2820	2.3752	1.3064	368.5	226.5	83.0	12S	176W
2018	101	-1169 Oct 26	21:47:06	28427	-39186	027	T+	-p	0.0685	2.7189	1.7456	320.4	210.5	97.2	9N	149E
2019	101	-1168 Apr 20	20:32:23	28418	-39180	032	P	t-	0.4805	2.0032	0.9499	349.1	204.1	-	7S	170E
2020	101	-1168 Oct 15	11:47:13	28409	-39174	037	P	a-	-0.6275	1.7113	0.7018	308.1	170.7	-	4N	60W
2021	102	-1167 Mar 11	17:12:52	28401	-39169	004	N	-a	-1.4303	0.2250	-0.7583	129.3	-	-	7N	137W
2022	102	-1167 Apr 10	03:06:40	28400	-39168	042	N	a-	1.2083	0.6426	-0.3610	216.8	-	-	2S	73E
2023	102	-1167 Sep 05	03:28:12	28392	-39163	009	N	-t	1.5486	0.0566	-1.0230	73.1	-	-	10S	66E
2024	102	-1167 Oct 04	19:52:51	28390	-39162	047	N	t-	-1.3827	0.3535	-0.7112	173.4	-	-	1S	179E
2025	102	-1166 Mar 01	08:33:24	28383	-39157	014	P	-a	-0.6837	1.5802	0.6262	285.7	157.5	-	11N	6W
2026	102	-1166 Aug 25	03:20:15	28374	-39151	019	P	-t	0.8592	1.3246	0.2389	315.6	116.7	-	14S	69E
2027	102	-1165 Feb 19	00:58:18	28364	-39145	024	T+	pp	0.0187	2.8076	1.8393	323.5	213.3	99.4	15N	108E
2028	102	-1165 Aug 14	05:45:22	28355	-39139	029	T+	pp	0.1089	2.6801	1.6366	356.7	226.7	100.9	18S	33E
2029	102	-1164 Feb 08	13:48:11	28346	-39133	034	P	a-	0.7657	1.4631	0.4432	300.5	144.3	-	19N	84W
2030	102	-1164 Aug 02	15:27:26	28337	-39127	039	P	a-	-0.6318	1.6910	0.7062	303.8	170.6	-	21S	113W
2031	102	-1164 Dec 29	01:38:08	28329	-39122	006	N	-t	-1.4517	0.2449	-0.8558	151.8	-	-	23N	95E
2032	102	-1163 Jan 27	19:30:11	28327	-39121	044	N	t-	1.5366	0.0771	-0.9997	84.7	-	-	22N	170W
2033	102	-1163 Jun 23	23:49:57	28320	-39116	011	N	-a	1.1924	0.6410	-0.3016	201.3	-	-	22S	119E
2034	102	-1163 Jul 23	06:54:47	28318	-39115	049	N	a-	-1.3170	0.4155	-0.5334	166.0	-	-	24S	14E
2035	102	-1163 Dec 18	00:52:30	28310	-39110	016	P	-t	-0.7722	1.4877	0.3952	325.8	145.2	-	22N	105E
2036	102	-1162 Jun 13	15:49:47	28301	-39104	021	P	-a	0.4804	1.9623	0.9903	314.1	191.9	-	22S	122W
2037	102	-1162 Dec 07	04:58:40	28292	-39098	026	T-	pp	-0.0536	2.7809	1.7388	349.6	222.9	101.0	21N	42E
2038	102	-1161 Jun 03	02:28:57	28283	-39092	031	T	pp	-0.2990	2.3221	1.2969	347.8	217.5	79.4	21S	78E
2039	102	-1161 Nov 26	16:15:04	28274	-39086	036	P	a-	0.6274	1.7015	0.7117	299.3	167.9	-	20N	128W
2040	102	-1160 May 22	06:01:05	28264	-39080	041	N	t-	-1.0965	0.8785	-0.1861	267.8	-	-	19S	25E

Cat Num	Canon Plate	Calendar Date	TD of Greatest Eclipse	ΔT s	Luna Num	Saros Num	Ecl. Type	QSE	Gamma	Pen. Mag.	Um. Mag.	Eclipse Phase Durations Pen. m	Par. m	Total m	Greatest in Zenith Lat.	Lng.
2041	103	-1160 Oct 16	20:48:47	28257	-39075	008	N	-a	-1.4115	0.2574	-0.7216	134.4	-	-	4N	164E
2042	103	-1160 Nov 15	07:33:33	28255	-39074	046	N	a-	1.2650	0.5227	-0.4494	184.7	-	-	17N	1E
2043	103	-1159 Apr 11	17:10:45	28248	-39069	013	P	-t	0.9605	1.1167	0.0745	285.4	64.9	-	3S	139W
2044	103	-1159 Oct 06	09:34:25	28238	-39063	018	P	-a	-0.7807	1.4375	0.4136	297.6	139.4	-	0N	27W
2045	103	-1158 Apr 01	01:44:23	28229	-39057	023	T+	pp	0.1593	2.5605	1.5707	334.7	217.0	95.8	0N	93E
2046	103	-1158 Sep 25	15:39:08	28220	-39051	028	T-	pp	-0.0850	2.7405	1.6641	367.6	230.5	102.3	4S	118W
2047	103	-1157 Mar 21	16:27:42	28211	-39045	033	P	a-	-0.5782	1.7731	0.8205	296.5	175.3	-	4N	127W
2048	103	-1157 Sep 14	15:46:36	28202	-39039	038	P	t-	0.6243	1.7603	0.6654	347.4	182.5	-	7S	120W
2049	103	-1156 Feb 10	00:12:32	28194	-39034	005	N	-a	1.3162	0.4338	-0.5483	174.3	-	-	19N	120E
2050	103	-1156 Mar 10	09:17:47	28192	-39033	043	N	a-	-1.2861	0.4760	-0.4802	177.7	-	-	7N	19W
2051	103	-1156 Aug 04	02:58:14	28185	-39028	010	N	-t	-1.5011	0.1159	-0.9087	98.5	-	-	22S	74E
2052	103	-1156 Sep 02	16:44:22	28183	-39027	048	N	t-	1.3056	0.4934	-0.5685	203.3	-	-	11S	134W
2053	103	-1155 Jan 29	11:10:40	28175	-39022	015	P	-a	0.6513	1.6829	0.6433	320.7	171.1	-	21N	46W
2054	103	-1155 Jul 24	14:43:07	28166	-39016	020	P	-a	-0.7043	1.5500	0.5808	291.0	155.9	-	23S	103W
2055	103	-1154 Jan 18	14:42:39	28157	-39010	025	T-	pp	-0.0564	2.7998	1.7098	374.3	233.6	104.1	22N	100W
2056	103	-1154 Jul 14	07:06:21	28148	-39004	030	T+	pp	0.0333	2.7691	1.8241	317.2	211.3	99.0	23S	10E
2057	103	-1153 Jan 07	14:00:19	28139	-38998	035	P	t-	-0.7492	1.5315	0.4360	331.4	152.5	-	23N	91W
2058	103	-1153 Jul 03	23:34:23	28129	-38992	040	P	a-	0.7782	1.4124	0.4474	282.2	139.7	-	23S	122E
2059	103	-1153 Nov 28	02:08:34	28122	-38987	007	N	-a	1.4552	0.2006	-0.8248	125.7	-	-	21N	83E
2060	103	-1153 Dec 27	16:39:56	28120	-38986	045	N	a-	-1.3976	0.3204	-0.7329	163.0	-	-	22N	132W
2061	104	-1152 May 23	23:45:49	28113	-38981	012	N	-t	-1.1169	0.8284	-0.2110	253.5	-	-	19S	118E
2062	104	-1152 Nov 16	15:00:24	28103	-38975	017	P	-a	0.7336	1.5018	0.5217	284.6	146.9	-	17N	111W
2063	104	-1151 May 13	01:48:56	28094	-38969	022	T	-t	-0.3704	2.2128	1.1445	364.1	219.3	60.4	15S	88E
2064	104	-1151 Nov 06	06:37:51	28085	-38963	027	T+	-p	0.0699	2.7174	1.7419	320.9	210.6	97.2	13N	15E
2065	104	-1150 May 02	03:04:31	28076	-38957	032	T	t-	0.3968	2.1547	1.1054	353.4	213.2	51.4	11S	70E
2066	104	-1150 Oct 26	20:20:47	28067	-38951	037	P	a-	-0.6250	1.7182	0.7040	309.7	171.4	-	8N	169E
2067	104	-1149 Mar 23	00:47:13	28059	-38946	004	N	-a	-1.4796	0.1317	-0.8458	99.4	-	-	2N	107E
2068	104	-1149 Apr 21	10:11:37	28058	-38945	042	N	a-	1.1339	0.7765	-0.2217	234.0	-	-	6S	36W
2069	104	-1149 Oct 16	03:56:53	28048	-38939	047	N	t-	-1.3749	0.3703	-0.6996	177.9	-	-	4N	56E
2070	104	-1148 Mar 11	16:29:43	28041	-38934	014	P	-a	-0.7301	1.4933	0.5428	280.2	148.5	-	7N	127W
2071	104	-1148 Sep 04	10:35:44	28032	-38928	019	P	-t	0.9000	1.2510	0.1628	309.1	97.3	-	11S	42W
2072	104	-1147 Mar 01	08:59:08	28022	-38922	024	T-	pp	-0.0219	2.8013	1.8340	324.1	213.8	99.6	12N	14W
2073	104	-1147 Aug 24	13:12:50	28013	-38916	029	T+	pp	0.1541	2.5972	1.5535	354.7	224.6	97.4	15S	81W
2074	104	-1146 Feb 18	21:37:17	28004	-38910	034	P	h-	0.7324	1.5242	0.5043	305.6	153.0	-	16N	157E
2075	104	-1146 Aug 13	23:11:41	27995	-38904	039	P	a-	-0.5811	1.7841	0.7992	307.5	178.1	-	19S	130E
2076	104	-1145 Jan 09	09:25:25	27987	-38899	006	N	-t	-1.4686	0.2130	-0.8859	142.0	-	-	22N	22W
2077	104	-1145 Feb 08	03:02:17	27986	-38898	044	N	t-	1.5076	0.1298	-0.9459	109.6	-	-	20N	76E
2078	104	-1145 Jul 05	07:22:19	27978	-38893	011	N	-a	1.2589	0.5202	-0.4249	183.7	-	-	23S	5E
2079	104	-1145 Aug 03	14:44:08	27977	-38892	049	N	a-	-1.2640	0.5139	-0.4370	182.8	-	-	22S	104W
2080	104	-1145 Dec 29	08:50:45	27969	-38887	016	P	-t	-0.7815	1.4686	0.3801	323.5	142.4	-	23N	15W
2081	105	-1144 Jun 23	23:04:32	27960	-38881	021	P	-a	0.5546	1.8284	0.8520	309.9	182.9	-	23S	129E
2082	105	-1144 Dec 17	13:25:21	27951	-38875	026	T-	pp	-0.0565	2.7733	1.7356	348.0	222.3	100.8	23N	85W
2083	105	-1143 Jun 13	09:12:05	27941	-38869	031	T-	pp	-0.2178	2.4732	1.4437	352.6	223.0	92.1	23S	24W
2084	105	-1143 Dec 07	01:03:16	27932	-38863	036	P	a-	0.6255	1.7037	0.7165	298.6	168.1	-	22N	99E
2085	105	-1142 Jun 02	12:19:16	27923	-38857	041	N*	t-	-1.0097	1.0386	-0.0275	286.8	-	-	21S	71W
2086	105	-1142 Oct 28	05:33:32	27915	-38852	008	N	-a	-1.4188	0.2456	-0.7365	131.8	-	-	8N	31E
2087	105	-1142 Nov 26	16:27:42	27914	-38851	046	N	a-	1.2652	0.5225	-0.4500	184.8	-	-	20N	133W
2088	105	-1141 Apr 22	23:53:24	27906	-38846	013	N	-t	1.0355	0.9766	-0.0606	269.7	-	-	7S	118E
2089	105	-1141 Oct 17	17:56:55	27897	-38840	018	P	-a	-0.7925	1.4186	0.3893	297.6	136.1	-	5N	155W
2090	105	-1140 Apr 11	09:01:41	27888	-38834	023	T+	pp	0.2264	2.4346	1.4503	331.3	213.7	89.2	4S	18W
2091	105	-1140 Oct 05	23:29:35	27879	-38828	028	T-	pp	-0.1046	2.7071	1.6255	368.7	230.4	101.3	1N	122E
2092	105	-1139 Apr 01	00:08:17	27870	-38822	033	P	a-	-0.5178	1.8821	0.9333	300.5	183.2	-	0S	116E
2093	105	-1139 Sep 24	23:18:52	27861	-38816	038	P	t-	0.5958	1.8140	0.7161	350.4	187.6	-	3S	125E
2094	105	-1138 Feb 20	08:15:50	27853	-38811	005	N	-a	1.3550	0.3624	-0.6192	160.8	-	-	16N	3W
2095	105	-1138 Mar 21	17:04:37	27851	-38810	043	N	a-	-1.2355	0.5682	-0.3868	192.6	-	-	3N	138W
2096	105	-1138 Aug 15	10:25:35	27844	-38805	010	Ne	-h	-1.5451	0.0351	-0.9893	54.4	-	-	19S	39W
2097	105	-1138 Sep 14	00:23:31	27842	-38804	048	N	h-	1.2701	0.5589	-0.5036	214.5	-	-	7S	109E
2098	105	-1137 Feb 09	19:00:07	27835	-38799	015	P	-a	0.6844	1.6220	0.5824	318.2	165.0	-	19N	164W
2099	105	-1137 Aug 04	22:26:00	27826	-38793	020	P	-a	-0.7557	1.4559	0.4864	284.5	144.4	-	21S	140E
2100	105	-1136 Jan 29	22:18:07	27816	-38787	025	T-	pp	-0.0287	2.8497	1.7615	374.8	234.2	105.0	21N	146E

Cat Num	Canon Plate	Calendar Date	TD of Greatest Eclipse	ΔT s	Luna Num	Saros Num	Ecl. Type	QSE	Gamma	Pen. Mag.	Um. Mag.	Eclipse Phase Durations Pen. m	Par. m	Total m	Greatest in Zenith Lat.	Lng.
2101	106	-1136 Jul 24	14:49:40	27807	-38781	030	T-	pp	-0.0257	2.7844	1.8369	317.5	211.2	99.0	22S	106W
2102	106	-1135 Jan 17	21:44:23	27798	-38775	035	P	t-	-0.7287	1.5668	0.4756	332.9	158.1	-	22N	153E
2103	106	-1135 Jul 14	07:00:37	27789	-38769	040	P	a-	0.7151	1.5306	0.5608	290.8	154.1	-	23S	10E
2104	106	-1135 Dec 08	10:44:38	27781	-38764	007	N	-a	1.4508	0.2067	-0.8150	127.1	-	-	23N	47W
2105	106	-1134 Jan 07	00:52:06	27780	-38763	045	N	a-	-1.3836	0.3433	-0.7046	167.6	-	-	22N	105E
2106	106	-1134 Jun 04	06:19:42	27772	-38758	012	N	-t	-1.2027	0.6727	-0.3701	232.9	-	-	22S	18E
2107	106	-1134 Jul 03	18:28:54	27771	-38757	050	Nb	t-	1.5265	0.0674	-0.9532	75.6	-	-	22S	163W
2108	106	-1134 Nov 27	23:52:07	27763	-38752	017	P	-a	0.7322	1.5037	0.5251	284.2	147.1	-	20N	116E
2109	106	-1133 May 24	08:05:53	27754	-38746	022	P	-t	-0.4592	2.0498	0.9814	358.4	209.6	-	19S	8W
2110	106	-1133 Nov 17	15:29:41	27745	-38740	027	T+	-p	0.0703	2.7174	1.7403	321.4	210.8	97.2	17N	119W
2111	106	-1132 May 12	09:36:55	27736	-38734	032	T	t-	0.3126	2.3074	1.2618	356.5	220.2	76.7	15S	31W
2112	106	-1132 Nov 06	04:56:00	27727	-38728	037	P	a-	-0.6246	1.7211	0.7028	311.1	171.7	-	13N	39E
2113	106	-1131 Apr 02	08:16:29	27719	-38723	004	Ne	-a	-1.5335	0.0301	-0.9421	47.8	-	-	2S	7W
2114	106	-1131 May 01	17:15:07	27717	-38722	042	N	a-	1.0569	0.9152	-0.0781	249.3	-	-	10S	144W
2115	106	-1131 Oct 26	12:05:06	27708	-38716	047	N	t-	-1.3712	0.3795	-0.6951	180.6	-	-	8N	68W
2116	106	-1130 Mar 23	00:16:48	27701	-38711	014	P	-a	-0.7843	1.3923	0.4450	273.6	136.4	-	3N	114E
2117	106	-1130 Sep 15	18:02:24	27692	-38705	019	P	-t	0.9319	1.1937	0.1032	303.7	78.1	-	7S	156W
2118	106	-1129 Mar 12	16:49:06	27682	-38699	024	T-	pp	-0.0711	2.7106	1.7440	324.5	213.8	98.9	8N	133W
2119	106	-1129 Sep 04	20:52:06	27673	-38693	029	T+	pp	0.1906	2.5302	1.4864	352.7	222.4	93.7	11S	163E
2120	106	-1128 Mar 01	05:15:00	27664	-38687	034	P	h-	0.6906	1.6009	0.5808	311.3	162.8	-	12N	41E
2121	107	-1128 Aug 24	07:06:12	27655	-38681	039	P	a-	-0.5382	1.8630	0.8778	310.2	183.6	-	16S	9E
2122	107	-1127 Jan 19	17:05:49	27648	-38676	006	N	-t	-1.4904	0.1719	-0.9248	128.0	-	-	21N	138W
2123	107	-1127 Feb 18	10:24:38	27646	-38675	044	N	t-	1.4714	0.1954	-0.8789	133.9	-	-	17N	36W
2124	107	-1127 Jul 15	14:57:49	27638	-38670	011	N	-a	1.3224	0.4053	-0.5429	164.2	-	-	22S	110W
2125	107	-1127 Aug 13	22:39:49	27637	-38669	049	N	a-	-1.2162	0.6027	-0.3505	196.2	-	-	19S	136E
2126	107	-1126 Jan 08	16:45:55	27629	-38664	016	P	-t	-0.7936	1.4440	0.3601	320.6	138.8	-	23N	134W
2127	107	-1126 Jul 05	06:20:12	27620	-38658	021	P	-a	0.6267	1.6985	0.7171	305.1	172.2	-	23S	19E
2128	107	-1126 Dec 28	21:49:15	27611	-38652	026	T-	pp	-0.0624	2.7599	1.7272	346.4	221.7	100.6	24N	149E
2129	107	-1125 Jun 24	15:56:01	27602	-38646	031	T-	pp	-0.1384	2.6214	1.5871	356.3	226.9	99.8	24S	126W
2130	107	-1125 Dec 18	09:49:59	27593	-38640	036	P	a-	0.6225	1.7077	0.7235	298.0	168.4	-	24N	33W
2131	107	-1124 Jun 12	18:39:23	27584	-38634	041	P	t-	-0.9241	1.1965	0.1287	303.0	87.0	-	23S	167W
2132	107	-1124 Nov 07	14:20:30	27576	-38629	008	N	-a	-1.4239	0.2376	-0.7475	130.0	-	-	12N	102W
2133	107	-1124 Dec 07	01:20:43	27575	-38628	046	N	a-	1.2651	0.5226	-0.4499	184.9	-	-	23N	93E
2134	107	-1123 May 03	06:35:53	27567	-38623	013	N	-t	1.1111	0.8355	-0.1971	252.1	-	-	11S	15E
2135	107	-1123 Oct 28	02:22:32	27558	-38617	018	P	-a	-0.8009	1.4055	0.3714	297.9	133.7	-	9N	77E
2136	107	-1122 Apr 22	16:15:22	27549	-38611	023	T	-p	0.2973	2.3019	1.3226	327.1	209.1	79.1	8S	129W
2137	107	-1122 Oct 17	07:26:03	27540	-38605	028	T-	pp	-0.1186	2.6838	1.5975	369.8	230.4	100.4	5N	1E
2138	107	-1121 Apr 12	07:43:23	27531	-38599	033	T	a-	-0.4526	2.0001	1.0543	304.3	190.3	34.3	5S	1W
2139	107	-1121 Oct 06	06:59:26	27522	-38593	038	P	t-	0.5741	1.8550	0.7548	352.6	191.2	-	1N	8E
2140	107	-1120 Mar 02	16:08:53	27514	-38588	005	N	-a	1.4014	0.2770	-0.7043	142.1	-	-	12N	123W
2141	108	-1120 Apr 01	00:44:09	27513	-38587	043	N	a-	-1.1795	0.6707	-0.2837	207.5	-	-	1S	105E
2142	108	-1120 Sep 24	08:13:09	27503	-38581	048	N	h-	1.2424	0.6098	-0.4529	222.4	-	-	3S	10W
2143	108	-1119 Feb 20	02:38:37	27496	-38576	015	P	-h	0.7258	1.5461	0.5066	314.5	156.2	-	15N	80E
2144	108	-1119 Aug 15	06:17:14	27487	-38570	020	P	-a	-0.8007	1.3736	0.4034	278.4	132.8	-	19S	21E
2145	108	-1118 Feb 09	05:42:52	27478	-38564	025	T+	pp	0.0074	2.8879	1.8018	375.1	234.7	105.4	18N	34E
2146	108	-1118 Aug 04	22:39:39	27469	-38558	030	T-	pp	-0.0790	2.6880	1.7375	317.5	210.6	97.7	20S	135E
2147	108	-1117 Jan 29	05:18:25	27460	-38552	035	P	t-	-0.6996	1.6178	0.5315	335.3	165.4	-	20N	39E
2148	108	-1117 Jul 25	14:32:32	27451	-38546	040	P	a-	0.6573	1.6393	0.6642	298.2	165.4	-	22S	104W
2149	108	-1117 Dec 19	19:18:41	27443	-38541	007	N	-a	1.4490	0.2078	-0.8096	126.9	-	-	25N	175W
2150	108	-1116 Jan 18	08:57:49	27441	-38540	045	N	a-	-1.3635	0.3772	-0.6647	174.4	-	-	21N	17W
2151	108	-1116 Jun 14	12:53:10	27434	-38535	012	N	-t	-1.2874	0.5192	-0.5275	208.5	-	-	24S	81W
2152	108	-1116 Jul 14	01:28:20	27432	-38534	050	N	t-	1.4635	0.1858	-0.8406	124.9	-	-	22S	91E
2153	108	-1116 Dec 08	08:43:36	27425	-38529	017	P	-a	0.7313	1.5043	0.5276	283.8	147.2	-	23N	18W
2154	108	-1115 Jun 03	14:22:11	27416	-38523	022	P	-t	-0.5485	1.8862	0.8175	351.3	197.1	-	21S	104W
2155	108	-1115 Nov 28	00:22:52	27407	-38517	027	T+	-p	0.0691	2.7203	1.7420	321.9	211.1	97.3	20N	107E
2156	108	-1114 May 23	16:07:23	27398	-38511	032	T+	pp	0.2259	2.4648	1.4224	358.6	225.3	91.5	18S	130W
2157	108	-1114 Nov 17	13:34:02	27389	-38505	037	P	a-	-0.6274	1.7178	0.6960	312.1	171.6	-	16N	92W
2158	108	-1113 May 13	00:17:25	27379	-38499	042	P	a-	0.9778	1.0581	0.0691	262.9	59.6	-	14S	108E
2159	108	-1113 Nov 06	20:16:23	27370	-38493	047	N	t-	-1.3706	0.3827	-0.6958	181.9	-	-	12N	168E
2160	108	-1112 Apr 02	07:58:55	27363	-38488	014	P	-a	-0.8427	1.2838	0.3391	265.8	121.0	-	1S	4W

Cat Num	Canon Plate	Calendar Date	TD of Greatest Eclipse	ΔT s	Luna Num	Saros Num	Ecl. Type	QSE	Gamma	Pen. Mag.	Um. Mag.	Eclipse Phase Durations Pen. m	Par. m	Total m	Greatest in Zenith Lat.	Lng.
2161	109	-1112 Sep 26	01:38:06	27354	-38482	019	P	-t	0.9567	1.1491	0.0568	299.2	58.2	-	2S	88E
2162	109	-1111 Mar 23	00:32:13	27345	-38476	024	T-	-p	-0.1254	2.6107	1.6446	324.5	213.2	96.7	3N	109E
2163	109	-1111 Sep 15	04:40:18	27336	-38470	029	T+	pp	0.2204	2.4757	1.4318	350.7	220.3	90.0	7S	44E
2164	109	-1110 Mar 12	12:43:31	27327	-38464	034	P	t-	0.6423	1.6896	0.6692	317.5	172.8	-	8N	73W
2165	109	-1110 Sep 04	15:09:55	27318	-38458	039	P	a-	-0.5026	1.9284	0.9428	312.0	187.6	-	12S	113W
2166	109	-1109 Jan 31	00:35:41	27310	-38453	006	N	-t	-1.5198	0.1166	-0.9773	105.9	-	-	19N	109E
2167	109	-1109 Mar 01	17:35:13	27308	-38452	044	N	t-	1.4266	0.2770	-0.7959	158.5	-	-	13N	145W
2168	109	-1109 Jul 26	22:40:31	27301	-38447	011	N	-a	1.3795	0.3023	-0.6494	143.5	-	-	21S	134E
2169	109	-1109 Aug 25	06:44:53	27299	-38446	049	N	a-	-1.1763	0.6772	-0.2786	206.4	-	-	16S	13E
2170	109	-1108 Jan 20	00:33:13	27292	-38441	016	P	-t	-0.8121	1.4076	0.3289	316.8	133.0	-	22N	109E
2171	109	-1108 Jul 15	13:40:43	27283	-38435	021	P	-a	0.6938	1.5783	0.5914	299.8	160.1	-	23S	92W
2172	109	-1107 Jan 08	06:07:40	27274	-38429	026	T-	pp	-0.0733	2.7372	1.7099	344.6	221.0	100.2	23N	24E
2173	109	-1107 Jul 04	22:42:59	27265	-38423	031	T-	pp	-0.0627	2.7629	1.7234	359.2	229.2	103.7	24S	132E
2174	109	-1107 Dec 28	18:32:20	27256	-38417	036	P	a-	0.6155	1.7189	0.7382	297.8	169.4	-	24N	163W
2175	109	-1106 Jun 24	01:00:53	27247	-38411	041	P	t-	-0.8395	1.3529	0.2827	317.1	126.3	-	24S	96E
2176	109	-1106 Nov 18	23:09:57	27239	-38406	008	N	-a	-1.4269	0.2333	-0.7541	129.2	-	-	16N	124E
2177	109	-1106 Dec 18	10:11:04	27238	-38405	046	N	a-	1.2633	0.5257	-0.4462	185.5	-	-	24N	40W
2178	109	-1105 May 14	13:16:24	27230	-38400	013	N	-t	1.1887	0.6911	-0.3374	231.6	-	-	14S	87W
2179	109	-1105 Jun 13	02:00:33	27229	-38399	051	Nb	t-	-1.5694	0.0050	-1.0479	21.5	-	-	24S	81E
2180	109	-1105 Nov 08	10:51:41	27221	-38394	018	P	-a	-0.8056	1.3991	0.3607	298.7	132.4	-	13N	52W
2181	110	-1104 May 02	23:26:51	27212	-38388	023	T	-p	0.3708	2.1649	1.1900	322.1	203.2	63.3	12S	121E
2182	110	-1104 Oct 27	15:28:20	27203	-38382	028	T-	pp	-0.1269	2.6706	1.5803	370.9	230.5	99.9	10N	121W
2183	110	-1103 Apr 22	15:13:16	27194	-38376	033	T	a-	-0.3827	2.1270	1.1838	307.9	196.6	60.7	9S	116W
2184	110	-1103 Oct 16	14:47:56	27185	-38370	038	P	t-	0.5586	1.8844	0.7823	354.1	193.6	-	6N	111W
2185	110	-1102 Mar 13	23:53:16	27177	-38365	005	N	-a	1.4542	0.1801	-0.8010	115.9	-	-	8N	119E
2186	110	-1102 Apr 12	08:16:37	27176	-38364	043	N	a-	-1.1182	0.7830	-0.1711	222.1	-	-	5S	11W
2187	110	-1102 Oct 05	16:12:20	27167	-38358	048	N	h-	1.2212	0.6486	-0.4139	227.8	-	-	2N	132W
2188	110	-1101 Mar 03	10:06:23	27159	-38353	015	P	-h	0.7751	1.4556	0.4162	309.4	144.1	-	12N	34W
2189	110	-1101 Aug 26	14:17:42	27150	-38347	020	P	-a	-0.8389	1.3039	0.3329	272.9	121.7	-	15S	100W
2190	110	-1100 Feb 20	12:58:55	27141	-38341	025	T+	pp	0.0505	2.8077	1.7238	375.1	234.8	105.0	15N	77W
2191	110	-1100 Aug 15	06:36:09	27132	-38335	030	T-	-p	-0.1269	2.6019	1.6481	317.2	209.6	95.4	18S	15E
2192	110	-1099 Feb 08	12:46:23	27123	-38329	035	P	t-	-0.6647	1.6791	0.5980	338.1	173.3	-	17N	74W
2193	110	-1099 Aug 04	22:09:45	27114	-38323	040	P	a-	0.6044	1.7392	0.7584	304.6	174.5	-	20S	141E
2194	110	-1099 Dec 30	03:49:03	27107	-38318	007	N	-a	1.4508	0.2022	-0.8104	124.8	-	-	25N	57E
2195	110	-1098 Jan 28	16:56:55	27105	-38317	045	N	a-	-1.3372	0.4222	-0.6134	182.9	-	-	19N	137W
2196	110	-1098 Jun 25	19:27:54	27098	-38312	012	N	-t	-1.3698	0.3702	-0.6808	179.2	-	-	25S	179E
2197	110	-1098 Jul 25	08:32:26	27096	-38311	050	N	t-	1.4048	0.2966	-0.7359	156.9	-	-	21S	15W
2198	110	-1098 Dec 19	17:31:58	27089	-38306	017	P	-a	0.7335	1.4991	0.5248	283.0	146.7	-	24N	150W
2199	110	-1097 Jun 14	20:41:47	27080	-38300	022	P	-t	-0.6348	1.7281	0.6587	343.0	181.8	-	23S	160E
2200	110	-1097 Dec 09	09:12:59	27071	-38294	027	T+	-p	0.0703	2.7184	1.7394	322.4	211.3	97.4	22N	26W
2201	111	-1096 Jun 02	22:42:11	27062	-38288	032	T+	pp	0.1417	2.6180	1.5784	359.6	228.3	100.0	20S	130E
2202	111	-1096 Nov 27	22:11:08	27053	-38282	037	P	a-	-0.6298	1.7148	0.6901	313.0	171.5	-	19N	138E
2203	111	-1095 May 23	07:19:50	27044	-38276	042	P	a-	0.8976	1.2035	0.2183	274.7	103.4	-	17S	1E
2204	111	-1095 Nov 17	04:29:23	27034	-38270	047	N	t-	-1.3717	0.3821	-0.6995	182.4	-	-	15N	43E
2205	111	-1094 Apr 13	15:33:56	27027	-38265	014	P	-a	-0.9072	1.1644	0.2218	256.4	99.5	-	6S	121W
2206	111	-1094 Oct 07	09:23:47	27018	-38259	019	P	-t	0.9735	1.1188	0.0251	295.8	38.8	-	2N	30W
2207	111	-1093 Apr 03	08:05:23	27009	-38253	024	T-	-p	-0.1874	2.4970	1.5307	324.0	211.8	92.4	1S	7W
2208	111	-1093 Sep 26	12:40:09	27000	-38247	029	T+	-p	0.2415	2.4369	1.3932	348.9	218.5	87.0	3S	78W
2209	111	-1092 Mar 22	20:01:47	26991	-38241	034	P	t-	0.5866	1.7922	0.7713	324.0	183.0	-	4N	175E
2210	111	-1092 Sep 14	23:22:40	26982	-38235	039	P	a-	-0.4740	1.9812	0.9952	313.3	190.4	-	8S	122E
2211	111	-1091 Feb 10	07:57:27	26974	-38230	006	Ne	-t	-1.5551	0.0501	-1.0406	69.9	-	-	16N	2W
2212	111	-1091 Mar 12	00:36:33	26973	-38229	044	N	t-	1.3747	0.3716	-0.6999	182.3	-	-	9N	107E
2213	111	-1091 Aug 06	06:28:49	26965	-38224	011	N	-a	1.4317	0.2085	-0.7472	120.4	-	-	19S	16E
2214	111	-1091 Sep 04	14:57:34	26964	-38223	049	N	a-	-1.1428	0.7402	0.2105	214.5	-	-	12S	111W
2215	111	-1090 Jan 30	08:13:51	26956	-38218	016	P	-t	-0.8367	1.3595	0.2864	311.8	124.6	-	20N	7W
2216	111	-1090 Jul 26	21:05:03	26948	-38212	021	P	-a	0.7567	1.4658	0.4730	294.2	146.3	-	21S	156E
2217	111	-1089 Jan 19	14:21:08	26939	-38206	026	T-	pp	-0.0893	2.7049	1.6835	342.8	220.3	99.5	22N	99W
2218	111	-1089 Jul 16	05:33:52	26929	-38200	031	T+	pp	0.0092	2.8638	1.8188	361.2	230.2	104.8	23S	28E
2219	111	-1088 Jan 09	03:09:49	26920	-38194	036	P	a-	0.6041	1.7378	0.7610	297.9	171.1	-	24N	67E
2220	111	-1088 Jul 04	07:27:49	26911	-38188	041	P	t-	-0.7591	1.5017	0.4291	328.8	152.2	-	25S	1W

Cat Num	Canon Plate	Calendar Date	TD of Greatest Eclipse	ΔT s	Luna Num	Saros Num	Ecl. Type	QSE	Gamma	Pen. Mag.	Um. Mag.	Eclipse Phase Durations Pen. m	Par. m	Total m	Greatest in Zenith Lat.	Lng.
2221	112	-1088 Nov 29	07:59:19	26904	-38183	008	N	-a	-1.4297	0.2291	-0.7602	128.4	-	-	19N	9W
2222	112	-1088 Dec 28	18:57:52	26903	-38182	046	N	a-	1.2591	0.5331	-0.4380	186.8	-	-	25N	171W
2223	112	-1087 May 24	19:57:56	26895	-38177	013	N	-t	1.2661	0.5472	-0.4777	208.1	-	-	17S	171E
2224	112	-1087 Jun 23	08:32:46	26894	-38176	051	N	t-	-1.4856	0.1580	-0.8937	119.5	-	-	25S	18W
2225	112	-1087 Nov 18	19:21:32	26886	-38171	018	P	-a	-0.8090	1.3947	0.3526	299.5	131.4	-	16N	179E
2226	112	-1086 May 14	06:37:19	26877	-38165	023	T	-a	0.4458	2.0254	1.0544	316.4	195.7	35.2	15S	11E
2227	112	-1086 Nov 07	23:32:47	26868	-38159	028	T-	pp	-0.1332	2.6608	1.5673	371.8	230.6	99.4	14N	116E
2228	112	-1085 May 03	22:40:07	26859	-38153	033	T	p-	-0.3100	2.2594	1.3182	310.9	201.9	76.4	12S	130E
2229	112	-1085 Oct 27	22:42:29	26850	-38147	038	P	t-	0.5479	1.9047	0.8015	354.9	195.1	-	10N	128E
2230	112	-1084 Mar 24	07:28:36	26843	-38142	005	Ne	-a	1.5136	0.0712	-0.9103	73.8	-	-	4N	3E
2231	112	-1084 Apr 22	15:43:46	26841	-38141	043	N	a-	-1.0528	0.9031	-0.0514	236.1	-	-	9S	125W
2232	112	-1084 Oct 16	00:18:41	26832	-38135	048	N	a-	1.2051	0.6780	-0.3840	231.6	-	-	6N	104E
2233	112	-1083 Mar 13	17:23:33	26825	-38130	015	P	-t	0.8321	1.3510	0.3115	302.7	127.0	-	8N	145W
2234	112	-1083 Sep 05	22:27:19	26816	-38124	020	P	-a	-0.8703	1.2469	0.2749	268.1	111.2	-	12S	136E
2235	112	-1082 Mar 02	20:02:19	26807	-38118	025	T+	pp	0.1035	2.7091	1.6277	374.6	234.1	103.0	11N	175E
2236	112	-1082 Aug 26	14:41:10	26798	-38112	030	T-	-p	-0.1674	2.5293	1.5719	316.9	208.4	92.6	15S	108W
2237	112	-1081 Feb 19	20:03:52	26789	-38106	035	P	t-	-0.6208	1.7569	0.6813	341.4	182.1	-	14N	175E
2238	112	-1081 Aug 16	05:54:07	26780	-38100	040	P	a-	0.5583	1.8268	0.8401	310.0	181.5	-	17S	24E
2239	112	-1080 Jan 10	12:14:04	26772	-38095	007	N	-a	1.4576	0.1872	-0.8202	119.7	-	-	25N	69W
2240	112	-1080 Feb 09	00:48:15	26771	-38094	045	N	a-	-1.3039	0.4801	-0.5489	193.1	-	-	17N	104E
2241	113	-1080 Jul 06	02:06:02	26763	-38089	012	N	-t	-1.4482	0.2286	-0.8271	143.1	-	-	25S	79E
2242	113	-1080 Aug 04	15:42:43	26762	-38088	050	N	t-	1.3516	0.3975	-0.6414	180.8	-	-	19S	124W
2243	113	-1080 Dec 30	02:17:41	26754	-38083	017	P	-a	0.7382	1.4891	0.5175	281.9	145.7	-	25N	79E
2244	113	-1079 Jun 25	03:02:41	26745	-38077	022	P	-t	-0.7200	1.5724	0.5019	333.4	162.7	-	24S	64E
2245	113	-1079 Dec 19	18:01:52	26737	-38071	027	T+	-p	0.0722	2.7150	1.7358	322.9	211.6	97.4	23N	159W
2246	113	-1078 Jun 14	05:18:48	26728	-38065	032	T+	pp	0.0578	2.7708	1.7334	359.4	229.6	104.1	22S	30E
2247	113	-1078 Dec 09	06:46:41	26719	-38059	037	P	a-	-0.6318	1.7122	0.6852	314.0	171.5	-	21N	9E
2248	113	-1077 Jun 03	14:24:28	26710	-38053	042	P	a-	0.8180	1.3479	0.3658	284.8	130.6	-	20S	107W
2249	113	-1077 Nov 28	12:42:38	26701	-38047	047	N	t-	-1.3734	0.3803	-0.7038	182.5	-	-	18N	81W
2250	113	-1076 Apr 23	23:05:55	26693	-38042	014	P	-a	-0.9740	1.0410	0.0999	245.7	67.9	-	10S	124E
2251	113	-1076 May 23	05:40:35	26692	-38041	052	Nb	a-	1.5114	0.0562	-0.8874	63.2	-	-	17S	24E
2252	113	-1076 Oct 17	17:15:31	26684	-38036	019	P	-t	0.9858	1.0967	0.0022	293.2	11.5	-	7N	150W
2253	113	-1075 Apr 13	15:33:30	26675	-38030	024	T-	-p	-0.2529	2.3771	1.4103	323.0	209.3	85.4	5S	122W
2254	113	-1075 Oct 06	20:48:03	26666	-38024	029	T+	-p	0.2565	2.4090	1.3658	347.3	217.0	84.5	2N	157E
2255	113	-1074 Apr 03	03:10:32	26657	-38018	034	P	t-	0.5240	1.9074	0.8856	330.5	192.9	-	0S	65E
2256	113	-1074 Sep 26	07:44:37	26649	-38012	039	T	a-	-0.4525	2.0207	1.0343	314.0	192.2	27.8	3S	6W
2257	113	-1073 Mar 23	07:27:29	26640	-38006	044	N	t-	1.3148	0.4809	-0.5893	205.5	-	-	5N	2E
2258	113	-1073 Aug 17	14:25:20	26632	-38001	011	N	-a	1.4771	0.1274	-0.8325	95.0	-	-	16S	105W
2259	113	-1073 Sep 15	23:18:50	26631	-38000	049	N	a-	-1.1164	0.7900	-0.1715	220.6	-	-	8S	121E
2260	113	-1072 Feb 10	15:45:37	26623	-37995	016	P	-t	-0.8689	1.2974	0.2303	305.5	112.4	-	17N	121W
2261	114	-1072 Aug 06	04:36:41	26614	-37989	021	P	-a	0.8126	1.3663	0.3674	288.7	131.4	-	19S	42E
2262	114	-1071 Jan 29	22:27:18	26605	-37983	026	T-	pp	-0.1119	2.6603	1.6452	340.8	219.3	98.4	20N	138E
2263	114	-1071 Jul 26	12:29:15	26596	-37977	031	T+	pp	0.0762	2.7437	1.6928	362.5	230.0	103.3	22S	76W
2264	114	-1070 Jan 19	11:41:03	26587	-37971	036	P	a-	0.5874	1.7663	0.7937	298.5	173.5	-	23N	61W
2265	114	-1070 Jul 15	13:59:53	26578	-37965	041	P	t-	-0.6830	1.6428	0.5673	338.6	171.3	-	24S	100W
2266	114	-1070 Dec 10	16:46:42	26571	-37960	008	N	-a	-1.4338	0.2223	-0.7683	126.8	-	-	21N	141W
2267	114	-1069 Jan 09	03:38:01	26570	-37959	046	N	a-	1.2502	0.5488	-0.4211	189.4	-	-	25N	59E
2268	114	-1069 Jun 05	02:41:25	26562	-37954	013	N	-t	1.3423	0.4059	-0.6158	180.9	-	-	20S	68E
2269	114	-1069 Jul 04	15:11:30	26561	-37953	051	N	t-	-1.4061	0.3036	-0.7474	163.0	-	-	25S	118W
2270	114	-1069 Nov 30	03:51:49	26553	-37948	018	P	-a	-0.8109	1.3927	0.3477	300.5	131.0	-	19N	51E
2271	114	-1068 May 24	13:48:02	26544	-37942	023	P	-a	0.5212	1.8853	0.9175	309.8	186.6	-	18S	99W
2272	114	-1068 Nov 18	07:39:36	26535	-37936	028	T-	pp	-0.1367	2.6555	1.5595	372.7	230.8	99.2	17N	7W
2273	114	-1067 May 14	06:04:35	26526	-37930	033	T-	p-	-0.2353	2.3958	1.4559	313.4	206.1	87.0	16S	17E
2274	114	-1067 Nov 07	06:42:34	26517	-37924	038	P	t-	0.5412	1.9171	0.8136	355.2	196.0	-	14N	7E
2275	114	-1066 May 03	23:04:58	26509	-37918	043	P	a-	-0.9831	1.0316	0.0761	249.5	60.4	-	13S	122E
2276	114	-1066 Oct 27	08:33:07	26500	-37912	048	N	a-	1.1946	0.6967	-0.3642	233.6	-	-	10N	21W
2277	114	-1065 Mar 25	00:31:18	26492	-37907	015	P	-t	0.8957	1.2345	0.1947	294.1	102.5	-	3N	105E
2278	114	-1065 Sep 17	06:45:25	26483	-37901	020	P	-a	-0.8951	1.2019	0.2287	264.1	102.0	-	7S	9E
2279	114	-1064 Mar 13	02:57:59	26474	-37895	025	T+	pp	0.1625	2.5996	1.5206	373.4	232.5	98.8	7N	69E
2280	114	-1064 Sep 05	22:53:51	26465	-37889	030	T-	-p	-0.2013	2.4690	1.5079	316.7	207.1	89.5	11S	127E

Cat Num	Canon Plate	Calendar Date	TD of Greatest Eclipse	ΔT s	Luna Num	Saros Num	Ecl. Type	QSE	Gamma	Pen. Mag.	Um. Mag.	Eclipse Phase Durations Pen. m	Par. m	Total m	Greatest in Zenith Lat.	Lng.
2281	115	-1063 Mar 02	03:14:46	26457	-37883	035	P	t-	-0.5707	1.8460	0.7760	344.8	190.8	-	11N	66E
2282	115	-1063 Aug 26	13:44:42	26448	-37877	040	P	a-	0.5180	1.9038	0.9110	314.7	187.2	-	14S	96W
2283	115	-1062 Jan 20	20:33:46	26440	-37872	007	N	-a	1.4694	0.1626	-0.8392	111.4	-	-	24N	166E
2284	115	-1062 Feb 19	08:32:50	26439	-37871	045	N	a-	-1.2640	0.5499	-0.4724	204.3	-	-	14N	14W
2285	115	-1062 Jul 17	08:48:57	26431	-37866	012	Ne	-t	-1.5218	0.0961	-0.9646	94.2	-	-	25S	22W
2286	115	-1062 Aug 15	23:00:01	26430	-37865	050	N	t-	1.3044	0.4875	-0.5580	199.5	-	-	16S	125E
2287	115	-1061 Jan 10	10:56:25	26422	-37860	017	P	-a	0.7492	1.4674	0.4989	280.2	143.4	-	24N	51W
2288	115	-1061 Jul 06	09:30:50	26413	-37854	022	P	-t	-0.7992	1.4277	0.3557	323.0	140.0	-	25S	34W
2289	115	-1061 Dec 31	02:45:04	26405	-37848	027	T+	-p	0.0790	2.7025	1.7234	323.4	211.8	97.4	24N	70E
2290	115	-1060 Jun 24	12:01:47	26396	-37842	032	T-	pp	-0.0219	2.8358	1.8002	358.3	229.3	104.6	24S	72W
2291	115	-1060 Dec 19	15:17:49	26387	-37836	037	P	a-	-0.6308	1.7150	0.6862	315.2	172.1	-	23N	119W
2292	115	-1059 Jun 13	21:32:25	26378	-37830	042	P	a-	0.7398	1.4901	0.5105	293.4	150.4	-	22S	145E
2293	115	-1059 Dec 08	20:53:19	26369	-37824	047	N	t-	-1.3737	0.3806	-0.7051	183.0	-	-	21N	156E
2294	115	-1058 May 05	06:32:06	26362	-37819	014	N	-a	-1.0456	0.9093	-0.0311	232.9	-	-	14S	10E
2295	115	-1058 Jun 03	13:02:23	26360	-37818	052	N	a-	1.4374	0.1916	-0.7511	115.1	-	-	19S	88W
2296	115	-1058 Oct 29	01:15:45	26353	-37813	019	N*	-t	0.9917	1.0860	-0.0086	291.6	-	-	11N	88E
2297	115	-1057 Apr 24	22:54:00	26344	-37807	024	T	-a	-0.3242	2.2470	1.2790	321.1	205.4	74.0	9S	126E
2298	115	-1057 Oct 18	05:04:40	26335	-37801	029	T+	-p	0.2653	2.3924	1.3501	345.8	215.8	83.0	6N	31E
2299	115	-1056 Apr 13	10:10:51	26326	-37795	034	T	t-	0.4553	2.0342	1.0111	336.9	202.1	16.7	5S	42W
2300	115	-1056 Oct 06	16:15:04	26317	-37789	039	T	a-	-0.4375	2.0484	1.0617	314.2	193.3	36.8	1N	136W
2301	116	-1055 Apr 02	14:09:22	26308	-37783	044	N	t-	1.2477	0.6033	-0.4656	227.8	-	-	0N	101W
2302	116	-1055 Aug 27	22:28:12	26301	-37778	011	N	-a	1.5170	0.0564	-0.9081	63.8	-	-	12S	133E
2303	116	-1055 Sep 26	07:47:34	26299	-37777	049	N	a-	-1.0963	0.8283	-0.1361	225.2	-	-	4S	8W
2304	116	-1054 Feb 20	23:10:32	26292	-37772	016	P	-t	-0.9073	1.2238	0.1630	297.9	95.2	-	13N	126E
2305	116	-1054 Aug 17	12:13:58	26283	-37766	021	P	-a	0.8627	1.2777	0.2722	283.4	115.1	-	16S	73W
2306	116	-1053 Feb 10	06:26:30	26274	-37760	026	T-	pp	-0.1410	2.6036	1.5949	338.5	218.0	96.6	17N	18E
2307	116	-1053 Aug 06	19:31:19	26265	-37754	031	T+	pp	0.1368	2.6356	1.5787	363.2	229.0	99.9	20S	177E
2308	116	-1052 Jan 30	20:06:28	26256	-37748	036	P	a-	0.5659	1.8037	0.8353	299.4	176.5	-	21N	172E
2309	116	-1052 Jul 25	20:38:30	26248	-37742	041	P	t-	-0.6120	1.7745	0.6961	346.6	185.7	-	23S	160E
2310	116	-1052 Dec 21	01:30:45	26240	-37737	008	N	-a	-1.4401	0.2111	-0.7804	124.0	-	-	22N	87E
2311	116	-1051 Jan 19	12:12:29	26239	-37736	046	N	a-	1.2372	0.5719	-0.3966	193.1	-	-	24N	70W
2312	116	-1051 Jun 15	09:29:14	26231	-37731	013	N	-t	1.4154	0.2705	-0.7486	148.9	-	-	21S	35W
2313	116	-1051 Jul 14	21:57:25	26230	-37730	051	N	t-	-1.3312	0.4408	-0.6097	193.4	-	-	25S	140E
2314	116	-1051 Dec 10	12:18:49	26222	-37725	018	P	-a	-0.8150	1.3863	0.3390	301.1	129.9	-	21N	76W
2315	116	-1050 Jun 04	21:00:16	26214	-37719	023	P	-a	0.5962	1.7464	0.7811	302.5	175.8	-	20S	152E
2316	116	-1050 Nov 29	15:45:33	26205	-37713	028	T-	pp	-0.1406	2.6491	1.5515	373.4	231.0	99.0	20N	129W
2317	116	-1049 May 25	13:29:02	26196	-37707	033	T-	p-	-0.1603	2.5333	1.5938	315.2	209.0	93.9	19S	96W
2318	116	-1049 Nov 18	14:44:07	26187	-37701	038	P	t-	0.5354	1.9275	0.8245	355.3	196.7	-	18N	115W
2319	116	-1048 May 14	06:23:49	26178	-37695	043	P	a-	-0.9119	1.1631	0.2060	261.9	97.8	-	17S	11E
2320	116	-1048 Nov 06	16:52:30	26169	-37689	048	N	a-	1.1874	0.7089	-0.3502	234.6	-	-	15N	147W
2321	117	-1047 Apr 04	07:28:45	26162	-37684	015	P	-t	0.9667	1.1045	0.0642	283.1	60.1	-	1S	1W
2322	117	-1047 Sep 27	15:12:18	26153	-37678	020	P	-a	-0.9134	1.1688	0.1947	261.0	94.3	-	3S	120W
2323	117	-1046 Mar 24	09:42:42	26144	-37672	025	T+	pp	0.2302	2.4742	1.3977	371.4	229.4	91.1	3N	34W
2324	117	-1046 Sep 17	07:15:14	26135	-37666	030	T-	-p	-0.2280	2.4220	1.4570	316.5	206.0	86.6	7S	0W
2325	117	-1045 Mar 13	10:15:52	26126	-37660	035	P	t-	-0.5117	1.9514	0.8871	348.4	199.6	-	7N	42W
2326	117	-1045 Sep 06	21:43:55	26117	-37654	040	P	a-	0.4854	1.9667	0.9678	318.7	191.5	-	10S	143E
2327	117	-1044 Feb 01	04:47:06	26110	-37649	007	N	-a	1.4873	0.1268	-0.8693	98.3	-	-	21N	42E
2328	117	-1044 Mar 01	16:10:21	26109	-37648	045	N	a-	-1.2172	0.6324	-0.3833	216.2	-	-	10N	130W
2329	117	-1044 Aug 26	06:23:43	26100	-37642	050	N	t-	1.2627	0.5673	-0.4848	214.5	-	-	13S	13E
2330	117	-1043 Jan 20	19:30:29	26092	-37637	017	P	-a	0.7644	1.4377	0.4725	277.9	140.0	-	23N	180W
2331	117	-1043 Jul 16	16:04:05	26084	-37631	022	P	-t	-0.8746	1.2903	0.2167	311.8	111.5	-	24S	132W
2332	117	-1042 Jan 10	11:23:11	26075	-37625	027	T+	-p	0.0895	2.6831	1.7044	323.8	212.0	97.2	23N	59W
2333	117	-1042 Jul 05	18:49:59	26066	-37619	032	T-	pp	-0.0981	2.6952	1.6609	356.4	227.6	102.0	24S	175W
2334	117	-1042 Dec 30	23:44:18	26057	-37613	037	P	a-	-0.6266	1.7233	0.6932	316.7	173.3	-	23N	114E
2335	117	-1041 Jun 25	04:45:24	26048	-37607	042	P	a-	0.6650	1.6265	0.6488	300.4	165.4	-	23S	36E
2336	117	-1041 Dec 20	05:00:07	26039	-37601	047	N	t-	-1.3713	0.3855	-0.7012	184.6	-	-	22N	34E
2337	117	-1040 May 15	13:57:58	26032	-37596	014	N	-a	-1.1173	0.7775	-0.1626	218.4	-	-	17S	103W
2338	117	-1040 Jun 13	20:27:44	26030	-37595	052	N	a-	1.3658	0.3228	-0.6197	147.5	-	-	21S	159E
2339	117	-1040 Nov 08	09:19:55	26023	-37590	019	N*	-t	0.9949	1.0798	-0.0141	290.5	-	-	15N	34W
2340	117	-1039 May 05	06:09:53	26014	-37584	024	T	-a	-0.3985	2.1116	1.1418	318.5	199.9	55.3	13S	14E

Cat Num	Canon Plate	Calendar Date	TD of Greatest Eclipse	ΔT s	Luna Num	Saros Num	Ecl. Type	QSE	Gamma	Pen. Mag.	Um. Mag.	Eclipse Phase Durations Pen. m	Par. m	Total m	Greatest in Zenith Lat.	Lng.
2341	118	-1039 Oct 28	13:27:15	26005	-37578	029	T+	-p	0.2700	2.3831	1.3422	344.4	214.9	82.0	10N	96W
2342	118	-1038 Apr 24	17:04:12	25997	-37572	034	T	t-	0.3815	2.1705	1.1456	342.9	210.4	58.4	9S	148W
2343	118	-1038 Oct 18	00:52:09	25988	-37566	039	T	a-	-0.4276	2.0665	1.0799	314.1	193.9	41.6	6N	93E
2344	118	-1037 Apr 13	20:43:24	25979	-37560	044	N	t-	1.1743	0.7376	-0.3305	248.9	-	-	4S	158E
2345	118	-1037 Oct 07	16:23:52	25970	-37554	049	N	a-	-1.0824	0.8551	-0.1119	228.3	-	-	1N	139W
2346	118	-1036 Mar 03	06:27:18	25963	-37549	016	P	-t	-0.9527	1.1374	0.0828	288.7	68.4	-	10N	15E
2347	118	-1036 Aug 27	19:58:26	25954	-37543	021	P	-a	0.9060	1.2016	0.1896	278.5	97.5	-	13S	169E
2348	118	-1035 Feb 20	14:18:18	25945	-37537	026	T-	pp	-0.1771	2.5342	1.5319	336.0	216.3	93.8	14N	102W
2349	118	-1035 Aug 17	02:40:40	25936	-37531	031	T+	pp	0.1904	2.5403	1.4773	363.6	227.3	95.0	17S	69E
2350	118	-1034 Feb 10	04:22:42	25927	-37525	036	P	a-	0.5363	1.8557	0.8917	300.9	180.4	-	18N	47E
2351	118	-1034 Aug 06	03:25:47	25919	-37519	041	P	t-	-0.5478	1.8939	0.8123	353.0	196.6	-	20S	57E
2352	118	-1033 Jan 01	10:09:48	25911	-37514	008	N	-a	-1.4503	0.1927	-0.7992	119.0	-	-	22N	43W
2353	118	-1033 Jan 30	20:38:31	25910	-37513	046	N	a-	1.2179	0.6066	-0.3604	198.3	-	-	21N	163E
2354	118	-1033 Jun 26	16:22:30	25902	-37508	013	N	-t	1.4849	0.1419	-0.8752	108.7	-	-	22S	139W
2355	118	-1033 Jul 26	04:52:29	25901	-37507	051	N	t-	-1.2622	0.5673	-0.4831	216.1	-	-	23S	35E
2356	118	-1033 Dec 21	20:42:05	25894	-37502	018	P	-a	-0.8215	1.3753	0.3263	301.3	128.0	-	23N	158E
2357	118	-1032 Jun 15	04:16:03	25885	-37496	023	P	-a	0.6690	1.6119	0.6485	294.6	163.2	-	22S	42E
2358	118	-1032 Dec 09	23:50:59	25876	-37490	028	T-	pp	-0.1446	2.6423	1.5439	374.0	231.2	98.8	22N	109E
2359	118	-1031 Jun 04	20:52:22	25867	-37484	033	T-	pp	-0.0842	2.6729	1.7333	316.4	210.9	98.0	21S	152E
2360	118	-1031 Nov 28	22:48:45	25858	-37478	038	P	t-	0.5317	1.9337	0.8320	355.0	197.1	-	20N	123E
2361	119	-1030 May 25	13:39:43	25850	-37472	043	P	a-	-0.8389	1.2983	0.3388	273.4	123.4	-	20S	100W
2362	119	-1030 Nov 18	01:15:57	25841	-37466	048	N	a-	1.1827	0.7163	-0.3404	234.8	-	-	18N	86E
2363	119	-1029 Apr 15	14:18:49	25833	-37461	015	N	-t	1.0425	0.9658	-0.0755	269.5	-	-	5S	106W
2364	119	-1029 Oct 08	23:46:52	25825	-37455	020	P	-a	-0.9259	1.1463	0.1713	258.7	88.6	-	1N	110E
2365	119	-1028 Apr 03	16:21:38	25816	-37449	025	T	pt	0.3025	2.3403	1.2660	368.3	224.8	78.4	1S	136W
2366	119	-1028 Sep 27	15:42:52	25807	-37443	030	T-	-p	-0.2493	2.3847	1.4161	316.5	205.0	84.0	2S	129W
2367	119	-1027 Mar 23	17:12:00	25798	-37437	035	T	t-	-0.4478	2.0660	1.0072	351.7	207.6	13.9	3N	148W
2368	119	-1027 Sep 17	05:49:58	25789	-37431	040	T	p-	0.4591	2.0181	1.0130	322.1	194.8	17.5	6S	19E
2369	119	-1026 Feb 11	12:52:12	25782	-37426	007	N	-a	1.5125	0.0775	-0.9125	76.9	-	-	19N	80W
2370	119	-1026 Mar 12	23:39:59	25781	-37425	045	N	a-	-1.1629	0.7288	-0.2804	228.8	-	-	6N	115E
2371	119	-1026 Sep 06	13:56:06	25772	-37419	050	N	t-	1.2285	0.6332	-0.4252	226.2	-	-	9S	102W
2372	119	-1025 Feb 01	03:56:03	25764	-37414	017	P	-a	0.7871	1.3943	0.4325	274.7	134.7	-	21N	53E
2373	119	-1025 Jul 27	22:46:49	25756	-37408	022	P	-t	-0.9422	1.1672	0.0917	300.4	73.8	-	23S	126E
2374	119	-1024 Jan 21	19:52:46	25747	-37402	027	T+	-p	0.1068	2.6508	1.6728	324.1	212.1	96.6	22N	173E
2375	119	-1024 Jul 16	01:47:20	25738	-37396	032	T-	pp	-0.1679	2.5666	1.5333	353.8	224.7	96.9	23S	80E
2376	119	-1023 Jan 10	08:03:23	25729	-37390	037	P	a-	-0.6172	1.7411	0.7102	318.7	175.4	-	23N	11W
2377	119	-1023 Jul 05	12:03:02	25721	-37384	042	P	a-	0.5929	1.7582	0.7818	306.1	177.0	-	23S	75W
2378	119	-1023 Dec 30	13:01:34	25712	-37378	047	N	t-	-1.3650	0.3971	-0.6897	187.5	-	-	23N	87W
2379	119	-1022 May 26	21:20:57	25704	-37373	014	N	-a	-1.1915	0.6416	-0.2991	201.3	-	-	20S	144E
2380	119	-1022 Jun 25	03:55:10	25703	-37372	052	N	a-	1.2953	0.4523	-0.4906	172.3	-	-	22S	47E
2381	120	-1022 Nov 19	17:27:58	25696	-37367	019	N*	-t	0.9955	1.0779	-0.0145	289.6	-	-	18N	158W
2382	120	-1021 May 16	13:21:16	25687	-37361	024	P	-a	-0.4758	1.9708	0.9986	314.8	192.6	-	17S	95W
2383	120	-1021 Nov 08	21:56:14	25678	-37355	029	T+	-p	0.2706	2.3810	1.3421	343.1	214.3	81.8	14N	135E
2384	120	-1020 May 04	23:51:22	25669	-37349	034	T	pp	0.3035	2.3147	1.2876	348.3	217.4	78.5	13S	108E
2385	120	-1020 Oct 28	09:35:08	25660	-37343	039	T	a-	-0.4222	2.0763	1.0901	313.8	194.0	44.0	10N	39W
2386	120	-1019 Apr 24	03:11:14	25652	-37337	044	N	t-	1.0957	0.8813	-0.1861	268.4	-	-	8S	59E
2387	120	-1019 Oct 18	01:06:28	25643	-37331	049	N	a-	-1.0742	0.8715	-0.0980	230.3	-	-	5N	88E
2388	120	-1018 Mar 14	13:36:34	25636	-37326	016	N*	-t	-1.0048	1.0387	-0.0096	277.6	-	-	6N	94W
2389	120	-1018 Sep 08	03:50:12	25627	-37320	021	P	-a	0.9422	1.1385	0.1199	274.3	78.6	-	9S	49E
2390	120	-1017 Mar 03	22:03:38	25618	-37314	026	T-	-p	-0.2192	2.4536	1.4577	333.1	214.0	89.6	10N	140E
2391	120	-1017 Aug 28	09:57:18	25609	-37308	031	T+	-p	0.2373	2.4573	1.3884	363.7	225.2	88.9	14S	42W
2392	120	-1016 Feb 21	12:32:45	25601	-37302	036	P	a-	0.5015	1.9175	0.9577	302.8	184.7	-	15N	77W
2393	120	-1016 Aug 16	10:22:07	25592	-37296	041	P	t-	-0.4908	2.0000	0.9154	358.0	204.8	-	18S	48W
2394	120	-1015 Jan 11	18:42:52	25585	-37291	008	N	-a	-1.4648	0.1661	-0.8260	111.0	-	-	22N	171W
2395	120	-1015 Feb 10	04:57:06	25583	-37290	046	N	a-	1.1931	0.6513	-0.3141	204.8	-	-	19N	37E
2396	120	-1015 Jul 06	23:21:46	25576	-37285	013	Ne	-t	1.5502	0.0212	-0.9942	42.3	-	-	22S	115E
2397	120	-1015 Aug 05	11:56:21	25574	-37284	051	N	t-	-1.1992	0.6829	-0.3676	233.8	-	-	21S	72W
2398	120	-1014 Jan 01	04:59:21	25567	-37279	018	P	-a	-0.8320	1.3566	0.3064	300.9	124.8	-	23N	33E
2399	120	-1014 Jun 26	11:36:32	25558	-37273	023	P	-a	0.7383	1.4841	0.5220	286.2	149.0	-	23S	69W
2400	120	-1014 Dec 21	07:50:44	25549	-37267	028	T-	pp	-0.1525	2.6278	1.5294	374.3	231.2	98.2	23N	11W

Cat Num	Canon Plate	Calendar Date	TD of Greatest Eclipse	ΔT s	Luna Num	Saros Num	Ecl. Type	QSE	Gamma	Pen. Mag.	Um. Mag.	Eclipse Phase Durations Pen. m	Par. m	Total m	Greatest in Zenith Lat.	Lng.
2401	121	-1013 Jun 16	04:19:04	25541	-37261	033	T-	pp	-0.0111	2.8075	1.8671	316.9	211.5	99.5	23S	40E
2402	121	-1013 Dec 10	06:51:51	25532	-37255	038	P	t-	0.5268	1.9415	0.8419	354.7	197.7	-	23N	2E
2403	121	-1012 Jun 04	20:55:08	25523	-37249	043	P	a-	-0.7660	1.4334	0.4709	283.7	143.0	-	22S	150E
2404	121	-1012 Nov 28	09:40:43	25514	-37243	048	N	a-	1.1785	0.7225	-0.3310	234.8	-	-	21N	41W
2405	121	-1011 Apr 25	21:01:20	25507	-37238	015	N	-t	1.1232	0.8185	-0.2241	252.7	-	-	9S	151E
2406	121	-1011 Oct 19	08:28:43	25498	-37232	020	P	-a	-0.9331	1.1335	0.1579	257.3	85.1	-	6N	23W
2407	121	-1010 Apr 14	22:50:51	25490	-37226	025	T	-t	0.3826	2.1924	1.1200	364.0	218.0	55.4	5S	124E
2408	121	-1010 Oct 09	00:18:52	25481	-37220	030	T-	-p	-0.2635	2.3604	1.3883	316.7	204.3	82.0	2N	100E
2409	121	-1009 Apr 04	00:00:30	25472	-37214	035	T	t-	-0.3766	2.1939	1.1405	354.7	214.8	58.6	2S	107E
2410	121	-1009 Sep 28	14:03:13	25463	-37208	040	T	p-	0.4395	2.0571	1.0461	324.9	197.2	32.7	2S	106W
2411	121	-1008 Feb 22	20:50:37	25456	-37203	007	Ne	-a	1.5437	0.0173	-0.9668	36.4	-	-	15N	159E
2412	121	-1008 Mar 23	07:04:02	25455	-37202	045	N	a-	-1.1029	0.8356	-0.1672	241.2	-	-	2N	2E
2413	121	-1008 Sep 16	21:36:04	25446	-37196	050	N	t-	1.2009	0.6870	-0.3777	235.4	-	-	5S	141E
2414	121	-1007 Feb 11	12:15:04	25439	-37191	017	P	-a	0.8155	1.3406	0.3823	270.7	127.6	-	18N	73W
2415	121	-1007 Aug 07	05:36:31	25430	-37185	022	N*	-t	-1.0043	1.0543	-0.0232	288.8	-	-	20S	23E
2416	121	-1006 Feb 01	04:15:41	25421	-37179	027	T+	-p	0.1294	2.6091	1.6319	324.4	212.1	95.6	20N	47E
2417	121	-1006 Jul 27	08:52:42	25412	-37173	032	T-	pp	-0.2322	2.4484	1.4157	350.7	221.0	89.6	22S	27W
2418	121	-1005 Jan 21	16:14:29	25404	-37167	037	P	a-	-0.6017	1.7696	0.7382	321.2	178.6	-	21N	134W
2419	121	-1005 Jul 16	19:28:37	25395	-37161	042	P	a-	0.5263	1.8798	0.9043	310.6	185.8	-	23S	173E
2420	121	-1004 Jan 10	20:56:40	25386	-37155	047	N	t-	-1.3537	0.4176	-0.6688	192.3	-	-	22N	154E
2421	122	-1004 Jun 06	04:45:11	25379	-37150	014	N	-a	-1.2645	0.5084	-0.4334	181.8	-	-	23S	32E
2422	122	-1004 Jul 05	11:27:15	25377	-37149	052	N	a-	1.2284	0.5757	-0.3682	192.0	-	-	23S	67W
2423	122	-1004 Nov 30	01:36:37	25370	-37144	019	N*	-t	0.9962	1.0756	-0.0147	288.7	-	-	21N	80E
2424	122	-1003 May 26	20:31:01	25362	-37138	024	P	-a	-0.5539	1.8291	0.8539	310.3	183.3	-	20S	155E
2425	122	-1003 Nov 19	06:27:28	25353	-37132	029	T+	-p	0.2699	2.3810	1.3447	341.8	213.8	81.9	18N	6E
2426	122	-1002 May 16	06:33:51	25344	-37126	034	T+	pp	0.2225	2.4648	1.4350	352.9	223.1	91.6	16S	5E
2427	122	-1002 Nov 08	18:22:14	25335	-37120	039	T	a-	-0.4197	2.0805	1.0952	313.2	193.9	45.0	14N	173W
2428	122	-1001 May 05	09:34:43	25327	-37114	044	N*	t-	1.0136	1.0320	-0.0352	286.2	-	-	12S	39W
2429	122	-1001 Oct 29	09:53:14	25318	-37108	049	N	a-	-1.0694	0.8812	-0.0903	231.6	-	-	9N	45W
2430	122	-1000 Mar 24	20:39:25	25311	-37103	016	N	-t	-1.0626	0.9294	-0.1128	264.4	-	-	1N	158E
2431	122	-1000 Sep 18	11:50:01	25302	-37097	021	P	-a	0.9711	1.0887	0.0637	271.0	57.9	-	5S	73W
2432	122	-0999 Mar 14	05:40:48	25293	-37091	026	T-	-p	-0.2691	2.3589	1.3692	329.8	210.9	83.1	6N	24E
2433	122	-0999 Sep 07	17:22:36	25284	-37085	031	T	-p	0.2764	2.3885	1.3136	363.7	223.1	82.3	10S	155W
2434	122	-0998 Mar 03	20:33:44	25276	-37079	036	T	a-	0.4585	1.9942	1.0386	304.9	189.3	29.0	11N	161E
2435	122	-0998 Aug 27	17:29:01	25267	-37073	041	T	t-	-0.4421	2.0910	1.0032	361.8	210.7	9.4	14S	156W
2436	122	-0997 Jan 23	03:07:35	25260	-37068	008	N	-a	-1.4858	0.1277	-0.8644	97.9	-	-	20N	62E
2437	122	-0997 Feb 21	13:05:53	25258	-37067	046	N	a-	1.1607	0.7100	-0.2539	212.9	-	-	15N	86W
2438	122	-0997 Aug 16	19:11:41	25250	-37061	051	N	t-	-1.1443	0.7839	-0.2669	247.3	-	-	18S	178E
2439	122	-0996 Jan 12	13:10:30	25242	-37056	018	P	-a	-0.8466	1.3302	0.2793	299.7	120.0	-	22N	90W
2440	122	-0996 Jul 06	19:01:42	25234	-37050	023	P	-a	0.8044	1.3624	0.4010	277.4	132.7	-	23S	179E
2441	123	-0996 Dec 31	15:47:08	25225	-37044	028	T-	pp	-0.1629	2.6084	1.5107	374.5	231.1	97.4	24N	130W
2442	123	-0995 Jun 26	11:47:30	25216	-37038	033	T+	pp	0.0604	2.7177	1.7758	316.8	211.2	98.6	24S	73W
2443	123	-0995 Dec 20	14:53:25	25208	-37032	038	P	t-	0.5198	1.9529	0.8563	354.4	198.6	-	24N	118W
2444	123	-0994 Jun 16	04:10:15	25199	-37026	043	P	a-	-0.6933	1.5687	0.6026	293.1	158.9	-	23S	40E
2445	123	-0994 Dec 09	18:06:28	25190	-37020	048	N	a-	1.1741	0.7287	-0.3211	234.8	-	-	23N	168W
2446	123	-0993 May 07	03:39:20	25183	-37015	015	N	-t	1.2066	0.6663	-0.3781	232.3	-	-	12S	49E
2447	123	-0993 Jun 05	15:41:34	25181	-37014	053	Nb	t-	-1.5151	0.0869	-0.9312	85.8	-	-	23S	133W
2448	123	-0993 Oct 30	17:14:48	25174	-37009	020	P	-a	-0.9375	1.1255	0.1496	256.3	82.9	-	10N	156W
2449	123	-0992 Apr 25	05:16:54	25165	-37003	025	P	-t	0.4653	2.0398	0.9692	358.3	208.8	-	9S	25E
2450	123	-0992 Oct 19	08:59:46	25157	-36997	030	T	-p	-0.2736	2.3435	1.3681	317.0	203.9	80.4	7N	32W
2451	123	-0991 Apr 14	06:44:32	25148	-36991	035	T	t-	-0.3006	2.3307	1.2824	357.0	220.8	79.0	6S	4E
2452	123	-0991 Oct 08	22:22:39	25139	-36985	040	T	p-	0.4255	2.0856	1.0691	327.3	199.0	39.8	3N	127E
2453	123	-0990 Apr 03	14:22:14	25131	-36979	045	N	a-	-1.0370	0.9535	-0.0433	253.3	-	-	2S	110W
2454	123	-0990 Sep 28	05:23:39	25122	-36973	050	N	t-	1.1798	0.7207	-0.3410	242.4			1S	22E
2455	123	-0989 Feb 22	20:24:37	25115	-36968	017	P	-a	0.8522	1.2716	0.3166	265.5	117.2	-	14N	163E
2456	123	-0989 Aug 18	12:38:00	25106	-36962	022	N	-t	-1.0573	0.9579	-0.1217	277.9	-	-	18S	84W
2457	123	-0988 Feb 12	12:29:23	25097	-36956	027	T+	-p	0.1595	2.5533	1.5769	324.4	211.6	93.9	17N	78W
2458	123	-0988 Aug 06	16:07:46	25089	-36950	032	T	-p	-0.2897	2.3427	1.3103	347.3	216.8	80.3	20S	137W
2459	123	-0987 Feb 01	00:16:18	25080	-36944	037	P	a-	-0.5793	1.8109	0.7793	324.3	182.8	-	19N	105E
2460	123	-0987 Jul 27	03:01:33	25071	-36938	042	T	a-	0.4647	1.9928	1.0175	314.1	192.6	20.1	21S	59E

Cat Num	Canon Plate	Calendar Date	TD of Greatest Eclipse	ΔT s	Luna Num	Saros Num	Ecl. Type	QSE	Gamma	Pen. Mag.	Um. Mag.	Eclipse Phase Durations Pen. m	Par. m	Total m	Greatest in Zenith Lat.	Lng.
2461	124	-0986 Jan 21	04:42:12	25063	-36932	047	N	t-	-1.3352	0.4511	-0.6343	199.5	-	-	21N	37E
2462	124	-0986 Jun 17	12:09:40	25056	-36927	014	N	-a	-1.3370	0.3763	-0.5675	158.6	-	-	24S	80W
2463	124	-0986 Jul 16	19:04:31	25054	-36926	052	N	a-	1.1653	0.6923	-0.2534	208.0	-	-	22S	178E
2464	124	-0986 Dec 11	09:45:47	25047	-36921	019	N*	-t	0.9972	1.0721	-0.0151	287.5	-	-	23N	43W
2465	124	-0985 Jun 07	03:39:17	25038	-36915	024	P	-a	-0.6322	1.6872	0.7084	304.7	171.6	-	22S	47E
2466	124	-0985 Nov 30	15:00:47	25030	-36909	029	T+	-p	0.2685	2.3819	1.3488	340.5	213.4	82.0	21N	123W
2467	124	-0984 May 26	13:13:32	25021	-36903	034	T+	pp	0.1399	2.6179	1.5850	356.6	227.2	99.8	19S	96W
2468	124	-0984 Nov 19	03:12:56	25012	-36897	039	T	a-	-0.4196	2.0799	1.0959	312.5	193.6	45.1	17N	54E
2469	124	-0983 May 15	15:53:33	25004	-36891	044	P	t-	0.9274	1.1901	0.1229	302.5	85.2	-	15S	136W
2470	124	-0983 Nov 08	18:44:41	24995	-36885	049	N	a-	-1.0688	0.8833	-0.0899	232.0	-	-	13N	179W
2471	124	-0982 Apr 05	03:36:57	24988	-36880	016	N	-t	-1.1257	0.8108	-0.2256	248.8	-	-	3S	51E
2472	124	-0982 Sep 29	19:56:29	24979	-36874	021	P	-a	0.9937	1.0502	0.0191	268.5	32.0	-	0S	163E
2473	124	-0981 Mar 25	13:12:57	24970	-36868	026	T	-p	-0.3242	2.2550	1.2711	325.9	206.8	73.6	2N	92W
2474	124	-0981 Sep 19	00:56:09	24962	-36862	031	T	-p	0.3083	2.3328	1.2524	363.7	221.1	75.6	5S	89E
2475	124	-0980 Mar 14	04:29:13	24953	-36856	036	T	a-	0.4109	2.0798	1.1279	307.2	193.9	51.5	7N	40E
2476	124	-0980 Sep 07	00:44:39	24944	-36850	041	T	t-	-0.4002	2.1693	1.0787	364.7	215.2	45.3	10S	93E
2477	124	-0979 Feb 02	11:24:38	24937	-36845	008	N	-a	-1.5123	0.0789	-0.9129	77.5	-	-	18N	63W
2478	124	-0979 Mar 03	21:07:10	24936	-36844	046	N	a-	1.1227	0.7790	-0.1837	221.8	-	-	11N	152E
2479	124	-0979 Aug 27	02:37:29	24927	-36838	051	N	h-	-1.0969	0.8710	-0.1802	257.7	-	-	15S	65E
2480	124	-0978 Jan 22	21:11:35	24920	-36833	018	P	-a	-0.8686	1.2899	0.2388	297.4	112.0	-	21N	150E
2481	125	-0978 Jul 18	02:34:21	24911	-36827	023	P	-a	0.8649	1.2513	0.2901	268.6	114.4	-	22S	65E
2482	125	-0977 Jan 11	23:34:51	24903	-36821	028	T-	pp	-0.1798	2.5767	1.4804	374.3	230.7	95.9	23N	113E
2483	125	-0977 Jul 07	19:21:12	24894	-36815	033	T+	-p	0.1276	2.5955	1.6516	316.3	209.9	95.8	24S	173E
2484	125	-0977 Dec 31	22:49:31	24885	-36809	038	P	t-	0.5079	1.9728	0.8799	354.4	200.2	-	24N	123E
2485	125	-0976 Jun 26	11:27:46	24877	-36803	043	P	a-	-0.6233	1.6992	0.7289	301.4	171.7	-	24S	70W
2486	125	-0976 Dec 20	02:30:03	24868	-36797	048	N	a-	1.1672	0.7391	-0.3063	235.2	-	-	25N	66E
2487	125	-0975 May 17	10:10:53	24861	-36792	015	N	-t	1.2939	0.5073	-0.5393	206.5	-	-	16S	51W
2488	125	-0975 Jun 15	22:28:08	24859	-36791	053	N	t-	-1.4416	0.2242	-0.7984	136.7	-	-	24S	124E
2489	125	-0975 Nov 10	02:05:51	24852	-36786	020	P	-a	-0.9383	1.1240	0.1483	255.9	82.4	-	14N	70E
2490	125	-0974 May 06	11:36:49	24844	-36780	025	P	-t	0.5525	1.8792	0.8099	351.0	196.4	-	13S	72W
2491	125	-0974 Oct 30	17:46:10	24835	-36774	030	T	-p	-0.2788	2.3354	1.3571	317.5	203.8	79.6	11N	166W
2492	125	-0973 Apr 25	13:24:18	24826	-36768	035	T-	pp	-0.2204	2.4756	1.4320	358.6	225.3	92.1	10S	98W
2493	125	-0973 Oct 20	06:48:27	24818	-36762	040	T	p-	0.4178	2.1023	1.0806	329.3	200.2	43.0	7N	1W
2494	125	-0972 Apr 13	21:36:22	24809	-36756	045	P	a-	-0.9666	1.0798	0.0886	264.7	67.2	-	6S	139E
2495	125	-0972 Oct 08	13:17:41	24800	-36750	050	N	t-	1.1643	0.7599	-0.3161	247.7	-	-	3N	98W
2496	125	-0971 Mar 05	04:27:44	24793	-36745	017	P	-a	0.8943	1.1927	0.2408	259.2	103.3	-	11N	41E
2497	125	-0971 Aug 28	19:48:31	24785	-36739	022	N	-t	-1.1035	0.8743	-0.2074	267.5	-	-	14S	167E
2498	125	-0970 Feb 22	20:33:56	24776	-36733	027	T+	-p	0.1968	2.4845	1.5089	324.2	210.8	90.9	14N	159E
2499	125	-0970 Aug 17	23:32:58	24767	-36727	032	T	-p	-0.3398	2.2507	1.2185	343.8	212.4	69.3	17S	111E
2500	125	-0969 Feb 12	08:08:53	24759	-36721	037	P	h-	-0.5499	1.8649	0.8332	327.9	187.8	-	16N	14W
2501	126	-0969 Aug 07	10:42:59	24750	-36715	042	T	a-	0.4096	2.0940	1.1187	316.6	197.6	50.6	19S	57W
2502	126	-0968 Feb 01	12:19:20	24741	-36709	047	N	t-	-1.3102	0.4962	-0.5879	208.7	-	-	19N	78W
2503	126	-0968 Jun 27	19:37:55	24734	-36704	014	N	-a	-1.4062	0.2506	-0.6957	131.2	-	-	25S	167E
2504	126	-0968 Jul 27	02:48:30	24733	-36703	052	N	a-	1.1076	0.7992	-0.1486	221.1	-	-	21S	61E
2505	126	-0968 Dec 21	17:52:05	24726	-36698	019	N*	-t	1.0010	1.0633	-0.0202	285.7	-	-	24N	165W
2506	126	-0967 Jun 17	10:47:18	24717	-36692	024	P	-a	-0.7098	1.5470	0.5639	298.1	157.3	-	24S	61W
2507	126	-0967 Dec 10	23:33:23	24708	-36686	029	T+	-p	0.2685	2.3799	1.3507	339.1	212.9	82.0	23N	109E
2508	126	-0966 Jun 06	19:52:04	24700	-36680	034	T+	pp	0.0570	2.7718	1.7352	359.3	229.6	104.2	21S	163E
2509	126	-0966 Nov 30	12:03:16	24691	-36674	039	T	p-	-0.4188	2.0804	1.0982	311.9	193.4	45.6	20N	80W
2510	126	-0965 May 26	22:11:59	24683	-36668	044	P	t-	0.8405	1.3498	0.2820	316.7	126.2	-	18S	128E
2511	126	-0965 Nov 20	03:37:47	24674	-36662	049	N	a-	-1.0693	0.8829	-0.0916	232.2	-	-	17N	46E
2512	126	-0964 Apr 15	10:29:50	24667	-36657	016	N	-t	-1.1930	0.6845	-0.3463	230.5	-	-	7S	55W
2513	126	-0964 Oct 10	04:09:32	24658	-36651	021	N*	-a	1.0103	1.0227	-0.0142	266.9	-	-	4N	38E
2514	126	-0963 Apr 04	20:38:42	24650	-36645	026	T	-p	-0.3854	2.1398	1.1616	321.2	201.5	58.8	2S	154E
2515	126	-0963 Sep 29	08:38:22	24641	-36639	031	T	-p	0.3322	2.2915	1.2058	363.9	219.4	69.5	1S	28W
2516	126	-0962 Mar 25	12:15:12	24632	-36633	036	T	p-	0.3550	2.1806	1.2322	309.6	198.5	67.1	2N	79W
2517	126	-0962 Sep 18	08:11:39	24624	-36627	041	T	t-	-0.3674	2.2309	1.1376	366.7	218.2	58.7	6S	21W
2518	126	-0961 Feb 13	19:32:15	24617	-36622	008	Ne	-a	-1.5457	0.0174	-0.9741	36.7	-	-	15N	174E
2519	126	-0961 Mar 15	04:59:15	24615	-36621	046	N	a-	1.0778	0.8609	-0.1008	231.6	-	-	7N	31E
2520	126	-0961 Sep 07	10:13:45	24607	-36615	051	N	h-	-1.0571	0.9443	-0.1072	265.5	-	-	11S	50W

Cat Num	Canon Plate	Calendar Date	TD of Greatest Eclipse	ΔT s	Luna Num	Saros Num	Ecl. Type	QSE	Gamma	Pen. Mag.	Um. Mag.	Eclipse Phase Durations Pen. m	Par. m	Total m	Greatest in Zenith Lat.	Lng.
2521	127	-0960 Feb 03	05:04:24	24599	-36610	018	P	-h	-0.8965	1.2386	0.1876	294.0	100.3	-	18N	31E
2522	127	-0960 Jul 28	10:13:34	24591	-36604	023	P	-a	0.9205	1.1496	0.1879	259.8	93.3	-	21S	51W
2523	127	-0959 Jan 22	07:16:03	24582	-36598	028	T-	pp	-0.2019	2.5350	1.4407	373.8	230.0	93.6	21N	3W
2524	127	-0959 Jul 18	02:58:09	24574	-36592	033	T+	-p	0.1920	2.4788	1.5322	315.3	207.8	91.1	23S	58E
2525	127	-0958 Jan 11	06:41:45	24565	-36586	038	P	t-	0.4923	1.9992	0.9107	354.5	202.2	-	24N	5E
2526	127	-0958 Jul 07	18:47:50	24556	-36580	043	P	a-	-0.5561	1.8250	0.8499	308.6	182.2	-	24S	180E
2527	127	-0958 Dec 31	10:50:16	24548	-36574	048	N	a-	1.1568	0.7556	-0.2847	236.5	-	-	25N	59W
2528	127	-0957 May 28	16:41:40	24541	-36569	015	N	-t	1.3804	0.3499	-0.6993	174.6	-	-	18S	150W
2529	127	-0957 Jun 27	05:17:36	24539	-36568	053	N	t-	-1.3708	0.3565	-0.6711	171.1	-	-	25S	21E
2530	127	-0957 Nov 21	10:58:45	24532	-36563	020	P	-a	-0.9377	1.1247	0.1496	255.6	82.7	-	17N	65W
2531	127	-0956 May 16	17:55:51	24524	-36557	025	P	-t	0.6404	1.7173	0.6491	342.3	180.7	-	16S	169W
2532	127	-0956 Nov 10	02:34:54	24515	-36551	030	T	-p	-0.2819	2.3309	1.3501	318.1	203.8	79.1	15N	61E
2533	127	-0955 May 05	20:02:50	24506	-36545	035	T-	pp	-0.1381	2.6243	1.5851	359.2	228.2	100.2	13S	160E
2534	127	-0955 Oct 30	15:17:56	24498	-36539	040	T	p-	0.4134	2.1128	1.0863	331.0	201.0	44.5	11N	130W
2535	127	-0954 Apr 25	04:46:39	24489	-36533	045	P	a-	-0.8917	1.2147	0.2286	275.4	105.5	-	10S	29E
2536	127	-0954 Oct 19	21:17:59	24481	-36527	050	N	t-	1.1542	0.7808	-0.3000	251.5	-	-	8N	140E
2537	127	-0953 Mar 16	12:22:24	24474	-36522	017	P	-a	0.9439	1.1002	0.1511	251.4	82.9	-	6N	80W
2538	127	-0953 Sep 09	03:09:53	24465	-36516	022	N	-t	-1.1414	0.8057	-0.2778	258.3	-	-	10S	55E
2539	127	-0952 Mar 05	04:29:10	24456	-36510	027	T+	-p	0.2415	2.4022	1.4271	323.7	209.3	86.3	10N	39E
2540	127	-0952 Aug 28	07:09:15	24448	-36504	032	T	-p	-0.3818	2.1736	1.1414	340.5	208.1	57.0	14S	5W
2541	128	-0951 Feb 22	15:49:57	24439	-36498	037	P	t-	-0.5119	1.9347	0.9030	332.1	193.8	-	13N	131W
2542	128	-0951 Aug 17	18:33:21	24431	-36492	042	T	a-	0.3611	2.1832	1.2075	318.5	201.2	64.9	16S	176W
2543	128	-0950 Feb 11	19:45:36	24422	-36486	047	N	t-	-1.2768	0.5566	-0.5258	220.0	-	-	16N	170E
2544	128	-0950 Jul 09	03:09:32	24415	-36481	014	N	-a	-1.4725	0.1306	-0.8189	96.0	-	-	25S	54E
2545	128	-0950 Aug 07	10:39:49	24414	-36480	052	N	a-	1.0558	0.8956	-0.0548	231.9	-	-	18S	58W
2546	128	-0949 Jan 02	01:54:36	24406	-36475	019	N*	-t	1.0087	1.0469	-0.0321	283.0	-	-	25N	75E
2547	128	-0949 Jun 28	17:57:27	24398	-36469	024	P	-a	-0.7847	1.4119	0.4240	290.7	139.8	-	25S	169W
2548	128	-0949 Dec 22	08:05:33	24389	-36463	029	T+	-p	0.2701	2.3749	1.3501	337.6	212.3	81.8	24N	19W
2549	128	-0948 Jun 17	02:29:39	24381	-36457	034	T-	pp	-0.0258	2.8312	1.7905	361.1	230.5	105.0	23S	62E
2550	128	-0948 Dec 10	20:54:20	24372	-36451	039	T	p-	-0.4183	2.0804	1.1004	311.2	193.3	46.0	22N	147E
2551	128	-0947 Jun 06	04:29:37	24364	-36445	044	P	t-	0.7525	1.5118	0.4432	329.2	154.4	-	21S	32E
2552	128	-0947 Nov 30	12:31:51	24355	-36439	049	N	a-	-1.0712	0.8798	-0.0955	232.2	-	-	19N	88W
2553	128	-0946 Apr 26	17:19:24	24348	-36434	016	N	-t	-1.2635	0.5524	-0.4732	208.8	-	-	11S	160W
2554	128	-0946 May 26	05:47:39	24347	-36433	054	N	t-	1.4859	0.1557	-0.8926	118.4	-	-	18S	12E
2555	128	-0946 Oct 21	12:28:00	24339	-36428	021	N*	-a	1.0216	1.0046	-0.0376	266.2	-	-	8N	88W
2556	128	-0945 Apr 16	04:01:26	24331	-36422	026	T	-a	-0.4499	2.0189	1.0457	315.9	194.9	32.3	7S	41E
2557	128	-0945 Oct 10	16:25:57	24322	-36416	031	T	-p	0.3512	2.2590	1.1687	364.1	218.0	63.8	3N	147W
2558	128	-0944 Apr 04	19:57:09	24314	-36410	036	T	p-	0.2955	2.2882	1.3427	311.8	202.6	78.5	2S	163E
2559	128	-0944 Sep 28	15:47:05	24305	-36404	041	T	p-	-0.3411	2.2803	1.1848	368.0	220.4	66.8	2S	137W
2560	128	-0943 Mar 25	12:42:49	24297	-36398	046	N	a-	1.0264	0.9549	-0.0062	242.0	-	-	3N	87W
2561	129	-0943 Sep 17	18:00:29	24288	-36392	051	N*	h-	-1.0247	1.0037	-0.0479	271.2	-	-	7S	169W
2562	129	-0942 Feb 13	12:45:43	24281	-36387	018	P	-h	-0.9330	1.1715	0.1207	288.9	81.4	-	15N	86W
2563	129	-0942 Aug 08	18:01:57	24273	-36381	023	P	-a	0.9691	1.0608	0.0984	251.6	68.2	-	18S	169W
2564	129	-0941 Feb 02	14:46:05	24264	-36375	028	T-	pp	-0.2329	2.4770	1.3851	372.8	228.6	89.7	19N	116W
2565	129	-0941 Jul 29	10:43:00	24255	-36369	033	T+	-p	0.2494	2.3750	1.4251	314.1	205.3	84.8	21S	59W
2566	129	-0940 Jan 22	14:26:36	24247	-36363	038	P	t-	0.4702	2.0374	0.9538	354.9	204.9	-	22N	112W
2567	129	-0940 Jul 18	02:11:29	24238	-36357	043	P	a-	-0.4927	1.9440	0.9636	315.0	190.6	-	23S	68E
2568	129	-0939 Jan 10	19:05:57	24230	-36351	048	N	a-	1.1420	0.7799	-0.2548	238.6	-	-	24N	177E
2569	129	-0939 Jun 07	23:09:51	24223	-36346	015	N	-t	1.4674	0.1917	-0.8606	131.5	-	-	20S	112E
2570	129	-0939 Jul 07	12:08:28	24221	-36345	053	N	t-	-1.3019	0.4856	-0.5473	198.2	-	-	25S	82W
2571	129	-0939 Dec 01	19:52:36	24214	-36340	020	P	-a	-0.9370	1.1255	0.1514	255.4	83.1	-	20N	161E
2572	129	-0938 May 28	00:12:24	24206	-36334	025	P	-t	0.7302	1.5522	0.4847	331.7	160.3	-	19S	96E
2573	129	-0938 Nov 21	11:26:22	24197	-36328	030	T	-p	-0.2825	2.3310	1.3481	318.7	204.0	79.0	18N	73W
2574	129	-0937 May 17	02:40:23	24189	-36322	035	T-	pp	-0.0539	2.7769	1.7416	358.8	229.5	104.2	17S	59E
2575	129	-0937 Nov 10	23:50:10	24180	-36316	040	T	p-	0.4120	2.1173	1.0868	332.5	201.6	44.7	15N	100E
2576	129	-0936 May 05	11:55:35	24172	-36310	045	P	a-	-0.8146	1.3539	0.3724	284.9	131.5	-	14S	80W
2577	129	-0936 Oct 30	05:23:10	24163	-36304	050	N	t-	1.1485	0.7934	-0.2916	254.0	-	-	12N	17E
2578	129	-0935 Mar 26	20:09:43	24156	-36299	017	P	-a	0.9996	0.9969	0.0502	241.9	48.5	-	2N	161E
2579	129	-0935 Apr 25	03:12:37	24155	-36298	055	Nb	a-	-1.5053	0.0681	-0.8771	69.5	-	-	11S	51E
2580	129	-0935 Sep 19	10:41:00	24147	-36293	022	N	-t	-1.1718	0.7507	-0.3344	250.4	-	-	6S	60W

Cat Num	Canon Plate	Calendar Date	TD of Greatest Eclipse	ΔT s	Luna Num	Saros Num	Ecl. Type	QSE	Gamma	Pen. Mag.	Um. Mag.	Eclipse Phase Durations Pen. m	Par. m	Total m	Greatest in Zenith Lat.	Lng.
2581	130	-0934 Mar 16	12:15:44	24139	-36287	027	T	-p	0.2928	2.3080	1.3332	322.6	206.9	79.1	6N	80W
2582	130	-0934 Sep 08	14:55:43	24130	-36281	032	T	-p	-0.4163	2.1102	1.0781	337.3	204.1	43.1	10S	123W
2583	130	-0933 Mar 05	23:21:26	24122	-36275	037	P	t-	-0.4667	2.0176	0.9858	336.5	200.1	-	9N	114E
2584	130	-0933 Aug 29	02:33:22	24113	-36269	042	T	p-	0.3200	2.2588	1.2825	319.7	203.8	73.7	13S	62E
2585	130	-0932 Feb 23	03:03:02	24105	-36263	047	N	t-	-1.2365	0.6297	-0.4508	232.7	-	-	12N	59E
2586	130	-0932 Jul 19	10:45:40	24098	-36258	014	Ne	-a	-1.5347	0.0182	-0.9350	36.2	-	-	24S	61W
2587	130	-0932 Aug 17	18:37:59	24096	-36257	052	P	a-	1.0097	0.9817	0.0284	240.7	36.6	-	16S	179W
2588	130	-0931 Jan 12	09:52:05	24089	-36252	019	N*	-t	1.0211	1.0215	-0.0524	279.2	-	-	24N	44W
2589	130	-0931 Jul 09	01:10:17	24081	-36246	024	P	-a	-0.8567	1.2826	0.2893	282.5	118.2	-	24S	82E
2590	130	-0930 Jan 01	16:32:17	24072	-36240	029	T	-p	0.2769	2.3599	1.3401	335.8	211.4	80.8	24N	146W
2591	130	-0930 Jun 28	09:09:30	24064	-36234	034	T-	pp	-0.1060	2.6864	1.6410	361.9	229.8	102.3	24S	38W
2592	130	-0930 Dec 22	05:41:47	24055	-36228	039	T	p-	-0.4141	2.0866	1.1094	310.8	193.5	47.8	23N	15E
2593	130	-0929 Jun 17	10:49:50	24047	-36222	044	P	t-	0.6663	1.6706	0.6007	339.8	175.3	-	22S	64W
2594	130	-0929 Dec 11	21:23:23	24038	-36216	049	N	a-	-1.0713	0.8799	-0.0957	232.4	-	-	21N	139E
2595	130	-0928 May 07	00:07:15	24031	-36211	016	N	-t	-1.3361	0.4169	-0.6040	182.9	-	-	15S	96E
2596	130	-0928 Jun 05	12:18:41	24030	-36210	054	N	t-	1.3999	0.3123	-0.7336	164.8	-	-	20S	87W
2597	130	-0928 Oct 31	20:50:45	24023	-36205	021	N	-a	1.0288	0.9939	-0.0531	266.1	-	-	13N	144E
2598	130	-0927 Apr 26	11:18:29	24014	-36199	026	P	-a	-0.5199	1.8882	0.9195	309.6	186.5	-	11S	71W
2599	130	-0927 Oct 21	00:21:12	24006	-36193	031	T	-p	0.3634	2.2387	1.1443	364.5	217.1	59.5	8N	92E
2600	130	-0926 Apr 16	03:31:51	23997	-36187	036	T+	p-	0.2296	2.4079	1.4649	313.7	206.2	87.5	6S	47E
2601	131	-0926 Oct 09	23:31:57	23989	-36181	041	T	p-	-0.3221	2.3159	1.2187	368.8	221.7	71.7	3N	105E
2602	131	-0925 Apr 05	20:18:45	23980	-36175	046	P	a-	0.9692	1.0599	0.0989	252.7	68.7	-	1S	157E
2603	131	-0925 Sep 29	01:57:48	23972	-36169	051	N*	a-	-0.9999	1.0493	-0.0023	275.1	-	-	2S	70E
2604	131	-0924 Feb 24	20:18:14	23965	-36164	018	P	-t	-0.9758	1.0930	0.0425	282.2	49.0	-	12N	160E
2605	131	-0924 Aug 19	01:57:19	23956	-36158	023	P	-a	1.0124	0.9820	0.0184	243.7	29.8	-	15S	71E
2606	131	-0923 Feb 12	22:08:44	23948	-36152	028	T-	-p	-0.2695	2.4084	1.3192	371.3	226.5	83.9	16N	132E
2607	131	-0923 Aug 08	18:33:21	23939	-36146	033	T	-p	0.3019	2.2805	1.3270	312.8	202.3	77.1	19S	178W
2608	131	-0922 Feb 01	22:04:15	23931	-36140	038	T	t-	0.4417	2.0870	1.0088	355.5	208.2	15.3	20N	133E
2609	131	-0922 Jul 29	09:40:35	23922	-36134	043	T	a-	-0.4346	2.0534	1.0673	320.4	197.4	39.2	22S	45W
2610	131	-0921 Jan 22	03:16:17	23914	-36128	048	N	a-	1.1225	0.8128	-0.2160	241.7	-	-	23N	55E
2611	131	-0921 Jun 19	05:40:06	23907	-36123	015	Ne	-t	1.5516	0.0389	-1.0168	60.3	-	-	22S	13E
2612	131	-0921 Jul 18	19:04:05	23905	-36122	053	N	t-	-1.2373	0.6070	-0.4316	220.0	-	-	24S	174E
2613	131	-0921 Dec 13	04:44:21	23898	-36117	020	P	-a	-0.9383	1.1225	0.1499	254.9	82.6	-	22N	28E
2614	131	-0920 Jun 07	06:31:43	23890	-36111	025	P	-t	0.8174	1.3919	0.3247	319.8	134.4	-	21S	0W
2615	131	-0920 Dec 01	20:16:55	23881	-36105	030	T	-p	-0.2835	2.3299	1.3455	319.4	204.2	78.9	21N	154E
2616	131	-0919 May 27	09:18:34	23873	-36099	035	T+	pp	0.0307	2.8176	1.7858	357.5	229.1	104.5	19S	42W
2617	131	-0919 Nov 21	08:23:25	23864	-36093	040	T	p-	0.4124	2.1185	1.0844	333.9	202.0	44.2	19N	29W
2618	131	-0918 May 16	19:03:53	23856	-36087	045	P	a-	-0.7360	1.4960	0.5185	293.3	151.2	-	17S	171E
2619	131	-0918 Nov 10	13:30:28	23848	-36081	050	N	t-	1.1452	0.8011	-0.2872	255.8	-	-	16N	106W
2620	131	-0917 Apr 07	03:50:22	23840	-36076	017	N	-a	1.0607	0.8838	-0.0610	230.5	-	-	2S	43E
2621	132	-0917 May 06	10:39:22	23839	-36075	055	N	a-	-1.4345	0.1969	-0.7461	116.6	-	-	15S	63W
2622	132	-0917 Sep 30	18:23:02	23832	-36070	022	N	-t	-1.1935	0.7114	-0.3748	244.3	-	-	2S	177W
2623	132	-0916 Mar 26	19:53:20	23824	-36064	027	T	-a	0.3509	2.2014	1.2264	320.9	203.4	67.8	2N	163E
2624	132	-0916 Sep 18	22:52:41	23815	-36058	032	T	-p	-0.4430	2.0612	1.0291	334.5	200.6	26.7	6S	116E
2625	132	-0915 Mar 16	06:41:42	23807	-36052	037	T	t-	-0.4133	2.1158	1.0838	341.2	206.6	45.1	5N	2E
2626	132	-0915 Sep 08	10:43:07	23798	-36046	042	T	p-	0.2864	2.3210	1.3438	320.4	205.5	79.3	9S	62W
2627	132	-0914 Mar 05	10:08:06	23790	-36040	047	N	t-	-1.1864	0.7206	-0.3578	246.9	-	-	9N	49W
2628	132	-0914 Aug 29	02:45:02	23781	-36034	052	P	a-	0.9708	1.0546	0.0981	247.7	67.3	-	12S	58E
2629	132	-0913 Jan 23	17:43:31	23774	-36029	019	N	-t	1.0392	0.9854	-0.0828	274.2	-	-	22N	163W
2630	132	-0913 Jul 20	08:26:35	23766	-36023	024	P	-a	-0.9246	1.1608	0.1618	273.8	90.3	-	24S	27W
2631	132	-0912 Jan 13	00:55:32	23757	-36017	029	T	-p	0.2873	2.3380	1.3236	333.8	210.3	79.2	23N	88E
2632	132	-0912 Jul 08	15:51:29	23749	-36011	034	T-	pp	-0.1836	2.5465	1.4962	361.8	227.7	96.3	24S	140W
2633	132	-0911 Jan 01	14:26:23	23740	-36005	039	T	p-	-0.4076	2.0969	1.1228	310.4	194.0	50.4	23N	116W
2634	132	-0911 Jun 27	17:11:41	23732	-35999	044	P	t-	0.5812	1.8275	0.7561	348.7	191.6	-	23S	160W
2635	132	-0911 Dec 22	06:12:49	23724	-35993	049	N	a-	-1.0702	0.8819	-0.0936	232.9	-	-	22N	6E
2636	132	-0910 May 18	06:55:07	23717	-35988	016	N	-t	-1.4092	0.2805	-0.7362	151.3	-	-	18S	7W
2637	132	-0910 Jun 16	18:53:44	23715	-35987	054	N	t-	1.3158	0.4657	-0.5782	197.8	-	-	22S	174E
2638	132	-0910 Nov 12	05:15:08	23708	-35982	021	N	-a	1.0337	0.9869	-0.0642	266.4	-	-	16N	17E
2639	132	-0909 May 07	18:35:30	23700	-35976	026	P	-a	-0.5906	1.7564	0.7916	302.6	176.5	-	14S	178E
2640	132	-0909 Nov 01	08:20:19	23691	-35970	031	T	-t	0.3722	2.2244	1.1266	365.0	216.5	56.1	12N	29W

Cat Num	Canon Plate	Calendar Date	TD of Greatest Eclipse	ΔT s	Luna Num	Saros Num	Ecl. Type	QSE	Gamma	Pen. Mag.	Um. Mag.	Eclipse Phase ---- Durations ---- Pen. m	Par. m	Total m	Greatest in Zenith Lat.	Lng.
2641	133	-0908 Apr 26	11:03:25	23683	-35964	036	T+	p-	0.1610	2.5328	1.5917	315.3	208.9	93.8	10S	68W
2642	133	-0908 Oct 20	07:23:27	23674	-35958	041	T	p-	-0.3082	2.3419	1.2437	369.2	222.5	74.9	7N	15W
2643	133	-0907 Apr 16	03:48:01	23666	-35952	046	P	a-	0.9069	1.1743	0.2129	263.3	99.5	-	6S	42E
2644	133	-0907 Oct 09	10:04:01	23658	-35946	051	P	a-	-0.9817	1.0823	0.0312	277.6	41.5	-	2N	54W
2645	133	-0906 Mar 07	03:37:51	23650	-35941	018	N	-t	-1.0277	0.9975	-0.0527	273.1	-	-	8N	48E
2646	133	-0906 Aug 30	10:02:56	23642	-35935	023	N	-a	1.0475	0.9181	-0.0468	236.9	-	-	12S	52W
2647	133	-0905 Feb 24	05:19:57	23634	-35929	028	T	-t	-0.3149	2.3236	1.2374	369.1	223.4	74.7	13N	23E
2648	133	-0905 Aug 20	02:31:31	23625	-35923	033	T	-p	0.3474	2.1990	1.2416	311.4	199.2	68.2	16S	61E
2649	133	-0904 Feb 13	05:33:26	23617	-35917	038	T	t-	0.4057	2.1501	1.0776	356.4	211.9	44.4	17N	20E
2650	133	-0904 Aug 08	17:15:34	23608	-35911	043	T	p-	-0.3825	2.1521	1.1600	325.1	202.7	58.8	19S	160W
2651	133	-0903 Feb 01	11:18:55	23600	-35905	048	N	a-	1.0957	0.8587	-0.1637	246.1	-	-	21N	67W
2652	133	-0903 Jul 29	02:03:46	23592	-35899	053	N	t-	-1.1765	0.7216	-0.3230	238.3	-	-	22S	68E
2653	133	-0903 Dec 23	13:34:06	23585	-35894	020	P	-a	-0.9414	1.1158	0.1451	254.0	81.3	-	23N	104W
2654	133	-0902 Jun 18	12:52:50	23576	-35888	025	P	-t	0.9031	1.2348	0.1674	306.3	98.6	-	22S	97W
2655	133	-0902 Dec 13	05:06:01	23568	-35882	030	T	-p	-0.2853	2.3272	1.3418	320.0	204.5	78.7	22N	21E
2656	133	-0901 Jun 07	15:59:41	23559	-35876	035	T+	pp	0.1139	2.6636	1.6347	355.1	227.1	101.3	22S	144W
2657	133	-0901 Dec 02	16:56:59	23551	-35870	040	T	p-	0.4139	2.1171	1.0801	335.1	202.4	43.2	21N	158W
2658	133	-0900 May 27	02:12:10	23542	-35864	045	P	a-	-0.6564	1.6404	0.6662	300.5	166.9	-	20S	63E
2659	133	-0900 Nov 20	21:40:24	23534	-35858	050	N	t-	1.1444	0.8040	-0.2871	256.8	-	-	19N	130E
2660	133	-0899 Apr 17	11:25:33	23527	-35853	017	N	-a	1.1261	0.7630	-0.1803	216.9	-	-	6S	73W
2661	134	-0899 May 16	18:03:46	23526	-35852	055	N	a-	-1.3616	0.3298	-0.6115	148.9	-	-	18S	176W
2662	134	-0899 Oct 11	02:13:31	23519	-35847	022	N	-t	-1.2093	0.6828	-0.4042	239.6	-	-	3N	63E
2663	134	-0898 Apr 07	03:22:41	23510	-35841	027	T	-a	0.4154	2.0833	1.1078	318.4	198.4	48.7	3S	49E
2664	134	-0898 Sep 30	06:59:04	23502	-35835	032	P	-p	-0.4631	2.0242	0.9925	332.1	197.7	-	1S	8W
2665	134	-0897 Mar 27	13:53:39	23493	-35829	037	T	p-	-0.3535	2.2257	1.1932	345.8	212.9	66.3	1N	109W
2666	134	-0897 Sep 19	19:00:27	23485	-35823	042	T+	p-	0.2584	2.3726	1.3947	320.8	206.6	83.2	5S	172E
2667	134	-0896 Mar 15	17:05:03	23477	-35817	047	N	t-	-1.1297	0.8236	-0.2528	261.5	-	-	5N	156W
2668	134	-0896 Sep 08	10:59:19	23468	-35811	052	P	a-	0.9380	1.1165	0.1566	253.5	84.4	-	8S	67W
2669	134	-0895 Feb 03	01:26:45	23461	-35806	019	N	-t	1.0645	0.9361	-0.1261	267.5	-	-	20N	81E
2670	134	-0895 Jul 30	15:47:51	23453	-35800	024	P	-a	-0.9875	1.0487	0.0434	264.7	47.7	-	22S	138W
2671	134	-0894 Jan 23	09:11:21	23444	-35794	029	T	-p	0.3047	2.3032	1.2945	331.5	208.7	76.2	22N	36W
2672	134	-0894 Jul 19	22:39:10	23436	-35788	034	T-	-p	-0.2557	2.4168	1.3612	361.0	224.3	86.9	23S	118E
2673	134	-0893 Jan 12	23:03:50	23428	-35782	039	T	p-	-0.3948	2.1188	1.1481	310.5	195.0	54.9	23N	115E
2674	134	-0893 Jul 08	23:40:22	23419	-35776	044	P	t-	0.5012	1.9753	0.9020	355.8	203.8	-	23S	102E
2675	134	-0892 Jan 02	14:57:15	23411	-35770	049	N	a-	-1.0652	0.8908	-0.0843	234.2	-	-	23N	125W
2676	134	-0892 May 28	13:43:13	23404	-35765	016	N	-t	-1.4828	0.1436	-0.8694	109.1	-	-	21S	111W
2677	134	-0892 Jun 27	01:32:48	23403	-35764	054	N	t-	1.2334	0.6162	-0.4262	223.5	-	-	23S	73E
2678	134	-0892 Nov 22	13:41:18	23396	-35759	021	N	-a	1.0367	0.9833	-0.0713	267.1	-	-	20N	110W
2679	134	-0891 May 18	01:49:58	23387	-35753	026	P	-a	-0.6644	1.6193	0.6579	294.5	164.0	-	18S	68E
2680	134	-0891 Nov 11	16:23:41	23379	-35747	031	T	-t	0.3769	2.2170	1.1166	365.5	216.2	54.1	16N	151W
2681	135	-0890 May 07	18:30:09	23370	-35741	036	T+	pp	0.0882	2.6658	1.7259	316.3	210.8	97.8	14S	178E
2682	135	-0890 Oct 31	15:22:26	23362	-35735	041	T	p-	-0.2999	2.3574	1.2587	369.2	222.8	76.7	11N	136W
2683	135	-0889 Apr 27	11:11:42	23354	-35729	046	P	a-	0.8407	1.2963	0.3340	273.7	122.7	-	10S	71W
2684	135	-0889 Oct 20	18:17:41	23345	-35723	051	P	a-	-0.9688	1.1057	0.0554	279.0	55.0	-	6N	179W
2685	135	-0888 Mar 17	10:49:04	23338	-35718	018	N	-t	-1.0855	0.8913	-0.1587	261.8	-	-	4N	62W
2686	135	-0888 Sep 09	18:16:20	23330	-35712	023	N	-a	1.0768	0.8651	-0.1013	230.9	-	-	8S	177W
2687	135	-0887 Mar 06	12:21:58	23322	-35706	028	T	-t	-0.3677	2.2251	1.1420	366.1	219.0	59.7	9N	84W
2688	135	-0887 Aug 30	10:36:45	23313	-35700	033	T	-p	0.3868	2.1288	1.1672	310.2	196.1	58.1	12S	62W
2689	135	-0886 Feb 23	12:55:06	23305	-35694	038	T	t-	0.3627	2.2259	1.1594	357.4	215.9	61.9	13N	92W
2690	135	-0886 Aug 20	00:57:26	23296	-35688	043	T	p-	-0.3367	2.2393	1.2409	329.1	206.8	70.4	16S	83E
2691	135	-0885 Feb 12	19:14:49	23288	-35682	048	N	a-	1.0626	0.9162	-0.0998	251.6	-	-	18N	173E
2692	135	-0885 Aug 09	09:11:02	23280	-35676	053	N	t-	-1.1223	0.8242	-0.2266	253.2	-	-	20S	40W
2693	135	-0884 Jan 03	22:19:25	23273	-35671	020	P	-a	-0.9483	1.1019	0.1335	252.6	78.1	-	23N	124E
2694	135	-0884 Jun 28	19:18:19	23264	-35665	025	P	-t	0.9852	1.0845	0.0166	291.4	31.7		23S	166E
2695	135	-0884 Dec 23	13:51:18	23256	-35659	030	T	-p	-0.2898	2.3192	1.3331	320.5	204.6	78.1	23N	110W
2696	135	-0883 Jun 17	22:44:43	23248	-35653	035	T+	pp	0.1944	2.5147	1.4882	351.8	223.5	94.6	23S	114E
2697	135	-0883 Dec 13	01:26:46	23239	-35647	040	T	p-	0.4131	2.1197	1.0805	336.4	203.0	43.4	23N	74E
2698	135	-0882 Jun 07	09:22:52	23231	-35641	045	P	a-	-0.5779	1.7832	0.8117	306.6	179.3	-	22S	46W
2699	135	-0882 Dec 02	05:49:01	23223	-35635	050	N	t-	1.1434	0.8068	-0.2861	257.8	-	-	22N	7E
2700	135	-0881 Apr 28	18:55:48	23216	-35630	017	N	-a	1.1956	0.6351	-0.3075	200.6	-	-	10S	172E

Cat Num	Canon Plate	Calendar Date	TD of Greatest Eclipse	ΔT s	Luna Num	Saros Num	Ecl. Type	QSE	Gamma	Pen. Mag.	Um. Mag.	Eclipse Phase Durations Pen. m	Par. m	Total m	Greatest in Zenith Lat.	Lng.
2701	136	-0881 May 28	01:27:49	23214	-35629	055	N	a-	-1.2880	0.4645	-0.4760	174.3	-	-	21S	72E
2702	136	-0881 Oct 22	10:11:12	23207	-35624	022	N	-t	-1.2198	0.6637	-0.4234	236.2	-	-	7N	58W
2703	136	-0880 Apr 17	10:44:38	23199	-35618	027	P	-a	0.4852	1.9558	0.9793	315.0	191.6	-	7S	64W
2704	136	-0880 Oct 10	15:14:44	23191	-35612	032	P	-p	-0.4763	1.9997	0.9686	330.1	195.6	-	3N	134W
2705	136	-0879 Apr 06	20:53:51	23182	-35606	037	T	pp	-0.2851	2.3516	1.3183	350.2	218.8	81.7	3S	144E
2706	136	-0879 Sep 30	03:27:21	23174	-35600	042	T+	p-	0.2381	2.4103	1.4317	320.9	207.3	85.6	1S	43E
2707	136	-0878 Mar 26	23:50:59	23166	-35594	047	N	t-	-1.0645	0.9423	-0.1323	276.5	-	-	0N	100E
2708	136	-0878 Sep 19	19:22:00	23157	-35588	052	P	a-	0.9125	1.1651	0.2018	257.9	95.2	-	4S	165E
2709	136	-0877 Feb 14	09:02:46	23150	-35583	019	N	-t	1.0961	0.8749	-0.1809	259.2	-	-	17N	34W
2710	136	-0877 Aug 10	23:15:17	23142	-35577	024	N	-a	-1.0441	0.9480	-0.0637	255.6	-	-	19S	109E
2711	136	-0876 Feb 03	17:22:11	23134	-35571	029	T	-p	0.3270	2.2593	1.2567	328.9	206.8	72.0	19N	159W
2712	136	-0876 Jul 30	05:30:24	23125	-35565	034	T	-t	-0.3240	2.2942	1.2332	359.6	219.9	73.3	21S	14E
2713	136	-0875 Jan 23	07:36:50	23117	-35559	039	T	p-	-0.3781	2.1476	1.1805	310.7	196.4	60.0	21N	14W
2714	136	-0875 Jul 19	06:14:08	23109	-35553	044	T	t-	0.4249	2.1164	1.0410	361.5	213.3	33.2	22S	3E
2715	136	-0874 Jan 12	23:35:41	23100	-35547	049	N	a-	-1.0559	0.9075	-0.0669	236.3	-	-	22N	105E
2716	136	-0874 Jun 08	20:34:39	23093	-35542	016	Ne	-t	-1.5543	0.0110	-0.9990	30.4	-	-	23S	145E
2717	136	-0874 Jul 08	08:19:15	23092	-35541	054	N	t-	1.1556	0.7584	-0.2830	243.7	-	-	23S	29W
2718	136	-0874 Dec 03	22:05:58	23085	-35536	021	N	-a	1.0402	0.9782	-0.0792	267.5	-	-	22N	123E
2719	136	-0873 May 29	09:06:28	23077	-35530	026	P	-a	-0.7370	1.4848	0.5261	285.8	149.4	-	21S	43W
2720	136	-0873 Nov 23	00:26:49	23068	-35524	031	T	-t	0.3814	2.2096	1.1074	366.0	216.0	52.2	19N	87E
2721	137	-0872 May 18	01:56:11	23060	-35518	036	T+	pp	0.0148	2.8003	1.8609	316.8	211.6	99.5	17S	65E
2722	137	-0872 Nov 10	23:25:06	23052	-35512	041	T	p-	-0.2944	2.3674	1.2691	368.9	223.0	77.8	15N	102E
2723	137	-0871 May 07	18:29:33	23043	-35506	046	P	a-	0.7704	1.4261	0.4623	283.6	142.0	-	13S	177E
2724	137	-0871 Oct 31	02:38:05	23035	-35500	051	P	a-	-0.9606	1.1200	0.0712	279.4	62.0	-	11N	54E
2725	137	-0870 Mar 28	17:48:55	23028	-35495	018	N	-t	-1.1513	0.7706	-0.2795	247.1	-	-	0S	170W
2726	137	-0870 Sep 21	02:39:04	23020	-35489	023	N	-a	1.0989	0.8253	-0.1425	226.2	-	-	3S	55E
2727	137	-0869 Mar 17	19:13:49	23012	-35483	028	T	-t	-0.4285	2.1121	1.0319	362.0	212.8	29.4	5N	170E
2728	137	-0869 Sep 10	18:50:58	23003	-35477	033	T	-a	0.4184	2.0729	1.1071	309.2	193.4	47.4	8S	173E
2729	137	-0868 Mar 05	20:08:07	22995	-35471	038	T	t-	0.3121	2.3158	1.2553	358.4	219.8	75.7	9N	158E
2730	137	-0868 Aug 30	08:46:03	22987	-35465	043	T	p-	-0.2976	2.3143	1.3095	332.6	210.1	78.0	13S	35W
2731	137	-0867 Feb 23	03:02:59	22978	-35459	048	N	a-	1.0225	0.9865	-0.0229	257.9	-	-	14N	55E
2732	137	-0867 Aug 19	16:24:54	22970	-35453	053	N	t-	-1.0743	0.9155	-0.1417	265.5	-	-	17S	150W
2733	137	-0866 Jan 14	06:58:44	22963	-35448	020	P	-a	-0.9602	1.0788	0.1130	250.3	72.0	-	22N	6W
2734	137	-0866 Jul 10	01:49:18	22955	-35442	025	N	-t	1.0625	0.9430	-0.1257	275.5	-	-	23S	68E
2735	137	-0865 Jan 03	22:32:24	22947	-35436	030	T	-p	-0.2969	2.3062	1.3199	320.9	204.5	77.0	23N	120E
2736	137	-0865 Jun 29	05:34:31	22938	-35430	035	T+	pp	0.2718	2.3715	1.3470	347.8	218.7	84.1	24S	11E
2737	137	-0865 Dec 24	09:53:23	22930	-35424	040	T	p-	0.4106	2.1252	1.0842	337.7	203.8	44.5	24N	52W
2738	137	-0864 Jun 17	16:36:15	22922	-35418	045	P	a-	-0.5010	1.9232	0.9539	311.5	189.0	-	24S	155W
2739	137	-0864 Dec 12	13:56:12	22913	-35412	050	N	t-	1.1414	0.8110	-0.2829	258.9	-	-	24N	114W
2740	137	-0863 May 09	02:21:39	22906	-35407	017	N	-a	1.2684	0.5016	-0.4409	180.8	-	-	13S	59E
2741	138	-0863 Jun 07	08:51:28	22905	-35406	055	N	a-	-1.2139	0.6003	-0.3399	195.4	-	-	23S	40W
2742	138	-0863 Nov 01	18:15:33	22898	-35401	022	N	-t	-1.2253	0.6532	-0.4334	234.1	-	-	11N	179E
2743	138	-0862 Apr 28	18:01:01	22890	-35395	027	P	-a	0.5586	1.8219	0.8438	310.5	182.7	-	11S	176W
2744	138	-0862 Oct 21	23:36:44	22882	-35389	032	P	-p	-0.4850	1.9832	0.9532	328.3	194.0	-	8N	99E
2745	138	-0861 Apr 18	03:48:10	22873	-35383	037	T-	pp	-0.2128	2.4849	1.4505	354.2	223.7	92.6	7S	38E
2746	138	-0861 Oct 11	12:01:11	22865	-35377	042	T+	p-	0.2234	2.4376	1.4584	320.8	207.6	87.2	4N	87W
2747	138	-0860 Apr 06	06:29:46	22857	-35371	047	N*	t-	-0.9934	1.0718	-0.0010	291.0	-	-	4S	2W
2748	138	-0860 Sep 30	03:51:26	22848	-35365	052	P	a-	0.8928	1.2029	0.2363	261.3	102.6	-	0N	36E
2749	138	-0859 Feb 24	16:30:46	22842	-35360	019	N	-t	1.1348	0.8006	-0.2488	248.7	-	-	13N	148W
2750	138	-0859 Aug 21	06:49:47	22833	-35354	024	N	-a	-1.0941	0.8595	-0.1588	246.8	-	-	16S	6W
2751	138	-0858 Feb 14	01:23:28	22825	-35348	029	T	-p	0.3580	2.1992	1.2027	325.9	204.0	65.0	16N	79E
2752	138	-0858 Aug 10	12:30:19	22817	-35342	034	T	-t	-0.3846	2.1859	1.1192	357.8	214.8	54.5	19S	92W
2753	138	-0857 Feb 03	16:01:25	22808	-35336	039	T	p-	-0.3539	2.1901	1.2268	311.3	198.2	66.3	19N	141W
2754	138	-0857 Jul 30	12:56:04	22800	-35330	044	T	t-	0.3548	2.2462	1.1683	365.8	220.2	64.5	21S	99W
2755	138	-0856 Jan 24	08:06:42	22792	-35324	049	N	a-	-1.0410	0.9345	-0.0389	239.4	-	-	20N	23W
2756	138	-0856 Jul 18	15:12:49	22784	-35318	054	N	t-	1.0821	0.8928	-0.1478	260.0	-	-	22S	134W
2757	138	-0856 Dec 14	06:28:20	22777	-35313	021	N	-a	1.0448	0.9708	-0.0886	267.6	-	-	24N	3W
2758	138	-0855 Jun 08	16:22:40	22768	-35307	026	P	-a	-0.8103	1.3494	0.3926	276.0	131.4	-	23S	153W
2759	138	-0855 Dec 03	08:31:17	22760	-35301	031	T	-t	0.3845	2.2044	1.1013	366.3	215.8	50.8	22N	35W
2760	138	-0854 May 29	09:20:28	22752	-35295	036	T-	pp	-0.0601	2.7173	1.7776	316.7	211.3	98.8	20S	48W

Cat Num	Canon Plate	Calendar Date	TD of Greatest Eclipse	ΔT s	Luna Num	Saros Num	Ecl. Type	QSE	Gamma	Pen. Mag.	Um. Mag.	Eclipse Phase Durations Pen. m	Par. m	Total m	Greatest in Zenith Lat.	Lng.
2761	139	-0854 Nov 22	07:30:50	22744	-35289	041	T	p-	-0.2908	2.3733	1.2762	368.3	222.9	78.5	18N	20W
2762	139	-0853 May 19	01:44:41	22735	-35283	046	P	a-	0.6983	1.5594	0.5934	292.9	158.1	-	17S	67E
2763	139	-0853 Nov 11	11:04:10	22727	-35277	051	P	a-	-0.9561	1.1271	0.0803	279.1	65.6	-	14N	74W
2764	139	-0852 Apr 08	00:40:39	22720	-35272	018	N	-t	-1.2226	0.6400	-0.4103	228.8	-	-	5S	85E
2765	139	-0852 May 07	13:07:29	22719	-35271	056	Nb	t-	1.5204	0.0781	-0.9417	81.6	-	-	13S	103W
2766	139	-0852 Oct 01	11:09:09	22712	-35266	023	N	-a	1.1154	0.7958	-0.1736	222.6	-	-	1N	74W
2767	139	-0851 Mar 28	01:57:43	22704	-35260	028	P	-t	-0.4957	1.9872	0.9102	356.8	204.5	-	0N	67E
2768	139	-0851 Sep 21	03:12:08	22695	-35254	033	T	-a	0.4437	2.0287	1.0586	308.5	191.0	35.6	4S	46E
2769	139	-0850 Mar 17	03:13:47	22687	-35248	038	T+	p-	0.2547	2.4181	1.3636	359.2	223.4	86.7	5N	49E
2770	139	-0850 Sep 10	16:42:17	22679	-35242	043	T-	p-	-0.2657	2.3761	1.3649	335.5	212.5	83.2	9S	156W
2771	139	-0849 Mar 06	10:44:51	22671	-35236	048	P	a-	0.9764	1.0677	0.0648	264.7	57.8	-	10N	63W
2772	139	-0849 Aug 30	23:45:28	22662	-35230	053	N	t-	-1.0324	0.9956	-0.0678	275.8	-	-	13S	99E
2773	139	-0848 Jan 25	15:31:28	22656	-35225	020	P	-a	-0.9776	1.0455	0.0824	247.2	61.8	-	20N	134W
2774	139	-0848 Jul 20	08:27:33	22647	-35219	025	N	-t	1.1338	0.8128	-0.2570	258.9	-	-	22S	32W
2775	139	-0847 Jan 14	07:05:49	22639	-35213	030	T	-p	-0.3102	2.2817	1.2956	321.0	204.1	74.8	22N	9W
2776	139	-0847 Jul 09	12:30:58	22631	-35207	035	T	-p	0.3447	2.2371	1.2140	343.1	212.7	69.0	23S	94W
2777	139	-0846 Jan 03	18:12:50	22623	-35201	040	T	p-	0.4031	2.1396	1.0974	339.3	205.1	47.8	24N	177W
2778	139	-0846 Jun 28	23:55:04	22614	-35195	045	T	a-	-0.4276	2.0571	1.0892	315.3	196.5	44.4	24S	94E
2779	139	-0846 Dec 23	21:57:32	22606	-35189	050	N	t-	1.1352	0.8224	-0.2718	260.9	-	-	25N	125E
2780	139	-0845 May 20	09:45:40	22599	-35184	017	N	-a	1.3422	0.3663	-0.5767	156.7	-	-	16S	54W
2781	140	-0845 Jun 18	16:18:02	22598	-35183	055	N	a-	-1.1420	0.7325	-0.2083	212.9	-	-	24S	152W
2782	140	-0845 Nov 13	02:24:54	22591	-35178	022	N	-t	-1.2272	0.6492	-0.4361	232.9	-	-	15N	56E
2783	140	-0844 May 09	01:10:54	22583	-35172	027	P	-a	0.6365	1.6801	0.6998	304.8	171.0	-	14S	75E
2784	140	-0844 Nov 01	08:06:04	22575	-35166	032	P	-p	-0.4884	1.9762	0.9477	326.9	193.1	-	12N	30W
2785	140	-0843 Apr 28	10:33:26	22566	-35160	037	T-	pp	-0.1341	2.6300	1.5940	357.5	227.5	100.2	11S	65W
2786	140	-0843 Oct 21	20:42:26	22558	-35154	042	T+	p-	0.2143	2.4544	1.4750	320.6	207.6	88.0	8N	141E
2787	140	-0842 Apr 17	12:59:55	22550	-35148	047	P	t-	-0.9152	1.2145	0.1432	305.2	91.7	-	8S	101W
2788	140	-0842 Oct 11	12:28:31	22542	-35142	052	P	a-	0.8796	1.2287	0.2589	263.7	107.1	-	5N	96W
2789	140	-0841 Mar 07	23:52:17	22535	-35137	019	N	-t	1.1796	0.7151	-0.3278	235.9	-	-	10N	100E
2790	140	-0841 Sep 01	14:30:07	22527	-35131	024	N	-a	-1.1385	0.7815	-0.2435	238.4	-	-	13S	123W
2791	140	-0840 Feb 25	09:19:38	22518	-35125	029	T	-p	0.3942	2.1297	1.1394	322.4	200.5	54.9	13N	41W
2792	140	-0840 Aug 20	19:36:08	22510	-35119	034	T	-t	-0.4397	2.0876	1.0153	355.7	209.1	20.2	16S	161E
2793	140	-0839 Feb 14	00:18:53	22502	-35113	039	T	p-	-0.3237	2.2436	1.2841	312.1	200.4	72.8	16N	94E
2794	140	-0839 Aug 09	19:45:55	22494	-35107	044	T	t-	0.2910	2.3646	1.2843	369.0	225.2	80.3	18S	158E
2795	140	-0838 Feb 03	16:30:02	22485	-35101	049	N	a-	-1.0203	0.9718	-0.0005	243.6	-	-	18N	150W
2796	140	-0838 Jul 29	22:16:04	22477	-35095	054	N*	t-	1.0155	1.0149	-0.0253	272.8	-	-	20S	120E
2797	140	-0838 Dec 25	14:44:59	22470	-35090	021	N	-a	1.0532	0.9561	-0.1048	266.8	-	-	25N	127W
2798	140	-0837 Jun 19	23:44:04	22462	-35084	026	P	-a	-0.8798	1.2210	0.2657	265.7	109.8	-	24S	96E
2799	140	-0837 Dec 14	16:32:33	22454	-35078	031	T	-t	0.3900	2.1944	1.0912	366.3	215.5	48.4	23N	155W
2800	140	-0836 Jun 08	16:45:07	22446	-35072	036	T-	-p	-0.1345	2.5811	1.6406	315.9	209.9	95.6	22S	160W
2801	141	-0836 Dec 02	15:37:28	22438	-35066	041	T	p-	-0.2875	2.3784	1.2833	367.6	222.9	79.2	21N	142W
2802	141	-0835 May 29	08:57:01	22429	-35060	046	P	a-	0.6244	1.6965	0.7275	301.5	171.8	-	19S	43W
2803	141	-0835 Nov 21	19:32:32	22421	-35054	051	P	a-	-0.9528	1.1318	0.0877	278.5	68.2	-	18N	158E
2804	141	-0834 Apr 19	07:23:02	22414	-35049	018	N	-t	-1.3001	0.4981	-0.5528	205.2	-	-	9S	18W
2805	141	-0834 May 18	19:53:46	22413	-35048	056	N	t-	1.4473	0.2137	-0.8090	133.8	-	-	16S	153E
2806	141	-0834 Oct 12	19:47:07	22406	-35043	023	N	-a	1.1260	0.7770	-0.1936	220.2	-	-	6N	154E
2807	141	-0833 Apr 08	08:33:46	22398	-35037	028	P	-t	-0.5690	1.8513	0.7771	350.1	193.5	-	4S	34W
2808	141	-0833 Oct 02	11:40:17	22390	-35031	033	T	-a	0.4629	1.9957	1.0214	308.1	189.2	21.8	1N	83W
2809	141	-0832 Mar 27	10:12:34	22381	-35025	038	T+	pp	0.1907	2.5325	1.4839	359.6	226.3	95.2	1N	58W
2810	141	-0832 Sep 21	00:45:45	22373	-35019	043	T-	p-	-0.2407	2.4251	1.4076	338.1	214.4	86.6	4S	81E
2811	141	-0831 Mar 16	18:17:58	22365	-35013	048	P	a-	0.9223	1.1638	0.1673	272.2	91.2	-	6N	178W
2812	141	-0831 Sep 10	07:14:32	22357	-35007	053	N*	t-	-0.9979	1.0620	-0.0075	284.0	-	-	9S	15W
2813	141	-0830 Feb 04	23:56:43	22350	-35002	020	P	-a	-1.0012	1.0006	0.0405	242.8	43.6	-	17N	98E
2814	141	-0830 Jul 31	15:14:16	22342	-34996	025	N	-t	1.1985	0.6947	-0.3764	241.9	-	-	20S	135W
2815	141	-0829 Jan 25	15:32:21	22334	-34990	030	T	-p	-0.3287	2.2477	1.2619	320.9	203.3	71.4	21N	136W
2816	141	-0829 Jul 20	19:35:30	22326	-34984	035	T	-t	0.4118	2.1135	1.0915	338.1	205.9	46.8	22S	159E
2817	141	-0828 Jan 15	02:26:40	22317	-34978	040	T	p-	0.3918	2.1607	1.1178	341.1	206.8	52.4	23N	60E
2818	141	-0828 Jul 09	07:18:01	22309	-34972	045	T	a-	-0.3572	2.1859	1.2190	318.2	202.2	66.6	24S	17W
2819	141	-0827 Jan 03	05:54:59	22301	-34966	050	N	t-	1.1264	0.8383	-0.2555	263.4	-	-	25N	6E
2820	141	-0827 May 30	17:08:14	22294	-34961	017	N	-a	1.4166	0.2305	-0.7137	126.1	-	-	19S	166W

Cat Num	Canon Plate	Calendar Date	TD of Greatest Eclipse	ΔT s	Luna Num	Saros Num	Ecl. Type	QSE	Gamma	Pen. Mag.	Um. Mag.	Eclipse Phase Durations Pen. m	Par. m	Total m	Greatest in Zenith Lat.	Lng.
2821	142	-0827 Jun 28	23:47:13	22293	-34960	055	N	a-	-1.0725	0.8606	-0.0813	227.7	-	-	25S	95E
2822	142	-0827 Nov 23	10:36:37	22286	-34955	022	N	-t	-1.2277	0.6473	-0.4360	232.0	-	-	18N	68W
2823	142	-0826 May 20	08:18:06	22278	-34949	027	P	-a	0.7155	1.5365	0.5534	297.8	156.2	-	17S	34W
2824	142	-0826 Nov 12	16:39:08	22270	-34943	032	P	-p	-0.4892	1.9737	0.9473	325.7	192.5	-	15N	160W
2825	142	-0825 May 09	17:15:05	22261	-34937	037	T-	pp	-0.0532	2.7795	1.7415	359.9	229.9	104.3	15S	168W
2826	142	-0825 Nov 02	05:27:24	22253	-34931	042	T+	p-	0.2082	2.4656	1.4862	320.2	207.5	88.5	12N	8E
2827	142	-0824 Apr 27	19:25:44	22245	-34925	047	P	t-	-0.8335	1.3638	0.2938	318.1	128.6	-	12S	160E
2828	142	-0824 Oct 21	21:10:42	22237	-34919	052	P	a-	0.8711	1.2458	0.2732	265.4	109.8	-	9N	132E
2829	142	-0823 Mar 18	07:04:58	22230	-34914	019	N	-t	1.2320	0.6157	-0.4207	220.0	-	-	5N	10W
2830	142	-0823 Apr 16	20:28:26	22229	-34913	057	Nb	t-	-1.5586	0.0252	-1.0287	48.4	-	-	8S	145E
2831	142	-0823 Sep 11	22:18:59	22222	-34908	024	N	-a	-1.1749	0.7180	-0.3135	231.1	-	-	9S	118E
2832	142	-0822 Mar 07	17:06:41	22214	-34902	029	T	-a	0.4386	2.0453	1.0610	318.2	195.9	37.2	9N	160W
2833	142	-0822 Sep 01	02:51:34	22206	-34896	034	P	-t	-0.4861	2.0053	0.9275	353.7	203.5	-	12S	50E
2834	142	-0821 Feb 25	08:27:18	22197	-34890	039	T	p-	-0.2859	2.3111	1.3553	313.1	202.9	79.4	12N	30W
2835	142	-0821 Aug 21	02:46:38	22189	-34884	044	T+	pp	0.2358	2.4671	1.3843	371.2	228.5	89.7	15S	51E
2836	142	-0820 Feb 15	00:44:20	22181	-34878	049	P	a-	-0.9929	1.0215	0.0503	248.8	49.1	-	15N	86E
2837	142	-0820 Aug 09	05:27:51	22173	-34872	054	P	h-	0.9547	1.1264	0.0863	283.1	69.0	-	18S	11E
2838	142	-0819 Jan 04	22:56:35	22166	-34867	021	N	-h	1.0650	0.9348	-0.1269	265.2	-	-	25N	111E
2839	142	-0819 Jun 30	07:08:01	22158	-34861	026	P	-a	-0.9479	1.0958	0.1411	254.6	81.3	-	25S	16W
2840	142	-0819 Dec 25	00:30:20	22150	-34855	031	T	-t	0.3981	2.1792	1.0768	366.0	214.9	44.7	24N	86E
2841	143	-0818 Jun 20	00:11:00	22142	-34849	036	T-	-p	-0.2078	2.4475	1.5054	314.6	207.4	89.9	24S	88E
2842	143	-0818 Dec 13	23:44:14	22134	-34843	041	T	p-	-0.2835	2.3843	1.2919	366.8	222.8	80.1	23N	96E
2843	143	-0817 Jun 09	16:08:08	22125	-34837	046	P	a-	0.5503	1.8342	0.8618	309.2	183.3	-	21S	152W
2844	143	-0817 Dec 03	04:02:50	22117	-34831	051	P	a-	-0.9504	1.1346	0.0937	277.8	70.3	-	20N	30E
2845	143	-0816 Apr 29	13:59:50	22110	-34826	018	N	-t	-1.3808	0.3504	-0.7015	175.0	-	-	13S	119W
2846	143	-0816 May 29	02:37:11	22109	-34825	056	N	t-	1.3723	0.3529	-0.6731	170.5	-	-	19S	51E
2847	143	-0816 Oct 23	04:30:47	22102	-34820	023	N	-a	1.1321	0.7663	-0.2054	218.8	-	-	10N	21E
2848	143	-0815 Apr 18	15:02:58	22094	-34814	028	P	-t	-0.6475	1.7059	0.6344	341.9	179.1	-	8S	134W
2849	143	-0815 Oct 12	20:14:50	22086	-34808	033	P	-a	0.4762	1.9732	0.9949	308.1	187.9	-	5N	146E
2850	143	-0814 Apr 07	17:06:23	22078	-34802	038	T+	pp	0.1220	2.6558	1.6129	359.5	228.3	101.0	3S	164W
2851	143	-0814 Oct 02	08:55:19	22070	-34796	043	T-	p-	-0.2217	2.4630	1.4395	340.4	215.8	88.9	0N	43W
2852	143	-0813 Mar 28	01:46:36	22062	-34790	048	P	a-	0.8636	1.2684	0.2779	279.6	115.3	-	2N	68E
2853	143	-0813 Sep 21	14:51:15	22053	-34784	053	P	t-	-0.9701	1.1158	0.0405	290.6	48.6	-	5S	131W
2854	143	-0812 Feb 16	08:14:04	22047	-34779	020	N	-a	-1.0314	0.9439	-0.0134	237.1	-	-	14N	27W
2855	143	-0812 Mar 16	16:31:39	22045	-34778	058	Nb	a-	1.5330	0.0218	-0.9321	39.6	-	-	6N	153W
2856	143	-0812 Aug 10	22:09:43	22039	-34773	025	N	-t	1.2561	0.5896	-0.4829	224.8	-	-	17S	120E
2857	143	-0811 Feb 04	23:49:49	22030	-34767	030	T	-p	-0.3539	2.2012	1.2158	320.5	201.9	66.0	18N	99E
2858	143	-0811 Jul 31	02:49:22	22022	-34761	035	P	-t	0.4717	2.0031	0.9818	332.9	198.7	-	20S	50E
2859	143	-0810 Jan 25	10:29:45	22014	-34755	040	T	p-	0.3728	2.1957	1.1525	343.3	209.2	59.1	21N	62W
2860	143	-0810 Jul 20	14:49:19	22006	-34749	045	T	p-	-0.2931	2.3032	1.3366	320.3	206.2	79.2	23S	131W
2861	144	-0809 Jan 14	13:44:07	21998	-34743	050	N	t-	1.1118	0.8646	-0.2282	267.1	-	-	24N	111W
2862	144	-0809 Jun 11	00:30:48	21991	-34738	017	Ne	-a	1.4905	0.0959	-0.8503	82.5	-	-	21S	82E
2863	144	-0809 Jul 10	07:20:32	21990	-34737	055	P	a-	-1.0064	0.9828	0.0391	240.3	42.9	-	25S	19W
2864	144	-0809 Dec 04	18:49:35	21983	-34732	022	N	-t	-1.2274	0.6464	-0.4343	231.1	-	-	20N	168E
2865	144	-0808 May 30	15:21:54	21975	-34726	027	P	-a	0.7960	1.3905	0.4040	289.5	137.1	-	20S	141W
2866	144	-0808 Nov 23	01:16:03	21967	-34720	032	P	-p	-0.4877	1.9751	0.9512	324.6	192.2	-	19N	70E
2867	144	-0807 May 19	23:50:05	21959	-34714	037	T+	pp	0.0322	2.8193	1.7787	361.4	230.7	105.0	18S	92E
2868	144	-0807 Nov 12	14:17:34	21950	-34708	042	T+	p-	0.2060	2.4694	1.4905	319.8	207.4	88.6	16N	126W
2869	144	-0806 May 09	01:46:42	21942	-34702	047	P	t-	-0.7477	1.5209	0.4518	329.9	155.7	-	15S	63E
2870	144	-0806 Nov 02	05:57:38	21934	-34696	052	P	a-	0.8667	1.2551	0.2799	266.5	111.2	-	13N	1W
2871	144	-0805 Mar 29	14:13:02	21927	-34691	019	N	-t	1.2887	0.5086	-0.5216	201.1	-	-	1N	119W
2872	144	-0805 Apr 28	03:05:52	21926	-34690	057	N	t-	-1.4784	0.1702	-0.8793	123.6	-	-	12S	43E
2873	144	-0805 Sep 23	06:14:50	21919	-34685	024	N	-a	-1.2045	0.6669	-0.3711	224.9	-	-	4S	3W
2874	144	-0804 Mar 18	00:48:05	21911	-34679	029	P	-a	0.4884	1.9510	0.9725	313.4	190.2	-	5N	82E
2875	144	-0804 Sep 11	10:13:52	21903	-34673	034	P	-t	-0.5262	1.9344	0.8511	351.8	198.0	-	8S	62W
2876	144	-0803 Mar 07	16:28:57	21895	-34667	039	T-	p-	-0.2422	2.3895	1.4372	314.2	205.3	85.5	8N	152W
2877	144	-0803 Aug 31	09:56:42	21887	-34661	044	T+	pp	0.1876	2.5567	1.4715	372.7	230.7	95.7	12S	58W
2878	144	-0802 Feb 25	08:49:33	21879	-34655	049	P	a-	-0.9585	1.0842	0.1140	255.1	73.5	-	11N	37W
2879	144	-0802 Aug 20	12:50:51	21871	-34649	054	P	h-	0.9017	1.2236	0.1836	291.0	98.9	-	15S	101W
2880	144	-0801 Jan 16	07:00:07	21864	-34644	021	N	-h	1.0829	0.9022	-0.1599	262.2	-	-	23N	10W

Cat Num	Canon Plate	Calendar Date	TD of Greatest Eclipse	ΔT s	Luna Num	Saros Num	Ecl. Type	QSE	Gamma	Pen. Mag.	Um. Mag.	Eclipse Phase Durations Pen. m	Par. m	Total m	Greatest in Zenith Lat.	Lng.
2881	145	-0801 Jul 11	14:37:59	21856	-34638	026	P	-a	-1.0115	0.9789	0.0244	243.2	34.3	-	24S	129W
2882	145	-0800 Jan 05	08:21:54	21848	-34632	031	T	-t	0.4111	2.1546	1.0536	365.3	213.7	37.6	24N	32W
2883	145	-0800 Jun 30	07:40:02	21840	-34626	036	T	-p	-0.2784	2.3190	1.3747	312.7	204.0	81.3	24S	25W
2884	145	-0800 Dec 24	07:47:01	21832	-34620	041	T	p-	-0.2758	2.3966	1.3078	366.0	223.1	81.6	23N	25W
2885	145	-0799 Jun 19	23:19:23	21823	-34614	046	P	a-	0.4770	1.9708	0.9943	316.1	192.9	-	23S	99E
2886	145	-0799 Dec 13	12:31:58	21815	-34608	051	P	a-	-0.9461	1.1405	0.1036	277.2	73.6	-	22N	98W
2887	145	-0798 May 10	20:30:55	21809	-34603	018	N	-t	-1.4648	0.1969	-0.8564	133.4	-	-	16S	141E
2888	145	-0798 Jun 09	09:19:29	21807	-34602	056	N	t-	1.2970	0.4931	-0.5369	199.7	-	-	21S	51W
2889	145	-0798 Nov 03	13:18:53	21801	-34597	023	N	-a	1.1352	0.7610	-0.2113	218.0	-	-	14N	112W
2890	145	-0797 Apr 29	21:27:38	21792	-34591	028	P	-t	-0.7297	1.5539	0.4847	331.9	160.3	-	12S	128E
2891	145	-0797 Oct 24	04:55:31	21784	-34585	033	P	-a	0.4843	1.9601	0.9782	308.4	187.1	-	9N	14E
2892	145	-0796 Apr 17	23:54:11	21776	-34579	038	T+	pp	0.0472	2.7904	1.7528	358.7	229.2	104.2	8S	92E
2893	145	-0796 Oct 12	17:11:53	21768	-34573	043	T-	p-	-0.2092	2.4888	1.4595	342.5	216.9	90.4	5N	169W
2894	145	-0795 Apr 07	09:08:25	21760	-34567	048	P	a-	0.7981	1.3858	0.4010	287.0	135.6	-	3S	45W
2895	145	-0795 Oct 01	22:36:08	21752	-34561	053	P	t-	-0.9495	1.1564	0.0757	295.6	66.2	-	1S	110E
2896	145	-0794 Feb 26	16:22:40	21745	-34556	020	N	-a	-1.0687	0.8740	-0.0806	229.8	-	-	11N	151W
2897	145	-0794 Mar 28	00:16:46	21744	-34555	058	N	a-	1.4784	0.1201	-0.8301	91.9	-	-	2N	89E
2898	145	-0794 Aug 22	05:16:03	21737	-34550	025	N	-t	1.3054	0.5000	-0.5740	208.5	-	-	14S	12E
2899	145	-0793 Feb 16	07:59:33	21729	-34544	030	T	-a	-0.3849	2.1440	1.1591	319.7	199.9	57.9	15N	25W
2900	145	-0793 Aug 11	10:11:32	21721	-34538	035	P	-t	0.5257	1.9037	0.8829	327.7	191.2	-	18S	62W
2901	146	-0792 Feb 05	18:25:25	21713	-34532	040	T	p-	0.3485	2.2403	1.1970	345.7	212.0	66.4	19N	179E
2902	146	-0792 Jul 30	22:26:57	21705	-34526	045	T-	p-	-0.2339	2.4119	1.4452	321.6	209.0	87.3	21S	114E
2903	146	-0791 Jan 24	21:25:18	21697	-34520	050	N	t-	1.0909	0.9022	-0.1890	272.0	-	-	22N	133E
2904	146	-0791 Jul 20	14:59:07	21689	-34514	055	P	a-	-0.9447	1.0973	0.1513	251.0	83.0	-	23S	134W
2905	146	-0791 Dec 15	03:01:46	21682	-34509	022	N	-t	-1.2285	0.6426	-0.4345	229.8	-	-	22N	45E
2906	146	-0790 Jun 10	22:26:02	21674	-34503	027	P	-a	0.8754	1.2469	0.2564	279.9	112.0	-	21S	112E
2907	146	-0790 Dec 04	09:52:19	21666	-34497	032	P	-p	-0.4875	1.9738	0.9532	323.3	191.8	-	21N	59W
2908	146	-0789 May 31	06:24:46	21658	-34491	037	T+	pp	0.1171	2.6651	1.6216	361.9	229.8	101.9	20S	8W
2909	146	-0789 Nov 23	23:09:09	21650	-34485	042	T+	p-	0.2048	2.4710	1.4931	319.3	207.2	88.7	19N	101E
2910	146	-0788 May 19	08:04:41	21642	-34479	047	P	t-	-0.6591	1.6832	0.6145	340.4	176.9	-	18S	33W
2911	146	-0788 Nov 12	14:47:39	21634	-34473	052	P	a-	0.8653	1.2587	0.2814	267.2	111.6	-	17N	135W
2912	146	-0787 Apr 08	21:14:15	21627	-34468	019	N	-t	1.3513	0.3907	-0.6336	177.4	-	-	3S	133E
2913	146	-0787 May 08	09:39:44	21626	-34467	057	N	t-	-1.3943	0.3225	-0.7230	167.1	-	-	16S	57W
2914	146	-0787 Oct 03	14:18:01	21619	-34462	024	N	-a	-1.2276	0.6275	-0.4165	219.9	-	-	0S	125W
2915	146	-0786 Mar 29	08:21:37	21611	-34456	029	P	-a	0.5454	1.8435	0.8704	307.7	182.8	-	1N	33W
2916	146	-0786 Sep 22	17:46:08	21603	-34450	034	P	-t	-0.5577	1.8791	0.7908	350.3	193.2	-	4S	177W
2917	146	-0785 Mar 19	00:22:21	21595	-34444	039	T-	p-	-0.1914	2.4811	1.5320	315.3	207.6	90.9	4N	87E
2918	146	-0785 Sep 11	17:16:40	21587	-34438	044	T+	pp	0.1470	2.6322	1.5449	373.5	232.0	99.4	8S	169W
2919	146	-0784 Mar 07	16:45:43	21579	-34432	049	P	a-	-0.9172	1.1596	0.1902	262.2	94.1	-	7N	158W
2920	146	-0784 Aug 30	20:23:51	21571	-34426	054	P	h-	0.8558	1.3079	0.2678	297.1	117.7	-	11S	144E
2921	147	-0783 Jan 26	14:54:40	21564	-34421	021	N	-h	1.1073	0.8575	-0.2045	257.5	-	-	22N	129W
2922	147	-0783 Jul 21	22:13:29	21556	-34415	026	N	-a	-1.0711	0.8699	-0.0851	231.5	-	-	23S	117E
2923	147	-0783 Aug 20	06:56:22	21554	-34414	064	N	a-	1.5166	0.0660	-0.9159	70.7	-	-	14S	14W
2924	147	-0782 Jan 15	16:07:18	21548	-34409	031	T	-t	0.4290	2.1207	1.0217	364.1	211.9	24.2	23N	149W
2925	147	-0782 Jul 11	15:12:49	21540	-34403	036	T	-p	-0.3455	2.1973	1.2502	310.4	199.6	69.4	24S	139W
2926	147	-0781 Jan 04	15:46:30	21532	-34397	041	T-	p-	-0.2648	2.4147	1.3302	365.2	223.5	83.7	23N	145W
2927	147	-0781 Jul 01	06:32:38	21524	-34391	046	T	p-	0.4061	2.1033	1.1221	322.1	200.6	52.0	23S	10W
2928	147	-0781 Dec 24	21:00:21	21516	-34385	051	P	a-	-0.9406	1.1484	0.1160	276.8	77.5	-	23N	135E
2929	147	-0780 May 21	02:57:48	21509	-34380	018	Ne	-t	-1.5509	0.0399	-1.0152	61.0	-	-	20S	43E
2930	147	-0780 Jun 19	16:00:40	21508	-34379	056	N	t-	1.2213	0.6343	-0.4000	224.4	-	-	22S	152W
2931	147	-0780 Nov 13	22:10:51	21501	-34374	023	N	-a	1.1354	0.7607	-0.2119	217.8	-	-	18N	114E
2932	147	-0780 Dec 13	09:08:42	21499	-34373	061	N	a-	-1.5529	0.0004	-0.9836	5.2	-	-	21N	48W
2933	147	-0779 May 10	03:49:05	21493	-34368	028	P	-t	-0.8145	1.3972	0.3300	320.1	135.3	-	16S	30E
2934	147	-0779 Nov 03	13:39:43	21485	-34362	033	P	-a	0.4892	1.9520	0.9676	309.0	186.7	-	13N	118W
2935	147	-0778 Apr 29	06:40:16	21477	-34356	038	T-	pp	-0.0299	2.8195	1.7868	357.1	228.8	104.4	12S	12W
2936	147	-0778 Oct 24	01:33:16	21469	-34350	043	T-	p-	-0.2013	2.5060	1.4716	344.3	217.8	91.2	9N	64E
2937	147	-0777 Apr 18	16:27:38	21461	-34344	048	P	a-	0.7297	1.5088	0.5292	293.8	152.3	-	7S	157W
2938	147	-0777 Oct 13	06:26:36	21453	-34338	053	P	t-	-0.9338	1.1877	0.1020	299.5	76.8	-	4N	9W
2939	147	-0776 Mar 09	00:23:47	21446	-34333	020	N	-a	-1.1122	0.7929	-0.1591	220.7	-	-	7N	87E
2940	147	-0776 Apr 07	07:56:37	21445	-34332	058	N	a-	1.4192	0.2270	-0.7199	124.9	-	-	2S	28W

Cat Num	Canon Plate	Calendar Date	TD of Greatest Eclipse	ΔT s	Luna Num	Saros Num	Ecl. Type	QSE	Gamma	Pen. Mag.	Um. Mag.	Eclipse Phase Durations Pen. m	Eclipse Phase Durations Par. m	Eclipse Phase Durations Total m	Greatest in Zenith Lat.	Greatest in Zenith Lng.
2941	148	-0776 Sep 01	12:32:43	21438	-34327	025	N	-t	1.3465	0.4252	-0.6501	193.3	-	-	10S	99W
2942	148	-0775 Feb 26	15:57:59	21430	-34321	030	T	-a	-0.4247	2.0709	1.0864	318.3	196.8	43.8	11N	146W
2943	148	-0775 Aug 21	17:45:10	21422	-34315	035	P	-h	0.5711	1.8205	0.7998	322.8	184.0	-	15S	177W
2944	148	-0774 Feb 16	02:09:11	21414	-34309	040	T	p-	0.3154	2.3010	1.2578	348.4	215.3	74.6	15N	62E
2945	148	-0774 Aug 11	06:14:12	21406	-34303	045	T-	p-	-0.1822	2.5072	1.5399	322.4	210.7	92.3	18S	4W
2946	148	-0773 Feb 05	04:56:20	21398	-34297	050	N	t-	1.0621	0.9540	-0.1352	278.5	-	-	19N	20E
2947	148	-0773 Jul 31	22:44:28	21390	-34291	055	P	a-	-0.8886	1.2015	0.2527	259.8	105.8	-	22S	109E
2948	148	-0773 Dec 26	11:12:04	21383	-34286	022	N	-t	-1.2316	0.6347	-0.4381	227.7	-	-	23N	78W
2949	148	-0772 Jun 21	05:28:20	21375	-34280	027	P	-a	0.9550	1.1032	0.1081	268.7	74.5	-	23S	6E
2950	148	-0772 Dec 14	18:29:45	21367	-34274	032	P	-p	-0.4871	1.9727	0.9559	322.0	191.5	-	23N	171E
2951	148	-0771 Jun 10	12:56:32	21359	-34268	037	T+	pp	0.2031	2.5089	1.4619	361.2	227.1	94.4	22S	107W
2952	148	-0771 Dec 04	08:01:46	21351	-34262	042	T+	p-	0.2043	2.4712	1.4948	318.8	207.0	88.7	22N	33W
2953	148	-0770 May 30	14:21:44	21343	-34256	047	P	t-	-0.5696	1.8473	0.7788	349.4	193.7	-	21S	129W
2954	148	-0770 Nov 23	23:39:56	21335	-34250	052	P	a-	0.8665	1.2574	0.2784	267.6	111.2	-	20N	91E
2955	148	-0769 Apr 20	04:12:50	21328	-34245	019	N	-t	1.4167	0.2678	-0.7509	147.9	-	-	7S	26E
2956	148	-0769 May 19	16:13:44	21327	-34244	057	N	t-	-1.3088	0.4775	-0.5643	199.7	-	-	19S	157W
2957	148	-0769 Oct 14	22:26:25	21320	-34239	024	N	-a	-1.2456	0.5974	-0.4524	216.2	-	-	4N	111E
2958	148	-0768 Apr 08	15:51:27	21312	-34233	029	P	-a	0.6063	1.7292	0.7613	301.2	173.7	-	4S	148W
2959	148	-0768 Oct 03	01:25:14	21304	-34227	034	P	-t	-0.5833	1.8345	0.7415	349.0	189.0	-	0N	66E
2960	148	-0767 Mar 29	08:08:40	21296	-34221	039	T-	p-	-0.1344	2.5843	1.6379	316.2	209.5	95.3	0S	32W
2961	149	-0767 Sep 22	00:46:46	21288	-34215	044	T+	pp	0.1142	2.6934	1.6041	373.9	232.7	101.5	4S	76E
2962	149	-0766 Mar 19	00:33:42	21280	-34209	049	P	a-	-0.8695	1.2467	0.2778	269.9	112.6	-	3N	83E
2963	149	-0766 Sep 11	04:07:19	21272	-34203	054	P	a-	0.8174	1.3783	0.3381	301.6	130.5	-	7S	26E
2964	149	-0765 Feb 06	22:39:14	21266	-34198	021	N	-t	1.1389	0.7993	-0.2624	250.7	-	-	19N	114E
2965	149	-0765 Aug 02	05:57:11	21258	-34192	026	N	-a	-1.1244	0.7725	-0.1833	220.0	-	-	21S	0E
2966	149	-0765 Aug 31	14:54:41	21256	-34191	064	N	a-	1.4739	0.1445	-0.8377	103.5	-	-	11S	135W
2967	149	-0764 Jan 26	23:43:50	21250	-34186	031	P	-t	0.4540	2.0735	0.9771	362.2	209.1	-	21N	97E
2968	149	-0764 Jul 21	22:50:21	21242	-34180	036	T	-a	-0.4084	2.0835	1.1331	307.8	194.6	52.5	23S	106E
2969	149	-0763 Jan 14	23:39:44	21234	-34174	041	T-	p-	-0.2478	2.4434	1.3638	364.5	224.2	86.7	22N	97E
2970	149	-0763 Jul 11	13:49:12	21226	-34168	046	T	p-	0.3386	2.2298	1.2434	327.3	206.8	70.8	23S	119W
2971	149	-0762 Jan 04	05:23:14	21218	-34162	051	P	a-	-0.9295	1.1661	0.1388	277.2	84.2	-	23N	10E
2972	149	-0762 Jun 30	22:44:30	21210	-34156	056	N	t-	1.1479	0.7713	-0.2679	245.2	-	-	23S	106E
2973	149	-0762 Nov 25	07:04:43	21203	-34151	023	N	-a	1.1349	0.7615	-0.2108	217.8	-	-	21N	20W
2974	149	-0762 Dec 24	17:54:45	21202	-34150	061	N	a-	-1.5469	0.0099	-0.9710	27.0	-	-	22N	180W
2975	149	-0761 May 21	10:08:26	21195	-34145	028	P	-t	-0.9010	1.2377	0.1721	306.3	99.9	-	19S	66W
2976	149	-0761 Nov 14	22:27:21	21187	-34139	033	P	-a	0.4910	1.9509	0.9629	309.7	186.8	-	17N	109E
2977	149	-0760 May 09	13:23:29	21179	-34133	038	T-	pp	-0.1107	2.6690	1.6409	354.6	226.9	101.4	15S	114W
2978	149	-0760 Nov 03	09:59:25	21171	-34127	043	T-	p-	-0.1982	2.5140	1.4749	345.9	218.4	91.6	13N	64W
2979	149	-0759 Apr 28	23:41:43	21163	-34121	048	P	a-	0.6559	1.6418	0.6669	300.3	166.8	-	11S	92E
2980	149	-0759 Oct 23	14:24:26	21155	-34115	053	P	t-	-0.9246	1.2068	0.1168	302.2	82.2	-	8N	130W
2981	150	-0758 Mar 20	08:17:06	21149	-34110	020	N	-a	-1.1621	0.7003	-0.2495	209.5	-	-	2N	34W
2982	150	-0758 Apr 18	15:31:17	21147	-34109	058	N	a-	1.3556	0.3423	-0.6018	151.5	-	-	6S	144W
2983	150	-0758 Sep 12	19:59:43	21141	-34104	025	N	-t	1.3799	0.3646	-0.7120	179.7	-	-	6S	147E
2984	150	-0757 Mar 09	23:48:34	21133	-34098	030	T	-a	-0.4703	1.9871	1.0027	316.2	192.6	8.0	7N	94E
2985	150	-0757 Sep 02	01:28:21	21125	-34092	035	P	-a	0.6095	1.7499	0.7294	318.3	177.3	-	11S	66E
2986	150	-0756 Feb 27	09:43:52	21117	-34086	040	T	pp	0.2755	2.3741	1.3310	351.2	218.7	82.6	12N	54W
2987	150	-0756 Aug 21	14:08:42	21109	-34080	045	T-	p-	-0.1359	2.5925	1.6244	322.7	211.7	95.5	15S	124W
2988	150	-0755 Feb 15	12:18:31	21101	-34074	050	N*	t-	1.0263	1.0185	-0.0684	285.9	-	-	16N	92W
2989	150	-0755 Aug 11	06:36:35	21093	-34068	055	P	a-	-0.8384	1.2952	0.3433	267.2	121.7	-	19S	10W
2990	150	-0754 Jan 05	19:17:23	21086	-34063	022	N	-t	-1.2393	0.6182	-0.4497	224.2	-	-	22N	161E
2991	150	-0754 Jul 02	12:34:03	21078	-34057	027	N	-a	1.0305	0.9671	-0.0330	256.5	-	-	23S	102W
2992	150	-0754 Dec 26	03:03:59	21070	-34051	032	P	-p	-0.4897	1.9658	0.9533	320.5	190.8	-	23N	43E
2993	150	-0753 Jun 21	19:29:40	21062	-34045	037	T	-t	0.2873	2.3564	1.3057	359.5	222.6	81.8	23S	154E
2994	150	-0753 Dec 15	16:52:28	21055	-34039	042	T+	p-	0.2021	2.4744	1.4998	318.3	206.9	88.9	23N	166W
2995	150	-0752 Jun 09	20:39:50	21047	-34033	047	P	t-	-0.4808	2.0104	0.9416	356.8	206.8	-	23S	136E
2996	150	-0752 Dec 04	08:31:17	21039	-34027	052	P	a-	0.8670	1.2570	0.2769	268.0	111.0	-	22N	42W
2997	150	-0751 Apr 30	11:06:58	21032	-34022	019	N	-t	1.4864	0.1373	-0.8762	106.6	-	-	11S	80W
2998	150	-0751 May 29	22:47:48	21031	-34021	057	N	t-	-1.2218	0.6354	-0.4032	226.1	-	-	21S	103E
2999	150	-0751 Oct 25	06:40:49	21024	-34016	024	N	-a	-1.2582	0.5769	-0.4780	213.7	-	-	9N	15W
3000	150	-0750 Apr 19	23:16:10	21016	-34010	029	P	-a	0.6723	1.6059	0.6424	293.7	162.3	-	8S	99E

Cat Num	Canon Plate	Calendar Date	TD of Greatest Eclipse	ΔT s	Luna Num	Saros Num	Ecl. Type	QSE	Gamma	Pen. Mag.	Um. Mag.	Eclipse Phase Durations Pen. m	Par. m	Total m	Greatest in Zenith Lat.	Lng.
3001	151	-0750 Oct 14	09:11:50	21008	-34004	034	P	-t	-0.6025	1.8014	0.7044	348.1	185.6	-	5N	52W
3002	151	-0749 Apr 09	15:48:46	21000	-33998	039	T-	pp	-0.0721	2.6975	1.7534	316.9	210.9	98.2	4S	149W
3003	151	-0749 Oct 03	08:27:00	20992	-33992	044	T+	pp	0.0893	2.7400	1.6491	373.9	232.9	102.7	1N	41W
3004	151	-0748 Mar 29	08:12:08	20984	-33986	049	P	a-	-0.8147	1.3474	0.3786	278.1	129.9	-	1S	34W
3005	151	-0748 Sep 21	12:01:20	20976	-33980	054	P	a-	0.7867	1.4346	0.3945	304.8	139.3	-	3S	94W
3006	151	-0747 Feb 17	06:13:50	20970	-33975	021	N	-t	1.1778	0.7275	-0.3336	241.5	-	-	16N	1W
3007	151	-0747 Aug 12	13:48:32	20962	-33969	026	N	-a	-1.1720	0.6858	-0.2714	208.9	-	-	19S	119W
3008	151	-0747 Sep 10	23:02:40	20961	-33968	064	N	a-	1.4383	0.2101	-0.7727	123.8	-	-	7S	101E
3009	151	-0746 Feb 06	07:11:52	20954	-33963	031	P	-t	0.4860	2.0132	0.9199	359.5	205.1	-	18N	16W
3010	151	-0746 Aug 02	06:34:20	20946	-33957	036	T	-a	-0.4658	1.9802	1.0260	305.1	189.1	24.0	21S	11W
3011	151	-0745 Jan 26	07:28:14	20938	-33951	041	T-	p-	-0.2260	2.4807	1.4065	363.8	225.1	90.0	20N	21W
3012	151	-0745 Jul 22	21:08:50	20930	-33945	046	T	p-	0.2743	2.3507	1.3586	331.8	211.7	82.6	22S	130E
3013	151	-0744 Jan 15	13:43:15	20922	-33939	051	P	a-	-0.9152	1.1895	0.1678	278.0	92.0	-	22N	116W
3014	151	-0744 Jul 11	05:30:51	20914	-33933	056	N	t-	1.0767	0.9046	-0.1398	263.0	-	-	22S	4E
3015	151	-0744 Dec 05	15:58:42	20908	-33928	023	N	-a	1.1345	0.7620	-0.2097	217.8	-	-	23N	154W
3016	151	-0743 Jan 04	02:37:01	20906	-33927	061	N	a-	-1.5378	0.0248	-0.9526	42.6	-	-	22N	50E
3017	151	-0743 May 31	16:28:12	20900	-33922	028	P	-t	-0.9871	1.0791	0.0147	290.5	29.9	-	22S	162W
3018	151	-0743 Nov 25	07:15:40	20892	-33916	033	P	-a	0.4921	1.9501	0.9598	310.5	186.9	-	20N	24W
3019	151	-0742 May 20	20:08:29	20884	-33910	038	T-	pp	-0.1908	2.5199	1.4959	351.2	223.5	95.0	18S	143E
3020	151	-0742 Nov 14	18:26:35	20876	-33904	043	T-	p-	-0.1965	2.5191	1.4759	347.4	219.0	91.8	16N	168E
3021	152	-0741 May 10	06:56:05	20868	-33898	048	P	a-	0.5813	1.7766	0.8059	305.9	178.6	-	15S	19W
3022	152	-0741 Nov 03	22:26:03	20860	-33892	053	P	t-	-0.9186	1.2196	0.1260	304.2	85.4	-	12N	108E
3023	152	-0740 Mar 30	16:02:13	20854	-33887	020	N	-a	-1.2187	0.5956	-0.3525	195.3	-	-	2S	152W
3024	152	-0740 Apr 28	23:00:56	20852	-33886	058	N	a-	1.2875	0.4661	-0.4758	174.6	-	-	10S	101E
3025	152	-0740 Sep 23	03:37:11	20846	-33881	025	N	-t	1.4054	0.3183	-0.7594	168.3	-	-	2S	31E
3026	152	-0739 Mar 20	07:28:28	20838	-33875	030	P	-a	-0.5243	1.8880	0.9035	313.3	186.7	-	3N	23W
3027	152	-0739 Sep 12	09:22:58	20830	-33869	035	P	-a	0.6394	1.6949	0.6745	314.3	171.5	-	7S	54W
3028	152	-0738 Mar 09	17:06:14	20822	-33863	040	T+	pp	0.2263	2.4644	1.4214	354.1	222.3	90.3	8N	166W
3029	152	-0738 Sep 01	22:13:51	20814	-33857	045	T-	p-	-0.0980	2.6627	1.6935	322.8	212.2	97.3	11S	114E
3030	152	-0737 Feb 26	19:30:20	20806	-33851	050	P	t-	0.9823	1.0980	0.0136	294.5	28.8	-	12N	159E
3031	152	-0737 Aug 22	14:35:40	20798	-33845	055	P	a-	-0.7942	1.3783	0.4227	273.4	133.5	-	16S	131W
3032	152	-0736 Jan 17	03:18:32	20792	-33840	022	N	-t	-1.2509	0.5941	-0.4683	219.5	-	-	21N	40E
3033	152	-0736 Jul 12	19:41:25	20784	-33834	027	N	-a	1.1034	0.8363	-0.1695	242.9	-	-	22S	151E
3034	152	-0736 Aug 11	05:48:53	20782	-33833	065	Nb	a-	-1.5394	0.0260	-0.9596	44.8	-	-	19S	0E
3035	152	-0735 Jan 05	11:34:52	20776	-33828	032	P	-a	-0.4959	1.9517	0.9442	318.6	189.7	-	23N	85W
3036	152	-0735 Jul 02	02:03:55	20768	-33822	037	T	-t	0.3697	2.2073	1.1523	356.8	216.3	61.0	23S	55E
3037	152	-0735 Dec 26	01:41:18	20760	-33816	042	T+	p-	0.1980	2.4806	1.5083	317.9	207.0	89.3	24N	62E
3038	152	-0734 Jun 21	03:00:35	20752	-33810	047	T	t-	-0.3935	2.1709	1.1016	362.7	216.8	51.3	24S	40E
3039	152	-0734 Dec 15	17:21:11	20744	-33804	052	P	a-	0.8668	1.2578	0.2769	268.5	111.2	-	24N	174W
3040	152	-0733 May 11	18:01:26	20738	-33799	019	Ne	-t	1.5565	0.0063	-1.0025	23.0	-	-	14S	175E
3041	153	-0733 Jun 10	05:25:27	20736	-33798	057	N	t-	-1.1364	0.7909	-0.2450	247.4	-	-	23S	3E
3042	153	-0733 Nov 05	14:58:51	20730	-33793	024	N	-a	-1.2671	0.5629	-0.4966	212.3	-	-	13N	141W
3043	153	-0732 Apr 30	06:37:39	20722	-33787	029	P	-a	0.7417	1.4765	0.5171	285.0	148.2	-	11S	14W
3044	153	-0732 Oct 24	17:04:20	20714	-33781	034	P	-t	-0.6164	1.7775	0.6771	347.6	183.0	-	9N	172W
3045	153	-0731 Apr 19	23:23:47	20706	-33775	039	T-	pp	-0.0054	2.8189	1.8766	317.1	211.5	99.4	8S	95E
3046	153	-0731 Oct 13	16:15:18	20698	-33769	044	T+	pp	0.0704	2.7751	1.6833	373.7	232.9	103.3	5N	160W
3047	153	-0730 Apr 09	15:43:51	20690	-33763	049	P	a-	-0.7547	1.4576	0.4884	286.3	145.5	-	5S	149W
3048	153	-0730 Oct 02	20:05:10	20683	-33757	054	P	a-	0.7633	1.4774	0.4377	306.8	145.3	-	1N	143E
3049	153	-0729 Feb 28	13:38:16	20676	-33752	021	N	-t	1.2243	0.6420	-0.4185	229.4	-	-	12N	114W
3050	153	-0729 Aug 23	21:47:53	20668	-33746	026	N	-a	-1.2137	0.6101	-0.3487	198.3	-	-	16S	120E
3051	153	-0729 Sep 22	07:19:06	20667	-33745	064	N	a-	1.4090	0.2642	-0.7192	137.8	-	-	2S	25W
3052	153	-0728 Feb 17	14:30:23	20660	-33740	031	P	-t	0.5258	1.9385	0.8486	355.9	199.6	-	15N	127W
3053	153	-0728 Aug 12	14:25:10	20652	-33734	036	P	-a	-0.5174	1.8877	0.9292	302.3	183.3	-	18S	129W
3054	153	-0727 Feb 05	15:07:03	20644	-33728	041	T-	p-	-0.1954	2.5340	1.4656	363.2	226.3	93.9	18N	136W
3055	153	-0727 Aug 02	04:34:53	20637	-33722	046	T+	pp	0.2160	2.4606	1.4625	335.5	215.4	90.4	20S	18E
3056	153	-0726 Jan 25	21:55:49	20629	-33716	051	P	a-	-0.8938	1.2259	0.2100	279.9	102.0	-	20N	121E
3057	153	-0726 Jul 22	12:22:13	20621	-33710	056	N*	t-	1.0101	1.0296	-0.0202	278.2	-	-	21S	100W
3058	153	-0726 Dec 17	00:50:34	20614	-33705	023	N	-a	1.1363	0.7582	-0.2124	217.2	-	-	24N	73E
3059	153	-0725 Jan 15	11:13:25	20613	-33704	061	N	a-	-1.5238	0.0486	-0.9250	59.4	-	-	21N	80W
3060	153	-0725 Jun 11	22:49:32	20606	-33699	028	N	-t	-1.0721	0.9227	-0.1409	272.6	-	-	23S	101E

Cat Num	Canon Plate	Calendar Date	TD of Greatest Eclipse	ΔT	Luna Num	Saros Num	Ecl. Type	QSE	Gamma	Pen. Mag.	Um. Mag.	Eclipse Phase Durations Pen.	Eclipse Phase Durations Par.	Eclipse Phase Durations Total	Greatest in Zenith Lat.	Greatest in Zenith Lng.
				s								m	m	m		
3061	154	-0725 Dec 06	16:04:30	20598	-33693	033	P	-a	0.4924	1.9505	0.9584	311.3	187.2	-	22N	157W
3062	154	-0724 May 31	02:52:11	20591	-33687	038	T	pp	-0.2731	2.3672	1.3467	346.8	218.5	84.0	21S	41E
3063	154	-0724 Nov 25	02:56:08	20583	-33681	043	T-	pp	-0.1976	2.5188	1.4722	348.8	219.5	91.8	19N	40E
3064	154	-0723 May 20	14:08:12	20575	-33675	048	P	a-	0.5036	1.9174	0.9501	310.8	188.6	-	18S	128W
3065	154	-0723 Nov 14	06:31:01	20567	-33669	053	P	t-	-0.9157	1.2264	0.1300	305.6	86.8	-	16N	15W
3066	154	-0722 Apr 10	23:41:15	20561	-33664	020	N	-a	-1.2801	0.4822	-0.4647	177.8	-	-	6S	91E
3067	154	-0722 May 10	06:28:12	20559	-33663	058	N	a-	1.2172	0.5944	-0.3460	194.5	-	-	14S	13W
3068	154	-0722 Oct 04	11:24:42	20553	-33658	025	N	-t	1.4237	0.2850	-0.7933	159.4	-	-	3N	88W
3069	154	-0722 Nov 03	05:33:53	20551	-33657	063	N	t-	-1.5822	0.0062	-1.0958	24.7	-	-	12N	0W
3070	154	-0721 Mar 31	15:01:21	20545	-33652	030	P	-a	-0.5833	1.7799	0.7951	309.5	179.0	-	1S	138W
3071	154	-0721 Sep 23	17:25:54	20537	-33646	035	P	-a	0.6633	1.6510	0.6307	310.9	166.5	-	2S	177W
3072	154	-0720 Mar 20	00:20:20	20529	-33640	040	T+	pp	0.1711	2.5657	1.5227	356.8	225.6	96.7	3N	83E
3073	154	-0720 Sep 12	06:26:53	20521	-33634	045	T-	p-	-0.0664	2.7212	1.7508	322.6	212.3	98.2	7S	12W
3074	154	-0719 Mar 09	02:32:11	20513	-33628	050	P	t-	0.9305	1.1917	0.1100	303.8	80.7	-	8N	51E
3075	154	-0719 Sep 01	22:43:06	20506	-33622	055	P	a-	-0.7572	1.4480	0.4886	278.4	142.2	-	12S	105E
3076	154	-0718 Jan 27	11:12:53	20499	-33617	022	N	-t	-1.2683	0.5592	-0.4973	212.7	-	-	19N	79W
3077	154	-0718 Jul 24	02:54:06	20491	-33611	027	N	-a	1.1708	0.7156	-0.2961	228.6	-	-	21S	42E
3078	154	-0718 Aug 22	13:34:08	20490	-33610	065	N	a-	-1.4999	0.1015	-0.8901	88.3	-	-	16S	117W
3079	154	-0717 Jan 16	19:59:49	20483	-33605	032	P	-a	-0.5077	1.9276	0.9252	316.4	188.0	-	22N	149E
3080	154	-0717 Jul 13	08:43:05	20475	-33599	037	T	-t	0.4471	2.0675	1.0080	353.3	208.5	14.6	23S	46W
3081	155	-0716 Jan 06	10:25:02	20468	-33593	042	T+	p-	0.1897	2.4946	1.5249	317.6	207.2	90.2	24N	69W
3082	155	-0716 Jul 01	09:25:03	20460	-33587	047	T	t-	-0.3089	2.3266	1.2563	367.1	224.2	77.2	24S	57W
3083	155	-0716 Dec 26	02:07:12	20452	-33581	052	P	a-	0.8640	1.2630	0.2819	269.4	112.3	-	25N	54E
3084	155	-0715 Jun 20	12:07:09	20444	-33575	057	N	t-	-1.0530	0.9428	-0.0909	265.0	-	-	24S	98W
3085	155	-0715 Nov 15	23:18:54	20438	-33570	024	N	-a	-1.2735	0.5531	-0.5102	211.4	-	-	16N	93E
3086	155	-0714 May 11	13:57:12	20430	-33564	029	P	-a	0.8131	1.3439	0.3878	275.3	130.6	-	15S	126W
3087	155	-0714 Nov 05	01:02:41	20422	-33558	034	P	-t	-0.6249	1.7633	0.6602	347.4	181.4	-	13N	67E
3088	155	-0713 May 01	06:53:55	20414	-33552	039	T+	pp	0.0653	2.7085	1.7673	316.9	211.2	98.6	12S	20W
3089	155	-0713 Oct 25	00:11:45	20406	-33546	044	T+	pp	0.0577	2.7986	1.7063	373.2	232.6	103.6	9N	79E
3090	155	-0712 Apr 19	23:08:01	20398	-33540	049	P	a-	-0.6893	1.5780	0.6079	294.6	159.8	-	9S	97E
3091	155	-0712 Oct 13	04:18:31	20391	-33534	054	P	a-	0.7466	1.5078	0.4687	307.8	149.2	-	6N	18E
3092	155	-0711 Mar 10	20:51:17	20384	-33529	021	N	-t	1.2791	0.5411	-0.5187	213.2	-	-	8N	136E
3093	155	-0711 Apr 09	10:11:53	20383	-33528	059	N	t-	-1.5042	0.1109	-0.9149	97.2	-	-	6S	68W
3094	155	-0711 Sep 03	05:56:33	20376	-33523	026	N	-a	-1.2483	0.5475	-0.4131	188.8	-	-	12S	3W
3095	155	-0711 Oct 02	15:45:11	20375	-33522	064	N	a-	1.3869	0.3050	-0.6789	147.1	-	-	2N	154W
3096	155	-0710 Feb 27	21:40:11	20368	-33517	031	P	-t	0.5728	1.8505	0.7642	351.3	192.3	-	11N	124E
3097	155	-0710 Aug 23	22:22:50	20361	-33511	036	P	-a	-0.5629	1.8064	0.8435	299.7	177.5	-	15S	110E
3098	155	-0709 Feb 16	22:40:19	20353	-33505	041	T-	pp	-0.1593	2.5970	1.5347	362.5	227.4	97.6	15N	109E
3099	155	-0709 Aug 13	12:06:15	20345	-33499	046	T+	pp	0.1629	2.5613	1.5568	338.7	218.1	95.5	17S	96W
3100	155	-0708 Feb 06	06:02:45	20337	-33493	051	P	a-	-0.8670	1.2719	0.2622	282.3	112.7	-	17N	2W
3101	156	-0708 Aug 01	19:18:03	20329	-33487	056	P	t-	0.9474	1.1475	0.0919	291.2	72.4	-	19S	156E
3102	156	-0708 Dec 27	09:39:42	20323	-33482	023	N	-a	1.1405	0.7496	-0.2196	216.1	-	-	25N	59W
3103	156	-0707 Jan 25	19:44:02	20322	-33481	061	N	a-	-1.5050	0.0811	-0.8885	76.3	-	-	19N	152E
3104	156	-0707 Jun 22	05:14:56	20315	-33476	028	N	-t	-1.1540	0.7723	-0.2909	252.8	-	-	25S	5E
3105	156	-0707 Dec 17	00:49:37	20307	-33470	033	P	-a	0.4955	1.9454	0.9519	311.9	187.2	-	24N	72E
3106	156	-0706 Jun 11	09:41:03	20299	-33464	038	T	-p	-0.3522	2.2205	1.2030	341.7	212.0	67.4	23S	63W
3107	156	-0706 Dec 06	11:23:41	20292	-33458	043	T-	pp	-0.1973	2.5208	1.4714	350.1	220.1	91.9	22N	88W
3108	156	-0705 May 31	21:21:59	20284	-33452	048	T	a-	0.4264	2.0575	1.0932	314.7	196.5	45.3	20S	122E
3109	156	-0705 Nov 25	14:36:42	20276	-33446	053	P	t-	-0.9135	1.2314	0.1329	306.7	87.9	-	19N	137W
3110	156	-0704 Apr 21	07:14:14	20269	-33441	020	N	-a	-1.3465	0.3601	-0.5861	155.6	-	-	10S	25W
3111	156	-0704 May 20	13:53:20	20268	-33440	058	N	a-	1.1446	0.7272	-0.2124	212.2	-	-	17S	126W
3112	156	-0704 Oct 14	19:21:09	20262	-33435	025	N	-t	1.4353	0.2638	-0.8146	153.3	-	-	7N	151E
3113	156	-0704 Nov 13	13:35:34	20260	-33434	063	N	t-	-1.5756	0.0182	-1.0838	42.1	-	-	15N	122W
3114	156	-0703 Apr 10	22:24:03	20254	-33429	030	P	-a	-0.6499	1.6582	0.6724	304.4	168.6	-	6S	109E
3115	156	-0703 Oct 04	01:39:46	20246	-33423	035	P	-a	0.6791	1.6218	0.6020	308.3	162.9	-	2N	57E
3116	156	-0702 Mar 31	07:23:37	20238	-33417	040	T+	pp	0.1079	2.6818	1.6385	359.2	228.3	101.6	1S	25W
3117	156	-0702 Sep 23	14:49:03	20231	-33411	045	T-	p-	-0.0420	2.7666	1.7951	322.4	212.2	98.5	3S	139W
3118	156	-0701 Mar 20	09:24:35	20223	-33405	050	P	t-	0.8709	1.2998	0.2207	313.5	112.6	-	4N	54W
3119	156	-0701 Sep 13	06:58:30	20215	-33399	055	P	a-	-0.7271	1.5053	0.5420	282.5	148.7	-	8S	21W
3120	156	-0700 Feb 07	19:00:18	20208	-33394	022	N	-t	-1.2921	0.5123	-0.5379	203.6	-	-	16N	164E

Cat Num	Canon Plate	Calendar Date	TD of Greatest Eclipse	ΔT s	Luna Num	Saros Num	Ecl. Type	QSE	Gamma	Pen. Mag.	Um. Mag.	Eclipse Phase Durations Pen. m	Par. m	Total m	Greatest in Zenith Lat.	Lng.
3121	157	-0700 Mar 08	09:55:49	20207	-33393	060	Nb	t-	1.5726	0.0056	-1.0603	22.9	-	-	9N	61W
3122	157	-0700 Aug 03	10:10:45	20201	-33388	027	N	-a	1.2339	0.6030	-0.4151	213.2	-	-	19S	68W
3123	157	-0700 Sep 01	21:26:18	20199	-33387	065	N	a-	-1.4664	0.1662	-0.8317	112.8	-	-	12S	123E
3124	157	-0699 Jan 27	04:19:35	20193	-33382	032	P	-a	-0.5246	1.8938	0.8970	313.7	185.6	-	20N	23E
3125	157	-0699 Jul 23	15:26:44	20185	-33376	037	P	-t	0.5202	1.9359	0.8715	349.1	199.2	-	21S	147W
3126	157	-0698 Jan 16	19:03:39	20177	-33370	042	T+	p-	0.1768	2.5167	1.5501	317.5	207.7	91.4	22N	162E
3127	157	-0698 Jul 12	15:55:53	20170	-33364	047	T-	pp	-0.2291	2.4737	1.4021	370.2	229.3	91.4	24S	155W
3128	157	-0697 Jan 06	10:49:38	20162	-33358	052	P	a-	0.8588	1.2725	0.2915	270.6	114.3	-	24N	76W
3129	157	-0697 Jul 01	18:53:48	20154	-33352	057	P	t-	-0.9722	1.0901	0.0581	279.5	57.0	-	25S	159E
3130	157	-0697 Nov 27	07:40:04	20147	-33347	024	N	-a	-1.2783	0.5459	-0.5207	210.9	-	-	19N	33W
3131	157	-0696 May 21	21:16:22	20140	-33341	029	P	-a	0.8853	1.2099	0.2566	264.5	108.0	-	18S	123E
3132	157	-0696 Nov 15	09:03:27	20132	-33335	034	P	-t	-0.6315	1.7521	0.6471	347.2	180.1	-	17N	54W
3133	157	-0695 May 11	14:21:08	20124	-33329	039	T+	-p	0.1387	2.5736	1.6328	316.0	209.8	95.4	16S	133W
3134	157	-0695 Nov 04	08:14:04	20116	-33323	044	T+	pp	0.0491	2.8142	1.7222	372.6	232.3	103.6	13N	43W
3135	157	-0694 May 01	06:27:43	20109	-33317	049	P	a-	-0.6205	1.7050	0.7335	302.4	172.5	-	13S	15W
3136	157	-0694 Oct 24	12:38:06	20101	-33311	054	P	a-	0.7340	1.5303	0.4923	308.2	151.8	-	10N	109W
3137	157	-0693 Mar 22	03:54:48	20094	-33306	021	N	-t	1.3406	0.4279	-0.6314	192.0	-	-	4N	28E
3138	157	-0693 Apr 20	17:09:29	20093	-33305	059	N	t-	-1.4388	0.2316	-0.7957	139.4	-	-	10S	175W
3139	157	-0693 Sep 14	14:13:30	20087	-33300	026	N	-a	-1.2769	0.4961	-0.4666	180.5	-	-	8S	129W
3140	157	-0693 Oct 14	00:18:26	20085	-33299	064	N	a-	1.3701	0.3359	-0.6481	153.7	-	-	6N	76E
3141	158	-0692 Mar 10	04:39:09	20079	-33294	031	P	-t	0.6286	1.7462	0.6637	345.3	182.3	-	7N	18E
3142	158	-0692 Sep 03	06:28:25	20071	-33288	036	P	-a	-0.6016	1.7377	0.7700	297.4	172.0	-	11S	13W
3143	158	-0691 Feb 27	06:03:50	20063	-33282	041	T-	pp	-0.1144	2.6762	1.6203	361.7	228.5	100.9	11N	3W
3144	158	-0691 Aug 23	19:45:24	20056	-33276	046	T+	pp	0.1170	2.6488	1.6379	341.4	220.0	98.7	14S	148E
3145	158	-0690 Feb 16	14:00:54	20048	-33270	051	P	a-	-0.8321	1.3328	0.3295	285.7	124.8	-	14N	122W
3146	158	-0690 Aug 13	02:21:49	20040	-33264	056	P	t-	0.8915	1.2531	0.1916	302.0	103.3	-	17S	48E
3147	158	-0689 Jan 07	18:24:31	20034	-33259	023	N	-a	1.1489	0.7333	-0.2339	214.0	-	-	25N	170E
3148	158	-0689 Feb 06	04:07:44	20032	-33258	061	N	a-	-1.4804	0.1242	-0.8414	93.8	-	-	17N	26E
3149	158	-0689 Jul 03	11:43:55	20026	-33253	028	N	-t	-1.2332	0.6269	-0.4362	230.7	-	-	25S	93W
3150	158	-0689 Dec 28	09:32:23	20018	-33247	033	P	-a	0.5003	1.9371	0.9427	312.4	187.0	-	24N	58W
3151	158	-0688 Jun 21	16:32:27	20010	-33241	038	T	-t	-0.4305	2.0756	1.0606	335.7	204.0	38.4	24S	166W
3152	158	-0688 Dec 16	19:49:12	20003	-33235	043	T-	pp	-0.1962	2.5239	1.4724	351.4	220.7	92.2	23N	146E
3153	158	-0687 Jun 11	04:36:57	19995	-33229	048	T	a-	0.3493	2.1979	1.2358	317.9	202.8	68.8	22S	12E
3154	158	-0687 Dec 05	22:42:47	19987	-33223	053	P	t-	-0.9118	1.2352	0.1356	307.6	88.8	-	21N	101E
3155	158	-0686 May 02	14:43:22	19981	-33218	020	N	-a	-1.4155	0.2333	-0.7127	126.9	-	-	14S	139W
3156	158	-0686 May 31	21:18:32	19979	-33217	058	N	a-	1.0720	0.8602	-0.0791	227.6	-	-	20S	122E
3157	158	-0686 Oct 26	03:24:23	19973	-33212	025	N	-t	1.4422	0.2510	-0.8271	149.3	-	-	11N	29E
3158	158	-0686 Nov 24	21:39:43	19972	-33211	063	N	t-	-1.5704	0.0274	-1.0737	51.6	-	-	18N	116E
3159	158	-0685 Apr 22	05:41:36	19965	-33206	030	P	-a	-0.7199	1.5305	0.5433	298.1	155.2	-	10S	3W
3160	158	-0685 Oct 15	10:01:00	19957	-33200	035	P	-a	0.6898	1.6019	0.5827	306.1	160.3	-	6N	70W
3161	159	-0684 Apr 10	14:18:17	19950	-33194	040	T+	pp	0.0384	2.8097	1.7658	361.1	230.1	104.5	5S	131W
3162	159	-0684 Oct 03	23:18:52	19942	-33188	045	T-	pp	-0.0235	2.8010	1.8283	322.1	211.9	98.6	1N	92E
3163	159	-0683 Mar 30	16:08:39	19934	-33182	050	P	t-	0.8043	1.4207	0.3442	323.2	138.1	-	0S	157W
3164	159	-0683 Sep 23	15:21:32	19926	-33176	055	P	a-	-0.7035	1.5506	0.5834	285.7	153.4	-	3S	148W
3165	159	-0682 Feb 18	02:39:50	19920	-33171	022	N	-t	-1.3229	0.4526	-0.5912	191.6	-	-	13N	47E
3166	159	-0682 Mar 19	16:58:18	19919	-33170	060	N	t-	1.5137	0.1108	-0.9494	100.6	-	-	5N	168W
3167	159	-0682 Aug 14	17:35:11	19912	-33165	027	N	-a	1.2896	0.5041	-0.5205	197.8	-	-	16S	180E
3168	159	-0682 Sep 13	05:26:36	19911	-33164	065	N	a-	-1.4399	0.2179	-0.7861	129.2	-	-	8S	1E
3169	159	-0681 Feb 07	12:32:16	19904	-33159	032	P	-a	-0.5479	1.8481	0.8570	310.4	182.3	-	17N	101W
3170	159	-0681 Aug 03	22:16:33	19897	-33153	037	P	-t	0.5874	1.8152	0.7457	344.5	188.7	-	19S	109E
3171	159	-0680 Jan 28	03:35:20	19889	-33147	042	T+	p-	0.1582	2.5493	1.5858	317.4	208.3	92.9	20N	33E
3172	159	-0680 Jul 22	22:34:06	19881	-33141	047	T-	pp	-0.1552	2.6101	1.5370	372.2	232.5	99.6	22S	105E
3173	159	-0679 Jan 16	19:23:58	19874	-33135	052	P	a-	0.8476	1.2928	0.3122	272.6	118.1	-	23N	155E
3174	159	-0679 Jul 12	01:48:08	19866	-33129	057	P	t-	-0.8966	1.2283	0.1976	291.2	102.6	-	24S	55E
3175	159	-0679 Dec 07	15:59:46	19859	-33124	024	N	-a	-1.2834	0.5378	-0.5312	210.2	-	-	21N	158W
3176	159	-0678 Jun 02	04:35:57	19852	-33118	029	P	-a	0.9577	1.0759	0.1247	252.5	76.5	-	20S	12E
3177	159	-0678 Nov 26	17:05:54	19844	-33112	034	P	-t	-0.6365	1.7435	0.6375	347.1	179.2	-	19N	176W
3178	159	-0677 May 22	21:45:50	19836	-33106	039	T+	-p	0.2139	2.4358	1.4948	314.4	207.3	89.3	18S	114E
3179	159	-0677 Nov 15	16:21:53	19829	-33100	044	T+	pp	0.0445	2.8222	1.7312	371.8	232.0	103.6	17N	166W
3180	159	-0676 May 11	13:40:13	19821	-33094	049	P	a-	-0.5465	1.8419	0.8683	310.0	184.0	-	16S	125W

Cat Num	Canon Plate	Calendar Date	TD of Greatest Eclipse	ΔT s	Luna Num	Saros Num	Ecl. Type	QSE	Gamma	Pen. Mag.	Um. Mag.	Eclipse Phase Durations Pen. m	Par. m	Total m	Greatest in Zenith Lat.	Lng.
3181	160	-0676 Nov 03	21:05:09	19813	-33088	054	P	a-	0.7268	1.5427	0.5062	308.0	153.2	-	14N	123E
3182	160	-0675 Apr 01	10:48:10	19807	-33083	021	N	-t	1.4091	0.3020	-0.7569	163.6	-	-	0S	77W
3183	160	-0675 Apr 30	23:59:18	19805	-33082	059	N	t-	-1.3679	0.3625	-0.6665	172.9	-	-	14S	81E
3184	160	-0675 Sep 24	22:38:53	19799	-33077	026	N	-a	-1.2990	0.4566	-0.5081	173.8	-	-	3S	102E
3185	160	-0675 Oct 24	08:58:29	19798	-33076	064	N	a-	1.3586	0.3569	-0.6271	157.8	-	-	11N	56W
3186	160	-0674 Mar 21	11:30:25	19791	-33071	031	P	-t	0.6905	1.6308	0.5520	337.9	169.2	-	3N	87W
3187	160	-0674 Sep 14	14:41:35	19784	-33065	036	P	-a	-0.6335	1.6816	0.7091	295.5	167.0	-	7S	138W
3188	160	-0673 Mar 10	13:21:49	19776	-33059	041	T-	pp	-0.0640	2.7655	1.7159	360.6	229.0	103.3	7N	115W
3189	160	-0673 Sep 04	03:30:04	19768	-33053	046	T+	pp	0.0765	2.7265	1.7090	343.7	221.4	100.6	10S	30E
3190	160	-0672 Feb 27	21:53:30	19760	-33047	051	P	a-	-0.7917	1.4038	0.4067	289.5	136.6	-	10N	118E
3191	160	-0672 Aug 23	09:32:19	19753	-33041	056	P	t-	0.8414	1.3481	0.2807	311.2	123.8	-	14S	61W
3192	160	-0671 Jan 18	03:02:42	19746	-33036	023	N	-a	1.1628	0.7067	-0.2585	210.6	-	-	23N	41E
3193	160	-0671 Feb 16	12:23:12	19745	-33035	061	N	a-	-1.4490	0.1797	-0.7817	112.1	-	-	13N	100W
3194	160	-0671 Jul 13	18:20:30	19739	-33030	028	N	-t	-1.3065	0.4925	-0.5708	206.8	-	-	24S	167E
3195	160	-0671 Aug 12	09:59:52	19737	-33029	066	Nb	t-	1.5751	0.0092	-1.0729	30.0	-	-	16S	68W
3196	160	-0670 Jan 07	18:08:41	19731	-33024	033	P	-a	0.5103	1.9190	0.9241	312.5	186.2	-	24N	173E
3197	160	-0670 Jul 02	23:30:55	19723	-33018	038	P	-t	-0.5038	1.9401	0.9271	329.2	194.8	-	24S	89E
3198	160	-0670 Dec 28	04:08:55	19715	-33012	043	T-	pp	-0.1908	2.5346	1.4816	352.8	221.6	93.0	24N	21E
3199	160	-0669 Jun 22	11:56:15	19708	-33006	048	T	p-	0.2749	2.3336	1.3731	320.1	207.3	82.3	23S	98W
3200	160	-0669 Dec 17	06:45:36	19700	-33000	053	P	t-	-0.9076	1.2430	0.1429	308.7	91.2	-	22N	19W
3201	161	-0668 May 12	22:07:14	19694	-32995	020	Ne	-a	-1.4885	0.0998	-0.8469	84.1	-	-	18S	108E
3202	161	-0668 Jun 11	04:43:12	19692	-32994	058	P	a-	0.9988	0.9948	0.0549	241.3	50.7	-	21S	9E
3203	161	-0668 Nov 05	11:34:23	19686	-32989	025	N	-t	1.4444	0.2464	-0.8306	147.7	-	-	15N	95W
3204	161	-0668 Dec 05	05:44:41	19685	-32988	063	N	t-	-1.5652	0.0361	-1.0633	59.0	-	-	20N	5W
3205	161	-0667 May 02	12:51:31	19678	-32983	030	P	-a	-0.7954	1.3929	0.4038	290.2	137.2	-	14S	112W
3206	161	-0667 Oct 25	18:29:57	19671	-32977	035	P	-a	0.6950	1.5918	0.5737	304.5	158.8	-	11N	161E
3207	161	-0666 Apr 21	21:04:40	19663	-32971	040	T-	pp	-0.0372	2.8122	1.7674	362.3	230.8	104.8	9S	125E
3208	161	-0666 Oct 15	07:57:00	19655	-32965	045	T-	pp	-0.0115	2.8236	1.8500	321.7	211.6	98.5	6N	40W
3209	161	-0665 Apr 10	22:44:33	19648	-32959	050	P	t-	0.7310	1.5541	0.4800	332.7	159.8	-	4S	101E
3210	161	-0665 Oct 04	23:51:57	19640	-32953	055	P	a-	-0.6863	1.5840	0.6130	288.2	156.6	-	1N	82E
3211	161	-0664 Feb 29	10:12:40	19634	-32948	022	N	-t	-1.3596	0.3819	-0.6552	176.3	-	-	9N	68W
3212	161	-0664 Mar 29	23:53:46	19632	-32947	060	N	t-	1.4481	0.2283	-0.8262	142.4	-	-	1N	85E
3213	161	-0664 Aug 25	01:05:56	19626	-32942	027	N	-a	1.3389	0.4170	-0.6143	182.3	-	-	12S	65E
3214	161	-0664 Sep 23	13:34:19	19625	-32941	065	N	a-	-1.4202	0.2570	-0.7531	140.4	-	-	4S	122W
3215	161	-0663 Feb 17	20:37:27	19618	-32936	032	P	-a	-0.5780	1.7899	0.8046	306.5	177.8	-	14N	137E
3216	161	-0663 Aug 14	05:13:56	19610	-32930	037	P	-t	0.6475	1.7075	0.6328	339.7	177.5	-	17S	4E
3217	161	-0662 Feb 07	12:00:48	19603	-32924	042	T+	p-	0.1344	2.5913	1.6310	317.5	209.1	94.6	18N	94W
3218	161	-0662 Aug 03	05:19:48	19595	-32918	047	T-	pp	-0.0869	2.7361	1.6614	373.2	234.2	104.0	20S	2E
3219	161	-0661 Jan 28	03:52:26	19587	-32912	052	P	a-	0.8323	1.3208	0.3407	275.2	123.2	-	21N	28E
3220	161	-0661 Jul 23	08:49:45	19580	-32906	057	P	h-	-0.8258	1.3576	0.3279	300.7	129.3	-	23S	51W
3221	162	-0661 Dec 19	00:16:45	19573	-32901	024	N	-h	-1.2902	0.5263	-0.5445	208.8	-	-	22N	78E
3222	162	-0660 Jun 12	11:56:41	19566	-32895	029	N	-a	1.0297	0.9432	-0.0067	239.2	-	-	22S	99W
3223	162	-0660 Jul 11	19:40:20	19564	-32894	067	Nb	a-	-1.5130	0.0681	-0.9050	71.7	-	-	25S	146E
3224	162	-0660 Dec 07	01:07:43	19558	-32889	034	P	-t	-0.6417	1.7340	0.6278	346.7	178.2	-	22N	64E
3225	162	-0659 Jun 02	05:10:26	19550	-32883	039	T	-p	0.2890	2.2983	1.3564	312.2	203.5	79.9	21S	2E
3226	162	-0659 Nov 26	00:30:37	19543	-32877	044	T+	pp	0.0401	2.8293	1.7401	370.9	231.6	103.6	20N	72E
3227	162	-0658 May 22	20:51:23	19535	-32871	049	T	a-	-0.4721	1.9797	1.0035	316.8	193.6	9.1	19S	126E
3228	162	-0658 Nov 15	05:36:02	19527	-32865	054	P	a-	0.7223	1.5499	0.5155	307.3	153.9	-	18N	6W
3229	162	-0657 Apr 12	17:33:17	19521	-32860	021	N	-t	1.4831	0.1662	-0.8926	123.1	-	-	5S	179E
3230	162	-0657 May 12	06:43:56	19520	-32859	059	N	t-	-1.2934	0.5004	-0.5309	201.3	-	-	17S	22W
3231	162	-0657 Oct 06	07:11:32	19513	-32854	026	N	-a	-1.3157	0.4269	-0.5396	168.5	-	-	1N	28W
3232	162	-0657 Nov 04	17:43:30	19512	-32853	064	N	a-	1.3509	0.3709	-0.6127	160.3	-	-	15N	172E
3233	162	-0656 Mar 31	18:12:54	19506	-32848	031	P	-t	0.7595	1.5023	0.4272	328.8	151.8	-	1S	170E
3234	162	-0656 Sep 24	23:02:50	19498	-32842	036	P	-a	-0.6587	1.6378	0.6606	294.1	162.7	-	3S	95E
3235	162	-0655 Mar 20	20:30:26	19490	-32836	041	T-	pp	-0.0049	2.8709	1.8276	359.1	229.0	104.3	3N	136E
3236	162	-0655 Sep 14	11:23:51	19483	-32830	046	T+	pp	0.0442	2.7890	1.7650	345.8	222.3	101.5	6S	90W
3237	162	-0654 Mar 10	05:38:03	19475	-32824	051	P	a-	-0.7435	1.4890	0.4982	293.9	148.7	-	6N	0W
3238	162	-0654 Sep 03	16:50:41	19468	-32818	056	P	t-	0.7979	1.4308	0.3575	318.8	138.4	-	10S	172W
3239	162	-0653 Jan 29	11:34:26	19461	-32813	023	N	-a	1.1823	0.6698	-0.2931	205.8	-	-	21N	88W
3240	162	-0653 Feb 27	20:31:24	19460	-32812	061	N	a-	-1.4115	0.2465	-0.7109	130.2	-	-	10N	137E

Cat Num	Canon Plate	Calendar Date	TD of Greatest Eclipse	ΔT s	Luna Num	Saros Num	Ecl. Type	QSE	Gamma	Pen. Mag.	Um. Mag.	Eclipse Phase Durations Pen. m	Par. m	Total m	Greatest in Zenith Lat.	Lng.
3241	163	-0653 Jul 25	01:04:03	19453	-32807	028	N	-t	-1.3747	0.3677	-0.6961	180.4	-	-	23S	66E
3242	163	-0653 Aug 23	16:59:48	19452	-32806	066	N	t-	1.5206	0.1107	-0.9744	103.0	-	-	13S	174W
3243	163	-0652 Jan 19	02:38:56	19446	-32801	033	P	-a	0.5247	1.8926	0.8976	312.2	184.9	-	22N	45E
3244	163	-0652 Jul 13	06:34:33	19438	-32795	038	P	-t	-0.5737	1.8110	0.7995	322.1	184.4	-	24S	18W
3245	163	-0651 Jan 07	12:23:33	19431	-32789	043	T-	pp	-0.1820	2.5512	1.4973	354.2	222.6	94.1	23N	102W
3246	163	-0651 Jul 02	19:19:41	19423	-32783	048	T+	p-	0.2033	2.4646	1.5050	321.5	210.4	90.9	24S	150E
3247	163	-0651 Dec 27	14:44:25	19415	-32777	053	P	t-	-0.9005	1.2560	0.1563	310.2	95.3	-	23N	139W
3248	163	-0650 Jun 22	12:11:03	19408	-32771	058	P	a-	0.9282	1.1251	0.1841	253.1	91.1	-	23S	103W
3249	163	-0650 Nov 16	19:49:08	19401	-32766	025	N	-t	1.4437	0.2468	-0.8285	147.3	-	-	19N	141E
3250	163	-0650 Dec 16	13:49:32	19400	-32765	063	N	t-	-1.5591	0.0460	-1.0508	66.5	-	-	22N	127W
3251	163	-0649 May 13	19:57:16	19394	-32760	030	P	-a	-0.8732	1.2514	0.2598	280.7	112.8	-	17S	139E
3252	163	-0649 Nov 06	03:04:06	19386	-32754	035	P	-a	0.6968	1.5878	0.5711	303.3	158.1	-	15N	32E
3253	163	-0648 May 02	03:45:08	19378	-32748	040	T-	pp	-0.1170	2.6664	1.6204	362.5	230.0	101.8	13S	23E
3254	163	-0648 Oct 25	16:40:37	19371	-32742	045	T-	pp	-0.0039	2.8378	1.8636	321.4	211.4	98.4	10N	172W
3255	163	-0647 Apr 21	05:14:14	19363	-32736	050	P	t-	0.6523	1.6973	0.6253	341.6	178.1	-	9S	2E
3256	163	-0647 Oct 15	08:28:55	19356	-32730	055	P	a-	-0.6749	1.6067	0.6321	290.0	158.8	-	5N	49W
3257	163	-0646 Mar 11	17:38:09	19349	-32725	022	N	-t	-1.4027	0.2994	-0.7311	156.6	-	-	5N	179E
3258	163	-0646 Apr 10	06:44:20	19348	-32724	060	N	t-	1.3774	0.3553	-0.6939	174.9	-	-	4S	19W
3259	163	-0646 Sep 05	08:43:37	19342	-32719	027	N	-a	1.3815	0.3422	-0.6957	167.0	-	-	9S	51W
3260	163	-0646 Oct 04	21:47:51	19340	-32718	065	N	a-	-1.4059	0.2862	-0.7298	148.4	-	-	0N	112E
3261	164	-0645 Mar 01	04:35:19	19334	-32713	032	P	-a	-0.6148	1.7195	0.7399	301.9	172.0	-	10N	16E
3262	164	-0645 Aug 25	12:19:19	19326	-32707	037	P	-t	0.7004	1.6131	0.5333	334.9	165.9	-	13S	104W
3263	164	-0644 Feb 18	20:16:34	19319	-32701	042	T+	p-	0.1024	2.6484	1.6914	317.6	209.9	96.4	14N	141E
3264	164	-0644 Aug 13	12:15:01	19311	-32695	047	T-	pp	-0.0260	2.8489	1.7722	373.4	234.6	105.6	17S	103W
3265	164	-0643 Feb 07	12:10:50	19303	-32689	052	P	a-	0.8091	1.3628	0.3835	278.8	130.1	-	18N	97W
3266	164	-0643 Aug 02	16:01:09	19296	-32683	057	P	h-	-0.7618	1.4748	0.4456	308.1	147.6	-	21S	159W
3267	164	-0643 Dec 29	08:27:48	19289	-32678	024	N	-h	-1.3009	0.5071	-0.5647	205.9	-	-	23N	45W
3268	164	-0642 Jun 23	19:20:59	19282	-32672	029	N	-a	1.0991	0.8155	-0.1336	225.0	-	-	23S	149E
3269	164	-0642 Jul 23	03:09:50	19281	-32671	067	N	a-	-1.4476	0.1880	-0.7847	117.5	-	-	23S	33E
3270	164	-0642 Dec 18	09:08:12	19274	-32666	034	P	-t	-0.6477	1.7226	0.6171	346.1	177.1	-	23N	57W
3271	164	-0641 Jun 13	12:33:47	19267	-32660	039	T	-a	0.3649	2.1599	1.2164	309.3	198.4	65.3	22S	110W
3272	164	-0641 Dec 07	08:42:05	19259	-32654	044	T+	pp	0.0374	2.8331	1.7465	369.8	231.2	103.6	22N	52W
3273	164	-0640 Jun 02	03:58:32	19251	-32648	049	T	p-	-0.3953	2.1222	1.1427	323.0	201.9	55.9	22S	18E
3274	164	-0640 Nov 25	14:10:23	19244	-32642	054	P	a-	0.7197	1.5532	0.5216	306.4	154.2	-	21N	135W
3275	164	-0639 Apr 23	00:10:14	19238	-32637	021	Ne	-t	1.5626	0.0204	-1.0386	43.8	-	-	9S	78E
3276	164	-0639 May 22	13:23:51	19236	-32636	059	N	t-	-1.2155	0.6448	-0.3892	226.2	-	-	20S	123W
3277	164	-0639 Oct 16	15:51:37	19230	-32631	026	N	-a	-1.3268	0.4073	-0.5610	164.8	-	-	5N	160W
3278	164	-0639 Nov 15	02:32:52	19229	-32630	064	N	a-	1.3466	0.3784	-0.6044	161.5	-	-	18N	39E
3279	164	-0638 Apr 12	00:50:03	19222	-32625	031	P	-t	0.8330	1.3656	0.2940	317.9	128.4	-	5S	68E
3280	164	-0638 Oct 06	07:29:39	19215	-32619	036	P	-a	-0.6789	1.6031	0.6212	293.1	159.1	-	2N	34W
3281	165	-0637 Apr 01	03:35:09	19207	-32613	041	T+	pp	0.0585	2.7693	1.7321	357.0	228.1	103.4	1S	28E
3282	165	-0637 Sep 25	19:23:25	19200	-32607	046	T+	pp	0.0174	2.8414	1.8110	347.7	223.0	101.9	2S	148E
3283	165	-0636 Mar 20	13:15:57	19192	-32601	051	P	a-	-0.6890	1.5860	0.6012	298.4	160.3	-	2N	117W
3284	165	-0636 Sep 14	00:16:57	19184	-32595	056	P	t-	0.7613	1.5009	0.4219	325.0	149.1	-	6S	75E
3285	165	-0635 Feb 08	19:58:04	19178	-32590	023	N	-a	1.2085	0.6206	-0.3400	199.1	-	-	18N	146E
3286	165	-0635 Mar 10	04:31:39	19177	-32589	061	N	a-	-1.3676	0.3252	-0.6284	148.2	-	-	6N	15E
3287	165	-0635 Aug 04	07:57:27	19170	-32584	028	N	-t	-1.4352	0.2571	-0.8075	152.1	-	-	21S	38W
3288	165	-0635 Sep 03	00:10:06	19169	-32583	066	N	t-	1.4737	0.1981	-0.8898	136.7	-	-	9S	77E
3289	165	-0634 Jan 29	10:59:59	19163	-32578	033	P	-a	0.5464	1.8529	0.8579	311.3	182.5	-	20N	81W
3290	165	-0634 Jul 24	13:47:57	19155	-32572	038	P	-t	-0.6366	1.6950	0.6846	315.1	173.4	-	22S	127W
3291	165	-0633 Jan 18	20:29:55	19148	-32566	043	T-	pp	-0.1669	2.5791	1.5247	355.7	223.8	95.8	22N	136E
3292	165	-0633 Jul 14	02:48:38	19140	-32560	048	T+	p-	0.1356	2.5886	1.6295	322.3	212.3	96.1	23S	37E
3293	165	-0632 Jan 07	22:37:18	19133	-32554	053	P	t-	-0.8885	1.2773	0.1787	312.3	101.7	-	23N	103E
3294	165	-0632 Jul 02	19:41:31	19125	-32548	058	P	a-	0.8594	1.2522	0.3092	263.5	116.1	-	23S	143E
3295	165	-0632 Nov 27	04:06:01	19119	-32543	025	N	-t	1.4419	0.2489	-0.8239	147.5	-	-	22N	16E
3296	165	-0632 Dec 26	21:50:06	19117	-32542	063	N	t-	-1.5490	0.0628	-1.0307	77.4	-	-	22N	113E
3297	165	-0631 May 24	02:58:23	19111	-32537	030	P	-a	-0.9538	1.1051	0.1104	269.2	75.3	-	20S	33E
3298	165	-0631 Nov 16	11:43:13	19104	-32531	035	P	-a	0.6954	1.5893	0.5747	302.4	158.0	-	18N	99W
3299	165	-0630 May 13	10:20:00	19096	-32525	040	T-	pp	-0.2005	2.5142	1.4664	361.7	227.4	94.8	17S	77W
3300	165	-0630 Nov 06	01:29:13	19088	-32519	045	T-	pp	-0.0002	2.8448	1.8703	321.1	211.1	98.3	14N	54E

Cat Num	Canon Plate	Calendar Date	TD of Greatest Eclipse	ΔT s	Luna Num	Saros Num	Ecl. Type	QSE	Gamma	Pen. Mag.	Um. Mag.	Eclipse Phase Durations Pen. m	Par. m	Total m	Greatest in Zenith Lat.	Lng.
3301	166	-0629 May 02	11:38:28	19081	-32513	050	P	t-	0.5689	1.8495	0.7794	349.6	193.7	-	12S	96W
3302	166	-0629 Oct 26	17:11:40	19073	-32507	055	P	a-	-0.6688	1.6196	0.6417	291.4	159.9	-	10N	179E
3303	166	-0628 Mar 22	00:56:00	19067	-32502	022	N	-t	-1.4525	0.2048	-0.8192	130.0	-	-	1N	67E
3304	166	-0628 Apr 20	13:28:34	19066	-32501	060	N	t-	1.3005	0.4938	-0.5502	202.7	-	-	8S	123W
3305	166	-0628 Sep 15	16:28:54	19059	-32496	027	N	-a	1.4169	0.2804	-0.7641	152.8	-	-	4S	169W
3306	166	-0628 Oct 15	06:08:23	19058	-32495	065	N	a-	-1.3980	0.3036	-0.7180	153.2	-	-	5N	15W
3307	166	-0627 Mar 11	12:26:04	19052	-32490	032	P	-a	-0.6582	1.6370	0.6631	296.5	164.5	-	6N	104W
3308	166	-0627 Sep 04	19:32:58	19044	-32484	037	P	-t	0.7459	1.5320	0.4472	330.5	154.2	-	9S	146E
3309	166	-0626 Mar 01	04:25:36	19037	-32478	042	T+	pp	0.0646	2.7161	1.7622	317.8	210.6	97.9	10N	17E
3310	166	-0626 Aug 24	19:20:01	19029	-32472	047	T+	pp	0.0274	2.8473	1.7689	373.1	234.2	105.3	14S	150E
3311	166	-0625 Feb 18	20:22:18	19022	-32466	052	P	a-	0.7811	1.4140	0.4353	282.9	137.9	-	15N	138E
3312	166	-0625 Aug 13	23:20:50	19014	-32460	057	P	h-	-0.7037	1.5812	0.5524	313.9	161.0	-	18S	90E
3313	166	-0624 Jan 09	16:33:32	19008	-32455	024	N	-h	-1.3150	0.4815	-0.5908	201.7	-	-	22N	166W
3314	166	-0624 Feb 08	07:39:48	19006	-32454	062	N	h-	1.5550	0.0224	-1.0129	43.7	-	-	19N	31W
3315	166	-0624 Jul 04	02:49:29	19000	-32449	029	N	-a	1.1651	0.6943	-0.2547	209.8	-	-	23S	36E
3316	166	-0624 Aug 02	10:47:04	18999	-32448	067	N	a-	-1.3878	0.2975	-0.6750	145.9	-	-	21S	82W
3317	166	-0624 Dec 28	17:03:16	18993	-32443	034	P	-t	-0.6578	1.7035	0.5993	345.0	175.1	-	23N	175W
3318	166	-0623 Jun 23	20:00:14	18985	-32437	039	T	-a	0.4377	2.0274	1.0816	305.9	192.2	41.8	23S	138E
3319	166	-0623 Dec 17	16:51:14	18977	-32431	044	T+	pp	0.0327	2.8400	1.7567	368.6	230.9	103.6	23N	174W
3320	166	-0622 Jun 13	11:06:24	18970	-32425	049	T	p-	-0.3200	2.2624	1.2790	328.3	208.6	75.0	23S	89W
3321	167	-0622 Dec 06	22:44:25	18962	-32419	054	P	a-	0.7165	1.5574	0.5292	305.5	154.7	-	23N	96E
3322	167	-0621 Jun 02	20:01:46	18955	-32413	059	N	t-	-1.1368	0.7909	-0.2464	247.9	-	-	22S	136E
3323	167	-0621 Oct 28	00:37:07	18949	-32408	026	N	-a	-1.3339	0.3950	-0.5747	162.5	-	-	10N	67E
3324	167	-0621 Nov 26	11:23:47	18947	-32407	064	N	a-	1.3434	0.3835	-0.5979	162.2	-	-	21N	95W
3325	167	-0620 Apr 22	07:19:16	18941	-32402	031	P	-t	0.9127	1.2177	0.1492	304.6	93.3	-	9S	31W
3326	167	-0620 Oct 16	16:03:53	18933	-32396	036	P	-a	-0.6925	1.5803	0.5941	292.7	156.5	-	6N	164W
3327	167	-0619 Apr 11	10:32:48	18926	-32390	041	T+	pp	0.1284	2.6381	1.6067	354.3	226.0	100.0	5S	79W
3328	167	-0619 Oct 06	03:30:41	18918	-32384	046	T-	pp	-0.0020	2.8727	1.8363	349.4	223.5	102.1	2N	24E
3329	167	-0618 Mar 31	20:47:27	18911	-32378	051	P	a-	-0.6285	1.6941	0.7149	303.0	171.1	-	2S	128E
3330	167	-0618 Sep 25	07:52:18	18903	-32372	056	P	t-	0.7328	1.5559	0.4716	329.9	156.7	-	1S	41W
3331	167	-0617 Feb 20	04:14:04	18897	-32367	023	N	-a	1.2409	0.5599	-0.3984	190.3	-	-	15N	20E
3332	167	-0617 Mar 21	12:24:31	18896	-32366	061	N	a-	-1.3176	0.4151	-0.5349	165.8	-	-	2N	105W
3333	167	-0617 Aug 15	14:59:18	18890	-32361	028	N	-t	-1.4892	0.1583	-0.9072	120.2	-	-	18S	145W
3334	167	-0617 Sep 14	07:29:50	18888	-32360	066	N	t-	1.4338	0.2726	-0.8179	159.3	-	-	5S	35W
3335	167	-0616 Feb 09	19:13:26	18882	-32355	033	P	-a	0.5736	1.8028	0.8080	310.0	179.2	-	17N	155E
3336	167	-0616 Aug 03	21:09:18	18874	-32349	038	P	-h	-0.6939	1.5896	0.5798	308.0	161.7	-	20S	122E
3337	167	-0615 Jan 29	04:27:32	18867	-32343	043	T-	pp	-0.1455	2.6185	1.5640	357.2	225.2	97.9	20N	16E
3338	167	-0615 Jul 24	10:24:22	18859	-32337	048	T+	pp	0.0730	2.7035	1.7443	322.5	213.2	98.8	22S	77W
3339	167	-0614 Jan 18	06:23:46	18852	-32331	053	P	t-	-0.8714	1.3078	0.2109	315.1	110.1	-	21N	14W
3340	167	-0614 Jul 14	03:15:51	18844	-32325	058	P	a-	0.7941	1.3732	0.4278	272.4	134.3	-	22S	29E
3341	168	-0614 Dec 08	12:23:46	18838	-32320	025	N	-t	1.4400	0.2508	-0.8188	147.5	-	-	24N	109W
3342	168	-0613 Jan 07	05:47:23	18837	-32319	063	N	t-	-1.5357	0.0852	-1.0043	89.7	-	-	22N	6W
3343	168	-0613 Jun 04	09:57:46	18831	-32314	030	N	-a	-1.0346	0.9586	-0.0397	255.8	-	-	22S	73W
3344	168	-0613 Nov 27	20:24:03	18823	-32308	035	P	-a	0.6931	1.5922	0.5801	301.5	158.2	-	21N	130E
3345	168	-0612 May 23	16:50:25	18815	-32302	040	T	-t	-0.2866	2.3573	1.3073	359.8	222.9	82.0	19S	176W
3346	168	-0612 Nov 16	10:20:37	18808	-32296	045	T+	pp	0.0014	2.8425	1.8681	320.7	210.9	98.2	18N	80W
3347	168	-0611 May 12	18:00:01	18800	-32290	050	P	t-	0.4830	2.0064	0.9377	356.6	206.4	-	16S	167E
3348	168	-0611 Nov 06	01:57:22	18793	-32284	055	P	a-	-0.6656	1.6270	0.6462	292.4	160.6	-	14N	46E
3349	168	-0610 Apr 02	08:08:34	18787	-32279	022	N	-t	-1.5071	0.1015	-0.9162	91.9	-	-	3S	43W
3350	168	-0610 May 01	20:11:15	18785	-32278	060	N	t-	1.2210	0.6374	-0.4020	226.2	-	-	12S	134E
3351	168	-0610 Sep 27	00:21:36	18779	-32273	027	N	-a	1.4456	0.2310	-0.8197	139.9	-	-	0S	71E
3352	168	-0610 Oct 26	14:33:31	18778	-32272	065	N	a-	-1.3943	0.3129	-0.7138	156.1	-	-	9N	143W
3353	168	-0609 Mar 22	20:09:33	18772	-32267	032	P	-a	-0.7081	1.5428	0.5741	290.1	154.9	-	1N	138E
3354	168	-0609 Sep 16	02:55:10	18764	-32261	037	P	-t	0.7839	1.4647	0.3752	326.5	143.0	-	5S	34E
3355	168	-0608 Mar 11	12:25:26	18757	-32255	042	T+	pp	0.0193	2.7979	1.8468	317.8	211.1	98.9	6N	105W
3356	168	-0608 Sep 04	02:36:09	18749	-32249	047	T+	pp	0.0718	2.7667	1.6864	372.4	233.2	103.9	10S	39E
3357	168	-0607 Mar 01	04:22:11	18742	-32243	052	P	a-	0.7443	1.4811	0.5030	288.0	147.1	-	11N	17E
3358	168	-0607 Aug 24	06:52:12	18734	-32237	057	P	a-	-0.6543	1.6717	0.6431	318.2	170.6	-	14S	24W
3359	168	-0606 Jan 20	00:30:56	18728	-32232	024	N	-t	-1.3345	0.4456	-0.6266	195.1	-	-	20N	74E
3360	168	-0606 Feb 18	15:27:17	18727	-32231	062	N	t-	1.5278	0.0724	-0.9629	78.5	-	-	15N	149W

Cat Num	Canon Plate	Calendar Date	TD of Greatest Eclipse	ΔT s	Luna Num	Saros Num	Ecl. Type	QSE	Gamma	Pen. Mag.	Um. Mag.	Eclipse Phase Durations Pen. m	Par. m	Total m	Greatest in Zenith Lat.	Lng.
3361	169	-0606 Jul 15	10:22:42	18720	-32226	029	N	-a	1.2275	0.5801	-0.3693	193.7	-	-	22S	78W
3362	169	-0606 Aug 13	18:32:09	18719	-32225	067	N	a-	-1.3340	0.3965	-0.5763	166.5	-	-	18S	161E
3363	169	-0605 Jan 09	00:53:54	18713	-32220	034	P	-t	-0.6712	1.6779	0.5759	343.3	172.3	-	23N	67E
3364	169	-0605 Jul 05	03:28:10	18705	-32214	039	P	-a	0.5086	1.8988	0.9502	301.8	184.7	-	23S	25E
3365	169	-0605 Dec 29	00:58:47	18698	-32208	044	T+	pp	0.0258	2.8506	1.7714	367.3	230.5	103.6	24N	65E
3366	169	-0604 Jun 23	18:12:41	18690	-32202	049	T-	pp	-0.2442	2.4038	1.4159	332.9	213.9	87.3	24S	163E
3367	169	-0604 Dec 17	07:18:47	18683	-32196	054	P	a-	0.7128	1.5622	0.5380	304.6	155.3	-	24N	32W
3368	169	-0603 Jun 13	02:38:36	18675	-32190	059	N	t-	-1.0574	0.9383	-0.1027	267.0	-	-	24S	36E
3369	169	-0603 Nov 07	09:26:42	18669	-32185	026	N	-a	-1.3380	0.3881	-0.5827	161.2	-	-	13N	66W
3370	169	-0603 Dec 06	20:14:54	18668	-32184	064	N	a-	1.3404	0.3882	-0.5915	162.8	-	-	23N	132E
3371	169	-0602 May 03	13:46:32	18662	-32179	031	P	-t	0.9940	1.0671	0.0015	289.2	9.7	-	13S	130W
3372	169	-0602 Oct 28	00:42:38	18654	-32173	036	P	-a	-0.7019	1.5652	0.5750	292.7	154.8	-	10N	64E
3373	169	-0601 Apr 22	17:27:36	18647	-32167	041	T+	pp	0.2017	2.5010	1.4751	350.7	222.7	93.6	9S	175E
3374	169	-0601 Oct 17	11:43:05	18639	-32161	046	T-	pp	-0.0164	2.8490	1.8070	351.1	224.0	102.0	7N	101W
3375	169	-0600 Apr 11	04:14:20	18632	-32155	051	P	a-	-0.5634	1.8109	0.8371	307.4	180.9	-	6S	14E
3376	169	-0600 Oct 05	15:34:51	18624	-32149	056	P	t-	0.7102	1.5998	0.5106	333.8	162.4	-	3N	158W
3377	169	-0599 Mar 02	12:20:32	18618	-32144	023	N	-a	1.2812	0.4850	-0.4712	178.5	-	-	11N	103W
3378	169	-0599 Mar 31	20:10:02	18617	-32143	061	N	a-	-1.2614	0.5166	-0.4303	183.0	-	-	3S	136E
3379	169	-0599 Aug 25	22:12:46	18611	-32138	028	N	-t	-1.5348	0.0752	-0.9911	83.3	-	-	15S	105E
3380	169	-0599 Sep 24	15:00:04	18609	-32137	066	N	t-	1.4016	0.3330	-0.7598	175.0	-	-	1S	149W
3381	170	-0598 Feb 20	03:16:58	18603	-32132	033	P	-a	0.6088	1.7383	0.7436	307.8	174.3	-	14N	33E
3382	170	-0598 Aug 15	04:40:06	18596	-32126	038	P	-h	-0.7444	1.4967	0.4874	301.1	149.9	-	17S	8E
3383	170	-0597 Feb 09	12:14:58	18588	-32120	043	T-	pp	-0.1163	2.6720	1.6177	358.8	226.8	100.2	17N	102W
3384	170	-0597 Aug 04	18:07:23	18581	-32114	048	T+	pp	0.0158	2.8088	1.8489	322.2	213.3	99.7	19S	166E
3385	170	-0596 Jan 29	14:00:06	18573	-32108	053	P	t-	-0.8463	1.3527	0.2581	318.9	121.0	-	19N	129W
3386	170	-0596 Jul 24	10:55:13	18566	-32102	058	P	a-	0.7332	1.4866	0.5382	280.1	148.2	-	21S	86W
3387	170	-0596 Dec 18	20:40:11	18560	-32097	025	N	-t	1.4399	0.2490	-0.8167	146.5	-	-	25N	127E
3388	170	-0595 Jan 17	13:37:39	18558	-32096	063	N	t-	-1.5161	0.1188	-0.9660	105.3	-	-	21N	124W
3389	170	-0595 Jun 14	16:55:52	18552	-32091	030	N	-a	-1.1153	0.8128	-0.1898	240.2	-	-	24S	178W
3390	170	-0595 Jul 14	02:24:19	18551	-32090	068	N	a-	1.4981	0.0988	-0.8809	86.8	-	-	22S	41E
3391	170	-0595 Dec 08	05:05:06	18545	-32085	035	P	-a	0.6915	1.5937	0.5848	300.6	158.3	-	23N	0W
3392	170	-0594 Jun 03	23:19:08	18537	-32079	040	T	-t	-0.3733	2.1995	1.1468	356.6	216.3	60.1	22S	85E
3393	170	-0594 Nov 27	19:14:40	18530	-32073	045	T+	pp	0.0011	2.8428	1.8690	320.4	210.8	98.1	20N	146E
3394	170	-0593 May 24	00:17:59	18522	-32067	050	T	t-	0.3938	2.1695	1.1019	362.3	216.8	51.4	19S	71E
3395	170	-0593 Nov 17	10:47:01	18515	-32061	055	P	a-	-0.6660	1.6274	0.6442	293.1	160.6	-	17N	87W
3396	170	-0592 May 12	02:50:56	18507	-32055	060	N	t-	1.1376	0.7883	-0.2468	246.8	-	-	15S	33E
3397	170	-0592 Oct 07	08:21:24	18501	-32050	027	N	-a	1.4675	0.1938	-0.8629	129.2	-	-	4N	51W
3398	170	-0592 Nov 05	23:02:56	18500	-32049	065	N	a-	-1.3949	0.3142	-0.7171	157.0	-	-	13N	89E
3399	170	-0591 Apr 02	03:46:55	18494	-32044	032	P	-a	-0.7637	1.4383	0.4746	282.6	142.7	-	3S	21E
3400	170	-0591 Sep 26	10:26:19	18486	-32038	037	P	-t	0.8141	1.4115	0.3175	323.2	132.9	-	1S	81W
3401	171	-0590 Mar 22	20:19:18	18479	-32032	042	T-	pp	-0.0309	2.7753	1.8267	317.6	211.2	98.9	2N	135E
3402	171	-0590 Sep 15	10:00:53	18471	-32026	047	T+	pp	0.1096	2.6980	1.6162	371.4	232.0	101.7	6S	74W
3403	171	-0589 Mar 12	12:14:59	18464	-32020	052	P	a-	0.7025	1.5578	0.5800	293.4	156.5	-	7N	103W
3404	171	-0589 Sep 04	14:32:47	18456	-32014	057	P	a-	-0.6118	1.7498	0.7212	321.3	177.8	-	11S	141W
3405	171	-0588 Jan 31	08:19:04	18450	-32009	024	N	-t	-1.3608	0.3971	-0.6747	185.5	-	-	18N	44W
3406	171	-0588 Feb 29	23:04:08	18449	-32008	062	N	t-	1.4929	0.1363	-0.8990	107.5	-	-	11N	95E
3407	171	-0588 Jul 25	18:02:20	18443	-32003	029	N	-a	1.2849	0.4753	-0.4752	176.9	-	-	20S	167E
3408	171	-0588 Aug 24	02:26:30	18441	-32002	067	N	a-	-1.2871	0.4828	-0.4906	181.8	-	-	15S	41E
3409	171	-0587 Jan 19	08:36:22	18435	-31997	034	P	-t	-0.6910	1.6401	0.5409	340.7	167.9	-	21N	49W
3410	171	-0587 Jul 15	11:01:32	18428	-31991	039	P	-a	0.5747	1.7793	0.8272	297.5	176.3	-	22S	89W
3411	171	-0586 Jan 08	09:00:11	18420	-31985	044	T+	pp	0.0133	2.8711	1.7966	365.9	230.1	103.7	23N	56W
3412	171	-0586 Jul 05	01:22:56	18413	-31979	049	T-	pp	-0.1726	2.5376	1.5447	336.7	217.7	95.1	24S	55E
3413	171	-0586 Dec 28	15:50:05	18406	-31973	054	P	a-	0.7061	1.5723	0.5526	304.0	156.6	-	24N	160W
3414	171	-0585 Jun 24	09:15:16	18398	-31967	059	P	t-	-0.9786	1.0850	0.0399	203.9	40.0		25S	64W
3415	171	-0585 Nov 18	18:19:31	18392	-31962	026	N	-a	-1.3398	0.3852	-0.5863	160.6	-	-	17N	160E
3416	171	-0585 Dec 18	05:04:47	18391	-31961	064	N	a-	1.3364	0.3944	-0.5830	163.7	-	-	25N	0E
3417	171	-0584 May 13	20:09:24	18385	-31956	031	N	-t	1.0784	0.9109	-0.1521	271.0	-	-	16S	133E
3418	171	-0584 Nov 07	09:25:57	18377	-31950	036	P	-a	-0.7071	1.5574	0.5636	293.1	153.8	-	14N	68W
3419	171	-0583 May 03	00:18:12	18370	-31944	041	T	pp	0.2792	2.3563	1.3354	346.3	217.8	82.9	13S	71E
3420	171	-0583 Oct 27	20:01:50	18362	-31938	046	T-	pp	-0.0245	2.8367	1.7896	352.7	224.5	102.0	11N	133E

Cat Num	Canon Plate	Calendar Date	TD of Greatest Eclipse	ΔT s	Luna Num	Saros Num	Ecl. Type	QSE	Gamma	Pen. Mag.	Um. Mag.	Eclipse Phase Durations Pen. m	Par. m	Total m	Greatest in Zenith Lat.	Lng.
3421	172	-0582 Apr 22	11:36:24	18355	-31932	051	P	a-	-0.4934	1.9369	0.9678	311.4	189.5	-	10S	98W
3422	172	-0582 Oct 16	23:24:02	18347	-31926	056	P	t-	0.6935	1.6326	0.5391	336.9	166.3	-	7N	83E
3423	172	-0581 Mar 13	20:19:31	18341	-31921	023	N	-a	1.3274	0.3991	-0.5551	163.4	-	-	7N	135E
3424	172	-0581 Apr 12	03:49:35	18340	-31920	061	N	a-	-1.2003	0.6273	-0.3169	199.3	-	-	7S	19E
3425	172	-0581 Sep 06	05:35:49	18334	-31915	028	Ne	-t	-1.5732	0.0052	-1.0622	22.0	-	-	11S	7W
3426	172	-0581 Oct 05	22:39:23	18333	-31914	066	N	t-	1.3760	0.3809	-0.7138	186.2	-	-	3N	94E
3427	172	-0580 Mar 02	11:10:26	18326	-31909	033	P	-a	0.6513	1.6600	0.6655	304.6	167.6	-	10N	87W
3428	172	-0580 Aug 25	12:20:21	18319	-31903	038	P	-a	-0.7879	1.4167	0.4076	294.7	138.3	-	14S	108W
3429	172	-0579 Feb 19	19:52:35	18312	-31897	043	T-	pp	-0.0799	2.7386	1.6845	360.3	228.3	102.3	13N	143E
3430	172	-0579 Aug 15	01:58:11	18304	-31891	048	T-	pp	-0.0351	2.7738	1.8129	321.7	212.7	99.2	17S	47E
3431	172	-0578 Feb 08	21:28:32	18297	-31885	053	P	t-	-0.8150	1.4089	0.3169	323.4	133.0	-	16N	118E
3432	172	-0578 Aug 04	18:40:50	18289	-31879	058	P	a-	0.6777	1.5901	0.6382	286.5	159.0	-	19S	156E
3433	172	-0578 Dec 30	04:54:47	18283	-31874	025	N	-t	1.4418	0.2431	-0.8180	144.2	-	-	25N	4E
3434	172	-0577 Jan 28	21:22:38	18282	-31873	063	N	t-	-1.4916	0.1611	-0.9186	121.7	-	-	19N	119E
3435	172	-0577 Jun 25	23:53:44	18276	-31868	030	N	-a	-1.1947	0.6695	-0.3379	222.2	-	-	25S	77E
3436	172	-0577 Jul 25	09:47:47	18274	-31867	068	N	a-	1.4410	0.2064	-0.7789	124.8	-	-	20S	71W
3437	172	-0577 Dec 19	13:45:23	18268	-31862	035	P	-a	0.6911	1.5925	0.5872	299.4	158.2	-	24N	130W
3438	172	-0576 Jun 14	05:47:17	18261	-31856	040	P	-t	-0.4598	2.0424	0.9867	352.2	207.3	-	23S	12W
3439	172	-0576 Dec 08	04:07:19	18253	-31850	045	T+	pp	0.0020	2.8405	1.8677	320.1	210.6	98.1	22N	13E
3440	172	-0575 Jun 03	06:37:20	18246	-31844	050	T	tp	0.3053	2.3315	1.2647	366.8	224.6	78.3	21S	25W
3441	173	-0575 Nov 27	19:36:50	18239	-31838	055	P	a-	-0.6668	1.6270	0.6417	293.7	160.6	-	20N	140E
3442	173	-0574 May 23	09:31:34	18231	-31832	060	N	t-	1.0537	0.9402	-0.0910	264.2	-	-	18S	69W
3443	173	-0574 Oct 18	16:26:07	18225	-31827	027	N	-a	1.4846	0.1652	-0.8969	120.0	-	-	9N	174W
3444	173	-0574 Nov 17	07:33:31	18224	-31826	065	N	a-	-1.3971	0.3121	-0.7231	157.1	-	-	16N	40W
3445	173	-0573 Apr 13	11:18:58	18218	-31821	032	P	-a	-0.8241	1.3252	0.3660	273.9	127.1	-	7S	94W
3446	173	-0573 Oct 07	18:05:34	18210	-31815	037	P	-t	0.8372	1.3712	0.2733	320.6	124.2	-	4N	162E
3447	173	-0572 Apr 02	04:03:37	18203	-31809	042	T-	pp	-0.0890	2.6677	1.7212	317.1	210.7	97.5	2S	16E
3448	173	-0572 Sep 25	17:37:16	18196	-31803	047	T+	pp	0.1383	2.6461	1.5629	370.3	230.6	99.5	1S	170E
3449	173	-0571 Mar 22	19:57:06	18188	-31797	052	P	a-	0.6525	1.6494	0.6716	299.4	166.4	-	2N	139E
3450	173	-0571 Sep 14	22:24:32	18181	-31791	057	P	a-	-0.5775	1.8126	0.7841	323.4	182.8	-	6S	99E
3451	173	-0570 Feb 10	15:57:04	18175	-31786	024	N	-t	-1.3943	0.3353	-0.7357	171.8	-	-	15N	159W
3452	173	-0570 Mar 12	06:30:25	18173	-31785	062	N	t-	1.4505	0.2142	-0.8211	134.3	-	-	7N	18W
3453	173	-0570 Aug 06	01:49:16	18167	-31780	029	N	-a	1.3366	0.3812	-0.5708	159.7	-	-	18S	49E
3454	173	-0570 Sep 04	10:30:12	18166	-31779	067	N	a-	-1.2473	0.5561	-0.4180	193.3	-	-	11S	81W
3455	173	-0569 Jan 30	16:12:10	18160	-31774	034	P	-t	-0.7161	1.5923	0.4964	337.2	162.0	-	19N	163W
3456	173	-0569 Jul 26	18:38:05	18152	-31768	039	P	-a	0.6374	1.6662	0.7101	292.8	166.8	-	21S	157E
3457	173	-0568 Jan 19	16:58:05	18145	-31762	044	T-	pp	-0.0026	2.8880	1.8189	364.4	229.7	103.8	22N	175W
3458	173	-0568 Jul 15	08:34:55	18138	-31756	049	T-	pp	-0.1037	2.6669	1.6684	339.7	220.4	99.8	23S	53W
3459	173	-0567 Jan 08	00:17:34	18130	-31750	054	P	a-	0.6957	1.5888	0.5741	303.7	158.6	-	24N	74E
3460	173	-0567 Jul 04	15:54:22	18123	-31744	059	P	t-	-0.9025	1.2271	0.1774	298.4	99.5	-	25S	164W
3461	174	-0567 Nov 29	03:13:54	18117	-31739	026	N	-a	-1.3401	0.3847	-0.5870	160.5	-	-	19N	25E
3462	174	-0567 Dec 28	13:52:25	18116	-31738	064	N	a-	1.3310	0.4029	-0.5719	165.0	-	-	25N	131W
3463	174	-0566 May 25	02:32:38	18109	-31733	031	N	-t	1.1626	0.7554	-0.3054	250.2	-	-	18S	36E
3464	174	-0566 Nov 18	18:10:36	18102	-31727	036	P	-a	-0.7106	1.5524	0.5556	293.7	153.2	-	17N	160E
3465	174	-0565 May 14	07:08:35	18095	-31721	041	T	-p	0.3576	2.2100	1.1938	340.9	211.4	66.0	16S	34W
3466	174	-0565 Nov 08	04:22:55	18087	-31715	046	T-	pp	-0.0301	2.8287	1.7770	354.2	225.0	102.0	15N	7E
3467	174	-0564 May 02	18:55:07	18080	-31709	051	T	a-	-0.4198	2.0698	1.1050	314.9	197.0	47.9	14S	150E
3468	174	-0564 Oct 27	07:18:49	18072	-31703	056	P	t-	0.6817	1.6562	0.5590	339.2	169.1	-	12N	38W
3469	174	-0563 Mar 24	04:10:03	18066	-31698	023	N	-a	1.3803	0.3013	-0.6514	143.4	-	-	3N	16E
3470	174	-0563 Apr 22	11:24:14	18065	-31697	061	N	a-	-1.1351	0.7459	-0.1963	214.6	-	-	11S	97W
3471	174	-0563 Oct 16	06:26:13	18058	-31691	066	N	t-	1.3559	0.4183	-0.6777	194.3	-	-	8N	24W
3472	174	-0562 Mar 13	18:54:07	18052	-31686	033	P	-a	0.7012	1.5685	0.5739	300.4	158.4	-	6N	155E
3473	174	-0562 Sep 05	20:11:14	18044	-31680	038	P	-a	-0.8234	1.3516	0.3425	289.1	127.7	-	10S	133E
3474	174	-0561 Mar 03	03:18:46	18037	-31674	043	T-	pp	-0.0349	2.8210	1.7674	361.7	229.5	104.0	9N	29E
3475	174	-0561 Aug 26	09:57:21	18030	-31668	048	T-	pp	-0.0794	2.6932	1.7310	320.9	211.8	97.9	13S	74W
3476	174	-0560 Feb 20	04:46:04	18022	-31662	053	P	t-	-0.7748	1.4812	0.3922	328.8	146.4	-	13N	8E
3477	174	-0560 Aug 15	02:33:41	18015	-31656	058	P	a-	0.6285	1.6824	0.7267	292.0	167.3	-	16S	37E
3478	174	-0559 Jan 09	13:03:41	18009	-31651	025	N	-t	1.4491	0.2271	-0.8287	139.1	-	-	24N	118W
3479	174	-0559 Feb 08	04:57:58	18008	-31650	063	N	t-	-1.4586	0.2188	-0.8552	140.7	-	-	16N	5E
3480	174	-0559 Jul 06	06:53:57	18001	-31645	030	N	-h	-1.2711	0.5320	-0.4807	201.7	-	-	25S	29W

Cat Num	Canon Plate	Calendar Date	TD of Greatest Eclipse	ΔT s	Luna Num	Saros Num	Ecl. Type	QSE	Gamma	Pen. Mag.	Um. Mag.	Eclipse Phase Durations Pen. m	Par. m	Total m	Greatest in Zenith Lat.	Lng.
3481	175	-0559 Aug 04	17:17:31	18000	-31644	068	N	h-	1.3894	0.3040	-0.6871	150.8	-	-	18S	176E
3482	175	-0559 Dec 29	22:23:12	17994	-31639	035	P	-a	0.6938	1.5855	0.5844	298.0	157.5	-	24N	101E
3483	175	-0558 Jun 25	12:15:47	17987	-31633	040	P	-t	-0.5451	1.8876	0.8285	346.6	195.9	-	24S	110W
3484	175	-0558 Dec 19	12:59:22	17979	-31627	045	T+	pp	0.0037	2.8367	1.8654	319.8	210.6	98.1	24N	120W
3485	175	-0557 Jun 14	12:56:46	17972	-31621	050	T+	pp	0.2163	2.4946	1.4282	369.9	230.1	93.4	23S	121W
3486	175	-0557 Dec 09	04:26:52	17965	-31615	055	P	a-	-0.6685	1.6246	0.6378	294.2	160.5	-	22N	7E
3487	175	-0556 Jun 02	16:11:56	17957	-31609	060	P	t-	0.9684	1.0951	0.0672	279.4	61.1	-	20S	170W
3488	175	-0556 Oct 29	00:36:46	17951	-31604	027	N	-a	1.4960	0.1467	-0.9203	113.7	-	-	13N	62E
3489	175	-0556 Nov 27	16:05:44	17950	-31603	065	N	a-	-1.4014	0.3060	-0.7326	156.2	-	-	19N	169W
3490	175	-0555 Apr 23	18:47:12	17944	-31598	032	P	-a	-0.8882	1.2056	0.2504	264.0	106.7	-	11S	152E
3491	175	-0555 Oct 18	01:51:10	17936	-31592	037	P	-t	0.8547	1.3407	0.2394	318.6	116.9	-	8N	44E
3492	175	-0554 Apr 13	11:43:42	17929	-31586	042	T-	-p	-0.1506	2.5539	1.6090	316.2	209.4	94.4	7S	101W
3493	175	-0554 Oct 07	01:22:13	17922	-31580	047	T+	pp	0.1606	2.6058	1.5217	369.2	229.4	97.4	3N	52E
3494	175	-0553 Apr 03	03:31:45	17914	-31574	052	P	a-	0.5971	1.7512	0.7732	305.5	176.1	-	2S	23E
3495	175	-0553 Sep 26	06:25:23	17907	-31568	057	P	a-	-0.5500	1.8630	0.8346	324.7	186.4	-	2S	23W
3496	175	-0552 Feb 21	23:24:53	17901	-31563	024	N	-t	-1.4350	0.2601	-0.8100	152.7	-	-	12N	87E
3497	175	-0552 Mar 22	13:46:39	17900	-31562	062	N	t-	1.4007	0.3058	-0.7298	159.6	-	-	3N	129W
3498	175	-0552 Aug 16	09:44:13	17894	-31557	029	N	-a	1.3819	0.2990	-0.6549	142.5	-	-	15S	71W
3499	175	-0552 Sep 14	18:43:09	17892	-31556	067	N	a-	-1.2149	0.6161	-0.3589	201.9	-	-	7S	154E
3500	175	-0551 Feb 09	23:37:28	17886	-31551	034	P	-t	-0.7495	1.5291	0.4370	332.5	153.4	-	16N	85E
3501	176	-0551 Aug 06	02:22:10	17879	-31545	039	P	-a	0.6933	1.5660	0.6054	288.2	156.9	-	18S	40E
3502	176	-0550 Jan 30	00:48:18	17872	-31539	044	T-	pp	-0.0252	2.8436	1.7803	362.7	229.2	103.6	19N	67E
3503	176	-0550 Jul 26	15:51:20	17864	-31533	049	T-	pp	-0.0395	2.7876	1.7833	342.1	221.9	102.1	21S	163W
3504	176	-0549 Jan 19	08:39:30	17857	-31527	054	P	a-	0.6802	1.6144	0.6051	303.9	161.6	-	22N	52W
3505	176	-0549 Jul 15	22:36:32	17850	-31521	059	P	t-	-0.8295	1.3634	0.3089	311.0	129.1	-	24S	95E
3506	176	-0549 Dec 10	12:07:26	17844	-31516	026	N	-a	-1.3416	0.3818	-0.5897	160.0	-	-	21N	108W
3507	176	-0548 Jan 08	22:34:35	17842	-31515	064	N	a-	1.3210	0.4196	-0.5520	167.9	-	-	25N	98E
3508	176	-0548 Jun 04	08:54:40	17836	-31510	031	N	-t	1.2477	0.5983	-0.4608	225.6	-	-	21S	61W
3509	176	-0548 Nov 29	02:56:38	17829	-31504	036	P	-a	-0.7127	1.5500	0.5506	294.5	153.0	-	20N	28E
3510	176	-0547 May 24	13:58:32	17822	-31498	041	T	-t	0.4374	2.0616	1.0493	334.6	203.1	34.8	19S	137W
3511	176	-0547 Nov 18	12:46:21	17814	-31492	046	T-	pp	-0.0330	2.8254	1.7698	355.7	225.5	102.1	18N	120W
3512	176	-0546 May 14	02:12:20	17807	-31486	051	T	a-	-0.3442	2.2069	1.2457	317.8	203.0	69.9	17S	39E
3513	176	-0546 Nov 07	15:18:48	17800	-31480	056	P	t-	0.6744	1.6710	0.5708	340.8	170.7	-	15N	159W
3514	176	-0545 Apr 04	11:52:54	17794	-31475	023	N	-a	1.4393	0.1925	-0.7590	115.9	-	-	2S	102W
3515	176	-0545 May 03	18:53:36	17792	-31474	061	N	a-	-1.0655	0.8729	-0.0678	229.0	-	-	15S	149E
3516	176	-0545 Oct 27	14:21:12	17785	-31468	066	N	t-	1.3418	0.4447	-0.6521	199.6	-	-	12N	145W
3517	176	-0544 Mar 24	02:28:40	17779	-31463	033	P	-a	0.7578	1.4649	0.4699	294.9	146.2	-	2N	39E
3518	176	-0544 Sep 16	04:11:56	17772	-31457	038	P	-a	-0.8519	1.2993	0.2902	284.2	118.0	-	6S	11E
3519	176	-0543 Mar 13	10:34:57	17764	-31451	043	T+	pp	0.0177	2.8524	1.7991	362.8	230.3	104.5	5N	82W
3520	176	-0543 Sep 05	18:05:26	17757	-31445	048	T-	-p	-0.1167	2.6255	1.6618	320.1	210.7	96.0	9S	163E
3521	177	-0542 Mar 02	11:55:56	17750	-31439	053	P	t-	-0.7281	1.5652	0.4794	334.5	159.7	-	9N	101W
3522	177	-0542 Aug 26	10:32:35	17743	-31433	058	P	a-	0.5845	1.7652	0.8053	296.6	174.0	-	13S	84W
3523	177	-0541 Jan 20	21:08:38	17736	-31428	025	N	-t	1.4603	0.2037	-0.8465	131.4	-	-	23N	121E
3524	177	-0541 Feb 19	12:27:33	17735	-31427	063	N	t-	-1.4200	0.2867	-0.7814	159.4	-	-	12N	109W
3525	177	-0541 Jul 17	13:57:05	17729	-31422	030	N	-h	-1.3440	0.4010	-0.6174	178.2	-	-	24S	135W
3526	177	-0541 Aug 16	00:52:44	17728	-31421	068	N	h-	1.3426	0.3930	-0.6043	170.8	-	-	15S	61E
3527	177	-0540 Jan 10	06:56:03	17722	-31416	035	P	-a	0.7011	1.5697	0.5733	296.0	155.9	-	24N	27W
3528	177	-0540 Jul 05	18:47:26	17715	-31410	040	P	-t	-0.6270	1.7392	0.6763	340.0	182.0	-	24S	152E
3529	177	-0540 Dec 29	21:47:00	17707	-31404	045	T+	pp	0.0093	2.8255	1.8561	319.5	210.6	98.1	24N	109E
3530	177	-0539 Jun 24	19:21:28	17700	-31398	050	T+	pp	0.1312	2.6508	1.5843	371.7	233.4	101.8	24S	142E
3531	177	-0539 Dec 19	13:13:01	17693	-31392	055	P	a-	-0.6674	1.6272	0.6394	295.0	161.0	-	23N	124W
3532	177	-0538 Jun 13	22:57:11	17685	-31386	060	P	t-	0.8855	1.2457	0.2207	291.9	108.0	-	22S	88E
3533	177	-0538 Nov 09	08:50:28	17679	-31381	027	N	-a	1.5044	0.1334	-0.9379	109.0	-	-	17N	62W
3534	177	-0538 Dec 09	00:36:08	17678	-31380	065	N	a-	-1.4049	0.3010	-0.7403	155.6	-	-	21N	64E
3535	177	-0537 May 05	02:11:03	17672	-31375	032	P	-a	-0.9565	1.0786	0.1268	252.5	77.1	-	15S	39E
3536	177	-0537 Oct 29	09:43:55	17665	-31369	037	P	-t	0.8663	1.3209	0.2169	317.4	111.7	-	12N	76W
3537	177	-0536 Apr 23	19:16:29	17657	-31363	042	T-	-p	-0.2184	2.4290	1.4851	314.6	207.0	88.7	11S	144E
3538	177	-0536 Oct 17	09:16:58	17650	-31357	047	T+	pp	0.1754	2.5788	1.4943	368.1	228.3	95.8	7N	68W
3539	177	-0535 Apr 13	10:57:01	17643	-31351	052	P	a-	0.5345	1.8664	0.8876	311.8	185.5	-	6S	90W
3540	177	-0535 Oct 06	14:36:48	17636	-31345	057	P	a-	-0.5302	1.8992	0.8713	325.3	188.6	-	2N	148W

Cat Num	Canon Plate	Calendar Date	TD of Greatest Eclipse	ΔT s	Luna Num	Saros Num	Ecl. Type	QSE	Gamma	Pen. Mag.	Um. Mag.	Eclipse Phase ---- Durations ---- Pen. m	Par. m	Total m	Greatest in Zenith Lat.	Lng.
3541	178	-0534 Mar 04	06:42:30	17630	-31340	024	N	-t	-1.4829	0.1718	-0.8972	125.3	-	-	8N	24W
3542	178	-0534 Apr 02	20:53:22	17628	-31339	062	N	t-	1.3440	0.4100	-0.6261	183.7	-	-	1S	122E
3543	178	-0534 Aug 27	17:46:12	17622	-31334	029	N	-a	1.4218	0.2269	-0.7292	124.9	-	-	11S	167E
3544	178	-0534 Sep 26	03:04:05	17621	-31333	067	N	a-	-1.1886	0.6648	-0.3111	208.3	-	-	3S	26E
3545	178	-0533 Feb 21	06:55:22	17615	-31328	034	P	-t	-0.7889	1.4548	0.3668	326.6	142.0	-	12N	26W
3546	178	-0533 Aug 17	10:11:30	17608	-31322	039	P	-a	0.7441	1.4751	0.5098	283.7	146.4	-	15S	79W
3547	178	-0532 Feb 10	08:31:35	17600	-31316	044	T-	pp	-0.0544	2.7869	1.7298	360.8	228.5	103.0	16N	50W
3548	178	-0532 Aug 05	23:12:29	17593	-31310	049	T+	pp	0.0198	2.8270	1.8164	344.1	222.6	102.4	19S	86E
3549	178	-0531 Jan 29	16:55:54	17586	-31304	054	P	a-	0.6596	1.6494	0.6458	304.5	165.4	-	20N	176W
3550	178	-0531 Jul 26	05:23:45	17579	-31298	059	P	t-	-0.7609	1.4919	0.4322	321.7	150.3	-	22S	7W
3551	178	-0531 Dec 20	20:58:38	17573	-31293	026	N	-a	-1.3451	0.3750	-0.5958	158.7	-	-	22N	119E
3552	178	-0530 Jan 19	07:11:40	17571	-31292	064	N	a-	1.3071	0.4435	-0.5249	171.9	-	-	23N	31W
3553	178	-0530 Jun 15	15:20:58	17565	-31287	031	N	-t	1.3294	0.4476	-0.6101	197.5	-	-	22S	158W
3554	178	-0530 Jul 15	06:08:51	17564	-31286	069	N	t-	-1.5124	0.1200	-0.9536	106.9	-	-	24S	19W
3555	178	-0530 Dec 10	11:41:22	17558	-31281	036	P	-a	-0.7153	1.5463	0.5448	295.1	152.7	-	22N	103W
3556	178	-0529 Jun 04	20:49:51	17551	-31275	041	P	-t	0.5169	1.9139	0.9052	327.3	193.0	-	21S	119E
3557	178	-0529 Nov 29	21:09:48	17544	-31269	046	T-	pp	-0.0351	2.8233	1.7645	357.0	226.1	102.2	21N	114E
3558	178	-0528 May 24	09:29:11	17536	-31263	051	T-	p-	-0.2677	2.3457	1.3875	319.9	207.6	83.4	20S	71W
3559	178	-0528 Nov 17	23:20:21	17529	-31257	056	P	t-	0.6689	1.6822	0.5798	342.1	172.0	-	19N	80E
3560	178	-0527 Apr 14	19:28:57	17523	-31252	023	Ne	-a	1.5033	0.0747	-0.8762	73.1	-	-	6S	142E
3561	179	-0527 May 14	02:20:49	17522	-31251	061	P	a-	-0.9942	1.0034	0.0635	242.2	54.4	-	18S	35E
3562	179	-0527 Nov 06	22:21:34	17515	-31245	066	N	t-	1.3319	0.4628	-0.6339	203.0	-	-	16N	94E
3563	179	-0526 Apr 04	09:54:21	17508	-31240	033	P	-a	0.8209	1.3495	0.3538	287.9	129.5	-	2S	74W
3564	179	-0526 Sep 27	12:21:37	17501	-31234	038	P	-a	-0.8738	1.2590	0.2501	280.2	109.9	-	2S	114W
3565	179	-0525 Mar 24	17:40:25	17494	-31228	043	T+	pp	0.0782	2.7412	1.6882	363.3	230.3	103.3	1N	170E
3566	179	-0525 Sep 17	02:22:22	17487	-31222	048	T-	-p	-0.1470	2.5709	1.6054	319.3	209.5	94.0	5S	37E
3567	179	-0524 Mar 12	18:53:29	17479	-31216	053	P	t-	-0.6714	1.6676	0.5851	340.8	173.4	-	5N	152E
3568	179	-0524 Sep 05	18:40:22	17472	-31210	058	P	a-	0.5483	1.8338	0.8696	300.3	179.0	-	9S	152E
3569	179	-0523 Jan 31	05:06:01	17466	-31205	025	N	-t	1.4782	0.1678	-0.8762	119.1	-	-	20N	1E
3570	179	-0523 Mar 01	19:47:32	17465	-31204	063	N	t-	-1.3729	0.3701	-0.6919	179.0	-	-	9N	139E
3571	179	-0523 Jul 27	21:04:52	17459	-31199	030	N	-h	-1.4118	0.2796	-0.7448	151.2	-	-	22S	118E
3572	179	-0523 Aug 26	08:35:01	17458	-31198	068	N	h-	1.3023	0.4701	-0.5336	186.3	-	-	12S	56W
3573	179	-0522 Jan 20	15:23:38	17452	-31193	035	P	-a	0.7133	1.5448	0.5533	293.4	153.3	-	22N	154W
3574	179	-0522 Jul 17	01:23:12	17444	-31187	040	P	-t	-0.7048	1.5985	0.5315	332.6	165.5	-	23S	52E
3575	179	-0521 Jan 10	06:31:20	17437	-31181	045	T+	pp	0.0175	2.8092	1.8420	319.2	210.6	98.2	23N	22W
3576	179	-0521 Jul 06	01:48:34	17430	-31175	050	T+	pp	0.0477	2.8042	1.7374	372.4	234.9	105.7	24S	45E
3577	179	-0521 Dec 30	21:56:48	17423	-31169	055	P	a-	-0.6649	1.6320	0.6436	296.0	161.8	-	23N	105E
3578	179	-0520 Jun 24	05:45:32	17416	-31163	060	P	t-	0.8039	1.3943	0.3716	302.5	136.6	-	23S	15W
3579	179	-0520 Nov 19	17:06:24	17409	-31158	027	N	-h	1.5102	0.1245	-0.9503	105.7	-	-	20N	173E
3580	179	-0520 Dec 19	09:03:50	17408	-31157	065	N	h-	-1.4070	0.2981	-0.7452	155.4	-	-	22N	63W
3581	180	-0519 May 15	09:33:45	17402	-31152	032	P	-a	-1.0261	0.9494	0.0004	239.6	4.3	-	18S	73W
3582	180	-0519 Jun 13	16:58:32	17401	-31151	070	Nb	a-	1.5034	0.0832	-0.8849	79.0	-	-	21S	176E
3583	180	-0519 Nov 08	17:41:07	17395	-31146	037	P	-t	0.8742	1.3074	0.2014	316.5	108.0	-	16N	164E
3584	180	-0518 May 05	02:46:09	17388	-31140	042	T	-p	-0.2884	2.3002	1.3567	312.5	203.5	79.9	14S	29E
3585	180	-0518 Oct 28	17:17:19	17381	-31134	047	T+	pp	0.1860	2.5591	1.4748	367.1	227.4	94.5	12N	170E
3586	180	-0517 Apr 24	18:16:43	17373	-31128	052	T	a-	0.4681	1.9889	1.0089	317.8	194.1	14.5	10S	158E
3587	180	-0517 Oct 17	22:55:30	17366	-31122	057	P	a-	-0.5160	1.9248	0.8977	325.3	190.0	-	7N	86E
3588	180	-0516 Mar 14	13:48:42	17360	-31117	024	Ne	-t	-1.5386	0.0690	-0.9989	80.3	-	-	3N	132W
3589	180	-0516 Apr 13	03:50:34	17359	-31116	062	N	t-	1.2805	0.5269	-0.5101	206.6	-	-	6S	15E
3590	180	-0516 Sep 07	01:57:13	17353	-31111	029	N	-a	1.4545	0.1680	-0.7905	108.1	-	-	7S	42E
3591	180	-0516 Oct 06	11:33:32	17352	-31110	067	N	a-	-1.1692	0.7009	-0.2759	212.8	-	-	2N	103W
3592	180	-0515 Mar 03	14:03:00	17346	-31105	034	P	-t	-0.8362	1.3658	0.2820	319.1	126.0	-	8N	135W
3593	180	-0515 Aug 27	18:08:42	17338	-31099	039	P	-a	0.7878	1.3974	0.4270	279.5	135.9	-	12S	160E
3594	180	-0514 Feb 20	16:06:46	17331	-31093	044	T-	pp	-0.0908	2.7169	1.6661	358.7	227.4	101.7	13N	165W
3595	180	-0514 Aug 17	06:40:42	17324	-31087	049	T+	pp	0.0721	2.7342	1.7171	345.6	222.6	101.3	16S	27W
3596	180	-0513 Feb 10	01:04:34	17317	-31081	054	P	a-	0.6320	1.6971	0.6995	305.8	170.2	-	17N	61E
3597	180	-0513 Aug 06	12:16:33	17310	-31075	059	P	t-	-0.6974	1.6111	0.5460	330.9	166.2	-	20S	111W
3598	180	-0512 Jan 01	05:46:22	17304	-31070	026	N	-a	-1.3515	0.3628	-0.6069	156.3	-	-	22N	13W
3599	180	-0512 Jan 30	15:41:34	17302	-31069	064	N	a-	1.2875	0.4778	-0.4870	177.5	-	-	21N	159W
3600	180	-0512 Jun 25	21:50:22	17296	-31064	031	N	-t	1.4085	0.3020	-0.7548	164.1	-	-	22S	104E

Cat Num	Canon Plate	Calendar Date	TD of Greatest Eclipse	ΔT s	Luna Num	Saros Num	Ecl. Type	QSE	Gamma	Pen. Mag.	Um. Mag.	Eclipse Phase Durations Pen. m	Par. m	Total m	Greatest in Zenith Lat.	Lng.
3601	181	-0512 Jul 25	12:46:13	17295	-31063	069	N	t-	-1.4403	0.2532	-0.8224	153.4	-	-	23S	119W
3602	181	-0512 Dec 20	20:23:10	17289	-31058	036	P	-a	-0.7196	1.5391	0.5361	295.5	152.1	-	23N	127E
3603	181	-0511 Jun 15	03:44:34	17282	-31052	041	P	-t	0.5943	1.7704	0.7647	319.3	181.2	-	23S	14E
3604	181	-0511 Dec 10	05:32:55	17275	-31046	046	T-	pp	-0.0362	2.8225	1.7612	358.3	226.6	102.4	23N	12W
3605	181	-0510 Jun 04	16:46:12	17268	-31040	051	T-	p-	-0.1907	2.4858	1.5299	321.3	210.9	92.2	22S	178E
3606	181	-0510 Nov 29	07:24:10	17260	-31034	056	P	t-	0.6655	1.6891	0.5854	343.0	172.9	-	21N	41W
3607	181	-0509 May 25	09:45:33	17253	-31028	061	P	a-	-0.9209	1.1378	0.1980	254.2	94.4	-	20S	77W
3608	181	-0509 Nov 18	06:26:56	17246	-31022	066	N	t-	1.3250	0.4751	-0.6209	205.1	-	-	19N	28W
3609	181	-0508 Apr 14	17:11:18	17240	-31017	033	P	-a	0.8903	1.2228	0.2259	279.0	105.8	-	6S	174E
3610	181	-0508 Oct 07	20:40:46	17233	-31011	038	P	-a	-0.8890	1.2311	0.2224	277.1	103.7	-	3N	120E
3611	181	-0507 Apr 04	00:37:41	17226	-31005	043	T+	pp	0.1448	2.6190	1.5661	363.2	229.2	99.7	3S	64E
3612	181	-0507 Sep 27	10:46:53	17218	-30999	048	T-	-p	-0.1713	2.5271	1.5600	318.5	208.4	92.0	1S	91W
3613	181	-0506 Mar 24	01:44:46	17211	-30993	053	P	t-	-0.6093	1.7799	0.7007	347.0	186.2	-	1N	47E
3614	181	-0506 Sep 17	02:55:15	17204	-30987	058	P	a-	0.5183	1.8910	0.9224	303.5	182.9	-	4S	27E
3615	181	-0505 Feb 11	12:57:12	17198	-30982	025	N	-t	1.5016	0.1215	-0.9161	101.3	-	-	17N	117W
3616	181	-0505 Mar 13	03:00:44	17197	-30981	063	N	t-	-1.3192	0.4655	-0.5904	198.2	-	-	5N	29E
3617	181	-0505 Aug 08	04:17:39	17191	-30976	030	N	-h	-1.4745	0.1678	-0.8629	118.9	-	-	20S	8E
3618	181	-0505 Sep 06	16:24:10	17190	-30975	068	N	h-	1.2682	0.5360	-0.4740	198.6	-	-	8S	175W
3619	181	-0504 Jan 31	23:44:46	17184	-30970	035	P	-a	0.7317	1.5085	0.5222	290.1	149.3	-	20N	80E
3620	181	-0504 Jul 27	08:05:25	17176	-30964	040	P	-t	-0.7768	1.4686	0.3972	324.6	146.4	-	22S	49W
3621	182	-0503 Jan 20	15:07:50	17169	-30958	045	T+	pp	0.0327	2.7802	1.8155	318.9	210.5	98.1	21N	152W
3622	182	-0503 Jul 16	08:24:17	17162	-30952	050	T-	pp	-0.0296	2.8377	1.7702	372.0	234.7	105.9	23S	54W
3623	182	-0502 Jan 10	06:34:14	17155	-30946	055	P	a-	-0.6574	1.6459	0.6573	297.4	163.5	-	22N	24W
3624	182	-0502 Jul 05	12:39:56	17148	-30940	060	P	h-	0.7258	1.5367	0.5159	311.1	156.8	-	23S	119W
3625	182	-0502 Dec 01	01:22:09	17142	-30935	027	N	-h	1.5157	0.1159	-0.9617	102.4	-	-	23N	49E
3626	182	-0502 Dec 30	17:26:31	17141	-30934	065	N	h-	-1.4057	0.3013	-0.7436	156.7	-	-	22N	171E
3627	182	-0501 May 26	16:54:48	17135	-30929	032	N	-a	-1.0977	0.8170	-0.1300	224.9	-	-	21S	175E
3628	182	-0501 Jun 25	00:14:49	17133	-30928	070	N	a-	1.4275	0.2215	-0.7448	126.9	-	-	22S	66E
3629	182	-0501 Nov 20	01:41:38	17127	-30923	037	P	-t	0.8791	1.2989	0.1918	315.9	105.5	-	19N	43E
3630	182	-0500 May 15	10:10:36	17119	-30917	042	T	-a	-0.3627	2.1641	1.2203	309.6	198.6	65.8	18S	83W
3631	182	-0500 Nov 08	01:25:12	17110	-30911	047	T+	pp	0.1913	2.5490	1.4656	366.0	226.6	93.8	15N	47E
3632	182	-0499 May 05	01:29:45	17102	-30905	052	T	p-	0.3969	2.1205	1.1385	323.5	201.9	55.2	14S	47E
3633	182	-0499 Oct 28	07:21:33	17093	-30899	057	P	a-	-0.5069	1.9409	0.9149	324.9	190.6	-	11N	42W
3634	182	-0498 Apr 24	10:40:33	17084	-30893	062	N	t-	1.2117	0.6538	-0.3843	228.1	-	-	10S	89W
3635	182	-0498 Sep 18	10:16:18	17077	-30888	029	N	-a	1.4812	0.1203	-0.8407	91.8	-	-	3S	84W
3636	182	-0498 Oct 17	20:10:28	17075	-30887	067	N	a-	-1.1554	0.7264	-0.2510	215.7	-	-	6N	126E
3637	182	-0497 Mar 14	21:02:14	17068	-30882	034	P	-t	-0.8906	1.2637	0.1843	309.8	103.3	-	4N	118E
3638	182	-0497 Sep 08	02:12:37	17059	-30876	039	P	-a	0.8256	1.3307	0.3551	275.8	125.5	-	8S	37E
3639	182	-0496 Mar 02	23:35:06	17051	-30870	044	T-	pp	-0.1340	2.6343	1.5901	356.1	225.8	99.2	9N	81E
3640	182	-0496 Aug 27	14:15:23	17042	-30864	049	T+	pp	0.1183	2.6529	1.6291	346.8	222.2	99.1	12S	143W
3641	183	-0495 Feb 20	09:06:27	17033	-30858	054	P	a-	0.5979	1.7565	0.7649	307.4	175.6	-	14N	61W
3642	183	-0495 Aug 16	19:17:01	17025	-30852	059	P	t-	-0.6405	1.7183	0.6477	338.5	178.4	-	16S	142E
3643	183	-0494 Jan 11	14:29:48	17018	-30847	026	N	-a	-1.3612	0.3442	-0.6242	152.6	-	-	21N	144W
3644	183	-0494 Feb 10	00:05:32	17016	-30846	064	N	a-	1.2631	0.5207	-0.4405	184.3	-	-	18N	74E
3645	183	-0494 Jul 07	04:25:22	17009	-30841	031	N	-t	1.4834	0.1644	-0.8918	122.3	-	-	22S	4E
3646	183	-0494 Aug 05	19:30:53	17008	-30840	069	N	t-	-1.3732	0.3773	-0.7004	185.1	-	-	20S	139E
3647	183	-0493 Jan 01	05:00:53	17000	-30835	036	P	-a	-0.7268	1.5265	0.5223	295.6	150.8	-	23N	3W
3648	183	-0493 Jun 26	10:43:43	16992	-30829	041	P	-t	0.6687	1.6326	0.6293	310.6	167.5	-	23S	91W
3649	183	-0493 Dec 21	13:51:29	16983	-30823	046	T-	pp	-0.0406	2.8154	1.7522	359.5	227.1	102.5	24N	137W
3650	183	-0492 Jun 15	00:05:46	16974	-30817	051	T-	p-	-0.1151	2.6237	1.6695	322.0	212.9	97.4	23S	68E
3651	183	-0492 Dec 09	15:26:16	16966	-30811	056	P	t-	0.6610	1.6977	0.5935	343.8	174.0	-	23N	162W
3652	183	-0491 Jun 04	17:10:51	16957	-30805	061	P	a-	-0.8480	1.2719	0.3315	264.9	119.9	-	23S	171E
3653	183	-0491 Nov 28	14:33:06	16949	-30799	066	N	t-	1.3183	0.4867	-0.6078	206.9	-	-	22N	150W
3654	183	-0490 Apr 26	00:21:38	16941	-30794	033	P	-a	0.9640	1.0882	0.0897	268.2	68.2	-	10S	65E
3655	183	-0490 Oct 19	05:07:52	16933	-30788	038	P	-a	-0.8984	1.2134	0.2054	274.8	99.6	-	7N	9W
3656	183	-0489 Apr 15	07:24:17	16924	-30782	043	T+	pp	0.2193	2.4824	1.4292	362.1	226.5	92.3	7S	40W
3657	183	-0489 Oct 08	19:20:03	16916	-30776	048	T-	-p	-0.1886	2.4960	1.5274	317.9	207.5	90.4	4N	138E
3658	183	-0488 Apr 03	08:25:26	16907	-30770	053	P	t-	-0.5385	1.9082	0.8321	353.1	198.3	-	3S	55W
3659	183	-0488 Sep 27	11:18:29	16898	-30764	058	P	a-	0.4956	1.9349	0.9619	306.0	185.7	-	0S	101W
3660	183	-0487 Feb 21	20:39:48	16891	-30759	025	N	-t	1.5327	0.0612	-0.9698	72.0	-	-	14N	125E

Cat Num	Canon Plate	Calendar Date	TD of Greatest Eclipse	ΔT s	Luna Num	Saros Num	Ecl. Type	QSE	Gamma	Pen. Mag.	Um. Mag.	Eclipse Phase ---- Durations ---- Pen. m	Par. m	Total m	Greatest in Zenith Lat.	Lng.
3661	184	-0487 Mar 23	10:05:34	16890	-30758	063	N	t-	-1.2577	0.5752	-0.4747	217.1	-	-	0N	79W
3662	184	-0487 Aug 18	11:37:33	16883	-30753	030	N	-h	-1.5305	0.0684	-0.9689	76.9	-	-	17S	103W
3663	184	-0487 Sep 17	00:21:24	16881	-30752	068	N	h-	1.2412	0.5887	-0.4277	208.0	-	-	4S	64E
3664	184	-0486 Feb 11	07:59:42	16874	-30747	035	P	-a	0.7559	1.4614	0.4805	286.1	143.8	-	17N	45W
3665	184	-0486 Aug 07	14:52:37	16866	-30741	040	P	-t	-0.8442	1.3472	0.2713	316.0	123.5	-	19S	152W
3666	184	-0485 Jan 31	23:39:21	16857	-30735	045	T+	pp	0.0520	2.7434	1.7813	318.6	210.5	97.9	19N	80E
3667	184	-0485 Jul 27	15:05:43	16849	-30729	050	T-	pp	-0.1028	2.7038	1.6353	370.8	233.1	103.1	21S	156W
3668	184	-0484 Jan 21	15:05:25	16840	-30723	055	P	a-	-0.6454	1.6679	0.6793	299.3	166.0	-	21N	153W
3669	184	-0484 Jul 15	19:40:47	16832	-30717	060	P	h-	0.6518	1.6716	0.6523	318.0	171.9	-	22S	135E
3670	184	-0484 Dec 11	09:36:53	16825	-30712	027	N	-h	1.5213	0.1066	-0.9731	98.6	-	-	24N	75W
3671	184	-0483 Jan 10	01:43:36	16823	-30711	065	N	h-	-1.4005	0.3113	-0.7346	159.7	-	-	22N	47E
3672	184	-0483 Jun 06	00:17:31	16816	-30706	032	N	-a	-1.1681	0.6870	-0.2585	208.6	-	-	23S	64E
3673	184	-0483 Jul 05	07:36:37	16815	-30705	070	N	a-	1.3554	0.3532	-0.6118	157.8	-	-	22S	45W
3674	184	-0483 Nov 30	09:42:28	16808	-30700	037	P	-t	0.8837	1.2906	0.1831	315.3	103.3	-	22N	78W
3675	184	-0482 May 26	17:34:51	16799	-30694	042	T	-a	-0.4368	2.0284	1.0838	306.0	192.4	42.4	20S	164E
3676	184	-0482 Nov 19	09:36:17	16791	-30688	047	T+	pp	0.1946	2.5422	1.4603	364.9	226.0	93.3	19N	77W
3677	184	-0481 May 16	08:38:03	16782	-30682	052	T	p-	0.3226	2.2581	1.2738	328.6	208.6	74.5	17S	61W
3678	184	-0481 Nov 08	15:53:00	16774	-30676	057	P	a-	-0.5015	1.9500	0.9256	324.1	190.8	-	15N	171W
3679	184	-0480 May 04	17:24:01	16765	-30670	062	N	t-	1.1379	0.7899	-0.2498	248.2	-	-	13S	168E
3680	184	-0480 Sep 28	18:43:37	16758	-30665	029	N	-a	1.5017	0.0839	-0.8796	77.0	-	-	1N	147E
3681	185	-0480 Oct 28	04:53:04	16757	-30664	067	N	a-	-1.1461	0.7437	-0.2339	217.6	-	-	10N	6W
3682	185	-0479 Mar 25	03:52:38	16750	-30659	034	P	-t	-0.9522	1.1486	0.0735	298.4	66.2	-	0S	13E
3683	185	-0479 Sep 18	10:25:09	16741	-30653	039	P	-a	0.8560	1.2776	0.2968	272.8	116.0	-	3S	88W
3684	185	-0478 Mar 14	06:55:55	16733	-30647	044	T-	pp	-0.1841	2.5392	1.5015	353.2	223.5	95.0	5N	31W
3685	185	-0478 Sep 07	21:57:11	16724	-30641	049	T+	-p	0.1576	2.5841	1.5537	347.9	221.5	96.4	8S	100E
3686	185	-0477 Mar 03	17:00:44	16716	-30635	054	P	a-	0.5569	1.8287	0.8432	309.5	181.5	-	10N	179E
3687	185	-0477 Aug 28	02:25:35	16708	-30629	059	P	t-	-0.5908	1.8122	0.7362	344.8	187.7	-	13S	34E
3688	185	-0476 Jan 22	23:05:49	16701	-30624	026	N	-a	-1.3771	0.3142	-0.6525	146.2	-	-	20N	87E
3689	185	-0476 Feb 21	08:19:53	16699	-30623	064	N	a-	1.2310	0.5778	-0.3799	192.8	-	-	14N	51W
3690	185	-0476 Jul 17	11:07:25	16692	-30618	031	Ne	-t	1.5527	0.0370	-1.0190	58.5	-	-	21S	97W
3691	185	-0476 Aug 16	02:25:34	16691	-30617	069	N	t-	-1.3136	0.4880	-0.5920	208.3	-	-	17S	34E
3692	185	-0475 Jan 11	13:32:46	16684	-30612	036	P	-a	-0.7381	1.5061	0.5013	295.2	148.6	-	22N	131W
3693	185	-0475 Jul 06	17:48:55	16676	-30606	041	P	-t	0.7391	1.5025	0.5011	301.5	152.0	-	23S	162E
3694	185	-0475 Dec 31	22:06:02	16667	-30600	046	T-	pp	-0.0475	2.8033	1.7388	360.6	227.7	102.5	24N	100E
3695	185	-0474 Jun 26	07:28:35	16659	-30594	051	T-	pp	-0.0417	2.7578	1.8046	321.9	213.6	99.8	24S	44W
3696	185	-0474 Dec 20	23:27:14	16650	-30588	056	P	t-	0.6555	1.7075	0.6036	344.6	175.3	-	24N	78E
3697	185	-0473 Jun 16	00:35:06	16642	-30582	061	P	a-	-0.7742	1.4078	0.4662	274.6	139.4	-	24S	59E
3698	185	-0473 Dec 09	22:41:44	16634	-30576	066	N	t-	1.3128	0.4954	-0.5967	208.1	-	-	24N	88E
3699	185	-0472 May 06	07:25:39	16627	-30571	033	N	-a	1.0418	0.9467	-0.0540	255.0	-	-	14S	43W
3700	185	-0472 Jun 04	16:30:04	16625	-30570	071	Nb	a-	-1.5485	0.0029	-0.9699	15.0	-	-	23S	179E
3701	186	-0472 Oct 29	13:41:34	16618	-30565	038	P	-a	-0.9033	1.2040	0.1970	273.2	97.4	-	11N	139W
3702	186	-0471 Apr 25	14:04:56	16610	-30559	043	T	-t	0.2979	2.3385	1.2848	360.0	222.2	79.6	11S	143W
3703	186	-0471 Oct 19	03:59:38	16602	-30553	048	T-	-p	-0.2008	2.4744	1.5045	317.4	206.8	89.1	8N	7E
3704	186	-0470 Apr 14	15:01:14	16593	-30547	053	P	t-	-0.4633	2.0447	0.9716	358.5	208.8	-	8S	156W
3705	186	-0470 Oct 08	19:47:23	16585	-30541	058	P	a-	0.4780	1.9694	0.9922	308.1	187.8	-	4N	130E
3706	186	-0469 Apr 03	17:05:26	16577	-30535	063	N	t-	-1.1910	0.6946	-0.3494	234.8	-	-	4S	173E
3707	186	-0469 Sep 28	08:25:24	16568	-30529	068	N	h-	1.2202	0.6305	-0.3922	215.3	-	-	1N	59W
3708	186	-0468 Feb 22	16:05:52	16562	-30524	035	P	-a	0.7881	1.3995	0.4240	281.0	136.0	-	13N	168W
3709	186	-0468 Aug 17	21:49:01	16553	-30518	040	P	-t	-0.9036	1.2405	0.1600	307.6	96.5	-	16S	103E
3710	186	-0467 Feb 11	08:02:00	16545	-30512	045	T+	-p	0.0791	2.6923	1.7329	318.1	210.2	97.2	16N	47W
3711	186	-0467 Aug 06	21:57:12	16537	-30506	050	T-	pp	-0.1684	2.5840	1.5145	369.0	230.5	97.9	19S	100E
3712	186	-0466 Jan 31	23:27:57	16528	-30500	055	P	a-	-0.6267	1.7021	0.7138	301.8	169.6	-	18N	81E
3713	186	-0466 Jul 27	02:50:32	16520	-30494	060	P	a-	0.5839	1.7959	0.7775	323.3	183.2	-	21S	27E
3714	186	-0466 Dec 22	17:47:51	16513	-30489	027	N	-h	1.5294	0.0925	-0.9884	92.2	-	-	25N	163E
3715	186	-0465 Jan 21	09:52:13	16512	-30488	065	N	h-	-1.3894	0.3320	-0.7145	165.1	-	-	20N	76W
3716	186	-0465 Jun 17	07:40:20	16505	-30483	032	N	-a	-1.2388	0.5569	-0.3879	189.9	-	-	25S	48W
3717	186	-0465 Jul 16	15:03:17	16504	-30482	070	N	a-	1.2863	0.4798	-0.4847	181.2	-	-	21S	157W
3718	186	-0465 Dec 11	17:43:49	16497	-30477	037	P	-t	0.8880	1.2824	0.1756	314.5	101.2	-	24N	162E
3719	186	-0464 Jun 06	00:56:59	16488	-30471	042	P	-a	-0.5126	1.8902	0.9439	301.5	184.4	-	23S	52E
3720	186	-0464 Nov 29	17:50:24	16480	-30465	047	T+	pp	0.1963	2.5380	1.4584	363.7	225.5	93.0	21N	159E

Cat Num	Canon Plate	Calendar Date	TD of Greatest Eclipse	ΔT s	Luna Num	Saros Num	Ecl. Type	QSE	Gamma	Pen. Mag.	Um. Mag.	Pen. m	Par. m	Total m	Lat.	Lng.
3721	187	-0463 May 26	15:42:49	16472	-30459	052	T+	pp	0.2460	2.4002	1.4129	333.2	214.0	87.2	20S	169W
3722	187	-0463 Nov 19	00:29:39	16464	-30453	057	P	a-	-0.4994	1.9529	0.9307	323.1	190.6	-	18N	59E
3723	187	-0462 May 16	00:01:49	16456	-30447	062	N	t-	1.0601	0.9338	-0.1079	266.8	-	-	16S	66E
3724	187	-0462 Oct 10	03:17:56	16449	-30442	029	N	-a	1.5170	0.0572	-0.9087	63.7	-	-	6N	16E
3725	187	-0462 Nov 08	13:41:02	16447	-30441	067	N	a-	-1.1409	0.7532	-0.2244	218.4	-	-	14N	139W
3726	187	-0461 Apr 05	10:36:35	16441	-30436	034	N*	-t	-1.0191	1.0236	-0.0472	284.8	-	-	4S	90W
3727	187	-0461 Sep 29	18:44:34	16432	-30430	039	P	-a	0.8799	1.2363	0.2503	270.6	107.5	-	1N	145E
3728	187	-0460 Mar 24	14:10:02	16424	-30424	044	T-	pp	-0.2405	2.4323	1.4011	349.6	220.2	88.3	0N	142W
3729	187	-0460 Sep 18	05:46:39	16416	-30418	049	T+	-p	0.1896	2.5288	1.4918	348.9	220.7	93.4	4S	19W
3730	187	-0459 Mar 14	00:49:13	16408	-30412	054	P	a-	0.5105	1.9111	0.9313	311.6	187.3	-	5N	59E
3731	187	-0459 Sep 07	09:41:36	16400	-30406	059	P	t-	-0.5476	1.8942	0.8129	350.0	195.0	-	9S	77W
3732	187	-0458 Feb 02	07:35:34	16393	-30401	026	N	-a	-1.3980	0.2750	-0.6900	137.4	-	-	17N	41W
3733	187	-0458 Mar 03	16:28:09	16391	-30400	064	N	a-	1.1940	0.6440	-0.3103	202.0	-	-	10N	174W
3734	187	-0458 Aug 27	09:29:39	16383	-30394	069	N	t-	-1.2607	0.5861	-0.4962	226.0	-	-	14S	73W
3735	187	-0457 Jan 22	21:57:27	16376	-30389	036	P	-a	-0.7548	1.4755	0.4703	294.1	145.1	-	20N	103E
3736	187	-0457 Jul 18	01:00:32	16368	-30383	041	P	-h	0.8050	1.3808	0.3807	292.0	134.5	-	22S	53E
3737	187	-0456 Jan 12	06:13:11	16360	-30377	046	T-	pp	-0.0598	2.7810	1.7159	361.5	228.1	102.4	23N	22W
3738	187	-0456 Jul 06	14:56:35	16352	-30371	051	T+	pp	0.0278	2.7830	1.8303	321.3	213.3	99.8	23S	156W
3739	187	-0456 Dec 31	07:21:55	16344	-30365	056	P	t-	0.6452	1.7259	0.6231	345.8	177.6	-	24N	41W
3740	187	-0455 Jun 26	08:02:56	16336	-30359	061	P	a-	-0.7037	1.5383	0.5947	282.9	154.5	-	24S	54W
3741	188	-0455 Dec 20	06:48:18	16328	-30353	066	N	t-	1.3054	0.5075	-0.5815	209.8	-	-	25N	34W
3742	188	-0454 May 17	14:24:22	16321	-30348	033	N	-h	1.1227	0.7996	-0.2038	238.9	-	-	17S	150W
3743	188	-0454 Jun 15	23:40:04	16320	-30347	071	N	h-	-1.4798	0.1307	-0.8459	99.6	-	-	25S	71E
3744	188	-0454 Nov 09	22:20:25	16313	-30342	038	P	-a	-0.9047	1.2008	0.1951	272.1	96.7	-	15N	90E
3745	188	-0453 May 06	20:38:09	16305	-30336	043	T	-t	0.3816	2.1853	1.1306	356.7	215.6	57.0	14S	117E
3746	188	-0453 Oct 30	12:45:41	16297	-30330	048	T-	-p	-0.2080	2.4617	1.4907	317.1	206.3	88.3	12N	126W
3747	188	-0452 Apr 24	21:29:03	16289	-30324	053	T	t-	-0.3810	2.1943	1.1239	363.4	217.9	56.2	11S	105E
3748	188	-0452 Oct 19	04:23:53	16280	-30318	058	T	p-	0.4669	1.9917	1.0105	309.8	189.2	15.3	9N	1W
3749	188	-0451 Apr 13	23:59:40	16272	-30312	063	N	t-	-1.1184	0.8251	-0.2132	251.5	-	-	8S	68E
3750	188	-0451 Oct 08	16:35:47	16264	-30306	068	N	h-	1.2048	0.6616	-0.3671	220.9	-	-	5N	176E
3751	188	-0450 Mar 05	00:05:59	16258	-30301	035	P	-a	0.8260	1.3273	0.3570	275.0	125.8	-	9N	70E
3752	188	-0450 Aug 29	04:53:05	16249	-30295	040	P	-t	-0.9565	1.1457	0.0607	299.3	60.3	-	13S	5W
3753	188	-0449 Feb 22	16:17:28	16241	-30289	045	T+	-p	0.1122	2.6302	1.6735	317.6	209.8	96.0	12N	172W
3754	188	-0449 Aug 18	04:56:55	16233	-30283	050	T-	pp	-0.2278	2.4756	1.4048	366.6	227.2	90.8	16S	6W
3755	188	-0448 Feb 12	07:42:47	16225	-30277	055	P	a-	-0.6022	1.7468	0.7589	304.8	174.0	-	15N	44W
3756	188	-0448 Aug 06	10:09:03	16217	-30271	060	P	a-	0.5220	1.9090	0.8913	327.3	191.6	-	18S	84W
3757	188	-0447 Jan 02	01:53:37	16211	-30266	027	N	-t	1.5410	0.0714	-1.0101	81.3	-	-	25N	41E
3758	188	-0447 Jan 31	17:52:27	16209	-30265	065	N	t-	-1.3721	0.3637	-0.6829	172.8	-	-	18N	163E
3759	188	-0447 Jun 27	15:07:55	16203	-30260	032	N	-a	-1.3061	0.4334	-0.5113	169.4	-	-	25S	160W
3760	188	-0447 Jul 26	22:38:05	16201	-30259	070	N	a-	1.2229	0.5959	-0.3684	199.2	-	-	20S	88E
3761	189	-0447 Dec 22	01:42:47	16195	-30254	037	P	-t	0.8946	1.2697	0.1642	313.2	98.0	-	24N	42E
3762	189	-0446 Jun 17	08:20:03	16187	-30248	042	P	-a	-0.5873	1.7543	0.8058	296.3	174.8	-	24S	59W
3763	189	-0446 Dec 11	02:04:39	16179	-30242	047	T+	pp	0.1986	2.5322	1.4556	362.4	224.9	92.7	23N	35E
3764	189	-0445 Jun 06	22:46:02	16171	-30236	052	T+	pp	0.1686	2.5441	1.5531	337.0	218.2	95.7	22S	84E
3765	189	-0445 Nov 30	09:07:28	16163	-30230	057	P	a-	-0.4976	1.9548	0.9353	322.0	190.4	-	21N	71W
3766	189	-0444 May 26	06:36:23	16155	-30224	062	P	t-	0.9799	1.0823	0.0380	283.8	46.9	-	19S	34W
3767	189	-0444 Oct 20	11:59:08	16148	-30219	029	N	-a	1.5271	0.0396	-0.9284	53.1	-	-	10N	116W
3768	189	-0444 Nov 18	22:31:49	16147	-30218	067	N	a-	-1.1376	0.7589	-0.2181	218.8	-	-	17N	87E
3769	189	-0443 Apr 15	17:14:29	16140	-30213	034	N	-t	-1.0910	0.8898	-0.1771	268.6	-	-	9S	168E
3770	189	-0443 Oct 10	03:10:31	16132	-30207	039	P	-a	0.8982	1.2052	0.2143	269.0	100.2	-	5N	17E
3771	189	-0442 Apr 04	21:18:27	16124	-30201	044	T	-p	-0.3027	2.3151	1.2900	345.3	215.8	78.2	4S	108E
3772	189	-0442 Sep 29	13:43:49	16116	-30195	049	T+	-p	0.2145	2.4862	1.4430	349.9	220.1	90.6	0N	141W
3773	189	-0441 Mar 25	08:28:55	16108	-30189	054	T	a-	0.4557	2.0087	1.0346	314.0	193.3	28.1	1N	58W
3774	189	-0441 Sep 18	17:07:07	16100	-30183	059	P	t-	-0.5123	1.9615	0.8750	354.1	200.3	-	5S	170E
3775	189	-0440 Feb 13	15:56:27	16093	-30178	026	N	-a	-1.4263	0.2221	-0.7410	124.2	-	-	14N	168W
3776	189	-0440 Mar 14	00:27:01	16092	-30177	064	N	a-	1.1490	0.7249	-0.2261	212.4	-	-	6N	64E
3777	189	-0440 Sep 06	16:44:40	16084	-30171	069	N	t-	-1.2157	0.6697	-0.4147	239.4	-	-	10S	176E
3778	189	-0439 Feb 02	06:13:22	16078	-30166	036	P	-a	-0.7784	1.4324	0.4271	292.1	139.6	-	18N	22W
3779	189	-0439 Jul 28	08:20:41	16070	-30160	041	P	-h	0.8650	1.2703	0.2713	282.6	115.0	-	20S	58W
3780	189	-0438 Jan 22	14:13:58	16062	-30154	046	T-	pp	-0.0767	2.7500	1.6848	362.4	228.5	102.1	21N	143W

Cat Num	Canon Plate	Calendar Date	TD of Greatest Eclipse	ΔT s	Luna Num	Saros Num	Ecl. Type	QSE	Gamma	Pen. Mag.	Um. Mag.	Eclipse Phase Durations Pen. m	Par. m	Total m	Greatest in Zenith Lat.	Lng.
3781	190	-0438 Jul 17	22:28:37	16054	-30148	051	T+	pp	0.0946	2.6607	1.7077	320.2	212.0	97.8	22S	90E
3782	190	-0437 Jan 11	15:12:59	16046	-30142	056	P	t-	0.6319	1.7495	0.6485	347.2	180.5	-	23N	158W
3783	190	-0437 Jul 07	15:32:32	16038	-30136	061	P	a-	-0.6348	1.6659	0.7198	290.3	166.7	-	24S	167W
3784	190	-0437 Dec 31	14:52:27	16030	-30130	066	N	h-	1.2952	0.5242	-0.5608	212.3	-	-	25N	155W
3785	190	-0436 May 27	21:19:39	16024	-30125	033	N	-h	1.2054	0.6496	-0.3571	219.5	-	-	19S	105E
3786	190	-0436 Jun 26	06:51:03	16022	-30124	071	N	h-	-1.4124	0.2566	-0.7243	138.6	-	-	25S	37W
3787	190	-0436 Nov 20	07:03:11	16016	-30119	038	P	-a	-0.9037	1.2016	0.1979	271.4	97.1	-	18N	41W
3788	190	-0435 May 17	03:08:27	16008	-30113	043	P	-t	0.4672	2.0289	0.9728	352.0	206.5	-	17S	18E
3789	190	-0435 Nov 09	21:34:46	16000	-30107	048	T-	-p	-0.2130	2.4528	1.4811	316.8	205.9	87.7	16N	100E
3790	190	-0434 May 06	03:54:57	15992	-30101	053	T	tp	-0.2967	2.3478	1.2799	367.2	225.1	80.0	15S	6E
3791	190	-0434 Oct 30	13:04:33	15984	-30095	058	T	p-	0.4597	2.0069	1.0220	311.2	190.2	22.1	13N	133W
3792	190	-0433 Apr 25	06:49:35	15976	-30089	063	N	t-	-1.0409	0.9646	-0.0684	266.8	-	-	12S	37W
3793	190	-0433 Oct 20	00:52:16	15968	-30083	068	N	h-	1.1951	0.6823	-0.3520	224.8	-	-	9N	50E
3794	190	-0432 Mar 15	07:57:54	15962	-30078	035	P	-a	0.8713	1.2416	0.2764	267.7	111.8	-	5N	50W
3795	190	-0432 Sep 08	12:07:00	15954	-30072	040	N*	-t	-1.0010	1.0662	-0.0232	291.8	-	-	9S	115W
3796	190	-0431 Mar 05	00:23:38	15946	-30066	045	T+	-p	0.1532	2.5538	1.5995	316.8	208.9	93.7	8N	64E
3797	190	-0431 Aug 28	12:08:33	15938	-30060	050	T	pp	-0.2782	2.3839	1.3117	364.1	223.6	82.4	12S	115W
3798	190	-0430 Feb 22	15:48:02	15931	-30054	055	P	a-	-0.5703	1.8052	0.8176	308.4	179.3	-	12N	167W
3799	190	-0430 Aug 17	17:37:02	15923	-30048	060	P	a-	0.4667	2.0103	0.9929	330.3	197.9	-	15S	163E
3800	190	-0429 Jan 13	09:52:55	15916	-30043	027	N	-t	1.5573	0.0416	-1.0400	62.4	-	-	24N	78W
3801	191	-0429 Feb 12	01:42:45	15915	-30042	065	N	t-	-1.3477	0.4086	-0.6380	183.0	-	-	14N	45E
3802	191	-0429 Jul 08	22:38:41	15908	-30037	032	N	-a	-1.3714	0.3139	-0.6313	145.6	-	-	25S	87E
3803	191	-0429 Aug 07	06:20:12	15907	-30036	070	N	a-	1.1647	0.7031	-0.2616	213.6	-	-	17S	28W
3804	191	-0428 Jan 02	09:37:20	15901	-30031	037	P	-t	0.9047	1.2500	0.1467	311.2	92.9	-	24N	76W
3805	191	-0428 Jun 27	15:44:26	15893	-30025	042	P	-a	-0.6606	1.6214	0.6699	290.4	163.3	-	24S	171W
3806	191	-0428 Dec 21	10:18:49	15885	-30019	047	T+	pp	0.2020	2.5239	1.4512	360.9	224.3	92.3	24N	88W
3807	191	-0427 Jun 17	05:48:19	15877	-30013	052	T+	pp	0.0914	2.6879	1.6927	340.0	221.0	100.7	23S	22W
3808	191	-0427 Dec 10	17:46:32	15869	-30007	057	P	a-	-0.4960	1.9560	0.9399	320.9	190.2	-	22N	159E
3809	191	-0426 Jun 06	13:08:38	15862	-30001	062	P	t-	0.8981	1.2338	0.1866	299.1	102.0	-	21S	133W
3810	191	-0426 Oct 31	20:46:04	15855	-29996	029	N	-a	1.5330	0.0298	-0.9401	46.2	-	-	14N	111E
3811	191	-0426 Nov 30	07:25:15	15854	-29995	067	N	a-	-1.1366	0.7604	-0.2158	218.7	-	-	20N	47W
3812	191	-0425 Apr 26	23:46:37	15847	-29990	034	N	-t	-1.1676	0.7473	-0.3157	249.1	-	-	13S	68E
3813	191	-0425 Oct 21	11:42:52	15840	-29984	039	P	-a	0.9107	1.1846	0.1890	268.2	94.6	-	10N	113W
3814	191	-0424 Apr 15	04:22:11	15832	-29978	044	T	-p	-0.3698	2.1890	1.1698	340.1	209.9	62.2	8S	0E
3815	191	-0424 Oct 09	21:46:59	15824	-29972	049	T+	-p	0.2336	2.4541	1.4049	351.0	219.5	88.0	5N	97E
3816	191	-0423 Apr 04	16:04:29	15816	-29966	054	T	a-	0.3967	2.1143	1.1453	316.3	198.7	55.5	3S	174W
3817	191	-0423 Sep 29	00:40:26	15809	-29960	059	P	t-	-0.4836	2.0166	0.9254	357.4	204.4	-	0S	54E
3818	191	-0422 Feb 24	00:10:19	15802	-29955	026	N	-a	-1.4600	0.1594	-0.8020	105.9	-	-	10N	67E
3819	191	-0422 Mar 25	08:20:04	15801	-29954	064	N	a-	1.0994	0.8144	-0.1338	222.9	-	-	2N	57W
3820	191	-0422 Sep 18	00:08:38	15793	-29948	069	N	t-	-1.1774	0.7411	-0.3452	249.9	-	-	5S	63E
3821	192	-0421 Feb 13	14:20:42	15787	-29943	036	P	-a	-0.8081	1.3778	0.3725	289.2	131.8	-	14N	145W
3822	192	-0421 Aug 08	15:49:35	15779	-29937	041	P	-a	0.9183	1.1722	0.1737	273.5	93.0	-	17S	171W
3823	192	-0420 Feb 02	22:03:43	15771	-29931	046	T-	pp	-0.1017	2.7041	1.6392	363.0	228.6	101.2	18N	99E
3824	192	-0420 Jul 28	06:08:44	15764	-29925	051	T+	-p	0.1547	2.5507	1.5970	318.7	210.1	94.3	20S	26W
3825	192	-0419 Jan 21	22:55:20	15756	-29919	056	P	t-	0.6120	1.7847	0.6860	349.0	184.6	-	21N	86E
3826	192	-0419 Jul 17	23:07:23	15748	-29913	061	P	a-	-0.5707	1.7851	0.8359	296.6	176.4	-	23S	79E
3827	192	-0418 Jan 10	22:51:19	15741	-29907	066	N	h-	1.2803	0.5493	-0.5312	216.1	-	-	24N	86E
3828	192	-0418 Jun 08	04:12:57	15734	-29902	033	N	-h	1.2881	0.4997	-0.5108	196.2	-	-	21S	1E
3829	192	-0418 Jul 07	14:04:09	15733	-29901	071	N	h-	-1.3476	0.3781	-0.6077	167.0	-	-	25S	146W
3830	192	-0418 Dec 01	15:47:52	15726	-29896	038	P	-a	-0.9018	1.2039	0.2026	270.7	98.0	-	21N	173W
3831	192	-0417 May 28	09:33:18	15719	-29890	043	P	-t	0.5564	1.8662	0.8083	345.7	194.3	-	20S	80W
3832	192	-0417 Nov 21	06:28:03	15711	-29884	048	T-	-p	-0.2146	2.4501	1.4781	316.6	205.8	87.5	19N	34W
3833	192	-0416 May 16	10:16:37	15703	-29878	053	T-	pp	-0.2087	2.5082	1.4423	370.0	230.4	94.4	18S	91W
3834	192	-0416 Nov 09	21:49:34	15696	-29872	058	T	p-	0.4566	2.0141	1.0259	312.3	190.7	24.1	16N	95E
3835	192	-0415 May 05	13:37:08	15688	-29866	063	P	t-	-0.9603	1.1100	0.0818	280.6	67.2	-	15S	141W
3836	192	-0415 Oct 30	09:14:17	15680	-29860	068	N	h-	1.1908	0.6927	-0.3466	227.3	-	-	13N	77W
3837	192	-0414 Mar 26	15:43:59	15674	-29855	035	P	-a	0.9220	1.1462	0.1858	259.2	92.7	-	1N	168W
3838	192	-0414 Sep 19	19:28:59	15666	-29849	040	N	-t	-1.0387	0.9992	-0.0944	284.9	-	-	4S	132E
3839	192	-0413 Mar 16	08:23:00	15659	-29843	045	T+	-p	0.1998	2.4672	1.5152	315.7	207.4	90.1	4N	58W
3840	192	-0413 Sep 08	19:30:02	15651	-29837	050	T	-p	-0.3213	2.3055	1.2321	361.6	219.9	73.0	8S	133E

Cat Num	Canon Plate	Calendar Date	TD of Greatest Eclipse	ΔT	Luna Num	Saros Num	Ecl. Type	QSE	Gamma	Pen. Mag.	Um. Mag.	Eclipse Phase Durations Pen.	Par.	Total	Greatest in Zenith Lat.	Lng.
				s								m	m	m		
3841	193	-0412 Mar 04	23:43:55	15643	-29831	055	P	a-	-0.5312	1.8768	0.8895	312.4	185.3	-	8N	73E
3842	193	-0412 Aug 28	01:15:33	15636	-29825	060	T	a-	0.4191	2.0976	1.0804	332.3	202.4	43.4	12S	46E
3843	193	-0411 Jan 23	17:44:41	15630	-29820	027	Ne	-t	1.5791	0.0013	-1.0798	11.0	-	-	22N	163E
3844	193	-0411 Feb 22	09:23:50	15628	-29819	065	N	t-	-1.3162	0.4662	-0.5803	195.0	-	-	11N	72W
3845	193	-0411 Jul 19	06:15:21	15622	-29814	032	N	-a	-1.4320	0.2033	-0.7431	118.3	-	-	24S	28W
3846	193	-0411 Aug 17	14:10:30	15621	-29813	070	N	a-	1.1125	0.7991	-0.1663	225.0	-	-	14S	147W
3847	193	-0410 Jan 12	17:26:31	15614	-29808	037	P	-t	0.9193	1.2219	0.1214	308.3	84.7	-	23N	167E
3848	193	-0410 Jul 08	23:12:12	15607	-29802	042	P	-a	-0.7306	1.4947	0.5397	283.9	149.9	-	24S	77E
3849	193	-0409 Jan 01	18:29:02	15599	-29796	047	T+	pp	0.2093	2.5084	1.4402	359.3	223.5	91.4	24N	150E
3850	193	-0409 Jun 28	12:50:51	15592	-29790	052	T+	pp	0.0152	2.8300	1.8301	342.3	222.5	102.7	24S	129W
3851	193	-0409 Dec 22	02:23:37	15584	-29784	057	P	a-	-0.4921	1.9612	0.9490	319.9	190.3	-	23N	30E
3852	193	-0408 Jun 16	19:41:05	15576	-29778	062	P	t-	0.8168	1.3847	0.3341	312.6	133.9	-	23S	128E
3853	193	-0408 Nov 11	05:36:07	15570	-29773	029	N	-a	1.5368	0.0236	-0.9478	41.1	-	-	18N	22W
3854	193	-0408 Dec 10	16:17:01	15569	-29772	067	N	a-	-1.1340	0.7644	-0.2102	218.9	-	-	22N	180E
3855	193	-0407 May 07	06:15:50	15563	-29767	034	N	-t	-1.2466	0.6006	-0.4591	225.9	-	-	16S	31W
3856	193	-0407 Oct 31	20:20:06	15555	-29761	039	P	-a	0.9192	1.1713	0.1714	268.0	90.6	-	14N	116E
3857	193	-0406 Apr 26	11:21:26	15547	-29755	044	T	-t	-0.4415	2.0547	1.0409	334.1	202.4	31.7	12S	107W
3858	193	-0406 Oct 21	05:56:35	15540	-29749	049	T+	-p	0.2467	2.4328	1.3782	352.2	219.3	86.1	9N	27W
3859	193	-0405 Apr 15	23:33:23	15532	-29743	054	T	a-	0.3312	2.2321	1.2679	318.3	203.6	72.4	7S	71E
3860	193	-0405 Oct 10	08:22:37	15525	-29737	059	P	t-	-0.4625	2.0575	0.9619	360.0	207.2	-	4N	63W
3861	194	-0404 Mar 06	08:13:48	15518	-29732	026	N	-a	-1.5019	0.0817	-0.8781	76.4	-	-	6N	56W
3862	194	-0404 Apr 04	16:04:16	15517	-29731	064	N	a-	1.0425	0.9176	-0.0281	233.9	-	-	3S	175W
3863	194	-0404 Sep 28	07:43:58	15510	-29725	069	N	t-	-1.1475	0.7968	-0.2912	257.4	-	-	1S	53W
3864	194	-0403 Feb 23	22:18:43	15503	-29720	036	P	-a	-0.8447	1.3107	0.3054	285.2	121.0	-	11N	94E
3865	194	-0403 Aug 18	23:27:10	15496	-29714	041	P	-a	0.9653	1.0858	0.0876	264.8	66.6	-	14S	73E
3866	194	-0402 Feb 13	05:45:30	15488	-29708	046	T-	pp	-0.1325	2.6474	1.5829	363.4	228.3	99.6	15N	17W
3867	194	-0402 Aug 08	13:54:55	15481	-29702	051	T+	-p	0.2101	2.4497	1.4949	317.0	207.7	89.3	18S	144W
3868	194	-0401 Feb 02	06:30:34	15473	-29696	056	P	t-	0.5860	1.8309	0.7352	351.2	189.5	-	19N	29W
3869	194	-0401 Jul 29	06:46:08	15466	-29690	061	P	a-	-0.5100	1.8983	0.9454	302.0	184.3	-	21S	36W
3870	194	-0400 Jan 22	06:45:29	15458	-29684	066	N	h-	1.2607	0.5826	-0.4927	221.0	-	-	22N	33W
3871	194	-0400 Jun 18	11:06:01	15452	-29679	033	N	-h	1.3698	0.3519	-0.6629	167.8	-	-	22S	103W
3872	194	-0400 Jul 17	21:20:54	15451	-29678	071	N	h-	-1.2860	0.4938	-0.4974	189.7	-	-	24S	104E
3873	194	-0400 Dec 12	00:32:05	15445	-29673	038	P	-a	-0.9010	1.2038	0.2055	269.9	98.4	-	22N	56E
3874	194	-0399 Jun 07	15:58:54	15437	-29667	043	P	-t	0.6442	1.7062	0.6462	338.1	179.0	-	22S	177W
3875	194	-0399 Dec 01	15:21:58	15430	-29661	048	T-	-p	-0.2157	2.4479	1.4761	316.5	205.7	87.3	21N	168W
3876	194	-0398 May 27	16:38:05	15422	-29655	053	T-	pp	-0.1199	2.6702	1.6061	371.5	233.7	102.7	20S	173E
3877	194	-0398 Nov 21	06:36:25	15415	-29649	058	T	p-	0.4555	2.0177	1.0267	313.3	191.1	24.4	20N	38W
3878	194	-0397 May 16	20:23:42	15407	-29643	063	P	t-	-0.8778	1.2592	0.2355	292.7	111.2	-	18S	116E
3879	194	-0397 Nov 10	17:39:09	15400	-29637	068	N	h-	1.1892	0.6979	-0.3459	228.9	-	-	17N	156E
3880	194	-0396 Apr 05	23:22:55	15394	-29632	035	P	-a	0.9792	1.0389	0.0829	249.0	62.7	-	3S	75E
3881	195	-0396 Sep 30	03:01:04	15386	-29626	040	N	-t	-1.0682	0.9469	-0.1503	279.3	-	-	0S	18E
3882	195	-0395 Mar 26	16:14:14	15379	-29620	045	T+	-p	0.2533	2.3680	1.4178	314.2	205.1	84.4	0N	177W
3883	195	-0395 Sep 19	03:02:04	15371	-29614	050	T	-p	-0.3565	2.2413	1.1669	359.1	216.4	63.2	4S	18E
3884	195	-0394 Mar 16	07:30:08	15364	-29608	055	P	a-	-0.4847	1.9621	0.9748	316.8	191.6	-	4N	46W
3885	195	-0394 Sep 08	09:04:29	15356	-29602	060	T	a-	0.3790	2.1711	1.1541	333.5	205.5	58.5	8S	73W
3886	195	-0393 Mar 05	16:52:34	15349	-29596	065	N	t-	-1.2758	0.5404	-0.5059	209.1	-	-	7N	174E
3887	195	-0393 Jul 30	13:57:23	15343	-29591	032	N	-a	-1.4886	0.1003	-0.8477	83.9	-	-	22S	144W
3888	195	-0393 Aug 28	22:09:41	15342	-29590	070	N	a-	1.0672	0.8828	-0.0836	234.1	-	-	11S	91E
3889	195	-0392 Jan 24	01:08:40	15335	-29585	037	P	-t	0.9397	1.1826	0.0857	304.1	71.5	-	21N	51E
3890	195	-0392 Jul 19	06:43:43	15328	-29579	042	P	-a	-0.7969	1.3750	0.4159	277.0	134.3	-	23S	37W
3891	195	-0391 Jan 12	02:35:16	15321	-29573	047	T+	pp	0.2207	2.4848	1.4218	357.3	222.5	90.0	23N	28E
3892	195	-0391 Jul 08	19:55:44	15313	-29567	052	T-	pp	-0.0581	2.7540	1.7488	343.8	222.8	102.1	23S	125E
3893	195	-0390 Jan 01	10:59:28	15306	-29561	057	P	a-	-0.4863	1.9696	0.9617	319.0	190.7	-	23N	99W
3894	195	-0390 Jun 28	02:12:52	15298	-29555	062	P	t-	0.7353	1.5362	0.4817	324.6	157.6	-	23S	30E
3895	195	-0390 Nov 22	14:29:35	15292	-29550	029	N	-a	1.5382	0.0217	-0.9508	39.4	-	-	21N	156W
3896	195	-0390 Dec 22	01:08:45	15291	-29549	067	N	a-	-1.1313	0.7684	-0.2044	219.1	-	-	23N	47E
3897	195	-0389 May 18	12:42:56	15285	-29544	034	N	-t	-1.3277	0.4503	-0.6063	197.9	-	-	19S	129W
3898	195	-0389 Jun 17	03:09:25	15284	-29543	072	N	t-	1.4986	0.1426	-0.9259	116.1	-	-	22S	15E
3899	195	-0389 Nov 12	05:00:52	15277	-29538	039	P	-a	0.9241	1.1642	0.1605	268.4	88.0	-	17N	15W
3900	195	-0388 May 06	18:18:57	15270	-29532	044	P	-t	-0.5156	1.9163	0.9075	327.0	193.0	-	16S	147E

Cat Num	Canon Plate	Calendar Date	TD of Greatest Eclipse	ΔT	Luna Num	Saros Num	Ecl. Type	QSE	Gamma	Pen. Mag.	Um. Mag.	Eclipse Phase Durations Pen.	Par.	Total	Greatest in Zenith Lat.	Lng.
				s								m	m	m		
3901	196	-0388 Oct 31	14:10:40	15263	-29526	049	T+	-p	0.2555	2.4192	1.3596	353.4	219.2	84.7	13N	152W
3902	196	-0387 Apr 26	07:00:12	15255	-29520	054	T+	p-	0.2635	2.3543	1.3943	319.9	207.6	83.8	11S	42W
3903	196	-0387 Oct 20	16:09:54	15248	-29514	059	P	t-	-0.4458	2.0900	0.9908	362.0	209.3	-	8N	178E
3904	196	-0386 Apr 15	23:44:19	15241	-29508	064	P	a-	0.9822	1.0272	0.0835	244.6	62.2	-	7S	68E
3905	196	-0386 Oct 09	15:27:42	15233	-29502	069	N	t-	-1.1237	0.8411	-0.2481	262.9	-	-	3N	170W
3906	196	-0385 Mar 07	06:05:49	15227	-29497	036	P	-a	-0.8894	1.2286	0.2232	279.5	105.0	-	7N	25W
3907	196	-0385 Aug 30	07:14:57	15220	-29491	041	P	-a	1.0048	1.0133	0.0151	257.0	27.9	-	10S	45W
3908	196	-0384 Feb 24	13:15:13	15212	-29485	046	T-	pp	-0.1724	2.5737	1.5099	363.3	227.5	96.6	11N	131W
3909	196	-0384 Aug 18	21:50:18	15205	-29479	051	T+	-p	0.2581	2.3624	1.4060	315.2	205.0	83.6	15S	96E
3910	196	-0383 Feb 12	13:55:19	15198	-29473	056	P	t-	0.5515	1.8925	0.8002	354.0	195.5	-	16N	141W
3911	196	-0383 Aug 08	14:32:35	15191	-29467	061	T	a-	-0.4562	1.9991	1.0422	306.5	190.3	30.4	18S	154W
3912	196	-0382 Feb 01	14:32:26	15183	-29461	066	N	h-	1.2347	0.6276	-0.4421	227.5	-	-	20N	150W
3913	196	-0382 Jun 29	17:58:40	15177	-29456	033	N	-h	1.4503	0.2066	-0.8131	130.9	-	-	22S	153E
3914	196	-0382 Jul 29	04:41:11	15176	-29455	071	N	h-	-1.2281	0.6030	-0.3941	208.3	-	-	22S	6W
3915	196	-0382 Dec 23	09:15:28	15170	-29450	038	P	-a	-0.9014	1.2013	0.2064	268.8	98.4	-	23N	75W
3916	196	-0381 Jun 18	22:23:07	15162	-29444	043	P	-t	0.7319	1.5466	0.4839	328.9	159.3	-	23S	86E
3917	196	-0381 Dec 13	00:15:56	15155	-29438	048	T-	-p	-0.2169	2.4455	1.4744	316.4	205.6	87.2	23N	59E
3918	196	-0380 Jun 06	22:59:28	15148	-29432	053	T-	pp	-0.0308	2.8332	1.7703	371.8	235.0	106.2	22S	76E
3919	196	-0380 Dec 01	15:24:54	15141	-29426	058	T	p-	0.4565	2.0169	1.0236	314.1	191.3	23.1	22N	170W
3920	196	-0379 May 27	03:10:26	15133	-29420	063	P	h-	-0.7940	1.4109	0.3911	303.2	139.5	-	21S	13E
3921	197	-0379 Nov 21	02:06:22	15126	-29414	068	N	h-	1.1904	0.6978	-0.3499	229.8	-	-	20N	28E
3922	197	-0378 Apr 17	06:57:36	15120	-29409	035	N	-a	1.0404	0.9248	-0.0273	237.1	-	-	7S	41W
3923	197	-0378 May 16	14:33:05	15119	-29408	073	Nb	a-	-1.4958	0.0966	-0.8704	84.9	-	-	19S	158W
3924	197	-0378 Oct 11	10:41:00	15113	-29403	040	N	-t	-1.0911	0.9066	-0.1939	274.7	-	-	4N	99W
3925	197	-0377 Apr 06	23:57:38	15105	-29397	045	T	-p	0.3132	2.2574	1.3087	312.1	201.9	75.5	4S	64E
3926	197	-0377 Sep 30	10:44:01	15098	-29391	050	T	-p	-0.3842	2.1910	1.1158	356.8	213.3	53.4	0N	100W
3927	197	-0376 Mar 26	15:07:40	15091	-29385	055	T	p-	-0.4319	2.0593	1.0716	321.3	197.9	40.4	1S	162W
3928	197	-0376 Sep 18	17:02:50	15084	-29379	060	T	a-	0.3460	2.2316	1.2147	334.1	207.7	67.5	3S	166E
3929	197	-0375 Mar 16	00:12:15	15076	-29373	065	N	t-	-1.2287	0.6267	-0.4195	223.9	-	-	3N	62E
3930	197	-0375 Aug 09	21:47:02	15070	-29368	032	Ne	-a	-1.5392	0.0085	-0.9416	24.7	-	-	19S	97E
3931	197	-0375 Sep 08	06:17:37	15069	-29367	070	N	a-	1.0288	0.9538	-0.0138	241.1	-	-	7S	32W
3932	197	-0374 Feb 03	08:43:36	15063	-29362	037	P	-t	0.9660	1.1323	0.0393	298.6	48.7	-	19N	63W
3933	197	-0374 Jul 30	14:19:32	15056	-29356	042	P	-a	-0.8591	1.2632	0.2996	269.8	116.2	-	21S	151W
3934	197	-0373 Jan 23	10:35:49	15049	-29350	047	T+	pp	0.2376	2.4509	1.3935	355.1	221.1	87.8	21N	92W
3935	197	-0373 Jul 20	03:04:02	15041	-29344	052	T-	pp	-0.1278	2.6290	1.6181	344.7	222.0	99.0	22S	17E
3936	197	-0372 Jan 12	19:29:05	15034	-29338	057	P	a-	-0.4747	1.9885	0.9856	318.4	191.7	-	22N	133E
3937	197	-0372 Jul 08	08:48:47	15027	-29332	062	P	t-	0.6574	1.6812	0.6226	334.9	175.5	-	23S	70W
3938	197	-0372 Dec 02	23:23:18	15021	-29327	029	N	-a	1.5399	0.0188	-0.9542	36.8	-	-	23N	70E
3939	197	-0371 Jan 01	09:56:06	15020	-29326	067	N	a-	-1.1247	0.7794	-0.1912	220.2	-	-	22N	85W
3940	197	-0371 May 28	19:10:08	15014	-29321	034	N	-t	-1.4086	0.3005	-0.7535	163.4	-	-	22S	133E
3941	198	-0371 Jun 27	09:32:39	15013	-29320	072	N	t-	1.4147	0.2969	-0.7721	165.1	-	-	22S	82W
3942	198	-0371 Nov 22	13:42:59	15007	-29315	039	P	-a	0.9275	1.1596	0.1526	268.8	86.1	-	20N	146W
3943	198	-0370 May 18	01:15:00	14999	-29309	044	P	-t	-0.5918	1.7741	0.7698	319.0	181.5	-	19S	42E
3944	198	-0370 Nov 11	22:28:45	14992	-29303	049	T+	-p	0.2601	2.4129	1.3490	354.7	219.4	84.0	17N	82E
3945	198	-0369 May 07	14:21:30	14985	-29297	054	T+	p-	0.1904	2.4865	1.5303	321.1	210.7	92.1	15S	154W
3946	198	-0369 Nov 01	00:04:45	14978	-29291	059	T	t-	-0.4355	2.1103	1.0081	363.4	210.6	14.8	13N	58E
3947	198	-0368 Apr 26	07:17:52	14971	-29285	064	P	a-	0.9165	1.1471	0.2047	255.1	95.9	-	11S	48W
3948	198	-0368 Oct 19	23:20:11	14964	-29279	069	N	t-	-1.1064	0.8732	-0.2167	266.6	-	-	8N	70E
3949	198	-0367 Mar 17	13:43:49	14958	-29274	036	P	-a	-0.9408	1.1345	0.1288	272.3	81.1	-	2N	141W
3950	198	-0367 Sep 09	15:12:35	14950	-29268	041	N	-a	1.0371	0.9540	-0.0442	250.2	-	-	6S	167W
3951	198	-0366 Mar 06	20:36:11	14943	-29262	046	T-	-p	-0.2188	2.4881	1.4252	362.8	225.9	91.6	7N	117E
3952	198	-0366 Aug 30	05:52:34	14936	-29256	051	T	-p	0.3006	2.2854	1.3271	313.4	202.3	77.0	11S	26W
3953	198	-0365 Feb 23	21:12:36	14929	-29250	056	P	t-	0.5109	1.9653	0.8766	357.0	201.8	-	12N	108E
3954	198	-0365 Aug 19	22:25:01	14922	-29244	061	T	p-	-0.4080	2.0899	1.1285	310.3	195.1	51.8	15S	87E
3955	198	-0364 Feb 12	22:12:10	14915	-29238	066	N	h-	1.2021	0.6842	-0.3795	235.3	-	-	16N	94E
3956	198	-0364 Jul 10	00:54:43	14909	-29233	033	Ne	-h	1.5266	0.0693	-0.9557	77.1	-	-	22S	48E
3957	198	-0364 Aug 08	12:07:55	14908	-29232	071	N	h-	-1.1762	0.7013	-0.3020	223.5	-	-	19S	119W
3958	198	-0363 Jan 02	17:55:51	14902	-29227	038	P	-a	-0.9045	1.1936	0.2026	267.3	97.3	-	23N	155E
3959	198	-0363 Jun 29	04:50:17	14894	-29221	043	P	-t	0.8164	1.3929	0.3274	318.5	134.4	-	23S	11W
3960	198	-0363 Dec 23	09:07:05	14887	-29215	048	T-	-p	-0.2203	2.4387	1.4686	316.2	205.5	86.9	23N	74W

Cat Num	Canon Plate	Calendar Date	TD of Greatest Eclipse	ΔT	Luna Num	Saros Num	Ecl. Type	QSE	Gamma	Pen. Mag.	Um. Mag.	Eclipse Phase Durations Pen.	Par.	Total	Greatest in Zenith Lat.	Lng.
				s								m	m	m		
3961	199	-0362 Jun 18	05:23:59	14880	-29209	053	T+	pp	0.0561	2.7863	1.7243	371.0	234.4	105.5	23S	20W
3962	199	-0362 Dec 13	00:11:24	14873	-29203	058	T	p-	0.4563	2.0182	1.0232	315.0	191.7	22.9	24N	58E
3963	199	-0361 Jun 07	09:58:28	14866	-29197	063	P	h-	-0.7101	1.5633	0.5469	312.1	160.5	-	23S	90W
3964	199	-0361 Dec 02	10:33:18	14859	-29191	068	N	h-	1.1920	0.6963	-0.3545	230.4	-	-	23N	98W
3965	199	-0360 Apr 27	14:27:25	14853	-29186	035	N	-a	1.1059	0.8029	-0.1459	223.3	-	-	11S	156W
3966	199	-0360 May 26	21:48:05	14852	-29185	073	N	a-	-1.4185	0.2368	-0.7271	130.8	-	-	22S	92E
3967	199	-0360 Oct 21	18:27:46	14846	-29180	040	N	-t	-1.1082	0.8765	-0.2267	271.2	-	-	9N	142E
3968	199	-0359 Apr 17	07:34:51	14839	-29174	045	T	-a	0.3780	2.1380	1.1902	309.3	197.3	61.7	8S	52W
3969	199	-0359 Oct 10	18:35:59	14832	-29168	050	T	-p	-0.4041	2.1546	1.0791	354.8	210.7	44.6	5N	141E
3970	199	-0358 Apr 06	22:35:36	14825	-29162	055	T	p-	-0.3718	2.1698	1.1815	325.9	204.1	62.4	5S	83E
3971	199	-0358 Sep 30	01:11:14	14818	-29156	060	T	p-	0.3205	2.2783	1.2616	334.3	209.0	73.1	1N	42E
3972	199	-0357 Mar 27	07:20:30	14810	-29150	065	N	t-	-1.1732	0.7286	-0.3176	239.8	-	-	1S	47W
3973	199	-0357 Sep 19	14:34:53	14803	-29144	070	P	a-	0.9976	1.0118	0.0428	246.4	45.1	-	3S	159W
3974	199	-0356 Feb 14	16:09:01	14798	-29139	037	N*	-t	1.0003	1.0672	-0.0214	291.3	-	-	16N	176W
3975	199	-0356 Aug 09	22:01:56	14791	-29133	042	P	-a	-0.9154	1.1624	0.1938	262.7	95.1	-	18S	92E
3976	199	-0355 Feb 02	18:30:51	14783	-29127	047	T+	-p	0.2602	2.4065	1.3551	352.6	219.4	84.6	18N	148E
3977	199	-0355 Jul 30	10:15:53	14776	-29121	052	T-	-p	-0.1936	2.5113	1.4943	345.0	220.2	93.6	20S	92W
3978	199	-0354 Jan 23	03:55:05	14769	-29115	057	T	a-	-0.4593	2.0140	1.0163	318.1	193.1	19.4	20N	7E
3979	199	-0354 Jul 19	15:27:26	14762	-29109	062	P	t-	0.5819	1.8220	0.7588	343.7	189.7	-	22S	170W
3980	199	-0354 Dec 14	08:16:48	14756	-29104	029	N	-a	1.5416	0.0157	-0.9576	33.6	-	-	25N	63W
3981	200	-0353 Jan 12	18:39:29	14755	-29103	067	N	a-	-1.1151	0.7956	-0.1724	221.8	-	-	21N	144E
3982	200	-0353 Jun 09	01:37:48	14749	-29098	034	N	-t	-1.4892	0.1515	-0.9003	117.3	-	-	24S	35E
3983	200	-0353 Jul 08	15:59:40	14748	-29097	072	N	t-	1.3329	0.4475	-0.6224	199.8	-	-	22S	179W
3984	200	-0353 Dec 03	22:25:46	14742	-29092	039	P	-a	0.9299	1.1566	0.1469	269.5	84.8	-	23N	83E
3985	200	-0352 May 28	08:12:29	14735	-29086	044	P	-t	-0.6677	1.6329	0.6326	310.0	167.7	-	21S	64W
3986	200	-0352 Nov 22	06:47:05	14728	-29080	049	T+	-p	0.2638	2.4079	1.3404	356.0	219.7	83.3	20N	43W
3987	200	-0351 May 17	21:43:30	14721	-29074	054	T+	p-	0.1172	2.6193	1.6661	321.6	212.7	97.3	18S	94E
3988	200	-0351 Nov 11	08:02:50	14714	-29068	059	T	t-	-0.4279	2.1255	1.0210	364.5	211.6	23.8	16N	62W
3989	200	-0350 May 07	14:47:45	14707	-29062	064	P	a-	0.8479	1.2727	0.3311	265.1	119.8	-	14S	162W
3990	200	-0350 Oct 31	07:19:19	14700	-29056	069	N	t-	-1.0937	0.8965	-0.1935	269.0	-	-	12N	51W
3991	200	-0349 Mar 28	21:11:44	14694	-29051	036	P	-a	-0.9996	1.0269	0.0207	263.1	33.1	-	2S	105E
3992	200	-0349 Sep 20	23:20:20	14688	-29045	041	N	-a	1.0618	0.9088	-0.0896	244.6	-	-	2S	70E
3993	200	-0348 Mar 17	03:44:43	14680	-29039	046	T	-p	-0.2746	2.3855	1.3233	361.5	223.2	83.4	3N	8E
3994	200	-0348 Sep 09	14:04:56	14673	-29033	051	T	-p	0.3347	2.2237	1.2634	311.8	199.7	70.6	7S	151W
3995	200	-0347 Mar 06	04:19:28	14667	-29027	056	P	t-	0.4616	2.0537	0.9688	360.2	208.6	-	8N	0W
3996	200	-0347 Aug 30	06:24:49	14660	-29021	061	T	p-	-0.3664	2.1685	1.2024	313.6	198.8	63.6	11S	35W
3997	200	-0346 Feb 23	05:44:02	14653	-29015	066	N	h-	1.1624	0.7541	-0.3034	244.3	-	-	13N	20W
3998	200	-0346 Aug 19	19:40:19	14646	-29009	071	N	h-	-1.1298	0.7897	-0.2200	236.2	-	-	16S	127E
3999	200	-0345 Jan 14	02:31:12	14640	-29004	038	P	-a	-0.9126	1.1767	0.1901	265.1	94.2	-	21N	26E
4000	200	-0345 Jul 10	11:19:10	14633	-28998	043	P	-t	0.8989	1.2433	0.1745	306.7	100.5	-	22S	109W
4001	201	-0344 Jan 03	17:55:10	14626	-28992	048	T-	-p	-0.2263	2.4269	1.4583	315.9	205.3	86.4	23N	154E
4002	201	-0344 Jun 28	11:51:56	14619	-28986	053	T+	pp	0.1408	2.6305	1.5691	369.0	232.0	100.8	24S	118W
4003	201	-0344 Dec 23	08:55:26	14612	-28980	058	T	p-	0.4546	2.0220	1.0257	316.0	192.3	24.1	24N	72W
4004	201	-0343 Jun 17	16:50:00	14605	-28974	063	P	h-	-0.6277	1.7129	0.6994	319.4	176.5	-	24S	167E
4005	201	-0343 Dec 12	18:59:43	14598	-28968	068	N	h-	1.1940	0.6939	-0.3595	230.9	-	-	24N	135E
4006	201	-0342 May 08	21:53:46	14592	-28963	035	N	-a	1.1746	0.6755	-0.2705	207.0	-	-	15S	91E
4007	201	-0342 Jun 07	05:02:57	14591	-28962	073	N	a-	-1.3403	0.3790	-0.5822	162.8	-	-	24S	17W
4008	201	-0342 Nov 02	02:21:05	14585	-28957	040	N	-t	-1.1200	0.8558	-0.2494	268.6	-	-	13N	23E
4009	201	-0341 Apr 28	15:06:33	14578	-28951	045	T	-a	0.4470	2.0111	1.0637	305.7	191.4	37.2	12S	167W
4010	201	-0341 Oct 22	02:35:49	14572	-28945	050	T	-p	-0.4184	2.1283	1.0528	353.0	208.7	36.6	9N	19E
4011	201	-0340 Apr 17	05:56:21	14565	-28939	055	T	p-	-0.3064	2.2903	1.3010	330.3	209.8	77.4	9S	29W
4012	201	-0340 Oct 10	09:28:13	14558	-28933	060	T	p-	0.3013	2.3133	1.2970	334.1	209.7	76.7	5N	84W
4013	201	-0339 Apr 06	14:21:18	14551	-28927	065	N	t-	-1.1119	0.8410	-0.2054	255.5	-	-	6S	154W
4014	201	-0339 Sep 29	22:59:07	14544	-28921	070	P	a-	0.9719	1.0596	0.0893	250.6	64.7	-	2N	74E
4015	201	-0338 Feb 24	23:26:47	14538	-28916	037	N	-t	1.0410	0.9902	-0.0937	282.2	-	-	12N	74E
4016	201	-0338 Aug 21	05:50:30	14531	-28910	042	P	-a	-0.9662	1.0718	0.0979	255.7	68.7	-	15S	26W
4017	201	-0337 Feb 14	02:17:04	14524	-28904	047	T	-p	0.2907	2.3472	1.3023	349.6	216.9	79.5	15N	31E
4018	201	-0337 Aug 10	17:33:53	14517	-28898	052	T-	-p	-0.2533	2.4049	1.3815	344.9	217.7	86.2	18S	158E
4019	201	-0336 Feb 03	12:13:02	14511	-28892	057	T	a-	-0.4366	2.0529	1.0606	318.2	195.2	36.9	18N	118W
4020	201	-0336 Jul 29	22:13:02	14504	-28886	062	P	t-	0.5126	1.9516	0.8838	351.0	200.6	-	20S	88E

Cat Num	Canon Plate	Calendar Date	TD of Greatest Eclipse	ΔT s	Luna Num	Saros Num	Ecl. Type	QSE	Gamma	Pen. Mag.	Um. Mag.	Eclipse Phase Durations Pen. m	Par. m	Total m	Greatest in Zenith Lat.	Lng.
4021	202	-0336 Dec 24	17:06:19	14498	-28881	029	N	-a	1.5469	0.0059	-0.9671	20.7	-	-	25N	165E
4022	202	-0335 Jan 23	03:15:31	14497	-28880	067	N	a-	-1.0996	0.8226	-0.1425	224.7	-	-	20N	15E
4023	202	-0335 Jun 19	08:09:37	14491	-28875	034	Ne	-t	-1.5668	0.0083	-1.0417	27.7	-	-	25S	63W
4024	202	-0335 Jul 18	22:34:06	14490	-28874	072	N	t-	1.2560	0.5891	-0.4819	226.0	-	-	21S	82E
4025	202	-0335 Dec 14	07:07:05	14484	-28869	039	P	-a	0.9331	1.1518	0.1399	269.9	83.0	-	24N	47W
4026	202	-0334 Jun 08	15:10:15	14477	-28863	044	P	-t	-0.7442	1.4908	0.4938	300.0	150.9	-	23S	169W
4027	202	-0334 Dec 03	15:07:01	14471	-28857	049	T+	-p	0.2656	2.4062	1.3358	357.2	220.0	83.1	22N	168W
4028	202	-0333 May 29	05:03:18	14464	-28851	054	T+	pp	0.0413	2.7573	1.8065	321.5	213.5	99.8	21S	18W
4029	202	-0333 Nov 22	16:04:36	14457	-28845	059	T	t-	-0.4234	2.1344	1.0285	365.2	212.2	27.6	19N	177E
4030	202	-0332 May 17	22:14:03	14450	-28839	064	P	a-	0.7762	1.4043	0.4627	274.4	139.0	-	18S	85E
4031	202	-0332 Nov 10	15:25:20	14443	-28833	069	N	t-	-1.0858	0.9107	-0.1787	270.2	-	-	15N	174W
4032	202	-0331 Apr 08	04:31:43	14437	-28828	036	N	-a	-1.0638	0.9096	-0.0975	251.6	-	-	6S	7W
4033	202	-0331 May 07	14:03:46	14436	-28827	074	Nb	a-	1.5502	0.0008	-0.9742	7.8	-	-	14S	152W
4034	202	-0331 Oct 01	07:36:21	14431	-28822	041	N	-a	1.0806	0.8744	-0.1240	240.1	-	-	2N	56W
4035	202	-0330 Mar 28	10:45:05	14424	-28816	046	T	-t	-0.3363	2.2720	1.2104	359.4	219.1	70.2	1S	100W
4036	202	-0330 Sep 20	22:24:32	14417	-28810	051	T	-p	0.3630	2.1730	1.2105	310.4	197.3	64.1	3S	83E
4037	202	-0329 Mar 17	11:17:24	14410	-28804	056	T	t-	0.4050	2.1556	1.0747	363.4	215.2	44.3	4N	106W
4038	202	-0329 Sep 10	14:31:50	14403	-28798	061	T	p-	-0.3314	2.2351	1.2642	316.3	201.7	71.2	7S	158W
4039	202	-0328 Mar 05	13:08:44	14396	-28792	066	N	h-	1.1160	0.8360	-0.2152	254.0	-	-	9N	133W
4040	202	-0328 Aug 30	03:19:32	14390	-28786	071	N	h-	-1.0896	0.8667	-0.1494	246.7	-	-	12S	11E
4041	203	-0327 Jan 24	11:00:41	14384	-28781	038	P	-a	-0.9259	1.1500	0.1680	262.0	88.7	-	19N	101W
4042	203	-0327 Jul 20	17:54:25	14377	-28775	043	P	-t	0.9753	1.1048	0.0324	294.1	44.2	-	21S	152E
4043	203	-0326 Jan 14	02:37:53	14370	-28769	048	T-	-p	-0.2367	2.4070	1.4401	315.5	204.9	85.3	22N	24E
4044	203	-0326 Jul 09	18:24:59	14364	-28763	053	T+	pp	0.2218	2.4819	1.4207	366.1	228.1	92.3	23S	143E
4045	203	-0325 Jan 03	17:34:37	14357	-28757	058	T	p-	0.4497	2.0313	1.0342	317.1	193.3	27.8	24N	158E
4046	203	-0325 Jun 28	23:46:26	14350	-28751	063	P	h-	-0.5486	1.8569	0.8457	325.1	188.6	-	24S	62E
4047	203	-0325 Dec 24	03:21:03	14343	-28745	068	N	h-	1.1931	0.6966	-0.3586	232.1	-	-	25N	10E
4048	203	-0324 May 19	05:18:17	14338	-28740	035	N	-a	1.2448	0.5455	-0.3985	188.1	-	-	17S	21W
4049	203	-0324 Jun 17	12:21:15	14336	-28739	073	N	a-	-1.2642	0.5177	-0.4416	187.2	-	-	25S	127W
4050	203	-0324 Nov 12	10:19:00	14331	-28734	040	N	-t	-1.1276	0.8426	-0.2638	266.9	-	-	16N	98W
4051	203	-0323 May 08	22:33:43	14324	-28728	045	P	-a	0.5195	1.8783	0.9306	301.3	183.6	-	15S	80E
4052	203	-0323 Nov 01	10:42:41	14317	-28722	050	T	-p	-0.4274	2.1114	1.0366	351.5	207.2	30.6	13N	104W
4053	203	-0322 Apr 28	13:09:23	14310	-28716	055	T-	pp	-0.2357	2.4208	1.4299	334.3	214.7	88.4	13S	139W
4054	203	-0322 Oct 21	17:53:48	14304	-28710	060	T	p-	0.2885	2.3364	1.3208	333.5	210.0	78.8	10N	148E
4055	203	-0321 Apr 17	21:10:46	14297	-28704	065	N	t-	-1.0423	0.9690	-0.0780	271.4	-	-	10S	101E
4056	203	-0321 Oct 11	07:32:24	14290	-28698	070	P	a-	0.9535	1.0940	0.1225	253.4	75.3	-	6N	57W
4057	203	-0320 Mar 07	06:35:02	14285	-28693	037	N	-t	1.0892	0.8992	-0.1798	270.8	-	-	8N	35W
4058	203	-0320 Aug 31	13:46:13	14278	-28687	042	P	-a	-1.0103	0.9937	0.0143	249.2	26.6	-	11S	147W
4059	203	-0319 Feb 24	09:57:04	14271	-28681	047	T	-p	0.3270	2.2773	1.2389	346.2	213.8	72.1	11N	85W
4060	203	-0319 Aug 21	00:57:54	14264	-28675	052	T	-p	-0.3069	2.3099	1.2800	344.6	214.7	76.8	14S	46E
4061	204	-0318 Feb 13	20:25:32	14258	-28669	057	T	a-	-0.4089	2.1009	1.1142	318.5	197.6	49.7	14N	117E
4062	204	-0318 Aug 10	05:03:13	14251	-28663	062	T	t-	0.4474	2.0737	1.0009	357.1	209.3	4.9	17S	16W
4063	204	-0317 Feb 03	11:46:02	14244	-28657	067	N	a-	-1.0797	0.8576	-0.1045	228.4	-	-	17N	113W
4064	204	-0317 Jul 30	05:15:36	14238	-28651	072	N	t-	1.1839	0.7220	-0.3503	246.8	-	-	19S	19W
4065	204	-0317 Dec 25	15:44:46	14232	-28646	039	P	-a	0.9386	1.1425	0.1288	270.0	80.0	-	25N	176W
4066	204	-0316 Jun 18	22:12:43	14225	-28640	044	P	-h	-0.8177	1.3546	0.3604	289.4	131.1	-	24S	85E
4067	204	-0316 Dec 13	23:24:16	14218	-28634	049	T+	-p	0.2690	2.4009	1.3283	358.2	220.3	82.6	23N	68E
4068	204	-0315 Jun 08	12:25:00	14212	-28628	054	T-	pp	-0.0333	2.7711	1.8220	320.8	213.2	99.8	23S	129W
4069	204	-0315 Dec 03	00:06:05	14205	-28622	059	T	t-	-0.4189	2.1432	1.0366	365.8	212.7	31.2	21N	56E
4070	204	-0314 May 29	05:38:56	14198	-28616	064	P	a-	0.7034	1.5380	0.5959	282.9	154.7	-	20S	27W
4071	204	-0314 Nov 21	23:34:17	14192	-28610	069	N	h-	-1.0798	0.9209	-0.1671	270.8	-	-	18N	63E
4072	204	-0313 Apr 19	11:41:31	14186	-28605	036	N	-h	-1.1349	0.7797	-0.2287	236.9	-	-	10S	117W
4073	204	-0313 May 18	21:14:43	14185	-28604	074	N	h-	1.4829	0.1255	-0.8518	97.8	-	-	17S	99E
4074	204	-0313 Oct 12	16:01:16	14179	-28599	041	N	-a	1.0927	0.8520	-0.1462	236.9	-	-	7N	176E
4075	204	-0312 Apr 07	17:34:51	14173	-28593	046	T	-t	-0.4057	2.1444	1.0833	356.3	213.1	46.2	6S	156E
4076	204	-0312 Oct 01	06:52:17	14166	-28587	051	T	-p	0.3845	2.1345	1.1700	309.2	195.4	58.3	2N	46W
4077	204	-0311 Mar 27	18:06:29	14159	-28581	056	T	t-	0.3410	2.2711	1.1941	366.3	221.4	68.6	1S	149E
4078	204	-0311 Sep 20	22:46:51	14153	-28575	061	T	p-	-0.3037	2.2885	1.3127	318.6	203.8	76.3	3S	76E
4079	204	-0310 Mar 16	20:24:55	14146	-28569	066	N	h-	1.0618	0.9322	-0.1125	264.4	-	-	4N	116E
4080	204	-0310 Sep 10	11:05:38	14139	-28563	071	N	h-	-1.0558	0.9320	-0.0908	255.4	-	-	8S	108W

Cat Num	Canon Plate	Calendar Date	TD of Greatest Eclipse	ΔT s	Luna Num	Saros Num	Ecl. Type	QSE	Gamma	Pen. Mag.	Um. Mag.	Eclipse Phase Durations Pen. m	Par. m	Total m	Greatest in Zenith Lat.	Lng.
4081	205	-0309 Feb 04	19:23:44	14134	-28558	038	P	-a	-0.9448	1.1128	0.1356	258.0	79.9	-	17N	132E
4082	205	-0309 Aug 01	00:35:07	14127	-28552	043	N	-t	1.0465	0.9761	-0.1001	280.8	-	-	19S	51E
4083	205	-0308 Jan 25	11:14:05	14121	-28546	048	T-	-p	-0.2521	2.3776	1.4127	314.9	204.3	83.5	20N	106W
4084	205	-0308 Jul 20	01:05:15	14114	-28540	053	T	pp	0.2971	2.3438	1.2823	362.4	222.8	79.5	22S	43E
4085	205	-0307 Jan 14	02:09:02	14107	-28534	058	T	p-	0.4419	2.0460	1.0483	318.5	194.6	33.0	23N	30E
4086	205	-0307 Jul 09	06:47:43	14101	-28528	063	P	a-	-0.4725	1.9957	0.9864	329.6	198.0	-	24S	43W
4087	205	-0306 Jan 03	11:38:53	14094	-28522	068	N	h-	1.1901	0.7028	-0.3538	233.7	-	-	25N	114W
4088	205	-0306 May 30	12:41:55	14089	-28517	035	N	-a	1.3158	0.4145	-0.5280	165.8	-	-	20S	133W
4089	205	-0306 Jun 28	19:42:11	14087	-28516	073	N	a-	-1.1898	0.6536	-0.3044	206.9	-	-	25S	122E
4090	205	-0306 Nov 23	18:20:07	14082	-28511	040	N	-t	-1.1328	0.8331	-0.2736	265.7	-	-	19N	142E
4091	205	-0305 May 20	05:56:59	14075	-28505	045	P	-a	0.5948	1.7405	0.7919	295.8	173.8	-	18S	32W
4092	205	-0305 Nov 12	18:55:06	14069	-28499	050	T	-p	-0.4327	2.1010	1.0276	350.1	206.2	26.6	16N	132E
4093	205	-0304 May 08	20:17:51	14062	-28493	055	T-	pp	-0.1620	2.5573	1.5642	337.7	218.6	96.2	16S	112E
4094	205	-0304 Nov 01	02:24:23	14055	-28487	060	T	p-	0.2792	2.3529	1.3384	332.8	210.0	80.2	14N	19E
4095	205	-0303 Apr 28	03:55:20	14049	-28481	065	P	t-	-0.9692	1.1037	0.0557	286.4	56.7	-	13S	2W
4096	205	-0303 Oct 21	16:11:33	14042	-28475	070	P	a-	0.9399	1.1193	0.1469	255.3	82.1	-	10N	172E
4097	205	-0302 Mar 18	13:35:10	14037	-28470	037	N	-t	1.1441	0.7959	-0.2782	256.7	-	-	4N	142W
4098	205	-0302 Sep 11	21:49:04	14030	-28464	042	N	-a	-1.0480	0.9274	-0.0576	243.3	-	-	7S	91E
4099	205	-0301 Mar 07	17:28:37	14024	-28458	047	T	-p	0.3711	2.1930	1.1612	342.1	209.7	60.7	7N	160E
4100	205	-0301 Sep 01	08:29:56	14017	-28452	052	T	-a	-0.3532	2.2283	1.1918	344.2	211.5	65.6	11S	69W
4101	206	-0300 Feb 25	04:28:32	14010	-28446	057	T	a-	-0.3726	2.1647	1.1837	319.1	200.6	61.6	11N	5W
4102	206	-0300 Aug 20	12:03:15	14004	-28440	062	T	t-	0.3904	2.1807	1.1031	362.0	215.7	51.3	14S	122W
4103	206	-0299 Feb 13	20:08:15	13997	-28434	067	N	a-	-1.0530	0.9051	-0.0540	233.3	-	-	14N	120E
4104	206	-0299 Aug 09	12:05:24	13991	-28428	072	N	t-	1.1175	0.8446	-0.2293	263.4	-	-	17S	123W
4105	206	-0298 Jan 05	00:17:48	13985	-28423	039	P	-a	0.9476	1.1266	0.1118	269.4	74.9	-	24N	56E
4106	206	-0298 Jun 30	05:18:17	13979	-28417	044	P	-h	-0.8896	1.2215	0.2295	277.8	106.3	-	25S	22W
4107	206	-0298 Dec 25	07:38:25	13972	-28411	049	T	-p	0.2743	2.3920	1.3181	359.1	220.5	81.8	24N	55W
4108	206	-0297 Jun 19	19:47:02	13966	-28405	054	T-	pp	-0.1081	2.6334	1.6853	319.4	211.8	97.4	24S	120E
4109	206	-0297 Dec 14	08:07:52	13959	-28399	059	T	t-	-0.4145	2.1510	1.0447	366.1	213.3	34.5	23N	64W
4110	206	-0296 Jun 08	13:02:48	13952	-28393	064	P	a-	0.6302	1.6731	0.7297	290.6	167.7	-	22S	139W
4111	206	-0296 Dec 02	07:45:27	13946	-28387	069	N	h-	-1.0750	0.9287	-0.1572	271.0	-	-	21N	60W
4112	206	-0295 Apr 29	18:45:37	13940	-28382	036	N	-h	-1.2096	0.6436	-0.3666	219.1	-	-	14S	135E
4113	206	-0295 May 29	04:23:26	13939	-28381	074	N	h-	1.4143	0.2528	-0.7274	137.7	-	-	19S	10W
4114	206	-0295 Oct 23	00:33:28	13934	-28376	041	N	-a	1.0998	0.8388	-0.1590	234.8	-	-	11N	46E
4115	206	-0294 Apr 19	00:16:45	13927	-28370	046	P	-t	-0.4807	2.0065	0.9456	351.8	204.8	-	10S	53E
4116	206	-0294 Oct 12	15:27:02	13921	-28364	051	T	-p	0.4004	2.1064	1.1398	308.4	193.8	53.3	6N	176W
4117	206	-0293 Apr 08	00:48:33	13914	-28358	056	T+	pp	0.2707	2.3982	1.3249	368.9	226.7	84.7	5S	47E
4118	206	-0293 Oct 02	07:08:43	13908	-28352	061	T	p-	-0.2822	2.3304	1.3498	320.5	205.4	79.7	1N	51W
4119	206	-0292 Mar 27	03:35:05	13901	-28346	066	P	h-	1.0016	1.0395	0.0010	274.8	7.5	-	0N	6E
4120	206	-0292 Sep 20	18:59:25	13895	-28340	071	N	h-	-1.0290	0.9845	-0.0447	262.2	-	-	4S	132E
4121	207	-0291 Feb 15	03:40:08	13889	-28335	038	P	-a	-0.9696	1.0648	0.0925	252.9	66.3	-	13N	7E
4122	207	-0291 Aug 11	07:22:49	13883	-28329	043	N	-t	1.1114	0.8591	-0.2210	267.0	-	-	16S	52W
4123	207	-0290 Feb 04	19:43:14	13876	-28323	048	T	-p	-0.2734	2.3375	1.3747	314.2	203.3	80.8	17N	126E
4124	207	-0290 Jul 31	07:53:26	13870	-28317	053	T	-t	0.3662	2.2172	1.1553	358.2	216.6	61.5	20S	60W
4125	207	-0289 Jan 25	10:34:31	13863	-28311	058	T	p-	0.4273	2.0728	1.0750	320.3	196.6	40.9	21N	97W
4126	207	-0289 Jul 20	13:56:49	13857	-28305	063	T	a-	-0.4018	2.1247	1.1168	332.9	204.8	51.9	22S	151W
4127	207	-0288 Jan 14	19:48:28	13850	-28299	068	N	t-	1.1814	0.7191	-0.3383	236.7	-	-	23N	124E
4128	207	-0288 Jun 09	20:06:30	13845	-28294	035	N	-a	1.3863	0.2848	-0.6570	139.0	-	-	21S	115E
4129	207	-0288 Jul 09	03:08:49	13844	-28293	073	N	a-	-1.1192	0.7828	-0.1746	223.0	-	-	24S	10E
4130	207	-0288 Dec 04	02:21:11	13838	-28288	040	N	-t	-1.1380	0.8235	-0.2828	264.2	-	-	21N	21E
4131	207	-0287 May 30	13:18:35	13832	-28282	045	P	-a	0.6710	1.6016	0.6515	289.4	161.6	-	20S	144W
4132	207	-0287 Nov 23	03:12:00	13826	-28276	050	T	-p	-0.4348	2.0962	1.0248	348.8	205.5	25.1	19N	7E
4133	207	-0286 May 20	03:19:29	13819	-28270	055	T-	pp	-0.0836	2.7024	1.7066	340.5	221.4	101.1	19S	6E
4134	207	-0286 Nov 12	11:01:41	13813	-28264	060	T	p-	0.2749	2.3599	1.3471	331.9	209.7	80.0	17N	111W
4135	207	-0285 May 09	10:31:36	13806	-28258	065	P	t-	-0.8903	1.2492	0.1999	300.8	105.5	-	17S	102W
4136	207	-0285 Nov 02	00:57:20	13800	-28252	070	P	a-	0.9318	1.1347	0.1616	256.4	85.8	-	14N	40E
4137	207	-0284 Mar 28	20:27:03	13794	-28247	037	N	-t	1.2057	0.6804	-0.3888	239.4	-	-	1S	113E
4138	207	-0284 Sep 22	05:59:57	13788	-28241	042	N	-a	-1.0785	0.8743	-0.1163	238.5	-	-	3S	33W
4139	207	-0283 Mar 18	00:54:28	13781	-28235	047	T	-t	0.4208	2.0985	1.0732	337.3	204.5	42.0	3N	47E
4140	207	-0283 Sep 11	16:07:33	13775	-28229	052	T	-a	-0.3939	2.1570	1.1139	343.8	208.2	52.0	7S	175E

Cat Num	Canon Plate	Calendar Date	TD of Greatest Eclipse	ΔT s	Luna Num	Saros Num	Ecl. Type	QSE	Gamma	Pen. Mag.	Um. Mag.	Eclipse Phase Durations Pen. m	Par. m	Total m	Greatest in Zenith Lat.	Lng.
4141	208	-0282 Mar 07	12:26:14	13768	-28223	057	T	p-	-0.3312	2.2378	1.2624	319.8	203.5	71.6	7N	126W
4142	208	-0282 Aug 31	19:09:38	13762	-28217	062	T	p-	0.3389	2.2778	1.1952	366.0	220.6	68.5	10S	130E
4143	208	-0281 Feb 25	04:22:31	13756	-28211	067	P	a-	-1.0199	0.9643	0.0083	239.1	19.8	-	10N	5W
4144	208	-0281 Aug 20	19:04:22	13749	-28205	072	N	t-	1.0578	0.9551	-0.1205	276.6	-	-	13S	131E
4145	208	-0280 Jan 16	08:44:17	13744	-28200	039	P	-a	0.9613	1.1018	0.0862	268.0	66.2	-	23N	70W
4146	208	-0280 Jul 10	12:30:42	13737	-28194	044	P	-h	-0.9565	1.0980	0.1078	266.0	73.8	-	24S	130W
4147	208	-0279 Jan 04	15:46:02	13731	-28188	049	T	-p	0.2841	2.3743	1.2995	359.7	220.3	80.1	24N	177W
4148	208	-0279 Jun 30	03:13:49	13725	-28182	054	T-	-p	-0.1794	2.5024	1.5548	317.4	209.4	92.6	24S	8E
4149	208	-0279 Dec 24	16:05:58	13718	-28176	059	T	t-	-0.4072	2.1639	1.0586	366.6	214.2	39.3	23N	177E
4150	208	-0278 Jun 19	20:26:49	13712	-28170	064	P	a-	0.5573	1.8078	0.8625	297.5	178.5	-	23S	109E
4151	208	-0278 Dec 13	15:56:41	13705	-28164	069	N	h-	-1.0695	0.9374	-0.1456	271.2	-	-	22N	177E
4152	208	-0277 May 11	01:42:39	13700	-28159	036	N	-h	-1.2888	0.4993	-0.5131	196.5	-	-	18S	29E
4153	208	-0277 Jun 09	11:29:51	13699	-28158	074	N	h-	1.3443	0.3831	-0.6007	168.2	-	-	21S	117W
4154	208	-0277 Nov 03	09:11:34	13694	-28153	041	N	-a	1.1028	0.8329	-0.1641	233.5	-	-	15N	84W
4155	208	-0276 Apr 29	06:51:03	13687	-28147	046	P	-t	-0.5612	1.8589	0.7978	345.9	193.5	-	14S	47W
4156	208	-0276 Oct 23	00:08:53	13681	-28141	051	T	-p	0.4107	2.0884	1.1200	307.9	192.8	49.7	11N	52E
4157	208	-0275 Apr 18	07:23:53	13674	-28135	056	T+	pp	0.1947	2.5359	1.4661	370.7	230.9	95.8	9S	54W
4158	208	-0275 Oct 12	15:37:12	13668	-28129	061	T-	p-	-0.2668	2.3611	1.3758	322.2	206.5	81.9	6N	180W
4159	208	-0274 Apr 07	10:38:43	13662	-28123	066	P	h-	0.9349	1.1587	0.1263	285.1	82.9	-	4S	102W
4160	208	-0274 Oct 02	03:00:27	13655	-28117	071	N*	h-	-1.0091	1.0243	-0.0112	267.6	-	-	1N	10E
4161	209	-0273 Feb 26	11:47:47	13650	-28112	038	P	-a	-1.0019	1.0030	0.0357	246.4	41.4	-	9N	116W
4162	209	-0273 Aug 22	14:19:01	13644	-28106	043	N	-t	1.1687	0.7559	-0.3282	253.5	-	-	13S	158W
4163	209	-0272 Feb 16	04:04:45	13637	-28100	048	T	-p	-0.3007	2.2863	1.3257	313.2	201.9	76.8	14N	0E
4164	209	-0272 Aug 10	14:50:35	13631	-28094	053	T	-t	0.4286	2.1030	1.0405	353.6	209.8	32.5	17S	165W
4165	209	-0271 Feb 04	18:53:04	13625	-28088	058	T	p-	0.4077	2.1087	1.1108	322.4	199.0	49.4	18N	138E
4166	209	-0271 Jul 30	21:13:33	13618	-28082	063	T	a-	-0.3365	2.2440	1.2370	335.2	209.8	70.9	20S	99E
4167	209	-0270 Jan 25	03:51:37	13612	-28076	068	N	t-	1.1683	0.7433	-0.3143	240.8	-	-	21N	3E
4168	209	-0270 Jun 21	03:31:47	13607	-28071	035	N	-a	1.4562	0.1565	-0.7852	104.2	-	-	22S	3E
4169	209	-0270 Jul 20	10:39:55	13605	-28070	073	N	a-	-1.0519	0.9063	-0.0510	236.4	-	-	23S	103W
4170	209	-0270 Dec 15	10:22:56	13600	-28065	040	N	-t	-1.1425	0.8145	-0.2905	262.8	-	-	22N	99W
4171	209	-0269 Jun 10	20:39:26	13594	-28059	045	P	-a	0.7469	1.4634	0.5110	282.0	146.6	-	22S	105E
4172	209	-0269 Dec 04	11:29:58	13588	-28053	050	T	-p	-0.4365	2.0917	1.0230	347.5	204.9	24.2	22N	117W
4173	209	-0268 May 30	10:19:37	13581	-28047	055	T-	pp	-0.0051	2.8481	1.8490	342.6	222.8	102.9	21S	100W
4174	209	-0268 Nov 22	19:41:38	13575	-28041	060	T	p-	0.2727	2.3630	1.3523	330.9	209.4	81.0	20N	119E
4175	209	-0267 May 19	17:04:41	13569	-28035	065	P	t-	-0.8091	1.3990	0.3481	314.0	136.5	-	20S	158E
4176	209	-0267 Nov 12	09:46:16	13562	-28029	070	P	a-	0.9263	1.1448	0.1714	257.0	88.2	-	18N	93W
4177	209	-0266 Apr 09	03:12:33	13557	-28024	037	N	-t	1.2725	0.5555	-0.5090	218.3	-	-	5S	9E
4178	209	-0266 May 08	18:09:15	13556	-28023	075	Nb	t-	-1.5751	0.0028	-1.0667	16.4	-	-	17S	142E
4179	209	-0266 Oct 03	14:17:30	13551	-28018	042	N	-a	-1.1031	0.8320	-0.1642	234.6	-	-	1N	159W
4180	209	-0265 Mar 29	08:11:43	13544	-28012	047	P	-t	0.4786	1.9893	0.9705	331.6	197.6	-	1S	65W
4181	210	-0265 Sep 22	23:53:54	13538	-28006	052	T	-a	-0.4265	2.1003	1.0508	343.5	205.3	35.5	2S	57E
4182	210	-0264 Mar 17	20:15:16	13532	-28000	057	T	p-	-0.2819	2.3256	1.3555	320.6	206.5	80.6	2N	115E
4183	210	-0264 Sep 11	02:25:28	13525	-27994	062	T	p-	0.2954	2.3600	1.2726	369.1	224.2	78.7	6S	19E
4184	210	-0263 Mar 07	12:28:13	13519	-27988	067	P	a-	-0.9799	1.0363	0.0831	245.9	62.1	-	6N	128W
4185	210	-0263 Aug 31	02:13:47	13513	-27982	072	N*	t-	1.0057	1.0515	-0.0258	286.9	-	-	10S	23E
4186	210	-0262 Jan 26	17:03:07	13508	-27977	039	P	-a	0.9807	1.0666	0.0505	265.5	51.1	-	21N	165E
4187	210	-0262 Jul 21	19:48:21	13501	-27971	044	N	-a	-1.0198	0.9812	-0.0079	253.7	-	-	23S	120E
4188	210	-0261 Jan 15	23:47:52	13495	-27965	049	T	-p	0.2982	2.3487	1.2737	360.0	219.9	77.5	22N	63E
4189	210	-0261 Jul 11	10:44:03	13489	-27959	054	T-	-p	-0.2481	2.3765	1.4285	315.0	206.1	85.4	23S	105W
4190	210	-0260 Jan 05	00:00:02	13482	-27953	059	T	t-	-0.3965	2.1827	1.0792	367.0	215.4	45.4	23N	58E
4191	210	-0260 Jun 30	03:53:01	13476	-27947	064	P	a-	0.4865	1.9391	0.9911	303.4	187.3	-	23S	3W
4192	210	-0260 Dec 24	00:07:39	13470	-27941	069	N	h-	-1.0628	0.9478	-0.1316	271.5	-	-	23N	55E
4193	210	-0259 May 21	08:35:22	13465	-27936	036	N	-h	-1.3702	0.3514	-0.6637	167.9	-	-	21S	75W
4194	210	-0259 Jun 19	18:35:16	13464	-27935	074	N	h-	1.2741	0.5139	-0.4739	193.2	-	-	22S	136E
4195	210	-0259 Nov 13	17:54:25	13458	-27930	041	N	-a	1.1026	0.8326	-0.1632	232.8	-	-	18N	144E
4196	210	-0258 May 10	13:20:17	13452	-27924	046	P	-t	-0.6452	1.7050	0.6435	338.4	178.8	-	17S	146W
4197	210	-0258 Nov 03	08:55:38	13446	-27918	051	T	-p	0.4168	2.0780	1.1081	307.6	192.1	47.3	14N	81W
4198	210	-0257 Apr 29	13:54:02	13440	-27912	056	T+	pp	0.1141	2.6821	1.6157	371.7	233.7	102.9	13S	154W
4199	210	-0257 Oct 24	00:11:26	13433	-27906	061	T-	p-	-0.2567	2.3817	1.3921	323.7	207.4	83.2	10N	50E
4200	210	-0256 Apr 17	17:38:51	13427	-27900	066	P	h-	0.8644	1.2853	0.2586	294.7	116.0	-	8S	151E

Cat Num	Canon Plate	Calendar Date	TD of Greatest Eclipse	ΔT s	Luna Num	Saros Num	Ecl. Type	QSE	Gamma	Pen. Mag.	Um. Mag.	Eclipse Phase Durations Pen. m	Par. m	Total m	Greatest in Zenith Lat.	Lng.
4201	211	-0256 Oct 12	11:06:48	13421	-27894	071	P	h-	-0.9942	1.0546	0.0132	271.9	26.9	-	5N	113W
4202	211	-0255 Mar 08	19:49:00	13416	-27889	038	N	-a	-1.0400	0.9309	-0.0318	238.5	-	-	5N	122E
4203	211	-0255 Apr 07	04:28:52	13415	-27888	076	Nb	a-	1.5485	0.0038	-0.9710	17.2	-	-	4S	10W
4204	211	-0255 Sep 01	21:24:15	13409	-27883	043	N	-t	1.2183	0.6668	-0.4211	240.5	-	-	9S	94E
4205	211	-0254 Feb 26	12:17:26	13403	-27877	048	T	-p	-0.3354	2.2216	1.2631	311.8	199.8	70.6	10N	125W
4206	211	-0254 Aug 21	21:57:32	13397	-27871	053	P	-t	0.4835	2.0027	0.9394	349.0	202.6	-	13S	87E
4207	211	-0253 Feb 16	03:01:19	13391	-27865	058	T	p-	0.3804	2.1589	1.1611	324.8	202.1	58.7	14N	15E
4208	211	-0253 Aug 11	04:40:22	13385	-27859	063	T	p-	-0.2787	2.3497	1.3434	336.5	213.1	81.9	17S	14W
4209	211	-0252 Feb 05	11:43:40	13378	-27853	068	N	t-	1.1471	0.7822	-0.2754	246.8	-	-	18N	115W
4210	211	-0252 Jul 01	11:01:10	13373	-27848	035	Ne	-a	1.5228	0.0347	-0.9076	49.5	-	-	22S	110W
4211	211	-0252 Jul 30	18:19:20	13372	-27847	073	P	a-	-0.9909	1.0184	0.0607	247.1	53.8	-	21S	142E
4212	211	-0252 Dec 25	18:21:48	13367	-27842	040	N	-t	-1.1490	0.8015	-0.3015	260.7	-	-	23N	141E
4213	211	-0251 Jun 21	03:59:44	13361	-27836	045	P	-a	0.8227	1.3258	0.3705	273.6	127.8	-	23S	5W
4214	211	-0251 Dec 14	19:49:11	13355	-27830	050	T	-p	-0.4378	2.0874	1.0222	346.0	204.3	23.7	23N	118E
4215	211	-0250 Jun 10	17:16:05	13348	-27824	055	T+	pp	0.0752	2.7214	1.7184	343.8	222.9	101.8	23S	155E
4216	211	-0250 Dec 04	04:24:15	13342	-27818	060	T+	p-	0.2717	2.3635	1.3555	329.7	209.0	81.1	22N	12W
4217	211	-0249 May 30	23:32:31	13336	-27812	065	P	t-	-0.7239	1.5564	0.5033	326.1	160.6	-	22S	60E
4218	211	-0249 Nov 23	18:39:02	13330	-27806	070	P	a-	0.9238	1.1493	0.1761	257.2	89.2	-	21N	133E
4219	211	-0248 Apr 19	09:52:45	13325	-27801	037	N	-t	1.3438	0.4224	-0.6376	192.2	-	-	9S	93W
4220	211	-0248 May 19	00:30:24	13323	-27800	075	N	t-	-1.4899	0.1584	-0.9095	122.2	-	-	20S	45E
4221	212	-0248 Oct 13	22:41:01	13318	-27795	042	N	-a	-1.1223	0.7993	-0.2021	231.6	-	-	6N	73E
4222	212	-0247 Apr 08	15:25:25	13312	-27789	047	P	-t	0.5402	1.8732	0.8605	325.1	189.2	-	5S	175W
4223	212	-0247 Oct 03	07:46:31	13306	-27783	052	P	-a	-0.4532	2.0545	0.9988	343.5	202.7	-	2N	63W
4224	212	-0246 Mar 29	03:58:25	13300	-27777	057	T-	p-	-0.2270	2.4237	1.4588	321.3	209.1	88.1	2S	3W
4225	212	-0246 Sep 22	09:49:03	13294	-27771	062	T+	pp	0.2584	2.4302	1.3383	371.6	226.9	85.6	2S	93W
4226	212	-0245 Mar 18	20:26:41	13287	-27765	067	P	a-	-0.9340	1.1192	0.1686	253.1	87.4	-	2N	111E
4227	212	-0245 Sep 11	09:33:24	13281	-27759	072	P	t-	0.9610	1.1343	0.0555	295.0	57.2	-	6S	89W
4228	212	-0244 Feb 07	01:12:47	13276	-27754	039	P	-a	1.0070	1.0185	0.0021	261.6	10.6	-	18N	42E
4229	212	-0244 Aug 01	03:15:32	13270	-27748	044	N	-a	-1.0763	0.8771	-0.1112	241.7	-	-	20S	8E
4230	212	-0243 Jan 26	07:41:14	13264	-27742	049	T	-p	0.3187	2.3109	1.2360	359.9	218.8	73.1	20N	56W
4231	212	-0243 Jul 21	18:19:53	13258	-27736	054	T	-p	-0.3126	2.2585	1.3098	312.2	202.0	75.6	22S	141E
4232	212	-0242 Jan 15	07:47:53	13252	-27730	059	T	t-	-0.3804	2.2109	1.1099	367.7	217.2	53.0	22N	58W
4233	212	-0242 Jul 11	11:22:22	13245	-27724	064	T	a-	0.4183	2.0658	1.1147	308.5	194.4	49.2	23S	115W
4234	212	-0241 Jan 04	08:13:50	13239	-27718	069	N	h-	-1.0515	0.9663	-0.1086	272.7	-	-	22N	66W
4235	212	-0241 Jun 01	15:24:03	13234	-27713	036	N	-h	-1.4535	0.2003	-0.8183	129.0	-	-	23S	178W
4236	212	-0241 Jul 01	01:41:50	13233	-27712	074	N	h-	1.2053	0.6426	-0.3500	214.3	-	-	23S	29E
4237	212	-0241 Nov 25	02:40:10	13228	-27707	041	N	-a	1.1010	0.8347	-0.1593	232.4	-	-	21N	13E
4238	212	-0240 May 20	19:45:32	13222	-27701	046	P	-t	-0.7317	1.5469	0.4845	329.1	159.4	-	20S	117E
4239	212	-0240 Nov 13	17:45:53	13216	-27695	051	T	-p	0.4202	2.0722	1.1012	307.5	191.7	45.8	18N	146E
4240	212	-0239 May 09	20:20:54	13210	-27689	056	T+	pp	0.0301	2.8347	1.7713	371.7	234.9	106.1	16S	108E
4241	213	-0239 Nov 03	08:51:00	13203	-27683	061	T-	p-	-0.2514	2.3934	1.3998	325.0	208.0	83.9	14N	81W
4242	213	-0238 Apr 29	00:33:17	13197	-27677	066	P	h-	0.7878	1.4232	0.4019	303.8	141.1	-	12S	46E
4243	213	-0238 Oct 23	19:20:15	13191	-27671	071	P	h-	-0.9857	1.0730	0.0261	274.9	37.8	-	9N	122E
4244	213	-0237 Mar 20	03:42:06	13186	-27666	038	N	-a	-1.0853	0.8453	-0.1128	228.7	-	-	1N	2E
4245	213	-0237 Apr 18	11:55:04	13185	-27665	076	N	a-	1.4822	0.1230	-0.8468	95.6	-	-	8S	124W
4246	213	-0237 Sep 13	04:39:20	13180	-27660	043	N	-t	1.2598	0.5925	-0.4991	228.6	-	-	5S	16W
4247	213	-0236 Mar 08	20:21:46	13174	-27654	048	T	-p	-0.3769	2.1444	1.1878	309.9	196.9	61.2	6N	113E
4248	213	-0236 Sep 01	05:15:24	13168	-27648	053	P	-t	0.5303	1.9173	0.8532	344.6	195.7	-	10S	24W
4249	213	-0235 Feb 26	11:01:52	13162	-27642	058	T	p-	0.3474	2.2194	1.2218	327.5	205.4	67.8	10N	106W
4250	213	-0235 Aug 21	12:14:55	13156	-27636	063	T-	p-	-0.2264	2.4456	1.4395	337.2	215.4	89.0	14S	128W
4251	213	-0234 Feb 15	19:27:50	13150	-27630	068	N	t-	1.1203	0.8312	-0.2260	253.8	-	-	15N	128E
4252	213	-0234 Aug 11	02:05:02	13143	-27624	073	P	a-	-0.9348	1.1217	0.1632	256.0	86.8	-	18S	24E
4253	213	-0233 Jan 06	02:16:29	13138	-27619	040	N	-t	-1.1592	0.7814	-0.3187	257.5	-	-	22N	23E
4254	213	-0233 Jul 02	11:22:06	13132	-27613	045	P	-a	0.8961	1.1928	0.2340	264.3	103.8	-	23S	116W
4255	213	-0233 Dec 26	04:06:15	13126	-27607	050	T	-p	-0.4416	2.0785	1.0175	344.4	203.5	21.0	23N	6W
4256	213	-0232 Jun 21	00:13:34	13120	-27601	055	T+	pp	0.1540	2.5790	1.5717	344.2	221.5	97.5	23S	50E
4257	213	-0232 Dec 14	13:05:15	13114	-27595	060	T+	p-	0.2689	2.3671	1.3622	328.7	208.7	81.6	24N	142W
4258	213	-0231 Jun 10	06:00:41	13108	-27589	065	P	t-	-0.6394	1.7127	0.6570	336.7	179.4	-	23S	37W
4259	213	-0231 Dec 04	03:32:05	13102	-27583	070	P	a-	0.9217	1.1529	0.1802	257.3	90.2	-	23N	0W
4260	213	-0230 Apr 30	16:27:36	13097	-27578	037	N	-t	1.4194	0.2815	-0.7743	158.4	-	-	12S	167E

Cat Num	Canon Plate	Calendar Date	TD of Greatest Eclipse	ΔT s	Luna Num	Saros Num	Ecl. Type	QSE	Gamma	Pen. Mag.	Um. Mag.	Eclipse Phase Durations Pen. m	Par. m	Total m	Greatest in Zenith Lat.	Lng.
4261	214	-0230 May 30	06:49:26	13096	-27577	075	N	t-	-1.4022	0.3187	-0.7480	170.6	-	-	22S	50W
4262	214	-0230 Oct 25	07:10:31	13091	-27572	042	N	-a	-1.1361	0.7764	-0.2299	229.6	-	-	10N	56W
4263	214	-0229 Apr 19	22:32:56	13085	-27566	047	P	-t	0.6075	1.7468	0.7398	317.5	178.6	-	9S	76E
4264	214	-0229 Oct 14	15:46:26	13079	-27560	052	P	-a	-0.4730	2.0210	0.9597	343.7	200.6	-	6N	175E
4265	214	-0228 Apr 08	11:34:41	13073	-27554	057	T-	p-	-0.1658	2.5336	1.5733	321.8	211.3	93.9	6S	119W
4266	214	-0228 Oct 02	17:22:38	13067	-27548	062	T+	pp	0.2300	2.4843	1.3883	373.5	228.7	89.9	3N	152E
4267	214	-0227 Mar 29	04:17:04	13061	-27542	067	P	a-	-0.8816	1.2142	0.2659	260.9	108.3	-	2S	9W
4268	214	-0227 Sep 21	17:03:05	13055	-27536	072	P	t-	0.9238	1.2033	0.1232	301.1	84.3	-	1S	157E
4269	214	-0226 Feb 17	09:13:30	13049	-27531	039	N	-a	1.0399	0.9581	-0.0583	256.2	-	-	14N	79W
4270	214	-0226 Aug 12	10:50:16	13043	-27525	044	N	-a	-1.1276	0.7828	-0.2052	229.8	-	-	18S	107W
4271	214	-0225 Feb 06	15:25:08	13037	-27519	049	T	-t	0.3464	2.2597	1.1856	359.2	217.1	66.1	17N	172W
4272	214	-0225 Aug 02	02:01:59	13031	-27513	054	T	-p	-0.3721	2.1499	1.1998	309.2	197.4	62.9	20S	25E
4273	214	-0224 Jan 26	15:29:33	13025	-27507	059	T	p-	-0.3590	2.2486	1.1506	368.4	219.3	61.2	19N	174W
4274	214	-0224 Jul 21	18:55:36	13019	-27501	064	T	p-	0.3539	2.1859	1.2310	312.9	200.1	67.4	21S	131E
4275	214	-0223 Jan 14	16:16:46	13013	-27495	069	N	h-	-1.0368	0.9908	-0.0792	274.4	-	-	21N	173E
4276	214	-0223 Jun 11	22:11:42	13008	-27490	036	Ne	-t	-1.5362	0.0503	-0.9720	65.8	-	-	24S	80E
4277	214	-0223 Jul 11	08:50:31	13007	-27489	074	N	t-	1.1388	0.7672	-0.2305	232.3	-	-	22S	79W
4278	214	-0223 Dec 05	11:27:58	13002	-27484	041	N	-a	1.0982	0.8388	-0.1531	232.3	-	-	23N	119W
4279	214	-0222 Jun 01	02:07:33	12996	-27478	046	P	-t	-0.8199	1.3856	0.3219	317.9	133.4	-	22S	20E
4280	214	-0222 Nov 25	02:38:57	12990	-27472	051	T	-p	0.4214	2.0703	1.0987	307.5	191.6	45.3	21N	12E
4281	215	-0221 May 21	02:46:14	12984	-27466	056	T-	pp	-0.0557	2.7865	1.7256	370.7	234.3	105.5	19S	11E
4282	215	-0221 Nov 14	17:33:03	12978	-27460	061	T-	p-	-0.2488	2.4000	1.4029	326.2	208.5	84.2	17N	148E
4283	215	-0220 May 09	07:27:25	12972	-27454	066	P	h-	0.7097	1.5640	0.5475	311.8	160.4	-	16S	59W
4284	215	-0220 Nov 03	03:37:42	12966	-27448	071	P	h-	-0.9809	1.0843	0.0325	277.2	42.2	-	13N	3W
4285	215	-0219 Mar 30	11:29:13	12961	-27443	038	N	-a	-1.1359	0.7504	-0.2035	217.1	-	-	3S	117W
4286	215	-0219 Apr 28	19:18:20	12960	-27442	076	N	a-	1.4127	0.2482	-0.7172	133.7	-	-	12S	124E
4287	215	-0219 Sep 23	12:02:43	12955	-27437	043	N	-t	1.2943	0.5309	-0.5642	217.9	-	-	0S	129W
4288	215	-0218 Mar 20	04:17:50	12949	-27431	048	T	-a	-0.4251	2.0552	1.1002	307.5	193.0	46.0	1N	8W
4289	215	-0218 Sep 12	12:43:55	12943	-27425	053	P	-t	0.5687	1.8471	0.7823	340.5	189.2	-	5S	138W
4290	215	-0217 Mar 09	18:50:25	12937	-27419	058	T	p-	0.3054	2.2964	1.2988	330.5	209.2	76.8	6N	135E
4291	215	-0217 Sep 01	20:00:56	12931	-27413	063	T-	p-	-0.1829	2.5253	1.5194	337.4	216.7	93.4	10S	114E
4292	215	-0216 Feb 27	03:00:15	12925	-27407	068	N	t-	1.0849	0.8958	-0.1610	262.5	-	-	11N	13E
4293	215	-0216 Aug 21	09:59:23	12919	-27401	073	P	a-	-0.8856	1.2127	0.2529	263.2	106.5	-	15S	95W
4294	215	-0215 Jan 16	10:05:16	12914	-27396	040	N	-t	-1.1740	0.7524	-0.3442	252.9	-	-	21N	94W
4295	215	-0215 Jul 12	18:47:07	12909	-27390	045	P	-a	0.9665	1.0657	0.1028	254.2	70.2	-	22S	132E
4296	215	-0214 Jan 05	12:20:59	12903	-27384	050	T	-p	-0.4480	2.0643	1.0081	342.4	202.4	14.3	23N	129W
4297	215	-0214 Jul 02	07:09:55	12897	-27378	055	T+	-p	0.2326	2.4373	1.4251	343.8	218.8	89.5	23S	54W
4298	215	-0214 Dec 25	21:46:18	12891	-27372	060	T+	p-	0.2655	2.3713	1.3702	327.6	208.5	82.1	24N	88E
4299	215	-0213 Jun 21	12:27:52	12885	-27366	065	P	t-	-0.5546	1.8698	0.8111	345.9	194.5	-	24S	134W
4300	215	-0213 Dec 15	12:25:04	12879	-27360	070	P	a-	0.9196	1.1562	0.1846	257.4	91.2	-	24N	133W
4301	216	-0212 May 10	23:00:42	12874	-27355	037	N	-t	1.4965	0.1382	-0.9137	112.1	-	-	15S	67E
4302	216	-0212 Jun 09	13:09:57	12873	-27354	075	N	t-	-1.3148	0.4786	-0.5872	205.8	-	-	24S	146W
4303	216	-0212 Nov 04	15:44:57	12868	-27349	042	N	-a	-1.1451	0.7621	-0.2486	228.7	-	-	14N	175E
4304	216	-0211 Apr 30	05:38:18	12862	-27343	047	P	-t	0.6775	1.6156	0.6140	308.8	165.5	-	13S	32W
4305	216	-0211 Oct 24	23:51:25	12856	-27337	052	P	-a	-0.4879	1.9963	0.9297	344.1	199.1	-	11N	53E
4306	216	-0210 Apr 19	19:07:03	12850	-27331	057	T-	pp	-0.1008	2.6509	1.6948	321.9	212.8	97.8	10S	126E
4307	216	-0210 Oct 14	01:03:18	12844	-27325	062	T+	pp	0.2072	2.5280	1.4284	374.9	229.9	92.8	7N	35E
4308	216	-0209 Apr 09	12:00:38	12838	-27319	067	P	a-	-0.8234	1.3201	0.3735	268.8	126.5	-	7S	127W
4309	216	-0209 Oct 03	00:42:59	12832	-27313	072	P	t-	0.8938	1.2588	0.1776	305.6	100.2	-	3N	40E
4310	216	-0208 Feb 28	17:04:34	12827	-27308	039	N	-a	1.0801	0.8843	-0.1322	248.8	-	-	10N	162E
4311	216	-0208 Aug 22	18:34:35	12821	-27302	044	N	-a	-1.1720	0.7013	-0.2866	218.6	-	-	14S	136E
4312	216	-0208 Sep 21	06:02:33	12820	-27301	082	N	a-	1.5431	0.0391	-0.9860	57.0	-	-	1S	39W
4313	216	-0207 Feb 16	22:59:13	12816	-27296	049	T	-t	0.3816	2.1946	1.1214	358.0	214.4	54.8	14N	73E
4314	216	-0207 Aug 12	09:51:43	12810	-27290	054	T	-a	-0.4257	2.0525	1.1007	306.1	192.4	46.0	17S	94W
4315	216	-0206 Feb 05	23:01:45	12804	-27284	059	T	p-	-0.3297	2.3006	1.2062	369.4	221.9	70.3	17N	72E
4316	216	-0206 Aug 02	02:34:13	12798	-27278	064	T	p-	0.2943	2.2974	1.3382	316.5	204.5	78.7	19S	15E
4317	216	-0205 Jan 26	00:12:42	12792	-27272	069	N*	h-	-1.0154	1.0273	-0.0372	277.2	-	-	19N	54E
4318	216	-0205 Jul 22	16:03:28	12786	-27266	074	N	t-	1.0763	0.8848	-0.1186	247.7	-	-	20S	173E
4319	216	-0205 Dec 16	20:14:13	12781	-27261	041	N	-a	1.0976	0.8385	-0.1507	231.6	-	-	25N	110E
4320	216	-0204 Jun 11	08:29:19	12775	-27255	046	P	-t	-0.9078	1.2252	0.1598	304.9	96.4	-	24S	76W

Cat Num	Canon Plate	Calendar Date	TD of Greatest Eclipse	ΔT s	Luna Num	Saros Num	Ecl. Type	QSE	Gamma	Pen. Mag.	Um. Mag.	Eclipse Phase ---- Durations ---- Pen. m	Par. m	Total m	Greatest in Zenith Lat.	Lng.
4321	217	-0204 Dec 05	11:33:13	12769	-27249	051	T	-p	0.4221	2.0692	1.0974	307.6	191.6	45.0	23N	121W
4322	217	-0203 May 31	09:09:57	12763	-27243	056	T-	pp	-0.1435	2.6243	1.5656	368.4	231.9	100.7	22S	86W
4323	217	-0203 Nov 25	02:18:04	12758	-27237	061	T-	p-	-0.2489	2.4013	1.4011	327.3	208.9	84.2	20N	16E
4324	217	-0202 May 20	14:18:45	12752	-27231	066	P	h-	0.6278	1.7121	0.7001	318.9	176.4	-	19S	163W
4325	217	-0202 Nov 14	11:59:07	12746	-27225	071	P	h-	-0.9801	1.0878	0.0317	278.6	41.9	-	17N	129W
4326	217	-0201 Apr 10	19:08:52	12741	-27220	038	N	-a	-1.1929	0.6439	-0.3062	202.8	-	-	8S	126E
4327	217	-0201 May 10	02:37:30	12740	-27219	076	N	a-	1.3390	0.3816	-0.5799	163.2	-	-	15S	12E
4328	217	-0201 Oct 04	19:36:11	12735	-27214	043	N	-t	1.3207	0.4841	-0.6140	209.1	-	-	4N	116E
4329	217	-0200 Mar 30	12:06:45	12729	-27208	048	T	-a	-0.4790	1.9556	1.0019	304.4	187.8	6.6	3S	127W
4330	217	-0200 Sep 22	20:21:52	12723	-27202	053	P	-t	0.6002	1.7896	0.7242	336.7	183.4	-	1S	106E
4331	217	-0199 Mar 20	02:31:27	12717	-27196	058	T+	p-	0.2579	2.3837	1.3859	333.5	212.8	84.9	2N	18E
4332	217	-0199 Sep 12	03:55:36	12712	-27190	063	T-	p-	-0.1457	2.5935	1.5876	337.2	217.3	96.2	6S	7W
4333	217	-0198 Mar 09	10:22:54	12706	-27184	068	N	t-	1.0422	0.9740	-0.0823	272.2	-	-	7N	99W
4334	217	-0198 Sep 01	18:01:27	12700	-27178	073	P	a-	-0.8423	1.2928	0.3315	268.9	120.4	-	11S	143E
4335	217	-0197 Jan 27	17:47:49	12695	-27173	040	N	-t	-1.1942	0.7133	-0.3791	246.7	-	-	18N	150E
4336	217	-0197 Jul 24	02:16:48	12689	-27167	045	N	-a	1.0324	0.9471	-0.0204	243.5	-	-	20S	19E
4337	217	-0197 Aug 22	10:35:52	12688	-27166	083	Nb	a-	-1.5282	0.0326	-0.9254	48.5	-	-	15S	105W
4338	217	-0196 Jan 16	20:30:01	12683	-27161	050	P	-p	-0.4598	2.0398	0.9889	340.0	200.6	-	21N	109E
4339	217	-0196 Jul 12	14:10:33	12677	-27155	055	T	-a	0.3066	2.3042	1.2866	342.6	214.7	77.6	23S	160W
4340	217	-0195 Jan 05	06:23:26	12672	-27149	060	T+	p-	0.2585	2.3821	1.3852	326.6	208.5	83.1	23N	41W
4341	218	-0195 Jul 01	18:56:57	12666	-27143	065	P	t-	-0.4717	2.0237	0.9616	353.7	206.5	-	24S	128E
4342	218	-0195 Dec 25	21:15:34	12660	-27137	070	P	a-	0.9156	1.1628	0.1926	257.8	93.0	-	25N	95E
4343	218	-0194 Jun 20	19:32:15	12654	-27131	075	N	t-	-1.2281	0.6374	-0.4280	233.6	-	-	25S	118E
4344	218	-0194 Nov 16	00:22:07	12649	-27126	042	N	-a	-1.1518	0.7520	-0.2627	228.2	-	-	17N	45E
4345	218	-0193 May 11	12:40:18	12643	-27120	047	P	-t	0.7513	1.4778	0.4810	298.9	149.1	-	16S	139W
4346	218	-0193 Nov 05	08:01:50	12638	-27114	052	P	-a	-0.4977	1.9805	0.9094	344.8	198.1	-	14N	71W
4347	218	-0192 Apr 30	02:35:01	12632	-27108	057	T-	pp	-0.0313	2.7766	1.8241	321.5	213.4	99.8	14S	13E
4348	218	-0192 Oct 24	08:51:03	12626	-27102	062	T+	pp	0.1902	2.5606	1.4580	376.0	230.8	94.7	11N	83W
4349	218	-0191 Apr 19	19:38:06	12620	-27096	067	P	a-	-0.7603	1.4352	0.4899	276.7	142.5	-	11S	117E
4350	218	-0191 Oct 13	08:32:42	12614	-27090	072	P	t-	0.8711	1.3008	0.2190	308.6	110.4	-	7N	79W
4351	218	-0190 Mar 11	00:44:45	12609	-27085	039	N	-a	1.1284	0.7958	-0.2209	239.0	-	-	6N	45E
4352	218	-0190 Apr 09	11:10:50	12609	-27084	077	Nb	a-	-1.5269	0.0466	-0.9344	59.9	-	-	7S	115W
4353	218	-0190 Sep 03	02:28:11	12604	-27079	044	N	-a	-1.2096	0.6323	-0.3558	208.4	-	-	10S	16E
4354	218	-0190 Oct 02	14:08:32	12603	-27078	082	N	a-	1.5148	0.0909	-0.9340	86.2	-	-	3N	162W
4355	218	-0189 Feb 28	06:22:59	12598	-27073	049	T	-t	0.4246	2.1152	1.0432	356.0	210.4	33.6	10N	39W
4356	218	-0189 Aug 23	17:49:16	12592	-27067	054	T	-a	-0.4730	1.9669	1.0130	303.1	187.5	16.9	14S	146E
4357	218	-0188 Feb 17	06:26:15	12586	-27061	059	T	p-	-0.2938	2.3644	1.2740	370.4	224.8	79.0	13N	40W
4358	218	-0188 Aug 12	10:19:03	12581	-27055	064	T+	p-	0.2403	2.3987	1.4349	319.6	207.7	86.2	16S	102W
4359	218	-0187 Feb 05	08:03:58	12575	-27049	069	P	h-	-0.9893	1.0721	0.0136	280.7	27.9	-	16N	64W
4360	218	-0187 Aug 01	23:19:19	12569	-27043	074	N	t-	1.0167	0.9971	-0.0123	261.1	-	-	18S	63E
4361	219	-0187 Dec 27	04:59:55	12564	-27038	041	N	-a	1.0981	0.8359	-0.1500	230.7	-	-	25N	21W
4362	219	-0186 Jun 22	14:51:38	12558	-27032	046	N*	-t	-0.9948	1.0666	-0.0008	289.9	-	-	25S	171W
4363	219	-0186 Dec 16	20:26:04	12553	-27026	051	T	-p	0.4239	2.0658	1.0942	307.6	191.5	44.3	24N	106E
4364	219	-0185 Jun 11	15:35:58	12547	-27020	056	T-	pp	-0.2299	2.4647	1.4078	365.0	227.6	91.3	23S	177E
4365	219	-0185 Dec 06	11:02:24	12541	-27014	061	T-	p-	-0.2488	2.4027	1.4000	328.3	209.3	84.3	22N	115W
4366	219	-0184 May 30	21:12:42	12535	-27008	066	P	h-	0.5469	1.8585	0.8504	324.6	188.8	-	21S	92E
4367	219	-0184 Nov 24	20:20:37	12530	-27002	071	P	h-	-0.9799	1.0900	0.0303	279.9	41.0	-	19N	105E
4368	219	-0183 Apr 21	02:44:46	12525	-26997	038	N	-a	-1.2534	0.5313	-0.4156	185.9	-	-	12S	10E
4369	219	-0183 May 20	09:56:45	12524	-26996	076	N	a-	1.2643	0.5169	-0.4412	186.9	-	-	18S	99W
4370	219	-0183 Oct 15	03:17:24	12519	-26991	043	N	-t	1.3408	0.4484	-0.6522	202.0	-	-	8N	0W
4371	219	-0182 Apr 10	19:46:59	12513	-26985	048	P	-a	-0.5400	1.8434	0.8905	300.4	180.9	-	7S	116E
4372	219	-0182 Oct 04	04:10:42	12507	-26979	053	P	-t	0.6235	1.7470	0.6813	333.6	178.7	-	3N	13W
4373	219	-0181 Mar 31	10:01:36	12502	-26973	058	T+	pp	0.2019	2.4866	1.4882	336.5	216.4	92.0	2S	97W
4374	219	-0181 Sep 23	12:01:18	12496	-26967	063	T-	p-	-0.1168	2.6466	1.6407	336.7	217.5	97.7	2S	130W
4375	219	-0180 Mar 19	17:34:12	12490	-26961	068	P	t-	0.9908	1.0682	0.0122	283.0	26.7	-	3N	151E
4376	219	-0180 Sep 12	02:13:27	12484	-26955	073	P	a-	-0.8069	1.3587	0.3957	273.4	130.1	-	7S	18E
4377	219	-0179 Feb 07	01:22:56	12480	-26950	040	N	-t	-1.2204	0.6629	-0.4249	238.4	-	-	15N	36E
4378	219	-0179 Aug 03	09:50:23	12474	-26944	045	N	-a	1.0944	0.8361	-0.1366	232.4	-	-	18S	95W
4379	219	-0179 Sep 01	18:37:09	12473	-26943	083	N	a-	-1.4899	0.1051	-0.8574	86.6	-	-	11S	133E
4380	219	-0178 Jan 27	04:34:54	12468	-26938	050	P	-t	-0.4758	2.0075	0.9625	337.2	198.4	-	19N	13W

Cat Num	Canon Plate	Calendar Date	TD of Greatest Eclipse	ΔT s	Luna Num	Saros Num	Ecl. Type	QSE	Gamma	Pen. Mag.	Um. Mag.	Eclipse Phase Durations Pen. m	Par. m	Total m	Greatest in Zenith Lat.	Lng.
4381	220	-0178 Jul 23	21:13:32	12463	-26932	055	T	-a	0.3774	2.1772	1.1538	340.9	209.5	59.6	21S	94E
4382	220	-0177 Jan 16	14:56:27	12457	-26926	060	T+	p-	0.2475	2.4000	1.4077	325.8	208.8	84.7	22N	169W
4383	220	-0177 Jul 13	01:29:07	12451	-26920	065	T	t-	-0.3917	2.1723	1.1065	360.1	215.8	52.2	23S	30E
4384	220	-0176 Jan 06	06:03:22	12445	-26914	070	P	a-	0.9098	1.1727	0.2043	258.4	95.6	-	24N	37W
4385	220	-0176 Jul 01	01:58:44	12440	-26908	075	N	t-	-1.1437	0.7923	-0.2731	256.1	-	-	25S	22E
4386	220	-0176 Nov 26	09:00:50	12435	-26903	042	N	-a	-1.1565	0.7450	-0.2731	228.1	-	-	20N	85W
4387	220	-0175 May 21	19:43:03	12429	-26897	047	P	-t	0.8252	1.3400	0.3476	287.9	128.9	-	19S	114E
4388	220	-0175 Nov 15	16:15:09	12423	-26891	052	P	-a	-0.5045	1.9701	0.8949	345.7	197.5	-	18N	165E
4389	220	-0174 May 11	09:59:53	12418	-26885	057	T+	pp	0.0414	2.7565	1.8071	320.6	213.1	99.6	17S	100W
4390	220	-0174 Nov 04	16:44:45	12412	-26879	062	T+	pp	0.1782	2.5839	1.4789	376.7	231.4	96.0	15N	157E
4391	220	-0173 May 01	03:10:54	12406	-26873	067	P	a-	-0.6935	1.5575	0.6128	284.3	156.5	-	14S	2E
4392	220	-0173 Oct 24	16:29:53	12401	-26867	072	P	t-	0.8538	1.3325	0.2506	310.6	117.3	-	11N	161E
4393	220	-0172 Mar 21	08:15:20	12396	-26862	039	N	-a	1.1835	0.6949	-0.3222	226.4	-	-	2N	69W
4394	220	-0172 Apr 19	18:34:28	12395	-26861	077	N	a-	-1.4686	0.1541	-0.8280	108.3	-	-	11S	132E
4395	220	-0172 Sep 13	10:31:52	12390	-26856	044	N	-a	-1.2398	0.5771	-0.4113	199.7	-	-	6S	107W
4396	220	-0172 Oct 12	22:23:19	12389	-26855	082	N	a-	1.4930	0.1306	-0.8937	102.6	-	-	8N	72E
4397	220	-0171 Mar 10	13:36:30	12384	-26850	049	P	-t	0.4754	2.0213	0.9505	353.0	204.8	-	6N	149W
4398	220	-0171 Sep 03	01:54:23	12379	-26844	054	P	-a	-0.5141	1.8925	0.9362	300.3	182.6	-	10S	23E
4399	220	-0170 Feb 27	13:41:04	12373	-26838	059	T-	pp	-0.2496	2.4433	1.3573	371.4	227.8	87.5	9N	150W
4400	220	-0170 Aug 23	18:10:56	12367	-26832	064	T+	p-	0.1926	2.4889	1.5200	322.2	210.2	91.3	13S	139E
4401	221	-0169 Feb 16	15:45:33	12362	-26826	069	P	h-	-0.9542	1.1333	0.0811	285.4	67.1	-	13N	179E
4402	221	-0169 Aug 13	06:42:12	12356	-26820	074	P	t-	0.9634	1.0981	0.0823	272.4	65.9	-	15S	49W
4403	221	-0168 Jan 07	13:41:22	12351	-26815	041	N	-a	1.1030	0.8251	-0.1572	228.9	-	-	24N	151W
4404	221	-0168 Jul 02	21:16:27	12346	-26809	046	N	-t	-1.0790	0.9132	-0.1566	273.0	-	-	25S	92E
4405	221	-0168 Dec 27	05:16:35	12340	-26803	051	T	-p	0.4278	2.0583	1.0873	307.4	191.3	42.8	24N	26W
4406	221	-0167 Jun 21	22:04:14	12334	-26797	056	T	pp	-0.3154	2.3071	1.2517	360.6	221.5	76.0	24S	79E
4407	221	-0167 Dec 16	19:46:18	12329	-26791	061	T-	p-	-0.2491	2.4033	1.3986	329.3	209.8	84.4	23N	115E
4408	221	-0166 Jun 11	04:06:18	12323	-26785	066	T	a-	0.4645	2.0081	1.0034	329.3	198.8	9.1	22S	12W
4409	221	-0166 Dec 06	04:43:36	12317	-26779	071	P	t-	-0.9816	1.0884	0.0257	280.7	37.9	-	21N	21W
4410	221	-0165 May 02	10:15:39	12313	-26774	038	N	-a	-1.3184	0.4107	-0.5335	165.1	-	-	15S	104W
4411	221	-0165 May 31	17:15:02	12312	-26773	076	N	a-	1.1880	0.6557	-0.2998	207.0	-	-	20S	151E
4412	221	-0165 Oct 26	11:06:14	12307	-26768	043	N	-t	1.3549	0.4235	-0.6791	196.8	-	-	13N	119W
4413	221	-0164 Apr 21	03:22:03	12301	-26762	048	P	-a	-0.6049	1.7240	0.7715	295.5	172.2	-	11S	0W
4414	221	-0164 Oct 14	12:08:40	12296	-26756	053	P	-t	0.6404	1.7160	0.6504	331.0	175.1	-	8N	134W
4415	221	-0163 Apr 10	17:24:33	12290	-26750	058	T+	pp	0.1409	2.5990	1.5997	339.2	219.4	97.6	7S	150E
4416	221	-0163 Oct 03	20:14:56	12284	-26744	063	T-	p-	-0.0938	2.6888	1.6830	336.0	217.3	98.6	3N	105E
4417	221	-0162 Mar 31	00:36:58	12279	-26738	068	P	t-	0.9330	1.1741	0.1185	294.1	82.0	-	2S	44E
4418	221	-0162 Sep 23	10:33:11	12273	-26732	073	P	a-	-0.7780	1.4126	0.4479	276.8	137.1	-	2S	108W
4419	221	-0161 Feb 18	08:49:27	12268	-26727	040	N	-t	-1.2535	0.5996	-0.4831	227.5	-	-	12N	77W
4420	221	-0161 Aug 14	17:31:05	12263	-26721	045	N	-a	1.1495	0.7376	-0.2405	221.3	-	-	15S	149E
4421	222	-0161 Sep 13	02:46:53	12262	-26720	083	N	a-	-1.4590	0.1641	-0.8028	107.9	-	-	7S	9E
4422	222	-0160 Feb 07	12:32:51	12257	-26715	050	P	-t	-0.4980	1.9637	0.9248	333.9	195.3	-	16N	133W
4423	222	-0160 Aug 03	04:21:58	12252	-26709	055	T	-a	0.4426	2.0606	1.0311	338.7	203.4	27.9	18S	14W
4424	222	-0159 Jan 26	23:23:16	12246	-26703	060	T+	p-	0.2309	2.4279	1.4405	325.1	209.3	86.8	20N	64E
4425	222	-0159 Jul 23	08:06:45	12240	-26697	065	T	tp	-0.3167	2.3121	1.2423	365.2	222.7	75.2	21S	70W
4426	222	-0158 Jan 16	14:45:01	12235	-26691	070	P	a-	0.8988	1.1916	0.2254	259.8	100.1	-	23N	167W
4427	222	-0158 Jul 12	08:30:19	12229	-26685	075	N	t-	-1.0624	0.9416	-0.1241	274.6	-	-	24S	76W
4428	222	-0158 Dec 07	17:39:19	12224	-26680	042	N	-a	-1.1609	0.7383	-0.2826	228.0	-	-	21N	145E
4429	222	-0157 Jun 02	02:45:49	12219	-26674	047	P	-h	0.8998	1.2012	0.2124	275.6	102.4	-	21S	8E
4430	222	-0157 Nov 27	00:29:42	12213	-26668	052	P	-a	-0.5096	1.9624	0.8841	346.5	197.2	-	20N	41E
4431	222	-0156 May 21	17:23:21	12208	-26662	057	T+	pp	0.1157	2.6190	1.6719	319.0	211.6	97.0	19S	148E
4432	222	-0156 Nov 15	00:43:42	12202	-26656	062	T+	pp	0.1708	2.5983	1.4919	377.2	231.7	96.7	18N	37E
4433	222	-0155 May 11	10:38:15	12196	-26650	067	P	a-	-0.6223	1.6880	0.7435	291.6	169.0	-	17S	111W
4434	222	-0155 Nov 04	00:34:55	12191	-26644	072	P	h-	0.8422	1.3537	0.2722	311.6	121.5	-	15N	38E
4435	222	-0154 Apr 01	15:36:03	12186	-26639	039	N	-h	1.2457	0.5812	-0.4365	210.1	-	-	2S	178E
4436	222	-0154 May 01	01:50:19	12185	-26638	077	N	h-	-1.4048	0.2719	-0.7116	142.8	-	-	15S	21E
4437	222	-0154 Sep 24	18:44:55	12181	-26633	044	N	-a	-1.2635	0.5338	-0.4549	192.3	-	-	2S	129E
4438	222	-0154 Oct 24	06:46:34	12180	-26632	082	N	a-	1.4769	0.1597	-0.8638	112.8	-	-	12N	55W
4439	222	-0153 Mar 21	20:39:26	12175	-26627	049	P	-t	0.5344	1.9124	0.8429	348.9	197.1	-	2N	103E
4440	222	-0153 Sep 14	10:08:20	12169	-26621	054	P	-a	-0.5483	1.8311	0.8722	297.9	178.1	-	6S	102W

Cat Num	Canon Plate	Calendar Date	TD of Greatest Eclipse	ΔT s	Luna Num	Saros Num	Ecl. Type	QSE	Gamma	Pen. Mag.	Um. Mag.	Eclipse Phase Durations Pen. m	Par. m	Total m	Greatest in Zenith Lat.	Lng.
4441	223	-0152 Mar 09	20:48:37	12164	-26615	059	T-	pp	-0.1989	2.5342	1.4526	372.3	230.5	94.7	5N	102E
4442	223	-0152 Sep 03	02:08:53	12158	-26609	064	T+	p-	0.1503	2.5691	1.5951	324.3	212.0	94.6	9S	18E
4443	223	-0151 Feb 26	23:22:10	12153	-26603	069	P	h-	-0.9140	1.2039	0.1582	290.6	92.5	-	9N	64E
4444	223	-0151 Aug 23	14:10:24	12147	-26597	074	P	t-	0.9149	1.1904	0.1681	282.2	93.4	-	12S	162W
4445	223	-0150 Jan 17	22:18:32	12143	-26592	041	N	-a	1.1119	0.8069	-0.1715	226.2	-	-	22N	80E
4446	223	-0150 Jul 14	03:45:01	12137	-26586	046	N	-t	-1.1597	0.7666	-0.3059	254.2	-	-	24S	5W
4447	223	-0149 Jan 07	14:02:52	12131	-26580	051	T	-a	0.4354	2.0439	1.0739	307.1	190.7	39.6	23N	157W
4448	223	-0149 Jul 03	04:38:27	12126	-26574	056	T	-t	-0.3966	2.1577	1.1032	355.2	213.8	51.0	24S	19W
4449	223	-0149 Dec 28	04:25:19	12120	-26568	061	T-	p-	-0.2456	2.4104	1.4043	330.5	210.5	85.0	23N	15W
4450	223	-0148 Jun 21	11:05:52	12115	-26562	066	T	a-	0.3856	2.1513	1.1494	332.8	206.3	58.1	23S	117W
4451	223	-0148 Dec 16	13:03:46	12109	-26556	071	P	t-	-0.9813	1.0901	0.0251	281.8	37.6	-	23N	146W
4452	223	-0147 May 12	17:43:38	12105	-26551	038	N	-a	-1.3860	0.2855	-0.6566	139.1	-	-	19S	142E
4453	223	-0147 Jun 11	00:34:33	12104	-26550	076	N	a-	1.1117	0.7946	-0.1589	224.1	-	-	22S	40E
4454	223	-0147 Nov 05	19:00:56	12099	-26545	043	N	-t	1.3645	0.4066	-0.6973	193.0	-	-	16N	122E
4455	223	-0146 May 02	10:50:40	12093	-26539	048	P	-a	-0.6751	1.5953	0.6425	289.4	160.8	-	15S	114W
4456	223	-0146 Oct 25	20:15:14	12088	-26533	053	P	-t	0.6508	1.6966	0.6315	329.0	172.6	-	12N	103E
4457	223	-0145 Apr 22	00:38:01	12082	-26527	058	T+	pp	0.0728	2.7247	1.7242	341.5	221.7	101.4	11S	40E
4458	223	-0145 Oct 15	04:38:38	12077	-26521	063	T-	p-	-0.0782	2.7172	1.7118	335.2	217.0	98.9	7N	22W
4459	223	-0144 Apr 10	07:30:07	12071	-26515	068	P	t-	0.8678	1.2936	0.2380	305.4	114.5	-	6S	62W
4460	223	-0144 Oct 03	19:01:15	12066	-26509	073	P	a-	-0.7556	1.4545	0.4880	279.3	142.1	-	2N	123E
4461	224	-0143 Feb 28	16:08:07	12061	-26504	040	N	-t	-1.2932	0.5242	-0.5533	213.6	-	-	8N	172E
4462	224	-0143 Aug 25	01:17:57	12056	-26498	045	N	-a	1.1990	0.6497	-0.3342	210.4	-	-	11S	31E
4463	224	-0143 Sep 23	11:04:15	12055	-26497	083	N	a-	-1.4344	0.2115	-0.7600	122.2	-	-	3S	117W
4464	224	-0142 Feb 17	20:23:30	12050	-26492	050	P	-t	-0.5273	1.9067	0.8744	329.8	191.0	-	12N	109E
4465	224	-0142 Aug 14	11:35:38	12045	-26486	055	P	-a	0.5027	1.9536	0.9178	336.2	196.5	-	15S	123W
4466	224	-0141 Feb 07	07:44:19	12039	-26480	060	T+	p-	0.2087	2.4660	1.4837	324.5	210.0	89.4	17N	61W
4467	224	-0141 Aug 03	14:50:16	12033	-26474	065	T-	pp	-0.2464	2.4431	1.3691	369.3	227.8	88.5	19S	172W
4468	224	-0140 Jan 27	23:21:10	12028	-26468	070	P	a-	0.8834	1.2189	0.2549	261.9	106.0	-	20N	64E
4469	224	-0140 Jul 22	15:09:06	12022	-26462	075	P	t-	-0.9859	1.0823	0.0161	289.6	31.1	-	22S	177W
4470	224	-0140 Dec 18	02:16:49	12018	-26457	042	N	-a	-1.1656	0.7309	-0.2923	227.8	-	-	22N	16E
4471	224	-0139 Jun 12	09:50:18	12012	-26451	047	P	-h	0.9739	1.0636	0.0780	262.2	63.0	-	22S	99W
4472	224	-0139 Dec 07	08:44:35	12007	-26445	052	P	-h	-0.5138	1.9560	0.8751	347.3	196.9	-	22N	82W
4473	224	-0138 Jun 02	00:46:22	12001	-26439	057	T+	-p	0.1907	2.4804	1.5351	316.8	209.0	91.7	22S	37E
4474	224	-0138 Nov 26	08:44:12	11996	-26433	062	T+	pp	0.1644	2.6104	1.5033	377.5	231.9	97.3	21N	83W
4475	224	-0137 May 22	18:03:37	11990	-26427	067	P	a-	-0.5496	1.8219	0.8767	298.2	179.6	-	20S	137E
4476	224	-0137 Nov 15	08:44:45	11985	-26421	072	P	h-	0.8339	1.3684	0.2881	311.9	124.4	-	19N	85W
4477	224	-0136 Apr 11	22:48:55	11980	-26416	039	N	-h	1.3133	0.4575	-0.5611	189.3	-	-	6S	68E
4478	224	-0136 May 11	09:02:06	11979	-26415	077	N	h-	-1.3381	0.3954	-0.5903	170.9	-	-	18S	88W
4479	224	-0136 Oct 05	03:05:50	11975	-26410	044	N	-a	-1.2817	0.5005	-0.4885	186.4	-	-	2N	2E
4480	224	-0136 Nov 03	15:15:12	11974	-26409	082	N	a-	1.4645	0.1819	-0.8404	119.7	-	-	16N	177E
4481	225	-0135 Apr 01	03:33:22	11969	-26404	049	P	-t	0.6002	1.7911	0.7227	343.4	186.7	-	3S	2W
4482	225	-0135 Sep 24	18:30:13	11964	-26398	054	P	-a	-0.5762	1.7812	0.8197	295.8	174.2	-	1S	131E
4483	225	-0134 Mar 21	03:45:27	11958	-26392	059	T-	pp	-0.1389	2.6419	1.5649	372.9	232.8	100.7	1N	4W
4484	225	-0134 Sep 14	10:15:22	11953	-26386	064	T+	pp	0.1156	2.6355	1.6562	326.2	213.2	96.7	5S	105W
4485	225	-0133 Mar 10	06:49:26	11947	-26380	069	P	h-	-0.8652	1.2902	0.2510	296.6	114.6	-	5N	50W
4486	225	-0133 Sep 03	21:46:42	11942	-26374	074	P	t-	0.8738	1.2692	0.2402	290.2	110.8	-	8S	83E
4487	225	-0132 Jan 29	06:48:46	11937	-26369	041	N	-a	1.1270	0.7770	-0.1972	222.0	-	-	20N	48W
4488	225	-0132 Feb 27	16:56:58	11936	-26368	079	N	a-	-1.5274	0.0488	-0.9383	61.0	-	-	8N	159E
4489	225	-0132 Jul 24	10:19:56	11932	-26363	046	N	-t	-1.2348	0.6303	-0.4452	233.9	-	-	22S	104W
4490	225	-0131 Jan 17	22:44:32	11926	-26357	051	T	-a	0.4468	2.0223	1.0537	306.5	189.8	34.0	22N	72E
4491	225	-0131 Jul 13	11:16:53	11921	-26351	056	P	-t	-0.4751	2.0134	0.9595	349.0	204.4	-	23S	119W
4492	225	-0130 Jan 07	13:01:18	11916	-26345	061	T-	p-	-0.2402	2.4207	1.4136	331.7	211.3	85.9	23N	143W
4493	225	-0130 Jul 02	18:08:36	11910	-26339	066	T	a-	0.3081	2.2925	1.2928	335.3	211.9	77.3	23S	137E
4494	225	-0130 Dec 27	21:20:49	11905	-26333	071	P	t-	-0.9792	1.0947	0.0281	283.1	39.9	-	23N	91E
4495	225	-0129 May 24	01:09:34	11900	-26328	038	N	-a	-1.4555	0.1573	-0.7834	104.4	-	-	21S	30E
4496	225	-0129 Jun 22	07:56:30	11899	-26327	076	N	a-	1.0367	0.9316	-0.0205	238.6	-	-	23S	71W
4497	225	-0129 Nov 17	03:01:11	11895	-26322	043	N	-t	1.3700	0.3968	-0.7076	190.7	-	-	20N	1E
4498	225	-0128 May 12	18:16:06	11889	-26316	048	P	-a	-0.7474	1.4632	0.5095	282.3	146.5	-	18S	134E
4499	225	-0128 Nov 05	04:27:48	11884	-26310	053	P	-t	0.6572	1.6843	0.6203	327.3	170.9	-	16N	21W
4500	225	-0127 May 02	07:46:21	11878	-26304	058	T+	pp	0.0012	2.8569	1.8546	343.3	223.0	102.9	14S	68W

Cat Num	Canon Plate	Calendar Date	TD of Greatest Eclipse	ΔT s	Luna Num	Saros Num	Ecl. Type	QSE	Gamma	Pen. Mag.	Um. Mag.	Eclipse Phase Durations Pen. m	Par. m	Total m	Greatest in Zenith Lat.	Lng.
4501	226	-0127 Oct 25	13:09:00	11873	-26298	063	T-	p-	-0.0676	2.7363	1.7316	334.3	216.6	99.0	11N	151W
4502	226	-0126 Apr 21	14:14:55	11867	-26292	068	P	t-	0.7965	1.4246	0.3689	316.5	140.2	-	10S	165W
4503	226	-0126 Oct 15	03:36:39	11862	-26286	073	P	a-	-0.7394	1.4852	0.5171	281.1	145.5	-	7N	7W
4504	226	-0125 Mar 11	23:18:18	11857	-26281	040	N	-t	-1.3398	0.4360	-0.6362	195.8	-	-	4N	63E
4505	226	-0125 Apr 10	15:04:46	11857	-26280	078	Nb	t-	1.5505	0.0507	-1.0242	69.9	-	-	5S	176W
4506	226	-0125 Sep 05	09:12:19	11852	-26275	045	N	-a	1.2417	0.5742	-0.4155	200.1	-	-	7S	89W
4507	226	-0125 Oct 04	19:28:33	11851	-26274	083	N	a-	-1.4160	0.2476	-0.7283	132.2	-	-	2N	115E
4508	226	-0124 Feb 29	04:06:47	11847	-26269	050	P	-t	-0.5634	1.8372	0.8114	325.1	185.5	-	9N	9W
4509	226	-0124 Aug 24	18:56:49	11841	-26263	055	P	-a	0.5556	1.8598	0.8176	333.7	189.4	-	12S	125E
4510	226	-0123 Feb 17	15:57:45	11836	-26257	060	T+	p-	0.1799	2.5164	1.5393	324.0	210.9	92.2	13N	174E
4511	226	-0123 Aug 13	21:40:31	11830	-26251	065	T-	pp	-0.1821	2.5633	1.4850	372.3	231.3	96.8	16S	85E
4512	226	-0122 Feb 07	07:49:32	11825	-26245	070	P	a-	0.8619	1.2569	0.2955	264.7	113.5	-	17N	63W
4513	226	-0122 Aug 02	21:56:16	11820	-26239	075	P	t-	-0.9155	1.2119	0.1448	301.7	91.3	-	20S	81E
4514	226	-0122 Dec 29	10:49:20	11815	-26234	042	N	-a	-1.1735	0.7172	-0.3077	226.7	-	-	22N	112W
4515	226	-0121 Jun 23	16:58:25	11810	-26228	047	N	-h	1.0455	0.9309	-0.0522	247.8	-	-	23S	154E
4516	226	-0121 Dec 18	16:57:23	11804	-26222	052	P	-h	-0.5185	1.9481	0.8656	347.9	196.6	-	23N	155E
4517	226	-0120 Jun 12	08:10:00	11799	-26216	057	T+	-p	0.2655	2.3426	1.3984	314.0	205.2	83.3	23S	75W
4518	226	-0120 Dec 06	16:46:18	11793	-26210	062	T+	pp	0.1594	2.6195	1.5124	377.6	232.1	97.7	23N	156E
4519	226	-0119 Jun 02	01:26:17	11788	-26204	067	T	a-	-0.4751	1.9591	1.0126	304.2	188.7	16.8	22S	26E
4520	226	-0119 Nov 25	16:59:21	11783	-26198	072	P	h-	0.8286	1.3772	0.2987	311.7	126.2	-	21N	152E
4521	227	-0118 Apr 23	05:51:52	11778	-26193	039	N	-h	1.3877	0.3216	-0.6983	161.4	-	-	10S	39W
4522	227	-0118 May 22	16:07:23	11777	-26192	077	N	h-	-1.2668	0.5275	-0.4608	195.7	-	-	21S	165E
4523	227	-0118 Oct 16	11:35:43	11773	-26187	044	N	-a	-1.2935	0.4790	-0.5101	182.3	-	-	7N	127W
4524	227	-0118 Nov 14	23:50:14	11772	-26186	082	N	a-	1.4566	0.1955	-0.8250	123.6	-	-	19N	48E
4525	227	-0117 Apr 12	10:18:30	11767	-26181	049	P	-t	0.6723	1.6583	0.5908	336.4	172.9	-	7S	105W
4526	227	-0117 Oct 06	02:59:37	11762	-26175	054	P	-a	-0.5979	1.7428	0.7787	294.2	170.9	-	3N	2E
4527	227	-0116 Mar 31	10:36:25	11756	-26169	059	T-	pp	-0.0739	2.7590	1.6864	372.9	234.3	104.6	3S	109W
4528	227	-0116 Sep 24	18:28:42	11751	-26163	064	T+	pp	0.0873	2.6900	1.7054	327.9	214.2	98.0	0S	130E
4529	227	-0115 Mar 20	14:11:16	11746	-26157	069	P	h-	-0.8108	1.3866	0.3539	302.6	133.6	-	1N	162W
4530	227	-0115 Sep 14	05:28:55	11740	-26151	074	P	t-	0.8382	1.3379	0.3023	297.1	123.6	-	4S	35W
4531	227	-0114 Feb 08	15:13:23	11736	-26146	041	N	-a	1.1473	0.7376	-0.2322	216.6	-	-	17N	174W
4532	227	-0114 Mar 10	00:51:25	11735	-26145	079	N	a-	-1.4834	0.1265	-0.8547	97.3	-	-	4N	39E
4533	227	-0114 Aug 04	17:01:44	11731	-26140	046	N	-t	-1.3040	0.5048	-0.5739	212.1	-	-	20S	155E
4534	227	-0113 Jan 29	07:18:26	11725	-26134	051	T	-a	0.4647	1.9887	1.0215	305.6	188.2	21.8	19N	56W
4535	227	-0113 Jul 24	18:04:26	11720	-26128	056	P	-t	-0.5468	1.8816	0.8280	342.4	193.9	-	21S	139E
4536	227	-0112 Jan 18	21:29:54	11714	-26122	061	T-	p-	-0.2290	2.4417	1.4339	333.1	212.5	87.5	21N	90E
4537	227	-0112 Jul 13	01:18:25	11709	-26116	066	T+	p-	0.2353	2.4251	1.4271	336.8	215.6	88.6	22S	29E
4538	227	-0111 Jan 07	05:31:43	11704	-26110	071	P	t-	-0.9726	1.1074	0.0398	285.2	47.5	-	22N	32W
4539	227	-0111 Jun 03	08:35:04	11699	-26105	038	Ne	-a	-1.5255	0.0284	-0.9116	44.8	-	-	24S	82W
4540	227	-0111 Jul 02	15:22:15	11698	-26104	076	P	a-	0.9640	1.0646	0.1132	250.9	72.9	-	23S	178E
4541	228	-0111 Nov 27	11:03:39	11694	-26099	043	N	-t	1.3737	0.3898	-0.7142	189.0	-	-	22N	119W
4542	228	-0110 May 24	01:36:16	11688	-26093	048	P	-a	-0.8234	1.3244	0.3692	273.7	127.6	-	21S	23E
4543	228	-0110 Nov 16	12:46:28	11683	-26087	053	P	-t	0.6594	1.6795	0.6172	326.0	170.1	-	19N	146W
4544	228	-0109 May 13	14:47:43	11678	-26081	058	T-	pp	-0.0752	2.7223	1.7175	344.4	223.1	101.8	18S	175W
4545	228	-0109 Nov 05	21:45:46	11672	-26075	063	T-	p-	-0.0615	2.7469	1.7432	333.3	216.1	99.0	15N	79E
4546	228	-0108 May 01	20:52:28	11667	-26069	068	P	t-	0.7195	1.5661	0.5098	327.2	161.6	-	14S	94E
4547	228	-0108 Oct 25	12:19:09	11662	-26063	073	P	a-	-0.7287	1.5055	0.5360	282.2	147.6	-	11N	139W
4548	228	-0107 Mar 22	06:20:20	11657	-26058	040	N	-t	-1.3931	0.3354	-0.7313	172.8	-	-	1S	45W
4549	228	-0107 Apr 20	21:38:22	11656	-26057	078	N	t-	1.4755	0.1868	-0.8850	132.4	-	-	9S	84E
4550	228	-0107 Sep 15	17:13:18	11652	-26052	045	N	-a	1.2784	0.5101	-0.4856	190.5	-	-	3S	149E
4551	228	-0107 Oct 15	03:59:52	11651	-26051	083	N	a-	-1.4036	0.2725	-0.7077	138.7	-	-	6N	14W
4552	228	-0106 Mar 11	11:43:08	11647	-26046	050	P	-t	-0.6060	1.7556	0.7365	319.5	178.5	-	4N	124W
4553	228	-0106 Sep 05	02:24:51	11641	-26040	055	P	-h	0.6017	1.7785	0.7297	331.3	182.2	-	8S	12E
4554	228	-0105 Mar 01	00:03:51	11636	-26034	060	T+	p-	0.1444	2.5788	1.6069	323.6	211.8	95.0	9N	51E
4555	228	-0105 Aug 25	04:39:15	11631	-26028	065	T-	pp	-0.1251	2.6702	1.5874	374.5	233.6	101.6	12S	21W
4556	228	-0104 Feb 18	16:11:42	11625	-26022	070	P	a-	0.8356	1.3039	0.3450	268.1	121.9	-	14N	170E
4557	228	-0104 Aug 13	04:51:11	11620	-26016	075	P	t-	-0.8506	1.3315	0.2634	311.5	120.8	-	17S	24W
4558	228	-0103 Jan 08	19:17:49	11616	-26011	042	N	-a	-1.1840	0.6986	-0.3276	224.9	-	-	22N	122E
4559	228	-0103 Feb 07	07:27:02	11615	-26010	080	N	a-	1.5540	0.0019	-0.9889	12.0	-	-	18N	59W
4560	228	-0103 Jul 04	00:11:02	11610	-26005	047	N	-a	1.1140	0.8043	-0.1769	232.5	-	-	22S	45E

Cat Num	Canon Plate	Calendar Date	TD of Greatest Eclipse	ΔT s	Luna Num	Saros Num	Ecl. Type	QSE	Gamma	Pen. Mag.	Um. Mag.	Eclipse Phase Durations Pen. m	Par. m	Total m	Greatest in Zenith Lat.	Lng.
4561	229	-0103 Aug 02	10:16:17	11609	-26004	085	Nb	a-	-1.5379	0.0436	-0.9714	60.3	-	-	20S	105W
4562	229	-0103 Dec 29	01:06:13	11605	-25999	052	P	-t	-0.5262	1.9345	0.8509	348.1	195.9	-	23N	33E
4563	229	-0102 Jun 23	15:35:33	11600	-25993	057	T	-p	0.3391	2.2073	1.2635	310.5	200.3	70.9	23S	174E
4564	229	-0102 Dec 18	00:46:42	11594	-25987	062	T+	pp	0.1529	2.6309	1.5248	377.5	232.4	98.4	24N	37E
4565	229	-0101 Jun 13	08:49:48	11589	-25981	067	T	a-	-0.4015	2.0951	1.1468	309.4	196.2	55.2	24S	86W
4566	229	-0101 Dec 07	01:14:01	11584	-25975	072	P	h-	0.8226	1.3867	0.3109	311.4	128.2	-	23N	28E
4567	229	-0100 May 03	12:49:33	11579	-25970	039	N	-h	1.4651	0.1806	-0.8411	123.0	-	-	14S	145W
4568	229	-0100 Jun 01	23:11:44	11578	-25969	077	N	h-	-1.1954	0.6602	-0.3313	217.0	-	-	23S	58E
4569	229	-0100 Oct 26	20:12:08	11574	-25964	044	N	-a	-1.3006	0.4659	-0.5230	179.6	-	-	11N	103E
4570	229	-0100 Nov 25	08:28:03	11573	-25963	082	N	a-	1.4505	0.2056	-0.8126	126.2	-	-	22N	82W
4571	229	-0099 Apr 22	16:55:24	11569	-25958	049	P	-t	0.7505	1.5145	0.4477	327.5	154.2	-	11S	154E
4572	229	-0099 Oct 16	11:36:14	11563	-25952	054	P	-a	-0.6138	1.7149	0.7483	293.0	168.4	-	7N	129W
4573	229	-0098 Apr 11	17:18:59	11558	-25946	059	T-	pp	-0.0016	2.8895	1.8212	372.2	234.8	106.1	7S	149E
4574	229	-0098 Oct 06	02:50:04	11553	-25940	064	T+	pp	0.0660	2.7317	1.7419	329.4	214.9	98.8	4N	3E
4575	229	-0097 Mar 31	21:24:58	11547	-25934	069	P	h-	-0.7487	1.4975	0.4711	309.0	150.9	-	4S	88E
4576	229	-0097 Sep 25	13:20:18	11542	-25928	074	P	t-	0.8107	1.3915	0.3496	302.6	132.4	-	0N	154W
4577	229	-0096 Feb 19	23:30:26	11538	-25923	041	N	-a	1.1744	0.6855	-0.2799	209.4	-	-	13N	60E
4578	229	-0096 Mar 20	08:38:39	11537	-25922	079	N	a-	-1.4328	0.2165	-0.7591	125.7	-	-	0S	80W
4579	229	-0096 Aug 14	23:50:54	11532	-25917	046	N	-t	-1.3671	0.3908	-0.6912	188.8	-	-	17S	52E
4580	229	-0096 Sep 13	16:42:16	11532	-25916	084	Nb	t-	1.5681	0.0231	-1.0613	47.2	-	-	4S	156E
4581	230	-0095 Feb 08	15:45:57	11527	-25911	051	P	-a	0.4879	1.9452	0.9798	304.2	185.9	-	16N	176E
4582	230	-0095 Aug 04	00:59:20	11522	-25905	056	P	-t	-0.6134	1.7594	0.7057	335.4	182.4	-	19S	34E
4583	230	-0094 Jan 29	05:51:34	11517	-25899	061	T-	p-	-0.2129	2.4715	1.4633	334.6	213.9	89.6	19N	36W
4584	230	-0094 Jul 24	08:34:17	11511	-25893	066	T+	p-	0.1667	2.5505	1.5538	337.4	218.0	95.5	21S	80W
4585	230	-0093 Jan 18	13:36:32	11506	-25887	071	P	t-	-0.9618	1.1275	0.0595	288.0	58.1	-	20N	153W
4586	230	-0093 Jul 13	22:52:38	11501	-25881	076	P	a-	0.8947	1.1917	0.2404	261.3	104.1	-	22S	65E
4587	230	-0093 Dec 08	19:07:08	11496	-25876	043	N	-t	1.3768	0.3834	-0.7194	187.3	-	-	24N	120E
4588	230	-0092 Jun 03	08:55:57	11491	-25870	048	P	-a	-0.8993	1.1864	0.2290	263.9	102.8	-	23S	88W
4589	230	-0092 Nov 26	21:08:40	11486	-25864	053	P	-h	0.6596	1.6779	0.6180	324.8	169.7	-	22N	88E
4590	230	-0091 May 23	21:44:36	11481	-25858	058	T-	pp	-0.1544	2.5785	1.5709	344.6	221.8	97.5	20S	80E
4591	230	-0091 Nov 16	06:26:52	11475	-25852	063	T-	p-	-0.0585	2.7518	1.7495	332.4	215.7	98.9	18N	52W
4592	230	-0090 May 13	03:24:45	11470	-25846	068	P	t-	0.6385	1.7152	0.6580	337.1	179.6	-	17S	5W
4593	230	-0090 Nov 05	21:06:05	11465	-25840	073	P	a-	-0.7218	1.5186	0.5481	282.9	148.9	-	15N	88E
4594	230	-0089 Apr 02	13:15:17	11461	-25835	040	N	-t	-1.4522	0.2243	-0.8371	142.2	-	-	5S	150W
4595	230	-0089 May 02	04:07:43	11460	-25834	078	N	t-	1.3963	0.3307	-0.7383	173.7	-	-	13S	15W
4596	230	-0089 Sep 27	01:22:14	11455	-25829	045	N	-a	1.3079	0.4589	-0.5427	182.4	-	-	1N	25E
4597	230	-0089 Oct 26	12:36:51	11454	-25828	083	N	a-	-1.3961	0.2882	-0.6960	142.9	-	-	10N	144W
4598	230	-0088 Mar 21	19:13:00	11450	-25823	050	P	-t	-0.6547	1.6630	0.6503	313.0	169.6	-	0N	121E
4599	230	-0088 Sep 15	09:59:46	11445	-25817	055	P	-h	0.6412	1.7093	0.6542	329.1	175.2	-	4S	103W
4600	230	-0087 Mar 11	08:02:33	11440	-25811	060	T+	p-	0.1023	2.6535	1.6867	323.2	212.6	97.4	5N	70W
4601	231	-0087 Sep 04	11:46:44	11434	-25805	065	T-	pp	-0.0756	2.7632	1.6761	376.1	234.8	104.2	9S	129W
4602	231	-0086 Mar 01	00:23:32	11429	-25799	070	P	a-	0.8011	1.3660	0.4096	272.5	131.6	-	10N	46E
4603	231	-0086 Aug 24	11:57:02	11424	-25793	075	P	t-	-0.7938	1.4364	0.3671	319.2	140.2	-	13S	131W
4604	231	-0085 Jan 20	03:38:16	11419	-25788	042	N	-a	-1.2003	0.6691	-0.3579	221.5	-	-	20N	3W
4605	231	-0085 Feb 18	15:37:22	11419	-25787	080	N	a-	1.5309	0.0440	-0.9465	58.1	-	-	14N	178E
4606	231	-0085 Jul 15	07:29:49	11414	-25782	047	N	-a	1.1783	0.6857	-0.2941	216.5	-	-	21S	65W
4607	231	-0085 Aug 13	17:40:08	11413	-25781	085	N	a-	-1.4777	0.1538	-0.8607	111.7	-	-	17S	144E
4608	231	-0084 Jan 09	09:08:56	11409	-25776	052	P	-t	-0.5382	1.9127	0.8288	347.9	194.5	-	22N	87W
4609	231	-0084 Jul 03	23:04:37	11404	-25770	057	T	-a	0.4099	2.0775	1.1335	306.5	194.3	52.6	23S	61E
4610	231	-0084 Dec 28	08:45:39	11398	-25764	062	T+	pp	0.1452	2.6442	1.5398	377.3	232.6	99.1	24N	82W
4611	231	-0083 Jun 23	16:11:50	11393	-25758	067	T	p-	-0.3273	2.2327	1.2817	314.0	202.3	73.3	24S	164E
4612	231	-0083 Dec 17	09:30:45	11388	-25752	072	P	a-	0.8178	1.3940	0.3215	310.9	129.8	-	24N	95W
4613	231	-0082 May 14	19:40:04	11384	-25747	039	Ne	-t	1.5465	0.0323	-0.9917	52.9	-	-	17S	111E
4614	231	-0082 Jun 13	06:12:46	11383	-25746	077	N	t-	-1.1223	0.7963	-0.1990	236.1			24S	47W
4615	231	-0082 Nov 07	04:54:46	11379	-25741	044	N	-a	-1.3034	0.4605	-0.5278	178.2	-	-	14N	29W
4616	231	-0082 Dec 06	17:08:06	11378	-25740	082	N	a-	1.4457	0.2129	-0.8025	127.9	-	-	24N	149E
4617	231	-0081 May 03	23:26:13	11373	-25735	049	P	-t	0.8328	1.3633	0.2969	316.5	128.7	-	14S	55E
4618	231	-0081 Oct 27	20:19:10	11368	-25729	054	P	-a	-0.6242	1.6969	0.7280	292.3	166.6	-	12N	100E
4619	231	-0080 Apr 21	23:57:43	11363	-25723	059	T+	pp	0.0744	2.7538	1.6896	370.7	233.8	104.6	11S	47E
4620	231	-0080 Oct 16	11:16:30	11358	-25717	064	T+	pp	0.0497	2.7642	1.7694	330.8	215.4	99.2	8N	125W

Cat Num	Canon Plate	Calendar Date	TD of Greatest Eclipse	ΔT s	Luna Num	Saros Num	Ecl. Type	QSE	Gamma	Pen. Mag.	Um. Mag.	Eclipse Phase Durations Pen. m	Par. m	Total m	Greatest in Zenith Lat.	Lng.
4621	232	-0079 Apr 11	04:34:56	11352	-25711	069	P	a-	-0.6824	1.6161	0.5957	314.9	165.8	-	8S	22W
4622	232	-0079 Oct 05	21:17:30	11347	-25705	074	P	t-	0.7887	1.4351	0.3870	307.2	138.9	-	5N	85E
4623	232	-0078 Mar 02	07:39:19	11343	-25700	041	N	-a	1.2085	0.6207	-0.3403	200.1	-	-	9N	63W
4624	232	-0078 Mar 31	16:18:47	11342	-25699	079	N	a-	-1.3758	0.3184	-0.6518	150.5	-	-	4S	163E
4625	232	-0078 Aug 26	06:49:10	11338	-25694	046	N	-t	-1.4225	0.2907	-0.7945	164.4	-	-	13S	54W
4626	232	-0078 Sep 25	00:09:28	11337	-25693	084	N	t-	1.5356	0.0852	-1.0041	90.3	-	-	1N	43E
4627	232	-0077 Feb 20	00:04:33	11332	-25688	051	P	-a	0.5184	1.8886	0.9247	302.3	182.7	-	13N	51E
4628	232	-0077 Aug 15	08:04:21	11327	-25682	056	P	-t	-0.6722	1.6516	0.5978	328.4	170.4	-	16S	73W
4629	232	-0076 Feb 09	14:03:32	11322	-25676	061	T-	pp	-0.1894	2.5147	1.5064	336.3	215.6	92.3	15N	159W
4630	232	-0076 Aug 03	15:59:09	11317	-25670	066	T+	pp	0.1046	2.6640	1.6681	337.4	219.1	99.3	18S	168E
4631	232	-0075 Jan 28	21:32:19	11312	-25664	071	P	t-	-0.9442	1.1597	0.0918	291.8	72.0	-	18N	88E
4632	232	-0075 Jul 24	06:27:59	11307	-25658	076	P	a-	0.8291	1.3123	0.3605	270.0	125.1	-	20S	49W
4633	232	-0075 Dec 19	03:09:50	11302	-25653	043	N	-t	1.3808	0.3752	-0.7256	185.2	-	-	25N	0W
4634	232	-0074 Jun 14	16:13:31	11297	-25647	048	P	-a	-0.9763	1.0465	0.0862	252.5	64.5	-	24S	162E
4635	232	-0074 Jul 13	23:31:40	11296	-25646	086	Nb	a-	1.5358	0.0136	-0.9345	31.3	-	-	21S	54E
4636	232	-0074 Dec 08	05:32:02	11292	-25641	053	P	-h	0.6595	1.6765	0.6196	323.5	169.4	-	23N	37W
4637	232	-0073 Jun 04	04:38:01	11287	-25635	058	T-	-p	-0.2355	2.4314	1.4205	343.9	218.9	89.3	22S	24W
4638	232	-0073 Nov 27	15:11:43	11282	-25629	063	T-	p-	-0.0577	2.7522	1.7519	331.3	215.2	98.8	21N	177E
4639	232	-0072 May 23	09:52:19	11276	-25623	068	P	t-	0.5541	1.8707	0.8123	346.1	194.7	-	20S	103W
4640	232	-0072 Nov 16	05:57:20	11271	-25617	073	P	a-	-0.7185	1.5251	0.5538	283.2	149.4	-	18N	45W
4641	233	-0071 Apr 12	20:04:09	11267	-25612	040	Ne	-t	-1.5162	0.1041	-0.9521	97.6	-	-	9S	106E
4642	233	-0071 May 12	10:32:34	11266	-25611	078	N	t-	1.3125	0.4831	-0.5833	206.6	-	-	16S	113W
4643	233	-0071 Oct 07	09:37:53	11262	-25606	045	N	-a	1.3311	0.4193	-0.5881	175.7	-	-	6N	100W
4644	233	-0071 Nov 05	21:18:58	11261	-25605	083	N	a-	-1.3936	0.2947	-0.6933	144.7	-	-	14N	84E
4645	233	-0070 Apr 02	02:35:39	11257	-25600	050	P	-t	-0.7100	1.5584	0.5521	305.4	158.3	-	4S	9E
4646	233	-0070 Sep 26	17:42:43	11251	-25594	055	P	-h	0.6730	1.6540	0.5926	327.4	168.9	-	1N	139E
4647	233	-0069 Mar 22	15:54:42	11246	-25588	060	T+	pp	0.0542	2.7392	1.7773	322.6	213.1	99.1	1N	170E
4648	233	-0069 Sep 15	19:02:34	11241	-25582	065	T-	pp	-0.0332	2.8431	1.7519	377.2	235.4	105.3	4S	120E
4649	233	-0068 Mar 11	08:29:01	11236	-25576	070	P	a-	0.7615	1.4375	0.4833	277.2	141.5	-	5N	77W
4650	233	-0068 Sep 03	19:12:21	11231	-25570	075	P	t-	-0.7439	1.5284	0.4581	325.1	154.0	-	9S	119E
4651	233	-0067 Jan 30	11:52:04	11226	-25565	042	N	-a	-1.2211	0.6311	-0.3964	216.6	-	-	17N	127W
4652	233	-0067 Feb 28	23:39:55	11226	-25564	080	N	a-	1.5025	0.0960	-0.8942	85.7	-	-	10N	56E
4653	233	-0067 Jul 25	14:54:32	11221	-25559	047	N	-a	1.2382	0.5753	-0.4036	199.8	-	-	19S	176W
4654	233	-0067 Aug 24	01:12:34	11220	-25558	085	N	a-	-1.4235	0.2531	-0.7611	141.5	-	-	14S	29E
4655	233	-0066 Jan 19	17:05:12	11216	-25553	052	P	-t	-0.5546	1.8824	0.7988	347.2	192.4	-	20N	154E
4656	233	-0066 Jul 15	06:38:24	11211	-25547	057	T	-a	0.4767	1.9554	1.0105	302.2	187.5	15.3	22S	53W
4657	233	-0065 Jan 08	16:38:18	11206	-25541	062	T+	pp	0.1325	2.6662	1.5643	377.1	233.0	100.2	23N	160E
4658	233	-0065 Jul 04	23:37:55	11201	-25535	067	T-	p-	-0.2570	2.3632	1.4090	317.7	207.0	84.6	24S	52E
4659	233	-0065 Dec 28	17:44:40	11196	-25529	072	P	a-	0.8105	1.4053	0.3368	310.7	132.3	-	24N	142E
4660	233	-0064 Jun 23	13:14:32	11190	-25523	077	N	t-	-1.0505	0.9303	-0.0694	252.9	-	-	25S	153W
4661	234	-0064 Nov 17	13:40:41	11186	-25518	044	N	-a	-1.3040	0.4588	-0.5285	177.5	-	-	18N	161W
4662	234	-0064 Dec 17	01:47:22	11185	-25517	082	N	a-	1.4400	0.2216	-0.7904	129.9	-	-	25N	20E
4663	234	-0063 May 14	05:51:46	11181	-25512	049	P	-t	0.9184	1.2063	0.1399	303.4	90.5	-	17S	43W
4664	234	-0063 Nov 07	05:07:05	11176	-25506	054	P	-a	-0.6307	1.6858	0.7151	292.0	165.5	-	15N	33W
4665	234	-0062 May 03	06:29:39	11171	-25500	059	T+	pp	0.1564	2.6014	1.5409	368.1	231.3	99.5	15S	52W
4666	234	-0062 Oct 27	19:49:52	11166	-25494	064	T+	pp	0.0396	2.7849	1.7855	332.1	215.9	99.4	12N	106E
4667	234	-0061 Apr 22	11:39:00	11160	-25488	069	P	a-	-0.6101	1.7458	0.7311	320.5	179.1	-	12S	129W
4668	234	-0061 Oct 17	05:21:30	11155	-25482	074	P	t-	0.7730	1.4668	0.4129	310.8	143.3	-	9N	37W
4669	234	-0060 Mar 12	15:40:39	11151	-25477	041	N	-a	1.2492	0.5440	-0.4128	188.4	-	-	5N	175E
4670	234	-0060 Apr 10	23:53:20	11150	-25476	079	N	a-	-1.3136	0.4301	-0.5353	172.4	-	-	8S	48E
4671	234	-0060 Sep 05	13:56:45	11146	-25471	046	N	-t	-1.4701	0.2050	-0.8835	139.2	-	-	9S	162W
4672	234	-0060 Oct 05	07:45:02	11145	-25470	084	N	t-	1.5099	0.1346	-0.9592	113.2	-	-	5N	73W
4673	234	-0059 Mar 02	08:15:01	11141	-25465	051	P	-a	0.5553	1.8199	0.8577	299.7	178.3	-	9N	73W
4674	234	-0059 Aug 25	15:18:07	11136	-25459	056	P	-t	-0.7246	1.5557	0.5015	321.6	158.1	-	13S	177E
4675	234	-0058 Feb 19	22:07:06	11131	-25453	061	T-	pp	-0.1597	2.5691	1.5608	338.0	217.4	95.2	12N	79E
4676	234	-0058 Aug 14	23:32:16	11126	-25447	066	T+	pp	0.0484	2.7669	1.7714	336.9	219.3	101.0	15S	54E
4677	234	-0057 Feb 09	05:19:04	11120	-25441	071	P	t-	-0.9199	1.2040	0.1365	296.6	87.5	-	15N	29W
4678	234	-0057 Aug 04	14:10:32	11115	-25435	076	P	a-	0.7690	1.4230	0.4702	277.3	140.4	-	18S	166W
4679	234	-0057 Dec 30	11:10:48	11111	-25430	043	N	-t	1.3866	0.3631	-0.7350	182.0	-	-	25N	120W
4680	234	-0056 Jun 24	23:31:52	11106	-25424	048	N	-a	-1.0518	0.9096	-0.0541	239.6	-	-	25S	53E

Cat Num	Canon Plate	Calendar Date	TD of Greatest Eclipse	ΔT s	Luna Num	Saros Num	Ecl. Type	QSE	Gamma	Pen. Mag.	Um. Mag.	Eclipse Phase Durations Pen. m	Par. m	Total m	Greatest in Zenith Lat.	Lng.
4681	235	-0056 Jul 24	07:06:57	11105	-25423	086	N	a-	1.4760	0.1250	-0.8264	94.0	-	-	19S	60W
4682	235	-0056 Dec 18	13:55:43	11101	-25418	053	P	-h	0.6600	1.6737	0.6205	322.1	169.1	-	24N	162W
4683	235	-0055 Jun 14	11:30:00	11096	-25412	058	T	-a	-0.3168	2.2842	1.2693	342.3	214.3	75.8	24S	128W
4684	235	-0055 Dec 07	23:57:04	11091	-25406	063	T-	p-	-0.0571	2.7521	1.7543	330.3	214.8	98.7	23N	46E
4685	235	-0054 Jun 03	16:17:18	11086	-25400	068	P	t-	0.4680	2.0296	0.9694	353.9	207.1	-	22S	160E
4686	235	-0054 Nov 27	14:50:14	11080	-25394	073	P	a-	-0.7166	1.5288	0.5572	283.4	149.8	-	20N	178W
4687	235	-0053 May 23	16:56:42	11075	-25388	078	N	t-	1.2273	0.6384	-0.4260	233.6	-	-	19S	150E
4688	235	-0053 Oct 18	17:58:59	11071	-25383	045	N	-a	1.3493	0.3886	-0.6242	170.3	-	-	10N	133E
4689	235	-0053 Nov 17	06:02:57	11070	-25382	083	N	a-	-1.3931	0.2972	-0.6939	145.7	-	-	17N	47W
4690	235	-0052 Apr 12	09:53:46	11066	-25377	050	P	-t	-0.7698	1.4456	0.4452	296.6	144.0	-	8S	102W
4691	235	-0052 Oct 07	01:32:43	11061	-25371	055	P	-h	0.6983	1.6105	0.5433	326.1	163.4	-	5N	20E
4692	235	-0051 Apr 01	23:38:59	11056	-25365	060	T-	pp	-0.0009	2.8348	1.8774	321.8	213.2	99.7	4S	52E
4693	235	-0051 Sep 26	02:27:51	11051	-25359	065	T+	pp	0.0014	2.9033	1.8083	377.9	235.5	105.5	0N	7E
4694	235	-0050 Mar 22	16:24:44	11046	-25353	070	P	a-	0.7140	1.5237	0.5715	282.6	151.9	-	1N	162E
4695	235	-0050 Sep 15	02:39:25	11041	-25347	075	P	t-	-0.7029	1.6041	0.5328	329.5	163.8	-	5S	5E
4696	235	-0049 Feb 10	19:55:00	11036	-25342	042	N	-a	-1.2500	0.5783	-0.4494	209.1	-	-	14N	111E
4697	235	-0049 Mar 12	07:31:05	11035	-25341	080	N	a-	1.4658	0.1633	-0.8268	111.4	-	-	6N	63W
4698	235	-0049 Aug 05	22:28:04	11031	-25336	047	N	-a	1.2916	0.4771	-0.5013	183.2	-	-	17S	70E
4699	235	-0049 Sep 04	08:56:23	11030	-25335	085	N	a-	-1.3774	0.3375	-0.6765	161.6	-	-	10S	88W
4700	235	-0048 Jan 31	00:53:19	11026	-25330	052	P	-t	-0.5769	1.8411	0.7583	345.8	189.2	-	18N	37E
4701	236	-0048 Jul 25	14:16:35	11021	-25324	057	P	-a	0.5399	1.8403	0.8939	297.6	179.9	-	20S	168W
4702	236	-0047 Jan 19	00:27:01	11016	-25318	062	T+	pp	0.1164	2.6941	1.5954	376.8	233.4	101.4	21N	43E
4703	236	-0047 Jul 15	07:05:20	11011	-25312	067	T-	p-	-0.1887	2.4905	1.5325	320.9	210.5	92.1	23S	60W
4704	236	-0046 Jan 08	01:56:02	11006	-25306	072	P	a-	0.8001	1.4220	0.3583	310.7	135.6	-	23N	20E
4705	236	-0046 Jul 04	20:15:49	11001	-25300	077	P	t-	-0.9789	1.0641	0.0595	267.9	56.1	-	24S	102E
4706	236	-0046 Nov 28	22:30:03	10997	-25295	044	N	-a	-1.3026	0.4607	-0.5252	177.5	-	-	20N	67E
4707	236	-0046 Dec 28	10:25:45	10996	-25294	082	N	a-	1.4331	0.2322	-0.7757	132.4	-	-	25N	109W
4708	236	-0045 May 25	12:14:35	10992	-25289	049	N*	-t	1.0057	1.0463	-0.0204	287.7	-	-	20S	139W
4709	236	-0045 Nov 18	13:57:39	10987	-25283	054	P	-a	-0.6351	1.6785	0.7063	291.9	164.8	-	18N	166W
4710	236	-0044 May 13	13:00:48	10981	-25277	059	T+	pp	0.2398	2.4468	1.3898	364.5	226.9	89.9	18S	151W
4711	236	-0044 Nov 07	04:26:31	10976	-25271	064	T+	pp	0.0329	2.7994	1.7957	333.4	216.4	99.6	16N	24W
4712	236	-0043 May 02	18:40:06	10971	-25265	069	P	a-	-0.5343	1.8823	0.8730	325.4	190.3	-	15S	124E
4713	236	-0043 Oct 27	13:30:22	10966	-25259	074	P	t-	0.7619	1.4898	0.4306	313.7	146.4	-	13N	161W
4714	236	-0042 Mar 23	23:34:46	10962	-25254	041	N	-a	1.2959	0.4562	-0.4967	173.6	-	-	1N	55E
4715	236	-0042 Apr 22	07:23:08	10961	-25253	079	N	a-	-1.2470	0.5500	-0.4108	192.1	-	-	12S	66W
4716	236	-0042 Sep 16	21:14:06	10957	-25248	046	N	-t	-1.5100	0.1333	-0.9582	113.0	-	-	5S	87E
4717	236	-0042 Oct 16	15:27:47	10956	-25247	084	N	t-	1.4899	0.1734	-0.9243	128.3	-	-	9N	170E
4718	236	-0041 Mar 13	16:16:04	10952	-25242	051	P	-a	0.5999	1.7374	0.7765	296.4	172.3	-	5N	165E
4719	236	-0041 Sep 05	22:43:31	10947	-25236	056	P	-t	-0.7683	1.4758	0.4212	315.3	146.3	-	9S	65E
4720	236	-0040 Mar 02	06:00:35	10942	-25230	061	T-	pp	-0.1223	2.6378	1.6294	339.9	219.2	98.1	8N	41W
4721	237	-0040 Aug 25	07:14:21	10937	-25224	066	T-	pp	-0.0012	2.8533	1.8580	335.9	218.9	101.2	12S	63W
4722	237	-0039 Feb 19	12:55:29	10932	-25218	071	P	t-	-0.8878	1.2626	0.1958	302.5	104.1	-	11N	144W
4723	237	-0039 Aug 14	22:00:16	10927	-25212	076	P	a-	0.7145	1.5238	0.5696	283.2	151.9	-	15S	76E
4724	237	-0038 Jan 09	19:06:08	10923	-25207	043	N	-t	1.3970	0.3423	-0.7523	176.6	-	-	24N	122E
4725	237	-0038 Jul 06	06:51:17	10918	-25201	048	N	-a	-1.1257	0.7762	-0.1916	225.3	-	-	24S	57W
4726	237	-0038 Aug 04	14:47:55	10917	-25200	086	N	a-	1.4211	0.2278	-0.7276	126.0	-	-	17S	176W
4727	237	-0038 Dec 29	22:17:13	10913	-25195	053	P	-h	0.6631	1.6658	0.6170	320.3	168.2	-	24N	73E
4728	237	-0037 Jun 25	18:21:42	10908	-25189	058	T	-h	-0.3972	2.1388	1.1195	339.7	208.1	53.2	24S	129E
4729	237	-0037 Dec 19	08:41:57	10903	-25183	063	T-	p-	-0.0554	2.7536	1.7588	329.3	214.4	98.6	24N	85W
4730	237	-0036 Jun 13	22:41:22	10897	-25177	068	T	t-	0.3813	2.1897	1.1274	360.5	217.1	56.7	23S	64E
4731	237	-0036 Dec 07	23:44:48	10892	-25171	073	P	a-	-0.7161	1.5295	0.5581	283.4	149.9	-	22N	48E
4732	237	-0035 Jun 02	23:18:12	10887	-25165	078	N	t-	1.1391	0.7994	-0.2632	256.8	-	-	21S	54E
4733	237	-0035 Oct 29	02:26:14	10883	-25160	045	N	-a	1.3618	0.3681	-0.6498	166.7	-	-	14N	5E
4734	237	-0035 Nov 27	14:50:06	10882	-25159	083	N	a-	-1.3958	0.2936	-0.7002	145.3	-	-	20N	179W
4735	237	-0034 Apr 23	17:06:58	10878	-25154	050	P	-h	-0.8346	1.3239	0.3290	286.5	125.6	-	12S	148E
4736	237	-0034 Oct 18	09:29:26	10873	-25148	055	P	-h	0.7173	1.5784	0.5058	325.4	159.0	-	10N	100W
4737	237	-0033 Apr 13	07:18:16	10868	-25142	060	T-	pp	-0.0607	2.7230	1.7696	320.6	212.6	98.9	8S	64W
4738	237	-0033 Oct 07	10:01:57	10863	-25136	065	T+	pp	0.0288	2.8548	1.7563	378.3	235.4	105.1	5N	108W
4739	237	-0032 Apr 02	00:14:45	10858	-25130	070	P	a-	0.6622	1.6180	0.6674	288.1	161.9	-	3S	43E
4740	237	-0032 Sep 25	10:15:03	10853	-25124	075	P	t-	-0.6682	1.6682	0.5960	332.7	171.1	-	1S	110W

Cat Num	Canon Plate	Calendar Date	TD of Greatest Eclipse	ΔT s	Luna Num	Saros Num	Ecl. Type	QSE	Gamma	Pen. Mag.	Um. Mag.	Eclipse Phase Durations Pen. m	Par. m	Total m	Greatest in Zenith Lat.	Lng.
4741	238	-0031 Feb 21	03:49:48	10849	-25119	042	N	-a	-1.2843	0.5153	-0.5125	199.2	-	-	10N	8W
4742	238	-0031 Mar 22	15:14:30	10848	-25118	080	N	a-	1.4238	0.2404	-0.7498	134.6	-	-	2N	179E
4743	238	-0031 Aug 16	06:09:23	10844	-25113	047	N	-a	1.3391	0.3899	-0.5885	166.5	-	-	14S	47W
4744	238	-0031 Sep 14	16:49:26	10843	-25112	085	N	a-	-1.3382	0.4094	-0.6045	176.2	-	-	6S	152E
4745	238	-0030 Feb 10	08:31:13	10839	-25107	052	P	-t	-0.6068	1.7857	0.7041	343.5	184.3	-	15N	78W
4746	238	-0030 Aug 05	22:01:46	10834	-25101	057	P	-a	0.5972	1.7361	0.7878	293.0	171.9	-	17S	75E
4747	238	-0029 Jan 30	08:07:01	10829	-25095	062	T+	pp	0.0933	2.7346	1.6396	376.4	233.9	102.9	19N	72W
4748	238	-0029 Jul 26	14:38:10	10824	-25089	067	T-	pp	-0.1254	2.6089	1.6465	323.3	212.9	96.7	21S	174W
4749	238	-0028 Jan 19	10:01:13	10819	-25083	072	P	a-	0.7841	1.4487	0.3903	311.3	140.5	-	22N	101W
4750	238	-0028 Jul 15	03:20:42	10814	-25077	077	P	t-	-0.9111	1.1912	0.1812	280.9	96.6	-	23S	5W
4751	238	-0028 Dec 09	07:19:38	10810	-25072	044	N	-a	-1.3014	0.4619	-0.5221	177.3	-	-	22N	65W
4752	238	-0027 Jan 07	19:00:02	10809	-25071	082	N	a-	1.4227	0.2492	-0.7543	136.4	-	-	24N	123E
4753	238	-0027 Jun 04	18:33:47	10805	-25066	049	N	-t	1.0949	0.8829	-0.1846	269.1	-	-	21S	125E
4754	238	-0027 Nov 28	22:50:38	10800	-25060	054	P	-a	-0.6374	1.6749	0.7017	292.0	164.5	-	21N	61E
4755	238	-0026 May 24	19:28:54	10795	-25054	059	T	pt	0.3257	2.2876	1.2336	359.7	220.6	73.6	20S	111E
4756	238	-0026 Nov 18	13:06:38	10790	-25048	064	T+	pp	0.0299	2.8067	1.7994	334.7	216.8	99.7	19N	155W
4757	238	-0025 May 14	01:38:36	10785	-25042	069	T	a-	-0.4554	2.0245	1.0200	329.6	199.7	22.2	18S	18E
4758	238	-0025 Nov 07	21:44:44	10780	-25036	074	P	t-	0.7565	1.5020	0.4382	315.8	147.9	-	17N	75E
4759	238	-0024 Apr 03	07:22:00	10776	-25031	041	N	-a	1.3486	0.3578	-0.5915	155.0	-	-	3S	64W
4760	238	-0024 May 02	14:48:42	10775	-25030	079	N	a-	-1.1763	0.6778	-0.2791	209.9	-	-	16S	179W
4761	239	-0024 Sep 27	04:40:15	10771	-25025	046	N	-t	-1.5428	0.0746	-1.0197	85.0	-	-	0S	26W
4762	239	-0024 Oct 26	23:17:15	10770	-25024	084	N	t-	1.4753	0.2019	-0.8992	138.4	-	-	13N	52E
4763	239	-0023 Mar 24	00:09:25	10766	-25019	051	P	-a	0.6505	1.6440	0.6843	292.1	164.5	-	0N	45E
4764	239	-0023 Sep 16	06:18:50	10761	-25013	056	P	-t	-0.8046	1.4093	0.3543	309.6	135.2	-	4S	51W
4765	239	-0022 Mar 13	13:43:41	10756	-25007	061	T-	pp	-0.0772	2.7207	1.7121	341.6	220.9	100.6	4N	158W
4766	239	-0022 Sep 05	15:06:11	10751	-25001	066	T-	pp	-0.0435	2.7758	1.7804	334.8	218.0	100.5	8S	178E
4767	239	-0021 Mar 02	20:22:17	10746	-24995	071	P	t-	-0.8485	1.3343	0.2683	309.2	120.9	-	7N	103E
4768	239	-0021 Aug 26	05:57:19	10741	-24989	076	P	a-	0.6660	1.6137	0.6577	288.1	160.8	-	11S	45W
4769	239	-0020 Jan 21	02:56:51	10737	-24984	043	N	-t	1.4112	0.3140	-0.7764	169.2	-	-	22N	5E
4770	239	-0020 Feb 19	20:49:27	10736	-24983	081	Nb	t-	-1.5749	0.0141	-1.0771	37.1	-	-	11N	96E
4771	239	-0020 Jul 16	14:14:06	10732	-24978	048	N	-a	-1.1958	0.6497	-0.3226	209.5	-	-	23S	168W
4772	239	-0020 Aug 14	22:34:46	10731	-24977	086	N	a-	1.3713	0.3213	-0.6384	148.7	-	-	14S	66E
4773	239	-0019 Jan 09	06:35:52	10727	-24972	053	P	-h	0.6692	1.6521	0.6083	318.2	166.8	-	23N	51W
4774	239	-0019 Jul 06	01:13:43	10722	-24966	058	P	-h	-0.4762	1.9964	0.9721	336.2	200.0	-	24S	26E
4775	239	-0019 Dec 29	17:24:40	10717	-24960	063	T-	p-	-0.0514	2.7592	1.7679	328.2	214.1	98.7	23N	145E
4776	239	-0018 Jun 25	05:06:46	10712	-24954	068	T	pp	0.2960	2.3475	1.2826	365.8	224.6	80.2	23S	33W
4777	239	-0018 Dec 19	08:36:48	10707	-24948	073	P	a-	-0.7137	1.5337	0.5629	283.6	150.4	-	23N	84W
4778	239	-0017 Jun 14	05:43:03	10702	-24942	078	N	t-	1.0524	0.9577	-0.1036	276.0	-	-	22S	42W
4779	239	-0017 Nov 09	10:57:12	10698	-24937	045	N	-a	1.3711	0.3534	-0.6690	164.2	-	-	18N	123W
4780	239	-0017 Dec 08	23:36:28	10697	-24936	083	N	a-	-1.3982	0.2904	-0.7057	144.9	-	-	21N	49E
4781	240	-0016 May 04	00:17:18	10693	-24931	050	P	-h	-0.9024	1.1969	0.2071	275.1	101.1	-	16S	39E
4782	240	-0016 Oct 28	17:31:20	10688	-24925	055	P	-h	0.7313	1.5551	0.4777	325.1	155.6	-	14N	138E
4783	240	-0015 Apr 23	14:51:41	10683	-24919	060	T-	pp	-0.1259	2.6016	1.6518	318.9	211.1	96.3	12S	179W
4784	240	-0015 Oct 17	17:44:37	10678	-24913	065	T+	pp	0.0489	2.8194	1.7181	378.6	235.1	104.5	9N	135E
4785	240	-0014 Apr 13	07:55:02	10673	-24907	070	P	a-	0.6026	1.7268	0.7773	293.9	171.8	-	7S	74W
4786	240	-0014 Oct 06	18:02:13	10668	-24901	075	P	t-	-0.6423	1.7160	0.6433	334.7	175.9	-	4N	132E
4787	240	-0013 Mar 04	11:33:13	10664	-24896	042	N	-a	-1.3268	0.4373	-0.5906	185.5	-	-	6N	126W
4788	240	-0013 Apr 02	22:47:39	10663	-24895	080	N	a-	1.3745	0.3312	-0.6595	157.2	-	-	3S	64E
4789	240	-0013 Aug 27	13:59:32	10659	-24890	047	N	-a	1.3801	0.3147	-0.6639	150.2	-	-	10S	166W
4790	240	-0013 Sep 26	00:52:16	10658	-24889	085	N	a-	-1.3061	0.4683	-0.5455	186.8	-	-	1S	30E
4791	240	-0012 Feb 21	15:59:42	10654	-24884	052	P	-t	-0.6436	1.7174	0.6372	340.1	177.7	-	11N	168E
4792	240	-0012 Aug 16	05:53:14	10649	-24878	057	P	-a	0.6492	1.6418	0.6911	288.4	163.5	-	14S	43W
4793	240	-0011 Feb 09	15:40:17	10645	-24872	062	T+	pp	0.0643	2.7856	1.6950	375.8	234.4	104.2	15N	174E
4794	240	-0011 Aug 05	22:14:21	10640	-24866	067	T-	pp	-0.0656	2.7210	1.7539	325.3	214.4	99.3	18S	72E
4795	240	-0010 Jan 29	18:02:01	10635	-24860	072	P	a-	0.7637	1.4832	0.4306	312.4	146.2	-	19N	138E
4796	240	-0010 Jul 26	10:28:25	10630	-24854	077	P	t-	-0.8465	1.3127	0.2969	292.3	122.0	-	21S	112W
4797	240	-0010 Dec 20	16:08:29	10626	-24849	044	N	-a	-1.3012	0.4610	-0.5205	176.7	-	-	22N	163E
4798	240	-0009 Jan 19	03:29:54	10625	-24848	082	N	a-	1.4082	0.2732	-0.7254	142.0	-	-	22N	4W
4799	240	-0009 Jun 16	00:54:24	10621	-24843	049	N	-t	1.1822	0.7233	-0.3452	247.7	-	-	22S	30E
4800	240	-0009 Dec 10	07:44:05	10616	-24837	054	P	-a	-0.6388	1.6726	0.6987	292.1	164.3	-	22N	73W

Cat Num	Canon Plate	Calendar Date	TD of Greatest Eclipse	ΔT s	Luna Num	Saros Num	Ecl. Type	QSE	Gamma	Pen. Mag.	Um. Mag.	Eclipse Phase ---- Durations ---- Pen. m	Par. m	Total m	Greatest in Zenith Lat.	Lng.
4801	241	-0008 Jun 04	01:58:18	10611	-24831	059	T	-t	0.4111	2.1294	1.0782	353.7	212.2	44.8	22S	13E
4802	241	-0008 Nov 28	21:47:33	10606	-24825	064	T+	pp	0.0282	2.8115	1.8009	335.9	217.3	99.9	21N	75E
4803	241	-0007 May 24	08:37:10	10601	-24819	069	T	a-	-0.3757	2.1687	1.1685	332.8	207.1	61.3	21S	87W
4804	241	-0007 Nov 18	06:00:55	10596	-24813	074	P	t-	0.7529	1.5106	0.4428	317.6	148.9	-	20N	49W
4805	241	-0006 Apr 14	15:02:35	10592	-24808	041	N	-a	1.4069	0.2493	-0.6970	130.5	-	-	7S	179E
4806	241	-0006 May 13	22:11:34	10591	-24807	079	N	a-	-1.1026	0.8114	-0.1422	225.9	-	-	19S	69E
4807	241	-0006 Oct 08	12:16:05	10587	-24802	046	N	-t	-1.5681	0.0294	-1.0672	53.6	-	-	4N	142W
4808	241	-0006 Nov 07	07:11:56	10586	-24801	084	N	t-	1.4648	0.2225	-0.8812	145.2	-	-	17N	68W
4809	241	-0005 Apr 04	07:54:39	10582	-24796	051	P	-a	0.7076	1.5390	0.5799	286.8	154.3	-	4S	73W
4810	241	-0005 Sep 27	14:04:02	10577	-24790	056	P	-t	-0.8337	1.3561	0.3009	304.7	125.2	-	0S	168W
4811	241	-0004 Mar 23	21:17:09	10572	-24784	061	T-	pp	-0.0248	2.8170	1.8080	343.2	222.2	102.2	0S	87E
4812	241	-0004 Sep 15	23:07:42	10567	-24778	066	T-	pp	-0.0784	2.7117	1.7163	333.5	216.9	99.2	4S	56E
4813	241	-0003 Mar 13	03:37:06	10562	-24772	071	P	t-	-0.8000	1.4228	0.3577	316.8	138.1	-	3N	8W
4814	241	-0003 Sep 05	14:03:00	10558	-24766	076	P	a-	0.6245	1.6909	0.7329	291.9	167.5	-	7S	167W
4815	241	-0002 Jan 31	10:39:52	10554	-24761	043	N	-t	1.4320	0.2736	-0.8121	158.1	-	-	19N	111W
4816	241	-0002 Mar 02	04:02:22	10553	-24760	081	N	t-	-1.5311	0.0925	-0.9948	94.4	-	-	7N	14W
4817	241	-0002 Jul 27	21:41:06	10549	-24755	048	N	-a	-1.2618	0.5313	-0.4462	192.5	-	-	21S	80E
4818	241	-0002 Aug 26	06:29:14	10548	-24754	086	N	a-	1.3280	0.4031	-0.5612	165.7	-	-	10S	54W
4819	241	-0001 Jan 20	14:48:33	10544	-24749	053	P	-h	0.6811	1.6275	0.5893	315.4	164.2	-	21N	174W
4820	241	-0001 Jul 17	08:09:12	10539	-24743	058	P	-h	-0.5514	1.8612	0.8315	332.0	190.3	-	23S	78W
4821	242	0000 Jan 10	02:04:40	10534	-24737	063	T-	p-	-0.0445	2.7699	1.7825	327.2	213.8	98.8	22N	16E
4822	242	0000 Jul 05	11:33:06	10529	-24731	068	T+	pp	0.2117	2.5037	1.4360	369.9	230.1	93.9	23S	129W
4823	242	0000 Dec 29	17:27:44	10524	-24725	073	P	a-	-0.7104	1.5392	0.5695	283.9	151.2	-	23N	143E
4824	242	0001 Jun 24	12:08:47	10519	-24719	078	P	t-	0.9653	1.1170	0.0567	292.5	57.9	-	23S	139W
4825	242	0001 Nov 19	19:31:30	10515	-24714	045	N	-a	1.3768	0.3449	-0.6815	163.0	-	-	21N	108E
4826	242	0001 Dec 19	08:21:56	10514	-24713	083	N	a-	-1.4007	0.2866	-0.7112	144.4	-	-	22N	81W
4827	242	0002 May 15	07:24:42	10510	-24708	050	P	-h	-0.9732	1.0647	0.0795	262.0	63.5	-	19S	69W
4828	242	0002 Nov 09	01:38:13	10505	-24702	055	P	-h	0.7405	1.5403	0.4587	325.2	153.3	-	17N	16E
4829	242	0003 May 04	22:22:05	10501	-24696	060	T-	-p	-0.1939	2.4752	1.5284	316.7	208.7	91.3	15S	67E
4830	242	0003 Oct 29	01:32:40	10496	-24690	065	T+	pp	0.0645	2.7919	1.6883	378.7	234.8	103.9	13N	17E
4831	242	0004 Apr 23	15:31:17	10491	-24684	070	P	a-	0.5399	1.8414	0.8926	299.4	180.8	-	11S	170E
4832	242	0004 Oct 17	01:57:00	10486	-24678	075	P	h-	-0.6219	1.7535	0.6806	335.9	179.4	-	8N	12E
4833	242	0005 Mar 14	19:06:49	10482	-24673	042	N	-h	-1.3763	0.3466	-0.6814	167.1	-	-	2N	119E
4834	242	0005 Apr 13	06:12:33	10481	-24672	080	N	h-	1.3192	0.4330	-0.5585	178.6	-	-	7S	49W
4835	242	0005 Sep 06	21:58:22	10477	-24667	047	N	-a	1.4149	0.2513	-0.7279	134.7	-	-	6S	73E
4836	242	0005 Oct 06	09:04:02	10476	-24666	085	N	a-	-1.2805	0.5151	-0.4985	194.5	-	-	3N	94W
4837	242	0006 Mar 03	23:17:23	10472	-24661	052	P	-t	-0.6887	1.6337	0.5552	335.6	168.4	-	7N	58E
4838	242	0006 Aug 27	13:53:00	10467	-24655	057	P	-a	0.6944	1.5603	0.6069	284.1	155.3	-	10S	165W
4839	242	0007 Feb 20	23:03:27	10462	-24649	062	T+	pp	0.0268	2.8521	1.7661	375.1	234.7	105.2	12N	62E
4840	242	0007 Aug 17	05:58:18	10457	-24643	067	T-	pp	-0.0129	2.8203	1.8479	326.9	215.2	100.2	15S	45W
4841	243	0008 Feb 10	01:55:27	10453	-24637	072	P	a-	0.7366	1.5298	0.4834	314.1	153.2	-	16N	19E
4842	243	0008 Aug 05	17:40:45	10448	-24631	077	P	t-	-0.7865	1.4259	0.4039	302.2	140.5	-	19S	139E
4843	243	0008 Dec 31	00:54:48	10444	-24626	044	N	-a	-1.3035	0.4553	-0.5234	175.3	-	-	22N	32E
4844	243	0009 Jan 29	11:53:50	10443	-24625	082	N	a-	1.3887	0.3065	-0.6869	149.4	-	-	20N	130W
4845	243	0009 Jun 26	07:15:25	10439	-24620	049	N	-t	1.2681	0.5664	-0.5037	222.8	-	-	23S	66W
4846	243	0009 Dec 20	16:35:45	10434	-24614	054	P	-a	-0.6419	1.6670	0.6930	292.1	164.0	-	23N	155E
4847	243	0010 Jun 15	08:28:25	10429	-24608	059	P	-t	0.4965	1.9717	0.9226	346.6	201.7	-	23S	85W
4848	243	0010 Dec 10	06:28:52	10424	-24602	064	T+	pp	0.0276	2.8139	1.8008	337.1	217.8	100.0	23N	55W
4849	243	0011 Jun 04	15:36:12	10419	-24596	069	T	a-	-0.2951	2.3147	1.3183	335.2	212.7	79.8	22S	168E
4850	243	0011 Nov 29	14:18:40	10414	-24590	074	P	t-	0.7514	1.5151	0.4439	319.0	149.5	-	22N	173W
4851	243	0012 Apr 24	22:38:24	10410	-24585	041	N	-a	1.4693	0.1335	-0.8101	96.4	-	-	11S	64E
4852	243	0012 May 24	05:33:04	10410	-24584	079	N	a-	-1.0272	0.9484	-0.0025	240.1	-	-	21S	42W
4853	243	0012 Nov 17	15:11:20	10405	-24578	084	N	t-	1.4580	0.2359	-0.8697	149.5	-	-	20N	172E
4854	243	0013 Apr 14	15:31:46	10401	-24573	051	P	-a	0.7709	1.4226	0.4638	280.2	140.8	-	0S	171E
4855	243	0013 Oct 07	21:59:19	10396	-24567	056	P	-t	-0.8553	1.3164	0.2612	300.7	117.0	-	4N	71E
4856	243	0014 Apr 04	04:41:23	10391	-24561	061	T+	pp	0.0341	2.8002	1.7906	344.4	222.9	102.4	5S	26W
4857	243	0014 Sep 27	07:18:39	10386	-24555	066	T-	pp	-0.1062	2.6608	1.6652	332.2	215.7	97.7	1N	68W
4858	243	0015 Mar 24	10:43:00	10381	-24549	071	P	t-	-0.7449	1.5236	0.4593	324.7	154.4	-	1S	116W
4859	243	0015 Sep 16	22:17:02	10376	-24543	076	P	a-	0.5899	1.7555	0.7953	294.9	172.5	-	3S	68E
4860	243	0016 Feb 11	18:17:06	10372	-24538	043	N	-t	1.4576	0.2239	-0.8567	143.2	-	-	16N	134E

Cat Num	Canon Plate	Calendar Date	TD of Greatest Eclipse	ΔT s	Luna Num	Saros Num	Ecl. Type	QSE	Gamma	Pen. Mag.	Um. Mag.	Eclipse Phase Durations Pen. m	Par. m	Total m	Greatest in Zenith Lat.	Lng.
4861	244	0016 Mar 12	11:08:18	10372	-24537	081	N	t-	-1.4812	0.1821	-0.9012	131.2	-	-	3N	122W
4862	244	0016 Aug 07	05:12:17	10367	-24532	048	N	-a	-1.3235	0.4207	-0.5622	173.8	-	-	19S	34W
4863	244	0016 Sep 05	14:29:41	10367	-24531	086	N	a-	1.2900	0.4753	-0.4939	179.3	-	-	6S	175W
4864	244	0017 Jan 30	22:56:51	10363	-24526	053	P	-h	0.6974	1.5946	0.5622	312.0	160.6	-	19N	64E
4865	244	0017 Jul 27	15:08:05	10358	-24520	058	P	-h	-0.6228	1.7330	0.6976	327.2	178.9	-	21S	177E
4866	244	0018 Jan 20	10:38:27	10353	-24514	063	T-	pp	-0.0319	2.7909	1.8078	326.2	213.6	99.0	21N	113W
4867	244	0018 Jul 16	18:04:17	10348	-24508	068	T+	pp	0.1315	2.6524	1.5814	372.7	233.6	101.7	22S	132E
4868	244	0019 Jan 10	02:13:15	10343	-24502	073	P	a-	-0.7026	1.5528	0.5844	284.7	152.8	-	22N	12E
4869	244	0019 Jul 05	18:40:32	10338	-24496	078	P	t-	0.8822	1.2693	0.2096	306.0	108.8	-	23S	123E
4870	244	0019 Dec 01	04:05:20	10334	-24491	045	N	-a	1.3824	0.3363	-0.6935	161.7	-	-	23N	20W
4871	244	0019 Dec 30	17:02:45	10334	-24490	083	N	a-	-1.4001	0.2883	-0.7107	145.2	-	-	22N	149E
4872	244	0020 May 25	14:32:20	10330	-24485	050	N	-h	-1.0447	0.9315	-0.0497	247.5	-	-	22S	177W
4873	244	0020 Nov 19	09:47:34	10325	-24479	055	P	-h	0.7471	1.5300	0.4448	325.5	151.6	-	20N	107W
4874	244	0021 May 15	05:47:44	10320	-24473	060	T-	-p	-0.2664	2.3411	1.3967	313.8	205.1	83.2	19S	46W
4875	244	0021 Nov 08	09:27:47	10315	-24467	065	T+	pp	0.0745	2.7744	1.6692	378.8	234.6	103.5	16N	102W
4876	244	0022 May 04	23:00:32	10310	-24461	070	T	a-	0.4715	1.9669	1.0181	304.8	189.1	20.1	15S	56E
4877	244	0022 Oct 28	10:00:48	10305	-24455	075	P	h-	-0.6081	1.7787	0.7063	336.4	181.5	-	12N	110W
4878	244	0023 Mar 26	02:29:29	10301	-24450	042	N	-h	-1.4336	0.2416	-0.7867	141.4	-	-	2S	7E
4879	244	0023 Apr 24	13:29:16	10301	-24449	080	N	h-	1.2579	0.5461	-0.4467	199.2	-	-	11S	160W
4880	244	0023 Sep 18	06:07:14	10297	-24444	047	N	-a	1.4424	0.2011	-0.7787	120.8	-	-	2S	51W
4881	245	0023 Oct 17	17:25:13	10296	-24443	085	N	a-	-1.2617	0.5494	-0.4636	199.6	-	-	7N	139E
4882	245	0024 Mar 14	06:25:48	10292	-24438	052	P	-t	-0.7407	1.5375	0.4608	329.6	156.0	-	3N	51W
4883	245	0024 Sep 06	21:59:30	10287	-24432	057	P	-a	0.7339	1.4893	0.5329	280.1	147.2	-	6S	72E
4884	245	0025 Mar 03	06:19:48	10282	-24426	062	T-	pp	-0.0166	2.8684	1.7873	374.0	234.5	105.5	8N	49W
4885	245	0025 Aug 27	13:47:51	10277	-24420	067	T+	pp	0.0341	2.7841	1.8063	328.2	215.4	100.0	11S	164W
4886	245	0026 Feb 20	09:41:53	10273	-24414	072	P	a-	0.7030	1.5882	0.5482	316.3	161.0	-	12N	98W
4887	245	0026 Aug 17	00:58:51	10268	-24408	077	P	t-	-0.7322	1.5288	0.5005	310.7	154.6	-	15S	29E
4888	245	0027 Jan 11	09:37:50	10264	-24403	044	N	-a	-1.3087	0.4442	-0.5313	173.0	-	-	21N	99W
4889	245	0027 Feb 09	20:12:05	10263	-24402	082	N	a-	1.3644	0.3484	-0.6397	158.1	-	-	17N	104E
4890	245	0027 Jul 07	13:40:37	10259	-24397	049	N	-t	1.3501	0.4167	-0.6551	194.0	-	-	22S	162W
4891	245	0027 Aug 06	05:04:30	10258	-24396	087	N	t-	-1.5179	0.1084	-0.9625	101.0	-	-	19S	32W
4892	245	0028 Jan 01	01:24:22	10254	-24391	054	P	-a	-0.6472	1.6572	0.6833	291.9	163.2	-	23N	23E
4893	245	0028 Jun 25	15:03:17	10249	-24385	059	P	-t	0.5784	1.8204	0.7731	338.5	189.1	-	23S	176E
4894	245	0028 Dec 20	15:07:37	10244	-24379	064	T+	pp	0.0255	2.8189	1.8036	338.2	218.4	100.2	24N	176E
4895	245	0029 Jun 14	22:37:10	10240	-24373	069	T-	p-	-0.2151	2.4598	1.4666	336.6	216.6	91.2	24S	62E
4896	245	0029 Dec 09	22:35:32	10235	-24367	074	P	t-	0.7499	1.5192	0.4454	320.3	150.1	-	24N	63E
4897	245	0030 May 06	06:10:02	10231	-24362	041	Ne	-a	1.5348	0.0122	-0.9295	29.4	-	-	14S	51W
4898	245	0030 Jun 04	12:55:00	10230	-24361	079	P	a-	-0.9517	1.0859	0.1371	252.5	79.9	-	23S	153W
4899	245	0030 Nov 28	23:11:31	10225	-24355	084	N	t-	1.4521	0.2472	-0.8594	153.0	-	-	23N	52E
4900	245	0031 Apr 25	23:02:48	10221	-24350	051	P	-a	0.8386	1.2986	0.3395	272.2	123.0	-	12S	56E
4901	246	0031 Oct 19	06:03:52	10216	-24344	056	P	-t	-0.8698	1.2896	0.2349	297.6	111.1	-	9N	51W
4902	246	0032 Apr 14	11:56:36	10212	-24338	061	T+	pp	0.0997	2.6803	1.6697	345.1	222.7	100.6	9S	137W
4903	246	0032 Oct 07	15:38:15	10207	-24332	066	T-	pp	-0.1274	2.6220	1.6263	331.0	214.6	96.3	5N	165E
4904	246	0033 Apr 03	17:37:53	10202	-24326	071	P	t-	-0.6813	1.6399	0.5764	332.8	170.1	-	5S	139E
4905	246	0033 Sep 27	06:39:46	10197	-24320	076	P	a-	0.5624	1.8071	0.8446	297.2	176.2	-	1N	60W
4906	246	0034 Feb 22	01:44:01	10193	-24315	043	N	-t	1.4918	0.1584	-0.9166	120.7	-	-	12N	21E
4907	246	0034 Mar 23	18:02:46	10192	-24314	081	N	t-	-1.4215	0.2895	-0.7897	163.6	-	-	1S	133E
4908	246	0034 Aug 18	12:50:05	10188	-24309	048	N	-a	-1.3791	0.3217	-0.6671	154.1	-	-	15S	149W
4909	246	0034 Sep 16	22:38:50	10188	-24308	086	N	a-	1.2595	0.5338	-0.4403	189.5	-	-	2S	61E
4910	246	0035 Feb 11	06:57:48	10184	-24303	053	P	-h	0.7207	1.5488	0.5227	307.7	155.4	-	15N	57W
4911	246	0035 Aug 07	22:11:37	10179	-24297	058	P	-h	-0.6894	1.6139	0.5725	321.9	165.9	-	18S	71E
4912	246	0036 Jan 31	19:07:10	10174	-24291	063	T-	pp	-0.0146	2.8203	1.8418	325.2	213.4	99.2	18N	120E
4913	246	0036 Jul 27	00:39:47	10169	-24285	068	T+	pp	0.0551	2.7944	1.7200	374.6	235.4	105.5	20S	33E
4914	246	0037 Jan 20	10:54:28	10164	-24279	073	P	a-	-0.6917	1.5720	0.6053	285.8	155.1	-	20N	118W
4915	246	0037 Jul 16	01:16:00	10160	-24273	078	P	t-	0.8011	1.4179	0.3584	317.4	138.8	-	21S	24E
4916	246	0037 Dec 11	12:39:56	10156	-24268	045	N	-a	1.3868	0.3298	-0.7029	160.8	-	-	25N	148W
4917	246	0038 Jan 10	01:39:57	10155	-24267	083	N	a-	-1.3974	0.2938	-0.7061	146.9	-	-	21N	20E
4918	246	0038 Jun 05	21:40:23	10151	-24262	050	N	-h	-1.1165	0.7980	-0.1798	231.4	-	-	24S	76E
4919	246	0038 Jul 05	07:09:37	10150	-24261	088	Nb	h-	1.5077	0.0952	-0.9125	88.4	-	-	22S	65W
4920	246	0038 Nov 30	17:57:58	10146	-24256	055	P	-h	0.7523	1.5219	0.4340	325.9	150.3	-	22N	131E

Cat Num	Canon Plate	Calendar Date	TD of Greatest Eclipse	ΔT s	Luna Num	Saros Num	Ecl. Type	QSE	Gamma	Pen. Mag.	Um. Mag.	Eclipse Phase Durations Pen. m	Par. m	Total m	Greatest in Zenith Lat.	Lng.
4921	247	0039 May 26	13:13:25	10141	-24250	060	T	-p	-0.3391	2.2067	1.2641	310.3	200.2	71.0	21S	158W
4922	247	0039 Nov 19	17:26:37	10137	-24244	065	T+	pp	0.0817	2.7615	1.6555	378.6	234.3	103.1	19N	138E
4923	247	0040 May 15	06:26:43	10132	-24238	070	T	a-	0.4009	2.0967	1.1473	309.7	196.3	55.3	18S	56W
4924	247	0040 Nov 07	18:10:06	10127	-24232	075	P	a-	-0.5980	1.7967	0.7251	336.4	182.8	-	16N	127E
4925	247	0041 Apr 05	09:43:36	10123	-24227	042	N	-h	-1.4966	0.1263	-0.9026	103.6	-	-	7S	103W
4926	247	0041 May 04	20:40:03	10122	-24226	080	N	h-	1.1924	0.6673	-0.3273	218.4	-	-	14S	91E
4927	247	0041 Sep 28	14:24:35	10118	-24221	047	N	-a	1.4635	0.1626	-0.8178	108.8	-	-	3N	176W
4928	247	0041 Oct 28	01:53:07	10117	-24220	085	N	a-	-1.2479	0.5743	-0.4380	202.9	-	-	12N	11E
4929	247	0042 Mar 25	13:23:08	10113	-24215	052	P	-t	-0.8009	1.4262	0.3512	321.9	138.7	-	2S	157W
4930	247	0042 Sep 18	06:15:02	10109	-24209	057	P	-a	0.7659	1.4322	0.4727	276.8	139.9	-	2S	53W
4931	247	0043 Mar 14	13:26:49	10104	-24203	062	T-	pp	-0.0678	2.7719	1.6958	372.6	233.8	104.4	3N	157W
4932	247	0043 Sep 07	21:44:44	10099	-24197	067	T+	pp	0.0744	2.7130	1.7296	329.3	215.3	99.0	7S	75E
4933	247	0044 Mar 02	17:20:47	10094	-24191	072	P	a-	0.6624	1.6594	0.6260	319.0	169.4	-	8N	146E
4934	247	0044 Aug 27	08:23:54	10090	-24185	077	P	t-	-0.6843	1.6199	0.5850	317.9	165.4	-	12S	84W
4935	247	0045 Jan 21	18:14:53	10086	-24180	044	N	-a	-1.3193	0.4229	-0.5490	168.8	-	-	19N	132E
4936	247	0045 Feb 20	04:21:53	10085	-24179	082	N	a-	1.3326	0.4040	-0.5786	168.7	-	-	13N	19W
4937	247	0045 Jul 17	20:09:29	10081	-24174	049	N	-t	1.4287	0.2737	-0.8003	159.3	-	-	21S	100E
4938	247	0045 Aug 16	11:57:07	10080	-24173	087	N	t-	-1.4587	0.2194	-0.8563	142.6	-	-	16S	136W
4939	247	0046 Jan 11	10:08:27	10076	-24168	054	P	-a	-0.6563	1.6402	0.6670	291.3	161.9	-	22N	107W
4940	247	0046 Jul 06	21:43:01	10071	-24162	059	P	-t	0.6571	1.6755	0.6295	329.7	174.3	-	23S	76E
4941	248	0046 Dec 31	23:42:42	10067	-24156	064	T+	pp	0.0210	2.8279	1.8111	339.4	219.0	100.5	23N	48E
4942	248	0047 Jun 26	05:42:20	10062	-24150	069	T-	pp	-0.1377	2.6006	1.6100	337.1	218.9	97.8	24S	44W
4943	248	0047 Dec 21	06:51:24	10057	-24144	074	P	t-	0.7484	1.5229	0.4473	321.5	150.7	-	24N	61W
4944	248	0048 Jun 14	20:16:49	10052	-24138	079	P	a-	-0.8754	1.2251	0.2777	263.6	111.3	-	24S	96E
4945	248	0048 Dec 09	07:13:58	10048	-24132	084	N	t-	1.4478	0.2553	-0.8516	155.4	-	-	24N	68W
4946	248	0049 May 06	06:27:57	10044	-24127	051	P	-a	0.9106	1.1669	0.2070	262.6	98.1	-	15S	56W
4947	248	0049 Oct 29	14:16:32	10039	-24121	056	P	-h	-0.8788	1.2725	0.2186	295.3	107.1	-	12N	175W
4948	248	0050 Apr 25	19:03:44	10034	-24115	061	T+	-p	0.1712	2.5498	1.5378	345.1	221.2	96.0	12S	115E
4949	248	0050 Oct 19	00:06:10	10029	-24109	066	T-	-p	-0.1426	2.5940	1.5985	329.8	213.6	95.1	9N	37E
4950	248	0051 Apr 15	00:25:55	10025	-24103	071	P	t-	-0.6122	1.7664	0.7034	340.8	184.4	-	9S	35E
4951	248	0051 Oct 08	15:08:47	10020	-24097	076	P	a-	0.5398	1.8496	0.8849	299.1	178.9	-	6N	172E
4952	248	0052 Mar 04	09:05:07	10016	-24092	043	N	-t	1.5311	0.0835	-0.9858	87.9	-	-	8N	90W
4953	248	0052 Apr 03	00:51:56	10015	-24091	081	N	t-	-1.3568	0.4062	-0.6690	191.4	-	-	6S	29E
4954	248	0052 Aug 28	20:33:55	10011	-24086	048	N	-a	-1.4292	0.2328	-0.7621	132.7	-	-	12S	94E
4955	248	0052 Sep 27	06:54:24	10010	-24085	086	N	a-	1.2345	0.5823	-0.3969	197.6	-	-	2N	64W
4956	248	0053 Feb 21	14:51:48	10006	-24080	053	P	-h	0.7505	1.4908	0.4711	302.5	148.3	-	12N	177W
4957	248	0053 Aug 18	05:21:12	10002	-24074	058	P	-h	-0.7500	1.5057	0.4581	316.4	151.6	-	15S	38W
4958	248	0054 Feb 11	03:28:21	9997	-24068	063	T+	pp	0.0096	2.8272	1.8533	324.1	213.2	99.2	15N	6W
4959	248	0054 Aug 07	07:23:25	9992	-24062	068	T-	pp	-0.0147	2.8704	1.7924	375.5	235.8	106.2	17S	68W
4960	248	0055 Jan 31	19:27:30	9987	-24056	073	P	a-	-0.6739	1.6036	0.6388	287.6	158.6	-	17N	114E
4961	249	0055 Jul 27	08:01:10	9983	-24050	078	P	t-	0.7267	1.5545	0.4949	326.3	159.4	-	20S	78W
4962	249	0055 Dec 22	21:11:25	9979	-24045	045	N	-a	1.3935	0.3186	-0.7163	158.7	-	-	25N	85E
4963	249	0056 Jan 21	10:10:15	9978	-24044	083	N	a-	-1.3896	0.3083	-0.6920	150.7	-	-	19N	108W
4964	249	0056 Jun 16	04:50:14	9974	-24039	050	N	-a	-1.1878	0.6659	-0.3092	213.4	-	-	25S	32W
4965	249	0056 Jul 15	14:15:13	9973	-24038	088	N	a-	1.4325	0.2324	-0.7736	135.9	-	-	21S	172W
4966	249	0056 Dec 11	02:08:06	9969	-24033	055	P	-t	0.7573	1.5137	0.4239	326.1	149.1	-	24N	9E
4967	249	0057 Jun 05	20:37:27	9964	-24027	060	T	-a	-0.4139	2.0689	1.1273	306.0	194.0	51.4	23S	91E
4968	249	0057 Nov 30	01:28:26	9960	-24021	065	T+	pp	0.0867	2.7524	1.6464	378.4	234.1	102.8	22N	17E
4969	249	0058 May 26	13:48:35	9955	-24015	070	T	p-	0.3270	2.2331	1.2824	314.2	202.5	73.4	20S	167W
4970	249	0058 Nov 19	02:25:57	9950	-24009	075	P	a-	-0.5923	1.8064	0.7365	335.9	183.4	-	19N	2E
4971	249	0059 May 16	03:45:04	9945	-24003	080	N	t-	1.1228	0.7960	-0.2008	236.4	-	-	17S	16W
4972	249	0059 Oct 09	22:50:07	9942	-23998	047	N	-a	1.4790	0.1346	-0.8464	99.0	-	-	7N	56E
4973	249	0059 Nov 08	10:27:17	9941	-23997	085	N	a-	-1.2384	0.5911	-0.4200	204.9	-	-	15N	118W
4974	249	0060 Apr 04	20:12:30	9937	-23992	052	P	-t	-0.8670	1.3040	0.2307	312.3	114.6	-	6S	99E
4975	249	0060 Sep 28	14:37:59	9932	-23986	057	P	-a	0.7918	1.3862	0.4235	274.1	133.4	-	2N	180E
4976	249	0061 Mar 24	20:25:56	9927	-23980	062	T-	pp	-0.1262	2.6622	1.5911	370.5	232.2	101.4	1S	97E
4977	249	0061 Sep 18	05:48:54	9923	-23974	067	T+	-p	0.1081	2.6542	1.6650	330.3	215.1	97.5	3S	47W
4978	249	0062 Mar 14	00:53:05	9918	-23968	072	P	a-	0.6151	1.7429	0.7160	322.0	177.9	-	4N	31E
4979	249	0062 Sep 07	15:56:02	9913	-23962	077	P	t-	-0.6428	1.6993	0.6579	324.0	173.8	-	7S	162E
4980	249	0063 Feb 02	02:46:05	9909	-23957	044	N	-a	-1.3352	0.3919	-0.5762	162.7	-	-	16N	4E

Cat Num	Canon Plate	Calendar Date	TD of Greatest Eclipse	ΔT s	Luna Num	Saros Num	Ecl. Type	QSE	Gamma	Pen. Mag.	Um. Mag.	Eclipse Phase Durations Pen. m	Par. m	Total m	Greatest in Zenith Lat.	Lng.
4981	250	0063 Mar 03	12:25:21	9908	-23956	082	N	a-	1.2950	0.4701	-0.5070	180.2	-	-	9N	141W
4982	250	0063 Jul 29	02:46:05	9905	-23951	049	N	-t	1.5007	0.1426	-0.9336	116.4	-	-	18S	0E
4983	250	0063 Aug 27	18:58:15	9904	-23950	087	N	t-	-1.4061	0.3185	-0.7620	170.7	-	-	12S	117E
4984	250	0064 Jan 22	18:47:36	9900	-23945	054	P	-a	-0.6692	1.6161	0.6437	290.3	159.9	-	19N	123E
4985	250	0064 Jul 17	04:28:44	9895	-23939	059	P	-t	0.7315	1.5385	0.4935	320.2	157.3	-	21S	25W
4986	250	0065 Jan 11	08:12:39	9890	-23933	064	T+	pp	0.0130	2.8432	1.8252	340.5	219.6	100.8	22N	79W
4987	250	0065 Jul 06	12:52:09	9886	-23927	069	T-	pp	-0.0633	2.7361	1.7475	336.7	219.8	101.1	23S	152W
4988	250	0065 Dec 31	15:01:36	9881	-23921	074	P	t-	0.7428	1.5337	0.4568	323.1	152.5	-	24N	178E
4989	250	0066 Jun 26	03:41:56	9876	-23915	079	P	a-	-0.8015	1.3604	0.4137	273.0	133.0	-	24S	15W
4990	250	0066 Dec 20	15:13:55	9871	-23909	084	N	t-	1.4417	0.2660	-0.8402	158.5	-	-	25N	173E
4991	250	0067 May 17	13:49:02	9868	-23904	051	P	-a	0.9854	1.0303	0.0690	251.2	57.9	-	18S	168W
4992	250	0067 Jun 15	20:57:27	9867	-23903	089	Nb	a-	-1.5189	0.0432	-0.9019	55.6	-	-	25S	85E
4993	250	0067 Nov 09	22:34:28	9863	-23898	056	P	-h	-0.8844	1.2617	0.2092	293.4	104.6	-	16N	60E
4994	250	0068 May 06	02:04:08	9858	-23892	061	T+	-p	0.2473	2.4112	1.3973	344.2	218.4	87.6	16S	9E
4995	250	0068 Oct 29	08:41:07	9853	-23886	066	T-	-p	-0.1525	2.5756	1.5806	328.8	212.8	94.1	13N	93W
4996	250	0069 Apr 25	07:03:32	9849	-23880	071	P	t-	-0.5351	1.9077	0.8449	348.5	197.6	-	13S	66W
4997	250	0069 Oct 18	23:46:12	9844	-23874	076	P	a-	0.5243	1.8791	0.9123	300.3	180.7	-	10N	41E
4998	250	0070 Apr 14	07:31:58	9839	-23868	081	N	t-	-1.2844	0.5372	-0.5341	216.9	-	-	10S	73W
4999	250	0070 Sep 09	04:25:28	9835	-23863	048	N	-a	-1.4723	0.1570	-0.8441	110.2	-	-	8S	25W
5000	250	0070 Oct 08	15:17:47	9835	-23862	086	N	a-	1.2164	0.6178	-0.3662	203.4	-	-	6N	168E
5001	251	0071 Mar 04	22:38:14	9831	-23857	053	P	-h	0.7872	1.4201	0.4069	296.3	138.7	-	8N	65E
5002	251	0071 Aug 29	12:38:10	9826	-23851	058	P	-h	-0.8036	1.4105	0.3566	311.1	136.2	-	11S	148W
5003	251	0072 Feb 22	11:43:32	9821	-23845	063	T+	pp	0.0393	2.7703	1.8012	323.0	212.8	98.9	11N	131W
5004	251	0072 Aug 17	14:12:55	9817	-23839	068	T-	pp	-0.0795	2.7532	1.6715	375.7	235.1	104.4	14S	172W
5005	251	0073 Feb 11	03:54:58	9812	-23833	073	P	a-	-0.6518	1.6432	0.6804	289.8	162.7	-	14N	14W
5006	251	0073 Aug 06	14:53:14	9807	-23827	078	P	t-	0.6568	1.6829	0.6230	333.6	174.7	-	17S	178E
5007	251	0074 Jan 02	05:39:16	9803	-23822	045	N	-a	1.4024	0.3032	-0.7334	155.6	-	-	25N	42W
5008	251	0074 Jan 31	18:33:21	9802	-23821	083	N	a-	-1.3768	0.3319	-0.6686	156.5	-	-	16N	126E
5009	251	0074 Jun 27	12:03:48	9798	-23816	050	N	-a	-1.2566	0.5384	-0.4345	193.6	-	-	25S	140W
5010	251	0074 Jul 26	21:28:30	9798	-23815	088	N	a-	1.3622	0.3608	-0.6439	166.7	-	-	19S	79E
5011	251	0074 Dec 22	10:16:17	9794	-23810	055	P	-t	0.7635	1.5029	0.4119	326.1	147.5	-	24N	112W
5012	251	0075 Jun 17	04:03:28	9789	-23804	060	P	-a	-0.4871	1.9344	0.9933	301.1	186.5	-	24S	21W
5013	251	0075 Dec 11	09:30:03	9784	-23798	065	T+	pp	0.0920	2.7421	1.6369	378.0	233.9	102.5	23N	102W
5014	251	0076 Jun 05	21:10:05	9780	-23792	070	T+	p-	0.2531	2.3696	1.4171	317.9	207.4	85.2	22S	82E
5015	251	0076 Nov 29	10:44:11	9775	-23786	075	P	a-	-0.5879	1.8133	0.7456	335.1	183.8	-	21N	122W
5016	251	0077 May 26	10:45:32	9770	-23780	080	N	t-	1.0503	0.9306	-0.0692	253.2	-	-	20S	122W
5017	251	0077 Oct 20	07:23:24	9766	-23775	047	N	-a	1.4889	0.1164	-0.8649	92.0	-	-	11N	74W
5018	251	0077 Nov 18	19:05:53	9766	-23774	085	N	a-	-1.2320	0.6020	-0.4073	205.9	-	-	18N	112E
5019	251	0078 Apr 16	02:53:06	9762	-23769	052	P	-t	-0.9397	1.1698	0.0980	300.5	76.2	-	10S	3W
5020	251	0078 Oct 09	23:08:40	9757	-23763	057	P	-a	0.8115	1.3515	0.3859	272.0	128.1	-	7N	50E
5021	252	0079 Apr 05	03:17:40	9752	-23757	062	T-	pp	-0.1912	2.5405	1.4743	367.7	229.5	95.7	5S	8W
5022	252	0079 Sep 29	14:01:11	9748	-23751	067	T+	-p	0.1346	2.6084	1.6136	331.3	214.8	96.0	2N	172W
5023	252	0080 Mar 24	08:17:51	9743	-23745	072	P	a-	0.5607	1.8395	0.8191	325.1	186.4	-	0S	82W
5024	252	0080 Sep 17	23:35:36	9738	-23739	077	P	t-	-0.6082	1.7661	0.7182	329.1	180.3	-	3S	46E
5025	252	0081 Feb 12	11:10:06	9734	-23734	044	N	-a	-1.3572	0.3496	-0.6146	154.0	-	-	13N	123W
5026	252	0081 Mar 13	20:20:42	9734	-23733	082	N	a-	1.2503	0.5494	-0.4223	192.7	-	-	5N	99E
5027	252	0081 Aug 08	09:29:38	9730	-23728	049	Ne	-t	1.5667	0.0230	-1.0558	47.3	-	-	16S	101W
5028	252	0081 Sep 07	02:08:15	9729	-23727	087	N	t-	-1.3606	0.4043	-0.6809	191.3	-	-	8S	8E
5029	252	0082 Feb 02	03:18:35	9725	-23722	054	P	-a	-0.6886	1.5799	0.6087	288.7	156.6	-	17N	5W
5030	252	0082 Jul 28	11:22:28	9720	-23716	059	P	-t	0.7998	1.4128	0.3684	310.4	138.2	-	19S	129W
5031	252	0083 Jan 22	16:36:21	9716	-23710	064	T+	pp	0.0008	2.8659	1.8472	341.6	220.3	101.1	20N	155E
5032	252	0083 Jul 17	20:07:38	9711	-23704	069	T+	pp	0.0073	2.8381	1.8510	335.7	219.5	101.8	22S	99E
5033	252	0084 Jan 11	23:07:35	9706	-23698	074	P	t-	0.7345	1.5492	0.4718	325.0	154.9	-	23N	57E
5034	252	0084 Jul 06	11:09:43	9701	-23692	079	P	a-	-0.7294	1.4926	0.5459	281.1	149.5	-	24S	127W
5035	252	0084 Dec 30	23:12:01	9697	-23686	084	N	t-	1.4338	0.2799	-0.8249	162.3	-	-	25N	54E
5036	252	0085 May 27	21:05:37	9693	-23681	051	N	-a	1.0632	0.8886	-0.0748	237.6	-	-	20S	83E
5037	252	0085 Jun 26	04:21:20	9692	-23680	089	N	a-	-1.4500	0.1707	-0.7766	109.4	-	-	25S	26W
5038	252	0085 Nov 20	06:58:27	9688	-23675	056	P	-h	-0.8858	1.2580	0.2075	292.1	103.9	-	19N	67W
5039	252	0086 May 17	08:59:30	9684	-23669	061	T	-h	0.3265	2.2670	1.2507	342.3	213.8	73.6	19S	96W
5040	252	0086 Nov 09	17:21:11	9679	-23663	066	T-	-p	-0.1588	2.5636	1.5695	327.8	212.1	93.5	17N	137E

Cat Num	Canon Plate	Calendar Date	TD of Greatest Eclipse	ΔT s	Luna Num	Saros Num	Ecl. Type	QSE	Gamma	Pen. Mag.	Um. Mag.	Eclipse Phase Durations Pen. m	Par. m	Total m	Greatest in Zenith Lat.	Lng.
5041	253	0087 May 06	13:37:09	9674	-23657	071	P	t-	-0.4551	2.0546	0.9917	355.4	208.7	-	16S	166W
5042	253	0087 Oct 30	08:28:59	9669	-23651	076	P	a-	0.5134	1.9000	0.9314	301.2	182.0	-	14N	91W
5043	253	0088 Apr 24	14:07:46	9665	-23645	081	N	t-	-1.2075	0.6765	-0.3912	239.7	-	-	13S	173W
5044	253	0088 Sep 19	12:23:25	9661	-23640	048	N	-a	-1.5096	0.0917	-0.9156	85.1	-	-	3S	146W
5045	253	0088 Oct 18	23:46:51	9660	-23639	086	N	a-	1.2034	0.6441	-0.3446	207.8	-	-	11N	40E
5046	253	0089 Mar 15	06:18:11	9656	-23634	053	P	-h	0.8302	1.3379	0.3312	288.9	126.2	-	3N	51W
5047	253	0089 Sep 08	20:02:49	9652	-23628	058	P	-h	-0.8504	1.3279	0.2677	306.0	119.9	-	7S	99E
5048	253	0090 Mar 04	19:49:38	9647	-23622	063	T+	pp	0.0773	2.6982	1.7338	321.7	212.1	98.0	7N	106E
5049	253	0090 Aug 28	21:12:52	9642	-23616	068	T-	pp	-0.1360	2.6515	1.5661	375.4	233.6	100.9	11S	82E
5050	253	0091 Feb 22	12:13:30	9638	-23610	073	P	a-	-0.6221	1.6966	0.7358	292.7	167.9	-	10N	139W
5051	253	0091 Aug 17	21:55:26	9633	-23604	078	P	t-	0.5941	1.7983	0.7378	339.2	186.1	-	14S	72E
5052	253	0092 Jan 13	14:00:41	9629	-23599	045	N	-a	1.4160	0.2787	-0.7591	150.0	-	-	23N	166W
5053	253	0092 Feb 12	02:47:35	9628	-23598	083	N	a-	-1.3574	0.3675	-0.6331	164.6	-	-	13N	2E
5054	253	0092 Jul 07	19:21:59	9624	-23593	050	N	-a	-1.3227	0.4165	-0.5548	171.7	-	-	24S	110E
5055	253	0092 Aug 06	04:50:08	9624	-23592	088	N	a-	1.2972	0.4797	-0.5242	189.4	-	-	16S	32W
5056	253	0093 Jan 01	18:19:53	9620	-23587	055	P	-t	0.7729	1.4858	0.3943	325.5	145.0	-	24N	127E
5057	253	0093 Jun 27	11:30:09	9615	-23581	060	P	-a	-0.5600	1.8008	0.8593	295.5	177.5	-	24S	133W
5058	253	0093 Dec 21	17:31:38	9610	-23575	065	T+	pp	0.0978	2.7306	1.6271	377.4	233.6	102.2	24N	138E
5059	253	0094 Jun 17	04:30:25	9606	-23569	070	T+	pp	0.1786	2.5077	1.5526	321.1	211.1	93.2	23S	29W
5060	253	0094 Dec 10	19:04:29	9601	-23563	075	P	a-	-0.5842	1.8186	0.7540	334.2	184.0	-	23N	113E
5061	254	0095 Jun 06	17:43:22	9596	-23557	080	P	t-	0.9761	1.0685	0.0653	268.5	58.8	-	22S	133E
5062	254	0095 Oct 31	16:04:01	9592	-23552	047	N	-a	1.4941	0.1070	-0.8744	88.2	-	-	15N	155E
5063	254	0095 Nov 30	03:48:43	9592	-23551	085	N	a-	-1.2282	0.6079	-0.3992	206.2	-	-	20N	19W
5064	254	0096 Apr 26	09:26:31	9588	-23546	052	N*	-t	-1.0177	1.0262	-0.0444	285.9	-	-	14S	103W
5065	254	0096 Oct 20	07:46:06	9583	-23540	057	P	-a	0.8257	1.3270	0.3583	270.6	124.0	-	11N	80W
5066	254	0097 Apr 15	10:03:20	9578	-23534	062	T-	pp	-0.2619	2.4082	1.3469	363.9	225.3	86.1	9S	111W
5067	254	0097 Oct 09	22:20:04	9574	-23528	067	T+	-p	0.1548	2.5741	1.5739	332.3	214.5	94.6	6N	62E
5068	254	0098 Apr 04	15:36:48	9569	-23522	072	P	a-	0.5004	1.9469	0.9329	328.2	194.4	-	5S	167E
5069	254	0098 Sep 29	07:22:36	9564	-23516	077	P	t-	-0.5804	1.8203	0.7661	333.4	185.2	-	1N	73W
5070	254	0099 Feb 23	19:27:49	9561	-23511	044	N	-a	-1.3845	0.2976	-0.6628	142.5	-	-	9N	112E
5071	254	0099 Mar 25	04:10:45	9560	-23510	082	N	a-	1.2008	0.6377	-0.3289	205.1	-	-	0N	21W
5072	254	0099 Sep 18	09:25:31	9555	-23504	087	N	t-	-1.3209	0.4794	-0.6102	207.3	-	-	4S	103W
5073	254	0100 Feb 13	11:43:02	9551	-23499	054	P	-a	-0.7131	1.5343	0.5642	286.4	152.1	-	13N	132W
5074	254	0100 Aug 07	18:24:40	9547	-23493	059	P	-t	0.8618	1.2989	0.2548	300.6	116.5	-	16S	124E
5075	254	0101 Feb 02	00:51:22	9542	-23487	064	T-	pp	-0.0181	2.8344	1.8153	342.7	220.9	101.3	17N	31E
5076	254	0101 Jul 28	03:30:19	9537	-23481	069	T+	pp	0.0728	2.7174	1.7313	334.2	218.3	100.3	20S	12W
5077	254	0102 Jan 22	07:05:05	9533	-23475	074	P	t-	0.7199	1.5761	0.4987	327.5	159.0	-	21N	63W
5078	254	0102 Jul 17	18:43:31	9528	-23469	079	P	a-	-0.6619	1.6167	0.6695	287.8	162.1	-	22S	119E
5079	254	0103 Jan 11	07:03:37	9523	-23463	084	N	t-	1.4206	0.3031	-0.7996	168.4	-	-	23N	63W
5080	254	0103 Jun 08	04:21:27	9519	-23458	051	N	-a	1.1408	0.7476	-0.2185	221.8	-	-	22S	27W
5081	255	0103 Jul 07	11:49:00	9519	-23457	089	N	a-	-1.3844	0.2925	-0.6578	141.9	-	-	25S	137W
5082	255	0103 Dec 01	15:25:15	9515	-23452	056	P	-h	-0.8856	1.2570	0.2091	291.0	104.0	-	21N	167E
5083	255	0104 May 27	15:49:27	9510	-23446	061	T	-h	0.4092	2.1168	1.0976	339.3	207.1	48.4	21S	161E
5084	255	0104 Nov 20	02:06:11	9505	-23440	066	T-	-p	-0.1616	2.5579	1.5650	326.8	211.6	93.1	20N	5E
5085	255	0105 May 16	20:03:52	9501	-23434	071	T	t-	-0.3700	2.2110	1.1477	361.4	218.2	60.7	19S	97E
5086	255	0105 Nov 09	17:17:35	9496	-23428	076	P	a-	0.5072	1.9121	0.9421	301.8	182.6	-	17N	137E
5087	255	0106 May 05	20:37:37	9492	-23422	081	N	t-	-1.1249	0.8263	-0.2381	260.3	-	-	17S	88E
5088	255	0106 Sep 30	20:29:48	9488	-23417	048	N	-a	-1.5395	0.0398	-0.9737	56.6	-	-	1N	91E
5089	255	0106 Oct 30	08:22:39	9487	-23416	086	N	a-	1.1964	0.6590	-0.3340	210.6	-	-	15N	90W
5090	255	0107 Mar 26	13:51:50	9483	-23411	053	P	-h	0.8794	1.2445	0.2441	280.3	109.4	-	1S	166W
5091	255	0107 Sep 20	03:34:19	9478	-23405	058	P	-h	-0.8908	1.2568	0.1906	301.3	102.6	-	3S	15W
5092	255	0108 Mar 15	03:50:07	9474	-23399	063	T+	pp	0.1205	2.6167	1.6568	320.2	211.1	96.2	3N	15W
5093	255	0108 Sep 08	04:20:54	9469	-23393	068	T-	pp	-0.1859	2.5618	1.4727	374.7	231.5	96.0	7S	26W
5094	255	0109 Mar 04	20:24:13	9464	-23387	073	P	a-	-0.5863	1.7614	0.8025	295.9	173.6	-	0N	97E
5095	255	0109 Aug 28	05:07:04	9460	-23381	078	P	t-	0.5383	1.9011	0.8399	343.4	194.5	-	10S	38W
5096	255	0110 Jan 23	22:15:49	9456	-23376	045	N	-a	1.4340	0.2460	-0.7925	141.8	-	-	21N	70E
5097	255	0110 Feb 22	10:53:24	9455	-23375	083	N	a-	-1.3321	0.4139	-0.5866	174.4	-	-	10N	120W
5098	255	0110 Jul 19	02:46:36	9451	-23370	050	N	-a	-1.3840	0.3034	-0.6669	147.6	-	-	23S	1W
5099	255	0110 Aug 17	12:21:04	9450	-23369	088	N	a-	1.2387	0.5866	-0.4167	206.6	-	-	13S	146W
5100	255	0111 Jan 13	02:17:53	9447	-23364	055	P	-t	0.7865	1.4607	0.3695	324.2	141.1	-	23N	8E

Cat Num	Canon Plate	Calendar Date	TD of Greatest Eclipse	ΔT s	Luna Num	Saros Num	Ecl. Type	QSE	Gamma	Pen. Mag.	Um. Mag.	Eclipse Phase Durations Pen. m	Par. m	Total m	Greatest in Zenith Lat.	Lng.
5101	256	0111 Jul 08	19:01:45	9442	-23358	060	P	-a	-0.6290	1.6747	0.7323	289.6	167.3	-	24S	114E
5102	256	0112 Jan 02	01:30:20	9437	-23352	065	T+	pp	0.1064	2.7137	1.6126	376.6	233.3	101.7	23N	19E
5103	256	0112 Jun 27	11:51:18	9433	-23346	070	T+	pp	0.1050	2.6443	1.6860	323.6	213.7	98.0	24S	139W
5104	256	0112 Dec 21	03:24:24	9428	-23340	075	P	a-	-0.5793	1.8257	0.7647	333.3	184.5	-	23N	11W
5105	256	0113 Jun 17	00:39:51	9423	-23334	080	P	t-	0.9012	1.2080	0.2008	282.4	101.5	-	23S	28E
5106	256	0113 Nov 11	00:49:12	9420	-23329	047	N	-a	1.4962	0.1030	-0.8781	86.3	-	-	19N	23E
5107	256	0113 Dec 10	12:31:29	9419	-23328	085	N	a-	-1.2238	0.6146	-0.3898	206.5	-	-	22N	149W
5108	256	0114 May 07	15:53:58	9415	-23323	052	N	-t	-1.0998	0.8749	-0.1947	268.3	-	-	17S	159E
5109	256	0114 Oct 31	16:29:23	9410	-23317	057	P	-a	0.8354	1.3105	0.3393	269.8	121.1	-	15N	148E
5110	256	0115 Apr 26	16:44:15	9406	-23311	062	T	-t	-0.3370	2.2682	1.2113	359.2	219.4	70.5	13S	147E
5111	256	0115 Oct 21	06:44:43	9401	-23305	067	T+	-p	0.1697	2.5494	1.5439	333.4	214.4	93.4	10N	65W
5112	256	0116 Apr 14	22:50:23	9396	-23299	072	T	a-	0.4345	2.0648	1.0567	331.0	201.6	36.9	9S	56E
5113	256	0116 Oct 09	15:17:08	9392	-23293	077	P	t-	-0.5594	1.8617	0.8016	336.9	188.8	-	6N	167E
5114	256	0117 Mar 06	03:36:01	9388	-23288	044	N	-a	-1.4199	0.2306	-0.7259	126.0	-	-	5N	12W
5115	256	0117 Apr 04	11:52:04	9387	-23287	082	N	a-	1.1435	0.7404	-0.2213	218.0	-	-	4S	138W
5116	256	0117 Sep 28	16:53:03	9382	-23281	087	N	t-	-1.2892	0.5397	-0.5542	219.0	-	-	1N	144E
5117	256	0118 Feb 23	19:58:14	9379	-23276	054	P	-a	-0.7452	1.4748	0.5060	283.3	145.6	-	9N	103E
5118	256	0118 Aug 19	01:36:55	9374	-23270	059	P	-t	0.9165	1.1985	0.1545	291.1	91.8	-	13S	15E
5119	256	0119 Feb 13	08:57:54	9369	-23264	064	T-	pp	-0.0433	2.7883	1.7689	343.7	221.4	101.3	14N	91W
5120	256	0119 Aug 08	11:01:20	9365	-23258	069	T+	pp	0.1323	2.6079	1.6224	332.2	216.3	97.1	17S	126W
5121	257	0120 Feb 02	14:56:03	9360	-23252	074	P	t-	0.7004	1.6115	0.5347	330.5	164.0	-	18N	179E
5122	257	0120 Jul 28	02:21:13	9355	-23246	079	P	a-	-0.5974	1.7356	0.7874	293.6	172.2	-	20S	5E
5123	257	0121 Jan 21	14:51:16	9351	-23240	084	N	t-	1.4040	0.3321	-0.7678	175.7	-	-	21N	180W
5124	257	0121 Jun 18	11:35:45	9347	-23235	051	N	-a	1.2185	0.6066	-0.3628	203.4	-	-	22S	136W
5125	257	0121 Jul 17	19:19:08	9346	-23234	089	N	a-	-1.3213	0.4101	-0.5437	166.5	-	-	23S	110E
5126	257	0121 Dec 11	23:53:41	9342	-23229	056	P	-h	-0.8849	1.2567	0.2121	289.9	104.4	-	22N	40E
5127	257	0122 Jun 07	22:37:31	9338	-23223	061	P	-t	0.4923	1.9660	0.9434	335.2	198.3	-	22S	58E
5128	257	0122 Dec 01	10:53:36	9333	-23217	066	T-	-p	-0.1626	2.5552	1.5640	325.9	211.2	92.9	22N	126W
5129	257	0123 May 28	02:29:02	9328	-23211	071	T	pp	-0.2837	2.3698	1.3057	366.4	225.6	82.7	21S	0W
5130	257	0123 Nov 21	02:08:23	9324	-23205	076	P	a-	0.5030	1.9204	0.9493	302.3	183.1	-	20N	4E
5131	257	0124 May 16	03:06:14	9319	-23199	081	N	t-	-1.0404	0.9798	-0.0816	278.4	-	-	20S	10W
5132	257	0124 Nov 09	17:01:56	9314	-23193	086	N	a-	1.1928	0.6677	-0.3292	212.5	-	-	18N	140E
5133	257	0125 Apr 05	21:18:32	9311	-23188	053	P	-h	0.9350	1.1394	0.1450	270.0	85.2	-	5S	80E
5134	257	0125 Sep 30	11:14:22	9306	-23182	058	P	-h	-0.9234	1.1999	0.1278	297.4	85.0	-	2N	132W
5135	257	0126 Mar 26	11:42:21	9301	-23176	063	T+	-p	0.1710	2.5218	1.5662	318.4	209.4	92.9	1S	135W
5136	257	0126 Sep 19	11:39:11	9297	-23170	068	T-	pp	-0.2273	2.4876	1.3951	373.8	229.3	90.6	2S	137W
5137	257	0127 Mar 16	04:25:48	9292	-23164	073	P	a-	-0.5431	1.8397	0.8825	299.7	179.8	-	2N	25W
5138	257	0127 Sep 08	12:30:19	9287	-23158	078	P	t-	0.4912	1.9879	0.9261	346.4	200.6	-	6S	150W
5139	257	0128 Feb 04	06:22:08	9284	-23153	045	N	-a	1.4585	0.2013	-0.8376	129.2	-	-	18N	52W
5140	257	0128 Mar 04	18:48:56	9283	-23152	083	N	a-	-1.2993	0.4741	-0.5263	186.2	-	-	5N	119E
5141	258	0128 Jul 29	10:16:54	9279	-23147	050	N	-a	-1.4415	0.1976	-0.7722	120.0	-	-	21S	114W
5142	258	0128 Aug 27	20:01:05	9278	-23146	088	N	a-	1.1866	0.6821	-0.3210	219.9	-	-	10S	98E
5143	258	0129 Jan 23	10:08:52	9274	-23141	055	P	-t	0.8054	1.4256	0.3353	322.1	135.4	-	20N	109W
5144	258	0129 Jul 19	02:36:35	9270	-23135	060	P	-a	-0.6956	1.5533	0.6093	283.1	155.6	-	22S	0E
5145	258	0130 Jan 12	09:24:15	9265	-23129	065	T+	pp	0.1193	2.6883	1.5905	375.6	232.8	101.0	22N	99W
5146	258	0130 Jul 08	19:14:06	9260	-23123	070	T+	pp	0.0334	2.7775	1.8154	325.4	215.1	100.3	23S	110E
5147	258	0131 Jan 01	11:43:30	9256	-23117	075	P	a-	-0.5727	1.8357	0.7791	332.5	185.3	-	23N	135W
5148	258	0131 Jun 28	07:35:26	9251	-23111	080	P	t-	0.8262	1.3478	0.3361	295.0	129.2	-	23S	76W
5149	258	0131 Nov 22	09:38:19	9247	-23106	047	N	-a	1.4959	0.1033	-0.8772	86.3	-	-	22N	109W
5150	258	0131 Dec 21	21:14:54	9247	-23105	085	N	a-	-1.2194	0.6210	-0.3801	206.8	-	-	22N	80E
5151	258	0132 May 17	22:17:24	9243	-23100	052	N	-t	-1.1847	0.7189	-0.3501	247.2	-	-	20S	62E
5152	258	0132 Nov 11	01:17:25	9238	-23094	057	P	-a	0.8409	1.3016	0.3280	269.5	119.4	-	18N	16E
5153	258	0133 May 06	23:20:42	9234	-23088	062	T	-t	-0.4163	2.1205	1.0679	353.3	211.5	41.8	17S	47E
5154	258	0133 Oct 31	15:15:03	9229	-23082	067	T+	-p	0.1791	2.5346	1.5243	334.6	214.5	92.6	14N	166E
5155	258	0134 Apr 26	06:00:42	9224	-23076	072	T	a-	0.3648	2.1898	1.1873	333.3	207.7	64.1	13S	53W
5156	258	0134 Oct 20	23:16:55	9220	-23070	077	P	t-	-0.5435	1.8937	0.8281	339.8	191.5	-	10N	46E
5157	258	0135 Mar 17	11:38:17	9216	-23065	044	N	-a	-1.4604	0.1545	-0.7984	103.7	-	-	1N	134W
5158	258	0135 Apr 15	19:30:01	9215	-23064	082	N	a-	1.0829	0.8494	-0.1079	230.2	-	-	8S	106E
5159	258	0135 Oct 10	00:28:11	9210	-23058	087	N	t-	-1.2635	0.5888	-0.5088	228.1	-	-	5N	29E
5160	258	0136 Mar 06	04:05:42	9207	-23053	054	P	-a	-0.7833	1.4044	0.4367	279.2	137.0	-	5N	20W

Cat Num	Canon Plate	Calendar Date	TD of Greatest Eclipse	ΔT s	Luna Num	Saros Num	Ecl. Type	QSE	Gamma	Pen. Mag.	Um. Mag.	Eclipse Phase Durations Pen. m	Par. m	Total m	Greatest in Zenith Lat.	Lng.
5161	259	0136 Aug 29	08:58:32	9202	-23047	059	P	-t	0.9642	1.1110	0.0669	282.1	61.0	-	9S	97W
5162	259	0137 Feb 23	16:54:38	9197	-23041	064	T-	pp	-0.0757	2.7288	1.7093	344.5	221.7	100.7	10N	149E
5163	259	0137 Aug 18	18:41:26	9193	-23035	069	T+	pp	0.1847	2.5116	1.5263	330.1	213.9	92.8	14S	118E
5164	259	0138 Feb 12	22:35:33	9188	-23029	074	P	t-	0.6724	1.6625	0.5865	334.3	170.7	-	14N	64E
5165	259	0138 Aug 08	10:07:24	9184	-23023	079	P	a-	-0.5400	1.8418	0.8920	298.2	179.9	-	17S	113W
5166	259	0139 Feb 01	22:30:28	9179	-23017	084	N	t-	1.3808	0.3729	-0.7236	185.2	-	-	19N	65E
5167	259	0139 Jun 29	18:51:03	9175	-23012	051	N	-a	1.2945	0.4692	-0.5041	182.0	-	-	22S	115E
5168	259	0139 Jul 29	02:54:05	9174	-23011	089	N	a-	-1.2626	0.5199	-0.4379	186.1	-	-	21S	4W
5169	259	0139 Dec 23	08:21:15	9170	-23006	056	P	-a	-0.8854	1.2538	0.2130	288.5	104.3	-	23N	86W
5170	259	0140 Jun 18	05:23:41	9166	-23000	061	P	-t	0.5756	1.8152	0.7886	329.9	186.9	-	23S	44W
5171	259	0140 Dec 11	19:42:12	9161	-22994	066	T-	-p	-0.1633	2.5528	1.5639	325.0	210.8	92.8	23N	102E
5172	259	0141 Jun 07	08:50:20	9157	-22988	071	T-	pp	-0.1944	2.5341	1.4688	370.2	231.1	96.1	23S	96W
5173	259	0141 Dec 01	11:02:25	9152	-22982	076	P	a-	0.5016	1.9234	0.9516	302.5	183.3	-	23N	129W
5174	259	0142 May 27	09:32:47	9147	-22976	081	P	t-	-0.9533	1.1384	0.0797	294.3	68.4	-	22S	107W
5175	259	0142 Nov 21	01:44:36	9143	-22970	086	N	a-	1.1925	0.6700	-0.3304	213.5	-	-	21N	9E
5176	259	0143 Apr 17	04:40:57	9139	-22965	053	P	-h	0.9949	1.0267	0.0380	258.3	44.1	-	9S	32W
5177	259	0143 Oct 11	19:01:54	9134	-22959	058	P	-h	-0.9491	1.1556	0.0780	294.4	67.0	-	6N	110E
5178	259	0144 Apr 05	19:28:30	9130	-22953	063	T+	-p	0.2270	2.4170	1.4653	316.1	207.0	87.7	6S	107E
5179	259	0144 Sep 29	19:06:02	9125	-22947	068	T-	pp	-0.2618	2.4259	1.3301	372.8	227.1	84.9	2N	109E
5180	259	0145 Mar 26	12:20:04	9120	-22941	073	P	a-	-0.4942	1.9288	0.9730	303.6	186.1	-	2S	146W
5181	260	0145 Sep 18	20:03:29	9116	-22935	078	P	t-	0.4511	2.0618	0.9994	348.4	205.0	-	2S	96E
5182	260	0146 Feb 14	14:19:19	9112	-22930	045	N	-a	1.4897	0.1442	-0.8948	110.3	-	-	15N	172W
5183	260	0146 Mar 16	02:35:11	9111	-22929	083	N	a-	-1.2595	0.5471	-0.4534	199.3	-	-	1N	1E
5184	260	0146 Aug 09	17:55:43	9107	-22924	050	N	-a	-1.4930	0.1032	-0.8666	87.2	-	-	18S	130E
5185	260	0146 Sep 08	03:51:17	9107	-22923	088	N	a-	1.1418	0.7644	-0.2386	230.2	-	-	5S	21W
5186	260	0147 Feb 03	17:52:16	9103	-22918	055	P	-t	0.8303	1.3793	0.2902	318.8	127.0	-	18N	135E
5187	260	0147 Jul 30	10:16:55	9098	-22912	060	P	-a	-0.7577	1.4404	0.4944	276.5	142.7	-	20S	115W
5188	260	0148 Jan 23	17:12:45	9094	-22906	065	T+	pp	0.1372	2.6536	1.5596	374.4	232.0	99.8	20N	144E
5189	260	0148 Jul 19	02:40:16	9089	-22900	070	T-	pp	-0.0349	2.7769	1.8105	326.6	215.5	100.3	21S	2W
5190	260	0149 Jan 11	19:57:43	9084	-22894	075	P	a-	-0.5613	1.8540	0.8024	331.9	186.8	-	21N	101E
5191	260	0149 Jul 08	14:32:46	9080	-22888	080	P	t-	0.7533	1.4842	0.4674	306.2	149.9	-	22S	180E
5192	260	0149 Dec 02	18:29:12	9076	-22883	047	N	-a	1.4948	0.1046	-0.8747	86.7	-	-	24N	119E
5193	260	0150 Jan 01	05:55:06	9075	-22882	085	N	a-	-1.2118	0.6331	-0.3643	207.8	-	-	22N	49W
5194	260	0150 May 29	04:38:46	9071	-22877	052	N	-t	-1.2706	0.5612	-0.5076	221.8	-	-	23S	34W
5195	260	0150 Jun 27	19:18:39	9071	-22876	090	Nb	t-	1.5618	0.0232	-1.0383	46.9	-	-	22S	108E
5196	260	0150 Nov 22	10:07:33	9067	-22871	057	P	-a	0.8449	1.2952	0.3198	269.4	118.1	-	21N	117W
5197	260	0151 May 18	05:55:48	9062	-22865	062	P	-t	-0.4975	1.9696	0.9208	346.2	201.4	-	20S	53W
5198	260	0151 Nov 11	23:49:40	9057	-22859	067	T+	-p	0.1847	2.5265	1.5118	335.8	214.7	92.1	18N	37E
5199	260	0152 May 06	13:06:28	9053	-22853	072	T	a-	0.2901	2.3242	1.3269	335.1	212.7	80.6	16S	160W
5200	260	0152 Oct 31	07:23:25	9048	-22847	077	P	t-	-0.5336	1.9145	0.8439	342.1	193.2	-	14N	76W
5201	261	0153 Mar 27	19:31:47	9044	-22842	044	Ne	-a	-1.5084	0.0647	-0.8849	67.6	-	-	4S	106E
5202	261	0153 Apr 26	03:01:22	9044	-22841	082	P	a-	1.0158	0.9705	0.0170	242.3	28.7	-	12S	8W
5203	261	0153 Oct 20	08:12:17	9039	-22835	087	N	t-	-1.2447	0.6248	-0.4761	234.5	-	-	9N	88W
5204	261	0154 Mar 17	12:03:13	9035	-22830	054	P	-a	-0.8293	1.3194	0.3526	273.8	124.9	-	1N	141W
5205	261	0154 Sep 09	16:31:39	9031	-22824	059	N*	-t	1.0034	1.0391	-0.0052	274.0	-	-	5S	149E
5206	261	0155 Mar 07	00:42:30	9026	-22818	064	T-	pp	-0.1147	2.6573	1.6377	345.1	221.6	99.2	6N	30E
5207	261	0155 Aug 30	02:29:41	9021	-22812	069	T+	-p	0.2312	2.4263	1.4409	327.8	211.2	87.7	10S	0E
5208	261	0156 Feb 24	06:07:20	9017	-22806	074	P	t-	0.6385	1.7240	0.6492	338.5	178.2	-	11N	50W
5209	261	0156 Aug 18	17:59:30	9012	-22800	079	P	a-	-0.4873	1.9394	0.9876	301.9	186.0	-	14S	128E
5210	261	0157 Feb 12	06:02:07	9008	-22794	084	N	t-	1.3510	0.4256	-0.6670	196.6	-	-	15N	48W
5211	261	0157 Jul 10	02:07:35	9004	-22789	051	N	-a	1.3687	0.3354	-0.6425	156.5	-	-	21S	6E
5212	261	0157 Aug 08	10:34:05	9003	-22788	089	N	a-	-1.2082	0.6219	-0.3403	202.1	-	-	18S	120W
5213	261	0158 Jan 02	16:47:31	8999	-22783	056	P	-a	-0.8879	1.2470	0.2107	286.7	103.4	-	22N	148E
5214	261	0158 Jun 29	12:11:02	8994	-22777	061	P	-t	0.6571	1.6678	0.6369	323.5	172.9	-	23S	146W
5215	261	0158 Dec 23	04:29:10	8990	-22771	066	T-	-p	-0.1659	2.5466	1.5604	324.1	210.4	92.6	24N	29W
5216	261	0159 Jun 18	15:13:26	8985	-22765	071	T-	pp	-0.1069	2.6956	1.6287	372.8	234.5	103.5	24S	168E
5217	261	0159 Dec 12	19:56:04	8981	-22759	076	P	a-	0.4999	1.9265	0.9545	302.8	183.5	-	24N	98E
5218	261	0160 Jun 06	15:59:27	8976	-22753	081	P	t-	-0.8651	1.2990	0.2425	308.1	116.4	-	23S	156E
5219	261	0160 Dec 01	10:28:20	8971	-22747	086	N	a-	1.1937	0.6693	-0.3340	214.0	-	-	23N	122W
5220	261	0161 Apr 27	11:58:38	8968	-22742	053	N	-h	1.0592	0.9060	-0.0774	244.7	-	-	13S	143W

Cat Num	Canon Plate	Calendar Date	TD of Greatest Eclipse	ΔT s	Luna Num	Saros Num	Ecl. Type	QSE	Gamma	Pen. Mag.	Um. Mag.	Eclipse Phase Durations Pen. m	Par. m	Total m	Greatest in Zenith Lat.	Lng.
5221	262	0161 Oct 22	02:55:49	8963	-22736	058	P	-t	-0.9691	1.1214	0.0389	292.1	47.7	-	10N	10W
5222	262	0162 Apr 17	03:08:02	8958	-22730	063	T	-p	0.2891	2.3014	1.3531	313.2	203.5	79.6	10S	10W
5223	262	0162 Oct 11	02:42:39	8954	-22724	068	T	-p	-0.2884	2.3785	1.2801	371.9	225.1	79.7	6N	6W
5224	262	0163 Apr 06	20:06:21	8949	-22718	073	T	a-	-0.4388	2.0300	1.0752	307.6	192.3	40.3	6S	96E
5225	262	0163 Sep 30	03:46:43	8945	-22712	078	T	h-	0.4183	2.1220	1.0592	349.7	208.1	38.6	3N	22W
5226	262	0164 Feb 25	22:06:19	8941	-22707	045	N	-h	1.5282	0.0736	-0.9655	79.5	-	-	11N	70E
5227	262	0164 Mar 26	10:11:32	8940	-22706	083	N	h-	-1.2126	0.6334	-0.3675	213.3	-	-	3S	115W
5228	262	0164 Aug 20	01:42:21	8936	-22701	050	Ne	-a	-1.5391	0.0188	-0.9514	37.4	-	-	15S	13E
5229	262	0164 Sep 18	11:51:25	8935	-22700	088	N	a-	1.1041	0.8336	-0.1695	237.9	-	-	1S	143W
5230	262	0165 Feb 14	01:25:02	8932	-22695	055	P	-t	0.8633	1.3179	0.2307	314.1	114.4	-	14N	21E
5231	262	0165 Aug 09	18:03:16	8927	-22689	060	P	-a	-0.8148	1.3370	0.3883	269.8	128.5	-	17S	128E
5232	262	0166 Feb 03	00:54:18	8922	-22683	065	T+	pp	0.1612	2.6075	1.5178	372.8	231.0	97.9	17N	28E
5233	262	0166 Jul 30	10:10:38	8918	-22677	070	T-	pp	-0.0989	2.6619	1.6906	327.4	215.0	98.4	19S	115W
5234	262	0167 Jan 23	04:07:59	8913	-22671	075	P	a-	-0.5458	1.8797	0.8335	331.7	188.8	-	19N	21W
5235	262	0167 Jul 19	21:32:33	8909	-22665	080	P	t-	0.6832	1.6156	0.5934	316.0	166.1	-	21S	74E
5236	262	0167 Dec 14	03:21:33	8905	-22660	047	N	-a	1.4933	0.1067	-0.8711	87.4	-	-	25N	14W
5237	262	0168 Jan 12	14:33:16	8904	-22659	085	N	a-	-1.2022	0.6486	-0.3446	209.4	-	-	21N	178W
5238	262	0168 Jun 08	10:57:46	8900	-22654	052	N	-t	-1.3577	0.4014	-0.6675	190.5	-	-	24S	129W
5239	262	0168 Jul 08	01:47:33	8899	-22653	090	N	t-	1.4833	0.1689	-0.8958	125.2	-	-	22S	10E
5240	262	0168 Dec 02	19:00:20	8896	-22648	057	P	-a	0.8466	1.2928	0.3159	269.6	117.6	-	23N	110E
5241	263	0169 May 28	12:29:57	8891	-22642	062	P	-t	-0.5803	1.8161	0.7707	337.9	188.7	-	22S	152W
5242	263	0169 Nov 22	08:26:30	8886	-22636	067	T+	-p	0.1877	2.5229	1.5044	337.0	215.1	91.9	20N	92W
5243	263	0170 May 17	20:11:58	8882	-22630	072	T+	p-	0.2142	2.4613	1.4686	336.2	216.4	91.2	19S	92E
5244	263	0170 Nov 11	15:33:27	8877	-22624	077	P	t-	-0.5269	1.9289	0.8539	344.1	194.5	-	17N	161E
5245	263	0171 May 07	10:30:52	8873	-22618	082	P	a-	0.9470	1.0952	0.1451	253.3	82.1	-	15S	122W
5246	263	0171 Oct 31	16:01:23	8868	-22612	087	N	t-	-1.2299	0.6534	-0.4502	239.3	-	-	13N	153E
5247	263	0172 Mar 27	19:53:27	8864	-22607	054	P	-a	-0.8808	1.2247	0.2584	267.2	108.6	-	3S	100E
5248	263	0172 Sep 20	00:14:49	8860	-22601	059	N	-t	1.0351	0.9810	-0.0633	267.1	-	-	1S	31E
5249	263	0173 Mar 17	08:18:36	8855	-22595	064	T-	-p	-0.1624	2.5699	1.5501	345.4	220.9	96.2	2N	85W
5250	263	0173 Sep 09	10:28:23	8850	-22589	069	T+	-p	0.2696	2.3561	1.3704	325.5	208.5	82.3	6S	121W
5251	263	0174 Mar 06	13:26:59	8846	-22583	074	P	t-	0.5955	1.8024	0.7288	343.3	186.6	-	7N	161W
5252	263	0174 Aug 30	02:00:28	8841	-22577	079	T	a-	-0.4421	2.0236	1.0695	304.9	190.5	38.5	10S	7E
5253	263	0175 Feb 23	13:24:09	8837	-22571	084	N	t-	1.3133	0.4927	-0.5957	209.8	-	-	11N	160W
5254	263	0175 Jul 21	09:27:54	8833	-22566	051	N	-a	1.4385	0.2097	-0.7732	125.7	-	-	20S	105W
5255	263	0175 Aug 19	18:21:01	8832	-22565	089	N	a-	-1.1601	0.7127	-0.2544	215.0	-	-	15S	122E
5256	263	0176 Jan 14	01:09:46	8828	-22560	056	P	-a	-0.8943	1.2327	0.2016	284.3	101.0	-	21N	23E
5257	263	0176 Jul 09	18:58:00	8823	-22554	061	P	-t	0.7375	1.5227	0.4868	316.0	155.3	-	22S	113E
5258	263	0177 Jan 02	13:14:26	8819	-22548	066	T-	-p	-0.1701	2.5373	1.5543	323.1	209.9	92.2	23N	160W
5259	263	0177 Jun 28	21:36:29	8814	-22542	071	T-	pp	-0.0198	2.8563	1.7874	374.2	235.9	106.5	24S	72E
5260	263	0177 Dec 23	04:48:32	8810	-22536	076	P	a-	0.4978	1.9304	0.9585	303.0	183.9	-	24N	35W
5261	264	0178 Jun 17	22:27:50	8805	-22530	081	P	t-	-0.7776	1.4587	0.4040	319.8	146.3	-	24S	59E
5262	264	0178 Dec 12	19:12:35	8800	-22524	086	N	a-	1.1964	0.6655	-0.3403	214.2	-	-	24N	108E
5263	264	0179 May 08	19:13:37	8797	-22519	053	N	-h	1.1264	0.7803	-0.1984	229.0	-	-	16S	107E
5264	264	0179 Jun 07	04:36:23	8796	-22518	091	N	h-	-1.4922	0.1221	-0.8822	99.7	-	-	24S	34W
5265	264	0179 Nov 02	10:55:05	8792	-22513	058	P	-t	-0.9839	1.0963	0.0094	290.5	23.6	-	14N	130W
5266	264	0180 Apr 27	10:43:18	8787	-22507	063	T	-p	0.3549	2.1792	1.2339	309.7	198.9	67.4	13S	125W
5267	264	0180 Oct 21	10:27:05	8783	-22501	068	T	-p	-0.3088	2.3422	1.2415	371.1	223.4	75.0	11N	123W
5268	264	0181 Apr 17	03:45:05	8778	-22495	073	T	p-	-0.3774	2.1423	1.1881	311.6	198.2	61.7	10S	20W
5269	264	0181 Oct 10	11:39:46	8774	-22489	078	T	a-	0.3929	2.1688	1.1059	350.3	210.2	50.7	7N	141W
5270	264	0182 Apr 06	17:39:41	8769	-22483	083	N	h-	-1.1597	0.7308	-0.2708	227.7	-	-	7S	132E
5271	264	0182 Sep 29	20:00:09	8764	-22477	088	N	a-	1.0728	0.8909	-0.1121	243.8	-	-	3N	94E
5272	264	0183 Feb 25	08:49:05	8761	-22472	055	P	-t	0.9027	1.2445	0.1593	308.0	96.2	-	10N	91W
5273	264	0183 Aug 21	01:56:50	8756	-22466	060	P	-a	-0.8658	1.2449	0.2933	263.3	113.3	-	14S	8E
5274	264	0184 Feb 14	08:28:06	8751	-22460	065	T+	pp	0.1918	2.5487	1.4639	370.9	229.5	94.8	14N	86W
5275	264	0184 Aug 09	17:45:41	8747	-22454	070	T-	-p	-0.1584	2.5555	1.5789	327.8	213.8	94.9	16S	131E
5276	264	0185 Feb 02	12:11:36	8742	-22448	075	P	a-	-0.5240	1.9167	0.8766	331.8	191.6	-	16N	142W
5277	264	0185 Jul 30	04:37:03	8738	-22442	080	P	t-	0.6174	1.7392	0.7111	324.5	178.9	-	19S	32W
5278	264	0185 Dec 24	12:11:07	8734	-22437	047	N	-a	1.4945	0.1033	-0.8724	85.9	-	-	25N	145W
5279	264	0186 Jan 22	23:04:27	8733	-22436	085	N	a-	-1.1863	0.6754	-0.3132	212.4	-	-	19N	54E
5280	264	0186 Jun 19	17:18:41	8729	-22431	052	N	-t	-1.4429	0.2454	-0.8241	151.1	-	-	25S	136E

Cat Num	Canon Plate	Calendar Date	TD of Greatest Eclipse	ΔT s	Luna Num	Saros Num	Ecl. Type	QSE	Gamma	Pen. Mag.	Um. Mag.	Eclipse Phase Durations Pen. m	Par. m	Total m	Greatest in Zenith Lat.	Lng.
5281	265	0186 Jul 19	08:22:09	8728	-22430	090	N	t-	1.4088	0.3073	-0.7611	167.2	-	-	20S	89W
5282	265	0186 Dec 14	03:52:26	8725	-22425	057	P	-a	0.8493	1.2884	0.3105	269.7	116.8	-	24N	22W
5283	265	0187 Jun 08	19:04:52	8720	-22419	062	P	-t	-0.6630	1.6628	0.6204	328.3	173.2	-	23S	109E
5284	265	0187 Dec 03	17:04:34	8715	-22413	067	T+	-p	0.1893	2.5217	1.4999	338.3	215.5	91.8	22N	139E
5285	265	0188 May 28	03:15:54	8711	-22407	072	T+	pp	0.1357	2.6032	1.6145	336.6	218.8	98.0	21S	14W
5286	265	0188 Nov 21	23:47:05	8706	-22401	077	P	t-	-0.5240	1.9361	0.8575	345.6	195.2	-	20N	37E
5287	265	0189 May 17	17:55:32	8702	-22395	082	P	a-	0.8735	1.2287	0.2813	263.7	111.9	-	18S	126E
5288	265	0189 Nov 10	23:57:48	8697	-22389	087	N	t-	-1.2208	0.6710	-0.4345	242.3	-	-	16N	34E
5289	265	0190 Apr 08	03:34:57	8693	-22384	054	P	-a	-0.9390	1.1178	0.1517	259.0	84.7	-	8S	17W
5290	265	0190 Oct 01	08:08:00	8689	-22378	059	N	-h	1.0594	0.9363	-0.1080	261.4	-	-	4N	88W
5291	265	0191 Mar 28	15:46:07	8684	-22372	064	T-	-p	-0.2164	2.4711	1.4509	345.1	219.3	91.0	3S	161E
5292	265	0191 Sep 20	18:35:47	8679	-22366	069	T	-p	0.3014	2.2979	1.3118	323.5	206.0	76.9	1S	116E
5293	265	0192 Mar 16	20:37:36	8675	-22360	074	P	t-	0.5455	1.8935	0.8213	348.3	195.3	-	2N	89E
5294	265	0192 Sep 09	10:08:08	8670	-22354	079	T	p-	-0.4024	2.0978	1.1410	307.4	194.0	53.6	6S	117W
5295	265	0193 Mar 05	20:38:25	8666	-22348	084	N	t-	1.2688	0.5721	-0.5118	224.0	-	-	7N	90E
5296	265	0193 Jul 31	16:51:54	8662	-22343	051	N	-a	1.5044	0.0916	-0.8968	84.3	-	-	17S	144E
5297	265	0193 Aug 30	02:14:16	8661	-22342	089	N	a-	-1.1176	0.7931	-0.1791	225.8	-	-	11S	3E
5298	265	0194 Jan 24	09:26:58	8657	-22337	056	P	-a	-0.9053	1.2096	0.1840	281.2	96.5	-	19N	102W
5299	265	0194 Jul 21	01:49:29	8653	-22331	061	P	-t	0.8131	1.3867	0.3457	307.7	134.1	-	20S	9E
5300	265	0195 Jan 13	21:55:51	8648	-22325	066	T-	-p	-0.1777	2.5215	1.5420	322.0	209.4	91.6	21N	70E
5301	266	0195 Jul 10	04:03:03	8643	-22319	071	T+	pp	0.0641	2.7762	1.7051	374.4	235.6	105.5	23S	24W
5302	266	0196 Jan 03	13:37:39	8639	-22313	076	P	a-	0.4931	1.9387	0.9675	303.5	184.6	-	24N	166W
5303	266	0196 Jun 28	05:00:01	8634	-22307	081	P	t-	-0.6924	1.6143	0.5610	329.4	167.8	-	24S	39W
5304	266	0196 Dec 23	03:53:36	8630	-22301	086	N	a-	1.1971	0.6653	-0.3423	214.8	-	-	25N	22W
5305	266	0197 May 19	02:26:09	8626	-22296	053	N	-a	1.1963	0.6500	-0.3245	210.9	-	-	18S	2W
5306	266	0197 Jun 17	11:34:20	8625	-22295	091	N	a-	-1.4105	0.2703	-0.7308	145.8	-	-	25S	139W
5307	266	0197 Nov 12	18:58:42	8621	-22290	058	N*	-t	-0.9947	1.0784	-0.0122	289.5	-	-	17N	108E
5308	266	0198 May 08	18:14:34	8617	-22284	063	T	-a	0.4244	2.0505	1.1075	305.5	193.0	47.5	16S	121E
5309	266	0198 Nov 01	18:17:50	8612	-22278	068	T	-p	-0.3241	2.3148	1.2126	370.3	221.9	71.0	14N	118E
5310	266	0199 Apr 28	11:17:53	8607	-22272	073	T	p-	-0.3115	2.2632	1.3090	315.4	203.5	76.1	14S	135W
5311	266	0199 Oct 21	19:42:04	8603	-22266	078	T	a-	0.3744	2.2027	1.1400	350.4	211.4	57.5	11N	97E
5312	266	0200 Apr 17	00:57:37	8598	-22260	083	N	h-	-1.0994	0.8420	-0.1606	242.6	-	-	11S	21E
5313	266	0200 Oct 10	04:18:39	8594	-22254	088	N	a-	1.0489	0.9348	-0.0681	247.9	-	-	7N	32W
5314	266	0201 Mar 07	16:02:06	8590	-22249	055	P	-t	0.9505	1.1556	0.0727	299.8	65.8	-	6N	159E
5315	266	0201 Aug 31	09:58:02	8585	-22243	060	P	-a	-0.9106	1.1644	0.2095	257.3	96.9	-	10S	113W
5316	266	0202 Feb 24	15:53:20	8580	-22237	065	T+	pp	0.2301	2.4759	1.3963	368.4	227.3	90.1	10N	161E
5317	266	0202 Aug 21	01:27:30	8576	-22231	070	T-	-p	-0.2115	2.4608	1.4786	327.9	212.1	90.0	13S	14E
5318	266	0203 Feb 13	20:10:07	8571	-22225	075	P	a-	-0.4969	1.9632	0.9293	332.2	194.9	-	13N	97E
5319	266	0203 Aug 10	11:45:01	8567	-22219	080	P	t-	0.5553	1.8562	0.8221	331.9	189.3	-	16S	140W
5320	266	0204 Jan 04	20:59:22	8563	-22214	047	N	-a	1.4974	0.0968	-0.8765	83.1	-	-	24N	84E
5321	267	0204 Feb 03	07:31:42	8562	-22213	085	N	a-	-1.1668	0.7087	-0.2750	216.1	-	-	16N	73W
5322	267	0204 Jun 29	23:41:02	8558	-22208	052	Ne	-t	-1.5265	0.0924	-0.9779	94.0	-	-	25S	40E
5323	267	0204 Jul 29	15:00:30	8557	-22207	090	N	t-	1.3371	0.4409	-0.6314	198.3	-	-	18S	171E
5324	267	0204 Dec 24	12:43:10	8554	-22202	057	P	-a	0.8527	1.2824	0.3040	269.7	115.8	-	24N	154W
5325	267	0205 Jun 19	01:42:04	8549	-22196	062	P	-t	-0.7444	1.5122	0.4723	317.7	154.3	-	24S	9E
5326	267	0205 Dec 14	01:41:42	8544	-22190	067	T+	-p	0.1909	2.5200	1.4955	339.5	216.0	91.8	24N	10E
5327	267	0206 Jun 08	10:22:12	8540	-22184	072	T+	pp	0.0584	2.7432	1.7581	336.1	219.8	101.3	23S	121W
5328	267	0206 Dec 03	07:59:57	8535	-22178	077	P	t-	-0.5211	1.9430	0.8615	347.0	196.0	-	22N	86W
5329	267	0207 May 29	01:20:47	8531	-22172	082	P	a-	0.8000	1.3624	0.4171	272.9	133.4	-	21S	14E
5330	267	0207 Nov 22	07:56:46	8526	-22166	087	N	t-	-1.2136	0.6848	-0.4218	244.5	-	-	19N	86W
5331	267	0208 Apr 18	11:08:28	8522	-22161	054	P	-a	-1.0031	1.0003	0.0340	248.8	40.8	-	12S	132W
5332	267	0208 May 17	18:36:22	8521	-22160	092	Nb	a-	1.5180	0.0452	-0.9007	56.8	-	-	18S	115E
5333	267	0208 Oct 11	16:10:24	8518	-22155	059	N	-h	1.0771	0.9037	-0.1403	256.8	-	-	8N	150E
5334	267	0209 Apr 07	23:02:53	8513	-22149	064	T	-p	-0.2784	2.3577	1.3367	344.1	216.5	82.5	7S	50E
5335	267	0209 Oct 01	02:52:37	8508	-22143	069	T	-p	0.3259	2.2531	1.2667	321.6	203.8	72.1	3N	10W
5336	267	0210 Mar 28	03:36:38	8504	-22137	074	P	t-	0.4864	2.0012	0.9302	353.6	204.2	-	2S	17W
5337	267	0210 Sep 20	18:25:10	8499	-22131	079	T	p-	-0.3705	2.1576	1.1982	309.2	196.5	62.4	2S	118E
5338	267	0211 Mar 17	03:43:20	8495	-22125	084	N	t-	1.2164	0.6660	-0.4133	238.9	-	-	3N	18W
5339	267	0211 Sep 10	10:14:18	8490	-22119	089	N	a-	-1.0814	0.8623	-0.1153	234.5	-	-	7S	119W
5340	267	0212 Feb 04	17:38:20	8486	-22114	056	P	-a	-0.9219	1.1763	0.1566	277.0	89.1	-	16N	135E

Cat Num	Canon Plate	Calendar Date	TD of Greatest Eclipse	ΔT (s)	Luna Num	Saros Num	Ecl. Type	QSE	Gamma	Pen. Mag.	Um. Mag.	Eclipse Phase Durations Pen. (m)	Par. (m)	Total (m)	Greatest in Zenith Lat.	Lng.
5341	268	0212 Jul 31	08:43:57	8481	-22108	061	P	-t	0.8847	1.2581	0.2114	298.7	107.3	-	18S	95W
5342	268	0213 Jan 24	06:31:27	8477	-22102	066	T-	-p	-0.1906	2.4959	1.5204	320.8	208.7	90.5	19N	59W
5343	268	0213 Jul 20	10:33:28	8472	-22096	071	T+	pp	0.1446	2.6299	1.5560	373.7	233.6	100.8	21S	122W
5344	268	0214 Jan 13	22:22:45	8468	-22090	076	P	a-	0.4855	1.9521	0.9818	304.2	185.6	-	22N	63E
5345	268	0214 Jul 09	11:36:52	8463	-22084	081	P	t-	-0.6100	1.7651	0.7128	337.2	184.0	-	23S	139W
5346	268	0215 Jan 03	12:31:27	8458	-22078	086	N	a-	1.1960	0.6680	-0.3410	215.8	-	-	24N	150W
5347	268	0215 May 30	09:38:48	8455	-22073	053	N	-a	1.2666	0.5192	-0.4517	190.2	-	-	20S	111W
5348	268	0215 Jun 28	18:35:59	8454	-22072	091	N	a-	-1.3306	0.4156	-0.5829	177.6	-	-	25S	116E
5349	268	0215 Nov 24	03:05:30	8450	-22067	058	N*	-t	-1.0022	1.0661	-0.0275	289.0	-	-	20N	13W
5350	268	0216 May 19	01:42:04	8445	-22061	063	P	-a	0.4971	1.9161	0.9749	300.4	185.4	-	19S	8E
5351	268	0216 Nov 12	02:14:52	8441	-22055	068	T	-p	-0.3344	2.2963	1.1932	369.6	220.9	68.1	18N	2W
5352	268	0217 May 08	18:45:21	8436	-22049	073	T-	p-	-0.2419	2.3913	1.4366	318.7	208.0	86.6	17S	112E
5353	268	0217 Nov 01	03:51:19	8432	-22043	078	T	a-	0.3609	2.2270	1.1651	350.1	212.0	61.7	15N	26W
5354	268	0218 Apr 28	08:09:17	8427	-22037	083	N	t-	-1.0348	0.9613	-0.0428	257.0	-	-	15S	89W
5355	268	0218 Oct 21	12:44:34	8422	-22031	088	N	a-	1.0308	0.9678	-0.0347	250.6	-	-	12N	160W
5356	268	0219 Mar 18	23:06:50	8419	-22026	055	N*	-t	1.0046	1.0552	-0.0254	289.7	-	-	2N	51E
5357	268	0219 Sep 11	18:06:04	8414	-22020	060	P	-a	-0.9496	1.0945	0.1361	251.7	78.9	-	6S	123E
5358	268	0220 Mar 06	23:10:55	8409	-22014	065	T	pp	0.2751	2.3906	1.3164	365.4	224.3	83.0	6N	51E
5359	268	0220 Aug 31	09:15:47	8405	-22008	070	T-	-p	-0.2587	2.3773	1.3892	327.8	210.1	84.2	9S	104W
5360	268	0221 Feb 24	03:59:16	8400	-22002	075	P	a-	-0.4613	2.0253	0.9979	333.0	198.8	-	9N	21W
5361	269	0221 Aug 20	19:00:25	8396	-21996	080	P	t-	0.5000	1.9609	0.9204	338.2	197.4	-	12S	110E
5362	269	0222 Jan 15	05:42:03	8392	-21991	047	N	-a	1.5053	0.0810	-0.8894	76.0	-	-	23N	47W
5363	269	0222 Feb 13	15:50:36	8391	-21990	085	N	a-	-1.1400	0.7553	-0.2234	221.3	-	-	12N	161E
5364	269	0222 Aug 09	21:46:20	8386	-21984	090	N	t-	1.2713	0.5637	-0.5126	222.1	-	-	15S	69E
5365	269	0223 Jan 04	21:29:18	8382	-21979	057	P	-a	0.8598	1.2695	0.2909	269.1	113.7	-	24N	75E
5366	269	0223 Jun 30	08:23:17	8378	-21973	062	P	-t	-0.8231	1.3666	0.3288	306.1	131.2	-	24S	91W
5367	269	0223 Dec 25	10:16:37	8373	-21967	067	T+	-p	0.1937	2.5160	1.4894	340.6	216.4	91.6	24N	118W
5368	269	0224 Jun 18	17:28:31	8369	-21961	072	T-	pp	-0.0198	2.8128	1.8305	334.9	219.4	101.8	24S	132E
5369	269	0224 Dec 13	16:13:39	8364	-21955	077	P	t-	-0.5194	1.9471	0.8635	348.1	196.6	-	23N	151E
5370	269	0225 Jun 08	08:44:11	8359	-21949	082	P	a-	0.7244	1.5004	0.5564	281.3	150.6	-	22S	97W
5371	269	0225 Dec 02	15:58:28	8355	-21943	087	N	t-	-1.2085	0.6945	-0.4125	246.0	-	-	21N	154E
5372	269	0226 Apr 29	18:35:22	8351	-21938	054	N	-a	-1.0721	0.8742	-0.0929	236.3	-	-	15S	114E
5373	269	0226 May 29	02:00:34	8350	-21937	092	N	a-	1.4498	0.1708	-0.7760	109.4	-	-	20S	3E
5374	269	0226 Oct 23	00:21:58	8346	-21932	059	N	-h	1.0884	0.8828	-0.1606	253.6	-	-	12N	26E
5375	269	0227 Apr 19	06:12:14	8342	-21926	064	T	-h	-0.3455	2.2351	1.2132	342.3	212.3	68.8	11S	59W
5376	269	0227 Oct 12	11:16:57	8337	-21920	069	T	-p	0.3447	2.2188	1.2320	320.0	201.9	67.8	8N	138W
5377	269	0228 Apr 07	10:28:01	8332	-21914	074	T	t-	0.4216	2.1195	1.0499	358.7	212.5	36.3	6S	122W
5378	269	0228 Oct 01	02:49:12	8328	-21908	079	T	p-	-0.3448	2.2062	1.2439	310.7	198.4	68.1	3N	10W
5379	269	0229 Mar 27	10:40:07	8323	-21902	084	N	t-	1.1570	0.7727	-0.3020	254.0	-	-	1S	124W
5380	269	0229 Sep 20	18:21:52	8319	-21896	089	N	a-	-1.0522	0.9187	-0.0642	241.5	-	-	2S	118E
5381	270	0230 Feb 15	01:43:28	8315	-21891	056	P	-a	-0.9438	1.1329	0.1193	271.9	77.9	-	12N	13E
5382	270	0230 Aug 11	15:44:30	8310	-21885	061	P	-t	0.9503	1.1406	0.0882	289.3	70.7	-	15S	159E
5383	270	0231 Feb 04	15:00:59	8306	-21879	066	T-	-p	-0.2087	2.4606	1.4892	319.4	207.8	88.8	16N	174E
5384	270	0231 Jul 31	17:10:42	8301	-21873	071	T+	pp	0.2192	2.4944	1.4178	372.1	230.2	92.7	19S	138E
5385	270	0232 Jan 25	07:01:26	8296	-21867	076	T	a-	0.4729	1.9745	1.0055	305.2	187.3	11.0	20N	67W
5386	270	0232 Jul 19	18:19:51	8292	-21861	081	P	t-	-0.5317	1.9084	0.8567	343.3	196.2	-	22S	120E
5387	270	0233 Jan 13	21:03:10	8287	-21855	086	N	a-	1.1909	0.6777	-0.3323	217.8	-	-	23N	82E
5388	270	0233 Jun 09	16:52:15	8283	-21850	053	N	-a	1.3365	0.3894	-0.5785	166.2	-	-	22S	140E
5389	270	0233 Jul 09	01:43:24	8283	-21849	091	N	a-	-1.2544	0.5544	-0.4420	201.6	-	-	24S	9E
5390	270	0233 Dec 04	11:12:06	8279	-21844	058	N*	-t	-1.0091	1.0546	-0.0410	288.4	-	-	22N	135W
5391	270	0234 May 30	09:08:40	8274	-21838	063	P	-a	0.5705	1.7809	0.8406	294.6	176.1	-	21S	104W
5392	270	0234 Nov 23	10:16:03	8269	-21832	068	T	-p	-0.3410	2.2844	1.1812	369.0	220.1	66.2	20N	122W
5393	270	0235 May 20	02:08:10	8265	-21826	073	T-	pp	-0.1688	2.5259	1.5700	321.6	211.6	94.0	20S	1E
5394	270	0235 Nov 12	12:07:20	8260	-21820	078	T	a-	0.3524	2.2419	1.1813	349.4	212.2	64.2	18N	151W
5395	270	0236 May 08	15:13:02	8256	-21814	083	P	t-	-0.9647	1.0907	0.0847	271.1	66.9	-	18S	164E
5396	270	0236 Oct 31	21:18:31	8251	-21808	088	N	a-	1.0185	0.9900	-0.0118	252.2	-	-	15N	71E
5397	270	0237 Mar 29	05:59:54	8247	-21803	055	N	-t	1.0676	0.9385	-0.1397	276.6	-	-	2S	54W
5398	270	0237 Sep 22	02:22:53	8242	-21797	060	P	-a	-0.9814	1.0381	0.0760	247.0	59.5	-	2S	3W
5399	270	0238 Mar 18	06:20:14	8238	-21791	065	T	-p	0.3274	2.2919	1.2232	361.6	220.1	72.1	1N	58W
5400	270	0238 Sep 11	17:10:37	8233	-21785	070	T	-p	-0.2995	2.3054	1.3113	327.8	208.0	77.5	5S	136E

Cat Num	Canon Plate	Calendar Date	TD of Greatest Eclipse	ΔT s	Luna Num	Saros Num	Ecl. Type	QSE	Gamma	Pen. Mag.	Um. Mag.	Pen. m	Par. m	Total m	Lat.	Lng.
5401	271	0239 Mar 07	11:42:56	8229	-21779	075	T	a-	-0.4203	2.0973	1.0764	333.9	202.9	42.5	5N	138W
5402	271	0239 Sep 01	02:21:20	8224	-21773	080	T	t-	0.4501	2.0557	1.0088	343.6	203.9	15.1	9S	1W
5403	271	0240 Jan 26	14:20:22	8220	-21768	047	N	-a	1.5168	0.0582	-0.9092	64.4	-	-	20N	176W
5404	271	0240 Feb 25	00:03:13	8219	-21767	085	N	a-	-1.1079	0.8117	-0.1618	227.4	-	-	9N	37E
5405	271	0240 Aug 20	04:37:52	8215	-21761	090	N	t-	1.2099	0.6784	-0.4021	241.4	-	-	12S	35W
5406	271	0241 Jan 15	06:11:26	8211	-21756	057	P	-a	0.8699	1.2507	0.2724	268.1	110.5	-	22N	55W
5407	271	0241 Jul 10	15:10:04	8206	-21750	062	P	-t	-0.8980	1.2284	0.1922	293.7	102.0	-	23S	167E
5408	271	0242 Jan 04	18:46:22	8202	-21744	067	T+	-p	0.2001	2.5051	1.4769	341.6	216.8	91.2	23N	115E
5409	271	0242 Jun 30	00:40:21	8197	-21738	072	T-	pp	-0.0943	2.6748	1.6948	333.0	217.9	99.6	24S	24E
5410	271	0242 Dec 25	00:23:45	8192	-21732	077	P	t-	-0.5150	1.9558	0.8707	349.4	197.6	-	23N	29E
5411	271	0243 Jun 19	16:09:20	8188	-21726	082	P	a-	0.6500	1.6366	0.6934	288.6	164.4	-	23S	151E
5412	271	0243 Dec 13	23:59:41	8183	-21720	087	N	t-	-1.2027	0.7049	-0.4016	247.5	-	-	22N	34E
5413	271	0244 May 10	01:56:36	8179	-21715	054	N	-a	-1.1449	0.7411	-0.2273	221.3	-	-	19S	3E
5414	271	0244 Jun 08	09:23:11	8179	-21714	092	N	a-	1.3804	0.2989	-0.6495	143.4	-	-	22S	108W
5415	271	0244 Nov 02	08:40:26	8175	-21709	059	N	-h	1.0945	0.8709	-0.1714	251.3	-	-	16N	100W
5416	271	0245 Apr 29	13:11:59	8170	-21703	064	T	-h	-0.4193	2.1003	1.0770	339.5	206.1	43.3	15S	165W
5417	271	0245 Oct 22	19:49:31	8165	-21697	069	T	-p	0.3572	2.1960	1.2091	318.7	200.5	64.8	12N	93E
5418	271	0246 Apr 18	17:10:07	8161	-21691	074	T	t-	0.3496	2.2511	1.1824	363.4	220.0	66.6	10S	136E
5419	271	0246 Oct 12	11:20:31	8156	-21685	079	T	p-	-0.3253	2.2434	1.2785	311.9	199.7	71.9	7N	139W
5420	271	0247 Apr 07	17:29:14	8152	-21679	084	N	t-	1.0906	0.8921	-0.1780	269.0	-	-	6S	132E
5421	272	0247 Oct 02	02:36:48	8147	-21673	089	N	a-	-1.0295	0.9631	-0.0253	247.0	-	-	2N	7W
5422	272	0248 Feb 26	09:40:27	8143	-21668	056	P	-a	-0.9733	1.0757	0.0684	265.4	59.3	-	8N	108W
5423	272	0248 Aug 21	22:50:28	8138	-21662	061	N*	-t	1.0104	1.0334	-0.0250	279.7	-	-	11S	51E
5424	272	0249 Feb 14	23:23:23	8134	-21656	066	T-	-p	-0.2332	2.4136	1.4464	317.8	206.5	86.3	13N	47E
5425	272	0249 Aug 10	23:54:59	8129	-21650	071	T	pp	0.2879	2.3699	1.2902	369.8	225.8	81.1	16S	36E
5426	272	0250 Feb 04	15:33:17	8125	-21644	076	T	p-	0.4551	2.0065	1.0390	306.6	189.4	29.1	17N	165E
5427	272	0250 Jul 31	01:10:55	8120	-21638	081	P	t-	-0.4591	2.0414	0.9900	348.0	205.3	-	19S	17E
5428	272	0251 Jan 25	05:29:35	8115	-21632	086	N	a-	1.1825	0.6936	-0.3171	220.7	-	-	20N	44W
5429	272	0251 Jun 21	00:07:12	8111	-21627	053	N	-a	1.4057	0.2612	-0.7044	137.3	-	-	22S	32E
5430	272	0251 Jul 20	08:55:40	8111	-21626	091	N	a-	-1.1810	0.6883	-0.3065	220.8	-	-	22S	99W
5431	272	0251 Dec 15	19:19:16	8107	-21621	058	N*	-t	-1.0149	1.0446	-0.0525	287.9	-	-	23N	104E
5432	272	0252 Jun 09	16:34:22	8102	-21615	063	P	-a	0.6446	1.6448	0.7050	287.9	164.9	-	22S	144E
5433	272	0252 Dec 03	18:19:33	8098	-21609	068	T	-p	-0.3462	2.2744	1.1719	368.2	219.4	64.6	22N	117E
5434	272	0253 May 30	09:27:57	8093	-21603	073	T-	pp	-0.0938	2.6646	1.7068	323.9	214.1	98.6	22S	110W
5435	272	0253 Nov 22	20:27:42	8088	-21597	078	T	p-	0.3467	2.2513	1.1927	348.5	212.2	65.7	21N	84E
5436	272	0254 May 19	22:13:01	8084	-21591	083	P	t-	-0.8921	1.2253	0.2169	284.3	105.4	-	21S	58E
5437	272	0254 Nov 12	05:56:12	8079	-21585	088	P	a-	1.0088	1.0072	0.0064	253.2	18.0	-	19N	59W
5438	272	0255 Apr 09	12:46:05	8075	-21580	055	N	-t	1.1356	0.8126	-0.2634	260.7	-	-	6S	157W
5439	272	0255 Oct 03	10:46:26	8071	-21574	060	P	-a	-1.0076	0.9917	0.0260	243.0	35.1	-	3N	130W
5440	272	0256 Mar 28	13:21:07	8066	-21568	065	T	-t	0.3871	2.1796	1.1163	356.9	214.4	53.9	3S	165W
5441	273	0256 Sep 22	01:12:44	8061	-21562	070	T	-p	-0.3336	2.2458	1.2458	327.8	206.0	70.6	1S	14E
5442	273	0257 Mar 17	19:17:48	8057	-21556	075	T	a-	-0.3712	2.1841	1.1697	334.9	207.1	61.4	1N	106E
5443	273	0257 Sep 11	09:50:39	8052	-21550	080	T	p-	0.4077	2.1367	1.0834	348.0	208.9	45.4	4S	115W
5444	273	0258 Feb 05	22:50:15	8048	-21545	047	Ne	-a	1.5356	0.0220	-0.9420	39.7	-	-	17N	56E
5445	273	0258 Mar 07	08:06:48	8047	-21544	085	N	a-	-1.0679	0.8826	-0.0860	234.7	-	-	4N	86W
5446	273	0258 Aug 31	11:39:32	8043	-21538	090	N	t-	1.1564	0.7787	-0.3060	256.6	-	-	8S	142W
5447	273	0259 Jan 26	14:46:56	8039	-21533	057	P	-a	0.8855	1.2219	0.2441	266.3	105.2	-	20N	176E
5448	273	0259 Jul 21	22:02:33	8034	-21527	062	P	-t	-0.9690	1.0975	0.0625	280.6	59.1	-	22S	64E
5449	273	0260 Jan 16	03:11:11	8030	-21521	067	T+	-p	0.2100	2.4875	1.4582	342.5	216.9	90.3	21N	11W
5450	273	0260 Jul 10	07:55:34	8025	-21515	072	T-	pp	-0.1671	2.5403	1.5621	330.4	215.2	94.9	23S	85W
5451	273	0261 Jan 04	08:29:58	8020	-21509	077	P	t-	-0.5081	1.9688	0.8831	350.8	198.9	-	22N	92W
5452	273	0261 Jun 29	23:35:54	8016	-21503	082	P	a-	0.5764	1.7717	0.8285	294.8	175.5	-	23S	39E
5453	273	0261 Dec 24	08:00:34	8011	-21497	087	N	t-	-1.1962	0.7161	-0.3891	249.0	-	-	23N	86W
5454	273	0262 May 21	09:13:33	8007	-21492	054	N	-a	-1.2203	0.6037	-0.3666	203.2	-	-	21S	107W
5455	273	0262 Jun 19	16:45:31	8006	-21491	092	N	a-	1.3112	0.4270	-0.5236	169.7	-	-	22S	141E
5456	273	0262 Nov 13	17:04:33	8003	-21486	059	N	-a	1.0970	0.8656	-0.1751	249.9	-	-	19N	134E
5457	273	0263 May 10	20:06:35	7998	-21480	064	P	-t	-0.4964	1.9598	0.9346	335.5	197.8	-	18S	90E
5458	273	0263 Nov 03	04:28:18	7993	-21474	069	T	-p	0.3650	2.1814	1.1947	317.6	199.4	62.6	15N	37W
5459	273	0264 Apr 28	23:44:49	7989	-21468	074	T+	pp	0.2719	2.3933	1.3254	367.5	226.4	84.7	14S	36E
5460	273	0264 Oct 22	19:58:27	7984	-21462	079	T	p-	-0.3114	2.2702	1.3027	312.8	200.6	74.3	11N	90E

Column groups: Eclipse Phase Durations (Pen., Par., Total); Greatest in Zenith (Lat., Lng.).

Cat Num	Canon Plate	Calendar Date	TD of Greatest Eclipse	ΔT	Luna Num	Saros Num	Ecl. Type	QSE	Gamma	Pen. Mag.	Um. Mag.	Eclipse Phase Durations Pen.	Par.	Total	Greatest in Zenith Lat.	Lng.
				s								m	m	m		
5461	274	0265 Apr 18	00:12:24	7979	-21456	084	N*	t-	1.0186	1.0221	-0.0437	283.2	-	-	10S	30E
5462	274	0265 Oct 12	10:58:04	7975	-21450	089	P	a-	-1.0125	0.9968	0.0033	251.3	12.8	-	7N	134W
5463	274	0266 Mar 08	17:30:47	7971	-21445	056	P	-a	-1.0088	1.0073	0.0063	257.5	18.1	-	4N	133E
5464	274	0266 Sep 02	06:04:50	7966	-21439	061	N	-t	1.0628	0.9402	-0.1240	270.5	-	-	7S	59W
5465	274	0267 Feb 26	07:38:57	7961	-21433	066	T-	-p	-0.2635	2.3557	1.3928	316.0	204.8	82.6	9N	78W
5466	274	0267 Aug 22	06:47:02	7957	-21427	071	T	-t	0.3501	2.2574	1.1746	367.1	220.7	65.5	12S	68W
5467	274	0268 Feb 15	23:57:16	7952	-21421	076	T	p-	0.4311	2.0497	1.0836	308.4	192.1	42.2	13N	38E
5468	274	0268 Aug 10	08:10:56	7948	-21415	081	T	t-	-0.3933	2.1622	1.1108	351.3	212.0	52.2	16S	89W
5469	274	0269 Feb 04	13:46:02	7943	-21409	086	N	a-	1.1670	0.7221	-0.2888	225.2	-	-	17N	169W
5470	274	0269 Jul 01	07:26:01	7939	-21404	053	N	-a	1.4722	0.1383	-0.8256	100.7	-	-	22S	78W
5471	274	0269 Jul 30	16:16:33	7938	-21403	091	N	a-	-1.1137	0.8113	-0.1825	235.9	-	-	20S	150E
5472	274	0269 Dec 26	03:22:47	7934	-21398	058	N*	-t	-1.0229	1.0303	-0.0673	286.8	-	-	23N	16W
5473	274	0270 Jun 21	00:01:05	7930	-21392	063	P	-a	0.7179	1.5104	0.5703	280.4	151.5	-	23S	32E
5474	274	0270 Dec 15	02:22:52	7925	-21386	068	T	-p	-0.3516	2.2636	1.1627	367.3	218.7	63.0	23N	3W
5475	274	0271 Jun 10	16:45:36	7920	-21380	073	T-	pp	-0.0177	2.8055	1.8453	325.5	215.4	100.6	23S	140E
5476	274	0271 Dec 04	04:51:46	7916	-21374	078	T	p-	0.3434	2.2561	1.2000	347.4	211.9	66.7	23N	42W
5477	274	0272 May 30	05:06:03	7911	-21368	083	P	t-	-0.8147	1.3688	0.3574	296.9	132.9	-	23S	45W
5478	274	0272 Nov 22	14:39:36	7906	-21362	088	P	a-	1.0035	1.0162	0.0170	253.3	29.1	-	22N	170E
5479	274	0273 Apr 19	19:22:51	7903	-21357	055	N	-t	1.2102	0.6747	-0.3993	240.8	-	-	10S	102E
5480	274	0273 Oct 13	19:17:34	7898	-21351	060	N	-a	-1.0274	0.9571	-0.0121	240.0	-	-	7N	101E
5481	275	0274 Apr 08	20:15:43	7893	-21345	065	P	-t	0.4523	2.0572	0.9993	351.2	207.1	-	7S	89E
5482	275	0274 Oct 03	09:22:12	7889	-21339	070	T	-a	-0.3606	2.1993	1.1933	328.0	204.2	63.8	4N	110W
5483	275	0275 Mar 29	02:47:37	7884	-21333	075	T	a-	-0.3170	2.2804	1.2724	335.8	211.0	75.0	3S	8W
5484	275	0275 Sep 22	17:25:49	7879	-21327	080	T	p-	0.3710	2.2073	1.1478	351.9	212.8	59.2	0S	129E
5485	275	0276 Mar 17	16:04:41	7875	-21321	085	N	a-	-1.0228	0.9629	-0.0008	242.3	-	-	0N	153E
5486	275	0276 Sep 10	18:48:49	7870	-21315	090	N	t-	1.1088	0.8681	-0.2206	268.9	-	-	4S	109E
5487	275	0277 Feb 05	23:15:03	7866	-21310	057	P	-a	0.9069	1.1823	0.2052	263.6	97.2	-	17N	49E
5488	275	0277 Aug 01	05:03:37	7861	-21304	062	N	-t	-1.0338	0.9783	-0.0559	267.4	-	-	19S	42W
5489	275	0278 Jan 26	11:28:31	7857	-21298	067	T+	-p	0.2255	2.4594	1.4294	343.2	216.8	88.7	19N	135W
5490	275	0278 Jul 21	15:17:39	7852	-21292	072	T-	-p	-0.2349	2.4154	1.4384	327.4	211.7	87.9	21S	164E
5491	275	0279 Jan 15	16:29:20	7847	-21286	077	P	t-	-0.4961	1.9910	0.9051	352.5	201.0	-	21N	149E
5492	275	0279 Jul 11	07:06:47	7843	-21280	082	P	a-	0.5060	1.9010	0.9573	300.1	184.3	-	22S	74W
5493	275	0280 Jan 04	15:57:05	7838	-21274	087	N	t-	-1.1860	0.7337	-0.3694	251.3	-	-	22N	155E
5494	275	0280 May 31	16:26:06	7834	-21269	054	N	-a	-1.2984	0.4618	-0.5110	180.8	-	-	23S	145E
5495	275	0280 Jun 30	00:08:08	7833	-21268	092	N	a-	1.2426	0.5544	-0.3991	191.5	-	-	22S	30E
5496	275	0280 Nov 24	01:33:04	7830	-21263	059	N	-a	1.0967	0.8650	-0.1735	248.9	-	-	22N	7E
5497	275	0281 May 21	02:54:54	7825	-21257	064	P	-t	-0.5777	1.8119	0.7843	330.1	186.7	-	21S	13W
5498	275	0281 Nov 13	13:11:53	7820	-21251	069	T	-p	0.3695	2.1729	1.1868	316.6	198.7	61.4	19N	169W
5499	275	0282 May 10	06:13:40	7816	-21245	074	T+	pp	0.1896	2.5441	1.4766	370.7	231.3	96.5	17S	62W
5500	275	0282 Nov 03	04:42:37	7811	-21239	079	T	p-	-0.3025	2.2875	1.3179	313.5	201.2	75.8	15N	42W
5501	276	0283 Apr 29	06:49:36	7806	-21233	084	P	t-	0.9408	1.1627	0.1011	296.7	76.8	-	13S	71W
5502	276	0283 Oct 23	19:26:01	7802	-21227	089	P	a-	-1.0017	1.0192	0.0207	254.5	32.2	-	11N	98E
5503	276	0284 Mar 19	01:13:41	7798	-21222	056	N	-a	-1.0511	0.9265	-0.0682	248.0	-	-	0S	16E
5504	276	0284 Sep 12	13:26:43	7793	-21216	061	N	-t	1.1079	0.8605	-0.2098	261.9	-	-	3S	171W
5505	276	0285 Mar 08	15:45:39	7788	-21210	066	T	-p	-0.3015	2.2839	1.3252	313.8	202.4	77.0	4N	159E
5506	276	0285 Sep 01	13:48:43	7784	-21204	071	T	-t	0.4042	2.1596	1.0737	364.2	215.2	44.0	8S	175W
5507	276	0286 Feb 26	08:13:46	7779	-21198	076	T	p-	0.4015	2.1033	1.1387	310.4	195.2	53.5	9N	87W
5508	276	0286 Aug 21	15:19:30	7774	-21192	081	T	t-	-0.3336	2.2718	1.2203	353.6	216.7	70.8	13S	163E
5509	276	0287 Feb 15	21:55:26	7770	-21186	086	N	a-	1.1467	0.7595	-0.2516	230.8	-	-	14N	68E
5510	276	0287 Jul 12	14:48:45	7766	-21181	053	Ne	-a	1.5359	0.0208	-0.9419	39.4	-	-	21S	171E
5511	276	0287 Aug 10	23:44:14	7765	-21180	091	N	a-	-1.0511	0.9258	-0.0673	248.2	-	-	17S	37E
5512	276	0288 Jan 06	11:23:04	7761	-21175	058	N*	-t	-1.0332	1.0113	-0.0861	285.1	-	-	22N	136W
5513	276	0288 Jul 01	07:28:44	7757	-21169	063	P	-a	0.7903	1.3781	0.4369	272.1	135.4	-	23S	80W
5514	276	0288 Dec 25	10:25:39	7752	-21163	068	T	-p	-0.3579	2.2510	1.1524	366.2	218.0	61.1	23N	123W
5515	276	0289 Jun 21	00:03:20	7747	-21157	073	T+	pp	0.0577	2.7336	1.7703	326.5	215.5	100.1	24S	31E
5516	276	0289 Dec 14	13:15:32	7743	-21151	078	T	p-	0.3392	2.2621	1.2093	346.2	211.8	67.8	24N	167W
5517	276	0290 Jun 10	11:58:47	7738	-21145	083	P	t-	-0.7378	1.5117	0.4967	308.2	154.0	-	24S	149W
5518	276	0290 Dec 03	23:24:25	7733	-21139	088	P	a-	0.9992	1.0230	0.0259	253.3	35.8	-	23N	39E
5519	276	0291 May 01	01:54:20	7729	-21134	055	N	-t	1.2885	0.5300	-0.5422	216.4	-	-	14S	3E
5520	276	0291 May 30	16:42:48	7729	-21133	093	Nb	t-	-1.5493	0.0457	-1.0149	65.7	-	-	23S	140E

Cat Num	Canon Plate	Calendar Date	TD of Greatest Eclipse	ΔT s	Luna Num	Saros Num	Ecl. Type	QSE	Gamma	Pen. Mag.	Um. Mag.	Eclipse Phase Durations Pen. m	Par. m	Total m	Greatest in Zenith Lat.	Lng.
5521	277	0291 Oct 25	03:54:20	7725	-21128	060	N	-a	-1.0426	0.9310	-0.0416	237.8	-	-	11N	29W
5522	277	0292 Apr 19	03:04:17	7720	-21122	065	P	-t	0.5230	1.9249	0.8721	344.4	197.6	-	11S	14W
5523	277	0292 Oct 13	17:38:27	7715	-21116	070	T	-a	-0.3816	2.1636	1.1520	328.4	202.8	57.5	8N	125E
5524	277	0293 Apr 08	10:09:32	7711	-21110	075	T-	p-	-0.2551	2.3909	1.3890	336.6	214.5	85.7	7S	120W
5525	277	0293 Oct 03	01:10:04	7706	-21104	080	T	p-	0.3421	2.2632	1.1977	355.1	215.7	67.5	4N	12E
5526	277	0294 Mar 28	23:54:39	7701	-21098	085	P	a-	-0.9706	1.0563	0.0972	250.6	67.7	-	4S	34E
5527	277	0294 Sep 22	02:07:28	7697	-21092	090	N	t-	1.0685	0.9441	-0.1485	278.7	-	-	0N	2W
5528	277	0295 Feb 17	07:34:50	7693	-21087	057	P	-a	0.9350	1.1304	0.1541	259.7	85.0	-	13N	77W
5529	277	0295 Aug 12	12:13:01	7688	-21081	062	N	-t	-1.0926	0.8701	-0.1636	254.2	-	-	16S	150W
5530	277	0296 Feb 06	19:37:21	7683	-21075	067	T+	-p	0.2472	2.4196	1.3892	343.5	216.3	86.0	16N	102E
5531	277	0296 Jul 31	22:45:33	7679	-21069	072	T	-p	-0.2985	2.2983	1.3219	324.0	207.4	78.4	19S	52E
5532	277	0297 Jan 26	00:22:08	7674	-21063	077	P	t-	-0.4791	2.0218	0.9366	354.4	203.7	-	18N	30E
5533	277	0297 Jul 21	14:41:54	7669	-21057	082	T	a-	0.4391	2.0245	1.0796	304.4	191.3	41.2	20S	172E
5534	277	0298 Jan 14	23:49:19	7665	-21051	087	N	t-	-1.1720	0.7580	-0.3423	254.5	-	-	20N	38E
5535	277	0298 Jun 11	23:37:34	7661	-21046	054	N	-a	-1.3764	0.3202	-0.6557	153.2	-	-	25S	36E
5536	277	0298 Jul 11	07:33:40	7660	-21045	092	N	a-	1.1766	0.6772	-0.2798	209.7	-	-	21S	81W
5537	277	0298 Dec 05	10:05:11	7656	-21040	059	N	-a	1.0946	0.8675	-0.1683	248.3	-	-	24N	121W
5538	277	0299 Jun 01	09:39:17	7651	-21034	064	P	-t	-0.6611	1.6602	0.6298	323.3	172.2	-	23S	114W
5539	277	0299 Nov 24	21:59:39	7647	-21028	069	T	-p	0.3711	2.1696	1.1845	315.9	198.2	61.0	21N	60E
5540	277	0300 May 20	12:38:15	7642	-21022	074	T+	pp	0.1038	2.7015	1.6341	373.0	234.6	103.7	20S	159W
5541	278	0300 Nov 13	13:30:59	7637	-21016	079	T	p-	-0.2975	2.2978	1.3262	314.0	201.5	76.5	18N	174W
5542	278	0301 May 09	13:23:23	7633	-21010	084	P	t-	0.8594	1.3102	0.2524	309.0	118.5	-	17S	170W
5543	278	0301 Nov 03	03:58:53	7628	-21004	089	P	a-	-0.9953	1.0331	0.0301	256.7	38.9	-	14N	31W
5544	278	0302 Mar 30	08:50:58	7624	-20999	056	N	-a	-1.0986	0.8363	-0.1524	236.8	-	-	5S	100W
5545	278	0302 Sep 23	20:55:33	7619	-20993	061	N	-t	1.1466	0.7923	-0.2836	254.0	-	-	1N	76E
5546	278	0303 Mar 19	23:45:52	7615	-20987	066	T	-p	-0.3453	2.2016	1.2468	311.2	199.4	68.8	0N	37E
5547	278	0303 Sep 12	20:59:20	7610	-20981	071	P	-t	0.4510	2.0753	0.9864	361.3	209.6	-	4S	76E
5548	278	0304 Mar 08	16:20:17	7605	-20975	076	T	p-	0.3638	2.1717	1.2086	312.8	198.7	64.3	5N	150E
5549	278	0304 Aug 31	22:38:34	7601	-20969	081	T	p-	-0.2817	2.3673	1.3155	355.0	220.0	81.6	9S	52E
5550	278	0305 Feb 26	05:53:39	7596	-20963	086	N	a-	1.1180	0.8122	-0.1990	238.2	-	-	10N	53W
5551	278	0305 Aug 21	07:22:06	7591	-20957	091	P	a-	-0.9958	1.0270	0.0342	257.8	41.8	-	13S	78W
5552	278	0306 Jan 16	19:15:47	7587	-20952	058	N	-t	-1.0490	0.9817	-0.1148	282.0	-	-	20N	106E
5553	278	0306 Jul 12	15:00:40	7583	-20946	063	P	-a	0.8591	1.2526	0.3099	263.4	116.2	-	21S	167E
5554	278	0307 Jan 05	18:25:38	7578	-20940	068	T	-p	-0.3666	2.2335	1.1380	364.9	217.0	58.4	22N	118E
5555	278	0307 Jul 02	07:20:25	7573	-20934	073	T+	-p	0.1328	2.5977	1.6306	326.8	214.5	96.9	23S	79W
5556	278	0307 Dec 25	21:40:21	7569	-20928	078	T	p-	0.3352	2.2676	1.2187	345.0	211.6	69.0	24N	68E
5557	278	0308 Jun 20	18:48:20	7564	-20922	083	P	t-	-0.6593	1.6577	0.6386	318.5	171.2	-	24S	109E
5558	278	0308 Dec 14	08:10:56	7559	-20916	088	P	a-	0.9957	1.0281	0.0335	253.1	40.5	-	24N	92W
5559	278	0309 May 11	08:19:25	7555	-20911	055	N	-t	1.3715	0.3770	-0.6936	185.1	-	-	17S	95W
5560	278	0309 Jun 09	23:06:16	7554	-20910	093	N	t-	-1.4661	0.1991	-0.8632	135.6	-	-	24S	43E
5561	279	0309 Nov 04	12:37:24	7551	-20905	060	N	-a	-1.0526	0.9141	-0.0615	236.5	-	-	15N	161W
5562	279	0310 Apr 30	09:49:20	7546	-20899	065	P	-t	0.5974	1.7859	0.7380	336.3	185.6	-	14S	117W
5563	279	0310 Oct 25	01:59:52	7541	-20893	070	T	-a	-0.3976	2.1369	1.1199	329.0	201.7	51.7	12N	2W
5564	279	0311 Apr 19	17:28:26	7536	-20887	075	T-	p-	-0.1896	2.5080	1.5119	336.9	217.2	93.5	11S	129E
5565	279	0311 Oct 14	09:00:15	7532	-20881	080	T	p-	0.3188	2.3088	1.2376	357.8	218.0	73.0	8N	107W
5566	279	0312 Apr 08	07:38:25	7527	-20875	085	P	a-	-0.9128	1.1602	0.2054	258.9	96.7	-	8S	83W
5567	279	0312 Oct 02	09:34:39	7522	-20869	090	N*	t-	1.0348	1.0077	-0.0885	286.3	-	-	5N	115W
5568	279	0313 Feb 27	15:46:26	7518	-20864	057	P	-a	0.9693	1.0669	0.0914	254.6	66.2	-	9N	159E
5569	279	0313 Aug 22	19:32:44	7514	-20858	062	N	-t	-1.1438	0.7761	-0.2573	241.6	-	-	13S	99E
5570	279	0314 Feb 17	03:36:44	7509	-20852	067	T	-p	0.2761	2.3669	1.3362	343.6	215.2	81.8	12N	19W
5571	279	0314 Aug 12	06:22:48	7504	-20846	072	T	-p	-0.3552	2.1942	1.2180	320.5	202.8	66.6	16S	64W
5572	279	0315 Feb 06	08:06:11	7500	-20840	077	P	t-	-0.4553	2.0650	0.9807	356.6	207.1	-	15N	86W
5573	279	0315 Aug 01	22:22:27	7495	-20834	082	T	p-	0.3765	2.1402	1.1936	308.0	196.7	62.0	18S	56E
5574	279	0316 Jan 26	07:35:14	7490	-20828	087	N	t-	-1.1524	0.7922	-0.3046	250.0			10N	79W
5575	279	0316 Jun 22	06:47:58	7486	-20823	054	N	-a	-1.4545	0.1788	-0.8010	116.4	-	-	25S	71W
5576	279	0316 Jul 21	15:02:38	7486	-20822	092	N	a-	1.1136	0.7949	-0.1662	225.2	-	-	20S	166E
5577	279	0316 Dec 15	18:37:07	7482	-20817	059	N	-a	1.0934	0.8679	-0.1644	247.5	-	-	25N	112E
5578	279	0317 Jun 11	16:20:48	7477	-20811	064	P	-t	-0.7458	1.5065	0.4727	315.0	153.5	-	24S	145E
5579	279	0317 Dec 05	06:49:18	7472	-20805	069	T	-p	0.3718	2.1674	1.1839	315.1	197.9	60.8	23N	73W
5580	279	0318 May 31	18:59:58	7467	-20799	074	T+	pp	0.0160	2.8628	1.7951	374.1	236.0	106.6	22S	105E

Cat Num	Canon Plate	Calendar Date	TD of Greatest Eclipse	ΔT s	Luna Num	Saros Num	Ecl. Type	QSE	Gamma	Pen. Mag.	Um. Mag.	Eclipse Phase Durations Pen. m	Par. m	Total m	Greatest in Zenith Lat.	Lng.
5581	280	0318 Nov 24	22:22:06	7463	-20793	079	T	p-	-0.2947	2.3036	1.3305	314.5	201.8	76.9	21N	53E
5582	280	0319 May 20	19:54:21	7458	-20787	084	P	t-	0.7744	1.4643	0.4100	320.0	147.1	-	19S	91E
5583	280	0319 Nov 14	12:36:16	7453	-20781	089	P	a-	-0.9934	1.0387	0.0315	258.2	39.9	-	18N	161W
5584	280	0320 Apr 09	16:20:28	7449	-20776	056	N	-a	-1.1532	0.7333	-0.2497	223.1	-	-	9S	146E
5585	280	0320 May 09	02:06:07	7449	-20775	094	N	a-	1.4856	0.1345	-0.8706	104.5	-	-	16S	2W
5586	280	0320 Oct 04	04:32:57	7445	-20770	061	N	-t	1.1776	0.7382	-0.3432	247.3	-	-	6N	40W
5587	280	0321 Mar 30	07:38:17	7440	-20764	066	T	-p	-0.3959	2.1068	1.1557	307.9	195.3	56.3	4S	83W
5588	280	0321 Sep 23	04:19:55	7435	-20758	071	P	-t	0.4898	2.0054	0.9137	358.5	204.4	-	0N	36W
5589	280	0322 Mar 20	00:19:18	7431	-20752	076	T	p-	0.3203	2.2511	1.2891	315.3	202.4	73.9	1N	28E
5590	280	0322 Sep 12	06:08:06	7426	-20746	081	T-	p-	-0.2373	2.4489	1.3967	355.7	222.0	88.5	5S	62W
5591	280	0323 Mar 09	13:43:57	7421	-20740	086	N	h-	1.0838	0.8750	-0.1363	246.5	-	-	5N	172W
5592	280	0323 Sep 01	15:07:36	7416	-20734	091	P	a-	-0.9461	1.1182	0.1255	265.5	78.8	-	9S	164E
5593	280	0324 Jan 28	03:03:12	7413	-20729	058	N	-t	-1.0686	0.9450	-0.1501	278.0	-	-	17N	11W
5594	280	0324 Jul 22	22:36:23	7408	-20723	063	P	-a	0.9244	1.1340	0.1890	254.2	92.3	-	19S	52E
5595	280	0325 Jan 16	02:20:38	7403	-20717	068	T	-p	-0.3796	2.2076	1.1159	363.1	215.5	53.9	21N	1W
5596	280	0325 Jul 12	14:40:46	7398	-20711	073	T+	-p	0.2043	2.4686	1.4973	326.4	212.5	91.2	22S	171E
5597	280	0326 Jan 05	06:02:03	7394	-20705	078	T	p-	0.3282	2.2780	1.2337	343.8	211.7	70.8	23N	57W
5598	280	0326 Jul 02	01:39:34	7389	-20699	083	P	t-	-0.5829	1.8002	0.7766	327.6	185.1	-	24S	6E
5599	280	0326 Dec 25	16:55:15	7384	-20693	088	P	a-	0.9904	1.0365	0.0449	253.1	46.7	-	25N	138E
5600	280	0327 May 22	14:42:20	7380	-20688	055	N	-t	1.4554	0.2223	-0.8471	144.1	-	-	19S	169E
5601	281	0327 Jun 21	05:30:30	7380	-20687	093	N	t-	-1.3836	0.3516	-0.7128	178.0	-	-	25S	53W
5602	281	0327 Nov 15	21:23:53	7376	-20682	060	N	-a	-1.0597	0.9024	-0.0759	235.7	-	-	18N	67E
5603	281	0328 May 10	16:29:40	7371	-20676	065	P	-t	0.6762	1.6391	0.5957	326.8	170.3	-	17S	142E
5604	281	0328 Nov 04	10:26:55	7366	-20670	070	T	-a	-0.4083	2.1198	1.0979	329.9	201.1	47.1	16N	129W
5605	281	0329 Apr 30	00:42:02	7361	-20664	075	T-	pp	-0.1190	2.6349	1.6442	336.8	219.0	98.8	15S	19E
5606	281	0329 Oct 24	16:57:19	7357	-20658	080	T	p-	0.3020	2.3423	1.2660	360.1	219.6	76.6	12N	133E
5607	281	0330 Apr 19	15:16:16	7352	-20652	085	P	a-	-0.8498	1.2739	0.3228	267.2	119.1	-	12S	161E
5608	281	0330 Oct 13	17:11:45	7347	-20646	090	N*	t-	1.0093	1.0560	-0.0433	291.9	-	-	9N	129E
5609	281	0331 Mar 10	23:48:46	7343	-20641	057	P	-a	1.0110	0.9902	0.0154	247.8	27.5	-	5N	37E
5610	281	0331 Sep 03	03:01:45	7339	-20635	062	N	-t	-1.1882	0.6945	-0.3388	229.6	-	-	9S	15W
5611	281	0332 Feb 28	11:26:32	7334	-20629	067	T	-a	0.3121	2.3008	1.2701	343.1	213.5	75.4	8N	137W
5612	281	0332 Aug 22	14:07:55	7329	-20623	072	T	-p	-0.4061	2.1009	1.1246	316.9	197.9	51.8	12S	179E
5613	281	0333 Feb 16	15:40:44	7324	-20617	077	T	t-	-0.4241	2.1215	1.0387	359.2	211.2	31.9	12N	159E
5614	281	0333 Aug 12	06:09:39	7320	-20611	082	T	p-	0.3193	2.2463	1.2974	310.8	200.8	74.2	15S	61W
5615	281	0334 Feb 05	15:15:09	7315	-20605	087	N	t-	-1.1273	0.8361	-0.2566	264.2	-	-	15N	166E
5616	281	0334 Jul 03	13:58:40	7311	-20600	054	Ne	-a	-1.5312	0.0402	-0.9439	56.1	-	-	25S	179W
5617	281	0334 Aug 01	22:35:06	7310	-20599	092	N	a-	1.0539	0.9067	-0.0589	238.5	-	-	17S	52E
5618	281	0334 Dec 27	03:09:05	7306	-20594	059	N	-a	1.0930	0.8668	-0.1616	246.4	-	-	25N	15W
5619	281	0335 Jun 22	23:01:25	7302	-20588	064	P	-t	-0.8303	1.3534	0.3159	305.2	128.9	-	24S	45E
5620	281	0335 Dec 16	15:39:41	7297	-20582	069	T	-p	0.3722	2.1658	1.1842	314.4	197.6	60.8	24N	156E
5621	282	0336 Jun 11	01:19:49	7292	-20576	074	T-	pp	-0.0732	2.7580	1.6897	374.0	235.5	105.3	23S	9E
5622	282	0336 Dec 05	07:14:42	7287	-20570	079	T	p-	-0.2934	2.3065	1.3324	314.8	202.0	77.2	22N	80W
5623	282	0337 May 31	02:25:45	7283	-20564	084	P	t-	0.6889	1.6198	0.5685	329.3	168.6	-	21S	8W
5624	282	0337 Nov 24	21:14:46	7278	-20558	089	P	a-	-0.9930	1.0414	0.0306	259.4	39.4	-	20N	69E
5625	282	0338 Apr 20	23:46:47	7274	-20553	056	N	-a	-1.2110	0.6245	-0.3532	207.2	-	-	13S	33E
5626	282	0338 May 20	09:07:53	7273	-20552	094	N	a-	1.4078	0.2749	-0.7256	146.8	-	-	19S	108W
5627	282	0338 Oct 15	12:17:29	7269	-20547	061	N	-t	1.2024	0.6952	-0.3911	241.8	-	-	10N	158W
5628	282	0339 Apr 10	15:24:31	7265	-20541	066	T	-a	-0.4520	2.0022	1.0544	304.1	190.0	34.3	8S	159E
5629	282	0339 Oct 04	11:49:32	7260	-20535	071	P	-t	0.5214	1.9487	0.8545	356.0	199.6	-	5N	150W
5630	282	0340 Mar 30	08:09:22	7255	-20529	076	T+	p-	0.2695	2.3438	1.3827	317.9	206.0	82.4	4S	91W
5631	282	0340 Sep 22	13:48:44	7250	-20523	081	T-	p-	-0.2015	2.5147	1.4623	355.9	223.2	92.7	0S	179W
5632	282	0341 Mar 19	21:22:21	7246	-20517	086	N	h-	1.0408	0.9540	-0.0575	256.1	-	-	1N	72E
5633	282	0341 Sep 11	23:04:32	7241	-20511	091	P	a-	-0.9051	1.1935	0.2007	271.3	98.3	-	5S	43E
5634	282	0342 Feb 07	10:41:18	7237	-20506	058	N	-t	-1.0950	0.8957	-0.1975	272.1	-	-	14N	126W
5635	282	0342 Aug 03	06:17:24	7232	-20500	063	P	-a	0.9852	1.0238	0.0762	244.7	59.5	-	17S	64W
5636	282	0343 Jan 27	10:10:21	7227	-20494	068	T	-p	-0.3973	2.1729	1.0856	360.9	213.5	46.7	18N	118W
5637	282	0343 Jul 23	22:03:33	7223	-20488	073	T	-p	0.2729	2.3452	1.3690	325.6	209.4	82.7	20S	60E
5638	282	0344 Jan 16	14:20:47	7218	-20482	078	T	p-	0.3178	2.2946	1.2555	342.7	212.1	73.3	21N	178E
5639	282	0344 Jul 12	08:30:50	7213	-20476	083	P	t-	-0.5072	1.9417	0.9131	335.6	196.6	-	23S	97W
5640	282	0345 Jan 05	01:38:25	7209	-20470	088	P	a-	0.9835	1.0473	0.0592	253.4	53.5	-	24N	8E

Cat Num	Canon Plate	Calendar Date	TD of Greatest Eclipse	ΔT s	Luna Num	Saros Num	Ecl. Type	QSE	Gamma	Pen. Mag.	Um. Mag.	Eclipse Phase Durations Pen. m	Par. m	Total m	Greatest in Zenith Lat.	Lng.
5641	283	0345 Jun 01	21:03:03	7205	-20465	055	Ne	-t	1.5406	0.0656	-1.0030	79.4	-	-	21S	73E
5642	283	0345 Jul 01	11:55:58	7204	-20464	093	N	t-	-1.3016	0.5033	-0.5635	210.4	-	-	25S	149W
5643	283	0345 Nov 26	06:13:23	7200	-20459	060	N	-a	-1.0644	0.8951	-0.0855	235.4	-	-	20N	65W
5644	283	0346 May 21	23:10:10	7195	-20453	065	P	-t	0.7555	1.4915	0.4522	316.0	151.3	-	20S	41E
5645	283	0346 Nov 15	18:58:00	7190	-20447	070	T	-a	-0.4148	2.1101	1.0838	330.9	200.9	43.8	19N	103E
5646	283	0347 May 11	07:53:39	7186	-20441	075	T-	pp	-0.0456	2.7671	1.7813	336.0	219.7	101.5	18S	90W
5647	283	0347 Nov 05	00:59:39	7181	-20435	080	T	p-	0.2899	2.3668	1.2859	362.0	220.8	78.9	16N	11E
5648	283	0348 Apr 29	22:50:04	7176	-20429	085	P	a-	-0.7830	1.3948	0.4470	275.2	137.4	-	16S	46E
5649	283	0348 Oct 24	00:56:02	7171	-20423	090	N*	t-	0.9890	1.0946	-0.0072	296.1	-	-	13N	12E
5650	283	0349 Mar 21	07:42:07	7167	-20418	057	N	-a	1.0597	0.9005	-0.0739	239.2	-	-	1N	83W
5651	283	0349 Apr 19	15:55:44	7167	-20417	095	N	a-	-1.4957	0.0884	-0.8620	79.3	-	-	13S	150E
5652	283	0349 Sep 13	10:42:02	7163	-20412	062	N	-h	-1.2247	0.6274	-0.4058	219.0	-	-	5S	131W
5653	283	0350 Mar 10	19:06:45	7158	-20406	067	T	-a	0.3554	2.2213	1.1905	342.1	210.7	65.2	4N	106E
5654	283	0350 Sep 02	22:01:41	7153	-20400	072	T	-a	-0.4507	2.0192	1.0425	313.4	193.0	30.9	8S	59E
5655	283	0351 Feb 27	23:05:34	7148	-20394	077	T	t-	-0.3851	2.1922	1.1110	362.1	215.7	53.0	8N	47E
5656	283	0351 Aug 23	14:03:56	7144	-20388	082	T+	p-	0.2677	2.3422	1.3908	313.0	203.8	82.2	11S	179E
5657	283	0352 Feb 16	22:45:16	7139	-20382	087	N	t-	-1.0938	0.8953	-0.1928	271.2	-	-	11N	52E
5658	283	0352 Aug 12	06:13:36	7134	-20376	092	P	a-	0.9996	1.0090	0.0384	249.8	43.4	-	14S	63W
5659	283	0353 Jan 06	11:37:48	7130	-20371	059	N	-a	1.0961	0.8586	-0.1652	244.4	-	-	24N	142W
5660	283	0353 Jul 03	05:43:02	7125	-20365	064	P	-t	-0.9128	1.2041	0.1623	293.9	94.8	-	24S	56W
5661	284	0353 Dec 27	00:27:46	7121	-20359	069	T	-p	0.3751	2.1592	1.1800	313.6	197.1	60.1	24N	24E
5662	284	0354 Jun 22	07:41:06	7116	-20353	074	T-	pp	-0.1613	2.5969	1.5275	372.8	233.1	99.6	24S	86W
5663	284	0354 Dec 16	16:07:47	7111	-20347	079	T	p-	-0.2924	2.3087	1.3340	315.2	202.2	77.4	23N	148E
5664	284	0355 Jun 11	08:56:17	7106	-20341	084	P	t-	0.6013	1.7792	0.7305	337.4	185.6	-	23S	106W
5665	284	0355 Dec 06	05:55:51	7102	-20335	089	P	a-	-0.9951	1.0390	0.0251	260.0	35.8	-	22N	61W
5666	284	0356 May 01	07:07:42	7098	-20330	056	N	-a	-1.2742	0.5061	-0.4667	188.0	-	-	16S	79W
5667	284	0356 May 30	16:07:42	7097	-20329	094	N	a-	1.3268	0.4216	-0.5748	178.5	-	-	21S	146E
5668	284	0356 Oct 25	20:09:25	7093	-20324	061	N	-t	1.2205	0.6642	-0.4266	237.7	-	-	14N	83E
5669	284	0357 Apr 20	23:04:20	7088	-20318	066	P	-a	-0.5138	1.8874	0.9424	299.5	183.2	-	12S	43E
5670	284	0357 Oct 14	19:29:06	7084	-20312	071	P	-t	0.5453	1.9059	0.8097	353.9	195.7	-	9N	94E
5671	284	0358 Apr 10	15:52:56	7079	-20306	076	T+	p-	0.2138	2.4458	1.4851	320.3	209.3	89.5	8S	151E
5672	284	0358 Oct 03	21:38:14	7074	-20300	081	T-	p-	-0.1720	2.5689	1.5164	355.6	223.9	95.5	4N	62E
5673	284	0359 Mar 31	04:53:11	7069	-20294	086	P	h-	0.9923	1.0432	0.0312	266.2	40.8	-	3S	43W
5674	284	0359 Sep 23	07:09:36	7064	-20288	091	P	a-	-0.8701	1.2579	0.2649	275.7	111.5	-	1S	80W
5675	284	0360 Feb 18	18:10:48	7060	-20283	058	N	-t	-1.1281	0.8338	-0.2570	264.2	-	-	11N	121E
5676	284	0360 Aug 13	14:04:29	7056	-20277	063	N	-a	1.0409	0.9232	-0.0276	235.2	-	-	14S	179E
5677	284	0360 Sep 11	22:17:15	7055	-20276	101	N	a-	-1.5179	0.0515	-0.9065	60.5	-	-	6S	54E
5678	284	0361 Feb 06	17:53:12	7051	-20271	068	T	-t	-0.4212	2.1266	1.0442	358.1	210.8	34.0	15N	125E
5679	284	0361 Aug 03	05:31:47	7046	-20265	073	T	-a	0.3363	2.2315	1.2500	324.4	205.6	71.1	17S	53W
5680	284	0362 Jan 26	22:33:23	7041	-20259	078	T	p-	0.3014	2.3218	1.2882	341.8	212.8	76.8	19N	55E
5681	285	0362 Jul 23	15:27:04	7037	-20253	083	T	t-	-0.4363	2.0746	1.0406	342.4	205.6	32.0	21S	158E
5682	285	0363 Jan 16	10:17:04	7032	-20247	088	P	a-	0.9727	1.0651	0.0809	254.3	62.2	-	22N	121W
5683	285	0363 Jul 12	18:24:05	7027	-20241	093	N	t-	-1.2215	0.6517	-0.4180	236.5	-	-	23S	114E
5684	285	0363 Dec 07	15:03:53	7023	-20236	060	N	-a	-1.0680	0.8893	-0.0930	235.2	-	-	22N	162E
5685	285	0364 Jun 01	05:49:18	7018	-20230	065	P	-t	0.8362	1.3415	0.3059	303.7	126.9	-	21S	59W
5686	285	0364 Nov 26	03:31:03	7014	-20224	070	T	-a	-0.4192	2.1041	1.0739	332.0	200.9	41.4	21N	26W
5687	285	0365 May 21	15:02:52	7009	-20218	075	T+	pp	0.0306	2.7924	1.8108	334.5	219.2	101.6	20S	162E
5688	285	0365 Nov 15	09:06:45	7004	-20212	080	T	p-	0.2825	2.3823	1.2976	363.6	221.7	80.3	19N	111W
5689	285	0366 May 11	06:19:37	6999	-20206	085	P	a-	-0.7122	1.5234	0.5783	282.7	153.0	-	19S	68W
5690	285	0366 Nov 04	08:47:09	6995	-20200	090	P	t-	0.9740	1.1230	0.0193	299.1	34.3	-	17N	107W
5691	285	0367 Apr 01	15:26:47	6991	-20195	057	N	-a	1.1151	0.7989	-0.1753	228.3	-	-	4S	159E
5692	285	0367 Apr 30	23:29:00	6990	-20194	095	N	a-	-1.4341	0.2013	-0.7488	118.6	-	-	16S	35E
5693	285	0367 Sep 24	18:32:06	6986	-20189	062	N	-h	-1.2542	0.5733	-0.4599	209.8	-	-	0S	110E
5694	285	0368 Mar 21	02:35:36	6981	-20183	067	T	-a	0.4073	2.1262	1.0953	340.3	206.6	47.7	0S	8W
5695	285	0368 Sep 13	06:04:43	6976	-20177	072	P	-a	-0.4882	1.9507	0.9733	310.2	188.4	-	4S	63W
5696	285	0369 Mar 10	06:20:35	6972	-20171	077	T	p-	-0.3386	2.2767	1.1973	365.0	220.4	68.7	4N	63W
5697	285	0369 Sep 02	22:05:42	6967	-20165	082	T+	p-	0.2226	2.4265	1.4722	314.8	206.0	87.6	8S	57E
5698	285	0370 Feb 27	06:08:40	6962	-20159	087	N	t-	-1.0544	0.9653	-0.1181	278.8	-	-	7N	60W
5699	285	0370 Aug 23	13:58:02	6957	-20153	092	P	a-	0.9507	1.1014	0.1254	259.3	77.7	-	11S	179E
5700	285	0371 Jan 17	20:04:09	6953	-20148	059	N	-a	1.1020	0.8454	-0.1734	241.9	-	-	22N	92E

Cat Num	Canon Plate	Calendar Date	TD of Greatest Eclipse	ΔT s	Luna Num	Saros Num	Ecl. Type	QSE	Gamma	Pen. Mag.	Um. Mag.	Eclipse Phase Durations Pen. m	Par. m	Total m	Greatest in Zenith Lat.	Lng.
5701	286	0371 Jul 14	12:25:32	6949	-20142	064	P	-t	-0.9935	1.0583	0.0119	281.0	26.3	-	23S	157W
5702	286	0372 Jan 07	09:14:17	6944	-20136	069	T	-p	0.3798	2.1492	1.1728	312.6	196.6	59.0	23N	107W
5703	286	0372 Jul 02	14:04:31	6939	-20130	074	T-	pp	-0.2478	2.4390	1.3683	370.3	228.8	88.9	23S	178E
5704	286	0372 Dec 27	00:58:10	6934	-20124	079	T	p-	-0.2893	2.3144	1.3396	315.7	202.6	77.9	23N	16E
5705	286	0373 Jun 21	15:31:05	6930	-20118	084	P	t-	0.5162	1.9343	0.8878	343.7	198.5	-	23S	156E
5706	286	0373 Dec 16	14:35:11	6925	-20112	089	P	a-	-0.9961	1.0384	0.0220	260.8	33.6	-	23N	170E
5707	286	0374 May 12	14:27:24	6921	-20107	056	N	-a	-1.3388	0.3854	-0.5831	165.3	-	-	20S	170E
5708	286	0374 Jun 10	23:10:19	6920	-20106	094	N	a-	1.2468	0.5666	-0.4262	203.2	-	-	22S	40E
5709	286	0374 Nov 06	04:05:36	6916	-20101	061	N	-t	1.2348	0.6398	-0.4548	234.5	-	-	17N	36W
5710	286	0375 May 02	06:39:57	6911	-20095	066	P	-a	-0.5794	1.7659	0.8231	294.0	174.8	-	16S	73W
5711	286	0375 Oct 26	03:16:29	6906	-20089	071	P	-t	0.5630	1.8742	0.7765	352.2	192.6	-	13N	24W
5712	286	0376 Apr 20	23:27:30	6902	-20083	076	T+	pp	0.1510	2.5610	1.6003	322.6	212.1	95.1	12S	36E
5713	286	0376 Oct 14	05:38:29	6897	-20077	081	T-	p-	-0.1505	2.6081	1.5559	355.1	224.0	97.1	8N	59W
5714	286	0377 Apr 10	12:13:33	6892	-20071	086	P	t-	0.9360	1.1469	0.1341	277.0	83.6	-	7S	154W
5715	286	0377 Oct 03	15:24:57	6887	-20065	091	P	a-	-0.8428	1.3080	0.3148	278.8	120.4	-	4N	155E
5716	286	0378 Mar 01	01:30:03	6883	-20060	058	N	-t	-1.1690	0.7574	-0.3307	253.8	-	-	6N	9E
5717	286	0378 Aug 24	21:59:16	6879	-20054	063	N	-a	1.0902	0.8346	-0.1199	226.1	-	-	10S	59E
5718	286	0378 Sep 23	06:34:46	6878	-20053	101	N	a-	-1.4867	0.1099	-0.8505	87.9	-	-	1S	71W
5719	286	0379 Feb 18	01:28:50	6874	-20048	068	P	-t	-0.4515	2.0683	0.9913	354.7	207.1	-	11N	10E
5720	286	0379 Aug 14	13:04:11	6869	-20042	073	T	-a	0.3954	2.1259	1.1387	322.9	201.1	55.0	14S	167W
5721	287	0380 Feb 07	06:41:29	6864	-20036	078	T	p-	0.2806	2.3569	1.3293	341.0	213.8	80.8	16N	67W
5722	287	0380 Aug 02	22:26:39	6859	-20030	083	T	p-	-0.3691	2.2008	1.1610	348.2	212.8	61.5	18S	53E
5723	287	0381 Jan 26	18:50:52	6855	-20024	088	P	a-	0.9577	1.0905	0.1106	255.9	72.3	-	19N	110E
5724	287	0381 Jul 23	00:56:52	6850	-20018	093	N	t-	-1.1451	0.7935	-0.2792	257.7	-	-	21S	15E
5725	287	0381 Dec 17	23:54:43	6846	-20013	060	N	-a	-1.0710	0.8846	-0.0991	235.2	-	-	23N	30E
5726	287	0382 Jun 12	12:30:29	6841	-20007	065	P	-t	0.9160	1.1934	0.1611	290.1	93.7	-	22S	160W
5727	287	0382 Dec 07	12:04:55	6836	-20001	070	T	-a	-0.4221	2.1003	1.0668	333.2	201.0	39.5	23N	154W
5728	287	0383 Jun 01	22:12:39	6832	-19995	075	T+	pp	0.1073	2.6499	1.6721	332.3	217.5	99.1	22S	54E
5729	287	0383 Nov 26	17:15:55	6827	-19989	080	T	p-	0.2770	2.3940	1.3060	364.9	222.5	81.3	22N	127E
5730	287	0384 May 21	13:46:34	6822	-19983	085	P	a-	-0.6387	1.6573	0.7142	289.5	166.2	-	21S	180E
5731	287	0384 Nov 14	16:43:27	6817	-19977	090	P	t-	0.9629	1.1441	0.0391	301.2	48.7	-	20N	134E
5732	287	0385 Apr 11	23:03:56	6813	-19972	057	N	-a	1.1760	0.6872	-0.2872	214.7	-	-	8S	43E
5733	287	0385 May 11	06:58:30	6812	-19971	095	N	a-	-1.3696	0.3199	-0.6305	148.1	-	-	19S	78W
5734	287	0385 Oct 05	02:31:49	6808	-19966	062	N	-h	-1.2768	0.5317	-0.5012	202.2	-	-	4N	12W
5735	287	0386 Apr 01	09:55:38	6804	-19960	067	P	-h	0.4656	2.0195	0.9881	337.7	200.9	-	4S	119W
5736	287	0386 Sep 24	14:17:03	6799	-19954	072	P	-a	-0.5186	1.8954	0.9173	307.4	184.3	-	0N	172E
5737	287	0387 Mar 21	13:25:12	6794	-19948	077	T	pp	-0.2838	2.3763	1.2989	367.9	225.1	81.7	1S	171W
5738	287	0387 Sep 14	06:15:23	6789	-19942	082	T+	p-	0.1840	2.4989	1.5415	316.2	207.5	91.1	3S	67W
5739	287	0388 Mar 09	13:22:21	6785	-19936	087	N*	t-	-1.0063	1.0509	-0.0275	287.6	-	-	3N	170W
5740	287	0388 Sep 02	21:50:16	6780	-19930	092	P	a-	0.9085	1.1817	0.2001	267.3	97.3	-	7S	60E
5741	288	0389 Jan 28	04:23:33	6776	-19925	059	N	-a	1.1144	0.8198	-0.1936	237.8	-	-	19N	33W
5742	288	0389 Jul 24	19:12:39	6771	-19919	064	N	-t	-1.0696	0.9212	-0.1302	267.1	-	-	21S	101E
5743	288	0390 Jan 17	17:56:15	6766	-19913	069	T	-p	0.3891	2.1305	1.1573	311.4	195.7	56.5	21N	123E
5744	288	0390 Jul 13	20:31:14	6761	-19907	074	T	-t	-0.3315	2.2861	1.2137	366.8	222.7	71.7	22S	81E
5745	288	0391 Jan 07	09:46:04	6757	-19901	079	T	p-	-0.2842	2.3237	1.3490	316.3	203.1	78.8	22N	116W
5746	288	0391 Jul 02	22:08:30	6752	-19895	084	T	t-	0.4318	2.0883	1.0435	348.8	208.6	33.5	23S	56E
5747	288	0391 Dec 27	23:13:03	6747	-19889	089	P	a-	-0.9968	1.0382	0.0198	261.7	32.0	-	22N	41E
5748	288	0392 May 22	21:43:32	6743	-19884	056	N	-a	-1.4069	0.2586	-0.7062	136.5	-	-	22S	60E
5749	288	0392 Jun 21	06:13:49	6742	-19883	094	N	a-	1.1661	0.7131	-0.2768	223.7	-	-	23S	66W
5750	288	0392 Nov 16	12:07:33	6738	-19878	061	N	-t	1.2440	0.6245	-0.4732	232.5	-	-	20N	157W
5751	288	0393 May 12	14:11:45	6734	-19872	066	P	-a	-0.6484	1.6386	0.6974	287.6	164.2	-	19S	173E
5752	288	0393 Nov 05	11:10:42	6729	-19866	071	P	-t	0.5755	1.8517	0.7531	350.8	190.2	-	17N	143W
5753	288	0394 May 02	06:57:39	6724	-19860	076	T+	pp	0.0850	2.6825	1.7213	324.5	214.2	98.9	15S	78W
5754	288	0394 Oct 25	13:47:02	6719	-19854	081	T-	p-	-0.1349	2.6365	1.5848	354.3	223.9	98.0	12N	178E
5755	288	0395 Apr 21	19:26:17	6714	-19848	086	P	t-	0.8742	1.2608	0.2470	287.9	112.1	-	11S	96E
5756	288	0395 Oct 14	23:48:01	6710	-19842	091	P	a-	-0.8213	1.3475	0.3542	281.0	126.6	-	8N	28E
5757	288	0396 Mar 11	08:40:35	6706	-19837	058	N	-t	-1.2165	0.6686	-0.4166	240.5	-	-	2N	100W
5758	288	0396 Sep 04	06:01:12	6701	-19831	063	N	-a	1.1333	0.7573	-0.2010	217.4	-	-	6S	63W
5759	288	0396 Oct 03	14:59:50	6700	-19830	101	N	a-	-1.4618	0.1567	-0.8060	104.5	-	-	3N	161E
5760	288	0397 Feb 28	08:56:09	6696	-19825	068	P	-t	-0.4891	1.9965	0.9251	350.6	202.1	-	7N	103W

Cat Num	Canon Plate	Calendar Date	TD of Greatest Eclipse	ΔT s	Luna Num	Saros Num	Ecl. Type	QSE	Gamma	Pen. Mag.	Um. Mag.	Eclipse Phase Durations Pen. m	Par. m	Total m	Greatest in Zenith Lat.	Lng.
5761	289	0397 Aug 24	20:43:52	6691	-19819	073	T	-a	0.4476	2.0332	1.0401	321.3	196.3	30.5	11S	77E
5762	289	0398 Feb 17	14:42:35	6686	-19813	078	T+	p-	0.2533	2.4039	1.3825	340.3	215.0	85.2	12N	171E
5763	289	0398 Aug 14	05:31:29	6682	-19807	083	T	p-	-0.3069	2.3178	1.2721	353.1	218.3	77.1	15S	54W
5764	289	0399 Feb 07	03:18:16	6677	-19801	088	P	a-	0.9372	1.1259	0.1504	258.3	83.7	-	16N	17W
5765	289	0399 Aug 03	07:35:32	6672	-19795	093	N	t-	-1.0732	0.9271	-0.1490	275.2	-	-	19S	85W
5766	289	0399 Dec 29	08:42:19	6668	-19790	060	N	-a	-1.0766	0.8746	-0.1099	234.6	-	-	22N	101W
5767	289	0400 Jun 22	19:13:43	6663	-19784	065	P	-t	0.9949	1.0474	0.0177	275.0	31.6	-	23S	99E
5768	289	0400 Dec 17	20:37:21	6658	-19778	070	T	-a	-0.4257	2.0951	1.0590	334.2	201.2	37.3	23N	79E
5769	289	0401 Jun 12	05:23:31	6654	-19772	075	T+	pp	0.1840	2.5076	1.5329	329.4	214.5	93.6	23S	54W
5770	289	0401 Dec 07	01:25:34	6649	-19766	080	T	p-	0.2725	2.4035	1.3132	366.1	223.1	82.1	23N	5E
5771	289	0402 Jun 01	21:12:31	6644	-19760	085	P	a-	-0.5640	1.7936	0.8519	295.6	177.2	-	23S	68E
5772	289	0402 Nov 26	00:44:37	6639	-19754	090	P	t-	0.9555	1.1578	0.0524	302.5	56.2	-	22N	14E
5773	289	0403 Apr 23	06:32:23	6635	-19749	057	N	-a	1.2433	0.5640	-0.4110	197.5	-	-	11S	71W
5774	289	0403 May 22	14:22:19	6634	-19748	095	N	a-	-1.3008	0.4464	-0.5048	173.3	-	-	22S	170E
5775	289	0403 Oct 16	10:41:10	6630	-19743	062	N	-h	-1.2926	0.5025	-0.5300	196.5	-	-	8N	135W
5776	289	0404 Apr 11	17:05:42	6626	-19737	067	P	-t	0.5310	1.8997	0.8678	333.9	193.1	-	8S	131E
5777	289	0404 Oct 04	22:38:21	6621	-19731	072	P	-a	-0.5422	1.8523	0.8735	305.0	180.8	-	5N	46E
5778	289	0405 Mar 31	20:20:38	6616	-19725	077	T-	pp	-0.2216	2.4895	1.4139	370.5	229.3	92.1	5S	83E
5779	289	0405 Sep 24	14:33:21	6611	-19719	082	T+	p-	0.1523	2.5586	1.5979	317.3	208.6	93.5	1N	167E
5780	289	0406 Mar 20	20:29:58	6607	-19713	087	P	t-	-0.9525	1.1471	0.0739	296.5	66.0	-	1S	82E
5781	290	0406 Sep 14	05:48:01	6602	-19707	092	P	a-	0.8713	1.2527	0.2654	274.1	111.3	-	3S	61W
5782	290	0407 Feb 08	12:38:57	6598	-19702	059	N	-a	1.1310	0.7864	-0.2212	232.7	-	-	16N	157W
5783	290	0407 Aug 05	02:03:47	6593	-19696	064	N	-t	-1.1417	0.7917	-0.2650	251.8	-	-	18S	2W
5784	290	0408 Jan 29	02:32:45	6588	-19690	069	T	-p	0.4032	2.1029	1.1332	310.0	194.3	52.3	18N	6W
5785	290	0408 Jul 24	03:03:41	6583	-19684	074	T	-t	-0.4106	2.1421	1.0677	362.5	215.0	42.3	20S	17W
5786	290	0409 Jan 17	18:28:20	6579	-19678	079	T	p-	-0.2746	2.3410	1.3669	317.0	204.0	80.4	20N	114E
5787	290	0409 Jul 13	04:53:15	6574	-19672	084	T	t-	0.3525	2.2330	1.1896	352.3	215.8	66.6	22S	45W
5788	290	0410 Jan 07	07:45:13	6569	-19666	089	P	a-	-0.9934	1.0452	0.0254	263.2	36.3	-	21N	87W
5789	290	0410 Jun 03	05:01:28	6565	-19661	056	N	-a	-1.4742	0.1336	-0.8282	98.9	-	-	24S	49W
5790	290	0410 Jul 02	13:23:17	6564	-19660	094	N	a-	1.0890	0.8533	-0.1342	240.3	-	-	22S	174W
5791	290	0410 Nov 27	20:11:18	6560	-19655	061	N	-t	1.2516	0.6118	-0.4882	230.9	-	-	23N	82E
5792	290	0411 May 23	21:39:55	6555	-19649	066	P	-a	-0.7205	1.5056	0.5655	280.1	151.0	-	22S	61E
5793	290	0411 Nov 16	19:10:51	6551	-19643	071	P	-t	0.5835	1.8370	0.7382	349.6	188.6	-	20N	96E
5794	290	0412 May 12	14:21:02	6546	-19637	076	T+	pp	0.0135	2.8142	1.8519	325.9	215.5	100.6	18S	170E
5795	290	0412 Nov 04	22:03:21	6541	-19631	081	T-	p-	-0.1249	2.6543	1.6037	353.3	223.6	98.4	16N	53E
5796	290	0413 May 02	02:30:17	6536	-19625	086	P	t-	0.8058	1.3871	0.3717	298.8	135.4	-	15S	11W
5797	290	0413 Oct 25	08:19:57	6531	-19619	091	P	a-	-0.8063	1.3749	0.3819	282.2	130.6	-	12N	101W
5798	290	0414 Mar 22	15:41:20	6527	-19614	058	N	-t	-1.2715	0.5664	-0.5159	223.5	-	-	2S	153E
5799	290	0414 Sep 15	14:10:10	6523	-19608	063	N	-a	1.1706	0.6910	-0.2714	209.5	-	-	2S	173E
5800	290	0414 Oct 14	23:32:16	6522	-19607	101	N	a-	-1.4429	0.1927	-0.7723	115.5	-	-	7N	31E
5801	291	0415 Mar 11	16:16:10	6518	-19602	068	P	-t	-0.5332	1.9126	0.8471	345.5	195.8	-	3N	146E
5802	291	0415 Sep 05	04:29:29	6513	-19596	073	P	-a	0.4939	1.9513	0.9519	319.7	191.4	-	6S	41W
5803	291	0416 Feb 28	22:36:15	6508	-19590	078	T+	p-	0.2190	2.4637	1.4487	339.7	216.3	89.8	8N	52E
5804	291	0416 Aug 24	12:42:36	6503	-19584	083	T-	pp	-0.2507	2.4241	1.3722	357.1	222.5	87.0	11S	163W
5805	291	0417 Feb 17	11:39:28	6499	-19578	088	P	a-	0.9112	1.1712	0.2003	261.4	95.8	-	12N	143W
5806	291	0417 Aug 13	14:21:02	6494	-19572	093	N*	t-	-1.0065	1.0513	-0.0283	289.7	-	-	16S	173E
5807	291	0418 Jan 08	17:26:39	6490	-19567	060	N	-a	-1.0848	0.8598	-0.1251	233.4	-	-	21N	128E
5808	291	0418 Jul 04	02:02:17	6485	-19561	065	N	-t	1.0700	0.9083	-0.1191	258.9	-	-	22S	3W
5809	291	0418 Dec 29	05:07:46	6480	-19555	070	T	-a	-0.4302	2.0879	1.0497	335.1	201.2	34.5	23N	48W
5810	291	0419 Jun 23	12:36:11	6475	-19549	075	T+	-p	0.2600	2.3669	1.3947	325.8	210.3	84.8	23S	162W
5811	291	0419 Dec 18	09:34:45	6471	-19543	080	T+	p-	0.2681	2.4124	1.3204	367.2	223.8	82.9	24N	117W
5812	291	0420 Jun 12	04:38:37	6466	-19537	085	P	a-	-0.4894	1.9302	0.9892	300.9	186.3	-	24S	44W
5813	291	0420 Dec 06	08:46:28	6461	-19531	090	P	t-	0.9486	1.1704	0.0653	303.5	62.6	-	24N	106W
5814	291	0421 May 03	13:55:13	6457	-19526	057	N	-a	1.3143	0.4343	-0.5420	176.1	-	-	15S	177E
5815	291	0421 Jun 01	21:45:05	6456	-19525	095	N	a-	-1.2316	0.5743	-0.3785	194.6	-	-	24S	59E
5816	291	0421 Oct 26	18:58:40	6452	-19520	062	N	-a	-1.3024	0.4841	-0.5477	192.6	-	-	12N	99E
5817	291	0422 Apr 23	00:08:11	6447	-19514	067	P	-t	0.6018	1.7703	0.7374	328.8	182.6	-	12S	24E
5818	291	0422 Oct 16	07:06:52	6443	-19508	072	P	-a	-0.5604	1.8191	0.8398	303.0	178.0	-	9N	83W
5819	291	0423 Apr 12	03:07:14	6438	-19502	077	T-	pp	-0.1525	2.6154	1.5416	372.6	232.7	99.9	9S	20W
5820	291	0423 Oct 05	22:59:03	6433	-19496	082	T+	p-	0.1272	2.6064	1.6425	318.2	209.3	94.9	5N	39E

Cat Num	Canon Plate	Calendar Date	TD of Greatest Eclipse	ΔT s	Luna Num	Saros Num	Ecl. Type	QSE	Gamma	Pen. Mag.	Um. Mag.	Eclipse Phase Durations Pen. m	Par. m	Total m	Greatest in Zenith Lat.	Lng.
5821	292	0424 Mar 31	03:27:25	6428	-19490	087	P	t-	-0.8896	1.2599	0.1918	305.9	104.3	-	5S	25W
5822	292	0424 Sep 24	13:54:44	6423	-19484	092	P	a-	0.8422	1.3091	0.3160	279.6	120.9	-	2N	176E
5823	292	0425 Feb 18	20:46:11	6419	-19479	059	N	-a	1.1550	0.7395	-0.2622	225.8	-	-	12N	80E
5824	292	0425 Aug 15	09:01:38	6415	-19473	064	N	-t	-1.2071	0.6744	-0.3878	236.0	-	-	15S	108W
5825	292	0426 Feb 08	11:02:26	6410	-19467	069	T	-p	0.4233	2.0642	1.0981	308.1	192.4	45.4	15N	134W
5826	292	0426 Aug 04	09:42:58	6405	-19461	074	P	-t	-0.4840	2.0085	0.9318	357.4	206.0	-	18S	118W
5827	292	0427 Jan 29	03:05:30	6400	-19455	079	T-	p-	-0.2611	2.3653	1.3921	317.9	205.1	82.5	18N	16W
5828	292	0427 Jul 24	11:42:43	6395	-19449	084	T	t-	0.2761	2.3729	1.3303	354.8	221.0	83.4	20S	148W
5829	292	0428 Jan 18	16:13:16	6390	-19443	089	P	a-	-0.9874	1.0567	0.0359	265.2	43.2	-	20N	147E
5830	292	0428 Jun 13	12:18:43	6387	-19438	056	Ne	-a	-1.5427	0.0066	-0.9526	22.2	-	-	25S	159W
5831	292	0428 Jul 12	20:36:32	6386	-19437	094	P	a-	1.0138	0.9904	0.0047	254.3	15.7	-	21S	78E
5832	292	0428 Dec 08	04:16:29	6382	-19432	061	N	-t	1.2575	0.6018	-0.4998	229.6	-	-	24N	38W
5833	292	0429 Jun 03	05:07:01	6377	-19426	066	P	-a	-0.7936	1.3714	0.4316	271.6	134.7	-	23S	52W
5834	292	0429 Nov 27	03:15:32	6372	-19420	071	P	-t	0.5886	1.8275	0.7293	348.6	187.5	-	22N	25W
5835	292	0430 May 23	21:41:13	6367	-19414	076	T-	pp	-0.0599	2.7298	1.7659	326.7	215.7	100.1	21S	60E
5836	292	0430 Nov 16	06:24:52	6363	-19408	081	T-	p-	-0.1184	2.6655	1.6164	352.2	223.1	98.6	19N	73W
5837	292	0431 May 13	09:29:00	6358	-19402	086	P	t-	0.7338	1.5202	0.5028	309.2	154.9	-	18S	117W
5838	292	0431 Nov 05	16:57:27	6353	-19396	091	P	a-	-0.7958	1.3940	0.4014	282.8	133.2	-	15N	129E
5839	292	0432 Apr 01	22:33:01	6349	-19391	058	N	-t	-1.3330	0.4520	-0.6273	201.7	-	-	6S	49E
5840	292	0432 May 01	14:00:49	6348	-19390	096	Nb	t-	1.5387	0.0671	-0.9973	79.6	-	-	14S	175E
5841	293	0432 Sep 25	22:27:10	6344	-19385	063	N	-a	1.2010	0.6373	-0.3292	202.7	-	-	3N	47E
5842	293	0432 Oct 25	08:11:22	6343	-19384	101	N	a-	-1.4296	0.2182	-0.7489	122.7	-	-	11N	100W
5843	293	0433 Mar 21	23:28:42	6339	-19379	068	P	-t	-0.5839	1.8166	0.7569	339.5	187.6	-	1S	36E
5844	293	0433 Sep 15	12:22:19	6335	-19373	073	P	-a	0.5334	1.8820	0.8764	318.3	186.7	-	2S	161W
5845	293	0434 Mar 11	06:23:02	6330	-19367	078	T+	p-	0.1781	2.5354	1.5269	339.1	217.6	94.1	4N	67W
5846	293	0434 Sep 04	20:01:17	6325	-19361	083	T-	pp	-0.2015	2.5174	1.4594	360.5	225.5	93.4	7S	86E
5847	293	0435 Feb 28	19:52:28	6320	-19355	088	P	a-	0.8783	1.2293	0.2630	265.3	108.6	-	8N	92E
5848	293	0435 Aug 24	21:14:38	6315	-19349	093	P	t-	-0.9461	1.1639	0.0808	301.6	69.6	-	12S	68E
5849	293	0436 Jan 20	02:05:05	6311	-19344	060	N	-a	-1.0973	0.8370	-0.1480	231.3	-	-	19N	1W
5850	293	0436 Feb 18	12:09:44	6310	-19343	098	Nb	a-	1.5443	0.0056	-0.9573	20.1	-	-	13N	152W
5851	293	0436 Jul 14	08:56:32	6306	-19338	065	N	-t	1.1410	0.7773	-0.2485	241.8	-	-	21S	107W
5852	293	0437 Jan 08	13:32:12	6302	-19332	070	T	-a	-0.4386	2.0731	1.0334	335.7	200.8	28.5	22N	174W
5853	293	0437 Jul 03	19:53:18	6297	-19326	075	T	-p	0.3328	2.2323	1.2620	321.6	205.1	72.2	23S	88E
5854	293	0437 Dec 28	17:41:27	6292	-19320	080	T+	p-	0.2627	2.4228	1.3299	368.1	224.5	84.0	24N	122E
5855	293	0438 Jun 23	12:05:03	6287	-19314	085	T	a-	-0.4149	2.0668	1.1259	305.5	193.7	51.1	24S	155W
5856	293	0438 Dec 17	16:49:55	6283	-19308	090	P	t-	0.9427	1.1806	0.0767	304.1	67.7	-	24N	134E
5857	293	0439 May 14	21:11:34	6278	-19303	057	N	-a	1.3895	0.2972	-0.6808	148.1	-	-	18S	67E
5858	293	0439 Jun 13	05:05:00	6278	-19302	095	N	a-	-1.1607	0.7055	-0.2495	213.4	-	-	25S	51W
5859	293	0439 Nov 07	03:23:48	6274	-19297	062	N	-a	-1.3072	0.4747	-0.5559	190.2	-	-	16N	28W
5860	293	0439 Dec 06	18:28:44	6273	-19296	100	N	a-	1.5720	0.0057	-1.0584	22.5	-	-	24N	107E
5861	294	0440 May 03	07:01:41	6269	-19291	067	P	-t	0.6790	1.6292	0.5951	322.1	168.5	-	15S	80W
5862	294	0440 Oct 26	15:43:29	6264	-19285	072	P	-a	-0.5725	1.7972	0.8174	301.4	175.9	-	13N	147E
5863	294	0441 Apr 22	09:46:51	6259	-19279	077	T-	pp	-0.0778	2.7516	1.6793	374.1	235.1	104.8	13S	121W
5864	294	0441 Oct 16	07:31:19	6255	-19273	082	T+	p-	0.1076	2.6439	1.6768	319.0	209.8	95.9	10N	90W
5865	294	0442 Apr 11	10:20:22	6250	-19267	087	P	t-	-0.8222	1.3810	0.3179	314.9	131.6	-	9S	129W
5866	294	0442 Oct 05	22:07:34	6245	-19261	092	P	a-	0.8187	1.3550	0.3563	284.1	127.9	-	6N	51E
5867	294	0443 Mar 02	04:47:16	6241	-19256	059	N	-a	1.1846	0.6821	-0.3135	217.2	-	-	8N	42W
5868	294	0443 Mar 31	15:53:13	6240	-19255	097	N	a-	-1.5144	0.0871	-0.9287	84.8	-	-	6S	148E
5869	294	0443 Aug 26	16:05:12	6236	-19250	064	N	-t	-1.2670	0.5671	-0.5006	219.5	-	-	11S	145E
5870	294	0444 Feb 19	19:25:21	6231	-19244	069	T	-a	0.4493	2.0144	1.0521	305.9	189.9	33.5	11N	99E
5871	294	0444 Aug 14	16:30:29	6227	-19238	074	P	-t	-0.5509	1.8870	0.8079	352.0	196.0	-	15S	139E
5872	294	0445 Feb 08	11:33:57	6222	-19232	079	T-	p-	-0.2406	2.4025	1.4302	319.0	206.6	85.4	14N	143W
5873	294	0445 Aug 03	18:41:53	6217	-19226	084	T+	p-	0.2065	2.5002	1.4583	356.1	224.3	93.2	17S	106E
5874	294	0446 Jan 29	00:33:14	6212	-19220	089	P	a-	-0.9752	1.0793	0.0578	268.2	54.8	-	17N	21E
5875	294	0446 Jul 24	03:56:14	6207	-19214	094	P	a-	0.9428	1.1200	0.1357	265.7	82.0	-	19S	33W
5876	294	0446 Dec 19	12:20:01	6203	-19209	061	N	-t	1.2645	0.5893	-0.5130	227.7	-	-	25N	159W
5877	294	0447 Jun 14	12:33:07	6199	-19203	066	P	-a	-0.8677	1.2356	0.2955	262.0	113.7	-	24S	163W
5878	294	0447 Dec 08	11:21:40	6194	-19197	071	P	-t	0.5926	1.8195	0.7226	347.5	186.5	-	24N	146W
5879	294	0448 Jun 03	04:56:53	6189	-19191	076	T-	-p	-0.1365	2.5905	1.6242	326.9	214.6	96.8	23S	50W
5880	294	0448 Nov 26	14:51:16	6184	-19185	081	T-	p-	-0.1150	2.6705	1.6237	351.0	222.6	98.7	21N	161E

Cat Num	Canon Plate	Calendar Date	TD of Greatest Eclipse	ΔT s	Luna Num	Saros Num	Ecl. Type	QSE	Gamma	Pen. Mag.	Um. Mag.	Eclipse Phase Durations Pen. m	Par. m	Total m	Greatest in Zenith Lat.	Lng.
5881	295	0449 May 23	16:21:54	6179	-19179	086	P	t-	0.6578	1.6610	0.6411	319.0	171.6	-	20S	139E
5882	295	0449 Nov 16	01:40:11	6175	-19173	091	P	a-	-0.7890	1.4060	0.4144	283.0	134.8	-	18N	2W
5883	295	0450 Apr 13	05:16:32	6171	-19168	058	N	-t	-1.4006	0.3265	-0.7500	173.3	-	-	11S	54W
5884	295	0450 May 12	20:31:53	6170	-19167	096	N	t-	1.4616	0.2085	-0.8559	138.7	-	-	17S	76E
5885	295	0450 Oct 07	06:51:37	6166	-19162	063	N	-a	1.2254	0.5946	-0.3760	197.1	-	-	7N	80W
5886	295	0450 Nov 05	16:56:38	6165	-19161	101	N	a-	-1.4212	0.2344	-0.7345	127.0	-	-	15N	128E
5887	295	0451 Apr 02	06:33:15	6161	-19156	068	P	-t	-0.6418	1.7075	0.6537	332.3	177.0	-	6S	72W
5888	295	0451 Sep 26	20:22:09	6156	-19150	073	P	-a	0.5664	1.8247	0.8129	317.3	182.3	-	2N	78E
5889	295	0452 Mar 21	14:02:47	6151	-19144	078	T+	p-	0.1304	2.6198	1.6176	338.4	218.8	97.8	1S	177E
5890	295	0452 Sep 15	03:27:27	6147	-19138	083	T-	pp	-0.1592	2.5982	1.5340	363.3	227.8	97.6	3S	28W
5891	295	0453 Mar 11	03:58:27	6142	-19132	088	P	a-	0.8392	1.2988	0.3371	269.9	121.5	-	4N	31W
5892	295	0453 Sep 04	04:17:10	6137	-19126	093	P	t-	-0.8927	1.2637	0.1770	311.2	101.6	-	8S	39W
5893	295	0454 Jan 30	10:38:22	6133	-19121	060	N	-a	-1.1136	0.8070	-0.1779	228.3	-	-	16N	130W
5894	295	0454 Feb 28	20:27:21	6132	-19120	098	N	a-	1.5172	0.0544	-0.9067	62.3	-	-	9N	83E
5895	295	0454 Jul 25	15:56:55	6128	-19115	065	N	-t	1.2076	0.6543	-0.3700	223.7	-	-	19S	147E
5896	295	0455 Jan 19	21:51:55	6123	-19109	070	T	-a	-0.4502	2.0524	1.0116	336.1	200.1	16.9	20N	61E
5897	295	0455 Jul 15	03:14:53	6119	-19103	075	T	-p	0.4025	2.1038	1.1349	317.0	198.8	53.9	21S	23W
5898	295	0456 Jan 09	01:42:52	6114	-19097	080	T+	pp	0.2531	2.4403	1.3474	369.1	225.5	85.7	22N	3E
5899	295	0456 Jul 03	19:34:25	6109	-19091	085	T	p-	-0.3426	2.1997	1.2582	309.1	199.5	70.1	23S	92E
5900	295	0456 Dec 28	00:50:43	6104	-19085	090	P	t-	0.9344	1.1947	0.0929	305.0	74.2	-	24N	14E
5901	296	0457 May 25	04:24:55	6100	-19080	057	N	-a	1.4663	0.1575	-0.8229	109.6	-	-	20S	42W
5902	296	0457 Jun 23	12:26:22	6100	-19079	095	N	a-	-1.0911	0.8345	-0.1233	229.8	-	-	25S	162W
5903	296	0457 Nov 17	11:53:00	6096	-19074	062	N	-a	-1.3095	0.4696	-0.5592	188.7	-	-	18N	155W
5904	296	0457 Dec 17	02:49:24	6095	-19073	100	N	a-	1.5642	0.0185	-1.0423	40.5	-	-	25N	17W
5905	296	0458 May 14	13:50:08	6091	-19068	067	P	-t	0.7592	1.4830	0.4471	313.9	150.0	-	18S	176E
5906	296	0458 Nov 07	00:25:45	6086	-19062	072	P	-a	-0.5804	1.7827	0.8029	300.3	174.5	-	16N	16E
5907	296	0459 May 03	16:18:09	6081	-19056	077	T+	pp	0.0033	2.8877	1.8167	374.6	236.0	106.5	16S	139E
5908	296	0459 Oct 27	16:10:32	6076	-19050	082	T+	p-	0.0941	2.6702	1.7002	319.7	210.1	96.4	13N	139E
5909	296	0460 Apr 21	17:05:19	6072	-19044	087	P	t-	-0.7478	1.5153	0.4569	323.7	154.1	-	13S	128E
5910	296	0460 Oct 16	06:27:57	6067	-19038	092	P	a-	0.8021	1.3882	0.3841	287.6	132.6	-	10N	75W
5911	296	0461 Mar 12	12:39:38	6063	-19033	059	N	-a	1.2219	0.6105	-0.3790	206.1	-	-	4N	161W
5912	296	0461 Apr 10	23:11:45	6062	-19032	097	N	a-	-1.4503	0.2016	-0.8081	127.1	-	-	10S	37E
5913	296	0461 Sep 05	23:17:33	6058	-19027	064	N	-t	-1.3191	0.4745	-0.5988	203.3	-	-	7S	36E
5914	296	0461 Oct 05	14:59:50	6057	-19026	102	Nb	t-	1.5718	0.0033	-1.0551	17.3	-	-	7N	157E
5915	296	0462 Mar 02	03:40:50	6053	-19021	069	P	-a	0.4820	1.9527	0.9941	303.2	186.4	-	7N	26W
5916	296	0462 Aug 25	23:25:23	6049	-19015	074	P	-t	-0.6119	1.7763	0.6947	346.3	185.2	-	11S	34E
5917	296	0463 Feb 19	19:55:45	6044	-19009	079	T-	p-	-0.2149	2.4490	1.4779	320.3	208.2	88.5	11N	90E
5918	296	0463 Aug 15	01:48:36	6039	-19003	084	T+	pp	0.1421	2.6183	1.5767	356.7	226.3	99.0	14S	1W
5919	296	0464 Feb 09	08:45:14	6034	-18997	089	P	a-	-0.9576	1.1119	0.0900	272.1	68.4	-	14N	102W
5920	296	0464 Aug 03	11:22:45	6029	-18991	094	P	a-	0.8765	1.2413	0.2578	275.1	110.7	-	17S	145W
5921	297	0464 Dec 29	20:21:47	6025	-18986	061	N	-t	1.2725	0.5744	-0.5277	225.4	-	-	24N	82E
5922	297	0465 Jun 24	20:00:37	6021	-18980	066	P	-a	-0.9402	1.1030	0.1619	251.4	85.8	-	25S	85E
5923	297	0465 Dec 18	19:28:13	6016	-18974	071	P	-t	0.5967	1.8108	0.7160	346.3	185.6	-	24N	93E
5924	297	0466 Jun 14	12:12:10	6011	-18968	076	T-	-p	-0.2126	2.4523	1.4831	326.3	212.3	90.5	24S	159W
5925	297	0466 Dec 07	23:20:07	6006	-18962	081	T-	p-	-0.1130	2.6729	1.6288	349.7	222.1	98.6	23N	34E
5926	297	0467 Jun 03	23:10:33	6001	-18956	086	P	t-	0.5792	1.8068	0.7839	328.1	185.9	-	22S	36E
5927	297	0467 Nov 27	10:26:19	5997	-18950	091	P	a-	-0.7846	1.4134	0.4230	282.8	135.7	-	21N	134W
5928	297	0468 Apr 23	11:53:18	5993	-18945	058	N	-t	-1.4732	0.1919	-0.8819	134.4	-	-	14S	155W
5929	297	0468 May 23	02:58:54	5992	-18944	096	N	t-	1.3810	0.3565	-0.7083	179.3	-	-	20S	21W
5930	297	0468 Oct 17	15:22:37	5988	-18939	063	N	-a	1.2442	0.5620	-0.4125	192.6	-	-	11N	151E
5931	297	0468 Nov 16	01:45:15	5987	-18938	101	N	a-	-1.4158	0.2451	-0.7253	129.8	-	-	18N	4W
5932	297	0469 Apr 12	13:32:05	5983	-18933	068	P	-t	-0.7049	1.5888	0.5406	323.8	163.5	-	10S	178W
5933	297	0469 Oct 07	04:29:25	5978	-18927	073	P	-a	0.5925	1.7797	0.7619	316.6	178.5	-	7N	46W
5934	297	0470 Apr 01	21:36:16	5974	-18921	078	T+	pp	0.0768	2.7151	1.7190	337.5	219.4	100.4	5S	62E
5935	297	0470 Sep 26	11:00:41	5969	-18915	083	T-	pp	-0.1234	2.6668	1.5967	365.6	229.3	100.3	2N	142W
5936	297	0471 Mar 22	11:56:40	5964	-18909	088	P	a-	0.7934	1.3805	0.4231	275.0	134.2	-	0S	152W
5937	297	0471 Sep 15	11:29:32	5959	-18903	093	P	t-	-0.8472	1.3489	0.2588	318.8	121.4	-	3S	149W
5938	297	0472 Feb 10	19:01:59	5955	-18898	060	N	-a	-1.1372	0.7634	-0.2211	223.6	-	-	13N	103E
5939	297	0472 Mar 11	04:34:10	5955	-18897	098	N	a-	1.4822	0.1178	-0.8415	91.3	-	-	5N	41W
5940	297	0472 Aug 04	23:06:07	5951	-18892	065	N	-t	1.2677	0.5436	-0.4798	205.4	-	-	16S	39E

Cat Num	Canon Plate	Calendar Date	TD of Greatest Eclipse	ΔT s	Luna Num	Saros Num	Ecl. Type	QSE	Gamma	Pen. Mag.	Um. Mag.	Eclipse Phase Durations Pen. m	Par. m	Total m	Greatest in Zenith Lat.	Lng.
5941	298	0472 Sep 03	12:30:19	5950	-18891	103	N	t-	-1.5287	0.0828	-0.9764	86.6	-	-	8S	163W
5942	298	0473 Jan 30	06:02:55	5946	-18886	070	P	-a	-0.4680	2.0201	0.9787	336.0	198.7	-	17N	62W
5943	298	0473 Jul 25	10:42:42	5941	-18880	075	T	-a	0.4675	1.9840	1.0159	312.0	191.8	19.1	19S	135W
5944	298	0474 Jan 19	09:38:27	5936	-18874	080	T+	pp	0.2394	2.4653	1.3728	370.2	226.7	88.1	20N	116W
5945	298	0474 Jul 15	03:06:47	5932	-18868	085	T	p-	-0.2731	2.3280	1.3853	312.1	203.9	82.0	22S	21W
5946	298	0475 Jan 08	08:49:33	5927	-18862	090	P	t-	0.9242	1.2121	0.1130	306.1	81.6	-	23N	105W
5947	298	0475 Jun 05	11:32:25	5923	-18857	057	Ne	-a	1.5465	0.0118	-0.9715	30.5	-	-	21S	149W
5948	298	0475 Jul 04	19:46:13	5922	-18856	095	P	a-	-1.0210	0.9650	0.0037	244.5	13.5	-	24S	88E
5949	298	0475 Nov 28	20:27:36	5918	-18851	062	N	-a	-1.3082	0.4709	-0.5557	188.3	-	-	21N	76E
5950	298	0475 Dec 28	11:11:04	5917	-18850	100	N	a-	1.5564	0.0307	-1.0262	51.9	-	-	25N	142W
5951	298	0476 May 24	20:32:20	5913	-18845	067	P	-t	0.8431	1.3302	0.2921	303.6	124.5	-	20S	75E
5952	298	0476 Nov 17	09:12:48	5908	-18839	072	P	-a	-0.5846	1.7749	0.7954	299.4	173.6	-	19N	116W
5953	298	0477 May 13	22:45:17	5904	-18833	077	T+	pp	0.0875	2.7327	1.6627	374.1	235.2	104.6	19S	42E
5954	298	0477 Nov 07	00:54:50	5899	-18827	082	T+	p-	0.0854	2.6876	1.7149	320.3	210.4	96.7	17N	7E
5955	298	0478 May 02	23:47:14	5894	-18821	087	P	t-	-0.6698	1.6560	0.6022	331.6	172.5	-	16S	26E
5956	298	0478 Oct 27	14:52:48	5889	-18815	092	P	a-	0.7899	1.4132	0.4039	290.5	135.9	-	14N	157E
5957	298	0479 Mar 23	20:26:33	5886	-18810	059	N	-a	1.2643	0.5298	-0.4539	192.7	-	-	0S	80E
5958	298	0479 Apr 22	06:26:26	5885	-18809	097	N	a-	-1.3821	0.3238	-0.6800	158.5	-	-	14S	73W
5959	298	0479 Sep 17	06:37:00	5881	-18804	064	N	-t	-1.3647	0.3935	-0.6852	187.1	-	-	3S	76W
5960	298	0479 Oct 16	22:54:35	5880	-18803	102	N	t-	1.5537	0.0393	-1.0249	59.5	-	-	11N	37E
5961	299	0480 Mar 12	11:47:44	5876	-18798	069	P	-a	0.5220	1.8775	0.9225	299.8	181.8	-	3N	150W
5962	299	0480 Sep 05	06:30:36	5871	-18792	074	P	-t	-0.6646	1.6808	0.5967	340.8	174.3	-	7S	74W
5963	299	0481 Mar 02	04:08:15	5866	-18786	079	T-	p-	-0.1818	2.5093	1.5392	321.7	210.0	91.9	6N	34W
5964	299	0481 Aug 25	09:05:22	5862	-18780	084	T+	pp	0.0851	2.7227	1.6812	356.5	227.1	102.0	10S	112W
5965	299	0482 Feb 19	16:47:26	5857	-18774	089	P	a-	-0.9326	1.1578	0.1357	277.2	83.7	-	10N	136E
5966	299	0482 Aug 14	18:57:45	5852	-18768	094	P	a-	0.8163	1.3516	0.3684	282.6	129.8	-	13S	100E
5967	299	0483 Jan 10	04:18:09	5848	-18763	061	N	-t	1.2846	0.5520	-0.5493	221.6	-	-	23N	37W
5968	299	0483 Jul 06	03:28:49	5844	-18757	066	P	-a	-1.0120	0.9721	0.0293	239.6	37.2	-	24S	27W
5969	299	0483 Aug 04	10:31:02	5843	-18756	104	N	a-	1.4920	0.0940	-0.8541	81.3	-	-	16S	133W
5970	299	0483 Dec 30	03:33:13	5839	-18751	071	P	-t	0.6026	1.7984	0.7067	344.8	184.4	-	24N	27W
5971	299	0484 Jun 24	19:26:05	5834	-18745	076	T	-p	-0.2892	2.3137	1.3407	324.9	208.7	80.4	24S	93E
5972	299	0484 Dec 18	07:49:08	5829	-18739	081	T-	p-	-0.1102	2.6762	1.6356	348.3	221.6	98.7	24N	93W
5973	299	0485 Jun 14	05:56:53	5825	-18733	086	P	t-	0.4992	1.9553	0.9289	336.3	197.9	-	23S	66W
5974	299	0485 Dec 07	19:15:13	5820	-18727	091	P	a-	-0.7819	1.4174	0.4289	282.4	136.3	-	22N	94E
5975	299	0486 May 04	18:23:37	5816	-18722	058	Ne	-t	-1.5503	0.0493	-1.0222	68.9	-	-	18S	107E
5976	299	0486 Jun 03	09:21:48	5815	-18721	096	N	t-	1.2971	0.5110	-0.5546	211.9	-	-	21S	117W
5977	299	0486 Oct 28	23:59:43	5811	-18716	063	N	-a	1.2580	0.5388	-0.4396	189.5	-	-	15N	20E
5978	299	0486 Nov 27	10:37:44	5810	-18715	101	N	a-	-1.4138	0.2493	-0.7222	130.9	-	-	20N	137W
5979	299	0487 Apr 23	20:25:16	5806	-18710	068	P	-t	-0.7733	1.4605	0.4178	313.8	146.1	-	14S	77E
5980	299	0487 Oct 18	12:43:16	5802	-18704	073	P	-a	0.6122	1.7466	0.7229	316.3	175.5	-	11N	170W
5981	300	0488 Apr 12	05:03:16	5797	-18698	078	T+	pp	0.0170	2.8218	1.8317	336.2	219.5	101.6	9S	52W
5982	300	0488 Oct 06	18:42:13	5792	-18692	083	T-	pp	-0.0953	2.7213	1.6455	367.6	230.4	101.9	6N	101E
5983	300	0489 Apr 01	19:48:58	5787	-18686	088	P	a-	0.7426	1.4719	0.5185	280.3	146.3	-	5S	88E
5984	300	0489 Sep 25	18:49:56	5783	-18680	093	P	t-	-0.8081	1.4222	0.3290	324.8	135.4	-	1N	100E
5985	300	0490 Feb 21	03:18:45	5779	-18675	060	N	-a	-1.1660	0.7104	-0.2736	217.3	-	-	9N	22W
5986	300	0490 Mar 22	12:34:42	5778	-18674	098	N	a-	1.4425	0.1898	-0.7680	115.3	-	-	0N	162W
5987	300	0490 Aug 16	06:23:31	5774	-18669	065	N	-t	1.3217	0.4442	-0.5785	186.8	-	-	12S	72W
5988	300	0490 Sep 14	19:57:05	5773	-18668	103	N	t-	-1.4826	0.1675	-0.8922	121.9	-	-	4S	84E
5989	300	0491 Feb 10	14:06:05	5769	-18663	070	P	-a	-0.4915	1.9771	0.9355	335.3	196.4	-	13N	177E
5990	300	0491 Aug 05	18:16:35	5765	-18657	075	P	-a	0.5281	1.8727	0.9048	306.8	184.2	-	16S	111E
5991	300	0492 Jan 30	17:26:12	5760	-18651	080	T+	pp	0.2197	2.5008	1.4095	371.3	228.2	91.1	17N	127E
5992	300	0492 Jul 25	10:44:18	5755	-18645	085	T-	p-	-0.2078	2.4487	1.5041	314.4	207.0	89.6	20S	136W
5993	300	0493 Jan 18	16:41:33	5750	-18639	090	P	t-	0.9082	1.2397	0.1441	308.0	91.6	-	21N	137E
5994	300	0493 Jul 15	03:10:25	5746	-18633	095	P	a-	-0.9550	1.0881	0.1227	257.1	76.8	-	23S	23W
5995	300	0493 Dec 09	05:03:29	5742	-18628	062	N	-a	-1.3062	0.4732	-0.5506	188.0	-	-	22N	53W
5996	300	0494 Jan 07	19:29:20	5741	-18627	100	N	a-	1.5456	0.0483	-1.0041	64.8	-	-	24N	94E
5997	300	0494 Jun 05	03:11:03	5737	-18622	067	P	-t	0.9285	1.1747	0.1340	291.4	86.5	-	22S	25W
5998	300	0494 Nov 28	18:02:34	5732	-18616	072	P	-a	-0.5869	1.7704	0.7915	298.7	173.1	-	21N	111E
5999	300	0495 May 25	05:07:40	5727	-18610	077	T+	pp	0.1752	2.5715	1.5022	372.4	232.6	98.2	21S	55W
6000	300	0495 Nov 18	09:43:17	5723	-18604	082	T+	p-	0.0800	2.6985	1.7236	320.9	210.6	96.9	20N	125W

Cat Num	Canon Plate	Calendar Date	TD of Greatest Eclipse	ΔT	Luna Num	Saros Num	Ecl. Type	QSE	Gamma	Pen. Mag.	Um. Mag.	Pen.	Par.	Total	Lat.	Lng.
				s								m	m	m		
6001	301	0496 May 13	06:24:02	5718	-18598	087	P	t-	-0.5867	1.8065	0.7568	338.6	187.9	-	19S	74W
6002	301	0496 Nov 06	23:23:50	5713	-18592	092	P	a-	0.7834	1.4275	0.4135	292.6	137.7	-	18N	29E
6003	301	0497 Apr 03	04:06:04	5709	-18587	059	N	-a	1.3134	0.4369	-0.5412	175.8	-	-	5S	36W
6004	301	0497 May 02	13:36:41	5709	-18586	097	N	a-	-1.3090	0.4552	-0.5432	184.8	-	-	17S	178E
6005	301	0497 Sep 27	14:04:31	5705	-18581	064	N	-t	-1.4031	0.3256	-0.7583	171.8	-	-	1N	171E
6006	301	0497 Oct 27	06:55:24	5704	-18580	102	N	t-	1.5411	0.0652	-1.0043	76.8	-	-	15N	84W
6007	301	0498 Mar 23	19:47:39	5700	-18575	069	P	-a	0.5681	1.7911	0.8395	295.8	175.9	-	1S	89E
6008	301	0498 Sep 16	13:45:04	5695	-18569	074	P	-t	-0.7100	1.5988	0.5123	335.6	163.6	-	3S	176E
6009	301	0499 Mar 13	12:12:02	5691	-18563	079	T-	pp	-0.1419	2.5820	1.6129	323.2	211.8	95.2	2N	157W
6010	301	0499 Sep 05	16:31:35	5686	-18557	084	T+	pp	0.0350	2.8146	1.7731	355.8	227.0	103.2	6S	135E
6011	301	0500 Mar 02	00:40:27	5681	-18551	089	P	h-	-0.9011	1.2156	0.1934	283.2	99.4	-	6N	16E
6012	301	0500 Aug 25	02:41:11	5676	-18545	094	P	a-	0.7623	1.4506	0.4675	288.5	143.6	-	10S	17W
6013	301	0501 Jan 20	12:08:52	5672	-18540	061	N	-t	1.3008	0.5214	-0.5784	216.0	-	-	21N	154W
6014	301	0501 Jul 16	11:01:18	5667	-18534	066	N	-a	-1.0801	0.8482	-0.0968	227.0	-	-	22S	141W
6015	301	0501 Aug 14	18:18:54	5667	-18533	104	N	a-	1.4366	0.1966	-0.7533	116.4	-	-	13S	109E
6016	301	0502 Jan 09	11:36:08	5663	-18528	071	P	-t	0.6109	1.7814	0.6933	342.9	182.7	-	22N	147W
6017	301	0502 Jul 06	02:40:41	5658	-18522	076	T	-a	-0.3643	2.1778	1.2008	322.9	203.8	64.9	23S	16W
6018	301	0502 Dec 29	16:17:46	5653	-18516	081	T-	p-	-0.1063	2.6813	1.6447	347.0	221.2	98.8	23N	141E
6019	301	0503 Jun 25	12:42:21	5648	-18510	086	T	t-	0.4191	2.1042	1.0738	343.4	207.8	42.8	23S	167W
6020	301	0503 Dec 19	04:03:22	5644	-18504	091	P	a-	-0.7784	1.4227	0.4364	282.0	137.0	-	23N	37W
6021	302	0504 Jun 13	15:44:17	5639	-18498	096	N	t-	1.2125	0.6668	-0.3999	238.7	-	-	22S	147E
6022	302	0504 Nov 08	08:41:40	5635	-18493	063	N	-a	1.2677	0.5226	-0.4591	187.4	-	-	19N	111W
6023	302	0504 Dec 07	19:30:38	5634	-18492	101	N	a-	-1.4123	0.2524	-0.7197	131.8	-	-	22N	90E
6024	302	0505 May 04	03:15:22	5630	-18487	068	P	-t	-0.8448	1.3269	0.2892	302.5	123.6	-	17S	27W
6025	302	0505 Oct 28	21:01:51	5625	-18481	073	P	-a	0.6272	1.7217	0.6926	316.4	173.2	-	15N	64E
6026	302	0506 Apr 23	12:25:55	5620	-18475	078	T-	pp	-0.0474	2.7633	1.7786	334.4	218.8	101.1	13S	164W
6027	302	0506 Oct 18	02:30:34	5616	-18469	083	T-	pp	-0.0733	2.7643	1.6833	369.3	231.2	102.9	10N	18W
6028	302	0507 Apr 13	03:32:37	5611	-18463	088	P	a-	0.6842	1.5772	0.6275	285.9	158.0	-	9S	30W
6029	302	0507 Oct 07	02:20:32	5606	-18457	093	P	t-	-0.7769	1.4809	0.3847	329.4	145.1	-	5N	14W
6030	302	0508 Mar 03	11:24:46	5602	-18452	060	N	-a	-1.2031	0.6420	-0.3416	208.5	-	-	5N	145W
6031	302	0508 Apr 01	20:25:03	5601	-18451	098	N	a-	1.3949	0.2765	-0.6802	138.3	-	-	4S	78E
6032	302	0508 Aug 26	13:51:10	5597	-18446	065	N	-h	1.3683	0.3585	-0.6639	168.6	-	-	9S	175E
6033	302	0508 Sep 25	03:35:20	5597	-18445	103	N	h-	-1.4447	0.2374	-0.8228	143.9	-	-	0N	32W
6034	302	0509 Feb 20	21:58:17	5593	-18440	070	P	-a	-0.5232	1.9190	0.8772	333.9	192.8	-	10N	57E
6035	302	0509 Aug 16	01:58:44	5588	-18434	075	P	-a	0.5824	1.7732	0.8051	301.7	176.4	-	13S	6W
6036	302	0510 Feb 10	01:06:17	5583	-18428	080	T+	pp	0.1942	2.5468	1.4571	372.4	229.8	94.5	14N	11E
6037	302	0510 Aug 05	18:25:38	5578	-18422	085	T-	p-	-0.1460	2.5633	1.6163	316.1	209.2	94.5	17S	108E
6038	302	0511 Jan 30	00:29:38	5573	-18416	090	P	t-	0.8886	1.2736	0.1822	310.3	102.3	-	18N	20E
6039	302	0511 Jul 26	10:36:11	5569	-18410	095	P	a-	-0.8912	1.2075	0.2376	268.3	105.3	-	20S	135W
6040	302	0511 Dec 20	13:40:30	5565	-18405	062	N	-a	-1.3042	0.4752	-0.5453	187.7	-	-	22N	179E
6041	303	0512 Jan 19	03:44:14	5564	-18404	100	N	a-	1.5313	0.0719	-0.9753	78.5	-	-	22N	29W
6042	303	0512 Jun 15	09:47:02	5560	-18399	067	N*	-t	1.0151	1.0175	-0.0264	276.8	-	-	23S	124W
6043	303	0512 Dec 09	02:54:28	5555	-18393	072	P	-a	-0.5877	1.7684	0.7907	298.2	172.8	-	23N	21W
6044	303	0513 Jun 04	11:29:13	5550	-18387	077	T+	pp	0.2636	2.4091	1.3401	369.5	227.8	86.4	22S	150W
6045	303	0513 Nov 28	18:33:24	5546	-18381	082	T+	p-	0.0764	2.7060	1.7292	321.5	210.8	97.0	22N	103E
6046	303	0514 May 24	13:00:56	5541	-18375	087	P	t-	-0.5027	1.9586	0.9129	344.5	200.3	-	22S	174W
6047	303	0514 Nov 18	07:57:20	5536	-18369	092	P	a-	0.7797	1.4365	0.4182	294.4	138.8	-	21N	99W
6048	303	0515 Apr 14	11:39:47	5532	-18364	059	N	-a	1.3677	0.3345	-0.6382	154.7	-	-	9S	152W
6049	303	0515 May 13	20:43:57	5531	-18363	097	N	a-	-1.2324	0.5933	-0.4001	207.2	-	-	20S	70E
6050	303	0515 Oct 08	21:39:41	5527	-18358	064	N	-t	-1.4348	0.2700	-0.8188	157.7	-	-	6N	56E
6051	303	0515 Nov 07	15:00:56	5527	-18357	102	N	t-	1.5328	0.0829	-0.9913	86.8	-	-	18N	154E
6052	303	0516 Apr 03	03:39:59	5523	-18352	069	P	-a	0.6205	1.6935	0.7449	290.9	168.3	-	5S	31W
6053	303	0516 Sep 26	21:09:37	5518	-18346	074	P	-t	-0.7473	1.5314	0.4428	330.9	153.6	-	2N	64E
6054	303	0517 Mar 23	20:06:37	5513	-18340	079	T-	pp	-0.0949	2.6679	1.6995	324.6	213.5	98.0	2S	83E
6055	303	0517 Sep 16	00:08:53	5508	-18334	084	T-	pp	-0.0065	2.8670	1.8254	354.8	226.5	103.2	2S	19E
6056	303	0518 Mar 13	08:23:11	5504	-18328	089	P	h-	-0.8620	1.2876	0.2652	290.1	115.5	-	2N	101W
6057	303	0518 Sep 05	10:33:20	5499	-18322	094	P	a-	0.7150	1.5376	0.5543	293.2	153.9	-	6S	136W
6058	303	0519 Jan 31	19:52:03	5495	-18317	061	N	-t	1.3228	0.4800	-0.6176	208.1	-	-	18N	90E
6059	303	0519 Jul 27	18:37:31	5490	-18311	066	N	-a	-1.1451	0.7305	-0.2174	213.6	-	-	20S	105E
6060	303	0519 Aug 26	02:14:07	5489	-18310	104	N	a-	1.3870	0.2888	-0.6634	139.9	-	-	9S	11W

Cat Num	Canon Plate	Calendar Date	TD of Greatest Eclipse	ΔT s	Luna Num	Saros Num	Ecl. Type	QSE	Gamma	Pen. Mag.	Um. Mag.	Eclipse Phase Durations Pen. m	Par. m	Total m	Greatest in Zenith Lat.	Lng.
6061	304	0520 Jan 20	19:32:49	5485	-18305	071	P	-t	0.6247	1.7540	0.6703	340.3	180.0	-	20N	94E
6062	304	0520 Jul 16	09:57:30	5481	-18299	076	T	-a	-0.4368	2.0473	1.0655	320.2	197.5	38.7	22S	126W
6063	304	0521 Jan 09	00:43:42	5476	-18293	081	T-	p-	-0.0994	2.6917	1.6598	345.5	220.9	99.1	22N	15E
6064	304	0521 Jul 05	19:28:10	5471	-18287	086	T	p-	0.3402	2.2513	1.2164	349.5	215.6	70.1	23S	91E
6065	304	0521 Dec 29	12:50:36	5466	-18281	091	P	a-	-0.7739	1.4296	0.4460	281.8	138.1	-	22N	168W
6066	304	0522 Jun 24	22:06:12	5462	-18275	096	N	t-	1.1271	0.8243	-0.2440	261.6	-	-	23S	51E
6067	304	0522 Nov 19	17:27:50	5458	-18270	063	N	-a	1.2737	0.5132	-0.4716	186.3	-	-	21N	118E
6068	304	0522 Dec 19	04:24:18	5457	-18269	101	N	a-	-1.4120	0.2531	-0.7193	132.0	-	-	22N	43W
6069	304	0523 May 15	10:00:51	5453	-18264	068	P	-t	-0.9205	1.1856	0.1526	289.2	91.3	-	20S	129W
6070	304	0523 Nov 09	05:26:03	5448	-18258	073	P	-a	0.6368	1.7065	0.6725	317.0	171.7	-	18N	62W
6071	304	0524 May 03	19:44:18	5444	-18252	078	T-	pp	-0.1162	2.6345	1.6548	332.1	217.1	98.4	16S	85E
6072	304	0524 Oct 28	10:24:53	5439	-18246	083	T-	pp	-0.0569	2.7968	1.7112	370.8	231.8	103.5	14N	137W
6073	304	0525 Apr 23	11:12:15	5434	-18240	088	P	a-	0.6221	1.6895	0.7429	291.3	168.6	-	12S	146W
6074	304	0525 Oct 17	09:59:06	5429	-18234	093	P	t-	-0.7519	1.5280	0.4294	332.8	152.2	-	9N	130W
6075	304	0526 Mar 14	19:23:31	5425	-18229	060	N	-a	-1.2455	0.5642	-0.4191	197.5	-	-	1N	94E
6076	304	0526 Apr 13	04:09:35	5425	-18228	098	N	a-	1.3433	0.3709	-0.5850	159.0	-	-	8S	40W
6077	304	0526 Sep 06	21:27:13	5421	-18223	065	N	-h	1.4087	0.2843	-0.7379	150.7	-	-	5S	60E
6078	304	0526 Oct 06	11:21:56	5420	-18222	103	N	h-	-1.4128	0.2961	-0.7643	159.4	-	-	5N	150W
6079	304	0527 Mar 04	05:41:46	5416	-18217	070	P	-h	-0.5610	1.8496	0.8079	331.8	187.8	-	5N	60W
6080	304	0527 Aug 27	09:48:54	5411	-18211	075	P	-a	0.6303	1.6857	0.7170	296.8	168.5	-	9S	125W
6081	305	0528 Feb 21	08:35:33	5406	-18205	080	T+	pp	0.1605	2.6077	1.5199	373.5	231.6	98.2	10N	102W
6082	305	0528 Aug 16	02:14:41	5402	-18199	085	T-	pp	-0.0909	2.6657	1.7160	317.3	210.5	97.3	14S	11W
6083	305	0529 Feb 09	08:09:09	5397	-18193	090	P	t-	0.8621	1.3199	0.2331	313.4	114.6	-	15N	95W
6084	305	0529 Aug 05	18:07:29	5392	-18187	095	P	a-	-0.8326	1.3176	0.3426	277.9	124.9	-	17S	112E
6085	305	0529 Dec 30	22:14:57	5388	-18182	062	N	-a	-1.3046	0.4724	-0.5441	186.5	-	-	22N	51E
6086	305	0530 Jan 29	11:53:03	5388	-18181	100	N	a-	1.5117	0.1050	-0.9366	94.2	-	-	19N	151W
6087	305	0530 Jun 26	16:23:02	5384	-18176	067	N	-t	1.1001	0.8632	-0.1842	260.0	-	-	23S	137E
6088	305	0530 Dec 20	11:45:39	5379	-18170	072	P	-a	-0.5893	1.7645	0.7885	297.5	172.4	-	23N	153W
6089	305	0531 Jun 15	17:48:24	5374	-18164	077	T	-t	0.3536	2.2440	1.1749	365.2	220.8	65.8	23S	115E
6090	305	0531 Dec 10	03:25:05	5369	-18158	082	T+	p-	0.0746	2.7102	1.7319	322.0	211.1	97.2	23N	30W
6091	305	0532 Jun 03	19:36:27	5365	-18152	087	T	t-	-0.4168	2.1145	1.0721	349.2	210.1	42.8	23S	87E
6092	305	0532 Nov 28	16:33:14	5360	-18146	092	P	a-	0.7789	1.4398	0.4178	295.7	139.1	-	23N	132E
6093	305	0533 Apr 24	19:08:13	5356	-18141	059	N	-a	1.4266	0.2241	-0.7438	127.5	-	-	12S	95E
6094	305	0533 May 24	03:49:59	5355	-18140	097	N	a-	-1.1538	0.7354	-0.2537	226.5	-	-	22S	37W
6095	305	0533 Oct 19	05:23:14	5351	-18135	064	N	-t	-1.4587	0.2283	-0.8649	145.9	-	-	10N	61W
6096	305	0533 Nov 17	23:11:15	5351	-18134	102	N	t-	1.5291	0.0916	-0.9867	91.4	-	-	21N	31E
6097	305	0534 Apr 14	11:25:38	5347	-18129	069	P	-a	0.6786	1.5855	0.6397	285.1	158.7	-	9S	149W
6098	305	0534 Oct 08	04:43:19	5342	-18123	074	P	-t	-0.7773	1.4772	0.3868	326.7	144.6	-	6N	51W
6099	305	0535 Apr 04	03:53:12	5337	-18117	079	T-	pp	-0.0418	2.7651	1.7971	325.8	214.7	99.8	6S	36W
6100	305	0535 Sep 27	07:56:26	5332	-18111	084	T-	pp	-0.0406	2.8044	1.7627	353.6	225.6	102.5	2N	99W
6101	306	0536 Mar 23	15:55:39	5328	-18105	089	P	t-	-0.8153	1.3734	0.3508	297.7	131.7	-	3S	144E
6102	306	0536 Sep 15	18:35:11	5323	-18099	094	P	a-	0.6749	1.6114	0.6277	296.7	161.4	-	2S	102E
6103	306	0537 Feb 11	03:27:52	5319	-18094	061	N	-t	1.3504	0.4279	-0.6671	197.4	-	-	14N	25W
6104	306	0537 Aug 07	02:18:44	5314	-18088	066	N	-a	-1.2056	0.6212	-0.3300	199.5	-	-	17S	11W
6105	306	0537 Sep 05	10:16:09	5314	-18087	104	N	a-	1.3429	0.3710	-0.5836	157.4	-	-	5S	133W
6106	306	0538 Jan 31	03:25:09	5310	-18082	071	P	-t	0.6425	1.7187	0.6399	337.2	176.4	-	17N	25W
6107	306	0538 Jul 27	17:17:25	5305	-18076	076	P	-a	-0.5058	1.9232	0.9363	317.0	190.2	-	20S	124E
6108	306	0539 Jan 20	09:05:32	5300	-18070	081	T-	p-	-0.0884	2.7093	1.6825	344.1	220.6	99.5	20N	111W
6109	306	0539 Jul 17	02:15:12	5295	-18064	086	T+	pp	0.2631	2.3952	1.3554	354.6	221.7	85.7	21S	11W
6110	306	0540 Jan 09	21:34:15	5291	-18058	091	P	a-	-0.7661	1.4423	0.4620	281.9	140.0	-	21N	61E
6111	306	0540 Jul 05	04:31:16	5286	-18052	096	N	t-	1.0439	0.9779	-0.0922	280.8	-	-	22S	46W
6112	306	0540 Nov 30	02:14:49	5282	-18047	063	N	-a	1.2789	0.5048	-0.4825	185.3	-	-	24N	14W
6113	306	0540 Dec 29	13:14:13	5281	-18046	101	N	a-	-1.4089	0.2586	-0.7135	133.5	-	-	22N	175W
6114	306	0541 May 25	16:46:38	5277	-18041	068	P	-t	-0.9967	1.0436	0.0149	274.3	28.9	-	22S	129E
6115	306	0541 Nov 19	13:52:58	5273	-18035	073	P	-a	0.6438	1.6959	0.6576	317.7	170.7	-	21N	171E
6116	306	0542 May 15	02:59:39	5268	-18029	078	T-	pp	-0.1884	2.4998	1.5246	329.1	214.2	93.1	19S	25W
6117	306	0542 Nov 08	18:24:22	5263	-18023	083	T-	pp	-0.0452	2.8203	1.7307	372.1	232.3	103.8	17N	102E
6118	306	0543 May 04	18:45:44	5259	-18017	088	P	a-	0.5545	1.8123	0.8682	296.6	178.3	-	16S	99E
6119	306	0543 Oct 28	17:46:25	5254	-18011	093	P	t-	-0.7338	1.5620	0.4616	335.1	156.9	-	13N	112E
6120	306	0544 Mar 25	03:10:43	5250	-18006	060	N	-a	-1.2964	0.4706	-0.5125	182.6	-	-	4S	25W

Cat Num	Canon Plate	Calendar Date	TD of Greatest Eclipse	ΔT s	Luna Num	Saros Num	Ecl. Type	QSE	Gamma	Pen. Mag.	Um. Mag.	Eclipse Phase Durations Pen. m	Par. m	Total m	Greatest in Zenith Lat.	Lng.
6121	307	0544 Apr 23	11:45:05	5249	-18005	098	N	a-	1.2848	0.4781	-0.4775	179.0	-	-	12S	155W
6122	307	0544 Sep 17	05:14:51	5245	-18000	065	N	-h	1.4406	0.2256	-0.7965	134.5	-	-	0S	59W
6123	307	0544 Oct 16	19:19:38	5244	-17999	103	N	h-	-1.3888	0.3401	-0.7202	169.7	-	-	9N	89E
6124	307	0545 Mar 14	13:14:23	5241	-17994	070	P	-h	-0.6067	1.7657	0.7240	328.7	181.0	-	1N	175W
6125	307	0545 Sep 06	17:47:10	5236	-17988	075	P	-a	0.6719	1.6097	0.6402	292.2	160.9	-	5S	114E
6126	307	0546 Mar 03	15:56:24	5231	-17982	080	T+	pp	0.1203	2.6804	1.5949	374.5	233.3	101.7	6N	146E
6127	307	0546 Aug 27	10:09:38	5226	-17976	085	T-	pp	-0.0411	2.7588	1.8059	318.2	211.1	98.6	10S	131W
6128	307	0547 Feb 20	15:41:42	5222	-17970	090	P	t-	0.8290	1.3780	0.2962	317.2	127.8	-	11N	150E
6129	307	0547 Aug 17	01:42:51	5217	-17964	095	P	a-	-0.7779	1.4206	0.4402	286.3	139.9	-	14S	3W
6130	307	0548 Jan 11	06:47:52	5213	-17959	062	N	-a	-1.3071	0.4655	-0.5465	184.6	-	-	20N	77W
6131	307	0548 Feb 09	19:56:47	5212	-17958	100	N	a-	1.4871	0.1472	-0.8884	110.7	-	-	15N	87E
6132	307	0548 Jul 06	22:59:59	5208	-17953	067	N	-t	1.1833	0.7126	-0.3387	240.7	-	-	22S	37E
6133	307	0548 Dec 30	20:35:05	5204	-17947	072	P	-a	-0.5927	1.7573	0.7833	296.7	171.8	-	22N	75E
6134	307	0549 Jun 26	00:10:45	5199	-17941	077	T	-t	0.4409	2.0839	1.0144	359.8	211.7	19.9	23S	19E
6135	307	0549 Dec 20	12:16:10	5194	-17935	082	T+	pp	0.0726	2.7142	1.7350	322.6	211.4	97.3	24N	162W
6136	307	0550 Jun 15	02:13:49	5190	-17929	087	T	t-	-0.3314	2.2698	1.2303	352.7	217.6	72.4	24S	13W
6137	307	0550 Dec 10	01:09:23	5185	-17923	092	P	a-	0.7790	1.4413	0.4161	296.9	139.2	-	24N	3E
6138	307	0551 May 06	02:33:23	5181	-17918	059	N	-a	1.4885	0.1083	-0.8553	89.2	-	-	15S	18W
6139	307	0551 Jun 04	10:56:21	5180	-17917	097	N	a-	-1.0744	0.8790	-0.1062	242.9	-	-	24S	144W
6140	307	0551 Oct 30	13:12:52	5176	-17912	064	N	-t	-1.4777	0.1955	-0.9016	135.7	-	-	13N	180W
6141	308	0551 Nov 29	07:22:40	5176	-17911	102	N	t-	1.5266	0.0978	-0.9837	94.7	-	-	23N	91W
6142	308	0552 Apr 24	19:05:07	5172	-17906	069	P	-a	0.7419	1.4682	0.5244	278.1	146.3	-	13S	95E
6143	308	0552 Oct 18	12:26:42	5167	-17900	074	P	-t	-0.8000	1.4364	0.3446	323.4	137.2	-	10N	168W
6144	308	0553 Apr 14	11:31:59	5162	-17894	079	T+	pp	0.0174	2.8099	1.8418	326.7	215.4	100.4	10S	152W
6145	308	0553 Oct 07	15:53:05	5158	-17888	084	T-	pp	-0.0681	2.7538	1.7124	352.3	224.6	101.4	6N	140E
6146	308	0554 Apr 03	23:18:17	5153	-17882	089	P	t-	-0.7612	1.4729	0.4498	305.8	147.3	-	7S	32E
6147	308	0554 Sep 27	02:46:01	5148	-17876	094	P	a-	0.6416	1.6726	0.6884	299.3	166.9	-	3N	23W
6148	308	0555 Feb 22	10:52:57	5144	-17871	061	N	-t	1.3863	0.3607	-0.7313	182.3	-	-	11N	137W
6149	308	0555 Mar 24	03:43:32	5144	-17870	099	Nb	t-	-1.5487	0.0532	-1.0202	71.0	-	-	3S	34W
6150	308	0555 Aug 18	10:05:54	5140	-17865	066	N	-a	-1.2609	0.5216	-0.4335	185.0	-	-	14S	129W
6151	308	0555 Sep 16	18:26:52	5139	-17864	104	N	a-	1.3058	0.4403	-0.5171	170.4	-	-	1S	103E
6152	308	0556 Feb 11	11:09:19	5135	-17859	071	P	-t	0.6675	1.6701	0.5967	333.1	171.2	-	14N	141W
6153	308	0556 Aug 07	00:41:59	5130	-17853	076	P	-a	-0.5701	1.8082	0.8156	313.5	181.8	-	17S	12E
6154	308	0557 Jan 30	17:21:29	5126	-17847	081	T-	pp	-0.0718	2.7370	1.7158	342.7	220.4	100.1	17N	125E
6155	308	0557 Jul 27	09:06:13	5121	-17841	086	T+	pp	0.1900	2.5321	1.4871	358.7	226.1	95.3	19S	114W
6156	308	0558 Jan 20	06:14:47	5116	-17835	091	P	a-	-0.7554	1.4601	0.4833	282.4	142.5	-	19N	69W
6157	308	0558 Jul 16	10:57:54	5112	-17829	096	P	t-	0.9616	1.1301	0.0577	297.4	59.0	-	21S	143W
6158	308	0558 Dec 11	11:03:49	5108	-17824	063	N	-a	1.2824	0.4995	-0.4899	184.9	-	-	25N	145W
6159	308	0559 Jan 09	22:02:29	5107	-17823	101	N	a-	-1.4049	0.2656	-0.7059	135.3	-	-	20N	53E
6160	308	0559 Jun 05	23:31:22	5103	-17818	068	N	-t	-1.0745	0.8988	-0.1260	257.4	-	-	24S	27E
6161	309	0559 Nov 30	22:22:15	5098	-17812	073	P	-a	0.6480	1.6902	0.6480	318.6	170.2	-	23N	44E
6162	309	0560 May 25	10:13:38	5094	-17806	078	T-	-p	-0.2624	2.3619	1.3906	325.4	210.1	84.5	22S	134W
6163	309	0560 Nov 19	02:28:07	5089	-17800	083	T-	pp	-0.0373	2.8366	1.7435	373.2	232.7	104.0	20N	19W
6164	309	0561 May 15	02:17:07	5084	-17794	088	P	a-	0.4850	1.9387	0.9967	301.3	186.7	-	19S	15W
6165	309	0561 Nov 08	01:38:42	5080	-17788	093	P	t-	-0.7195	1.5890	0.4873	336.7	160.4	-	17N	7W
6166	309	0562 Apr 05	10:51:17	5076	-17783	060	N	-a	-1.3520	0.3686	-0.6147	163.6	-	-	8S	142W
6167	309	0562 May 04	19:16:30	5075	-17782	098	N	a-	1.2233	0.5911	-0.3648	197.3	-	-	15S	91E
6168	309	0562 Sep 28	13:11:14	5071	-17777	065	N	-a	1.4663	0.1785	-0.8435	119.8	-	-	4N	180W
6169	309	0562 Oct 28	03:24:31	5070	-17776	103	N	a-	-1.3698	0.3747	-0.6852	177.0	-	-	13N	33W
6170	309	0563 Mar 25	20:36:29	5067	-17771	070	P	-t	-0.6602	1.6676	0.6260	324.3	171.6	-	3S	73E
6171	309	0563 Sep 18	01:54:29	5062	-17765	075	P	-a	0.7065	1.5466	0.5761	288.1	153.9	-	1S	9W
6172	309	0564 Mar 13	23:06:18	5057	-17759	080	T+	pp	0.0713	2.7691	1.6860	375.3	234.8	104.5	2N	37E
6173	309	0564 Sep 06	18:12:55	5053	-17753	085	T+	pp	0.0017	2.8328	1.8764	318.8	211.3	98.9	6S	107E
6174	309	0565 Mar 02	23:04:59	5048	-17747	090	P	t-	0.7879	1.4508	0.3743	321.7	141.7	-	7N	38E
6175	309	0565 Aug 27	09:25:59	5043	-17741	095	P	a-	-0.7303	1.5109	0.5248	293.4	151.2	-	10S	120W
6176	309	0566 Jan 21	15:15:54	5039	-17736	062	N	-a	-1.3141	0.4503	-0.5568	181.1	-	-	18N	156E
6177	309	0566 Feb 20	03:53:05	5039	-17735	100	N	a-	1.4558	0.2014	-0.8280	128.2	-	-	12N	33W
6178	309	0566 Jul 18	05:38:45	5035	-17730	067	N	-t	1.2637	0.5673	-0.4884	218.6	-	-	20S	63W
6179	309	0566 Aug 16	19:12:38	5034	-17729	105	Nb	t-	-1.5296	0.0708	-0.9678	78.6	-	-	15S	94E
6180	309	0567 Jan 11	05:21:17	5030	-17724	072	P	-a	-0.5990	1.7445	0.7730	295.6	170.9	-	21N	56W

Cat Num	Canon Plate	Calendar Date	TD of Greatest Eclipse	ΔT s	Luna Num	Saros Num	Ecl. Type	QSE	Gamma	Pen. Mag.	Um. Mag.	Eclipse Phase Durations Pen. m	Par. m	Total m	Greatest in Zenith Lat.	Lng.
6181	310	0567 Jul 07	06:34:55	5025	-17718	077	P	-t	0.5263	1.9277	0.8573	353.2	200.3	-	22S	77W
6182	310	0567 Dec 31	21:04:38	5021	-17712	082	T+	pp	0.0689	2.7213	1.7416	323.1	211.7	97.6	23N	67E
6183	310	0568 Jun 25	08:53:42	5016	-17706	087	T-	p-	-0.2473	2.4230	1.3859	355.0	222.8	88.3	24S	113W
6184	310	0568 Dec 20	09:45:04	5011	-17700	092	P	a-	0.7798	1.4411	0.4134	297.9	139.2	-	24N	125W
6185	310	0569 Jun 14	18:03:30	5007	-17694	097	P	a-	-0.9946	1.0238	0.0418	257.2	46.2	-	24S	109E
6186	310	0569 Nov 09	21:08:04	5003	-17689	064	N	-t	-1.4916	0.1715	-0.9288	127.6	-	-	17N	61E
6187	310	0569 Dec 09	15:35:23	5002	-17688	102	N	t-	1.5258	0.1007	-0.9835	96.4	-	-	25N	146E
6188	310	0570 May 06	02:39:45	4998	-17683	069	P	-a	0.8093	1.3438	0.4016	269.9	130.5	-	16S	20W
6189	310	0570 Oct 29	20:18:13	4994	-17677	074	P	-t	-0.8163	1.4068	0.3141	320.7	131.4	-	14N	73E
6190	310	0571 Apr 25	19:02:19	4989	-17671	079	T+	pp	0.0830	2.6897	1.7212	327.2	215.3	99.2	14S	94E
6191	310	0571 Oct 18	23:59:27	4984	-17665	084	T-	pp	-0.0884	2.7164	1.6754	351.0	223.5	100.4	11N	17E
6192	310	0572 Apr 14	06:32:05	4980	-17659	089	P	t-	-0.7009	1.5839	0.5601	314.1	162.1	-	11S	78W
6193	310	0572 Oct 07	11:05:09	4975	-17653	094	P	a-	0.6148	1.7220	0.7374	301.1	170.9	-	7N	149W
6194	310	0573 Mar 04	18:10:09	4971	-17648	061	N	-t	1.4279	0.2825	-0.8062	162.4	-	-	7N	112E
6195	310	0573 Apr 03	10:40:47	4970	-17647	099	N	t-	-1.4914	0.1577	-0.9143	121.3	-	-	7S	140W
6196	310	0573 Aug 28	17:59:54	4966	-17642	066	N	-a	-1.3104	0.4329	-0.5263	170.3	-	-	10S	111E
6197	310	0573 Sep 27	02:44:59	4966	-17641	104	N	a-	1.2751	0.4981	-0.4621	180.3	-	-	4N	24W
6198	310	0574 Feb 21	18:47:51	4962	-17636	071	P	-t	0.6978	1.6116	0.5440	328.3	164.5	-	10N	103E
6199	310	0574 Aug 18	08:10:53	4957	-17630	076	P	-a	-0.6298	1.7016	0.7030	309.7	172.5	-	13S	101W
6200	310	0575 Feb 11	01:32:10	4952	-17624	081	T-	pp	-0.0499	2.7742	1.7589	341.2	220.2	100.7	13N	2E
6201	311	0575 Aug 07	16:01:52	4948	-17618	086	T+	pp	0.1212	2.6610	1.6104	362.0	229.1	101.1	16S	141E
6202	311	0576 Jan 31	14:48:13	4943	-17612	091	P	a-	-0.7385	1.4893	0.5163	283.5	146.3	-	16N	162E
6203	311	0576 Jul 26	17:31:30	4939	-17606	096	P	t-	0.8844	1.2729	0.1980	311.0	107.1	-	18S	118E
6204	311	0576 Dec 21	19:50:46	4935	-17601	063	N	-a	1.2876	0.4907	-0.5003	183.9	-	-	25N	84E
6205	311	0577 Jan 20	06:44:41	4934	-17600	101	N	a-	-1.3962	0.2812	-0.6894	139.1	-	-	18N	77W
6206	311	0577 Jun 16	06:19:01	4930	-17595	068	N	-t	-1.1505	0.7577	-0.2639	238.8	-	-	25S	75W
6207	311	0577 Dec 11	06:50:38	4925	-17589	073	P	-a	0.6524	1.6837	0.6384	319.5	169.7	-	24N	83W
6208	311	0578 Jun 05	17:27:34	4921	-17583	078	T	-p	-0.3375	2.2225	1.2545	320.9	204.7	71.3	23S	118E
6209	311	0578 Nov 30	10:34:02	4916	-17577	083	T-	pp	-0.0317	2.8481	1.7524	374.1	233.0	104.1	22N	140W
6210	311	0579 May 26	09:43:51	4911	-17571	088	T	a-	0.4114	2.0731	1.1325	305.6	194.0	52.3	21S	127W
6211	311	0579 Nov 19	09:37:50	4907	-17565	093	P	t-	-0.7103	1.6061	0.5039	337.6	162.5	-	20N	127W
6212	311	0580 Apr 15	18:21:52	4903	-17560	060	N	-a	-1.4151	0.2533	-0.7306	137.5	-	-	12S	104E
6213	311	0580 May 15	02:41:40	4902	-17559	098	N	a-	1.1570	0.7131	-0.2437	214.6	-	-	18S	21W
6214	311	0580 Oct 08	21:17:59	4898	-17554	065	N	-a	1.4847	0.1446	-0.8771	107.8	-	-	8N	57E
6215	311	0580 Nov 07	11:37:22	4898	-17553	103	N	a-	-1.3563	0.3990	-0.6599	181.7	-	-	16N	157W
6216	311	0581 Apr 05	03:48:48	4894	-17548	070	P	-t	-0.7208	1.5565	0.5147	318.6	158.9	-	7S	37W
6217	311	0581 Sep 28	10:10:51	4889	-17542	075	P	-a	0.7344	1.4960	0.5244	284.6	147.7	-	4N	135W
6218	311	0582 Mar 25	06:07:37	4884	-17536	080	T+	pp	0.0155	2.8702	1.7896	375.6	235.6	106.0	3S	70W
6219	311	0582 Sep 18	02:23:05	4880	-17530	085	T+	pp	0.0385	2.7671	1.8070	319.2	211.2	98.5	1S	17W
6220	311	0583 Mar 14	06:21:40	4875	-17524	090	P	t-	0.7406	1.5347	0.4638	326.5	155.3	-	2N	73W
6221	312	0583 Sep 07	17:15:19	4871	-17518	095	P	a-	-0.6887	1.5903	0.5982	299.4	160.0	-	6S	121E
6222	312	0584 Feb 01	23:38:54	4867	-17513	062	N	-a	-1.3256	0.4264	-0.5754	175.9	-	-	15N	29E
6223	312	0584 Mar 02	11:42:34	4866	-17512	100	N	a-	1.4183	0.2671	-0.7558	146.1	-	-	7N	152W
6224	312	0584 Jul 28	12:22:13	4862	-17507	067	N	-t	1.3390	0.4314	-0.6290	193.7	-	-	17S	165W
6225	312	0584 Aug 27	02:28:31	4861	-17506	105	N	t-	-1.4809	0.1633	-0.8815	119.0	-	-	11S	17W
6226	312	0585 Jan 21	14:03:47	4857	-17501	072	P	-a	-0.6083	1.7261	0.7573	294.3	169.5	-	19N	173E
6227	312	0585 Jul 17	13:04:06	4853	-17495	077	P	-t	0.6076	1.7791	0.7076	345.7	186.6	-	21S	175W
6228	312	0586 Jan 11	05:49:30	4848	-17489	082	T+	pp	0.0626	2.7329	1.7531	323.7	212.1	97.9	22N	64W
6229	312	0586 Jul 06	15:38:20	4844	-17483	087	T-	pp	-0.1662	2.5707	1.5357	356.1	226.1	97.6	23S	146E
6230	312	0586 Dec 31	18:16:58	4839	-17477	092	P	a-	0.7779	1.4455	0.4158	299.2	140.0	-	24N	108E
6231	312	0587 Jun 26	01:13:11	4834	-17471	097	P	a-	-0.9157	1.1672	0.1879	269.3	95.7	-	24S	2E
6232	312	0587 Nov 21	05:07:11	4831	-17466	064	N	-t	-1.5021	0.1535	-0.9492	121.1	-	-	19N	59W
6233	312	0587 Dec 20	23:46:02	4830	-17465	102	N	t-	1.5237	0.1054	-0.9807	98.8	-	-	25N	24E
6234	312	0588 May 16	10:10:43	4826	-17460	069	P	-a	0.8796	1.2143	0.2731	260.4	109.7	-	19S	134W
6235	312	0588 Nov 09	04:15:35	4821	-17454	074	P	-t	-0.8283	1.3849	0.2921	318.5	126.9	-	17N	47W
6236	312	0589 May 06	02:27:11	4817	-17448	079	T+	-p	0.1525	2.5627	1.5932	327.1	214.2	95.6	17S	18W
6237	312	0589 Oct 29	08:13:54	4812	-17442	084	T-	pp	-0.1026	2.6901	1.6498	349.6	222.6	99.4	14N	107W
6238	312	0590 Apr 25	13:36:38	4807	-17436	089	P	t-	-0.6336	1.7078	0.6829	322.4	176.1	-	14S	175E
6239	312	0590 Oct 18	19:32:46	4803	-17430	094	P	a-	0.5946	1.7592	0.7743	302.2	173.7	-	11N	83E
6240	312	0591 Mar 16	01:16:48	4799	-17425	061	N	-t	1.4777	0.1894	-0.8958	134.0	-	-	2N	4E

Cat Num	Canon Plate	Calendar Date	TD of Greatest Eclipse	ΔT s	Luna Num	Saros Num	Ecl. Type	QSE	Gamma	Pen. Mag.	Um. Mag.	Eclipse Phase Durations Pen. m	Par. m	Total m	Greatest in Zenith Lat.	Lng.
6241	313	0591 Apr 14	17:27:28	4798	-17424	099	N	t-	-1.4258	0.2773	-0.7934	159.3	-	-	11S	117E
6242	313	0591 Sep 09	02:01:43	4794	-17419	066	N	-a	-1.3534	0.3561	-0.6075	156.0	-	-	6S	11W
6243	313	0591 Oct 08	11:11:50	4794	-17418	104	N	a-	1.2514	0.5431	-0.4200	187.6	-	-	8N	152W
6244	313	0592 Mar 04	02:16:44	4790	-17413	071	P	-t	0.7367	1.5372	0.4757	322.1	155.1	-	6N	11W
6245	313	0592 Aug 28	15:47:05	4785	-17407	076	P	-a	-0.6828	1.6075	0.6026	306.0	162.9	-	9S	143E
6246	313	0593 Feb 21	09:36:13	4781	-17401	081	T-	pp	-0.0214	2.8233	1.8141	339.7	219.9	101.2	10N	120W
6247	313	0593 Aug 17	23:02:30	4776	-17395	086	T+	pp	0.0572	2.7814	1.7251	364.5	230.9	104.1	13S	35E
6248	313	0594 Feb 10	23:16:48	4771	-17389	091	P	a-	-0.7172	1.5263	0.5573	285.1	150.9	-	13N	35E
6249	313	0594 Aug 07	00:10:03	4767	-17383	096	P	t-	0.8107	1.4095	0.3318	322.5	135.9	-	16S	18E
6250	313	0595 Jan 02	04:36:03	4763	-17378	063	N	-a	1.2938	0.4799	-0.5122	182.5	-	-	24N	47W
6251	313	0595 Jan 31	15:21:36	4762	-17377	101	N	a-	-1.3838	0.3033	-0.6661	144.3	-	-	15N	153E
6252	313	0595 Jun 27	13:08:17	4758	-17372	068	N	-t	-1.2258	0.6181	-0.4007	217.8	-	-	25S	177W
6253	313	0595 Jul 27	01:33:13	4758	-17371	106	Nb	t-	1.5221	0.0900	-0.9596	90.2	-	-	18S	3W
6254	313	0595 Dec 22	15:18:33	4754	-17366	073	P	-a	0.6565	1.6774	0.6296	320.3	169.2	-	24N	151E
6255	313	0596 Jun 16	00:43:12	4749	-17360	078	T	-a	-0.4119	2.0847	1.1194	315.8	198.0	50.9	24S	8E
6256	313	0596 Dec 10	18:39:58	4744	-17354	083	T-	pp	-0.0264	2.8587	1.7612	375.0	233.4	104.3	23N	99E
6257	313	0597 Jun 05	17:11:22	4740	-17348	088	T	p-	0.3381	2.2072	1.2673	309.2	199.8	71.2	23S	121E
6258	313	0597 Nov 29	17:39:59	4735	-17342	093	P	t-	-0.7030	1.6194	0.5175	338.1	164.2	-	22N	113E
6259	313	0598 Apr 27	01:46:02	4731	-17337	060	N	-a	-1.4824	0.1303	-0.8546	100.1	-	-	16S	8W
6260	313	0598 May 26	10:03:29	4731	-17336	098	N	a-	1.0884	0.8398	-0.1184	230.5	-	-	20S	132W
6261	314	0598 Oct 20	05:32:42	4727	-17331	065	N	-a	1.4976	0.1206	-0.9006	98.3	-	-	13N	67W
6262	314	0598 Nov 18	19:55:27	4726	-17330	103	N	a-	-1.3461	0.4168	-0.6405	184.8	-	-	19N	78E
6263	314	0599 Apr 16	10:52:10	4722	-17325	070	P	-t	-0.7879	1.4335	0.3913	311.1	141.6	-	11S	144W
6264	314	0599 Oct 09	18:35:43	4718	-17319	075	P	-a	0.7556	1.4576	0.4850	281.8	142.7	-	8N	97E
6265	314	0600 Apr 04	12:58:45	4713	-17313	080	T-	pp	-0.0484	2.8085	1.7303	375.4	235.6	105.6	7S	175W
6266	314	0600 Sep 28	10:42:03	4708	-17307	085	T+	-p	0.0679	2.7152	1.7514	319.6	210.9	97.7	3N	143W
6267	314	0601 Mar 24	13:29:56	4704	-17301	090	P	t-	0.6858	1.6326	0.5672	331.7	168.6	-	2S	179E
6268	314	0601 Sep 18	01:11:32	4699	-17295	095	P	a-	-0.6534	1.6581	0.6600	304.5	166.8	-	2S	0E
6269	314	0602 Feb 12	07:55:24	4695	-17290	062	N	-a	-1.3431	0.3917	-0.6046	168.5	-	-	12N	95W
6270	314	0602 Mar 13	19:24:44	4695	-17289	100	N	a-	1.3739	0.3452	-0.6711	164.1	-	-	3N	91E
6271	314	0602 Aug 08	19:10:29	4691	-17284	067	N	-t	1.4094	0.3048	-0.7605	165.3	-	-	14S	92E
6272	314	0602 Sep 07	09:51:15	4690	-17283	105	N	t-	-1.4381	0.2450	-0.8062	145.3	-	-	7S	129W
6273	314	0603 Feb 01	22:39:19	4686	-17278	072	P	-a	-0.6239	1.6959	0.7302	292.5	167.1	-	16N	44E
6274	314	0603 Jul 28	19:38:19	4682	-17272	077	P	-t	0.6849	1.6380	0.5650	337.3	170.6	-	18S	86E
6275	314	0604 Jan 22	14:28:53	4677	-17266	082	T+	pp	0.0520	2.7523	1.7727	324.4	212.7	98.4	19N	166E
6276	314	0604 Jul 16	22:28:47	4673	-17260	087	T-	pp	-0.0887	2.7121	1.6786	356.3	227.7	102.5	21S	43E
6277	314	0605 Jan 11	02:44:30	4668	-17254	092	P	a-	0.7732	1.4548	0.4237	300.9	141.6	-	22N	19W
6278	314	0605 Jul 06	08:27:08	4663	-17248	097	P	a-	-0.8393	1.3064	0.3292	279.4	123.7	-	24S	107W
6279	314	0605 Dec 01	13:09:47	4660	-17243	064	N	-t	-1.5094	0.1410	-0.9634	116.4	-	-	21N	179W
6280	314	0605 Dec 31	07:55:11	4659	-17242	102	N	t-	1.5209	0.1111	-0.9761	101.6	-	-	24N	97W
6281	315	0606 May 27	17:37:56	4655	-17237	069	P	-a	0.9530	1.0794	0.1385	249.3	79.6	-	21S	114E
6282	315	0606 Jun 26	00:11:01	4654	-17236	107	Nb	a-	-1.5278	0.0256	-0.9169	42.6	-	-	25S	17E
6283	315	0606 Nov 20	12:19:35	4650	-17231	074	P	-t	-0.8354	1.3717	0.2792	316.9	124.1	-	20N	168W
6284	315	0607 May 17	09:46:22	4646	-17225	079	T+	-p	0.2260	2.4287	1.4576	326.3	211.8	89.0	20S	129W
6285	315	0607 Nov 09	16:35:38	4641	-17219	084	T-	pp	-0.1117	2.6726	1.6336	348.3	221.7	98.7	18N	127E
6286	315	0608 May 05	20:34:34	4637	-17213	089	P	t-	-0.5617	1.8407	0.8142	330.3	188.6	-	18S	69E
6287	315	0608 Oct 29	04:07:37	4632	-17207	094	P	a-	0.5800	1.7862	0.8012	302.9	175.5	-	15N	47W
6288	315	0609 Mar 26	08:16:40	4628	-17202	061	Ne	-t	1.5328	0.0866	-0.9951	91.3	-	-	2S	103W
6289	315	0609 Apr 25	00:08:34	4627	-17201	099	N	t-	-1.3555	0.4058	-0.6637	190.7	-	-	15S	16E
6290	315	0609 Sep 19	10:09:43	4624	-17196	066	N	-a	-1.3914	0.2887	-0.6794	141.7	-	-	2S	134W
6291	315	0609 Oct 18	19:44:15	4623	-17195	104	N	a-	1.2323	0.5795	-0.3863	193.2	-	-	12N	79E
6292	315	0610 Mar 15	09:40:14	4619	-17190	071	P	-t	0.7808	1.4533	0.3978	314.9	143.2	-	2N	123W
6293	315	0610 Sep 08	23:29:23	4615	-17184	076	P	-a	-0.7301	1.5240	0.5127	302.4	152.9	-	5S	26E
6294	315	0611 Mar 04	17:32:31	4610	-17178	081	T+	pp	0.0143	2.8331	1.8302	338.1	219.5	101.2	5N	119E
6295	315	0611 Aug 29	06:10:27	4605	-17172	086	T-	pp	-0.0001	2.8891	1.8269	366.3	231.7	104.9	9S	74W
6296	315	0612 Feb 22	07:37:09	4601	-17166	091	P	a-	-0.6888	1.5763	0.6115	287.3	156.5	-	9N	92W
6297	315	0612 Aug 17	06:57:52	4596	-17160	096	P	t-	0.7443	1.5328	0.4524	331.8	155.8	-	12S	85W
6298	315	0613 Jan 12	13:15:27	4592	-17155	063	N	-a	1.3047	0.4602	-0.5326	179.5	-	-	22N	176W
6299	315	0613 Feb 10	23:50:12	4592	-17154	101	N	a-	-1.3651	0.3370	-0.6311	151.8	-	-	12N	25E
6300	315	0613 Jul 07	20:04:08	4588	-17149	068	N	-t	-1.2967	0.4869	-0.5296	195.0	-	-	24S	79E

Cat Num	Canon Plate	Calendar Date	TD of Greatest Eclipse	ΔT s	Luna Num	Saros Num	Ecl. Type	QSE	Gamma	Pen. Mag.	Um. Mag.	Eclipse Phase Durations Pen. m	Eclipse Phase Durations Par. m	Eclipse Phase Durations Total m	Greatest in Zenith Lat.	Greatest in Zenith Lng.
6301	316	0613 Aug 06	08:29:10	4587	-17148	106	N	t-	1.4514	0.2195	-0.8298	138.8	-	-	15S	108W
6302	316	0614 Jan 01	23:42:34	4583	-17143	073	P	-a	0.6634	1.6658	0.6160	320.7	168.2	-	23N	26E
6303	316	0614 Jun 27	08:00:24	4579	-17137	078	P	-a	-0.4858	1.9481	0.9848	310.0	189.9	-	24S	101W
6304	316	0614 Dec 22	02:45:31	4574	-17131	083	T-	pp	-0.0210	2.8690	1.7705	375.6	233.7	104.4	24N	22W
6305	316	0615 Jun 17	00:37:32	4570	-17125	088	T+	p-	0.2633	2.3446	1.4046	312.2	204.5	83.5	23S	9E
6306	316	0615 Dec 11	01:44:34	4565	-17119	093	P	t-	-0.6971	1.6296	0.5288	338.3	165.5	-	23N	8W
6307	316	0616 Jun 05	17:22:01	4560	-17113	098	P	a-	1.0173	0.9713	0.0109	245.2	23.2	-	22S	117E
6308	316	0616 Oct 30	13:56:41	4557	-17108	065	N	-a	1.5043	0.1080	-0.9124	92.8	-	-	16N	166E
6309	316	0616 Nov 29	04:19:10	4556	-17107	103	N	a-	-1.3394	0.4281	-0.6271	186.5	-	-	21N	48W
6310	316	0617 Apr 26	17:47:20	4552	-17102	070	P	-t	-0.8606	1.3004	0.2574	301.8	117.6	-	15S	111E
6311	316	0617 Oct 20	03:07:47	4548	-17096	075	P	-a	0.7714	1.4291	0.4555	279.6	138.7	-	12N	32W
6312	316	0618 Apr 15	19:42:39	4543	-17090	080	T-	pp	-0.1182	2.6793	1.6035	374.4	234.3	102.6	11S	83E
6313	316	0618 Oct 09	19:07:54	4538	-17084	085	T+	-p	0.0911	2.6744	1.7068	320.1	210.6	96.8	7N	89E
6314	316	0619 Apr 04	20:30:48	4534	-17078	090	P	t-	0.6240	1.7431	0.6832	336.8	181.1	-	6S	72E
6315	316	0619 Sep 29	09:15:02	4529	-17072	095	P	a-	-0.6248	1.7136	0.7094	308.8	172.0	-	3N	122W
6316	316	0620 Feb 23	16:05:46	4526	-17067	062	N	-a	-1.3661	0.3465	-0.6442	158.5	-	-	8N	141E
6317	316	0620 Mar 24	03:00:44	4525	-17066	100	N	a-	1.3235	0.4345	-0.5755	181.7	-	-	1S	25W
6318	316	0620 Aug 19	02:05:18	4521	-17061	067	N	-t	1.4737	0.1893	-0.8809	132.1	-	-	11S	12W
6319	316	0620 Sep 17	17:20:56	4520	-17060	105	N	t-	-1.4012	0.3157	-0.7417	164.6	-	-	2S	117E
6320	316	0621 Feb 12	07:08:56	4516	-17055	072	P	-a	-0.6446	1.6564	0.6938	290.1	163.8	-	12N	84W
6321	317	0621 Aug 08	02:20:49	4512	-17049	077	P	-t	0.7556	1.5091	0.4345	328.6	152.6	-	15S	16W
6322	317	0622 Feb 01	23:02:22	4507	-17043	082	T+	pp	0.0369	2.7797	1.8006	325.0	213.2	98.9	16N	37E
6323	317	0622 Jul 28	05:25:43	4503	-17037	087	T-	pp	-0.0157	2.8455	1.8131	355.5	227.9	104.1	19S	62W
6324	317	0623 Jan 22	11:05:47	4498	-17031	092	P	a-	0.7644	1.4716	0.4395	303.0	144.2	-	20N	144W
6325	317	0623 Jul 17	15:46:39	4494	-17025	097	P	a-	-0.7668	1.4387	0.4630	287.8	143.3	-	22S	143E
6326	317	0623 Dec 12	21:11:51	4490	-17020	064	N	-t	-1.5166	0.1281	-0.9771	111.1	-	-	22N	61E
6327	317	0624 Jan 11	15:57:41	4489	-17019	102	N	t-	1.5136	0.1247	-0.9628	107.8	-	-	23N	142E
6328	317	0624 Jun 07	01:04:30	4485	-17014	069	P	-a	1.0266	0.9445	0.0032	236.8	12.4	-	22S	2E
6329	317	0624 Jul 06	07:40:28	4485	-17013	107	N	a-	-1.4581	0.1535	-0.7893	103.2	-	-	24S	96W
6330	317	0624 Nov 30	20:26:50	4481	-17008	074	P	-t	-0.8398	1.3630	0.2717	315.5	122.4	-	22N	70E
6331	317	0625 May 27	17:01:46	4476	-17002	079	T	-a	0.3019	2.2905	1.3173	324.7	208.0	78.3	22S	122E
6332	317	0625 Nov 20	01:02:11	4472	-16996	084	T-	pp	-0.1175	2.6611	1.6238	347.0	221.0	98.1	20N	0E
6333	317	0626 May 17	03:25:41	4467	-16990	089	P	t-	-0.4850	1.9823	0.9539	337.8	199.8	-	20S	35W
6334	317	0626 Nov 09	12:48:58	4463	-16984	094	P	a-	0.5703	1.8038	0.8191	303.0	176.6	-	18N	177W
6335	317	0627 May 06	06:40:15	4458	-16978	099	N	t-	-1.2776	0.5484	-0.5204	218.9	-	-	18S	83W
6336	317	0627 Sep 30	18:26:31	4454	-16973	066	N	-a	-1.4220	0.2350	-0.7378	128.8	-	-	3N	100E
6337	317	0627 Oct 30	04:24:39	4454	-16972	104	N	a-	1.2200	0.6034	-0.3649	196.9	-	-	16N	52W
6338	317	0628 Mar 25	16:54:58	4450	-16967	071	P	-t	0.8324	1.3554	0.3061	306.2	127.0	-	2S	126E
6339	317	0628 Sep 19	07:19:11	4445	-16961	076	P	-a	-0.7702	1.4536	0.4359	299.2	143.1	-	1S	93W
6340	317	0629 Mar 15	01:22:20	4441	-16955	081	T+	pp	0.0562	2.7533	1.7565	336.2	218.7	100.5	1N	0E
6341	318	0629 Sep 08	13:25:57	4436	-16949	086	T-	pp	-0.0505	2.7997	1.7317	367.7	231.8	104.2	5S	176E
6342	318	0630 Mar 04	15:51:20	4432	-16943	091	P	a-	-0.6551	1.6361	0.6754	290.0	162.7	-	5N	143E
6343	318	0630 Aug 28	13:52:42	4427	-16937	096	P	t-	0.6831	1.6466	0.5631	339.5	170.7	-	8S	170E
6344	318	0631 Jan 23	21:50:58	4423	-16932	063	N	-a	1.3184	0.4352	-0.5579	175.4	-	-	20N	55E
6345	318	0631 Feb 22	08:12:23	4423	-16931	101	N	a-	-1.3416	0.3793	-0.5874	160.6	-	-	8N	102W
6346	318	0631 Jul 19	03:05:01	4419	-16926	068	N	-t	-1.3643	0.3619	-0.6528	169.4	-	-	22S	27W
6347	318	0631 Aug 17	15:33:22	4418	-16925	106	N	t-	1.3860	0.3395	-0.7097	170.3	-	-	12S	145E
6348	318	0632 Jan 13	08:02:02	4414	-16920	073	P	-a	0.6733	1.6482	0.5971	320.7	166.6	-	22N	99W
6349	318	0632 Jul 07	15:22:21	4410	-16914	078	P	-a	-0.5564	1.8177	0.8559	303.8	180.6	-	23S	149E
6350	318	0633 Jan 01	10:48:22	4405	-16908	083	T-	pp	-0.0135	2.8831	1.7843	376.2	234.1	104.6	23N	142W
6351	318	0633 Jun 27	08:05:29	4401	-16902	088	T+	p-	0.1899	2.4795	1.5390	314.5	207.9	91.5	23S	103W
6352	318	0633 Dec 21	09:48:43	4396	-16896	093	P	t-	-0.6904	1.6410	0.5420	338.4	166.9	-	23N	128W
6353	318	0634 Jun 17	00:39:45	4392	-16890	098	P	a-	0.9459	1.1037	0.1406	258.6	82.0	-	23S	8E
6354	318	0634 Nov 10	22:26:12	4388	-16885	065	N	-a	1.5072	0.1020	-0.9171	90.0	-	-	20N	38E
6355	318	0634 Dec 10	12:44:10	4387	-16884	103	N	a-	-1.3332	0.4381	-0.6143	187.8	-	-	22N	173W
6356	318	0635 May 08	00:34:27	4383	-16879	070	P	-t	-0.9389	1.1574	0.1133	290.2	79.9	-	18S	8E
6357	318	0635 Oct 31	11:46:53	4379	-16873	075	P	-a	0.7819	1.4101	0.4359	278.0	135.8	-	16N	162W
6358	318	0636 Apr 26	02:19:09	4374	-16867	080	T-	pp	-0.1939	2.5393	1.4656	372.5	231.6	96.0	14S	18W
6359	318	0636 Oct 20	03:40:23	4370	-16861	085	T+	-p	0.1088	2.6438	1.6727	320.5	210.4	96.0	12N	40W
6360	318	0637 Apr 15	03:25:41	4365	-16855	090	P	t-	0.5564	1.8644	0.8099	341.7	192.5	-	10S	33W

Cat Num	Canon Plate	Calendar Date	TD of Greatest Eclipse	ΔT s	Luna Num	Saros Num	Ecl. Type	QSE	Gamma	Pen. Mag.	Um. Mag.	Eclipse Phase Durations Pen. m	Par. m	Total m	Greatest in Zenith Lat.	Lng.
6361	319	0637 Oct 09	17:26:00	4361	-16849	095	P	a-	-0.6029	1.7568	0.7468	312.4	175.9	-	7N	114E
6362	319	0638 Mar 06	00:08:14	4357	-16844	062	N	-a	-1.3962	0.2885	-0.6964	144.9	-	-	3N	19E
6363	319	0638 Apr 04	10:29:18	4356	-16843	100	N	a-	1.2660	0.5368	-0.4669	199.1	-	-	5S	138W
6364	319	0638 Aug 30	09:07:10	4353	-16838	067	N	-t	1.5314	0.0859	-0.9895	90.1	-	-	7S	119W
6365	319	0638 Sep 29	00:58:43	4352	-16837	105	N	t-	-1.3712	0.3739	-0.6897	178.9	-	-	2N	1E
6366	319	0639 Feb 23	15:30:42	4348	-16832	072	P	-a	-0.6720	1.6043	0.6449	287.0	159.1	-	8N	149E
6367	319	0639 Aug 19	09:11:47	4344	-16826	077	P	-t	0.8194	1.3930	0.3166	319.8	132.5	-	12S	120W
6368	319	0640 Feb 13	07:27:28	4339	-16820	082	T+	pp	0.0153	2.8190	1.8405	325.7	213.8	99.3	13N	90W
6369	319	0640 Aug 07	12:31:44	4335	-16814	087	T+	pp	0.0507	2.7809	1.7492	354.1	226.9	103.2	16S	169W
6370	319	0641 Feb 01	19:20:41	4330	-16808	092	P	a-	0.7512	1.4961	0.4634	305.6	148.0	-	17N	92E
6371	319	0641 Jul 27	23:11:29	4326	-16802	097	P	a-	-0.6977	1.5649	0.5901	294.7	158.2	-	19S	31E
6372	319	0641 Dec 23	05:14:18	4322	-16797	064	N	-t	-1.5232	0.1160	-0.9893	105.9	-	-	22N	60W
6373	319	0642 Jan 21	23:56:00	4321	-16796	102	N	t-	1.5033	0.1435	-0.9438	115.6	-	-	21N	23E
6374	319	0642 Jun 18	08:30:02	4317	-16791	069	N	-a	1.1008	0.8089	-0.1334	222.5	-	-	22S	110W
6375	319	0642 Jul 17	15:12:29	4317	-16790	107	N	a-	-1.3904	0.2783	-0.6653	137.3	-	-	22S	151E
6376	319	0642 Dec 12	04:36:58	4313	-16785	074	P	-t	-0.8428	1.3565	0.2672	314.2	121.2	-	23N	52W
6377	319	0643 Jun 08	00:13:30	4308	-16779	079	T	-a	0.3802	2.1483	1.1722	322.2	202.6	60.7	23S	14E
6378	319	0643 Dec 01	09:33:27	4304	-16773	084	T-	pp	-0.1204	2.6548	1.6198	345.7	220.3	97.8	22N	127W
6379	319	0644 May 27	10:13:17	4299	-16767	089	T	t-	-0.4060	2.1286	1.0977	344.5	209.3	48.9	22S	137W
6380	319	0644 Nov 19	21:34:01	4295	-16761	094	P	a-	0.5632	1.8166	0.8325	303.0	177.2	-	21N	51E
6381	320	0645 May 16	13:09:28	4290	-16755	099	N	t-	-1.1974	0.6953	-0.3730	243.2	-	-	21S	179E
6382	320	0645 Oct 11	02:49:50	4287	-16750	066	N	-a	-1.4472	0.1910	-0.7862	116.9	-	-	7N	27W
6383	320	0645 Nov 09	13:09:24	4286	-16749	104	N	a-	1.2114	0.6203	-0.3503	199.5	-	-	19N	177E
6384	320	0646 Apr 06	00:04:06	4282	-16744	071	P	-t	0.8895	1.2476	0.2044	296.0	105.1	-	6S	18E
6385	320	0646 Sep 30	15:15:57	4278	-16738	076	P	-a	-0.8039	1.3949	0.3709	296.5	133.7	-	3N	147E
6386	320	0647 Mar 26	09:05:04	4273	-16732	081	T+	pp	0.1048	2.6610	1.6703	334.0	217.4	98.7	3S	117W
6387	320	0647 Sep 19	20:50:01	4269	-16726	086	T-	pp	-0.0935	2.7236	1.6499	368.7	231.4	102.5	1S	64E
6388	320	0648 Mar 14	23:56:21	4264	-16720	091	P	a-	-0.6132	1.7109	0.7541	293.2	169.5	-	1N	21E
6389	320	0648 Sep 07	20:58:42	4260	-16714	096	P	t-	0.6304	1.7447	0.6583	345.4	181.6	-	4S	62E
6390	320	0649 Feb 03	06:18:35	4256	-16709	063	N	-a	1.3386	0.3983	-0.5949	168.8	-	-	17N	72W
6391	320	0649 Mar 04	16:25:34	4255	-16708	101	N	a-	-1.3112	0.4345	-0.5308	171.2	-	-	4N	134E
6392	320	0649 Jul 29	10:13:10	4251	-16703	068	N	-t	-1.4270	0.2462	-0.7672	140.7	-	-	20S	134W
6393	320	0649 Aug 27	22:47:19	4251	-16702	106	N	t-	1.3270	0.4478	-0.6015	193.0	-	-	8S	35E
6394	320	0650 Jan 23	16:14:37	4247	-16697	073	P	-a	0.6885	1.6209	0.5690	320.1	163.9	-	19N	138E
6395	320	0650 Jul 18	22:48:17	4243	-16691	078	P	-a	-0.6247	1.6920	0.7310	297.0	170.1	-	21S	37E
6396	320	0651 Jan 12	18:45:55	4238	-16685	083	T-	pp	-0.0019	2.9040	1.8058	376.6	234.4	104.8	21N	99E
6397	320	0651 Jul 08	15:34:40	4234	-16679	088	T+	pp	0.1175	2.6130	1.6712	316.2	210.2	96.4	22S	145E
6398	320	0652 Jan 01	17:51:58	4229	-16673	093	P	t-	-0.6823	1.6545	0.5582	338.5	168.7	-	22N	112E
6399	320	0652 Jun 27	07:56:56	4225	-16667	098	P	a-	0.8746	1.2363	0.2698	270.6	111.9	-	23S	102W
6400	320	0652 Nov 21	07:01:04	4221	-16662	065	N	-a	1.5069	0.1017	-0.9158	89.6	-	-	22N	90W
6401	321	0652 Dec 20	21:10:19	4220	-16661	103	N	a-	-1.3269	0.4478	-0.6011	189.0	-	-	22N	61E
6402	321	0653 May 18	07:16:09	4216	-16656	070	N*	-t	-1.0205	1.0084	-0.0371	276.1	-	-	21S	93W
6403	321	0653 Nov 10	20:32:02	4212	-16650	075	P	-a	0.7881	1.3990	0.4243	276.9	134.1	-	19N	66E
6404	321	0654 May 07	08:49:12	4207	-16644	080	T	pp	-0.2749	2.3897	1.3179	369.4	227.0	84.2	18S	116W
6405	321	0654 Oct 31	12:19:22	4203	-16638	085	T+	-p	0.1208	2.6234	1.6489	321.0	210.2	95.3	15N	171W
6406	321	0655 Apr 26	10:15:24	4198	-16632	090	P	t-	0.4834	1.9959	0.9466	346.1	202.5	-	14S	137W
6407	321	0655 Oct 21	01:43:16	4194	-16626	095	P	a-	-0.5868	1.7890	0.7734	315.3	178.6	-	11N	12W
6408	321	0656 Mar 16	08:04:06	4190	-16621	062	N	-a	-1.4322	0.2196	-0.7597	126.7	-	-	1S	102W
6409	321	0656 Apr 14	17:53:08	4190	-16620	100	N	a-	1.2035	0.6485	-0.3492	215.5	-	-	9S	109E
6410	321	0656 Oct 09	08:43:47	4185	-16614	105	N	t-	-1.3473	0.4208	-0.6486	189.7	-	-	6N	116W
6411	321	0657 Mar 05	23:46:12	4181	-16609	072	P	-a	-0.7048	1.5425	0.5864	283.3	153.1	-	4N	24E
6412	321	0657 Aug 29	16:11:01	4177	-16603	077	P	-t	0.8767	1.2888	0.2106	310.9	109.6	-	8S	134E
6413	321	0658 Feb 23	15:45:20	4172	-16597	082	T-	pp	-0.0118	2.8251	1.8473	326.4	214.3	99.6	9N	145E
6414	321	0658 Aug 18	19:46:17	4168	-16591	087	T+	pp	0.1108	2.6703	1.6392	352.2	225.1	100.5	12S	81E
6415	321	0659 Feb 13	03:25:32	4164	-16585	092	P	h-	0.7305	1.5342	0.5011	309.1	153.5	-	13N	30W
6416	321	0659 Aug 08	06:44:20	4159	-16579	097	P	a-	-0.6346	1.6806	0.7063	300.2	169.3	-	16S	83W
6417	321	0660 Jan 03	13:12:46	4155	-16574	064	N	-t	-1.5327	0.0983	-1.0062	97.7	-	-	21N	179W
6418	321	0660 Feb 02	07:45:06	4155	-16573	102	N	t-	1.4862	0.1745	-0.9121	127.3	-	-	17N	95W
6419	321	0660 Jun 28	15:57:32	4151	-16568	069	N	-a	1.1732	0.6769	-0.2670	206.6	-	-	22S	138E
6420	321	0660 Jul 27	22:50:41	4150	-16567	107	N	a-	-1.3273	0.3947	-0.5504	161.7	-	-	20S	36E

Cat Num	Canon Plate	Calendar Date	TD of Greatest Eclipse	ΔT s	Luna Num	Saros Num	Ecl. Type	QSE	Gamma	Pen. Mag.	Um. Mag.	Eclipse Phase Durations Pen. m	Par. m	Total m	Greatest in Zenith Lat.	Lng.
6421	322	0660 Dec 22	12:45:46	4146	-16562	074	P	-t	-0.8473	1.3469	0.2603	312.6	119.5	-	23N	173W
6422	322	0661 Jun 18	07:24:37	4142	-16556	079	T	-a	0.4580	2.0071	1.0277	318.9	195.6	25.5	23S	94W
6423	322	0661 Dec 11	18:06:47	4138	-16550	084	T-	pp	-0.1221	2.6501	1.6180	344.3	219.7	97.6	23N	105E
6424	322	0662 Jun 07	16:55:29	4133	-16544	089	T	pp	-0.3234	2.2816	1.2478	350.5	217.2	74.3	23S	122E
6425	322	0662 Dec 01	06:23:27	4129	-16538	094	P	a-	0.5595	1.8228	0.8398	302.7	177.5	-	23N	81W
6426	322	0663 May 27	19:32:50	4124	-16532	099	N	t-	-1.1126	0.8508	-0.2174	265.1	-	-	23S	82E
6427	322	0663 Oct 22	11:20:57	4120	-16527	066	N	-a	-1.4659	0.1588	-0.8228	107.2	-	-	11N	156W
6428	322	0663 Nov 20	21:59:12	4120	-16526	104	N	a-	1.2073	0.6287	-0.3437	200.9	-	-	22N	44E
6429	322	0664 Apr 16	07:06:19	4116	-16521	071	P	-t	0.9527	1.1287	0.0913	284.0	71.1	-	10S	90W
6430	322	0664 Oct 10	23:20:38	4112	-16515	076	P	-a	-0.8303	1.3496	0.3196	294.5	125.4	-	7N	24E
6431	322	0665 Apr 05	16:42:03	4107	-16509	081	T+	pp	0.1591	2.5585	1.5736	331.4	215.4	95.3	7S	127E
6432	322	0665 Sep 30	04:20:51	4103	-16503	086	T-	pp	-0.1304	2.6587	1.5795	369.5	230.7	100.2	4N	51W
6433	322	0666 Mar 26	07:55:36	4098	-16497	091	P	a-	-0.5664	1.7949	0.8419	296.6	176.4	-	4S	101W
6434	322	0666 Sep 19	04:13:04	4094	-16491	096	P	t-	0.5841	1.8311	0.7419	350.1	189.9	-	0S	48W
6435	322	0667 Feb 14	14:38:46	4090	-16486	063	N	-a	1.3642	0.3512	-0.6419	159.6	-	-	13N	162E
6436	322	0667 Mar 16	00:30:20	4089	-16485	101	N	a-	-1.2744	0.5013	-0.4628	183.0	-	-	0S	11E
6437	322	0667 Aug 09	17:28:51	4086	-16480	068	N	-t	-1.4843	0.1406	-0.8718	107.0	-	-	17S	116E
6438	322	0667 Sep 08	06:11:10	4085	-16479	106	N	t-	1.2749	0.5435	-0.5061	210.1	-	-	4S	77W
6439	322	0668 Feb 04	00:20:07	4081	-16474	073	P	-a	0.7087	1.5840	0.5316	318.8	159.8	-	16N	16E
6440	322	0668 Jul 29	06:20:47	4077	-16468	078	P	-a	-0.6879	1.5759	0.6153	290.1	158.6	-	19S	77W
6441	323	0669 Jan 23	02:37:33	4072	-16462	083	T+	pp	0.0143	2.8808	1.7837	376.9	234.6	105.0	19N	19W
6442	323	0669 Jul 18	23:08:32	4068	-16456	088	T+	pp	0.0490	2.7396	1.7959	317.2	211.3	98.9	21S	31E
6443	323	0670 Jan 12	01:51:41	4064	-16450	093	P	t-	-0.6708	1.6740	0.5810	338.9	171.3	-	21N	8W
6444	323	0670 Jul 08	15:14:59	4059	-16444	098	P	a-	0.8044	1.3670	0.3966	281.4	133.5	-	22S	149E
6445	323	0670 Dec 02	15:39:13	4055	-16439	065	N	-a	1.5049	0.1043	-0.9109	90.4	-	-	24N	140E
6446	323	0671 Jan 01	05:35:08	4055	-16438	103	N	a-	-1.3188	0.4607	-0.5841	190.7	-	-	22N	65W
6447	323	0671 May 29	13:53:18	4051	-16433	070	N	-t	-1.1050	0.8544	-0.1931	259.1	-	-	23S	167E
6448	323	0671 Nov 22	05:20:49	4047	-16427	075	P	-a	0.7918	1.3922	0.4175	276.2	133.0	-	22N	66W
6449	323	0672 May 17	15:15:17	4042	-16421	080	T	-t	-0.3592	2.2344	1.1641	365.0	220.3	63.9	20S	146E
6450	323	0672 Nov 10	21:03:25	4038	-16415	085	T+	-p	0.1288	2.6103	1.6328	321.6	210.2	94.8	19N	58E
6451	323	0673 May 06	17:01:08	4033	-16409	090	T	t-	0.4061	2.1352	1.0907	349.9	211.0	47.6	17S	120E
6452	323	0673 Oct 31	10:06:07	4029	-16403	095	P	a-	-0.5760	1.8116	0.7908	317.7	180.6	-	15N	138W
6453	323	0674 Mar 27	15:52:43	4025	-16398	062	N	-a	-1.4746	0.1390	-0.8348	101.2	-	-	5S	139E
6454	323	0674 Apr 26	01:11:35	4025	-16397	100	N	a-	1.1354	0.7707	-0.2214	231.0	-	-	13S	2W
6455	323	0674 Oct 20	16:36:37	4020	-16391	105	N	t-	-1.3301	0.4550	-0.6198	197.4	-	-	10N	124E
6456	323	0675 Mar 17	07:51:56	4016	-16386	072	P	-a	-0.7459	1.4657	0.5126	278.5	144.8	-	0S	99W
6457	323	0675 Sep 09	23:21:02	4012	-16380	077	P	-t	0.9254	1.2004	0.1203	302.7	83.8	-	3S	25E
6458	323	0676 Mar 05	23:54:07	4008	-16374	082	T-	pp	-0.0460	2.7620	1.7849	327.0	214.6	99.4	4N	21E
6459	323	0676 Aug 29	03:10:50	4003	-16368	087	T+	pp	0.1636	2.5733	1.5424	350.0	222.7	96.5	8S	31W
6460	323	0677 Feb 23	11:22:19	3999	-16362	092	P	h-	0.7040	1.5829	0.5497	313.1	160.0	-	9N	150W
6461	324	0677 Aug 18	14:24:52	3994	-16356	097	P	a-	-0.5771	1.7861	0.8119	304.5	177.9	-	13S	161E
6462	324	0678 Jan 13	21:08:25	3991	-16351	064	N	-t	-1.5442	0.0764	-1.0266	86.3	-	-	20N	63E
6463	324	0678 Feb 12	15:27:08	3990	-16350	102	N	t-	1.4638	0.2151	-0.8702	141.1	-	-	14N	149E
6464	324	0678 Jul 09	23:25:34	3986	-16345	069	N	-a	1.2448	0.5467	-0.3995	188.4	-	-	21S	26E
6465	324	0678 Aug 08	06:33:10	3986	-16344	107	N	a-	-1.2677	0.5051	-0.4419	181.0	-	-	17S	81W
6466	324	0679 Jan 02	20:55:00	3982	-16339	074	P	-t	-0.8519	1.3366	0.2535	310.8	117.8	-	22N	65E
6467	324	0679 Jun 29	14:35:07	3977	-16333	079	P	-a	0.5354	1.8672	0.8838	314.7	186.7	-	23S	158E
6468	324	0679 Dec 23	02:40:26	3973	-16327	084	T-	pp	-0.1243	2.6443	1.6158	342.9	219.1	97.3	23N	23W
6469	324	0680 Jun 17	23:37:13	3969	-16321	089	T-	pp	-0.2413	2.4340	1.3968	355.4	223.3	89.2	24S	22E
6470	324	0680 Dec 11	15:14:04	3964	-16315	094	P	a-	0.5568	1.8270	0.8455	302.2	177.7	-	24N	147E
6471	324	0681 Jun 07	01:55:18	3960	-16309	099	N*	t-	-1.0266	1.0086	-0.0597	284.2	-	-	24S	13W
6472	324	0681 Nov 01	19:56:27	3956	-16304	066	N	-a	-1.4810	0.1332	-0.8525	98.7	-	-	14N	74E
6473	324	0681 Dec 01	06:50:31	3956	-16303	104	N	a-	1.2047	0.6342	-0.3397	201.9	-	-	24N	88W
6474	324	0682 Apr 27	14:04:47	3952	-16298	071	N*	-t	1.0196	1.0031	-0.0287	270.1	-	-	14S	165E
6475	324	0682 Oct 22	07:31:05	3947	-16292	076	P	-a	-0.8513	1.3140	0.2781	293.1	118.1	-	11N	100W
6476	324	0683 Apr 17	00:12:10	3943	-16286	081	T+	-p	0.2200	2.4439	1.4645	328.3	212.5	89.5	11S	13E
6477	324	0683 Oct 11	12:00:16	3939	-16280	086	T-	pp	-0.1597	2.6075	1.5231	370.2	230.0	97.7	8N	167W
6478	324	0684 Apr 05	15:47:00	3934	-16274	091	P	a-	-0.5127	1.8918	0.9422	300.2	183.3	-	8S	140E
6479	324	0684 Sep 29	11:37:19	3930	-16268	096	P	t-	0.5455	1.9032	0.8115	353.7	196.0	-	4N	161W
6480	324	0685 Feb 24	22:49:22	3926	-16263	063	N	-a	1.3972	0.2906	-0.7023	146.4	-	-	9N	38E

Cat Num	Canon Plate	Calendar Date	TD of Greatest Eclipse	ΔT s	Luna Num	Saros Num	Ecl. Type	QSE	Gamma	Pen. Mag.	Um. Mag.	Eclipse Phase Durations Pen. m	Par. m	Total m	Greatest in Zenith Lat.	Lng.
6481	325	0685 Mar 26	08:26:20	3926	-16262	101	N	a-	-1.2310	0.5805	-0.3826	195.7	-	-	4S	110W
6482	325	0685 Aug 20	00:54:09	3922	-16257	068	Ne	-h	-1.5348	0.0477	-0.9641	62.6	-	-	13S	4E
6483	325	0685 Sep 18	13:46:00	3921	-16256	106	N	h-	1.2307	0.6248	-0.4250	222.9	-	-	0N	167E
6484	325	0686 Feb 14	08:15:54	3917	-16251	073	P	-h	0.7362	1.5336	0.4811	316.5	153.7	-	13N	103W
6485	325	0686 Aug 09	13:59:05	3913	-16245	078	P	-a	-0.7469	1.4677	0.5068	283.0	146.1	-	16S	168E
6486	325	0687 Feb 03	10:21:39	3909	-16239	083	T+	pp	0.0365	2.8391	1.7438	376.9	234.8	104.8	16N	135W
6487	325	0687 Jul 30	06:46:23	3904	-16233	088	T-	pp	-0.0161	2.8012	1.8550	317.8	211.6	99.3	18S	84W
6488	325	0688 Jan 23	09:46:28	3900	-16227	093	P	t-	-0.6545	1.7018	0.6130	339.6	174.7	-	18N	127W
6489	325	0688 Jul 18	22:35:37	3896	-16221	098	P	a-	0.7369	1.4932	0.5183	290.9	150.2	-	20S	38E
6490	325	0688 Dec 13	00:20:13	3892	-16216	065	N	-a	1.5017	0.1087	-0.9038	91.9	-	-	25N	11E
6491	325	0689 Jan 11	13:58:39	3891	-16215	103	N	a-	-1.3085	0.4772	-0.5629	193.0	-	-	20N	169E
6492	325	0689 Jun 08	20:26:39	3888	-16210	070	N	-t	-1.1912	0.6972	-0.3526	238.5	-	-	24S	69E
6493	325	0689 Jul 08	09:04:39	3887	-16209	108	Nb	t-	1.5561	0.0165	-1.0111	37.9	-	-	21S	120W
6494	325	0689 Dec 02	14:12:58	3883	-16204	075	P	-a	0.7932	1.3896	0.4152	275.7	132.5	-	23N	161E
6495	325	0690 May 28	21:38:06	3879	-16198	080	T	-t	-0.4462	2.0740	1.0050	359.3	211.1	11.7	22S	50E
6496	325	0690 Nov 22	05:51:18	3874	-16192	085	T+	-p	0.1331	2.6038	1.6236	322.2	210.3	94.6	21N	74W
6497	325	0691 May 17	23:43:54	3870	-16186	090	T	t-	0.3254	2.2810	1.2410	352.8	217.9	73.7	20S	19E
6498	325	0691 Nov 11	18:33:40	3866	-16180	095	P	a-	-0.5697	1.8255	0.7998	319.7	181.8	-	18N	94E
6499	325	0692 Apr 06	23:36:07	3862	-16175	062	Ne	-a	-1.5217	0.0499	-0.9187	60.9	-	-	9S	22E
6500	325	0692 May 06	08:27:53	3861	-16174	100	N	a-	1.0645	0.8982	-0.0888	245.0	-	-	16S	112W
6501	326	0692 Oct 31	00:33:45	3857	-16168	105	N	t-	-1.3168	0.4818	-0.5979	203.3	-	-	14N	4E
6502	326	0693 Mar 27	15:51:57	3853	-16163	072	P	-a	-0.7917	1.3801	0.4298	272.8	134.3	-	5S	139E
6503	326	0693 Sep 20	06:40:41	3849	-16157	077	P	-t	0.9666	1.1256	0.0438	295.2	51.0	-	1N	86W
6504	326	0694 Mar 17	07:54:00	3845	-16151	082	T-	-p	-0.0871	2.6863	1.7097	327.3	214.6	98.5	0N	100W
6505	326	0694 Sep 09	10:45:10	3840	-16145	087	T+	pp	0.2093	2.4895	1.4587	347.7	220.1	91.6	4S	146W
6506	326	0695 Mar 06	19:08:00	3836	-16139	092	P	t-	0.6691	1.6471	0.6138	318.0	167.9	-	5N	92E
6507	326	0695 Aug 29	22:14:46	3832	-16133	097	P	a-	-0.5268	1.8785	0.9040	307.8	184.3	-	9S	42E
6508	326	0696 Jan 25	04:56:04	3828	-16128	064	N	-t	-1.5616	0.0434	-1.0574	65.2	-	-	17N	54W
6509	326	0696 Feb 23	22:57:48	3827	-16127	102	N	t-	1.4326	0.2715	-0.8123	157.9	-	-	10N	36E
6510	326	0696 Jul 20	06:58:24	3824	-16122	069	N	-a	1.3120	0.4250	-0.5242	168.4	-	-	19S	88W
6511	326	0696 Aug 18	14:23:54	3823	-16121	107	N	a-	-1.2148	0.6034	-0.3461	196.0	-	-	13S	161E
6512	326	0697 Jan 13	05:00:08	3819	-16116	074	P	-t	-0.8601	1.3195	0.2405	308.4	114.7	-	20N	56W
6513	326	0697 Jul 09	21:45:56	3815	-16110	079	P	-a	0.6114	1.7299	0.7420	309.7	175.8	-	22S	50E
6514	326	0698 Jan 02	11:12:53	3811	-16104	084	T-	pp	-0.1281	2.6353	1.6108	341.4	218.5	97.0	23N	150W
6515	326	0698 Jun 29	06:17:01	3806	-16098	089	T-	pp	-0.1585	2.5877	1.5467	359.3	227.7	98.6	23S	78W
6516	326	0698 Dec 23	00:04:48	3802	-16092	094	P	a-	0.5537	1.8316	0.8521	301.8	177.9	-	24N	16E
6517	326	0699 Jun 18	08:15:36	3798	-16086	099	P	t-	-0.9385	1.1707	0.1016	301.1	77.8	-	24S	109W
6518	326	0699 Nov 13	04:37:54	3794	-16081	066	N	-a	-1.4911	0.1165	-0.8729	92.7	-	-	17N	56W
6519	326	0699 Dec 12	15:43:47	3793	-16080	104	N	a-	1.2041	0.6358	-0.3391	202.3	-	-	25N	139E
6520	326	0700 May 07	20:59:22	3790	-16075	071	N	-t	1.0904	0.8706	-0.1560	254.0	-	-	17S	60E
6521	327	0700 Nov 01	15:46:52	3785	-16069	076	P	-a	-0.8673	1.2874	0.2461	292.2	111.9	-	15N	136E
6522	327	0701 Apr 27	07:38:50	3781	-16063	081	T	-p	0.2846	2.3227	1.3485	324.6	208.5	80.9	15S	100W
6523	327	0701 Oct 21	19:46:39	3777	-16057	086	T-	pp	-0.1828	2.5674	1.4783	370.8	229.3	95.4	12N	76E
6524	327	0702 Apr 16	23:32:06	3772	-16051	091	T	a-	-0.4535	1.9989	1.0523	303.7	189.8	33.6	12S	22E
6525	327	0702 Oct 10	19:10:42	3768	-16045	096	P	t-	0.5139	1.9623	0.8684	356.3	200.5	-	8N	85E
6526	327	0703 Mar 08	06:51:47	3764	-16040	063	N	-a	1.4362	0.2190	-0.7738	128.3	-	-	5N	84W
6527	327	0703 Apr 06	16:14:43	3764	-16039	101	N	a-	-1.1820	0.6702	-0.2923	208.9	-	-	9S	132E
6528	327	0703 Sep 29	21:30:42	3759	-16033	106	N	h-	1.1934	0.6934	-0.3566	232.5	-	-	5N	50E
6529	327	0704 Feb 25	16:02:28	3756	-16028	073	P	-h	0.7706	1.4704	0.4181	313.1	145.1	-	9N	139E
6530	327	0704 Aug 19	21:46:01	3751	-16022	078	P	-a	-0.7994	1.3717	0.4102	276.2	133.1	-	12S	50E
6531	327	0705 Feb 13	17:58:13	3747	-16016	083	T+	pp	0.0649	2.7859	1.6928	376.8	234.7	104.2	12N	110E
6532	327	0705 Aug 09	14:29:23	3743	-16010	088	T-	pp	-0.0769	2.6912	1.7421	317.9	211.0	98.0	15S	159E
6533	327	0706 Feb 02	17:35:35	3738	-16004	093	P	t-	-0.6328	1.7392	0.6550	340.7	179.0	-	15N	116E
6534	327	0706 Jul 30	05:59:52	3734	-15998	098	P	a-	0.6726	1.6136	0.6336	299.3	163.5	-	17S	74W
6535	327	0706 Dec 24	08:59:57	3731	-15993	065	N	-a	1.5002	0.1097	-0.8994	92.1	-	-	25N	118W
6536	327	0707 Jan 22	22:16:17	3730	-15992	103	N	a-	-1.2927	0.5034	-0.5314	196.9	-	-	18N	45E
6537	327	0707 Jun 20	02:58:47	3726	-15987	070	N	-t	-1.2775	0.5405	-0.5122	213.9	-	-	25S	29W
6538	327	0707 Jul 19	15:57:32	3726	-15986	108	N	t-	1.4877	0.1446	-0.8881	111.6	-	-	19S	137E
6539	327	0707 Dec 13	23:05:35	3722	-15981	075	P	-a	0.7948	1.3863	0.4127	275.2	132.1	-	24N	29E
6540	327	0708 Jun 08	04:00:34	3718	-15975	080	P	-t	-0.5335	1.9135	0.8453	352.3	199.3	-	24S	45W

Cat Num	Canon Plate	Calendar Date	TD of Greatest Eclipse	ΔT s	Luna Num	Saros Num	Ecl. Type	QSE	Gamma	Pen. Mag.	Um. Mag.	Eclipse Phase Durations Pen. m	Par. m	Total m	Greatest in Zenith Lat.	Lng.
6541	328	0708 Dec 02	14:40:25	3713	-15969	085	T+	-p	0.1362	2.5991	1.6168	322.9	210.5	94.4	23N	154E
6542	328	0709 May 28	06:26:03	3709	-15963	090	T+	p-	0.2431	2.4301	1.3941	354.8	222.9	88.9	22S	82W
6543	328	0709 Nov 22	03:04:52	3705	-15957	095	P	a-	-0.5670	1.8328	0.8028	321.3	182.6	-	20N	33W
6544	328	0710 May 17	15:39:32	3700	-15951	100	P	a-	0.9885	1.0352	0.0528	258.1	51.8	-	19S	139E
6545	328	0710 Nov 11	08:37:48	3696	-15945	105	N	t-	-1.3094	0.4976	-0.5863	207.0	-	-	17N	117W
6546	328	0711 Apr 07	23:43:02	3693	-15940	072	P	-a	-0.8454	1.2804	0.3325	265.7	119.9	-	9S	20E
6547	328	0711 Oct 01	14:11:14	3688	-15934	077	N*	-t	0.9994	1.0661	-0.0173	288.8	-	-	5N	160E
6548	328	0712 Mar 27	15:44:25	3684	-15928	082	T-	-p	-0.1357	2.5969	1.6206	327.4	214.0	96.3	4S	140E
6549	328	0712 Sep 19	18:30:19	3680	-15922	087	T+	-p	0.2470	2.4203	1.3896	345.3	217.4	86.5	0N	96E
6550	328	0713 Mar 17	02:45:03	3675	-15916	092	P	t-	0.6279	1.7226	0.6893	323.2	176.3	-	1N	24W
6551	328	0713 Sep 09	06:12:00	3671	-15910	097	P	a-	-0.4820	1.9610	0.9859	310.4	189.1	-	5S	79W
6552	328	0714 Feb 04	12:38:47	3667	-15905	064	Ne	-t	-1.5827	0.0032	-1.0947	17.9	-	-	14N	171W
6553	328	0714 Mar 06	06:20:38	3667	-15904	102	N	t-	1.3954	0.3387	-0.7432	175.5	-	-	6N	77W
6554	328	0714 Jul 31	14:33:59	3663	-15899	069	N	-a	1.3764	0.3085	-0.6441	145.4	-	-	16S	157E
6555	328	0714 Aug 29	22:20:25	3662	-15898	107	N	a-	-1.1668	0.6929	-0.2594	208.3	-	-	9S	40E
6556	328	0715 Jan 24	13:01:07	3659	-15893	074	P	-t	-0.8723	1.2947	0.2205	305.2	109.9	-	18N	176W
6557	328	0715 Jul 21	04:59:07	3655	-15887	079	P	-a	0.6845	1.5984	0.6054	303.9	162.9	-	20S	59W
6558	328	0716 Jan 13	19:42:46	3650	-15881	084	T-	pp	-0.1345	2.6211	1.6013	339.8	217.9	96.5	21N	83E
6559	328	0716 Jul 09	12:59:00	3646	-15875	089	T-	pp	-0.0782	2.7372	1.6920	362.3	230.4	103.7	22S	179W
6560	328	0717 Jan 02	08:52:59	3642	-15869	094	P	a-	0.5484	1.8402	0.8630	301.6	178.5	-	23N	116W
6561	329	0717 Jun 28	14:38:36	3637	-15863	099	P	t-	-0.8523	1.3293	0.2593	315.4	121.5	-	24S	156E
6562	329	0717 Nov 23	13:21:42	3634	-15858	066	N	-a	-1.4991	0.1034	-0.8893	87.7	-	-	20N	173E
6563	329	0717 Dec 23	00:35:34	3633	-15857	104	N	a-	1.2028	0.6385	-0.3370	202.9	-	-	25N	7E
6564	329	0718 May 19	03:51:17	3630	-15852	071	N	-t	1.1639	0.7333	-0.2885	235.3	-	-	19S	44W
6565	329	0718 Nov 13	00:07:06	3625	-15846	076	P	-a	-0.8788	1.2687	0.2226	291.9	107.1	-	18N	10E
6566	329	0719 May 08	15:01:04	3621	-15840	081	T	-p	0.3534	2.1942	1.2245	320.1	203.3	67.6	18S	149E
6567	329	0719 Nov 02	03:39:07	3617	-15834	086	T-	pp	-0.2003	2.5374	1.4442	371.4	228.7	93.3	16N	43W
6568	329	0720 Apr 27	07:11:17	3613	-15828	091	T	a-	-0.3893	2.1155	1.1714	307.1	195.7	58.7	15S	94W
6569	329	0720 Oct 21	02:53:28	3608	-15822	096	P	t-	0.4900	2.0071	0.9114	358.0	203.6	-	12N	32W
6570	329	0721 Mar 18	14:44:16	3605	-15817	063	N	-a	1.4826	0.1337	-0.8590	101.3	-	-	1N	157E
6571	329	0721 Apr 16	23:54:57	3604	-15816	101	N	a-	-1.1268	0.7714	-0.1909	222.3	-	-	12S	15E
6572	329	0721 Oct 10	05:25:14	3600	-15810	106	N	a-	1.1632	0.7487	-0.3012	239.6	-	-	9N	70W
6573	329	0722 Mar 07	23:38:28	3596	-15805	073	P	-t	0.8127	1.3929	0.3409	308.4	133.0	-	4N	24E
6574	329	0722 Aug 31	05:40:33	3592	-15799	078	P	-a	-0.8462	1.2864	0.3239	269.7	119.6	-	9S	70W
6575	329	0723 Feb 25	01:24:07	3588	-15793	083	T+	pp	0.1017	2.7170	1.6265	376.2	234.2	102.8	8N	2W
6576	329	0723 Aug 20	22:18:59	3583	-15787	088	T-	-p	-0.1319	2.5920	1.6394	317.7	209.9	95.4	12S	41E
6577	329	0724 Feb 14	01:18:04	3579	-15781	093	P	t-	-0.6051	1.7875	0.7084	342.1	184.1	-	12N	1W
6578	329	0724 Aug 09	13:28:17	3575	-15775	098	P	a-	0.6127	1.7263	0.7409	306.7	174.2	-	14S	173E
6579	329	0725 Jan 03	17:38:46	3571	-15770	065	N	-a	1.5002	0.1079	-0.8975	91.0	-	-	24N	113E
6580	329	0725 Feb 02	06:29:58	3571	-15769	103	N	a-	-1.2731	0.5366	-0.4926	201.8	-	-	15N	79W
6581	330	0725 Jun 30	09:30:44	3567	-15764	070	N	-t	-1.3626	0.3858	-0.6700	183.9	-	-	24S	127W
6582	330	0725 Jul 29	22:53:00	3566	-15763	108	N	t-	1.4218	0.2682	-0.7699	151.0	-	-	17S	32E
6583	330	0725 Dec 24	07:58:42	3563	-15758	075	P	-a	0.7960	1.3834	0.4110	274.7	131.8	-	24N	104W
6584	330	0726 Jun 19	10:21:50	3559	-15752	080	P	-t	-0.6216	1.7515	0.6839	343.7	184.2	-	24S	141W
6585	330	0726 Dec 13	23:31:01	3554	-15746	085	T+	-p	0.1377	2.5973	1.6132	323.6	210.7	94.4	24N	22E
6586	330	0727 Jun 08	13:08:30	3550	-15740	090	T+	pp	0.1598	2.5810	1.5486	355.7	226.2	98.2	23S	177E
6587	330	0727 Dec 03	11:36:54	3546	-15734	095	P	a-	-0.5657	1.8369	0.8033	322.8	183.1	-	22N	161W
6588	330	0728 May 27	22:51:52	3542	-15728	100	P	a-	0.9121	1.1733	0.1950	269.6	97.3	-	21S	31E
6589	330	0728 Nov 21	16:44:17	3537	-15722	105	N	t-	-1.3041	0.5092	-0.5784	209.8	-	-	20N	121E
6590	330	0729 Apr 18	07:28:45	3534	-15717	072	P	-a	-0.9032	1.1733	0.2274	257.4	100.7	-	13S	98W
6591	330	0729 Oct 11	21:49:56	3530	-15711	077	N*	-t	1.0260	1.0180	-0.0666	283.2	-	-	9N	44E
6592	330	0730 Apr 07	23:26:39	3525	-15705	082	T-	-p	-0.1904	2.4966	1.5203	327.1	212.7	92.3	8S	23E
6593	330	0730 Oct 01	02:25:09	3521	-15699	087	T	-p	0.2772	2.3647	1.3342	343.1	214.9	81.5	5N	24W
6594	330	0731 Mar 28	10:09:54	3517	-15693	092	P	t-	0.5778	1.8146	0.7811	329.0	185.4	-	4S	137W
6595	330	0731 Sep 20	14:19:20	3513	-15687	097	T	a-	-0.4453	2.0286	1.0528	312.1	192.6	34.2	0S	158E
6596	330	0732 Mar 16	13:32:02	3509	-15681	102	N	t-	1.3497	0.4217	-0.6583	194.4	-	-	1N	174E
6597	330	0732 Aug 10	22:15:39	3505	-15676	069	N	-a	1.4353	0.2024	-0.7542	119.2	-	-	13S	41E
6598	330	0732 Sep 09	06:25:00	3504	-15675	107	N	a-	-1.1257	0.7699	-0.1855	218.1	-	-	5S	82W
6599	330	0733 Feb 03	20:55:25	3501	-15670	074	P	-t	-0.8903	1.2591	0.1902	301.0	102.3	-	14N	65E
6600	330	0733 Jul 31	12:15:44	3497	-15664	079	P	-a	0.7535	1.4746	0.4761	297.7	147.9	-	17S	169W

Cat Num	Canon Plate	Calendar Date	TD of Greatest Eclipse	ΔT s	Luna Num	Saros Num	Ecl. Type	QSE	Gamma	Pen. Mag.	Um. Mag.	Eclipse Phase Durations Pen. m	Par. m	Total m	Greatest in Zenith Lat.	Lng.
6601	331	0734 Jan 24	04:07:54	3492	-15658	084	T-	pp	-0.1456	2.5982	1.5835	338.1	217.1	95.7	18N	44W
6602	331	0734 Jul 20	19:41:42	3488	-15652	089	T+	pp	0.0006	2.8821	1.8322	364.3	231.6	105.3	20S	80E
6603	331	0735 Jan 13	17:38:29	3484	-15646	094	P	a-	0.5408	1.8526	0.8783	301.5	179.4	-	21N	113E
6604	331	0735 Jul 09	21:03:32	3480	-15640	099	P	t-	-0.7675	1.4856	0.4143	327.7	150.0	-	23S	59E
6605	331	0735 Dec 04	22:07:43	3476	-15635	066	N	-a	-1.5049	0.0943	-0.9012	84.0	-	-	21N	42E
6606	331	0736 Jan 03	09:25:09	3476	-15634	104	N	a-	1.2004	0.6431	-0.3326	203.8	-	-	24N	125W
6607	331	0736 May 29	10:42:35	3472	-15629	071	N	-t	1.2382	0.5947	-0.4228	213.9	-	-	21S	147W
6608	331	0736 Jun 27	22:44:48	3471	-15628	109	N	t-	-1.4922	0.1422	-0.9019	112.6	-	-	25S	33E
6609	331	0736 Nov 23	08:31:02	3468	-15623	076	P	-a	-0.8861	1.2575	0.2071	292.1	103.8	-	21N	116W
6610	331	0737 May 18	22:20:50	3464	-15617	081	T	-a	0.4250	2.0609	1.0952	314.9	196.6	45.8	20S	38E
6611	331	0737 Nov 12	11:36:54	3459	-15611	086	T-	-p	-0.2130	2.5159	1.4191	372.0	228.3	91.6	19N	163W
6612	331	0738 May 08	14:46:04	3455	-15605	091	T	p-	-0.3214	2.2390	1.2969	310.1	200.9	74.2	18S	151E
6613	331	0738 Nov 01	10:43:51	3451	-15599	096	P	t-	0.4715	2.0416	0.9447	359.2	205.8	-	16N	150W
6614	331	0739 Mar 29	22:27:20	3448	-15594	063	Ne	-a	1.5359	0.0360	-0.9569	53.2	-	-	3S	39E
6615	331	0739 Apr 28	07:28:09	3447	-15593	101	N	a-	-1.0662	0.8826	-0.0799	235.6	-	-	16S	99W
6616	331	0739 Oct 21	13:28:49	3443	-15587	106	N	a-	1.1394	0.7922	-0.2574	244.6	-	-	13N	168E
6617	331	0740 Mar 18	07:05:14	3439	-15582	073	P	-t	0.8617	1.3029	0.2511	302.2	116.0	-	0N	90W
6618	331	0740 Sep 10	13:43:12	3435	-15576	078	P	-a	-0.8867	1.2126	0.2490	263.6	105.8	-	4S	168E
6619	331	0741 Mar 07	08:41:47	3431	-15570	083	T+	pp	0.1452	2.6359	1.5481	375.3	233.1	100.0	4N	113W
6620	331	0741 Aug 31	06:15:24	3427	-15564	088	T-	-p	-0.1810	2.5039	1.5475	317.3	208.5	91.8	8S	80W
6621	332	0742 Feb 24	08:51:59	3422	-15558	093	P	t-	-0.5698	1.8496	0.7760	344.1	190.1	-	8N	115W
6622	332	0742 Aug 20	21:02:12	3418	-15552	098	P	a-	0.5580	1.8296	0.8384	313.0	182.8	-	11S	59E
6623	332	0743 Jan 15	02:13:37	3415	-15547	065	N	-a	1.5041	0.0986	-0.9024	86.8	-	-	22N	15W
6624	332	0743 Feb 13	14:36:05	3414	-15546	103	N	a-	-1.2466	0.5823	-0.4409	208.4	-	-	11N	159E
6625	332	0743 Jul 11	16:04:57	3411	-15541	070	N	-t	-1.4449	0.2366	-0.8229	146.4	-	-	23S	134E
6626	332	0743 Aug 10	05:54:17	3410	-15540	108	N	t-	1.3608	0.3830	-0.6608	179.3	-	-	14S	74W
6627	332	0744 Jan 04	16:47:56	3406	-15535	075	P	-a	0.8009	1.3735	0.4028	273.8	130.5	-	23N	125E
6628	332	0744 Jun 29	16:46:15	3402	-15529	080	P	-t	-0.7073	1.5942	0.5266	334.0	165.7	-	24S	123E
6629	332	0744 Dec 24	08:20:06	3398	-15523	085	T+	-p	0.1405	2.5927	1.6074	324.2	211.0	94.3	24N	110W
6630	332	0745 Jun 18	19:51:37	3394	-15517	090	T+	pp	0.0762	2.7330	1.7036	355.7	227.9	103.1	24S	76E
6631	332	0745 Dec 13	20:09:45	3390	-15511	095	P	a-	-0.5656	1.8386	0.8019	324.0	183.5	-	23N	71E
6632	332	0746 Jun 08	06:02:24	3386	-15505	100	P	a-	0.8329	1.3169	0.3421	279.9	125.8	-	22S	77W
6633	332	0746 Dec 03	00:53:43	3381	-15499	105	N	t-	-1.3017	0.5150	-0.5753	211.4	-	-	22N	1W
6634	332	0747 Apr 29	15:06:35	3378	-15494	072	P	-a	-0.9676	1.0545	0.1101	247.3	71.3	-	16S	146E
6635	332	0747 May 28	21:51:50	3377	-15493	110	Nb	a-	1.5242	0.0320	-0.9101	47.7	-	-	20S	45E
6636	332	0747 Oct 23	05:38:53	3374	-15488	077	N	-t	1.0447	0.9841	-0.1013	279.0	-	-	14N	75W
6637	332	0748 Apr 18	07:01:00	3370	-15482	082	T-	-p	-0.2510	2.3857	1.4090	326.2	210.5	85.7	12S	92W
6638	332	0748 Oct 11	10:28:52	3365	-15476	087	T	-p	0.3009	2.3211	1.2908	341.0	212.6	77.0	9N	147W
6639	332	0749 Apr 07	17:26:53	3361	-15470	092	P	t-	0.5220	1.9173	0.8834	334.8	194.2	-	8S	113E
6640	332	0749 Sep 30	22:34:35	3357	-15464	097	T	a-	-0.4148	2.0851	1.1085	313.4	195.2	48.1	4N	33E
6641	333	0750 Mar 27	20:34:13	3353	-15458	102	N	t-	1.2966	0.5181	-0.5598	213.7	-	-	3S	67E
6642	333	0750 Aug 22	06:02:05	3350	-15453	069	N	-a	1.4900	0.1043	-0.8567	86.6	-	-	9S	77W
6643	333	0750 Sep 20	14:36:38	3349	-15452	107	N	a-	-1.0904	0.8363	-0.1224	226.0	-	-	1S	153E
6644	333	0751 Feb 15	04:44:01	3345	-15447	074	P	-t	-0.9135	1.2135	0.1504	295.8	91.2	-	11N	53W
6645	333	0751 Aug 11	19:37:11	3341	-15441	079	P	-a	0.8176	1.3600	0.3556	291.1	130.6	-	14S	80E
6646	333	0752 Feb 04	12:27:31	3337	-15435	084	T-	pp	-0.1619	2.5656	1.5563	336.2	216.1	94.5	15N	169W
6647	333	0752 Jul 31	02:29:52	3333	-15429	089	T+	pp	0.0741	2.7497	1.6949	365.4	231.4	103.9	18S	23W
6648	333	0753 Jan 24	02:19:22	3329	-15423	094	P	a-	0.5294	1.8721	0.9009	301.8	180.8	-	19N	17W
6649	333	0753 Jul 20	03:32:38	3325	-15417	099	P	t-	-0.6857	1.6365	0.5635	337.9	170.8	-	21S	38W
6650	333	0753 Dec 15	06:53:21	3321	-15412	066	N	-a	-1.5106	0.0849	-0.9129	80.0	-	-	22N	89W
6651	333	0754 Jan 13	18:10:39	3321	-15411	104	N	a-	1.1951	0.6526	-0.3229	205.4	-	-	22N	105E
6652	333	0754 Jun 09	17:34:32	3317	-15406	071	N	-t	1.3124	0.4568	-0.5569	189.1	-	-	22S	110E
6653	333	0754 Jul 09	05:26:24	3316	-15405	109	N	t-	-1.4101	0.2919	-0.7505	158.6	-	-	24S	67W
6654	333	0754 Dec 04	16:55:33	3313	-15400	076	P	-a	-0.8924	1.2476	0.1936	292.4	100.9	-	22N	119E
6655	333	0755 May 30	05:39:02	3309	-15394	081	P	-a	0.4985	1.9242	0.9620	308.9	188.4	-	22S	72W
6656	333	0755 Nov 23	19:38:22	3305	-15388	086	T-	-p	-0.2222	2.5005	1.4010	372.5	228.0	90.4	21N	77E
6657	333	0756 May 18	22:17:22	3301	-15382	091	T-	p-	-0.2504	2.3687	1.4279	312.7	205.2	85.1	21S	38E
6658	333	0756 Nov 11	18:40:07	3296	-15376	096	P	t-	0.4577	2.0673	0.9699	359.8	207.2	-	19N	90E
6659	333	0757 May 08	14:55:18	3292	-15370	101	P	a-	-1.0014	1.0020	0.0387	248.6	43.5	-	19S	148E
6660	333	0757 Oct 31	21:41:14	3288	-15364	106	N	a-	1.1221	0.8235	-0.2253	247.9	-	-	17N	44E

Cat Num	Canon Plate	Calendar Date	TD of Greatest Eclipse	ΔT s	Luna Num	Saros Num	Ecl. Type	QSE	Gamma	Pen. Mag.	Um. Mag.	Eclipse Phase Durations Pen. m	Par. m	Total m	Greatest in Zenith Lat.	Lng.
6661	334	0758 Mar 29	14:19:50	3285	-15359	073	P	-t	0.9195	1.1967	0.1453	293.8	89.8	-	4S	160E
6662	334	0758 Sep 21	21:54:44	3281	-15353	078	P	-a	-0.9203	1.1518	0.1866	258.4	92.3	-	0S	43E
6663	334	0759 Mar 18	15:48:35	3277	-15347	083	T+	pp	0.1971	2.5391	1.4544	373.7	231.1	95.0	0S	138E
6664	334	0759 Sep 11	14:19:21	3272	-15341	088	T-	-p	-0.2234	2.4281	1.4677	316.9	206.8	87.6	4S	158E
6665	334	0760 Mar 06	16:18:27	3268	-15335	093	P	t-	-0.5277	1.9239	0.8560	346.2	196.5	-	4N	132E
6666	334	0760 Aug 31	04:42:27	3264	-15329	098	P	a-	0.5094	1.9218	0.9246	318.4	189.6	-	7S	58W
6667	334	0761 Jan 25	10:45:27	3261	-15324	065	N	-a	1.5111	0.0834	-0.9129	79.6	-	-	19N	143W
6668	334	0761 Feb 23	22:37:07	3260	-15323	103	N	a-	-1.2151	0.6369	-0.3801	215.8	-	-	7N	37E
6669	334	0761 Jul 21	22:40:15	3257	-15318	070	Ne	-t	-1.5252	0.0914	-0.9721	92.4	-	-	21S	35E
6670	334	0761 Aug 20	12:59:01	3256	-15317	108	N	t-	1.3031	0.4919	-0.5578	202.0	-	-	10S	179E
6671	334	0762 Jan 15	01:35:12	3253	-15312	075	P	-a	0.8077	1.3601	0.3915	272.6	128.9	-	21N	7W
6672	334	0762 Jul 10	23:13:12	3248	-15306	080	P	-t	-0.7911	1.4406	0.3728	322.9	142.7	-	23S	26E
6673	334	0763 Jan 04	17:06:25	3244	-15300	085	T+	-p	0.1450	2.5848	1.5987	324.8	211.2	94.2	23N	119E
6674	334	0763 Jun 30	02:38:41	3240	-15294	090	T-	pp	-0.0052	2.8620	1.8352	354.6	227.9	104.3	23S	26W
6675	334	0763 Dec 25	04:40:18	3236	-15288	095	P	a-	-0.5643	1.8423	0.8030	325.4	184.1	-	23N	55W
6676	334	0764 Jun 18	13:15:57	3232	-15282	100	P	a-	0.7555	1.4575	0.4856	288.6	146.2	-	23S	174E
6677	334	0764 Dec 13	09:01:39	3228	-15276	105	N	t-	-1.2984	0.5220	-0.5704	213.2	-	-	22N	122W
6678	334	0765 May 09	22:41:10	3225	-15271	072	N	-a	-1.0343	0.9315	-0.0120	235.7	-	-	19S	32E
6679	334	0765 Jun 08	05:18:46	3224	-15270	110	N	a-	1.4527	0.1626	-0.7784	106.1	-	-	22S	67W
6680	334	0765 Nov 02	13:34:44	3221	-15265	077	N	-t	1.0583	0.9594	-0.1264	275.6	-	-	17N	166E
6681	335	0766 Apr 29	14:26:51	3216	-15259	082	T	-a	-0.3178	2.2635	1.2860	324.7	206.9	75.1	16S	155E
6682	335	0766 Oct 22	18:41:39	3212	-15253	087	T	-p	0.3179	2.2896	1.2599	339.1	210.8	73.4	13N	89E
6683	335	0767 Apr 19	00:33:10	3208	-15247	092	T	t-	0.4581	2.0348	1.0002	340.8	203.0	2.4	12S	5E
6684	335	0767 Oct 12	06:58:38	3204	-15241	097	T	a-	-0.3912	2.1288	1.1514	314.2	196.9	55.9	8N	95W
6685	335	0768 Apr 07	03:26:13	3200	-15235	102	N	t-	1.2353	0.6296	-0.4464	233.2	-	-	7S	38W
6686	335	0768 Sep 01	13:56:35	3197	-15230	069	Ne	-a	1.5377	0.0190	-0.9466	37.3	-	-	5S	163E
6687	335	0768 Sep 30	22:57:01	3196	-15229	107	N	a-	-1.0625	0.8892	-0.0728	232.0	-	-	4N	27E
6688	335	0769 Feb 25	12:24:44	3193	-15224	074	P	-t	-0.9436	1.1553	0.0983	289.2	74.1	-	7N	170W
6689	335	0769 Aug 22	03:03:05	3189	-15218	079	P	-a	0.8767	1.2546	0.2440	284.3	110.3	-	10S	33W
6690	335	0770 Feb 14	20:40:58	3185	-15212	084	T-	-p	-0.1839	2.5223	1.5188	334.1	214.8	92.6	11N	67E
6691	335	0770 Aug 11	09:22:06	3180	-15206	089	T+	pp	0.1431	2.6256	1.5655	365.8	230.1	99.9	14S	127W
6692	335	0771 Feb 04	10:53:52	3176	-15200	094	P	a-	0.5127	1.9010	0.9332	302.4	182.8	-	16N	146W
6693	335	0771 Jul 31	10:07:28	3172	-15194	099	P	t-	-0.6085	1.7792	0.7043	346.3	186.4	-	18S	138W
6694	335	0771 Dec 26	15:38:15	3169	-15189	066	N	-a	-1.5164	0.0754	-0.9243	75.7	-	-	22N	140E
6695	335	0772 Jan 25	02:51:28	3168	-15188	104	N	a-	1.1870	0.6673	-0.3076	207.6	-	-	19N	26W
6696	335	0772 Jun 20	00:28:23	3165	-15183	071	N	-t	1.3856	0.3207	-0.6896	159.8	-	-	22S	6E
6697	335	0772 Jul 19	12:13:23	3164	-15182	109	N	t-	-1.3311	0.4363	-0.6048	190.7	-	-	22S	169W
6698	335	0772 Dec 15	01:20:14	3161	-15177	076	P	-a	-0.8978	1.2394	0.1824	292.7	98.4	-	23N	7W
6699	335	0773 Jun 09	12:57:21	3157	-15171	081	P	-a	0.5723	1.7873	0.8280	302.1	178.5	-	23S	178E
6700	335	0773 Dec 04	03:42:30	3153	-15165	086	T-	-p	-0.2288	2.4894	1.3879	373.0	227.8	89.4	23N	44W
6701	336	0774 May 30	05:44:59	3149	-15159	091	T-	p-	-0.1764	2.5041	1.5640	314.8	208.4	92.6	22S	74W
6702	336	0774 Nov 23	02:42:09	3145	-15153	096	P	t-	0.4481	2.0848	0.9875	359.9	208.1	-	22N	30W
6703	336	0775 May 19	22:17:38	3141	-15147	101	P	a-	-0.9333	1.1276	0.1630	260.9	88.1	-	21S	37E
6704	336	0775 Nov 12	05:59:14	3137	-15141	106	N	a-	1.1088	0.8474	-0.2002	250.0	-	-	20N	80W
6705	336	0776 Apr 08	21:26:01	3133	-15136	073	P	-t	0.9831	1.0800	0.0286	283.4	40.6	-	8S	52E
6706	336	0776 Oct 02	06:14:18	3129	-15130	078	P	-a	-0.9474	1.1027	0.1361	253.9	79.2	-	4N	83W
6707	336	0777 Mar 28	22:47:16	3125	-15124	083	T+	pp	0.2556	2.4301	1.3485	371.4	228.0	86.9	4S	32E
6708	336	0777 Sep 21	22:29:48	3121	-15118	088	T-	-p	-0.2601	2.3629	1.3983	316.5	205.0	83.0	1N	34E
6709	336	0778 Mar 17	23:36:51	3117	-15112	093	P	t-	-0.4782	2.0117	0.9498	348.6	203.0	-	1S	20E
6710	336	0778 Sep 11	12:29:30	3113	-15106	098	P	a-	0.4670	2.0028	0.9993	323.1	195.0	-	3S	176W
6711	336	0779 Feb 05	19:09:30	3110	-15101	065	N	-a	1.5249	0.0555	-0.9359	64.9	-	-	16N	90E
6712	336	0779 Mar 07	06:28:42	3109	-15100	103	N	a-	-1.1750	0.7072	-0.3034	224.8	-	-	3N	82W
6713	336	0779 Aug 31	20:12:04	3105	-15094	108	N	t-	1.2523	0.5882	-0.4676	219.8	-	-	6S	69E
6714	336	0780 Jan 26	10:16:16	3102	-15089	075	P	-a	0.8200	1.3363	0.3700	270.8	125.7	-	19N	137W
6715	336	0780 Jul 21	05:45:40	3098	-15083	080	P	-t	-0.8704	1.2954	0.2270	311.0	113.6	-	21S	72W
6716	336	0781 Jan 15	01:47:56	3094	-15077	085	T+	-p	0.1533	2.5699	1.5833	325.4	211.4	93.8	21N	11W
6717	336	0781 Jul 10	09:29:49	3090	-15071	090	T-	pp	-0.0842	2.7159	1.6913	352.7	226.4	102.3	22S	129W
6718	336	0782 Jan 04	13:07:50	3086	-15065	095	P	a-	-0.5613	1.8487	0.8076	326.8	185.0	-	22N	178E
6719	336	0782 Jun 29	20:30:03	3082	-15059	100	P	a-	0.6776	1.5993	0.6296	296.1	162.3	-	23S	66E
6720	336	0782 Dec 24	17:09:39	3078	-15053	105	N	t-	-1.2956	0.5279	-0.5658	214.7	-	-	22N	116E

Cat Num	Canon Plate	Calendar Date	TD of Greatest Eclipse	ΔT s	Luna Num	Saros Num	Ecl. Type	QSE	Gamma	Pen. Mag.	Um. Mag.	Eclipse Phase Durations Pen. m	Par. m	Total m	Greatest in Zenith Lat.	Lng.
6721	337	0783 May 21	06:10:31	3074	-15048	072	N	-a	-1.1053	0.8011	-0.1422	221.7	-	-	22S	81W
6722	337	0783 Jun 19	12:44:53	3074	-15047	110	N	a-	1.3803	0.2953	-0.6453	141.1	-	-	22S	179W
6723	337	0783 Nov 13	21:37:37	3070	-15042	077	N	-t	1.0667	0.9438	-0.1417	273.2	-	-	20N	45E
6724	337	0784 May 09	21:47:09	3066	-15036	082	T	-a	-0.3885	2.1345	1.1555	322.3	201.9	58.0	19S	44E
6725	337	0784 Nov 02	03:02:29	3062	-15030	087	T	-p	0.3292	2.2684	1.2396	337.5	209.3	70.8	16N	37W
6726	337	0785 Apr 29	07:32:14	3058	-15024	092	T	t-	0.3891	2.1619	1.1263	346.3	210.9	55.1	15S	101W
6727	337	0785 Oct 22	15:29:49	3054	-15018	097	T	p-	-0.3732	2.1621	1.1841	314.6	198.0	60.9	12N	137E
6728	337	0786 Apr 18	10:10:53	3050	-15012	102	N	t-	1.1680	0.7521	-0.3220	252.0	-	-	11S	140W
6729	337	0786 Oct 12	07:24:09	3046	-15006	107	N	a-	-1.0406	0.9310	-0.0343	236.7	-	-	8N	101W
6730	337	0787 Mar 08	19:58:14	3043	-15001	074	P	-t	-0.9799	1.0854	0.0347	281.2	44.3	-	2N	76E
6731	337	0787 Sep 02	10:36:12	3039	-14995	079	P	-a	0.9288	1.1623	0.1453	277.8	86.6	-	6S	147W
6732	337	0788 Feb 26	04:48:17	3035	-14989	084	T-	-p	-0.2114	2.4689	1.4712	331.8	213.2	89.9	7N	56W
6733	337	0788 Aug 21	16:20:45	3031	-14983	089	T+	pp	0.2061	2.5127	1.4473	365.6	227.8	93.6	11S	128E
6734	337	0789 Feb 14	19:22:01	3027	-14977	094	P	a-	0.4906	1.9396	0.9754	303.3	185.3	-	12N	86E
6735	337	0789 Aug 10	16:49:37	3023	-14971	099	P	t-	-0.5370	1.9114	0.8344	353.0	198.3	-	15S	121E
6736	337	0790 Jan 06	00:19:09	3020	-14966	066	N	-a	-1.5250	0.0601	-0.9409	67.9	-	-	21N	10E
6737	337	0790 Feb 04	11:24:47	3019	-14965	104	N	a-	1.1732	0.6922	-0.2819	211.3	-	-	16N	154W
6738	337	0790 Jul 01	07:25:17	3016	-14960	071	N	-t	1.4568	0.1886	-0.8190	123.5	-	-	22S	98W
6739	337	0790 Jul 30	19:07:33	3015	-14959	109	N	t-	-1.2566	0.5725	-0.4677	215.0	-	-	19S	87E
6740	337	0790 Dec 26	09:42:21	3012	-14954	076	P	-a	-0.9044	1.2283	0.1691	292.8	95.2	-	22N	132W
6741	338	0791 Jun 20	20:17:03	3008	-14948	081	P	-a	0.6455	1.6520	0.6948	294.5	166.7	-	23S	68E
6742	338	0791 Dec 15	11:45:36	3004	-14942	086	T-	-p	-0.2357	2.4772	1.3746	373.3	227.7	88.4	23N	164W
6743	338	0792 Jun 09	13:12:03	3000	-14936	091	T-	pp	-0.1020	2.6406	1.7005	316.3	210.6	97.3	23S	174E
6744	338	0792 Dec 03	10:47:42	2996	-14930	096	T	t-	0.4413	2.0967	1.0004	359.7	208.7	3.1	23N	151W
6745	338	0793 May 30	05:34:46	2992	-14924	101	P	a-	-0.8615	1.2604	0.2938	272.7	116.4	-	23S	73W
6746	338	0793 Nov 22	14:23:41	2988	-14918	106	N	a-	1.1000	0.8626	-0.1832	251.0	-	-	22N	154E
6747	338	0794 Apr 20	04:21:44	2985	-14913	073	N	-t	1.0539	0.9501	-0.1015	270.2	-	-	12S	53W
6748	338	0794 Oct 13	14:42:25	2981	-14907	078	P	-a	-0.9682	1.0653	0.0972	250.3	67.2	-	8N	149E
6749	338	0795 Apr 09	05:35:52	2977	-14901	083	T	-t	0.3224	2.3061	1.2275	368.1	223.2	73.5	8S	71W
6750	338	0795 Oct 03	06:48:40	2973	-14895	088	T	-p	-0.2895	2.3110	1.3421	316.2	203.5	78.5	5N	93W
6751	338	0796 Mar 28	06:48:48	2969	-14889	093	T	t-	-0.4225	2.1110	1.0550	350.9	209.2	37.5	5S	89W
6752	338	0796 Sep 21	20:22:35	2965	-14883	098	T	p-	0.4303	2.0732	1.0635	327.1	199.3	38.4	1N	64E
6753	338	0797 Feb 16	03:29:04	2962	-14878	065	Ne	-a	1.5430	0.0196	-0.9666	38.5	-	-	12N	35W
6754	338	0797 Mar 17	14:15:23	2961	-14877	103	N	a-	-1.1301	0.7865	-0.2178	234.1	-	-	1S	160E
6755	338	0797 Sep 11	03:30:50	2957	-14871	108	N	t-	1.2065	0.6753	-0.3866	234.5	-	-	2S	42W
6756	338	0798 Feb 05	18:52:02	2954	-14866	075	P	-a	0.8368	1.3042	0.3405	268.3	121.1	-	15N	94E
6757	338	0798 Aug 01	12:23:42	2950	-14860	080	P	-t	-0.9452	1.1585	0.0892	298.2	72.6	-	18S	172W
6758	338	0799 Jan 26	10:24:05	2946	-14854	085	T+	-p	0.1656	2.5475	1.5608	325.8	211.5	93.0	18N	140W
6759	338	0799 Jul 21	16:27:53	2942	-14848	090	T-	pp	-0.1584	2.5789	1.5559	350.1	223.6	97.5	20S	126E
6760	338	0800 Jan 15	21:29:08	2938	-14842	095	P	a-	-0.5539	1.8630	0.8205	328.6	186.6	-	20N	53E
6761	339	0800 Jul 10	03:50:05	2934	-14836	100	P	a-	0.6037	1.7340	0.7660	302.2	174.5	-	21S	45W
6762	339	0801 Jan 04	01:13:24	2930	-14830	105	N	t-	-1.2895	0.5393	-0.5550	217.2	-	-	21N	4W
6763	339	0801 May 31	13:37:25	2927	-14825	072	N	-a	-1.1781	0.6678	-0.2758	205.4	-	-	24S	167E
6764	339	0801 Jun 29	20:12:23	2926	-14824	110	N	a-	1.3087	0.4267	-0.5141	167.5	-	-	22S	69E
6765	339	0801 Nov 24	05:45:07	2923	-14819	077	N	-t	1.0719	0.9336	-0.1508	271.4	-	-	22N	77W
6766	339	0802 May 21	05:01:30	2919	-14813	082	T	-a	-0.4636	1.9978	1.0168	318.9	195.0	19.9	21S	65W
6767	339	0802 Nov 13	11:29:30	2915	-14807	087	T	-p	0.3360	2.2552	1.2277	336.0	208.2	69.1	19N	164W
6768	339	0803 May 10	14:22:49	2911	-14801	092	T	pp	0.3139	2.3006	1.2637	351.5	218.0	76.2	18S	155E
6769	339	0803 Nov 03	00:08:20	2907	-14795	097	T	p-	-0.3608	2.1850	1.2066	314.7	198.6	63.9	16N	6E
6770	339	0804 Apr 28	16:47:44	2904	-14789	102	N	t-	1.0944	0.8864	-0.1861	270.0	-	-	15S	119E
6771	339	0804 Oct 22	15:58:05	2900	-14783	107	N	a-	-1.0245	0.9622	-0.0063	240.1	-	-	12N	129E
6772	339	0805 Mar 19	03:24:04	2896	-14778	074	N*	-t	-1.0232	1.0029	-0.0415	271.4	-	-	2S	37W
6773	339	0805 Sep 12	18:15:42	2893	-14772	079	P	-a	0.9746	1.0815	0.0580	271.6	55.6	-	2S	96E
6774	339	0806 Mar 08	12:46:53	2889	-14766	084	T-	-p	-0.2470	2.4005	1.4089	329.2	211.0	85.7	3N	177W
6775	339	0806 Sep 01	23:26:04	2885	-14760	089	T+	-p	0.2630	2.4111	1.3402	365.0	224.9	85.3	7S	20E
6776	339	0807 Feb 26	03:42:39	2881	-14754	094	T	a-	0.4621	1.9901	1.0295	304.7	188.5	25.4	8N	40W
6777	339	0807 Aug 21	23:39:45	2877	-14748	099	P	t-	-0.4715	2.0328	0.9535	358.2	207.2	-	11S	17E
6778	339	0808 Jan 17	08:55:34	2874	-14743	066	N	-a	-1.5369	0.0388	-0.9631	54.8	-	-	19N	119W
6779	339	0808 Feb 15	19:51:09	2873	-14742	104	N	a-	1.1543	0.7264	-0.2468	216.0	-	-	12N	78E
6780	339	0808 Jul 11	14:27:06	2870	-14737	071	Ne	-t	1.5247	0.0629	-0.9426	71.8	-	-	20S	156E

Cat Num	Canon Plate	Calendar Date	TD of Greatest Eclipse	ΔT s	Luna Num	Saros Num	Ecl. Type	QSE	Gamma	Pen. Mag.	Um. Mag.	Eclipse Phase ---- Durations ---- Pen. m	Par. m	Total m	Greatest in Zenith Lat.	Lng.
6781	340	0808 Aug 10	02:09:44	2869	-14736	109	N	t-	-1.1873	0.6994	-0.3402	233.9	-	-	16S	19W
6782	340	0809 Jan 05	18:01:57	2866	-14731	076	P	-a	-0.9123	1.2147	0.1538	292.5	91.2	-	21N	104E
6783	340	0809 Jul 01	03:37:57	2862	-14725	081	P	-a	0.7181	1.5179	0.5623	286.2	152.9	-	22S	42W
6784	340	0809 Dec 25	19:48:48	2858	-14719	086	T-	-p	-0.2422	2.4654	1.3624	373.5	227.5	87.5	23N	76E
6785	340	0810 Jun 20	20:38:25	2854	-14713	091	T-	pp	-0.0274	2.7779	1.8371	317.1	211.6	99.4	24S	62E
6786	340	0810 Dec 14	18:54:29	2850	-14707	096	T	t-	0.4350	2.1075	1.0128	359.4	209.1	18.5	24N	88E
6787	340	0811 Jun 10	12:50:09	2847	-14701	101	P	a-	-0.7888	1.3950	0.4258	283.5	137.8	-	24S	179E
6788	340	0811 Dec 03	22:51:00	2843	-14695	106	N	a-	1.0931	0.8741	-0.1694	251.5	-	-	24N	27E
6789	340	0812 Apr 30	11:11:23	2839	-14690	073	N	-t	1.1289	0.8127	-0.2393	254.2	-	-	15S	157W
6790	340	0812 Oct 23	23:16:18	2836	-14684	078	P	-a	-0.9846	1.0358	0.0664	247.4	55.7	-	12N	19E
6791	340	0813 Apr 19	12:18:18	2832	-14678	083	T	-t	0.3940	2.1731	1.0974	363.7	216.7	50.3	12S	173W
6792	340	0813 Oct 13	15:14:00	2828	-14672	088	T	-p	-0.3133	2.2694	1.2965	316.0	202.1	74.3	9N	140E
6793	340	0814 Apr 08	13:52:16	2824	-14666	093	T	t-	-0.3589	2.2246	1.1745	353.0	215.1	64.2	9S	164E
6794	340	0814 Oct 03	04:23:30	2820	-14660	098	T	p-	0.4009	2.1303	1.1144	330.5	202.6	50.8	6N	57W
6795	340	0815 Mar 28	21:53:31	2816	-14654	103	N	a-	-1.0774	0.8800	-0.1181	244.1	-	-	5S	44E
6796	340	0815 Sep 22	10:58:23	2813	-14648	108	N	t-	1.1683	0.7485	-0.3194	246.1	-	-	2N	156W
6797	340	0816 Feb 17	03:19:25	2809	-14643	075	P	-a	0.8606	1.2592	0.2982	264.8	114.0	-	12N	34W
6798	340	0816 Aug 11	19:10:41	2805	-14637	080	N*	-t	-1.0130	1.0346	-0.0357	285.2	-	-	15S	85E
6799	340	0817 Feb 05	18:53:17	2802	-14631	085	T+	-p	0.1832	2.5150	1.5285	326.1	211.3	91.8	15N	93E
6800	340	0817 Jul 31	23:31:43	2798	-14625	090	T-	pp	-0.2288	2.4491	1.4274	346.8	219.8	89.9	18S	20E
6801	341	0818 Jan 26	05:45:02	2794	-14619	095	P	a-	-0.5428	1.8839	0.8405	330.6	188.8	-	17N	71W
6802	341	0818 Jul 21	11:13:59	2790	-14613	100	P	a-	0.5321	1.8649	0.8979	307.2	184.2	-	19S	156W
6803	341	0819 Jan 15	09:12:36	2786	-14607	105	N	t-	-1.2802	0.5564	-0.5377	220.6	-	-	19N	124W
6804	341	0819 Jun 11	21:02:15	2783	-14602	072	N	-a	-1.2522	0.5323	-0.4123	186.2	-	-	25S	56E
6805	341	0819 Jul 11	03:42:30	2782	-14601	110	N	a-	1.2391	0.5550	-0.3868	188.6	-	-	21S	44W
6806	341	0819 Dec 05	13:57:15	2779	-14596	077	N	-t	1.0742	0.9286	-0.1541	270.1	-	-	24N	161E
6807	341	0820 May 31	12:12:24	2775	-14590	082	P	-a	-0.5405	1.8579	0.8745	314.5	186.1	-	23S	173W
6808	341	0820 Nov 23	20:01:23	2772	-14584	087	T	-p	0.3396	2.2478	1.2220	334.6	207.4	68.2	22N	69E
6809	341	0821 May 20	21:08:36	2768	-14578	092	T+	pp	0.2354	2.4455	1.4068	356.0	223.7	90.0	21S	53E
6810	341	0821 Nov 13	08:52:22	2764	-14572	097	T	p-	-0.3529	2.1995	1.2211	314.6	198.9	65.7	19N	125W
6811	341	0822 May 09	23:18:04	2760	-14566	102	N*	t-	1.0154	1.0307	-0.0404	286.9	-	-	18S	21E
6812	341	0822 Nov 03	00:38:02	2756	-14560	107	P	a-	-1.0137	0.9835	0.0120	242.6	23.9	-	15N	1W
6813	341	0823 Mar 30	10:43:23	2753	-14555	074	N	-t	-1.0722	0.9097	-0.1284	259.8	-	-	6S	149W
6814	341	0823 Sep 24	02:02:13	2749	-14549	079	N*	-a	1.0138	1.0128	-0.0171	266.1	-	-	3N	22W
6815	341	0824 Mar 18	20:39:25	2746	-14543	084	T	-p	-0.2883	2.3219	1.3360	326.1	208.2	79.7	1S	63E
6816	341	0824 Sep 12	06:39:37	2742	-14537	089	T	-p	0.3125	2.3230	1.2466	364.2	221.6	75.2	2S	90W
6817	341	0825 Mar 08	11:55:53	2738	-14531	094	T	a-	0.4275	2.0519	1.0949	306.3	191.9	44.7	4N	165W
6818	341	0825 Sep 01	06:38:15	2734	-14525	099	T	t-	-0.4126	2.1421	1.0605	362.3	213.9	40.0	7S	89W
6819	341	0826 Jan 27	17:25:25	2731	-14520	066	Ne	-a	-1.5533	0.0089	-0.9935	26.4	-	-	16N	114E
6820	341	0826 Feb 26	04:08:47	2730	-14519	104	N	a-	1.1290	0.7722	-0.1999	222.1	-	-	8N	47W
6821	342	0826 Aug 21	09:21:41	2727	-14513	109	N	t-	-1.1250	0.8134	-0.2259	248.6	-	-	12S	129W
6822	342	0827 Jan 17	02:14:19	2724	-14508	076	P	-a	-0.9251	1.1918	0.1298	291.4	84.3	-	19N	19W
6823	342	0827 Jul 12	11:03:38	2720	-14502	081	P	-a	0.7871	1.3909	0.4362	277.5	137.0	-	21S	154W
6824	342	0828 Jan 06	03:47:56	2716	-14496	086	T-	-p	-0.2513	2.4486	1.3460	373.4	227.1	86.1	22N	43W
6825	342	0828 Jul 01	04:05:59	2712	-14490	091	T+	pp	0.0461	2.7441	1.8020	317.3	211.6	99.1	23S	50W
6826	342	0828 Dec 25	03:01:14	2708	-14484	096	T	t-	0.4284	2.1184	1.0261	358.9	209.6	26.2	24N	33W
6827	342	0829 Jun 20	20:03:13	2705	-14478	101	P	a-	-0.7151	1.5318	0.5594	293.4	155.2	-	24S	70E
6828	342	0829 Dec 14	07:21:21	2701	-14472	106	N	a-	1.0879	0.8822	-0.1583	251.5	-	-	25N	99W
6829	342	0830 May 11	17:51:21	2698	-14467	073	N	-t	1.2104	0.6635	-0.3893	233.7	-	-	18S	102E
6830	342	0830 Jun 10	06:33:54	2697	-14466	111	Nb	t-	-1.5398	0.0457	-0.9806	63.1	-	-	25S	88W
6831	342	0830 Nov 04	07:57:58	2694	-14461	078	P	-a	-0.9950	1.0173	0.0467	245.5	46.8	-	16N	112W
6832	342	0831 Apr 30	18:53:15	2690	-14455	083	P	-t	0.4716	2.0295	0.9565	358.0	207.8	-	16S	87E
6833	342	0831 Oct 24	23:46:04	2686	-14449	088	T	-p	-0.3310	2.2389	1.2620	316.1	201.0	70.7	13N	11E
6834	342	0832 Apr 18	20:51:19	2683	-14443	093	T	t-	-0.2910	2.3464	1.3019	354.6	220.1	80.6	13S	58E
6835	342	0832 Oct 13	12:31:07	2679	-14437	098	T	p-	0.3781	2.1753	1.1533	333.5	205.1	58.2	10N	180E
6836	342	0833 Apr 08	05:26:40	2675	-14431	103	N	a-	-1.0197	0.9828	-0.0092	254.0	-	-	10S	71W
6837	342	0833 Oct 02	18:32:17	2671	-14425	108	N	t-	1.1356	0.8114	-0.2622	255.6	-	-	6N	90E
6838	342	0834 Feb 27	11:40:48	2668	-14420	075	P	-a	0.8895	1.2048	0.2465	260.5	104.4	-	8N	160W
6839	342	0834 Aug 23	02:06:02	2665	-14414	080	N	-t	-1.0744	0.9226	-0.1491	272.1	-	-	11S	20W
6840	342	0835 Feb 17	03:14:18	2661	-14408	085	T+	-p	0.2072	2.4707	1.4845	326.3	210.8	89.7	11N	34W

Cat Num	Canon Plate	Calendar Date	TD of Greatest Eclipse	ΔT s	Luna Num	Saros Num	Ecl. Type	QSE	Gamma	Pen. Mag.	Um. Mag.	Eclipse Phase Durations Pen. m	Eclipse Phase Durations Par. m	Eclipse Phase Durations Total m	Greatest in Zenith Lat.	Greatest in Zenith Lng.
6841	343	0835 Aug 12	06:45:13	2657	-14402	090	T	-p	-0.2922	2.3323	1.3114	343.1	215.3	79.9	14S	89W
6842	343	0836 Feb 06	13:52:42	2653	-14396	095	P	h-	-0.5253	1.9161	0.8722	333.1	191.8	-	14N	167E
6843	343	0836 Jul 31	18:44:36	2650	-14390	100	T	a-	0.4654	1.9871	1.0207	311.2	191.7	21.7	17S	91E
6844	343	0837 Jan 25	17:04:56	2646	-14384	105	N	t-	-1.2655	0.5830	-0.5103	225.5	-	-	17N	118E
6845	343	0837 Jun 22	04:27:12	2643	-14379	072	N	-a	-1.3259	0.3978	-0.5484	163.3	-	-	25S	56W
6846	343	0837 Jul 21	11:16:30	2642	-14378	110	N	a-	1.1723	0.6782	-0.2651	206.0	-	-	19S	157W
6847	343	0837 Dec 15	22:10:08	2639	-14373	077	N	-t	1.0763	0.9235	-0.1568	268.7	-	-	25N	39E
6848	343	0838 Jun 11	19:19:00	2635	-14367	082	P	-a	-0.6199	1.7137	0.7270	309.0	174.6	-	24S	81E
6849	343	0838 Dec 05	04:36:36	2632	-14361	087	T	-p	0.3412	2.2436	1.2203	333.4	206.8	67.8	23N	59W
6850	343	0839 Jun 01	03:49:08	2628	-14355	092	T+	pp	0.1534	2.5971	1.5561	359.7	228.0	99.1	22S	47W
6851	343	0839 Nov 24	17:39:54	2624	-14349	097	T	p-	-0.3476	2.2092	1.2311	314.4	199.1	66.9	21N	104E
6852	343	0840 May 20	05:43:53	2620	-14343	102	P	t-	0.9323	1.1827	0.1125	302.3	81.7	-	20S	76W
6853	343	0840 Nov 13	09:23:17	2617	-14337	107	P	a-	-1.0074	0.9964	0.0224	244.2	32.6	-	18N	133W
6854	343	0841 Apr 09	17:55:20	2614	-14332	074	N	-t	-1.1279	0.8044	-0.2275	245.8	-	-	10S	102E
6855	343	0841 Oct 04	09:55:45	2610	-14326	079	N	-a	1.0463	0.9563	-0.0799	261.3	-	-	7N	141W
6856	343	0842 Mar 30	04:24:21	2606	-14320	084	T	-p	-0.3367	2.2301	1.2500	322.5	204.5	70.7	6S	55W
6857	343	0842 Sep 23	14:01:24	2603	-14314	089	T	-t	0.3546	2.2485	1.1669	363.3	218.3	63.7	2N	158E
6858	343	0843 Mar 19	20:00:50	2599	-14308	094	T	p-	0.3858	2.1265	1.1730	308.2	195.7	58.9	1S	72E
6859	343	0843 Sep 12	13:47:07	2595	-14302	099	T	t-	-0.3617	2.2366	1.1527	365.2	218.8	61.5	3S	163E
6860	343	0844 Mar 08	12:19:11	2591	-14296	104	N	a-	1.0985	0.8278	-0.1435	229.0	-	-	4N	171W
6861	344	0844 Aug 31	16:41:53	2588	-14290	109	N	h-	-1.0684	0.9173	-0.1220	260.4	-	-	8S	120E
6862	344	0845 Jan 27	10:21:40	2585	-14285	076	P	-h	-0.9412	1.1625	0.1000	289.5	74.5	-	16N	141W
6863	344	0845 Jul 22	18:33:10	2581	-14279	081	P	-a	0.8531	1.2696	0.3152	268.3	118.3	-	19S	93E
6864	344	0846 Jan 16	11:43:00	2577	-14273	086	T-	-p	-0.2638	2.4251	1.3237	373.0	226.5	84.1	20N	162W
6865	344	0846 Jul 12	11:35:24	2574	-14267	091	T+	-p	0.1179	2.6133	1.6693	317.0	210.5	96.5	21S	162W
6866	344	0847 Jan 05	11:06:14	2570	-14261	096	T	t-	0.4197	2.1327	1.0437	358.4	210.3	33.7	23N	153W
6867	344	0847 Jul 02	03:17:32	2566	-14255	101	P	a-	-0.6429	1.6662	0.6900	302.2	169.3	-	24S	38W
6868	344	0847 Dec 25	15:50:18	2563	-14249	106	N	a-	1.0809	0.8933	-0.1436	251.9	-	-	24N	134E
6869	344	0848 May 22	00:28:20	2560	-14244	073	N	-t	1.2933	0.5121	-0.5418	208.9	-	-	20S	3E
6870	344	0848 Jun 20	13:19:57	2559	-14243	111	N	t-	-1.4660	0.1831	-0.8468	125.2	-	-	25S	170E
6871	344	0848 Nov 14	16:43:54	2556	-14238	078	P	-a	-1.0022	1.0046	0.0333	244.2	39.6	-	19N	117E
6872	344	0849 May 11	01:23:11	2552	-14232	083	P	-t	0.5532	1.8784	0.8079	350.8	196.1	-	18S	11W
6873	344	0849 Nov 04	08:23:41	2549	-14226	088	T	-p	-0.3439	2.2172	1.2366	316.4	200.3	67.8	17N	119W
6874	344	0850 Apr 30	03:44:32	2545	-14220	093	T-	pp	-0.2175	2.4786	1.4394	355.7	224.0	92.1	16S	47W
6875	344	0850 Oct 24	20:45:26	2541	-14214	098	T	p-	0.3612	2.2091	1.1815	336.0	207.0	62.9	14N	55E
6876	344	0851 Apr 19	12:52:52	2538	-14208	103	P	a-	-0.9554	1.0980	0.1116	264.0	74.6	-	13S	176E
6877	344	0851 Oct 14	02:15:28	2534	-14202	108	N	t-	1.1107	0.8597	-0.2193	262.7	-	-	10N	27W
6878	344	0852 Mar 09	19:53:32	2531	-14197	075	P	-a	0.9258	1.1368	0.1812	254.9	90.4	-	3N	75E
6879	344	0852 Sep 02	09:10:46	2527	-14191	080	N	-t	-1.1287	0.8236	-0.2493	259.3	-	-	7S	127W
6880	344	0853 Feb 27	11:26:50	2524	-14185	085	T+	-p	0.2379	2.4142	1.4284	326.1	209.9	86.5	7N	158W
6881	345	0853 Aug 22	14:06:49	2520	-14179	090	T	-p	-0.3500	2.2260	1.2057	339.2	210.3	67.2	11S	159E
6882	345	0854 Feb 16	21:51:13	2516	-14173	095	P	t-	-0.5012	1.9604	0.9164	336.0	195.6	-	10N	46E
6883	345	0854 Aug 12	02:21:32	2513	-14167	100	T	a-	0.4033	2.1010	1.1347	314.2	197.3	53.4	14S	24W
6884	345	0855 Feb 06	00:50:13	2509	-14161	105	N	t-	-1.2454	0.6191	-0.4728	231.8	-	-	13N	2E
6885	345	0855 Jul 03	11:52:35	2506	-14156	072	N	-a	-1.3988	0.2653	-0.6832	135.3	-	-	24S	167W
6886	345	0855 Aug 01	18:55:01	2506	-14155	110	N	a-	1.1094	0.7948	-0.1506	220.5	-	-	16S	87E
6887	345	0855 Dec 27	06:23:14	2503	-14150	077	N	-t	1.0789	0.9172	-0.1599	267.1	-	-	24N	84W
6888	345	0856 Jun 22	02:25:03	2499	-14144	082	P	-a	-0.6989	1.5706	0.5803	302.3	160.3	-	24S	26W
6889	345	0856 Dec 15	13:14:16	2495	-14138	087	T	-p	0.3417	2.2413	1.2209	332.1	206.3	67.7	24N	172E
6890	345	0857 Jun 11	10:26:01	2492	-14132	092	T+	pp	0.0691	2.7531	1.7094	362.4	230.8	104.1	23S	147W
6891	345	0857 Dec 05	02:30:48	2488	-14126	097	T	p-	-0.3447	2.2140	1.2367	314.0	199.0	67.5	23N	29W
6892	345	0858 May 31	12:06:43	2484	-14120	102	P	t-	0.8462	1.3402	0.2708	316.3	123.9	-	22S	172W
6893	345	0858 Nov 24	18:11:30	2481	-14114	107	P	a-	-1.0038	1.0040	0.0277	245.4	36.3	-	21N	95E
6894	345	0859 Apr 21	01:02:34	2478	-14109	074	N	-t	-1.1880	0.6912	-0.3348	229.4	-	-	14S	6W
6895	345	0859 May 20	13:38:04	2477	-14108	112	Nb	t-	1.5700	0.0007	-1.0461	8.1	-	-	19S	165E
6896	345	0859 Oct 15	17:56:26	2474	-14103	079	N	-a	1.0723	0.9116	-0.1307	257.4	-	-	11N	97E
6897	345	0860 Apr 09	12:04:18	2471	-14097	084	T	-p	-0.3899	2.1298	1.1552	318.4	199.9	57.4	10S	171W
6898	345	0860 Oct 03	21:30:19	2467	-14091	089	T	-t	0.3903	2.1855	1.0990	362.5	215.0	50.2	6N	45E
6899	345	0861 Mar 30	03:58:57	2463	-14085	094	T	p-	0.3383	2.2122	1.2617	310.1	199.4	70.4	5S	49W
6900	345	0861 Sep 22	21:05:29	2460	-14079	099	T	t-	-0.3183	2.3173	1.2312	367.3	222.2	73.5	1N	52E

Cat Num	Canon Plate	Calendar Date	TD of Greatest Eclipse	ΔT s	Luna Num	Saros Num	Ecl. Type	QSE	Gamma	Pen. Mag.	Um. Mag.	Eclipse Phase Durations Pen. m	Par. m	Total m	Greatest in Zenith Lat.	Lng.
6901	346	0862 Mar 19	20:18:53	2456	-14073	104	N	a-	1.0598	0.8983	-0.0721	237.4	-	-	0S	68E
6902	346	0862 Sep 12	00:13:30	2453	-14067	109	N*	h-	-1.0201	1.0059	-0.0333	269.3	-	-	4S	6E
6903	346	0863 Feb 07	18:19:18	2450	-14062	076	P	-h	-0.9644	1.1199	0.0573	286.2	56.9	-	13N	99E
6904	346	0863 Aug 03	02:09:34	2446	-14056	081	P	-a	0.9140	1.1579	0.2033	259.0	96.4	-	16S	21W
6905	346	0864 Jan 27	19:30:38	2443	-14050	086	T	-p	-0.2821	2.3906	1.2910	372.2	225.4	80.8	17N	81E
6906	346	0864 Jul 22	19:08:57	2439	-14044	091	T+	-p	0.1860	2.4897	1.5431	316.1	208.5	91.8	19S	84E
6907	346	0865 Jan 15	19:08:25	2435	-14038	096	T	t-	0.4083	2.1517	1.0667	358.0	211.3	41.3	21N	87E
6908	346	0865 Jul 12	10:31:09	2432	-14032	101	P	a-	-0.5710	1.8004	0.8198	310.2	181.1	-	22S	147W
6909	346	0866 Jan 05	00:19:44	2428	-14026	106	N	a-	1.0736	0.9044	-0.1283	252.1	-	-	23N	8E
6910	346	0866 Jun 02	06:59:17	2425	-14021	073	N	-t	1.3794	0.3549	-0.7006	177.0	-	-	22S	95W
6911	346	0866 Jul 01	20:03:56	2425	-14020	111	N	t-	-1.3912	0.3224	-0.7116	164.7	-	-	24S	70E
6912	346	0866 Nov 26	01:34:40	2422	-14015	078	P	-a	-1.0058	0.9982	0.0264	243.4	35.3	-	21N	16W
6913	346	0867 May 22	07:49:10	2418	-14009	083	P	-t	0.6379	1.7219	0.6535	342.1	180.9	-	21S	108W
6914	346	0867 Nov 15	17:06:44	2415	-14003	088	T	-p	-0.3516	2.2047	1.2206	316.9	199.9	65.9	19N	110E
6915	346	0868 May 10	10:35:28	2411	-13997	093	T-	pp	-0.1411	2.6163	1.5822	356.1	226.7	99.5	19S	150W
6916	346	0868 Nov 04	05:04:36	2408	-13991	098	T	p-	0.3491	2.2339	1.2010	338.2	208.4	65.9	17N	70W
6917	346	0869 Apr 29	20:16:05	2404	-13985	103	P	a-	-0.8876	1.2197	0.2386	273.4	106.9	-	17S	64E
6918	346	0869 Oct 24	10:04:36	2400	-13979	108	N	t-	1.0909	0.8985	-0.1854	268.4	-	-	14N	145W
6919	346	0870 Mar 21	03:58:24	2398	-13974	075	P	-a	0.9687	1.0569	0.1037	247.9	69.2	-	1S	48W
6920	346	0870 Sep 13	16:25:26	2394	-13968	080	N	-t	-1.1754	0.7387	-0.3356	247.3	-	-	3S	123E
6921	347	0871 Mar 10	19:30:43	2390	-13962	085	T	-p	0.2753	2.3455	1.3601	325.6	208.4	81.5	3N	80E
6922	347	0871 Sep 02	21:38:28	2387	-13956	090	T	-p	-0.4002	2.1336	1.1137	335.3	205.1	51.3	7S	45E
6923	347	0872 Feb 28	05:40:08	2383	-13950	095	P	t-	-0.4700	2.0177	0.9736	339.3	200.2	-	6N	72W
6924	347	0872 Aug 22	10:06:58	2380	-13944	100	T	a-	0.3477	2.2031	1.2364	316.4	201.5	68.4	10S	142W
6925	347	0873 Feb 16	08:26:20	2376	-13938	105	N	t-	-1.2181	0.6683	-0.4218	239.9	-	-	10N	113W
6926	347	0873 Jul 13	19:19:38	2373	-13933	072	N	-a	-1.4698	0.1364	-0.8151	98.4	-	-	23S	81E
6927	347	0873 Aug 12	02:38:51	2373	-13932	110	N	a-	1.0508	0.9036	-0.0444	232.7	-	-	13S	30W
6928	347	0874 Jan 06	14:34:28	2370	-13927	077	N	-t	1.0837	0.9064	-0.1668	264.9	-	-	23N	154E
6929	347	0874 Jul 03	09:30:34	2366	-13921	082	P	-a	-0.7776	1.4285	0.4338	294.6	142.3	-	24S	132W
6930	347	0874 Dec 26	21:50:44	2363	-13915	087	T	-p	0.3440	2.2353	1.2185	330.7	205.6	67.3	24N	44E
6931	347	0875 Jun 22	17:01:48	2359	-13909	092	T-	pp	-0.0154	2.8533	1.8064	364.1	231.8	105.4	24S	114E
6932	347	0875 Dec 16	11:22:45	2356	-13903	097	T	p-	-0.3421	2.2182	1.2420	313.7	199.0	68.0	23N	161W
6933	347	0876 Jun 10	18:27:49	2352	-13897	102	P	t-	0.7583	1.5013	0.4322	328.6	152.7	-	23S	92E
6934	347	0876 Dec 05	03:02:18	2349	-13891	107	P	a-	-1.0027	1.0071	0.0290	246.1	37.1	-	22N	37W
6935	347	0877 May 01	08:04:53	2346	-13886	074	N	-t	-1.2528	0.5693	-0.4510	209.7	-	-	18S	113W
6936	347	0877 May 30	20:15:13	2345	-13885	112	N	t-	1.4857	0.1535	-0.8896	116.7	-	-	21S	65E
6937	347	0877 Oct 26	02:03:46	2342	-13880	079	N	-a	1.0920	0.8784	-0.1696	254.7	-	-	15N	25W
6938	347	0878 Apr 20	19:36:55	2339	-13874	084	T	-a	-0.4498	2.0172	1.0476	313.5	193.8	32.8	14S	75E
6939	347	0878 Oct 15	05:08:15	2335	-13868	089	T	-t	0.4180	2.1369	1.0457	361.8	212.2	34.7	11N	71W
6940	347	0879 Apr 10	11:50:07	2332	-13862	094	T	p-	0.2849	2.3088	1.3612	312.1	203.0	79.9	9S	168W
6941	348	0879 Oct 04	04:33:34	2328	-13856	099	T	p-	-0.2823	2.3843	1.2963	368.8	224.6	81.1	6N	62W
6942	348	0880 Mar 30	04:11:45	2325	-13850	104	P	a-	1.0161	0.9784	0.0084	246.2	20.4	-	5S	52W
6943	348	0880 Sep 22	07:54:42	2322	-13844	109	P	a-	-0.9785	1.0823	0.0430	276.2	48.6	-	1N	111W
6944	348	0881 Feb 18	02:09:41	2319	-13839	076	P	-h	-0.9928	1.0677	0.0053	281.8	17.5	-	9N	19W
6945	348	0881 Aug 13	09:51:22	2315	-13833	081	P	-a	0.9707	1.0542	0.0988	249.5	68.0	-	13S	138W
6946	348	0882 Feb 07	03:12:24	2312	-13827	086	T	-p	-0.3049	2.3475	1.2502	371.1	223.9	76.2	14N	34W
6947	348	0882 Aug 03	02:47:08	2308	-13821	091	T+	-p	0.2497	2.3744	1.4247	314.9	205.7	85.0	16S	31W
6948	348	0883 Jan 27	03:04:51	2305	-13815	096	T	t-	0.3916	2.1800	1.0995	357.7	212.8	49.8	18N	33W
6949	348	0883 Jul 23	17:49:11	2301	-13809	101	P	a-	-0.5035	1.9267	0.9412	317.0	190.5	-	20S	103E
6950	348	0884 Jan 16	08:45:25	2298	-13803	106	N	a-	1.0630	0.9215	-0.1064	253.0	-	-	21N	118W
6951	348	0884 Jun 12	13:29:13	2295	-13798	073	N	-t	1.4652	0.1984	-0.8591	134.6	-	-	22S	167E
6952	348	0884 Jul 12	02:49:55	2294	-13797	111	N	t-	-1.3183	0.4584	-0.5801	194.7	-	-	23S	32W
6953	348	0884 Dec 06	10:26:38	2292	-13792	078	P	-a	-1.0085	0.9931	0.0213	242.8	31.7	-	22N	148W
6954	348	0885 Jun 01	14:13:12	2288	-13786	083	P	-t	0.7240	1.5630	0.4964	331.8	161.6	-	22S	155E
6955	348	0885 Nov 26	01:52:31	2285	-13780	088	T	-p	-0.3571	2.1962	1.2092	317.5	199.7	64.5	22N	21W
6956	348	0886 May 21	17:22:29	2281	-13774	093	T-	pp	-0.0603	2.7622	1.7327	355.6	227.9	103.6	21S	108E
6957	348	0886 Nov 15	13:28:41	2278	-13768	098	T	p-	0.3417	2.2499	1.2123	340.0	209.4	67.6	20N	164E
6958	348	0887 May 11	03:34:50	2274	-13762	103	P	a-	-0.8151	1.3503	0.3739	282.2	130.8	-	20S	46W
6959	348	0887 Nov 04	17:59:53	2271	-13756	108	N	t-	1.0765	0.9273	-0.1611	272.7	-	-	18N	95E
6960	348	0888 Mar 31	11:55:10	2268	-13751	075	P	-a	1.0182	0.9649	0.0139	239.2	25.6	-	5S	168W

Cat Num	Canon Plate	Calendar Date	TD of Greatest Eclipse	ΔT s	Luna Num	Saros Num	Ecl. Type	QSE	Gamma	Pen. Mag.	Um. Mag.	Eclipse Phase Durations Pen. m	Par. m	Total m	Greatest in Zenith Lat.	Lng.
6961	349	0888 Apr 29	19:18:38	2267	-13750	113	Nb	a-	-1.5015	0.0751	-0.8702	72.8	-	-	17S	78E
6962	349	0888 Sep 23	23:50:47	2265	-13745	080	N	-t	-1.2137	0.6689	-0.4066	236.5	-	-	1N	10E
6963	349	0889 Mar 21	03:24:49	2261	-13739	085	T	-p	0.3202	2.2630	1.2778	324.7	206.1	73.9	2S	40W
6964	349	0889 Sep 13	05:19:21	2258	-13733	090	T	-p	-0.4438	2.0537	1.0339	331.5	200.0	28.6	3S	72W
6965	349	0890 Mar 10	13:19:07	2254	-13727	095	T	t-	-0.4313	2.0887	1.0447	343.0	205.2	33.4	2N	172E
6966	349	0890 Sep 02	18:00:35	2251	-13721	100	T	p-	0.2983	2.2942	1.3268	318.0	204.4	77.7	6S	98E
6967	349	0891 Feb 27	15:53:21	2248	-13715	105	N	t-	-1.1834	0.7308	-0.3572	249.4	-	-	6N	134E
6968	349	0891 Jul 25	02:50:16	2245	-13710	072	Ne	-a	-1.5377	0.0137	-0.9413	31.6	-	-	21S	32W
6969	349	0891 Aug 23	10:29:38	2244	-13709	110	P	a-	0.9978	1.0023	0.0513	242.8	49.0	-	9S	148W
6970	349	0892 Jan 17	22:43:39	2241	-13704	077	N	-t	1.0911	0.8904	-0.1782	262.1	-	-	21N	32E
6971	349	0892 Jul 13	16:36:46	2238	-13698	082	P	-a	-0.8546	1.2894	0.2900	285.7	119.2	-	22S	121E
6972	349	0893 Jan 06	06:26:50	2234	-13692	087	T	-p	0.3474	2.2271	1.2142	329.2	205.0	66.6	22N	85W
6973	349	0893 Jul 02	23:37:17	2231	-13686	092	T-	pp	-0.0998	2.7003	1.6499	364.7	231.2	103.0	23S	16E
6974	349	0893 Dec 26	20:14:13	2228	-13680	097	T	p-	-0.3391	2.2230	1.2484	313.3	199.1	68.7	23N	67E
6975	349	0894 Jun 22	00:49:05	2224	-13674	102	P	t-	0.6701	1.6632	0.5940	339.2	174.5	-	23S	3W
6976	349	0894 Dec 16	11:53:12	2221	-13668	107	P	a-	-1.0018	1.0093	0.0298	246.7	37.7	-	23N	169W
6977	349	0895 May 12	15:05:21	2218	-13663	074	N	-t	-1.3197	0.4439	-0.5712	186.6	-	-	20S	141E
6978	349	0895 Jun 11	02:54:24	2218	-13662	112	N	t-	1.4018	0.3058	-0.7339	161.9	-	-	22S	35W
6979	349	0895 Nov 06	10:15:12	2215	-13657	079	N	-a	1.1078	0.8521	-0.2011	252.6	-	-	18N	148W
6980	349	0896 May 01	03:06:49	2211	-13651	084	P	-a	-0.5129	1.8992	0.9342	307.9	186.4	-	17S	39W
6981	350	0896 Oct 25	12:52:55	2208	-13645	089	T	-t	0.4402	2.0984	1.0030	361.3	209.8	9.1	14N	172E
6982	350	0897 Apr 20	19:34:20	2205	-13639	094	T+	p-	0.2254	2.4168	1.4716	313.9	206.2	87.7	13S	75E
6983	350	0897 Oct 14	12:11:10	2201	-13633	099	T-	p-	-0.2536	2.4378	1.3482	369.6	226.1	86.1	10N	177W
6984	350	0898 Apr 10	11:55:02	2198	-13627	104	P	a-	0.9647	1.0725	0.1027	255.8	70.4	-	9S	169W
6985	350	0898 Oct 03	15:47:13	2195	-13621	109	P	a-	-0.9452	1.1433	0.1041	281.2	74.7	-	5N	130E
6986	350	0899 Mar 01	09:48:18	2192	-13616	076	N	-t	-1.0300	0.9993	-0.0627	275.3	-	-	5N	135W
6987	350	0899 Aug 24	17:42:20	2188	-13610	081	P	-a	1.0204	0.9636	0.0071	240.6	18.5	-	9S	103E
6988	350	0900 Feb 18	10:45:18	2185	-13604	086	T	-p	-0.3349	2.2910	1.1967	369.3	221.6	68.9	10N	149W
6989	350	0900 Aug 13	10:30:10	2182	-13598	091	T	-p	0.3091	2.2671	1.3139	313.3	202.4	76.1	13S	148W
6990	350	0901 Feb 06	10:56:34	2178	-13592	096	T	t-	0.3704	2.2164	1.1410	357.5	214.5	58.4	14N	151W
6991	350	0901 Aug 03	01:09:31	2175	-13586	101	T	a-	-0.4388	2.0482	1.0573	323.0	198.2	36.5	17S	7W
6992	350	0902 Jan 26	17:07:20	2172	-13580	106	N	a-	1.0483	0.9458	-0.0769	254.6	-	-	18N	116E
6993	350	0902 Jun 23	19:56:14	2169	-13575	073	Ne	-t	1.5524	0.0397	-1.0203	61.2	-	-	22S	70E
6994	350	0902 Jul 23	09:37:09	2168	-13574	111	N	t-	-1.2466	0.5924	-0.4510	219.5	-	-	20S	134W
6995	350	0902 Dec 17	19:20:33	2165	-13569	078	P	-a	-1.0103	0.9897	0.0184	242.4	29.4	-	23N	79E
6996	350	0903 Jun 12	20:36:58	2162	-13563	083	P	-t	0.8107	1.4032	0.3380	319.8	136.5	-	23S	59E
6997	350	0903 Dec 07	10:40:03	2159	-13557	088	T	-p	-0.3608	2.1906	1.2011	318.2	199.7	63.5	23N	152W
6998	350	0904 Jun 01	00:10:31	2155	-13551	093	T+	pp	0.0207	2.8327	1.8073	354.2	227.7	104.2	23S	5E
6999	350	0904 Nov 25	21:55:42	2152	-13545	098	T	p-	0.3375	2.2599	1.2180	341.7	210.2	68.5	22N	38E
7000	350	0905 May 21	10:51:32	2149	-13539	103	P	a-	-0.7399	1.4862	0.5141	290.2	149.7	-	22S	155W
7001	351	0905 Nov 15	02:00:01	2146	-13533	108	N	t-	1.0662	0.9481	-0.1441	275.9	-	-	21N	25W
7002	351	0906 Apr 11	19:45:25	2143	-13528	075	N	-a	1.0731	0.8633	-0.0859	228.8	-	-	9S	73E
7003	351	0906 May 11	02:52:46	2142	-13527	113	N	a-	-1.4355	0.1951	-0.7479	115.9	-	-	20S	36W
7004	351	0906 Oct 05	07:26:04	2139	-13522	080	N	-t	-1.2449	0.6123	-0.4642	227.1	-	-	5N	105W
7005	351	0907 Apr 01	11:10:21	2136	-13516	085	T	-a	0.3718	2.1684	1.1831	323.1	202.7	62.1	6S	158W
7006	351	0907 Sep 24	13:11:11	2133	-13510	090	P	-a	-0.4793	1.9884	0.9688	328.0	195.3	-	2N	169E
7007	351	0908 Mar 20	20:48:35	2130	-13504	095	T	p-	-0.3853	2.1730	1.1291	346.8	210.5	55.5	2S	58E
7008	351	0908 Sep 13	02:02:20	2126	-13498	100	T+	p-	0.2550	2.3741	1.4057	319.1	206.5	84.0	2S	23W
7009	351	0909 Mar 09	23:10:48	2123	-13492	105	N	t-	-1.1411	0.8073	-0.2782	260.3	-	-	1N	23E
7010	351	0909 Sep 02	18:27:38	2120	-13486	110	P	a-	0.9507	1.0904	0.1360	251.2	78.9	-	5S	91E
7011	351	0910 Jan 28	06:46:30	2117	-13481	077	N	-t	1.1043	0.8636	-0.1998	257.8	-	-	18N	88W
7012	351	0910 Jul 24	23:45:48	2114	-13475	082	P	-a	-0.9284	1.1568	0.1520	275.9	88.4	-	20S	14E
7013	351	0911 Jan 17	14:58:46	2110	-13469	087	T	-p	0.3551	2.2106	1.2022	327.5	203.9	64.8	20N	148E
7014	351	0911 Jul 14	06:14:57	2107	-13463	092	T-	pp	-0.1816	2.5522	1.4978	364.4	228.9	96.8	21S	84W
7015	351	0912 Jan 07	05:02:56	2104	-13457	097	T	p-	-0.3334	2.2324	1.2599	313.1	199.4	70.0	22N	64W
7016	351	0912 Jul 02	07:12:37	2100	-13451	102	P	t-	0.5831	1.8231	0.7535	348.0	191.2	-	22S	99W
7017	351	0912 Dec 26	20:43:53	2097	-13445	107	P	a-	-1.0012	1.0109	0.0305	247.3	38.2	-	22N	59E
7018	351	0913 May 22	22:02:04	2094	-13440	074	N	-t	-1.3904	0.3118	-0.6985	157.6	-	-	23S	37E
7019	351	0913 Jun 21	09:33:25	2094	-13439	112	N	t-	1.3161	0.4617	-0.5752	195.4	-	-	22S	135W
7020	351	0913 Nov 16	18:32:25	2091	-13434	079	N	-a	1.1181	0.8354	-0.2225	251.5	-	-	21N	87E

Cat Num	Canon Plate	Calendar Date	TD of Greatest Eclipse	ΔT s	Luna Num	Saros Num	Ecl. Type	QSE	Gamma	Pen. Mag.	Um. Mag.	Eclipse Phase Durations Pen. m	Par. m	Total m	Greatest in Zenith Lat.	Lng.
7021	352	0914 May 12	10:31:56	2088	-13428	084	P	-a	-0.5808	1.7724	0.8115	301.3	177.0	-	20S	151W
7022	352	0914 Nov 05	20:44:43	2085	-13422	089	P	-t	0.4561	2.0708	0.9720	361.0	207.9	-	18N	54E
7023	352	0915 May 02	03:13:23	2081	-13416	094	T+	p-	0.1613	2.5334	1.5900	315.5	208.8	93.6	16S	41W
7024	352	0915 Oct 25	19:58:25	2078	-13410	099	T-	p-	-0.2320	2.4781	1.3874	370.0	227.0	89.2	14N	65E
7025	352	0916 Apr 20	19:32:10	2075	-13404	104	P	a-	0.9090	1.1749	0.2049	265.4	98.2	-	12S	75E
7026	352	0916 Oct 13	23:48:01	2072	-13398	109	P	a-	-0.9178	1.1935	0.1546	284.8	90.1	-	9N	8E
7027	352	0917 Mar 11	17:18:47	2069	-13393	076	N	-t	-1.0729	0.9202	-0.1412	267.0	-	-	1N	111E
7028	352	0917 Sep 04	01:40:03	2066	-13387	081	N	-a	1.0648	0.8829	-0.0751	232.0	-	-	5S	18W
7029	352	0918 Feb 28	18:09:14	2062	-13381	086	T	-t	-0.3721	2.2212	1.1301	366.9	218.3	57.4	6N	99E
7030	352	0918 Aug 24	18:19:51	2059	-13375	091	T	-p	0.3626	2.1710	1.2138	311.7	198.7	65.0	9S	93E
7031	352	0919 Feb 17	18:40:56	2056	-13369	096	T	t-	0.3425	2.2647	1.1949	357.5	216.7	67.4	11N	92E
7032	352	0919 Aug 14	08:35:22	2053	-13363	101	T	p-	-0.3793	2.1603	1.1637	328.2	204.3	59.8	14S	120W
7033	352	0920 Feb 07	01:22:57	2050	-13357	106	N	a-	1.0279	0.9805	-0.0366	257.2	-	-	15N	8W
7034	352	0920 Aug 02	16:29:26	2046	-13351	111	N	t-	-1.1794	0.7184	-0.3302	239.7	-	-	18S	122E
7035	352	0920 Dec 28	04:12:44	2044	-13346	078	P	-a	-1.0135	0.9833	0.0129	241.7	24.6	-	22N	53W
7036	352	0921 Jun 23	03:00:54	2040	-13340	083	P	-t	0.8972	1.2440	0.1798	306.0	101.8	-	23S	37W
7037	352	0921 Dec 17	19:27:42	2037	-13334	088	T	-p	-0.3639	2.1859	1.1944	318.9	199.8	62.7	23N	77E
7038	352	0922 Jun 12	06:58:18	2034	-13328	093	T+	pp	0.1026	2.6807	1.6589	351.9	225.9	101.5	23S	97W
7039	352	0922 Dec 07	06:23:41	2031	-13322	098	T	p-	0.3348	2.2664	1.2210	343.2	210.9	69.0	24N	89W
7040	352	0923 Jun 01	18:07:03	2028	-13316	103	P	a-	-0.6630	1.6256	0.6571	297.3	165.0	-	23S	96E
7041	353	0923 Nov 26	10:04:30	2024	-13310	108	N	t-	1.0602	0.9605	-0.1346	278.0	-	-	23N	145W
7042	353	0924 Apr 22	03:28:25	2022	-13305	075	N	-a	1.1338	0.7511	-0.1967	216.1	-	-	13S	44W
7043	353	0924 May 21	10:22:51	2021	-13304	113	N	a-	-1.3660	0.3217	-0.6195	146.9	-	-	23S	149W
7044	353	0924 Oct 15	15:10:51	2019	-13299	080	N	-t	-1.2689	0.5686	-0.5086	219.3	-	-	10N	138E
7045	353	0925 Apr 11	18:46:56	2015	-13293	085	T	-a	0.4300	2.0616	1.0760	320.8	197.9	41.5	10S	87E
7046	353	0925 Oct 04	21:12:36	2012	-13287	090	P	-a	-0.5078	1.9360	0.9165	324.9	191.2	-	6N	47E
7047	353	0926 Apr 01	04:06:53	2009	-13281	095	T	p-	-0.3309	2.2730	1.2291	350.8	215.9	71.7	7S	53W
7048	353	0926 Sep 24	10:13:22	2006	-13275	100	T+	p-	0.2190	2.4406	1.4711	319.8	207.9	88.1	2N	147W
7049	353	0927 Mar 21	06:19:21	2003	-13269	105	N	t-	-1.0915	0.8970	-0.1860	271.9	-	-	3S	85W
7050	353	0927 Sep 14	02:32:30	1999	-13263	110	P	a-	0.9095	1.1678	0.2098	258.1	97.1	-	1S	32W
7051	353	0928 Feb 08	14:45:03	1997	-13258	077	N	-t	1.1216	0.8291	-0.2287	252.4	-	-	14N	152E
7052	353	0928 Aug 04	06:58:24	1994	-13252	082	P	-a	-0.9982	1.0315	0.0212	265.4	33.7	-	17S	95W
7053	353	0929 Jan 27	23:27:27	1990	-13246	087	T	-p	0.3661	2.1878	1.1844	325.5	202.7	62.2	17N	21E
7054	353	0929 Jul 24	12:54:35	1987	-13240	092	T-	pp	-0.2611	2.4085	1.3498	363.1	225.1	86.3	19S	176E
7055	353	0930 Jan 17	13:48:41	1984	-13234	097	T	p-	-0.3250	2.2465	1.2764	313.0	199.9	71.8	19N	165E
7056	353	0930 Jul 13	13:40:18	1981	-13228	102	P	t-	0.4989	1.9780	0.9077	355.2	204.1	-	21S	164E
7057	353	0931 Jan 07	05:30:34	1978	-13222	107	P	a-	-0.9976	1.0178	0.0368	248.4	42.0	-	21N	72W
7058	353	0931 Jun 03	05:00:15	1975	-13217	074	N	-t	-1.4606	0.1810	-0.8252	121.0	-	-	24S	68W
7059	353	0931 Jul 02	16:18:07	1975	-13216	112	N	t-	1.2335	0.6120	-0.4225	220.9	-	-	22S	124E
7060	353	0931 Nov 28	02:51:41	1972	-13211	079	N	-a	1.1266	0.8219	-0.2400	250.7	-	-	23N	37W
7061	354	0932 May 22	17:55:33	1969	-13205	084	P	-a	-0.6506	1.6426	0.6853	293.9	165.8	-	22S	98E
7062	354	0932 Nov 16	04:40:50	1966	-13199	089	P	-t	0.4684	2.0498	0.9481	360.8	206.5	-	21N	65W
7063	354	0933 May 12	10:47:29	1963	-13193	094	T+	pp	0.0929	2.6583	1.7161	316.6	210.7	97.6	19S	155W
7064	354	0933 Nov 05	03:53:12	1959	-13187	099	T-	p-	-0.2162	2.5072	1.4160	369.9	227.5	91.2	17N	54W
7065	354	0934 May 02	03:00:10	1956	-13181	104	P	a-	0.8463	1.2903	0.3196	275.4	121.0	-	16S	38W
7066	354	0934 Oct 25	07:58:55	1953	-13175	109	P	a-	-0.8978	1.2298	0.1915	287.0	99.4	-	13N	115W
7067	354	0935 Mar 23	00:37:35	1951	-13170	076	N	-t	-1.1244	0.8255	-0.2353	256.0	-	-	4S	0W
7068	354	0935 Sep 15	09:46:31	1947	-13164	081	N	-a	1.1025	0.8147	-0.1450	224.3	-	-	0S	141W
7069	354	0935 Oct 14	19:25:58	1947	-13163	119	N	a-	-1.5283	0.0450	-0.9379	57.9	-	-	9N	73E
7070	354	0936 Mar 11	01:23:54	1944	-13158	086	T	-t	-0.4167	2.1375	1.0498	363.8	213.8	36.4	1N	11W
7071	354	0936 Sep 04	02:16:16	1941	-13152	091	T	-a	0.4102	2.0859	1.1243	310.0	194.8	51.0	5S	27W
7072	354	0937 Feb 28	02:18:08	1938	-13146	096	T	p-	0.3079	2.3251	1.2612	357.6	219.2	76.1	6N	23W
7073	354	0937 Aug 24	16:05:45	1935	-13140	101	T	p-	-0.3243	2.2643	1.2616	332.7	209.2	73.3	10S	127E
7074	354	0938 Feb 17	09:33:20	1932	-13134	106	P	a-	1.0023	1.0244	0.0133	260.6	26.3	-	11N	131W
7075	354	0938 Aug 13	23:26:21	1929	-13128	111	N	t-	-1.1163	0.8370	-0.2172	256.7	-	-	14S	17E
7076	354	0939 Jan 08	13:02:53	1926	-13123	078	P	-a	-1.0187	0.9733	0.0041	240.6	13.8	-	21N	175E
7077	354	0939 Jul 04	09:28:30	1923	-13117	083	P	-t	0.9807	1.0905	0.0269	290.9	40.2	-	22S	133W
7078	354	0939 Dec 29	04:14:47	1920	-13111	088	T	-p	-0.3669	2.1811	1.1881	319.6	200.0	61.9	23N	54W
7079	354	0940 Jun 22	13:48:38	1917	-13105	093	T+	pp	0.1834	2.5307	1.5122	348.7	222.6	95.4	23S	161E
7080	354	0940 Dec 17	14:51:55	1914	-13099	098	T	p-	0.3331	2.2712	1.2227	344.6	211.6	69.4	24N	145E

Cat Num	Canon Plate	Calendar Date	TD of Greatest Eclipse	ΔT	Luna Num	Saros Num	Ecl. Type	QSE	Gamma	Pen. Mag.	Um. Mag.	Eclipse Phase ---- Durations ---- Pen.	Par.	Total	Greatest in Zenith Lat.	Lng.
				s								m	m	m		
7081	355	0941 Jun 12	01:23:11	1911	-13093	103	P	a-	-0.5857	1.7657	0.8003	303.4	177.3	-	24S	13W
7082	355	0941 Dec 06	18:10:10	1908	-13087	108	N	t-	1.0551	0.9711	-0.1263	279.8	-	-	24N	94E
7083	355	0942 May 03	11:05:37	1905	-13082	075	N	-a	1.1993	0.6305	-0.3165	200.6	-	-	16S	159W
7084	355	0942 Jun 01	17:51:08	1905	-13081	113	N	a-	-1.2946	0.4523	-0.4880	172.0	-	-	24S	99E
7085	355	0942 Oct 26	23:04:38	1902	-13076	080	N	-t	-1.2864	0.5366	-0.5409	213.2	-	-	13N	18E
7086	355	0943 Apr 23	02:16:28	1899	-13070	085	P	-a	0.4936	1.9454	0.9590	317.6	191.5	-	13S	27W
7087	355	0943 Oct 16	05:22:35	1896	-13064	090	P	-a	-0.5302	1.8947	0.8756	322.1	187.7	-	10N	76W
7088	355	0944 Apr 11	11:16:37	1893	-13058	095	T-	pp	-0.2701	2.3847	1.3406	354.6	221.0	84.2	10S	162W
7089	355	0944 Oct 04	18:32:45	1890	-13052	100	T+	p-	0.1896	2.4952	1.5246	320.1	208.8	90.8	7N	86E
7090	355	0945 Mar 31	13:17:13	1887	-13046	105	N*	t-	-1.0333	1.0026	-0.0779	284.3	-	-	7S	169E
7091	355	0945 Sep 24	10:45:43	1884	-13040	110	P	a-	0.8754	1.2324	0.2706	263.7	109.4	-	3N	156W
7092	355	0946 Feb 18	22:35:45	1881	-13035	077	N	-t	1.1459	0.7814	-0.2704	245.2	-	-	11N	33E
7093	355	0946 Aug 15	14:16:45	1878	-13029	082	N	-a	-1.0624	0.9168	-0.0996	254.4	-	-	14S	155E
7094	355	0947 Feb 08	07:48:45	1875	-13023	087	T	-p	0.3842	2.1521	1.1539	323.2	200.8	57.3	14N	105W
7095	355	0947 Aug 04	19:39:58	1872	-13017	092	T	-t	-0.3354	2.2745	1.2112	361.1	219.9	70.6	16S	74E
7096	355	0948 Jan 28	22:29:43	1869	-13011	097	T	p-	-0.3123	2.2684	1.3011	313.1	200.8	74.3	17N	34E
7097	355	0948 Jul 23	20:11:39	1866	-13005	102	T	t-	0.4170	2.1288	1.0574	361.0	214.0	39.0	19S	66E
7098	355	0949 Jan 17	14:14:23	1863	-12999	107	P	a-	-0.9921	1.0280	0.0470	249.8	47.5	-	19N	157E
7099	355	0949 Jun 13	11:57:25	1860	-12994	074	Ne	-t	-1.5324	0.0473	-0.9551	62.3	-	-	25S	172W
7100	355	0949 Jul 12	23:05:38	1860	-12993	112	N	t-	1.1515	0.7615	-0.2711	241.9	-	-	20S	22E
7101	356	0949 Dec 08	11:13:05	1857	-12988	079	N	-a	1.1325	0.8129	-0.2524	250.5	-	-	24N	161W
7102	356	0950 Jun 03	01:16:29	1854	-12982	084	P	-a	-0.7231	1.5081	0.5536	285.4	151.8	-	24S	12W
7103	356	0950 Nov 27	12:41:45	1851	-12976	089	P	-t	0.4766	2.0359	0.9321	360.8	205.5	-	23N	175E
7104	356	0951 May 23	18:18:29	1848	-12970	094	T+	pp	0.0221	2.7879	1.8464	317.2	211.6	99.4	21S	92E
7105	356	0951 Nov 16	11:53:47	1845	-12964	099	T-	p-	-0.2046	2.5286	1.4374	369.7	227.7	92.6	20N	174W
7106	356	0952 May 12	10:23:59	1842	-12958	104	P	a-	0.7807	1.4113	0.4394	284.9	139.8	-	19S	150W
7107	356	0952 Nov 04	16:16:46	1839	-12952	109	P	a-	-0.8826	1.2573	0.2199	288.4	105.8	-	16N	120E
7108	356	0953 Apr 02	07:47:11	1837	-12947	076	N	-t	-1.1824	0.7187	-0.3415	241.9	-	-	8S	109W
7109	356	0953 Sep 25	18:00:31	1834	-12941	081	N	-a	1.1343	0.7571	-0.2045	217.3	-	-	4N	95E
7110	356	0953 Oct 25	03:56:18	1833	-12940	119	N	a-	-1.5085	0.0814	-0.9018	77.4	-	-	12N	55W
7111	356	0954 Mar 22	08:29:52	1831	-12935	086	P	-t	-0.4686	2.0404	0.9564	359.6	207.7	-	3S	119W
7112	356	0954 Sep 15	10:20:13	1828	-12929	091	T	-a	0.4511	2.0131	1.0471	308.4	190.9	32.1	1S	149W
7113	356	0955 Mar 11	09:47:24	1825	-12923	096	T+	p-	0.2660	2.3990	1.3412	357.8	221.7	84.3	2N	137W
7114	356	0955 Sep 04	23:43:41	1822	-12917	101	T	p-	-0.2763	2.3555	1.3465	336.5	212.9	82.0	6S	11E
7115	356	0956 Feb 28	17:36:42	1819	-12911	106	P	a-	0.9704	1.0799	0.0748	264.8	61.7	-	7N	107E
7116	356	0956 Aug 24	06:28:59	1816	-12905	111	N	t-	-1.0582	0.9463	-0.1135	271.0	-	-	10S	89W
7117	356	0957 Jan 18	21:48:41	1813	-12900	078	N	-a	-1.0274	0.9564	-0.0112	238.9	-	-	18N	44E
7118	356	0957 Jul 14	15:59:57	1810	-12894	083	N	-t	1.0609	0.9432	-0.1201	274.3	-	-	20S	128E
7119	356	0958 Jan 08	12:57:28	1807	-12888	088	T	-p	-0.3733	2.1700	1.1760	320.1	199.8	60.2	21N	176E
7120	356	0958 Jul 03	20:42:11	1804	-12882	093	T+	pp	0.2625	2.3844	1.3684	344.6	217.9	85.5	22S	58E
7121	357	0958 Dec 28	23:17:44	1801	-12876	098	T	p-	0.3302	2.2777	1.2269	346.0	212.3	70.1	23N	20E
7122	357	0959 Jun 23	08:40:46	1798	-12870	103	P	a-	-0.5088	1.9056	0.9426	308.4	187.2	-	24S	123W
7123	357	0959 Dec 18	02:15:57	1795	-12864	108	N	t-	1.0504	0.9804	-0.1185	281.4	-	-	25N	26W
7124	357	0960 May 13	18:37:52	1793	-12859	075	N	-a	1.2685	0.5034	-0.4433	181.7	-	-	19S	87E
7125	357	0960 Jun 12	01:18:14	1792	-12858	113	N	a-	-1.2222	0.5849	-0.3550	193.0	-	-	25S	13W
7126	357	0960 Nov 06	07:06:47	1790	-12853	080	N	-t	-1.2978	0.5156	-0.5617	209.0	-	-	17N	102W
7127	357	0961 May 03	09:36:58	1787	-12847	085	P	-a	0.5637	1.8173	0.8297	313.3	182.8	-	17S	138W
7128	357	0961 Oct 26	13:41:49	1784	-12841	090	P	-a	-0.5457	1.8660	0.8474	319.8	185.0	-	14N	158E
7129	357	0962 Apr 22	18:16:53	1781	-12835	095	T-	pp	-0.2025	2.5089	1.4644	358.0	225.4	93.9	14S	92E
7130	357	0962 Oct 16	03:00:36	1778	-12829	100	T+	p-	0.1670	2.5372	1.5655	320.3	209.2	92.5	11N	41W
7131	357	0963 Apr 11	20:07:31	1775	-12823	105	P	t-	-0.9689	1.1195	0.0416	296.7	50.2	-	11S	65E
7132	357	0963 Oct 05	19:06:28	1772	-12817	110	P	a-	0.8480	1.2846	0.3190	268.1	118.0	-	8N	77E
7133	357	0964 Mar 01	06:21:13	1770	-12812	077	N	-t	1.1752	0.7245	-0.3210	236.4	-	-	6N	84W
7134	357	0964 Aug 25	21:39:04	1767	-12806	082	N	-a	-1.1223	0.8101	-0.2126	243.0	-	-	10S	43E
7135	357	0965 Feb 18	16:05:35	1764	-12800	087	T	-p	0.4066	2.1081	1.1155	320.5	198.4	50.1	10N	130E
7136	357	0965 Aug 15	02:30:04	1761	-12794	092	T	-t	-0.4053	2.1487	1.0806	358.4	213.6	45.6	13S	29W
7137	357	0966 Feb 08	07:04:04	1758	-12788	097	T	p-	-0.2940	2.3006	1.3362	313.4	202.0	77.5	13N	95W
7138	357	0966 Aug 04	02:50:13	1755	-12782	102	T	t-	0.3406	2.2697	1.1969	365.3	221.2	69.0	16S	34W
7139	357	0967 Jan 28	22:51:32	1752	-12776	107	P	a-	-0.9814	1.0474	0.0666	252.1	56.5	-	16N	28E
7140	357	0967 Jul 24	06:00:42	1749	-12770	112	N	t-	1.0745	0.9020	-0.1291	258.7	-	-	18S	82W

Cat Num	Canon Plate	Calendar Date	TD of Greatest Eclipse	ΔT s	Luna Num	Saros Num	Ecl. Type	QSE	Gamma	Pen. Mag.	Um. Mag.	Eclipse Phase Durations Pen. m	Par. m	Total m	Greatest in Zenith Lat.	Lng.
7141	358	0967 Dec 19	19:32:11	1747	-12765	079	N	-a	1.1396	0.8011	-0.2669	249.8	-	-	25N	75E
7142	358	0968 Jun 13	08:38:46	1744	-12759	084	P	-a	-0.7952	1.3748	0.4226	276.1	135.1	-	24S	123W
7143	358	0968 Dec 07	20:43:51	1741	-12753	089	P	-t	0.4836	2.0237	0.9185	360.7	204.6	-	24N	55E
7144	358	0969 Jun 03	01:45:37	1738	-12747	094	T-	pp	-0.0520	2.7330	1.7915	317.3	211.6	99.0	23S	20W
7145	358	0969 Nov 26	19:59:45	1735	-12741	099	T-	p-	-0.1968	2.5426	1.4521	369.1	227.7	93.4	22N	65E
7146	358	0970 May 23	17:41:28	1733	-12735	104	P	a-	0.7102	1.5416	0.5678	294.2	156.2	-	21S	101E
7147	358	0970 Nov 16	00:41:26	1730	-12729	109	P	a-	-0.8722	1.2758	0.2397	289.0	109.8	-	19N	6W
7148	358	0971 Apr 13	14:46:11	1727	-12724	076	N	-t	-1.2481	0.5979	-0.4619	223.8	-	-	12S	145E
7149	358	0971 May 13	04:03:35	1727	-12723	114	Nb	t-	1.5318	0.0619	-0.9674	73.4	-	-	18S	55W
7150	358	0971 Oct 07	02:23:49	1724	-12718	081	N	-a	1.1592	0.7124	-0.2511	211.7	-	-	8N	32W
7151	358	0971 Nov 05	12:34:29	1724	-12717	119	N	a-	-1.4945	0.1072	-0.8761	88.5	-	-	16N	175E
7152	358	0972 Apr 01	15:26:56	1721	-12712	086	P	-t	-0.5277	1.9302	0.8499	354.4	199.6	-	7S	136E
7153	358	0972 Sep 25	18:30:55	1719	-12706	091	P	-a	0.4860	1.9515	0.9807	307.1	187.3	-	4N	87E
7154	358	0973 Mar 21	17:09:47	1716	-12700	096	T+	p-	0.2176	2.4846	1.4330	357.8	224.1	91.5	2S	111E
7155	358	0973 Sep 15	07:27:56	1713	-12694	101	T-	p-	-0.2346	2.4353	1.4199	339.7	215.8	87.9	1S	107W
7156	358	0974 Mar 11	01:32:38	1710	-12688	106	P	a-	0.9317	1.1478	0.1489	269.7	85.9	-	3N	13W
7157	358	0974 Sep 04	13:38:51	1707	-12682	111	N*	t-	-1.0065	1.0441	-0.0215	282.8	-	-	6S	162E
7158	358	0975 Jan 30	06:30:05	1705	-12677	078	N	-a	-1.0398	0.9327	-0.0330	236.4	-	-	15N	87W
7159	358	0975 Jul 25	22:37:10	1702	-12671	083	N	-t	1.1367	0.8042	-0.2591	256.4	-	-	18S	29E
7160	358	0976 Jan 19	21:36:18	1699	-12665	088	T	-a	-0.3823	2.1539	1.1591	320.4	199.5	57.7	19N	47E
7161	359	0976 Jul 14	03:41:09	1696	-12659	093	T	-p	0.3379	2.2450	1.2312	339.9	212.0	70.9	21S	47W
7162	359	0977 Jan 08	07:40:11	1693	-12653	098	T	p-	0.3252	2.2877	1.2352	347.5	213.3	71.4	22N	105W
7163	359	0977 Jul 03	16:00:39	1691	-12647	103	T	a-	-0.4331	2.0437	1.0824	312.5	195.1	42.6	23S	128E
7164	359	0977 Dec 28	10:20:09	1688	-12641	108	N	t-	1.0448	0.9910	-0.1086	283.0	-	-	24N	146W
7165	359	0978 May 25	02:06:44	1685	-12636	075	N	-a	1.3399	0.3725	-0.5745	158.5	-	-	20S	25W
7166	359	0978 Jun 23	08:46:20	1685	-12635	113	N	a-	-1.1506	0.7164	-0.2237	210.8	-	-	25S	124W
7167	359	0978 Nov 17	15:14:24	1682	-12630	080	N	-t	-1.3051	0.5017	-0.5748	205.9	-	-	19N	136E
7168	359	0979 May 14	16:52:28	1680	-12624	085	P	-a	0.6369	1.6838	0.6945	307.8	171.7	-	19S	113E
7169	359	0979 Nov 06	22:08:13	1677	-12618	090	P	-a	-0.5557	1.8473	0.8296	318.0	183.1	-	17N	32E
7170	359	0980 May 03	01:09:21	1674	-12612	095	T-	pp	-0.1291	2.6438	1.5985	360.9	228.8	100.7	17S	12W
7171	359	0980 Oct 26	11:35:12	1671	-12606	100	T+	p-	0.1498	2.5694	1.5967	320.3	209.5	93.6	15N	171W
7172	359	0981 Apr 22	02:48:48	1668	-12600	105	P	t-	-0.8971	1.2500	0.1745	309.0	100.9	-	15S	36W
7173	359	0981 Oct 16	03:34:43	1666	-12594	110	P	a-	0.8268	1.3253	0.3560	271.6	124.1	-	12N	51W
7174	359	0982 Mar 12	13:56:56	1663	-12589	077	N	-t	1.2130	0.6518	-0.3873	224.8	-	-	2N	160E
7175	359	0982 Sep 06	05:09:09	1660	-12583	082	N	-a	-1.1749	0.7168	-0.3124	231.8	-	-	6S	71W
7176	359	0983 Mar 02	00:14:18	1658	-12577	087	T	-p	0.4367	2.0502	1.0631	317.3	195.2	37.6	6N	7E
7177	359	0983 Aug 26	09:26:42	1655	-12571	092	P	-t	-0.4691	2.0340	0.9610	355.3	206.4	-	9S	135W
7178	359	0984 Feb 19	15:31:52	1652	-12565	097	T-	p-	-0.2700	2.3430	1.3817	313.9	203.4	81.3	9N	137E
7179	359	0984 Aug 14	09:35:35	1649	-12559	102	T+	pp	0.2692	2.4015	1.3272	368.4	226.3	84.7	13S	137W
7180	359	0985 Feb 08	07:22:35	1646	-12553	107	P	a-	-0.9665	1.0746	0.0943	255.1	67.1	-	12N	100W
7181	360	0985 Aug 03	13:01:16	1644	-12547	112	P	t-	1.0007	1.0368	0.0068	272.5	19.5	-	15S	172E
7182	360	0985 Dec 30	03:50:38	1641	-12542	079	N	-h	1.1466	0.7892	-0.2807	249.0	-	-	24N	49W
7183	360	0986 Jun 24	16:01:32	1639	-12536	084	P	-a	-0.8673	1.2416	0.2910	265.8	114.0	-	24S	127E
7184	360	0986 Dec 19	04:46:27	1636	-12530	089	P	-t	0.4900	2.0120	0.9064	360.4	203.9	-	24N	64W
7185	360	0987 Jun 14	09:12:40	1633	-12524	094	T-	-p	-0.1261	2.5975	1.6553	316.7	210.4	96.2	24S	132W
7186	360	0987 Dec 08	04:09:24	1630	-12518	099	T-	p-	-0.1911	2.5523	1.4632	368.4	227.6	93.9	23N	56W
7187	360	0988 Jun 03	00:55:42	1627	-12512	104	P	a-	0.6377	1.6757	0.6996	302.9	170.3	-	22S	8W
7188	360	0988 Nov 26	09:10:40	1625	-12506	109	P	a-	-0.8647	1.2885	0.2543	289.0	112.5	-	21N	133W
7189	360	0989 Apr 23	21:37:46	1622	-12501	076	N	-t	-1.3188	0.4681	-0.5914	200.9	-	-	16S	41E
7190	360	0989 May 23	10:53:23	1622	-12500	114	N	t-	1.4607	0.1935	-0.8379	128.7	-	-	20S	158W
7191	360	0989 Oct 17	10:53:56	1620	-12495	081	N	-a	1.1786	0.6779	-0.2876	207.1	-	-	13N	161W
7192	360	0989 Nov 15	21:17:18	1619	-12494	119	N	a-	-1.4842	0.1259	-0.8571	95.6	-	-	19N	45E
7193	360	0990 Apr 12	22:15:53	1617	-12489	086	P	-t	-0.5933	1.8081	0.7313	347.8	189.0	-	11S	32E
7194	360	0990 Oct 07	02:49:28	1614	-12483	091	P	-a	0.5138	1.9028	0.9274	306.1	184.2	-	8N	39W
7195	360	0991 Apr 02	00:25:28	1611	-12477	096	T+	pp	0.1630	2.5819	1.5363	357.6	226.0	97.4	7S	1E
7196	360	0991 Sep 26	15:18:46	1609	-12471	101	T-	pp	-0.1991	2.5035	1.4818	342.6	218.0	92.0	3N	134E
7197	360	0992 Mar 21	09:21:55	1606	-12465	106	P	a-	0.8867	1.2273	0.2344	275.3	106.2	-	2S	132W
7198	360	0992 Sep 14	20:56:23	1603	-12459	111	P	t-	-0.9615	1.1297	0.0584	292.6	58.4	-	2S	51E
7199	360	0993 Feb 09	15:03:57	1601	-12454	078	N	-a	-1.0584	0.8974	-0.0662	232.8	-	-	12N	144E
7200	360	0993 Mar 11	00:08:28	1600	-12453	116	Nb	a-	1.5408	0.0098	-0.9489	26.5	-	-	3N	7E

Cat Num	Canon Plate	Calendar Date	TD of Greatest Eclipse	ΔT s	Luna Num	Saros Num	Ecl. Type	QSE	Gamma	Pen. Mag.	Um. Mag.	Eclipse Phase ---- Durations ---- Pen. m	Par. m	Total m	Greatest in Zenith Lat.	Lng.
7201	361	0993 Aug 05	05:21:17	1598	-12448	083	N	-t	1.2070	0.6753	-0.3885	237.6	-	-	15S	73W
7202	361	0994 Jan 30	06:07:59	1595	-12442	088	T	-a	-0.3967	2.1275	1.1325	320.4	198.7	53.2	16N	81W
7203	361	0994 Jul 25	10:46:43	1593	-12436	093	T	-t	0.4086	2.1143	1.1021	334.7	205.1	49.0	18S	154W
7204	361	0995 Jan 19	15:56:11	1590	-12430	098	T	p-	0.3157	2.3058	1.2522	349.0	214.6	73.7	19N	131E
7205	361	0995 Jul 14	23:25:10	1587	-12424	103	T	a-	-0.3607	2.1760	1.2159	315.7	201.0	66.0	21S	16E
7206	361	0996 Jan 08	18:21:43	1585	-12418	108	N*	t-	1.0376	1.0043	-0.0952	284.9	-	-	22N	94E
7207	361	0996 Jun 04	09:31:24	1582	-12413	075	N	-a	1.4143	0.2367	-0.7114	128.2	-	-	22S	137W
7208	361	0996 Jul 03	16:14:11	1582	-12412	113	N	a-	-1.0788	0.8487	-0.0924	226.4	-	-	24S	124E
7209	361	0996 Nov 27	23:27:59	1580	-12407	080	N	-t	-1.3084	0.4950	-0.5800	204.0	-	-	21N	13E
7210	361	0997 May 25	00:01:22	1577	-12401	085	P	-a	0.7145	1.5426	0.5510	301.0	157.0	-	21S	5E
7211	361	0997 Nov 17	06:41:06	1574	-12395	090	P	-a	-0.5614	1.8362	0.8198	316.5	181.8	-	20N	97W
7212	361	0998 May 14	07:54:10	1571	-12389	095	T-	pp	-0.0501	2.7894	1.7430	363.1	231.1	104.7	20S	113W
7213	361	0998 Nov 06	20:16:42	1569	-12383	100	T+	p-	0.1379	2.5915	1.6181	320.2	209.5	94.2	18N	59E
7214	361	0999 May 03	09:24:58	1566	-12377	105	P	t-	-0.8206	1.3892	0.3159	320.6	132.9	-	18S	136W
7215	361	0999 Oct 27	12:08:14	1563	-12371	110	P	a-	0.8103	1.3574	0.3844	274.4	128.6	-	15N	180W
7216	361	1000 Mar 22	21:27:53	1561	-12366	077	N	-t	1.2555	0.5707	-0.4620	211.0	-	-	2S	46E
7217	361	1000 Apr 21	10:47:39	1561	-12365	115	Nb	t-	-1.5354	0.0665	-0.9847	77.8	-	-	15S	157W
7218	361	1000 Sep 16	12:44:57	1558	-12360	082	N	-a	-1.2220	0.6336	-0.4021	220.8	-	-	1S	174E
7219	361	1000 Oct 16	01:35:42	1558	-12359	120	N	a-	1.5463	0.0276	-0.9862	47.1	-	-	12N	22W
7220	361	1001 Mar 12	08:16:18	1556	-12354	087	P	-a	0.4728	1.9809	0.9995	313.6	191.1	-	2N	115W
7221	362	1001 Sep 05	16:30:33	1553	-12348	092	P	-t	-0.5266	1.9311	0.8531	351.9	198.6	-	5S	118E
7222	362	1002 Mar 01	23:51:54	1550	-12342	097	T-	p-	-0.2396	2.3972	1.4390	314.6	205.1	85.4	5N	11E
7223	362	1002 Aug 25	16:30:31	1548	-12336	102	T+	pp	0.2051	2.5200	1.4440	370.4	229.7	94.0	9S	119E
7224	362	1003 Feb 19	15:44:36	1545	-12330	107	P	a-	-0.9446	1.1145	0.1347	259.2	79.9	-	8N	133E
7225	362	1003 Aug 14	20:11:58	1542	-12324	112	P	h-	0.9340	1.1589	0.1296	283.3	83.4	-	12S	64E
7226	362	1004 Jan 10	12:03:58	1540	-12319	079	N	-h	1.1574	0.7701	-0.3012	247.1	-	-	22N	171W
7227	362	1004 Jul 04	23:26:40	1538	-12313	084	P	-a	-0.9379	1.1116	0.1619	254.6	86.5	-	23S	16E
7228	362	1004 Dec 29	12:47:12	1535	-12307	089	P	-t	0.4980	1.9973	0.8920	360.0	202.9	-	23N	177E
7229	362	1005 Jun 24	16:38:42	1532	-12301	094	T-	-p	-0.2011	2.4605	1.5170	315.5	208.1	90.6	24S	117E
7230	362	1005 Dec 18	12:19:45	1530	-12295	099	T-	p-	-0.1856	2.5612	1.4745	367.6	227.4	94.5	23N	178W
7231	362	1006 Jun 14	08:06:25	1527	-12289	104	P	a-	0.5629	1.8145	0.8354	310.9	182.4	-	23S	116W
7232	362	1006 Dec 07	17:43:42	1524	-12283	109	P	a-	-0.8595	1.2968	0.2651	288.6	114.4	-	22N	99E
7233	362	1007 May 05	04:21:00	1522	-12278	076	N	-t	-1.3951	0.3281	-0.7315	170.8	-	-	19S	60W
7234	362	1007 Jun 03	17:38:22	1522	-12277	114	N	t-	1.3862	0.3314	-0.7024	167.0	-	-	22S	101E
7235	362	1007 Oct 28	19:31:01	1520	-12272	081	N	-a	1.1928	0.6527	-0.3145	203.7	-	-	16N	69E
7236	362	1007 Nov 27	06:04:31	1519	-12271	119	N	a-	-1.4772	0.1385	-0.8438	99.9	-	-	21N	87W
7237	362	1008 Apr 23	04:57:48	1517	-12266	086	P	-t	-0.6646	1.6754	0.6021	339.7	175.1	-	15S	69W
7238	362	1008 Oct 17	11:14:53	1514	-12260	091	P	-a	0.5357	1.8649	0.8849	305.5	181.5	-	12N	166W
7239	362	1009 Apr 12	07:33:32	1512	-12254	096	T+	pp	0.1011	2.6924	1.6529	357.1	227.4	101.7	11S	107W
7240	362	1009 Oct 06	23:17:13	1509	-12248	101	T-	pp	-0.1707	2.5588	1.5308	345.1	219.7	94.7	7N	14E
7241	363	1010 Apr 01	17:04:25	1506	-12242	106	P	a-	0.8352	1.3188	0.3319	281.2	124.2	-	6S	111E
7242	363	1010 Sep 26	04:21:48	1504	-12236	111	P	t-	-0.9230	1.2030	0.1262	300.6	85.2	-	2N	61W
7243	363	1011 Feb 20	23:31:12	1502	-12231	078	N	-a	-1.0827	0.8518	-0.1096	227.8	-	-	8N	17E
7244	363	1011 Mar 22	08:13:07	1501	-12230	116	N	a-	1.5000	0.0829	-0.8721	76.6	-	-	1S	115W
7245	363	1011 Aug 16	12:13:51	1499	-12225	083	N	-t	1.2709	0.5584	-0.5060	218.1	-	-	11S	177W
7246	363	1012 Feb 10	14:34:00	1496	-12219	088	T	-a	-0.4153	2.0935	1.0984	320.2	197.5	46.5	12N	152E
7247	363	1012 Aug 04	17:58:16	1494	-12213	093	P	-t	0.4752	1.9914	0.9806	329.1	197.2	-	15S	97E
7248	363	1013 Jan 30	00:06:26	1491	-12207	098	T	p-	0.3023	2.3306	1.2763	350.7	216.2	76.7	16N	9E
7249	363	1013 Jul 25	06:54:14	1489	-12201	103	T	p-	-0.2916	2.3024	1.3429	318.0	205.5	79.5	19S	96W
7250	363	1014 Jan 19	02:16:49	1486	-12195	108	N*	t-	1.0254	1.0261	-0.0726	287.5	-	-	20N	24W
7251	363	1014 Jun 15	16:55:18	1484	-12190	075	Ne	-a	1.4887	0.1009	-0.8488	84.9	-	-	22S	112E
7252	363	1014 Jul 14	23:45:44	1483	-12189	113	P	a-	-1.0101	0.9755	0.0329	239.7	39.3	-	22S	11E
7253	363	1014 Dec 09	07:43:47	1481	-12184	080	N	-t	-1.3102	0.4906	-0.5822	202.6	-	-	22N	110W
7254	363	1015 Jun 05	07:07:30	1479	-12178	085	P	-a	0.7935	1.3991	0.4048	292.7	138.1	-	22S	101W
7255	363	1015 Nov 28	15:17:02	1476	-12172	090	P	-a	-0.5651	1.8284	0.8138	315.1	180.8	-	22N	135E
7256	363	1016 May 24	14:34:02	1474	-12166	095	T+	pp	0.0320	2.8235	1.7755	364.4	231.8	105.2	22S	146E
7257	363	1016 Nov 17	05:02:51	1471	-12160	100	T+	p-	0.1298	2.6066	1.6328	320.0	209.5	94.6	21N	73W
7258	363	1017 May 13	15:53:23	1468	-12154	105	P	t-	-0.7377	1.5403	0.4689	331.5	158.2	-	20S	126E
7259	363	1017 Nov 06	20:48:25	1466	-12148	110	P	a-	0.7998	1.3784	0.4021	276.4	131.3	-	19N	50E
7260	363	1018 Apr 03	04:50:06	1464	-12143	077	N	-t	1.3052	0.4762	-0.5501	193.6	-	-	6S	65W

Cat Num	Canon Plate	Calendar Date	TD of Greatest Eclipse	ΔT s	Luna Num	Saros Num	Ecl. Type	QSE	Gamma	Pen. Mag.	Um. Mag.	Eclipse Phase Durations Pen. m	Par. m	Total m	Greatest in Zenith Lat.	Lng.
7261	364	1018 May 02	17:36:53	1463	-12142	115	N	t-	-1.4601	0.2021	-0.8441	133.4	-	-	18S	100E
7262	364	1018 Sep 27	20:29:00	1461	-12137	082	N	-a	-1.2613	0.5647	-0.4774	210.8	-	-	3N	57E
7263	364	1018 Oct 27	09:53:57	1461	-12136	120	N	a-	1.5341	0.0527	-0.9666	65.2	-	-	16N	147W
7264	364	1019 Mar 23	16:10:23	1459	-12131	087	P	-a	0.5161	1.8988	0.9229	309.2	185.8	-	3S	125E
7265	364	1019 Sep 16	23:43:14	1456	-12125	092	P	-t	-0.5760	1.8429	0.7600	348.7	190.8	-	1S	9E
7266	364	1020 Mar 12	08:04:38	1454	-12119	097	T-	p-	-0.2030	2.4628	1.5077	315.4	206.8	89.5	1N	113W
7267	364	1020 Sep 04	23:33:25	1451	-12113	102	T+	pp	0.1470	2.6274	1.5496	371.6	231.8	99.7	5S	12E
7268	364	1021 Mar 01	23:59:24	1448	-12107	107	P	a-	-0.9175	1.1639	0.1848	264.0	93.1	-	4N	9E
7269	364	1021 Aug 25	03:30:44	1446	-12101	112	P	h-	0.8727	1.2712	0.2423	292.1	111.9	-	8S	47W
7270	364	1022 Jan 20	20:12:07	1444	-12096	079	N	-h	1.1715	0.7446	-0.3274	244.3	-	-	20N	67E
7271	364	1022 Jul 16	06:55:23	1441	-12090	084	P	-a	-1.0059	0.9866	0.0374	242.8	42.2	-	22S	96W
7272	364	1022 Aug 14	14:51:42	1441	-12089	122	Nb	a-	1.5452	0.0087	-0.9639	25.6	-	-	11S	143E
7273	364	1023 Jan 09	20:45:36	1439	-12084	089	P	-t	0.5079	1.9787	0.8744	359.1	201.6	-	22N	58E
7274	364	1023 Jul 06	00:06:25	1436	-12078	094	T	-p	-0.2742	2.3274	1.3820	313.7	204.6	82.0	23S	5E
7275	364	1023 Dec 29	20:30:03	1434	-12072	099	T-	p-	-0.1793	2.5712	1.4873	366.6	227.3	95.1	23N	60E
7276	364	1024 Jun 24	15:16:44	1431	-12066	104	P	a-	0.4885	1.9528	0.9702	318.1	192.6	-	23S	137E
7277	364	1024 Dec 18	02:18:03	1429	-12060	109	P	a-	-0.8549	1.3037	0.2750	288.1	116.0	-	23N	28W
7278	364	1025 May 15	10:57:37	1427	-12055	076	N	-t	-1.4754	0.1809	-0.8790	128.8	-	-	21S	160W
7279	364	1025 Jun 14	00:19:56	1426	-12054	114	N	t-	1.3095	0.4737	-0.5631	197.7	-	-	22S	1E
7280	364	1025 Nov 08	04:13:54	1424	-12049	081	N	-a	1.2025	0.6356	-0.3331	201.3	-	-	19N	61W
7281	365	1025 Dec 07	14:53:57	1424	-12048	119	N	a-	-1.4718	0.1477	-0.8335	102.9	-	-	22N	141E
7282	365	1026 May 04	11:34:12	1422	-12043	086	P	-t	-0.7406	1.5344	0.4643	330.1	157.1	-	18S	169W
7283	365	1026 Oct 28	19:46:13	1419	-12037	091	P	-a	0.5522	1.8367	0.8524	305.2	179.5	-	16N	65E
7284	365	1027 Apr 23	14:37:01	1417	-12031	096	T+	pp	0.0345	2.8117	1.7780	356.0	227.9	103.9	14S	146E
7285	365	1027 Oct 18	07:22:25	1414	-12025	101	T-	pp	-0.1486	2.6025	1.5685	347.3	220.9	96.6	11N	108W
7286	365	1028 Apr 12	00:40:17	1412	-12019	106	P	a-	0.7775	1.4218	0.4405	287.3	140.4	-	10S	4W
7287	365	1028 Oct 06	11:54:42	1409	-12013	111	P	t-	-0.8911	1.2642	0.1821	307.1	101.7	-	7N	176W
7288	365	1029 Mar 03	07:49:56	1407	-12008	078	N	-a	-1.1137	0.7937	-0.1654	221.3	-	-	4N	109W
7289	365	1029 Apr 01	16:09:21	1407	-12007	116	N	a-	1.4523	0.1685	-0.7830	108.2	-	-	5S	124E
7290	365	1029 Aug 26	19:15:46	1405	-12002	083	N	-t	1.3273	0.4553	-0.6099	198.4	-	-	7S	76E
7291	365	1030 Feb 20	22:49:33	1402	-11996	088	T	-a	-0.4416	2.0450	1.0500	319.5	195.4	33.7	8N	27E
7292	365	1030 Aug 16	01:18:59	1400	-11990	093	P	-h	0.5349	1.8815	0.8715	323.4	189.0	-	12S	14W
7293	365	1031 Feb 10	08:07:50	1397	-11984	098	T	p-	0.2828	2.3666	1.3120	352.6	218.2	80.6	13N	112W
7294	365	1031 Aug 05	14:28:58	1395	-11978	103	T-	p-	-0.2270	2.4210	1.4616	319.6	208.6	88.0	16S	150E
7295	365	1032 Jan 30	10:06:26	1392	-11972	108	N*	t-	1.0092	1.0552	-0.0421	290.9	-	-	17N	142W
7296	365	1032 Jul 25	07:19:48	1390	-11966	113	P	a-	-0.9439	1.0982	0.1534	251.2	83.6	-	19S	103W
7297	365	1032 Dec 19	16:02:06	1388	-11961	080	N	-t	-1.3107	0.4883	-0.5817	201.5	-	-	22N	126E
7298	365	1033 Jun 15	14:08:41	1385	-11955	085	P	-a	0.8753	1.2507	0.2530	282.7	112.1	-	23S	154E
7299	365	1033 Dec 08	23:57:10	1383	-11949	090	P	-a	-0.5660	1.8256	0.8133	313.9	180.3	-	23N	6E
7300	365	1034 Jun 04	21:09:34	1380	-11943	095	T+	pp	0.1168	2.6688	1.6189	364.6	230.9	102.1	23S	48E
7301	366	1034 Nov 28	13:52:33	1378	-11937	100	T+	p-	0.1246	2.6161	1.6422	319.7	209.4	94.8	22N	156E
7302	366	1035 May 24	22:19:50	1375	-11931	105	P	t-	-0.6530	1.6949	0.6251	341.1	178.0	-	22S	29E
7303	366	1035 Nov 18	05:32:39	1373	-11925	110	P	a-	0.7934	1.3918	0.4124	277.9	133.0	-	21N	81W
7304	366	1036 Apr 13	12:07:45	1371	-11920	077	N	-t	1.3593	0.3738	-0.6463	172.4	-	-	10S	176W
7305	366	1036 May 13	00:23:32	1371	-11919	115	N	t-	-1.3814	0.3440	-0.6974	171.1	-	-	21S	2W
7306	366	1036 Oct 08	04:18:46	1368	-11914	082	N	-a	-1.2952	0.5058	-0.5425	201.5	-	-	7N	62W
7307	366	1036 Nov 06	18:16:32	1368	-11913	120	N	a-	1.5260	0.0701	-0.9544	75.4	-	-	19N	87E
7308	366	1037 Apr 02	23:58:30	1366	-11908	087	P	-a	0.5649	1.8067	0.8360	304.1	179.1	-	7S	7E
7309	366	1037 Sep 27	07:04:05	1364	-11902	092	P	-t	-0.6185	1.7672	0.6796	345.6	183.2	-	3N	103W
7310	366	1038 Mar 23	16:08:34	1361	-11896	097	T-	p-	-0.1590	2.5422	1.5899	316.2	208.6	93.3	3S	125E
7311	366	1038 Sep 16	06:47:20	1359	-11890	102	T+	pp	0.0974	2.7193	1.6398	372.2	232.9	102.8	1S	98W
7312	366	1039 Mar 13	08:04:49	1356	-11884	107	P	a-	-0.8831	1.2268	0.2482	269.8	107.2	-	0N	114W
7313	366	1039 Sep 05	10:59:14	1354	-11878	112	P	a-	0.8183	1.3708	0.3423	298.9	130.6	-	4S	160W
7314	366	1040 Feb 01	04:12:21	1352	-11873	079	N	-h	1.1915	0.7080	-0.3641	239.7	-	-	17N	53W
7315	366	1040 Jul 26	14:28:59	1350	-11867	084	N	-a	-1.0703	0.8686	-0.0809	230.3	-	-	19S	150E
7316	366	1040 Aug 24	22:35:38	1349	-11866	122	N	a-	1.4885	0.1128	-0.8599	91.2	-	-	8S	26E
7317	366	1041 Jan 20	04:37:49	1347	-11861	089	P	-t	0.5226	1.9507	0.8483	357.8	199.5	-	19N	60W
7318	366	1041 Jul 16	07:35:26	1345	-11855	094	T	-p	-0.3458	2.1972	1.2491	311.4	200.1	69.4	21S	107W
7319	366	1042 Jan 09	04:38:02	1342	-11849	099	T-	p-	-0.1706	2.5853	1.5053	365.6	227.3	95.9	21N	61W
7320	366	1042 Jul 05	22:26:38	1340	-11843	104	T	a-	0.4145	2.0908	1.1039	324.4	201.0	48.5	22S	30E

Cat Num	Canon Plate	Calendar Date	TD of Greatest Eclipse	ΔT s	Luna Num	Saros Num	Ecl. Type	QSE	Gamma	Pen. Mag.	Um. Mag.	Eclipse Phase Durations Pen. m	Par. m	Total m	Greatest in Zenith Lat.	Lng.
7321	367	1042 Dec 29	10:51:45	1338	-11837	109	P	a-	-0.8490	1.3126	0.2877	287.7	118.1	-	22N	156W
7322	367	1043 May 26	17:28:50	1336	-11832	076	Ne	-t	-1.5590	0.0279	-1.0329	51.4	-	-	23S	102E
7323	367	1043 Jun 25	06:59:58	1335	-11831	114	N	t-	1.2317	0.6181	-0.4222	223.6	-	-	22S	99W
7324	367	1043 Nov 19	13:02:16	1333	-11826	081	N	-a	1.2084	0.6255	-0.3444	199.9	-	-	22N	167E
7325	367	1043 Dec 18	23:45:22	1333	-11825	119	N	a-	-1.4678	0.1543	-0.8254	104.9	-	-	22N	10E
7326	367	1044 May 14	18:04:40	1331	-11820	086	P	-t	-0.8213	1.3848	0.3177	318.4	132.7	-	21S	93E
7327	367	1044 Nov 08	04:23:18	1328	-11814	091	P	-a	0.5636	1.8178	0.8295	305.3	178.1	-	19N	64W
7328	367	1045 May 03	21:35:07	1326	-11808	096	T-	pp	-0.0376	2.8033	1.7750	354.2	227.3	103.8	17S	40E
7329	367	1045 Oct 28	15:33:52	1324	-11802	101	T-	pp	-0.1325	2.6348	1.5951	349.2	221.9	97.7	15N	128E
7330	367	1046 Apr 23	08:10:38	1321	-11796	106	P	a-	0.7145	1.5347	0.5587	293.3	154.8	-	14S	118W
7331	367	1046 Oct 17	19:35:27	1319	-11790	111	P	t-	-0.8660	1.3128	0.2258	312.2	112.7	-	11N	68E
7332	367	1047 Mar 14	16:02:02	1317	-11785	078	N	-a	-1.1500	0.7260	-0.2310	213.2	-	-	1S	127E
7333	367	1047 Apr 13	00:00:20	1317	-11784	116	N	a-	1.4005	0.2620	-0.6862	133.6	-	-	9S	5E
7334	367	1047 Sep 07	02:25:55	1315	-11779	083	N	-t	1.3774	0.3638	-0.7022	178.5	-	-	3S	32W
7335	367	1047 Oct 06	19:05:28	1314	-11778	121	N	t-	-1.5573	0.0467	-1.0450	67.2	-	-	6N	76E
7336	367	1048 Mar 03	06:58:13	1312	-11773	088	P	-a	-0.4731	1.9873	0.9924	318.3	192.6	-	4N	96W
7337	367	1048 Aug 26	08:47:40	1310	-11767	093	P	-a	0.5886	1.7827	0.7732	317.7	180.5	-	8S	127W
7338	367	1049 Feb 20	16:00:09	1308	-11761	098	T+	pp	0.2564	2.4150	1.3604	354.7	220.5	85.3	9N	129E
7339	367	1049 Aug 15	22:10:32	1305	-11755	103	T-	p-	-0.1675	2.5304	1.5706	320.6	210.5	93.5	12S	33E
7340	367	1050 Feb 09	17:47:23	1303	-11749	108	P	t-	0.9861	1.0966	0.0013	295.3	8.9	-	13N	103E
7341	368	1050 Aug 05	14:59:51	1301	-11743	113	P	a-	-0.8826	1.2120	0.2645	261.0	108.2	-	16S	142E
7342	368	1050 Dec 31	00:18:08	1299	-11738	080	N	-t	-1.3134	0.4814	-0.5849	199.5	-	-	21N	3E
7343	368	1051 Jun 26	21:10:32	1296	-11732	085	P	-a	0.9553	1.1059	0.1042	271.3	73.7	-	22S	48E
7344	368	1051 Dec 20	08:37:46	1294	-11726	090	P	-a	-0.5669	1.8224	0.8130	312.7	179.9	-	23N	123W
7345	368	1052 Jun 15	03:41:44	1292	-11720	095	T+	pp	0.2035	2.5110	1.4587	363.7	228.1	94.6	23S	50W
7346	368	1052 Dec 08	22:44:34	1289	-11714	100	T+	p-	0.1212	2.6221	1.6486	319.5	209.3	94.9	24N	24E
7347	368	1053 Jun 04	04:42:18	1287	-11708	105	P	t-	-0.5650	1.8558	0.7872	349.5	194.3	-	24S	66W
7348	368	1053 Nov 28	14:20:20	1285	-11702	110	P	a-	0.7901	1.3991	0.4172	279.0	133.9	-	23N	148E
7349	368	1054 Apr 24	19:18:25	1283	-11697	077	N	-t	1.4195	0.2604	-0.7538	144.7	-	-	14S	75E
7350	368	1054 May 24	07:06:50	1282	-11696	115	N	t-	-1.2989	0.4932	-0.5438	201.2	-	-	23S	103W
7351	368	1054 Oct 19	12:16:56	1280	-11691	082	N	-a	-1.3213	0.4607	-0.5935	194.0	-	-	11N	178E
7352	368	1054 Nov 18	02:44:30	1280	-11690	120	N	a-	1.5231	0.0778	-0.9513	79.7	-	-	22N	39W
7353	368	1055 Apr 14	07:40:05	1278	-11685	087	P	-a	0.6197	1.7035	0.7378	298.1	170.7	-	11S	110W
7354	368	1055 Oct 08	14:32:40	1276	-11679	092	P	-t	-0.6543	1.7038	0.6118	342.8	176.0	-	8N	144E
7355	368	1056 Apr 03	00:05:48	1273	-11673	097	T-	p-	-0.1093	2.6320	1.6822	317.0	210.0	96.5	8S	4E
7356	368	1056 Sep 26	14:10:44	1271	-11667	102	T+	pp	0.0551	2.7978	1.7167	372.2	233.2	104.3	4N	150E
7357	368	1057 Mar 23	16:01:00	1269	-11661	107	P	a-	-0.8419	1.3021	0.3240	276.2	121.5	-	4S	126E
7358	368	1057 Sep 15	18:37:43	1267	-11655	112	P	a-	0.7712	1.4572	0.4287	304.1	143.8	-	0N	84E
7359	368	1058 Feb 11	12:04:59	1265	-11650	079	N	-t	1.2167	0.6617	-0.4102	233.3	-	-	13N	172W
7360	368	1058 Aug 06	22:08:18	1262	-11644	084	N	-a	-1.1301	0.7591	-0.1912	217.5	-	-	16S	35E
7361	369	1058 Sep 05	06:27:56	1262	-11643	122	N	a-	1.4381	0.2055	-0.7676	121.8	-	-	4S	93W
7362	369	1059 Jan 31	12:24:30	1260	-11638	089	P	-t	0.5417	1.9144	0.8144	355.8	196.7	-	16N	177W
7363	369	1059 Jul 27	15:08:45	1258	-11632	094	T	-a	-0.4135	2.0746	1.1234	308.5	194.7	50.9	18S	139E
7364	369	1060 Jan 20	12:43:26	1256	-11626	099	T-	pp	-0.1590	2.6044	1.5287	364.5	227.3	97.0	19N	178E
7365	369	1060 Jul 16	05:37:00	1253	-11620	104	T	p-	0.3417	2.2267	1.2352	330.0	207.8	70.2	20S	78W
7366	369	1061 Jan 08	19:24:07	1251	-11614	109	P	a-	-0.8414	1.3246	0.3038	287.5	120.7	-	20N	77E
7367	369	1061 Jul 05	13:39:56	1249	-11608	114	N	t-	1.1542	0.7624	-0.2818	245.7	-	-	21S	161E
7368	369	1061 Nov 29	21:52:57	1247	-11603	081	N	-a	1.2125	0.6183	-0.3523	198.9	-	-	24N	35E
7369	369	1061 Dec 29	08:34:39	1247	-11602	119	N	a-	-1.4621	0.1638	-0.8139	107.8	-	-	21N	122W
7370	369	1062 May 26	00:32:48	1245	-11597	086	P	-t	-0.9039	1.2319	0.1674	304.9	98.4	-	23S	4W
7371	369	1062 Nov 19	13:04:07	1242	-11591	091	P	-a	0.5717	1.8048	0.8129	305.6	177.2	-	21N	166E
7372	369	1063 May 15	04:31:01	1240	-11585	096	T-	pp	-0.1122	2.6637	1.6405	351.6	225.5	100.9	20S	64W
7373	369	1063 Nov 08	23:49:19	1238	-11579	101	T-	pp	-0.1205	2.6595	1.6146	351.1	222.8	98.5	18N	4E
7374	369	1064 May 03	15:36:35	1236	-11573	106	P	a-	0.6470	1.6562	0.6850	298.9	167.5	-	17S	130E
7375	369	1064 Oct 28	03:23:09	1233	-11567	111	P	t-	-0.8469	1.3502	0.2587	316.2	120.3	-	14N	49W
7376	369	1065 Mar 25	00:03:44	1232	-11562	078	N	-a	-1.1947	0.6431	-0.3121	202.5	-	-	5S	5E
7377	369	1065 Apr 23	07:42:38	1231	-11561	116	N	a-	1.3416	0.3688	-0.5768	156.7	-	-	13S	111W
7378	369	1065 Sep 17	09:47:23	1229	-11556	083	N	-t	1.4189	0.2881	-0.7788	159.7	-	-	1N	144W
7379	369	1065 Oct 17	02:42:50	1229	-11555	121	N	t-	-1.5292	0.0992	-0.9943	97.5	-	-	10N	39W
7380	369	1066 Mar 14	14:55:47	1227	-11550	088	P	-a	-0.5131	1.9138	0.9190	316.4	188.4	-	0S	143E

Cat Num	Canon Plate	Calendar Date	TD of Greatest Eclipse	ΔT s	Luna Num	Saros Num	Ecl. Type	QSE	Gamma	Pen. Mag.	Um. Mag.	Eclipse Phase ---- Durations ---- Pen. m	Par. m	Total m	Greatest in Zenith Lat.	Lng.
7381	370	1066 Sep 06	16:26:32	1225	-11544	093	P	-a	0.6349	1.6977	0.6885	312.3	172.2	-	4S	117E
7382	370	1067 Mar 03	23:42:08	1223	-11538	098	T+	pp	0.2223	2.4774	1.4232	356.9	223.0	90.5	5N	12E
7383	370	1067 Aug 27	05:59:57	1220	-11532	103	T-	p-	-0.1142	2.6286	1.6680	321.1	211.7	96.7	8S	85W
7384	370	1068 Feb 21	01:20:57	1218	-11526	108	P	t-	0.9572	1.1484	0.0555	300.5	57.8	-	10N	11W
7385	370	1068 Aug 15	22:43:39	1216	-11520	113	P	a-	-0.8248	1.3197	0.3689	269.4	125.9	-	13S	25E
7386	370	1069 Jan 10	08:34:17	1214	-11515	080	N	-t	-1.3166	0.4734	-0.5888	197.3	-	-	20N	121W
7387	370	1069 Jul 07	04:10:43	1212	-11509	085	N	-a	1.0352	0.9616	-0.0445	258.1	-	-	21S	57W
7388	370	1069 Dec 30	17:18:20	1210	-11503	090	P	-a	-0.5684	1.8179	0.8119	311.4	179.3	-	22N	107E
7389	370	1070 Jun 26	10:13:14	1207	-11497	095	T	-t	0.2897	2.3541	1.2990	361.7	223.4	81.4	23S	148W
7390	370	1070 Dec 20	07:37:34	1205	-11491	100	T+	p-	0.1190	2.6258	1.6531	319.2	209.3	95.0	24N	109W
7391	370	1071 Jun 15	11:05:23	1203	-11485	105	P	t-	-0.4770	2.0168	0.9491	356.5	207.1	-	24S	162W
7392	370	1071 Dec 09	23:09:02	1201	-11479	110	P	a-	0.7883	1.4035	0.4194	279.9	134.4	-	24N	17E
7393	370	1072 May 05	02:26:49	1199	-11474	077	N	-t	1.4820	0.1429	-0.8658	107.8	-	-	17S	32W
7394	370	1072 Jun 03	13:50:39	1199	-11473	115	N	t-	-1.2156	0.6440	-0.3891	225.7	-	-	24S	156E
7395	370	1072 Oct 29	20:20:03	1197	-11468	082	N	-a	-1.3427	0.4243	-0.6354	187.6	-	-	15N	57E
7396	370	1072 Nov 28	11:14:15	1196	-11467	120	N	a-	1.5223	0.0815	-0.9518	81.8	-	-	24N	166W
7397	370	1073 Apr 24	15:15:43	1195	-11462	087	P	-a	0.6800	1.5906	0.6295	291.2	160.2	-	14S	135E
7398	370	1073 Oct 18	22:09:44	1193	-11456	092	P	-t	-0.6827	1.6537	0.5577	340.4	169.7	-	12N	29E
7399	370	1074 Apr 14	07:55:33	1190	-11450	097	T-	pp	-0.0536	2.7331	1.7855	317.5	211.1	98.6	11S	114W
7400	370	1074 Oct 07	21:44:00	1188	-11444	102	T+	pp	0.0205	2.8619	1.7795	371.9	233.1	104.7	8N	36E
7401	371	1075 Apr 03	23:48:09	1186	-11438	107	P	a-	-0.7940	1.3900	0.4119	283.3	135.6	-	8S	8E
7402	371	1075 Sep 27	02:26:29	1184	-11432	112	P	a-	0.7320	1.5291	0.5007	307.9	153.2	-	4N	35W
7403	371	1076 Feb 22	19:47:44	1182	-11427	079	N	-t	1.2491	0.6018	-0.4695	224.3	-	-	9N	72E
7404	371	1076 Aug 17	05:53:26	1180	-11421	084	N	-a	-1.1855	0.6583	-0.2933	204.4	-	-	13S	83W
7405	371	1076 Sep 15	14:28:39	1179	-11420	122	N	a-	1.3941	0.2866	-0.6871	142.5	-	-	1N	145E
7406	371	1077 Feb 10	20:03:13	1178	-11415	089	P	-t	0.5673	1.8658	0.7690	353.1	192.7	-	13N	68E
7407	371	1077 Aug 06	22:46:01	1176	-11409	094	T	-a	-0.4775	1.9590	1.0041	305.4	188.4	9.6	15S	24E
7408	371	1078 Jan 30	20:42:20	1173	-11403	099	T-	pp	-0.1414	2.6342	1.5635	363.4	227.6	98.5	16N	58E
7409	371	1078 Jul 27	12:50:23	1171	-11397	104	T+	p-	0.2721	2.3570	1.3602	334.7	213.2	83.3	18S	173E
7410	371	1079 Jan 20	03:53:25	1169	-11391	109	P	a-	-0.8304	1.3424	0.3263	287.7	124.4	-	18N	50W
7411	371	1079 Jul 16	20:20:34	1167	-11385	114	N	t-	1.0775	0.9052	-0.1432	264.9	-	-	20S	61E
7412	371	1079 Dec 11	06:46:01	1165	-11380	081	N	-a	1.2149	0.6140	-0.3570	198.3	-	-	25N	97W
7413	371	1080 Jan 09	17:22:50	1165	-11379	119	N	a-	-1.4555	0.1747	-0.8005	111.1	-	-	20N	107E
7414	371	1080 Jun 05	06:58:39	1163	-11374	086	P	-t	-0.9885	1.0755	0.0133	289.1	28.3	-	24S	100W
7415	371	1080 Nov 29	21:48:25	1161	-11368	091	P	-a	0.5763	1.7981	0.8029	306.2	176.7	-	23N	36E
7416	371	1081 May 25	11:23:33	1159	-11362	096	T-	pp	-0.1905	2.5177	1.4992	348.2	222.2	94.7	22S	167W
7417	371	1081 Nov 19	08:09:51	1157	-11356	101	T-	pp	-0.1135	2.6747	1.6251	352.7	223.5	99.0	21N	120W
7418	371	1082 May 14	22:59:40	1155	-11350	106	P	a-	0.5763	1.7838	0.8169	304.1	178.5	-	20S	19E
7419	371	1082 Nov 08	11:16:05	1152	-11344	111	P	t-	-0.8323	1.3788	0.2835	319.3	125.6	-	18N	168W
7420	371	1083 Apr 05	07:59:31	1151	-11339	078	N	-a	-1.2441	0.5516	-0.4019	189.4	-	-	9S	115W
7421	372	1083 May 04	15:21:46	1150	-11338	116	N	a-	1.2799	0.4808	-0.4625	176.9	-	-	16S	133E
7422	372	1083 Sep 28	17:18:17	1148	-11333	083	N	-t	1.4534	0.2252	-0.8424	141.6	-	-	6N	102E
7423	372	1083 Oct 28	10:27:33	1148	-11332	121	N	t-	-1.5062	0.1421	-0.9530	116.2	-	-	14N	156W
7424	372	1084 Mar 24	22:45:33	1146	-11327	088	P	-a	-0.5588	1.8300	0.8352	313.7	182.8	-	5S	24E
7425	372	1084 Sep 17	00:14:00	1144	-11321	093	P	-a	0.6746	1.6246	0.6155	307.3	164.2	-	1N	1W
7426	372	1085 Mar 14	07:14:33	1142	-11315	098	T+	pp	0.1811	2.5528	1.4988	359.1	225.6	95.5	0N	102W
7427	372	1085 Sep 06	13:57:40	1140	-11309	103	T-	p-	-0.0677	2.7144	1.7526	321.2	212.1	98.3	4S	154E
7428	372	1086 Mar 03	08:43:45	1138	-11303	108	P	t-	0.9199	1.2155	0.1254	306.9	86.1	-	5N	123W
7429	372	1086 Aug 27	06:35:39	1136	-11297	113	P	a-	-0.7742	1.4144	0.4600	276.3	138.8	-	9S	94W
7430	372	1087 Jan 21	16:45:37	1134	-11292	080	N	-t	-1.3237	0.4579	-0.5995	193.5	-	-	17N	117E
7431	372	1087 Jul 18	11:13:13	1132	-11286	085	N	-a	1.1118	0.8234	-0.1877	243.4	-	-	19S	163W
7432	372	1087 Aug 16	21:27:44	1132	-11285	123	Nb	a-	-1.5358	0.0343	-0.9548	51.8	-	-	13S	44E
7433	372	1088 Jan 11	01:56:01	1130	-11280	090	P	-a	-0.5727	1.8081	0.8059	309.9	178.4	-	20N	21W
7434	372	1088 Jul 06	16:45:07	1128	-11274	095	T	-t	0.3746	2.2000	1.1416	358.5	216.7	59.2	22S	114E
7435	372	1088 Dec 30	16:29:04	1126	-11268	100	T+	p-	0.1152	2.6321	1.6608	319.0	209.3	95.2	23N	119E
7436	372	1089 Jun 25	17:27:45	1124	-11262	105	T	t-	-0.3880	2.1799	1.1127	362.2	217.1	53.8	24S	103E
7437	372	1089 Dec 20	07:58:18	1122	-11256	110	P	a-	0.7874	1.4060	0.4203	280.7	134.7	-	24N	114W
7438	372	1090 May 16	09:30:58	1120	-11251	077	Ne	-t	1.5484	0.0186	-0.9850	39.2	-	-	19S	139W
7439	372	1090 Jun 14	20:34:35	1120	-11250	115	N	t-	-1.1311	0.7974	-0.2323	246.3	-	-	25S	56E
7440	372	1090 Nov 10	04:28:53	1118	-11245	082	N	-a	-1.3586	0.3976	-0.6670	182.7	-	-	18N	66W

Cat Num	Canon Plate	Calendar Date	TD of Greatest Eclipse	ΔT s	Luna Num	Saros Num	Ecl. Type	QSE	Gamma	Pen. Mag.	Um. Mag.	Eclipse Phase Durations Pen. m	Par. m	Total m	Greatest in Zenith Lat.	Lng.
7441	373	1090 Dec 09	19:45:28	1117	-11244	120	N	a-	1.5233	0.0813	-0.9555	82.0	-	-	25N	67E
7442	373	1091 May 05	22:46:58	1116	-11239	087	P	-a	0.7444	1.4704	0.5133	283.1	147.0	-	17S	22E
7443	373	1091 Oct 30	05:54:30	1114	-11233	092	P	-t	-0.7043	1.6159	0.5163	338.6	164.5	-	15N	88W
7444	373	1092 Apr 24	15:38:18	1112	-11227	097	T+	pp	0.0080	2.8160	1.8702	317.7	211.6	99.3	15S	129E
7445	373	1092 Oct 18	05:26:35	1110	-11221	102	T-	pp	-0.0070	2.8872	1.8037	371.4	232.6	104.6	12N	81W
7446	373	1093 Apr 14	07:26:59	1108	-11215	107	P	a-	-0.7400	1.4892	0.5109	290.7	149.2	-	12S	108W
7447	373	1093 Oct 07	10:25:02	1105	-11209	112	P	a-	0.6997	1.5883	0.5601	310.6	160.0	-	9N	155W
7448	373	1094 Mar 05	03:20:50	1104	-11204	079	N	-t	1.2887	0.5288	-0.5416	212.2	-	-	5N	42W
7449	373	1094 Apr 03	17:48:02	1103	-11203	117	Nb	t-	-1.5448	0.0414	-0.9944	60.1	-	-	9S	97E
7450	373	1094 Aug 28	13:46:15	1102	-11198	084	N	-a	-1.2350	0.5683	-0.3850	191.5	-	-	9S	158E
7451	373	1094 Sep 26	22:38:25	1101	-11197	122	N	a-	1.3568	0.3555	-0.6191	157.4	-	-	5N	22E
7452	373	1095 Feb 22	03:35:10	1100	-11192	089	P	-t	0.5987	1.8065	0.7131	349.6	187.3	-	9N	45W
7453	373	1095 Aug 18	06:28:04	1098	-11186	094	P	-a	-0.5372	1.8516	0.8925	302.0	181.5	-	12S	92W
7454	373	1096 Feb 11	04:36:49	1096	-11180	099	T-	pp	-0.1193	2.6720	1.6068	362.3	227.8	100.1	12N	61W
7455	373	1096 Aug 06	20:07:07	1094	-11174	104	T+	pp	0.2060	2.4812	1.4788	338.6	217.2	91.9	15S	64E
7456	373	1097 Jan 30	12:17:27	1092	-11168	109	P	a-	-0.8144	1.3691	0.3581	288.5	129.3	-	15N	176W
7457	373	1097 Jul 27	03:04:22	1090	-11162	114	N*	t-	1.0037	1.0430	-0.0101	281.3	-	-	17S	41W
7458	373	1097 Dec 21	15:38:35	1088	-11157	081	N	-a	1.2181	0.6081	-0.3629	197.5	-	-	25N	131E
7459	373	1098 Jan 20	02:06:11	1088	-11156	119	N	a-	-1.4449	0.1928	-0.7796	116.3	-	-	17N	24W
7460	373	1098 Jun 16	13:25:58	1086	-11151	086	N	-t	-1.0720	0.9213	-0.1390	271.4	-	-	25S	163E
7461	374	1098 Dec 11	06:32:29	1084	-11145	091	P	-a	0.5808	1.7911	0.7932	306.8	176.3	-	24N	94W
7462	374	1099 Jun 05	18:17:13	1082	-11139	096	T-	pp	-0.2690	2.3716	1.3572	344.0	217.4	84.5	23S	90E
7463	374	1099 Nov 30	16:32:22	1080	-11133	101	T-	pp	-0.1086	2.6858	1.6322	354.2	224.1	99.4	23N	115E
7464	374	1100 May 25	06:19:17	1078	-11127	106	P	a-	0.5017	1.9186	0.9556	308.8	188.0	-	22S	91W
7465	374	1100 Nov 18	19:14:28	1076	-11121	111	P	t-	-0.8223	1.3988	0.3003	321.6	129.2	-	20N	73E
7466	374	1101 Apr 15	15:46:10	1074	-11116	078	N	-a	-1.3010	0.4467	-0.5056	172.3	-	-	13S	127E
7467	374	1101 May 14	22:54:41	1074	-11115	116	N	a-	1.2128	0.6032	-0.3385	195.7	-	-	19S	20E
7468	374	1101 Oct 09	01:00:25	1072	-11110	083	N	-t	1.4795	0.1776	-0.8906	126.0	-	-	10N	14W
7469	374	1101 Nov 07	18:20:53	1072	-11109	121	N	t-	-1.4895	0.1732	-0.9228	127.7	-	-	17N	86E
7470	374	1102 Apr 05	06:24:14	1070	-11104	088	P	-a	-0.6129	1.7308	0.7357	310.0	175.0	-	9S	92W
7471	374	1102 Sep 28	08:12:27	1068	-11098	093	P	-a	0.7063	1.5664	0.5575	302.9	157.3	-	5N	122W
7472	374	1103 Mar 25	14:36:57	1066	-11092	098	T+	pp	0.1325	2.6418	1.5882	361.1	228.1	99.9	4S	146E
7473	374	1103 Sep 17	22:02:58	1064	-11086	103	T-	pp	-0.0272	2.7894	1.8262	321.1	212.1	98.9	0N	32E
7474	374	1104 Mar 13	15:58:54	1062	-11080	108	P	t-	0.8762	1.2941	0.2070	313.8	109.3	-	1N	127E
7475	374	1104 Sep 06	14:33:30	1060	-11074	113	P	a-	-0.7288	1.4997	0.5414	282.3	148.9	-	5S	145E
7476	374	1105 Feb 01	00:52:56	1059	-11069	080	N	-t	-1.3351	0.4344	-0.6175	188.2	-	-	14N	5W
7477	374	1105 Mar 02	16:23:09	1058	-11068	118	Nb	t-	1.5583	0.0341	-1.0362	56.3	-	-	6N	122E
7478	374	1105 Jul 28	18:17:13	1057	-11063	085	N	-h	1.1862	0.6896	-0.3269	226.8	-	-	16S	91E
7479	374	1105 Aug 27	05:02:14	1056	-11062	123	N	h-	-1.4880	0.1251	-0.8699	98.6	-	-	9S	71W
7480	374	1106 Jan 21	10:31:12	1055	-11057	090	P	-a	-0.5798	1.7929	0.7951	308.1	177.1	-	18N	150W
7481	375	1106 Jul 17	23:19:32	1053	-11051	095	P	-t	0.4570	2.0506	0.9887	354.3	208.1	-	20S	16E
7482	375	1107 Jan 11	01:17:57	1051	-11045	100	T+	p-	0.1092	2.6423	1.6726	318.8	209.4	95.5	21N	12W
7483	375	1107 Jul 06	23:54:10	1049	-11039	105	T	t-	-0.3018	2.3381	1.2710	366.4	224.5	78.9	22S	7E
7484	375	1107 Dec 31	16:45:56	1047	-11033	110	P	a-	0.7857	1.4097	0.4227	281.6	135.3	-	23N	115E
7485	375	1108 Jun 25	03:20:28	1045	-11027	115	N	t-	-1.0469	0.9505	-0.0762	263.7	-	-	24S	45W
7486	375	1108 Nov 20	12:41:08	1043	-11022	082	N	-a	-1.3711	0.3769	-0.6921	178.9	-	-	20N	172E
7487	375	1108 Dec 20	04:16:00	1043	-11021	120	N	a-	1.5246	0.0804	-0.9593	81.9	-	-	25N	59W
7488	375	1109 May 16	06:14:47	1041	-11016	087	P	-a	0.8119	1.3447	0.3912	274.0	130.5	-	20S	90W
7489	375	1109 Nov 09	13:45:00	1039	-11010	092	P	-t	-0.7211	1.5865	0.4841	337.2	160.2	-	18N	155E
7490	375	1110 May 05	23:15:37	1037	-11004	097	T+	pp	0.0740	2.6943	1.7496	317.5	211.2	98.3	18S	14E
7491	375	1110 Oct 29	13:18:02	1035	-10998	102	T-	pp	-0.0276	2.8496	1.7657	370.6	232.1	104.2	16N	161E
7492	375	1111 Apr 25	14:58:17	1033	-10992	107	P	a-	-0.6802	1.5991	0.6202	298.2	162.0	-	16S	138E
7493	375	1111 Oct 18	18:32:06	1032	-10986	112	P	a-	0.6737	1.6358	0.6080	312.4	165.0	-	13N	82E
7494	375	1112 Mar 15	10:43:34	1030	-10981	079	N	-t	1.3358	0.4419	-0.6275	195.9	-	-	1N	154W
7495	375	1112 Apr 14	01:00:10	1030	-10980	117	N	t-	-1.4900	0.1421	-0.8941	110.7	-	-	13S	12W
7496	375	1112 Sep 07	21:46:49	1028	-10975	084	N	-a	-1.2786	0.4893	-0.4660	179.0	-	-	5S	37E
7497	375	1112 Oct 07	06:57:02	1028	-10974	122	N	a-	1.3264	0.4118	-0.5637	168.2	-	-	9N	104W
7498	375	1113 Mar 04	10:56:15	1026	-10969	089	P	-t	0.6388	1.7309	0.6414	344.9	179.7	-	4N	157W
7499	375	1113 Aug 28	14:16:50	1024	-10963	094	P	-a	-0.5908	1.7555	0.7919	298.6	174.2	-	8S	150E
7500	375	1114 Feb 21	12:23:23	1022	-10957	099	T-	pp	-0.0900	2.7228	1.6635	361.1	228.1	101.7	8N	178W

Cat Num	Canon Plate	Calendar Date	TD of Greatest Eclipse	ΔT s	Luna Num	Saros Num	Ecl. Type	QSE	Gamma	Pen. Mag.	Um. Mag.	Eclipse Phase Durations Pen. m	Par. m	Total m	Greatest in Zenith Lat.	Lng.
7501	376	1114 Aug 18	03:29:04	1020	-10951	104	T+	pp	0.1451	2.5959	1.5875	341.9	220.1	97.3	11S	48W
7502	376	1115 Feb 10	20:36:02	1018	-10945	109	P	a-	-0.7934	1.4049	0.3994	289.9	135.4	-	11N	59E
7503	376	1115 Aug 07	09:52:28	1017	-10939	114	P	t-	0.9337	1.1739	0.1158	295.4	81.4	-	14S	143W
7504	376	1116 Jan 02	00:31:03	1015	-10934	081	N	-a	1.2216	0.6015	-0.3690	196.6	-	-	24N	1W
7505	376	1116 Jan 31	10:45:59	1015	-10933	119	N	a-	-1.4315	0.2157	-0.7536	122.5	-	-	14N	154W
7506	376	1116 Jun 26	19:52:58	1013	-10928	086	N	-t	-1.1559	0.7667	-0.2921	250.9	-	-	24S	67E
7507	376	1116 Dec 21	15:17:52	1011	-10922	091	P	-a	0.5839	1.7867	0.7865	307.5	176.2	-	24N	136E
7508	376	1117 Jun 16	01:10:59	1009	-10916	096	T	-p	-0.3485	2.2239	1.2133	338.8	211.1	68.5	24S	14W
7509	376	1117 Dec 11	00:56:11	1007	-10910	101	T-	pp	-0.1054	2.6932	1.6363	355.7	224.7	99.7	23N	10W
7510	376	1118 Jun 05	13:39:05	1005	-10904	106	T	a-	0.4266	2.0549	1.0950	312.7	195.7	45.5	23S	159E
7511	376	1118 Nov 30	03:16:19	1004	-10898	111	P	t-	-0.8150	1.4133	0.3124	323.3	131.6	-	22N	47W
7512	376	1119 Apr 26	23:28:02	1002	-10893	078	N	-a	-1.3613	0.3356	-0.6158	151.2	-	-	17S	11E
7513	376	1119 May 26	06:25:57	1002	-10892	116	N	a-	1.1443	0.7283	-0.2124	212.3	-	-	21S	93W
7514	376	1119 Oct 20	08:50:17	1000	-10887	083	N	-t	1.5000	0.1401	-0.9283	112.0	-	-	14N	133W
7515	376	1119 Nov 19	02:18:55	1000	-10886	121	N	t-	-1.4760	0.1981	-0.8981	136.2	-	-	20N	33W
7516	376	1120 Apr 15	13:55:59	998	-10881	088	P	-a	-0.6720	1.6228	0.6270	305.2	165.0	-	13S	154E
7517	376	1120 Oct 08	16:19:42	996	-10875	093	P	-a	0.7314	1.5204	0.5115	299.1	151.3	-	9N	116E
7518	376	1121 Apr 04	21:48:29	994	-10869	098	T+	pp	0.0758	2.7458	1.6924	363.0	230.1	103.3	8S	37E
7519	376	1121 Sep 28	06:17:30	993	-10863	103	T+	pp	0.0058	2.8295	1.8648	320.8	211.9	98.9	4N	93W
7520	376	1122 Mar 24	23:03:27	991	-10857	108	P	t-	0.8240	1.3883	0.3044	321.4	130.7	-	3S	20E
7521	377	1122 Sep 17	22:39:18	989	-10851	113	P	a-	-0.6902	1.5725	0.6101	287.1	156.5	-	0S	22E
7522	377	1123 Feb 12	08:53:07	987	-10846	080	N	-t	-1.3525	0.3994	-0.6467	180.2	-	-	10N	126W
7523	377	1123 Mar 13	23:48:00	987	-10845	118	N	t-	1.5141	0.1123	-0.9522	101.1	-	-	2N	10E
7524	377	1123 Aug 09	01:26:18	985	-10840	085	N	-h	1.2552	0.5660	-0.4563	209.0	-	-	13S	17W
7525	377	1123 Sep 07	12:43:57	985	-10839	123	N	h-	-1.4464	0.2045	-0.7966	125.7	-	-	5S	172E
7526	377	1124 Feb 01	19:00:48	984	-10834	090	P	-a	-0.5920	1.7681	0.7751	305.8	175.1	-	15N	83E
7527	377	1124 Jul 28	05:56:15	982	-10828	095	P	-t	0.5367	1.9064	0.8406	349.2	197.5	-	17S	84W
7528	377	1125 Jan 21	10:02:47	980	-10822	100	T+	p-	0.0998	2.6585	1.6908	318.6	209.6	96.0	18N	143W
7529	377	1125 Jul 17	06:23:50	978	-10816	105	T-	pp	-0.2180	2.4920	1.4247	369.3	229.5	93.0	20S	91W
7530	377	1126 Jan 11	01:29:55	976	-10810	110	P	a-	0.7818	1.4172	0.4296	282.7	136.5	-	22N	15W
7531	377	1126 Jul 06	10:10:16	974	-10804	115	P	t-	-0.9647	1.0999	0.0757	278.1	64.4	-	23S	148W
7532	377	1126 Dec 01	20:57:03	973	-10799	082	N	-h	-1.3794	0.3635	-0.7093	176.5	-	-	22N	48E
7533	377	1126 Dec 31	12:45:19	973	-10798	120	N	h-	1.5261	0.0787	-0.9632	81.3	-	-	24N	175E
7534	377	1127 May 27	13:39:51	971	-10793	087	P	-a	0.8822	1.2143	0.2637	263.6	108.9	-	21S	158E
7535	377	1127 Nov 20	21:40:54	969	-10787	092	P	-t	-0.7332	1.5654	0.4608	336.1	156.9	-	21N	36E
7536	377	1128 May 16	06:48:04	967	-10781	097	T+	-p	0.1439	2.5658	1.6217	316.7	209.9	95.0	20S	99W
7537	377	1128 Nov 08	21:17:28	966	-10775	102	T-	pp	-0.0425	2.8224	1.7384	369.8	231.5	103.7	19N	41E
7538	377	1129 May 05	22:21:39	964	-10769	107	P	a-	-0.6146	1.7200	0.7401	305.7	174.1	-	19S	27E
7539	377	1129 Oct 29	02:48:21	962	-10763	112	P	a-	0.6544	1.6709	0.6437	313.4	168.2	-	16N	42W
7540	377	1130 Mar 26	17:57:04	960	-10758	079	N	-t	1.3894	0.3430	-0.7254	174.6	-	-	4S	96E
7541	378	1130 Apr 25	08:04:26	960	-10757	117	N	t-	-1.4296	0.2535	-0.7835	146.8	-	-	16S	119W
7542	378	1130 Sep 19	05:54:50	959	-10752	084	N	-a	-1.3164	0.4211	-0.5364	167.1	-	-	0S	86W
7543	378	1130 Oct 18	15:22:48	958	-10751	122	N	a-	1.3014	0.4580	-0.5182	176.3	-	-	13N	129E
7544	378	1131 Mar 15	18:10:31	957	-10746	089	P	-t	0.6846	1.6449	0.5594	339.2	170.1	-	0N	94E
7545	378	1131 Sep 08	22:12:05	955	-10740	094	P	-a	-0.6385	1.6704	0.7020	295.3	166.8	-	4S	30E
7546	378	1132 Mar 03	20:03:48	953	-10734	099	T-	pp	-0.0547	2.7844	1.7312	359.8	228.2	103.0	4N	66E
7547	378	1132 Aug 28	10:55:54	951	-10728	104	T+	pp	0.0890	2.7019	1.6873	344.6	222.1	100.6	7S	160W
7548	378	1133 Feb 21	04:48:14	950	-10722	109	P	a-	-0.7663	1.4517	0.4520	291.8	142.5	-	7N	65W
7549	378	1133 Aug 17	16:46:45	948	-10716	114	P	t-	0.8689	1.2955	0.2323	307.3	113.7	-	11S	112E
7550	378	1134 Jan 12	09:18:53	946	-10711	081	N	-a	1.2292	0.5871	-0.3826	194.6	-	-	22N	132W
7551	378	1134 Feb 10	19:17:59	946	-10710	119	N	a-	-1.4116	0.2505	-0.7155	131.4	-	-	11N	78E
7552	378	1134 Jul 08	02:25:23	944	-10705	086	N	-t	-1.2357	0.6196	-0.4382	228.3	-	-	23S	31W
7553	378	1135 Jan 02	00:00:14	943	-10699	091	P	-a	0.5894	1.7774	0.7755	308.0	175.7	-	23N	6E
7554	378	1135 Jun 27	08:07:45	941	-10693	096	T	-t	-0.4264	2.0793	1.0717	332.9	203.3	41.4	24S	117W
7555	378	1135 Dec 22	09:18:55	939	-10687	101	T-	pp	-0.1019	2.7011	1.6414	357.0	225.3	100.0	23N	135W
7556	378	1136 Jun 15	20:58:24	937	-10681	106	T	a-	0.3500	2.1941	1.2368	315.8	201.9	68.6	23S	49E
7557	378	1136 Dec 10	11:19:58	936	-10675	111	P	t-	-0.8096	1.4241	0.3216	324.6	133.5	-	23N	167W
7558	378	1137 May 07	07:01:23	934	-10670	078	N	-a	-1.4281	0.2128	-0.7384	121.9	-	-	20S	103W
7559	378	1137 Jun 05	13:52:59	934	-10669	116	N	a-	1.0723	0.8603	-0.0800	227.7	-	-	22S	155E
7560	378	1137 Oct 30	16:50:45	932	-10664	083	N	-t	1.5128	0.1164	-0.9519	102.1	-	-	17N	107E

Cat Num	Canon Plate	Calendar Date	TD of Greatest Eclipse	ΔT s	Luna Num	Saros Num	Ecl. Type	QSE	Gamma	Pen. Mag.	Um. Mag.	Eclipse Phase Durations Pen. m	Par. m	Total m	Greatest in Zenith Lat.	Lng.
7561	379	1137 Nov 29	10:23:17	932	-10663	121	N	t-	-1.4668	0.2147	-0.8810	141.3	-	-	21N	154W
7562	379	1138 Apr 26	21:18:25	931	-10658	088	P	-a	-0.7379	1.5023	0.5055	299.0	151.6	-	16S	43E
7563	379	1138 Oct 20	00:35:55	929	-10652	093	P	-a	0.7500	1.4861	0.4775	296.0	146.5	-	13N	9W
7564	379	1139 Apr 16	04:51:10	927	-10646	098	T+	pp	0.0124	2.8620	1.8088	364.3	231.3	105.0	12S	69W
7565	379	1139 Oct 09	14:39:55	925	-10640	103	T+	pp	0.0327	2.7810	1.8147	320.5	211.4	98.5	9N	140E
7566	379	1140 Apr 04	05:59:42	924	-10634	108	P	t-	0.7647	1.4955	0.4147	329.2	150.0	-	8S	86W
7567	379	1140 Sep 28	06:51:51	922	-10628	113	P	a-	-0.6576	1.6345	0.6678	291.2	162.4	-	4N	102W
7568	379	1141 Feb 22	16:48:14	920	-10623	080	N	-t	-1.3748	0.3554	-0.6845	170.0	-	-	6N	115E
7569	379	1141 Mar 24	07:06:11	920	-10622	118	N	t-	1.4633	0.2024	-0.8561	134.1	-	-	3S	101W
7570	379	1141 Aug 19	08:39:40	919	-10617	085	N	-h	1.3195	0.4511	-0.5773	189.5	-	-	10S	126W
7571	379	1141 Sep 17	20:32:05	918	-10616	123	N	h-	-1.4106	0.2734	-0.7340	145.1	-	-	1S	54E
7572	379	1142 Feb 12	03:24:57	917	-10611	090	P	-a	-0.6093	1.7339	0.7459	303.0	172.3	-	11N	44W
7573	379	1142 Aug 08	12:39:04	915	-10605	095	P	-t	0.6106	1.7729	0.7029	343.5	185.2	-	14S	175E
7574	379	1143 Feb 01	18:43:18	913	-10599	100	T+	p-	0.0870	2.6808	1.7154	318.5	209.9	96.7	15N	87E
7575	379	1143 Jul 28	12:59:07	912	-10593	105	T-	pp	-0.1382	2.6385	1.5707	370.9	232.6	101.0	18S	170E
7576	379	1144 Jan 22	10:09:46	910	-10587	110	P	a-	0.7750	1.4299	0.4418	284.2	138.5	-	19N	145W
7577	379	1144 Jul 16	17:05:15	908	-10581	115	P	h-	-0.8858	1.2438	0.2217	290.0	107.6	-	21S	109E
7578	379	1144 Dec 12	05:12:49	907	-10576	082	N	-h	-1.3876	0.3500	-0.7258	174.0	-	-	22N	75W
7579	379	1145 Jan 10	21:09:35	906	-10575	120	N	h-	1.5241	0.0833	-0.9604	83.9	-	-	22N	50E
7580	379	1145 Jun 06	21:03:47	905	-10570	087	P	-a	0.9538	1.0817	0.1333	251.9	78.8	-	22S	47E
7581	380	1145 Dec 01	05:40:07	903	-10564	092	P	-t	-0.7425	1.5490	0.4429	335.2	154.3	-	22N	83W
7582	380	1146 May 27	14:17:43	902	-10558	097	T+	-p	0.2160	2.4334	1.4893	315.2	207.4	89.1	22S	148E
7583	380	1146 Nov 20	05:21:53	900	-10552	102	T-	pp	-0.0538	2.8013	1.7179	368.9	230.9	103.2	21N	79W
7584	380	1147 May 17	05:39:50	898	-10546	107	P	a-	-0.5454	1.8480	0.8664	312.8	184.9	-	21S	83W
7585	380	1147 Nov 09	11:11:45	896	-10540	112	P	a-	0.6404	1.6961	0.6699	313.8	170.4	-	19N	168W
7586	380	1148 Apr 06	00:59:23	895	-10535	079	N	-t	1.4511	0.2293	-0.8381	144.5	-	-	8S	10W
7587	380	1148 May 05	14:59:11	895	-10534	117	N	t-	-1.3620	0.3779	-0.6601	177.8	-	-	19S	137E
7588	380	1148 Sep 29	14:11:31	893	-10529	084	N	-a	-1.3475	0.3652	-0.5947	156.4	-	-	4N	149E
7589	380	1148 Oct 28	23:56:59	893	-10528	122	N	a-	1.2830	0.4920	-0.4848	181.8	-	-	17N	0E
7590	380	1149 Mar 26	01:14:23	892	-10523	089	P	-t	0.7387	1.5435	0.4622	331.9	156.9	-	4S	14W
7591	380	1149 Sep 19	06:15:22	890	-10517	094	P	-a	-0.6793	1.5981	0.6247	292.4	159.6	-	0N	92W
7592	380	1150 Mar 15	03:35:35	888	-10511	099	T-	pp	-0.0115	2.8605	1.8136	358.2	228.1	103.8	0S	48W
7593	380	1150 Sep 08	18:30:27	887	-10505	104	T+	pp	0.0401	2.7950	1.7740	346.9	223.3	102.1	3S	85E
7594	380	1151 Mar 04	12:54:24	885	-10499	109	P	a-	-0.7334	1.5091	0.5151	294.3	150.2	-	3N	172E
7595	380	1151 Aug 28	23:46:23	883	-10493	114	P	t-	0.8086	1.4088	0.3403	317.5	135.7	-	7S	6E
7596	380	1152 Jan 23	18:04:09	882	-10488	081	N	-a	1.2392	0.5682	-0.4002	191.8	-	-	19N	97E
7597	380	1152 Feb 22	03:44:59	881	-10487	119	N	a-	-1.3878	0.2925	-0.6699	141.2	-	-	7N	50W
7598	380	1152 Jul 18	09:01:03	880	-10482	086	N	-t	-1.3134	0.4768	-0.5803	202.6	-	-	21S	130W
7599	380	1152 Aug 17	00:06:26	880	-10481	124	Nb	t-	1.5433	0.0650	-1.0123	79.1	-	-	10S	2E
7600	380	1153 Jan 12	08:39:57	878	-10476	091	P	-a	0.5965	1.7649	0.7619	308.3	175.0	-	21N	123W
7601	381	1153 Jul 07	15:07:34	877	-10470	096	P	-t	-0.5027	1.9381	0.9331	326.2	194.0	-	22S	138E
7602	381	1154 Jan 01	17:39:57	875	-10464	101	T-	pp	-0.0975	2.7102	1.6484	358.3	226.0	100.4	22N	101E
7603	381	1154 Jun 27	04:20:08	873	-10458	106	T	p-	0.2750	2.3309	1.3754	318.1	206.5	82.2	23S	61W
7604	381	1154 Dec 21	19:22:51	872	-10452	111	P	t-	-0.8036	1.4356	0.3322	325.9	135.6	-	23N	74E
7605	381	1155 May 18	14:32:01	870	-10447	078	Ne	-a	-1.4969	0.0869	-0.8647	79.0	-	-	22S	144E
7606	381	1155 Jun 16	21:20:48	870	-10446	116	P	a-	1.0007	0.9917	0.0511	241.2	48.9	-	23S	43E
7607	381	1155 Nov 11	00:57:26	869	-10441	083	N	-t	1.5216	0.1001	-0.9675	94.5	-	-	21N	14W
7608	381	1155 Dec 10	18:29:45	868	-10440	121	N	t-	-1.4587	0.2291	-0.8655	145.5	-	-	22N	86E
7609	381	1156 May 07	04:33:43	867	-10435	088	P	-a	-0.8089	1.3728	0.3747	291.3	133.6	-	19S	66W
7610	381	1156 Oct 30	08:59:49	865	-10429	093	P	-a	0.7631	1.4617	0.4537	293.6	142.9	-	17N	135W
7611	381	1157 Apr 26	11:44:52	864	-10423	098	T-	pp	-0.0579	2.7787	1.7253	365.1	231.6	104.4	16S	174W
7612	381	1157 Oct 19	23:10:14	862	-10417	103	T+	pp	0.0532	2.7440	1.7763	320.1	211.0	97.9	13N	12E
7613	381	1158 Apr 15	12:47:06	860	-10411	108	P	t-	0.6979	1.6165	0.5388	337.1	167.7	-	11S	171E
7614	381	1158 Oct 09	15:12:42	859	-10405	113	P	a-	-0.6320	1.6837	0.7127	294.5	166.8	-	8N	132E
7615	381	1159 Mar 06	00:35:43	857	-10400	080	N	-t	-1.4037	0.2992	-0.7344	156.0	-	-	2N	3W
7616	381	1159 Apr 04	14:16:56	857	-10399	118	N	t-	1.4057	0.3050	-0.7475	162.5	-	-	7S	150E
7617	381	1159 Aug 30	15:58:49	856	-10394	085	N	-h	1.3780	0.3469	-0.6878	168.6	-	-	6S	123E
7618	381	1159 Sep 29	04:27:05	855	-10393	123	N	h-	-1.3809	0.3310	-0.6826	159.5	-	-	3N	66W
7619	381	1160 Feb 23	11:42:25	854	-10388	090	P	-a	-0.6327	1.6884	0.7055	299.6	168.3	-	7N	169W
7620	381	1160 Aug 18	19:27:29	852	-10382	095	P	-t	0.6791	1.6493	0.5749	337.2	171.4	-	10S	72E

Cat Num	Canon Plate	Calendar Date	TD of Greatest Eclipse	ΔT s	Luna Num	Saros Num	Ecl. Type	QSE	Gamma	Pen. Mag.	Um. Mag.	Eclipse Phase Durations Pen. m	Par. m	Total m	Greatest in Zenith Lat.	Lng.
7621	382	1161 Feb 12	03:16:14	851	-10376	100	T+	p-	0.0677	2.7150	1.7521	318.5	210.2	97.4	11N	42W
7622	382	1161 Aug 07	19:41:17	849	-10370	105	T-	pp	-0.0637	2.7758	1.7071	371.6	234.1	104.9	15S	69E
7623	382	1162 Feb 01	18:43:30	848	-10364	110	P	a-	0.7638	1.4504	0.4622	286.2	141.6	-	16N	87E
7624	382	1162 Jul 28	00:06:36	846	-10358	115	P	h-	-0.8107	1.3808	0.3602	299.7	134.0	-	18S	3E
7625	382	1162 Dec 23	13:28:13	845	-10353	082	N	-h	-1.3954	0.3369	-0.7412	171.4	-	-	22N	163E
7626	382	1163 Jan 22	05:29:14	844	-10352	120	N	h-	1.5192	0.0929	-0.9520	88.8	-	-	20N	75W
7627	382	1163 Jun 18	04:27:57	843	-10347	087	P	-a	1.0257	0.9490	0.0023	238.8	10.6	-	22S	63W
7628	382	1163 Jul 17	11:51:17	843	-10346	125	Nb	a-	-1.5011	0.0863	-0.8798	80.1	-	-	22S	173W
7629	382	1163 Dec 12	13:42:18	841	-10341	092	P	-t	-0.7493	1.5368	0.4301	334.5	152.4	-	23N	158E
7630	382	1164 Jun 06	21:43:27	840	-10335	097	T	-p	0.2912	2.2957	1.3510	313.1	203.7	79.5	23S	37E
7631	382	1164 Nov 30	13:32:27	838	-10329	102	T-	pp	-0.0607	2.7880	1.7058	367.8	230.3	102.8	23N	159E
7632	382	1165 May 27	12:52:52	837	-10323	107	P	a-	-0.4727	1.9823	0.9986	319.5	194.5	-	23S	169E
7633	382	1165 Nov 19	19:41:18	835	-10317	112	P	a-	0.6307	1.7131	0.6884	313.6	171.7	-	22N	65E
7634	382	1166 Apr 17	07:53:54	834	-10312	079	Ne	-t	1.5180	0.1061	-0.9604	99.5	-	-	11S	115W
7635	382	1166 May 16	21:48:48	833	-10311	117	N	t-	-1.2909	0.5092	-0.5302	204.5	-	-	22S	35E
7636	382	1166 Oct 10	22:35:56	832	-10306	084	N	-a	-1.3726	0.3203	-0.6420	147.1	-	-	8N	22E
7637	382	1166 Nov 09	08:37:08	832	-10305	122	N	a-	1.2696	0.5169	-0.4603	185.6	-	-	20N	130W
7638	382	1167 Apr 06	08:12:22	830	-10300	089	P	-t	0.7979	1.4327	0.3557	323.3	139.8	-	8S	119W
7639	382	1167 Sep 30	14:24:46	829	-10294	094	P	-a	-0.7146	1.5358	0.5574	289.8	152.7	-	5N	145E
7640	382	1168 Mar 25	11:01:50	827	-10288	099	T+	pp	0.0372	2.8101	1.7695	356.3	227.4	103.4	5S	161W
7641	383	1168 Sep 19	02:11:28	826	-10282	104	T-	pp	-0.0028	2.8668	1.8393	348.8	224.0	102.6	1N	32W
7642	383	1169 Mar 14	20:52:04	824	-10276	109	P	a-	-0.6927	1.5808	0.5927	297.4	158.8	-	1S	52E
7643	383	1169 Sep 08	06:54:28	822	-10270	114	P	t-	0.7554	1.5090	0.4352	325.9	151.5	-	3S	102W
7644	383	1170 Feb 03	02:42:29	821	-10265	081	N	-a	1.2550	0.5384	-0.4285	187.3	-	-	16N	33W
7645	383	1170 Mar 04	12:03:29	821	-10264	119	N	a-	-1.3569	0.3475	-0.6115	152.9	-	-	3N	175W
7646	383	1170 Jul 29	15:44:25	820	-10259	086	N	-t	-1.3850	0.3451	-0.7115	174.1	-	-	18S	129E
7647	383	1170 Aug 28	07:00:31	819	-10258	124	N	t-	1.4807	0.1811	-0.8983	130.5	-	-	6S	103W
7648	383	1171 Jan 23	17:13:11	818	-10253	091	P	-a	0.6086	1.7431	0.7392	308.1	173.5	-	18N	109E
7649	383	1171 Jul 18	22:13:16	816	-10247	096	P	-t	-0.5751	1.8042	0.8013	319.0	183.3	-	20S	31E
7650	383	1172 Jan 13	01:56:16	815	-10241	101	T-	pp	-0.0899	2.7249	1.6616	359.6	226.7	101.0	20N	23W
7651	383	1172 Jul 07	11:43:06	813	-10235	106	T+	p-	0.2002	2.4674	1.5132	319.8	209.8	91.1	22S	171W
7652	383	1173 Jan 01	03:24:43	812	-10229	111	P	t-	-0.7967	1.4482	0.3448	327.1	138.0	-	22N	46W
7653	383	1173 Jun 27	04:47:42	810	-10223	116	P	a-	0.9282	1.1252	0.1837	253.4	91.1	-	22S	68W
7654	383	1173 Nov 21	09:11:11	809	-10218	083	N	-t	1.5255	0.0922	-0.9741	90.5	-	-	23N	137W
7655	383	1173 Dec 21	02:37:51	809	-10217	121	N	t-	-1.4510	0.2421	-0.8504	149.1	-	-	22N	35W
7656	383	1174 May 18	11:42:18	807	-10212	088	P	-a	-0.8845	1.2351	0.2350	281.8	108.3	-	22S	173W
7657	383	1174 Nov 10	17:31:32	806	-10206	093	P	-a	0.7709	1.4471	0.4398	291.7	140.6	-	20N	97E
7658	383	1175 May 07	18:31:15	804	-10200	098	T-	pp	-0.1334	2.6403	1.5865	364.9	230.4	100.9	19S	84E
7659	383	1175 Oct 31	07:47:11	803	-10194	103	T+	pp	0.0685	2.7167	1.7476	319.8	210.5	97.4	16N	117W
7660	383	1176 Apr 25	19:28:08	801	-10188	108	P	t-	0.6255	1.7479	0.6732	344.6	183.3	-	15S	70E
7661	384	1176 Oct 19	23:40:14	800	-10182	113	P	a-	-0.6126	1.7214	0.7462	297.1	169.9	-	12N	5E
7662	384	1177 Mar 16	08:16:15	798	-10177	080	N	-t	-1.4388	0.2316	-0.7955	137.5	-	-	2S	119W
7663	384	1177 Apr 14	21:21:09	798	-10176	118	N	t-	1.3418	0.4194	-0.6274	187.7	-	-	11S	43E
7664	384	1177 Sep 09	23:24:39	797	-10171	085	N	-h	1.4299	0.2548	-0.7862	146.4	-	-	1S	10E
7665	384	1177 Oct 09	12:29:21	797	-10170	123	N	h-	-1.3577	0.3766	-0.6431	170.1	-	-	8N	173E
7666	384	1178 Mar 05	19:53:46	795	-10165	090	P	-a	-0.6616	1.6328	0.6551	295.6	163.2	-	3N	67E
7667	384	1178 Aug 30	02:23:05	794	-10159	095	P	-t	0.7413	1.5374	0.4587	330.7	156.1	-	6S	33W
7668	384	1179 Feb 23	11:43:09	792	-10153	100	T+	pp	0.0435	2.7581	1.7978	318.5	210.6	98.1	7N	169W
7669	384	1179 Aug 19	02:31:51	791	-10147	105	T+	pp	0.0045	2.8847	1.8151	371.4	234.1	105.7	11S	34W
7670	384	1180 Feb 13	03:10:08	789	-10141	110	P	a-	0.7474	1.4805	0.4925	288.8	145.8	-	12N	40W
7671	384	1180 Aug 07	07:15:04	788	-10135	115	P	h-	-0.7403	1.5093	0.4899	307.6	152.7	-	15S	104W
7672	384	1181 Jan 02	21:40:36	786	-10130	082	N	-h	-1.4048	0.3204	-0.7593	167.9	-	-	21N	40E
7673	384	1181 Feb 01	13:41:16	786	-10129	120	N	h-	1.5091	0.1119	-0.9337	97.6	-	-	16N	162E
7674	384	1181 Jun 28	11:53:59	785	-10124	087	N	-a	1.0961	0.8193	-0.1263	224.6	-	-	22S	175W
7675	384	1181 Jul 27	19:19:16	785	-10123	125	N	a-	-1.4322	0.2124	-0.7529	123.8	-	-	19S	75E
7676	384	1181 Dec 22	21:43:01	783	-10118	092	P	-t	-0.7570	1.5225	0.4161	333.5	150.3	-	23N	39E
7677	384	1182 Jun 18	05:09:25	782	-10112	097	T	-a	0.3657	2.1596	1.2136	310.3	198.7	65.1	23S	74W
7678	384	1182 Dec 11	21:45:32	780	-10106	102	T-	pp	-0.0657	2.7779	1.6977	366.7	229.8	102.5	23N	37E
7679	384	1183 Jun 07	20:01:59	779	-10100	107	T	p-	-0.3976	2.1216	1.1352	325.6	202.7	54.8	24S	62E
7680	384	1183 Dec 01	04:15:19	777	-10094	112	P	a-	0.6242	1.7241	0.7013	313.1	172.4	-	23N	62W

Cat Num	Canon Plate	Calendar Date	TD of Greatest Eclipse	ΔT s	Luna Num	Saros Num	Ecl. Type	QSE	Gamma	Pen. Mag.	Um. Mag.	Eclipse Phase Durations Pen. m	Par. m	Total m	Greatest in Zenith Lat.	Lng.
7681	385	1184 May 27	04:31:35	776	-10088	117	N	t-	-1.2148	0.6497	-0.3916	228.7	-	-	23S	66W
7682	385	1184 Oct 21	07:08:36	775	-10083	084	N	-a	-1.3915	0.2868	-0.6778	139.7	-	-	12N	107W
7683	385	1184 Nov 19	17:23:21	774	-10082	122	N	a-	1.2607	0.5332	-0.4440	187.9	-	-	22N	99E
7684	385	1185 Apr 16	15:00:13	773	-10077	089	P	-t	0.8654	1.3067	0.2338	312.4	115.3	-	12S	138E
7685	385	1185 Oct 10	22:42:44	772	-10071	094	P	-a	-0.7424	1.4873	0.5038	287.8	146.7	-	9N	19E
7686	385	1186 Apr 05	18:20:26	770	-10065	099	T+	pp	0.0932	2.7041	1.6699	354.0	226.1	101.6	9S	88E
7687	385	1186 Sep 30	09:59:19	769	-10059	104	T-	pp	-0.0392	2.8031	1.7692	350.5	224.4	102.2	5N	150W
7688	385	1187 Mar 26	04:44:00	767	-10053	109	P	a-	-0.6465	1.6627	0.6803	300.7	167.3	-	5S	67W
7689	385	1187 Sep 19	14:09:42	766	-10047	114	P	t-	0.7084	1.5980	0.5189	332.9	163.5	-	2N	148E
7690	385	1188 Feb 14	11:15:29	764	-10042	081	N	-a	1.2752	0.5006	-0.4647	181.4	-	-	12N	161W
7691	385	1188 Mar 14	20:15:07	764	-10041	119	N	a-	-1.3205	0.4125	-0.5431	165.4	-	-	2S	61E
7692	385	1188 Aug 08	22:33:11	763	-10036	086	N	-t	-1.4527	0.2210	-0.8357	140.5	-	-	15S	26E
7693	385	1188 Sep 07	14:02:23	763	-10035	124	N	t-	1.4234	0.2874	-0.7943	162.8	-	-	2S	151E
7694	385	1189 Feb 03	01:41:20	761	-10030	091	P	-a	0.6245	1.7142	0.7098	307.5	171.3	-	15N	18W
7695	385	1189 Jul 29	05:24:59	760	-10024	096	P	-h	-0.6435	1.6779	0.6766	311.3	171.5	-	18S	77W
7696	385	1190 Jan 23	10:07:00	759	-10018	101	T-	pp	-0.0782	2.7469	1.6828	360.8	227.6	101.7	18N	145W
7697	385	1190 Jul 18	19:11:31	757	-10012	106	T+	p-	0.1293	2.5972	1.6436	320.6	211.8	96.3	20S	76E
7698	385	1191 Jan 12	11:23:18	756	-10006	111	P	t-	-0.7869	1.4658	0.3631	328.6	141.3	-	20N	165W
7699	385	1191 Jul 08	12:16:23	754	-10000	116	P	a-	0.8572	1.2564	0.3133	264.2	116.9	-	21S	180E
7700	385	1191 Dec 02	17:28:32	753	-9995	083	N	-t	1.5273	0.0880	-0.9766	88.2	-	-	24N	100E
7701	386	1192 Jan 01	10:45:09	753	-9994	121	N	t-	-1.4420	0.2573	-0.8325	153.1	-	-	21N	156W
7702	386	1192 May 28	18:46:26	752	-9989	088	P	-a	-0.9629	1.0924	0.0898	270.3	68.6	-	23S	81E
7703	386	1192 Nov 21	02:08:07	750	-9983	093	P	-a	0.7752	1.4385	0.4325	290.3	139.3	-	22N	32W
7704	386	1193 May 18	01:10:42	749	-9977	098	T-	pp	-0.2140	2.4929	1.4382	363.8	227.6	93.2	21S	16W
7705	386	1193 Nov 10	16:30:23	747	-9971	103	T+	-p	0.0789	2.6982	1.7279	319.6	210.2	96.9	19N	112E
7706	386	1194 May 07	02:03:17	746	-9965	108	P	t-	0.5479	1.8889	0.8169	351.5	196.8	-	18S	29W
7707	386	1194 Oct 31	08:13:30	744	-9959	113	P	a-	-0.5981	1.7500	0.7709	299.3	172.2	-	16N	124W
7708	386	1195 Mar 27	15:49:48	743	-9954	080	N	-t	-1.4800	0.1526	-0.8681	111.9	-	-	7S	126E
7709	386	1195 Apr 26	04:20:15	743	-9953	118	N	t-	1.2725	0.5438	-0.4974	210.3	-	-	14S	63W
7710	386	1195 Sep 21	06:58:05	742	-9948	085	N	-h	1.4749	0.1755	-0.8719	122.9	-	-	3N	104W
7711	386	1195 Oct 20	20:38:39	741	-9947	123	N	h-	-1.3406	0.4110	-0.6146	178.0	-	-	12N	50E
7712	386	1196 Mar 16	03:56:40	740	-9942	090	P	-a	-0.6982	1.5628	0.5904	290.7	156.2	-	2S	55W
7713	386	1196 Sep 09	09:26:43	739	-9936	095	P	-t	0.7966	1.4383	0.3550	324.3	139.7	-	2S	140W
7714	386	1197 Mar 05	20:01:37	737	-9930	100	T+	pp	0.0119	2.8148	1.8571	318.4	210.9	98.7	3N	65E
7715	386	1197 Aug 29	09:31:59	736	-9924	105	T+	pp	0.0655	2.7734	1.7026	370.6	233.1	104.3	7S	140W
7716	386	1198 Feb 23	11:28:24	734	-9918	110	P	a-	0.7245	1.5223	0.5345	292.1	151.3	-	8N	166W
7717	386	1198 Aug 18	14:32:52	733	-9912	115	P	a-	-0.6764	1.6262	0.6075	313.7	166.3	-	12S	145E
7718	386	1199 Jan 14	05:50:03	732	-9907	082	N	-h	-1.4158	0.3006	-0.7799	163.3	-	-	19N	81W
7719	386	1199 Feb 12	21:46:45	732	-9906	120	N	h-	1.4944	0.1390	-0.9070	108.9	-	-	13N	41E
7720	386	1199 Jul 09	19:21:31	730	-9901	087	N	-a	1.1656	0.6915	-0.2538	208.7	-	-	20S	74E
7721	387	1199 Aug 08	02:52:18	730	-9900	125	N	a-	-1.3665	0.3327	-0.6322	152.8	-	-	16S	39W
7722	387	1200 Jan 03	05:44:10	729	-9895	092	P	-t	-0.7645	1.5084	0.4030	332.4	148.2	-	21N	81W
7723	387	1200 Jun 28	12:34:22	728	-9889	097	T	-a	0.4407	2.0230	1.0751	306.7	192.3	40.3	22S	175E
7724	387	1200 Dec 22	06:00:34	726	-9883	102	T-	pp	-0.0701	2.7684	1.6910	365.5	229.3	102.2	23N	86W
7725	387	1201 Jun 18	03:08:28	725	-9877	107	T	p-	-0.3208	2.2641	1.2744	331.0	209.6	74.8	24S	44W
7726	387	1201 Dec 11	12:52:37	723	-9871	112	P	a-	0.6196	1.7313	0.7110	312.4	172.9	-	24N	169E
7727	387	1202 Jun 07	11:11:59	722	-9865	117	N	t-	-1.1369	0.7939	-0.2497	250.1	-	-	24S	166W
7728	387	1202 Nov 01	15:46:21	721	-9860	084	N	-a	-1.4068	0.2598	-0.7070	133.3	-	-	15N	123E
7729	387	1202 Dec 01	02:11:52	721	-9859	122	N	a-	1.2534	0.5463	-0.4306	189.7	-	-	24N	32W
7730	387	1203 Apr 27	21:44:13	719	-9854	089	P	-t	0.9364	1.1745	0.1056	299.8	78.8	-	15S	36E
7731	387	1203 Oct 22	07:06:23	718	-9848	094	P	-a	-0.7653	1.4479	0.4595	286.3	141.3	-	13N	107W
7732	387	1204 Apr 16	01:33:07	717	-9842	099	T+	pp	0.1551	2.5875	1.5593	351.1	223.8	97.6	12S	21W
7733	387	1204 Oct 10	17:54:16	715	-9836	104	T-	pp	-0.0691	2.7514	1.7113	352.0	224.5	101.4	10N	91E
7734	387	1205 Apr 05	12:28:31	714	-9830	109	P	a-	-0.5935	1.7572	0.7804	304.2	175.8	-	9S	176E
7735	387	1205 Sep 29	21:33:51	713	-9824	114	P	t-	0.6688	1.6732	0.5891	338.6	172.6	-	6N	36E
7736	387	1206 Feb 24	19:39:20	711	-9819	081	N	-a	1.3030	0.4487	-0.5149	172.7	-	-	8N	72E
7737	387	1206 Mar 26	04:18:08	711	-9818	119	N	a-	-1.2770	0.4907	-0.4617	178.9	-	-	6S	61W
7738	387	1206 Aug 20	05:32:45	710	-9813	086	N	-t	-1.5124	0.1115	-0.9453	100.6	-	-	12S	79W
7739	387	1206 Sep 18	21:15:39	710	-9812	124	N	t-	1.3744	0.3783	-0.7054	185.1	-	-	2N	41E
7740	387	1207 Feb 14	10:01:10	709	-9807	091	P	-a	0.6469	1.6732	0.6685	306.3	167.9	-	11N	143W

Cat Num	Canon Plate	Calendar Date	TD of Greatest Eclipse	ΔT s	Luna Num	Saros Num	Ecl. Type	QSE	Gamma	Pen. Mag.	Um. Mag.	Eclipse Phase Durations Pen. m	Par. m	Total m	Greatest in Zenith Lat.	Lng.
7741	388	1207 Aug 09	12:43:42	707	-9801	096	P	-h	-0.7071	1.5606	0.5604	303.5	158.5	-	15S	173E
7742	388	1208 Feb 03	18:10:32	706	-9795	101	T-	pp	-0.0610	2.7785	1.7140	362.0	228.4	102.6	14N	94E
7743	388	1208 Jul 29	02:44:05	704	-9789	106	T+	pp	0.0611	2.7223	1.7688	320.9	212.8	99.0	17S	37W
7744	388	1209 Jan 22	19:16:05	703	-9783	111	P	t-	-0.7725	1.4916	0.3903	330.5	146.0	-	17N	77E
7745	388	1209 Jul 18	19:47:15	702	-9777	116	P	a-	0.7880	1.3845	0.4391	273.6	136.0	-	19S	67E
7746	388	1209 Dec 13	01:50:03	701	-9772	083	N	-t	1.5268	0.0877	-0.9744	87.7	-	-	25N	25W
7747	388	1210 Jan 11	18:51:00	700	-9771	121	N	t-	-1.4309	0.2758	-0.8105	157.8	-	-	19N	83E
7748	388	1210 Jun 09	01:46:22	699	-9766	088	N	-h	-1.0435	0.9459	-0.0596	256.5	-	-	24S	24W
7749	388	1210 Dec 02	10:48:53	698	-9760	093	P	-a	0.7770	1.4344	0.4301	289.2	138.6	-	24N	161W
7750	388	1211 May 29	07:45:32	697	-9754	098	T	-t	-0.2977	2.3398	1.2840	361.5	222.9	79.8	23S	114W
7751	388	1211 Nov 22	01:18:37	695	-9748	103	T+	-p	0.0853	2.6867	1.7157	319.4	209.9	96.6	22N	19W
7752	388	1212 May 17	08:32:49	694	-9742	108	P	t-	0.4652	2.0394	0.9699	357.6	208.5	-	20S	126W
7753	388	1212 Nov 10	16:52:49	692	-9736	113	P	a-	-0.5890	1.7685	0.7856	300.9	173.7	-	19N	106E
7754	388	1213 Apr 06	23:17:18	691	-9731	080	Ne	-t	-1.5270	0.0632	-0.9512	72.2	-	-	11S	13E
7755	388	1213 May 06	11:15:13	691	-9730	118	N	t-	1.1983	0.6773	-0.3587	230.7	-	-	17S	167W
7756	388	1213 Oct 01	14:38:31	690	-9725	085	N	-h	1.5134	0.1080	-0.9457	97.4	-	-	7N	140E
7757	388	1213 Oct 31	04:53:28	690	-9724	123	N	h-	-1.3285	0.4361	-0.5951	183.6	-	-	15N	74W
7758	388	1214 Mar 27	11:53:30	689	-9719	090	P	-a	-0.7405	1.4827	0.5152	285.0	147.4	-	6S	175W
7759	388	1214 Sep 20	16:39:22	687	-9713	095	P	-t	0.8443	1.3529	0.2654	318.3	122.6	-	2N	111E
7760	388	1215 Mar 17	04:13:33	686	-9707	100	T-	pp	-0.0249	2.7896	1.8344	318.3	211.1	98.7	1S	59W
7761	389	1215 Sep 09	16:40:52	685	-9701	105	T+	pp	0.1199	2.6742	1.6022	369.2	231.5	101.4	3S	111E
7762	389	1216 Mar 05	19:38:20	683	-9695	110	P	a-	0.6956	1.5752	0.5878	296.0	157.7	-	4N	71E
7763	389	1216 Aug 28	21:59:36	682	-9689	115	P	a-	-0.6191	1.7311	0.7130	318.4	176.5	-	8S	33E
7764	389	1217 Jan 24	13:51:51	681	-9684	082	N	-t	-1.4322	0.2706	-0.8101	155.7	-	-	16N	158E
7765	389	1217 Feb 23	05:41:09	681	-9683	120	N	t-	1.4720	0.1803	-0.8659	123.9	-	-	8N	79W
7766	389	1217 Jul 20	02:53:32	680	-9678	087	N	-a	1.2316	0.5706	-0.3750	191.7	-	-	18S	40W
7767	389	1217 Aug 18	10:33:44	679	-9677	125	N	a-	-1.3072	0.4417	-0.5234	173.9	-	-	12S	155W
7768	389	1218 Jan 13	13:40:34	678	-9672	092	P	-t	-0.7754	1.4875	0.3839	330.6	145.0	-	19N	161E
7769	389	1218 Jul 09	20:01:09	677	-9666	097	P	-a	0.5137	1.8903	0.9398	302.6	184.6	-	21S	64E
7770	389	1219 Jan 02	14:13:50	676	-9660	102	T-	pp	-0.0764	2.7551	1.6811	364.1	228.7	101.9	22N	152E
7771	389	1219 Jun 29	10:13:47	674	-9654	107	T-	pp	-0.2440	2.4069	1.4133	335.6	215.1	87.6	23S	150W
7772	389	1219 Dec 22	21:30:57	673	-9648	112	P	a-	0.6153	1.7376	0.7203	311.6	173.3	-	24N	41E
7773	389	1220 Jun 17	17:47:16	672	-9642	117	N	t-	-1.0553	0.9449	-0.1015	269.5	-	-	24S	96E
7774	389	1220 Nov 12	00:31:04	671	-9637	084	N	-a	-1.4166	0.2427	-0.7260	129.1	-	-	18N	8W
7775	389	1220 Dec 11	11:03:49	670	-9636	122	N	a-	1.2491	0.5538	-0.4222	190.6	-	-	25N	164W
7776	389	1221 May 08	04:20:55	669	-9631	089	N*	-t	1.0129	1.0322	-0.0330	284.7	-	-	18S	63W
7777	389	1221 Nov 01	15:36:50	668	-9625	094	P	-a	-0.7819	1.4196	0.4266	285.4	137.2	-	16N	125E
7778	389	1222 Apr 27	08:40:32	667	-9619	099	T+	pp	0.2221	2.4617	1.4394	347.5	220.4	90.8	16S	128W
7779	389	1222 Oct 22	01:56:29	665	-9613	104	T-	pp	-0.0919	2.7126	1.6665	353.4	224.6	100.5	13N	30W
7780	389	1223 Apr 16	20:07:47	664	-9607	109	P	a-	-0.5354	1.8613	0.8897	307.7	183.8	-	13S	60E
7781	390	1223 Oct 11	05:05:11	663	-9601	114	P	t-	0.6353	1.7372	0.6482	343.3	179.5	-	10N	77W
7782	390	1224 Mar 07	03:57:05	662	-9596	081	N	-a	1.3356	0.3878	-0.5739	161.6	-	-	4N	53W
7783	390	1224 Apr 05	12:15:00	662	-9595	119	N	a-	-1.2286	0.5780	-0.3714	192.4	-	-	10S	179E
7784	390	1224 Aug 30	12:40:13	660	-9590	086	Ne	-t	-1.5663	0.0127	-1.0445	34.1	-	-	8S	173E
7785	390	1224 Sep 29	04:37:55	660	-9589	124	N	t-	1.3317	0.4577	-0.6281	201.9	-	-	6N	70W
7786	390	1225 Feb 24	18:12:46	659	-9584	091	P	-a	0.6756	1.6207	0.6159	304.3	163.0	-	7N	93E
7787	390	1225 Aug 19	20:10:52	658	-9578	096	P	-a	-0.7646	1.4547	0.4552	295.7	144.7	-	11S	61E
7788	390	1226 Feb 14	02:06:28	657	-9572	101	T-	pp	-0.0381	2.8205	1.7561	363.1	229.3	103.4	11N	25W
7789	390	1226 Aug 09	10:22:59	655	-9566	106	T-	pp	-0.0021	2.8309	1.8770	320.6	212.8	99.6	14S	152W
7790	390	1227 Feb 03	03:02:49	654	-9560	111	P	t-	-0.7531	1.5261	0.4269	333.0	151.9	-	14N	39W
7791	390	1227 Jul 30	03:22:22	653	-9554	116	P	a-	0.7224	1.5062	0.5580	281.8	150.7	-	16S	47W
7792	390	1227 Dec 24	10:11:43	652	-9549	083	N	-h	1.5267	0.0862	-0.9727	86.7	-	-	25N	149W
7793	390	1228 Jan 23	02:52:19	651	-9548	121	N	h-	-1.4157	0.3017	-0.7805	164.1	-	-	17N	37W
7794	390	1228 Jun 19	08:43:27	650	-9543	088	N	-h	-1.1254	0.7974	-0.2116	240.2	-	-	25S	128W
7795	390	1228 Jul 18	18:28:45	650	-9542	126	N	h-	1.4996	0.0982	-0.8858	87.1	-	-	18S	86E
7796	390	1228 Dec 12	19:31:51	649	-9537	093	P	-a	0.7777	1.4319	0.4299	288.1	138.2	-	24N	70E
7797	390	1229 Jun 08	14:17:10	648	-9531	098	T	-t	-0.3836	2.1831	1.1256	358.0	216.0	56.1	24S	148E
7798	390	1229 Dec 02	10:09:23	647	-9525	103	T+	-p	0.0900	2.6784	1.7070	319.2	209.7	96.3	23N	151W
7799	390	1230 May 28	15:00:09	645	-9519	108	T	t-	0.3801	2.1945	1.1271	362.6	217.9	56.8	22S	137E
7800	390	1230 Nov 22	01:36:20	644	-9513	113	P	a-	-0.5834	1.7804	0.7943	302.3	174.7	-	21N	24W

Cat Num	Canon Plate	Calendar Date	TD of Greatest Eclipse	ΔT s	Luna Num	Saros Num	Ecl. Type	QSE	Gamma	Pen. Mag.	Um. Mag.	Eclipse Phase Durations Pen. m	Par. m	Total m	Greatest in Zenith Lat.	Lng.
7801	391	1231 May 17	18:06:29	643	-9507	118	N	t-	1.1199	0.8187	-0.2123	249.0	-	-	20S	90E
7802	391	1231 Oct 12	22:26:32	642	-9502	085	N	-h	1.5451	0.0529	-1.0068	68.8	-	-	12N	22E
7803	391	1231 Nov 11	13:14:04	642	-9501	123	N	h-	-1.3215	0.4514	-0.5849	187.3	-	-	18N	161E
7804	391	1232 Apr 06	19:42:43	641	-9496	090	P	-a	-0.7898	1.3899	0.4273	278.1	135.8	-	10S	67E
7805	391	1232 Oct 01	00:01:20	639	-9490	095	P	-t	0.8838	1.2824	0.1907	312.9	105.1	-	7N	1W
7806	391	1233 Mar 27	12:15:38	638	-9484	100	T-	pp	-0.0700	2.7058	1.7529	318.0	210.8	98.0	6S	180E
7807	391	1233 Sep 20	00:00:52	637	-9478	105	T+	pp	0.1657	2.5909	1.5176	367.7	229.4	97.5	2N	0E
7808	391	1234 Mar 17	03:39:29	636	-9472	110	P	a-	0.6601	1.6402	0.6531	300.5	165.0	-	1S	50W
7809	391	1234 Sep 09	05:35:20	634	-9466	115	P	a-	-0.5682	1.8243	0.8064	321.9	184.1	-	3S	82W
7810	391	1235 Feb 04	21:47:54	633	-9461	082	N	-t	-1.4525	0.2333	-0.8472	145.4	-	-	13N	39E
7811	391	1235 Mar 06	13:27:40	633	-9460	120	N	t-	1.4439	0.2318	-0.8143	140.3	-	-	4N	164E
7812	391	1235 Jul 31	10:29:18	632	-9455	087	N	-a	1.2947	0.4553	-0.4911	173.0	-	-	15S	154W
7813	391	1235 Aug 29	18:21:53	632	-9454	125	N	a-	-1.2528	0.5419	-0.4237	190.3	-	-	8S	87E
7814	391	1236 Jan 24	21:33:27	631	-9449	092	P	-t	-0.7894	1.4605	0.3594	328.3	140.8	-	17N	43E
7815	391	1236 Jul 20	03:28:54	630	-9443	097	P	-a	0.5855	1.7602	0.8065	297.7	175.3	-	19S	48W
7816	391	1237 Jan 12	22:26:14	628	-9437	102	T-	pp	-0.0842	2.7387	1.6689	362.7	228.1	101.5	20N	29E
7817	391	1237 Jul 09	17:19:45	627	-9431	107	T-	pp	-0.1686	2.5476	1.5495	339.4	219.1	95.8	22S	104E
7818	391	1238 Jan 02	06:08:17	626	-9425	112	P	a-	0.6096	1.7463	0.7326	310.9	173.9	-	23N	87W
7819	391	1238 Jun 29	00:23:45	625	-9419	117	P	t-	-0.9751	1.0937	0.0442	286.4	50.7	-	24S	3W
7820	391	1238 Nov 23	09:19:23	624	-9414	084	N	-a	-1.4237	0.2306	-0.7396	126.1	-	-	20N	139W
7821	392	1238 Dec 22	19:55:49	623	-9413	122	N	a-	1.2449	0.5610	-0.4139	191.5	-	-	25N	65E
7822	392	1239 May 19	10:55:10	622	-9408	089	N	-t	1.0917	0.8859	-0.1758	267.1	-	-	20S	162W
7823	392	1239 Nov 13	00:11:32	621	-9402	094	P	-a	-0.7947	1.3983	0.4010	284.9	133.8	-	19N	3W
7824	392	1240 May 07	15:44:15	620	-9396	099	T	-p	0.2929	2.3290	1.3122	343.0	215.6	80.2	19S	125E
7825	392	1240 Nov 01	10:04:27	619	-9390	104	T-	pp	-0.1096	2.6829	1.6314	354.8	224.7	99.6	17N	152W
7826	392	1241 Apr 27	03:40:38	618	-9384	109	T	a-	-0.4710	1.9770	1.0102	311.1	191.2	15.3	16S	54W
7827	392	1241 Oct 21	12:45:10	616	-9378	114	P	t-	0.6089	1.7877	0.6944	347.0	184.5	-	14N	167E
7828	392	1242 Mar 18	12:05:32	615	-9373	081	N	-a	1.3760	0.3129	-0.6473	146.4	-	-	1S	177W
7829	392	1242 Apr 16	20:04:30	615	-9372	119	N	a-	-1.1740	0.6770	-0.2699	206.0	-	-	14S	60E
7830	392	1242 Oct 10	12:09:48	614	-9366	124	N	t-	1.2959	0.5241	-0.5633	214.5	-	-	10N	176E
7831	392	1243 Mar 08	02:14:43	613	-9361	091	P	-a	0.7116	1.5546	0.5498	301.3	156.1	-	3N	28W
7832	392	1243 Aug 31	03:46:51	612	-9355	096	P	-a	-0.8158	1.3606	0.3616	288.1	130.4	-	7S	54W
7833	392	1244 Feb 25	09:51:50	611	-9349	101	T-	pp	-0.0073	2.8769	1.8129	364.1	230.1	104.1	7N	142W
7834	392	1244 Aug 19	18:08:09	609	-9343	106	T-	pp	-0.0604	2.7243	1.7694	320.0	212.0	98.7	10S	91E
7835	392	1245 Feb 13	10:41:36	608	-9337	111	P	t-	-0.7272	1.5722	0.4757	336.0	159.1	-	10N	154W
7836	392	1245 Aug 09	11:02:11	607	-9331	116	P	a-	0.6610	1.6206	0.6691	288.8	162.3	-	13S	163W
7837	392	1246 Jan 03	18:33:05	606	-9326	083	N	-h	1.5278	0.0823	-0.9727	84.4	-	-	23N	87E
7838	392	1246 Feb 02	10:48:40	606	-9325	121	N	h-	-1.3956	0.3361	-0.7413	172.1	-	-	13N	156W
7839	392	1246 Jun 30	15:39:39	605	-9320	088	N	-h	-1.2069	0.6501	-0.3631	221.1	-	-	24S	129E
7840	392	1246 Jul 30	01:46:44	605	-9319	126	N	h-	1.4376	0.2144	-0.7746	128.0	-	-	16S	24W
7841	393	1246 Dec 24	04:16:48	604	-9314	093	P	-a	0.7778	1.4305	0.4311	287.1	138.1	-	24N	60W
7842	393	1247 Jun 19	20:45:31	602	-9308	098	P	-t	-0.4714	2.0230	0.9636	353.1	206.5	-	24S	51E
7843	393	1247 Dec 13	19:03:01	601	-9302	103	T+	-p	0.0924	2.6739	1.7025	319.0	209.6	96.2	24N	77E
7844	393	1248 Jun 07	21:24:58	600	-9296	108	T	tp	0.2922	2.3548	1.2892	366.5	225.2	81.0	23S	41E
7845	393	1248 Dec 02	10:22:46	599	-9290	113	P	a-	-0.5810	1.7864	0.7974	303.4	175.3	-	22N	155W
7846	393	1249 May 28	00:56:09	598	-9284	118	N	t-	1.0389	0.9650	-0.0615	265.1	-	-	22S	12W
7847	393	1249 Oct 23	06:21:04	597	-9279	085	N	-h	1.5706	0.0089	-1.0564	28.5	-	-	15N	97W
7848	393	1249 Nov 21	21:38:08	597	-9278	123	N	h-	-1.3179	0.4603	-0.5806	189.8	-	-	20N	35E
7849	393	1250 Apr 18	03:27:09	596	-9273	090	P	-a	-0.8433	1.2895	0.3313	270.1	121.1	-	14S	50W
7850	393	1250 Oct 12	07:30:04	594	-9267	095	P	-t	0.9176	1.2223	0.1269	308.0	86.6	-	11N	114W
7851	393	1251 Apr 07	20:11:51	593	-9261	100	T-	-p	-0.1198	2.6134	1.6625	317.5	210.1	96.0	10S	59E
7852	393	1251 Oct 01	07:29:58	592	-9255	105	T+	pp	0.2045	2.5202	1.4459	365.9	227.2	93.1	6N	113W
7853	393	1252 Mar 27	11:30:34	591	-9249	110	P	a-	0.6170	1.7193	0.7323	305.5	172.9	-	5S	169W
7854	393	1252 Sep 19	13:21:02	590	-9243	115	P	a-	-0.5247	1.9040	0.8863	324.5	189.7	-	1N	160E
7855	393	1253 Feb 15	05:33:59	589	-9238	082	N	-t	-1.4801	0.1824	-0.8974	129.4	-	-	9N	78W
7856	393	1253 Mar 16	21:02:30	589	-9237	120	N	t-	1.4071	0.2992	-0.7467	158.9	-	-	0S	49E
7857	393	1253 Aug 10	18:11:53	588	-9232	087	N	-a	1.3524	0.3501	-0.5976	153.2	-	-	12S	90E
7858	393	1253 Sep 09	02:19:22	587	-9231	125	N	a-	-1.2055	0.6290	-0.3374	202.9	-	-	4S	33W
7859	393	1254 Feb 04	05:18:13	586	-9226	092	P	-t	-0.8099	1.4213	0.3232	324.9	134.2	-	13N	73W
7860	393	1254 Jul 31	11:01:35	585	-9220	097	P	-a	0.6526	1.6390	0.6816	292.6	164.8	-	16S	162W

Cat Num	Canon Plate	Calendar Date	TD of Greatest Eclipse	ΔT s	Luna Num	Saros Num	Ecl. Type	QSE	Gamma	Pen. Mag.	Um. Mag.	Eclipse Phase Durations Pen. m	Par. m	Total m	Greatest in Zenith Lat.	Lng.
7861	394	1255 Jan 24	06:34:23	584	-9214	102	T-	pp	-0.0960	2.7147	1.6495	361.0	227.4	101.0	17N	92W
7862	394	1255 Jul 21	00:25:59	583	-9208	107	T-	pp	-0.0945	2.6862	1.6831	342.5	221.8	100.7	19S	3W
7863	394	1256 Jan 13	14:44:04	582	-9202	112	P	a-	0.6022	1.7578	0.7483	310.4	174.9	-	21N	145E
7864	394	1256 Jul 09	06:59:02	581	-9196	117	P	t-	-0.8945	1.2434	0.1902	301.5	103.3	-	22S	101W
7865	394	1256 Dec 03	18:11:29	580	-9191	084	N	-a	-1.4280	0.2232	-0.7482	124.2	-	-	22N	89E
7866	394	1257 Jan 02	04:47:00	580	-9190	122	N	a-	1.2402	0.5688	-0.4043	192.4	-	-	23N	67W
7867	394	1257 May 29	17:25:28	579	-9185	089	N	-t	1.1736	0.7340	-0.3247	246.2	-	-	22S	101E
7868	394	1257 Nov 23	08:51:11	578	-9179	094	P	-a	-0.8029	1.3853	0.3842	284.9	131.6	-	21N	133W
7869	394	1258 May 18	22:45:17	576	-9173	099	T	-p	0.3669	2.1906	1.1788	337.8	209.4	63.4	21S	20E
7870	394	1258 Nov 12	18:16:59	575	-9167	104	T-	-p	-0.1227	2.6615	1.6049	356.2	224.8	98.9	20N	85E
7871	394	1259 May 08	11:10:18	574	-9161	109	T	a-	-0.4034	2.0988	1.1365	314.1	197.6	53.8	19S	167W
7872	394	1259 Nov 01	20:31:53	573	-9155	114	P	t-	0.5883	1.8274	0.7303	349.8	188.3	-	17N	50E
7873	394	1260 Mar 28	20:06:05	572	-9150	081	N	-a	1.4225	0.2270	-0.7318	125.8	-	-	5S	62E
7874	394	1260 Apr 27	03:47:29	572	-9149	119	N	a-	-1.1141	0.7858	-0.1590	219.4	-	-	17S	56W
7875	394	1260 Oct 20	19:50:54	571	-9143	124	N	t-	1.2670	0.5778	-0.5110	223.8	-	-	14N	60E
7876	394	1261 Mar 18	10:07:56	570	-9138	091	P	-a	0.7540	1.4769	0.4720	297.3	146.9	-	1S	148W
7877	394	1261 Sep 10	11:32:15	569	-9132	096	P	-a	-0.8598	1.2796	0.2808	281.1	115.9	-	3S	172W
7878	394	1262 Mar 07	17:28:17	568	-9126	101	T+	pp	0.0302	2.8346	1.7710	364.9	230.6	104.2	2N	103E
7879	394	1262 Aug 31	02:01:40	566	-9120	106	T-	pp	-0.1120	2.6303	1.6741	319.1	210.8	96.5	6S	29W
7880	394	1263 Feb 24	18:12:42	565	-9114	111	P	t-	-0.6950	1.6299	0.5364	339.5	167.4	-	6N	92E
7881	395	1263 Aug 20	18:47:02	564	-9108	116	P	a-	0.6041	1.7269	0.7717	294.9	171.6	-	10S	80E
7882	395	1264 Jan 15	02:52:06	563	-9103	083	N	-h	1.5314	0.0734	-0.9771	79.5	-	-	21N	37W
7883	395	1264 Feb 13	18:38:57	563	-9102	121	N	h-	-1.3699	0.3805	-0.6915	181.7	-	-	10N	86E
7884	395	1264 Jul 10	22:36:10	562	-9097	088	N	-h	-1.2870	0.5053	-0.5124	198.6	-	-	22S	25E
7885	395	1264 Aug 09	09:08:46	562	-9096	126	N	h-	1.3792	0.3244	-0.6701	156.5	-	-	12S	135W
7886	395	1265 Jan 03	12:59:20	561	-9091	093	P	-a	0.7803	1.4241	0.4280	285.8	137.3	-	23N	170E
7887	395	1265 Jun 30	03:14:06	560	-9085	098	P	-t	-0.5583	1.8647	0.8028	347.0	194.4	-	23S	45W
7888	395	1265 Dec 24	03:55:57	559	-9079	103	T+	-p	0.0958	2.6675	1.6966	318.9	209.5	96.0	23N	55W
7889	395	1266 Jun 19	03:50:18	558	-9073	108	T+	pp	0.2044	2.5153	1.4511	369.2	230.3	94.8	23S	55W
7890	395	1266 Dec 13	19:09:37	557	-9067	113	P	a-	-0.5792	1.7909	0.7995	304.4	175.8	-	23N	75E
7891	395	1267 Jun 08	07:45:08	556	-9061	118	P	t-	0.9560	1.1151	0.0926	279.3	71.1	-	22S	114W
7892	395	1267 Dec 03	06:05:09	554	-9055	123	N	h-	-1.3174	0.4632	-0.5817	191.1	-	-	22N	90W
7893	395	1268 Apr 28	11:03:47	554	-9050	090	P	-a	-0.9038	1.1765	0.2223	260.6	100.6	-	17S	165W
7894	395	1268 Oct 22	15:08:29	552	-9044	095	P	-t	0.9433	1.1769	0.0781	304.1	68.4	-	15N	131E
7895	395	1269 Apr 18	03:59:31	551	-9038	100	T-	-p	-0.1768	2.5079	1.5585	316.5	208.6	92.4	13S	58W
7896	395	1269 Oct 11	15:09:34	550	-9032	105	T+	pp	0.2354	2.4639	1.3888	364.2	225.0	88.8	10N	131E
7897	395	1270 Apr 07	19:13:13	549	-9026	110	P	a-	0.5674	1.8103	0.8231	310.8	181.1	-	9S	74E
7898	395	1270 Sep 30	21:16:50	548	-9020	115	P	a-	-0.4886	1.9703	0.9527	326.1	193.7	-	5N	40E
7899	395	1271 Feb 26	13:12:42	547	-9015	082	N	-t	-1.5127	0.1219	-0.9568	106.5	-	-	5N	167E
7900	395	1271 Mar 28	04:28:45	547	-9014	120	N	t-	1.3642	0.3779	-0.6679	177.8	-	-	4S	64W
7901	396	1271 Aug 22	01:59:09	546	-9009	087	N	-a	1.4063	0.2520	-0.6976	131.1	-	-	8S	28W
7902	396	1271 Sep 20	10:24:01	546	-9008	125	N	a-	-1.1637	0.7062	-0.2613	213.0	-	-	1N	156W
7903	396	1272 Feb 15	12:57:53	545	-9003	092	P	-t	-0.8346	1.3742	0.2798	320.6	125.6	-	9N	171E
7904	396	1272 Aug 10	18:37:46	544	-8997	097	P	-a	0.7160	1.5249	0.5632	287.0	152.9	-	12S	83E
7905	396	1273 Feb 03	14:37:42	543	-8991	102	T-	pp	-0.1123	2.6821	1.6223	359.2	226.5	100.1	14N	147E
7906	396	1273 Jul 31	07:35:52	542	-8985	107	T-	pp	-0.0244	2.8175	1.8090	344.8	223.3	102.8	16S	111W
7907	396	1274 Jan 23	23:16:22	541	-8979	112	P	a-	0.5916	1.7749	0.7699	310.0	176.3	-	18N	17E
7908	396	1274 Jul 20	13:37:14	540	-8973	117	P	t-	-0.8167	1.3882	0.3311	314.5	133.8	-	20S	159E
7909	396	1274 Dec 15	03:03:39	539	-8968	084	N	-a	-1.4323	0.2157	-0.7565	122.3	-	-	22N	43W
7910	396	1275 Jan 13	13:35:00	539	-8967	122	N	a-	1.2330	0.5809	-0.3902	194.0	-	-	21N	162E
7911	396	1275 Jun 09	23:56:24	538	-8962	089	N	-t	1.2550	0.5833	-0.4726	222.1	-	-	22S	3E
7912	396	1275 Jul 09	14:25:50	538	-8961	127	Nb	t-	-1.5731	0.0064	-1.0631	25.0	-	-	23S	147E
7913	396	1275 Dec 04	17:32:19	537	-8956	094	P	-a	-0.8093	1.3752	0.3707	285.2	129.8	-	22N	98E
7914	396	1276 May 29	05:44:05	536	-8950	099	T	-t	0.4435	2.0477	1.0405	331.5	201.4	31.4	22S	84W
7915	396	1276 Nov 23	02:33:30	535	-8944	104	T-	-p	-0.1316	2.6473	1.5863	357.5	225.0	98.3	22N	39W
7916	396	1277 May 18	18:36:13	534	-8938	109	T	a-	-0.3321	2.2277	1.2691	316.6	203.1	72.3	21S	82E
7917	396	1277 Nov 12	04:24:08	532	-8932	114	P	t-	0.5727	1.8577	0.7574	352.0	191.0	-	20N	67W
7918	396	1278 Apr 09	03:58:00	532	-8927	081	N	-a	1.4755	0.1292	-0.8286	95.9	-	-	9S	57W
7919	396	1278 May 08	11:25:14	531	-8926	119	N	a-	-1.0501	0.9025	-0.0408	232.3	-	-	20S	171W
7920	396	1278 Nov 01	03:41:09	530	-8920	124	N	t-	1.2453	0.6181	-0.4715	230.3	-	-	18N	57W

Cat Num	Canon Plate	Calendar Date	TD of Greatest Eclipse	ΔT s	Luna Num	Saros Num	Ecl. Type	QSE	Gamma	Pen. Mag.	Um. Mag.	Eclipse Phase Durations Pen. m	Par. m	Total m	Greatest in Zenith Lat.	Lng.
7921	397	1279 Mar 29	17:50:49	530	-8915	091	P	-a	0.8041	1.3850	0.3799	292.0	134.1	-	6S	96E
7922	397	1279 Sep 21	19:26:50	528	-8909	096	P	-a	-0.8971	1.2112	0.2123	274.7	101.5	-	1N	69E
7923	397	1280 Mar 18	00:53:37	527	-8903	101	T+	pp	0.0761	2.7500	1.6872	365.4	230.6	103.2	2S	10W
7924	397	1280 Sep 10	10:02:47	526	-8897	106	T-	-p	-0.1577	2.5474	1.5895	318.0	209.2	93.5	2S	150W
7925	397	1281 Mar 07	01:33:37	525	-8891	111	P	t-	-0.6542	1.7029	0.6128	343.6	176.7	-	2N	19W
7926	397	1281 Aug 31	02:38:37	524	-8885	116	P	a-	0.5531	1.8227	0.8632	300.0	178.9	-	6S	38W
7927	397	1282 Jan 25	11:08:16	523	-8880	083	N	-h	1.5382	0.0585	-0.9870	70.8	-	-	18N	161W
7928	397	1282 Feb 24	02:23:04	523	-8879	121	N	h-	-1.3383	0.4357	-0.6307	192.7	-	-	6N	31W
7929	397	1282 Jul 22	05:33:18	522	-8874	088	N	-h	-1.3653	0.3642	-0.6585	171.6	-	-	20S	79W
7930	397	1282 Aug 20	16:34:27	522	-8873	126	N	h-	1.3244	0.4279	-0.5725	178.9	-	-	9S	113E
7931	397	1283 Jan 14	21:40:31	521	-8868	093	P	-a	0.7846	1.4145	0.4220	284.3	136.2	-	20N	41E
7932	397	1283 Jul 11	09:42:53	520	-8862	098	P	-t	-0.6444	1.7081	0.6435	339.5	179.1	-	22S	142W
7933	397	1284 Jan 04	12:48:12	519	-8856	103	T+	-p	0.0995	2.6603	1.6903	318.7	209.4	95.9	22N	173E
7934	397	1284 Jun 29	10:16:00	518	-8850	108	T+	pp	0.1164	2.6763	1.6131	370.7	233.4	102.7	23S	151W
7935	397	1284 Dec 24	03:56:38	517	-8844	113	P	a-	-0.5781	1.7938	0.8005	305.3	176.2	-	23N	56W
7936	397	1285 Jun 18	14:36:09	516	-8838	118	P	h-	0.8735	1.2647	0.2457	291.3	112.8	-	23S	143E
7937	397	1285 Dec 13	14:31:36	515	-8832	123	N	h-	-1.3172	0.4654	-0.5829	192.2	-	-	22N	144E
7938	397	1286 May 09	18:38:06	514	-8827	090	P	-a	-0.9665	1.0596	0.1088	249.9	71.4	-	20S	81E
7939	397	1286 Nov 02	22:53:19	513	-8821	095	P	-t	0.9638	1.1407	0.0391	300.8	48.7	-	18N	15E
7940	397	1287 Apr 29	11:41:31	512	-8815	100	T-	-p	-0.2383	2.3946	1.4463	315.0	206.3	86.5	17S	174W
7941	398	1287 Oct 22	22:57:41	511	-8809	105	T+	pp	0.2599	2.4193	1.3436	362.5	223.0	84.8	14N	14E
7942	398	1288 Apr 18	02:46:56	510	-8803	110	P	a-	0.5112	1.9137	0.9260	316.3	189.2	-	13S	40W
7943	398	1288 Oct 11	05:22:09	509	-8797	115	T	a-	-0.4598	2.0229	1.0056	327.1	196.5	11.6	10N	82W
7944	398	1289 Mar 08	20:39:13	508	-8792	082	Ne	-t	-1.5540	0.0455	-1.0319	65.6	-	-	0N	54E
7945	398	1289 Apr 07	11:43:12	508	-8791	120	N	t-	1.3125	0.4727	-0.5732	197.6	-	-	9S	173W
7946	398	1289 Sep 01	09:55:12	507	-8786	087	N	-a	1.4531	0.1673	-0.7846	107.6	-	-	4S	148W
7947	398	1289 Sep 30	18:38:31	507	-8785	125	N	a-	-1.1299	0.7689	-0.1998	220.5	-	-	5N	80E
7948	398	1290 Feb 25	20:28:11	506	-8780	092	P	-t	-0.8665	1.3135	0.2232	315.0	113.0	-	5N	58E
7949	398	1290 Aug 22	02:19:31	505	-8774	097	P	-a	0.7740	1.4207	0.4544	281.4	140.0	-	9S	33W
7950	398	1291 Feb 14	22:34:42	504	-8768	102	T-	pp	-0.1344	2.6386	1.5845	357.1	225.5	98.6	10N	27E
7951	398	1291 Aug 11	14:49:05	503	-8762	107	T+	pp	0.0418	2.7884	1.7741	346.5	223.8	102.6	13S	141E
7952	398	1292 Feb 04	07:43:14	502	-8756	112	P	a-	0.5759	1.8012	0.8012	310.0	178.4	-	15N	110W
7953	398	1292 Jul 30	20:17:43	501	-8750	117	P	t-	-0.7409	1.5294	0.4680	325.8	156.1	-	17S	59E
7954	398	1292 Dec 25	11:56:44	501	-8745	084	N	-a	-1.4360	0.2090	-0.7635	120.5	-	-	22N	176W
7955	398	1293 Jan 23	22:19:22	500	-8744	122	N	a-	1.2230	0.5981	-0.3706	196.3	-	-	18N	31E
7956	398	1293 Jun 20	06:27:25	500	-8739	089	N	-t	1.3364	0.4329	-0.6208	193.6	-	-	22S	94W
7957	398	1293 Jul 19	20:55:18	499	-8738	127	N	t-	-1.4918	0.1560	-0.9143	121.2	-	-	21S	50E
7958	398	1293 Dec 15	02:14:39	499	-8733	094	P	-a	-0.8142	1.3676	0.3602	285.5	128.5	-	23N	31W
7959	398	1294 Jun 09	12:43:31	498	-8727	099	P	-t	0.5205	1.9045	0.9014	324.4	191.6	-	23S	171E
7960	398	1294 Dec 04	10:52:56	497	-8721	104	T-	-p	-0.1372	2.6389	1.5741	358.8	225.3	98.0	23N	162W
7961	399	1295 May 30	01:59:34	496	-8715	109	T-	p-	-0.2580	2.3620	1.4067	318.6	207.4	84.6	23S	29W
7962	399	1295 Nov 23	12:21:47	495	-8709	114	P	t-	0.5616	1.8794	0.7765	353.7	192.9	-	22N	174E
7963	399	1296 Apr 19	11:43:02	494	-8704	081	Ne	-a	1.5337	0.0221	-0.9349	40.1	-	-	12S	174W
7964	399	1296 May 18	18:58:30	494	-8703	119	P	a-	-0.9826	1.0258	0.0835	244.5	62.3	-	22S	76E
7965	399	1296 Nov 11	11:38:19	493	-8697	124	N	t-	1.2282	0.6497	-0.4402	235.0	-	-	21N	176W
7966	399	1297 Apr 09	01:24:32	492	-8692	091	P	-a	0.8611	1.2807	0.2751	285.1	116.2	-	10S	19W
7967	399	1297 Oct 02	03:30:53	491	-8686	096	P	-a	-0.9276	1.1553	0.1565	269.2	87.6	-	6N	53W
7968	399	1298 Mar 29	08:10:17	490	-8680	101	T+	pp	0.1286	2.6533	1.5911	365.4	229.9	100.6	6S	120W
7969	399	1298 Sep 21	18:11:08	489	-8674	106	T-	-p	-0.1975	2.4751	1.5155	316.9	207.5	90.0	2N	87E
7970	399	1299 Mar 18	08:46:35	488	-8668	111	P	t-	-0.6068	1.7881	0.7016	348.0	186.2	-	2S	128W
7971	399	1299 Sep 11	10:36:37	487	-8662	116	P	a-	0.5076	1.9082	0.9444	304.4	184.7	-	1S	159W
7972	399	1300 Feb 05	19:17:56	486	-8657	083	N	-h	1.5506	0.0329	-1.0070	53.0	-	-	15N	77E
7973	399	1300 Mar 06	09:58:08	486	-8656	121	N	h-	-1.2987	0.5054	-0.5550	205.4	-	-	1N	146W
7974	399	1300 Aug 01	12:33:39	485	-8651	088	N	-h	-1.4397	0.2303	-0.7978	138.8	-	-	17S	175E
7975	399	1300 Aug 31	00:06:34	485	-8650	126	N	h-	1.2753	0.5211	-0.4854	196.5	-	-	5S	1W
7976	399	1301 Jan 25	06:16:32	484	-8645	093	P	-a	0.7937	1.3956	0.4072	282.1	133.8	-	18N	88W
7977	399	1301 Jul 21	16:15:37	483	-8639	098	P	-t	-0.7266	1.5588	0.4911	330.9	160.6	-	20S	120E
7978	399	1302 Jan 14	21:35:54	482	-8633	103	T+	-p	0.1072	2.6455	1.6768	318.5	209.3	95.6	20N	41E
7979	399	1302 Jul 10	16:46:06	481	-8627	108	T+	pp	0.0313	2.8320	1.7694	371.0	234.5	105.9	21S	112E
7980	399	1303 Jan 04	12:41:28	480	-8621	113	P	a-	-0.5756	1.7992	0.8044	306.4	176.9	-	21N	174E

Cat Num	Canon Plate	Calendar Date	TD of Greatest Eclipse	ΔT s	Luna Num	Saros Num	Ecl. Type	QSE	Gamma	Pen. Mag.	Um. Mag.	Eclipse Phase Durations Pen. m	Eclipse Phase Durations Par. m	Eclipse Phase Durations Total m	Greatest in Zenith Lat.	Greatest in Zenith Lng.
7981	400	1303 Jun 29	21:28:03	479	-8615	118	P	h-	0.7904	1.4157	0.3998	301.7	140.2	-	22S	41E
7982	400	1303 Dec 24	22:58:34	478	-8609	123	N	h-	-1.3179	0.4655	-0.5856	192.9	-	-	22N	18E
7983	400	1304 May 20	02:07:02	478	-8604	090	N	-a	-1.0344	0.9336	-0.0142	237.3	-	-	22S	31W
7984	400	1304 Jun 18	09:25:47	477	-8603	128	Nb	a-	1.4896	0.1061	-0.8572	88.5	-	-	22S	139W
7985	400	1304 Nov 13	06:45:34	477	-8598	095	P	-t	0.9780	1.1157	0.0118	298.5	26.9	-	21N	103W
7986	400	1305 May 09	19:16:38	476	-8592	100	T	-p	-0.3054	2.2711	1.3234	312.8	202.8	77.1	20S	72E
7987	400	1305 Nov 02	06:55:19	475	-8586	105	T	-p	0.2773	2.3874	1.3116	361.0	221.3	81.5	17N	106W
7988	400	1306 Apr 29	10:13:32	474	-8580	110	T	a-	0.4499	2.0267	1.0381	321.7	196.9	29.9	16S	152W
7989	400	1306 Oct 22	13:35:34	473	-8574	115	T	a-	-0.4369	2.0648	1.0478	327.6	198.4	33.4	13N	154E
7990	400	1307 Apr 18	18:50:03	472	-8568	120	N	t-	1.2553	0.5777	-0.4683	216.9	-	-	12S	79E
7991	400	1307 Sep 12	17:57:27	471	-8563	087	N	-a	1.4952	0.0914	-0.8630	80.2	-	-	0N	90E
7992	400	1307 Oct 12	03:00:20	471	-8562	125	N	a-	-1.1017	0.8211	-0.1488	226.4	-	-	9N	46W
7993	400	1308 Mar 08	03:50:42	470	-8557	092	P	-t	-0.9050	1.2406	0.1548	307.9	95.0	-	1N	54W
7994	400	1308 Sep 01	10:07:04	469	-8551	097	P	-a	0.8267	1.3266	0.3552	275.9	125.9	-	5S	151W
7995	400	1309 Feb 25	06:25:40	468	-8545	102	T-	pp	-0.1623	2.5844	1.5364	354.8	224.0	96.5	6N	91W
7996	400	1309 Aug 21	22:08:01	467	-8539	107	T+	pp	0.1025	2.6801	1.6598	347.7	223.4	100.5	9S	30E
7997	400	1310 Feb 14	16:04:18	466	-8533	112	P	a-	0.5548	1.8373	0.8425	310.5	181.3	-	11N	125E
7998	400	1310 Aug 11	03:04:21	466	-8527	117	P	t-	-0.6706	1.6608	0.5948	335.3	172.8	-	14S	43W
7999	400	1311 Jan 05	20:47:32	465	-8522	084	N	-a	-1.4416	0.1988	-0.7736	117.7	-	-	20N	53E
8000	400	1311 Feb 04	06:58:29	465	-8521	122	N	a-	1.2089	0.6226	-0.3435	199.5	-	-	15N	99W
8001	401	1311 Jul 01	13:00:36	464	-8516	089	N	-t	1.4161	0.2856	-0.7662	159.0	-	-	21S	168E
8002	401	1311 Jul 31	03:29:38	464	-8515	127	N	t-	-1.4136	0.3000	-0.7713	165.8	-	-	18S	49W
8003	401	1311 Dec 26	10:55:55	463	-8510	094	P	-a	-0.8194	1.3594	0.3496	285.9	127.1	-	22N	161W
8004	401	1312 Jun 19	19:43:46	462	-8504	099	P	-t	0.5971	1.7620	0.7624	316.4	179.9	-	23S	66E
8005	401	1312 Dec 14	19:11:58	461	-8498	104	T-	-p	-0.1424	2.6309	1.5630	360.0	225.7	97.7	23N	74E
8006	401	1313 Jun 09	09:22:06	460	-8492	109	T-	p-	-0.1828	2.4987	1.5459	319.9	210.5	92.7	24S	139W
8007	401	1313 Dec 03	20:22:21	459	-8486	114	P	t-	0.5534	1.8952	0.7906	354.9	194.3	-	24N	55E
8008	401	1314 May 30	02:28:18	458	-8480	119	P	a-	-0.9123	1.1547	0.2128	255.9	97.7	-	24S	36W
8009	401	1314 Nov 22	19:41:06	457	-8474	124	N	t-	1.2152	0.6734	-0.4162	238.2	-	-	23N	64E
8010	401	1315 Apr 20	08:49:00	457	-8469	091	P	-a	0.9245	1.1647	0.1583	276.4	90.0	-	13S	131W
8011	401	1315 Oct 13	11:44:11	456	-8463	096	P	-a	-0.9511	1.1121	0.1132	264.6	74.7	-	10N	177W
8012	401	1316 Apr 08	15:14:51	455	-8457	101	T+	pp	0.1902	2.5400	1.4785	364.7	228.2	95.4	10S	133E
8013	401	1316 Oct 02	02:28:20	454	-8451	106	T-	-p	-0.2301	2.4164	1.4547	315.8	205.8	86.5	6N	38W
8014	401	1317 Mar 28	15:49:55	453	-8445	111	P	t-	-0.5516	1.8876	0.8048	352.6	195.8	-	7S	125E
8015	401	1317 Sep 21	18:41:57	452	-8439	116	T	a-	0.4690	1.9815	1.0130	308.0	189.3	17.1	3N	79E
8016	401	1318 Mar 17	17:27:00	451	-8433	121	N	h-	-1.2533	0.5856	-0.4687	218.5	-	-	3S	101E
8017	401	1318 Aug 12	19:38:00	450	-8428	088	N	-h	-1.5096	0.1051	-0.9288	95.2	-	-	14S	69E
8018	401	1318 Sep 11	07:44:46	450	-8427	126	N	h-	1.2321	0.6037	-0.4092	210.8	-	-	1S	116W
8019	401	1319 Feb 05	14:49:06	449	-8422	093	P	-a	0.8062	1.3706	0.3866	279.5	130.5	-	14N	144E
8020	401	1319 Aug 01	22:50:32	449	-8416	098	P	-t	-0.8064	1.4142	0.3431	321.2	137.5	-	17S	21E
8021	402	1320 Jan 26	06:20:48	448	-8410	103	T+	-p	0.1171	2.6265	1.6594	318.3	209.2	95.2	17N	89W
8022	402	1320 Jul 20	23:20:21	447	-8404	108	T-	pp	-0.0510	2.7958	1.7335	370.2	233.9	105.3	19S	13E
8023	402	1321 Jan 14	21:22:17	446	-8398	113	P	a-	-0.5704	1.8092	0.8135	307.6	178.0	-	19N	44E
8024	402	1321 Jul 10	04:25:23	445	-8392	118	P	h-	0.7105	1.5609	0.5475	310.2	159.8	-	21S	64W
8025	402	1322 Jan 04	07:22:09	444	-8386	123	N	h-	-1.3164	0.4693	-0.5839	194.2	-	-	21N	107W
8026	402	1322 May 31	09:35:18	443	-8381	090	N	-a	-1.1028	0.8068	-0.1387	223.1	-	-	24S	143W
8027	402	1322 Jun 29	16:45:29	443	-8380	128	N	a-	1.4140	0.2438	-0.7173	132.1	-	-	21S	111E
8028	402	1322 Nov 24	14:41:01	442	-8375	095	N*	-t	0.9897	1.0951	-0.0102	296.5	-	-	23N	139E
8029	402	1323 May 21	02:48:09	442	-8369	100	T	-a	-0.3753	2.1430	1.1953	310.0	198.0	62.6	22S	41W
8030	402	1323 Nov 13	14:59:35	441	-8363	105	T	-p	0.2897	2.3643	1.2890	359.6	220.0	79.0	20N	134E
8031	402	1324 May 09	17:31:33	440	-8357	110	T	p-	0.3823	2.1514	1.1614	327.0	204.1	59.4	19S	98E
8032	402	1324 Nov 01	21:57:32	439	-8351	115	T	a-	-0.4205	2.0946	1.0783	327.6	199.5	42.3	17N	29E
8033	402	1325 Apr 29	01:47:03	438	-8345	120	N	t-	1.1908	0.6965	-0.3501	236.1	-	-	16S	26W
8034	402	1325 Sep 23	02:08:36	437	-8340	087	Ne	-a	1.5302	0.0286	-0.9286	45.1	-	-	5N	33W
8035	402	1325 Oct 22	11:30:16	437	-8339	125	N	a-	-1.0801	0.8613	-0.1097	230.6	-	-	13N	174W
8036	402	1326 Mar 19	11:03:40	436	-8334	092	P	-t	-0.9512	1.1535	0.0722	299.0	65.6	-	3S	163W
8037	402	1326 Sep 12	18:02:10	435	-8328	097	P	-a	0.8726	1.2451	0.2684	270.7	111.1	-	0S	89E
8038	402	1327 Mar 08	14:08:34	435	-8322	102	T-	pp	-0.1975	2.5166	1.4749	352.1	222.1	93.1	2N	152E
8039	402	1327 Sep 02	05:31:42	434	-8316	107	T+	-p	0.1582	2.5811	1.5545	348.5	222.3	96.8	5S	82W
8040	402	1328 Feb 26	00:18:42	433	-8310	112	P	a-	0.5278	1.8841	0.8948	311.2	184.7	-	7N	0E

Cat Num	Canon Plate	Calendar Date	TD of Greatest Eclipse	ΔT s	Luna Num	Saros Num	Ecl. Type	QSE	Gamma	Pen. Mag.	Um. Mag.	Eclipse Phase Durations Pen. m	Par. m	Total m	Greatest in Zenith Lat.	Lng.
8041	403	1328 Aug 21	09:56:15	432	-8304	117	P	t-	-0.6050	1.7834	0.7127	343.3	185.8	-	10S	147W
8042	403	1329 Jan 16	05:35:00	431	-8299	084	N	-a	-1.4497	0.1837	-0.7882	113.4	-	-	18N	79W
8043	403	1329 Feb 14	15:30:42	431	-8298	122	N	a-	1.1896	0.6565	-0.3066	204.0	-	-	11N	133E
8044	403	1329 Jul 11	19:37:15	430	-8293	089	N	-t	1.4929	0.1439	-0.9065	114.0	-	-	20S	69E
8045	403	1329 Aug 10	10:10:57	430	-8292	127	N	t-	-1.3403	0.4352	-0.6374	197.1	-	-	14S	150W
8046	403	1330 Jan 05	19:35:32	430	-8287	094	P	-a	-0.8250	1.3500	0.3383	286.1	125.6	-	21N	70E
8047	403	1330 Jul 01	02:46:07	429	-8281	099	P	-t	0.6729	1.6215	0.6250	307.6	166.1	-	22S	39W
8048	403	1330 Dec 26	03:30:41	428	-8275	104	T-	-p	-0.1471	2.6236	1.5532	361.1	226.0	97.5	23N	49W
8049	403	1331 Jun 20	16:44:34	427	-8269	109	T-	pp	-0.1072	2.6363	1.6856	320.6	212.4	97.6	23S	111E
8050	403	1331 Dec 15	04:24:50	426	-8263	114	P	t-	0.5467	1.9081	0.8025	355.8	195.5	-	24N	64W
8051	403	1332 Jun 09	09:55:11	425	-8257	119	P	a-	-0.8396	1.2881	0.3459	266.5	122.3	-	24S	147W
8052	403	1332 Dec 03	03:48:29	424	-8251	124	N	t-	1.2052	0.6911	-0.3975	240.4	-	-	24N	57W
8053	403	1333 Apr 30	16:06:09	424	-8246	091	P	-a	0.9930	1.0397	0.0322	265.7	41.4	-	16S	119E
8054	403	1333 Oct 23	20:04:51	423	-8240	096	P	-a	-0.9693	1.0787	0.0800	260.8	62.9	-	13N	57E
8055	403	1334 Apr 19	22:12:11	422	-8234	101	T+	-p	0.2569	2.4173	1.3562	363.3	225.1	86.7	14S	28E
8056	403	1334 Oct 13	10:52:55	421	-8228	106	T-	-p	-0.2565	2.3690	1.4053	314.9	204.2	83.1	10N	165W
8057	403	1335 Apr 08	22:45:02	420	-8222	111	P	t-	-0.4893	2.0000	0.9209	357.1	205.0	-	11S	20E
8058	403	1335 Oct 03	02:54:06	419	-8216	116	T	p-	0.4364	2.0437	1.0705	311.1	192.8	39.1	7N	45W
8059	403	1336 Mar 28	00:47:28	418	-8210	121	N	h-	-1.2002	0.6799	-0.3682	232.3	-	-	7S	10W
8060	403	1336 Sep 21	15:30:34	418	-8204	126	N	h-	1.1952	0.6745	-0.3447	222.3	-	-	4N	126E
8061	404	1337 Feb 15	23:13:25	417	-8199	093	P	-a	0.8261	1.3317	0.3523	275.9	125.0	-	10N	17E
8062	404	1337 Aug 12	05:32:46	416	-8193	098	P	-t	-0.8797	1.2814	0.2066	311.0	109.0	-	13S	80W
8063	404	1338 Feb 05	14:59:11	415	-8187	103	T+	-p	0.1328	2.5967	1.6315	318.0	208.9	94.5	13N	141E
8064	404	1338 Aug 01	06:00:25	414	-8181	108	T-	pp	-0.1292	2.6524	1.5901	368.4	231.8	101.4	16S	87W
8065	404	1339 Jan 26	05:58:09	414	-8175	113	P	a-	-0.5615	1.8259	0.8295	309.2	179.7	-	16N	84W
8066	404	1339 Jul 21	11:26:41	413	-8169	118	P	h-	0.6327	1.7028	0.6914	317.1	174.9	-	18S	169W
8067	404	1340 Jan 15	15:41:50	412	-8163	123	N	h-	-1.3126	0.4771	-0.5776	196.3	-	-	18N	129E
8068	404	1340 Jun 10	17:00:17	411	-8158	090	N	-a	-1.1742	0.6750	-0.2688	206.5	-	-	25S	107E
8069	404	1340 Jul 10	00:06:41	411	-8157	128	N	a-	1.3388	0.3810	-0.5785	162.6	-	-	20S	1E
8070	404	1340 Dec 04	22:41:46	410	-8152	095	N*	-t	0.9970	1.0820	-0.0240	295.1	-	-	24N	20E
8071	404	1341 May 31	10:15:31	409	-8146	100	T	-a	-0.4482	2.0095	1.0612	306.4	191.6	36.5	23S	153W
8072	404	1341 Nov 23	23:09:48	409	-8140	105	T	-p	0.2980	2.3486	1.2744	358.2	218.9	77.2	22N	12E
8073	404	1342 May 21	00:44:45	408	-8134	110	T	p-	0.3114	2.2824	1.2905	331.8	210.3	76.6	21S	10W
8074	404	1342 Nov 13	06:26:34	407	-8128	115	T	a-	-0.4091	2.1150	1.0997	327.2	200.1	47.3	20N	98W
8075	404	1343 May 10	08:36:45	406	-8122	120	N	t-	1.1208	0.8253	-0.2221	254.5	-	-	18S	129W
8076	404	1343 Nov 02	20:06:56	405	-8116	125	N	a-	-1.0639	0.8917	-0.0804	233.7	-	-	16N	56E
8077	404	1344 Mar 29	18:09:24	405	-8111	092	N*	-t	-1.0035	1.0552	-0.0213	288.3	-	-	8S	90E
8078	404	1344 Sep 23	02:03:46	404	-8105	097	P	-a	0.9124	1.1748	0.1928	266.0	95.4	-	4N	32W
8079	404	1345 Mar 18	21:44:41	403	-8099	102	T-	pp	-0.2390	2.4373	1.4020	349.0	219.5	88.1	3S	37E
8080	404	1345 Sep 12	13:02:48	402	-8093	107	T+	-p	0.2067	2.4953	1.4623	349.0	220.8	92.0	1S	164E
8081	405	1346 Mar 08	08:26:57	401	-8087	112	P	a-	0.4951	1.9412	0.9574	312.3	188.5	-	2N	123W
8082	405	1346 Sep 01	16:54:27	400	-8081	117	P	t-	-0.5451	1.8959	0.8204	350.0	195.9	-	6S	107E
8083	405	1347 Jan 27	14:17:28	400	-8076	084	N	-a	-1.4618	0.1611	-0.8100	106.5	-	-	15N	151E
8084	405	1347 Feb 25	23:56:18	400	-8075	122	N	a-	1.1650	0.7002	-0.2600	209.6	-	-	7N	6E
8085	405	1347 Jul 23	02:19:29	399	-8070	089	Ne	-t	1.5654	0.0105	-1.0388	31.0	-	-	17S	32W
8086	405	1347 Aug 21	16:59:59	399	-8069	127	N	t-	-1.2725	0.5603	-0.5137	220.8	-	-	11S	107E
8087	405	1348 Jan 17	04:09:50	398	-8064	094	P	-a	-0.8346	1.3330	0.3200	285.7	122.8	-	18N	58W
8088	405	1348 Jul 11	09:52:30	397	-8058	099	P	-h	0.7460	1.4861	0.4919	298.1	150.0	-	20S	145W
8089	405	1349 Jan 05	11:45:40	397	-8052	104	T-	-p	-0.1541	2.6116	1.5395	362.1	226.3	97.1	21N	172W
8090	405	1349 Jul 01	00:09:20	396	-8046	109	T-	pp	-0.0330	2.7719	1.8225	320.6	213.1	99.8	23S	0E
8091	405	1349 Dec 25	12:25:48	395	-8040	114	P	t-	0.5389	1.9224	0.8167	356.6	196.8	-	23N	177E
8092	405	1350 Jun 20	17:21:47	394	-8034	119	P	a-	-0.7670	1.4219	0.4788	276.0	141.2	-	24S	102E
8093	405	1350 Dec 14	11:59:03	393	-8028	124	N	h-	1.1975	0.7044	-0.3824	241.8	-	-	25N	178W
8094	405	1351 May 11	23:14:20	393	-8023	091	N	-h	1.0676	0.9036	-0.1054	252.2	-	-	19S	12E
8095	405	1351 Jun 10	08:50:44	392	-8022	129	Nb	h-	-1.5474	0.0079	-0.9709	24.8	-	-	25S	131W
8096	405	1351 Nov 04	04:34:08	392	-8017	096	P	-a	-0.9811	1.0568	0.0584	258.1	53.8	-	17N	70W
8097	405	1352 Apr 30	04:59:28	391	-8011	101	T	-t	0.3309	2.2814	1.2205	360.7	220.2	71.9	17S	74W
8098	405	1352 Oct 23	19:25:34	390	-8005	106	T	-p	-0.2765	2.3333	1.3677	314.2	202.9	80.2	14N	66E
8099	405	1353 Apr 19	05:32:00	389	-7999	111	T	t-	-0.4201	2.1251	1.0498	361.4	213.5	36.4	14S	82W
8100	405	1353 Oct 13	11:14:11	389	-7993	116	T	p-	0.4109	2.0928	1.1148	313.6	195.4	49.3	11N	171W

Cat Num	Canon Plate	Calendar Date	TD of Greatest Eclipse	ΔT s	Luna Num	Saros Num	Ecl. Type	QSE	Gamma	Pen. Mag.	Um. Mag.	Eclipse Phase Durations Pen. m	Par. m	Total m	Greatest in Zenith Lat.	Lng.
8101	406	1354 Apr 08	08:02:44	388	-7987	121	N	h-	-1.1418	0.7840	-0.2580	245.7	-	-	11S	120W
8102	406	1354 Oct 02	23:21:57	387	-7981	126	N	h-	1.1637	0.7355	-0.2900	231.8	-	-	8N	7E
8103	406	1355 Feb 27	07:33:13	386	-7976	093	P	-a	0.8503	1.2849	0.3103	271.6	117.8	-	6N	108W
8104	406	1355 Aug 23	12:20:17	386	-7970	098	P	-t	-0.9483	1.1575	0.0788	300.2	68.6	-	10S	177E
8105	406	1356 Feb 16	23:31:22	385	-7964	103	T+	-p	0.1535	2.5578	1.5946	317.6	208.5	93.3	10N	12E
8106	406	1356 Aug 11	12:47:57	384	-7958	108	T-	pp	-0.2017	2.5196	1.4568	365.9	228.4	94.4	13S	170E
8107	406	1357 Feb 05	14:27:27	383	-7952	113	P	a-	-0.5479	1.8509	0.8542	311.1	182.1	-	13N	148E
8108	406	1357 Jul 31	18:35:47	382	-7946	118	P	a-	0.5603	1.8348	0.8249	322.5	186.1	-	16S	84E
8109	406	1358 Jan 25	23:54:42	382	-7940	123	N	t-	-1.3040	0.4933	-0.5625	199.9	-	-	15N	6E
8110	406	1358 Jun 22	00:27:21	381	-7935	090	N	-a	-1.2441	0.5462	-0.3966	187.9	-	-	25S	5W
8111	406	1358 Jul 21	07:33:39	381	-7934	128	N	a-	1.2675	0.5113	-0.4473	185.5	-	-	18S	111W
8112	406	1358 Dec 16	06:43:08	380	-7929	095	N*	-t	1.0039	1.0693	-0.0367	293.6	-	-	24N	99W
8113	406	1359 Jun 11	17:39:47	379	-7923	100	P	-a	-0.5233	1.8723	0.9228	301.9	183.5	-	24S	97E
8114	406	1359 Dec 05	07:24:19	379	-7917	105	T	-p	0.3034	2.3378	1.2653	356.9	218.0	76.0	24N	111W
8115	406	1360 May 31	07:51:57	378	-7911	110	T+	pp	0.2362	2.4218	1.4274	336.2	215.6	88.6	23S	117W
8116	406	1360 Nov 23	15:00:39	377	-7905	115	T	a-	-0.4013	2.1286	1.1148	326.6	200.4	50.4	22N	134E
8117	406	1361 May 20	15:19:11	376	-7899	120	N	t-	1.0453	0.9645	-0.0841	272.0	-	-	21S	131E
8118	406	1361 Nov 13	04:49:55	375	-7893	125	N	a-	-1.0525	0.9129	-0.0598	235.7	-	-	19N	74W
8119	406	1362 Apr 10	01:06:46	375	-7888	092	N	-t	-1.0626	0.9445	-0.1274	275.2	-	-	12S	16W
8120	406	1362 Oct 04	10:12:08	374	-7882	097	P	-a	0.9460	1.1158	0.1284	261.9	78.7	-	8N	155W
8121	407	1363 Mar 30	05:13:30	373	-7876	102	T	-p	-0.2874	2.3452	1.3164	345.4	216.1	80.6	7S	76W
8122	407	1363 Sep 23	20:40:23	372	-7870	107	T+	-p	0.2489	2.4211	1.3816	349.4	219.0	86.5	4N	49E
8123	407	1364 Mar 18	16:26:32	372	-7864	112	T	a-	0.4548	2.0125	1.0342	313.8	192.8	27.8	2S	116E
8124	407	1364 Sep 12	00:00:44	371	-7858	117	P	t-	-0.4920	1.9957	0.9153	355.4	203.7	-	2S	0W
8125	407	1365 Feb 06	22:54:26	370	-7853	084	N	-a	-1.4782	0.1304	-0.8395	96.2	-	-	11N	21E
8126	407	1365 Mar 08	08:14:22	370	-7852	122	N	a-	1.1345	0.7547	-0.2025	216.2	-	-	3N	119W
8127	407	1365 Aug 31	23:57:40	369	-7846	127	N	t-	-1.2107	0.6745	-0.4012	239.4	-	-	6S	2E
8128	407	1366 Jan 27	12:39:18	369	-7841	094	P	-a	-0.8476	1.3097	0.2958	284.8	118.8	-	15N	175E
8129	407	1366 Jul 22	17:04:01	368	-7835	099	P	-h	0.8156	1.3574	0.3652	288.1	131.3	-	18S	107E
8130	407	1367 Jan 16	19:57:34	367	-7829	104	T-	-p	-0.1629	2.5959	1.5227	362.9	226.5	96.5	19N	66E
8131	407	1367 Jul 12	07:35:23	366	-7823	109	T+	pp	0.0404	2.7579	1.8092	319.9	212.7	99.5	21S	111W
8132	407	1368 Jan 05	20:26:16	366	-7817	114	P	t-	0.5309	1.9368	0.8316	357.3	198.1	-	22N	58E
8133	407	1368 Jul 01	00:48:23	365	-7811	119	P	a-	-0.6948	1.5552	0.6105	284.5	156.4	-	23S	10W
8134	407	1368 Dec 24	20:09:43	364	-7805	124	N	h-	1.1894	0.7181	-0.3662	243.1	-	-	24N	60E
8135	407	1369 May 22	06:17:43	363	-7800	091	N	-h	1.1448	0.7628	-0.2483	236.0	-	-	21S	94W
8136	407	1369 Jun 20	16:01:26	363	-7799	129	N	h-	-1.4792	0.1346	-0.8474	101.7	-	-	25S	122E
8137	407	1369 Nov 14	13:09:02	363	-7794	096	P	-a	-0.9887	1.0426	0.0449	256.0	47.1	-	19N	161E
8138	407	1370 May 11	11:41:22	362	-7788	101	T	-t	0.4088	2.1386	1.0777	357.0	213.3	44.8	19S	175W
8139	407	1370 Nov 04	04:03:32	361	-7782	106	T	-p	-0.2920	2.3057	1.3383	313.6	201.8	77.6	17N	63W
8140	407	1371 Apr 30	12:13:03	360	-7776	111	T	t-	-0.3457	2.2598	1.1879	365.0	220.9	67.6	17S	177E
8141	408	1371 Oct 24	19:40:31	360	-7770	116	T	p-	0.3911	2.1314	1.1490	315.8	197.4	55.7	15N	62E
8142	408	1372 Apr 18	15:09:54	359	-7764	121	N	h-	-1.0757	0.9022	-0.1338	259.3	-	-	15S	133E
8143	408	1372 Oct 13	07:21:39	358	-7758	126	N	h-	1.1395	0.7830	-0.2487	239.1	-	-	12N	113W
8144	408	1373 Mar 09	15:44:12	358	-7753	093	P	-a	0.8822	1.2240	0.2542	266.1	107.3	-	2N	128E
8145	408	1373 Sep 02	19:16:42	357	-7747	098	N*	-t	-1.0091	1.0478	-0.0346	289.5	-	-	6S	72E
8146	408	1374 Feb 27	07:55:14	356	-7741	103	T+	-p	0.1809	2.5063	1.5452	317.0	207.8	91.3	5N	114W
8147	408	1374 Aug 22	19:44:13	355	-7735	108	T-	pp	-0.2675	2.3991	1.3358	362.9	224.2	84.8	9S	65E
8148	408	1375 Feb 16	22:49:15	355	-7729	113	P	a-	-0.5287	1.8861	0.8894	313.5	185.1	-	9N	22E
8149	408	1375 Aug 12	01:50:39	354	-7723	118	P	a-	0.4917	1.9602	0.9514	326.7	194.8	-	12S	26W
8150	408	1376 Feb 06	08:01:19	353	-7717	123	N	t-	-1.2911	0.5172	-0.5390	204.9	-	-	12N	116W
8151	408	1376 Jul 02	07:54:03	352	-7712	090	N	-a	-1.3147	0.4166	-0.5258	166.0	-	-	24S	116W
8152	408	1376 Jul 31	15:04:49	352	-7711	128	N	a-	1.1989	0.6370	-0.3213	204.0	-	-	15S	136E
8153	408	1376 Dec 26	14:45:06	352	-7706	095	N*	-t	1.0104	1.0569	-0.0482	292.1	-	-	24N	142E
8154	408	1377 Jun 22	01:02:58	351	-7700	100	P	-a	-0.5989	1.7346	0.7831	296.5	173.5	-	24S	14W
8155	408	1377 Dec 15	15:42:20	350	-7694	105	T	-p	0.3069	2.3302	1.2601	355.5	217.3	75.2	24N	126E
8156	408	1378 Jun 11	14:55:52	349	-7688	110	T+	pp	0.1591	2.5646	1.5673	339.8	219.6	96.7	23S	137E
8157	408	1378 Dec 04	23:38:54	349	-7682	115	T	a-	-0.3963	2.1369	1.1249	325.8	200.4	52.2	23N	6E
8158	408	1379 May 31	21:57:14	348	-7676	120	P	t-	0.9665	1.1098	0.0597	288.2	58.8	-	22S	32E
8159	408	1379 Nov 24	13:37:20	347	-7670	125	N	a-	-1.0449	0.9270	-0.0461	236.9	-	-	21N	155E
8160	408	1380 Apr 20	07:57:16	347	-7665	092	N	-t	-1.1271	0.8236	-0.2435	259.4	-	-	15S	119W

Cat Num	Canon Plate	Calendar Date	TD of Greatest Eclipse	ΔT s	Luna Num	Saros Num	Ecl. Type	QSE	Gamma	Pen. Mag.	Um. Mag.	Eclipse Phase Durations Pen. m	Par. m	Total m	Greatest in Zenith Lat.	Lng.
8161	409	1380 Oct 14	18:27:37	346	-7659	097	P	-a	0.9731	1.0688	0.0759	258.6	61.1	-	12N	80E
8162	409	1381 Apr 09	12:36:50	345	-7653	102	T	-p	-0.3410	2.2437	1.2211	341.1	211.7	69.5	11S	172E
8163	409	1381 Oct 04	04:24:45	344	-7647	107	T	-p	0.2847	2.3587	1.3129	349.8	217.2	80.5	8N	68W
8164	409	1382 Mar 30	00:20:27	344	-7641	112	T	a-	0.4091	2.0937	1.1207	315.3	197.0	50.9	6S	3W
8165	409	1382 Sep 23	07:15:12	343	-7635	117	P	t-	-0.4458	2.0828	0.9977	359.9	209.6	-	3N	110W
8166	409	1383 Feb 18	07:23:57	342	-7630	084	N	-a	-1.5005	0.0889	-0.8799	79.8	-	-	7N	107W
8167	409	1383 Mar 19	16:24:19	342	-7629	122	N	a-	1.0974	0.8213	-0.1330	223.9	-	-	2S	117E
8168	409	1383 Sep 12	07:04:45	342	-7623	127	N	t-	-1.1559	0.7760	-0.3013	253.9	-	-	2S	106W
8169	409	1384 Feb 07	21:01:03	341	-7618	094	P	-a	-0.8661	1.2760	0.2615	283.1	112.6	-	12N	50E
8170	409	1384 Aug 02	00:22:35	340	-7612	099	P	-a	0.8800	1.2387	0.2478	277.9	109.7	-	15S	3W
8171	409	1385 Jan 27	04:01:25	339	-7606	104	T-	-p	-0.1775	2.5694	1.4957	363.5	226.5	95.3	16N	55W
8172	409	1385 Jul 22	15:06:49	339	-7600	109	T+	pp	0.1096	2.6310	1.6824	318.7	211.3	97.1	18S	136E
8173	409	1386 Jan 16	04:22:29	338	-7594	114	P	t-	0.5199	1.9563	0.8526	358.1	199.9	-	20N	61W
8174	409	1386 Jul 12	08:15:58	337	-7588	119	P	a-	-0.6236	1.6869	0.7399	292.1	168.9	-	21S	121W
8175	409	1387 Jan 05	04:20:25	337	-7582	124	N	h-	1.1807	0.7323	-0.3488	244.5	-	-	23N	61W
8176	409	1387 Jun 02	13:14:54	336	-7577	091	N	-h	1.2257	0.6157	-0.3979	216.1	-	-	22S	162E
8177	409	1387 Jul 01	23:09:54	336	-7576	129	N	h-	-1.4100	0.2635	-0.7222	141.3	-	-	24S	15E
8178	409	1387 Nov 25	21:50:13	335	-7571	096	P	-a	-0.9921	1.0358	0.0392	254.7	44.0	-	21N	32E
8179	409	1388 May 21	18:15:06	335	-7565	101	P	-t	0.4926	1.9850	0.9237	351.9	203.7	-	21S	87E
8180	409	1388 Nov 14	12:48:27	334	-7559	106	T	-p	-0.3020	2.2882	1.3193	313.2	201.1	75.9	20N	166E
8181	410	1389 May 10	18:48:43	333	-7553	111	T-	pp	-0.2665	2.4036	1.3350	367.9	226.8	85.6	20S	78E
8182	410	1389 Nov 04	04:12:46	332	-7547	116	T	p-	0.3767	2.1601	1.1733	317.6	198.9	59.6	18N	66W
8183	410	1390 Apr 29	22:13:51	332	-7541	121	N*	h-	-1.0059	1.0275	-0.0029	271.9	-	-	18S	26E
8184	410	1390 Oct 24	15:27:20	331	-7535	126	N	h-	1.1210	0.8200	-0.2176	244.9	-	-	16N	125E
8185	410	1391 Mar 20	23:49:01	330	-7530	093	P	-a	0.9197	1.1528	0.1878	259.6	92.9	-	2S	6E
8186	410	1391 Sep 14	02:19:56	330	-7524	098	N	-t	-1.0641	0.9489	-0.1374	278.7	-	-	1S	35W
8187	410	1392 Mar 09	16:12:14	329	-7518	103	T+	-p	0.2141	2.4444	1.4854	316.1	206.7	88.4	1N	121E
8188	410	1392 Sep 02	02:49:46	328	-7512	108	T	-p	-0.3265	2.2910	1.2271	359.5	219.4	72.3	5S	42W
8189	410	1393 Feb 27	07:02:14	328	-7506	113	P	a-	-0.5028	1.9337	0.9370	316.3	188.9	-	5N	102W
8190	410	1393 Aug 22	09:15:18	327	-7500	118	T	a-	0.4300	2.0730	1.0650	329.7	201.1	39.2	9S	138W
8191	410	1394 Feb 16	15:59:20	326	-7494	123	N	t-	-1.2718	0.5528	-0.5036	211.7	-	-	8N	124E
8192	410	1394 Jul 13	15:23:50	326	-7489	090	N	-a	-1.3829	0.2915	-0.6511	140.4	-	-	22S	132E
8193	410	1394 Aug 11	22:42:10	325	-7488	128	N	a-	1.1347	0.7549	-0.2035	219.0	-	-	12S	21E
8194	410	1395 Jan 06	22:44:41	325	-7483	095	N*	-t	1.0191	1.0401	-0.0633	290.0	-	-	22N	23E
8195	410	1395 Jul 03	08:26:04	324	-7477	100	P	-a	-0.6743	1.5974	0.6434	290.2	161.2	-	23S	124W
8196	410	1395 Dec 27	00:00:15	324	-7471	105	T	-p	0.3111	2.3209	1.2539	354.1	216.5	74.4	23N	3E
8197	410	1396 Jun 21	21:56:41	323	-7465	110	T+	pp	0.0803	2.7111	1.7102	342.7	222.3	101.4	23S	33E
8198	410	1396 Dec 15	08:19:17	322	-7459	115	T	a-	-0.3923	2.1430	1.1335	324.9	200.3	53.7	23N	123W
8199	410	1397 Jun 11	04:31:36	321	-7453	120	P	t-	0.8850	1.2603	0.2082	303.0	107.8	-	23S	67W
8200	410	1397 Dec 04	22:27:28	321	-7447	125	N	a-	-1.0394	0.9371	-0.0359	237.8	-	-	22N	23E
8201	411	1398 May 01	14:41:55	320	-7442	092	N	-t	-1.1967	0.6937	-0.3691	240.5	-	-	19S	140E
8202	411	1398 Oct 26	02:50:02	319	-7436	097	P	-a	0.9943	1.0325	0.0346	256.1	41.6	-	16N	45W
8203	411	1399 Apr 20	19:52:49	319	-7430	102	T	-p	-0.4017	2.1292	1.1127	336.0	205.8	51.3	15S	62E
8204	411	1399 Oct 15	12:16:35	318	-7424	107	T	-p	0.3135	2.3089	1.2570	350.3	215.6	74.5	12N	173E
8205	411	1400 Apr 09	08:06:51	317	-7418	112	T	a-	0.3562	2.1881	1.2202	316.9	201.2	66.5	10S	121W
8206	411	1400 Oct 03	14:38:36	317	-7412	117	T	t-	-0.4071	2.1561	1.0665	363.4	214.1	41.7	7N	138E
8207	411	1401 Feb 28	15:45:52	316	-7407	084	Ne	-a	-1.5288	0.0363	-0.9312	51.3	-	-	3N	127E
8208	411	1401 Mar 30	00:26:41	316	-7406	122	N	a-	1.0541	0.8993	-0.0523	232.3	-	-	6S	5W
8209	411	1401 Sep 22	14:22:16	315	-7400	127	N	t-	-1.1084	0.8639	-0.2150	265.2	-	-	2N	143E
8210	411	1402 Feb 18	05:16:30	315	-7395	094	P	-a	-0.8892	1.2339	0.2190	280.5	104.0	-	8N	75W
8211	411	1402 Aug 13	07:47:03	314	-7389	099	P	-a	0.9402	1.1276	0.1376	267.5	82.8	-	11S	115W
8212	411	1403 Feb 07	12:00:03	313	-7383	104	T-	-p	-0.1958	2.5359	1.4622	363.8	226.2	93.6	12N	175W
8213	411	1403 Aug 02	22:42:18	313	-7377	109	T+	-p	0.1755	2.5103	1.5612	317.1	209.1	92.7	15S	22E
8214	411	1404 Jan 27	12:13:43	312	-7371	114	P	t-	0.5045	1.9835	0.8819	359.1	202.2	-	17N	178W
8215	411	1404 Jul 22	15:46:36	311	-7365	119	P	a-	-0.5553	1.8137	0.8639	298.6	178.9	-	19S	126E
8216	411	1405 Jan 15	12:28:24	311	-7359	124	N	h-	1.1692	0.7514	-0.3257	246.4	-	-	20N	178E
8217	411	1405 Jun 12	20:10:11	310	-7354	091	N	-h	1.3072	0.4678	-0.5489	191.8	-	-	22S	59E
8218	411	1405 Jul 12	06:20:21	310	-7353	129	N	h-	-1.3426	0.3894	-0.6007	170.4	-	-	22S	92W
8219	411	1405 Dec 06	06:33:13	309	-7348	096	P	-a	-0.9946	1.0304	0.0353	253.5	41.7	-	22N	98W
8220	411	1406 Jun 02	00:46:25	309	-7342	101	P	-t	0.5777	1.8292	0.7671	345.3	191.2	-	23S	11W

Cat Num	Canon Plate	Calendar Date	TD of Greatest Eclipse	ΔT s	Luna Num	Saros Num	Ecl. Type	QSE	Gamma	Pen. Mag.	Um. Mag.	Eclipse Phase Durations Pen. m	Par. m	Total m	Greatest in Zenith Lat.	Lng.
8221	412	1406 Nov 25	21:36:51	308	-7336	106	T	-p	-0.3091	2.2757	1.3057	313.0	200.6	74.6	22N	35E
8222	412	1407 May 22	01:19:29	307	-7330	111	T-	pp	-0.1827	2.5558	1.4902	369.9	231.2	97.2	22S	20W
8223	412	1407 Nov 15	12:50:10	307	-7324	116	T	p-	0.3670	2.1799	1.1892	319.2	199.9	62.1	21N	165E
8224	412	1408 May 10	05:12:17	306	-7318	121	P	h-	-0.9307	1.1628	0.1378	283.7	86.1	-	21S	78W
8225	412	1408 Nov 03	23:39:10	305	-7312	126	N	h-	1.1080	0.8465	-0.1965	249.2	-	-	19N	2E
8226	412	1409 Mar 31	07:45:13	305	-7307	093	P	-a	0.9647	1.0679	0.1074	251.5	71.0	-	7S	114W
8227	412	1409 Sep 24	09:33:48	304	-7301	098	N	-t	-1.1102	0.8662	-0.2238	268.9	-	-	3N	144W
8228	412	1410 Mar 21	00:20:45	303	-7295	103	T+	-p	0.2544	2.3696	1.4125	315.0	205.0	83.9	3S	3W
8229	412	1410 Sep 13	10:04:27	303	-7289	108	T	-p	-0.3788	2.1956	1.1308	355.9	214.2	56.6	1S	151W
8230	412	1411 Mar 10	15:06:37	302	-7283	113	P	a-	-0.4704	1.9931	0.9965	319.5	193.3	-	0N	136E
8231	412	1411 Sep 02	16:48:09	301	-7277	118	T	a-	0.3740	2.1755	1.1679	331.9	205.8	60.8	5S	108E
8232	412	1412 Feb 27	23:47:40	301	-7271	123	N	t-	-1.2454	0.6012	-0.4551	220.3	-	-	4N	7E
8233	412	1412 Jul 23	22:56:30	300	-7266	090	N	-a	-1.4488	0.1710	-0.7726	108.7	-	-	19S	19E
8234	412	1412 Aug 22	06:26:36	300	-7265	128	N	a-	1.0758	0.8633	-0.0957	231.2	-	-	8S	96W
8235	412	1413 Jan 17	06:41:56	300	-7260	095	N*	-t	1.0298	1.0192	-0.0818	287.3	-	-	20N	95W
8236	412	1413 Jul 13	15:50:27	299	-7254	100	P	-a	-0.7479	1.4639	0.5068	283.2	146.5	-	21S	125E
8237	412	1414 Jan 06	08:17:53	298	-7248	105	T	-p	0.3163	2.3096	1.2463	352.5	215.7	73.3	22N	120W
8238	412	1414 Jul 03	04:57:15	298	-7242	110	T+	pp	0.0020	2.8567	1.8518	344.8	223.7	103.1	22S	72W
8239	412	1414 Dec 26	17:01:07	297	-7236	115	T	a-	-0.3889	2.1477	1.1411	324.0	200.2	55.0	22N	107E
8240	412	1415 Jun 22	11:03:06	296	-7230	120	P	t-	0.8015	1.4148	0.3602	316.4	139.0	-	23S	164W
8241	413	1415 Dec 16	07:19:54	296	-7224	125	N	a-	-1.0357	0.9437	-0.0290	238.2	-	-	22N	109W
8242	413	1416 May 11	21:22:21	295	-7219	092	N	-t	-1.2700	0.5571	-0.5015	217.7	-	-	21S	39E
8243	413	1416 Jun 10	11:56:50	295	-7218	130	Nb	t-	1.5632	0.0240	-1.0441	48.1	-	-	22S	178W
8244	413	1416 Nov 05	11:17:37	294	-7213	097	P	-a	1.0107	1.0048	0.0019	254.3	9.9	-	19N	172W
8245	413	1417 May 01	03:05:22	294	-7207	102	P	-t	-0.4661	2.0081	0.9973	330.1	198.6	-	18S	46W
8246	413	1417 Oct 25	20:15:07	293	-7201	107	T	-p	0.3364	2.2698	1.2122	350.8	214.2	68.9	16N	53E
8247	413	1418 Apr 20	15:48:18	292	-7195	112	T	p-	0.2987	2.2912	1.3281	318.3	205.0	78.1	14S	123E
8248	413	1418 Oct 14	22:09:21	292	-7189	117	T	t-	-0.3746	2.2179	1.1241	366.2	217.5	56.0	11N	25E
8249	413	1419 Apr 10	08:21:39	291	-7183	122	P	a-	1.0048	0.9886	0.0393	241.3	43.0	-	10S	124W
8250	413	1419 Oct 03	21:50:05	291	-7177	127	N	t-	-1.0687	0.9374	-0.1428	273.8	-	-	6N	31E
8251	413	1420 Feb 29	13:20:56	290	-7172	094	P	-a	-0.9205	1.1765	0.1614	276.6	90.3	-	4N	163E
8252	413	1420 Aug 23	15:21:06	289	-7166	099	P	-a	0.9934	1.0299	0.0404	257.5	45.3	-	7S	130E
8253	413	1421 Feb 17	19:48:27	289	-7160	104	T-	-p	-0.2216	2.4882	1.4150	363.8	225.4	90.6	8N	67E
8254	413	1421 Aug 13	06:24:23	288	-7154	109	T+	-p	0.2363	2.3993	1.4492	315.1	206.2	86.6	12S	94W
8255	413	1422 Feb 06	19:58:01	287	-7148	114	P	t-	0.4834	2.0208	0.9219	360.4	205.2	-	13N	66E
8256	413	1422 Aug 02	23:20:42	287	-7142	119	P	a-	-0.4902	1.9348	0.9817	304.3	187.1	-	16S	12E
8257	413	1423 Jan 26	20:33:05	286	-7136	124	N	h-	1.1544	0.7763	-0.2962	248.9	-	-	17N	57E
8258	413	1423 Jun 24	03:00:25	286	-7131	091	N	-h	1.3912	0.3153	-0.7050	160.4	-	-	22S	43W
8259	413	1423 Jul 23	13:30:12	285	-7130	129	N	h-	-1.2754	0.5152	-0.4798	194.6	-	-	20S	160E
8260	413	1423 Dec 17	15:20:01	285	-7125	096	P	-a	-0.9947	1.0293	0.0362	252.7	42.1	-	22N	132E
8261	414	1424 Jun 12	07:13:02	284	-7119	101	P	-t	0.6657	1.6684	0.6051	337.1	174.8	-	23S	107W
8262	414	1424 Dec 06	06:28:42	284	-7113	106	T	-p	-0.3134	2.2682	1.2975	312.9	200.3	73.7	23N	97W
8263	414	1425 Jun 01	07:48:18	283	-7107	111	T-	pp	-0.0970	2.7118	1.6488	370.9	233.8	103.8	23S	116W
8264	414	1425 Nov 25	21:31:52	282	-7101	116	T	p-	0.3617	2.1914	1.1971	320.5	200.7	63.3	23N	36E
8265	414	1426 May 21	12:09:04	282	-7095	121	P	h-	-0.8528	1.3032	0.2833	294.4	120.4	-	23S	178E
8266	414	1426 Nov 15	07:55:22	281	-7089	126	N	h-	1.0992	0.8652	-0.1830	252.5	-	-	22N	121W
8267	414	1427 Apr 11	15:36:23	281	-7084	093	P	-a	1.0144	0.9746	0.0184	242.1	29.7	-	10S	127E
8268	414	1427 May 10	23:39:25	280	-7083	131	Nb	a-	-1.5427	0.0112	-0.9572	29.1	-	-	21S	5E
8269	414	1427 Oct 05	16:55:33	280	-7078	098	N	-t	-1.1500	0.7951	-0.2985	259.7	-	-	7N	104E
8270	414	1428 Mar 31	08:21:15	279	-7072	103	T	-p	0.3013	2.2826	1.3273	313.4	202.5	77.2	7S	124W
8271	414	1428 Sep 23	17:29:54	279	-7066	108	T	-t	-0.4230	2.1148	1.0492	352.5	209.1	35.6	3N	96E
8272	414	1429 Mar 20	23:02:11	278	-7060	113	T	p-	-0.4312	2.0650	1.0683	323.0	198.1	39.5	4S	17E
8273	414	1429 Sep 13	00:30:32	277	-7054	118	T	a-	0.3250	2.2652	1.2579	333.2	209.1	72.8	0S	8W
8274	414	1430 Mar 10	07:26:31	277	-7048	123	N	t-	-1.2119	0.6625	-0.3935	230.5	-	-	0S	109W
8275	414	1430 Aug 04	06:34:42	276	-7043	090	Ne	-a	-1.5105	0.0585	-0.8865	64.2	-	-	16S	96W
8276	414	1430 Sep 02	14:18:56	276	-7042	128	P	a-	1.0229	0.9608	0.0008	241.1	6.3	-	4S	145E
8277	414	1431 Jan 28	14:33:23	276	-7037	095	N	-t	1.0455	0.9890	-0.1089	283.4	-	-	17N	147E
8278	414	1431 Jul 24	23:16:42	275	-7031	100	P	-a	-0.8195	1.3344	0.3735	275.4	128.6	-	19S	14E
8279	414	1432 Jan 17	16:32:40	274	-7025	105	T	-p	0.3247	2.2918	1.2331	350.6	214.6	71.5	19N	117E
8280	414	1432 Jul 13	11:58:17	274	-7019	110	T-	pp	-0.0751	2.7250	1.7154	346.1	223.7	101.9	20S	177W

Cat Num	Canon Plate	Calendar Date	TD of Greatest Eclipse	ΔT s	Luna Num	Saros Num	Ecl. Type	QSE	Gamma	Pen. Mag.	Um. Mag.	Eclipse Phase ---- Durations ---- Pen. m	Par. m	Total m	Greatest in Zenith Lat.	Lng.
8281	415	1433 Jan 06	01:40:46	273	-7013	115	T	p-	-0.3829	2.1570	1.1538	323.1	200.3	57.1	21N	22W
8282	415	1433 Jul 02	17:34:56	273	-7007	120	P	t-	0.7183	1.5689	0.5115	328.2	162.2	-	22S	98E
8283	415	1433 Dec 26	16:12:40	272	-7001	125	N	a-	-1.0319	0.9502	-0.0215	238.7	-	-	22N	119E
8284	415	1434 May 23	03:58:56	271	-6996	092	N	-t	-1.3466	0.4146	-0.6401	189.8	-	-	23S	60W
8285	415	1434 Jun 21	18:20:00	271	-6995	130	N	t-	1.4767	0.1822	-0.8850	130.6	-	-	22S	86E
8286	415	1434 Nov 16	19:50:42	271	-6990	097	N	-a	1.0223	0.9858	-0.0216	253.4	-	-	22N	60E
8287	415	1435 May 12	10:13:11	270	-6984	102	P	-t	-0.5356	1.8779	0.8725	323.2	189.4	-	21S	153W
8288	415	1435 Nov 06	04:20:16	270	-6978	107	T	-a	0.3529	2.2421	1.1792	351.5	213.2	64.2	19N	68W
8289	415	1436 Apr 30	23:23:12	269	-6972	112	T+	p-	0.2351	2.4058	1.4470	319.6	208.4	87.2	17S	9E
8290	415	1436 Oct 25	05:49:24	268	-6966	117	T	p-	-0.3498	2.2654	1.1678	368.4	219.8	64.1	15N	90W
8291	415	1437 Apr 20	16:10:28	268	-6960	122	P	a-	0.9505	1.0873	0.1399	250.4	80.0	-	14S	118E
8292	415	1437 Oct 14	05:26:54	267	-6954	127	N	t-	-1.0354	0.9990	-0.0824	280.4	-	-	10N	84W
8293	415	1438 Mar 11	21:17:55	267	-6949	094	P	-a	-0.9572	1.1092	0.0939	271.4	69.8	-	1S	43E
8294	415	1438 Sep 03	23:02:41	266	-6943	099	N	-a	1.0411	0.9422	-0.0470	247.9	-	-	3S	14E
8295	415	1439 Mar 01	03:28:54	265	-6937	104	T-	-p	-0.2536	2.4292	1.3568	363.4	224.1	86.2	4N	48W
8296	415	1439 Aug 24	14:12:06	265	-6931	109	T	-p	0.2925	2.2968	1.3451	312.9	202.9	78.7	8S	148E
8297	415	1440 Feb 18	03:35:32	264	-6925	114	P	t-	0.4566	2.0685	0.9728	361.9	208.8	-	9N	49W
8298	415	1440 Aug 13	07:00:03	263	-6919	119	T	a-	-0.4299	2.0475	1.0904	309.1	193.5	44.1	12S	103W
8299	415	1441 Feb 06	04:31:34	263	-6913	124	N	h-	1.1337	0.8117	-0.2557	252.7	-	-	14N	63W
8300	415	1441 Jul 04	09:52:12	262	-6908	091	N	-t	1.4727	0.1679	-0.8565	119.2	-	-	21S	146W
8301	416	1441 Aug 02	20:45:02	262	-6907	129	N	t-	-1.2127	0.6328	-0.3675	214.1	-	-	16S	51E
8302	416	1441 Dec 28	00:06:08	262	-6902	096	P	-a	-0.9956	1.0263	0.0358	251.8	41.8	-	22N	1E
8303	416	1442 Jun 23	13:39:06	261	-6896	101	P	-t	0.7535	1.5081	0.4432	327.3	153.7	-	22S	157E
8304	416	1442 Dec 17	15:21:21	261	-6890	106	T	-p	-0.3170	2.2617	1.2906	312.8	200.0	73.1	23N	131E
8305	416	1443 Jun 12	14:15:52	260	-6884	111	T-	pp	-0.0098	2.8706	1.8097	370.7	234.6	106.2	24S	147E
8306	416	1443 Dec 07	06:15:30	259	-6878	116	T	p-	0.3582	2.1994	1.2020	321.7	201.3	64.1	24N	94W
8307	416	1444 May 31	19:03:18	259	-6872	121	P	h-	-0.7715	1.4502	0.4347	303.9	145.3	-	24S	75E
8308	416	1444 Nov 25	16:15:51	258	-6866	126	N	h-	1.0945	0.8762	-0.1764	254.8	-	-	23N	115E
8309	416	1445 Apr 21	23:20:33	258	-6861	093	N	-a	1.0703	0.8700	-0.0823	230.7	-	-	14S	11E
8310	416	1445 May 21	07:02:27	258	-6860	131	N	a-	-1.4705	0.1417	-0.8227	101.9	-	-	23S	106W
8311	416	1445 Oct 16	00:26:27	257	-6855	098	N	-t	-1.1823	0.7374	-0.3593	251.8	-	-	11N	9W
8312	416	1446 Apr 11	16:14:02	256	-6849	103	T	-p	0.3545	2.1841	1.2303	311.3	199.1	67.0	11S	117E
8313	416	1446 Oct 05	01:05:29	256	-6843	108	P	-t	-0.4596	2.0480	0.9818	349.2	204.4	-	8N	19W
8314	416	1447 Apr 01	06:47:26	255	-6837	113	T	p-	-0.3842	2.1513	1.1546	326.8	203.2	58.1	8S	101W
8315	416	1447 Sep 24	08:22:29	255	-6831	118	T	p-	0.2832	2.3420	1.3348	334.0	211.3	80.5	4N	127W
8316	416	1448 Mar 20	14:55:10	254	-6825	123	N	t-	-1.1709	0.7375	-0.3180	242.1	-	-	5S	138E
8317	416	1448 Sep 12	22:19:32	253	-6819	128	P	a-	0.9764	1.0468	0.0854	249.0	63.3	-	0N	24E
8318	416	1449 Feb 07	22:19:16	253	-6814	095	N	-t	1.0658	0.9497	-0.1445	278.3	-	-	13N	30E
8319	416	1449 Aug 04	06:47:06	252	-6808	100	P	-a	-0.8872	1.2124	0.2473	267.0	106.8	-	16S	99W
8320	416	1450 Jan 28	00:45:08	252	-6802	105	T	-p	0.3360	2.2686	1.2148	348.5	213.2	69.0	16N	6W
8321	417	1450 Jul 24	19:00:10	251	-6796	110	T-	pp	-0.1507	2.5888	1.5742	346.6	222.4	97.8	18S	77E
8322	417	1451 Jan 17	10:19:23	251	-6790	115	T	p-	-0.3754	2.1687	1.1695	322.3	200.6	59.5	18N	151W
8323	417	1451 Jul 14	00:07:29	250	-6784	120	P	t-	0.6357	1.7221	0.6615	338.5	180.3	-	20S	0E
8324	417	1452 Jan 07	01:03:46	249	-6778	125	N	a-	-1.0269	0.9588	-0.0117	239.4	-	-	20N	13W
8325	417	1452 Jun 02	10:34:09	249	-6773	092	N	-t	-1.4244	0.2701	-0.7812	154.7	-	-	24S	158W
8326	417	1452 Jul 02	00:45:25	249	-6772	130	N	t-	1.3910	0.3392	-0.7275	175.5	-	-	21S	10W
8327	417	1452 Nov 27	04:26:52	248	-6767	097	N	-a	1.0311	0.9717	-0.0397	252.8	-	-	24N	68W
8328	417	1453 May 22	17:19:57	248	-6761	102	P	-t	-0.6067	1.7450	0.7446	315.4	178.2	-	23S	100E
8329	417	1453 Nov 16	12:28:47	247	-6755	107	T	-a	0.3661	2.2202	1.1526	352.3	212.5	59.9	21N	171E
8330	417	1454 May 12	06:55:11	247	-6749	112	T+	p-	0.1683	2.5265	1.5715	320.4	210.9	93.7	20S	104W
8331	417	1454 Nov 05	13:35:42	246	-6743	117	T	p-	-0.3300	2.3033	1.2025	370.0	221.6	69.6	18N	153E
8332	417	1455 May 01	23:51:27	245	-6737	122	P	a-	0.8897	1.1981	0.2522	259.7	105.8	-	17S	2E
8333	417	1455 Oct 25	13:13:24	245	-6731	127	N*	t-	-1.0093	1.0475	-0.0348	285.1	-	-	14N	159E
8334	417	1456 Mar 22	05:03:32	244	-6726	094	P	-a	-1.0027	1.0258	0.0103	264.2	23.5	-	5S	74W
8335	417	1456 Sep 14	06:54:33	244	-6720	099	N	-a	1.0812	0.8686	-0.1207	239.1	-	-	1N	105W
8336	417	1457 Mar 11	10:57:52	243	-6714	104	T	-p	-0.2944	2.3538	1.2823	362.4	221.9	79.1	0S	162W
8337	417	1457 Sep 03	22:08:21	243	-6708	109	T	-p	0.3420	2.2069	1.2534	310.6	199.3	69.5	4S	28E
8338	417	1458 Feb 28	11:04:53	242	-6702	114	T	t-	0.4229	2.1284	1.0365	363.6	212.8	31.3	5N	162W
8339	417	1458 Aug 24	14:43:49	241	-6696	119	T	p-	-0.3737	2.1528	1.1913	313.2	198.7	62.2	8S	140E
8340	417	1459 Feb 17	12:25:19	241	-6690	124	N	h-	1.1084	0.8552	-0.2065	257.2	-	-	10N	178E

Cat Num	Canon Plate	Calendar Date	TD of Greatest Eclipse	ΔT s	Luna Num	Saros Num	Ecl. Type	QSE	Gamma	Pen. Mag.	Um. Mag.	Eclipse Phase Durations Pen. m	Par. m	Total m	Greatest in Zenith Lat.	Lng.
8341	418	1459 Jul 15	16:42:43	240	-6685	091	Ne	-t	1.5537	0.0217	-1.0073	43.6	-	-	19S	111E
8342	418	1459 Aug 14	04:02:35	240	-6684	129	N	t-	-1.1531	0.7452	-0.2609	230.9	-	-	13S	59W
8343	418	1460 Jan 08	08:51:51	240	-6679	096	P	-a	-0.9976	1.0211	0.0335	250.6	40.3	-	20N	129W
8344	418	1460 Jul 03	20:04:26	239	-6673	101	P	-t	0.8412	1.3481	0.2813	315.7	125.7	-	21S	61E
8345	418	1460 Dec 28	00:14:45	239	-6667	106	T	-p	-0.3200	2.2561	1.2851	312.7	199.9	72.5	22N	1W
8346	418	1461 Jun 22	20:44:43	238	-6661	111	T+	pp	0.0770	2.7463	1.6874	369.4	233.5	104.6	23S	50E
8347	418	1461 Dec 17	15:00:00	238	-6655	116	T	p-	0.3559	2.2049	1.2048	322.9	201.9	64.6	24N	136E
8348	418	1462 Jun 12	01:58:55	237	-6649	121	P	h-	-0.6901	1.5974	0.5860	312.0	164.2	-	24S	29W
8349	418	1462 Dec 07	00:38:21	236	-6643	126	N	h-	1.0920	0.8826	-0.1738	256.5	-	-	24N	10W
8350	418	1463 May 03	06:59:35	236	-6638	093	N	-a	1.1308	0.7572	-0.1917	217.3	-	-	17S	105W
8351	418	1463 Jun 01	14:23:22	236	-6637	131	N	a-	-1.3957	0.2771	-0.6837	140.3	-	-	24S	144E
8352	418	1463 Oct 27	08:05:16	235	-6632	098	N	-t	-1.2081	0.6913	-0.4081	245.0	-	-	15N	124W
8353	418	1464 Apr 22	00:00:10	235	-6626	103	T	-a	0.4130	2.0764	1.1235	308.5	194.6	50.8	15S	0E
8354	418	1464 Oct 15	08:50:47	234	-6620	108	P	-t	-0.4889	1.9945	0.9280	346.2	200.2	-	12N	135W
8355	418	1465 Apr 11	14:24:28	234	-6614	113	T	p-	-0.3312	2.2489	1.2517	330.6	208.2	72.0	12S	144E
8356	418	1465 Oct 04	16:24:28	233	-6608	118	T+	p-	0.2490	2.4047	1.3976	334.3	212.7	85.5	8N	111E
8357	418	1466 Mar 31	22:14:13	233	-6602	123	N	t-	-1.1226	0.8259	-0.2293	254.6	-	-	9S	27E
8358	418	1466 Sep 24	06:28:03	232	-6596	128	P	a-	0.9362	1.1213	0.1585	255.4	85.3	-	5N	99W
8359	418	1467 Feb 19	05:57:51	232	-6591	095	N	-t	1.0924	0.8988	-0.1911	271.5	-	-	9N	85W
8360	418	1467 Aug 15	14:22:17	231	-6585	100	P	-a	-0.9504	1.0988	0.1289	258.4	78.6	-	12S	146E
8361	419	1468 Feb 08	08:50:31	230	-6579	105	T	-p	0.3539	2.2330	1.1847	346.0	211.3	64.5	12N	128W
8362	419	1468 Aug 04	02:06:04	230	-6573	110	T-	-p	-0.2221	2.4606	1.4404	346.4	220.0	90.8	15S	29W
8363	419	1469 Jan 27	18:53:18	229	-6567	115	T	p-	-0.3632	2.1888	1.1940	321.8	201.3	63.1	15N	81E
8364	419	1469 Jul 24	06:42:46	229	-6561	120	P	t-	0.5556	1.8708	0.8066	347.3	194.6	-	17S	99W
8365	419	1470 Jan 17	09:51:49	228	-6555	125	P	a-	-1.0194	0.9718	0.0030	240.5	11.9	-	18N	144W
8366	419	1470 Jun 13	17:08:38	228	-6550	092	N	-t	-1.5032	0.1241	-0.9242	105.9	-	-	25S	104E
8367	419	1470 Jul 13	07:13:12	228	-6549	130	N	t-	1.3060	0.4950	-0.5715	208.7	-	-	19S	106W
8368	419	1470 Dec 08	13:05:59	227	-6544	097	N	-a	1.0369	0.9627	-0.0521	252.8	-	-	24N	164E
8369	419	1471 Jun 03	00:23:10	227	-6538	102	P	-t	-0.6815	1.6054	0.6094	306.4	164.3	-	24S	5W
8370	419	1471 Nov 27	20:42:22	226	-6532	107	T	-a	0.3745	2.2070	1.1352	353.2	212.1	56.8	23N	48E
8371	419	1472 May 22	14:23:05	226	-6526	112	T+	pp	0.0973	2.6551	1.7033	320.7	212.5	98.0	22S	144E
8372	419	1472 Nov 15	21:28:20	225	-6520	117	T	p-	-0.3154	2.3313	1.2280	371.2	222.8	73.2	21N	36E
8373	419	1473 May 12	07:28:21	224	-6514	122	P	a-	0.8256	1.3153	0.3703	268.7	126.1	-	19S	112W
8374	419	1473 Nov 04	21:08:23	224	-6508	127	P	t-	-0.9890	1.0849	0.0023	288.5	11.7	-	17N	40E
8375	419	1474 Apr 02	12:41:25	223	-6503	094	N	-a	-1.0536	0.9326	-0.0831	255.2	-	-	9S	171E
8376	419	1474 Sep 25	14:54:00	223	-6497	099	N	-a	1.1158	0.8051	-0.1843	231.1	-	-	5N	134E
8377	419	1475 Mar 22	18:18:39	222	-6491	104	T	-t	-0.3413	2.2673	1.1969	360.8	218.7	68.2	4S	87E
8378	419	1475 Sep 15	06:11:44	222	-6485	109	T	-p	0.3855	2.1282	1.1726	308.5	195.7	58.8	1N	94W
8379	419	1476 Mar 10	18:25:12	221	-6479	114	T	t-	0.3817	2.2019	1.1140	365.5	217.2	53.9	1N	87E
8380	419	1476 Sep 03	22:34:58	221	-6473	119	T	p-	-0.3243	2.2458	1.2797	316.7	202.7	73.1	4S	21E
8381	420	1477 Feb 27	20:11:49	220	-6467	124	N	h-	1.0767	0.9104	-0.1454	262.7	-	-	6N	61E
8382	420	1477 Aug 24	11:25:47	220	-6461	129	N	t-	-1.0986	0.8482	-0.1639	244.9	-	-	9S	170W
8383	420	1478 Jan 18	17:33:33	219	-6456	096	P	-a	-1.0033	1.0090	0.0249	248.8	34.7	-	17N	101E
8384	420	1478 Jul 15	02:32:38	219	-6450	101	P	-t	0.9258	1.1940	0.1249	302.7	85.8	-	19S	36W
8385	420	1479 Jan 08	09:05:26	218	-6444	106	T	-p	-0.3253	2.2462	1.2757	312.6	199.6	71.6	20N	133W
8386	420	1479 Jul 04	03:14:45	218	-6438	111	T+	pp	0.1634	2.5872	1.5297	366.9	230.6	98.8	22S	47W
8387	420	1479 Dec 28	23:43:31	217	-6432	116	T	p-	0.3535	2.2104	1.2083	324.0	202.5	65.2	23N	7E
8388	420	1480 Jun 22	08:55:27	216	-6426	121	P	h-	-0.6085	1.7455	0.7375	318.9	179.1	-	24S	132W
8389	420	1480 Dec 17	09:00:49	216	-6420	126	N	h-	1.0903	0.8873	-0.1722	258.0	-	-	24N	134W
8390	420	1481 May 13	14:33:56	215	-6415	093	N	-a	1.1952	0.6376	-0.3085	201.5	-	-	19S	142E
8391	420	1481 Jun 11	21:44:03	215	-6414	131	N	a-	-1.3201	0.4144	-0.5435	168.9	-	-	25S	35E
8392	420	1481 Nov 06	15:52:48	215	-6409	098	N	-t	-1.2265	0.6587	-0.4429	240.0	-	-	18N	119E
8393	420	1482 May 03	07:39:00	214	-6403	103	T	-a	0.4772	1.9582	1.0059	305.0	188.5	11.5	18S	115W
8394	420	1482 Oct 26	16:45:28	214	-6397	108	P	-t	-0.5111	1.9538	0.8871	343.6	196.7	-	15N	106E
8395	420	1483 Apr 22	21:52:19	213	-6391	113	T-	p-	-0.2713	2.3591	1.3612	334.4	212.9	83.2	15S	32E
8396	420	1483 Oct 16	00:36:00	213	-6385	118	T+	p-	0.2218	2.4546	1.4476	334.2	213.5	88.7	12N	12W
8397	420	1484 Apr 11	05:22:47	212	-6379	123	N	t-	-1.0662	0.9292	-0.1257	267.9	-	-	13S	81W
8398	420	1484 Oct 04	14:45:35	212	-6373	128	P	a-	0.9031	1.1830	0.2185	260.3	99.2	-	9N	136E
8399	420	1485 Mar 01	13:29:44	211	-6368	095	N	-t	1.1247	0.8371	-0.2481	263.1	-	-	5N	162E
8400	420	1485 Aug 25	22:01:58	211	-6362	100	P	-a	-1.0093	0.9933	0.0183	249.4	30.1	-	8S	31E

Cat Num	Canon Plate	Calendar Date	TD of Greatest Eclipse	ΔT s	Luna Num	Saros Num	Ecl. Type	QSE	Gamma	Pen. Mag.	Um. Mag.	Eclipse Phase ---- Durations ---- Pen. m	Par. m	Total m	Greatest in Zenith Lat.	Lng.
8401	421	1486 Feb 18	16:51:51	210	-6356	105	T	-p	0.3760	2.1895	1.1471	343.1	209.0	58.2	9N	111E
8402	421	1486 Aug 15	09:15:15	210	-6350	110	T	-p	-0.2900	2.3389	1.3130	345.7	216.5	80.5	11S	137W
8403	421	1487 Feb 08	03:22:49	209	-6344	115	T	p-	-0.3469	2.2163	1.2262	321.3	202.3	67.3	12N	46W
8404	421	1487 Aug 04	13:21:13	209	-6338	120	P	t-	0.4783	2.0146	0.9466	354.6	205.8	-	14S	161E
8405	421	1488 Jan 28	18:35:29	208	-6332	125	P	a-	-1.0083	0.9911	0.0243	242.2	33.8	-	15N	85E
8406	421	1488 Jul 23	13:46:42	208	-6326	130	N	t-	1.2247	0.6444	-0.4223	234.4	-	-	17S	155E
8407	421	1488 Dec 18	21:43:56	207	-6321	097	N	-a	1.0434	0.9523	-0.0655	252.6	-	-	24N	36E
8408	421	1489 Jun 13	07:28:45	207	-6315	102	P	-h	-0.7554	1.4678	0.4757	296.6	147.8	-	24S	111W
8409	421	1489 Dec 08	04:57:05	206	-6309	107	T	-a	0.3817	2.1954	1.1202	354.1	211.8	53.9	24N	74W
8410	421	1490 Jun 02	21:49:13	206	-6303	112	T+	pp	0.0243	2.7877	1.8386	320.5	213.1	99.8	23S	33E
8411	421	1490 Nov 27	05:25:13	205	-6297	117	T	p-	-0.3042	2.3528	1.2476	372.1	223.7	75.7	22N	82W
8412	421	1491 May 23	15:00:00	205	-6291	122	P	a-	0.7569	1.4411	0.4965	277.4	143.5	-	21S	135E
8413	421	1491 Nov 16	05:10:39	204	-6285	127	P	t-	-0.9741	1.1121	0.0298	290.6	41.7	-	20N	80W
8414	421	1492 Apr 12	20:07:44	204	-6280	094	N	-a	-1.1130	0.8239	-0.1924	243.4	-	-	13S	58E
8415	421	1492 May 12	06:21:14	204	-6279	132	Nb	a-	1.5337	0.0347	-0.9474	52.0	-	-	19S	95W
8416	421	1492 Oct 05	23:04:19	203	-6274	099	N	-a	1.1424	0.7565	-0.2332	224.5	-	-	10N	11E
8417	421	1493 Apr 02	01:28:34	203	-6268	104	T	-t	-0.3964	2.1656	1.0963	358.3	214.0	49.5	9S	21W
8418	421	1493 Sep 25	14:23:01	202	-6262	109	T	-p	0.4226	2.0613	1.1033	306.4	192.2	46.4	5N	142E
8419	421	1494 Mar 22	01:37:17	202	-6256	114	T	t-	0.3334	2.2884	1.2046	367.4	221.7	70.1	4S	22W
8420	421	1494 Sep 15	06:32:25	201	-6250	119	T	p-	-0.2805	2.3287	1.3576	319.6	205.8	80.5	0N	99W
8421	422	1495 Mar 11	03:50:59	201	-6244	124	N	h-	1.0379	0.9785	-0.0711	269.3	-	-	1N	55W
8422	422	1495 Sep 04	18:54:17	200	-6238	129	N	t-	-1.0490	0.9425	-0.0759	256.9	-	-	5S	76E
8423	422	1496 Jan 30	02:12:22	200	-6233	096	P	-a	-1.0120	0.9911	0.0107	246.4	22.8	-	14N	29W
8424	422	1496 Jul 25	09:03:59	199	-6227	101	N*	-t	1.0074	1.0457	-0.0260	288.3	-	-	17S	134W
8425	422	1497 Jan 18	17:53:14	199	-6221	106	T	-p	-0.3330	2.2315	1.2619	312.3	199.2	70.2	18N	96E
8426	422	1497 Jul 14	09:49:55	198	-6215	111	T+	pp	0.2462	2.4347	1.3783	363.5	226.1	88.8	20S	145W
8427	422	1498 Jan 08	08:25:32	198	-6209	116	T	p-	0.3505	2.2168	1.2130	325.1	203.2	65.9	21N	123W
8428	422	1498 Jul 03	15:54:47	197	-6203	121	P	a-	-0.5279	1.8917	0.8867	324.4	190.8	-	23S	123E
8429	422	1498 Dec 28	17:22:57	197	-6197	126	N	t-	1.0890	0.8911	-0.1710	259.3	-	-	24N	102E
8430	422	1499 May 24	22:05:32	196	-6192	093	N	-a	1.2621	0.5136	-0.4301	182.8	-	-	21S	29E
8431	422	1499 Jun 23	05:05:35	196	-6191	131	N	a-	-1.2445	0.5520	-0.4037	191.8	-	-	24S	75W
8432	422	1499 Nov 17	23:45:52	196	-6186	098	N	-t	-1.2408	0.6331	-0.4700	235.8	-	-	20N	2E
8433	422	1500 May 13	15:12:36	195	-6180	103	P	-a	0.5457	1.8326	0.8802	300.5	180.6	-	20S	132E
8434	422	1500 Nov 06	00:48:33	195	-6174	108	P	-t	-0.5272	1.9239	0.8577	341.4	193.9	-	18N	15W
8435	422	1501 May 03	05:13:26	194	-6168	113	T-	pp	-0.2064	2.4787	1.4797	337.8	217.0	91.9	18S	79W
8436	422	1501 Oct 26	08:54:58	194	-6162	118	T+	p-	0.1999	2.4944	1.4878	333.8	213.9	90.8	16N	137W
8437	422	1502 Apr 22	12:22:44	193	-6156	123	N*	t-	-1.0034	1.0445	-0.0103	281.3	-	-	16S	174E
8438	422	1502 Oct 15	23:10:40	193	-6150	128	P	a-	0.8759	1.2336	0.2675	264.1	108.8	-	13N	9E
8439	422	1503 Mar 12	20:51:38	192	-6145	095	N	-t	1.1652	0.7603	-0.3200	251.9	-	-	1N	50E
8440	422	1503 Sep 06	05:48:25	192	-6139	100	N	-a	-1.0620	0.8993	-0.0811	240.6	-	-	4S	87W
8441	423	1504 Mar 01	00:44:47	191	-6133	105	T	-p	0.4057	2.1318	1.0956	339.7	205.8	47.6	4N	8W
8442	423	1504 Aug 25	16:30:40	191	-6127	110	T	-a	-0.3518	2.2287	1.1966	344.6	212.3	66.5	8S	113E
8443	423	1505 Feb 18	11:45:03	190	-6121	115	T	p-	-0.3240	2.2559	1.2708	321.2	203.7	72.4	8N	172W
8444	423	1505 Aug 14	20:05:59	190	-6115	120	T	t-	0.4063	2.1487	1.0766	360.6	214.4	44.7	11S	60E
8445	423	1506 Feb 08	03:14:07	189	-6109	125	P	a-	-0.9931	1.0177	0.0532	244.6	49.8	-	11N	44W
8446	423	1506 Aug 03	20:24:30	189	-6103	130	N	t-	1.1456	0.7897	-0.2776	255.4	-	-	14S	55E
8447	423	1506 Dec 30	06:22:25	189	-6098	097	N	-a	1.0490	0.9433	-0.0769	252.4	-	-	23N	93W
8448	423	1507 Jun 24	14:33:59	188	-6092	102	P	-h	-0.8306	1.3283	0.3394	285.5	127.0	-	24S	143E
8449	423	1507 Dec 19	13:13:01	188	-6086	107	T	-a	0.3872	2.1868	1.1089	354.9	211.7	51.6	24N	163E
8450	423	1508 Jun 13	05:14:02	187	-6080	112	T-	pp	-0.0503	2.7389	1.7918	319.6	212.6	99.4	24S	78W
8451	423	1508 Dec 07	13:26:13	187	-6074	117	T	p-	-0.2962	2.3681	1.2618	372.7	224.3	77.4	23N	158E
8452	423	1509 Jun 02	22:29:18	186	-6068	122	P	a-	0.6866	1.5702	0.6253	285.4	158.0	-	22S	23E
8453	423	1509 Nov 26	13:18:02	186	-6062	127	P	h-	-0.9625	1.1329	0.0514	291.9	54.5	-	22N	159E
8454	423	1510 Apr 24	03:27:17	185	-6057	094	N	-h	-1.1769	0.7070	-0.3100	229.0	-	-	17S	52W
8455	423	1510 May 23	13:38:07	185	-6056	132	N	h-	1.4701	0.1523	-0.8316	108.1	-	-	21S	156E
8456	423	1510 Oct 17	07:22:00	185	-6051	099	N	-a	1.1637	0.7175	-0.2724	219.0	-	-	14N	114W
8457	423	1511 Apr 13	08:29:06	184	-6045	104	P	-t	-0.4585	2.0510	0.9828	354.7	207.5	-	13S	127W
8458	423	1511 Oct 06	22:42:06	184	-6039	109	T	-a	0.4533	2.0061	1.0457	304.7	189.0	31.4	9N	17E
8459	423	1512 Apr 01	08:40:58	183	-6033	114	T	pp	0.2778	2.3883	1.3087	369.1	225.9	82.9	8S	129W
8460	423	1512 Sep 25	14:37:20	183	-6027	119	T-	p-	-0.2434	2.3993	1.4232	322.1	208.1	85.6	4N	139E

Cat Num	Canon Plate	Calendar Date	TD of Greatest Eclipse	ΔT s	Luna Num	Saros Num	Ecl. Type	QSE	Gamma	Pen. Mag.	Um. Mag.	Eclipse Phase Durations Pen. m	Par. m	Total m	Greatest in Zenith Lat.	Lng.
8461	424	1513 Mar 21	11:22:52	182	-6021	124	P	h-	0.9922	1.0591	0.0158	276.5	29.8	-	3S	169W
8462	424	1513 Sep 15	02:30:22	182	-6015	129	N*	t-	-1.0060	1.0245	-0.0003	266.8	-	-	0S	39W
8463	424	1514 Feb 09	10:45:15	182	-6010	096	N	-a	-1.0259	0.9636	-0.0127	243.0	-	-	10N	157W
8464	424	1514 Aug 05	15:39:55	181	-6004	101	N	-t	1.0845	0.9055	-0.1690	272.6	-	-	13S	127E
8465	424	1515 Jan 30	02:35:56	181	-5998	106	T	-p	-0.3447	2.2094	1.2410	311.9	198.5	67.8	15N	35W
8466	424	1515 Jul 25	16:29:57	180	-5992	111	T	pp	0.3252	2.2893	1.2336	359.2	220.1	73.4	17S	115E
8467	424	1516 Jan 19	17:02:13	180	-5986	116	T	p-	0.3438	2.2297	1.2248	326.4	204.2	67.6	19N	109E
8468	424	1516 Jul 13	22:58:24	179	-5980	121	T	a-	-0.4501	2.0335	1.0308	328.7	199.7	27.4	21S	18E
8469	424	1517 Jan 08	01:41:58	179	-5974	126	N	t-	1.0862	0.8971	-0.1668	260.9	-	-	22N	22W
8470	424	1517 Jun 04	05:34:12	178	-5969	093	N	-a	1.3316	0.3854	-0.5567	160.1	-	-	22S	83W
8471	424	1517 Jul 03	12:28:24	178	-5968	131	N	a-	-1.1693	0.6892	-0.2648	210.9	-	-	23S	175E
8472	424	1517 Nov 28	07:44:09	178	-5963	098	N	-t	-1.2509	0.6150	-0.4889	232.8	-	-	22N	117W
8473	424	1518 May 24	22:40:54	177	-5957	103	P	-a	0.6180	1.7002	0.7473	295.0	170.5	-	22S	20E
8474	424	1518 Nov 17	08:59:11	177	-5951	108	P	-t	-0.5379	1.9040	0.8385	339.5	191.9	-	21N	137W
8475	424	1519 May 14	12:25:26	177	-5945	113	T-	pp	-0.1349	2.6107	1.6101	340.8	220.4	98.3	21S	173E
8476	424	1519 Nov 06	17:22:27	176	-5939	118	T+	p-	0.1846	2.5223	1.5162	333.2	213.9	92.1	19N	96E
8477	424	1520 May 02	19:14:16	176	-5933	123	P	t-	-0.9344	1.1712	0.1162	294.6	81.5	-	19S	71E
8478	424	1520 Oct 26	07:43:02	175	-5927	128	P	a-	0.8549	1.2729	0.3053	266.9	115.5	-	16N	119W
8479	424	1521 Mar 23	04:07:01	175	-5922	095	N	-t	1.2110	0.6737	-0.4015	238.5	-	-	4S	60W
8480	424	1521 Sep 16	13:41:13	174	-5916	100	N	-a	-1.1087	0.8166	-0.1695	232.1	-	-	0N	154E
8481	425	1521 Oct 15	23:41:27	174	-5915	138	N	a-	1.5291	0.0386	-0.9345	53.1	-	-	14N	1E
8482	425	1522 Mar 12	08:32:37	174	-5910	105	T	-t	0.4404	2.0648	1.0350	335.8	201.8	29.3	0N	125W
8483	425	1522 Sep 05	23:50:39	173	-5904	110	T	-a	-0.4089	2.1270	1.0887	343.1	207.4	46.4	3S	2E
8484	425	1523 Mar 01	20:02:00	173	-5898	115	T	p-	-0.2961	2.3045	1.3244	321.2	205.3	77.7	4N	63E
8485	425	1523 Aug 26	02:57:01	173	-5892	120	T	t-	0.3395	2.2735	1.1971	365.5	221.0	68.9	7S	44W
8486	425	1524 Feb 19	11:45:16	172	-5886	125	P	a-	-0.9719	1.0554	0.0933	248.0	65.6	-	7N	173W
8487	425	1524 Aug 14	03:11:11	172	-5880	130	N	t-	1.0728	0.9238	-0.1443	272.1	-	-	10S	47W
8488	425	1525 Jan 09	14:56:50	171	-5875	097	N	-a	1.0577	0.9283	-0.0938	251.7	-	-	21N	140E
8489	425	1525 Jul 04	21:43:23	171	-5869	102	P	-h	-0.9029	1.1941	0.2080	273.8	101.0	-	23S	36E
8490	425	1525 Dec 29	21:26:19	170	-5863	107	T	-a	0.3944	2.1746	1.0947	355.6	211.4	48.4	23N	41E
8491	425	1526 Jun 24	12:39:44	170	-5857	112	T-	pp	-0.1248	2.6016	1.6559	318.2	211.0	96.4	23S	172E
8492	425	1526 Dec 18	21:27:49	169	-5851	117	T	p-	-0.2887	2.3819	1.2754	373.1	224.9	79.0	23N	39E
8493	425	1527 Jun 14	05:54:52	169	-5845	122	P	a-	0.6134	1.7051	0.7593	293.0	170.6	-	23S	88W
8494	425	1527 Dec 07	21:30:05	169	-5839	127	P	h-	-0.9540	1.1478	0.0678	292.6	62.3	-	23N	37E
8495	425	1528 May 04	10:37:14	168	-5834	094	N	-h	-1.2477	0.5778	-0.4406	210.4	-	-	20S	160W
8496	425	1528 Jun 02	20:49:34	168	-5833	132	N	h-	1.4025	0.2775	-0.7088	144.9	-	-	22S	48E
8497	425	1528 Oct 27	15:48:36	168	-5828	099	N	-a	1.1786	0.6902	-0.2998	214.8	-	-	17N	119E
8498	425	1529 Apr 23	15:20:28	167	-5822	104	P	-t	-0.5277	1.9238	0.8563	349.9	198.6	-	16S	130E
8499	425	1529 Oct 17	07:09:32	167	-5816	109	T	-a	0.4775	1.9630	1.0001	303.3	186.3	1.7	13N	111W
8500	425	1530 Apr 12	15:36:23	166	-5810	114	T+	pp	0.2151	2.5012	1.4260	370.5	229.6	92.9	12S	126E
8501	426	1530 Oct 06	22:49:14	166	-5804	119	T-	p-	-0.2126	2.4583	1.4771	324.3	209.9	89.0	9N	15E
8502	426	1531 Apr 01	18:47:57	166	-5798	124	P	h-	0.9400	1.1518	0.1149	284.2	79.0	-	7S	79E
8503	426	1531 Sep 26	10:13:23	165	-5792	129	P	t-	-0.9696	1.0946	0.0634	275.0	58.4	-	4N	155W
8504	426	1532 Feb 20	19:11:57	165	-5787	096	N	-a	-1.0452	0.9260	-0.0461	238.5	-	-	6N	76E
8505	426	1532 Aug 15	22:22:09	164	-5781	101	N	-t	1.1559	0.7761	-0.3016	256.0	-	-	10S	25E
8506	426	1533 Feb 09	11:13:52	164	-5775	106	T	-p	-0.3602	2.1804	1.2134	311.2	197.6	64.5	11N	164W
8507	426	1533 Aug 04	23:16:00	164	-5769	111	T	-t	0.3999	2.1520	1.0967	354.1	212.8	49.4	14S	13E
8508	426	1534 Jan 30	01:34:32	163	-5763	116	T	p-	0.3341	2.2478	1.2421	327.8	205.4	70.0	15N	19W
8509	426	1534 Jul 25	06:07:01	163	-5757	121	T	a-	-0.3754	2.1695	1.1688	331.9	206.4	61.1	18S	90W
8510	426	1535 Jan 19	09:56:17	162	-5751	126	N	t-	1.0800	0.9092	-0.1560	263.1	-	-	19N	145W
8511	426	1535 Jun 15	13:01:28	162	-5746	093	N	-a	1.4023	0.2550	-0.6860	131.7	-	-	22S	166E
8512	426	1535 Jul 14	19:54:01	162	-5745	131	N	a-	-1.0956	0.8239	-0.1292	226.9	-	-	21S	64E
8513	426	1535 Dec 09	15:45:20	162	-5740	098	N	-t	-1.2588	0.6006	-0.5034	230.2	-	-	22N	124E
8514	426	1536 Jun 04	06:06:33	161	-5734	103	P	-a	0.6920	1.5650	0.6109	288.5	157.9	-	23S	91W
8515	426	1536 Nov 27	17:14:01	161	-5728	108	P	-t	-0.5457	1.8889	0.8249	337.8	190.3	-	22N	100E
8516	426	1537 May 24	19:33:13	160	-5722	113	T-	pp	-0.0608	2.7478	1.7451	343.3	222.7	102.1	22S	67E
8517	426	1537 Nov 17	01:55:51	160	-5716	118	T+	p-	0.1738	2.5416	1.5365	332.5	213.8	92.9	21N	31W
8518	426	1538 May 14	01:58:00	159	-5710	123	P	t-	-0.8594	1.3090	0.2535	307.5	118.3	-	21S	30W
8519	426	1538 Nov 06	16:22:02	159	-5704	128	P	a-	0.8392	1.3024	0.3334	268.9	120.1	-	20N	111E
8520	426	1539 Apr 03	11:13:11	159	-5699	095	N	-t	1.2644	0.5732	-0.4968	221.4	-	-	8S	167W

Cat Num	Canon Plate	Calendar Date	TD of Greatest Eclipse	ΔT s	Luna Num	Saros Num	Ecl. Type	QSE	Gamma	Pen. Mag.	Um. Mag.	Eclipse Phase Durations Pen. m	Par. m	Total m	Greatest in Zenith Lat.	Lng.
8521	427	1539 Sep 27	21:41:55	158	-5693	100	N	-a	-1.1485	0.7464	-0.2455	224.4	-	-	4N	33E
8522	427	1539 Oct 27	08:11:00	158	-5692	138	N	a-	1.5126	0.0709	-0.9063	71.9	-	-	17N	127W
8523	427	1540 Mar 22	16:11:08	158	-5687	105	P	-t	0.4838	1.9821	0.9587	331.0	196.4	-	4S	119E
8524	427	1540 Sep 16	07:18:56	157	-5681	110	P	-a	-0.4584	2.0394	0.9947	341.6	202.4	-	1N	111W
8525	427	1541 Mar 12	04:11:19	157	-5675	115	T-	p-	-0.2612	2.3661	1.3912	321.3	207.1	83.3	1S	60W
8526	427	1541 Sep 05	09:54:46	157	-5669	120	T	pp	0.2784	2.3879	1.3072	369.3	225.8	82.7	3S	149W
8527	427	1542 Mar 01	20:10:00	156	-5663	125	P	a-	-0.9455	1.1026	0.1430	252.1	80.7	-	3N	60E
8528	427	1542 Aug 25	10:04:53	156	-5657	130	N*	t-	1.0046	1.0495	-0.0196	285.8	-	-	7S	151W
8529	427	1543 Jan 20	23:27:44	155	-5652	097	N	-a	1.0685	0.9092	-0.1143	250.3	-	-	19N	13E
8530	427	1543 Jul 16	04:55:02	155	-5646	102	P	-h	-0.9742	1.0623	0.0784	261.0	62.9	-	21S	71W
8531	427	1544 Jan 10	05:37:51	155	-5640	107	T	-h	0.4024	2.1605	1.0793	356.0	211.0	44.6	21N	81W
8532	427	1544 Jul 04	20:06:57	154	-5634	112	T-	-p	-0.1982	2.4665	1.5216	316.1	208.3	90.9	22S	60E
8533	427	1544 Dec 29	05:29:10	154	-5628	117	T	p-	-0.2809	2.3961	1.2900	373.4	225.5	80.7	22N	80W
8534	427	1545 Jun 24	13:21:01	153	-5622	122	P	a-	0.5409	1.8389	0.8916	299.6	181.0	-	22S	161E
8535	427	1545 Dec 18	05:45:01	153	-5616	127	P	h-	-0.9470	1.1596	0.0818	292.9	68.2	-	22N	85W
8536	427	1546 May 15	17:40:52	153	-5611	094	N	-h	-1.3223	0.4416	-0.5783	187.1	-	-	22S	94E
8537	427	1546 Jun 14	03:57:38	153	-5610	132	N	h-	1.3329	0.4067	-0.5824	174.0	-	-	22S	59W
8538	427	1546 Nov 08	00:21:33	152	-5605	099	N	-a	1.1891	0.6709	-0.3188	211.7	-	-	20N	9W
8539	427	1547 May 04	22:04:32	152	-5599	104	P	-t	-0.6022	1.7866	0.7198	343.7	186.8	-	19S	28E
8540	427	1547 Oct 28	15:43:40	152	-5593	109	P	-a	0.4960	1.9303	0.9651	302.2	184.2	-	17N	121E
8541	428	1548 Apr 22	22:24:51	151	-5587	114	T+	pp	0.1463	2.6255	1.5543	371.4	232.4	100.3	15S	24E
8542	428	1548 Oct 17	07:08:29	151	-5581	119	T-	p-	-0.1885	2.5051	1.5189	326.1	211.2	91.4	13N	110W
8543	428	1549 Apr 12	02:06:54	150	-5575	124	P	h-	0.8818	1.2555	0.2247	292.0	108.4	-	11S	31W
8544	428	1549 Oct 06	18:02:25	150	-5569	129	P	t-	-0.9388	1.1543	0.1168	281.9	78.9	-	8N	87E
8545	428	1550 Mar 03	03:31:35	150	-5564	096	N	-a	-1.0707	0.8771	-0.0907	232.7	-	-	2N	50W
8546	428	1550 Apr 01	12:54:38	150	-5563	134	Nb	a-	1.5506	0.0022	-0.9770	12.9	-	-	7S	168E
8547	428	1550 Aug 27	05:11:29	149	-5558	101	N	-t	1.2210	0.6583	-0.4226	238.8	-	-	6S	78W
8548	428	1551 Feb 20	19:43:12	149	-5552	106	T	-p	-0.3827	2.1382	1.1729	310.3	196.0	58.9	7N	68E
8549	428	1551 Aug 16	06:09:51	149	-5546	111	P	-t	0.4688	2.0256	0.9702	348.7	204.5	-	10S	92W
8550	428	1552 Feb 10	09:58:45	148	-5540	116	T	p-	0.3183	2.2769	1.2708	329.5	207.2	73.5	12N	145W
8551	428	1552 Aug 04	13:22:37	148	-5534	121	T	a-	-0.3055	2.2970	1.2978	334.1	211.3	77.5	15S	161E
8552	428	1553 Jan 29	18:03:50	147	-5528	126	N	t-	1.0690	0.9296	-0.1363	266.2	-	-	16N	94E
8553	428	1553 Jun 25	20:29:09	147	-5523	093	N	-a	1.4726	0.1258	-0.8148	93.6	-	-	21S	54E
8554	428	1553 Jul 25	03:24:11	147	-5522	131	P	a-	-1.0253	0.9528	0.0001	240.2	2.5	-	18S	49W
8555	428	1553 Dec 19	23:49:19	147	-5517	098	N	-t	-1.2643	0.5901	-0.5132	228.2	-	-	22N	4E
8556	428	1554 Jun 15	13:27:59	146	-5511	103	P	-a	0.7688	1.4250	0.4691	280.7	141.8	-	23S	159E
8557	428	1554 Dec 09	01:34:26	146	-5505	108	P	-t	-0.5496	1.8808	0.8188	336.3	189.3	-	23N	24W
8558	428	1555 Jun 05	02:34:54	146	-5499	113	T+	pp	0.0172	2.8291	1.8238	345.0	223.7	103.1	23S	38W
8559	428	1555 Nov 28	10:34:50	145	-5493	118	T+	p-	0.1671	2.5534	1.5495	331.7	213.5	93.3	23N	160W
8560	428	1556 May 24	08:36:04	145	-5487	123	P	t-	-0.7802	1.4547	0.3983	319.6	145.4	-	23S	129W
8561	429	1556 Nov 17	01:06:54	145	-5481	128	P	a-	0.8286	1.3224	0.3523	270.2	123.0	-	22N	19W
8562	429	1557 Apr 13	18:14:10	144	-5476	095	N	-t	1.3222	0.4644	-0.6004	200.7	-	-	11S	87E
8563	429	1557 May 13	09:23:04	144	-5475	133	Nb	t-	-1.5370	0.0734	-0.9975	83.9	-	-	22S	142W
8564	429	1557 Oct 08	05:47:56	144	-5470	100	N	-a	-1.1834	0.6853	-0.3122	217.2	-	-	8N	89W
8565	429	1557 Nov 06	16:45:01	144	-5469	138	N	a-	1.5003	0.0955	-0.8857	83.5	-	-	20N	105E
8566	429	1558 Apr 02	23:45:10	144	-5464	105	P	-t	0.5316	1.8911	0.8741	325.6	189.8	-	8S	5E
8567	429	1558 Sep 27	14:53:07	143	-5458	110	P	-a	-0.5022	1.9621	0.9112	340.1	197.2	-	5N	135E
8568	429	1559 Mar 23	12:13:18	143	-5452	115	T-	p-	-0.2197	2.4397	1.4698	321.5	209.0	88.5	5S	178E
8569	429	1559 Sep 16	17:01:00	142	-5446	120	T+	pp	0.2244	2.4891	1.4041	372.2	229.3	91.3	1N	103E
8570	429	1560 Mar 12	04:26:27	142	-5440	125	P	a-	-0.9127	1.1617	0.2045	257.1	95.8	-	1S	65W
8571	429	1560 Sep 04	17:08:50	142	-5434	130	P	t-	0.9440	1.1611	0.0910	296.6	72.8	-	3S	102E
8572	429	1561 Jan 31	07:51:02	141	-5429	097	N	-a	1.0850	0.8795	-0.1450	247.8	-	-	15N	113W
8573	429	1561 Jul 26	12:13:48	141	-5423	102	N	-a	-1.0402	0.9402	-0.0420	247.8	-	-	18S	179E
8574	429	1562 Jan 20	13:43:40	141	-5417	107	T	-t	0.4147	2.1384	1.0565	356.1	210.1	38.0	18N	158E
8575	429	1562 Jul 16	03:36:19	140	-5411	112	T-	-p	-0.2701	2.3345	1.3896	313.4	204.6	82.5	20S	52W
8576	429	1563 Jan 09	13:28:27	140	-5405	117	T-	p-	-0.2712	2.4131	1.3084	373.5	226.2	82.6	20N	161E
8577	429	1563 Jul 05	20:46:26	140	-5399	122	T	a-	0.4678	1.9741	1.0245	305.6	189.7	23.4	21S	50E
8578	429	1563 Dec 29	13:59:52	139	-5393	127	P	h-	-0.9392	1.1722	0.0975	293.1	74.1	-	22N	152E
8579	429	1564 May 26	00:37:06	139	-5388	094	N	-h	-1.4018	0.2969	-0.7252	156.1	-	-	24S	9W
8580	429	1564 Jun 24	11:03:08	139	-5387	132	N	h-	1.2616	0.5393	-0.4534	198.6	-	-	22S	165W

Cat Num	Canon Plate	Calendar Date	TD of Greatest Eclipse	ΔT s	Luna Num	Saros Num	Ecl. Type	QSE	Gamma	Pen. Mag.	Um. Mag.	Eclipse Phase Durations Pen. m	Par. m	Total m	Greatest in Zenith Lat.	Lng.
8581	430	1564 Nov 18	09:01:33	139	-5382	099	N	-a	1.1947	0.6604	-0.3288	209.8	-	-	23N	138W
8582	430	1564 Dec 17	21:16:37	139	-5381	137	N	a-	-1.5587	0.0066	-1.0109	23.0	-	-	22N	42E
8583	430	1565 May 15	04:41:16	138	-5376	104	P	-t	-0.6820	1.6400	0.5736	335.7	171.1	-	22S	71W
8584	430	1565 Nov 08	00:23:49	138	-5370	109	P	-a	0.5096	1.9065	0.9391	301.4	182.5	-	20N	9W
8585	430	1566 May 04	05:06:57	138	-5364	114	T+	pp	0.0717	2.7603	1.6932	371.6	234.1	104.7	18S	77W
8586	430	1566 Oct 28	15:34:06	137	-5358	119	T-	p-	-0.1704	2.5409	1.5498	327.8	212.2	92.9	16N	123E
8587	430	1567 Apr 23	09:18:49	137	-5352	124	P	h-	0.8168	1.3716	0.3469	299.8	131.9	-	15S	140W
8588	430	1567 Oct 18	01:59:02	137	-5346	129	P	t-	-0.9149	1.2012	0.1576	287.4	91.4	-	12N	33W
8589	430	1568 Mar 13	11:44:28	136	-5341	096	N	-a	-1.1021	0.8171	-0.1462	225.5	-	-	2S	174W
8590	430	1568 Apr 11	20:39:00	136	-5340	134	N	a-	1.4970	0.0979	-0.8760	85.4	-	-	11S	51E
8591	430	1568 Sep 06	12:08:34	136	-5335	101	N	-t	1.2796	0.5524	-0.5317	221.2	-	-	1S	177E
8592	430	1569 Mar 03	04:06:08	136	-5329	106	T	-a	-0.4104	2.0865	1.1228	308.9	193.9	50.5	3N	59W
8593	430	1569 Aug 26	13:12:21	136	-5323	111	P	-t	0.5313	1.9111	0.8556	342.9	195.6	-	6S	162E
8594	430	1570 Feb 20	18:16:30	135	-5317	116	T	p-	0.2981	2.3142	1.3078	331.4	209.2	77.5	7N	90E
8595	430	1570 Aug 15	20:44:34	135	-5311	121	T-	p-	-0.2401	2.4165	1.4183	335.4	214.6	87.6	11S	50E
8596	430	1571 Feb 10	02:04:37	135	-5305	126	N	t-	1.0535	0.9582	-0.1079	270.2	-	-	12N	27W
8597	430	1571 Aug 05	10:59:42	134	-5299	131	P	a-	-0.9591	1.0743	0.1215	251.3	75.3	-	15S	163W
8598	430	1571 Dec 31	07:51:32	134	-5294	098	N	-t	-1.2712	0.5767	-0.5249	225.6	-	-	21N	115W
8599	430	1572 Jun 25	20:49:39	134	-5288	103	P	-a	0.8446	1.2872	0.3288	271.9	121.5	-	22S	49E
8600	430	1572 Dec 19	09:56:07	133	-5282	108	P	-h	-0.5525	1.8740	0.8148	334.8	188.5	-	23N	148W
8601	431	1573 Jun 15	09:34:00	133	-5276	113	T+	pp	0.0964	2.6854	1.6770	345.9	223.4	101.1	23S	143W
8602	431	1573 Dec 08	19:16:17	133	-5270	118	T+	p-	0.1620	2.5617	1.5597	330.8	213.2	93.5	24N	71E
8603	431	1574 Jun 04	15:09:13	132	-5264	123	P	t-	-0.6977	1.6068	0.5492	330.8	167.1	-	24S	133E
8604	431	1574 Nov 28	09:55:43	132	-5258	128	P	a-	0.8213	1.3363	0.3655	271.1	125.0	-	24N	150W
8605	431	1575 Apr 25	01:06:06	132	-5253	095	N	-t	1.3873	0.3425	-0.7173	173.8	-	-	15S	16W
8606	431	1575 May 24	15:51:37	132	-5252	133	N	t-	-1.4542	0.2241	-0.8444	144.3	-	-	24S	122E
8607	431	1575 Oct 19	14:02:17	132	-5247	100	N	-a	-1.2107	0.6380	-0.3652	211.4	-	-	12N	147E
8608	431	1575 Nov 18	01:25:13	131	-5246	138	N	a-	1.4936	0.1097	-0.8751	89.6	-	-	23N	24W
8609	431	1576 Apr 13	07:11:16	131	-5241	105	P	-t	0.5865	1.7871	0.7763	319.2	181.4	-	12S	108W
8610	431	1576 Oct 07	22:35:11	131	-5235	110	P	-a	-0.5385	1.8987	0.8417	338.9	192.4	-	9N	18E
8611	431	1577 Apr 02	20:08:13	131	-5229	115	T-	p-	-0.1719	2.5248	1.5598	321.7	210.7	93.1	9S	59E
8612	431	1577 Sep 27	00:16:07	130	-5223	120	T+	pp	0.1781	2.5760	1.4869	374.4	231.6	96.7	5N	7W
8613	431	1578 Mar 23	12:35:23	130	-5217	125	P	a-	-0.8738	1.2318	0.2770	262.7	110.4	-	6S	172E
8614	431	1578 Sep 16	00:21:28	130	-5211	130	P	t-	0.8897	1.2613	0.1901	305.2	103.5	-	2N	7W
8615	431	1579 Feb 11	16:08:35	129	-5206	097	N	-a	1.1054	0.8424	-0.1828	244.2	-	-	12N	122E
8616	431	1579 Aug 06	19:37:41	129	-5200	102	N	-a	-1.1029	0.8248	-0.1564	234.1	-	-	15S	68E
8617	431	1580 Jan 31	21:43:39	129	-5194	107	T	-t	0.4311	2.1083	1.0264	355.8	208.7	26.3	15N	39E
8618	431	1580 Jul 26	11:09:58	129	-5188	112	T	-p	-0.3387	2.2089	1.2635	310.3	200.0	70.8	17S	166W
8619	431	1581 Jan 19	21:24:57	128	-5182	117	T-	p-	-0.2589	2.4347	1.3320	373.6	227.0	84.9	18N	43E
8620	431	1581 Jul 16	04:13:15	128	-5176	122	T	p-	0.3965	2.1064	1.1541	310.7	196.9	56.5	19S	62W
8621	432	1582 Jan 08	22:14:15	128	-5170	127	P	a-	-0.9304	1.1866	0.1155	293.5	80.2	-	20N	30E
8622	432	1582 Jun 06	07:29:32	128	-5165	094	N	-t	-1.4830	0.1491	-0.8754	112.5	-	-	25S	112W
8623	432	1582 Jul 05	18:08:00	127	-5164	132	N	t-	1.1903	0.6722	-0.3245	219.8	-	-	20S	89E
8624	432	1582 Dec 09	17:45:08	127	-5159	099	N	-a	1.1977	0.6544	-0.3340	208.5	-	-	24N	93E
8625	432	1583 Jan 08	05:55:04	127	-5158	137	N	a-	-1.5513	0.0186	-0.9958	38.4	-	-	21N	87W
8626	432	1583 Jun 05	11:12:22	127	-5153	104	P	-t	-0.7655	1.4867	0.4203	325.9	150.3	-	23S	168W
8627	432	1583 Nov 29	09:08:53	127	-5147	109	P	-a	0.5191	1.8900	0.9208	301.0	181.3	-	22N	140W
8628	432	1584 May 24	11:45:10	126	-5141	114	T-	pp	-0.0065	2.8780	1.8145	370.9	234.5	106.1	21S	177W
8629	432	1584 Nov 18	00:04:27	126	-5135	119	T-	p-	-0.1567	2.5683	1.5727	329.3	213.1	94.0	19N	4W
8630	432	1585 May 13	16:27:07	126	-5129	124	P	h-	0.7477	1.4955	0.4765	307.1	151.1	-	18S	113E
8631	432	1585 Nov 07	10:01:47	126	-5123	129	P	t-	-0.8965	1.2379	0.1885	291.8	99.8	-	16N	154W
8632	432	1586 Apr 03	19:49:08	125	-5118	096	N	-a	-1.1405	0.7445	-0.2146	216.4	-	-	7S	64E
8633	432	1586 May 03	04:16:35	125	-5117	134	N	a-	1.4373	0.2049	-0.7639	121.9	-	-	14S	64W
8634	432	1586 Sep 27	19:14:33	125	-5112	101	N	-t	1.3306	0.4604	-0.6270	203.8	-	-	3N	69E
8635	432	1587 Mar 24	12:19:48	125	-5106	106	T	-a	-0.4456	2.0211	1.0591	307.1	190.8	35.7	2S	177E
8636	432	1587 Sep 16	20:25:05	125	-5100	111	P	-t	0.5857	1.8114	0.7555	337.3	186.6	-	2S	53E
8637	432	1588 Mar 13	02:23:35	124	-5094	116	T+	p-	0.2698	2.3661	1.3596	333.5	211.6	82.5	3N	33W
8638	432	1588 Sep 05	04:15:59	124	-5088	121	T-	p-	-0.1818	2.5231	1.5254	336.0	216.6	93.8	7S	64W
8639	432	1589 Mar 02	09:56:49	124	-5082	126	N	t-	1.0319	0.9978	-0.0681	275.3	-	-	8N	145W
8640	432	1589 Aug 25	18:40:24	124	-5076	131	P	a-	-0.8969	1.1887	0.2352	260.7	102.9	-	12S	81E

Cat Num	Canon Plate	Calendar Date	TD of Greatest Eclipse	ΔT s	Luna Num	Saros Num	Ecl. Type	QSE	Gamma	Pen. Mag.	Um. Mag.	Eclipse Phase ---- Durations ---- Pen. m	Par. m	Total m	Greatest in Zenith Lat.	Lng.
8641	433	1590 Jan 20	15:53:21	123	-5071	098	N	-t	-1.2783	0.5624	-0.5370	222.7	-	-	19N	125E
8642	433	1590 Jul 17	04:09:55	123	-5065	103	P	-a	0.9206	1.1493	0.1879	261.7	93.9	-	20S	61W
8643	433	1591 Jan 09	18:19:22	123	-5059	108	P	-h	-0.5552	1.8674	0.8116	333.3	187.7	-	22N	88E
8644	433	1591 Jul 06	16:29:29	123	-5053	113	T+	pp	0.1776	2.5381	1.5260	345.9	221.6	95.7	23S	114E
8645	433	1591 Dec 30	04:00:35	122	-5047	118	T+	p-	0.1588	2.5665	1.5668	329.8	212.9	93.7	23N	59W
8646	433	1592 Jun 24	21:40:12	122	-5041	123	P	t-	-0.6135	1.7620	0.7029	340.7	184.6	-	24S	36E
8647	433	1592 Dec 18	18:46:58	122	-5035	128	P	a-	0.8160	1.3460	0.3749	271.7	126.4	-	24N	78E
8648	433	1593 May 15	07:55:13	122	-5030	095	N	-t	1.4548	0.2162	-0.8388	139.2	-	-	18S	119W
8649	433	1593 Jun 13	22:19:51	122	-5029	133	N	t-	-1.3703	0.3770	-0.6895	184.3	-	-	25S	25E
8650	433	1593 Nov 08	22:22:02	121	-5024	100	N	-a	-1.2329	0.6000	-0.4086	206.6	-	-	16N	22E
8651	433	1593 Dec 08	10:08:29	121	-5023	138	N	a-	1.4899	0.1180	-0.8701	93.1	-	-	24N	154W
8652	433	1594 May 04	14:32:42	121	-5018	105	P	-t	0.6461	1.6748	0.6699	311.9	171.0	-	15S	142E
8653	433	1594 Oct 29	06:23:44	121	-5012	110	P	-a	-0.5687	1.8462	0.7833	337.9	188.0	-	13N	99W
8654	433	1595 Apr 24	03:56:44	121	-5006	115	T-	p-	-0.1184	2.6206	1.6602	321.7	212.0	96.7	13S	59W
8655	433	1595 Oct 18	07:40:06	121	-5000	120	T+	pp	0.1391	2.6497	1.5567	376.0	233.1	100.2	10N	118W
8656	433	1596 Apr 12	20:35:30	120	-4994	125	P	a-	-0.8278	1.3151	0.3623	268.9	124.8	-	10S	51E
8657	433	1596 Oct 06	07:45:10	120	-4988	130	P	t-	0.8436	1.3466	0.2743	311.8	122.6	-	6N	119W
8658	433	1597 Mar 04	00:16:53	120	-4983	097	N	-a	1.1329	0.7921	-0.2335	238.7	-	-	7N	0W
8659	433	1597 Aug 27	03:08:44	120	-4977	102	N	-a	-1.1604	0.7189	-0.2616	220.3	-	-	11S	46W
8660	433	1597 Sep 25	13:40:53	120	-4976	140	N	a-	1.5117	0.0922	-0.9238	86.7	-	-	2N	152E
8661	434	1598 Feb 21	05:35:27	119	-4971	107	P	-t	0.4537	2.0666	0.9851	355.0	206.5	-	11N	80W
8662	434	1598 Aug 16	18:47:58	119	-4965	112	T	-a	-0.4040	2.0897	1.1432	306.8	194.6	54.2	14S	80E
8663	434	1599 Feb 10	05:14:35	119	-4959	117	T-	p-	-0.2410	2.4661	1.3663	373.8	228.1	88.0	14N	74W
8664	434	1599 Aug 06	11:42:06	119	-4953	122	T	p-	0.3272	2.2353	1.2794	315.1	202.6	73.1	17S	174W
8665	434	1600 Jan 30	06:25:29	118	-4947	127	P	a-	-0.9183	1.2067	0.1399	294.2	87.7	-	17N	93W
8666	434	1600 Jul 26	01:13:18	118	-4941	132	N	t-	1.1200	0.8035	-0.1978	238.3	-	-	18S	17W
8667	434	1600 Dec 20	02:31:54	117	-4936	099	N	-a	1.1989	0.6515	-0.3356	207.6	-	-	25N	38W
8668	434	1601 Jan 18	14:32:25	117	-4935	137	N	a-	-1.5425	0.0329	-0.9777	50.9	-	-	19N	145E
8669	434	1601 Jun 15	17:39:34	117	-4930	104	P	-t	-0.8517	1.3287	0.2621	314.1	121.7	-	24S	96E
8670	434	1601 Dec 09	17:58:33	116	-4924	109	P	-a	0.5249	1.8800	0.9092	300.8	180.6	-	23N	89E
8671	434	1602 Jun 04	18:18:12	116	-4918	114	T-	pp	-0.0895	2.7239	1.6640	369.2	233.2	104.0	23S	85E
8672	434	1602 Nov 29	08:40:26	115	-4912	119	T-	p-	-0.1482	2.5860	1.5861	330.7	213.7	94.6	21N	132W
8673	434	1603 May 24	23:30:55	115	-4906	124	P	a-	0.6737	1.6287	0.6150	313.9	167.4	-	20S	7E
8674	434	1603 Nov 18	18:10:30	114	-4900	129	P	t-	-0.8837	1.2641	0.2093	295.3	105.2	-	18N	84E
8675	434	1604 Apr 14	03:47:06	114	-4895	096	N	-a	-1.1848	0.6613	-0.2938	205.2	-	-	11S	57W
8676	434	1604 May 13	11:49:46	114	-4894	134	N	a-	1.3732	0.3201	-0.6441	150.2	-	-	17S	178W
8677	434	1604 Oct 08	02:29:43	113	-4889	101	N	-t	1.3742	0.3821	-0.7084	187.1	-	-	7N	40W
8678	434	1605 Apr 03	20:27:02	113	-4883	106	P	-a	-0.4858	1.9467	0.9861	304.8	186.9	-	6S	54E
8679	434	1605 Sep 27	03:46:02	112	-4877	111	P	-t	0.6341	1.7229	0.6666	331.8	177.4	-	2N	58W
8680	434	1606 Mar 24	10:23:18	112	-4871	116	T+	p-	0.2363	2.4276	1.4211	335.8	214.1	87.5	1S	154W
8681	435	1606 Sep 16	11:55:21	111	-4865	121	T-	p-	-0.1295	2.6189	1.6216	336.0	217.7	97.5	3S	180W
8682	435	1607 Mar 13	17:38:53	111	-4859	126	N*	t-	1.0028	1.0510	-0.0145	281.6	-	-	4N	98E
8683	435	1607 Sep 06	02:28:42	110	-4853	131	P	a-	-0.8407	1.2923	0.3378	268.3	121.3	-	8S	37W
8684	435	1608 Jan 31	23:50:22	110	-4848	098	N	-t	-1.2894	0.5405	-0.5558	218.3	-	-	16N	6E
8685	435	1608 Jul 27	11:33:02	109	-4842	103	P	-a	0.9936	1.0172	0.0520	250.7	50.5	-	18S	172W
8686	435	1608 Aug 25	19:20:07	109	-4841	141	Nb	a-	-1.5413	0.0062	-0.9471	21.1	-	-	12S	71E
8687	435	1609 Jan 20	02:39:41	108	-4836	108	P	-h	-0.5607	1.8552	0.8034	331.4	186.5	-	20N	37W
8688	435	1609 Jul 16	23:25:49	108	-4830	113	T+	-p	0.2570	2.3945	1.3783	345.0	218.3	86.3	21S	10E
8689	435	1610 Jan 09	12:44:36	107	-4824	118	T+	p-	0.1549	2.5721	1.5752	328.8	212.6	93.9	22N	171E
8690	435	1610 Jul 06	04:08:03	107	-4818	123	P	t-	-0.5273	1.9212	0.8600	349.4	199.1	-	23S	61W
8691	435	1610 Dec 30	03:39:54	106	-4812	128	P	a-	0.8124	1.3526	0.3817	272.1	127.3	-	24N	54W
8692	435	1611 May 26	14:38:02	106	-4807	095	Ne	-t	1.5270	0.0816	-0.9690	86.2	-	-	20S	140E
8693	435	1611 Jun 25	04:45:08	105	-4806	133	N	t-	-1.2836	0.5352	-0.5295	216.0	-	-	25S	71W
8694	435	1611 Nov 20	06:48:14	105	-4801	100	N	-a	-1.2493	0.5724	-0.4411	203.1	-	-	19N	105W
8695	435	1611 Dec 19	18:54:22	105	-4800	138	N	a-	1.4891	0.1208	-0.8700	94.4	-	-	25N	76E
8696	435	1612 May 14	21:48:17	104	-4795	105	P	-t	0.7111	1.5527	0.5534	303.4	157.9	-	18S	33E
8697	435	1612 Nov 08	14:19:56	104	-4789	110	P	-h	-0.5915	1.8071	0.7387	337.3	184.5	-	16N	142E
8698	435	1613 May 04	11:39:03	103	-4783	115	T-	pp	-0.0593	2.7271	1.7709	321.4	212.8	99.0	16S	175W
8699	435	1613 Oct 28	15:11:40	102	-4777	120	T+	pp	0.1064	2.7115	1.6149	377.2	234.1	102.3	13N	128E
8700	435	1614 Apr 24	04:28:48	102	-4771	125	P	a-	-0.7766	1.4083	0.4573	275.4	138.3	-	14S	68W

Cat Num	Canon Plate	Calendar Date	TD of Greatest Eclipse	ΔT s	Luna Num	Saros Num	Ecl. Type	QSE	Gamma	Pen. Mag.	Um. Mag.	Eclipse Phase Durations Pen. m	Par. m	Total m	Greatest in Zenith Lat.	Lng.
8701	436	1614 Oct 17	15:18:17	101	-4765	130	P	t-	0.8043	1.4191	0.3458	316.8	135.9	-	10N	127E
8702	436	1615 Mar 15	08:16:13	101	-4760	097	N	-a	1.1667	0.7300	-0.2956	231.4	-	-	3N	121W
8703	436	1615 Apr 13	19:38:31	100	-4759	135	Nb	a-	-1.5327	0.0399	-0.9488	55.6	-	-	10S	65E
8704	436	1615 Sep 07	10:47:18	100	-4754	102	N	-a	-1.2123	0.6234	-0.3568	206.5	-	-	7S	161W
8705	436	1615 Oct 06	21:30:17	100	-4753	140	N	a-	1.4676	0.1730	-0.8429	117.5	-	-	6N	34E
8706	436	1616 Mar 03	13:19:28	99	-4748	107	P	-t	0.4820	2.0142	0.9335	353.5	203.4	-	7N	164E
8707	436	1616 Aug 27	02:31:49	99	-4742	112	T	-a	-0.4643	1.9799	1.0318	303.1	188.6	26.4	10S	37W
8708	436	1617 Feb 20	12:58:54	98	-4736	117	T-	pp	-0.2185	2.5056	1.4091	373.9	229.3	91.4	11N	169E
8709	436	1617 Aug 16	19:15:10	97	-4730	122	T+	p-	0.2621	2.3568	1.3970	318.8	207.0	83.7	13S	72E
8710	436	1618 Feb 09	14:33:24	97	-4724	127	P	a-	-0.9027	1.2327	0.1709	295.4	96.3	-	14N	145E
8711	436	1618 Aug 06	08:19:44	96	-4718	132	N	t-	1.0513	0.9321	-0.0743	254.4	-	-	16S	124W
8712	436	1618 Dec 31	11:19:57	95	-4713	099	N	-a	1.1997	0.6492	-0.3361	206.8	-	-	24N	168W
8713	436	1619 Jan 29	23:07:09	95	-4712	137	N	a-	-1.5311	0.0518	-0.9547	63.5	-	-	16N	17E
8714	436	1619 Jun 27	00:04:48	95	-4707	104	P	-t	-0.9389	1.1690	0.1017	300.2	77.7	-	24S	0W
8715	436	1619 Dec 21	02:49:28	94	-4701	109	P	-a	0.5298	1.8716	0.8997	300.7	180.0	-	24N	42W
8716	436	1620 Jun 15	00:50:49	93	-4695	114	T-	pp	-0.1735	2.5683	1.5114	366.5	230.2	97.9	24S	12W
8717	436	1620 Dec 09	17:19:13	93	-4689	119	T-	p-	-0.1423	2.5987	1.5951	332.0	214.3	95.1	23N	99E
8718	436	1621 Jun 04	06:32:45	92	-4683	124	P	a-	0.5970	1.7670	0.7582	319.9	180.9	-	22S	98W
8719	436	1621 Nov 29	02:23:01	91	-4677	129	P	t-	-0.8746	1.2832	0.2237	298.1	108.8	-	21N	38W
8720	436	1622 Apr 25	11:37:48	91	-4672	096	N	-a	-1.2352	0.5669	-0.3845	191.4	-	-	14S	175W
8721	437	1622 May 24	19:18:16	90	-4671	134	N	a-	1.3046	0.4440	-0.5161	174.3	-	-	20S	70E
8722	437	1622 Oct 19	09:54:38	90	-4666	101	N	-t	1.4097	0.3184	-0.7750	171.9	-	-	11N	152W
8723	437	1623 Apr 15	04:23:33	89	-4660	106	P	-a	-0.5345	1.8566	0.8972	301.6	181.4	-	10S	66W
8724	437	1623 Oct 08	11:18:42	88	-4654	111	P	-t	0.6734	1.6508	0.5941	326.8	169.1	-	6N	173W
8725	437	1624 Apr 03	18:12:08	88	-4648	116	T+	pp	0.1945	2.5044	1.4977	338.1	216.8	92.5	5S	88E
8726	437	1624 Sep 26	19:44:24	87	-4642	121	T-	pp	-0.0845	2.7014	1.7042	335.6	218.0	99.5	2N	62E
8727	437	1625 Mar 24	01:11:28	86	-4636	126	P	t-	0.9665	1.1171	0.0524	289.0	55.0	-	1S	16W
8728	437	1625 Sep 16	10:24:16	86	-4630	131	P	a-	-0.7904	1.3855	0.4295	274.6	134.8	-	3S	157W
8729	437	1626 Feb 11	07:43:56	85	-4625	098	N	-t	-1.3035	0.5128	-0.5797	212.7	-	-	13N	112W
8730	437	1626 Aug 07	18:56:24	84	-4619	103	N	-a	1.0655	0.8874	-0.0820	238.2	-	-	15S	77E
8731	437	1626 Sep 06	03:05:57	84	-4618	141	N	a-	-1.4897	0.1028	-0.8544	85.7	-	-	8S	46W
8732	437	1627 Jan 31	10:59:17	84	-4613	108	P	-h	-0.5675	1.8404	0.7932	329.4	185.2	-	17N	161W
8733	437	1627 Jul 28	06:21:40	83	-4607	113	T	-h	0.3356	2.2527	1.2317	343.3	213.5	71.4	19S	94W
8734	437	1628 Jan 20	21:27:26	82	-4601	118	T+	p-	0.1499	2.5798	1.5861	327.9	212.4	94.2	20N	41E
8735	437	1628 Jul 16	10:36:57	81	-4595	123	T	t-	-0.4424	2.0783	1.0147	356.8	210.5	19.9	22S	158W
8736	437	1629 Jan 09	12:32:47	81	-4589	128	P	a-	0.8092	1.3582	0.3877	272.4	128.2	-	23N	174E
8737	437	1629 Jul 05	11:11:37	80	-4583	133	N	t-	-1.1969	0.6937	-0.3696	241.7	-	-	24S	167W
8738	437	1629 Nov 30	15:17:32	79	-4578	100	N	-a	-1.2625	0.5505	-0.4676	200.4	-	-	21N	129E
8739	437	1629 Dec 30	03:40:31	79	-4577	138	N	a-	1.4892	0.1219	-0.8713	95.0	-	-	25N	54W
8740	437	1630 May 26	05:01:22	79	-4572	105	P	-h	0.7788	1.4259	0.4318	293.7	141.7	-	20S	76W
8741	438	1630 Nov 19	22:20:56	78	-4566	110	P	-h	-0.6100	1.7757	0.7024	336.9	181.4	-	19N	22E
8742	438	1631 May 15	19:15:41	77	-4560	115	T+	pp	0.0052	2.8245	1.8721	320.7	213.0	99.8	19S	70E
8743	438	1631 Nov 08	22:51:38	76	-4554	120	T+	pp	0.0805	2.7605	1.6608	378.1	234.7	103.5	17N	13E
8744	438	1632 May 04	12:15:05	76	-4548	125	P	a-	-0.7196	1.5122	0.5625	282.1	151.1	-	17S	175E
8745	438	1632 Oct 27	23:00:32	75	-4542	130	P	t-	0.7718	1.4790	0.4051	320.5	145.4	-	14N	11E
8746	438	1633 Mar 25	16:05:28	74	-4537	097	N	-a	1.2080	0.6543	-0.3715	221.4	-	-	1S	121E
8747	438	1633 Apr 24	03:17:46	74	-4536	135	N	a-	-1.4852	0.1271	-0.8619	98.9	-	-	14S	50W
8748	438	1633 Sep 17	18:34:53	74	-4531	102	N	-a	-1.2576	0.5404	-0.4399	193.3	-	-	3S	81E
8749	438	1633 Oct 17	05:29:19	73	-4530	140	N	a-	1.4306	0.2407	-0.7750	137.3	-	-	11N	86W
8750	438	1634 Mar 14	20:53:01	73	-4525	107	P	-t	0.5184	1.9468	0.8674	351.1	198.8	-	3N	50E
8751	438	1634 Sep 07	10:21:50	72	-4519	112	P	-a	-0.5196	1.8794	0.9293	299.3	182.2	-	7S	155W
8752	438	1635 Mar 03	20:34:50	71	-4513	117	T-	pp	-0.1889	2.5580	1.4653	374.1	230.7	95.3	7N	55E
8753	438	1635 Aug 28	02:53:01	71	-4507	122	T+	p-	0.2013	2.4706	1.5064	321.9	210.2	90.8	10S	43W
8754	438	1636 Feb 20	22:34:50	70	-4501	127	P	a-	-0.8810	1.2699	0.2135	297.3	106.7	-	10N	25E
8755	438	1636 Aug 16	15:29:55	69	-4495	132	P	t-	0.9861	1.0545	0.0426	268.4	47.8	-	13S	128E
8756	438	1637 Jan 10	20:08:48	68	-4490	099	N	-a	1.2007	0.6462	-0.3369	205.9	-	-	23N	61E
8757	438	1637 Feb 09	07:38:45	68	-4489	137	N	a-	-1.5164	0.0763	-0.9256	76.7	-	-	13N	111W
8758	438	1637 Jul 07	06:28:09	68	-4484	104	N*	-t	-1.0270	1.0079	-0.0604	283.9	-	-	24S	96W
8759	438	1637 Dec 31	11:42:25	67	-4478	109	P	-a	0.5331	1.8660	0.8934	300.7	179.7	-	24N	174W
8760	438	1638 Jun 26	07:21:47	66	-4472	114	T-	pp	-0.2595	2.4091	1.3548	362.6	225.3	86.8	24S	110W

Cat Num	Canon Plate	Calendar Date	TD of Greatest Eclipse	ΔT s	Luna Num	Saros Num	Ecl. Type	QSE	Gamma	Pen. Mag.	Um. Mag.	Eclipse Phase Durations Pen. m	Par. m	Total m	Greatest in Zenith Lat.	Lng.
8761	439	1638 Dec 21	02:00:39	66	-4466	119	T-	pp	-0.1393	2.6058	1.5989	333.3	214.8	95.4	23N	30W
8762	439	1639 Jun 15	13:32:42	65	-4460	124	P	a-	0.5176	1.9103	0.9059	325.0	192.1	-	23S	157E
8763	439	1639 Dec 10	10:39:50	64	-4454	129	P	t-	-0.8695	1.2946	0.2308	300.1	110.7	-	22N	161W
8764	439	1640 May 05	19:23:35	63	-4449	096	N	-a	-1.2899	0.4649	-0.4833	174.8	-	-	18S	68E
8765	439	1640 Jun 04	02:45:05	63	-4448	134	N	a-	1.2340	0.5718	-0.3848	194.8	-	-	21S	42W
8766	439	1640 Oct 29	17:27:12	63	-4443	101	N	-t	1.4390	0.2657	-0.8301	157.8	-	-	15N	94E
8767	439	1641 Apr 25	12:14:36	62	-4437	106	P	-a	-0.5876	1.7588	0.8003	297.8	174.6	-	14S	176E
8768	439	1641 Oct 18	19:00:25	61	-4431	111	P	-t	0.7062	1.5909	0.5340	322.2	161.4	-	11N	71E
8769	439	1642 Apr 15	01:52:19	60	-4425	116	T+	pp	0.1463	2.5930	1.5860	340.4	219.3	96.9	10S	28W
8770	439	1642 Oct 08	03:42:19	60	-4419	121	T-	pp	-0.0462	2.7719	1.7746	335.0	217.9	100.3	6N	58W
8771	439	1643 Apr 04	08:33:34	59	-4413	126	P	t-	0.9222	1.1982	0.1341	297.3	87.2	-	5S	127W
8772	439	1643 Sep 27	18:28:14	58	-4407	131	P	a-	-0.7468	1.4663	0.5085	279.6	144.7	-	1N	81E
8773	439	1644 Feb 22	15:29:27	58	-4402	098	N	-t	-1.3240	0.4730	-0.6153	204.5	-	-	9N	131E
8774	439	1644 Aug 18	02:25:18	57	-4396	103	N	-a	1.1319	0.7679	-0.2062	225.2	-	-	12S	36W
8775	439	1644 Sep 16	10:59:45	57	-4395	141	N	a-	-1.4448	0.1873	-0.7742	115.0	-	-	4S	166W
8776	439	1645 Feb 10	19:13:47	56	-4390	108	P	-h	-0.5789	1.8168	0.7748	326.8	183.1	-	13N	75E
8777	439	1645 Aug 07	13:19:26	55	-4384	113	T	-h	0.4115	2.1160	1.0899	340.8	207.1	46.7	16S	162E
8778	439	1646 Jan 31	06:07:04	55	-4378	118	T+	p-	0.1417	2.5929	1.6030	327.0	212.3	94.7	17N	88W
8779	439	1646 Jul 27	17:06:36	54	-4372	123	T	t-	-0.3586	2.2335	1.1671	362.8	219.4	64.2	19S	105E
8780	439	1647 Jan 20	21:23:09	53	-4366	128	P	a-	0.8039	1.3674	0.3979	273.0	129.7	-	21N	43E
8781	440	1647 Jul 16	17:39:13	53	-4360	133	N	t-	-1.1100	0.8527	-0.2097	263.3	-	-	22S	97E
8782	440	1647 Dec 11	23:51:13	52	-4355	100	N	-a	-1.2717	0.5357	-0.4864	198.7	-	-	22N	1E
8783	440	1648 Jan 10	12:26:15	52	-4354	138	N	a-	1.4897	0.1219	-0.8730	95.3	-	-	23N	176E
8784	440	1648 Jun 05	12:11:39	51	-4349	105	P	-h	0.8497	1.2935	0.3041	282.8	120.8	-	22S	177E
8785	440	1648 Nov 30	06:26:42	51	-4343	110	P	-h	-0.6238	1.7525	0.6748	336.9	179.0	-	21N	99W
8786	440	1649 May 26	02:48:36	50	-4337	115	T+	pp	0.0732	2.6981	1.7489	319.6	212.2	98.6	21S	43W
8787	440	1649 Nov 19	06:38:54	49	-4331	120	T+	pp	0.0608	2.7980	1.6957	378.6	234.9	104.1	20N	103W
8788	440	1650 May 15	19:54:19	48	-4325	125	P	a-	-0.6570	1.6265	0.6778	288.8	163.1	-	20S	60E
8789	440	1650 Nov 08	06:52:25	48	-4319	130	P	t-	0.7464	1.5257	0.4515	323.1	152.0	-	17N	107W
8790	440	1651 Apr 05	23:45:28	47	-4314	097	N	-a	1.2558	0.5668	-0.4593	208.5	-	-	5S	5E
8791	440	1651 May 05	10:48:41	47	-4313	135	N	a-	-1.4321	0.2249	-0.7647	130.8	-	-	18S	163W
8792	440	1651 Sep 29	02:31:07	46	-4308	102	N	-a	-1.2966	0.4689	-0.5117	180.8	-	-	1N	39W
8793	440	1651 Oct 28	13:37:37	46	-4307	140	N	a-	1.4003	0.2963	-0.7192	151.1	-	-	14N	151E
8794	440	1652 Mar 25	04:17:30	46	-4302	107	P	-t	0.5617	1.8667	0.7887	347.9	192.6	-	2S	62W
8795	440	1652 Sep 17	18:19:29	45	-4296	112	P	-a	-0.5687	1.7906	0.8381	295.6	175.7	-	2S	84E
8796	440	1653 Mar 14	04:04:31	44	-4290	117	T-	pp	-0.1537	2.6205	1.5320	374.1	232.1	98.9	2N	59W
8797	440	1653 Sep 07	10:35:26	44	-4284	122	T+	pp	0.1448	2.5766	1.6076	324.4	212.6	95.4	6S	159W
8798	440	1654 Mar 03	06:31:30	43	-4278	127	P	a-	-0.8544	1.3156	0.2650	299.8	117.6	-	6N	95W
8799	440	1654 Aug 27	22:44:01	42	-4272	132	P	t-	0.9246	1.1704	0.1526	280.5	89.3	-	9S	19E
8800	440	1655 Jan 22	04:54:43	42	-4267	099	N	-a	1.2045	0.6379	-0.3425	204.3	-	-	21N	70W
8801	441	1655 Feb 20	16:03:56	42	-4266	137	N	a-	-1.4962	0.1109	-0.8860	91.8	-	-	9N	122E
8802	441	1655 Jul 18	12:53:09	41	-4261	104	N	-t	-1.1131	0.8506	-0.2192	265.4	-	-	22S	169E
8803	441	1656 Jan 11	20:33:34	40	-4255	109	P	-a	0.5380	1.8571	0.8843	300.5	179.2	-	22N	54E
8804	441	1656 Jul 06	13:55:39	40	-4249	114	T	-t	-0.3437	2.2535	1.2014	357.6	218.6	69.0	23S	152E
8805	441	1656 Dec 31	10:41:02	39	-4243	119	T-	pp	-0.1359	2.6134	1.6037	334.5	215.4	95.7	23N	159W
8806	441	1657 Jun 25	20:33:50	38	-4237	124	T	a-	0.4380	2.0545	1.0540	329.1	201.0	35.9	23S	52E
8807	441	1657 Dec 20	18:58:00	38	-4231	129	P	t-	-0.8661	1.3026	0.2353	301.9	112.0	-	23N	75E
8808	441	1658 May 17	03:01:54	37	-4226	096	N	-a	-1.3510	0.3513	-0.5941	153.4	-	-	21S	47W
8809	441	1658 Jun 15	10:08:00	37	-4225	134	N	a-	1.1593	0.7073	-0.2463	213.1	-	-	22S	152W
8810	441	1658 Nov 10	01:09:24	37	-4220	101	N	-t	1.4609	0.2267	-0.8712	146.2	-	-	18N	21W
8811	441	1658 Dec 09	20:10:47	37	-4219	139	N	t-	-1.5784	0.0141	-1.0897	37.2	-	-	22N	56E
8812	441	1659 May 06	19:56:13	36	-4214	106	P	-a	-0.6484	1.6470	0.6890	292.8	165.3	-	17S	60E
8813	441	1659 Oct 30	02:52:52	35	-4208	111	P	-t	0.7310	1.5454	0.4885	318.4	155.1	-	14N	47W
8814	441	1660 Apr 25	09:22:01	35	-4202	116	T+	pp	0.0901	2.6965	1.6890	342.5	221.4	100.5	13S	141W
8815	441	1660 Oct 18	11:50:08	34	-4196	121	T-	pp	-0.0151	2.8289	1.8316	334.2	217.5	100.5	10N	179E
8816	441	1661 Apr 14	15:46:19	34	-4190	126	P	t-	0.8708	1.2919	0.2286	306.2	112.5	-	9S	124E
8817	441	1661 Oct 08	02:39:03	33	-4184	131	P	a-	-0.7089	1.5370	0.5771	283.7	152.4	-	5N	43W
8818	441	1662 Mar 04	23:10:09	32	-4179	098	N	-t	-1.3487	0.4254	-0.6581	194.3	-	-	5N	15E
8819	441	1662 Aug 29	09:57:01	32	-4173	103	N	-a	1.1951	0.6546	-0.3247	211.1	-	-	8S	149W
8820	441	1662 Sep 27	18:59:20	32	-4172	141	N	a-	-1.4052	0.2625	-0.7037	135.6	-	-	1N	73E

Cat Num	Canon Plate	Calendar Date	TD of Greatest Eclipse	ΔT s	Luna Num	Saros Num	Ecl. Type	QSE	Gamma	Pen. Mag.	Um. Mag.	Eclipse Phase Durations Pen. m	Par. m	Total m	Greatest in Zenith Lat.	Lng.
8821	442	1663 Feb 22	03:23:34	31	-4167	108	P	-h	-0.5950	1.7845	0.7481	323.8	180.2	-	10N	47W
8822	442	1663 Aug 18	20:20:08	31	-4161	113	P	-h	0.4840	1.9858	0.9543	337.6	199.4	-	13S	56E
8823	442	1664 Feb 11	14:43:15	30	-4155	118	T+	p-	0.1303	2.6117	1.6258	326.1	212.3	95.4	14N	143E
8824	442	1664 Aug 06	23:40:04	29	-4149	123	T	pp	-0.2779	2.3830	1.3135	367.6	226.0	83.5	17S	6E
8825	442	1665 Jan 31	06:10:20	29	-4143	128	P	a-	0.7963	1.3807	0.4124	273.9	131.8	-	18N	89W
8826	442	1665 Jul 27	00:11:35	28	-4137	133	N*	t-	-1.0260	1.0063	-0.0554	281.0	-	-	20S	1W
8827	442	1665 Dec 22	08:25:41	28	-4132	100	N	-a	-1.2796	0.5230	-0.5028	197.3	-	-	22N	126W
8828	442	1666 Jan 20	21:09:12	28	-4131	138	N	a-	1.4886	0.1246	-0.8717	96.6	-	-	21N	46E
8829	442	1666 Jun 16	19:20:41	27	-4126	105	P	-h	0.9222	1.1584	0.1732	270.5	92.6	-	23S	70E
8830	442	1666 Dec 11	14:35:36	27	-4120	110	P	-h	-0.6344	1.7350	0.6536	337.0	177.2	-	23N	140E
8831	442	1667 Jun 06	10:18:29	26	-4114	115	T+	pp	0.1438	2.5671	1.6205	317.8	210.4	95.2	23S	155W
8832	442	1667 Nov 30	14:31:26	26	-4108	120	T+	pp	0.0455	2.8269	1.7228	379.0	235.1	104.5	22N	140E
8833	442	1668 May 26	03:28:47	25	-4102	125	P	a-	-0.5909	1.7477	0.7993	295.2	173.8	-	22S	53W
8834	442	1668 Nov 18	14:52:18	25	-4096	130	P	h-	0.7274	1.5606	0.4865	324.6	156.6	-	20N	133E
8835	442	1669 Apr 16	07:15:32	24	-4091	097	N	-h	1.3108	0.4660	-0.5604	191.5	-	-	9S	108W
8836	442	1669 May 15	18:11:57	24	-4090	135	N	h-	-1.3734	0.3332	-0.6575	158.2	-	-	20S	86E
8837	442	1669 Oct 09	10:35:58	24	-4085	102	N	-a	-1.3293	0.4092	-0.5719	169.5	-	-	5N	161W
8838	442	1669 Nov 07	21:54:08	23	-4084	140	N	a-	1.3760	0.3406	-0.6745	160.9	-	-	18N	27E
8839	442	1670 Apr 05	11:31:33	23	-4079	107	P	-t	0.6128	1.7720	0.6956	343.4	184.1	-	6S	172W
8840	442	1670 Sep 29	02:24:55	23	-4073	112	P	-a	-0.6114	1.7135	0.7583	292.2	169.3	-	2N	38W
8841	443	1671 Mar 25	11:23:35	22	-4067	117	T-	pp	-0.1095	2.6993	1.6154	374.0	233.4	102.3	2S	169W
8842	443	1671 Sep 18	18:25:15	22	-4061	122	T+	pp	0.0951	2.6705	1.6963	326.5	214.2	98.1	2S	82E
8843	443	1672 Mar 13	14:20:48	21	-4055	127	P	a-	-0.8210	1.3738	0.3293	303.0	129.5	-	2N	147E
8844	443	1672 Sep 07	06:03:27	21	-4049	132	P	t-	0.8681	1.2770	0.2531	291.0	113.7	-	5S	92W
8845	443	1673 Feb 01	13:38:12	20	-4044	099	N	-a	1.2108	0.6248	-0.3525	202.0	-	-	18N	160E
8846	443	1673 Mar 03	00:24:02	20	-4043	137	N	a-	-1.4715	0.1536	-0.8381	107.2	-	-	5N	4W
8847	443	1673 Jul 28	19:20:13	20	-4038	104	N	-t	-1.1971	0.6975	-0.3740	244.2	-	-	20S	72E
8848	443	1674 Jan 22	05:23:56	19	-4032	109	P	-a	0.5433	1.8472	0.8745	300.4	178.6	-	20N	78W
8849	443	1674 Jul 17	20:30:10	19	-4026	114	T	-t	-0.4280	2.0980	1.0478	351.6	209.9	35.2	21S	54E
8850	443	1675 Jan 11	19:21:44	18	-4020	119	T-	pp	-0.1334	2.6192	1.6073	335.7	216.0	96.0	22N	72E
8851	443	1675 Jul 07	03:36:11	18	-4014	124	T	a-	0.3582	2.1991	1.2020	332.3	207.9	66.2	22S	53W
8852	443	1676 Jan 01	03:16:18	18	-4008	129	P	t-	-0.8634	1.3091	0.2389	303.4	113.1	-	22N	48W
8853	443	1676 May 27	10:37:28	17	-4003	096	N	-a	-1.4146	0.2337	-0.7095	126.3	-	-	23S	160W
8854	443	1676 Jun 25	17:32:06	17	-4002	134	N	a-	1.0849	0.8427	-0.1086	228.9	-	-	22S	97E
8855	443	1676 Nov 20	08:58:24	17	-3997	101	N	-t	1.4777	0.1967	-0.9028	136.5	-	-	21N	138W
8856	443	1676 Dec 20	04:11:44	17	-3996	139	N	t-	-1.5710	0.0285	-1.0768	52.9	-	-	22N	63W
8857	443	1677 May 17	03:32:58	16	-3991	106	P	-a	-0.7127	1.5290	0.5710	286.8	153.7	-	20S	54W
8858	443	1677 Nov 09	10:52:40	16	-3985	111	P	-t	0.7504	1.5096	0.4529	315.1	149.7	-	18N	167W
8859	443	1678 May 06	16:44:20	16	-3979	116	T+	pp	0.0285	2.8099	1.8015	344.3	223.0	102.5	17S	108E
8860	443	1678 Oct 29	20:06:20	15	-3973	121	T+	pp	0.0093	2.8395	1.8422	333.2	216.9	100.2	14N	54E
8861	444	1679 Apr 25	22:48:29	15	-3967	126	P	t-	0.8115	1.4006	0.3379	315.6	134.9	-	13S	18E
8862	444	1679 Oct 19	10:58:53	14	-3961	131	P	a-	-0.6784	1.5940	0.6321	286.9	158.0	-	9N	168W
8863	444	1680 Mar 15	06:41:52	14	-3956	098	N	-t	-1.3804	0.3645	-0.7139	180.3	-	-	1N	99W
8864	444	1680 Apr 13	23:18:00	14	-3955	136	Nb	t-	1.5494	0.0560	-1.0254	73.6	-	-	8S	11E
8865	444	1680 Sep 08	17:35:19	14	-3950	103	N	-a	1.2521	0.5527	-0.4320	196.8	-	-	4S	95E
8866	444	1680 Oct 08	03:06:25	14	-3949	141	N	a-	-1.3719	0.3258	-0.6450	150.6	-	-	5N	49W
8867	444	1681 Mar 04	11:26:33	13	-3944	108	P	-h	-0.6173	1.7406	0.7101	320.1	176.2	-	6N	169W
8868	444	1681 Aug 29	03:25:58	13	-3938	113	P	-h	0.5511	1.8655	0.8282	334.0	190.6	-	9S	51W
8869	444	1682 Feb 21	23:13:12	13	-3932	118	T+	p-	0.1132	2.6408	1.6594	325.3	212.5	96.4	10N	15E
8870	444	1682 Aug 18	06:17:12	13	-3926	123	T-	pp	-0.2006	2.5267	1.4538	371.2	230.7	95.0	13S	93W
8871	444	1683 Feb 11	14:52:35	12	-3920	128	P	a-	0.7848	1.4011	0.4345	275.2	134.9	-	15N	141E
8872	444	1683 Aug 07	06:48:53	12	-3914	133	P	t-	-0.9452	1.1546	0.0931	295.7	73.7	-	17S	101W
8873	444	1684 Jan 02	17:00:34	12	-3909	100	N	-a	-1.2865	0.5119	-0.5169	196.1	-	-	22N	106E
8874	444	1684 Feb 01	05:47:50	12	-3908	138	N	a-	1.4849	0.1319	-0.8653	99.6	-	-	19N	83W
8875	444	1684 Jun 27	02:30:11	11	-3903	105	P	-h	0.9948	1.0233	0.0416	257.0	46.1	-	22S	37W
8876	444	1684 Dec 21	22:47:25	11	-3897	110	P	-h	-0.6415	1.7234	0.6389	337.4	176.0	-	23N	18E
8877	444	1685 Jun 16	17:45:43	11	-3891	115	T+	-p	0.2172	2.4314	1.4869	315.4	207.5	89.0	23S	94E
8878	444	1685 Dec 10	22:29:22	11	-3885	120	T+	pp	0.0346	2.8476	1.7422	379.1	235.1	104.7	23N	21E
8879	444	1686 Jun 06	10:58:28	10	-3879	125	P	a-	-0.5211	1.8759	0.9273	301.3	183.4	-	23S	165W
8880	444	1686 Nov 29	22:59:47	10	-3873	130	P	h-	0.7134	1.5860	0.5126	325.4	159.7	-	22N	12E

Cat Num	Canon Plate	Calendar Date	TD of Greatest Eclipse	ΔT s	Luna Num	Saros Num	Ecl. Type	QSE	Gamma	Pen. Mag.	Um. Mag.	Eclipse Phase Durations Pen. m	Par. m	Total m	Greatest in Zenith Lat.	Lng.
8881	445	1687 Apr 27	14:36:00	10	-3868	097	N	-h	1.3727	0.3527	-0.6741	169.0	-	-	13S	141E
8882	445	1687 May 27	01:28:13	10	-3867	135	N	h-	-1.3095	0.4512	-0.5410	182.7	-	-	23S	23W
8883	445	1687 Oct 20	18:50:03	10	-3862	102	N	-a	-1.3555	0.3615	-0.6201	159.6	-	-	9N	74E
8884	445	1687 Nov 19	06:18:50	10	-3861	140	N	a-	1.3575	0.3742	-0.6401	167.6	-	-	21N	99W
8885	445	1688 Apr 15	18:37:21	10	-3856	107	P	-t	0.6703	1.6657	0.5908	337.7	173.0	-	10S	81E
8886	445	1688 Oct 09	10:37:18	9	-3850	112	P	-a	-0.6485	1.6468	0.6888	289.1	163.2	-	6N	162W
8887	445	1689 Apr 04	18:36:37	9	-3844	117	T-	pp	-0.0599	2.7881	1.7089	373.6	234.2	104.7	6S	82E
8888	445	1689 Sep 29	02:21:15	9	-3838	122	T+	pp	0.0509	2.7541	1.7747	328.2	215.2	99.5	3N	38W
8889	445	1690 Mar 24	22:03:03	9	-3832	127	P	a-	-0.7809	1.4443	0.4062	306.7	141.7	-	3S	30E
8890	445	1690 Sep 18	13:28:36	9	-3826	132	P	t-	0.8169	1.3742	0.3440	300.0	131.2	-	1S	156E
8891	445	1691 Feb 12	22:16:33	9	-3821	099	N	-a	1.2219	0.6027	-0.3712	198.4	-	-	15N	30E
8892	445	1691 Mar 14	08:36:44	9	-3820	137	N	a-	-1.4403	0.2083	-0.7780	123.8	-	-	1N	127W
8893	445	1691 Aug 09	01:52:19	9	-3815	104	N	-t	-1.2766	0.5526	-0.5210	220.6	-	-	17S	26W
8894	445	1692 Feb 02	14:08:39	8	-3809	109	P	-a	0.5536	1.8281	0.8560	299.8	177.5	-	17N	152E
8895	445	1692 Jul 28	03:10:54	8	-3803	114	P	-t	-0.5080	1.9505	0.9016	344.7	199.6	-	19S	46W
8896	445	1693 Jan 22	03:58:35	8	-3797	119	T-	pp	-0.1278	2.6302	1.6167	336.9	216.7	96.6	19N	57W
8897	445	1693 Jul 17	10:41:02	8	-3791	124	T	p-	0.2793	2.3426	1.3482	334.4	213.0	82.4	21S	159W
8898	445	1694 Jan 11	11:33:05	8	-3785	129	P	t-	-0.8598	1.3167	0.2444	305.0	114.6	-	21N	171W
8899	445	1694 Jun 07	18:08:12	8	-3780	096	N	-a	-1.4823	0.1085	-0.8331	87.0	-	-	24S	88E
8900	445	1694 Jul 07	00:55:25	8	-3779	134	P	a-	1.0089	0.9813	0.0317	242.8	39.0	-	22S	13W
8901	446	1694 Dec 01	16:54:29	8	-3774	101	N	-t	1.4889	0.1765	-0.9238	129.5	-	-	23N	104E
8902	446	1694 Dec 31	12:13:59	8	-3773	139	N	t-	-1.5645	0.0405	-1.0653	63.1	-	-	22N	177E
8903	446	1695 May 28	11:01:57	8	-3768	106	P	-a	-0.7830	1.4003	0.4417	279.3	138.2	-	22S	166W
8904	446	1695 Nov 20	19:02:06	8	-3762	111	P	-t	0.7629	1.4863	0.4303	312.6	146.0	-	21N	71E
8905	446	1696 May 16	23:58:19	8	-3756	116	T-	pp	-0.0392	2.7910	1.7813	345.5	223.6	102.7	20S	1W
8906	446	1696 Nov 09	04:30:21	8	-3750	121	T+	pp	0.0281	2.8049	1.8078	332.3	216.3	99.8	17N	71W
8907	446	1697 May 06	05:42:53	8	-3744	126	P	t-	0.7461	1.5203	0.4581	324.9	154.5	-	16S	87W
8908	446	1697 Oct 29	19:25:59	8	-3738	131	P	a-	-0.6537	1.6404	0.6764	289.3	162.1	-	13N	65E
8909	446	1698 Mar 26	14:06:38	8	-3733	098	N	-t	-1.4182	0.2924	-0.7806	162.0	-	-	4S	149E
8910	446	1698 Apr 25	06:11:31	8	-3732	136	N	t-	1.4876	0.1675	-0.9101	125.8	-	-	12S	93W
8911	446	1698 Sep 20	01:18:39	8	-3727	103	N	-a	1.3043	0.4598	-0.5308	181.8	-	-	0N	22W
8912	446	1698 Oct 19	11:20:30	8	-3726	141	N	a-	-1.3447	0.3783	-0.5973	161.9	-	-	9N	173W
8913	446	1699 Mar 15	19:23:52	8	-3721	108	P	-h	-0.6453	1.6862	0.6619	315.7	171.0	-	1N	71E
8914	446	1699 Sep 09	10:37:10	8	-3715	113	P	-h	0.6130	1.7549	0.7117	330.0	180.7	-	5S	160W
8915	446	1700 Mar 05	07:37:27	8	-3709	118	T+	p-	0.0909	2.6795	1.7026	324.5	212.7	97.5	6N	111W
8916	446	1700 Aug 29	13:01:24	8	-3703	123	T-	pp	-0.1291	2.6597	1.5833	373.8	233.7	101.7	9S	165E
8917	446	1701 Feb 22	23:29:56	8	-3697	128	P	a-	0.7695	1.4283	0.4634	277.0	138.8	-	11N	11E
8918	446	1701 Aug 18	13:32:17	8	-3691	133	P	t-	-0.8683	1.2957	0.2342	307.8	114.3	-	14S	158E
8919	446	1702 Jan 14	01:33:37	8	-3686	100	N	-a	-1.2943	0.4988	-0.5324	194.5	-	-	20N	21W
8920	446	1702 Feb 12	14:21:25	8	-3685	138	N	a-	1.4778	0.1451	-0.8527	104.6	-	-	15N	149E
8921	447	1702 Jul 09	09:41:33	8	-3680	105	N	-h	1.0664	0.8904	-0.0883	242.1	-	-	21S	144W
8922	447	1703 Jan 03	06:58:07	8	-3674	110	P	-h	-0.6493	1.7103	0.6236	337.6	174.7	-	22N	103W
8923	447	1703 Jun 29	01:12:47	9	-3668	115	T	-p	0.2910	2.2953	1.3522	312.4	203.4	79.5	23S	18W
8924	447	1703 Dec 23	06:29:54	9	-3662	120	T+	pp	0.0258	2.8640	1.7582	379.1	235.1	104.7	23N	98W
8925	447	1704 Jun 17	18:25:46	9	-3656	125	T	a-	-0.4494	2.0078	1.0583	306.8	191.7	35.7	24S	84E
8926	447	1704 Dec 11	07:11:30	9	-3650	130	P	a-	0.7023	1.6057	0.5336	325.7	162.0	-	24N	110W
8927	447	1705 May 08	21:47:46	9	-3645	097	N	-h	1.4404	0.2288	-0.7989	138.1	-	-	16S	33E
8928	447	1705 Jun 07	08:39:03	9	-3644	135	N	h-	-1.2419	0.5763	-0.4180	204.8	-	-	24S	130W
8929	447	1705 Nov 01	03:12:43	9	-3639	102	N	-a	-1.3753	0.3252	-0.6568	151.6	-	-	13N	52W
8930	447	1705 Nov 30	14:50:04	9	-3638	140	N	a-	1.3439	0.3987	-0.6145	172.0	-	-	23N	134E
8931	447	1706 Apr 28	01:31:45	9	-3633	107	P	-t	0.7365	1.5436	0.4702	330.2	157.6	-	13S	23W
8932	447	1706 Oct 21	18:58:21	9	-3627	112	P	-a	-0.6785	1.5933	0.6323	286.4	157.8	-	10N	72E
8933	447	1707 Apr 17	01:39:36	9	-3621	117	T-	pp	-0.0018	2.8922	1.8178	372.7	234.4	105.7	10S	25W
8934	447	1707 Oct 11	10:24:59	9	-3615	122	T+	pp	0.0139	2.8249	1.8400	329.7	215.8	100.1	7N	159W
8935	447	1708 Apr 05	05:37:46	9	-3609	127	P	a-	-0.7337	1.5276	0.4959	310.9	154.0	-	7S	84W
8936	447	1708 Sep 29	21:01:05	9	-3603	132	P	t-	0.7725	1.4590	0.4224	307.5	144.1	-	3N	42E
8937	447	1709 Feb 24	06:50:38	9	-3598	099	N	-a	1.2369	0.5733	-0.3970	193.7	-	-	11N	99W
8938	447	1709 Mar 25	16:43:34	9	-3597	137	N	a-	-1.4038	0.2725	-0.7084	140.3	-	-	3S	110E
8939	447	1709 Aug 20	08:27:54	9	-3592	104	N	-t	-1.3529	0.4137	-0.6622	193.5	-	-	14S	126W
8940	447	1709 Sep 19	00:30:48	9	-3591	142	Nb	t-	1.5439	0.0642	-1.0136	78.2	-	-	0S	10W

Cat Num	Canon Plate	Calendar Date	TD of Greatest Eclipse	ΔT s	Luna Num	Saros Num	Ecl. Type	QSE	Gamma	Pen. Mag.	Um. Mag.	Eclipse Phase Durations Pen. m	Par. m	Total m	Greatest in Zenith Lat.	Lng.
8941	448	1710 Feb 13	22:50:15	9	-3586	109	P	-a	0.5663	1.8043	0.8330	299.1	176.0	-	14N	21E
8942	448	1710 Aug 09	09:55:48	9	-3580	114	P	-t	-0.5854	1.8080	0.7601	337.1	187.4	-	16S	147W
8943	448	1711 Feb 03	12:31:30	9	-3574	119	T-	pp	-0.1197	2.6458	1.6310	338.1	217.5	97.3	17N	176E
8944	448	1711 Jul 29	17:50:25	9	-3568	124	T+	p-	0.2032	2.4811	1.4889	335.7	216.4	92.2	19S	94E
8945	448	1712 Jan 23	19:47:09	9	-3562	129	P	t-	-0.8544	1.3274	0.2535	306.7	116.8	-	19N	66E
8946	448	1712 Jul 18	08:21:48	9	-3556	134	P	a-	0.9350	1.1162	0.1678	254.7	87.9	-	20S	124W
8947	448	1712 Dec 13	00:53:52	9	-3551	101	N	-t	1.4977	0.1604	-0.9401	123.4	-	-	25N	15W
8948	448	1713 Jan 11	20:14:55	9	-3550	139	N	t-	-1.5569	0.0545	-1.0513	73.0	-	-	20N	58E
8949	448	1713 Jun 08	18:28:15	9	-3545	106	P	-a	-0.8551	1.2685	0.3089	270.6	118.1	-	24S	83E
8950	448	1713 Dec 02	03:17:09	9	-3539	111	P	-h	0.7716	1.4698	0.4149	310.5	143.4	-	23N	52W
8951	448	1714 May 29	07:05:03	10	-3533	116	T-	pp	-0.1120	2.6582	1.6468	346.1	223.1	100.3	22S	107W
8952	448	1714 Nov 21	13:01:44	10	-3527	121	T+	pp	0.0414	2.7803	1.7836	331.3	215.6	99.4	20N	161E
8953	448	1715 May 18	12:28:52	10	-3521	126	P	t-	0.6739	1.6526	0.5906	334.1	172.1	-	19S	172E
8954	448	1715 Nov 11	04:00:14	10	-3515	131	P	a-	-0.6347	1.6760	0.7102	291.1	165.1	-	17N	64W
8955	448	1716 Apr 06	21:22:33	10	-3510	098	N	-t	-1.4634	0.2068	-0.8606	136.9	-	-	8S	40E
8956	448	1716 May 06	12:57:42	10	-3509	136	N	t-	1.4192	0.2912	-0.7828	163.7	-	-	15S	165E
8957	448	1716 Oct 01	09:10:11	10	-3504	103	N	-a	1.3494	0.3802	-0.6163	167.3	-	-	5N	141W
8958	448	1716 Oct 30	19:42:19	10	-3503	141	N	a-	-1.3238	0.4189	-0.5613	170.2	-	-	13N	61E
8959	448	1717 Mar 27	03:13:41	10	-3498	108	P	-h	-0.6800	1.6192	0.6013	310.4	164.1	-	3S	47W
8960	448	1717 Sep 20	17:54:25	10	-3492	113	P	-h	0.6690	1.6553	0.6059	325.9	170.2	-	0S	90E
8961	449	1718 Mar 16	15:54:36	10	-3486	118	T+	p-	0.0622	2.7298	1.7576	323.8	212.9	98.5	2N	124E
8962	449	1718 Sep 09	19:52:29	10	-3480	123	T-	pp	-0.0635	2.7818	1.7017	375.5	235.3	105.0	5S	61E
8963	449	1719 Mar 06	07:59:13	10	-3474	128	P	a-	0.7478	1.4671	0.5043	279.5	143.9	-	7N	117W
8964	449	1719 Aug 29	20:24:03	10	-3468	133	P	t-	-0.7974	1.4259	0.3641	317.5	139.5	-	10S	54E
8965	449	1720 Jan 25	10:04:09	10	-3463	100	N	-a	-1.3033	0.4832	-0.5498	192.3	-	-	18N	148W
8966	449	1720 Feb 23	22:48:32	10	-3462	138	N	a-	1.4666	0.1659	-0.8322	111.9	-	-	11N	22E
8967	449	1720 Jul 19	16:55:34	10	-3457	105	N	-a	1.1366	0.7604	-0.2159	225.9	-	-	20S	107E
8968	449	1720 Aug 18	02:31:47	10	-3456	143	N	a-	-1.4936	0.1213	-0.8866	99.2	-	-	15S	37W
8969	449	1721 Jan 13	15:08:19	10	-3451	110	P	-t	-0.6568	1.6972	0.6089	337.6	173.4	-	21N	135E
8970	449	1721 Jul 09	08:39:47	10	-3445	115	T	-p	0.3650	2.1590	1.2167	308.7	198.1	65.2	22S	129W
8971	449	1722 Jan 02	14:32:53	10	-3439	120	T+	pp	0.0187	2.8768	1.7714	378.9	235.0	104.8	23N	143E
8972	449	1722 Jun 29	01:49:16	10	-3433	125	T	p-	-0.3750	2.1452	1.1942	311.8	198.8	62.7	24S	27W
8973	449	1722 Dec 22	15:28:30	10	-3427	130	P	a-	0.6947	1.6187	0.5485	325.5	163.4	-	24N	128E
8974	449	1723 May 20	04:51:53	10	-3422	097	Ne	-t	1.5130	0.0963	-0.9326	91.0	-	-	19S	73W
8975	449	1723 Jun 18	15:45:35	10	-3421	135	N	t-	-1.1715	0.7068	-0.2902	224.8	-	-	25S	124E
8976	449	1723 Nov 12	11:42:32	10	-3416	102	N	-a	-1.3902	0.2982	-0.6842	145.2	-	-	16N	179W
8977	449	1723 Dec 11	23:25:51	10	-3415	140	N	a-	1.3333	0.4173	-0.5944	175.2	-	-	24N	7E
8978	449	1724 May 08	08:19:43	10	-3410	107	P	-t	0.8072	1.4131	0.3411	321.0	137.1	-	16S	126W
8979	449	1724 Nov 01	03:26:08	10	-3404	112	P	-a	-0.7028	1.5502	0.5863	284.3	153.1	-	14N	55W
8980	449	1725 Apr 27	08:37:09	10	-3398	117	T+	pp	0.0615	2.7803	1.7107	371.3	233.7	104.8	14S	130W
8981	450	1725 Oct 21	18:34:41	10	-3392	122	T-	pp	-0.0177	2.8206	1.8302	331.0	216.1	100.1	11N	78E
8982	450	1726 Apr 16	13:06:31	10	-3386	127	P	a-	-0.6806	1.6220	0.5966	315.2	165.7	-	11S	163E
8983	450	1726 Oct 11	04:40:33	10	-3380	132	P	t-	0.7341	1.5326	0.4897	314.0	154.0	-	8N	74W
8984	450	1727 Mar 07	15:16:35	10	-3375	099	N	-a	1.2592	0.5304	-0.4359	186.7	-	-	6N	134E
8985	450	1727 Apr 06	00:41:54	10	-3374	137	N	a-	-1.3596	0.3508	-0.6247	157.4	-	-	7S	10W
8986	450	1727 Aug 31	15:12:00	10	-3369	104	N	-t	-1.4223	0.2878	-0.7907	163.4	-	-	10S	133E
8987	450	1727 Sep 30	07:40:13	10	-3368	142	N	t-	1.4949	0.1565	-0.9259	121.5	-	-	4N	118W
8988	450	1728 Feb 25	07:24:24	10	-3363	109	P	-a	0.5855	1.7687	0.7984	297.8	173.5	-	10N	107W
8989	450	1728 Aug 19	16:48:40	10	-3357	114	P	-t	-0.6569	1.6763	0.6291	329.0	173.8	-	13S	109E
8990	450	1729 Feb 13	20:57:43	10	-3351	119	T-	pp	-0.1065	2.6705	1.6548	339.4	218.4	98.2	13N	49E
8991	450	1729 Aug 09	01:05:17	11	-3345	124	T+	pp	0.1305	2.6136	1.6232	336.1	218.4	98.0	16S	15W
8992	450	1730 Feb 03	03:55:45	11	-3339	129	P	t-	-0.8451	1.3449	0.2702	308.9	120.6	-	16N	56W
8993	450	1730 Jul 29	15:49:46	11	-3333	134	P	a-	0.8619	1.2503	0.3022	265.1	115.5	-	18S	124E
8994	450	1730 Dec 24	08:58:03	11	-3328	101	N	-t	1.5032	0.1500	-0.9498	119.3	-	-	25N	134W
8995	450	1731 Jan 23	04:14:13	11	-3327	139	N	t-	-1.5477	0.0709	-1.0338	83.2	-	-	18N	61W
8996	450	1731 Jun 20	01:49:42	11	-3322	106	P	-a	-0.9307	1.1306	0.1692	260.2	89.4	-	24S	27W
8997	450	1731 Dec 13	11:38:06	11	-3316	111	P	-h	0.7764	1.4601	0.4069	308.8	141.7	-	24N	176W
8998	450	1732 Jun 08	14:06:11	11	-3310	116	T-	-p	-0.1888	2.5185	1.5049	345.9	221.2	94.6	23S	148E
8999	450	1732 Dec 01	21:39:42	11	-3304	121	T+	pp	0.0502	2.7638	1.7678	330.4	215.1	99.0	22N	33E
9000	450	1733 May 28	19:08:10	11	-3298	126	P	t-	0.5966	1.7945	0.7324	342.8	187.6	-	21S	72E

Cat Num	Canon Plate	Calendar Date	TD of Greatest Eclipse	ΔT s	Luna Num	Saros Num	Ecl. Type	QSE	Gamma	Pen. Mag.	Um. Mag.	Eclipse Phase Durations Pen. m	Par. m	Total m	Greatest in Zenith Lat.	Lng.
9001	451	1733 Nov 21	12:40:54	11	-3292	131	P	a-	-0.6211	1.7020	0.7344	292.5	167.2	-	19N	166E
9002	451	1734 Apr 18	04:32:27	11	-3287	098	N	-t	-1.5136	0.1117	-0.9501	101.1	-	-	12S	69W
9003	451	1734 May 17	19:38:11	11	-3286	136	N	t-	1.3452	0.4252	-0.6453	195.0	-	-	18S	65E
9004	451	1734 Oct 12	17:08:05	11	-3281	103	N	-a	1.3883	0.3117	-0.6908	153.1	-	-	9N	99E
9005	451	1734 Nov 11	04:10:31	11	-3280	141	N	a-	-1.3088	0.4488	-0.5360	176.2	-	-	16N	66W
9006	451	1735 Apr 07	10:56:51	11	-3275	108	P	-h	-0.7210	1.5408	0.5292	304.2	155.3	-	7S	164W
9007	451	1735 Oct 02	01:19:13	11	-3269	113	P	-h	0.7178	1.5689	0.5134	322.1	159.4	-	4N	23W
9008	451	1736 Mar 27	00:05:53	11	-3263	118	T+	pp	0.0283	2.7897	1.8222	323.0	213.0	99.2	3S	0W
9009	451	1736 Sep 20	02:51:00	11	-3257	123	T-	pp	-0.0042	2.8925	1.8087	376.4	235.8	106.0	1S	44W
9010	451	1737 Mar 16	16:22:12	11	-3251	128	P	a-	0.7211	1.5151	0.5542	282.5	149.9	-	2N	117E
9011	451	1737 Sep 09	03:24:12	11	-3245	133	P	t-	-0.7325	1.5453	0.4830	325.3	157.4	-	6S	52W
9012	451	1738 Feb 04	18:29:02	11	-3240	100	N	-a	-1.3164	0.4597	-0.5747	188.6	-	-	15N	86E
9013	451	1738 Mar 06	07:07:15	11	-3239	138	N	a-	1.4492	0.1978	-0.8003	122.2	-	-	7N	103W
9014	451	1738 Jul 31	00:13:25	11	-3234	105	N	-a	1.2043	0.6353	-0.3392	208.4	-	-	17S	2W
9015	451	1738 Aug 29	09:51:16	11	-3233	143	N	a-	-1.4268	0.2433	-0.7636	138.5	-	-	11S	147W
9016	451	1739 Jan 24	23:14:14	11	-3228	110	P	-t	-0.6674	1.6783	0.5892	337.2	171.5	-	19N	15E
9017	451	1739 Jul 20	16:09:12	11	-3222	115	T	-a	0.4373	2.0265	1.0842	304.5	191.7	42.3	20S	119E
9018	451	1740 Jan 13	22:33:33	12	-3216	120	T+	pp	0.0096	2.8928	1.7887	378.5	235.0	104.9	21N	24E
9019	451	1740 Jul 09	09:13:16	12	-3210	125	T	p-	-0.3014	2.2813	1.3281	316.1	204.5	78.0	23S	137W
9020	451	1741 Jan 01	23:47:06	12	-3204	130	P	a-	0.6880	1.6297	0.5621	325.1	164.6	-	24N	4E
9021	452	1741 Jun 28	22:47:45	12	-3198	135	N	t-	-1.0981	0.8430	-0.1571	243.1	-	-	24S	19E
9022	452	1741 Nov 22	20:19:31	12	-3193	102	N	-a	-1.3999	0.2804	-0.7021	140.8	-	-	19N	52E
9023	452	1741 Dec 22	08:05:41	12	-3192	140	N	a-	1.3256	0.4305	-0.5792	177.1	-	-	25N	122W
9024	452	1742 May 19	14:58:37	12	-3187	107	P	-t	0.8845	1.2706	0.1998	309.6	107.2	-	19S	135E
9025	452	1742 Nov 12	12:01:43	12	-3181	112	P	-a	-0.7209	1.5183	0.5517	282.7	149.4	-	17N	176E
9026	452	1743 May 08	15:26:08	12	-3175	117	T+	pp	0.1323	2.6480	1.5829	369.0	231.8	101.1	17S	128E
9027	452	1743 Nov 02	02:52:32	12	-3169	122	T-	pp	-0.0421	2.7785	1.7829	332.3	216.4	99.7	15N	47W
9028	452	1744 Apr 26	20:29:07	12	-3163	127	P	a-	-0.6212	1.7278	0.7086	319.6	176.8	-	14S	52E
9029	452	1744 Oct 21	12:26:28	12	-3157	132	P	t-	0.7016	1.5954	0.5462	319.4	161.7	-	12N	169E
9030	452	1745 Mar 17	23:37:12	12	-3152	099	N	-a	1.2862	0.4789	-0.4836	177.9	-	-	2N	8E
9031	452	1745 Apr 16	08:34:55	12	-3151	137	N	a-	-1.3106	0.4382	-0.5321	173.9	-	-	11S	129W
9032	452	1745 Sep 10	22:02:28	12	-3146	104	N	-t	-1.4863	0.1716	-0.9095	127.5	-	-	6S	29E
9033	452	1745 Oct 10	14:56:57	12	-3145	142	N	t-	1.4515	0.2385	-0.8485	149.2	-	-	8N	132E
9034	452	1746 Mar 07	15:52:17	12	-3140	109	P	-a	0.6096	1.7239	0.7548	296.0	170.2	-	6N	125E
9035	452	1746 Aug 30	23:48:16	12	-3134	114	P	-t	-0.7237	1.5537	0.5069	320.5	158.7	-	9S	3E
9036	452	1747 Feb 25	05:17:43	12	-3128	119	T-	pp	-0.0887	2.7034	1.6871	340.7	219.4	99.3	9N	76W
9037	452	1747 Aug 20	08:26:59	12	-3122	124	T+	pp	0.0627	2.7375	1.7484	335.8	219.1	100.8	13S	126W
9038	452	1748 Feb 14	11:58:06	13	-3116	129	P	t-	-0.8312	1.3706	0.2956	311.7	126.0	-	12N	176W
9039	452	1748 Aug 08	23:23:28	13	-3110	134	P	a-	0.7929	1.3769	0.4287	273.8	134.9	-	15S	10E
9040	452	1749 Jan 03	17:03:06	13	-3105	101	N	-t	1.5085	0.1396	-0.9589	115.0	-	-	24N	106E
9041	453	1749 Feb 02	12:09:42	13	-3104	139	N	t-	-1.5351	0.0931	-1.0100	95.2	-	-	15N	179W
9042	453	1749 Jun 30	09:09:16	13	-3099	106	P	-a	-1.0073	0.9913	0.0276	248.2	36.9	-	24S	136W
9043	453	1749 Jul 29	16:31:02	13	-3098	144	Nb	a-	1.5083	0.0642	-0.8842	67.7	-	-	17S	113E
9044	453	1749 Dec 23	20:02:11	13	-3093	111	P	-h	0.7796	1.4532	0.4023	307.3	140.6	-	24N	60E
9045	453	1750 Jun 19	21:03:03	13	-3087	116	T-	-p	-0.2684	2.3736	1.3573	344.7	217.7	84.6	24S	45E
9046	453	1750 Dec 13	06:21:46	13	-3081	121	T+	pp	0.0561	2.7524	1.7576	329.5	214.5	98.7	23N	97W
9047	453	1751 Jun 09	01:41:59	13	-3075	126	P	t-	0.5150	1.9445	0.8819	350.7	200.8	-	22S	26W
9048	453	1751 Dec 02	21:26:52	13	-3069	131	P	a-	-0.6116	1.7202	0.7512	293.4	168.5	-	21N	36E
9049	453	1752 Apr 28	11:35:10	13	-3064	098	Ne	-t	-1.5698	0.0059	-1.0505	23.4	-	-	16S	175W
9050	453	1752 May 28	02:13:58	13	-3063	136	N	t-	1.2666	0.5677	-0.4996	221.9	-	-	20S	34W
9051	453	1752 Oct 23	01:12:56	13	-3058	103	N	-a	1.4211	0.2545	-0.7539	139.7	-	-	13N	22W
9052	453	1752 Nov 21	12:43:58	13	-3057	141	N	a-	-1.2983	0.4701	-0.5189	180.5	-	-	19N	166E
9053	453	1753 Apr 17	18:33:18	13	-3052	108	P	-h	-0.7683	1.4508	0.4454	297.0	143.9	-	11S	81E
9054	453	1753 Oct 12	08:51:29	13	-3046	113	P	-h	0.7597	1.4949	0.4333	318.5	148.7	-	8N	136W
9055	453	1754 Apr 07	08:08:01	13	-3040	118	T-	pp	-0.0140	2.8135	1.8507	322.0	212.8	99.4	7S	121W
9056	453	1754 Oct 01	09:58:28	14	-3034	123	T+	pp	0.0478	2.8144	1.7268	376.8	235.5	105.2	3N	152W
9057	453	1755 Mar 28	00:36:25	14	-3028	128	P	a-	0.6871	1.5764	0.6176	286.2	156.8	-	2S	7W
9058	453	1755 Sep 20	10:34:08	14	-3022	133	P	t-	-0.6745	1.6521	0.5890	331.3	170.5	-	2S	160W
9059	453	1756 Feb 16	02:47:53	14	-3017	100	N	-a	-1.3339	0.4282	-0.6071	183.2	-	-	11N	39W
9060	453	1756 Mar 16	15:17:47	14	-3016	138	N	a-	1.4259	0.2406	-0.7575	134.6	-	-	3N	133E

Cat Num	Canon Plate	Calendar Date	TD of Greatest Eclipse	ΔT s	Luna Num	Saros Num	Ecl. Type	QSE	Gamma	Pen. Mag.	Um. Mag.	Eclipse Phase Durations Pen. m	Par. m	Total m	Greatest in Zenith Lat.	Lng.
9061	454	1756 Aug 10	07:36:47	14	-3011	105	N	-a	1.2682	0.5174	-0.4558	189.6	-	-	14S	113W
9062	454	1756 Sep 08	17:19:28	14	-3010	143	N	a-	-1.3658	0.3549	-0.6514	165.0	-	-	7S	100E
9063	454	1757 Feb 04	07:16:54	14	-3005	110	P	-t	-0.6801	1.6551	0.5658	336.5	169.0	-	16N	106W
9064	454	1757 Jul 30	23:39:48	14	-2999	115	P	-a	0.5087	1.8957	0.9529	299.6	183.9	-	18S	6E
9065	454	1758 Jan 24	06:34:14	14	-2993	120	T+	pp	0.0003	2.9089	1.8067	378.0	234.9	104.9	19N	95W
9066	454	1758 Jul 20	16:36:44	14	-2987	125	T-	p-	-0.2279	2.4175	1.4615	319.7	208.9	88.3	21S	112E
9067	454	1759 Jan 13	08:06:35	14	-2981	130	P	a-	0.6812	1.6404	0.5762	324.6	165.9	-	22N	119W
9068	454	1759 Jul 10	05:48:57	14	-2975	135	N	t-	-1.0246	0.9797	-0.0240	259.5	-	-	23S	86W
9069	454	1759 Dec 04	05:01:57	14	-2970	102	N	-a	-1.4059	0.2693	-0.7129	137.9	-	-	21N	77W
9070	454	1760 Jan 02	16:47:25	14	-2969	140	N	a-	1.3191	0.4411	-0.5661	178.6	-	-	24N	109E
9071	454	1760 May 29	21:33:40	14	-2964	107	P	-t	0.9647	1.1231	0.0532	296.0	56.6	-	21S	36E
9072	454	1760 Nov 22	20:41:36	15	-2958	112	P	-a	-0.7355	1.4928	0.5236	281.5	146.2	-	20N	47E
9073	454	1761 May 18	22:11:58	15	-2952	117	T+	pp	0.2063	2.5099	1.4493	365.9	228.5	94.1	20S	26E
9074	454	1761 Nov 12	11:15:47	15	-2946	122	T-	-p	-0.0614	2.7457	1.7449	333.5	216.6	99.3	18N	173W
9075	454	1762 May 08	03:45:37	15	-2940	127	P	a-	-0.5557	1.8449	0.8317	323.8	186.9	-	18S	58W
9076	454	1762 Nov 01	20:19:49	15	-2934	132	P	t-	0.6759	1.6455	0.5905	323.8	167.4	-	15N	51E
9077	454	1763 Mar 29	07:49:13	15	-2929	099	N	-a	1.3206	0.4140	-0.5447	166.2	-	-	2S	115W
9078	454	1763 Apr 27	16:20:38	15	-2928	137	N	a-	-1.2552	0.5374	-0.4281	190.1	-	-	15S	114E
9079	454	1763 Sep 22	05:02:59	15	-2923	104	N	-t	-1.5421	0.0706	-1.0131	82.5	-	-	2S	77W
9080	454	1763 Oct 21	22:22:45	15	-2922	142	N	t-	1.4153	0.3071	-0.7842	168.6	-	-	12N	20E
9081	455	1764 Mar 18	00:11:05	15	-2917	109	P	-a	0.6410	1.6655	0.6976	293.5	165.5	-	1N	0W
9082	455	1764 Sep 10	06:58:27	15	-2911	114	P	-t	-0.7824	1.4458	0.3992	312.2	142.8	-	5S	105W
9083	455	1765 Mar 07	13:28:56	15	-2905	119	T-	pp	-0.0643	2.7483	1.7318	342.0	220.5	100.5	5N	161E
9084	455	1765 Aug 30	15:55:32	15	-2899	124	T-	pp	-0.0004	2.8513	1.8629	334.9	218.9	101.4	9S	121E
9085	455	1766 Feb 24	19:52:51	15	-2893	129	P	t-	-0.8116	1.4065	0.3317	315.2	133.0	-	9N	65E
9086	455	1766 Aug 20	07:01:43	15	-2887	134	P	a-	0.7272	1.4978	0.5489	281.3	149.6	-	12S	105W
9087	455	1767 Jan 15	01:08:11	15	-2882	101	N	-t	1.5141	0.1283	-0.9680	110.2	-	-	23N	14W
9088	455	1767 Feb 13	19:59:03	15	-2881	139	N	t-	-1.5174	0.1244	-0.9764	109.6	-	-	12N	63E
9089	455	1767 Jul 11	16:27:13	16	-2876	106	N	-a	-1.0845	0.8511	-0.1155	234.2	-	-	23S	115E
9090	455	1767 Aug 10	00:02:55	16	-2875	144	N	a-	1.4447	0.1826	-0.7690	113.2	-	-	14S	0E
9091	455	1768 Jan 04	04:29:31	16	-2870	111	P	-h	0.7810	1.4490	0.4011	305.9	140.1	-	24N	66W
9092	455	1768 Jun 30	03:56:19	16	-2864	116	T	-h	-0.3499	2.2256	1.2062	342.6	212.5	68.0	24S	58W
9093	455	1768 Dec 23	15:07:13	16	-2858	121	T+	pp	0.0597	2.7449	1.7517	328.6	214.1	98.4	24N	133E
9094	455	1769 Jun 19	08:11:40	16	-2852	126	T	t-	0.4302	2.1005	1.0372	357.6	211.9	31.5	23S	123W
9095	455	1769 Dec 13	06:17:04	16	-2846	131	P	a-	-0.6058	1.7312	0.7611	294.0	169.4	-	23N	96W
9096	455	1770 Jun 08	08:45:13	16	-2840	136	N	t-	1.1836	0.7186	-0.3457	245.4	-	-	22S	132W
9097	455	1770 Nov 03	09:24:30	16	-2835	103	N	-a	1.4478	0.2085	-0.8057	127.5	-	-	17N	145W
9098	455	1770 Dec 02	21:22:32	16	-2834	141	N	a-	-1.2926	0.4825	-0.5104	183.2	-	-	21N	37E
9099	455	1771 Apr 29	02:04:10	16	-2829	108	P	-h	-0.8211	1.3509	0.3515	288.5	129.3	-	15S	32W
9100	455	1771 Oct 23	16:30:23	16	-2823	113	P	-h	0.7955	1.4323	0.3649	315.3	138.3	-	12N	108E
9101	456	1772 Apr 17	16:04:45	16	-2817	118	T-	pp	-0.0612	2.7248	1.7663	320.9	212.3	98.6	11S	119E
9102	456	1772 Oct 11	17:14:41	16	-2811	123	T+	pp	0.0928	2.7337	1.6425	376.9	234.7	103.3	8N	98E
9103	456	1773 Apr 07	08:43:30	16	-2805	128	P	a-	0.6476	1.6479	0.6909	290.2	164.1	-	6S	130W
9104	456	1773 Sep 30	17:53:06	16	-2799	133	P	t-	-0.6232	1.7467	0.6829	336.0	180.3	-	3N	89E
9105	456	1774 Feb 26	10:58:53	16	-2794	100	N	-a	-1.3568	0.3865	-0.6494	175.3	-	-	7N	162W
9106	456	1774 Mar 27	23:19:02	16	-2793	138	N	a-	1.3960	0.2955	-0.7026	148.9	-	-	2S	12E
9107	456	1774 Aug 21	15:06:39	16	-2788	105	N	-a	1.3271	0.4089	-0.5635	169.8	-	-	11S	134E
9108	456	1774 Sep 20	00:57:04	16	-2787	143	N	a-	-1.3118	0.4539	-0.5520	184.2	-	-	2S	15W
9109	456	1775 Feb 15	15:10:58	16	-2782	110	P	-t	-0.6990	1.6202	0.5314	334.8	165.0	-	12N	136E
9110	456	1775 Aug 11	07:15:46	16	-2776	115	P	-a	0.5755	1.7737	0.8298	294.4	175.2	-	15S	108W
9111	456	1776 Feb 04	14:29:50	16	-2770	120	T-	pp	-0.0128	2.8846	1.7851	377.4	234.7	104.9	16N	146E
9112	456	1776 Jul 31	00:02:02	17	-2764	125	T-	pp	-0.1566	2.5500	1.5907	322.7	212.2	94.9	18S	1E
9113	456	1777 Jan 23	16:24:09	17	-2758	130	P	a-	0.6723	1.6547	0.5944	324.3	167.5	-	20N	117E
9114	456	1777 Jul 20	12:48:46	17	-2752	135	P	t-	-0.9508	1.1173	0.1093	274.3	75.8	-	21S	169E
9115	456	1777 Dec 14	13:48:58	17	-2747	102	N	-a	-1.4087	0.2638	-0.7179	136.2	-	-	22N	152E
9116	456	1778 Jan 13	01:29:24	17	-2746	140	N	a-	1.3124	0.4519	-0.5522	180.0	-	-	23N	20W
9117	456	1778 Jun 10	04:01:09	17	-2741	107	N	-t	1.0501	0.9660	-0.1034	279.3	-	-	22S	60W
9118	456	1778 Dec 04	05:27:52	17	-2735	112	P	-a	-0.7448	1.4769	0.5054	280.9	144.1	-	22N	84W
9119	456	1779 May 30	04:52:10	17	-2729	117	T	pp	0.2853	2.3629	1.3064	361.7	223.4	82.1	22S	74W
9120	456	1779 Nov 23	19:44:47	17	-2723	122	T-	-p	-0.0752	2.7227	1.7172	334.8	216.8	98.9	20N	61E

Cat Num	Canon Plate	Calendar Date	TD of Greatest Eclipse	ΔT s	Luna Num	Saros Num	Ecl. Type	QSE	Gamma	Pen. Mag.	Um. Mag.	Eclipse Phase ---- Durations ---- Pen. m	Par. m	Total m	Greatest in Zenith Lat.	Lng.
9121	457	1780 May 18	10:58:18	17	-2717	127	P	a-	-0.4862	1.9696	0.9620	327.5	195.8	-	20S	166W
9122	457	1780 Nov 12	04:19:52	17	-2711	132	P	t-	0.6565	1.6839	0.6232	327.5	171.5	-	18N	69W
9123	457	1781 Apr 08	15:55:06	17	-2706	099	N	-a	1.3602	0.3394	-0.6156	151.3	-	-	6S	122E
9124	457	1781 May 08	00:01:08	17	-2705	137	N	a-	-1.1949	0.6457	-0.3152	205.6	-	-	18S	2W
9125	457	1781 Nov 01	05:56:11	17	-2699	142	N	t-	1.3850	0.3646	-0.7308	183.0	-	-	16N	94W
9126	457	1782 Mar 29	08:23:07	17	-2694	109	P	-a	0.6778	1.5974	0.6307	290.3	159.4	-	3S	124W
9127	457	1782 Sep 21	14:17:33	17	-2688	114	P	-t	-0.8348	1.3498	0.3031	304.1	125.9	-	1S	144E
9128	457	1783 Mar 18	21:31:30	17	-2682	119	T-	pp	-0.0333	2.8053	1.7886	343.3	221.5	101.5	1N	39E
9129	457	1783 Sep 10	23:33:06	17	-2676	124	T-	pp	-0.0569	2.7474	1.7595	333.6	217.9	100.4	5S	6E
9130	457	1784 Mar 07	03:39:43	17	-2670	129	P	t-	-0.7858	1.4535	0.3792	319.4	141.5	-	4N	52W
9131	457	1784 Aug 30	14:45:57	17	-2664	134	P	a-	0.6662	1.6104	0.6602	287.4	161.1	-	8S	138E
9132	457	1785 Jan 25	09:10:57	17	-2659	101	N	-t	1.5218	0.1127	-0.9808	103.2	-	-	20N	134W
9133	457	1785 Feb 24	03:42:22	17	-2658	139	N	t-	-1.4947	0.1647	-0.9331	125.5	-	-	8N	53W
9134	457	1785 Jul 21	23:45:46	17	-2653	106	N	-a	-1.1605	0.7134	-0.2568	218.2	-	-	21S	5E
9135	457	1785 Aug 20	07:38:39	17	-2652	144	N	a-	1.3843	0.2953	-0.6598	142.9	-	-	11S	114W
9136	457	1786 Jan 14	12:55:56	17	-2647	111	P	-h	0.7838	1.4421	0.3978	304.2	139.1	-	22N	169E
9137	457	1786 Jul 11	10:47:19	17	-2641	116	T	-t	-0.4324	2.0761	1.0530	339.3	205.2	36.3	22S	160W
9138	457	1787 Jan 03	23:53:37	17	-2635	121	T+	pp	0.0632	2.7375	1.7465	327.6	213.6	98.2	23N	3E
9139	457	1787 Jun 30	14:39:32	17	-2629	126	T	t-	0.3440	2.2592	1.1947	363.5	220.8	68.6	23S	141E
9140	457	1787 Dec 24	15:08:29	16	-2623	131	P	a-	-0.6010	1.7405	0.7696	294.5	170.1	-	23N	133E
9141	458	1788 Jun 18	15:15:25	16	-2617	136	N	t-	1.0985	0.8734	-0.1883	265.8	-	-	22S	131E
9142	458	1788 Nov 13	17:42:36	16	-2612	103	N	-a	1.4689	0.1725	-0.8471	116.8	-	-	20N	90E
9143	458	1788 Dec 13	06:04:27	16	-2611	141	N	a-	-1.2898	0.4894	-0.5069	185.0	-	-	22N	92W
9144	458	1789 May 09	09:28:39	16	-2606	108	P	-h	-0.8799	1.2401	0.2465	278.7	109.6	-	18S	143W
9145	458	1789 Nov 03	00:17:06	16	-2600	113	P	-h	0.8242	1.3824	0.3094	312.7	128.7	-	16N	8W
9146	458	1790 Apr 28	23:53:39	16	-2594	118	T-	pp	-0.1156	2.6230	1.6685	319.4	211.1	96.6	15S	1E
9147	458	1790 Oct 23	00:40:41	16	-2588	123	T+	pp	0.1296	2.6678	1.5734	376.7	233.6	100.9	12N	14W
9148	458	1791 Apr 18	16:41:28	16	-2582	128	P	a-	0.6008	1.7331	0.7776	294.7	171.8	-	10S	110E
9149	458	1791 Oct 12	01:23:20	16	-2576	133	P	t-	-0.5802	1.8259	0.7615	339.4	187.4	-	7N	24W
9150	458	1792 Mar 08	19:02:23	16	-2571	100	N	-a	-1.3848	0.3351	-0.7011	164.5	-	-	3N	77E
9151	458	1792 Apr 07	07:12:01	16	-2570	138	N	a-	1.3602	0.3611	-0.6370	164.1	-	-	6S	107W
9152	458	1792 Aug 31	22:42:42	16	-2565	105	N	-a	1.3816	0.3086	-0.6634	148.5	-	-	7S	19E
9153	458	1792 Sep 30	08:43:05	16	-2564	143	N	a-	-1.2636	0.5423	-0.4634	199.0	-	-	2N	133W
9154	458	1793 Feb 25	22:59:34	16	-2559	110	P	-t	-0.7218	1.5777	0.4899	332.5	159.7	-	8N	18E
9155	458	1793 Aug 21	14:55:22	16	-2553	115	P	-a	0.6392	1.6576	0.7120	288.8	165.5	-	11S	137E
9156	458	1794 Feb 14	22:21:16	15	-2547	120	T-	pp	-0.0300	2.8514	1.7553	376.5	234.5	104.8	13N	28E
9157	458	1794 Aug 11	07:29:21	15	-2541	125	T-	pp	-0.0874	2.6790	1.7157	324.9	214.3	98.8	15S	111W
9158	458	1795 Feb 04	00:39:45	15	-2535	130	P	a-	0.6610	1.6732	0.6175	324.1	169.7	-	17N	6W
9159	458	1795 Jul 31	19:50:34	15	-2529	135	P	t-	-0.8791	1.2513	0.2387	287.3	110.4	-	19S	64E
9160	458	1795 Dec 25	22:37:34	15	-2524	102	N	-a	-1.4110	0.2591	-0.7215	134.9	-	-	22N	21E
9161	459	1796 Jan 24	10:09:20	15	-2523	140	N	a-	1.3035	0.4664	-0.5341	182.0	-	-	20N	149W
9162	459	1796 Jun 20	10:28:08	15	-2518	107	N	-t	1.1354	0.8095	-0.2597	259.9	-	-	22S	157W
9163	459	1796 Dec 14	14:16:42	15	-2512	112	P	-a	-0.7519	1.4648	0.4915	280.5	142.5	-	23N	145E
9164	459	1797 Jun 09	11:30:23	14	-2506	117	T	-t	0.3666	2.2119	1.1592	356.4	216.5	62.2	23S	173W
9165	459	1797 Dec 04	04:17:57	14	-2500	122	T-	-p	-0.0850	2.7070	1.6971	336.0	217.0	98.6	22N	67W
9166	459	1798 May 29	18:07:26	14	-2494	127	T	a-	-0.4129	2.1015	1.0990	330.7	203.4	48.0	22S	87E
9167	459	1798 Nov 23	12:25:24	14	-2488	132	P	t-	0.6421	1.7128	0.6472	330.4	174.5	-	21N	170E
9168	459	1799 Apr 19	23:52:13	14	-2483	099	N	-a	1.4072	0.2513	-0.7003	131.1	-	-	10S	2E
9169	459	1799 May 19	07:36:00	14	-2482	137	N	a-	-1.1294	0.7639	-0.1930	220.3	-	-	21S	115W
9170	459	1799 Nov 12	13:38:20	14	-2476	142	N	t-	1.3618	0.4090	-0.6897	193.4	-	-	19N	151E
9171	459	1800 Apr 09	16:26:01	13	-2471	109	P	-a	0.7221	1.5157	0.5500	286.0	151.0	-	7S	114E
9172	459	1800 Oct 02	21:46:40	13	-2465	114	P	-t	-0.8797	1.2673	0.2205	296.5	108.3	-	3N	31E
9173	459	1801 Mar 30	05:24:05	13	-2459	119	T+	pp	0.0052	2.8569	1.8400	344.5	222.3	102.1	4S	80W
9174	459	1801 Sep 22	07:18:55	13	-2453	124	T-	pp	-0.1074	2.6548	1.6669	332.1	216.4	98.2	1S	111W
9175	459	1802 Mar 19	11:15:28	13	-2447	129	P	t-	-0.7517	1.5156	0.4422	324.4	151.7	-	0N	167W
9176	459	1802 Sep 11	22:36:46	13	-2441	134	P	a-	0.6105	1.7134	0.7615	292.6	170.0	-	4S	20E
9177	459	1803 Feb 06	17:10:42	13	-2436	101	N	-t	1.5322	0.0918	-0.9980	93.0	-	-	17N	107E
9178	459	1803 Mar 08	11:17:40	13	-2435	139	N	t-	-1.4653	0.2167	-0.8776	143.2	-	-	4N	167W
9179	459	1803 Aug 03	07:05:09	12	-2430	106	N	-a	-1.2349	0.5790	-0.3953	200.0	-	-	19S	104W
9180	459	1803 Sep 01	15:19:10	12	-2429	144	N	a-	1.3282	0.4003	-0.5589	165.1	-	-	7S	130E

Cat Num	Canon Plate	Calendar Date	TD of Greatest Eclipse	ΔT s	Luna Num	Saros Num	Ecl. Type	QSE	Gamma	Pen. Mag.	Um. Mag.	Eclipse Phase Durations Pen. m	Eclipse Phase Durations Par. m	Eclipse Phase Durations Total m	Greatest in Zenith Lat.	Greatest in Zenith Lng.
9181	460	1804 Jan 26	21:21:21	12	-2424	111	P	-h	0.7882	1.4319	0.3918	302.3	137.8	-	20N	43E
9182	460	1804 Jul 22	17:38:00	12	-2418	116	P	-t	-0.5141	1.9284	0.9011	335.0	195.8	-	21S	97E
9183	460	1805 Jan 15	08:40:58	12	-2412	121	T+	pp	0.0663	2.7305	1.7420	326.7	213.2	98.1	21N	128W
9184	460	1805 Jul 11	21:04:52	12	-2406	126	T+	pp	0.2561	2.4214	1.3554	368.1	227.6	87.6	22S	45E
9185	460	1806 Jan 05	00:02:04	12	-2400	131	P	a-	-0.5981	1.7459	0.7748	294.8	170.5	-	22N	1E
9186	460	1806 Jun 30	21:44:38	12	-2394	136	N*	t-	1.0115	1.0320	-0.0276	283.6	-	-	22S	35E
9187	460	1806 Nov 26	02:05:38	12	-2389	103	N	-a	1.4853	0.1450	-0.8797	107.8	-	-	22N	35W
9188	460	1806 Dec 25	14:48:04	12	-2388	141	N	a-	-1.2891	0.4923	-0.5071	186.0	-	-	22N	138E
9189	460	1807 May 21	16:49:32	12	-2383	108	P	-h	-0.9423	1.1229	0.1346	267.4	82.0	-	21S	107E
9190	460	1807 Nov 15	08:09:59	12	-2377	113	P	-h	0.8473	1.3426	0.2646	310.7	120.1	-	19N	126W
9191	460	1808 May 10	07:38:16	12	-2371	118	T-	-p	-0.1736	2.5145	1.5637	317.5	209.2	92.8	18S	116W
9192	460	1808 Nov 03	08:13:38	12	-2365	123	T+	pp	0.1607	2.6121	1.5148	376.3	232.3	98.1	15N	127W
9193	460	1809 Apr 30	00:33:02	12	-2359	128	P	a-	0.5490	1.8275	0.8733	299.3	179.4	-	14S	9W
9194	460	1809 Oct 23	09:02:46	12	-2353	133	P	t-	-0.5440	1.8924	0.8275	341.8	192.6	-	11N	139W
9195	460	1810 Mar 21	02:54:49	12	-2348	100	N	-a	-1.4209	0.2689	-0.7674	148.7	-	-	1S	42W
9196	460	1810 Apr 19	14:54:06	12	-2347	138	N	a-	1.3165	0.4414	-0.5570	180.6	-	-	10S	137E
9197	460	1810 Sep 13	06:27:11	12	-2342	105	N	-a	1.4299	0.2202	-0.7520	126.2	-	-	3S	98W
9198	460	1810 Oct 12	16:39:20	12	-2341	143	N	a-	-1.2230	0.6167	-0.3889	210.1	-	-	6N	107E
9199	460	1811 Mar 10	06:37:39	12	-2336	110	P	-t	-0.7527	1.5205	0.4340	329.0	151.9	-	4N	97W
9200	460	1811 Sep 02	22:41:52	12	-2330	115	P	-a	0.6972	1.5522	0.6045	283.3	155.1	-	7S	19E
9201	461	1812 Feb 27	06:04:51	12	-2324	120	T-	pp	-0.0538	2.8057	1.7135	375.5	234.1	104.3	9N	88W
9202	461	1812 Aug 22	15:01:22	12	-2318	125	T-	pp	-0.0228	2.7998	1.8320	326.7	215.4	100.4	12S	135E
9203	461	1813 Feb 15	08:50:46	12	-2312	130	P	a-	0.6453	1.6993	0.6488	324.2	172.5	-	13N	129W
9204	461	1813 Aug 12	02:52:49	12	-2306	135	P	t-	-0.8085	1.3834	0.3655	299.0	134.5	-	16S	42W
9205	461	1814 Jan 06	07:28:19	12	-2301	102	N	-a	-1.4118	0.2568	-0.7223	134.0	-	-	21N	111W
9206	461	1814 Feb 04	18:47:01	12	-2300	140	N	a-	1.2926	0.4843	-0.5122	184.5	-	-	17N	82E
9207	461	1814 Jul 02	16:51:30	12	-2295	107	N	-t	1.2225	0.6498	-0.4196	236.7	-	-	22S	108E
9208	461	1814 Dec 26	23:08:27	12	-2289	112	P	-a	-0.7564	1.4573	0.4825	280.3	141.5	-	23N	13E
9209	461	1815 Jun 21	18:06:32	12	-2283	117	T	-t	0.4498	2.0575	1.0081	349.9	207.4	14.7	23S	89E
9210	461	1815 Dec 16	12:55:06	12	-2277	122	T-	-p	-0.0906	2.6987	1.6850	337.2	217.4	98.4	23N	165E
9211	461	1816 Jun 10	01:14:25	12	-2271	127	T	a-	-0.3368	2.2387	1.2410	333.2	209.5	71.3	23S	19W
9212	461	1816 Dec 04	20:35:14	12	-2265	132	P	t-	0.6319	1.7338	0.6636	332.8	176.7	-	23N	49E
9213	461	1817 May 01	07:44:01	12	-2260	099	N	-a	1.4587	0.1554	-0.7931	103.9	-	-	14S	116W
9214	461	1817 May 30	15:07:30	12	-2259	137	N	a-	-1.0607	0.8880	-0.0652	233.9	-	-	23S	132E
9215	461	1817 Nov 23	21:26:57	12	-2253	142	N	t-	1.3435	0.4441	-0.6577	201.1	-	-	22N	35E
9216	461	1818 Apr 21	00:20:29	12	-2248	109	P	-a	0.7728	1.4222	0.4572	280.7	140.0	-	11S	5W
9217	461	1818 Oct 14	05:25:27	12	-2242	114	P	-t	-0.9175	1.1981	0.1512	289.7	90.3	-	7N	84W
9218	461	1819 Apr 10	13:07:53	12	-2236	119	T+	pp	0.0502	2.7745	1.7574	345.4	222.7	101.8	8S	163E
9219	461	1819 Oct 03	15:13:28	12	-2230	124	T-	pp	-0.1510	2.5747	1.5868	330.4	214.5	95.3	4N	129E
9220	461	1820 Mar 29	18:42:43	12	-2224	129	P	t-	-0.7112	1.5895	0.5171	330.0	162.4	-	4S	80E
9221	462	1820 Sep 22	06:35:09	11	-2218	134	P	a-	0.5612	1.8049	0.8511	296.7	176.8	-	0N	101W
9222	462	1821 Feb 17	01:05:17	11	-2213	101	N	-t	1.5470	0.0626	-1.0230	76.8	-	-	13N	12W
9223	462	1821 Mar 18	18:44:59	11	-2212	139	N	t-	-1.4293	0.2810	-0.8095	161.8	-	-	0S	80E
9224	462	1821 Aug 13	14:26:43	11	-2207	106	N	-a	-1.3066	0.4497	-0.5292	179.2	-	-	16S	145E
9225	462	1821 Sep 11	23:04:56	11	-2206	144	N	a-	1.2766	0.4971	-0.4666	182.9	-	-	3S	12E
9226	462	1822 Feb 06	05:43:26	11	-2201	111	P	-a	0.7961	1.4151	0.3797	299.9	135.6	-	16N	82W
9227	462	1822 Aug 03	00:30:07	11	-2195	116	P	-t	-0.5938	1.7845	0.7526	329.8	184.2	-	18S	6W
9228	462	1823 Jan 26	17:24:54	11	-2189	121	T+	pp	0.0729	2.7169	1.7314	325.7	212.8	97.8	19N	102E
9229	462	1823 Jul 23	03:32:23	10	-2183	126	T+	pp	0.1699	2.5805	1.5124	371.6	232.3	98.6	20S	52W
9230	462	1824 Jan 16	08:54:09	10	-2177	131	P	a-	-0.5937	1.7538	0.7829	295.3	171.2	-	21N	131W
9231	462	1824 Jul 11	04:14:59	10	-2171	136	P	t-	0.9245	1.1908	0.1330	298.7	87.6	-	21S	63W
9232	462	1824 Dec 06	10:32:41	10	-2166	103	N	-a	1.4980	0.1241	-0.9053	100.3	-	-	24N	160W
9233	462	1825 Jan 04	23:31:53	10	-2165	141	N	a-	-1.2890	0.4937	-0.5081	186.8	-	-	21N	8E
9234	462	1825 Jun 01	00:06:18	10	-2160	108	P	-h	-1.0089	0.9983	0.0149	254.4	27.6	-	23S	2W
9235	462	1825 Nov 25	16:09:25	9	-2154	113	P	-h	0.8643	1.3137	0.2311	309.4	113.0	-	22N	114E
9236	462	1826 May 21	15:15:25	9	-2148	118	T-	-p	-0.2384	2.3941	1.4464	315.0	206.3	86.5	20S	130E
9237	462	1826 Nov 14	15:56:14	9	-2142	123	T+	pp	0.1840	2.5707	1.4709	375.9	231.2	95.6	18N	117E
9238	462	1827 May 11	08:16:57	9	-2136	128	P	a-	0.4910	1.9336	0.9801	304.0	186.8	-	17S	125W
9239	462	1827 Nov 03	16:51:55	8	-2130	133	P	h-	-0.5151	1.9457	0.8805	343.3	196.3	-	15N	103E
9240	462	1828 Mar 31	10:38:46	8	-2125	100	N	-h	-1.4630	0.1917	-0.8446	126.9	-	-	6S	159W

Cat Num	Canon Plate	Calendar Date	TD of Greatest Eclipse	ΔT s	Luna Num	Saros Num	Ecl. Type	QSE	Gamma	Pen. Mag.	Um. Mag.	Eclipse Phase Durations Pen. m	Par. m	Total m	Greatest in Zenith Lat.	Lng.
9241	463	1828 Apr 29	22:28:38	8	-2124	138	N	h-	1.2674	0.5318	-0.4671	197.2	-	-	13S	22E
9242	463	1828 Sep 23	14:19:43	8	-2119	105	N	-a	1.4724	0.1424	-0.8302	102.0	-	-	2N	143E
9243	463	1828 Oct 23	00:44:42	8	-2118	143	N	a-	-1.1889	0.6793	-0.3263	218.5	-	-	10N	15W
9244	463	1829 Mar 20	14:08:14	8	-2113	110	P	-t	-0.7891	1.4527	0.3680	324.5	141.5	-	1S	150E
9245	463	1829 Sep 13	06:33:27	8	-2107	115	P	-a	0.7509	1.4550	0.5046	277.7	143.9	-	3S	100W
9246	463	1830 Mar 09	13:42:57	7	-2101	120	T-	pp	-0.0824	2.7510	1.6632	374.2	233.4	103.3	4N	157E
9247	463	1830 Sep 02	22:37:55	7	-2095	125	T+	pp	0.0371	2.7760	1.8034	327.9	215.7	100.2	8S	20E
9248	463	1831 Feb 26	16:56:21	7	-2089	130	P	a-	0.6246	1.7346	0.6897	324.7	176.2	-	9N	109E
9249	463	1831 Aug 23	09:59:56	7	-2083	135	P	t-	-0.7428	1.5067	0.4833	309.1	152.4	-	12S	149W
9250	463	1832 Jan 17	16:18:15	7	-2078	102	N	-a	-1.4135	0.2527	-0.7245	132.8	-	-	19N	118E
9251	463	1832 Feb 16	03:20:41	7	-2077	140	N	a-	1.2783	0.5083	-0.4838	188.0	-	-	14N	46W
9252	463	1832 Jul 12	23:15:58	6	-2072	107	N	-t	1.3081	0.4929	-0.5769	209.3	-	-	21S	12E
9253	463	1832 Aug 11	14:14:52	6	-2071	145	Nb	t-	-1.5396	0.0662	-1.0000	79.1	-	-	17S	148E
9254	463	1833 Jan 06	07:59:45	6	-2066	112	P	-a	-0.7611	1.4491	0.4733	280.2	140.5	-	22N	118W
9255	463	1833 Jul 02	00:43:22	6	-2060	117	P	-t	0.5329	1.9036	0.8572	342.3	196.1	-	23S	10W
9256	463	1833 Dec 26	21:32:54	6	-2054	122	T-	-p	-0.0951	2.6920	1.6749	338.4	217.8	98.3	23N	37E
9257	463	1834 Jun 21	08:19:45	6	-2048	127	T-	p-	-0.2583	2.3806	1.3871	335.0	214.2	85.6	24S	125W
9258	463	1834 Dec 16	04:48:07	6	-2042	132	P	t-	0.6249	1.7487	0.6746	334.8	178.2	-	24N	73W
9259	463	1835 May 12	15:28:42	6	-2037	099	Ne	-a	1.5160	0.0489	-0.8969	58.8	-	-	17S	127E
9260	463	1835 Jun 10	22:35:53	6	-2036	137	P	a-	-0.9888	1.0184	0.0681	246.4	56.8	-	24S	21E
9261	464	1835 Dec 05	05:20:19	6	-2030	142	N	t-	1.3290	0.4717	-0.6322	206.9	-	-	23N	83W
9262	464	1836 May 01	08:06:48	5	-2025	109	P	-a	0.8299	1.3173	0.3528	274.0	125.2	-	14S	122W
9263	464	1836 Oct 24	13:14:42	5	-2019	114	P	-t	-0.9472	1.1435	0.0966	283.9	72.5	-	11N	158E
9264	464	1837 Apr 20	20:40:45	5	-2013	119	T+	pp	0.1033	2.6772	1.6597	346.0	222.5	100.2	12S	50E
9265	464	1837 Oct 13	23:17:06	5	-2007	124	T-	-p	-0.1878	2.5074	1.5192	328.6	212.6	92.0	8N	7E
9266	464	1838 Apr 10	01:58:48	5	-2001	129	P	t-	-0.6622	1.6788	0.6076	336.1	173.9	-	8S	30W
9267	464	1838 Oct 03	14:41:28	5	-1995	134	P	a-	0.5182	1.8849	0.9288	300.0	182.1	-	4N	137E
9268	464	1839 Feb 28	08:53:39	5	-1990	101	Ne	-t	1.5671	0.0232	-1.0576	46.8	-	-	9N	129W
9269	464	1839 Mar 30	02:03:05	5	-1989	139	N	t-	-1.3853	0.3595	-0.7269	181.4	-	-	5S	30W
9270	464	1839 Aug 24	21:52:11	5	-1984	106	N	-a	-1.3743	0.3281	-0.6560	155.5	-	-	12S	33E
9271	464	1839 Sep 23	06:57:39	5	-1983	144	N	a-	1.2311	0.5830	-0.3855	197.0	-	-	1N	107W
9272	464	1840 Feb 17	14:02:34	5	-1978	111	P	-a	0.8074	1.3917	0.3615	297.0	132.3	-	13N	153E
9273	464	1840 Aug 13	07:23:08	5	-1972	116	P	-t	-0.6716	1.6443	0.6074	323.5	170.0	-	15S	109W
9274	464	1841 Feb 06	02:07:13	5	-1966	121	T+	pp	0.0812	2.6999	1.7178	324.7	212.4	97.5	16N	28W
9275	464	1841 Aug 02	10:00:51	5	-1960	126	T+	pp	0.0846	2.7381	1.6678	373.8	235.0	104.6	18S	149W
9276	464	1842 Jan 26	17:44:27	5	-1954	131	P	a-	-0.5884	1.7634	0.7930	295.9	172.1	-	18N	97E
9277	464	1842 Jul 22	10:47:38	6	-1948	136	P	t-	0.8383	1.3480	0.2917	311.6	126.5	-	20S	161W
9278	464	1842 Dec 17	19:02:29	6	-1943	103	N	-a	1.5078	0.1082	-0.9253	94.1	-	-	25N	74E
9279	464	1843 Jan 16	08:14:24	6	-1942	141	N	a-	-1.2885	0.4955	-0.5083	187.7	-	-	20N	121W
9280	464	1843 Jun 12	07:22:16	6	-1937	108	N	-h	-1.0767	0.8717	-0.1073	240.0	-	-	24S	111W
9281	465	1843 Jul 11	16:49:54	6	-1936	146	Nb	h-	1.5468	0.0226	-0.9831	43.3	-	-	21S	108E
9282	465	1843 Dec 07	00:11:31	6	-1931	113	P	-t	0.8785	1.2896	0.2031	308.3	106.6	-	23N	5W
9283	465	1844 May 31	22:50:43	6	-1925	118	T	-p	-0.3050	2.2706	1.3255	312.0	202.4	77.1	22S	17E
9284	465	1844 Nov 24	23:44:52	6	-1919	123	T+	pp	0.2026	2.5375	1.4358	375.4	230.2	93.3	21N	1E
9285	465	1845 May 21	15:54:30	6	-1913	128	T	a-	0.4281	2.0488	1.0957	308.6	193.7	45.3	20S	121E
9286	465	1845 Nov 14	00:49:42	6	-1907	133	P	h-	-0.4924	1.9872	0.9221	344.1	198.9	-	18N	16W
9287	465	1846 Apr 11	18:11:36	6	-1902	100	N	-h	-1.5131	0.0998	-0.9366	92.6	-	-	10S	87E
9288	465	1846 May 11	05:53:43	6	-1901	138	N	h-	1.2111	0.6355	-0.3643	214.2	-	-	17S	89W
9289	465	1846 Oct 04	22:21:39	6	-1896	105	N	-a	1.5077	0.0779	-0.8954	75.7	-	-	6N	21E
9290	465	1846 Nov 03	08:59:23	6	-1895	143	N	a-	-1.1618	0.7287	-0.2766	224.5	-	-	14N	139W
9291	465	1847 Mar 31	21:27:03	6	-1890	110	P	-t	-0.8345	1.3684	0.2856	318.2	126.4	-	5S	39E
9292	465	1847 Sep 24	14:33:44	7	-1884	115	P	-a	0.7973	1.3714	0.4181	272.5	132.7	-	1N	139E
9293	465	1848 Mar 19	21:12:12	7	-1878	120	T-	pp	-0.1186	2.6823	1.5993	372.5	232.4	101.4	0N	44E
9294	465	1848 Sep 13	06:19:25	7	-1872	125	T+	pp	0.0922	2.6776	1.6997	328.8	215.4	98.6	4S	96W
9295	465	1849 Mar 09	00:55:49	7	-1866	130	P	a-	0.5980	1.7804	0.7414	325.6	180.6	-	5N	11W
9296	465	1849 Sep 02	17:10:17	7	-1860	135	P	t-	-0.6806	1.6238	0.5945	317.9	166.6	-	8S	103E
9297	465	1850 Jan 28	01:06:06	7	-1855	102	N	-a	-1.4175	0.2442	-0.7305	130.4	-	-	17N	13W
9298	465	1850 Feb 26	11:48:11	7	-1854	140	N	a-	1.2587	0.5420	-0.4453	192.8	-	-	10N	173W
9299	465	1850 Jul 24	05:40:16	7	-1849	107	N	-t	1.3933	0.3371	-0.7337	175.7	-	-	19S	84W
9300	465	1850 Aug 22	20:54:59	7	-1848	145	N	t-	-1.4679	0.1997	-0.8704	136.1	-	-	13S	47E

Cat Num	Canon Plate	Calendar Date	TD of Greatest Eclipse	ΔT s	Luna Num	Saros Num	Ecl. Type	QSE	Gamma	Pen. Mag.	Um. Mag.	Eclipse Phase Durations Pen. m	Par. m	Total m	Greatest in Zenith Lat.	Lng.
9301	466	1851 Jan 17	16:50:47	7	-1843	112	P	-a	-0.7660	1.4406	0.4642	280.0	139.5	-	20N	110E
9302	466	1851 Jul 13	07:21:41	7	-1837	117	P	-t	0.6154	1.7508	0.7069	333.6	182.3	-	21S	109W
9303	466	1852 Jan 07	06:10:44	7	-1831	122	T-	-p	-0.0991	2.6863	1.6663	339.6	218.2	98.3	22N	91W
9304	466	1852 Jul 01	15:26:13	7	-1825	127	T-	p-	-0.1799	2.5227	1.5330	335.9	217.4	94.6	23S	129E
9305	466	1852 Dec 26	13:03:06	7	-1819	132	P	t-	0.6203	1.7588	0.6815	336.5	179.3	-	24N	164E
9306	466	1853 Jun 21	06:01:48	7	-1813	137	P	a-	-0.9146	1.1535	0.2056	257.8	96.7	-	24S	90W
9307	466	1853 Dec 15	13:18:43	7	-1807	142	N	t-	1.3186	0.4917	-0.6138	211.0	-	-	24N	159E
9308	466	1854 May 12	15:46:09	7	-1802	109	P	-a	0.8919	1.2035	0.2389	266.0	105.0	-	17S	123E
9309	466	1854 Nov 04	21:12:48	7	-1796	114	P	-h	-0.9707	1.1004	0.0537	278.9	54.3	-	15N	38E
9310	466	1855 May 02	04:05:23	7	-1790	119	T+	-p	0.1625	2.5690	1.5509	346.1	221.5	96.5	15S	62W
9311	466	1855 Oct 25	07:29:40	7	-1784	124	T-	-p	-0.2177	2.4527	1.4643	327.0	210.8	88.7	12N	116W
9312	466	1856 Apr 20	09:06:46	7	-1778	129	P	t-	-0.6068	1.7799	0.7098	342.3	185.1	-	12S	137W
9313	466	1856 Oct 13	22:54:33	7	-1772	134	P	a-	0.4809	1.9545	0.9960	302.7	186.2	-	9N	13E
9314	466	1857 Apr 09	09:13:19	7	-1766	139	N	t-	-1.3347	0.4503	-0.6318	201.0	-	-	9S	138W
9315	466	1857 Sep 04	05:22:14	7	-1761	106	N	-a	-1.4376	0.2148	-0.7748	127.6	-	-	8S	80W
9316	466	1857 Oct 03	14:56:57	7	-1760	144	N	a-	1.1914	0.6585	-0.3150	208.4	-	-	5N	132E
9317	466	1858 Feb 27	22:14:20	7	-1755	111	P	-a	0.8252	1.3561	0.3316	293.0	126.9	-	9N	30E
9318	466	1858 Aug 24	14:20:46	7	-1749	116	P	-t	-0.7446	1.5130	0.4707	316.7	153.4	-	12S	146E
9319	466	1859 Feb 17	10:43:28	7	-1743	121	T+	pp	0.0950	2.6727	1.6944	323.6	211.9	96.9	12N	157W
9320	466	1859 Aug 13	16:34:26	7	-1737	126	T+	pp	0.0038	2.8877	1.8148	374.9	236.0	106.5	15S	113E
9321	467	1860 Feb 07	02:29:45	8	-1731	131	P	a-	-0.5790	1.7801	0.8106	296.8	173.7	-	15N	34W
9322	467	1860 Aug 01	17:25:03	8	-1725	136	P	t-	0.7551	1.5003	0.4450	322.2	152.3	-	17S	100E
9323	467	1860 Dec 28	03:33:54	8	-1720	103	N	-a	1.5159	0.0950	-0.9419	88.7	-	-	25N	53W
9324	467	1861 Jan 26	16:54:03	8	-1719	141	N	a-	-1.2864	0.5002	-0.5052	189.0	-	-	17N	109E
9325	467	1861 Jun 22	14:35:14	8	-1714	108	N	-a	-1.1476	0.7397	-0.2355	223.2	-	-	25S	142E
9326	467	1861 Jul 21	23:50:38	8	-1713	146	N	a-	1.4659	0.1695	-0.8334	116.6	-	-	19S	3E
9327	467	1861 Dec 17	08:18:43	8	-1708	113	P	-t	0.8880	1.2738	0.1840	307.8	101.9	-	24N	125W
9328	467	1862 Jun 12	06:21:03	8	-1702	118	T	-p	-0.3763	2.1387	1.1957	308.2	197.2	62.4	24S	95W
9329	467	1862 Dec 06	07:40:21	7	-1696	123	T+	pp	0.2157	2.5141	1.4112	374.9	229.3	91.5	23N	117W
9330	467	1863 Jun 01	23:26:14	7	-1690	128	T	p-	0.3605	2.1729	1.2195	312.9	199.9	66.1	22S	8E
9331	467	1863 Nov 25	08:56:11	7	-1684	133	P	a-	-0.4760	2.0170	0.9525	344.4	200.4	-	20N	137W
9332	467	1864 May 21	13:12:02	6	-1678	138	N	h-	1.1502	0.7480	-0.2531	230.6	-	-	19S	161E
9333	467	1864 Oct 15	06:30:47	6	-1673	105	N	-a	1.5379	0.0228	-0.9512	41.1	-	-	10N	102W
9334	467	1864 Nov 13	17:21:25	6	-1672	143	N	a-	-1.1402	0.7683	-0.2367	228.9	-	-	17N	96E
9335	467	1865 Apr 11	04:38:10	6	-1667	110	P	-t	-0.8856	1.2736	0.1929	310.4	105.4	-	9S	69W
9336	467	1865 Oct 04	22:40:01	5	-1661	115	P	-a	0.8386	1.2973	0.3408	267.6	121.2	-	5N	17E
9337	467	1866 Mar 31	04:33:35	5	-1655	120	T-	pp	-0.1614	2.6011	1.5232	370.4	230.7	98.2	4S	67W
9338	467	1866 Sep 24	14:07:14	4	-1649	125	T+	-p	0.1412	2.5904	1.6071	329.4	214.6	96.0	1N	146E
9339	467	1867 Mar 20	08:49:01	4	-1643	130	P	a-	0.5656	1.8366	0.8038	326.8	185.5	-	1N	130W
9340	467	1867 Sep 14	00:26:27	3	-1637	135	P	t-	-0.6239	1.7309	0.6956	325.5	177.8	-	4S	7W
9341	468	1868 Feb 08	09:49:41	3	-1632	102	N	-a	-1.4252	0.2288	-0.7433	126.2	-	-	14N	144W
9342	468	1868 Mar 08	20:09:48	3	-1631	140	N	a-	1.2339	0.5850	-0.3975	198.9	-	-	6N	61E
9343	468	1868 Aug 03	12:09:13	2	-1626	107	N	-t	1.4740	0.1894	-0.8825	133.5	-	-	16S	179E
9344	468	1868 Sep 02	03:41:27	2	-1625	145	N	t-	-1.4012	0.3242	-0.7499	171.9	-	-	9S	55W
9345	468	1869 Jan 28	01:38:27	2	-1620	112	P	-a	-0.7733	1.4272	0.4507	279.5	137.9	-	17N	21W
9346	468	1869 Jul 23	14:02:45	1	-1614	117	P	-t	0.6961	1.6017	0.5599	323.8	165.8	-	19S	151E
9347	468	1870 Jan 17	14:46:33	1	-1608	122	T-	-p	-0.1037	2.6789	1.6566	340.7	218.7	98.2	21N	141E
9348	468	1870 Jul 12	22:34:23	0	-1602	127	T-	pp	-0.1023	2.6636	1.6769	336.0	219.1	99.7	22S	23E
9349	468	1871 Jan 06	21:16:40	-0	-1596	132	P	t-	0.6154	1.7690	0.6893	338.0	180.5	-	23N	42E
9350	468	1871 Jul 02	13:27:48	-1	-1590	137	P	a-	-0.8401	1.2893	0.3432	267.9	122.4	-	24S	159E
9351	468	1871 Dec 26	21:19:33	-1	-1584	142	N	t-	1.3105	0.5068	-0.5995	213.9	-	-	25N	40E
9352	468	1872 May 22	23:18:24	-1	-1579	109	P	-a	0.9592	1.0801	0.1151	256.1	74.3	-	20S	10E
9353	468	1872 Nov 15	05:19:37	-2	-1573	114	P	-h	-0.9876	1.0691	0.0229	275.0	35.5	-	18N	83W
9354	468	1873 May 12	11:20:26	-2	-1567	119	T+	-p	0.2284	2.4484	1.4294	345.5	219.3	89.9	18S	171W
9355	468	1873 Nov 04	15:51:03	-2	-1561	124	T-	-p	-0.2408	2.4104	1.4217	325.4	209.1	85.7	15N	118E
9356	468	1874 May 01	16:03:17	-3	-1555	129	P	t-	-0.5426	1.8971	0.8282	348.7	196.2	-	16S	118E
9357	468	1874 Oct 25	07:16:22	-3	-1549	134	T	a-	0.4510	2.0108	1.0497	304.8	189.2	32.7	12N	113W
9358	468	1875 Apr 20	16:15:23	-3	-1543	139	N	t-	-1.2770	0.5540	-0.5238	220.2	-	-	13S	115E
9359	468	1875 Sep 15	12:57:27	-4	-1538	106	N	-a	-1.4955	0.1115	-0.8839	93.2	-	-	4S	165E
9360	468	1875 Oct 14	23:02:55	-4	-1537	144	N	a-	1.1577	0.7228	-0.2558	217.7	-	-	9N	10E

Cat Num	Canon Plate	Calendar Date	TD of Greatest Eclipse	ΔT s	Luna Num	Saros Num	Ecl. Type	QSE	Gamma	Pen. Mag.	Um. Mag.	Eclipse Phase Durations Pen. m	Par. m	Total m	Greatest in Zenith Lat.	Lng.
9361	469	1876 Mar 10	06:21:31	-4	-1532	111	P	-a	0.8474	1.3124	0.2937	288.4	119.7	-	5N	92W
9362	469	1876 Sep 03	21:22:33	-4	-1526	116	P	-t	-0.8130	1.3904	0.3425	309.2	133.9	-	8S	39E
9363	469	1877 Feb 27	19:15:38	-4	-1520	121	T+	-p	0.1125	2.6386	1.6644	322.5	211.3	96.2	8N	74E
9364	469	1877 Aug 23	23:11:35	-4	-1514	126	T-	pp	-0.0739	2.7606	1.6849	375.0	235.4	104.9	11S	13E
9365	469	1878 Feb 17	11:11:05	-5	-1508	131	P	a-	-0.5667	1.8021	0.8338	297.9	175.6	-	11N	164W
9366	469	1878 Aug 13	00:08:23	-5	-1502	136	P	t-	0.6756	1.6458	0.5912	330.8	171.1	-	14S	1W
9367	469	1879 Jan 08	12:04:10	-5	-1497	103	N	-a	1.5243	0.0812	-0.9587	82.3	-	-	24N	179W
9368	469	1879 Feb 07	01:28:35	-5	-1496	141	N	a-	-1.2809	0.5108	-0.4956	191.4	-	-	14N	19W
9369	469	1879 Jul 03	21:50:28	-5	-1491	108	N	-a	-1.2172	0.6103	-0.3618	204.6	-	-	24S	34E
9370	469	1879 Aug 02	06:57:44	-5	-1490	146	N	a-	1.3889	0.3098	-0.6909	155.0	-	-	17S	103W
9371	469	1879 Dec 28	16:26:22	-5	-1485	113	P	-t	0.8969	1.2587	0.1664	307.2	97.4	-	24N	114E
9372	469	1880 Jun 22	13:50:27	-5	-1479	118	T	-a	-0.4484	2.0057	1.0641	303.7	190.6	37.2	24S	153E
9373	469	1880 Dec 16	15:39:03	-5	-1473	123	T+	pp	0.2263	2.4949	1.3914	374.4	228.6	89.9	24N	124E
9374	469	1881 Jun 12	06:53:44	-5	-1467	128	T	p-	0.2898	2.3031	1.3488	316.8	205.2	79.9	23S	104W
9375	469	1881 Dec 05	17:08:33	-5	-1461	133	P	a-	-0.4640	2.0386	0.9751	344.2	201.4	-	22N	101E
9376	469	1882 Jun 01	20:21:58	-5	-1455	138	N	t-	1.0832	0.8717	-0.1311	246.8	-	-	21S	54E
9377	469	1882 Nov 25	01:51:01	-6	-1449	143	N	a-	-1.1243	0.7972	-0.2071	231.8	-	-	20N	31W
9378	469	1883 Apr 22	11:38:29	-6	-1444	110	P	-t	-0.9448	1.1639	0.0853	300.5	71.3	-	13S	175W
9379	469	1883 Oct 16	06:54:15	-6	-1438	115	P	-a	0.8732	1.2355	0.2757	263.3	110.1	-	10N	107W
9380	469	1884 Apr 10	11:46:34	-6	-1432	120	T-	pp	-0.2116	2.5064	1.4337	367.7	228.2	92.9	8S	176W
9381	470	1884 Oct 04	22:01:55	-6	-1426	125	T+	-p	0.1839	2.5151	1.5260	329.9	213.5	92.6	5N	27E
9382	470	1885 Mar 30	16:34:07	-6	-1420	130	P	a-	0.5257	1.9066	0.8802	328.4	191.0	-	4S	113E
9383	470	1885 Sep 24	07:48:12	-6	-1414	135	P	t-	-0.5725	1.8284	0.7867	332.0	186.7	-	0N	119W
9384	470	1886 Feb 18	18:29:01	-6	-1409	102	N	-a	-1.4365	0.2064	-0.7627	119.9	-	-	10N	86E
9385	470	1886 Mar 20	04:24:16	-6	-1408	140	N	a-	1.2029	0.6393	-0.3381	206.1	-	-	1N	64W
9386	470	1886 Aug 14	18:42:04	-6	-1403	107	Ne	-t	1.5511	0.0489	-1.0246	68.6	-	-	13S	80E
9387	470	1886 Sep 13	10:34:41	-6	-1402	145	N	t-	-1.3398	0.4390	-0.6392	198.4	-	-	5S	159W
9388	470	1887 Feb 08	10:22:04	-6	-1397	112	P	-a	-0.7836	1.4081	0.4317	278.7	135.6	-	14N	152W
9389	470	1887 Aug 03	20:48:53	-6	-1391	117	P	-t	0.7732	1.4595	0.4194	313.3	146.3	-	17S	49E
9390	470	1888 Jan 28	23:20:01	-6	-1385	122	T-	-p	-0.1095	2.6692	1.6452	341.8	219.1	98.1	18N	13E
9391	470	1888 Jul 23	05:44:47	-6	-1379	127	T-	pp	-0.0256	2.8030	1.8189	335.3	219.5	101.8	20S	85W
9392	470	1889 Jan 17	05:29:39	-6	-1373	132	P	t-	0.6106	1.7788	0.6972	339.5	181.7	-	21N	80W
9393	470	1889 Jul 12	20:53:52	-6	-1367	137	P	a-	-0.7654	1.4257	0.4807	276.8	141.7	-	23S	48E
9394	470	1890 Jan 06	05:21:26	-6	-1361	142	N	t-	1.3029	0.5209	-0.5854	216.6	-	-	24N	79W
9395	470	1890 Jun 03	06:44:42	-6	-1356	109	N	-a	1.0309	0.9492	-0.0168	244.1	-	-	21S	101W
9396	470	1890 Jul 02	14:08:36	-6	-1355	147	N	a-	-1.4871	0.1025	-0.8445	85.3	-	-	25S	149E
9397	470	1890 Nov 26	13:33:48	-6	-1350	114	P	-h	-0.9994	1.0470	0.0017	271.9	9.8	-	20N	154E
9398	470	1891 May 23	18:29:11	-6	-1344	119	T	-a	0.2988	2.3199	1.2996	344.1	215.7	79.1	20S	82E
9399	470	1891 Nov 16	00:18:46	-6	-1338	124	T-	-p	-0.2592	2.3766	1.3880	324.0	207.6	83.1	18N	8W
9400	470	1892 May 11	22:53:20	-6	-1332	129	P	t-	-0.4734	2.0236	0.9555	354.7	206.3	-	19S	16E
9401	471	1892 Nov 04	15:44:53	-6	-1326	134	T	p-	0.4267	2.0565	1.0930	306.5	191.5	44.1	16N	119E
9402	471	1893 Apr 30	23:08:58	-6	-1320	139	N	t-	-1.2119	0.6713	-0.4023	239.2	-	-	16S	12E
9403	471	1893 Sep 25	20:39:00	-6	-1315	106	Ne	-a	-1.5476	0.0190	-0.9825	38.9	-	-	0S	49E
9404	471	1893 Oct 25	07:16:16	-6	-1314	144	N	a-	1.1305	0.7753	-0.2084	225.0	-	-	13N	114W
9405	471	1894 Mar 21	14:20:27	-6	-1309	111	P	-h	0.8770	1.2549	0.2424	282.6	109.3	-	0N	147E
9406	471	1894 Sep 15	04:31:25	-6	-1303	116	P	-t	-0.8748	1.2797	0.2261	301.7	110.9	-	4S	69W
9407	471	1895 Mar 11	03:39:10	-6	-1297	121	T+	-p	0.1376	2.5904	1.6204	321.2	210.5	94.8	4N	52W
9408	471	1895 Sep 04	05:56:52	-6	-1291	126	T-	pp	-0.1449	2.6317	1.5530	374.4	233.5	100.5	7S	89W
9409	471	1896 Feb 28	19:45:40	-6	-1285	131	P	a-	-0.5488	1.8342	0.8673	299.5	178.2	-	7N	66E
9410	471	1896 Aug 23	06:57:19	-6	-1279	136	P	t-	0.5997	1.7849	0.7306	337.8	185.4	-	11S	104W
9411	471	1897 Jan 18	20:33:07	-6	-1274	103	N	-a	1.5332	0.0659	-0.9763	74.5	-	-	22N	55E
9412	471	1897 Feb 17	09:57:44	-6	-1273	141	N	a-	-1.2717	0.5281	-0.4790	194.8	-	-	11N	146W
9413	471	1897 Jul 14	05:05:19	-5	-1268	108	N	-a	-1.2879	0.4792	-0.4902	183.0	-	-	23S	75W
9414	471	1897 Aug 12	14:08:38	-5	-1267	146	N	a-	1.3135	0.4472	-0.5517	183.2	-	-	14S	148E
9415	471	1898 Jan 08	00:34:46	-5	-1262	113	P	-t	0.9046	1.2456	0.1515	306.7	93.3	-	23N	7W
9416	471	1898 Jul 03	21:17:20	-5	-1256	118	P	-a	-0.5228	1.8689	0.9280	298.4	182.3	-	23S	42E
9417	471	1898 Dec 27	23:41:52	-4	-1250	123	T+	pp	0.2339	2.4809	1.3777	373.8	228.0	88.8	24N	5E
9418	471	1899 Jun 23	14:17:53	-3	-1244	128	T+	p-	0.2169	2.4376	1.4820	320.1	209.5	89.5	23S	146E
9419	471	1899 Dec 17	01:25:45	-3	-1238	133	P	a-	-0.4551	2.0541	0.9922	343.7	202.0	-	23N	22W
9420	471	1900 Jun 13	03:27:37	-2	-1232	138	N*	t-	1.0134	1.0010	-0.0040	261.9	-	-	22S	52W

Cat Num	Canon Plate	Calendar Date	TD of Greatest Eclipse	ΔT s	Luna Num	Saros Num	Ecl. Type	QSE	Gamma	Pen. Mag.	Um. Mag.	Eclipse Phase Durations Pen. m	Par. m	Total m	Greatest in Zenith Lat.	Lng.
9421	472	1900 Dec 06	10:26:29	-1	-1226	143	N	a-	-1.1125	0.8183	-0.1851	233.7	-	-	21N	159W
9422	472	1901 May 03	18:30:38	-1	-1221	110	N*	-t	-1.0101	1.0431	-0.0334	288.2	-	-	17S	81E
9423	472	1901 Oct 27	15:15:18	-0	-1215	115	P	-a	0.9021	1.1841	0.2208	259.6	99.4	-	14N	127E
9424	472	1902 Apr 22	18:52:40	0	-1209	120	T-	pp	-0.2680	2.4002	1.3327	364.3	224.6	84.6	12S	76E
9425	472	1902 Oct 17	06:03:26	1	-1203	125	T+	-p	0.2201	2.4514	1.4566	330.4	212.3	88.8	9N	95W
9426	472	1903 Apr 12	00:12:59	2	-1197	130	P	a-	0.4798	1.9877	0.9677	330.1	196.5	-	8S	3W
9427	472	1903 Oct 06	15:17:33	2	-1191	135	P	t-	-0.5280	1.9133	0.8654	337.5	193.7	-	4N	128E
9428	472	1904 Mar 02	03:02:34	3	-1186	102	N	-a	-1.4528	0.1748	-0.7910	110.5	-	-	6N	43W
9429	472	1904 Mar 31	12:32:28	3	-1185	140	N	a-	1.1665	0.7036	-0.2688	214.2	-	-	3S	173E
9430	472	1904 Sep 24	17:34:44	3	-1179	145	N	t-	-1.2837	0.5440	-0.5384	219.1	-	-	1S	95E
9431	472	1905 Feb 19	19:00:02	4	-1174	112	P	-a	-0.7984	1.3809	0.4049	277.2	132.1	-	11N	78E
9432	472	1905 Aug 15	03:40:59	5	-1168	117	P	-t	0.8456	1.3259	0.2871	302.3	123.1	-	14S	54W
9433	472	1906 Feb 09	07:46:58	5	-1162	122	T-	-p	-0.1199	2.6507	1.6254	342.8	219.4	97.8	15N	113W
9434	472	1906 Aug 04	13:00:10	6	-1156	127	T+	pp	0.0477	2.7615	1.7793	333.9	218.7	101.2	17S	166E
9435	472	1907 Jan 29	13:38:00	6	-1150	132	P	t-	0.6027	1.7936	0.7110	341.1	183.5	-	19N	159E
9436	472	1907 Jul 25	04:22:27	7	-1144	137	P	a-	-0.6924	1.5595	0.6149	284.5	156.8	-	21S	64W
9437	472	1908 Jan 18	13:21:36	8	-1138	142	N	t-	1.2939	0.5370	-0.5685	219.5	-	-	22N	162E
9438	472	1908 Jun 14	14:06:32	8	-1133	109	N	-a	1.1053	0.8135	-0.1541	229.9	-	-	22S	149E
9439	472	1908 Jul 13	21:33:55	8	-1132	147	N	a-	-1.4185	0.2292	-0.7195	126.3	-	-	23S	38E
9440	472	1908 Dec 07	21:55:09	9	-1127	114	N*	-h	-1.0059	1.0344	-0.0096	269.7	-	-	22N	29E
9441	473	1909 Jun 04	01:28:51	10	-1121	119	T	-h	0.3755	2.1800	1.1580	341.6	210.2	60.4	22S	23W
9442	473	1909 Nov 27	08:54:41	10	-1115	124	T-	-p	-0.2712	2.3544	1.3660	322.8	206.5	81.2	21N	137W
9443	473	1910 May 24	05:34:16	11	-1109	129	T	t-	-0.3975	2.1625	1.0950	360.3	215.4	49.5	21S	85W
9444	473	1910 Nov 17	00:20:52	12	-1103	134	T	p-	0.4089	2.0905	1.1246	307.7	193.0	50.6	19N	9W
9445	473	1911 May 13	05:56:24	12	-1097	139	N	t-	-1.1413	0.7987	-0.2706	257.0	-	-	19S	90W
9446	473	1911 Nov 06	15:36:45	13	-1091	144	N	a-	1.1100	0.8154	-0.1733	230.6	-	-	17N	121E
9447	473	1912 Apr 01	22:14:16	13	-1086	111	P	-h	0.9116	1.1884	0.1820	275.8	95.2	-	4S	28E
9448	473	1912 Sep 26	11:44:50	14	-1080	116	P	-t	-0.9320	1.1779	0.1183	293.9	81.7	-	0N	178W
9449	473	1913 Mar 22	11:57:49	15	-1074	121	T+	-p	0.1671	2.5340	1.5683	319.7	209.4	92.8	0S	178W
9450	473	1913 Sep 15	12:48:19	15	-1068	126	T-	pp	-0.2109	2.5122	1.4304	373.0	230.5	93.5	3S	167E
9451	473	1914 Mar 12	04:13:08	16	-1062	131	P	a-	-0.5254	1.8764	0.9111	301.5	181.5	-	3N	61W
9452	473	1914 Sep 04	13:54:57	17	-1056	136	P	t-	0.5301	1.9127	0.8585	343.2	196.0	-	7S	151E
9453	473	1915 Jan 31	04:57:42	17	-1051	103	N	-a	1.5450	0.0451	-0.9989	62.0	-	-	19N	70W
9454	473	1915 Mar 01	18:19:32	17	-1050	141	N	a-	-1.2573	0.5548	-0.4528	199.8	-	-	7N	88E
9455	473	1915 Jul 26	12:24:39	18	-1045	108	N	-a	-1.3553	0.3545	-0.6129	158.7	-	-	21S	176E
9456	473	1915 Aug 24	21:27:17	18	-1044	146	N	a-	1.2435	0.5750	-0.4226	204.4	-	-	10S	38E
9457	473	1916 Jan 20	08:39:41	18	-1039	113	P	-t	0.9146	1.2277	0.1327	305.7	87.7	-	21N	127W
9458	473	1916 Jul 15	04:46:07	19	-1033	118	P	-a	-0.5956	1.7351	0.7944	292.4	172.5	-	22S	70W
9459	473	1917 Jan 08	07:44:48	19	-1027	123	T+	pp	0.2415	2.4663	1.3642	373.1	227.3	87.6	23N	114W
9460	473	1917 Jul 04	21:39:04	20	-1021	128	T+	pp	0.1419	2.5762	1.6185	322.9	212.8	96.0	23S	36E
9461	474	1917 Dec 28	09:46:32	20	-1015	133	T	a-	-0.4484	2.0652	1.0056	343.0	202.3	12.0	23N	146W
9462	474	1918 Jun 24	10:28:03	20	-1009	138	P	t-	0.9397	1.1376	0.1297	276.3	82.4	-	23S	157W
9463	474	1918 Dec 17	19:06:01	21	-1003	143	N	a-	-1.1035	0.8340	-0.1679	234.9	-	-	22N	73E
9464	474	1919 May 15	01:14:00	21	-998	110	N	-t	-1.0820	0.9103	-0.1644	273.1	-	-	20S	20W
9465	474	1919 Nov 07	23:44:28	21	-992	115	P	-a	0.9246	1.1445	0.1780	256.7	89.8	-	17N	0W
9466	474	1920 May 03	01:51:08	21	-986	120	T	-p	-0.3312	2.2818	1.2194	360.1	219.7	71.5	16S	29W
9467	474	1920 Oct 27	14:11:38	22	-980	125	T+	-p	0.2502	2.3992	1.3987	330.9	211.1	85.0	13N	143E
9468	474	1921 Apr 22	07:44:39	22	-974	130	T	a-	0.4269	2.0816	1.0678	331.9	202.0	40.1	12S	116W
9469	474	1921 Oct 16	22:53:59	22	-968	135	P	t-	-0.4902	1.9858	0.9317	342.1	199.1	-	9N	13E
9470	474	1922 Mar 13	11:28:48	23	-963	102	N	-a	-1.4752	0.1320	-0.8304	96.3	-	-	2N	170W
9471	474	1922 Apr 11	20:32:12	23	-962	140	N	a-	1.1228	0.7812	-0.1863	223.2	-	-	7S	53E
9472	474	1922 Oct 06	00:43:50	23	-956	145	N	t-	-1.2348	0.6358	-0.4508	235.2	-	-	4N	13W
9473	474	1923 Mar 03	03:32:09	23	-951	112	P	-a	-0.8175	1.3453	0.3701	275.2	127.1	-	6N	50W
9474	474	1923 Aug 26	10:39:52	23	-945	117	P	-t	0.9133	1.2013	0.1634	290.8	94.2	-	10S	160W
9475	474	1924 Feb 20	16:08:55	24	-939	122	T-	-p	-0.1338	2.6256	1.5995	343.6	219.7	97.1	11N	121E
9476	474	1924 Aug 14	20:20:30	24	-933	127	T+	pp	0.1175	2.6326	1.6519	332.0	216.9	98.2	14S	56E
9477	474	1925 Feb 08	21:42:22	24	-927	132	P	t-	0.5921	1.8134	0.7304	342.9	185.7	-	15N	38E
9478	474	1925 Aug 04	11:52:57	24	-921	137	P	a-	-0.6208	1.6909	0.7463	291.2	168.9	-	18S	177W
9479	474	1926 Jan 28	21:20:24	24	-915	142	N	t-	1.2836	0.5551	-0.5488	222.6	-	-	19N	43E
9480	474	1926 Jun 25	21:25:07	24	-910	109	N	-a	1.1814	0.6749	-0.2948	213.1	-	-	22S	40E

Cat Num	Canon Plate	Calendar Date	TD of Greatest Eclipse	ΔT s	Luna Num	Saros Num	Ecl. Type	QSE	Gamma	Pen. Mag.	Um. Mag.	Eclipse Phase Durations Pen. m	Par. m	Total m	Greatest in Zenith Lat.	Lng.
9481	475	1926 Jul 25	05:00:12	24	-909	147	N	a-	-1.3511	0.3542	-0.5970	155.5	-	-	21S	73W
9482	475	1926 Dec 19	06:20:07	24	-904	114	N*	-a	-1.0101	1.0257	-0.0163	268.0	-	-	22N	96W
9483	475	1927 Jun 15	08:24:41	24	-898	119	T	-t	0.4543	2.0365	1.0123	338.1	202.8	17.7	23S	126W
9484	475	1927 Dec 08	17:35:10	24	-892	124	T	-p	-0.2796	2.3389	1.3510	321.8	205.6	79.8	22N	94E
9485	475	1928 Jun 03	12:09:57	24	-886	129	T	tp	-0.3175	2.3092	1.2421	365.2	223.0	75.3	23S	177E
9486	475	1928 Nov 27	09:01:47	24	-880	134	T	p-	0.3952	2.1166	1.1486	308.7	194.2	54.8	22N	139W
9487	475	1929 May 23	12:37:45	24	-874	139	N	t-	-1.0650	0.9367	-0.1287	273.7	-	-	21S	169E
9488	475	1929 Nov 17	00:03:13	24	-868	144	N	a-	1.0947	0.8460	-0.1474	235.0	-	-	20N	5W
9489	475	1930 Apr 13	05:58:54	24	-863	111	P	-h	0.9545	1.1066	0.1064	267.3	73.4	-	8S	89W
9490	475	1930 Oct 07	19:07:10	24	-857	116	P	-t	-0.9811	1.0906	0.0252	286.6	38.3	-	5N	71E
9491	475	1931 Apr 02	20:07:55	24	-851	121	T+	-p	0.2043	2.4637	1.5021	317.9	207.8	89.6	5S	59E
9492	475	1931 Sep 26	19:48:29	24	-845	126	T-	pp	-0.2698	2.4058	1.3208	371.2	226.9	84.2	1N	61E
9493	475	1932 Mar 22	12:32:39	24	-839	131	P	a-	-0.4956	1.9303	0.9666	303.8	185.3	-	1S	173E
9494	475	1932 Sep 14	21:01:00	24	-833	136	P	t-	0.4664	2.0296	0.9752	347.2	204.0	-	3S	43E
9495	475	1933 Feb 10	13:17:33	24	-828	103	N	-a	1.5600	0.0182	-1.0270	39.6	-	-	16N	165E
9496	475	1933 Mar 12	02:33:03	24	-827	141	N	a-	-1.2369	0.5923	-0.4154	206.3	-	-	2N	36W
9497	475	1933 Aug 05	19:46:05	24	-822	108	N	-a	-1.4216	0.2322	-0.7338	129.5	-	-	18S	66E
9498	475	1933 Sep 04	04:52:20	24	-821	146	N	a-	1.1776	0.6955	-0.3013	221.4	-	-	6S	74W
9499	475	1934 Jan 30	16:42:42	24	-816	113	P	-t	0.9258	1.2073	0.1120	304.3	80.9	-	18N	113E
9500	475	1934 Jul 26	12:15:38	24	-810	118	P	-a	-0.6681	1.6025	0.6612	285.7	160.8	-	20S	178E
9501	476	1935 Jan 19	15:47:35	24	-804	123	T+	pp	0.2498	2.4502	1.3499	372.1	226.7	86.3	21N	126E
9502	476	1935 Jul 16	05:00:05	24	-798	128	T+	pp	0.0672	2.7146	1.7542	325.0	214.8	99.6	21S	73W
9503	476	1936 Jan 08	18:09:58	24	-792	133	T	a-	-0.4428	2.0740	1.0173	342.1	202.5	20.8	22N	89E
9504	476	1936 Jul 04	17:25:23	24	-786	138	P	t-	0.8642	1.2778	0.2668	289.5	116.3	-	22S	100E
9505	476	1936 Dec 28	03:49:09	24	-780	143	N	a-	-1.0970	0.8451	-0.1550	235.5	-	-	22N	57W
9506	476	1937 May 25	07:51:34	24	-775	110	N	-t	-1.1581	0.7697	-0.3033	254.7	-	-	22S	119W
9507	476	1937 Nov 18	08:19:26	24	-769	115	P	-a	0.9421	1.1141	0.1443	254.5	81.3	-	20N	129W
9508	476	1938 May 14	08:44:00	24	-763	120	T	-t	-0.3994	2.1540	1.0966	354.9	213.1	49.4	19S	132W
9509	476	1938 Nov 07	22:26:42	24	-757	125	T	-p	0.2738	2.3585	1.3525	331.5	210.2	81.4	17N	19E
9510	476	1939 May 03	15:11:43	24	-751	130	T	a-	0.3693	2.1842	1.1765	333.5	207.1	62.4	15S	131E
9511	476	1939 Oct 28	06:36:43	24	-745	135	P	t-	-0.4581	2.0477	0.9877	346.1	203.4	-	12N	103W
9512	476	1940 Mar 23	19:48:19	24	-740	102	N	-a	-1.5033	0.0788	-0.8803	74.6	-	-	3S	64E
9513	476	1940 Apr 22	04:26:25	25	-739	140	N	a-	1.0741	0.8683	-0.0945	232.5	-	-	11S	67W
9514	476	1940 Oct 16	08:01:17	25	-733	145	N	t-	-1.1924	0.7156	-0.3749	248.0	-	-	8N	124W
9515	476	1941 Mar 13	11:55:47	25	-728	112	P	-a	-0.8436	1.2970	0.3226	272.2	119.8	-	2N	177W
9516	476	1941 Sep 05	17:47:15	25	-722	117	P	-t	0.9746	1.0884	0.0511	279.3	53.4	-	6S	93E
9517	476	1942 Mar 03	00:21:54	25	-716	122	T-	-p	-0.1545	2.5879	1.5612	344.3	219.7	95.9	7N	2W
9518	476	1942 Aug 26	03:48:25	26	-710	127	T+	pp	0.1818	2.5142	1.5344	329.5	214.2	93.4	10S	57W
9519	476	1943 Feb 20	05:38:23	26	-704	132	P	t-	0.5751	1.8444	0.7616	345.1	189.0	-	12N	81W
9520	476	1943 Aug 15	19:28:46	26	-698	137	P	a-	-0.5533	1.8152	0.8697	296.7	178.4	-	15S	69E
9521	477	1944 Feb 09	05:14:57	26	-692	142	N	t-	1.2698	0.5792	-0.5223	226.7	-	-	16N	75W
9522	477	1944 Jul 06	04:40:01	27	-687	109	N	-a	1.2596	0.5328	-0.4398	192.7	-	-	21S	69W
9523	477	1944 Aug 04	12:26:51	27	-686	147	N	a-	-1.2842	0.4785	-0.4758	179.1	-	-	18S	175E
9524	477	1944 Dec 29	14:49:35	27	-681	114	N*	-a	-1.0114	1.0220	-0.0176	266.6	-	-	22N	138E
9525	477	1945 Jun 25	15:14:22	27	-675	119	P	-t	0.5370	1.8862	0.8593	333.3	192.7	-	23S	132E
9526	477	1945 Dec 19	02:20:47	27	-669	124	T	-p	-0.2845	2.3293	1.3424	320.9	204.9	78.9	23N	36W
9527	477	1946 Jun 14	18:39:17	28	-663	129	T-	pp	-0.2324	2.4654	1.3983	369.2	229.0	91.1	23S	80E
9528	477	1946 Dec 08	17:48:28	28	-657	134	T	p-	0.3864	2.1337	1.1639	309.5	194.9	57.3	23N	91E
9529	477	1947 Jun 03	19:15:44	28	-651	139	P	t-	-0.9849	1.0818	0.0202	288.9	34.7	-	23S	70E
9530	477	1947 Nov 28	08:34:29	28	-645	144	N	a-	1.0838	0.8683	-0.1297	238.3	-	-	22N	132W
9531	477	1948 Apr 23	13:39:18	28	-640	111	P	-a	1.0016	1.0171	0.0229	257.6	34.3	-	12S	155E
9532	477	1948 Oct 18	02:35:41	29	-634	116	N*	-t	-1.0245	1.0140	-0.0572	279.7	-	-	9N	42W
9533	477	1949 Apr 13	04:11:25	29	-628	121	T+	-p	0.2474	2.3825	1.4251	315.8	205.7	84.9	9S	62W
9534	477	1949 Oct 07	02:56:55	29	-622	126	T	-p	-0.3219	2.3118	1.2236	369.2	222.9	72.8	5N	47W
9535	477	1950 Apr 02	20:44:34	29	-616	131	T	a-	-0.4598	1.9951	1.0329	306.5	189.6	26.9	5S	50E
9536	477	1950 Sep 26	04:17:11	29	-610	136	T	t-	0.4101	2.1331	1.0783	350.1	209.8	44.3	1N	66W
9537	477	1951 Mar 23	10:37:33	30	-604	141	N	a-	-1.2099	0.6418	-0.3661	214.3	-	-	2S	158W
9538	477	1951 Aug 17	03:14:39	30	-599	108	N	-a	-1.4828	0.1195	-0.8456	93.6	-	-	15S	47W
9539	477	1951 Sep 15	12:27:06	30	-598	146	N	a-	1.1186	0.8034	-0.1928	234.6	-	-	2S	172E
9540	477	1952 Feb 11	00:39:48	30	-593	113	P	-t	0.9416	1.1781	0.0832	301.9	70.1	-	15N	6W

Cat Num	Canon Plate	Calendar Date	TD of Greatest Eclipse	ΔT s	Luna Num	Saros Num	Ecl. Type	QSE	Gamma	Pen. Mag.	Um. Mag.	Eclipse Phase Durations Pen. m	Eclipse Phase Durations Par. m	Eclipse Phase Durations Total m	Greatest in Zenith Lat.	Greatest in Zenith Lng.
9541	478	1952 Aug 05	19:47:55	30	-587	118	P	-a	-0.7383	1.4741	0.5317	278.4	147.2	-	18S	65E
9542	478	1953 Jan 29	23:47:49	30	-581	123	T+	-p	0.2606	2.4291	1.3314	371.0	225.8	84.5	18N	7E
9543	478	1953 Jul 26	12:21:10	30	-575	128	T-	pp	-0.0071	2.8265	1.8628	326.5	215.7	100.7	19S	176E
9544	478	1954 Jan 19	02:32:21	31	-569	133	T	a-	-0.4357	2.0853	1.0322	341.2	202.9	28.2	20N	35W
9545	478	1954 Jul 16	00:20:51	31	-563	138	P	t-	0.7876	1.4202	0.4054	301.6	140.9	-	21S	4W
9546	478	1955 Jan 08	12:33:20	31	-557	143	N	a-	-1.0906	0.8555	-0.1421	236.0	-	-	21N	173E
9547	478	1955 Jun 05	14:23:23	31	-552	110	N	-t	-1.2383	0.6218	-0.4498	232.3	-	-	24S	144E
9548	478	1955 Nov 29	17:00:00	31	-546	115	P	-a	0.9551	1.0917	0.1190	253.0	74.2	-	22N	102E
9549	478	1956 May 24	15:31:52	32	-540	120	P	-t	-0.4726	2.0174	0.9647	348.5	204.5	-	21S	126E
9550	478	1956 Nov 18	06:48:16	32	-534	125	T	-p	0.2917	2.3285	1.3172	332.2	209.5	78.4	20N	106W
9551	478	1957 May 13	22:31:28	32	-528	130	T	a-	0.3045	2.3001	1.2982	335.0	211.6	77.6	18S	21E
9552	478	1957 Nov 07	14:27:30	32	-522	135	T	t-	-0.4332	2.0963	1.0305	349.3	206.5	27.9	16N	139E
9553	478	1958 Apr 04	04:00:15	32	-517	102	Ne	-a	-1.5380	0.0135	-0.9422	31.0	-	-	7S	60W
9554	478	1958 May 03	12:13:29	32	-516	140	P	a-	1.0188	0.9676	0.0092	242.2	21.0	-	15S	176E
9555	478	1958 Oct 27	15:27:50	33	-510	145	N	t-	-1.1570	0.7825	-0.3118	257.9	-	-	12N	124E
9556	478	1959 Mar 24	20:11:57	33	-505	112	P	-a	-0.8757	1.2379	0.2643	268.2	109.5	-	2S	58E
9557	478	1959 Sep 17	01:03:37	33	-499	117	N	-t	1.0296	0.9874	-0.0496	268.0	-	-	2S	17W
9558	478	1960 Mar 13	08:28:21	33	-493	122	T-	-p	-0.1799	2.5415	1.5145	344.8	219.4	94.0	3N	125W
9559	478	1960 Sep 05	11:21:51	33	-487	127	T+	-p	0.2422	2.4031	1.4239	326.7	210.8	86.7	6S	171W
9560	478	1961 Mar 02	13:28:40	34	-481	132	P	t-	0.5540	1.8828	0.8006	347.6	192.9	-	8N	161E
9561	479	1961 Aug 26	03:08:51	34	-475	137	P	a-	-0.4894	1.9330	0.9863	301.4	186.0	-	11S	46W
9562	479	1962 Feb 19	13:03:42	34	-469	142	N	t-	1.2511	0.6120	-0.4865	231.9	-	-	12N	168E
9563	479	1962 Jul 17	11:54:49	34	-464	109	N	-a	1.3370	0.3924	-0.5835	168.3	-	-	20S	177W
9564	479	1962 Aug 15	19:57:30	34	-463	147	N	a-	-1.2210	0.5963	-0.3616	198.2	-	-	15S	62E
9565	479	1963 Jan 09	23:19:42	35	-458	114	N*	-a	-1.0128	1.0180	-0.0185	265.3	-	-	21N	12E
9566	479	1963 Jul 06	22:02:59	35	-452	119	P	-t	0.6197	1.7360	0.7060	327.2	179.9	-	22S	31E
9567	479	1963 Dec 30	11:07:25	35	-446	124	T	-p	-0.2889	2.3206	1.3350	320.0	204.3	78.1	23N	166W
9568	479	1964 Jun 25	01:06:50	35	-440	129	T-	pp	-0.1461	2.6238	1.5565	372.1	233.2	100.8	24S	16W
9569	479	1964 Dec 19	02:37:54	36	-434	134	T	p-	0.3801	2.1461	1.1748	310.1	195.5	58.9	24N	40W
9570	479	1965 Jun 14	01:49:26	36	-428	139	P	t-	-0.9005	1.2351	0.1767	302.7	100.3	-	24S	27W
9571	479	1965 Dec 08	17:10:32	36	-422	144	N	a-	1.0774	0.8820	-0.1201	240.7	-	-	24N	100E
9572	479	1966 May 04	21:12:06	37	-417	111	N	-a	1.0553	0.9157	-0.0728	246.0	-	-	15S	42E
9573	479	1966 Oct 29	10:12:53	37	-411	116	N	-t	-1.0599	0.9517	-0.1249	273.7	-	-	12N	157W
9574	479	1967 Apr 24	12:07:04	38	-405	121	T	-p	0.2972	2.2892	1.3356	313.3	202.8	77.9	12S	178E
9575	479	1967 Oct 18	10:15:48	38	-399	126	T	-t	-0.3653	2.2337	1.1426	367.1	218.9	59.8	9N	157W
9576	479	1968 Apr 13	04:48:01	39	-393	131	T	p-	-0.4173	2.0725	1.1116	309.5	194.1	48.5	9S	72W
9577	479	1968 Oct 06	11:42:35	39	-387	136	T	t-	0.3605	2.2242	1.1691	352.0	213.9	63.0	6N	179W
9578	479	1969 Apr 02	18:33:06	39	-381	141	N	a-	-1.1764	0.7033	-0.3047	223.7	-	-	6S	82E
9579	479	1969 Aug 27	10:48:15	40	-376	108	Ne	-a	-1.5407	0.0131	-0.9516	31.3	-	-	11S	161W
9580	479	1969 Sep 25	20:10:19	40	-375	146	N	a-	1.0655	0.9007	-0.0953	245.1	-	-	2N	55E
9581	480	1970 Feb 21	08:30:43	40	-370	113	P	-t	0.9619	1.1402	0.0463	298.5	52.7	-	11N	124W
9582	480	1970 Aug 17	03:24:06	41	-364	118	P	-a	-0.8053	1.3521	0.4079	270.6	131.3	-	14S	49W
9583	480	1971 Feb 10	07:45:21	41	-358	123	T	-p	0.2741	2.4026	1.3082	369.5	224.7	82.2	15N	112W
9584	480	1971 Aug 06	19:43:52	42	-352	128	T-	pp	-0.0794	2.6958	1.7283	327.3	215.5	99.4	17S	66E
9585	480	1972 Jan 30	10:54:05	42	-346	133	T	a-	-0.4273	2.0987	1.0497	340.3	203.4	34.8	17N	160W
9586	480	1972 Jul 26	07:16:22	43	-340	138	P	t-	0.7116	1.5618	0.5427	312.4	160.1	-	19S	108W
9587	480	1973 Jan 18	21:17:58	43	-334	143	N	a-	-1.0844	0.8655	-0.1293	236.5	-	-	19N	43E
9588	480	1973 Jun 15	20:50:41	44	-329	110	N	-t	-1.3216	0.4685	-0.6020	204.6	-	-	25S	48E
9589	480	1973 Jul 15	11:39:19	44	-328	148	Nb	t-	1.5177	0.1046	-0.9581	99.1	-	-	20S	174W
9590	480	1973 Dec 10	01:45:06	44	-323	115	P	-a	0.9644	1.0760	0.1007	252.0	68.5	-	24N	28W
9591	480	1974 Jun 04	22:16:44	45	-317	120	P	-t	-0.5488	1.8752	0.8269	341.0	193.6	-	23S	26E
9592	480	1974 Nov 29	15:14:07	45	-311	125	T	-p	0.3054	2.3057	1.2896	333.1	208.9	75.7	22N	129E
9593	480	1975 May 25	05:48:47	46	-305	130	T+	p-	0.2367	2.4218	1.4253	335.9	215.2	88.3	21S	88W
9594	480	1975 Nov 18	22:24:12	46	-299	135	T	p-	-0.4134	2.1352	1.0642	352.1	209.0	40.2	19N	20E
9595	480	1976 May 13	19:55:08	47	-293	140	P	a-	0.9585	1.0761	0.1217	251.8	75.4	-	18S	61E
9596	480	1976 Nov 06	23:01:59	47	-287	145	N	t-	-1.1275	0.8383	-0.2594	265.8	-	-	15N	11E
9597	480	1977 Apr 04	04:19:04	48	-282	112	P	-a	-0.9148	1.1657	0.1928	262.9	94.7	-	7S	64W
9598	480	1977 Sep 27	08:30:08	48	-276	117	N	-t	1.0768	0.9007	-0.1361	257.5	-	-	3N	130W
9599	480	1978 Mar 24	16:23:11	49	-270	122	T-	-p	-0.2140	2.4790	1.4518	344.9	218.5	90.7	2S	116E
9600	480	1978 Sep 16	19:05:01	49	-264	127	T	-p	0.2951	2.3060	1.3268	323.8	207.2	78.6	2S	73E

Cat Num	Canon Plate	Calendar Date	TD of Greatest Eclipse	ΔT s	Luna Num	Saros Num	Ecl. Type	QSE	Gamma	Pen. Mag.	Um. Mag.	Eclipse Phase Durations Pen. m	Par. m	Total m	Greatest in Zenith Lat.	Lng.
9601	481	1979 Mar 13	21:08:52	50	-258	132	P	t-	0.5253	1.9350	0.8538	350.6	197.7	-	3N	46E
9602	481	1979 Sep 06	10:55:02	50	-252	137	T	a-	-0.4305	2.0421	1.0936	305.1	191.9	44.4	7S	164W
9603	481	1980 Mar 01	20:46:03	51	-246	142	N	t-	1.2269	0.6545	-0.4405	238.5	-	-	8N	52E
9604	481	1980 Jul 27	19:08:59	51	-241	109	N	-a	1.4138	0.2535	-0.7264	137.5	-	-	18S	74E
9605	481	1980 Aug 26	03:31:20	51	-240	147	N	a-	-1.1608	0.7089	-0.2531	214.4	-	-	12S	52W
9606	481	1981 Jan 20	07:50:48	51	-235	114	N*	-a	-1.0141	1.0136	-0.0192	263.8	-	-	19N	115W
9607	481	1981 Jul 17	04:47:40	52	-229	119	P	-t	0.7045	1.5822	0.5486	319.6	163.2	-	21S	70W
9608	481	1982 Jan 09	19:56:44	52	-223	124	T	-p	-0.2916	2.3147	1.3310	319.1	203.8	77.7	22N	63E
9609	481	1982 Jul 06	07:31:47	53	-217	129	T-	pp	-0.0579	2.7860	1.7179	373.8	235.6	105.7	23S	112W
9610	481	1982 Dec 30	11:29:37	53	-211	134	T	p-	0.3758	2.1545	1.1822	310.6	195.9	60.0	24N	172W
9611	481	1983 Jun 25	08:23:11	53	-205	139	P	t-	-0.8151	1.3901	0.3348	314.7	134.6	-	24S	125W
9612	481	1983 Dec 20	01:49:57	54	-199	144	N	a-	1.0746	0.8890	-0.1167	242.3	-	-	24N	28W
9613	481	1984 May 15	04:41:03	54	-194	111	N	-a	1.1130	0.8071	-0.1760	232.5	-	-	18S	71W
9614	481	1984 Jun 13	14:26:39	54	-193	149	Nb	a-	-1.5239	0.0647	-0.9414	73.0	-	-	25S	143E
9615	481	1984 Nov 08	17:56:08	54	-188	116	N	-t	-1.0899	0.8992	-0.1825	268.4	-	-	16N	88E
9616	481	1985 May 04	19:57:19	55	-182	121	T	-p	0.3519	2.1870	1.2369	310.2	198.9	67.7	16S	60E
9617	481	1985 Oct 28	17:43:17	55	-176	126	T	-t	-0.4022	2.1673	1.0736	365.1	214.9	43.9	13N	91E
9618	481	1986 Apr 24	12:43:30	55	-170	131	T	p-	-0.3682	2.1620	1.2022	312.6	198.8	63.6	13S	169E
9619	481	1986 Oct 17	19:18:54	55	-164	136	T	h-	0.3188	2.3008	1.2455	353.2	216.8	73.7	10N	67E
9620	481	1987 Apr 14	02:19:50	55	-158	141	N	h-	-1.1364	0.7769	-0.2313	234.1	-	-	10S	35W
9621	482	1987 Oct 07	04:02:30	56	-152	146	N	a-	1.0189	0.9863	-0.0096	253.5	-	-	6N	64W
9622	482	1988 Mar 03	16:13:41	56	-147	113	N*	-t	0.9885	1.0907	-0.0017	293.8	-	-	7N	120E
9623	482	1988 Aug 27	11:05:29	56	-141	118	P	-a	-0.8681	1.2380	0.2915	262.5	113.0	-	11S	165W
9624	482	1989 Feb 20	15:36:18	56	-135	123	T	-p	0.2934	2.3651	1.2747	367.7	223.1	78.5	11N	130E
9625	482	1989 Aug 17	03:09:07	57	-129	128	T-	-p	-0.1490	2.5703	1.5984	327.5	214.3	95.8	14S	46W
9626	482	1990 Feb 09	19:12:02	57	-123	133	T	a-	-0.4148	2.1191	1.0750	339.6	204.3	42.3	14N	76E
9627	482	1990 Aug 06	14:13:16	57	-117	138	P	t-	0.6374	1.7005	0.6766	322.0	175.5	-	16S	148E
9628	482	1991 Jan 30	05:59:38	58	-111	143	N	a-	-1.0752	0.8807	-0.1106	237.5	-	-	17N	87W
9629	482	1991 Jun 27	03:15:41	58	-106	110	N	-t	-1.4063	0.3126	-0.7572	169.5	-	-	25S	48W
9630	482	1991 Jul 26	18:08:50	58	-105	148	N	t-	1.4369	0.2542	-0.8110	152.7	-	-	18S	89E
9631	482	1991 Dec 21	10:34:00	58	-100	115	P	-a	0.9709	1.0651	0.0876	251.5	64.1	-	24N	159W
9632	482	1992 Jun 15	04:57:57	59	-94	120	P	-t	-0.6288	1.7264	0.6822	332.2	179.8	-	24S	74W
9633	482	1992 Dec 09	23:45:05	59	-88	125	T	-p	0.3144	2.2915	1.2709	334.1	208.7	73.9	23N	2E
9634	482	1993 Jun 04	13:01:26	59	-82	130	T+	p-	0.1638	2.5532	1.5617	336.3	217.8	95.8	22S	164E
9635	482	1993 Nov 29	06:27:06	60	-76	135	T	p-	-0.3994	2.1633	1.0876	354.4	210.8	46.7	21N	99W
9636	482	1994 May 25	03:31:20	60	-70	140	P	a-	0.8933	1.1941	0.2432	261.2	104.6	-	20S	53W
9637	482	1994 Nov 18	06:44:54	61	-64	145	N	t-	-1.1047	0.8815	-0.2189	271.6	-	-	18N	105W
9638	482	1995 Apr 15	12:19:04	61	-59	112	P	-a	-0.9593	1.0836	0.1114	256.3	73.0	-	11S	175E
9639	482	1995 Oct 08	16:05:12	61	-53	117	N	-t	1.1179	0.8252	-0.2115	247.6	-	-	7N	116E
9640	482	1996 Apr 04	00:10:47	62	-47	122	T-	-p	-0.2534	2.4068	1.3795	344.7	217.1	85.8	6S	2W
9641	483	1996 Sep 27	02:55:24	62	-41	127	T	-p	0.3426	2.2188	1.2395	320.9	203.3	69.2	2N	46W
9642	483	1997 Mar 24	04:40:28	62	-35	132	P	t-	0.4899	1.9994	0.9195	353.9	203.1	-	1S	68W
9643	483	1997 Sep 16	18:47:42	63	-29	137	T	p-	-0.3768	2.1417	1.1909	308.2	196.4	61.5	3S	77E
9644	483	1998 Mar 13	04:21:08	63	-23	142	N	t-	1.1964	0.7086	-0.3824	246.4	-	-	4N	62W
9645	483	1998 Aug 08	02:25:57	63	-18	109	N	-a	1.4875	0.1206	-0.8637	96.4	-	-	15S	35W
9646	483	1998 Sep 06	11:11:11	63	-17	147	N	a-	-1.1057	0.8121	-0.1544	227.8	-	-	7S	168W
9647	483	1999 Jan 31	16:18:35	63	-12	114	N*	-a	-1.0189	1.0027	-0.0258	261.7	-	-	16N	119E
9648	483	1999 Jul 28	11:34:46	64	-6	119	P	-t	0.7862	1.4342	0.3966	310.9	142.5	-	18S	172W
9649	483	2000 Jan 21	04:44:34	64	0	124	T	-p	-0.2957	2.3060	1.3246	318.2	203.3	77.0	20N	68W
9650	483	2000 Jul 16	13:56:39	64	6	129	T+	pp	0.0302	2.8375	1.7684	374.4	236.0	106.4	21S	153E
9651	483	2001 Jan 09	20:21:40	64	12	134	T	p-	0.3720	2.1618	1.1889	311.0	196.3	61.0	22N	57E
9652	483	2001 Jul 05	14:56:23	64	18	139	P	t-	-0.7287	1.5475	0.4947	325.1	159.3	-	23S	137E
9653	483	2001 Dec 30	10:30:22	64	24	144	N	a-	1.0731	0.8933	-0.1155	243.5	-	-	24N	157W
9654	483	2002 May 26	12:04:26	64	29	111	N	-a	1.1758	0.6893	-0.2888	216.6	-	-	20S	179E
9655	483	2002 Jun 24	21:28:13	64	30	149	N	a-	-1.4439	0.2095	-0.7925	129.1	-	-	25S	39E
9656	483	2002 Nov 20	01:47:40	64	35	116	N	-t	-1.1126	0.8600	-0.2264	264.3	-	-	19N	30W
9657	483	2003 May 16	03:41:13	64	41	121	T	-a	0.4123	2.0747	1.1276	306.5	193.9	51.4	19S	56W
9658	483	2003 Nov 09	01:19:38	64	47	126	T	-t	-0.4319	2.1139	1.0178	363.2	211.4	22.0	16N	24W
9659	483	2004 May 04	20:31:17	65	53	131	T	p-	-0.3132	2.2627	1.3035	315.7	203.2	75.5	17S	51E
9660	483	2004 Oct 28	03:05:11	65	59	136	T	p-	0.2846	2.3637	1.3081	353.8	218.7	80.5	13N	50W

Cat Num	Canon Plate	Calendar Date	TD of Greatest Eclipse	ΔT s	Luna Num	Saros Num	Ecl. Type	QSE	Gamma	Pen. Mag.	Um. Mag.	Eclipse Phase Durations Pen. m	Par. m	Total m	Greatest in Zenith Lat.	Lng.
9661	484	2005 Apr 24	09:55:55	65	65	141	N	h-	-1.0885	0.8650	-0.1436	245.6	-	-	14S	150W
9662	484	2005 Oct 17	12:04:27	65	71	146	P	a-	0.9796	1.0585	0.0625	259.8	56.0	-	10N	175E
9663	484	2006 Mar 14	23:48:34	65	76	113	N*	-t	1.0210	1.0300	-0.0604	287.5	-	-	3N	6E
9664	484	2006 Sep 07	18:52:25	65	82	118	P	-a	-0.9262	1.1329	0.1837	254.4	91.1	-	7S	77E
9665	484	2007 Mar 03	23:21:59	65	88	123	T	-p	0.3175	2.3188	1.2328	365.4	221.1	73.4	7N	13E
9666	484	2007 Aug 28	10:38:27	66	94	128	T-	-p	-0.2145	2.4526	1.4758	327.3	212.2	90.0	10S	159W
9667	484	2008 Feb 21	03:27:09	66	100	133	T	a-	-0.3992	2.1451	1.1062	339.0	205.4	49.8	10N	48W
9668	484	2008 Aug 16	21:11:12	66	106	138	P	t-	0.5646	1.8366	0.8076	330.5	188.1	-	13S	43E
9669	484	2009 Feb 09	14:39:22	66	112	143	N	a-	-1.0640	0.8994	-0.0882	238.8	-	-	14N	144E
9670	484	2009 Jul 07	09:39:43	66	117	110	N	-t	-1.4915	0.1562	-0.9133	121.5	-	-	24S	143W
9671	484	2009 Aug 06	00:40:18	66	118	148	N	t-	1.3572	0.4019	-0.6661	189.8	-	-	16S	9W
9672	484	2009 Dec 31	19:23:46	67	123	115	P	-a	0.9765	1.0556	0.0763	251.1	60.0	-	24N	70E
9673	484	2010 Jun 26	11:39:34	67	129	120	P	-t	-0.7091	1.5773	0.5368	322.1	162.9	-	24S	174W
9674	484	2010 Dec 21	08:18:04	67	135	125	T	-p	0.3214	2.2807	1.2561	335.1	208.7	72.3	24N	125W
9675	484	2011 Jun 15	20:13:43	67	141	130	T+	pp	0.0897	2.6868	1.6999	336.1	219.3	100.2	23S	57E
9676	484	2011 Dec 10	14:32:56	68	147	135	T	p-	-0.3882	2.1860	1.1061	356.4	212.2	51.1	23N	140E
9677	484	2012 Jun 04	11:04:20	68	153	140	P	a-	0.8247	1.3183	0.3704	270.0	126.6	-	22S	166W
9678	484	2012 Nov 28	14:34:07	68	159	145	N	t-	-1.0869	0.9155	-0.1873	276.0	-	-	20N	139E
9679	484	2013 Apr 25	20:08:38	68	164	112	P	-a	-1.0121	0.9866	0.0148	247.7	27.0	-	14S	57E
9680	484	2013 May 25	04:11:06	68	165	150	Nb	a-	1.5350	0.0157	-0.9335	33.6	-	-	19S	63W
9681	485	2013 Oct 18	23:51:25	68	170	117	N	-h	1.1508	0.7649	-0.2718	239.1	-	-	11N	2W
9682	485	2014 Apr 15	07:46:48	69	176	122	T	-a	-0.3017	2.3182	1.2907	343.9	214.7	77.8	10S	116W
9683	485	2014 Oct 08	10:55:44	69	182	127	T	-p	0.3826	2.1456	1.1659	318.1	199.5	58.8	6N	167W
9684	485	2015 Apr 04	12:01:24	69	188	132	T	t-	0.4460	2.0792	1.0008	357.5	209.0	4.7	5S	179W
9685	485	2015 Sep 28	02:48:17	69	194	137	T	p-	-0.3296	2.2296	1.2764	310.7	199.9	71.9	2N	44W
9686	485	2016 Mar 23	11:48:21	70	200	142	N	t-	1.1591	0.7747	-0.3118	255.4	-	-	0S	175W
9687	485	2016 Sep 16	18:55:27	70	206	147	N	a-	-1.0548	0.9080	-0.0635	239.3	-	-	3S	75E
9688	485	2017 Feb 11	00:45:03	70	211	114	N	-a	-1.0254	0.9884	-0.0354	259.2	-	-	13N	8W
9689	485	2017 Aug 07	18:21:38	70	217	119	P	-t	0.8668	1.2886	0.2464	300.9	115.2	-	15S	86E
9690	485	2018 Jan 31	13:31:00	71	223	124	T	-p	-0.3014	2.2941	1.3155	317.2	202.7	76.1	17N	161E
9691	485	2018 Jul 27	20:22:54	71	229	129	T+	pp	0.1168	2.6792	1.6087	373.8	234.5	103.0	19S	56E
9692	485	2019 Jan 21	05:13:27	71	235	134	T	p-	0.3684	2.1684	1.1953	311.5	196.8	62.0	20N	75W
9693	485	2019 Jul 16	21:31:55	71	241	139	P	t-	-0.6430	1.7037	0.6531	333.7	177.9	-	22S	39E
9694	485	2020 Jan 10	19:11:11	72	247	144	N	a-	1.0726	0.8956	-0.1160	244.6	-	-	23N	74E
9695	485	2020 Jun 05	19:26:14	72	252	111	N	-a	1.2406	0.5683	-0.4053	198.2	-	-	21S	69E
9696	485	2020 Jul 05	04:31:12	72	253	149	N	a-	-1.3638	0.3546	-0.6436	165.0	-	-	24S	66W
9697	485	2020 Nov 30	09:44:01	72	258	116	N	-t	-1.1309	0.8285	-0.2620	261.0	-	-	21N	148W
9698	485	2021 May 26	11:19:53	72	264	121	T	-a	0.4774	1.9540	1.0095	302.0	187.4	14.5	21S	170W
9699	485	2021 Nov 19	09:04:06	73	270	126	P	-t	-0.4552	2.0720	0.9742	361.5	208.4	-	19N	139W
9700	485	2022 May 16	04:12:42	73	276	131	T-	p-	-0.2532	2.3726	1.4137	318.7	207.2	84.9	19S	64W
9701	486	2022 Nov 08	11:00:22	73	282	136	T+	p-	0.2570	2.4143	1.3589	353.9	219.8	85.0	17N	169W
9702	486	2023 May 05	17:24:05	73	288	141	N	h-	-1.0349	0.9636	-0.0457	257.5	-	-	17S	98E
9703	486	2023 Oct 28	20:15:18	74	294	146	P	a-	0.9471	1.1181	0.1220	264.6	77.4	-	14N	52E
9704	486	2024 Mar 25	07:13:59	74	299	113	N	-t	1.0609	0.9557	-0.1325	279.1	-	-	1S	106W
9705	486	2024 Sep 18	02:45:25	74	305	118	P	-a	-0.9792	1.0372	0.0848	246.3	62.8	-	3S	42W
9706	486	2025 Mar 14	06:59:56	75	311	123	T	-p	0.3484	2.2595	1.1784	362.6	218.3	65.4	3N	102W
9707	486	2025 Sep 07	18:12:58	75	317	128	T	-p	-0.2752	2.3440	1.3619	326.7	209.4	82.1	6S	87E
9708	486	2026 Mar 03	11:34:52	75	323	133	T	a-	-0.3765	2.1838	1.1507	338.6	207.2	58.3	6N	171W
9709	486	2026 Aug 28	04:14:04	75	329	138	P	t-	0.4964	1.9645	0.9299	337.8	198.1	-	9S	63W
9710	486	2027 Feb 20	23:14:06	76	335	143	N	a-	-1.0480	0.9266	-0.0569	241.0	-	-	10N	15E
9711	486	2027 Jul 18	16:04:09	76	340	110	Ne	-t	-1.5758	0.0014	-1.0680	11.8	-	-	22S	121E
9712	486	2027 Aug 17	07:14:59	76	341	148	N	t-	1.2797	0.5456	-0.5254	218.6	-	-	12S	108W
9713	486	2028 Jan 12	04:14:13	76	346	115	P	-a	0.9817	1.0468	0.0662	250.7	56.0	-	23N	61W
9714	486	2028 Jul 06	18:20:57	77	352	120	P	-t	-0.7903	1.4266	0.3892	310.6	141.5	-	23S	86E
9715	486	2028 Dec 31	16:53:15	77	358	125	T	-p	0.3258	2.2742	1.2463	336.2	208.8	71.3	23N	108E
9716	486	2029 Jun 26	03:23:22	77	364	130	T+	pp	0.0124	2.8266	1.8436	335.1	219.5	101.9	23S	50W
9717	486	2029 Dec 20	22:43:12	78	370	135	T	p-	-0.3811	2.2008	1.1174	358.0	213.3	53.7	23N	19E
9718	486	2030 Jun 15	18:34:34	78	376	140	P	a-	0.7534	1.4480	0.5025	278.2	144.4	-	23S	82E
9719	486	2030 Dec 09	22:28:51	78	382	145	N	t-	-1.0731	0.9416	-0.1628	279.2	-	-	22N	21E
9720	486	2031 May 07	03:52:02	78	387	112	N	-a	-1.0694	0.8814	-0.0904	237.3	-	-	18S	59W

Cat Num	Canon Plate	Calendar Date	TD of Greatest Eclipse	ΔT s	Luna Num	Saros Num	Ecl. Type	QSE	Gamma	Pen. Mag.	Um. Mag.	Eclipse Phase Durations Pen. m	Par. m	Total m	Greatest in Zenith Lat.	Lng.
9721	487	2031 Jun 05	11:45:17	78	388	150	N	a-	1.4731	0.1292	-0.8199	95.6	-	-	21S	176W
9722	487	2031 Oct 30	07:46:45	79	393	117	N	-h	1.1773	0.7161	-0.3204	231.8	-	-	15N	121W
9723	487	2032 Apr 25	15:14:51	79	399	122	T	-a	-0.3558	2.2192	1.1913	342.4	211.2	65.5	14S	131E
9724	487	2032 Oct 18	19:03:40	79	405	127	T	-p	0.4169	2.0830	1.1028	315.4	195.9	47.1	10N	71E
9725	487	2033 Apr 14	19:13:51	80	411	132	T	t-	0.3954	2.1711	1.0944	361.2	215.0	49.2	9S	72E
9726	487	2033 Oct 08	10:56:23	80	417	137	T	p-	-0.2889	2.3057	1.3497	312.6	202.4	78.8	6N	167W
9727	487	2034 Apr 03	19:06:59	80	423	142	N	t-	1.1144	0.8545	-0.2274	265.4	-	-	5S	75E
9728	487	2034 Sep 28	02:47:37	81	429	147	P	a-	-1.0110	0.9911	0.0144	248.7	26.7	-	1N	44W
9729	487	2035 Feb 22	09:06:12	81	434	114	N	-a	-1.0367	0.9652	-0.0535	255.7	-	-	9N	133W
9730	487	2035 Aug 19	01:12:15	81	440	119	P	-t	0.9433	1.1507	0.1037	289.8	76.5	-	12S	17W
9731	487	2036 Feb 11	22:13:06	82	446	124	T	-p	-0.3110	2.2751	1.2995	316.1	201.9	74.5	14N	31E
9732	487	2036 Aug 07	02:52:32	82	452	129	T+	pp	0.2004	2.5266	1.4544	372.1	231.3	95.3	16S	41W
9733	487	2037 Jan 31	14:01:38	82	458	134	T	p-	0.3619	2.1803	1.2074	312.1	197.5	63.7	18N	153E
9734	487	2037 Jul 27	04:09:53	83	464	139	P	t-	-0.5582	1.8584	0.8095	340.8	192.4	-	20S	60W
9735	487	2038 Jan 21	03:49:52	83	470	144	N	a-	1.0710	0.8996	-0.1140	245.8	-	-	21N	54W
9736	487	2038 Jun 17	02:45:02	83	475	111	N	-a	1.3082	0.4422	-0.5275	176.3	-	-	22S	41W
9737	487	2038 Jul 16	11:35:56	84	476	149	N	a-	-1.2837	0.4999	-0.4952	192.4	-	-	23S	172W
9738	487	2038 Dec 11	17:45:00	84	481	116	N	-t	-1.1448	0.8046	-0.2892	258.5	-	-	22N	93E
9739	487	2039 Jun 06	18:54:25	84	487	121	P	-a	0.5460	1.8272	0.8846	296.7	179.3	-	22S	77E
9740	487	2039 Nov 30	16:56:28	85	493	126	P	-t	-0.4721	2.0418	0.9426	360.1	206.0	-	21N	104E
9741	488	2040 May 26	11:46:22	85	499	131	T-	p-	-0.1872	2.4938	1.5348	321.4	210.7	92.2	21S	177W
9742	488	2040 Nov 18	19:04:40	85	505	136	T+	p-	0.2361	2.4525	1.3974	353.6	220.4	87.8	20N	70E
9743	488	2041 May 16	00:43:03	86	511	141	P	t-	-0.9746	1.0747	0.0645	269.7	58.5	-	20S	12W
9744	488	2041 Nov 08	04:35:05	86	517	146	P	a-	0.9212	1.1656	0.1696	268.0	90.3	-	18N	73W
9745	488	2042 Apr 05	14:30:11	86	522	113	N	-t	1.1080	0.8680	-0.2176	268.4	-	-	5S	144E
9746	488	2042 Sep 29	10:45:47	87	528	118	N	-a	-1.0261	0.9528	-0.0031	238.5	-	-	2N	163W
9747	488	2043 Mar 25	14:32:04	87	534	123	T	-t	0.3849	2.1900	1.1142	359.3	214.6	53.4	2S	144E
9748	488	2043 Sep 19	01:51:50	88	540	128	T	-a	-0.3316	2.2433	1.2556	325.8	206.0	71.7	2S	29W
9749	488	2044 Mar 13	19:38:33	88	546	133	T	a-	-0.3496	2.2303	1.2031	338.4	209.1	66.4	2N	68E
9750	488	2044 Sep 07	11:20:44	88	552	138	T	t-	0.4318	2.0860	1.0456	344.0	206.2	33.9	5S	171W
9751	488	2045 Mar 03	07:43:26	89	558	143	N	a-	-1.0274	0.9623	-0.0168	243.9	-	-	6N	113W
9752	488	2045 Aug 27	13:54:50	89	564	148	N	t-	1.2060	0.6825	-0.3919	241.7	-	-	9S	152E
9753	488	2046 Jan 22	13:02:37	90	569	115	P	-a	0.9885	1.0347	0.0532	250.0	50.4	-	21N	168E
9754	488	2046 Jul 18	01:06:05	90	575	120	P	-t	-0.8691	1.2807	0.2461	298.1	114.6	-	22S	14W
9755	488	2047 Jan 12	01:26:14	90	581	125	T	-p	0.3317	2.2649	1.2341	337.2	208.9	70.0	22N	19W
9756	488	2047 Jul 07	10:35:45	91	587	130	T-	pp	-0.0636	2.7310	1.7513	333.4	218.5	100.8	23S	157W
9757	488	2048 Jan 01	06:53:55	91	593	135	T	p-	-0.3745	2.2141	1.1280	359.4	214.3	55.9	23N	102W
9758	488	2048 Jun 26	02:02:28	92	599	140	P	a-	0.6796	1.5825	0.6388	285.7	159.2	-	23S	30W
9759	488	2048 Dec 20	06:27:48	92	605	145	N	t-	-1.0624	0.9617	-0.1436	281.6	-	-	22N	97W
9760	488	2049 May 17	11:26:39	92	610	112	N	-a	-1.1337	0.7638	-0.2085	224.3	-	-	21S	172W
9761	489	2049 Jun 15	19:14:12	92	611	150	N	a-	1.4068	0.2511	-0.6985	132.0	-	-	22S	72E
9762	489	2049 Nov 09	15:52:11	93	616	117	N	-h	1.1964	0.6808	-0.3553	226.1	-	-	18N	118E
9763	489	2050 May 06	22:32:02	94	622	122	T	-h	-0.4181	2.1052	1.0767	340.0	206.0	43.2	17S	21E
9764	489	2050 Oct 30	03:21:47	95	628	127	T	-p	0.4435	2.0345	1.0538	313.1	192.9	34.5	14N	54W
9765	489	2051 Apr 26	02:16:28	96	634	132	T	p-	0.3371	2.2773	1.2022	364.8	220.8	69.6	13S	34W
9766	489	2051 Oct 19	19:11:50	97	640	137	T-	p-	-0.2542	2.3708	1.4118	314.2	204.3	83.6	10N	69E
9767	489	2052 Apr 14	02:18:06	98	646	142	N	t-	1.0628	0.9466	-0.1305	276.0	-	-	9S	34W
9768	489	2052 Oct 08	10:45:58	99	652	147	P	a-	-0.9726	1.0642	0.0821	256.6	63.3	-	5N	164W
9769	489	2053 Mar 04	17:22:10	99	657	114	N	-a	-1.0530	0.9323	-0.0808	251.1	-	-	5N	102E
9770	489	2053 Aug 29	08:05:50	100	663	119	N*	-t	1.0164	1.0191	-0.0330	277.8	-	-	8S	121W
9771	489	2054 Feb 22	06:51:27	101	669	124	T	-p	-0.3242	2.2491	1.2769	314.7	200.9	72.1	10N	99W
9772	489	2054 Aug 18	09:26:30	102	675	129	T	pp	0.2806	2.3805	1.3062	369.5	226.5	82.9	13S	140W
9773	489	2055 Feb 11	22:46:17	103	681	134	T	p-	0.3526	2.1970	1.2246	312.9	198.4	66.0	14N	22E
9774	489	2055 Aug 07	10:53:18	104	687	139	P	t-	-0.4769	2.0069	0.9594	346.3	203.4	-	17S	161W
9775	489	2056 Feb 01	12:26:06	105	693	144	N	a-	1.0682	0.9056	-0.1096	247.2	-	-	18N	177E
9776	489	2056 Jun 27	10:03:09	106	698	111	N	-a	1.3769	0.3143	-0.6519	149.9	-	-	22S	149W
9777	489	2056 Jul 26	18:43:24	106	699	149	N	a-	-1.2048	0.6435	-0.3489	214.4	-	-	20S	81E
9778	489	2056 Dec 22	01:48:56	107	704	116	N	-t	-1.1559	0.7857	-0.3109	256.4	-	-	22N	27W
9779	489	2057 Jun 17	02:26:20	108	710	121	P	-a	0.6167	1.6967	0.7555	290.6	169.3	-	23S	36W
9780	489	2057 Dec 11	00:53:38	109	716	126	P	-t	-0.4853	2.0178	0.9181	358.8	204.0	-	23N	15W

Cat Num	Canon Plate	Calendar Date	TD of Greatest Eclipse	ΔT s	Luna Num	Saros Num	Ecl. Type	QSE	Gamma	Pen. Mag.	Um. Mag.	Eclipse Phase Durations Pen. m	Par. m	Total m	Greatest in Zenith Lat.	Lng.
9781	490	2058 Jun 06	19:15:48	110	722	131	T-	pp	-0.1181	2.6210	1.6611	323.6	213.4	97.3	23S	71E
9782	490	2058 Nov 30	03:16:18	111	728	136	T+	p-	0.2208	2.4802	1.4260	353.0	220.7	89.7	22N	52W
9783	490	2059 May 27	07:55:35	112	734	141	P	t-	-0.9097	1.1946	0.1829	281.7	97.2	-	22S	119W
9784	490	2059 Nov 19	13:01:36	113	740	146	P	a-	0.9004	1.2037	0.2079	270.5	99.2	-	20N	161E
9785	490	2060 Apr 15	21:37:04	114	745	113	N	-t	1.1621	0.7674	-0.3156	255.0	-	-	9S	37E
9786	490	2060 Oct 09	18:53:32	115	751	118	N	-a	-1.0670	0.8796	-0.0799	231.3	-	-	6N	74E
9787	490	2060 Nov 08	04:04:15	115	752	156	N	a-	1.5332	0.0266	-0.9375	43.6	-	-	18N	65W
9788	490	2061 Apr 04	21:54:05	116	757	123	T	-t	0.4300	2.1044	1.0341	355.0	209.6	29.9	6S	33E
9789	490	2061 Sep 29	09:38:13	117	763	128	T	-a	-0.3810	2.1556	1.1621	324.8	202.4	59.0	2N	146W
9790	490	2062 Mar 25	03:33:50	118	769	133	T	p-	-0.3150	2.2905	1.2695	338.3	211.3	74.7	2S	52W
9791	490	2062 Sep 18	18:34:02	119	775	138	T	p-	0.3735	2.1959	1.1496	349.3	212.4	59.5	1S	80E
9792	490	2063 Mar 14	16:05:49	120	781	143	P	a-	-1.0007	1.0088	0.0342	247.8	40.6	-	1N	121E
9793	490	2063 Sep 07	20:41:12	121	787	148	N	t-	1.1374	0.8101	-0.2678	260.4	-	-	5S	49E
9794	490	2064 Feb 02	21:48:57	122	792	115	P	-a	0.9969	1.0197	0.0377	249.0	42.5	-	18N	37E
9795	490	2064 Jul 28	07:52:48	123	798	120	P	-t	-0.9473	1.1361	0.1038	284.3	75.7	-	20S	116W
9796	490	2065 Jan 22	09:58:58	124	804	125	T	-p	0.3371	2.2561	1.2231	338.2	209.0	68.8	20N	146W
9797	490	2065 Jul 17	17:48:40	125	810	130	T-	pp	-0.1402	2.5890	1.6121	331.0	216.3	97.0	21S	95E
9798	490	2066 Jan 11	15:04:47	126	816	135	T	p-	-0.3687	2.2259	1.1378	360.7	215.2	57.9	21N	136E
9799	490	2066 Jul 07	09:30:29	127	822	140	P	a-	0.6055	1.7179	0.7753	292.3	171.3	-	22S	141W
9800	490	2066 Dec 31	14:30:10	128	828	145	N	t-	-1.0539	0.9773	-0.1281	283.3	-	-	22N	144E
9801	491	2067 May 28	18:56:08	129	833	112	N	-a	-1.2012	0.6403	-0.3329	208.5	-	-	23S	76E
9802	491	2067 Jun 27	02:41:06	129	834	150	N	a-	1.3394	0.3754	-0.5753	159.8	-	-	22S	39W
9803	491	2067 Nov 21	00:04:42	130	839	117	N	-h	1.2106	0.6544	-0.3811	221.5	-	-	21N	4W
9804	491	2068 May 17	05:42:17	131	845	122	P	-t	-0.4851	1.9826	0.9532	336.6	199.0	-	20S	86W
9805	491	2068 Nov 09	11:47:00	132	851	127	T	-p	0.4645	1.9962	1.0149	311.2	190.2	18.4	18N	180E
9806	491	2069 May 06	09:09:57	133	857	132	T+	pp	0.2717	2.3965	1.3229	368.1	226.2	84.3	17S	138W
9807	491	2069 Oct 30	03:35:06	134	863	137	T-	p-	-0.2263	2.4235	1.4616	315.4	205.6	86.8	14N	57W
9808	491	2070 Apr 25	09:21:24	135	869	142	N*	t-	1.0044	1.0515	-0.0209	286.9	-	-	12S	140W
9809	491	2070 Oct 19	18:51:12	137	875	147	P	a-	-0.9406	1.1258	0.1383	263.2	81.7	-	9N	74E
9810	491	2071 Mar 16	01:31:09	137	880	114	N	-a	-1.0756	0.8879	-0.1194	245.1	-	-	1N	20W
9811	491	2071 Sep 09	15:05:41	138	886	119	N	-t	1.0834	0.8989	-0.1586	265.2	-	-	4S	133E
9812	491	2072 Mar 04	15:23:07	140	892	124	T	-p	-0.3430	2.2127	1.2441	313.2	199.4	68.5	6N	133E
9813	491	2072 Aug 28	16:05:42	141	898	129	T	-t	0.3563	2.2428	1.1662	366.0	220.3	64.2	9S	119E
9814	491	2073 Feb 22	07:24:53	142	904	134	T	p-	0.3388	2.2218	1.2503	313.8	199.7	69.2	10N	107W
9815	491	2073 Aug 17	17:42:41	143	910	139	T	t-	-0.3998	2.1479	1.1013	350.5	211.6	50.1	13S	96E
9816	491	2074 Feb 11	20:55:58	144	916	144	N	a-	1.0611	0.9191	-0.0972	249.5	-	-	15N	50E
9817	491	2074 Jul 08	17:21:38	145	921	111	N	-a	1.4456	0.1870	-0.7765	116.6	-	-	21S	101E
9818	491	2074 Aug 07	01:56:03	145	922	149	N	a-	-1.1291	0.7813	-0.2091	232.2	-	-	17S	27W
9819	491	2075 Jan 02	09:55:03	146	927	116	N	-t	-1.1642	0.7714	-0.3271	254.9	-	-	22N	147W
9820	491	2075 Jun 28	09:55:35	147	933	121	P	-a	0.6897	1.5624	0.6220	283.4	157.0	-	23S	147W
9821	492	2075 Dec 22	08:55:55	148	939	126	P	-t	-0.4945	2.0008	0.9013	357.6	202.5	-	23N	134W
9822	492	2076 Jun 17	02:39:47	149	945	131	T-	pp	-0.0452	2.7554	1.7943	325.3	215.1	100.2	23S	39W
9823	492	2076 Dec 10	11:34:51	150	951	136	T+	p-	0.2102	2.4990	1.4460	352.2	220.6	90.8	23N	175W
9824	492	2077 Jun 06	14:59:52	151	957	141	P	t-	-0.8387	1.3257	0.3123	293.6	125.0	-	24S	135E
9825	492	2077 Nov 29	21:35:53	152	963	146	P	a-	0.8854	1.2309	0.2356	272.0	105.0	-	23N	34E
9826	492	2078 Apr 27	04:35:44	153	968	113	N	-t	1.2222	0.6558	-0.4246	238.2	-	-	13S	68W
9827	492	2078 Oct 21	03:08:03	154	974	118	N	-a	-1.1021	0.8171	-0.1462	224.8	-	-	10N	50W
9828	492	2078 Nov 19	12:40:04	154	975	156	N	a-	1.5147	0.0615	-0.9047	66.0	-	-	21N	166E
9829	492	2079 Apr 16	05:10:45	155	980	123	P	-t	0.4799	2.0100	0.9451	350.1	203.4	-	10S	77W
9830	492	2079 Oct 10	17:30:30	156	986	128	T	-a	-0.4246	2.0786	1.0791	323.8	198.7	42.4	7N	95E
9831	492	2080 Apr 04	11:23:38	157	992	133	T	p-	-0.2751	2.3607	1.3460	338.3	213.6	82.1	6S	170W
9832	492	2080 Sep 29	01:52:42	158	998	138	T	p-	0.3203	2.2967	1.2443	353.7	217.4	73.8	3N	30W
9833	492	2081 Mar 25	00:22:01	159	1004	143	P	a-	-0.9687	1.0652	0.0953	252.4	67.1	-	3S	4W
9834	492	2081 Sep 18	03:35:26	161	1010	148	N	t-	1.0747	0.9270	-0.1545	275.7	-	-	1S	55W
9835	492	2082 Feb 13	06:29:19	161	1015	115	P	-a	1.0101	0.9955	0.0134	247.2	25.5	-	14N	93W
9836	492	2082 Aug 08	14:46:42	163	1021	120	N*	-t	-1.0203	1.0011	-0.0294	269.8	-	-	17S	141E
9837	492	2083 Feb 02	18:26:46	164	1027	125	T	-p	0.3463	2.2400	1.2052	338.9	208.8	66.5	17N	88E
9838	492	2083 Jul 29	01:05:34	165	1033	130	T-	pp	-0.2143	2.4520	1.4773	328.0	212.9	90.4	19S	14W
9839	492	2084 Jan 22	23:13:00	166	1039	135	T	p-	-0.3610	2.2407	1.1513	362.0	216.3	60.5	19N	15E
9840	492	2084 Jul 17	16:58:51	167	1045	140	P	a-	0.5312	1.8540	0.9119	298.1	181.4	-	20S	107E

Cat Num	Canon Plate	Calendar Date	TD of Greatest Eclipse	ΔT s	Luna Num	Saros Num	Ecl. Type	QSE	Gamma	Pen. Mag.	Um. Mag.	Eclipse Phase Durations Pen. m	Par. m	Total m	Greatest in Zenith Lat.	Lng.
9841	493	2085 Jan 10	22:32:29	168	1051	145	N	t-	-1.0453	0.9927	-0.1119	284.9	-	-	21N	24E
9842	493	2085 Jun 08	02:17:36	169	1056	112	N	-a	-1.2745	0.5065	-0.4682	188.5	-	-	24S	34W
9843	493	2085 Jul 07	10:04:40	169	1057	150	N	a-	1.2694	0.5047	-0.4478	183.5	-	-	21S	149W
9844	493	2085 Dec 01	08:25:35	170	1062	117	N	-a	1.2189	0.6387	-0.3957	218.5	-	-	23N	128W
9845	493	2086 May 28	12:43:47	171	1068	122	P	-t	-0.5585	1.8486	0.8180	332.0	189.4	-	22S	169E
9846	493	2086 Nov 20	20:19:42	172	1074	127	P	-p	0.4799	1.9679	0.9865	309.5	188.1	-	20N	52E
9847	493	2087 May 17	15:55:20	173	1080	132	T+	pp	0.1999	2.5276	1.4554	371.0	230.6	95.1	19S	121E
9848	493	2087 Nov 10	12:05:33	174	1086	137	T-	p-	-0.2043	2.4654	1.5006	316.4	206.6	88.9	17N	175E
9849	493	2088 May 05	16:16:50	175	1092	142	P	t-	0.9387	1.1695	0.1019	297.9	77.1	-	16S	116E
9850	493	2088 Oct 30	03:03:20	177	1098	147	P	a-	-0.9147	1.1761	0.1831	268.5	93.6	-	13N	49W
9851	493	2089 Mar 26	09:34:14	178	1103	114	N	-a	-1.1038	0.8332	-0.1681	237.8	-	-	4S	142W
9852	493	2089 Sep 19	22:11:17	179	1109	119	N	-t	1.1447	0.7893	-0.2737	252.2	-	-	0N	26E
9853	493	2090 Mar 15	23:48:31	180	1115	124	T	-p	-0.3674	2.1659	1.2012	311.3	197.5	63.0	1N	6E
9854	493	2090 Sep 08	22:52:29	181	1121	129	T	-t	0.4257	2.1167	1.0377	362.0	213.1	31.9	5S	17E
9855	493	2091 Mar 05	15:58:22	182	1127	134	T	p-	0.3212	2.2537	1.2832	315.0	201.3	72.9	6N	124E
9856	493	2091 Aug 29	00:38:25	183	1133	139	T	t-	-0.3270	2.2810	1.2351	353.4	217.5	72.9	10S	8W
9857	493	2092 Feb 23	05:20:59	184	1139	144	N	a-	1.0509	0.9383	-0.0789	252.4	-	-	11N	76W
9858	493	2092 Jul 19	00:41:58	185	1144	111	Ne	-a	1.5131	0.0620	-0.8992	67.7	-	-	19S	8W
9859	493	2092 Aug 17	09:13:59	185	1145	149	N	a-	-1.0568	0.9131	-0.0757	246.7	-	-	14S	136W
9860	493	2093 Jan 12	18:00:03	186	1150	116	N	-t	-1.1733	0.7553	-0.3444	253.1	-	-	20N	93E
9861	494	2093 Jul 08	17:24:18	187	1156	121	P	-a	0.7632	1.4275	0.4872	275.3	141.9	-	22S	101E
9862	494	2094 Jan 01	17:00:06	188	1162	126	P	-t	-0.5024	1.9858	0.8871	356.5	201.2	-	22N	107E
9863	494	2094 Jun 28	10:01:57	190	1168	131	T+	pp	0.0288	2.7865	1.8234	326.5	215.7	100.6	23S	149W
9864	494	2094 Dec 21	19:56:32	191	1174	136	T+	p-	0.2016	2.5138	1.4627	351.2	220.5	91.6	24N	61E
9865	494	2095 Jun 17	22:00:11	192	1180	141	P	t-	-0.7653	1.4617	0.4459	304.7	146.9	-	24S	31E
9866	494	2095 Dec 11	06:15:02	193	1186	146	P	a-	0.8742	1.2510	0.2565	272.9	108.9	-	24N	95W
9867	494	2096 May 07	11:24:42	194	1191	113	N	-t	1.2896	0.5309	-0.5469	216.9	-	-	16S	171W
9868	494	2096 Jun 06	02:43:41	194	1192	151	Nb	t-	-1.5723	0.0047	-1.0584	21.2	-	-	24S	41W
9869	494	2096 Oct 31	11:30:23	195	1197	118	N	-a	-1.1307	0.7666	-0.2006	219.3	-	-	13N	175W
9870	494	2096 Nov 29	21:22:22	195	1198	156	N	a-	1.5017	0.0862	-0.8816	78.1	-	-	23N	37E
9871	494	2097 Apr 26	12:18:17	196	1203	123	P	-t	0.5377	1.9013	0.8420	344.0	195.2	-	13S	176E
9872	494	2097 Oct 21	01:30:55	197	1209	128	T	-a	-0.4608	2.0152	1.0097	323.1	195.2	15.2	11N	26W
9873	494	2098 Apr 15	19:04:48	198	1215	133	T-	p-	-0.2272	2.4454	1.4369	338.3	215.8	89.0	10S	74E
9874	494	2098 Oct 10	09:19:58	200	1221	138	T	pp	0.2749	2.3831	1.3246	357.4	221.0	82.7	7N	143W
9875	494	2099 Apr 05	08:30:56	201	1227	143	P	a-	-0.9304	1.1333	0.1680	257.7	88.1	-	7S	127W
9876	494	2099 Sep 29	10:36:38	202	1233	148	N*	t-	1.0174	1.0340	-0.0512	288.3	-	-	3N	161W
9877	494	2100 Feb 24	15:05:11	203	1238	115	N	-a	1.0267	0.9649	-0.0170	244.6	-	-	10N	138E
9878	494	2100 Aug 19	21:44:58	204	1244	120	N	-t	-1.0905	0.8716	-0.1575	254.2	-	-	13S	36E
9879	494	2101 Feb 14	02:50:00	205	1250	125	T	-a	0.3584	2.2184	1.1825	339.4	208.4	63.4	13N	38W
9880	494	2101 Aug 09	08:25:33	206	1256	130	T	-p	-0.2864	2.3189	1.3458	324.4	208.4	80.7	16S	124W
9881	495	2102 Feb 03	07:18:21	207	1262	135	T	p-	-0.3514	2.2585	1.1686	363.2	217.5	63.6	16N	105W
9882	495	2102 Jul 30	00:29:10	209	1268	140	T	a-	0.4586	1.9873	1.0451	303.0	189.5	31.3	18S	5W
9883	495	2103 Jan 23	06:34:00	210	1274	145	N*	t-	-1.0357	1.0095	-0.0937	286.5	-	-	19N	95W
9884	495	2103 Jun 20	09:36:11	211	1279	112	N	-a	-1.3492	0.3704	-0.6062	163.9	-	-	25S	143W
9885	495	2103 Jul 19	17:28:43	211	1280	150	N	a-	1.2002	0.6328	-0.3220	203.4	-	-	20S	100E
9886	495	2103 Dec 13	16:51:37	212	1285	117	N	-a	1.2239	0.6287	-0.4042	216.3	-	-	24N	106E
9887	495	2104 Jun 08	19:38:40	213	1291	122	P	-t	-0.6362	1.7069	0.6746	325.9	176.8	-	23S	66E
9888	495	2104 Dec 02	04:58:11	214	1297	127	P	-a	0.4910	1.9476	0.9661	308.2	186.5	-	22N	76W
9889	495	2105 May 28	22:34:06	215	1303	132	T+	pp	0.1227	2.6687	1.5976	373.1	233.9	102.3	21S	22E
9890	495	2105 Nov 21	20:42:00	216	1309	137	T-	p-	-0.1874	2.4976	1.5301	317.3	207.3	90.4	20N	47E
9891	495	2106 May 17	23:06:43	218	1315	142	P	t-	0.8677	1.2975	0.2345	308.4	114.5	-	19S	13E
9892	495	2106 Nov 11	11:22:14	219	1321	147	P	a-	-0.8947	1.2153	0.2171	272.8	101.7	-	16N	174W
9893	495	2107 Apr 07	17:30:11	220	1326	114	N	-a	-1.1382	0.7671	-0.2283	228.7	-	-	8S	99E
9894	495	2107 May 07	04:30:24	220	1327	152	Nb	a-	1.5588	0.0058	-1.0103	22.2	-	-	15S	67W
9895	495	2107 Oct 02	05:23:19	221	1332	119	N	-t	1.1997	0.6911	-0.3775	239.2	-	-	4N	83W
9896	495	2108 Mar 27	08:06:28	222	1338	124	T	-p	-0.3982	2.1076	1.1467	309.0	194.8	54.7	3S	119W
9897	495	2108 Sep 20	05:47:11	223	1344	129	P	-t	0.4884	2.0030	0.9213	357.6	205.1	-	1S	88W
9898	495	2109 Mar 17	00:22:28	224	1350	134	T	p-	0.2962	2.2989	1.3296	316.4	203.3	77.5	2N	2W
9899	495	2109 Sep 09	07:43:03	226	1356	139	T-	p-	-0.2608	2.4024	1.3568	355.2	221.6	85.6	6S	115W
9900	495	2110 Mar 06	13:37:20	227	1362	144	N	a-	1.0345	0.9686	-0.0491	256.4	-	-	7N	160E

Cat Num	Canon Plate	Calendar Date	TD of Greatest Eclipse	ΔT s	Luna Num	Saros Num	Ecl. Type	QSE	Gamma	Pen. Mag.	Um. Mag.	Eclipse Phase Durations Pen. m	Par. m	Total m	Greatest in Zenith Lat.	Lng.
9901	496	2110 Aug 29	16:38:48	228	1368	149	P	a-	-0.9893	1.0365	0.0488	258.6	49.8	-	10S	112E
9902	496	2111 Jan 25	02:03:07	229	1373	116	N	-t	-1.1833	0.7371	-0.3630	250.8	-	-	18N	27W
9903	496	2111 Jul 21	00:53:17	230	1379	121	P	-a	0.8362	1.2938	0.3530	266.3	123.2	-	20S	11W
9904	496	2112 Jan 14	01:06:36	231	1385	126	P	-t	-0.5087	1.9735	0.8764	355.3	200.1	-	21N	14W
9905	496	2112 Jul 09	17:19:51	232	1391	131	T+	pp	0.1055	2.6471	1.6814	326.9	215.1	98.4	22S	102E
9906	496	2113 Jan 02	04:22:59	234	1397	136	T+	p-	0.1964	2.5221	1.4735	350.1	220.2	92.1	23N	64W
9907	496	2113 Jun 29	04:55:26	235	1403	141	P	t-	-0.6887	1.6035	0.5849	315.1	165.2	-	24S	72W
9908	496	2113 Dec 22	14:58:42	236	1409	146	P	a-	0.8666	1.2644	0.2712	273.2	111.5	-	24N	136E
9909	496	2114 May 19	18:07:33	237	1414	113	N	-t	1.3611	0.3984	-0.6770	190.1	-	-	19S	89E
9910	496	2114 Jun 18	09:17:07	237	1415	151	N	t-	-1.4944	0.1478	-0.9157	117.4	-	-	25S	138W
9911	496	2114 Nov 12	19:59:33	238	1420	118	N	-a	-1.1533	0.7270	-0.2439	214.9	-	-	17N	58E
9912	496	2114 Dec 12	06:09:26	238	1421	156	N	a-	1.4932	0.1026	-0.8666	85.0	-	-	25N	93W
9913	496	2115 May 08	19:21:24	239	1426	123	P	-t	0.5996	1.7849	0.7311	336.9	185.0	-	17S	70E
9914	496	2115 Nov 02	09:36:34	241	1432	128	P	-a	-0.4919	1.9611	0.9498	322.4	191.9	-	14N	147W
9915	496	2116 Apr 27	02:41:18	242	1438	133	T-	p-	-0.1746	2.5388	1.5364	338.1	217.7	94.6	14S	40W
9916	496	2116 Oct 21	16:53:39	243	1444	138	T+	pp	0.2353	2.4586	1.3943	360.5	223.8	88.8	11N	104E
9917	496	2117 Apr 16	16:32:00	244	1450	143	P	a-	-0.8852	1.2139	0.2530	263.7	106.6	-	11S	113E
9918	496	2117 Oct 10	17:47:12	245	1456	148	P	t-	0.9675	1.1274	0.0387	298.5	48.4	-	8N	91E
9919	496	2118 Mar 07	23:33:04	246	1461	115	N	-a	1.0496	0.9228	-0.0588	240.8	-	-	6N	11E
9920	496	2118 Aug 31	04:51:45	247	1467	120	N	-t	-1.1543	0.7540	-0.2740	238.5	-	-	10S	71W
9921	497	2119 Feb 25	11:05:13	249	1473	125	T	-a	0.3765	2.1857	1.1490	339.5	207.4	58.1	10N	162W
9922	497	2119 Aug 20	15:51:55	250	1479	130	T	-p	-0.3538	2.1946	1.2227	320.3	203.1	67.3	13S	124E
9923	497	2120 Feb 14	15:17:20	251	1485	135	T	p-	-0.3371	2.2848	1.1950	364.6	219.1	67.9	13N	135E
9924	497	2120 Aug 09	08:01:32	252	1491	140	T	a-	0.3875	2.1182	1.1751	307.1	195.9	59.3	15S	118W
9925	497	2121 Feb 02	14:32:40	253	1497	145	N*	t-	-1.0235	1.0308	-0.0701	288.4	-	-	16N	146E
9926	497	2121 Jun 30	16:49:52	254	1502	112	N	-a	-1.4272	0.2286	-0.7505	131.0	-	-	25S	110E
9927	497	2121 Jul 30	00:52:44	255	1503	150	N	a-	1.1312	0.7610	-0.1968	220.8	-	-	17S	11W
9928	497	2121 Dec 24	01:22:13	256	1508	117	N	-a	1.2261	0.6238	-0.4071	214.8	-	-	25N	20W
9929	497	2122 Jun 20	02:27:47	257	1514	122	P	-t	-0.7177	1.5584	0.5240	318.3	160.2	-	24S	35W
9930	497	2122 Dec 13	13:42:41	258	1520	127	P	-a	0.4979	1.9349	0.9536	307.1	185.4	-	24N	154E
9931	497	2123 Jun 09	05:06:28	259	1526	132	T+	pp	0.0406	2.8189	1.7487	374.3	235.7	106.1	23S	76W
9932	497	2123 Dec 03	05:24:09	261	1532	137	T-	p-	-0.1755	2.5208	1.5507	318.0	207.7	91.4	22N	83W
9933	497	2124 May 28	05:50:59	262	1538	142	P	t-	0.7913	1.4356	0.3770	318.4	141.9	-	21S	87W
9934	497	2124 Nov 21	19:47:21	263	1544	147	P	a-	-0.8808	1.2434	0.2401	276.1	106.9	-	19N	61E
9935	497	2125 Apr 18	01:18:48	264	1549	114	N	-a	-1.1791	0.6890	-0.3005	217.5	-	-	12S	19W
9936	497	2125 May 17	11:46:31	264	1550	152	N	a-	1.4923	0.1249	-0.8854	100.9	-	-	18S	176W
9937	497	2125 Oct 12	12:43:06	265	1555	119	N	-t	1.2476	0.6061	-0.4679	226.6	-	-	9N	167E
9938	497	2126 Apr 07	16:17:55	266	1561	124	T	-a	-0.4346	2.0390	1.0817	306.3	191.5	41.6	7S	117E
9939	497	2126 Oct 01	12:49:48	268	1567	129	P	-t	0.5446	1.9011	0.8169	353.1	196.7	-	4N	166E
9940	497	2127 Mar 28	08:40:17	269	1573	134	T+	p-	0.2664	2.3531	1.3849	318.0	205.5	82.3	3S	128W
9941	498	2127 Sep 20	14:56:04	270	1579	139	T-	p-	-0.2007	2.5126	1.4672	356.2	224.2	93.5	1S	136E
9942	498	2128 Mar 16	21:46:08	271	1585	144	N*	a-	1.0128	1.0086	-0.0093	261.3	-	-	2N	37E
9943	498	2128 Sep 09	00:10:30	273	1591	149	P	a-	-0.9266	1.1513	0.1642	268.2	89.6	-	6S	2W
9944	498	2129 Feb 04	10:02:24	274	1596	116	N	-t	-1.1959	0.7140	-0.3858	247.6	-	-	15N	146W
9945	498	2129 Jul 31	08:24:31	275	1602	121	P	-a	0.9070	1.1644	0.2224	256.5	99.7	-	17S	123W
9946	498	2130 Jan 24	09:10:19	276	1608	126	P	-t	-0.5173	1.9565	0.8619	353.7	198.7	-	19N	134W
9947	498	2130 Jul 21	00:38:56	277	1614	131	T+	-p	0.1803	2.5114	1.5426	326.6	213.4	93.5	20S	7W
9948	498	2131 Jan 13	12:49:59	279	1620	136	T+	p-	0.1914	2.5296	1.4842	348.9	219.8	92.5	22N	171E
9949	498	2131 Jul 10	11:47:56	280	1626	141	P	t-	-0.6107	1.7483	0.7265	324.5	180.4	-	23S	174W
9950	498	2132 Jan 02	23:44:22	281	1632	146	P	a-	0.8604	1.2750	0.2835	273.4	113.6	-	24N	6E
9951	498	2132 May 30	00:42:50	282	1637	113	N	-t	1.4380	0.2563	-0.8169	154.4	-	-	21S	10W
9952	498	2132 Jun 28	15:45:35	282	1638	151	N	t-	-1.4127	0.2981	-0.7663	164.7	-	-	24S	126E
9953	498	2132 Nov 23	04:35:10	283	1643	118	N	-a	-1.1707	0.6968	-0.2776	211.4	-	-	19N	71W
9954	498	2132 Dec 22	14:59:51	283	1644	156	N	a-	1.4873	0.1138	-0.8564	89.5	-	-	25N	136E
9955	498	2133 May 19	02:16:19	284	1649	123	P	-t	0.6688	1.6553	0.6067	328.4	171.6	-	19S	34W
9956	498	2133 Nov 12	17:50:08	286	1655	128	P	-a	-0.5157	1.9203	0.9033	322.2	189.3	-	17N	90E
9957	498	2134 May 08	10:10:41	287	1661	133	T-	pp	-0.1152	2.6447	1.6482	337.6	219.0	98.8	17S	152W
9958	498	2134 Nov 02	00:34:20	288	1667	138	T+	pp	0.2022	2.5222	1.4521	363.1	225.9	93.0	15N	12W
9959	498	2135 Apr 28	00:26:35	289	1673	143	P	a-	-0.8344	1.3051	0.3483	269.9	123.2	-	15S	6W
9960	498	2135 Oct 22	01:06:05	291	1679	148	P	t-	0.9242	1.2085	0.1164	306.6	83.1	-	12N	20W

Cat Num	Canon Plate	Calendar Date	TD of Greatest Eclipse	ΔT s	Luna Num	Saros Num	Ecl. Type	QSE	Gamma	Pen. Mag.	Um. Mag.	Eclipse Phase Durations Pen. m	Par. m	Total m	Greatest in Zenith Lat.	Lng.
9961	499	2136 Mar 18	07:53:55	292	1684	115	N	-a	1.0778	0.8707	-0.1103	235.6	-	-	2N	115W
9962	499	2136 Apr 16	17:08:45	292	1685	153	Nb	a-	-1.5241	0.0396	-0.9174	53.3	-	-	12S	103E
9963	499	2136 Sep 10	12:05:06	293	1690	120	N	-t	-1.2134	0.6452	-0.3821	222.2	-	-	6S	180E
9964	499	2137 Mar 07	19:13:43	294	1696	125	T	-a	0.3992	2.1441	1.1069	339.3	205.8	50.0	5N	76E
9965	499	2137 Aug 30	23:24:05	296	1702	130	T	-p	-0.4171	2.0782	1.1069	315.9	197.1	48.3	9S	11E
9966	499	2138 Feb 24	23:09:56	297	1708	135	T	p-	-0.3178	2.3198	1.2306	366.1	221.1	73.1	9N	17E
9967	499	2138 Aug 20	15:38:46	298	1714	140	T	p-	0.3204	2.2421	1.2977	310.4	200.9	74.3	12S	127E
9968	499	2139 Feb 13	22:28:17	299	1720	145	N*	t-	-1.0082	1.0574	-0.0406	290.8	-	-	12N	27E
9969	499	2139 Jul 12	00:01:45	300	1725	112	Ne	-a	-1.5055	0.0866	-0.8957	81.9	-	-	23S	3E
9970	499	2139 Aug 10	08:18:10	301	1726	150	N	a-	1.0638	0.8865	-0.0749	236.0	-	-	15S	122W
9971	499	2140 Jan 04	09:55:16	302	1731	117	N	-a	1.2270	0.6209	-0.4075	213.6	-	-	24N	146W
9972	499	2140 Jun 30	09:13:17	303	1737	122	P	-t	-0.8012	1.4064	0.3695	308.9	138.2	-	24S	136W
9973	499	2140 Dec 23	22:29:48	304	1743	127	P	-a	0.5028	1.9255	0.9450	306.2	184.6	-	24N	24E
9974	499	2141 Jun 19	11:34:51	305	1749	132	T-	pp	-0.0446	2.8112	1.7415	374.4	235.9	106.1	23S	172W
9975	499	2141 Dec 13	14:10:16	307	1755	137	T-	p-	-0.1671	2.5374	1.5652	318.7	208.1	92.0	23N	147E
9976	499	2142 Jun 08	12:32:42	308	1761	142	P	t-	0.7118	1.5794	0.5247	327.2	163.2	-	22S	173E
9977	499	2142 Dec 03	04:16:11	309	1767	147	P	a-	-0.8704	1.2648	0.2569	278.9	110.6	-	21N	65W
9978	499	2143 Apr 29	09:01:22	310	1772	114	N	-a	-1.2256	0.6009	-0.3830	204.1	-	-	16S	135W
9979	499	2143 May 28	18:59:24	311	1773	152	N	a-	1.4219	0.2513	-0.7536	140.9	-	-	20S	76E
9980	499	2143 Oct 23	20:10:38	312	1778	119	N	-t	1.2885	0.5336	-0.5457	214.8	-	-	13N	54E
9981	500	2144 Apr 18	00:20:29	313	1784	124	T	-a	-0.4787	1.9563	1.0026	302.9	187.0	7.6	11S	4W
9982	500	2144 Oct 11	20:02:14	314	1790	129	P	-t	0.5930	1.8136	0.7269	348.7	188.3	-	8N	57E
9983	500	2145 Apr 07	16:48:09	315	1796	134	T+	p-	0.2285	2.4221	1.4550	319.7	207.9	87.3	7S	110E
9984	500	2145 Sep 30	22:19:51	317	1802	139	T-	pp	-0.1486	2.6083	1.5628	356.5	225.7	98.2	3N	24E
9985	500	2146 Mar 28	05:44:28	318	1808	144	P	a-	0.9832	1.0629	0.0448	267.6	48.6	-	2S	83W
9986	500	2146 Sep 20	07:51:28	319	1814	149	P	a-	-0.8707	1.2537	0.2669	275.9	112.3	-	2S	118W
9987	500	2147 Feb 15	17:57:14	320	1819	116	N	-t	-1.2114	0.6849	-0.4138	243.4	-	-	12N	95E
9988	500	2147 Aug 11	15:57:02	322	1825	121	P	-a	0.9765	1.0378	0.0941	245.8	66.0	-	14S	123E
9989	500	2147 Sep 09	23:11:33	322	1826	159	Nb	a-	-1.5372	0.0123	-0.9380	29.6	-	-	7S	13E
9990	500	2148 Feb 04	17:13:45	323	1831	126	P	-t	-0.5265	1.9380	0.8465	352.0	197.2	-	16N	106E
9991	500	2148 Jul 31	07:56:37	324	1837	131	T+	-p	0.2554	2.3755	1.4030	325.7	210.4	85.3	18S	116W
9992	500	2149 Jan 23	21:17:23	325	1843	136	T+	p-	0.1859	2.5379	1.4962	347.6	219.6	93.0	19N	45E
9993	500	2149 Jul 20	18:38:11	327	1849	141	P	t-	-0.5315	1.8957	0.8700	333.1	193.2	-	21S	84E
9994	500	2150 Jan 13	08:31:35	328	1855	146	P	a-	0.8552	1.2834	0.2941	273.3	115.3	-	22N	124W
9995	500	2150 Jun 10	07:14:48	329	1860	113	Ne	-t	1.5171	0.1101	-0.9612	102.4	-	-	22S	107W
9996	500	2150 Jul 09	22:13:22	329	1861	151	N	t-	-1.3300	0.4505	-0.6151	199.9	-	-	23S	29E
9997	500	2150 Dec 04	13:14:45	330	1866	118	N	-a	-1.1848	0.6727	-0.3051	208.7	-	-	21N	160E
9998	500	2151 Jan 02	23:51:26	330	1867	156	N	a-	1.4828	0.1225	-0.8484	92.8	-	-	24N	4E
9999	500	2151 May 30	09:09:10	331	1872	123	P	-t	0.7403	1.5216	0.4780	318.6	155.1	-	21S	136W
10000	500	2151 Nov 24	02:08:13	332	1878	128	P	-a	-0.5350	1.8875	0.8651	322.2	187.0	-	20N	34W
10001	501	2152 May 18	17:35:13	333	1884	133	T-	pp	-0.0511	2.7597	1.7688	336.7	219.6	101.2	20S	97E
10002	501	2152 Nov 12	08:21:47	334	1890	138	T+	pp	0.1753	2.5743	1.4989	365.4	227.5	95.8	18N	128W
10003	501	2153 May 08	08:14:39	335	1896	143	P	a-	-0.7781	1.4065	0.4535	276.3	138.3	-	18S	123W
10004	501	2153 Nov 01	08:34:03	336	1902	148	P	t-	0.8881	1.2763	0.1811	313.0	102.7	-	15N	132W
10005	501	2154 Mar 29	16:05:07	337	1907	115	N	-a	1.1135	0.8049	-0.1757	228.6	-	-	3S	122E
10006	501	2154 Apr 28	01:04:33	337	1908	153	N	a-	-1.4784	0.1228	-0.8328	93.2	-	-	15S	16W
10007	501	2154 Sep 21	19:29:06	338	1913	120	N	-t	-1.2646	0.5510	-0.4759	206.5	-	-	2S	68E
10008	501	2154 Oct 21	09:26:08	338	1914	158	Nb	t-	1.5560	0.0348	-1.0286	56.4	-	-	12N	145W
10009	501	2155 Mar 19	03:12:45	339	1919	125	T	-a	0.4293	2.0892	1.0517	338.5	203.4	35.5	1N	45W
10010	501	2155 Sep 11	07:03:11	340	1925	130	T	-a	-0.4752	1.9715	1.0003	311.4	190.7	2.6	5S	105W
10011	501	2156 Mar 07	06:54:14	341	1931	135	T	p-	-0.2922	2.3663	1.2782	367.7	223.5	79.0	5N	100W
10012	501	2156 Aug 30	23:20:37	342	1937	140	T+	p-	0.2569	2.3595	1.4132	313.0	204.6	83.9	8S	11E
10013	501	2157 Feb 24	06:16:36	343	1943	145	P	t-	-0.9868	1.0949	0.0005	294.2	5.6	-	9N	90W
10014	501	2157 Aug 20	15:46:36	344	1949	150	P	a-	0.9992	1.0070	0.0415	249.2	45.1	-	11S	125E
10015	501	2158 Jan 14	18:29:52	345	1954	117	N	-a	1.2275	0.6184	-0.4068	212.5	-	-	22N	87E
10016	501	2158 Jul 11	15:55:51	346	1960	122	P	-t	-0.8859	1.2525	0.2126	297.8	107.6	-	23S	124E
10017	501	2159 Jan 04	07:19:20	347	1966	127	P	-a	0.5061	1.9189	0.9394	305.4	183.9	-	23N	107W
10018	501	2159 Jun 30	18:00:09	348	1972	132	T-	pp	-0.1322	2.6504	1.5809	373.4	234.1	102.0	23S	92E
10019	501	2159 Dec 24	23:00:09	349	1978	137	T-	p-	-0.1619	2.5478	1.5737	319.2	208.4	92.4	23N	16E
10020	501	2160 Jun 18	19:10:10	350	1984	142	P	t-	0.6280	1.7313	0.6804	335.2	180.8	-	23S	74E

Cat Num	Canon Plate	Calendar Date	TD of Greatest Eclipse	ΔT s	Luna Num	Saros Num	Ecl. Type	QSE	Gamma	Pen. Mag.	Um. Mag.	Eclipse Phase Durations Pen. m	Par. m	Total m	Greatest in Zenith Lat.	Lng.
10021	502	2160 Dec 13	12:50:19	351	1990	147	P	a-	-0.8650	1.2768	0.2646	280.8	112.4	-	22N	167E
10022	502	2161 May 09	16:38:03	352	1995	114	N	-a	-1.2776	0.5027	-0.4757	187.7	-	-	19S	111E
10023	502	2161 Jun 08	02:08:54	352	1996	152	N	a-	1.3475	0.3853	-0.6146	171.5	-	-	22S	31W
10024	502	2161 Nov 03	03:45:49	353	2001	119	N	-t	1.3226	0.4736	-0.6107	204.1	-	-	16N	59W
10025	502	2162 Apr 29	08:17:08	354	2007	124	P	-a	-0.5280	1.8642	0.9137	298.9	181.3	-	15S	124W
10026	502	2162 Oct 23	03:23:58	356	2013	129	P	-t	0.6340	1.7395	0.6504	344.6	180.3	-	12N	54W
10027	502	2163 Apr 19	00:49:33	357	2019	134	T+	p-	0.1858	2.5000	1.5338	321.5	210.2	91.8	11S	11W
10028	502	2163 Oct 12	05:51:53	358	2025	139	T-	pp	-0.1026	2.6927	1.6471	356.3	226.3	100.9	7N	90W
10029	502	2164 Apr 07	13:34:34	359	2031	144	P	h-	0.9479	1.1278	0.1095	274.6	75.5	-	6S	159E
10030	502	2164 Sep 30	15:40:32	360	2037	149	P	a-	-0.8209	1.3449	0.3581	282.0	128.0	-	2N	124E
10031	502	2165 Feb 26	01:44:02	361	2042	116	N	-t	-1.2329	0.6447	-0.4524	237.1	-	-	8N	22W
10032	502	2165 Aug 21	23:34:21	362	2048	121	N	-a	1.0417	0.9193	-0.0269	234.5	-	-	11S	8E
10033	502	2165 Sep 20	07:05:16	362	2049	159	N	a-	-1.4858	0.1074	-0.8447	86.8	-	-	2S	106W
10034	502	2166 Feb 15	01:11:40	363	2054	126	P	-t	-0.5402	1.9109	0.8233	349.8	194.9	-	12N	13W
10035	502	2166 Aug 11	15:17:07	364	2060	131	T	-a	0.3273	2.2456	1.2688	324.1	206.3	73.3	15S	134E
10036	502	2167 Feb 04	05:41:32	365	2066	136	T+	p-	0.1772	2.5517	1.5143	346.4	219.4	93.8	16N	80W
10037	502	2167 Aug 01	01:28:59	366	2072	141	T	t-	-0.4536	2.0407	1.0108	340.5	203.6	16.6	18S	19W
10038	502	2168 Jan 24	17:17:30	367	2078	146	P	a-	0.8487	1.2940	0.3073	273.4	117.4	-	20N	105E
10039	502	2168 Jul 20	04:38:50	368	2084	151	N	t-	-1.2454	0.6065	-0.4607	228.9	-	-	22S	66W
10040	502	2168 Dec 14	21:59:24	369	2089	118	N	-a	-1.1945	0.6565	-0.3242	206.9	-	-	22N	31E
10041	503	2169 Jan 13	08:43:52	369	2090	156	N	a-	1.4794	0.1288	-0.8423	95.1	-	-	23N	127W
10042	503	2169 Jun 09	15:57:06	370	2095	123	P	-t	0.8158	1.3806	0.3416	307.2	133.5	-	22S	122E
10043	503	2169 Dec 04	10:31:58	371	2101	128	P	-a	-0.5488	1.8648	0.8375	322.5	185.3	-	22N	159W
10044	503	2170 May 30	00:55:17	372	2107	133	T+	pp	0.0174	2.8188	1.8330	335.3	219.3	101.7	22S	13W
10045	503	2170 Nov 23	16:16:21	373	2113	138	T+	pp	0.1554	2.6134	1.5331	367.2	228.6	97.7	21N	114E
10046	503	2171 May 19	15:56:55	374	2119	143	P	a-	-0.7166	1.5175	0.5680	282.7	151.9	-	21S	121E
10047	503	2171 Nov 12	16:10:02	376	2125	148	P	t-	0.8584	1.3323	0.2343	318.0	115.9	-	18N	115E
10048	503	2172 Apr 09	00:08:52	376	2130	115	N	-a	1.1548	0.7289	-0.2513	219.8	-	-	7S	0E
10049	503	2172 May 08	08:53:18	377	2131	153	N	a-	-1.4275	0.2157	-0.7389	122.7	-	-	19S	133W
10050	503	2172 Oct 02	03:01:41	378	2136	120	N	-h	-1.3098	0.4681	-0.5586	191.3	-	-	3N	46W
10051	503	2172 Oct 31	17:09:17	378	2137	158	N	h-	1.5197	0.1016	-0.9623	95.5	-	-	16N	100E
10052	503	2173 Mar 29	11:02:23	379	2142	125	P	-a	0.4662	2.0215	0.9839	337.1	199.9	-	3S	163W
10053	503	2173 Sep 21	14:50:18	380	2148	130	P	-a	-0.5272	1.8761	0.9047	306.9	184.0	-	1S	138E
10054	503	2174 Mar 18	14:30:50	381	2154	135	T-	pp	-0.2605	2.4236	1.3371	369.4	226.0	85.2	1N	146E
10055	503	2174 Sep 11	07:08:03	382	2160	140	T+	p-	0.1982	2.4683	1.5197	315.0	207.3	90.3	4S	106W
10056	503	2175 Mar 07	13:59:53	383	2166	145	P	t-	-0.9608	1.1406	0.0502	298.1	54.8	-	4N	154E
10057	503	2175 Aug 31	23:19:04	384	2172	150	P	a-	0.9385	1.1207	0.1506	260.6	84.8	-	8S	12E
10058	503	2176 Jan 26	03:03:39	385	2177	117	N	-a	1.2292	0.6133	-0.4082	210.9	-	-	20N	41W
10059	503	2176 Jul 21	22:36:52	386	2183	122	P	-t	-0.9707	1.0986	0.0553	284.6	56.3	-	21S	24E
10060	503	2177 Jan 14	16:08:55	387	2189	127	P	-a	0.5099	1.9112	0.9333	304.6	183.3	-	22N	122E
10061	504	2177 Jul 11	00:25:23	388	2195	132	T-	pp	-0.2199	2.4896	1.4199	371.2	230.4	93.0	22S	3W
10062	504	2178 Jan 04	07:50:14	390	2201	137	T-	p-	-0.1570	2.5574	1.5820	319.8	208.8	92.8	23N	115W
10063	504	2178 Jun 30	01:48:39	391	2207	142	P	t-	0.5438	1.8841	0.8364	341.8	194.6	-	23S	25W
10064	504	2178 Dec 24	21:26:20	392	2213	147	P	a-	-0.8614	1.2853	0.2694	282.5	113.6	-	23N	40E
10065	504	2179 May 21	00:09:10	393	2218	114	N	-a	-1.3346	0.3956	-0.5778	167.6	-	-	21S	2W
10066	504	2179 Jun 19	09:16:11	393	2219	152	N	a-	1.2701	0.5250	-0.4703	196.7	-	-	22S	137W
10067	504	2179 Nov 14	11:28:05	394	2224	119	N	-t	1.3504	0.4249	-0.6641	194.8	-	-	19N	175W
10068	504	2180 May 09	16:05:58	395	2230	124	P	-a	-0.5840	1.7599	0.8124	294.1	173.9	-	18S	119E
10069	504	2180 Nov 02	10:55:47	396	2236	129	P	-t	0.6669	1.6803	0.5891	341.0	173.2	-	16N	167W
10070	504	2181 Apr 29	08:40:08	397	2242	134	T+	pp	0.1345	2.5938	1.6281	323.2	212.3	95.9	15S	129W
10071	504	2181 Oct 22	13:35:21	398	2248	139	T-	pp	-0.0652	2.7613	1.7157	355.7	226.3	102.2	11N	154E
10072	504	2182 Apr 18	21:14:14	399	2254	144	P	h-	0.9049	1.2068	0.1883	282.5	98.3	-	10S	43E
10073	504	2182 Oct 11	23:38:06	400	2260	149	P	a-	-0.7776	1.4246	0.4376	286.8	139.4	-	7N	4E
10074	504	2183 Mar 09	09:24:16	401	2265	116	N	-t	-1.2592	0.5953	-0.4997	229.0	-	-	3N	137W
10075	504	2183 Sep 02	07:15:22	403	2271	121	N	-a	1.1036	0.8072	-0.1419	222.7	-	-	7S	107W
10076	504	2183 Oct 01	15:05:25	403	2272	159	N	a-	-1.4396	0.1933	-0.7610	115.5	-	-	2N	133E
10077	504	2184 Feb 26	09:05:35	404	2277	126	P	-t	-0.5579	1.8762	0.7930	347.1	192.0	-	8N	132W
10078	504	2184 Aug 21	22:38:35	405	2283	131	T	-a	0.3977	2.1190	1.1374	321.9	201.0	54.8	11S	23E
10079	504	2185 Feb 14	14:03:41	406	2289	136	T+	p-	0.1660	2.5697	1.5372	345.1	219.4	94.8	13N	154E
10080	504	2185 Aug 11	08:20:59	407	2295	141	T	t-	-0.3774	2.1828	1.1482	346.9	212.0	59.3	15S	122W

Cat Num	Canon Plate	Calendar Date	TD of Greatest Eclipse	ΔT s	Luna Num	Saros Num	Ecl. Type	QSE	Gamma	Pen. Mag.	Um. Mag.	Eclipse Phase Durations Pen. m	Par. m	Total m	Greatest in Zenith Lat.	Lng.
10081	505	2186 Feb 04	02:01:15	408	2301	146	P	a-	0.8402	1.3080	0.3244	273.7	120.1	-	17N	25W
10082	505	2186 Jul 31	11:07:14	409	2307	151	N	t-	-1.1629	0.7589	-0.3102	252.6	-	-	19S	163W
10083	505	2186 Dec 26	06:46:26	410	2312	118	N	-a	-1.2018	0.6443	-0.3389	205.7	-	-	22N	100W
10084	505	2187 Jan 24	17:35:19	410	2313	156	N	a-	1.4758	0.1353	-0.8356	97.5	-	-	21N	101E
10085	505	2187 Jun 20	22:44:13	411	2318	123	P	-t	0.8928	1.2372	0.2025	294.4	104.6	-	23S	21E
10086	505	2187 Dec 15	18:58:37	413	2324	128	P	-a	-0.5595	1.8473	0.8156	323.0	184.1	-	23N	76E
10087	505	2188 Jun 09	08:12:51	414	2330	133	T+	pp	0.0887	2.6856	1.7045	333.2	218.0	99.8	23S	122W
10088	505	2188 Dec 04	00:15:39	415	2336	138	T+	pp	0.1394	2.6448	1.5602	368.8	229.5	98.9	22N	5W
10089	505	2189 May 29	23:34:01	416	2342	143	P	a-	-0.6505	1.6375	0.6909	288.8	164.1	-	22S	8E
10090	505	2189 Nov 22	23:54:08	417	2348	148	P	t-	0.8349	1.3765	0.2764	321.7	125.0	-	21N	1W
10091	505	2190 Apr 20	08:03:22	418	2353	115	N	-a	1.2034	0.6397	-0.3402	208.3	-	-	11S	119W
10092	505	2190 May 19	16:35:26	418	2354	153	N	a-	-1.3714	0.3183	-0.6357	147.8	-	-	21S	112E
10093	505	2190 Oct 13	10:43:55	419	2359	120	N	-h	-1.3479	0.3979	-0.6285	176.9	-	-	7N	162W
10094	505	2190 Nov 12	01:00:53	419	2360	158	N	h-	1.4894	0.1572	-0.9067	117.9	-	-	19N	18W
10095	505	2191 Apr 09	18:42:01	420	2365	125	P	-a	0.5107	1.9399	0.9022	334.9	194.9	-	7S	82E
10096	505	2191 Oct 02	22:45:48	422	2371	130	P	-a	-0.5729	1.7925	0.8206	302.6	177.4	-	3N	18E
10097	505	2192 Mar 28	21:56:24	423	2377	135	T-	pp	-0.2202	2.4966	1.4120	371.1	228.8	91.6	4S	34E
10098	505	2192 Sep 21	15:02:17	424	2383	140	T+	p-	0.1453	2.5668	1.6154	316.5	209.1	94.4	0S	134E
10099	505	2193 Mar 17	21:34:32	425	2389	145	P	t-	-0.9274	1.1995	0.1138	303.0	81.6	-	0N	40E
10100	505	2193 Sep 11	06:57:00	426	2395	150	P	a-	0.8828	1.2256	0.2505	270.4	108.1	-	4S	104W
10101	506	2194 Feb 05	11:34:56	427	2400	117	N	-a	1.2338	0.6027	-0.4146	208.5	-	-	17N	168W
10102	506	2194 Aug 02	05:18:42	428	2406	122	N	-t	-1.0539	0.9479	-0.0993	269.6	-	-	19S	76W
10103	506	2195 Jan 26	00:58:49	429	2412	127	P	-a	0.5139	1.9028	0.9269	303.7	182.7	-	19N	10W
10104	506	2195 Jul 22	06:49:10	431	2418	132	T	pt	-0.3086	2.3271	1.2569	367.7	224.6	77.4	20S	99W
10105	506	2196 Jan 15	16:41:55	432	2424	137	T-	p-	-0.1537	2.5641	1.5877	320.3	209.1	93.1	21N	114E
10106	506	2196 Jul 10	08:26:05	433	2430	142	P	t-	0.4577	2.0407	0.9960	347.3	205.8	-	22S	123W
10107	506	2197 Jan 04	06:04:04	434	2436	147	P	a-	-0.8601	1.2894	0.2703	283.8	114.1	-	22N	88W
10108	506	2197 May 31	07:36:04	435	2441	114	N	-a	-1.3954	0.2817	-0.6871	142.4	-	-	23S	113W
10109	506	2197 Jun 29	16:22:52	435	2442	152	N	a-	1.1911	0.6680	-0.3233	217.8	-	-	22S	117E
10110	506	2197 Nov 24	19:17:17	436	2447	119	N	-t	1.3722	0.3870	-0.7061	187.0	-	-	22N	69E
10111	506	2198 May 20	23:50:32	437	2453	124	P	-a	-0.6436	1.6493	0.7041	288.4	164.8	-	21S	3E
10112	506	2198 Nov 13	18:34:28	439	2459	129	P	-t	0.6944	1.6306	0.5378	337.7	166.7	-	19N	79E
10113	506	2199 May 10	16:25:22	440	2465	134	T+	pp	0.0793	2.6951	1.7297	324.8	214.0	98.8	18S	115E
10114	506	2199 Nov 02	21:27:11	441	2471	139	T-	pp	-0.0340	2.8187	1.7731	354.9	226.0	102.7	15N	36E
10115	506	2200 Apr 30	04:44:34	442	2477	144	P	t-	0.8550	1.2987	0.2797	291.1	118.6	-	14S	70W
10116	506	2200 Oct 23	07:44:43	443	2483	149	P	a-	-0.7411	1.4917	0.5044	290.4	147.8	-	11N	118W
10117	506	2201 Mar 20	16:55:04	444	2488	116	N	-t	-1.2929	0.5323	-0.5602	217.8	-	-	1S	110E
10118	506	2201 Sep 13	15:03:17	445	2494	121	N	-a	1.1597	0.7059	-0.2464	210.7	-	-	3S	135E
10119	506	2201 Oct 12	23:14:05	446	2495	159	N	a-	-1.4001	0.2669	-0.6897	134.9	-	-	6N	10E
10120	506	2202 Mar 09	16:51:36	447	2500	126	P	-t	-0.5824	1.8288	0.7505	343.6	187.8	-	4N	111E
10121	507	2202 Sep 03	06:05:57	448	2506	131	T	-a	0.4622	2.0031	1.0163	319.5	194.9	19.6	7S	90W
10122	507	2203 Feb 26	22:20:39	449	2512	136	T+	p-	0.1500	2.5964	1.5692	343.9	219.5	96.1	9N	30E
10123	507	2203 Aug 23	15:15:12	450	2518	141	T	pp	-0.3039	2.3203	1.2806	352.2	218.6	78.1	12S	134E
10124	507	2204 Feb 16	10:41:26	451	2524	146	P	a-	0.8285	1.3276	0.3475	274.5	123.7	-	13N	155W
10125	507	2204 Aug 11	17:37:23	453	2530	151	N	t-	-1.0819	0.9088	-0.1627	272.6	-	-	16S	99E
10126	507	2205 Jan 06	15:34:56	454	2535	118	N	-a	-1.2079	0.6341	-0.3511	204.7	-	-	21N	129E
10127	507	2205 Feb 05	02:23:28	454	2536	156	N	a-	1.4699	0.1458	-0.8245	101.2	-	-	18N	30W
10128	507	2205 Jul 02	05:29:54	455	2541	123	P	-t	0.9713	1.0912	0.0603	279.8	58.0	-	22S	79W
10129	507	2205 Dec 27	03:28:47	456	2547	128	P	-a	-0.5665	1.8365	0.8008	323.8	183.4	-	23N	50W
10130	507	2206 Jun 21	15:28:26	457	2553	133	T+	pp	0.1626	2.5480	1.5711	330.5	215.5	95.4	23S	130E
10131	507	2206 Dec 16	08:19:07	458	2559	138	T+	pp	0.1275	2.6685	1.5802	370.2	230.3	99.8	23N	124W
10132	507	2207 Jun 11	07:07:39	460	2565	143	P	a-	-0.5811	1.7634	0.8193	294.5	174.8	-	24S	105W
10133	507	2207 Dec 05	07:45:24	461	2571	148	P	t-	0.8170	1.4102	0.3084	324.3	131.4	-	23N	117W
10134	507	2208 May 01	15:49:06	462	2576	115	N	-a	1.2583	0.5389	-0.4410	193.6	-	-	14S	124E
10135	507	2208 May 31	00:10:37	462	2577	153	N	a-	-1.3102	0.4305	-0.5233	170.4	-	-	23S	2W
10136	507	2208 Oct 24	18:36:02	463	2582	120	N	-h	-1.3789	0.3410	-0.6853	164.2	-	-	11N	80E
10137	507	2208 Nov 23	09:01:59	463	2583	158	N	h-	1.4659	0.2002	-0.8635	132.2	-	-	22N	137W
10138	507	2209 Apr 21	02:12:20	464	2588	125	P	-h	0.5617	1.8464	0.8086	331.7	188.1	-	11S	31W
10139	507	2209 Oct 14	06:49:53	465	2594	130	P	-a	-0.6119	1.7214	0.7487	298.6	171.1	-	7N	104W
10140	507	2210 Apr 10	05:13:49	467	2600	135	T-	pp	-0.1736	2.5812	1.4985	372.7	231.3	97.3	8S	76W

Cat Num	Canon Plate	Calendar Date	TD of Greatest Eclipse	ΔT s	Luna Num	Saros Num	Ecl. Type	QSE	Gamma	Pen. Mag.	Um. Mag.	Eclipse Phase ---- Durations ---- Pen. m	Par. m	Total m	Greatest in Zenith Lat.	Lng.
10141	508	2210 Oct 03	23:03:49	468	2606	140	T+	p-	0.0988	2.6537	1.6993	317.6	210.2	96.8	4N	13E
10142	508	2211 Mar 30	05:02:58	469	2612	145	P	t-	-0.8882	1.2689	0.1880	308.4	103.6	-	4S	73W
10143	508	2211 Sep 23	14:39:33	470	2618	150	P	a-	0.8314	1.3226	0.3421	278.9	124.9	-	1N	140E
10144	508	2212 Feb 17	20:03:16	471	2623	117	N	-a	1.2415	0.5862	-0.4264	205.2	-	-	13N	65E
10145	508	2212 Mar 18	09:06:46	471	2624	155	N	a-	-1.5477	0.0361	-0.9998	55.5	-	-	0S	133W
10146	508	2212 Aug 13	12:02:09	472	2629	122	N	-t	-1.1349	0.8015	-0.2499	252.6	-	-	15S	177W
10147	508	2213 Feb 06	09:44:33	474	2635	127	P	-a	0.5218	1.8872	0.9137	302.6	181.6	-	16N	141W
10148	508	2213 Aug 02	13:16:20	475	2641	132	T	-t	-0.3946	2.1698	1.0987	363.1	216.8	50.6	18S	165E
10149	508	2214 Jan 27	01:30:46	476	2647	137	T-	p-	-0.1480	2.5748	1.5979	320.9	209.5	93.6	18N	18W
10150	508	2214 Jul 22	15:06:54	477	2653	142	T	t-	0.3734	2.1941	1.1518	351.4	214.2	60.4	20S	137E
10151	508	2215 Jan 16	14:40:00	479	2659	147	P	a-	-0.8578	1.2950	0.2732	285.2	115.0	-	20N	144E
10152	508	2215 Jun 12	14:59:57	480	2664	114	N	-a	-1.4594	0.1624	-0.8024	108.9	-	-	25S	137E
10153	508	2215 Jul 11	23:30:24	480	2665	152	N	a-	1.1114	0.8125	-0.1753	235.8	-	-	21S	11E
10154	508	2215 Dec 07	03:12:05	481	2670	119	N	-t	1.3891	0.3578	-0.7388	180.7	-	-	24N	48W
10155	508	2216 Jun 01	07:27:36	482	2676	124	P	-a	-0.7097	1.5271	0.5839	281.6	153.0	-	23S	110W
10156	508	2216 Nov 25	02:22:56	483	2682	129	P	-t	0.7144	1.5946	0.5005	335.1	161.6	-	21N	37W
10157	508	2217 May 22	00:01:35	484	2688	134	T+	pp	0.0167	2.8100	1.8444	326.0	215.2	100.4	20S	1E
10158	508	2217 Nov 14	05:29:02	486	2694	139	T-	pp	-0.0100	2.8624	1.8171	353.9	225.5	102.6	18N	84W
10159	508	2218 May 11	12:05:34	487	2700	144	P	t-	0.7980	1.4036	0.3839	300.0	137.2	-	17S	180E
10160	508	2218 Nov 03	16:00:25	488	2706	149	P	a-	-0.7113	1.5465	0.5589	293.0	153.8	-	14N	118E
10161	509	2219 Apr 01	00:18:15	489	2711	116	N	-t	-1.3322	0.4587	-0.6309	203.5	-	-	5S	2W
10162	509	2219 Sep 24	22:55:52	490	2717	121	N	-a	1.2116	0.6125	-0.3435	198.5	-	-	2N	16E
10163	509	2219 Oct 24	07:29:32	491	2718	159	N	a-	-1.3663	0.3303	-0.6288	149.2	-	-	10N	114W
10164	509	2220 Mar 20	00:32:40	492	2723	126	P	-t	-0.6114	1.7728	0.6999	339.4	182.6	-	0S	4W
10165	509	2220 Sep 13	13:36:52	493	2729	131	P	-a	0.5228	1.8948	0.9023	316.6	188.0	-	3S	157E
10166	509	2221 Mar 09	06:32:19	494	2735	136	T+	p-	0.1290	2.6321	1.6106	342.7	219.6	97.5	5N	93W
10167	509	2221 Sep 02	22:13:42	495	2741	141	T-	pp	-0.2348	2.4498	1.4048	356.6	223.5	89.7	8S	29E
10168	509	2222 Feb 26	19:17:36	497	2747	146	P	a-	0.8137	1.3530	0.3767	275.6	128.0	-	9N	76E
10169	509	2222 Aug 23	00:12:08	498	2753	151	N*	t-	-1.0042	1.0525	-0.0214	289.4	-	-	12S	0E
10170	509	2223 Jan 18	00:22:32	499	2758	118	N	-a	-1.2144	0.6230	-0.3639	203.6	-	-	20N	1W
10171	509	2223 Feb 16	11:07:58	499	2759	156	N	a-	1.4616	0.1605	-0.8088	106.1	-	-	14N	161W
10172	509	2223 Jul 13	12:17:52	500	2764	123	N	-t	1.0485	0.9478	-0.0797	263.7	-	-	21S	179E
10173	509	2224 Jan 07	11:58:46	501	2770	128	P	-a	-0.5731	1.8260	0.7870	324.5	182.7	-	22N	176W
10174	509	2224 Jul 01	22:43:13	503	2776	133	T+	-p	0.2378	2.4081	1.4348	327.0	211.7	87.7	23S	22E
10175	509	2224 Dec 26	16:25:17	504	2782	138	T+	pp	0.1183	2.6868	1.5956	371.5	230.9	100.5	23N	116E
10176	509	2225 Jun 21	14:38:56	505	2788	143	P	a-	-0.5097	1.8936	0.9512	299.7	183.9	-	24S	143E
10177	509	2225 Dec 15	15:41:08	506	2794	148	P	t-	0.8026	1.4370	0.3343	326.2	136.2	-	24N	125E
10178	509	2226 May 12	23:26:43	507	2799	115	N	-a	1.3191	0.4276	-0.5527	174.8	-	-	17S	10E
10179	509	2226 Jun 11	07:41:30	508	2800	153	N	a-	-1.2459	0.5488	-0.4056	190.6	-	-	24S	113W
10180	509	2226 Nov 05	02:38:01	509	2805	120	N	-h	-1.4024	0.2977	-0.7283	153.5	-	-	14N	41W
10181	510	2226 Dec 04	17:10:21	509	2806	158	N	h-	1.4479	0.2330	-0.8300	141.8	-	-	24N	102E
10182	510	2227 May 02	09:32:27	510	2811	125	P	-t	0.6204	1.7388	0.7008	327.3	179.0	-	15S	141W
10183	510	2227 Oct 25	15:02:23	511	2817	130	P	-a	-0.6444	1.6621	0.6886	295.1	165.4	-	11N	133E
10184	510	2228 Apr 20	12:20:32	513	2823	135	T-	pp	-0.1184	2.6812	1.6007	374.0	233.6	102.1	12S	177E
10185	510	2228 Oct 14	07:13:09	514	2829	140	T+	pp	0.0586	2.7290	1.7713	318.5	210.8	98.0	8N	110W
10186	510	2229 Apr 09	12:21:26	515	2835	145	P	t-	-0.8403	1.3543	0.2786	314.6	124.2	-	8S	177E
10187	510	2229 Oct 03	22:29:31	516	2841	150	P	a-	0.7865	1.4077	0.4216	286.0	137.4	-	5N	22E
10188	510	2230 Feb 28	04:27:10	517	2846	117	N	-a	1.2537	0.5612	-0.4463	200.5	-	-	9N	61W
10189	510	2230 Mar 29	16:58:24	518	2847	155	N	a-	-1.5105	0.1012	-0.9285	91.9	-	-	5S	108E
10190	510	2230 Aug 24	18:48:09	519	2852	122	N	-t	-1.2127	0.6610	-0.3950	233.5	-	-	12S	81E
10191	510	2231 Feb 17	18:27:46	520	2858	127	P	-a	0.5321	1.8669	0.8961	301.3	180.3	-	13N	89E
10192	510	2231 Aug 13	19:45:54	521	2864	132	P	-t	-0.4786	2.0163	0.9439	357.4	206.9	-	15S	67E
10193	510	2232 Feb 07	10:17:44	522	2870	137	T-	p-	-0.1410	2.5876	1.6106	321.5	210.0	94.1	15N	149W
10194	510	2232 Aug 01	21:49:49	524	2876	142	T	t-	0.2898	2.3464	1.3061	354.3	220.4	81.1	18S	36E
10195	510	2233 Jan 26	23:15:04	525	2882	147	P	a-	-0.8553	1.3006	0.2767	286.6	116.0	-	18N	16E
10196	510	2233 Jun 22	22:22:33	526	2887	114	Ne	-a	-1.5248	0.0404	-0.9208	54.8	-	-	25S	27E
10197	510	2233 Jul 22	06:40:42	526	2888	152	N	a-	1.0328	0.9552	-0.0297	250.9	-	-	19S	97W
10198	510	2233 Dec 17	11:10:37	527	2893	119	N	-t	1.4027	0.3343	-0.7651	175.4	-	-	25N	166W
10199	510	2234 Jun 12	15:02:55	529	2899	124	P	-a	-0.7774	1.4022	0.4604	273.8	138.4	-	24S	137E
10200	510	2234 Dec 06	10:17:27	530	2905	129	P	-t	0.7300	1.5662	0.4715	332.8	157.3	-	23N	155W

Cat Num	Canon Plate	Calendar Date	TD of Greatest Eclipse	ΔT s	Luna Num	Saros Num	Ecl. Type	QSE	Gamma	Pen. Mag.	Um. Mag.	Eclipse Phase Durations Pen. m	Par. m	Total m	Greatest in Zenith Lat.	Lng.
10201	511	2235 Jun 02	07:32:48	531	2911	134	T-	pp	-0.0495	2.7501	1.7838	326.7	215.5	100.1	22S	111W
10202	511	2235 Nov 25	13:38:16	532	2917	139	T+	pp	0.0086	2.8647	1.8201	352.8	224.9	102.4	21N	154E
10203	511	2236 May 21	19:18:51	534	2923	144	P	t-	0.7352	1.5194	0.4987	309.1	154.2	-	20S	72E
10204	511	2236 Nov 14	00:24:12	535	2929	149	P	a-	-0.6879	1.5896	0.6019	294.8	158.2	-	17N	8W
10205	511	2237 Apr 11	07:30:45	536	2934	116	N	-t	-1.3795	0.3702	-0.7163	184.3	-	-	9S	111W
10206	511	2237 May 10	23:45:33	536	2935	154	N	t-	1.5296	0.0863	-0.9832	90.2	-	-	16S	5E
10207	511	2237 Oct 05	06:57:02	537	2940	121	N	-a	1.2563	0.5325	-0.4275	186.9	-	-	6N	105W
10208	511	2237 Nov 03	15:53:39	538	2941	159	N	a-	-1.3394	0.3809	-0.5807	159.5	-	-	14N	120E
10209	511	2238 Mar 31	08:05:25	539	2946	126	P	-t	-0.6477	1.7035	0.6362	334.3	175.6	-	5S	118W
10210	511	2238 Sep 24	21:14:02	540	2952	131	P	-a	0.5775	1.7975	0.7992	313.7	180.6	-	1N	42E
10211	511	2239 Mar 20	14:37:27	541	2958	136	T+	p-	0.1017	2.6791	1.6635	341.5	219.8	99.1	0N	145E
10212	511	2239 Sep 14	05:17:34	543	2964	141	T-	pp	-0.1708	2.5701	1.5194	360.2	227.0	97.0	4S	78W
10213	511	2240 Mar 09	03:46:50	544	2970	146	P	a-	0.7925	1.3898	0.4175	277.4	133.8	-	5N	52W
10214	511	2240 Sep 02	06:52:16	545	2976	151	P	t-	-0.9304	1.1894	0.1126	303.5	81.7	-	9S	101W
10215	511	2241 Jan 28	09:08:46	546	2981	118	N	-a	-1.2220	0.6096	-0.3784	202.1	-	-	17N	132W
10216	511	2241 Feb 26	19:46:41	547	2982	156	N	a-	1.4489	0.1833	-0.7848	113.2	-	-	10N	69E
10217	511	2241 Jul 23	19:08:10	548	2987	123	N	-t	1.1245	0.8071	-0.2177	245.9	-	-	19S	77E
10218	511	2242 Jan 17	20:28:26	549	2993	128	P	-a	-0.5794	1.8159	0.7742	325.2	182.2	-	20N	58E
10219	511	2242 Jul 13	05:59:11	550	2999	133	T	-p	0.3129	2.2688	1.2985	322.9	206.7	76.2	22S	86W
10220	511	2243 Jan 07	00:33:44	552	3005	138	T+	pp	0.1116	2.7004	1.6069	372.5	231.4	101.0	23N	5W
10221	512	2243 Jul 02	22:07:19	553	3011	143	T	a-	-0.4355	2.0292	1.0880	304.3	191.8	43.2	23S	32E
10222	512	2243 Dec 26	23:42:17	554	3017	148	P	t-	0.7922	1.4562	0.3533	327.5	139.5	-	24N	7E
10223	512	2244 May 23	06:56:57	555	3022	115	N	-a	1.3850	0.3070	-0.6741	150.3	-	-	19S	102W
10224	512	2244 Jun 21	15:07:43	555	3023	153	N	a-	-1.1783	0.6734	-0.2820	209.0	-	-	25S	136E
10225	512	2244 Nov 15	10:48:54	557	3028	120	N	-a	-1.4201	0.2651	-0.7605	144.8	-	-	17N	163W
10226	512	2244 Dec 15	01:25:20	557	3029	158	N	a-	1.4340	0.2578	-0.8040	148.4	-	-	25N	21W
10227	512	2245 May 12	16:43:42	558	3034	125	P	-t	0.6856	1.6194	0.5809	321.6	166.7	-	18S	111E
10228	512	2245 Nov 04	23:23:16	559	3040	130	P	-a	-0.6707	1.6143	0.6399	292.0	160.4	-	15N	8E
10229	512	2246 May 01	19:20:11	561	3046	135	T-	pp	-0.0575	2.7920	1.7136	374.9	235.2	105.2	15S	72E
10230	512	2246 Oct 25	15:28:38	562	3052	140	T+	pp	0.0236	2.7950	1.8339	319.2	211.1	98.6	12N	126E
10231	512	2247 Apr 20	19:34:24	563	3058	145	P	t-	-0.7870	1.4495	0.3789	320.9	142.4	-	12S	68E
10232	512	2247 Oct 15	06:25:30	565	3064	150	P	a-	0.7471	1.4829	0.4911	292.2	147.1	-	9N	98W
10233	512	2248 Mar 10	12:44:45	566	3069	117	N	-a	1.2717	0.5255	-0.4766	193.9	-	-	5N	174E
10234	512	2248 Apr 09	00:42:30	566	3070	155	N	a-	-1.4664	0.1790	-0.8443	120.9	-	-	9S	9W
10235	512	2248 Sep 04	01:38:56	567	3075	122	N	-t	-1.2857	0.5294	-0.5314	212.4	-	-	8S	22W
10236	512	2248 Oct 03	15:52:04	567	3076	160	Nb	t-	1.5365	0.0611	-0.9836	73.3	-	-	5N	121E
10237	512	2249 Feb 28	03:04:45	568	3081	127	P	-a	0.5480	1.8362	0.8685	299.5	178.2	-	9N	40W
10238	512	2249 Aug 24	02:22:14	570	3087	132	P	-t	-0.5572	1.8728	0.7990	350.9	195.3	-	12S	32W
10239	512	2250 Feb 17	18:58:32	571	3093	137	T-	p-	-0.1291	2.6094	1.6326	322.2	210.7	95.0	12N	81E
10240	512	2250 Aug 13	04:39:38	572	3099	142	T+	pp	0.2108	2.4907	1.4520	356.0	224.5	92.9	15S	66W
10241	513	2251 Feb 07	07:45:40	574	3105	147	P	a-	-0.8497	1.3117	0.2862	288.4	118.1	-	15N	111W
10242	513	2251 Aug 02	13:53:15	575	3111	152	P	a-	0.9548	1.0971	0.1145	263.8	75.6	-	17S	155E
10243	513	2251 Dec 28	19:12:49	576	3116	119	N	-t	1.4131	0.3163	-0.7853	171.1	-	-	25N	75E
10244	513	2252 Jun 22	22:33:27	577	3122	124	P	-a	-0.8494	1.2697	0.3286	264.6	119.3	-	24S	25E
10245	513	2252 Dec 16	18:18:41	579	3128	129	P	-t	0.7406	1.5468	0.4522	331.0	154.3	-	24N	87E
10246	513	2253 Jun 12	14:57:11	580	3134	134	T-	-p	-0.1210	2.6195	1.6520	326.9	214.8	97.6	23S	138E
10247	513	2253 Dec 05	21:55:59	581	3140	139	T+	pp	0.0213	2.8410	1.7974	351.6	224.2	102.0	22N	31E
10248	513	2254 Jun 02	02:24:40	583	3146	144	P	t-	0.6670	1.6452	0.6231	317.9	169.5	-	21S	34W
10249	513	2254 Nov 25	08:54:59	584	3152	149	P	a-	-0.6696	1.6231	0.6353	296.0	161.3	-	20N	135W
10250	513	2255 Apr 22	14:36:07	585	3157	116	N	-t	-1.4324	0.2717	-0.8116	159.2	-	-	13S	143E
10251	513	2255 May 22	06:33:13	585	3158	154	N	t-	1.4629	0.2082	-0.8604	138.7	-	-	19S	96W
10252	513	2255 Oct 16	15:04:25	587	3163	121	N	-a	1.2959	0.4621	-0.5021	175.7	-	-	10N	132E
10253	513	2255 Nov 15	00:24:34	587	3164	159	N	a-	-1.3181	0.4213	-0.5429	167.2	-	-	17N	7W
10254	513	2256 Apr 10	15:31:04	588	3169	126	P	-t	-0.6904	1.6221	0.5606	328.1	166.5	-	9S	130E
10255	513	2256 Oct 05	04:57:13	589	3175	131	P	-a	0.6264	1.7108	0.7064	310.9	173.0	-	5N	75W
10256	513	2257 Mar 30	22:36:42	591	3181	136	T+	pp	0.0687	2.7368	1.7272	340.2	220.0	100.4	4S	24E
10257	513	2257 Sep 24	12:27:34	592	3187	141	T-	pp	-0.1123	2.6802	1.6239	363.1	229.3	101.4	1N	174E
10258	513	2258 Mar 20	12:10:19	593	3193	146	P	a-	0.7663	1.4358	0.4676	279.8	140.3	-	1N	178W
10259	513	2258 Sep 13	13:39:53	595	3199	151	P	t-	-0.8623	1.3159	0.2362	315.1	116.3	-	4S	157E
10260	513	2259 Feb 08	17:51:42	596	3204	118	N	-a	-1.2320	0.5917	-0.3971	199.9	-	-	14N	98E

Cat Num	Canon Plate	Calendar Date	TD of Greatest Eclipse	ΔT s	Luna Num	Saros Num	Ecl. Type	QSE	Gamma	Pen. Mag.	Um. Mag.	Eclipse Phase Durations Pen. m	Par. m	Total m	Greatest in Zenith Lat.	Lng.
10261	514	2259 Mar 10	04:19:51	596	3205	156	N	a-	1.4323	0.2131	-0.7537	121.8	-	-	6N	59W
10262	514	2259 Aug 04	02:02:00	597	3210	123	N	-t	1.1981	0.6707	-0.3516	226.3	-	-	16S	27W
10263	514	2259 Sep 02	14:47:56	597	3211	161	Nb	t-	-1.5668	0.0099	-1.0435	30.2	-	-	9S	141E
10264	514	2260 Jan 29	04:55:22	599	3216	128	P	-a	-0.5871	1.8028	0.7589	325.7	181.3	-	18N	68W
10265	514	2260 Jul 23	13:17:09	600	3222	133	T	-p	0.3867	2.1322	1.1642	318.1	200.5	59.0	19S	165E
10266	514	2261 Jan 17	08:40:06	601	3228	138	T+	pp	0.1036	2.7157	1.6208	373.4	231.9	101.5	21N	125W
10267	514	2261 Jul 13	05:36:18	603	3234	143	T	p-	-0.3617	2.1643	1.2235	308.2	198.1	66.0	22S	80W
10268	514	2262 Jan 06	07:45:00	604	3240	148	P	t-	0.7832	1.4725	0.3702	328.3	142.2	-	23N	112W
10269	514	2262 Jun 03	14:20:47	605	3245	115	N	-a	1.4554	0.1784	-0.8039	116.3	-	-	21S	147E
10270	514	2262 Jul 02	22:31:30	605	3246	153	N	a-	-1.1089	0.8015	-0.1556	225.7	-	-	24S	26E
10271	514	2262 Nov 26	19:06:54	606	3251	120	N	-a	-1.4327	0.2415	-0.7833	138.1	-	-	20N	73E
10272	514	2262 Dec 26	09:43:59	607	3252	158	N	a-	1.4226	0.2778	-0.7821	153.3	-	-	25N	144W
10273	514	2263 May 23	23:46:25	608	3257	125	P	-t	0.7568	1.4891	0.4499	314.2	150.2	-	20S	5E
10274	514	2263 Nov 16	07:52:13	609	3263	130	P	-a	-0.6908	1.5778	0.6027	289.5	156.3	-	18N	119W
10275	514	2264 May 12	02:08:52	611	3269	135	T+	pp	0.0121	2.8742	1.7979	375.2	235.8	106.2	18S	31W
10276	514	2264 Nov 04	23:52:49	612	3275	140	T-	pp	-0.0039	2.8330	1.8685	319.8	211.2	98.6	16N	0E
10277	514	2265 May 01	02:38:34	613	3281	145	P	t-	-0.7260	1.5587	0.4934	327.4	159.3	-	16S	38W
10278	514	2265 Oct 25	14:28:59	615	3287	150	P	a-	0.7146	1.5454	0.5480	297.3	154.4	-	13N	141E
10279	514	2266 Mar 21	20:56:19	616	3292	117	N	-a	1.2951	0.4799	-0.5167	185.4	-	-	1N	51E
10280	514	2266 Apr 20	08:20:18	616	3293	155	N	a-	-1.4163	0.2677	-0.7494	145.9	-	-	13S	123W
10281	515	2266 Sep 15	08:35:37	617	3298	122	N	-t	-1.3530	0.4086	-0.6573	189.3	-	-	4S	127W
10282	515	2266 Oct 14	23:23:02	618	3299	160	N	t-	1.4969	0.1369	-0.9139	109.4	-	-	10N	8E
10283	515	2267 Mar 11	11:37:14	619	3304	127	P	-a	0.5679	1.7981	0.8336	297.4	175.5	-	4N	169W
10284	515	2267 Sep 04	09:02:59	620	3310	132	P	-t	-0.6322	1.7360	0.6605	343.6	181.7	-	8S	133W
10285	515	2268 Feb 29	03:35:27	621	3316	137	T-	p-	-0.1142	2.6365	1.6602	322.9	211.4	96.0	8N	48W
10286	515	2268 Aug 23	11:34:51	623	3322	142	T+	pp	0.1351	2.6290	1.5915	356.7	226.9	99.8	11S	170W
10287	515	2269 Feb 17	16:11:18	624	3328	147	P	a-	-0.8406	1.3289	0.3023	290.7	121.5	-	11N	123E
10288	515	2269 Aug 12	21:11:37	626	3334	152	P	a-	0.8806	1.2324	0.2517	274.5	109.6	-	14S	46E
10289	515	2270 Jan 08	03:16:20	627	3339	119	N	-t	1.4224	0.2999	-0.8031	167.1	-	-	23N	45W
10290	515	2270 Jul 04	06:03:42	628	3345	124	P	-a	-0.9214	1.1374	0.1965	254.2	94.0	-	24S	87W
10291	515	2270 Dec 28	02:22:42	630	3351	129	P	-t	0.7493	1.5304	0.4364	329.3	151.7	-	24N	33W
10292	515	2271 Jun 23	22:18:46	631	3357	134	T-	-p	-0.1943	2.4858	1.5166	326.4	212.9	92.3	24S	29E
10293	515	2271 Dec 17	06:18:50	632	3363	139	T+	pp	0.0303	2.8236	1.7815	350.4	223.5	101.7	23N	93W
10294	515	2272 Jun 12	09:23:36	634	3369	144	P	t-	0.5938	1.7804	0.7565	326.5	183.3	-	23S	138W
10295	515	2272 Dec 05	17:32:31	635	3375	149	P	a-	-0.6565	1.6470	0.6596	296.7	163.3	-	22N	97E
10296	515	2273 May 02	21:31:53	636	3380	116	N	-t	-1.4925	0.1598	-0.9205	123.2	-	-	17S	39E
10297	515	2273 Jun 01	13:13:38	637	3381	154	N	t-	1.3902	0.3413	-0.7267	175.7	-	-	21S	164E
10298	515	2273 Oct 26	23:19:49	638	3386	121	N	-a	1.3288	0.4037	-0.5647	165.5	-	-	14N	8E
10299	515	2273 Nov 25	09:01:53	638	3387	159	N	a-	-1.3020	0.4520	-0.5144	172.7	-	-	19N	136W
10300	515	2274 Apr 21	22:48:59	639	3392	126	P	-t	-0.7401	1.5280	0.4723	320.8	154.6	-	13S	20E
10301	516	2274 Oct 16	12:48:09	641	3398	131	P	-a	0.6683	1.6369	0.6264	308.3	165.5	-	10N	167E
10302	516	2275 Apr 11	06:28:07	642	3404	136	T+	pp	0.0283	2.8077	1.8043	338.8	219.9	101.3	8S	94W
10303	516	2275 Oct 05	19:44:07	643	3410	141	T-	pp	-0.0599	2.7794	1.7172	365.3	230.8	103.7	5N	64E
10304	516	2276 Mar 30	20:26:06	645	3416	146	P	a-	0.7334	1.4941	0.5301	282.8	147.8	-	3S	57E
10305	516	2276 Sep 23	20:35:24	646	3422	151	P	t-	-0.8004	1.4311	0.3482	324.7	138.9	-	0S	52E
10306	516	2277 Feb 19	02:29:15	647	3427	118	N	-a	-1.2459	0.5662	-0.4229	196.5	-	-	10N	32W
10307	516	2277 Mar 20	12:44:38	648	3428	156	N	a-	1.4094	0.2543	-0.7109	132.7	-	-	1N	174E
10308	516	2277 Aug 14	09:01:20	649	3433	123	N	-t	1.2678	0.5419	-0.4785	205.3	-	-	13S	132W
10309	516	2277 Sep 12	21:49:40	649	3434	161	N	t-	-1.4999	0.1328	-0.9207	109.1	-	-	5S	35E
10310	516	2278 Feb 08	13:19:10	650	3439	128	P	-a	-0.5965	1.7864	0.7409	326.0	180.2	-	14N	166E
10311	516	2278 Aug 03	20:37:40	652	3445	133	T	-a	0.4592	1.9983	1.0322	312.7	193.0	27.1	17S	55E
10312	516	2279 Jan 28	16:45:42	653	3451	138	T+	pp	0.0954	2.7312	1.6355	374.2	232.5	102.1	18N	114E
10313	516	2279 Jul 24	13:05:07	654	3457	143	T	p-	-0.2877	2.3002	1.3593	311.4	203.1	79.9	20S	168E
10314	516	2280 Jan 17	15:49:31	656	3463	148	P	t-	0.7749	1.4871	0.3860	329.0	144.8	-	21N	128E
10315	516	2280 Jun 13	21:38:07	657	3468	115	Ne	-a	1.5302	0.0420	-0.9420	57.3	-	-	22S	39E
10316	516	2280 Jul 13	05:52:36	657	3469	153	N	a-	-1.0376	0.9334	-0.0259	240.9	-	-	23S	84W
10317	516	2280 Dec 07	03:32:38	659	3474	120	N	-a	-1.4402	0.2273	-0.7964	133.7	-	-	21N	52W
10318	516	2281 Jan 05	18:06:46	659	3475	158	N	a-	1.4136	0.2931	-0.7644	156.7	-	-	24N	92E
10319	516	2281 Jun 03	06:42:43	660	3480	125	P	-t	0.8322	1.3513	0.3110	305.1	127.9	-	22S	98W
10320	516	2281 Nov 26	16:27:11	661	3486	130	P	-a	-0.7064	1.5494	0.5737	287.5	153.0	-	20N	113E

Cat Num	Canon Plate	Calendar Date	TD of Greatest Eclipse	ΔT s	Luna Num	Saros Num	Ecl. Type	QSE	Gamma	Pen. Mag.	Um. Mag.	Eclipse Phase Durations Pen. m	Par. m	Total m	Greatest in Zenith Lat.	Lng.
10321	517	2282 May 23	08:52:42	663	3492	135	T+	pp	0.0856	2.7385	1.6640	374.6	235.1	104.4	21S	131W
10322	517	2282 Nov 16	08:23:04	664	3498	140	T-	pp	-0.0260	2.7939	1.8261	320.3	211.2	98.4	19N	127W
10323	517	2283 May 12	09:37:20	666	3504	145	P	t-	-0.6596	1.6779	0.6177	333.7	174.4	-	19S	143W
10324	517	2283 Nov 05	22:38:43	667	3510	150	P	a-	0.6877	1.5976	0.5945	301.6	160.1	-	16N	19E
10325	517	2284 Apr 01	05:01:09	668	3515	117	N	-a	1.3247	0.4227	-0.5682	174.2	-	-	4S	71W
10326	517	2284 Apr 30	15:51:43	669	3516	155	N	a-	-1.3601	0.3677	-0.6431	168.7	-	-	16S	124E
10327	517	2284 Sep 25	15:39:31	670	3521	122	N	-t	-1.4139	0.2994	-0.7716	164.2	-	-	0N	126E
10328	517	2284 Oct 25	07:01:35	670	3522	160	N	t-	1.4633	0.2016	-0.8553	132.6	-	-	14N	107W
10329	517	2285 Mar 21	20:01:07	671	3527	127	P	-a	0.5954	1.7459	0.7847	294.6	171.7	-	0S	65E
10330	517	2285 Sep 14	15:53:25	673	3533	132	P	-t	-0.6997	1.6132	0.5357	336.1	166.9	-	4S	124E
10331	517	2286 Mar 11	12:04:37	674	3539	137	T-	pp	-0.0929	2.6753	1.6996	323.7	212.3	97.2	4N	176W
10332	517	2286 Sep 03	18:37:54	675	3545	142	T+	pp	0.0648	2.7576	1.7209	356.6	227.8	103.2	7S	83E
10333	517	2287 Mar 01	00:30:02	677	3551	147	P	a-	-0.8265	1.3553	0.3278	293.6	126.4	-	7N	2W
10334	517	2287 Aug 24	04:35:07	678	3557	152	P	a-	0.8094	1.3624	0.3829	283.4	132.2	-	10S	66W
10335	517	2288 Jan 19	11:19:30	680	3562	119	N	-t	1.4316	0.2832	-0.8203	162.8	-	-	22N	164W
10336	517	2288 Jul 14	13:31:30	681	3568	124	P	-a	-0.9954	1.0018	0.0604	242.3	53.1	-	22S	162E
10337	517	2288 Aug 12	20:15:31	681	3569	162	Nb	a-	1.4946	0.0875	-0.8572	78.4	-	-	13S	60E
10338	517	2289 Jan 07	10:30:56	682	3574	129	P	-t	0.7552	1.5188	0.4263	327.8	149.9	-	23N	153W
10339	517	2289 Jul 04	05:36:25	684	3580	134	T-	-p	-0.2703	2.3477	1.3760	325.1	209.6	83.3	23S	80W
10340	517	2289 Dec 27	14:46:24	685	3586	139	T+	pp	0.0364	2.8113	1.7713	349.1	222.9	101.4	23N	141E
10341	518	2290 Jun 23	16:17:39	687	3592	144	P	t-	0.5172	1.9221	0.8959	334.4	195.3	-	23S	119E
10342	518	2290 Dec 17	02:15:35	688	3598	149	P	a-	-0.6473	1.6636	0.6768	296.9	164.6	-	23N	32W
10343	518	2291 May 14	04:20:25	690	3603	116	Ne	-t	-1.5580	0.0379	-1.0391	60.6	-	-	20S	63W
10344	518	2291 Jun 12	19:47:41	690	3604	154	N	t-	1.3123	0.4841	-0.5835	206.7	-	-	22S	66E
10345	518	2291 Nov 07	07:41:44	691	3609	121	N	-a	1.3564	0.3553	-0.6174	156.4	-	-	18N	117W
10346	518	2291 Dec 06	17:45:10	691	3610	159	N	a-	-1.2908	0.4735	-0.4951	176.6	-	-	21N	94E
10347	518	2292 May 02	06:00:52	692	3615	126	P	-t	-0.7956	1.4233	0.3735	312.0	139.2	-	16S	88W
10348	518	2292 Oct 26	20:45:26	694	3621	131	P	-a	0.7041	1.5743	0.5577	306.0	158.3	-	13N	47E
10349	518	2293 Apr 21	14:13:28	695	3627	136	T-	pp	-0.0179	2.8237	1.8264	337.2	219.5	101.4	12S	149E
10350	518	2293 Oct 16	03:08:18	697	3633	141	T-	pp	-0.0144	2.8658	1.7979	367.1	231.6	104.6	9N	48W
10351	518	2294 Apr 11	04:36:13	698	3639	146	P	a-	0.6956	1.5613	0.6014	286.1	155.5	-	8S	66W
10352	518	2294 Oct 05	03:38:22	700	3645	151	P	t-	-0.7442	1.5356	0.4498	332.5	155.4	-	4N	54W
10353	518	2295 Mar 02	11:01:08	701	3650	118	N	-a	-1.2642	0.5329	-0.4564	191.7	-	-	6N	160W
10354	518	2295 Mar 31	21:02:51	701	3651	156	N	a-	1.3816	0.3044	-0.6593	144.6	-	-	3S	49E
10355	518	2295 Aug 25	16:07:04	702	3656	123	N	-t	1.3328	0.4218	-0.5971	182.5	-	-	9S	121E
10356	518	2295 Sep 24	05:00:02	703	3657	161	N	t-	-1.4387	0.2450	-0.8086	146.4	-	-	1S	74W
10357	518	2296 Feb 19	21:35:59	704	3662	128	P	-a	-0.6112	1.7600	0.7135	325.6	178.2	-	11N	42E
10358	518	2296 Aug 14	04:03:01	705	3668	133	P	-a	0.5282	1.8710	0.9060	306.8	184.5	-	14S	57W
10359	518	2297 Feb 08	00:46:04	707	3674	138	T+	pp	0.0832	2.7535	1.6578	374.9	233.1	102.8	15N	5W
10360	518	2297 Aug 03	20:37:10	708	3680	143	T-	p-	-0.2162	2.4317	1.4901	313.9	206.8	88.9	17S	55E
10361	519	2298 Jan 27	23:51:16	710	3686	148	P	t-	0.7640	1.5060	0.4070	329.8	148.0	-	19N	8E
10362	519	2298 Jul 24	13:14:16	711	3692	153	P	a-	-0.9673	1.0639	0.1017	254.5	70.0	-	21S	166E
10363	519	2298 Dec 18	12:03:40	713	3697	120	N	-a	-1.4439	0.2196	-0.8025	131.1	-	-	22N	179W
10364	519	2299 Jan 17	02:30:29	713	3698	158	N	a-	1.4048	0.3077	-0.7467	159.8	-	-	22N	32W
10365	519	2299 Jun 14	13:31:01	714	3703	125	P	-t	0.9131	1.2036	0.1618	293.7	94.6	-	22S	161E
10366	519	2299 Dec 08	01:09:08	716	3709	130	P	-a	-0.7167	1.5307	0.5547	286.0	150.6	-	22N	16W
10367	519	2300 Jun 03	15:28:06	717	3715	135	T+	pp	0.1655	2.5911	1.5183	373.1	232.8	99.0	22S	131E
10368	519	2300 Nov 27	17:00:33	719	3721	140	T-	-p	-0.0421	2.7660	1.7950	320.8	211.2	98.1	21N	105E
10369	519	2301 May 23	16:29:16	720	3727	145	P	t-	-0.5869	1.8089	0.7536	339.6	187.8	-	21S	115E
10370	519	2301 Nov 17	06:56:02	722	3733	150	P	a-	0.6678	1.6368	0.6283	305.1	164.1	-	19N	105W
10371	519	2302 Apr 13	12:59:53	723	3738	117	N	-a	1.3599	0.3552	-0.6300	160.1	-	-	8S	169E
10372	519	2302 May 12	23:17:54	723	3739	155	N	a-	-1.2986	0.4776	-0.5273	189.4	-	-	19S	12E
10373	519	2302 Oct 07	22:49:32	724	3744	122	N	-t	-1.4691	0.2007	-0.8754	136.1	-	-	4N	18E
10374	519	2302 Nov 06	14:46:26	724	3745	160	N	t-	1.4353	0.2560	-0.8068	149.3	-	-	17N	137E
10375	519	2303 Apr 03	04:19:40	726	3750	127	P	-a	0.6276	1.6851	0.7273	291.2	166.8	-	4S	61W
10376	519	2303 Sep 26	22:50:54	727	3756	132	P	-t	-0.7615	1.5007	0.4212	328.4	150.5	-	1N	19E
10377	519	2304 Mar 22	20:26:42	729	3762	137	T-	pp	-0.0661	2.7240	1.7491	324.6	213.1	98.4	1S	58E
10378	519	2304 Sep 15	01:48:59	730	3768	142	T+	pp	0.0003	2.8756	1.8394	355.7	227.6	103.9	3S	25W
10379	519	2305 Mar 12	08:41:35	732	3774	147	P	a-	-0.8072	1.3909	0.3629	297.2	132.7	-	3N	125W
10380	519	2305 Sep 04	12:05:50	733	3780	152	P	a-	0.7436	1.4828	0.5042	290.6	148.5	-	7S	179W

Cat Num	Canon Plate	Calendar Date	TD of Greatest Eclipse	ΔT s	Luna Num	Saros Num	Ecl. Type	QSE	Gamma	Pen. Mag.	Um. Mag.	Eclipse Phase Durations Pen. m	Par. m	Total m	Greatest in Zenith Lat.	Lng.
10381	520	2306 Jan 30	19:20:01	734	3785	119	N	-t	1.4427	0.2627	-0.8406	157.2	-	-	19N	77E
10382	520	2306 Mar 01	13:42:55	735	3786	157	N	t-	-1.5675	0.0238	-1.0597	47.5	-	-	7N	160E
10383	520	2306 Jul 26	21:01:27	736	3791	124	N	-a	-1.0676	0.8701	-0.0725	229.2	-	-	20S	50E
10384	520	2306 Aug 25	03:53:23	736	3792	162	N	a-	1.4297	0.2071	-0.7385	119.3	-	-	10S	55W
10385	520	2307 Jan 19	18:39:13	738	3797	129	P	-t	0.7616	1.5060	0.4156	326.1	148.0	-	21N	86E
10386	520	2307 Jul 16	12:51:48	739	3803	134	T	-a	-0.3475	2.2075	1.2329	323.0	204.9	69.1	22S	172E
10387	520	2308 Jan 08	23:16:24	741	3809	139	T+	pp	0.0413	2.8010	1.7637	347.8	222.3	101.1	22N	15E
10388	520	2308 Jul 04	23:07:47	742	3815	144	T	t-	0.4377	2.0694	1.0404	341.6	205.5	31.9	22S	17E
10389	520	2308 Dec 28	11:01:33	744	3821	149	P	a-	-0.6401	1.6763	0.6906	296.9	165.6	-	23N	162W
10390	520	2309 Jun 24	02:17:42	745	3827	154	N	t-	1.2308	0.6337	-0.4340	233.4	-	-	22S	31W
10391	520	2309 Nov 18	16:10:56	746	3832	121	N	-a	1.3782	0.3174	-0.6594	148.7	-	-	21N	116E
10392	520	2309 Dec 18	02:32:41	747	3833	159	N	a-	-1.2830	0.4888	-0.4815	179.3	-	-	22N	36W
10393	520	2310 May 14	13:06:26	748	3838	126	P	-t	-0.8566	1.3085	0.2643	301.8	118.7	-	19S	166E
10394	520	2310 Nov 08	04:49:17	749	3844	131	P	-a	0.7339	1.5227	0.5001	304.2	151.8	-	17N	73W
10395	520	2311 May 03	21:52:07	751	3850	136	T-	pp	-0.0706	2.7241	1.7326	335.3	218.5	100.3	16S	34E
10396	520	2311 Oct 28	10:40:16	752	3856	141	T+	pp	0.0241	2.8509	1.7773	368.6	231.9	104.5	13N	161W
10397	520	2312 Apr 22	12:37:09	754	3862	146	P	a-	0.6499	1.6433	0.6873	290.0	163.8	-	12S	174E
10398	520	2312 Oct 16	10:51:23	755	3868	151	P	t-	-0.6960	1.6257	0.5368	338.6	167.4	-	8N	163W
10399	520	2313 Mar 13	19:25:46	757	3873	118	N	-a	-1.2880	0.4890	-0.5001	184.8	-	-	2N	74E
10400	520	2313 Apr 12	05:12:28	757	3874	156	N	a-	1.3473	0.3667	-0.5956	158.0	-	-	7S	74W
10401	521	2313 Sep 05	23:20:24	758	3879	123	N	-t	1.3925	0.3118	-0.7061	158.0	-	-	5S	12E
10402	521	2313 Oct 05	12:20:12	758	3880	161	N	t-	-1.3843	0.3451	-0.7088	171.6	-	-	3N	176E
10403	521	2314 Mar 03	05:46:48	760	3885	128	P	-a	-0.6301	1.7256	0.6784	324.8	175.3	-	7N	81W
10404	521	2314 Aug 26	11:33:39	761	3891	133	P	-a	0.5935	1.7508	0.7866	300.7	175.1	-	10S	170W
10405	521	2315 Feb 20	08:42:48	763	3897	138	T+	pp	0.0684	2.7804	1.6853	375.4	233.6	103.6	11N	124W
10406	521	2315 Aug 16	04:10:27	764	3903	143	T-	p-	-0.1459	2.5616	1.6185	315.7	209.3	94.7	14S	58W
10407	521	2316 Feb 09	07:52:24	766	3909	148	P	t-	0.7523	1.5260	0.4299	330.5	151.4	-	16N	111W
10408	521	2316 Aug 04	20:36:17	767	3915	153	P	a-	-0.8979	1.1930	0.2273	266.6	103.1	-	18S	56E
10409	521	2316 Dec 29	20:38:45	769	3920	120	N	-a	-1.4452	0.2162	-0.8039	129.7	-	-	22N	54E
10410	521	2317 Jan 28	10:53:57	769	3921	158	N	a-	1.3953	0.3233	-0.7273	163.0	-	-	20N	157W
10411	521	2317 Jun 25	20:15:57	770	3926	125	P	-t	0.9955	1.0533	0.0097	280.2	23.7	-	22S	60E
10412	521	2317 Dec 19	09:55:14	772	3932	130	P	-a	-0.7237	1.5180	0.5419	284.9	149.0	-	23N	146W
10413	521	2318 Jun 14	22:00:42	773	3938	135	T+	pp	0.2477	2.4395	1.3681	370.4	228.8	88.9	23S	33E
10414	521	2318 Dec 09	01:41:59	775	3944	140	T-	-p	-0.0546	2.7445	1.7707	321.3	211.2	97.8	23N	24W
10415	521	2319 Jun 03	23:18:19	776	3950	145	P	t-	-0.5109	1.9459	0.8953	344.8	199.2	-	23S	13E
10416	521	2319 Nov 28	15:18:35	778	3956	150	P	a-	0.6528	1.6670	0.6533	308.0	167.1	-	22N	130E
10417	521	2320 Apr 23	20:50:31	779	3961	117	N	-a	1.4021	0.2749	-0.7048	141.4	-	-	11S	51E
10418	521	2320 May 23	06:38:12	780	3962	155	N	a-	-1.2312	0.5984	-0.4008	208.6	-	-	22S	97W
10419	521	2320 Oct 18	06:08:39	781	3967	122	N	-t	-1.5163	0.1166	-0.9645	104.8	-	-	8N	92W
10420	521	2320 Nov 16	22:38:58	781	3968	160	N	t-	1.4138	0.2982	-0.7701	161.1	-	-	20N	19E
10421	522	2321 Apr 13	12:29:34	782	3973	127	P	-a	0.6671	1.6109	0.6562	287.1	160.3	-	8S	176E
10422	522	2321 Oct 07	05:58:35	784	3979	132	P	-t	-0.8152	1.4033	0.3218	320.9	133.4	-	5N	89W
10423	522	2322 Apr 03	04:39:45	785	3985	137	T-	pp	-0.0323	2.7857	1.8116	325.4	213.9	99.4	5S	66W
10424	522	2322 Sep 26	09:09:54	787	3991	142	T-	pp	-0.0568	2.7718	1.7360	354.3	226.5	102.8	1N	136W
10425	522	2323 Mar 23	16:43:41	789	3997	147	P	h-	-0.7809	1.4392	0.4110	301.6	140.6	-	1S	114E
10426	522	2323 Sep 15	19:43:08	790	4003	152	P	a-	0.6823	1.5951	0.6167	296.5	160.9	-	3S	66E
10427	522	2324 Feb 11	03:17:27	791	4008	119	N	-t	1.4561	0.2377	-0.8646	149.9	-	-	16N	42W
10428	522	2324 Mar 11	21:29:30	792	4009	157	N	t-	-1.5451	0.0645	-1.0182	78.0	-	-	2N	43E
10429	522	2324 Aug 06	04:31:49	793	4014	124	N	-a	-1.1392	0.7395	-0.2050	214.5	-	-	18S	63W
10430	522	2324 Sep 04	11:36:05	793	4015	162	N	a-	1.3683	0.3205	-0.6266	146.8	-	-	6S	171W
10431	522	2325 Jan 30	02:46:50	795	4020	129	P	-t	0.7692	1.4906	0.4033	324.2	145.8	-	18N	35W
10432	522	2325 Jul 26	20:06:28	796	4026	134	T	-a	-0.4246	2.0679	1.0898	320.2	198.6	45.0	20S	64E
10433	522	2326 Jan 19	07:48:32	798	4032	139	T+	pp	0.0454	2.7918	1.7577	346.4	221.7	100.9	20N	111W
10434	522	2326 Jul 16	05:54:47	799	4038	144	T	p-	0.3565	2.2200	1.1878	347.8	214.0	65.9	21S	84W
10435	522	2327 Jan 08	19:50:18	801	4044	149	P	a-	-0.6349	1.6851	0.7008	296.7	166.3	-	22N	67E
10436	522	2327 Jul 05	08:44:32	802	4050	154	N	t-	1.1463	0.7888	-0.2792	256.7	-	-	22S	127W
10437	522	2327 Nov 30	00:45:50	804	4055	121	N	-a	1.3949	0.2887	-0.6920	142.6	-	-	23N	11W
10438	522	2327 Dec 29	11:23:32	804	4056	159	N	a-	-1.2784	0.4978	-0.4739	180.9	-	-	22N	168W
10439	522	2328 May 24	20:06:39	805	4061	126	P	-t	-0.9225	1.1848	0.1460	289.9	89.5	-	22S	61E
10440	522	2328 Nov 18	12:59:37	807	4067	131	P	-a	0.7574	1.4824	0.4541	302.9	146.1	-	20N	165E

Cat Num	Canon Plate	Calendar Date	TD of Greatest Eclipse	ΔT s	Luna Num	Saros Num	Ecl. Type	QSE	Gamma	Pen. Mag.	Um. Mag.	Eclipse Phase Durations Pen. m	Par. m	Total m	Greatest in Zenith Lat.	Lng.
10441	523	2329 May 14	05:26:11	808	4073	136	T-	pp	-0.1279	2.6162	1.6303	332.9	216.8	97.6	19S	79W
10442	523	2329 Nov 07	18:18:33	810	4079	141	T+	pp	0.0568	2.7935	1.7146	369.8	231.9	103.7	16N	85E
10443	523	2330 May 03	20:33:07	812	4085	146	P	a-	0.5998	1.7334	0.7810	293.9	171.8	-	15S	55E
10444	523	2330 Oct 27	18:12:55	813	4091	151	P	t-	-0.6542	1.7038	0.6121	343.5	176.4	-	12N	86E
10445	523	2331 Mar 25	03:42:19	815	4096	118	N	-a	-1.3179	0.4341	-0.5548	175.5	-	-	3S	51W
10446	523	2331 Apr 23	13:14:05	815	4097	156	N	a-	1.3070	0.4401	-0.5211	172.1	-	-	11S	165E
10447	523	2331 Sep 17	06:41:33	816	4102	123	N	-t	1.4465	0.2123	-0.8048	131.1	-	-	1S	99W
10448	523	2331 Oct 16	19:49:56	816	4103	161	N	t-	-1.3365	0.4329	-0.6213	190.2	-	-	8N	63E
10449	523	2332 Mar 13	13:48:34	818	4108	128	P	-a	-0.6557	1.6788	0.6313	323.2	170.9	-	2N	158E
10450	523	2332 Sep 05	19:11:20	819	4114	133	P	-a	0.6533	1.6412	0.6771	294.4	165.0	-	6S	75E
10451	523	2333 Mar 02	16:30:27	821	4120	138	T+	pp	0.0466	2.8198	1.7259	375.9	234.3	104.5	7N	119E
10452	523	2333 Aug 26	11:49:39	822	4126	143	T-	pp	-0.0807	2.6822	1.7372	317.0	210.8	97.9	10S	173W
10453	523	2334 Feb 19	15:48:18	824	4132	148	P	t-	0.7362	1.5538	0.4612	331.6	155.8	-	12N	130E
10454	523	2334 Aug 16	03:59:52	826	4138	153	P	a-	-0.8304	1.3190	0.3493	277.3	125.9	-	15S	55W
10455	523	2335 Jan 10	05:16:00	827	4143	120	N	-a	-1.4453	0.2147	-0.8029	128.9	-	-	21N	74W
10456	523	2335 Feb 08	19:15:18	827	4144	158	N	a-	1.3834	0.3429	-0.7034	166.9	-	-	16N	78E
10457	523	2335 Jul 07	02:55:54	829	4149	125	N	-t	1.0806	0.8983	-0.1476	263.8	-	-	22S	39W
10458	523	2335 Dec 30	18:45:24	830	4155	130	P	-a	-0.7278	1.5101	0.5345	284.0	148.0	-	22N	83E
10459	523	2336 Jun 25	04:27:14	832	4161	135	T	-t	0.3347	2.2794	1.2090	366.4	222.5	71.0	23S	62W
10460	523	2336 Dec 19	10:28:54	833	4167	140	T-	-p	-0.0624	2.7315	1.7552	321.9	211.3	97.6	23N	155W
10461	524	2337 Jun 14	06:03:17	835	4173	145	T	t-	-0.4307	2.0911	1.0448	349.2	208.8	34.0	24S	87W
10462	524	2337 Dec 08	23:45:53	837	4179	150	P	a-	0.6423	1.6887	0.6702	310.4	169.2	-	23N	5E
10463	524	2338 May 05	04:35:58	838	4184	117	N	-a	1.4491	0.1860	-0.7884	116.8	-	-	15S	66W
10464	524	2338 Jun 03	13:55:27	838	4185	155	N	a-	-1.1602	0.7259	-0.2680	225.8	-	-	23S	154E
10465	524	2338 Oct 29	13:35:05	839	4190	122	N	-t	-1.5569	0.0446	-1.0414	65.4	-	-	12N	156E
10466	524	2338 Nov 28	06:37:25	840	4191	160	N	t-	1.3976	0.3305	-0.7428	169.7	-	-	22N	99W
10467	524	2339 Apr 24	20:32:43	841	4196	127	P	-a	0.7125	1.5262	0.5746	282.1	151.9	-	12S	55E
10468	524	2339 Oct 18	13:14:22	843	4202	132	P	-t	-0.8625	1.3175	0.2339	313.8	115.1	-	9N	162E
10469	524	2340 Apr 13	12:45:17	844	4208	137	T+	pp	0.0074	2.8310	1.8576	326.1	214.5	99.9	9S	172E
10470	524	2340 Oct 06	16:39:55	846	4214	142	T-	pp	-0.1074	2.6788	1.6431	352.6	224.9	100.4	5N	111E
10471	524	2341 Apr 03	00:36:59	848	4220	147	P	h-	-0.7480	1.4997	0.4713	306.6	149.6	-	6S	5W
10472	524	2341 Sep 26	03:29:16	849	4226	152	P	a-	0.6276	1.6955	0.7172	301.1	170.2	-	2N	51W
10473	524	2342 Feb 21	11:09:44	850	4231	119	N	-t	1.4735	0.2049	-0.8958	139.6	-	-	12N	160W
10474	524	2342 Mar 23	05:08:18	851	4232	157	N	t-	-1.5166	0.1161	-0.9654	104.3	-	-	2S	73W
10475	524	2342 Aug 17	12:05:10	852	4237	124	N	-a	-1.2083	0.6140	-0.3328	198.2	-	-	15S	176W
10476	524	2342 Sep 15	19:24:22	852	4238	162	N	a-	1.3113	0.4261	-0.5229	167.6	-	-	2S	71E
10477	524	2343 Feb 10	10:51:39	854	4243	129	P	-t	0.7796	1.4698	0.3860	321.7	142.8	-	15N	156W
10478	524	2343 Aug 07	03:21:51	855	4249	134	P	-a	-0.5005	1.9307	0.9484	316.5	190.7	-	17S	45W
10479	524	2344 Jan 30	16:18:41	857	4255	139	T+	pp	0.0518	2.7782	1.7479	345.0	221.1	100.6	18N	122E
10480	524	2344 Jul 26	12:40:59	859	4261	144	T	pp	0.2752	2.3711	1.3351	353.2	220.7	83.8	19S	175E
10481	525	2345 Jan 19	04:38:50	860	4267	149	P	a-	-0.6290	1.6948	0.7125	296.6	167.1	-	20N	64W
10482	525	2345 Jul 15	15:10:59	862	4273	154	N	t-	1.0613	0.9453	-0.1235	276.8	-	-	20S	137E
10483	525	2345 Dec 10	09:24:58	863	4278	121	N	-a	1.4083	0.2658	-0.7184	137.4	-	-	24N	140W
10484	525	2346 Jan 08	20:14:46	863	4279	159	N	a-	-1.2743	0.5059	-0.4667	182.4	-	-	21N	61E
10485	525	2346 Jun 05	03:03:17	865	4284	126	P	-t	-0.9921	1.0545	0.0209	276.2	34.4	-	23S	42W
10486	525	2346 Nov 29	21:15:59	866	4290	131	P	-a	0.7756	1.4517	0.4180	302.1	141.3	-	22N	41E
10487	525	2347 May 25	12:53:35	868	4296	136	T-	pp	-0.1916	2.4966	1.5159	330.0	214.2	92.6	21S	170E
10488	525	2347 Nov 19	02:04:59	870	4302	141	T+	pp	0.0825	2.7489	1.6651	370.8	231.8	102.7	19N	31W
10489	525	2348 May 14	04:21:14	871	4308	146	P	a-	0.5425	1.8368	0.8878	298.0	179.7	-	18S	62W
10490	525	2348 Nov 07	01:44:25	873	4314	151	P	t-	-0.6202	1.7674	0.6733	347.2	183.0	-	16N	26W
10491	525	2349 Apr 04	11:49:55	874	4319	118	N	-a	-1.3547	0.3665	-0.6222	162.7	-	-	7S	173W
10492	525	2349 May 03	21:07:25	875	4320	156	N	a-	1.2602	0.5254	-0.4348	186.8	-	-	15S	46E
10493	525	2349 Sep 27	14:12:31	876	4325	123	N	-h	1.4935	0.1259	-0.8907	101.4	-	-	3N	148E
10494	525	2349 Oct 27	03:30:27	876	4326	161	N	h-	-1.2963	0.5069	-0.5476	203.8	-	-	11N	53W
10495	525	2350 Mar 24	21:42:58	878	4331	128	P	-h	-0.6865	1.6222	0.5747	320.9	165.1	-	2S	39E
10496	525	2350 Sep 17	02:54:54	879	4337	133	P	-a	0.7087	1.5396	0.5752	288.1	154.3	-	2S	41W
10497	525	2351 Mar 14	00:12:41	881	4343	138	T+	pp	0.0207	2.8665	1.7744	376.3	234.8	105.1	3N	3E
10498	525	2351 Sep 06	19:32:42	883	4349	143	T-	pp	-0.0188	2.7969	1.8495	317.7	211.4	99.1	6S	70E
10499	525	2352 Mar 01	23:39:18	884	4355	148	P	t-	0.7155	1.5898	0.5013	333.1	161.1	-	8N	12E
10500	525	2352 Aug 26	11:26:40	886	4361	153	P	a-	-0.7660	1.4394	0.4651	286.8	143.1	-	11S	167W

Cat Num	Canon Plate	Calendar Date	TD of Greatest Eclipse	ΔT s	Luna Num	Saros Num	Ecl. Type	QSE	Gamma	Pen. Mag.	Um. Mag.	Eclipse Phase Durations Pen. m	Par. m	Total m	Greatest in Zenith Lat.	Lng.
10501	526	2353 Jan 20	13:54:28	887	4366	120	N	-a	-1.4453	0.2132	-0.8012	128.0	-	-	19N	157E
10502	526	2353 Feb 19	03:33:51	888	4367	158	N	a-	1.3686	0.3676	-0.6738	171.7	-	-	13N	46W
10503	526	2353 Jul 17	09:35:20	889	4372	125	N	-t	1.1652	0.7444	-0.3043	244.7	-	-	20S	138W
10504	526	2354 Jan 10	03:35:57	891	4378	130	P	-a	-0.7321	1.5020	0.5271	283.2	147.0	-	21N	49W
10505	526	2354 Jul 06	10:53:56	892	4384	135	T	-t	0.4214	2.1198	1.0501	361.2	214.0	36.7	22S	158W
10506	526	2354 Dec 30	19:17:27	894	4390	140	T-	-p	-0.0685	2.7214	1.7430	322.5	211.4	97.5	23N	75E
10507	526	2355 Jun 25	12:46:19	895	4396	145	T	t-	-0.3479	2.2409	1.1986	352.7	216.4	68.0	24S	173E
10508	526	2355 Dec 20	08:16:43	897	4402	150	P	a-	0.6355	1.7033	0.6805	312.4	170.6	-	24N	121W
10509	526	2356 May 15	12:14:44	899	4407	117	N	-a	1.5017	0.0871	-0.8824	80.3	-	-	18S	180E
10510	526	2356 Jun 13	21:09:23	899	4408	155	N	a-	-1.0858	0.8601	-0.1291	241.4	-	-	24S	46E
10511	526	2356 Dec 08	14:40:36	900	4414	160	N	t-	1.3857	0.3546	-0.7232	176.0	-	-	24N	141E
10512	526	2357 May 05	04:27:44	902	4419	127	P	-a	0.7646	1.4292	0.4802	275.9	140.9	-	15S	64W
10513	526	2357 Oct 28	20:41:13	904	4425	132	P	-t	-0.9009	1.2478	0.1625	307.4	96.8	-	13N	50E
10514	526	2358 Apr 24	20:41:46	905	4431	137	T+	pp	0.0542	2.7449	1.7720	326.7	214.8	99.5	13S	53E
10515	526	2358 Oct 18	00:19:34	907	4437	142	T-	pp	-0.1509	2.5990	1.5634	350.7	222.9	97.2	9N	5W
10516	526	2359 Apr 14	08:20:06	909	4443	147	P	t-	-0.7077	1.5737	0.5452	312.4	159.6	-	10S	121W
10517	526	2359 Oct 07	11:23:43	910	4449	152	P	a-	0.5791	1.7845	0.8059	304.7	177.3	-	6N	170W
10518	526	2360 Mar 03	18:54:38	912	4454	119	N	-t	1.4964	0.1618	-0.9366	124.5	-	-	8N	84E
10519	526	2360 Apr 02	12:35:50	912	4455	157	N	t-	-1.4797	0.1831	-0.8970	130.4	-	-	6S	175E
10520	526	2360 Aug 27	19:41:56	913	4460	124	N	-a	-1.2741	0.4947	-0.4551	180.3	-	-	11S	69E
10521	527	2360 Sep 26	03:19:54	914	4461	162	N	a-	1.2602	0.5211	-0.4303	183.7	-	-	2N	49W
10522	527	2361 Feb 20	18:53:07	915	4466	129	P	-t	0.7931	1.4428	0.3633	318.7	138.7	-	11N	84E
10523	527	2361 Aug 17	10:38:41	917	4472	134	P	-a	-0.5740	1.7981	0.8112	312.1	181.2	-	14S	154W
10524	527	2362 Feb 10	00:47:27	918	4478	139	T+	pp	0.0599	2.7611	1.7352	343.5	220.5	100.3	15N	5W
10525	527	2362 Aug 06	19:27:11	920	4484	144	T+	pp	0.1948	2.5207	1.4807	357.5	225.7	94.9	16S	74E
10526	527	2363 Jan 30	13:27:21	922	4490	149	P	a-	-0.6231	1.7045	0.7245	296.4	167.9	-	17N	165E
10527	527	2363 Jul 26	21:36:02	923	4496	154	P	t-	0.9748	1.1044	0.0346	294.5	45.8	-	19S	41E
10528	527	2363 Dec 21	18:08:30	925	4501	121	N	-a	1.4179	0.2498	-0.7376	133.8	-	-	25N	91E
10529	527	2364 Jan 20	05:07:05	925	4502	159	N	a-	-1.2715	0.5114	-0.4618	183.4	-	-	19N	71W
10530	527	2364 Jun 15	09:57:15	926	4507	126	N	-t	-1.0645	0.9192	-0.1097	260.4	-	-	24S	145W
10531	527	2364 Dec 10	05:36:17	928	4513	131	P	-a	0.7898	1.4281	0.3895	301.8	137.4	-	24N	82W
10532	527	2365 Jun 04	20:18:36	930	4519	136	T-	-p	-0.2581	2.3722	1.3963	326.5	210.5	84.9	23S	59E
10533	527	2365 Nov 29	09:57:07	931	4525	141	T+	pp	0.1031	2.7133	1.6251	371.8	231.6	101.7	22N	148W
10534	527	2366 May 25	12:05:08	933	4531	146	T	a-	0.4817	1.9469	1.0007	302.0	186.9	3.9	20S	178W
10535	527	2366 Nov 18	09:23:06	935	4537	151	P	t-	-0.5917	1.8207	0.7245	350.0	188.0	-	19N	141W
10536	527	2367 Apr 15	19:48:55	936	4542	118	N	-a	-1.3977	0.2876	-0.7010	145.6	-	-	11S	66E
10537	527	2367 May 15	04:53:36	937	4543	156	N	a-	1.2082	0.6207	-0.3390	201.5	-	-	18S	70W
10538	527	2367 Oct 08	21:52:39	938	4548	123	N	-h	1.5336	0.0521	-0.9642	65.4	-	-	7N	32E
10539	527	2367 Nov 07	11:20:25	938	4549	161	N	h-	-1.2630	0.5678	-0.4867	213.8	-	-	15N	170W
10540	527	2368 Apr 04	05:25:31	940	4554	128	P	-h	-0.7261	1.5496	0.5021	317.3	156.6	-	6S	77W
10541	528	2368 Sep 27	10:47:20	941	4560	133	P	-a	0.7571	1.4511	0.4862	282.1	143.5	-	3N	160W
10542	528	2369 Mar 24	07:44:15	943	4566	138	T-	pp	-0.0135	2.8786	1.7885	376.4	235.1	105.4	1S	111W
10543	528	2369 Sep 17	03:22:09	945	4572	143	T+	pp	0.0374	2.7643	1.8139	318.1	211.3	98.8	2S	48W
10544	528	2370 Mar 13	07:22:52	946	4578	148	P	t-	0.6883	1.6373	0.5534	335.2	167.6	-	4N	104W
10545	528	2370 Sep 06	18:57:22	948	4584	153	P	a-	-0.7056	1.5529	0.5736	295.0	156.7	-	7S	80E
10546	528	2371 Jan 31	22:31:48	950	4589	120	N	-a	-1.4466	0.2089	-0.8019	126.3	-	-	16N	29E
10547	528	2371 Mar 02	11:47:04	950	4590	158	N	a-	1.3489	0.4010	-0.6350	178.0	-	-	9N	169W
10548	528	2371 Jul 28	16:11:29	951	4595	125	N	-t	1.2511	0.5884	-0.4635	221.7	-	-	18S	123E
10549	528	2371 Aug 27	05:10:45	952	4596	163	Nb	t-	-1.5155	0.0931	-0.9387	89.7	-	-	12S	73W
10550	528	2372 Jan 21	12:27:44	953	4601	130	P	-a	-0.7353	1.4953	0.5218	282.4	146.2	-	19N	180E
10551	528	2372 Jul 16	17:18:29	955	4607	135	P	-t	0.5095	1.9581	0.8887	354.5	202.8	-	21S	106E
10552	528	2373 Jan 10	04:07:33	956	4613	140	T-	-p	-0.0728	2.7144	1.7343	323.0	211.6	97.4	22N	56W
10553	528	2373 Jul 05	19:28:57	958	4619	145	T-	t-	-0.2642	2.3928	1.3540	355.1	222.1	85.7	23S	73E
10554	528	2373 Dec 30	16:50:27	960	4625	150	P	a-	0.6322	1.7113	0.6848	314.0	171.5	-	24N	112E
10555	528	2374 Jun 25	04:21:36	962	4631	155	P	a-	-1.0090	0.9989	0.0140	255.3	26.9	-	24S	61W
10556	528	2374 Dec 19	22:47:53	963	4637	160	N	t-	1.3775	0.3715	-0.7103	180.5	-	-	25N	21E
10557	528	2375 May 16	12:17:12	965	4642	127	P	-a	0.8213	1.3240	0.3774	268.7	126.9	-	18S	179E
10558	528	2375 Nov 09	04:16:34	966	4648	132	P	-t	-0.9330	1.1898	0.1030	301.8	77.6	-	16N	64W
10559	528	2376 May 05	04:29:55	968	4654	137	T+	-p	0.1074	2.6471	1.6744	326.9	214.4	97.8	16S	64W
10560	528	2376 Oct 28	08:09:00	970	4660	142	T-	pp	-0.1871	2.5326	1.4971	348.7	220.9	93.7	13N	122W

Cat Num	Canon Plate	Calendar Date	TD of Greatest Eclipse	ΔT s	Luna Num	Saros Num	Ecl. Type	QSE	Gamma	Pen. Mag.	Um. Mag.	Eclipse Phase Durations Pen. m	Par. m	Total m	Greatest in Zenith Lat.	Lng.
10561	529	2377 Apr 24	15:54:42	972	4666	147	P	t-	-0.6611	1.6595	0.6307	318.5	169.8	-	14S	125E
10562	529	2377 Oct 17	19:26:08	973	4672	152	P	a-	0.5368	1.8623	0.8833	307.4	182.7	-	10N	69E
10563	529	2378 Mar 15	02:32:23	975	4677	119	N	-t	1.5246	0.1087	-0.9871	102.5	-	-	4N	31W
10564	529	2378 Apr 13	19:55:16	975	4678	157	N	t-	-1.4366	0.2615	-0.8170	154.9	-	-	10S	65E
10565	529	2378 Sep 08	03:23:53	976	4683	124	N	-a	-1.3354	0.3840	-0.5691	160.9	-	-	7S	47W
10566	529	2378 Oct 07	11:22:21	977	4684	162	N	a-	1.2148	0.6057	-0.3484	196.5	-	-	7N	170W
10567	529	2379 Mar 04	02:48:34	978	4689	129	P	-t	0.8120	1.4056	0.3310	314.8	132.8	-	7N	35W
10568	529	2379 Aug 28	17:58:09	980	4695	134	P	-a	-0.6445	1.6712	0.6792	307.1	170.1	-	10S	95E
10569	529	2380 Feb 21	09:11:43	982	4701	139	T+	pp	0.0725	2.7355	1.7144	341.9	219.9	99.9	11N	131W
10570	529	2380 Aug 17	02:16:10	983	4707	144	T+	pp	0.1172	2.6655	1.6209	360.8	229.1	101.5	13S	29W
10571	529	2381 Feb 09	22:11:31	985	4713	149	P	a-	-0.6131	1.7214	0.7442	296.6	169.4	-	14N	34E
10572	529	2381 Aug 06	04:04:39	987	4719	154	P	t-	0.8908	1.2594	0.1881	309.5	104.5	-	16S	56W
10573	529	2382 Jan 01	02:54:07	988	4724	121	N	-a	1.4262	0.2361	-0.7541	130.5	-	-	24N	39W
10574	529	2382 Jan 30	13:57:10	989	4725	159	N	a-	-1.2668	0.5200	-0.4533	185.0	-	-	16N	157E
10575	529	2382 Jun 26	16:49:14	990	4730	126	N	-t	-1.1391	0.7802	-0.2444	242.3	-	-	24S	113E
10576	529	2382 Dec 21	14:00:40	992	4736	131	P	-a	0.8002	1.4114	0.3684	301.8	134.4	-	24N	153E
10577	529	2383 Jun 16	03:39:21	994	4742	136	T	-p	-0.3291	2.2398	1.2682	322.2	205.5	73.0	24S	50W
10578	529	2383 Dec 10	17:55:40	995	4748	141	T+	pp	0.1179	2.6881	1.5961	372.6	231.5	100.8	23N	93E
10579	529	2384 Jun 04	19:42:06	997	4754	146	T	a-	0.4149	2.0683	1.1244	305.7	193.6	50.8	22S	68E
10580	529	2384 Nov 28	17:11:03	999	4760	151	P	t-	-0.5705	1.8604	0.7626	351.9	191.5	-	21N	103E
10581	530	2385 Apr 26	03:39:11	1000	4765	118	N	-a	-1.4471	0.1969	-0.7917	121.8	-	-	15S	51W
10582	530	2385 May 25	12:32:57	1001	4766	156	N	a-	1.1510	0.7256	-0.2340	216.1	-	-	20S	175E
10583	530	2385 Nov 17	19:18:57	1002	4772	161	N	h-	-1.2358	0.6177	-0.4366	221.3	-	-	18N	71E
10584	530	2386 Apr 15	12:59:59	1004	4777	128	P	-h	-0.7715	1.4663	0.4189	312.6	145.3	-	11S	169E
10585	530	2386 Oct 08	18:46:46	1006	4783	133	P	-a	0.8002	1.3725	0.4066	276.4	132.5	-	7N	79E
10586	530	2387 Apr 04	15:07:51	1007	4789	138	T-	pp	-0.0541	2.8029	1.7153	376.2	235.0	104.9	6S	138E
10587	530	2387 Sep 28	11:16:56	1009	4795	143	T+	-p	0.0890	2.6712	1.7176	318.1	210.7	97.3	2N	167W
10588	530	2388 Mar 23	15:00:16	1011	4801	148	P	t-	0.6554	1.6951	0.6163	337.6	174.8	-	1S	141E
10589	530	2388 Sep 17	02:33:17	1013	4807	153	P	a-	-0.6501	1.6574	0.6727	302.2	167.5	-	3S	35W
10590	530	2389 Feb 11	07:06:06	1014	4812	120	N	-a	-1.4510	0.1989	-0.8078	122.9	-	-	13N	99W
10591	530	2389 Mar 12	19:54:34	1014	4813	158	N	a-	1.3239	0.4441	-0.5863	185.8	-	-	4N	68E
10592	530	2389 Aug 07	22:50:34	1016	4818	125	N	-t	1.3334	0.4392	-0.6162	194.9	-	-	15S	23E
10593	530	2389 Sep 06	12:15:03	1016	4819	163	N	t-	-1.4545	0.2079	-0.8294	133.3	-	-	8S	179W
10594	530	2390 Jan 31	21:17:34	1018	4824	130	P	-a	-0.7404	1.4853	0.5133	281.5	145.1	-	17N	48E
10595	530	2390 Jul 27	23:44:46	1019	4830	135	P	-t	0.5961	1.7992	0.7298	346.7	188.9	-	18S	10E
10596	530	2391 Jan 21	12:56:41	1021	4836	140	T-	-p	-0.0775	2.7063	1.7249	323.6	211.8	97.3	20N	173E
10597	530	2391 Jul 17	02:12:50	1023	4842	145	T-	pp	-0.1807	2.5444	1.5086	356.4	225.9	96.3	21S	27W
10598	530	2392 Jan 11	01:23:56	1025	4848	150	P	a-	0.6291	1.7186	0.6889	315.6	172.4	-	23N	15W
10599	530	2392 Jul 05	11:33:11	1026	4854	155	P	a-	-0.9306	1.1410	0.1598	267.6	88.7	-	24S	168W
10600	530	2392 Dec 30	06:57:20	1028	4860	160	N	t-	1.3717	0.3838	-0.7011	183.8	-	-	24N	100W
10601	531	2393 May 26	20:00:14	1030	4865	127	P	-a	0.8834	1.2091	0.2643	260.2	108.0	-	20S	64E
10602	531	2393 Nov 19	12:00:29	1031	4871	132	P	-t	-0.9583	1.1439	0.0560	297.0	57.5	-	19N	179W
10603	531	2394 May 16	12:10:00	1033	4877	137	T+	-p	0.1667	2.5384	1.5656	326.6	213.3	94.3	19S	179W
10604	531	2394 Nov 08	16:08:09	1035	4883	142	T-	pp	-0.2159	2.4796	1.4443	346.8	218.9	90.3	16N	118E
10605	531	2395 May 05	23:17:48	1037	4889	147	P	t-	-0.6059	1.7608	0.7316	325.0	180.5	-	17S	14E
10606	531	2395 Oct 29	03:37:44	1039	4895	152	P	a-	0.5017	1.9270	0.9474	309.4	186.6	-	14N	54W
10607	531	2396 Mar 25	10:01:24	1040	4900	119	Ne	-t	1.5593	0.0434	-1.0492	65.1	-	-	1S	144W
10608	531	2396 Apr 24	03:03:43	1040	4901	157	N	t-	-1.3852	0.3549	-0.7219	179.1	-	-	14S	43W
10609	531	2396 Sep 18	11:11:24	1042	4906	124	N	-a	-1.3918	0.2825	-0.6746	139.6	-	-	3S	164W
10610	531	2396 Oct 17	19:32:43	1042	4907	162	N	a-	1.1760	0.6783	-0.2786	206.6	-	-	11N	67E
10611	531	2397 Mar 14	10:38:04	1044	4912	129	P	-t	0.8363	1.3584	0.2891	310.0	124.6	-	3N	153W
10612	531	2397 Sep 08	01:22:03	1045	4918	134	P	-a	-0.7104	1.5531	0.5556	301.7	157.4	-	6S	16W
10613	531	2398 Mar 03	17:32:53	1047	4924	139	T+	pp	0.0885	2.7036	1.6877	340.2	219.1	99.2	7N	104E
10614	531	2398 Aug 28	09:06:27	1049	4930	144	T+	pp	0.0414	2.8069	1.7575	363.3	230.9	104.6	10S	132W
10615	531	2399 Feb 21	06:53:33	1051	4936	149	P	a-	-0.6013	1.7416	0.7675	297.0	171.1	-	10N	96W
10616	531	2399 Aug 17	10:35:26	1053	4942	154	P	t-	0.8080	1.4121	0.3391	322.3	137.2	-	13S	154W
10617	531	2400 Jan 12	11:40:47	1054	4947	121	N	-a	1.4330	0.2247	-0.7678	127.8	-	-	23N	169W
10618	531	2400 Feb 10	22:44:27	1054	4948	159	N	a-	-1.2606	0.5313	-0.4417	187.0	-	-	13N	26E
10619	531	2400 Jul 06	23:41:22	1056	4953	126	N	-t	-1.2140	0.6409	-0.3798	221.8	-	-	24S	11E
10620	531	2400 Aug 05	12:01:19	1056	4954	164	Nb	t-	1.5295	0.0755	-0.9722	82.8	-	-	15S	175W

Cat Num	Canon Plate	Calendar Date	TD of Greatest Eclipse	ΔT s	Luna Num	Saros Num	Ecl. Type	QSE	Gamma	Pen. Mag.	Um. Mag.	Eclipse Phase Durations Pen. m	Par. m	Total m	Greatest in Zenith Lat.	Lng.
10621	532	2400 Dec 31	22:26:34	1058	4959	131	P	-a	0.8085	1.3980	0.3512	301.9	132.0	-	24N	29E
10622	532	2401 Jun 26	10:59:50	1059	4965	136	T	-p	-0.4007	2.1066	1.1387	317.2	199.2	54.6	24S	160W
10623	532	2401 Dec 21	01:56:07	1061	4971	141	T+	pp	0.1306	2.6663	1.5711	373.3	231.4	99.9	24N	25W
10624	532	2402 Jun 16	03:16:59	1063	4977	146	T	p-	0.3462	2.1934	1.2513	309.1	199.2	69.4	23S	44W
10625	532	2402 Dec 10	01:04:27	1065	4983	151	P	t-	-0.5534	1.8923	0.7935	353.2	194.1	-	22N	14W
10626	532	2403 May 07	11:19:39	1066	4988	118	N	-a	-1.5036	0.0934	-0.8955	85.0	-	-	18S	167W
10627	532	2403 Jun 05	20:05:08	1067	4989	156	N	a-	1.0884	0.8406	-0.1195	230.4	-	-	21S	63E
10628	532	2403 Nov 29	03:25:44	1068	4995	161	N	a-	-1.2145	0.6565	-0.3973	226.6	-	-	20N	50W
10629	532	2404 Apr 25	20:22:55	1070	5000	128	P	-t	-0.8255	1.3670	0.3197	306.1	129.2	-	14S	58E
10630	532	2404 Oct 19	02:55:17	1072	5006	133	P	-a	0.8362	1.3070	0.3400	271.4	122.2	-	11N	43W
10631	532	2405 Apr 14	22:20:34	1074	5012	138	T-	pp	-0.1033	2.7113	1.6262	375.5	234.3	103.0	10S	29E
10632	532	2405 Oct 08	19:19:56	1075	5018	143	T+	-p	0.1336	2.5913	1.6341	318.1	209.7	95.1	6N	71E
10633	532	2406 Apr 03	22:29:30	1077	5024	148	P	t-	0.6153	1.7659	0.6924	340.4	182.5	-	5S	28E
10634	532	2406 Sep 28	10:14:02	1079	5030	153	P	a-	-0.5994	1.7532	0.7629	308.4	176.2	-	2N	151W
10635	532	2407 Feb 22	15:37:08	1081	5035	120	N	-a	-1.4584	0.1829	-0.8192	117.6	-	-	9N	133E
10636	532	2407 Mar 24	03:55:49	1081	5036	158	N	a-	1.2931	0.4974	-0.5268	194.8	-	-	0S	53W
10637	532	2407 Aug 19	05:30:19	1082	5041	125	N	-t	1.4137	0.2939	-0.7656	162.1	-	-	12S	77W
10638	532	2407 Sep 17	19:23:23	1083	5042	163	N	t-	-1.3973	0.3159	-0.7273	163.4	-	-	3S	73E
10639	532	2408 Feb 12	06:04:35	1084	5047	130	P	-a	-0.7479	1.4704	0.5006	280.2	143.5	-	13N	83W
10640	532	2408 Aug 07	06:12:58	1086	5053	135	P	-t	0.6809	1.6438	0.5740	337.6	171.8	-	16S	87W
10641	533	2409 Jan 31	21:44:38	1088	5059	140	T-	-p	-0.0827	2.6972	1.7150	324.2	212.0	97.3	17N	42E
10642	533	2409 Jul 27	08:59:05	1090	5065	145	T-	pp	-0.0983	2.6943	1.6613	356.7	227.9	102.2	19S	128W
10643	533	2410 Jan 21	09:56:47	1091	5071	150	P	a-	0.6263	1.7250	0.6926	317.0	173.2	-	20N	142W
10644	533	2410 Jul 16	18:45:58	1093	5077	155	P	a-	-0.8523	1.2829	0.3050	278.2	119.7	-	22S	85E
10645	533	2411 Jan 10	15:08:15	1095	5083	160	N	t-	1.3675	0.3928	-0.6946	186.2	-	-	23N	139E
10646	533	2411 Jun 07	03:37:47	1097	5088	127	P	-a	0.9498	1.0866	0.1431	250.0	80.8	-	22S	50W
10647	533	2411 Jul 06	10:22:42	1097	5089	165	Nb	a-	-1.5404	0.0027	-0.9404	14.0	-	-	24S	150W
10648	533	2411 Nov 30	19:52:22	1098	5094	132	P	-t	-0.9774	1.1091	0.0205	293.1	34.9	-	21N	64E
10649	533	2412 May 26	19:43:25	1100	5100	137	T+	-p	0.2306	2.4214	1.4479	325.8	211.1	88.2	21S	68E
10650	533	2412 Nov 19	00:15:25	1102	5106	142	T-	-p	-0.2386	2.4377	1.4030	345.0	217.1	87.1	19N	3W
10651	533	2413 May 16	06:33:36	1104	5112	147	P	t-	-0.5457	1.8718	0.8419	331.5	190.5	-	20S	95W
10652	533	2413 Nov 08	11:57:24	1106	5118	152	P	a-	0.4733	1.9794	0.9993	310.7	189.5	-	17N	179W
10653	533	2414 May 05	10:04:13	1108	5124	157	N	t-	-1.3275	0.4600	-0.6152	202.2	-	-	17S	148W
10654	533	2414 Sep 29	19:04:37	1109	5129	124	N	-a	-1.4432	0.1902	-0.7710	115.8	-	-	1N	77E
10655	533	2414 Oct 29	03:49:51	1109	5130	162	N	a-	1.1430	0.7403	-0.2195	214.7	-	-	14N	57W
10656	533	2415 Mar 25	18:20:36	1111	5135	129	P	-t	0.8667	1.2996	0.2360	304.0	113.1	-	1S	91E
10657	533	2415 Sep 19	08:51:06	1113	5141	134	P	-a	-0.7713	1.4444	0.4410	296.1	143.2	-	2S	129W
10658	533	2416 Mar 14	01:45:55	1115	5147	139	T+	pp	0.1118	2.6579	1.6477	338.3	218.2	98.1	3N	20W
10659	533	2416 Sep 07	16:02:33	1117	5153	144	T-	pp	-0.0288	2.8326	1.7780	364.9	231.5	104.9	6S	124E
10660	533	2417 Mar 03	15:29:02	1118	5159	149	P	a-	-0.5837	1.7721	0.8015	297.8	173.7	-	6N	135E
10661	534	2417 Aug 27	17:11:33	1120	5165	154	P	t-	0.7295	1.5573	0.4822	333.0	160.1	-	9S	107E
10662	534	2418 Jan 22	20:25:46	1122	5170	121	N	-a	1.4410	0.2109	-0.7835	124.3	-	-	21N	61E
10663	534	2418 Feb 21	07:26:24	1122	5171	159	N	a-	-1.2504	0.5496	-0.4228	190.0	-	-	9N	104W
10664	534	2418 Jul 18	06:34:23	1124	5176	126	N	-t	-1.2887	0.5020	-0.5153	198.1	-	-	22S	92W
10665	534	2418 Aug 16	18:46:36	1124	5177	164	N	t-	1.4488	0.2230	-0.8236	140.0	-	-	12S	84E
10666	534	2419 Jan 12	06:53:23	1126	5182	131	P	-a	0.8154	1.3869	0.3369	302.2	129.9	-	22N	97W
10667	534	2419 Jul 07	18:17:23	1127	5188	136	T	-a	-0.4754	1.9679	1.0032	311.4	191.3	8.6	23S	92E
10668	534	2420 Jan 01	10:00:56	1129	5194	141	T+	pp	0.1394	2.6514	1.5539	374.0	231.4	99.2	23N	145W
10669	534	2420 Jun 26	10:47:40	1131	5200	146	T	p-	0.2737	2.3259	1.3849	311.9	203.9	82.0	23S	156W
10670	534	2420 Dec 20	09:03:39	1133	5206	151	P	t-	-0.5406	1.9159	0.8168	353.9	195.9	-	23N	132W
10671	534	2421 Jun 16	03:32:46	1135	5212	156	P	a-	1.0225	0.9621	0.0011	244.0	7.3	-	22S	48W
10672	534	2421 Dec 09	11:40:03	1137	5218	161	N	a-	-1.1984	0.6857	-0.3671	230.3	-	-	22N	172W
10673	534	2422 May 07	03:37:48	1138	5223	128	P	-t	-0.8852	1.2576	0.2103	298.1	106.8	-	18S	51W
10674	534	2422 Oct 30	11:10:45	1140	5229	133	P	-a	0.8668	1.2513	0.2831	266.9	112.2	-	15N	167W
10675	534	2423 Apr 26	05:25:45	1142	5235	138	T-	pp	-0.1585	2.6086	1.5262	374.3	232.8	99.1	14S	77W
10676	534	2423 Oct 20	03:29:19	1144	5241	143	T+	-p	0.1723	2.5221	1.5612	317.9	208.6	92.3	10N	51W
10677	534	2424 Apr 14	05:51:07	1146	5247	148	P	t-	0.5686	1.8490	0.7810	343.5	190.6	-	9S	83W
10678	534	2424 Oct 08	18:01:43	1148	5253	153	P	a-	-0.5553	1.8371	0.8409	313.7	183.0	-	6N	91E
10679	534	2425 Mar 05	00:03:15	1149	5258	120	N	-a	-1.4701	0.1590	-0.8383	109.5	-	-	5N	6E
10680	534	2425 Apr 03	11:50:58	1149	5259	158	N	a-	1.2568	0.5609	-0.4572	204.7	-	-	4S	172W

Cat Num	Canon Plate	Calendar Date	TD of Greatest Eclipse	ΔT s	Luna Num	Saros Num	Ecl. Type	QSE	Gamma	Pen. Mag.	Um. Mag.	Eclipse Phase Durations Pen. m	Par. m	Total m	Greatest in Zenith Lat.	Lng.
10681	535	2425 Aug 29	12:14:31	1151	5264	125	N	-t	1.4893	0.1572	-0.9064	120.4	-	-	8S	179W
10682	535	2425 Sep 28	02:37:33	1151	5265	163	N	t-	-1.3452	0.4143	-0.6347	186.2	-	-	1N	36W
10683	535	2426 Feb 22	14:46:52	1153	5270	130	P	-a	-0.7594	1.4481	0.4805	278.5	140.9	-	9N	146E
10684	535	2426 Aug 18	12:46:04	1155	5276	135	P	-t	0.7614	1.4963	0.4258	327.5	151.4	-	12S	174E
10685	535	2427 Feb 12	06:28:08	1157	5282	140	T-	-p	-0.0913	2.6817	1.6991	324.7	212.2	97.1	14N	89W
10686	535	2427 Aug 07	15:48:57	1158	5288	145	T-	pp	-0.0180	2.8405	1.8096	356.0	228.3	104.4	16S	129E
10687	535	2428 Feb 01	18:26:27	1160	5294	150	P	a-	0.6218	1.7344	0.6999	318.6	174.4	-	18N	91E
10688	535	2428 Jul 27	02:01:20	1162	5300	155	P	a-	-0.7754	1.4228	0.4474	287.2	141.5	-	20S	24W
10689	535	2429 Jan 20	23:16:44	1164	5306	160	N	t-	1.3620	0.4038	-0.6854	189.1	-	-	21N	18E
10690	535	2429 Jun 17	11:11:33	1166	5311	127	P	-a	1.0191	0.9591	0.0164	238.3	27.8	-	22S	162W
10691	535	2429 Jul 16	17:51:15	1166	5312	165	N	a-	-1.4688	0.1337	-0.8085	96.5	-	-	23S	99E
10692	535	2429 Dec 11	03:51:44	1168	5317	132	N*	-t	-0.9904	1.0853	-0.0033	290.1	-	-	22N	55W
10693	535	2430 Jun 07	03:09:30	1169	5323	137	T	-p	0.3000	2.2946	1.3201	324.3	207.7	78.4	22S	43W
10694	535	2430 Nov 30	08:31:15	1171	5329	142	T-	-p	-0.2548	2.4076	1.3735	343.3	215.5	84.6	21N	126W
10695	535	2431 May 27	13:39:33	1173	5335	147	P	t-	-0.4783	1.9959	0.9651	338.0	200.2	-	22S	159E
10696	535	2431 Nov 19	20:25:36	1175	5341	152	T	a-	0.4513	2.0200	1.0394	311.6	191.4	29.6	20N	55E
10697	535	2432 May 15	16:54:29	1177	5347	157	N	t-	-1.2616	0.5802	-0.4934	224.6	-	-	20S	110E
10698	535	2432 Oct 10	03:05:14	1179	5352	124	N	-a	-1.4883	0.1097	-0.8560	88.8	-	-	6N	44W
10699	535	2432 Nov 08	12:15:14	1179	5353	162	N	a-	1.1169	0.7899	-0.1730	220.8	-	-	18N	177E
10700	535	2433 Apr 05	01:56:44	1181	5358	129	P	-t	0.9030	1.2300	0.1723	296.8	97.3	-	5S	23W
10701	536	2433 Sep 29	16:24:45	1182	5364	134	P	-a	-0.8273	1.3446	0.3352	290.4	127.2	-	2N	117E
10702	536	2434 Mar 25	09:54:38	1184	5370	139	T+	pp	0.1395	2.6043	1.5998	336.3	217.0	96.4	2S	142W
10703	536	2434 Sep 18	23:02:35	1186	5376	144	T-	pp	-0.0951	2.7137	1.6538	365.8	230.9	102.8	2S	18E
10704	536	2435 Mar 14	23:59:26	1188	5382	149	P	a-	-0.5619	1.8103	0.8434	298.9	176.7	-	2N	7E
10705	536	2435 Sep 07	23:52:36	1190	5388	154	P	t-	0.6545	1.6959	0.6186	341.9	177.3	-	5S	6E
10706	536	2436 Feb 03	05:08:57	1192	5393	121	N	-a	1.4502	0.1947	-0.8010	119.9	-	-	18N	68W
10707	536	2436 Mar 03	16:03:14	1192	5394	159	N	a-	-1.2367	0.5745	-0.3972	194.0	-	-	5N	126E
10708	536	2436 Jul 28	13:31:07	1194	5399	126	N	-t	-1.3610	0.3678	-0.6466	171.1	-	-	20S	165E
10709	536	2436 Aug 27	01:39:06	1194	5400	164	N	t-	1.3726	0.3624	-0.6832	175.7	-	-	9S	20W
10710	536	2437 Jan 22	15:17:33	1195	5405	131	P	-a	0.8239	1.3726	0.3201	302.2	127.3	-	20N	139E
10711	536	2437 Jul 18	01:38:01	1197	5411	136	P	-a	-0.5482	1.8329	0.8708	305.0	182.0	-	21S	18W
10712	536	2438 Jan 11	18:05:30	1199	5417	141	T+	pp	0.1483	2.6358	1.5366	374.5	231.3	98.5	22N	95E
10713	536	2438 Jul 07	18:17:19	1201	5423	146	T+	p-	0.2003	2.4604	1.5199	314.2	207.5	90.5	22S	92E
10714	536	2438 Dec 31	17:06:05	1203	5429	151	P	t-	-0.5300	1.9352	0.8365	354.4	197.3	-	23N	109E
10715	536	2439 Jun 27	10:55:58	1205	5435	156	P	a-	0.9531	1.0901	0.1276	256.9	78.1	-	22S	158W
10716	536	2439 Dec 20	19:59:36	1207	5441	161	N	a-	-1.1858	0.7079	-0.3434	232.7	-	-	22N	64E
10717	536	2440 May 17	10:41:49	1209	5446	128	P	-t	-0.9527	1.1339	0.0862	287.8	69.8	-	20S	156W
10718	536	2440 Nov 09	19:35:18	1210	5452	133	P	-a	0.8904	1.2086	0.2392	263.2	103.7	-	18N	67E
10719	536	2441 May 06	12:21:25	1212	5458	138	T-	pp	-0.2213	2.4921	1.4123	372.3	230.0	92.2	17S	179E
10720	536	2441 Oct 30	11:45:56	1214	5464	143	T+	-p	0.2048	2.4645	1.4997	317.8	207.4	89.3	14N	176W
10721	537	2442 Apr 25	13:05:12	1216	5470	148	P	t-	0.5147	1.9449	0.8825	346.6	198.5	-	13S	168E
10722	537	2442 Oct 20	01:55:57	1218	5476	153	P	a-	-0.5172	1.9101	0.9079	318.3	188.4	-	10N	28W
10723	537	2443 Mar 16	08:22:54	1220	5481	120	N	-a	-1.4877	0.1241	-0.8680	96.7	-	-	0N	119W
10724	537	2443 Apr 14	19:37:49	1220	5482	158	N	a-	1.2128	0.6385	-0.3734	215.7	-	-	8S	71E
10725	537	2443 Sep 09	19:02:39	1222	5487	125	Ne	-t	1.5609	0.0281	-1.0399	51.7	-	-	4S	78E
10726	537	2443 Oct 09	09:58:19	1222	5488	163	N	t-	-1.2987	0.5027	-0.5525	204.3	-	-	5N	147W
10727	537	2444 Mar 04	23:24:12	1224	5493	130	P	-a	-0.7754	1.4175	0.4526	276.3	137.3	-	5N	17E
10728	537	2444 Aug 28	19:24:50	1226	5499	135	P	-t	0.8375	1.3572	0.2857	316.7	126.5	-	8S	74E
10729	537	2445 Feb 22	15:07:08	1227	5505	140	T-	-p	-0.1033	2.6598	1.6771	325.2	212.4	96.7	10N	141E
10730	537	2445 Aug 17	22:44:46	1229	5511	145	T+	pp	0.0583	2.7658	1.7366	354.5	227.3	103.3	13S	25E
10731	537	2446 Feb 12	02:53:07	1231	5517	150	P	a-	0.6155	1.7467	0.7106	320.3	175.9	-	14N	35W
10732	537	2446 Aug 07	09:19:05	1233	5523	155	P	a-	-0.6995	1.5609	0.5875	294.9	158.2	-	17S	133W
10733	537	2447 Feb 01	07:24:27	1235	5529	160	N	t-	1.3562	0.4148	-0.6754	191.8	-	-	18N	103W
10734	537	2447 Jun 28	18:42:11	1237	5534	127	N	-a	1.0906	0.8279	-0.1147	224.6	-	-	22S	86E
10735	537	2447 Jul 28	01:20:15	1237	5535	165	N	a-	-1.3974	0.2645	-0.6775	134.0	-	-	20S	13W
10736	537	2447 Dec 22	11:56:10	1239	5540	132	N*	-t	-1.0001	1.0673	-0.0207	287.7	-	-	22N	174W
10737	537	2448 Jun 17	10:30:31	1241	5546	137	T	-a	0.3730	2.1615	1.1854	322.0	202.8	62.6	23S	152W
10738	537	2448 Dec 10	16:53:19	1243	5552	142	T-	-p	-0.2667	2.3852	1.3524	341.7	214.2	82.6	23N	110E
10739	537	2449 Jun 06	20:40:00	1245	5558	147	T	t-	-0.4068	2.1278	1.0956	344.0	208.7	48.3	23S	55E
10740	537	2449 Nov 30	04:59:05	1247	5564	152	T	a-	0.4334	2.0529	1.0721	312.1	192.8	39.5	22N	73W

Cat Num	Canon Plate	Calendar Date	TD of Greatest Eclipse	ΔT s	Luna Num	Saros Num	Ecl. Type	QSE	Gamma	Pen. Mag.	Um. Mag.	Eclipse Phase Durations Pen. m	Par. m	Total m	Greatest in Zenith Lat.	Lng.
10741	538	2450 May 26	23:38:32	1249	5570	157	N	t-	-1.1906	0.7097	-0.3626	245.5	-	-	22S	10E
10742	538	2450 Oct 21	11:12:18	1250	5575	124	N	-a	-1.5280	0.0391	-0.9312	53.5	-	-	9N	166W
10743	538	2450 Nov 19	20:46:40	1250	5576	162	N	a-	1.0959	0.8297	-0.1360	225.6	-	-	21N	49E
10744	538	2451 Apr 16	09:23:54	1252	5581	129	P	-t	0.9471	1.1462	0.0945	287.9	72.7	-	9S	135W
10745	538	2451 Oct 11	00:05:24	1254	5587	134	P	-a	-0.8768	1.2570	0.2413	284.9	109.8	-	6N	1E
10746	538	2452 Apr 04	17:54:43	1256	5593	139	T+	pp	0.1748	2.5366	1.5380	333.9	215.3	93.7	6S	97E
10747	538	2452 Sep 29	06:10:28	1258	5599	144	T-	pp	-0.1542	2.6079	1.5426	366.2	229.6	98.7	3N	90W
10748	538	2453 Mar 25	08:21:32	1260	5605	149	P	a-	-0.5329	1.8615	0.8982	300.5	180.5	-	3S	119W
10749	538	2453 Sep 18	06:42:17	1262	5611	154	P	t-	0.5863	1.8223	0.7426	349.0	190.2	-	1S	97W
10750	538	2454 Feb 13	13:48:10	1264	5616	121	N	-a	1.4626	0.1725	-0.8241	113.4	-	-	14N	162E
10751	538	2454 Mar 15	00:33:02	1264	5617	159	N	a-	-1.2176	0.6089	-0.3617	199.3	-	-	1N	1W
10752	538	2454 Aug 08	20:30:20	1265	5622	126	N	-t	-1.4320	0.2364	-0.7757	138.2	-	-	17S	60E
10753	538	2454 Sep 07	08:37:33	1266	5623	164	N	t-	1.2998	0.4955	-0.5495	202.2	-	-	5S	125W
10754	538	2455 Feb 02	23:40:14	1267	5628	131	P	-a	0.8331	1.3567	0.3023	302.0	124.4	-	17N	14E
10755	538	2455 Jul 29	08:58:50	1269	5634	136	P	-a	-0.6217	1.6971	0.7368	297.7	170.9	-	19S	127W
10756	538	2456 Jan 23	02:10:08	1271	5640	141	T+	pp	0.1568	2.6207	1.5206	374.9	231.3	97.9	20N	25W
10757	538	2456 Jul 18	01:45:42	1273	5646	146	T+	p-	0.1257	2.5973	1.6567	315.9	210.0	96.1	21S	19W
10758	538	2457 Jan 11	01:11:47	1275	5652	151	P	t-	-0.5215	1.9503	0.8527	354.5	198.3	-	21N	11W
10759	538	2457 Jul 07	18:16:32	1277	5658	156	P	a-	0.8822	1.2213	0.2567	268.8	109.1	-	22S	93E
10760	538	2457 Dec 31	04:23:20	1279	5664	161	N	a-	-1.1759	0.7250	-0.3242	234.3	-	-	22N	60W
10761	539	2458 May 28	17:39:26	1281	5669	128	N*	-t	-1.0244	1.0025	-0.0456	275.2	-	-	23S	100E
10762	539	2458 Nov 21	04:06:22	1283	5675	133	P	-a	0.9091	1.1749	0.2043	260.2	96.1	-	21N	60W
10763	539	2459 May 17	19:09:13	1285	5681	138	T	pt	-0.2904	2.3641	1.2869	369.4	225.7	80.8	20S	77E
10764	539	2459 Nov 10	20:09:31	1287	5687	143	T+	-p	0.2312	2.4179	1.4492	317.7	206.3	86.3	18N	59E
10765	539	2460 May 05	20:13:03	1289	5693	148	P	t-	0.4550	2.0519	0.9949	349.6	206.0	-	16S	61E
10766	539	2460 Oct 30	09:56:53	1291	5699	153	P	a-	-0.4853	1.9716	0.9636	322.2	192.7	-	14N	148W
10767	539	2461 Mar 26	16:36:11	1292	5704	120	N	-a	-1.5110	0.0788	-0.9080	77.1	-	-	4S	117E
10768	539	2461 Apr 25	03:19:08	1293	5705	158	N	a-	1.1634	0.7261	-0.2797	227.0	-	-	12S	45W
10769	539	2461 Oct 19	17:26:44	1295	5711	163	N	t-	-1.2586	0.5792	-0.4818	218.6	-	-	9N	100E
10770	539	2462 Mar 16	07:55:14	1296	5716	130	P	-a	-0.7966	1.3771	0.4150	273.3	132.2	-	1N	112W
10771	539	2462 Sep 09	02:10:01	1298	5722	135	P	-t	0.9084	1.2278	0.1551	305.3	94.9	-	4S	28W
10772	539	2463 Mar 05	23:39:34	1300	5728	140	T-	-p	-0.1201	2.6287	1.6462	325.5	212.5	96.1	6N	13E
10773	539	2463 Aug 29	05:47:14	1302	5734	145	T+	pp	0.1296	2.6342	1.6063	352.2	225.1	99.7	9S	81W
10774	539	2464 Feb 23	11:12:18	1304	5740	150	P	a-	0.6041	1.7683	0.7311	322.4	178.3	-	10N	159W
10775	539	2464 Aug 17	16:42:26	1306	5746	155	P	a-	-0.6277	1.6921	0.7201	301.2	170.9	-	13S	116E
10776	539	2465 Feb 11	15:26:51	1308	5752	160	N	t-	1.3470	0.4319	-0.6587	195.8	-	-	15N	137E
10777	539	2465 Jul 09	02:10:48	1310	5757	127	N	-a	1.1635	0.6943	-0.2489	208.7	-	-	21S	26W
10778	539	2465 Aug 07	08:51:22	1310	5758	165	N	a-	-1.3275	0.3929	-0.5495	161.3	-	-	17S	125W
10779	539	2466 Jan 01	20:04:17	1312	5763	132	N*	-t	-1.0070	1.0540	-0.0328	285.6	-	-	22N	65E
10780	539	2466 Jun 28	17:46:38	1314	5769	137	T	-a	0.4490	2.0231	1.0448	318.7	196.2	32.2	23S	100E
10781	540	2466 Dec 22	01:21:41	1316	5775	142	T	-p	-0.2740	2.3710	1.3398	340.3	213.2	81.3	23N	16W
10782	540	2467 Jun 18	03:31:37	1318	5781	147	T	p-	-0.3290	2.2713	1.2374	349.5	216.3	72.8	24S	47W
10783	540	2467 Dec 11	13:39:54	1320	5787	152	T	a-	0.4214	2.0751	1.0943	312.3	193.7	44.8	23N	159E
10784	540	2468 Jun 06	06:15:13	1322	5793	157	N	t-	-1.1139	0.8499	-0.2213	265.1	-	-	24S	89W
10785	540	2468 Nov 30	05:24:27	1324	5799	162	N	a-	1.0806	0.8593	-0.1092	229.2	-	-	23N	79W
10786	540	2469 Apr 26	16:45:33	1326	5804	129	P	-t	0.9960	1.0533	0.0077	277.6	20.9	-	13S	114E
10787	540	2469 Oct 21	07:52:38	1328	5810	134	P	-a	-0.9196	1.1815	0.1596	279.8	90.6	-	10N	116W
10788	540	2470 Apr 16	01:49:49	1330	5816	139	T+	-p	0.2150	2.4598	1.4670	331.3	213.2	89.8	10S	22W
10789	540	2470 Oct 10	13:23:39	1332	5822	144	T-	pp	-0.2084	2.5113	1.4405	366.2	227.5	93.0	7N	161E
10790	540	2471 Apr 05	16:38:08	1334	5828	149	P	a-	-0.4993	1.9215	0.9619	302.4	184.6	-	7S	116E
10791	540	2471 Sep 29	13:39:37	1336	5834	154	P	t-	0.5238	1.9383	0.8561	354.7	200.0	-	3N	158E
10792	540	2472 Feb 24	22:21:41	1338	5839	121	N	-a	1.4791	0.1425	-0.8547	103.6	-	-	11N	34E
10793	540	2472 Mar 25	08:55:10	1338	5840	159	N	a-	-1.1927	0.6540	-0.3155	205.9	-	-	3S	127W
10794	540	2472 Aug 19	03:36:26	1340	5845	126	N	-t	-1.4981	0.1141	-0.8961	96.8	-	-	14S	47W
10795	540	2472 Sep 17	15:45:51	1340	5846	164	N	t-	1.2338	0.6166	-0.4282	222.3	-	-	1S	127E
10796	540	2473 Feb 13	07:57:35	1342	5851	131	P	-a	0.8462	1.3335	0.2776	301.1	120.0	-	14N	110W
10797	540	2473 Aug 08	16:24:06	1344	5857	136	P	-a	-0.6919	1.5675	0.6087	289.9	158.2	-	16S	121E
10798	540	2474 Feb 02	10:11:05	1346	5863	141	T+	pp	0.1681	2.6001	1.4998	375.0	231.1	96.9	17N	144W
10799	540	2474 Jul 29	09:15:38	1348	5869	146	T+	pp	0.0523	2.7323	1.7910	316.9	211.3	98.9	18S	131W
10800	540	2475 Jan 22	09:16:58	1350	5875	151	P	t-	-0.5121	1.9665	0.8708	354.5	199.5	-	19N	131W

Cat Num	Canon Plate	Calendar Date	TD of Greatest Eclipse	ΔT s	Luna Num	Saros Num	Ecl. Type	QSE	Gamma	Pen. Mag.	Um. Mag.	Pen. m	Par. m	Total m	Lat.	Lng.
10801	541	2475 Jul 19	01:34:32	1352	5881	156	P	a-	0.8096	1.3559	0.3885	279.9	132.0	-	20S	16W
10802	541	2476 Jan 11	12:49:31	1354	5887	161	N	a-	-1.1672	0.7397	-0.3068	235.5	-	-	21N	175E
10803	541	2476 Jun 08	00:28:51	1355	5892	128	N	-t	-1.1019	0.8607	-0.1884	259.5	-	-	24S	2W
10804	541	2476 Dec 01	12:44:17	1357	5898	133	P	-a	0.9228	1.1503	0.1789	257.9	90.2	-	23N	172E
10805	541	2477 May 28	01:49:59	1359	5904	138	T	-t	-0.3652	2.2255	1.1507	365.3	219.6	61.5	22S	22W
10806	541	2477 Nov 21	04:40:34	1361	5910	143	T+	-p	0.2515	2.3825	1.4101	317.8	205.4	83.8	20N	68W
10807	541	2478 May 17	03:13:38	1364	5916	148	T	t-	0.3883	2.1715	1.1200	352.4	212.9	54.3	19S	43W
10808	541	2478 Nov 10	18:04:53	1366	5922	153	T	p-	-0.4598	2.0213	1.0074	325.5	195.9	13.3	17N	90E
10809	541	2479 Apr 07	00:42:37	1367	5927	120	Ne	-a	-1.5402	0.0223	-0.9591	41.1	-	-	8S	5W
10810	541	2479 May 06	10:53:14	1368	5928	158	N	a-	1.1070	0.8265	-0.1732	238.7	-	-	16S	158W
10811	541	2479 Oct 31	01:02:53	1370	5934	163	N	t-	-1.2253	0.6433	-0.4235	229.8	-	-	13N	14W
10812	541	2480 Mar 26	16:18:42	1371	5939	130	P	-a	-0.8242	1.3251	0.3659	269.5	125.1	-	4S	122E
10813	541	2480 Sep 19	09:03:33	1373	5945	135	P	-t	0.9726	1.1107	0.0366	293.8	46.8	-	0S	132W
10814	541	2481 Mar 16	08:06:00	1375	5951	140	T-	-p	-0.1415	2.5893	1.6072	325.8	212.3	95.0	1N	114W
10815	541	2481 Sep 08	12:56:08	1377	5957	145	T+	pp	0.1966	2.5108	1.4839	349.4	221.9	93.7	5S	171E
10816	541	2482 Mar 05	19:26:14	1380	5963	150	P	a-	0.5889	1.7964	0.7585	324.9	181.3	-	6N	77E
10817	541	2482 Aug 29	00:10:16	1382	5969	155	P	a-	-0.5588	1.8179	0.8468	306.3	180.9	-	10S	4E
10818	541	2483 Feb 22	23:24:23	1384	5975	160	N	t-	1.3340	0.4557	-0.6346	201.0	-	-	11N	18E
10819	541	2483 Jul 20	09:38:03	1385	5980	127	N	-a	1.2373	0.5595	-0.3848	190.1	-	-	19S	137W
10820	541	2483 Aug 18	16:25:11	1386	5981	165	N	a-	-1.2596	0.5181	-0.4253	183.0	-	-	14S	121E
10821	542	2484 Jan 13	04:14:53	1387	5986	132	N*	-t	-1.0124	1.0430	-0.0418	283.8	-	-	21N	56W
10822	542	2484 Jul 09	01:00:42	1389	5992	137	P	-a	0.5260	1.8831	0.9021	314.5	187.6	-	22S	8W
10823	542	2485 Jan 01	09:52:20	1391	5998	142	T	-p	-0.2800	2.3588	1.3298	338.8	212.3	80.2	23N	142W
10824	542	2485 Jun 28	10:20:45	1394	6004	147	T-	pp	-0.2499	2.4177	1.3816	354.2	222.4	87.8	23S	148W
10825	542	2485 Dec 21	22:24:27	1396	6010	152	T	a-	0.4123	2.0914	1.1110	312.2	194.2	48.3	24N	29E
10826	542	2486 Jun 17	12:46:55	1398	6016	157	N	t-	-1.0329	0.9982	-0.0722	283.1	-	-	24S	175E
10827	542	2486 Dec 11	14:07:10	1400	6022	162	N	a-	1.0694	0.8809	-0.0900	231.8	-	-	24N	152E
10828	542	2487 May 07	23:59:40	1401	6027	129	N	-t	1.0516	0.9483	-0.0913	265.1	-	-	16S	6E
10829	542	2487 Nov 01	15:47:25	1404	6033	134	P	-a	-0.9559	1.1180	0.0900	275.3	68.9	-	14N	125E
10830	542	2488 Apr 26	09:36:10	1406	6039	139	T+	-p	0.2634	2.3681	1.3812	328.1	210.3	83.7	14S	139W
10831	542	2488 Oct 20	20:46:19	1408	6045	144	T-	-p	-0.2544	2.4297	1.3535	365.9	225.2	86.4	11N	51E
10832	542	2489 Apr 16	00:46:37	1410	6051	149	T	a-	-0.4585	1.9945	1.0385	304.5	189.0	29.0	11S	6W
10833	542	2489 Oct 09	20:45:34	1412	6057	154	P	t-	0.4679	2.0422	0.9575	359.2	207.5	-	7N	51E
10834	542	2490 Mar 07	06:48:56	1414	6062	121	N	-a	1.5004	0.1035	-0.8940	88.8	-	-	6N	93W
10835	542	2490 Apr 05	17:09:28	1414	6063	159	N	a-	-1.1619	0.7100	-0.2584	213.7	-	-	8S	109E
10836	542	2490 Sep 28	23:02:21	1416	6069	164	N	t-	1.1733	0.7276	-0.3170	238.2	-	-	4N	17E
10837	542	2491 Feb 24	16:09:26	1418	6074	131	P	-a	0.8630	1.3030	0.2463	299.6	113.8	-	10N	127E
10838	542	2491 Aug 19	23:52:20	1420	6080	136	P	-a	-0.7602	1.4419	0.4837	281.6	143.5	-	13S	9E
10839	542	2492 Feb 13	18:09:20	1422	6086	141	T+	pp	0.1814	2.5754	1.4757	375.0	230.7	95.6	13N	97E
10840	542	2492 Aug 08	16:47:17	1424	6092	146	T-	pp	-0.0195	2.7933	1.8505	317.3	211.6	99.4	16S	116E
10841	543	2493 Feb 01	17:21:21	1426	6098	151	P	t-	-0.5016	1.9845	0.8915	354.5	200.8	-	16N	109E
10842	543	2493 Jul 29	08:53:05	1428	6104	156	P	a-	0.7379	1.4891	0.5184	289.8	150.0	-	18S	126W
10843	543	2494 Jan 21	21:17:45	1430	6110	161	N	a-	-1.1593	0.7525	-0.2908	236.4	-	-	19N	49E
10844	543	2494 Jun 19	07:12:41	1432	6115	128	N	-t	-1.1829	0.7127	-0.3375	240.4	-	-	24S	101W
10845	543	2494 Jul 18	19:45:14	1432	6116	166	Nb	t-	1.5603	0.0070	-1.0171	24.6	-	-	19S	71E
10846	543	2494 Dec 12	21:27:43	1434	6121	133	P	-a	0.9324	1.1330	0.1610	256.2	85.7	-	24N	42E
10847	543	2495 Jun 08	08:25:20	1436	6127	138	T	-t	-0.4447	2.0787	1.0060	360.0	211.2	12.9	23S	120W
10848	543	2495 Dec 02	13:16:56	1438	6133	143	T+	-p	0.2671	2.3557	1.3799	318.0	204.8	81.6	22N	164E
10849	543	2496 May 27	10:09:53	1440	6139	148	T	t-	0.3172	2.2994	1.2530	354.6	218.7	75.2	21S	147W
10850	543	2496 Nov 21	02:19:04	1442	6145	153	T	p-	-0.4399	2.0605	1.0411	328.4	198.4	31.1	20N	32W
10851	543	2497 May 16	18:23:08	1444	6151	158	N	a-	1.0463	0.9350	-0.0590	249.9	-	-	18S	90E
10852	543	2497 Nov 10	08:44:48	1447	6157	163	N	t-	-1.1969	0.6981	-0.3742	239.0	-	-	16N	129W
10853	543	2498 Apr 07	00:35:16	1448	6162	130	P	-a	-0.8576	1.2623	0.3058	264.8	115.3	-	8S	3W
10854	543	2498 Sep 30	16:05:36	1450	6160	135	N+	t	1.0300	1.0062	0.0694	282.4			4N	122E
10855	543	2499 Mar 27	16:22:54	1453	6174	140	T-	-p	-0.1705	2.5359	1.5542	325.9	211.9	93.1	3S	122E
10856	543	2499 Sep 19	20:14:17	1455	6180	145	T+	pp	0.2568	2.4000	1.3737	346.3	218.1	85.8	1S	61E
10857	543	2500 Mar 17	03:30:27	1457	6186	150	P	h-	0.5665	1.8378	0.7996	327.9	185.4	-	2N	44W
10858	543	2500 Sep 09	07:45:59	1459	6192	155	P	a-	-0.4957	1.9335	0.9628	310.3	188.5	-	6S	111W
10859	543	2501 Mar 06	07:13:33	1461	6198	160	N	t-	1.3146	0.4909	-0.5985	208.2	-	-	7N	99W
10860	543	2501 Jul 31	17:06:36	1463	6203	127	N	-a	1.3097	0.4276	-0.5184	168.6	-	-	17S	111E

Cat Num	Canon Plate	Calendar Date	TD of Greatest Eclipse	ΔT s	Luna Num	Saros Num	Ecl. Type	QSE	Gamma	Pen. Mag.	Um. Mag.	Eclipse Phase Durations Pen. m	Par. m	Total m	Greatest in Zenith Lat.	Lng.
10861	544	2501 Aug 30	00:04:13	1463	6204	165	N	a-	-1.1957	0.6361	-0.3089	200.4	-	-	10S	6E
10862	544	2502 Jan 24	12:26:40	1465	6209	132	N*	-t	-1.0173	1.0327	-0.0494	281.9	-	-	18N	178W
10863	544	2502 Jul 21	08:11:07	1467	6215	137	P	-a	0.6051	1.7398	0.7554	309.2	176.6	-	20S	115W
10864	544	2503 Jan 13	18:27:08	1469	6221	142	T	-p	-0.2833	2.3514	1.3250	337.4	211.6	79.6	21N	91E
10865	544	2503 Jul 10	17:04:26	1471	6227	147	T-	pp	-0.1673	2.5705	1.5317	358.1	227.0	97.7	22S	112E
10866	544	2504 Jan 03	07:13:07	1473	6233	152	T	p-	0.4063	2.1021	1.1225	312.0	194.5	50.5	23N	101W
10867	544	2504 Jun 28	19:14:33	1475	6239	157	P	t-	-0.9484	1.1530	0.0831	299.4	70.5	-	24S	79E
10868	544	2504 Dec 22	22:54:41	1478	6245	162	N	a-	1.0627	0.8943	-0.0787	233.5	-	-	24N	22E
10869	544	2505 May 19	07:09:44	1479	6250	129	N	-t	1.1112	0.8360	-0.1978	250.7	-	-	19S	102W
10870	544	2505 Nov 12	23:47:37	1481	6256	134	P	-a	-0.9869	1.0642	0.0301	271.4	40.3	-	17N	6E
10871	544	2506 May 08	17:18:30	1484	6262	139	T	-p	0.3158	2.2691	1.2877	324.5	206.6	75.1	17S	106E
10872	544	2506 Nov 02	04:15:24	1486	6268	144	T	-p	-0.2945	2.3586	1.2773	365.6	222.8	78.8	14N	62W
10873	544	2507 Apr 28	08:48:07	1488	6274	149	T	a-	-0.4117	2.0786	1.1261	306.8	193.6	51.0	14S	127W
10874	544	2507 Oct 22	04:00:56	1490	6280	154	T	t-	0.4193	2.1325	1.0454	362.6	213.0	34.7	11N	58W
10875	544	2508 Mar 18	15:08:38	1492	6285	121	N	-a	1.5273	0.0544	-0.9433	64.8	-	-	2N	142E
10876	544	2508 Apr 17	01:15:51	1492	6286	159	N	a-	-1.1252	0.7768	-0.1907	222.4	-	-	12S	13W
10877	544	2508 Oct 10	06:29:19	1494	6292	164	N	h-	1.1203	0.8248	-0.2198	250.4	-	-	8N	95W
10878	544	2509 Mar 08	00:13:07	1496	6297	131	P	-h	0.8858	1.2614	0.2042	297.0	104.6	-	6N	6E
10879	544	2509 Aug 31	07:27:37	1498	6303	136	P	-a	-0.8231	1.3263	0.3684	273.2	127.1	-	9S	105W
10880	544	2510 Feb 25	02:01:15	1500	6309	141	T+	pp	0.1998	2.5411	1.4426	374.6	230.1	93.7	9N	21W
10881	545	2510 Aug 21	00:21:36	1502	6315	146	T-	pp	-0.0891	2.6666	1.7219	317.2	210.9	97.7	12S	2E
10882	545	2511 Feb 14	01:22:52	1504	6321	151	P	t-	-0.4880	2.0077	0.9181	354.5	202.5	-	13N	11W
10883	545	2511 Aug 10	16:12:02	1507	6327	156	P	a-	0.6669	1.6215	0.6468	298.8	164.6	-	15S	125E
10884	545	2512 Feb 03	05:43:56	1509	6333	161	N	a-	-1.1489	0.7695	-0.2698	237.7	-	-	16N	77W
10885	545	2512 Jun 30	13:51:10	1511	6338	128	N	-t	-1.2672	0.5588	-0.4931	216.7	-	-	24S	160E
10886	545	2512 Jul 30	02:35:53	1511	6339	166	N	t-	1.4873	0.1429	-0.8850	110.6	-	-	17S	31W
10887	545	2512 Dec 24	06:15:46	1513	6344	133	P	-a	0.9388	1.1213	0.1490	255.0	82.5	-	24N	88W
10888	545	2513 Jun 19	14:55:56	1515	6350	138	P	-t	-0.5279	1.9251	0.8543	353.2	200.0	-	24S	143E
10889	545	2513 Dec 13	21:58:14	1517	6356	143	T	-p	0.2783	2.3367	1.3576	318.4	204.3	79.8	23N	35E
10890	545	2514 Jun 08	17:01:23	1519	6362	148	T+	p-	0.2410	2.4366	1.3951	356.2	223.3	89.1	23S	111E
10891	545	2514 Dec 03	10:39:16	1521	6368	153	T	p-	-0.4258	2.0891	1.0644	330.8	200.3	38.8	22N	156W
10892	545	2515 May 29	01:46:15	1523	6374	158	P	a-	0.9791	1.0557	0.0671	261.0	58.4	-	21S	21W
10893	545	2515 Nov 22	16:34:46	1526	6380	163	N	t-	-1.1754	0.7402	-0.3374	246.0	-	-	19N	114E
10894	545	2516 Apr 18	08:44:26	1527	6385	130	P	-a	-0.8975	1.1879	0.2341	258.9	102.0	-	12S	125W
10895	545	2516 Oct 11	23:17:01	1530	6391	135	N	-t	1.0802	0.9148	-0.1624	271.5	-	-	8N	14E
10896	545	2517 Apr 08	00:32:52	1532	6397	140	T-	-p	-0.2048	2.4726	1.4914	325.8	211.1	90.2	7S	1W
10897	545	2517 Oct 01	03:41:07	1534	6403	145	T	-p	0.3108	2.3007	1.2749	342.9	213.9	76.0	4N	51W
10898	545	2518 Mar 28	11:27:42	1536	6409	150	P	t-	0.5390	1.8883	0.8498	331.2	190.1	-	3S	164W
10899	545	2518 Sep 20	15:27:28	1538	6415	155	T	a-	-0.4370	2.0413	1.0707	313.4	194.4	39.5	1S	133E
10900	545	2519 Mar 17	14:56:26	1540	6421	160	N	t-	1.2905	0.5344	-0.5537	216.6	-	-	2N	145E
10901	546	2519 Aug 12	00:36:43	1542	6426	127	N	-a	1.3801	0.2995	-0.6489	143.1	-	-	14S	2W
10902	546	2519 Sep 10	07:48:23	1542	6427	165	N	a-	-1.1363	0.7462	-0.2007	214.8	-	-	6S	111W
10903	546	2520 Feb 04	20:36:27	1544	6432	132	N*	-t	-1.0242	1.0184	-0.0603	279.5	-	-	15N	60E
10904	546	2520 Jul 31	15:22:37	1546	6438	137	P	-a	0.6822	1.6002	0.6120	303.0	163.3	-	17S	138E
10905	546	2521 Jan 24	03:01:30	1549	6444	142	T	-p	-0.2872	2.3426	1.3195	336.0	210.8	78.9	19N	36W
10906	546	2521 Jul 20	23:47:33	1551	6450	147	T-	pp	-0.0849	2.7232	1.6814	361.1	229.9	103.3	20S	11E
10907	546	2522 Jan 13	16:02:19	1553	6456	152	T	p-	0.4007	2.1118	1.1333	311.8	194.7	52.4	22N	128E
10908	546	2522 Jul 10	01:40:39	1555	6462	157	P	t-	-0.8624	1.3106	0.2409	313.8	117.4	-	23S	17W
10909	546	2523 Jan 03	07:44:09	1557	6468	162	N	a-	1.0579	0.9041	-0.0706	234.9	-	-	24N	109W
10910	546	2523 May 30	14:13:13	1559	6473	129	N	-t	1.1767	0.7131	-0.3152	233.4	-	-	21S	153E
10911	546	2523 Nov 24	07:55:24	1561	6479	134	N*	-a	-1.0110	1.0229	-0.0168	268.4	-	-	19N	116W
10912	546	2524 May 19	00:54:02	1564	6485	139	T	-p	0.3744	2.1589	1.1826	320.3	201.7	61.8	19S	8W
10913	546	2524 Nov 12	11:52:58	1566	6491	144	T	-p	-0.3270	2.3013	1.2152	365.2	220.4	71.0	18N	176W
10914	546	2525 May 08	16:42:30	1568	6497	149	T	p-	-0.3589	2.1739	1.2244	309.2	198.0	66.0	18S	115E
10915	546	2525 Nov 01	11:26:31	1570	6503	154	T	t-	0.3790	2.2076	1.1183	365.0	216.9	54.7	15N	169W
10916	546	2526 Apr 28	09:13:29	1572	6509	159	N	a-	-1.0819	0.8560	-0.1107	232.0	-	-	15S	133W
10917	546	2526 Oct 21	14:05:41	1574	6515	164	N	h-	1.0739	0.9099	-0.1348	260.0	-	-	12N	151E
10918	546	2527 Mar 19	08:09:29	1576	6520	131	P	-h	0.9140	1.2096	0.1524	293.3	91.3	-	1N	114W
10919	546	2527 Sep 11	15:08:19	1578	6526	136	P	-a	-0.8822	1.2181	0.2599	264.5	108.3	-	5S	139E
10920	546	2528 Mar 07	09:46:47	1581	6532	141	T+	pp	0.2233	2.4970	1.4003	373.9	229.1	90.8	5N	137W

Cat Num	Canon Plate	Calendar Date	TD of Greatest Eclipse	ΔT s	Luna Num	Saros Num	Ecl. Type	QSE	Gamma	Pen. Mag.	Um. Mag.	Eclipse Phase Durations Pen. m	Par. m	Total m	Greatest in Zenith Lat.	Lng.
10921	547	2528 Aug 31	08:00:19	1583	6538	146	T-	-p	-0.1549	2.5471	1.5999	316.6	209.3	94.1	8S	113W
10922	547	2529 Feb 24	09:21:26	1585	6544	151	P	t-	-0.4712	2.0365	0.9508	354.6	204.5	-	9N	131W
10923	547	2529 Aug 20	23:32:24	1587	6550	156	P	a-	0.5976	1.7509	0.7718	306.8	176.6	-	11S	14E
10924	547	2530 Feb 13	14:09:12	1589	6556	161	N	a-	-1.1371	0.7891	-0.2459	239.3	-	-	12N	157E
10925	547	2530 Jul 11	20:27:10	1591	6561	128	N	-t	-1.3526	0.4032	-0.6507	187.3	-	-	23S	62E
10926	547	2530 Aug 10	09:26:46	1592	6562	166	N	t-	1.4146	0.2785	-0.7537	153.2	-	-	14S	134W
10927	547	2531 Jan 04	15:06:28	1593	6567	133	P	-a	0.9433	1.1130	0.1407	254.1	80.2	-	24N	141E
10928	547	2531 Jun 30	21:23:33	1596	6573	138	P	-t	-0.6134	1.7673	0.6981	344.9	185.7	-	24S	47E
10929	547	2531 Dec 25	06:43:03	1598	6579	143	T	-p	0.2865	2.3231	1.3412	318.8	204.0	78.5	24N	94W
10930	547	2532 Jun 18	23:51:43	1600	6585	148	T+	pp	0.1630	2.5775	1.5405	356.9	226.5	97.9	23S	9E
10931	547	2532 Dec 13	19:02:44	1602	6591	153	T	p-	-0.4148	2.1117	1.0822	332.9	201.8	43.7	23N	79E
10932	547	2533 Jun 08	09:07:46	1604	6597	158	P	a-	0.9093	1.1811	0.1975	271.2	98.1	-	22S	130W
10933	547	2533 Dec 03	00:30:10	1607	6603	163	N	t-	-1.1584	0.7739	-0.3083	251.5	-	-	21N	4W
10934	547	2534 Apr 29	16:45:47	1608	6608	130	P	-a	-0.9439	1.1015	0.1502	251.6	82.7	-	16S	114E
10935	547	2534 Oct 23	06:38:03	1611	6614	135	N	-t	1.1230	0.8370	-0.2416	261.5	-	-	12N	97W
10936	547	2535 Apr 19	08:33:01	1613	6620	140	T-	-p	-0.2470	2.3950	1.4141	325.3	209.6	85.6	11S	122W
10937	547	2535 Oct 12	11:18:38	1615	6626	145	T	-p	0.3567	2.2164	1.1908	339.5	209.6	65.0	8N	166W
10938	547	2536 Apr 07	19:13:18	1617	6632	150	P	t-	0.5027	1.9550	0.9165	335.1	195.6	-	7S	79E
10939	547	2536 Sep 30	23:18:44	1620	6638	155	T	a-	-0.3856	2.1357	1.1647	315.6	198.7	58.5	3N	15E
10940	547	2537 Mar 27	22:29:42	1622	6644	160	N	t-	1.2589	0.5915	-0.4949	226.8	-	-	2S	31E
10941	548	2537 Aug 22	08:09:08	1624	6649	127	N	-a	1.4483	0.1758	-0.7756	111.1	-	-	10S	115W
10942	548	2537 Sep 20	15:38:06	1624	6650	165	N	a-	-1.0812	0.8485	-0.1009	226.8	-	-	2S	131E
10943	548	2538 Feb 15	04:44:31	1626	6655	132	N*	-t	-1.0329	1.0003	-0.0744	276.6	-	-	12N	61W
10944	548	2538 Aug 11	22:33:32	1628	6661	137	P	-a	0.7588	1.4620	0.4693	295.7	146.7	-	14S	30E
10945	548	2539 Feb 04	11:35:58	1630	6667	142	T	-p	-0.2921	2.3317	1.3124	334.5	210.1	78.1	16N	164W
10946	548	2539 Aug 01	06:28:18	1633	6673	147	T-	pp	-0.0012	2.8786	1.8333	363.0	231.3	105.3	18S	88W
10947	548	2540 Jan 25	00:52:56	1635	6679	152	T	p-	0.3958	2.1200	1.1432	311.5	195.0	54.1	20N	4W
10948	548	2540 Jul 20	08:06:21	1637	6685	157	P	t-	-0.7756	1.4701	0.4001	326.4	147.7	-	21S	113W
10949	548	2541 Jan 13	16:35:03	1639	6691	162	N	a-	1.0544	0.9111	-0.0649	235.9	-	-	22N	120E
10950	548	2541 Jun 09	21:15:08	1641	6696	129	N	-t	1.2440	0.5870	-0.4361	213.4	-	-	22S	48E
10951	548	2541 Jul 09	09:38:29	1641	6697	167	Nb	t-	-1.5025	0.1243	-0.9219	105.7	-	-	24S	136W
10952	548	2541 Dec 04	16:08:21	1643	6702	134	N	-a	-1.0300	0.9907	-0.0543	266.2	-	-	21N	122E
10953	548	2542 May 30	08:25:22	1646	6708	139	T	-a	0.4375	2.0408	1.0695	315.3	195.5	39.4	21S	120W
10954	548	2542 Nov 23	19:37:00	1648	6714	144	T	-t	-0.3540	2.2542	1.1635	364.9	218.3	63.1	20N	69E
10955	548	2543 May 20	00:31:08	1650	6720	149	T	p-	-0.3011	2.2786	1.3317	311.4	202.1	77.5	20S	2W
10956	548	2543 Nov 12	19:00:51	1652	6726	154	T	t-	0.3451	2.2708	1.1796	366.7	219.8	65.9	18N	77E
10957	548	2544 May 08	17:03:20	1655	6732	159	N	a-	-1.0325	0.9464	-0.0198	242.2	-	-	18S	110E
10958	548	2544 Oct 31	21:52:27	1657	6738	164	N	h-	1.0350	0.9813	-0.0633	267.2	-	-	15N	34E
10959	548	2545 Mar 29	15:56:33	1659	6743	131	P	-h	0.9493	1.1448	0.0878	288.1	70.1	-	3S	129E
10960	548	2545 Sep 21	22:55:50	1661	6749	136	P	-a	-0.9362	1.1193	0.1604	255.9	86.1	-	1S	22E
10961	549	2546 Mar 18	17:24:17	1663	6755	141	T+	-p	0.2534	2.4406	1.3462	372.7	227.5	86.4	1N	108E
10962	549	2546 Sep 11	15:43:56	1665	6761	146	T-	-p	-0.2167	2.4353	1.4850	315.6	207.1	88.7	4S	130E
10963	549	2547 Mar 07	17:12:49	1668	6767	151	P	t-	-0.4480	2.0768	0.9958	355.1	207.1	-	5N	111E
10964	549	2547 Sep 01	06:56:14	1670	6773	156	P	a-	0.5317	1.8743	0.8902	313.7	186.4	-	8S	97W
10965	549	2548 Feb 24	22:29:44	1672	6779	161	N	a-	-1.1206	0.8168	-0.2131	241.8	-	-	8N	32E
10966	549	2548 Jul 22	03:01:34	1674	6784	128	N	-t	-1.4381	0.2475	-0.8088	149.3	-	-	21S	36W
10967	549	2548 Aug 20	16:20:01	1674	6785	166	N	t-	1.3441	0.4103	-0.6267	184.4	-	-	11S	122E
10968	549	2549 Jan 14	23:58:01	1676	6790	133	P	-a	0.9477	1.1048	0.1330	253.2	78.0	-	22N	10E
10969	549	2549 Jul 11	03:50:12	1678	6796	138	P	-t	-0.7000	1.6079	0.5398	335.1	167.5	-	23S	49W
10970	549	2550 Jan 04	15:31:00	1681	6802	143	T	-p	0.2921	2.3141	1.3299	319.4	203.9	77.6	23N	135E
10971	549	2550 Jun 30	06:38:43	1683	6808	148	T+	pp	0.0811	2.7258	1.6929	356.8	228.2	103.0	23S	91W
10972	549	2550 Dec 25	03:31:04	1685	6814	153	T	p-	-0.4084	2.1256	1.0918	334.7	202.8	46.1	23N	46W
10973	549	2551 Jun 19	16:25:17	1688	6820	158	P	a-	0.8350	1.3153	0.3361	280.7	125.0	-	23S	121E
10974	549	2551 Dec 14	08:31:21	1690	6826	163	N	t-	-1.1463	0.7981	-0.2882	255.6	-	-	22N	122W
10975	549	2552 May 10	00:40:14	1692	6831	130	P	-a	-0.9961	1.0046	0.0554	242.8	50.9	-	19S	4W
10976	549	2552 Jun 08	07:57:03	1692	6832	168	Nb	a-	1.5204	0.0407	-0.9049	53.8	-	-	21S	112W
10977	549	2552 Nov 02	14:09:22	1694	6837	135	N	-t	1.1581	0.7731	-0.3067	252.5	-	-	16N	150E
10978	549	2553 Apr 29	16:26:25	1696	6843	140	T	-p	-0.2942	2.3083	1.3276	324.4	207.4	78.8	15S	120E
10979	549	2553 Oct 22	19:04:11	1698	6849	145	T	-p	0.3969	2.1426	1.1172	336.2	205.3	52.0	12N	77E
10980	549	2554 Apr 19	02:51:33	1701	6855	150	P	t-	0.4610	2.0315	0.9929	339.2	201.4	-	11S	36W

Cat Num	Canon Plate	Calendar Date	TD of Greatest Eclipse	ΔT s	Luna Num	Saros Num	Ecl. Type	QSE	Gamma	Pen. Mag.	Um. Mag.	Eclipse Phase Durations Pen. m	Par. m	Total m	Greatest in Zenith Lat.	Lng.
10981	550	2554 Oct 12	07:17:08	1703	6861	155	T	p-	-0.3398	2.2201	1.2485	317.3	201.9	69.7	7N	105W
10982	550	2555 Apr 08	05:53:58	1705	6867	160	N	t-	1.2203	0.6615	-0.4230	238.4	-	-	6S	81W
10983	550	2555 Sep 02	15:45:57	1707	6872	127	Ne	-a	1.5126	0.0596	-0.8952	65.5	-	-	6S	130E
10984	550	2555 Oct 01	23:35:11	1708	6873	165	N	a-	-1.0323	0.9398	-0.0126	236.6	-	-	2N	11E
10985	550	2556 Feb 26	12:48:08	1709	6878	132	N	-t	-1.0459	0.9742	-0.0959	272.7	-	-	8N	178E
10986	550	2556 Aug 22	05:48:05	1712	6884	137	P	-a	0.8315	1.3309	0.3334	287.8	126.6	-	11S	79W
10987	550	2557 Feb 14	20:06:30	1714	6890	142	T	-p	-0.3008	2.3135	1.2986	332.7	209.0	76.6	12N	69E
10988	550	2557 Aug 11	13:11:57	1716	6896	147	T+	pp	0.0794	2.7370	1.6879	363.9	231.1	103.8	15S	171E
10989	550	2558 Feb 04	09:41:29	1719	6902	152	T	p-	0.3892	2.1311	1.1564	311.3	195.4	56.3	17N	135W
10990	550	2558 Jul 31	14:32:26	1721	6908	157	P	t-	-0.6888	1.6296	0.5590	337.3	170.3	-	19S	151E
10991	550	2559 Jan 25	01:25:29	1723	6914	162	N	a-	1.0511	0.9176	-0.0592	236.9	-	-	20N	11W
10992	550	2559 Jun 21	04:13:10	1725	6919	129	N	-t	1.3146	0.4550	-0.5634	189.5	-	-	22S	55W
10993	550	2559 Jul 20	16:18:32	1725	6920	167	N	t-	-1.4179	0.2780	-0.7654	155.4	-	-	22S	125E
10994	550	2559 Dec 16	00:26:02	1727	6925	134	N	-a	-1.0444	0.9666	-0.0832	264.6	-	-	22N	1W
10995	550	2560 Jun 09	15:52:08	1730	6931	139	P	-a	0.5047	1.9153	0.9482	309.5	187.8	-	22S	129E
10996	550	2560 Dec 04	03:28:39	1732	6937	144	T	-t	-0.3739	2.2195	1.1250	364.8	216.6	55.9	22N	48W
10997	550	2561 May 30	08:13:38	1734	6943	149	T-	p-	-0.2381	2.3931	1.4485	313.5	205.7	86.4	22S	117W
10998	550	2561 Nov 23	02:43:40	1737	6949	154	T	p-	0.3178	2.3216	1.2289	367.8	221.8	73.1	21N	37W
10999	550	2562 May 20	00:45:23	1739	6955	159	P	a-	-0.9772	1.0477	0.0816	252.6	62.7	-	21S	5W
11000	550	2562 Nov 12	05:48:45	1741	6961	164	N*	a-	1.0030	1.0401	-0.0045	272.5	-	-	19N	84W
11001	551	2563 Apr 09	23:33:33	1743	6966	131	P	-t	0.9919	1.0664	0.0097	281.2	23.7	-	7S	15E
11002	551	2563 Oct 03	06:51:07	1745	6972	136	P	-a	-0.9842	1.0318	0.0718	247.7	58.2	-	3N	98W
11003	551	2564 Mar 29	00:54:06	1748	6978	141	T	-p	0.2895	2.3731	1.2813	371.1	225.2	79.9	3S	5W
11004	551	2564 Sep 21	23:33:16	1750	6984	146	T	-p	-0.2732	2.3332	1.3795	314.4	204.3	81.6	0S	12E
11005	551	2565 Mar 18	00:59:22	1752	6990	151	T	t-	-0.4202	2.1252	1.0492	355.6	210.0	35.6	1N	6W
11006	551	2565 Sep 11	14:24:12	1755	6996	156	T	a-	0.4700	1.9902	1.0009	319.7	194.2	4.6	4S	150E
11007	551	2566 Mar 07	06:46:52	1757	7002	161	N	a-	-1.1006	0.8507	-0.1738	244.9	-	-	4N	92W
11008	551	2566 Aug 02	09:35:06	1759	7007	128	Ne	-t	-1.5233	0.0927	-0.9665	92.9	-	-	19S	134W
11009	551	2566 Aug 31	23:15:21	1759	7008	166	N	t-	1.2755	0.5387	-0.5034	209.7	-	-	7S	18E
11010	551	2567 Jan 26	08:49:56	1761	7013	133	P	-a	0.9521	1.0963	0.1254	252.3	75.8	-	20N	122W
11011	551	2567 Jul 22	10:17:26	1763	7019	138	P	-t	-0.7862	1.4493	0.3821	323.8	144.4	-	21S	145W
11012	551	2568 Jan 16	00:18:34	1766	7025	143	T	-p	0.2977	2.3047	1.3185	319.9	203.9	76.7	21N	5E
11013	551	2568 Jul 10	13:27:45	1768	7031	148	T-	pp	-0.0002	2.8724	1.8432	355.8	228.4	104.5	22S	167E
11014	551	2569 Jan 04	12:00:29	1770	7037	153	T	p-	-0.4031	2.1372	1.0997	336.4	203.8	48.0	22N	172W
11015	551	2569 Jun 29	23:42:27	1773	7043	158	P	a-	0.7595	1.4519	0.4767	289.1	145.4	-	22S	13E
11016	551	2569 Dec 24	16:35:18	1775	7049	163	N	t-	-1.1366	0.8176	-0.2721	258.9	-	-	22N	118E
11017	551	2570 May 21	08:28:02	1777	7054	130	N	-a	-1.0538	0.8977	-0.0497	232.2	-	-	21S	120W
11018	551	2570 Jun 19	15:30:31	1777	7055	168	N	a-	1.4525	0.1642	-0.7793	106.7	-	-	22S	135E
11019	551	2570 Nov 13	21:50:12	1779	7060	135	N	-t	1.1859	0.7227	-0.3581	245.0	-	-	19N	36E
11020	551	2571 May 11	00:09:06	1782	7066	140	T	-a	-0.3498	2.2064	1.2255	322.8	204.0	67.9	18S	4E
11021	552	2571 Nov 03	03:01:01	1784	7072	145	T	-p	0.4285	2.0843	1.0591	333.2	201.5	37.5	15N	42W
11022	552	2572 Apr 29	10:18:28	1786	7078	150	T	p-	0.4106	2.1240	1.0853	343.6	207.5	45.6	14S	148W
11023	552	2572 Oct 22	15:24:53	1789	7084	155	T	p-	-0.3012	2.2913	1.3190	318.4	204.1	76.9	11N	132E
11024	552	2573 Apr 18	13:08:28	1791	7090	160	N	t-	1.1738	0.7457	-0.3367	251.2	-	-	10S	170E
11025	552	2573 Oct 12	07:39:27	1793	7096	165	P	a-	-0.9892	1.0204	0.0649	244.7	55.0	-	7N	110W
11026	552	2574 Mar 08	20:47:08	1795	7101	132	N	-t	-1.0631	0.9400	-0.1249	267.8	-	-	4N	58E
11027	552	2574 Sep 02	13:03:42	1798	7107	137	P	-a	0.9023	1.2037	0.2008	278.9	100.5	-	7S	171E
11028	552	2575 Feb 26	04:34:54	1800	7113	142	T	-p	-0.3120	2.2906	1.2805	330.8	207.9	74.6	8N	58W
11029	552	2575 Aug 22	19:56:32	1802	7119	147	T+	pp	0.1584	2.5943	1.5408	364.0	229.3	98.8	11S	69E
11030	552	2576 Feb 15	18:27:33	1805	7125	152	T	p-	0.3803	2.1461	1.1739	311.2	196.1	59.0	13N	94E
11031	552	2576 Aug 10	21:01:26	1807	7131	157	P	t-	-0.6042	1.7854	0.7139	346.4	187.5	-	16S	54E
11032	552	2577 Feb 04	10:14:49	1809	7137	162	N	a-	1.0475	0.9244	-0.0529	238.0	-	-	17N	143W
11033	552	2577 Jul 01	11:11:08	1811	7142	129	N	-t	1.3860	0.3218	-0.6922	160.6	-	-	22S	159W
11034	552	2577 Jul 30	23:01:14	1812	7143	167	N	t-	-1.3341	0.4307	-0.6104	190.0	-	-	19S	24E
11035	552	2577 Dec 26	08:46:27	1814	7148	134	N	-a	-1.0560	0.9475	-0.1065	263.5	-	-	22N	124W
11036	552	2578 Jun 20	23:17:05	1816	7154	139	P	-a	0.5741	1.7859	0.8228	302.9	178.4	-	23S	19E
11037	552	2578 Dec 15	11:24:39	1818	7160	144	T	-t	-0.3903	2.1911	1.0933	364.7	215.1	48.8	23N	165W
11038	552	2579 Jun 10	15:51:26	1821	7166	149	T-	p-	-0.1708	2.5156	1.5728	315.2	208.5	92.9	23S	130E
11039	552	2579 Dec 04	10:34:13	1823	7172	154	T	p-	0.2963	2.3615	1.2678	368.5	223.1	77.8	22N	154W
11040	552	2580 May 30	08:21:38	1825	7178	159	P	a-	-0.9176	1.1573	0.1908	263.0	94.7	-	23S	118W

Cat Num	Canon Plate	Calendar Date	TD of Greatest Eclipse	ΔT s	Luna Num	Saros Num	Ecl. Type	QSE	Gamma	Pen. Mag.	Um. Mag.	Eclipse Phase Durations Pen. m	Par. m	Total m	Greatest in Zenith Lat.	Lng.
11041	553	2580 Nov 22	13:52:59	1828	7184	164	P	a-	0.9765	1.0883	0.0442	276.5	49.1	-	21N	155E
11042	553	2581 Apr 20	07:01:04	1830	7189	131	N	-t	1.0414	0.9755	-0.0809	272.2	-	-	11S	97W
11043	553	2581 Oct 13	14:54:30	1832	7195	136	N	-a	-1.0259	0.9559	-0.0054	240.1	-	-	7N	141E
11044	553	2582 Apr 09	08:13:59	1834	7201	141	T	-t	0.3337	2.2904	1.2017	368.7	221.9	69.7	7S	115W
11045	553	2582 Oct 03	07:29:03	1837	7207	146	T	-p	-0.3243	2.2415	1.2839	313.0	201.2	73.0	4N	107W
11046	553	2583 Mar 29	08:37:39	1839	7213	151	T	t-	-0.3849	2.1873	1.1167	356.3	213.4	53.6	4S	121W
11047	553	2583 Sep 22	21:58:06	1841	7219	156	T	p-	0.4136	2.0965	1.1015	324.9	200.5	47.9	0N	36E
11048	553	2584 Mar 17	14:56:25	1844	7225	161	N	a-	-1.0736	0.8975	-0.1213	249.3	-	-	0S	145E
11049	553	2584 Sep 11	06:16:14	1846	7231	166	N	t-	1.2116	0.6587	-0.3888	230.1	-	-	3S	88W
11050	553	2585 Feb 05	17:40:16	1848	7236	133	P	-a	0.9585	1.0839	0.1143	251.0	72.4	-	17N	106E
11051	553	2585 Aug 01	16:45:35	1850	7242	138	P	-t	-0.8717	1.2920	0.2253	310.8	113.3	-	18S	119E
11052	553	2586 Jan 26	09:06:34	1853	7248	143	T	-p	0.3029	2.2960	1.3083	320.5	203.9	75.8	19N	126W
11053	553	2586 Jul 21	20:16:47	1855	7254	148	T-	pp	-0.0826	2.7194	1.6935	353.8	226.9	102.6	20S	66E
11054	553	2587 Jan 15	20:31:07	1858	7260	153	T	p-	-0.3996	2.1452	1.1046	337.9	204.6	49.2	21N	62E
11055	553	2587 Jul 11	06:58:13	1860	7266	158	P	a-	0.6818	1.5926	0.6208	296.6	161.7	-	21S	95W
11056	553	2588 Jan 05	00:42:53	1862	7272	163	N	t-	-1.1301	0.8309	-0.2615	261.2	-	-	22N	2W
11057	553	2588 May 31	16:10:46	1864	7277	130	N	-a	-1.1157	0.7836	-0.1626	219.7	-	-	23S	125E
11058	553	2588 Jun 29	23:02:17	1865	7278	168	N	a-	1.3830	0.2909	-0.6511	140.2	-	-	22S	23E
11059	553	2588 Nov 24	05:39:09	1867	7283	135	N	-t	1.2078	0.6828	-0.3986	238.6	-	-	22N	81W
11060	553	2589 May 21	07:46:01	1869	7289	140	T	-a	-0.4096	2.0969	1.1156	320.5	199.4	50.5	21S	109W
11061	554	2589 Nov 13	11:06:01	1871	7295	145	T	-p	0.4544	2.0367	1.0118	330.5	198.1	16.9	19N	163W
11062	554	2590 May 10	17:36:56	1874	7301	150	T	p-	0.3540	2.2281	1.1891	347.9	213.5	65.8	18S	103E
11063	554	2590 Nov 02	23:40:12	1876	7307	155	T-	p-	-0.2685	2.3518	1.3786	319.1	205.7	81.9	15N	9E
11064	554	2591 Apr 29	20:14:12	1879	7313	160	N	t-	1.1201	0.8432	-0.2371	264.6	-	-	14S	64E
11065	554	2591 Oct 23	15:51:13	1881	7319	165	P	a-	-0.9525	1.0893	0.1306	251.2	77.4	-	11N	126E
11066	554	2592 Mar 19	04:38:54	1883	7324	132	N	-t	-1.0867	0.8938	-0.1655	261.3	-	-	1S	60W
11067	554	2592 Sep 12	20:25:19	1885	7330	137	P	-a	0.9671	1.0876	0.0791	269.7	64.3	-	3S	60E
11068	554	2593 Mar 08	12:57:32	1888	7336	142	T	-p	-0.3284	2.2579	1.2528	328.7	206.3	71.4	4N	176E
11069	554	2593 Sep 02	02:44:50	1890	7342	147	T+	pp	0.2335	2.4587	1.4006	363.2	226.2	90.3	7S	33W
11070	554	2594 Feb 26	03:08:54	1893	7348	152	T	p-	0.3674	2.1683	1.1989	311.3	197.0	62.5	9N	36W
11071	554	2594 Aug 22	03:34:22	1895	7354	157	P	t-	-0.5225	1.9360	0.8632	353.8	200.7	-	12S	45W
11072	554	2595 Feb 15	18:59:22	1897	7360	162	N	a-	1.0403	0.9376	-0.0397	239.7	-	-	13N	87E
11073	554	2595 Jul 12	18:08:13	1899	7365	129	N	-t	1.4585	0.1868	-0.8234	123.4	-	-	20S	98E
11074	554	2595 Aug 11	05:47:49	1900	7366	167	N	t-	-1.2521	0.5803	-0.4589	216.7	-	-	16S	77W
11075	554	2596 Jan 06	17:09:04	1902	7371	134	N	-a	-1.0652	0.9325	-0.1251	262.8	-	-	21N	112E
11076	554	2596 Jul 01	06:40:09	1904	7377	139	P	-a	0.6457	1.6529	0.6930	295.3	166.9	-	22S	91W
11077	554	2596 Dec 25	19:24:44	1907	7383	144	T	-t	-0.4030	2.1691	1.0688	364.6	213.9	42.3	23N	77E
11078	554	2597 Jun 20	23:25:26	1909	7389	149	T-	pp	-0.1005	2.6442	1.7025	316.4	210.5	97.3	23S	17E
11079	554	2597 Dec 14	18:32:07	1911	7395	154	T	p-	0.2805	2.3907	1.2965	368.7	223.9	80.9	23N	88E
11080	554	2598 Jun 10	15:50:01	1914	7401	159	P	a-	-0.8521	1.2778	0.3105	273.5	119.1	-	24S	131E
11081	555	2598 Dec 03	22:06:11	1916	7407	164	P	a-	0.9567	1.1243	0.0809	279.1	65.9	-	23N	34E
11082	555	2599 May 01	14:18:45	1918	7412	131	N	-t	1.0978	0.8717	-0.1843	260.8	-	-	14S	153E
11083	555	2599 Oct 24	23:06:08	1921	7418	136	N	-a	-1.0614	0.8917	-0.0711	233.2	-	-	11N	18E
11084	555	2600 Apr 20	15:25:41	1923	7424	141	T	-t	0.3845	2.1957	1.1101	365.5	217.3	53.2	11S	137E
11085	555	2600 Oct 14	15:31:59	1925	7430	146	T	-p	-0.3690	2.1614	1.1998	311.6	197.9	63.0	8N	132E
11086	555	2601 Apr 09	16:10:39	1928	7436	151	T	t-	-0.3445	2.2586	1.1937	357.1	216.8	67.3	8S	126E
11087	555	2601 Oct 04	05:36:14	1930	7442	156	T	p-	0.3614	2.1952	1.1943	329.5	205.6	64.4	5N	79W
11088	555	2602 Mar 29	23:01:38	1933	7448	161	N	a-	-1.0422	0.9520	-0.0609	254.2	-	-	4S	23E
11089	555	2602 Sep 23	13:21:50	1935	7454	166	N	t-	1.1517	0.7714	-0.2817	247.3	-	-	1N	165E
11090	555	2603 Feb 18	02:26:58	1937	7459	133	P	-a	0.9679	1.0657	0.0978	249.3	67.1	-	13N	25W
11091	555	2603 Aug 13	23:17:51	1940	7465	138	P	-t	-0.9539	1.1412	0.0746	296.6	66.6	-	15S	21E
11092	555	2604 Feb 07	17:51:11	1942	7471	143	T	-p	0.3106	2.2824	1.2936	320.9	203.7	74.5	16N	104E
11093	555	2604 Aug 02	03:10:29	1944	7477	148	T-	pp	-0.1619	2.5726	1.5493	351.0	224.1	97.5	18S	38W
11094	555	2605 Jan 27	04:58:58	1947	7483	153	T	p-	-0.3944	2.1561	1.1129	339.4	205.6	51.1	18N	64W
11095	555	2605 Jul 22	14:16:22	1949	7489	158	P	a-	0.6052	1.7318	0.7628	302.9	174.7	-	20S	156E
11096	555	2606 Jan 16	08:50:19	1952	7495	163	N	t-	-1.1237	0.8436	-0.2508	263.5	-	-	20N	123W
11097	555	2606 Jun 12	23:46:43	1954	7500	130	N	-a	-1.1830	0.6597	-0.2858	204.3	-	-	24S	12E
11098	555	2606 Jul 12	06:31:08	1954	7501	168	N	a-	1.3108	0.4229	-0.5181	166.9	-	-	21S	88W
11099	555	2606 Dec 06	13:36:27	1956	7506	135	N	-t	1.2234	0.6542	-0.4274	233.7	-	-	23N	161E
11100	555	2607 Jun 02	15:13:55	1959	7512	140	P	-a	-0.4765	1.9746	0.9923	317.3	193.0	-	23S	139E

Cat Num	Canon Plate	Calendar Date	TD of Greatest Eclipse	ΔT s	Luna Num	Saros Num	Ecl. Type	QSE	Gamma	Pen. Mag.	Um. Mag.	Eclipse Phase Durations Pen. m	Par. m	Total m	Greatest in Zenith Lat.	Lng.
11101	556	2607 Nov 25	19:20:19	1961	7518	145	P	-p	0.4734	2.0015	0.9772	328.1	195.4	-	21N	75E
11102	556	2608 May 22	00:45:18	1963	7524	150	T	pp	0.2896	2.3465	1.3069	352.1	219.1	80.7	20S	4W
11103	556	2608 Nov 14	08:04:33	1966	7530	155	T-	p-	-0.2426	2.3997	1.4256	319.5	206.7	85.1	18N	117W
11104	556	2609 May 11	03:10:50	1968	7536	160	N	t-	1.0591	0.9539	-0.1241	278.4	-	-	17S	40W
11105	556	2609 Nov 04	00:09:40	1971	7542	165	P	a-	-0.9216	1.1479	0.1857	256.6	91.6	-	14N	2E
11106	556	2610 Mar 31	12:25:16	1973	7547	132	N	-t	-1.1153	0.8384	-0.2149	253.5	-	-	5S	177W
11107	556	2610 Sep 25	03:50:47	1975	7553	137	N	-a	1.0277	0.9796	-0.0350	260.2	-	-	2N	52W
11108	556	2611 Mar 20	21:14:33	1978	7559	142	T	-p	-0.3502	2.2152	1.2156	326.2	204.3	66.6	0S	51E
11109	556	2611 Sep 14	09:37:36	1980	7565	147	T	-t	0.3042	2.3315	1.2685	361.7	221.9	77.8	3S	137W
11110	556	2612 Mar 09	11:45:52	1983	7571	152	T	p-	0.3508	2.1973	1.2308	311.6	198.3	66.7	5N	166W
11111	556	2612 Sep 02	10:12:37	1985	7577	157	T	t-	-0.4447	2.0795	1.0052	359.8	210.9	12.0	8S	145W
11112	556	2613 Feb 27	03:40:03	1988	7583	162	N	a-	1.0305	0.9555	-0.0216	241.9	-	-	9N	43W
11113	556	2613 Jul 24	01:08:17	1990	7588	129	Ne	-t	1.5292	0.0555	-0.9514	67.7	-	-	18S	7W
11114	556	2613 Aug 22	12:40:07	1990	7589	167	N	t-	-1.1735	0.7237	-0.3140	237.8	-	-	13S	180E
11115	556	2614 Jan 18	01:31:45	1992	7594	134	N	-a	-1.0735	0.9188	-0.1418	262.1	-	-	20N	13W
11116	556	2614 Jul 13	14:02:30	1994	7600	139	P	-a	0.7183	1.5184	0.5612	286.9	153.0	-	21S	159E
11117	556	2615 Jan 07	03:27:06	1997	7606	144	T	-t	-0.4133	2.1512	1.0490	364.6	213.0	36.0	22N	42W
11118	556	2615 Jul 03	06:57:20	1999	7612	149	T-	pp	-0.0286	2.7759	1.8347	317.0	211.5	99.3	23S	95W
11119	556	2615 Dec 27	02:33:58	2002	7618	154	T+	p-	0.2678	2.4139	1.3201	368.6	224.5	83.1	24N	30W
11120	556	2616 Jun 21	23:14:53	2004	7624	159	P	a-	-0.7845	1.4027	0.4340	283.3	138.6	-	24S	20E
11121	557	2616 Dec 15	06:25:32	2007	7630	164	P	a-	0.9418	1.1512	0.1090	280.7	76.0	-	24N	90W
11122	557	2617 May 12	21:27:24	2009	7635	131	N	-t	1.1607	0.7562	-0.2996	246.3	-	-	17S	46E
11123	557	2617 Nov 05	07:25:07	2011	7641	136	N	-a	-1.0911	0.8379	-0.1265	227.2	-	-	15N	107W
11124	557	2618 May 01	22:28:05	2014	7647	141	T	-t	0.4426	2.0873	1.0049	361.4	211.1	11.6	15S	31E
11125	557	2618 Oct 25	23:42:26	2016	7653	146	T	-a	-0.4073	2.0934	1.1276	310.4	194.8	51.5	12N	9E
11126	557	2619 Apr 20	23:33:42	2019	7659	151	T	p-	-0.2952	2.3461	1.2870	357.8	220.4	79.1	12S	15E
11127	557	2619 Oct 15	13:22:29	2021	7665	156	T	p-	0.3168	2.2802	1.2731	333.3	209.5	74.5	9N	164E
11128	557	2620 Apr 09	06:58:47	2024	7671	161	P	a-	-1.0036	1.0199	0.0130	260.0	26.0	-	8S	96W
11129	557	2620 Oct 03	20:33:57	2026	7677	166	N	t-	1.0976	0.8736	-0.1853	261.5	-	-	5N	57E
11130	557	2621 Feb 28	11:09:03	2028	7682	133	P	-a	0.9813	1.0401	0.0741	246.8	58.6	-	9N	155W
11131	557	2621 Aug 24	05:54:22	2031	7688	138	N	-t	-1.0327	0.9968	-0.0701	281.1	-	-	12S	79W
11132	557	2622 Feb 18	02:32:59	2033	7694	143	T	-p	0.3201	2.2652	1.2758	321.2	203.5	72.8	12N	26W
11133	557	2622 Aug 13	10:06:08	2036	7700	148	T-	pp	-0.2405	2.4273	1.4063	347.3	219.8	88.7	15S	141W
11134	557	2623 Feb 07	13:25:20	2038	7706	153	T	p-	-0.3887	2.1676	1.1224	341.0	206.7	53.2	15N	170E
11135	557	2623 Aug 02	21:35:48	2041	7712	158	P	a-	0.5288	1.8709	0.9040	308.2	185.1	-	17S	46E
11136	557	2624 Jan 27	16:56:50	2043	7718	163	N	t-	-1.1169	0.8567	-0.2389	265.7	-	-	18N	117E
11137	557	2624 Jun 23	07:19:46	2045	7723	130	N	-a	-1.2525	0.5322	-0.4131	186.0	-	-	25S	100W
11138	557	2624 Jul 22	14:00:56	2046	7724	168	N	a-	1.2392	0.5542	-0.3865	188.5	-	-	19S	160E
11139	557	2624 Dec 16	21:40:25	2048	7729	135	N	-t	1.2345	0.6336	-0.4476	230.0	-	-	24N	42E
11140	557	2625 Jun 12	22:36:39	2050	7735	140	P	-a	-0.5468	1.8464	0.8628	313.2	184.8	-	24S	30E
11141	558	2625 Dec 06	03:40:53	2053	7741	145	P	-p	0.4880	1.9745	0.9509	326.1	193.1	-	23N	49W
11142	558	2626 Jun 02	07:46:44	2055	7747	150	T+	pp	0.2203	2.4741	1.4338	355.8	223.9	91.7	22S	108W
11143	558	2626 Nov 25	16:35:27	2058	7753	155	T-	p-	-0.2223	2.4374	1.4626	319.7	207.3	87.3	20N	116E
11144	558	2627 May 22	09:59:12	2060	7759	160	P	t-	0.9914	1.0771	0.0011	292.1	8.0	-	19S	142W
11145	558	2627 Nov 15	08:35:40	2063	7765	165	P	a-	-0.8972	1.1943	0.2287	260.8	101.1	-	17N	124W
11146	558	2628 Apr 10	20:04:00	2065	7770	132	N	-t	-1.1504	0.7709	-0.2764	243.7	-	-	9S	68E
11147	558	2628 Oct 05	11:22:30	2067	7776	137	N	-a	1.0821	0.8828	-0.1379	250.7	-	-	6N	165W
11148	558	2629 Mar 31	05:24:37	2070	7782	142	T	-p	-0.3783	2.1609	1.1668	323.3	201.6	59.5	4S	72W
11149	558	2629 Sep 24	16:36:45	2072	7788	147	T	-t	0.3690	2.2151	1.1472	359.7	216.7	60.2	1N	117E
11150	558	2630 Mar 20	20:15:16	2075	7794	152	T	p-	0.3279	2.2378	1.2746	312.1	199.9	71.7	0N	67E
11151	558	2630 Sep 13	16:57:18	2077	7800	157	T	t-	-0.3718	2.2141	1.1381	364.3	218.5	59.0	4S	114E
11152	558	2631 Mar 10	12:13:33	2080	7806	162	P	a-	1.0154	0.9830	0.0065	245.0	17.7	-	5N	172W
11153	558	2631 Sep 02	19:39:18	2082	7812	167	N	t-	-1.0994	0.8591	-0.1775	254.8	-	-	9S	74E
11154	558	2632 Jan 29	09:52:07	2084	7817	134	N	-a	-1.0828	0.9028	-0.1600	261.1	-	-	17N	137W
11155	558	2632 Jul 23	21:26:01	2087	7823	139	P	-a	0.7902	1.3854	0.4302	277.6	136.4	-	19S	49E
11156	558	2633 Jan 17	11:31:11	2089	7829	144	T	-t	-0.4215	2.1367	1.0334	364.5	212.2	29.9	20N	162W
11157	558	2633 Jul 13	14:26:23	2092	7835	149	T+	pp	0.0456	2.7446	1.8033	317.1	211.5	99.1	22S	154E
11158	558	2634 Jan 06	10:40:40	2094	7841	154	T+	p-	0.2586	2.4303	1.3374	368.3	224.7	84.7	23N	150W
11159	558	2634 Jul 03	06:34:23	2097	7847	159	P	a-	-0.7131	1.5348	0.5640	292.7	155.3	-	24S	88W
11160	558	2634 Dec 26	14:51:05	2099	7853	164	P	a-	0.9309	1.1703	0.1298	281.5	82.4	-	24N	146E

Cat Num	Canon Plate	Calendar Date	TD of Greatest Eclipse	ΔT s	Luna Num	Saros Num	Ecl. Type	QSE	Gamma	Pen. Mag.	Um. Mag.	Eclipse Phase Durations Pen. m	Par. m	Total m	Greatest in Zenith Lat.	Lng.
11161	559	2635 May 24	04:26:35	2102	7858	131	N	-t	1.2304	0.6284	-0.4273	228.0	-	-	20S	58W
11162	559	2635 Jun 22	17:24:39	2102	7859	169	Nb	t-	-1.5336	0.0567	-0.9688	70.0	-	-	25S	108E
11163	559	2635 Nov 16	15:52:29	2104	7864	136	N	-a	-1.1144	0.7960	-0.1700	222.2	-	-	18N	127E
11164	559	2635 Dec 16	02:16:39	2105	7865	174	N	a-	1.5481	0.0105	-0.9760	28.1	-	-	25N	27W
11165	559	2636 May 12	05:23:36	2107	7870	141	P	-t	0.5064	1.9687	0.8894	356.1	202.9	-	18S	73W
11166	559	2636 Nov 05	07:58:56	2109	7876	146	T	-a	-0.4402	2.0352	1.0651	309.3	191.8	37.5	15N	115W
11167	559	2637 May 01	06:52:10	2112	7882	151	T-	p-	-0.2413	2.4421	1.3888	358.4	223.5	88.5	15S	95W
11168	559	2637 Oct 25	21:14:11	2114	7888	156	T	p-	0.2774	2.3556	1.3424	336.6	212.6	81.5	13N	46E
11169	559	2638 Apr 20	14:49:41	2117	7894	161	P	a-	-0.9590	1.0986	0.0978	266.4	70.4	-	12S	146E
11170	559	2638 Oct 15	03:52:17	2119	7900	166	N	t-	1.0489	0.9659	-0.0987	273.4	-	-	9N	53W
11171	559	2639 Mar 11	19:45:23	2121	7905	133	P	-a	0.9996	1.0054	0.0417	243.5	44.2	-	5N	75E
11172	559	2639 Sep 04	12:37:54	2124	7911	138	N	-t	-1.1060	0.8625	-0.2049	264.8	-	-	8S	180E
11173	559	2640 Feb 29	11:07:48	2126	7917	143	T	-p	0.3353	2.2376	1.2478	321.3	202.8	69.8	8N	155W
11174	559	2640 Aug 23	17:09:19	2129	7923	148	T	-p	-0.3138	2.2918	1.2726	343.0	214.5	76.1	11S	113E
11175	559	2641 Feb 17	21:46:04	2132	7929	153	T	p-	-0.3787	2.1866	1.1401	342.7	208.1	56.7	11N	45E
11176	559	2641 Aug 13	04:58:42	2134	7935	158	T	a-	0.4545	2.0065	1.0412	312.5	193.3	30.6	14S	64W
11177	559	2642 Feb 07	01:00:14	2137	7941	163	N	t-	-1.1078	0.8738	-0.2223	268.2	-	-	15N	3W
11178	559	2642 Jul 04	14:48:51	2139	7946	130	N	-a	-1.3251	0.3991	-0.5466	163.4	-	-	24S	148E
11179	559	2642 Aug 02	21:30:59	2139	7947	168	N	a-	1.1673	0.6864	-0.2548	207.0	-	-	16S	48E
11180	559	2642 Dec 28	05:49:29	2141	7952	135	N	-t	1.2419	0.6196	-0.4605	227.2	-	-	24N	78W
11181	560	2643 Jun 24	05:52:24	2144	7958	140	P	-a	-0.6220	1.7093	0.7237	307.8	173.8	-	24S	78W
11182	560	2643 Dec 17	12:08:59	2146	7964	145	P	-a	0.4971	1.9570	0.9347	324.3	191.5	-	24N	175W
11183	560	2644 Jun 12	14:40:34	2149	7970	150	T+	pp	0.1454	2.6121	1.5705	359.0	227.7	99.4	23S	149E
11184	560	2644 Dec 06	01:12:54	2151	7976	155	T-	p-	-0.2069	2.4659	1.4906	319.7	207.6	88.8	22N	12W
11185	560	2645 Jun 01	16:40:36	2154	7982	160	P	t-	0.9179	1.2111	0.1369	305.2	89.8	-	21S	119E
11186	560	2645 Nov 25	17:08:16	2157	7988	165	P	a-	-0.8783	1.2306	0.2617	264.1	107.7	-	20N	109E
11187	560	2646 Apr 22	03:35:31	2159	7993	132	N	-t	-1.1920	0.6915	-0.3496	231.6	-	-	13S	45W
11188	560	2646 Oct 16	19:00:12	2161	7999	137	N	-a	1.1308	0.7967	-0.2304	241.4	-	-	10N	80E
11189	560	2647 Apr 11	13:28:29	2164	8005	142	T	-p	-0.4123	2.0956	1.1071	320.0	198.2	48.5	9S	167E
11190	560	2647 Oct 05	23:42:33	2166	8011	147	T	-t	0.4277	2.1100	1.0370	357.4	210.9	31.3	5N	10E
11191	560	2648 Mar 31	04:38:32	2169	8017	152	T	p-	0.2996	2.2881	1.3281	312.9	201.8	76.9	4S	60W
11192	560	2648 Sep 23	23:50:09	2172	8023	157	T	t-	-0.3051	2.3373	1.2595	367.7	223.9	77.4	0N	10E
11193	560	2649 Mar 20	20:41:39	2174	8029	162	P	a-	0.9964	1.0174	0.0415	248.8	44.7	-	1N	61E
11194	560	2649 Sep 13	02:45:09	2177	8035	167	N	t-	-1.0296	0.9867	-0.0491	268.6	-	-	5S	33W
11195	560	2650 Feb 08	18:09:59	2179	8040	134	N	-a	-1.0934	0.8843	-0.1803	259.7	-	-	14N	100E
11196	560	2650 Aug 04	04:51:39	2181	8046	139	P	-a	0.8606	1.2555	0.3017	267.6	116.2	-	16S	62W
11197	560	2651 Jan 28	19:32:59	2184	8052	144	T	-t	-0.4314	2.1186	1.0151	364.1	211.2	20.2	18N	79E
11198	560	2651 Jul 24	21:56:01	2187	8058	149	T+	-p	0.1192	2.6099	1.6679	316.6	210.4	96.5	20S	42E
11199	560	2652 Jan 17	18:48:27	2189	8064	154	T+	p-	0.2501	2.4451	1.3539	367.7	224.9	86.0	21N	89E
11200	560	2652 Jul 13	13:52:27	2192	8070	159	P	a-	-0.6410	1.6683	0.6951	301.3	169.3	-	22S	163E
11201	561	2653 Jan 05	23:18:55	2194	8076	164	P	a-	0.9213	1.1867	0.1484	282.0	87.7	-	23N	20E
11202	561	2653 Jun 03	11:18:15	2196	8081	131	N	-t	1.3049	0.4919	-0.5641	204.8	-	-	21S	160W
11203	561	2653 Jul 03	00:17:37	2197	8082	169	N	t-	-1.4619	0.1895	-0.8384	126.9	-	-	24S	6E
11204	561	2653 Nov 27	00:26:25	2199	8087	136	N	-a	-1.1326	0.7633	-0.2042	218.2	-	-	20N	0W
11205	561	2653 Dec 26	11:00:19	2199	8088	174	N	a-	1.5371	0.0304	-0.9555	47.5	-	-	25N	156W
11206	561	2654 May 23	12:09:21	2202	8093	141	P	-t	0.5779	1.8359	0.7597	349.4	191.8	-	20S	174W
11207	561	2654 Nov 16	16:23:11	2204	8099	146	T	-a	-0.4662	1.9896	1.0153	308.6	189.2	18.5	18N	120E
11208	561	2655 May 12	14:01:47	2207	8105	151	T-	pp	-0.1797	2.5521	1.5046	358.6	226.1	96.0	18S	158E
11209	561	2655 Nov 06	05:13:36	2209	8111	156	T+	p-	0.2453	2.4175	1.3983	339.4	214.9	86.2	16N	73W
11210	561	2656 Apr 30	22:32:51	2212	8117	161	P	a-	-0.9076	1.1901	0.1952	273.3	97.8	-	16S	30E
11211	561	2656 Oct 25	11:18:45	2214	8123	166	N*	t-	1.0074	1.0448	-0.0254	283.0	-	-	13N	165W
11212	561	2657 Mar 22	04:15:37	2217	8128	133	N	-a	1.0230	0.9612	-0.0001	239.1	-	-	0N	53W
11213	561	2657 Sep 14	19:26:58	2219	8134	138	N	-t	-1.1750	0.7364	-0.3318	247.4	-	-	4S	77E
11214	561	2658 Mar 11	19:37:41	2222	8140	143	T	-p	0.3539	2.2034	1.2136	321.2	201.9	65.7	4N	77E
11215	561	2658 Sep 04	00:17:32	2224	8146	148	T	-p	-0.3839	2.1626	1.1447	338.1	208.0	57.7	7S	5E
11216	561	2659 Mar 01	06:01:08	2227	8152	153	T	p-	-0.3649	2.2123	1.1648	344.5	209.9	61.2	7N	78W
11217	561	2659 Aug 24	12:25:53	2230	8158	158	T	a-	0.3831	2.1370	1.1727	315.8	199.6	59.9	11S	176W
11218	561	2660 Feb 18	09:00:08	2232	8164	163	N	t-	-1.0960	0.8952	-0.2007	271.3	-	-	11N	123W
11219	561	2660 Jul 14	22:16:53	2234	8169	130	N	-a	-1.3979	0.2661	-0.6807	135.3	-	-	23S	37E
11220	561	2660 Aug 13	05:03:27	2235	8170	168	N	a-	1.0975	0.8149	-0.1272	222.7	-	-	13S	65W

Cat Num	Canon Plate	Calendar Date	TD of Greatest Eclipse	ΔT s	Luna Num	Saros Num	Ecl. Type	QSE	Gamma	Pen. Mag.	Um. Mag.	Eclipse Phase Durations Pen. m	Par. m	Total m	Greatest in Zenith Lat.	Lng.
11221	562	2661 Jan 07	14:01:35	2237	8175	135	N	-t	1.2474	0.6087	-0.4697	224.8	-	-	23N	160E
11222	562	2661 Jul 04	13:05:13	2240	8181	140	P	-a	-0.6987	1.5698	0.5818	301.2	160.0	-	23S	175E
11223	562	2661 Dec 27	20:41:10	2242	8187	145	P	-a	0.5035	1.9444	0.9238	322.8	190.2	-	24N	59E
11224	562	2662 Jun 23	21:28:04	2245	8193	150	T+	pp	0.0662	2.7583	1.7152	361.5	230.3	104.0	23S	48E
11225	562	2662 Dec 17	09:55:29	2247	8199	155	T-	p-	-0.1956	2.4867	1.5111	319.6	207.8	89.8	23N	141W
11226	562	2663 Jun 12	23:16:24	2250	8205	160	P	t-	0.8395	1.3542	0.2816	317.7	126.1	-	22S	21E
11227	562	2663 Dec 07	01:45:56	2253	8211	165	P	a-	-0.8640	1.2584	0.2865	266.7	112.4	-	22N	20W
11228	562	2664 May 02	11:00:10	2255	8216	132	N	-t	-1.2396	0.6009	-0.4340	216.8	-	-	17S	157W
11229	562	2664 Jun 01	00:26:03	2255	8217	170	Nb	t-	1.5565	0.0287	-1.0247	51.5	-	-	21S	3E
11230	562	2664 Oct 27	02:45:20	2257	8222	137	N	-a	1.1726	0.7231	-0.3102	232.7	-	-	14N	36W
11231	562	2665 Apr 21	21:25:02	2260	8228	142	T	-a	-0.4528	2.0184	1.0355	316.1	193.8	28.5	13S	48E
11232	562	2665 Oct 16	06:55:19	2263	8234	147	P	-t	0.4800	2.0165	0.9385	354.9	204.7	-	9N	98W
11233	562	2666 Apr 11	12:53:42	2265	8240	152	T+	p-	0.2646	2.3506	1.3938	313.8	203.9	82.3	8S	176E
11234	562	2666 Oct 05	06:52:02	2268	8246	157	T-	pp	-0.2457	2.4473	1.3675	370.1	227.7	88.3	5N	96W
11235	562	2667 Apr 01	04:59:45	2270	8252	162	P	a-	0.9700	1.0655	0.0903	253.8	65.7	-	3S	64W
11236	562	2667 Sep 24	10:00:39	2273	8258	167	P	h-	-0.9668	1.1018	0.0664	279.4	60.5	-	0S	142W
11237	562	2668 Feb 20	02:22:36	2275	8263	134	N	-h	-1.1072	0.8593	-0.2063	257.4	-	-	10N	23W
11238	562	2668 Aug 14	12:20:04	2278	8269	139	P	-a	0.9290	1.1295	0.1766	256.9	90.3	-	13S	174W
11239	562	2669 Feb 08	03:33:04	2281	8275	144	P	-t	-0.4424	2.0982	0.9951	363.5	210.0	-	15N	41W
11240	562	2669 Aug 04	05:25:20	2283	8281	149	T+	-p	0.1927	2.4758	1.5324	315.4	208.2	91.3	17S	70W
11241	563	2670 Jan 28	02:57:43	2286	8287	154	T+	p-	0.2422	2.4583	1.3696	367.1	225.1	87.3	18N	32W
11242	563	2670 Jul 24	21:06:28	2288	8293	159	P	a-	-0.5663	1.8070	0.8306	309.3	181.4	-	20S	55E
11243	563	2671 Jan 17	07:50:18	2291	8299	164	P	a-	0.9141	1.1985	0.1630	282.0	91.4	-	22N	106W
11244	563	2671 Jun 14	18:03:01	2293	8304	131	N	-t	1.3835	0.3478	-0.7088	175.0	-	-	22S	99E
11245	563	2671 Jul 14	07:06:49	2294	8305	169	N	t-	-1.3877	0.3272	-0.7036	165.3	-	-	23S	95W
11246	563	2671 Dec 08	09:06:19	2296	8310	136	N	-a	-1.1461	0.7392	-0.2295	215.1	-	-	21N	129W
11247	563	2672 Jan 06	19:46:59	2296	8311	174	N	a-	1.5285	0.0455	-0.9392	57.9	-	-	24N	74E
11248	563	2672 Jun 02	18:50:39	2299	8316	141	P	-t	0.6528	1.6971	0.6238	341.3	177.8	-	22S	87E
11249	563	2672 Nov 27	00:53:08	2301	8322	146	P	-a	-0.4868	1.9539	0.9755	308.1	187.1	-	21N	7W
11250	563	2673 May 22	21:07:24	2304	8328	151	T-	pp	-0.1140	2.6700	1.6281	358.4	227.8	101.2	21S	52E
11251	563	2673 Nov 16	13:18:11	2306	8334	156	T+	p-	0.2182	2.4702	1.4452	342.0	216.8	89.6	19N	166E
11252	563	2674 May 12	06:11:01	2309	8340	161	P	a-	-0.8512	1.2906	0.3014	280.2	119.3	-	19S	84W
11253	563	2674 Nov 05	18:52:17	2312	8346	166	P	t-	0.9718	1.1130	0.0373	291.1	46.8	-	17N	82E
11254	563	2675 Apr 02	12:37:33	2314	8351	133	N	-a	1.0534	0.9044	-0.0546	233.3	-	-	4S	178W
11255	563	2675 May 01	21:07:16	2314	8352	171	Nb	a-	-1.5178	0.0500	-0.9046	59.7	-	-	17S	52E
11256	563	2675 Sep 26	02:25:48	2317	8357	138	N	-t	-1.2363	0.6242	-0.4449	229.9	-	-	0N	28W
11257	563	2676 Mar 22	03:59:06	2319	8363	143	T	-a	0.3793	2.1567	1.1670	320.6	200.3	59.1	0S	48W
11258	563	2676 Sep 14	07:34:06	2322	8369	148	T	-t	-0.4478	2.0448	1.0280	332.9	200.9	26.2	3S	105W
11259	563	2677 Mar 11	14:08:06	2325	8375	153	T	p-	-0.3451	2.2490	1.2010	346.6	212.2	67.0	3N	160E
11260	563	2677 Sep 03	19:58:59	2327	8381	158	T	p-	0.3160	2.2598	1.2960	318.3	204.3	75.2	7S	70E
11261	564	2678 Feb 28	16:52:54	2330	8387	163	N	t-	-1.0788	0.9264	-0.1687	275.4	-	-	7N	119E
11262	564	2678 Jul 26	05:42:41	2332	8392	130	N	-a	-1.4721	0.1308	-0.8178	96.2	-	-	21S	73W
11263	564	2678 Aug 24	12:38:23	2333	8393	168	N	a-	1.0297	0.9402	-0.0036	236.2	-	-	10S	179W
11264	564	2679 Jan 18	22:16:02	2335	8398	135	N	-t	1.2515	0.5998	-0.4762	222.7	-	-	22N	38E
11265	564	2679 Jul 15	20:14:10	2337	8404	140	P	-a	-0.7779	1.4261	0.4350	293.3	142.0	-	22S	68E
11266	564	2680 Jan 08	05:16:52	2340	8410	145	P	-a	0.5078	1.9354	0.9170	321.3	189.2	-	23N	68W
11267	564	2680 Jul 04	04:11:07	2343	8416	150	T-	pp	-0.0161	2.8511	1.8060	363.0	231.3	105.3	23S	51W
11268	564	2680 Dec 27	18:43:05	2345	8422	155	T-	p-	-0.1879	2.5008	1.5253	319.4	207.9	90.4	23N	89E
11269	564	2681 Jun 23	05:46:34	2348	8428	160	P	t-	0.7562	1.5062	0.4350	329.1	153.2	-	23S	76W
11270	564	2681 Dec 17	10:29:13	2351	8434	165	P	a-	-0.8545	1.2772	0.3024	268.6	115.4	-	22N	149W
11271	564	2682 May 13	18:18:50	2353	8439	132	N	-t	-1.2927	0.5004	-0.5284	198.9	-	-	20S	94E
11272	564	2682 Jun 12	07:13:50	2353	8440	170	N	t-	1.4794	0.1679	-0.8809	122.4	-	-	22S	98W
11273	564	2682 Nov 07	10:37:27	2356	8445	137	N	-a	1.2077	0.6619	-0.3776	225.0	-	-	17N	154W
11274	564	2683 May 03	05:14:46	2358	8451	142	P	-a	-0.4996	1.9298	0.9523	311.5	188.3	-	16S	70W
11275	564	2683 Oct 27	14:16:32	2361	8457	147	P	-t	0.5247	1.9369	0.8540	352.5	198.6	-	13N	152E
11276	564	2684 Apr 21	21:02:50	2364	8463	152	T+	p-	0.2245	2.4226	1.4689	314.7	206.0	87.4	12S	54E
11277	564	2684 Oct 15	14:01:43	2366	8469	157	T-	pp	-0.1924	2.5462	1.4645	371.6	230.3	95.2	9N	156E
11278	564	2685 Apr 11	13:11:29	2369	8475	162	P	a-	0.9389	1.1222	0.1478	259.4	83.5	-	8S	173E
11279	564	2685 Oct 04	17:24:11	2372	8481	167	P	h-	-0.9096	1.2066	0.1716	288.2	95.5	-	4N	106E
11280	564	2686 Mar 02	10:28:53	2374	8486	134	N	-h	-1.1257	0.8258	-0.2405	253.9	-	-	6N	145W

Cat Num	Canon Plate	Calendar Date	TD of Greatest Eclipse	ΔT s	Luna Num	Saros Num	Ecl. Type	QSE	Gamma	Pen. Mag.	Um. Mag.	Eclipse Phase ---- Durations ---- Pen. m	Par. m	Total m	Greatest in Zenith Lat.	Lng.
11281	565	2686 Aug 25	19:52:22	2377	8492	139	P	-a	0.9946	1.0090	0.0563	245.6	51.8	-	9S	72E
11282	565	2687 Feb 19	11:28:02	2379	8498	144	P	-t	-0.4574	2.0701	0.9682	362.5	208.3	-	11N	159W
11283	565	2687 Aug 15	12:57:37	2382	8504	149	T+	-p	0.2634	2.3470	1.4016	313.7	205.0	83.4	14S	177E
11284	565	2688 Feb 08	11:03:30	2385	8510	154	T+	p-	0.2311	2.4771	1.3915	366.4	225.4	88.9	15N	153W
11285	565	2688 Aug 04	04:22:03	2387	8516	159	P	a-	-0.4936	1.9422	0.9620	316.3	191.4	-	17S	54W
11286	565	2689 Jan 27	16:21:23	2390	8522	164	P	a-	0.9063	1.2111	0.1790	282.0	95.3	-	19N	128E
11287	565	2689 Jun 25	00:41:31	2392	8527	131	N	-t	1.4660	0.1969	-0.8606	133.8	-	-	22S	1E
11288	565	2689 Jul 24	13:53:09	2393	8528	169	N	t-	-1.3117	0.4683	-0.5658	195.9	-	-	21S	164E
11289	565	2689 Dec 18	17:51:16	2395	8533	136	N	-a	-1.1558	0.7218	-0.2479	212.9	-	-	22N	101E
11290	565	2690 Jan 17	04:35:17	2395	8534	174	N	a-	1.5212	0.0581	-0.9249	65.3	-	-	22N	57W
11291	565	2690 Jun 14	01:25:03	2398	8539	141	P	-t	0.7329	1.5487	0.4780	331.4	159.2	-	23S	11W
11292	565	2690 Dec 08	09:29:22	2400	8545	146	P	-a	-0.5021	1.9276	0.9455	307.9	185.5	-	22N	135W
11293	565	2691 Jun 03	04:06:13	2403	8551	151	T-	pp	-0.0418	2.7998	1.7632	357.5	228.6	104.0	22S	52W
11294	565	2691 Nov 27	21:30:20	2406	8557	156	T+	pp	0.1981	2.5099	1.4793	344.2	218.2	91.8	21N	44E
11295	565	2692 May 22	13:43:08	2408	8563	161	P	a-	-0.7889	1.4021	0.4184	287.1	137.7	-	21S	163E
11296	565	2692 Nov 16	02:32:31	2411	8569	166	P	t-	0.9420	1.1702	0.0894	297.6	72.0	-	20N	32W
11297	565	2693 Apr 12	20:52:52	2413	8574	133	N	-a	1.0893	0.8373	-0.1193	226.0	-	-	8S	58E
11298	565	2693 May 12	05:00:40	2414	8575	171	N	a-	-1.4659	0.1435	-0.8076	100.1	-	-	20S	66W
11299	565	2693 Oct 06	09:32:29	2416	8580	138	N	-t	-1.2917	0.5232	-0.5469	212.2	-	-	4N	135W
11300	565	2694 Apr 02	12:12:44	2419	8586	143	T	-a	0.4106	2.0993	1.1098	319.7	198.0	49.0	5S	172W
11301	566	2694 Sep 25	14:58:03	2421	8592	148	P	-h	-0.5063	1.9370	0.9208	327.6	193.3	-	1N	144E
11302	566	2695 Mar 22	22:07:31	2424	8598	153	T	p-	-0.3199	2.2952	1.2470	348.9	214.8	73.3	1S	40E
11303	566	2695 Sep 15	03:38:20	2427	8604	158	T+	p-	0.2537	2.3741	1.4104	320.0	207.6	84.8	3S	46W
11304	566	2696 Mar 11	00:39:17	2430	8610	163	N	t-	-1.0567	0.9663	-0.1274	280.3	-	-	3N	2E
11305	566	2696 Sep 03	20:18:25	2432	8616	168	P	a-	0.9661	1.0579	0.1121	247.6	71.8	-	6S	65E
11306	566	2697 Jan 29	06:31:11	2435	8621	135	N	-t	1.2559	0.5902	-0.4828	220.4	-	-	19N	84W
11307	566	2697 Jul 26	03:21:15	2437	8627	140	P	-a	-0.8574	1.2820	0.2873	283.9	118.4	-	20S	38W
11308	566	2698 Jan 18	13:54:16	2440	8633	145	P	-a	0.5115	1.9274	0.9116	319.9	188.4	-	21N	164E
11309	566	2698 Jul 15	10:51:12	2443	8639	150	T-	pp	-0.1003	2.6979	1.6503	363.5	230.7	102.9	21S	151W
11310	566	2699 Jan 08	03:32:27	2445	8645	155	T-	p-	-0.1816	2.5121	1.5371	319.2	207.9	90.8	22N	42W
11311	566	2699 Jul 04	12:14:28	2448	8651	160	P	t-	0.6708	1.6625	0.5922	339.3	174.3	-	22S	172W
11312	566	2699 Dec 28	19:15:28	2451	8657	165	P	a-	-0.8478	1.2909	0.3137	270.1	117.5	-	22N	81E
11313	566	2700 May 25	01:32:20	2453	8662	132	N	-t	-1.3504	0.3916	-0.6315	176.9	-	-	22S	13W
11314	566	2700 Jun 23	13:59:43	2454	8663	170	N	t-	1.3996	0.3122	-0.7324	164.2	-	-	22S	161E
11315	566	2700 Nov 18	18:35:06	2456	8668	137	N	-a	1.2375	0.6101	-0.4352	218.0	-	-	20N	87E
11316	566	2701 May 14	12:58:21	2459	8674	142	P	-a	-0.5522	1.8308	0.8584	306.2	181.4	-	19S	175E
11317	566	2701 Nov 07	21:45:22	2461	8680	147	P	-t	0.5629	1.8693	0.7818	350.3	192.7	-	17N	40E
11318	566	2702 May 04	05:02:21	2464	8686	152	T+	p-	0.1763	2.5096	1.5588	315.7	208.0	92.1	16S	66W
11319	566	2702 Oct 27	21:21:47	2467	8692	157	T-	pp	-0.1470	2.6304	1.5467	372.5	231.8	99.4	13N	46E
11320	566	2703 Apr 23	21:12:57	2470	8698	162	P	a-	0.8998	1.1937	0.2199	266.0	101.0	-	12S	52E
11321	567	2703 Oct 17	00:58:02	2472	8704	167	P	h-	-0.8598	1.2979	0.2630	295.0	116.3	-	8N	8W
11322	567	2704 Mar 13	18:26:58	2475	8709	134	N	-h	-1.1501	0.7811	-0.2852	248.7	-	-	2N	95E
11323	567	2704 Sep 06	03:30:17	2477	8715	139	N	-a	1.0559	0.8968	-0.0562	233.9	-	-	5S	43W
11324	567	2704 Oct 05	11:58:46	2478	8716	177	Nb	a-	-1.5219	0.0542	-0.9236	63.8	-	-	3N	172W
11325	567	2705 Mar 02	19:18:45	2480	8721	144	P	-t	-0.4756	2.0358	0.9357	361.1	206.2	-	7N	83E
11326	567	2705 Aug 26	20:31:01	2483	8727	149	T	-p	0.3328	2.2210	1.2731	311.5	200.7	72.0	10S	63E
11327	567	2706 Feb 19	19:08:31	2485	8733	154	T+	p-	0.2189	2.4976	1.4159	365.6	225.7	90.7	12N	86E
11328	567	2706 Aug 16	11:37:00	2488	8739	159	T	a-	-0.4214	2.0768	1.0924	322.5	199.6	45.8	14S	162W
11329	567	2707 Feb 09	00:51:51	2491	8745	164	P	a-	0.8975	1.2254	0.1972	282.2	99.5	-	16N	1E
11330	567	2707 Jul 07	07:16:15	2493	8750	131	Ne	-t	1.5504	0.0428	-1.0160	63.4	-	-	21S	97W
11331	567	2707 Aug 05	20:39:14	2494	8751	169	N	t-	-1.2359	0.6093	-0.4287	221.3	-	-	18S	63E
11332	567	2707 Dec 31	02:40:18	2496	8756	136	N	-a	-1.1623	0.7102	-0.2601	211.4	-	-	22N	29W
11333	567	2708 Jan 29	13:24:03	2496	8757	174	N	a-	1.5145	0.0694	-0.9117	71.1	-	-	20N	172E
11334	567	2708 Jun 25	07:57:36	2499	8762	141	P	-t	0.8147	1.3974	0.3292	319.9	135.1	-	23S	100W
11335	567	2708 Dec 19	18:08:42	2502	8768	146	P	-a	-0.5144	1.9069	0.9212	307.9	184.2	-	23N	97E
11336	567	2709 Jun 14	11:03:25	2504	8774	151	T+	pp	0.0326	2.8141	1.7826	355.9	228.1	104.1	23S	155W
11337	567	2709 Dec 09	05:46:36	2507	8780	156	T+	pp	0.1821	2.5418	1.5060	346.2	219.3	93.4	23N	79W
11338	567	2710 Jun 03	21:10:23	2510	8786	161	P	a-	-0.7219	1.5225	0.5439	293.7	153.6	-	23S	53E
11339	567	2710 Nov 28	10:19:44	2513	8792	166	P	t-	0.9184	1.2159	0.1304	302.8	86.6	-	22N	148W
11340	567	2711 Apr 25	04:59:49	2515	8797	133	N	-a	1.1317	0.7584	-0.1961	216.9	-	-	12S	64W

Cat Num	Canon Plate	Calendar Date	TD of Greatest Eclipse	ΔT s	Luna Num	Saros Num	Ecl. Type	QSE	Gamma	Pen. Mag.	Um. Mag.	Eclipse Phase Durations Pen. m	Par. m	Total m	Greatest in Zenith Lat.	Lng.
11341	568	2711 May 24	12:47:50	2515	8798	171	N	a-	-1.4090	0.2463	-0.7018	129.8	-	-	22S	178E
11342	568	2711 Oct 18	16:49:35	2518	8803	138	N	-t	-1.3388	0.4372	-0.6339	195.2	-	-	8N	115E
11343	568	2712 Apr 13	20:16:52	2520	8809	143	T	-a	0.4490	2.0286	1.0393	318.2	194.7	30.0	9S	67E
11344	568	2712 Oct 06	22:32:13	2523	8815	148	P	-h	-0.5570	1.8438	0.8281	322.5	185.7	-	5N	30E
11345	568	2713 Apr 03	05:57:08	2526	8821	153	T	p-	-0.2874	2.3551	1.3067	351.4	217.8	80.1	5S	78W
11346	568	2713 Sep 26	11:24:41	2529	8827	158	T+	p-	0.1969	2.4786	1.5145	321.1	209.8	90.9	1N	163W
11347	568	2714 Mar 23	08:17:02	2531	8833	163	N*	t-	-1.0277	1.0184	-0.0734	286.3	-	-	2S	113W
11348	568	2714 Sep 16	04:03:38	2534	8839	168	P	a-	0.9068	1.1680	0.2195	257.3	99.1	-	2S	52W
11349	568	2715 Feb 10	14:44:02	2536	8844	135	N	-t	1.2626	0.5760	-0.4931	217.3	-	-	16N	153E
11350	568	2715 Aug 07	10:27:45	2539	8850	140	P	-a	-0.9365	1.1390	0.1402	273.1	84.7	-	17S	144W
11351	568	2716 Jan 30	22:32:36	2542	8856	145	P	-a	0.5151	1.9192	0.9066	318.4	187.6	-	18N	36E
11352	568	2716 Jul 26	17:28:56	2545	8862	150	T-	pp	-0.1855	2.5429	1.4926	363.0	228.4	96.4	19S	110E
11353	568	2717 Jan 19	12:23:43	2548	8868	155	T-	p-	-0.1768	2.5205	1.5464	318.9	207.9	91.2	20N	173W
11354	568	2717 Jul 15	18:39:40	2550	8874	160	P	t-	0.5828	1.8237	0.7541	348.2	191.3	-	21S	92E
11355	568	2718 Jan 09	04:04:42	2553	8880	165	P	a-	-0.8440	1.2988	0.3196	271.2	118.6	-	21N	49W
11356	568	2718 Jun 05	08:40:08	2555	8885	132	N	-t	-1.4132	0.2737	-0.7438	148.9	-	-	24S	119W
11357	568	2718 Jul 04	20:42:24	2556	8886	170	N	t-	1.3161	0.4636	-0.5772	196.5	-	-	22S	61E
11358	568	2718 Nov 30	02:39:40	2558	8891	137	N	-a	1.2608	0.5701	-0.4808	212.5	-	-	23N	33W
11359	568	2719 May 25	20:36:44	2561	8897	142	P	-a	-0.6097	1.7228	0.7552	300.1	172.7	-	21S	61E
11360	568	2719 Nov 19	05:21:16	2564	8903	147	P	-t	0.5947	1.8131	0.7212	348.3	187.3	-	20N	74W
11361	569	2720 May 14	12:56:50	2567	8909	152	T+	p-	0.1238	2.6047	1.6564	316.5	209.7	95.8	19S	176E
11362	569	2720 Nov 07	04:50:33	2569	8915	157	T-	pp	-0.1083	2.7023	1.6171	372.9	232.6	102.0	16N	66W
11363	569	2721 May 04	05:07:36	2572	8921	162	P	a-	0.8558	1.2743	0.3008	272.9	117.1	-	15S	67W
11364	569	2721 Oct 27	08:40:23	2575	8927	167	P	a-	-0.8162	1.3779	0.3431	300.2	130.9	-	12N	123W
11365	569	2722 Mar 25	02:17:14	2577	8932	134	N	-t	-1.1798	0.7264	-0.3398	241.7	-	-	3S	22W
11366	569	2722 Sep 17	11:14:28	2580	8938	139	N	-a	1.1120	0.7941	-0.1596	222.2	-	-	1S	159W
11367	569	2722 Oct 16	19:56:04	2580	8939	177	N	a-	-1.4756	0.1393	-0.8388	101.4	-	-	7N	69E
11368	569	2723 Mar 14	03:00:48	2583	8944	144	P	-t	-0.5004	1.9890	0.8913	358.9	202.9	-	3N	32W
11369	569	2723 Sep 07	04:10:10	2586	8950	149	T	-a	0.3969	2.1050	1.1539	308.9	195.8	56.2	6S	52W
11370	569	2724 Mar 02	03:07:53	2588	8956	154	T+	p-	0.2019	2.5265	1.4492	364.8	226.2	92.8	8N	33W
11371	569	2724 Aug 26	18:54:49	2591	8962	159	T	p-	-0.3524	2.2059	1.2168	327.9	206.3	67.5	10S	88E
11372	569	2725 Feb 19	09:18:57	2594	8968	164	P	a-	0.8854	1.2453	0.2216	282.7	104.8	-	12N	126W
11373	569	2725 Aug 16	03:25:48	2597	8974	169	N	t-	-1.1611	0.7487	-0.2935	242.9	-	-	15S	39W
11374	569	2726 Jan 10	11:31:30	2599	8979	136	N	-a	-1.1674	0.7010	-0.2695	210.1	-	-	21N	161W
11375	569	2726 Feb 08	22:10:40	2600	8980	174	N	a-	1.5059	0.0840	-0.8946	78.0	-	-	16N	42E
11376	569	2726 Jul 06	14:25:33	2602	8985	141	P	-t	0.9000	1.2399	0.1736	306.2	100.2	-	22S	156E
11377	569	2726 Dec 31	02:52:31	2605	8991	146	P	-a	-0.5225	1.8935	0.9048	308.2	183.4	-	23N	32W
11378	569	2727 Jun 25	17:57:08	2608	8997	151	T+	pp	0.1104	2.6689	1.6419	353.5	226.3	101.2	23S	103E
11379	569	2727 Dec 20	14:07:35	2610	9003	156	T+	pp	0.1710	2.5647	1.5242	348.0	220.3	94.5	23N	158E
11380	569	2728 Jun 14	04:34:21	2613	9009	161	P	a-	-0.6514	1.6496	0.6755	299.8	167.2	-	24S	57W
11381	570	2728 Dec 08	18:13:34	2616	9015	166	P	t-	0.9008	1.2503	0.1606	306.8	95.9	-	23N	95E
11382	570	2729 May 05	12:59:44	2618	9020	133	N	-a	1.1800	0.6688	-0.2837	205.6	-	-	15S	176E
11383	570	2729 Jun 03	20:29:16	2619	9021	171	N	a-	-1.3473	0.3582	-0.5872	154.7	-	-	24S	63E
11384	570	2729 Oct 29	00:15:40	2621	9026	138	N	-t	-1.3791	0.3637	-0.7084	178.9	-	-	12N	3E
11385	570	2730 Apr 25	04:13:03	2624	9032	143	P	-a	0.4934	1.9471	0.9579	316.1	190.3	-	13S	52W
11386	570	2730 Oct 18	06:15:28	2627	9038	148	P	-a	-0.6012	1.7626	0.7471	317.5	178.3	-	9N	86W
11387	570	2731 Apr 14	13:37:27	2630	9044	153	T-	pp	-0.2478	2.4277	1.3794	353.9	221.0	87.0	9S	167E
11388	570	2731 Oct 07	19:19:13	2632	9050	158	T+	p-	0.1462	2.5719	1.6071	321.7	211.2	94.8	6N	78E
11389	570	2732 Apr 02	15:47:26	2635	9056	163	N*	t-	-0.9928	1.0814	-0.0082	293.2	-	-	6S	135E
11390	570	2732 Sep 26	11:53:56	2638	9062	168	P	a-	0.8518	1.2704	0.3189	265.6	117.8	-	2N	170W
11391	570	2733 Feb 20	22:54:49	2640	9067	135	N	-t	1.2716	0.5573	-0.5075	213.3	-	-	12N	31E
11392	570	2733 Aug 17	17:35:07	2643	9073	140	N	-a	-1.0138	0.9994	-0.0040	260.8	-	-	14S	109E
11393	570	2734 Feb 10	07:08:47	2646	9079	145	P	-a	0.5210	1.9064	0.8976	316.8	186.5	-	15N	93W
11394	570	2734 Aug 07	00:06:17	2649	9085	150	T-	-t	-0.2703	2.3889	1.3353	361.3	224.2	84.8	17S	11E
11395	570	2735 Jan 30	21:13:45	2652	9091	155	T-	p-	-0.1706	2.5312	1.5584	318.6	208.0	91.7	18N	55E
11396	570	2735 Jul 27	01:06:30	2655	9097	160	P	t-	0.4956	1.9835	0.9141	355.6	204.7	-	19S	4W
11397	570	2736 Jan 20	12:52:56	2657	9103	165	P	a-	-0.8396	1.3076	0.3269	272.3	120.0	-	19N	180E
11398	570	2736 Jun 15	15:45:49	2660	9108	132	N	-t	-1.4783	0.1516	-0.8608	111.5	-	-	25S	135E
11399	570	2736 Jul 15	03:26:59	2660	9109	170	N	t-	1.2326	0.6151	-0.4222	222.2	-	-	20S	39W
11400	570	2736 Dec 10	10:49:08	2663	9114	137	N	-a	1.2799	0.5376	-0.5183	207.8	-	-	24N	153W

Cat Num	Canon Plate	Calendar Date	TD of Greatest Eclipse	ΔT s	Luna Num	Saros Num	Ecl. Type	QSE	Gamma	Pen. Mag.	Um. Mag.	Eclipse Phase Durations Pen. m	Par. m	Total m	Greatest in Zenith Lat.	Lng.
11401	571	2737 Jun 05	04:09:30	2665	9120	142	P	-a	-0.6726	1.6053	0.6421	292.9	161.9	-	23S	51W
11402	571	2737 Nov 29	13:04:48	2668	9126	147	P	-t	0.6201	1.7684	0.6727	346.6	182.6	-	22N	172E
11403	571	2738 May 25	20:43:19	2671	9132	152	T+	pp	0.0643	2.7129	1.7667	317.1	210.9	98.3	21S	60E
11404	571	2738 Nov 18	12:29:42	2674	9138	157	T-	pp	-0.0776	2.7593	1.6726	372.9	232.9	103.3	19N	180E
11405	571	2739 May 15	12:52:11	2677	9144	162	P	a-	0.8039	1.3694	0.3960	280.6	132.8	-	18S	178E
11406	571	2739 Nov 07	16:33:47	2680	9150	167	P	a-	-0.7807	1.4430	0.4083	304.0	141.0	-	15N	119E
11407	571	2740 Apr 04	09:58:17	2682	9155	134	N	-t	-1.2163	0.6593	-0.4063	232.3	-	-	7S	138W
11408	571	2740 Sep 27	19:04:39	2685	9161	139	N	-a	1.1636	0.7000	-0.2549	210.3	-	-	3N	83E
11409	571	2740 Oct 27	04:01:09	2685	9162	177	N	a-	-1.4349	0.2142	-0.7644	124.5	-	-	11N	53W
11410	571	2741 Mar 24	10:36:55	2688	9167	144	P	-t	-0.5301	1.9330	0.8383	356.2	198.7	-	2S	147W
11411	571	2741 Sep 17	11:52:55	2691	9173	149	T	-a	0.4576	1.9954	1.0408	306.0	190.2	29.9	2S	168W
11412	571	2742 Mar 13	11:02:37	2693	9179	154	T+	p-	0.1801	2.5639	1.4916	364.0	226.8	95.3	3N	152W
11413	571	2742 Sep 07	02:14:46	2696	9185	159	T	p-	-0.2857	2.3309	1.3365	332.5	211.5	80.9	6S	23W
11414	571	2743 Mar 02	17:43:04	2699	9191	164	P	a-	0.8700	1.2710	0.2523	283.7	111.0	-	8N	109E
11415	571	2743 Aug 27	10:15:10	2702	9197	169	N	t-	-1.0886	0.8839	-0.1628	261.4	-	-	11S	142W
11416	571	2744 Jan 21	20:23:11	2704	9202	136	N	-a	-1.1723	0.6919	-0.2785	208.9	-	-	19N	68E
11417	571	2744 Feb 20	06:54:26	2705	9203	174	N	a-	1.4949	0.1027	-0.8731	85.9	-	-	13N	89W
11418	571	2744 Jul 16	20:55:09	2707	9208	141	P	-t	0.9841	1.0848	0.0202	290.8	34.9	-	20S	59E
11419	571	2745 Jan 10	11:37:38	2710	9214	146	P	-a	-0.5292	1.8826	0.8912	308.6	182.8	-	21N	162W
11420	571	2745 Jul 06	00:50:21	2713	9220	151	T+	pp	0.1897	2.5214	1.4986	350.1	223.0	94.9	22S	0E
11421	572	2745 Dec 30	22:31:19	2716	9226	156	T+	pp	0.1627	2.5820	1.5372	349.6	221.1	95.3	23N	34E
11422	572	2746 Jun 25	11:55:59	2719	9232	161	P	a-	-0.5784	1.7815	0.8114	305.3	178.6	-	24S	167W
11423	572	2746 Dec 20	02:11:51	2722	9238	166	P	t-	0.8870	1.2774	0.1842	309.9	102.5	-	24N	23W
11424	572	2747 May 16	20:51:32	2724	9243	133	N	-a	1.2348	0.5675	-0.3835	191.4	-	-	18S	58E
11425	572	2747 Jun 15	04:06:13	2724	9244	171	N	a-	-1.2818	0.4774	-0.4659	176.4	-	-	24S	50W
11426	572	2747 Nov 09	07:52:28	2727	9249	138	N	-t	-1.4115	0.3048	-0.7681	164.4	-	-	15N	110W
11427	572	2747 Dec 09	01:24:50	2727	9250	176	N	t-	1.5632	0.0390	-1.0588	61.5	-	-	24N	13W
11428	572	2748 May 05	12:00:09	2730	9255	143	P	-a	0.5449	1.8527	0.8633	313.1	184.3	-	16S	169W
11429	572	2748 Oct 28	14:07:40	2733	9261	148	P	-a	-0.6389	1.6934	0.6781	312.9	171.3	-	13N	156E
11430	572	2749 Apr 24	21:07:32	2735	9267	153	T-	pp	-0.2004	2.5144	1.4663	356.5	224.1	93.4	13S	54E
11431	572	2749 Oct 18	03:21:40	2738	9273	158	T+	p-	0.1018	2.6540	1.6881	321.9	211.9	97.1	10N	43W
11432	572	2750 Apr 13	23:06:26	2741	9279	163	P	t-	-0.9488	1.1608	0.0737	301.2	66.4	-	10S	25E
11433	572	2750 Oct 07	19:51:22	2744	9285	168	P	a-	0.8030	1.3618	0.4069	272.5	131.4	-	6N	70E
11434	572	2751 Mar 04	07:00:48	2746	9290	135	N	-t	1.2850	0.5303	-0.5295	207.8	-	-	8N	90W
11435	572	2751 Aug 29	00:44:37	2749	9296	140	N	-a	-1.0883	0.8653	-0.1432	247.2	-	-	10S	1E
11436	572	2752 Feb 21	15:42:15	2752	9302	145	P	-a	0.5298	1.8881	0.8835	314.9	185.0	-	11N	139E
11437	572	2752 Aug 17	06:45:06	2755	9308	150	T	-t	-0.3533	2.2385	1.1814	358.7	218.2	66.0	13S	88W
11438	572	2753 Feb 10	06:03:22	2758	9314	155	T-	p-	-0.1639	2.5425	1.5715	318.4	208.1	92.2	14N	76W
11439	572	2753 Aug 06	07:32:44	2761	9320	160	T	t-	0.4075	2.1452	1.0757	361.6	215.2	44.6	16S	100W
11440	572	2754 Jan 30	21:42:08	2764	9326	165	P	a-	-0.8364	1.3141	0.3321	273.3	121.0	-	17N	49E
11441	573	2754 Jun 26	22:48:44	2766	9331	132	Ne	-t	-1.5463	0.0245	-0.9832	45.1	-	-	25S	31E
11442	573	2754 Jul 26	10:12:00	2767	9332	170	N	t-	1.1479	0.7689	-0.2655	243.8	-	-	18S	140W
11443	573	2754 Dec 21	19:03:15	2769	9337	137	N	-a	1.2944	0.5134	-0.5472	204.3	-	-	25N	85E
11444	573	2755 Jun 16	11:39:21	2772	9343	142	P	-a	-0.7384	1.4826	0.5233	284.8	148.6	-	24S	163W
11445	573	2755 Dec 10	20:54:39	2775	9349	147	P	-t	0.6402	1.7333	0.6342	345.3	178.6	-	23N	56E
11446	573	2756 Jun 05	04:26:06	2778	9355	152	T+	pp	0.0018	2.8267	1.8821	317.4	211.5	99.3	23S	55W
11447	573	2756 Nov 28	20:15:29	2781	9361	157	T-	pp	-0.0519	2.8069	1.7193	372.7	232.8	104.0	21N	64E
11448	573	2757 May 25	20:31:07	2784	9367	162	P	a-	0.7479	1.4723	0.4986	288.3	147.1	-	20S	63E
11449	573	2757 Nov 18	00:35:24	2786	9373	167	P	a-	-0.7511	1.4970	0.4627	306.7	148.4	-	18N	1W
11450	573	2758 Apr 15	17:28:58	2789	9378	134	N	-t	-1.2602	0.5783	-0.4867	219.8	-	-	11S	109E
11451	573	2758 May 15	07:26:11	2789	9379	172	Nb	t-	1.5528	0.0238	-1.0064	45.4	-	-	17S	100W
11452	573	2758 Oct 09	03:02:58	2792	9384	139	N	-a	1.2090	0.6176	-0.3388	199.0	-	-	7N	38W
11453	573	2758 Nov 07	12:15:27	2792	9385	177	N	a-	-1.4012	0.2763	-0.7029	140.3	-	-	15N	176W
11454	573	2759 Apr 04	18:03:23	2795	9390	144	P	-t	-0.5676	1.8626	0.7710	352.5	192.9	-	6S	101E
11455	573	2759 Sep 28	19:42:45	2798	9396	149	P	-a	0.5121	1.8973	0.9387	303.1	184.2	-	3N	74E
11456	573	2760 Mar 23	18:49:54	2801	9402	154	T+	pp	0.1517	2.6135	1.5465	363.2	227.6	98.0	1S	91E
11457	573	2760 Sep 17	09:40:35	2803	9408	159	T-	pp	-0.2246	2.4458	1.4459	336.3	215.5	89.5	2S	135W
11458	573	2761 Mar 13	02:01:39	2806	9414	164	P	a-	0.8496	1.3058	0.2923	285.2	118.5	-	4N	16W
11459	573	2761 Sep 06	17:07:18	2809	9420	169	N*	t-	-1.0189	1.0143	-0.0373	277.3	-	-	7S	115E
11460	573	2762 Feb 01	05:14:43	2812	9425	136	N	-a	-1.1775	0.6821	-0.2877	207.6	-	-	16N	64W

Cat Num	Canon Plate	Calendar Date	TD of Greatest Eclipse	ΔT s	Luna Num	Saros Num	Ecl. Type	QSE	Gamma	Pen. Mag.	Um. Mag.	Eclipse Phase Durations Pen. m	Par. m	Total m	Greatest in Zenith Lat.	Lng.
11461	574	2762 Mar 02	15:33:55	2812	9426	174	N	a-	1.4806	0.1274	-0.8452	95.2	-	-	9N	141E
11462	574	2762 Jul 28	03:24:16	2815	9431	141	N	-t	1.0681	0.9301	-0.1333	273.2	-	-	18S	37W
11463	574	2763 Jan 21	20:23:35	2817	9437	146	P	-a	-0.5347	1.8736	0.8801	309.0	182.3	-	19N	68E
11464	574	2763 Jul 17	07:43:39	2820	9443	151	T+	pp	0.2695	2.3731	1.3540	345.9	218.1	84.5	21S	102W
11465	574	2764 Jan 11	06:57:46	2823	9449	156	T+	pp	0.1577	2.5929	1.5446	351.1	221.8	95.8	22N	91W
11466	574	2764 Jul 05	19:15:49	2826	9455	161	P	a-	-0.5030	1.9180	0.9515	310.0	188.3	-	23S	84E
11467	574	2764 Dec 30	10:14:06	2829	9461	166	P	t-	0.8767	1.2977	0.2015	312.4	107.1	-	24N	142W
11468	574	2765 May 27	04:37:27	2832	9466	133	N	-a	1.2942	0.4579	-0.4918	173.9	-	-	20S	58W
11469	574	2765 Jun 25	11:39:32	2832	9467	171	N	a-	-1.2132	0.6024	-0.3393	195.6	-	-	25S	162W
11470	574	2765 Nov 19	15:38:17	2835	9472	138	N	-t	-1.4371	0.2580	-0.8155	151.5	-	-	18N	134E
11471	574	2765 Dec 19	09:20:56	2835	9473	176	N	t-	1.5466	0.0698	-1.0288	82.1	-	-	25N	130W
11472	574	2766 May 16	19:38:18	2837	9478	143	P	-a	0.6030	1.7462	0.7565	309.1	176.2	-	19S	77E
11473	574	2766 Nov 08	22:09:29	2840	9484	148	P	-a	-0.6694	1.6373	0.6222	308.9	165.1	-	16N	36E
11474	574	2767 May 06	04:28:53	2843	9490	153	T-	pp	-0.1468	2.6128	1.5648	358.9	226.9	98.8	17S	56W
11475	574	2767 Oct 29	11:31:52	2846	9496	158	T+	p-	0.0636	2.7247	1.7577	321.9	212.1	98.3	13N	165W
11476	574	2768 Apr 24	06:18:26	2849	9502	163	P	t-	-0.8992	1.2506	0.1661	309.6	98.5	-	14S	83W
11477	574	2768 Oct 18	03:55:36	2852	9508	168	P	a-	0.7601	1.4423	0.4838	278.3	141.8	-	10N	51W
11478	574	2769 Mar 14	15:02:24	2855	9513	135	N	-t	1.3025	0.4953	-0.5591	200.7	-	-	3N	149E
11479	574	2769 Sep 08	07:56:48	2858	9519	140	N	-a	-1.1596	0.7373	-0.2767	232.3	-	-	6S	107W
11480	574	2769 Oct 07	18:39:36	2858	9520	178	N	a-	1.5105	0.0826	-0.9101	80.0	-	-	7N	88E
11481	575	2770 Mar 04	00:11:40	2860	9525	145	P	-a	0.5427	1.8620	0.8621	312.6	183.0	-	7N	12E
11482	575	2770 Aug 28	13:27:15	2863	9531	150	T	-t	-0.4330	2.0943	1.0331	355.1	210.4	29.6	10S	171E
11483	575	2771 Feb 21	14:47:55	2866	9537	155	T-	p-	-0.1527	2.5621	1.5931	318.2	208.4	93.1	11N	153E
11484	575	2771 Aug 17	14:04:20	2869	9543	160	T	t-	0.3232	2.3001	1.2302	366.2	223.0	73.8	13S	162E
11485	575	2772 Feb 11	06:27:48	2872	9549	165	P	a-	-0.8304	1.3255	0.3429	274.6	123.0	-	14N	82W
11486	575	2772 Aug 05	17:00:33	2875	9555	170	N	t-	1.0649	0.9201	-0.1118	261.6	-	-	16S	118E
11487	575	2773 Jan 01	03:19:29	2878	9560	137	N	-a	1.3067	0.4928	-0.5717	201.3	-	-	24N	37W
11488	575	2773 Jun 26	19:05:54	2881	9566	142	P	-a	-0.8076	1.3539	0.3979	275.5	131.8	-	24S	87E
11489	575	2773 Dec 21	04:50:12	2884	9572	147	P	-t	0.6553	1.7068	0.6051	344.3	175.4	-	24N	61W
11490	575	2774 Jun 16	12:01:40	2886	9578	152	T-	pp	-0.0668	2.7069	1.7634	317.1	211.2	98.5	23S	168W
11491	575	2774 Dec 10	04:10:47	2889	9584	157	T-	pp	-0.0334	2.8411	1.7531	372.2	232.6	104.2	23N	53W
11492	575	2775 Jun 06	04:01:54	2892	9590	162	P	a-	0.6856	1.5871	0.6126	296.1	160.6	-	22S	48W
11493	575	2775 Nov 29	08:45:58	2895	9596	167	P	a-	-0.7280	1.5391	0.5053	308.5	153.6	-	21N	123W
11494	575	2776 Apr 26	00:50:25	2898	9601	134	N	-t	-1.3110	0.4849	-0.5794	203.5	-	-	15S	1W
11495	575	2776 May 25	14:39:01	2898	9602	172	N	t-	1.4962	0.1282	-0.9027	104.7	-	-	20S	152E
11496	575	2776 Oct 19	11:08:55	2901	9607	139	N	-a	1.2486	0.5458	-0.4124	188.3	-	-	11N	159W
11497	575	2776 Nov 17	20:37:54	2901	9608	177	N	a-	-1.3735	0.3274	-0.6523	151.7	-	-	18N	59E
11498	575	2777 Apr 15	01:22:23	2904	9613	144	P	-t	-0.6112	1.7808	0.6927	347.8	185.3	-	10S	9W
11499	575	2777 Oct 09	03:37:45	2907	9619	149	P	-a	0.5620	1.8080	0.8451	300.2	177.9	-	7N	46W
11500	575	2778 Apr 04	02:31:48	2910	9625	154	T+	pp	0.1179	2.6725	1.6113	362.3	228.2	100.5	5S	25W
11501	576	2778 Sep 28	17:10:59	2913	9631	159	T-	pp	-0.1682	2.5523	1.5464	339.6	218.4	95.2	2N	112E
11502	576	2779 Mar 24	10:14:32	2916	9637	164	P	a-	0.8238	1.3502	0.3424	287.3	126.9	-	1S	140W
11503	576	2779 Sep 18	00:04:53	2919	9643	169	P	t-	-0.9540	1.1361	0.0793	290.8	67.5	-	3S	10E
11504	576	2780 Feb 12	14:04:32	2921	9648	136	N	-a	-1.1840	0.6697	-0.2993	205.9	-	-	13N	164E
11505	576	2780 Mar 13	00:09:05	2922	9649	174	N	a-	1.4629	0.1581	-0.8111	105.6	-	-	4N	13E
11506	576	2780 Aug 07	09:56:19	2924	9654	141	N	-t	1.1498	0.7796	-0.2829	253.5	-	-	15S	135W
11507	576	2781 Feb 01	05:07:58	2927	9660	146	P	-a	-0.5411	1.8627	0.8674	309.4	181.8	-	17N	62W
11508	576	2781 Jul 27	14:38:59	2930	9666	151	T	-p	0.3484	2.2267	1.2108	340.8	211.8	68.4	19S	154E
11509	576	2782 Jan 21	15:23:04	2933	9672	156	T+	pp	0.1521	2.6047	1.5534	352.6	222.5	96.4	20N	144E
11510	576	2782 Jul 17	02:35:28	2936	9678	161	T	a-	-0.4266	2.0567	1.0932	314.0	196.2	45.2	21S	25W
11511	576	2783 Jan 10	18:18:02	2939	9684	166	P	t-	0.8680	1.3147	0.2164	314.4	110.9	-	23N	99E
11512	576	2783 Jun 07	12:17:27	2941	9689	133	N	-a	1.3582	0.3401	-0.6090	151.8	-	-	21S	172W
11513	576	2783 Jul 06	19:11:10	2942	9690	171	N	a-	-1.1430	0.7309	-0.2100	212.7	-	-	24S	86E
11514	576	2783 Nov 30	23:32:00	2944	9695	138	N	-t	-1.4570	0.2215	-0.8521	140.6	-	-	20N	17E
11515	576	2783 Dec 30	17:20:31	2945	9696	176	N	t-	1.5323	0.0959	-1.0027	95.9	-	-	24N	112E
11516	576	2784 May 27	03:08:54	2947	9701	143	P	-a	0.6667	1.6298	0.6392	304.0	165.7	-	21S	35W
11517	576	2784 Nov 19	06:20:13	2950	9707	148	P	-a	-0.6931	1.5936	0.5788	305.4	159.9	-	19N	86W
11518	576	2785 May 16	11:39:40	2953	9713	153	T-	pp	-0.0850	2.7262	1.6781	361.0	229.2	102.8	19S	163W
11519	576	2785 Nov 08	19:50:41	2956	9719	158	T+	pp	0.0321	2.7830	1.8148	321.7	212.1	98.7	17N	70E
11520	576	2786 May 05	13:19:52	2959	9725	163	P	t-	-0.8411	1.3559	0.2741	318.5	124.6	-	17S	171E

Cat Num	Canon Plate	Calendar Date	TD of Greatest Eclipse	ΔT s	Luna Num	Saros Num	Ecl. Type	QSE	Gamma	Pen. Mag.	Um. Mag.	Eclipse Phase Durations Pen. m	Par. m	Total m	Greatest in Zenith Lat.	Lng.
11521	577	2786 Oct 29	12:07:55	2962	9731	168	P	a-	0.7239	1.5106	0.5483	283.0	149.5	-	14N	174W
11522	577	2787 Mar 25	22:56:42	2965	9736	135	N	-t	1.3267	0.4481	-0.6005	191.0	-	-	1S	30E
11523	577	2787 Apr 24	13:54:20	2965	9737	173	N	t-	-1.5381	0.0689	-0.9971	79.8	-	-	14S	163E
11524	577	2787 Sep 19	15:14:14	2968	9742	140	N	-a	-1.2257	0.6188	-0.4009	216.3	-	-	2S	143E
11525	577	2787 Oct 19	02:29:28	2968	9743	178	N	a-	1.4715	0.1571	-0.8415	110.1	-	-	11N	29W
11526	577	2788 Mar 14	08:36:50	2971	9748	145	P	-a	0.5601	1.8278	0.8327	309.9	180.4	-	3N	114W
11527	577	2788 Sep 07	20:12:11	2974	9754	150	P	-t	-0.5098	1.9554	0.8901	350.6	200.9	-	6S	69E
11528	577	2789 Mar 03	23:29:51	2977	9760	155	T-	p-	-0.1391	2.5859	1.6193	318.1	208.8	94.1	6N	23E
11529	577	2789 Aug 27	20:38:53	2980	9766	160	T+	pp	0.2408	2.4518	1.3811	369.5	228.5	89.7	9S	63E
11530	577	2790 Feb 21	15:10:34	2983	9772	165	P	a-	-0.8225	1.3401	0.3571	276.2	125.5	-	10N	147E
11531	577	2790 Aug 16	23:52:50	2986	9778	170	P	t-	0.9836	1.0681	0.0383	276.5	46.3	-	12S	15E
11532	577	2791 Jan 12	11:38:09	2988	9783	137	N	-a	1.3165	0.4765	-0.5914	198.9	-	-	23N	160W
11533	577	2791 Feb 11	02:59:35	2989	9784	175	N	a-	-1.5670	0.0006	-1.0350	7.4	-	-	13N	30W
11534	577	2791 Jul 08	02:32:22	2991	9789	142	P	-a	-0.8773	1.2245	0.2713	265.2	110.6	-	23S	24W
11535	577	2792 Jan 01	12:48:24	2994	9795	147	P	-t	0.6682	1.6842	0.5804	343.4	172.6	-	24N	179W
11536	577	2792 Jun 26	19:36:07	2997	9801	152	T-	-p	-0.1364	2.5789	1.6360	316.3	209.9	95.5	23S	80E
11537	577	2792 Dec 20	12:11:32	3000	9807	157	T-	pp	-0.0186	2.8682	1.7803	371.6	232.3	104.2	23N	171W
11538	577	2793 Jun 16	11:27:20	3003	9813	162	P	a-	0.6193	1.7093	0.7336	303.6	172.9	-	23S	159W
11539	577	2793 Dec 09	17:03:55	3006	9819	167	P	a-	-0.7101	1.5715	0.5386	309.5	157.2	-	22N	114E
11540	577	2794 May 07	08:02:01	3009	9824	134	N	-t	-1.3690	0.3781	-0.6856	181.9	-	-	18S	109W
11541	578	2794 Jun 05	21:43:55	3009	9825	172	N	t-	1.4333	0.2441	-0.7880	143.4	-	-	21S	47E
11542	578	2794 Oct 30	19:23:28	3012	9830	139	N	-a	1.2815	0.4863	-0.4738	178.7	-	-	15N	77E
11543	578	2794 Nov 29	05:08:06	3012	9831	177	N	a-	-1.3518	0.3675	-0.6126	159.8	-	-	20N	68W
11544	578	2795 Apr 26	08:31:28	3015	9836	144	P	-t	-0.6630	1.6840	0.5996	341.7	175.0	-	14S	116W
11545	578	2795 Oct 20	11:41:11	3018	9842	149	P	-a	0.6046	1.7321	0.7647	297.5	171.8	-	11N	167W
11546	578	2796 Apr 14	10:05:56	3021	9848	154	T+	pp	0.0771	2.7444	1.6891	361.2	228.7	102.6	10S	139W
11547	578	2796 Oct 09	00:47:09	3024	9854	159	T-	pp	-0.1173	2.6487	1.6368	342.3	220.5	98.8	6N	2W
11548	578	2797 Apr 03	18:20:53	3027	9860	164	P	a-	0.7919	1.4059	0.4038	290.1	136.2	-	5S	98E
11549	578	2797 Sep 28	07:07:54	3030	9866	169	P	t-	-0.8938	1.2491	0.1870	302.3	102.3	-	1N	96W
11550	578	2798 Feb 22	22:50:07	3032	9871	136	N	-a	-1.1945	0.6498	-0.3178	203.2	-	-	9N	33E
11551	578	2798 Mar 24	08:36:37	3033	9872	174	N	a-	1.4388	0.2004	-0.7651	118.3	-	-	0S	115W
11552	578	2798 Aug 18	16:31:28	3035	9877	141	N	-t	1.2293	0.6336	-0.4284	231.4	-	-	12S	126E
11553	578	2799 Feb 12	13:50:02	3039	9883	146	P	-a	-0.5492	1.8484	0.8520	309.6	181.1	-	13N	168E
11554	578	2799 Aug 07	21:37:38	3042	9889	151	T	-t	0.4256	2.0836	1.0705	335.0	204.1	41.2	16S	50E
11555	578	2800 Feb 01	23:47:11	3045	9895	156	T+	pp	0.1461	2.6169	1.5632	353.9	223.2	97.0	17N	19E
11556	578	2800 Jul 27	09:56:36	3048	9901	161	T	a-	-0.3507	2.1949	1.2337	317.1	202.3	68.4	19S	134W
11557	578	2801 Jan 21	02:23:42	3051	9907	166	P	t-	0.8610	1.3284	0.2287	316.1	114.0	-	21N	21W
11558	578	2801 Jun 17	19:51:48	3053	9912	133	N	-a	1.4265	0.2146	-0.7343	122.1	-	-	22S	76E
11559	578	2801 Jul 17	02:40:12	3054	9913	171	N	a-	-1.0705	0.8637	-0.0768	228.0	-	-	22S	25W
11560	578	2801 Dec 11	07:34:19	3056	9918	138	N	-t	-1.4707	0.1962	-0.8771	132.3	-	-	22N	103W
11561	579	2802 Jan 10	01:25:15	3057	9919	176	N	t-	1.5217	0.1152	-0.9828	104.7	-	-	23N	7W
11562	579	2802 Jun 07	10:32:24	3059	9924	143	P	-a	0.7352	1.5047	0.5129	297.6	151.9	-	22S	145W
11563	579	2802 Nov 30	14:39:15	3062	9930	148	P	-a	-0.7109	1.5606	0.5464	302.5	155.7	-	21N	150E
11564	579	2803 May 27	18:42:59	3065	9936	153	T-	pp	-0.0179	2.8496	1.8012	362.5	230.6	104.8	21S	92E
11565	579	2803 Nov 20	04:17:09	3068	9942	158	T+	pp	0.0068	2.8300	1.8606	321.4	211.8	98.7	20N	55W
11566	579	2804 May 15	20:15:12	3071	9948	163	P	t-	-0.7776	1.4710	0.3919	327.3	146.4	-	20S	68E
11567	579	2804 Nov 08	20:26:02	3074	9954	168	P	a-	0.6928	1.5696	0.6034	287.0	155.7	-	17N	62E
11568	579	2805 Apr 05	06:45:50	3077	9959	135	N	-t	1.3557	0.3919	-0.6507	178.8	-	-	5S	87W
11569	579	2805 May 04	21:08:11	3078	9960	173	N	t-	-1.4821	0.1689	-0.8915	123.4	-	-	17S	55E
11570	579	2805 Sep 29	22:36:36	3080	9965	140	N	-a	-1.2869	0.5096	-0.5162	199.3	-	-	2N	32E
11571	579	2805 Oct 29	10:25:36	3081	9966	178	N	a-	1.4380	0.2215	-0.7830	130.5	-	-	15N	148W
11572	579	2806 Mar 25	16:54:41	3083	9971	145	P	-a	0.5840	1.7812	0.7913	306.6	176.7	-	1S	121E
11573	579	2806 Sep 19	03:03:31	3086	9977	150	P	-t	-0.5808	1.8275	0.7577	345.6	190.1	-	2S	34W
11574	579	2807 Mar 15	00:04:45	3089	9983	155	T	p	0.1195	2.6205	1.6566	318.1	209.3	95.3	2N	106W
11575	579	2807 Sep 08	03:21:14	3092	9989	160	T+	pp	0.1643	2.5926	1.5210	371.5	232.0	98.8	5S	38W
11576	579	2808 Mar 03	23:46:51	3095	9995	165	P	a-	-0.8098	1.3635	0.3805	278.4	129.3	-	6N	18E
11577	579	2808 Aug 27	06:51:56	3098	10001	170	P	t-	0.9066	1.2086	0.1804	288.7	98.1	-	9S	90W
11578	579	2809 Jan 22	19:56:13	3101	10006	137	N	-h	1.3263	0.4598	-0.6107	196.3	-	-	21N	77E
11579	579	2809 Feb 21	11:23:26	3101	10007	175	N	h-	-1.5610	0.0123	-1.0247	32.3	-	-	9N	156W
11580	579	2809 Jul 18	09:56:45	3104	10012	142	P	-a	-0.9494	1.0913	0.1402	253.5	80.9	-	22S	134W

Cat Num	Canon Plate	Calendar Date	TD of Greatest Eclipse	ΔT s	Luna Num	Saros Num	Ecl. Type	QSE	Gamma	Pen. Mag.	Um. Mag.	Eclipse Phase Durations Pen. m	Par. m	Total m	Greatest in Zenith Lat.	Lng.
11581	580	2810 Jan 11	20:50:27	3107	10018	147	P	-t	0.6779	1.6669	0.5620	342.6	170.5	-	22N	62E
11582	580	2810 Jul 08	03:06:03	3110	10024	152	T-	-p	-0.2100	2.4439	1.5009	314.9	207.5	89.7	23S	32W
11583	580	2810 Dec 31	20:18:19	3113	10030	157	T-	pp	-0.0081	2.8870	1.8000	370.8	231.9	104.1	23N	69E
11584	580	2811 Jun 27	18:46:47	3116	10036	162	P	a-	0.5485	1.8401	0.8626	310.9	183.9	-	23S	92E
11585	580	2811 Dec 21	01:29:09	3119	10042	167	P	a-	-0.6972	1.5945	0.5629	309.9	159.7	-	23N	10W
11586	580	2812 May 17	15:05:07	3122	10047	134	N	-t	-1.4327	0.2607	-0.8023	153.0	-	-	21S	146E
11587	580	2812 Jun 16	04:41:48	3122	10048	172	N	t-	1.3655	0.3693	-0.6642	175.0	-	-	22S	57W
11588	580	2812 Nov 10	03:44:37	3125	10053	139	N	-a	1.3094	0.4361	-0.5260	170.0	-	-	19N	48W
11589	580	2812 Dec 09	13:44:30	3125	10054	177	N	a-	-1.3347	0.3989	-0.5815	165.7	-	-	21N	165E
11590	580	2813 May 06	15:33:47	3128	10059	144	P	-t	-0.7203	1.5769	0.4963	334.4	161.9	-	17S	139E
11591	580	2813 Oct 30	19:50:33	3131	10065	149	P	-a	0.6418	1.6662	0.6941	295.1	165.9	-	15N	71E
11592	580	2814 Apr 25	17:33:04	3134	10071	154	T+	pp	0.0297	2.8283	1.7790	359.9	228.9	103.9	13S	109E
11593	580	2814 Oct 20	08:29:42	3137	10077	159	T-	pp	-0.0725	2.7340	1.7158	344.7	222.0	100.9	10N	118W
11594	580	2815 Apr 15	02:21:28	3140	10083	164	P	a-	0.7546	1.4715	0.4752	293.3	145.8	-	9S	22W
11595	580	2815 Oct 09	14:17:15	3143	10089	169	P	t-	-0.8391	1.3523	0.2848	312.0	124.8	-	6N	156E
11596	580	2816 Mar 05	07:31:10	3146	10094	136	N	-a	-1.2087	0.6229	-0.3431	199.5	-	-	5N	97W
11597	580	2816 Apr 03	16:58:39	3146	10095	174	N	a-	1.4103	0.2509	-0.7111	131.5	-	-	4S	120E
11598	580	2816 Aug 28	23:13:01	3149	10100	141	N	-t	1.3037	0.4969	-0.5648	207.2	-	-	8S	25E
11599	580	2816 Sep 27	14:42:10	3150	10101	179	Nb	t-	-1.5723	0.0135	-1.0670	36.2	-	-	1N	151E
11600	580	2817 Feb 22	22:27:41	3152	10106	146	P	-a	-0.5604	1.8283	0.8309	309.5	179.9	-	9N	39E
11601	581	2817 Aug 18	04:40:17	3155	10112	151	P	-t	0.5002	1.9456	0.9347	328.5	194.9	-	12S	56W
11602	581	2818 Feb 12	08:07:36	3158	10118	156	T+	pp	0.1380	2.6328	1.5773	355.3	224.1	97.8	14N	106W
11603	581	2818 Aug 07	17:20:37	3161	10124	161	T	p-	-0.2767	2.3297	1.3702	319.4	206.9	82.0	16S	115E
11604	581	2819 Feb 01	10:26:27	3165	10130	166	P	t-	0.8520	1.3452	0.2448	317.9	117.8	-	18N	140W
11605	581	2819 Jun 29	03:22:47	3167	10135	133	Ne	-a	1.4969	0.0856	-0.8637	78.2	-	-	22S	36W
11606	581	2819 Jul 28	10:10:35	3168	10136	171	P	a-	-0.9992	0.9947	0.0538	241.4	50.2	-	20S	137W
11607	581	2819 Dec 22	15:42:44	3170	10141	138	N	-t	-1.4797	0.1794	-0.8931	126.3	-	-	22N	137E
11608	581	2820 Jan 21	09:30:50	3171	10142	176	N	t-	1.5117	0.1328	-0.9639	112.1	-	-	21N	127W
11609	581	2820 Jun 17	17:49:32	3173	10147	143	P	-a	0.8082	1.3715	0.3781	289.6	133.5	-	23S	107E
11610	581	2820 Dec 10	23:05:01	3176	10153	148	P	-a	-0.7238	1.5367	0.5232	300.1	152.5	-	22N	25E
11611	581	2821 Jun 07	01:37:38	3179	10159	153	T+	pp	0.0556	2.7807	1.7316	363.3	230.9	104.4	23S	11W
11612	581	2821 Nov 30	12:51:18	3183	10165	158	T-	pp	-0.0124	2.8203	1.8499	321.1	211.5	98.5	22N	177E
11613	581	2822 May 27	03:00:29	3186	10171	163	P	t-	-0.7058	1.6015	0.5249	336.1	165.9	-	22S	32W
11614	581	2822 Nov 20	04:52:39	3189	10177	168	P	a-	0.6690	1.6152	0.6452	290.1	160.1	-	20N	64W
11615	581	2823 Apr 16	14:27:27	3191	10182	135	N	-t	1.3911	0.3237	-0.7127	162.8	-	-	9S	157E
11616	581	2823 May 16	04:14:42	3192	10183	173	N	t-	-1.4192	0.2815	-0.7734	157.0	-	-	20S	51W
11617	581	2823 Oct 11	06:05:12	3194	10188	140	N	-a	-1.3420	0.4116	-0.6203	181.6	-	-	6N	81W
11618	581	2823 Nov 09	18:28:33	3195	10189	178	N	a-	1.4106	0.2749	-0.7355	145.3	-	-	18N	91E
11619	581	2824 Apr 05	01:07:10	3198	10194	145	P	-a	0.6129	1.7255	0.7408	302.8	172.1	-	6S	3W
11620	581	2824 Sep 29	10:00:55	3201	10200	150	P	-t	-0.6461	1.7100	0.6356	340.2	178.0	-	2N	139W
11621	582	2825 Mar 25	16:34:39	3204	10206	155	T-	p-	-0.0956	2.6629	1.7017	318.1	209.9	96.5	2S	126E
11622	582	2825 Sep 18	10:09:06	3207	10212	160	T+	pp	0.0918	2.7262	1.6533	372.6	233.9	103.7	1S	140W
11623	582	2826 Mar 15	08:18:11	3210	10218	165	P	a-	-0.7934	1.3935	0.4107	281.1	134.1	-	1N	110W
11624	582	2826 Sep 07	13:57:35	3213	10224	170	P	h-	0.8337	1.3417	0.3147	298.6	126.6	-	5S	163E
11625	582	2827 Feb 03	04:12:29	3216	10229	137	N	-h	1.3371	0.4410	-0.6314	193.2	-	-	18N	46W
11626	582	2827 Mar 04	19:41:32	3216	10230	175	N	h-	-1.5510	0.0311	-1.0069	51.4	-	-	5N	80E
11627	582	2827 Jul 29	17:23:45	3219	10235	142	P	-a	-1.0197	0.9614	0.0119	240.7	23.9	-	19S	115E
11628	582	2827 Aug 28	01:05:38	3219	10236	180	Nb	a-	1.5211	0.0524	-0.9188	62.9	-	-	8S	3W
11629	582	2828 Jan 23	04:52:44	3222	10241	147	P	-t	0.6877	1.6493	0.5439	341.7	168.3	-	20N	57W
11630	582	2828 Jul 18	10:35:28	3225	10247	152	T	-p	-0.2838	2.3088	1.3651	312.8	203.9	80.6	21S	143W
11631	582	2829 Jan 11	04:27:25	3228	10253	157	T+	pp	0.0011	2.8992	1.8136	369.9	231.5	104.0	22N	52W
11632	582	2829 Jul 08	02:03:07	3231	10259	162	P	a-	0.4755	1.9752	0.9954	317.6	193.5	-	22S	16W
11633	582	2829 Dec 31	09:58:30	3234	10265	167	P	a-	-0.6872	1.6120	0.5822	309.9	161.4	-	22N	136W
11634	582	2830 May 28	21:58:39	3237	10270	134	N	-t	-1.5032	0.1312	-0.9315	110.1	-	-	23S	43E
11635	582	2830 Jun 27	11:33:31	3238	10271	172	N	t-	1.2930	0.5032	-0.5321	202.5	-	-	22S	159W
11636	582	2830 Nov 21	12:13:59	3240	10276	139	N	-a	1.3312	0.3973	-0.5668	162.8	-	-	21N	174W
11637	582	2830 Dec 20	22:26:34	3241	10277	177	N	a-	-1.3219	0.4223	-0.5579	169.8	-	-	22N	36E
11638	582	2831 May 17	22:27:44	3243	10282	144	P	-t	-0.7844	1.4574	0.3804	325.5	144.2	-	20S	36E
11639	582	2831 Nov 11	04:07:08	3246	10288	149	P	-a	0.6729	1.6115	0.6349	293.0	160.4	-	18N	53W
11640	582	2832 May 06	00:53:25	3250	10294	154	T-	pp	-0.0242	2.8354	1.7921	358.2	228.4	103.9	17S	0W

Cat Num	Canon Plate	Calendar Date	TD of Greatest Eclipse	ΔT s	Luna Num	Saros Num	Ecl. Type	QSE	Gamma	Pen. Mag.	Um. Mag.	Eclipse Phase Durations Pen. m	Par. m	Total m	Greatest in Zenith Lat.	Lng.
11641	583	2832 Oct 30	16:19:38	3253	10300	159	T-	pp	-0.0344	2.8072	1.7828	346.7	222.9	101.9	14N	125E
11642	583	2833 Apr 25	10:13:57	3256	10306	164	P	a-	0.7096	1.5511	0.5606	297.0	155.8	-	13S	140W
11643	583	2833 Oct 19	21:34:09	3259	10312	169	P	t-	-0.7907	1.4438	0.3709	320.2	140.9	-	10N	46E
11644	583	2834 Mar 16	16:06:06	3262	10317	136	N	-a	-1.2281	0.5863	-0.3778	194.3	-	-	0N	134E
11645	583	2834 Apr 15	01:12:29	3262	10318	174	N	a-	1.3749	0.3141	-0.6442	146.0	-	-	8S	4W
11646	583	2834 Sep 09	06:01:06	3265	10323	141	N	-t	1.3730	0.3698	-0.6920	180.6	-	-	4S	77W
11647	583	2834 Oct 08	21:43:36	3265	10324	179	N	t-	-1.5145	0.1208	-0.9621	107.4	-	-	5N	45E
11648	583	2835 Mar 06	06:59:29	3268	10329	146	P	-a	-0.5761	1.7998	0.8019	309.0	178.1	-	5N	89W
11649	583	2835 Aug 29	11:49:21	3271	10335	151	P	-t	0.5704	1.8160	0.8068	321.5	184.6	-	9S	163W
11650	583	2836 Feb 23	16:24:26	3274	10341	156	T+	pp	0.1276	2.6525	1.5958	356.6	225.0	98.7	10N	131E
11651	583	2836 Aug 18	00:46:58	3277	10347	161	T-	p-	-0.2040	2.4626	1.5043	320.9	210.1	90.7	13S	3E
11652	583	2837 Feb 11	18:28:15	3280	10353	166	P	t-	0.8424	1.3628	0.2625	319.7	121.8	-	14N	100E
11653	583	2837 Aug 07	17:40:59	3284	10359	171	P	a-	-0.9279	1.1261	0.1842	253.3	91.2	-	17S	110E
11654	583	2838 Jan 01	23:56:51	3286	10364	138	N	-t	-1.4851	0.1687	-0.9024	122.3	-	-	21N	15E
11655	583	2838 Jan 31	17:37:23	3287	10365	176	N	t-	1.5016	0.1502	-0.9442	118.8	-	-	19N	113E
11656	583	2838 Jun 29	01:01:07	3289	10370	143	P	-a	0.8850	1.2318	0.2362	279.9	108.1	-	22S	0W
11657	583	2838 Dec 22	07:37:16	3292	10376	148	P	-a	-0.7322	1.5205	0.5082	298.2	150.3	-	23N	101W
11658	583	2839 Jun 18	08:27:21	3296	10382	153	T+	pp	0.1326	2.6398	1.5898	363.3	229.8	100.8	23S	112W
11659	583	2839 Dec 11	21:30:26	3299	10388	158	T-	pp	-0.0278	2.7924	1.8211	320.8	211.2	98.3	23N	49E
11660	583	2840 Jun 06	09:42:03	3302	10394	163	P	t-	-0.6306	1.7383	0.6640	344.1	182.4	-	23S	132W
11661	584	2840 Nov 30	13:24:44	3305	10400	168	P	a-	0.6501	1.6517	0.6781	292.7	163.4	-	22N	169E
11662	584	2841 Apr 26	22:02:28	3308	10405	135	N	-t	1.4326	0.2445	-0.7856	141.9	-	-	13S	43E
11663	584	2841 May 26	11:15:35	3308	10406	173	N	t-	-1.3507	0.4045	-0.6451	185.4	-	-	22S	156W
11664	584	2841 Oct 21	13:40:22	3311	10411	140	N	-a	-1.3909	0.3250	-0.7133	163.4	-	-	10N	165E
11665	584	2841 Nov 20	02:38:13	3311	10412	178	N	a-	1.3890	0.3173	-0.6988	156.1	-	-	21N	30W
11666	584	2842 Apr 16	09:12:05	3314	10417	145	P	-a	0.6487	1.6572	0.6778	298.1	166.0	-	10S	124W
11667	584	2842 Oct 10	17:06:32	3317	10423	150	P	-t	-0.7043	1.6054	0.5264	334.7	165.2	-	6N	114E
11668	584	2843 Apr 06	00:55:44	3320	10429	155	T-	pp	-0.0642	2.7192	1.7608	318.2	210.5	97.7	6S	0E
11669	584	2843 Sep 29	17:07:01	3324	10435	160	T+	pp	0.0270	2.8459	1.7717	372.8	234.5	105.6	3N	115E
11670	584	2844 Mar 25	16:41:29	3327	10441	165	P	a-	-0.7707	1.4348	0.4524	284.4	140.1	-	3S	124E
11671	584	2844 Sep 17	21:10:30	3330	10447	170	P	h-	0.7655	1.4664	0.4403	306.7	146.5	-	1S	55E
11672	584	2845 Feb 13	12:25:13	3333	10452	137	N	-h	1.3502	0.4177	-0.6561	188.9	-	-	14N	169W
11673	584	2845 Mar 15	03:52:11	3333	10453	175	N	h-	-1.5358	0.0593	-0.9793	70.9	-	-	1N	43W
11674	584	2845 Aug 09	00:51:23	3336	10458	142	N	-a	-1.0901	0.8318	-0.1169	226.6	-	-	17S	3E
11675	584	2845 Sep 07	08:36:34	3336	10459	180	N	a-	1.4529	0.1771	-0.7934	113.9	-	-	4S	116W
11676	584	2846 Feb 02	12:54:05	3339	10464	147	P	-t	0.6979	1.6303	0.5253	340.5	165.9	-	17N	176W
11677	584	2846 Jul 29	18:03:14	3342	10470	152	T	-a	-0.3588	2.1719	1.2268	310.0	199.0	66.6	19S	105E
11678	584	2847 Jan 22	12:39:37	3345	10476	157	T+	pp	0.0084	2.8845	1.8011	368.8	231.1	103.9	20N	174W
11679	584	2847 Jul 19	09:16:19	3348	10482	162	T	p-	0.4007	2.1139	1.1312	323.6	201.7	53.9	20S	123W
11680	584	2848 Jan 11	18:31:29	3352	10488	167	P	a-	-0.6794	1.6252	0.5977	309.6	162.7	-	21N	97E
11681	585	2848 Jul 07	18:21:12	3355	10494	172	N	t-	1.2176	0.6427	-0.3949	226.5	-	-	21S	100E
11682	585	2848 Dec 01	20:49:50	3357	10499	139	N	-a	1.3481	0.3670	-0.5989	156.9	-	-	23N	59E
11683	585	2848 Dec 31	07:13:15	3358	10500	177	N	a-	-1.3127	0.4391	-0.5407	172.6	-	-	22N	94W
11684	585	2849 May 28	05:14:53	3361	10505	144	P	-t	-0.8542	1.3276	0.2543	314.7	120.0	-	22S	65W
11685	585	2849 Nov 21	12:30:21	3364	10511	149	P	-a	0.6981	1.5676	0.5864	291.4	155.7	-	21N	177W
11686	585	2850 May 17	08:08:10	3367	10517	154	T-	pp	-0.0838	2.7230	1.6857	356.0	227.2	102.3	20S	109W
11687	585	2850 Nov 11	00:15:59	3370	10523	159	T-	pp	-0.0023	2.8692	1.8387	348.5	223.5	102.2	17N	6E
11688	585	2851 May 06	18:01:05	3373	10529	164	P	a-	0.6594	1.6403	0.6554	300.8	165.5	-	16S	103E
11689	585	2851 Oct 31	04:58:44	3377	10535	169	P	t-	-0.7487	1.5235	0.4454	327.0	152.9	-	13N	65W
11690	585	2852 Mar 27	00:35:12	3379	10540	136	N	-a	-1.2522	0.5412	-0.4211	187.5	-	-	4S	6E
11691	585	2852 Apr 25	09:20:36	3380	10541	174	N	a-	1.3349	0.3856	-0.5692	160.5	-	-	12S	126W
11692	585	2852 Sep 19	12:56:17	3382	10546	141	N	-t	1.4368	0.2528	-0.8093	150.6	-	-	0N	178E
11693	585	2852 Oct 19	04:52:51	3383	10547	179	N	t-	-1.4622	0.2179	-0.8674	143.0	-	-	9N	63W
11694	585	2853 Mar 16	15:24:48	3386	10552	146	P	-a	-0.5965	1.7624	0.7643	308.1	175.4	-	1N	145E
11695	585	2853 Sep 08	19:05:00	3389	10558	151	P	-h	0.6356	1.6955	0.6877	314.2	173.3	-	4S	87E
11696	585	2854 Mar 06	00:33:07	3392	10564	156	T+	pp	0.1111	2.6830	1.6256	358.0	226.0	100.0	6N	8E
11697	585	2854 Aug 29	08:19:05	3395	10570	161	T-	p-	-0.1357	2.5877	1.6300	321.8	212.1	96.0	9S	110W
11698	585	2855 Feb 23	02:24:08	3398	10576	166	P	t-	0.8283	1.3882	0.2887	321.9	127.3	-	11N	19W
11699	585	2855 Aug 19	01:14:47	3402	10582	171	P	a-	-0.8594	1.2526	0.3091	263.6	116.1	-	13S	3W
11700	585	2856 Jan 13	08:12:54	3404	10587	138	N	-t	-1.4895	0.1596	-0.9093	118.7	-	-	20N	107W

Cat Num	Canon Plate	Calendar Date	TD of Greatest Eclipse	ΔT s	Luna Num	Saros Num	Ecl. Type	QSE	Gamma	Pen. Mag.	Um. Mag.	Eclipse Phase ---- Durations ---- Pen. m	Par. m	Total m	Greatest in Zenith Lat.	Lng.
11701	586	2856 Feb 12	01:40:49	3405	10588	176	N	t-	1.4886	0.1726	-0.9190	126.8	-	-	15N	8W
11702	586	2856 Jul 09	08:09:01	3408	10593	143	P	-a	0.9636	1.0889	0.0906	268.4	68.6	-	21S	106W
11703	586	2857 Jan 01	16:14:14	3411	10599	148	P	-a	-0.7373	1.5104	0.4998	296.6	148.8	-	22N	131E
11704	586	2857 Jun 28	15:09:08	3414	10605	153	T+	pp	0.2153	2.4889	1.4375	362.2	226.9	92.9	23S	148E
11705	586	2857 Dec 22	06:16:07	3417	10611	158	T-	pp	-0.0378	2.7743	1.8025	320.5	210.9	98.0	23N	81W
11706	586	2858 Jun 17	16:16:02	3420	10617	163	P	t-	-0.5494	1.8862	0.8140	351.6	196.7	-	24S	131E
11707	586	2858 Dec 11	22:03:11	3424	10623	168	P	a-	0.6370	1.6776	0.7006	294.7	165.7	-	24N	41E
11708	586	2859 May 08	05:30:26	3426	10628	135	N	-t	1.4800	0.1543	-0.8695	113.1	-	-	16S	68W
11709	586	2859 Jun 06	18:11:15	3427	10629	173	N	t-	-1.2771	0.5370	-0.5075	210.1	-	-	24S	101E
11710	586	2859 Nov 01	21:23:01	3429	10634	140	N	-a	-1.4327	0.2514	-0.7930	145.3	-	-	13N	50E
11711	586	2859 Dec 01	10:54:30	3430	10635	178	N	a-	1.3737	0.3481	-0.6734	163.8	-	-	23N	153W
11712	586	2860 Apr 26	17:11:09	3433	10640	145	P	-a	0.6900	1.5789	0.6047	292.8	158.3	-	13S	116E
11713	586	2860 Oct 21	00:18:17	3436	10646	150	P	-t	-0.7571	1.5109	0.4273	329.2	151.3	-	10N	6E
11714	586	2861 Apr 16	09:11:05	3439	10652	155	T-	pp	-0.0280	2.7842	1.8286	318.2	210.9	98.6	10S	123W
11715	586	2861 Oct 10	00:12:28	3442	10658	160	T-	pp	-0.0322	2.8371	1.7613	372.4	234.0	105.2	7N	8E
11716	586	2862 Apr 06	00:56:58	3446	10664	165	P	a-	-0.7422	1.4869	0.5050	288.4	147.2	-	7S	0E
11717	586	2862 Sep 29	04:32:10	3449	10670	170	P	a-	0.7036	1.5797	0.5542	313.0	160.9	-	3N	56W
11718	586	2863 Feb 24	20:33:14	3451	10675	137	N	-h	1.3665	0.3882	-0.6864	183.1	-	-	10N	70E
11719	586	2863 Mar 26	11:55:16	3452	10676	175	N	h-	-1.5153	0.0972	-0.9418	90.8	-	-	4S	164W
11720	586	2863 Aug 20	08:23:18	3455	10681	142	N	-a	-1.1571	0.7087	-0.2396	211.5	-	-	13S	110W
11721	587	2863 Sep 18	16:14:38	3455	10682	180	N	a-	1.3899	0.2926	-0.6776	144.5	-	-	0S	129E
11722	587	2864 Feb 13	20:52:07	3458	10687	147	P	-t	0.7108	1.6061	0.5022	338.9	162.8	-	14N	65E
11723	587	2864 Aug 09	01:33:03	3461	10693	152	T	-a	-0.4318	2.0389	1.0918	306.6	192.8	44.3	16S	7W
11724	587	2865 Feb 01	20:51:00	3464	10699	157	T+	pp	0.0170	2.8673	1.7868	367.7	230.6	103.7	17N	65E
11725	587	2865 Jul 29	16:27:39	3468	10705	162	T	p-	0.3250	2.2547	1.2685	329.0	208.5	73.9	18S	130E
11726	587	2866 Jan 22	03:06:00	3471	10711	167	P	a-	-0.6721	1.6372	0.6125	309.2	163.9	-	19N	30W
11727	587	2866 Jul 19	01:05:56	3474	10717	172	N	t-	1.1399	0.7866	-0.2538	247.9	-	-	20S	0W
11728	587	2866 Dec 13	05:31:19	3477	10722	139	N	-a	1.3610	0.3442	-0.6234	152.3	-	-	24N	70W
11729	587	2867 Jan 11	16:01:40	3477	10723	177	N	a-	-1.3045	0.4537	-0.5252	174.9	-	-	20N	135E
11730	587	2867 Jun 08	11:56:08	3480	10728	144	P	-t	-0.9288	1.1890	0.1189	301.8	83.6	-	24S	164W
11731	587	2867 Dec 02	21:00:11	3483	10734	149	P	-a	0.7178	1.5335	0.5479	290.3	151.7	-	23N	57E
11732	587	2868 May 27	15:16:52	3487	10740	154	T-	pp	-0.1491	2.6003	1.5686	353.1	224.9	98.4	22S	145E
11733	587	2868 Nov 21	08:18:53	3490	10746	159	T+	pp	0.0238	2.8326	1.7963	350.1	224.0	102.1	20N	114W
11734	587	2869 May 17	01:41:13	3493	10752	164	P	a-	0.6026	1.7419	0.7625	304.8	175.0	-	19S	11W
11735	587	2869 Nov 10	12:31:37	3496	10758	169	P	t-	-0.7140	1.5898	0.5067	332.4	161.8	-	17N	177W
11736	587	2870 Apr 07	08:55:29	3499	10763	136	N	-a	-1.2835	0.4828	-0.4775	178.2	-	-	8S	119W
11737	587	2870 May 06	17:19:39	3499	10764	174	N	a-	1.2875	0.4710	-0.4806	175.7	-	-	16S	114E
11738	587	2870 Sep 30	20:00:43	3502	10769	141	N	-t	1.4935	0.1491	-0.9136	116.5	-	-	5N	72E
11739	587	2870 Oct 30	12:12:36	3503	10770	179	N	t-	-1.4177	0.3007	-0.7869	166.7	-	-	13N	172W
11740	587	2871 Mar 27	23:42:41	3505	10775	146	P	-a	-0.6226	1.7147	0.7165	306.6	171.7	-	4S	20E
11741	588	2871 Sep 20	02:28:09	3509	10781	151	P	-h	0.6956	1.5850	0.5781	306.9	161.2	-	0S	24W
11742	588	2872 Mar 16	08:35:55	3512	10787	156	T+	pp	0.0905	2.7211	1.6633	359.3	227.1	101.3	1N	112W
11743	588	2872 Sep 08	15:55:59	3515	10793	161	T-	pp	-0.0708	2.7068	1.7491	322.0	213.0	98.8	5S	135E
11744	588	2873 Mar 05	10:15:36	3518	10799	166	P	t-	0.8106	1.4200	0.3219	324.5	133.8	-	6N	136W
11745	588	2873 Aug 29	08:50:19	3522	10805	171	P	a-	-0.7924	1.3766	0.4310	272.8	134.8	-	10S	117W
11746	588	2874 Jan 23	16:32:19	3524	10810	138	N	-t	-1.4918	0.1541	-0.9122	116.3	-	-	18N	129E
11747	588	2874 Feb 22	09:42:48	3525	10811	176	N	t-	1.4739	0.1979	-0.8901	135.1	-	-	11N	128W
11748	588	2874 Jul 20	15:14:11	3528	10816	143	N	-a	1.0433	0.9443	-0.0571	254.9	-	-	19S	148E
11749	588	2875 Jan 13	00:53:41	3531	10822	148	P	-a	-0.7410	1.5026	0.4941	295.2	147.7	-	21N	3E
11750	588	2875 Jul 09	21:48:44	3534	10828	153	T	-t	0.2988	2.3364	1.2834	360.1	222.2	79.5	22S	49E
11751	588	2876 Jan 02	15:04:42	3538	10834	158	T-	pp	-0.0454	2.7606	1.7885	320.2	210.7	97.8	23N	149E
11752	588	2876 Jun 27	22:47:35	3541	10840	163	P	t-	-0.4657	2.0389	0.9686	358.0	208.6	-	24S	34E
11753	588	2876 Dec 22	06:45:05	3544	10846	168	P	a-	0.6270	1.6974	0.7173	296.4	167.4	-	24N	87W
11754	588	2877 May 18	12:53:08	3547	10851	135	Ne	-t	1.5321	0.0556	-0.9623	68.2	-	-	18S	179W
11755	588	2877 Jun 17	01:03:49	3547	10852	173	N	t-	-1.1999	0.6762	-0.3635	231.7	-	-	24S	1W
11756	588	2877 Nov 12	05:12:17	3550	10857	140	N	-a	-1.4686	0.1885	-0.8619	127.1	-	-	17N	67W
11757	588	2877 Dec 11	19:15:31	3551	10858	178	N	a-	1.3625	0.3712	-0.6554	169.4	-	-	24N	84E
11758	588	2878 May 08	01:02:06	3553	10863	145	P	-a	0.7386	1.4872	0.5180	286.3	148.2	-	17S	1W
11759	588	2878 Nov 01	07:39:31	3557	10869	150	P	-t	-0.8017	1.4312	0.3432	324.0	137.6	-	14N	104W
11760	588	2879 Apr 27	17:18:01	3560	10875	155	T+	pp	0.0154	2.8060	1.8529	318.0	211.1	98.8	14S	115E

Cat Num	Canon Plate	Calendar Date	TD of Greatest Eclipse	ΔT s	Luna Num	Saros Num	Ecl. Type	QSE	Gamma	Pen. Mag.	Um. Mag.	Eclipse Phase Durations Pen. m	Par. m	Total m	Greatest in Zenith Lat.	Lng.
11761	589	2879 Oct 21	07:27:46	3563	10881	160	T-	pp	-0.0838	2.7431	1.6660	371.5	232.9	103.4	11N	101W
11762	589	2880 Apr 16	09:03:38	3566	10887	165	P	a-	-0.7070	1.5513	0.5697	293.0	155.2	-	11S	121W
11763	589	2880 Oct 09	12:03:07	3570	10893	170	P	a-	0.6484	1.6809	0.6557	317.8	171.6	-	7N	169W
11764	589	2881 Mar 07	04:34:07	3572	10898	137	N	-t	1.3878	0.3491	-0.7256	174.7	-	-	6N	51W
11765	589	2881 Apr 05	19:48:04	3573	10899	175	N	t-	-1.4873	0.1486	-0.8905	112.1	-	-	8S	78E
11766	589	2881 Aug 30	15:58:08	3576	10904	142	N	-a	-1.2220	0.5897	-0.3589	194.9	-	-	9S	136E
11767	589	2881 Sep 28	23:59:38	3576	10905	180	N	a-	1.3318	0.3992	-0.5711	166.7	-	-	4N	12E
11768	589	2882 Feb 24	04:46:33	3579	10910	147	P	-t	0.7267	1.5761	0.4740	336.8	158.9	-	10N	53W
11769	589	2882 Aug 20	09:04:14	3582	10916	152	P	-a	-0.5034	1.9088	0.9591	302.5	185.4	-	12S	120W
11770	589	2883 Feb 13	05:01:07	3586	10922	157	T+	pp	0.0274	2.8464	1.7695	366.4	230.2	103.5	13N	57W
11771	589	2883 Aug 09	23:39:02	3589	10928	162	T+	pp	0.2499	2.3944	1.4042	333.5	213.9	86.6	15S	22E
11772	589	2884 Feb 02	11:41:45	3592	10934	167	P	a-	-0.6649	1.6486	0.6274	308.8	165.0	-	16N	158W
11773	589	2884 Jul 29	07:47:55	3595	10940	172	N	t-	1.0605	0.9340	-0.1094	267.1	-	-	17S	100W
11774	589	2884 Dec 23	14:17:44	3598	10945	139	N	-a	1.3703	0.3280	-0.6410	148.9	-	-	25N	160E
11775	589	2885 Jan 22	00:52:23	3599	10946	177	N	a-	-1.2981	0.4648	-0.5128	176.6	-	-	18N	4E
11776	589	2885 Jun 18	18:33:10	3601	10951	144	N*	-t	-1.0070	1.0439	-0.0230	286.6	-	-	24S	98E
11777	589	2885 Dec 13	05:35:06	3605	10957	149	P	-a	0.7328	1.5082	0.5185	289.7	148.5	-	24N	70W
11778	589	2886 Jun 07	22:21:31	3608	10963	154	T-	pp	-0.2186	2.4700	1.4438	349.6	221.5	91.5	23S	40E
11779	589	2886 Dec 02	16:27:49	3611	10969	159	T+	pp	0.0441	2.7981	1.7563	351.6	224.3	101.8	22N	125E
11780	589	2887 May 28	09:17:42	3615	10975	164	P	a-	0.5420	1.8504	0.8761	308.6	183.5	-	21S	125W
11781	590	2887 Nov 21	20:10:22	3618	10981	169	P	t-	-0.6842	1.6467	0.5589	337.0	168.7	-	19N	69E
11782	590	2888 Apr 17	17:09:16	3621	10986	136	N	-a	-1.3200	0.4149	-0.5436	166.4	-	-	12S	117E
11783	590	2888 May 17	01:13:49	3621	10987	174	N	a-	1.2361	0.5640	-0.3847	190.3	-	-	18S	4W
11784	590	2888 Oct 11	03:14:10	3624	10992	141	N	-t	1.5434	0.0578	-1.0055	72.9	-	-	9N	37W
11785	590	2888 Nov 09	19:40:54	3625	10993	179	N	t-	-1.3793	0.3722	-0.7175	184.1	-	-	16N	76E
11786	590	2889 Apr 07	07:51:21	3627	10998	146	P	-a	-0.6557	1.6538	0.6556	304.2	166.4	-	8S	102W
11787	590	2889 Sep 30	09:59:39	3631	11004	151	P	-h	0.7495	1.4859	0.4796	299.7	148.7	-	4N	137W
11788	590	2890 Mar 27	16:28:28	3634	11010	156	T+	pp	0.0619	2.7734	1.7157	360.6	228.2	102.7	3S	129E
11789	590	2890 Sep 19	23:40:17	3637	11016	161	T-	pp	-0.0118	2.8153	1.8572	321.8	213.1	99.7	1S	19E
11790	590	2891 Mar 16	17:57:55	3641	11022	166	P	t-	0.7856	1.4649	0.3688	328.0	142.3	-	2N	108E
11791	590	2891 Sep 09	16:31:35	3644	11028	171	P	a-	-0.7303	1.4918	0.5435	280.5	148.9	-	5S	127E
11792	590	2892 Feb 04	00:50:55	3647	11033	138	N	-t	-1.4950	0.1464	-0.9164	113.1	-	-	15N	5E
11793	590	2892 Mar 04	17:39:19	3647	11034	176	N	t-	1.4542	0.2318	-0.8520	145.3	-	-	7N	113E
11794	590	2892 Jul 30	22:16:27	3650	11039	143	N	-a	1.1241	0.7979	-0.2074	238.9	-	-	17S	43E
11795	590	2892 Aug 29	07:53:48	3651	11040	181	N	a-	-1.4966	0.1023	-0.8789	88.5	-	-	10S	102W
11796	590	2893 Jan 23	09:34:59	3653	11045	148	P	-a	-0.7436	1.4964	0.4906	293.9	146.9	-	18N	126W
11797	590	2893 Jul 20	04:23:43	3657	11051	153	T	-t	0.3852	2.1790	1.1239	356.6	215.3	55.6	20S	49W
11798	590	2894 Jan 12	23:56:21	3660	11057	158	T-	pp	-0.0507	2.7508	1.7789	319.9	210.5	97.6	21N	18E
11799	590	2894 Jul 09	05:14:55	3663	11063	163	T	t-	-0.3782	2.1987	1.1298	363.3	218.2	57.4	22S	62W
11800	590	2895 Jan 02	15:31:21	3667	11069	168	P	a-	0.6211	1.7096	0.7268	297.7	168.5	-	23N	143E
11801	591	2895 Jun 28	07:53:55	3670	11075	173	N	t-	-1.1194	0.8218	-0.2135	250.7	-	-	24S	102W
11802	591	2895 Nov 23	13:07:29	3673	11080	140	N	-a	-1.4989	0.1358	-0.9203	108.9	-	-	19N	175E
11803	591	2895 Dec 23	03:40:24	3673	11081	178	N	a-	1.3551	0.3871	-0.6441	173.4	-	-	25N	41W
11804	591	2896 May 18	08:48:25	3676	11086	145	P	-a	0.7915	1.3877	0.4233	278.9	135.6	-	19S	117W
11805	591	2896 Nov 11	15:07:55	3679	11092	150	P	-t	-0.8401	1.3629	0.2708	319.3	123.6	-	17N	145E
11806	591	2897 May 08	01:17:49	3683	11098	155	T+	pp	0.0647	2.7144	1.7636	317.6	210.8	98.1	17S	5W
11807	591	2897 Oct 31	14:52:00	3686	11104	160	T-	pp	-0.1285	2.6618	1.5833	370.3	231.3	100.6	14N	148E
11808	591	2898 Apr 27	17:02:17	3689	11110	165	P	a-	-0.6660	1.6265	0.6452	298.1	163.6	-	15S	119E
11809	591	2898 Oct 20	19:43:17	3693	11116	170	P	a-	0.5999	1.7697	0.7448	321.5	179.7	-	11N	75E
11810	591	2899 Mar 18	12:27:24	3695	11121	137	N	-t	1.4146	0.2998	-0.7748	163.0	-	-	2N	169W
11811	591	2899 Apr 17	03:32:15	3696	11122	175	N	t-	-1.4532	0.2113	-0.8279	133.3	-	-	12S	38W
11812	591	2899 Sep 10	23:39:29	3699	11127	142	N	-a	-1.2819	0.4802	-0.4692	177.6	-	-	5S	21E
11813	591	2899 Oct 10	07:53:10	3699	11128	180	N	a-	1.2802	0.4941	-0.4766	183.4	-	-	8N	107W
11814	591	2900 Mar 07	12:35:27	3702	11133	147	P	-t	0.7471	1.5374	0.4377	333.0	153.6		6N	171W
11815	591	2900 Aug 31	16:38:06	3706	11139	152	P	-a	-0.5725	1.7836	0.8309	298.0	176.7	-	9S	127E
11816	591	2901 Feb 24	13:07:56	3709	11145	157	T+	pp	0.0412	2.8191	1.7463	365.0	229.6	103.2	9N	179W
11817	591	2901 Aug 21	06:51:31	3712	11151	162	T+	pp	0.1763	2.5318	1.5371	337.3	217.9	94.8	12S	86W
11818	591	2902 Feb 13	20:14:30	3716	11157	167	P	a-	-0.6548	1.6652	0.6480	308.5	166.7	-	13N	75E
11819	591	2902 Aug 10	14:30:31	3719	11163	172	P	t-	0.9817	1.0803	0.0332	283.9	43.9	-	14S	159E
11820	591	2903 Jan 04	23:07:17	3722	11168	139	N	-a	1.3775	0.3152	-0.6546	146.2	-	-	24N	30E

Cat Num	Canon Plate	Calendar Date	TD of Greatest Eclipse	ΔT s	Luna Num	Saros Num	Ecl. Type	QSE	Gamma	Pen. Mag.	Um. Mag.	Eclipse Phase Durations Pen. m	Eclipse Phase Durations Par. m	Eclipse Phase Durations Total m	Greatest in Zenith Lat.	Greatest in Zenith Lng.
11821	592	2903 Feb 03	09:41:43	3722	11169	177	N	a-	-1.2903	0.4781	-0.4977	178.6	-	-	15N	127W
11822	592	2903 Jul 01	01:07:21	3725	11174	144	N	-t	-1.0874	0.8950	-0.1692	269.0	-	-	24S	0E
11823	592	2903 Dec 25	14:13:38	3728	11180	149	P	-a	0.7445	1.4885	0.4951	289.3	145.9	-	24N	162E
11824	592	2904 Jun 19	05:22:46	3732	11186	154	T	pp	-0.2919	2.3330	1.3119	345.2	216.6	80.5	24S	64W
11825	592	2904 Dec 14	00:42:23	3735	11192	159	T+	pp	0.0592	2.7730	1.7261	353.1	224.6	101.5	23N	3E
11826	592	2905 Jun 08	16:47:10	3738	11198	164	T	a-	0.4748	1.9715	1.0018	312.3	191.4	6.5	22S	124E
11827	592	2905 Dec 03	03:57:28	3742	11204	169	P	t-	-0.6616	1.6904	0.5984	340.5	173.7	-	21N	47W
11828	592	2906 Apr 30	01:14:13	3745	11209	136	N	-a	-1.3639	0.3335	-0.6232	150.4	-	-	16S	4W
11829	592	2906 May 29	09:00:23	3745	11210	174	N	a-	1.1780	0.6692	-0.2770	205.0	-	-	20S	120W
11830	592	2906 Nov 22	03:18:35	3749	11216	179	N	t-	-1.3479	0.4308	-0.6606	196.9	-	-	19N	38W
11831	592	2907 Apr 19	15:51:24	3751	11221	146	P	-a	-0.6955	1.5808	0.5826	301.0	159.1	-	12S	137E
11832	592	2907 Oct 12	17:40:09	3755	11227	151	P	-a	0.7968	1.3988	0.3929	292.8	135.9	-	8N	107E
11833	592	2908 Apr 08	00:13:40	3758	11233	156	T+	pp	0.0281	2.8353	1.7778	361.8	229.2	103.8	7S	13E
11834	592	2908 Oct 01	07:29:50	3761	11239	161	T+	pp	0.0432	2.7579	1.7990	321.2	212.5	99.1	3N	99W
11835	592	2909 Mar 28	01:34:29	3765	11245	166	P	t-	0.7561	1.5179	0.4242	331.8	151.4	-	2S	7W
11836	592	2909 Sep 21	00:16:53	3768	11251	171	P	a-	-0.6720	1.6003	0.6488	287.2	160.1	-	1S	10E
11837	592	2910 Feb 15	09:08:12	3771	11256	138	N	-t	-1.4998	0.1355	-0.9233	108.5	-	-	11N	119W
11838	592	2910 Mar 17	01:30:26	3772	11257	176	N	t-	1.4298	0.2742	-0.8049	157.0	-	-	3N	5W
11839	592	2910 Aug 12	05:19:18	3774	11262	143	N	-h	1.2032	0.6549	-0.3546	220.6	-	-	14S	63W
11840	592	2910 Sep 10	15:19:19	3775	11263	181	N	h-	-1.4381	0.2122	-0.7739	126.7	-	-	6S	146E
11841	593	2911 Feb 04	18:15:58	3778	11268	148	P	-a	-0.7469	1.4888	0.4860	292.5	146.0	-	15N	105E
11842	593	2911 Aug 01	10:59:11	3781	11274	153	P	-t	0.4704	2.0238	0.9662	352.0	206.1	-	17S	147W
11843	593	2912 Jan 25	08:47:30	3785	11280	158	T-	-p	-0.0564	2.7401	1.7687	319.6	210.3	97.4	19N	114W
11844	593	2912 Jul 20	11:43:06	3788	11286	163	T	tp	-0.2912	2.3578	1.2900	367.4	225.5	81.2	21S	158W
11845	593	2913 Jan 14	00:18:24	3791	11292	168	P	a-	0.6164	1.7194	0.7343	298.8	169.4	-	22N	13E
11846	593	2913 Jul 09	14:42:35	3795	11298	173	N	t-	-1.0366	0.9716	-0.0597	267.4	-	-	23S	157E
11847	593	2913 Dec 04	21:09:04	3797	11303	140	N	-a	-1.5232	0.0940	-0.9675	91.3	-	-	21N	56E
11848	593	2914 Jan 03	12:08:10	3798	11304	178	N	a-	1.3509	0.3969	-0.6384	176.1	-	-	24N	166W
11849	593	2914 May 30	16:28:18	3801	11309	145	P	-a	0.8499	1.2784	0.3183	270.2	119.2	-	21S	129E
11850	593	2914 Nov 23	22:44:30	3804	11315	150	P	-t	-0.8712	1.3076	0.2118	315.3	110.4	-	20N	31E
11851	593	2915 May 20	09:10:06	3808	11321	155	T+	-p	0.1200	2.6119	1.6631	317.0	210.0	96.0	20S	122W
11852	593	2915 Nov 12	22:26:46	3811	11327	160	T-	pp	-0.1646	2.5961	1.5165	368.9	229.6	97.4	18N	35E
11853	593	2916 May 09	00:51:40	3814	11333	165	P	a-	-0.6179	1.7148	0.7335	303.5	172.3	-	18S	2E
11854	593	2916 Nov 01	03:32:55	3818	11339	170	P	a-	0.5585	1.8456	0.8208	324.1	185.6	-	15N	42W
11855	593	2917 Mar 29	20:11:52	3821	11344	137	N	-t	1.4477	0.2389	-0.8350	146.6	-	-	2S	75E
11856	593	2917 Apr 28	11:05:50	3821	11345	175	N	t-	-1.4114	0.2878	-0.7514	155.0	-	-	15S	152W
11857	593	2917 Sep 22	07:26:14	3824	11350	142	N	-a	-1.3377	0.3786	-0.5721	159.1	-	-	1S	97W
11858	593	2917 Oct 21	15:55:02	3825	11351	180	N	a-	1.2349	0.5776	-0.3938	196.3	-	-	12N	133E
11859	593	2918 Mar 18	20:16:49	3827	11356	147	P	-t	0.7735	1.4874	0.3907	330.0	146.1	-	1N	74E
11860	593	2918 Sep 12	00:16:22	3831	11362	152	P	-a	-0.6373	1.6665	0.7101	293.1	166.9	-	5S	12E
11861	594	2919 Mar 07	21:11:15	3834	11368	157	T+	pp	0.0585	2.7849	1.7170	363.4	229.0	102.8	5N	61E
11862	594	2919 Sep 01	14:05:41	3838	11374	162	T+	pp	0.1051	2.6649	1.6652	340.4	220.6	99.8	8S	165E
11863	594	2920 Feb 25	04:45:31	3841	11380	167	P	a-	-0.6427	1.6852	0.6723	308.4	168.7	-	9N	53W
11864	594	2920 Aug 20	21:13:30	3844	11386	172	P	t-	0.9038	1.2254	0.1743	298.7	98.8	-	11S	58E
11865	594	2921 Jan 15	07:59:27	3847	11391	139	N	-a	1.3825	0.3062	-0.6643	144.3	-	-	22N	102W
11866	594	2921 Feb 13	18:30:10	3848	11392	177	N	a-	-1.2821	0.4921	-0.4816	180.7	-	-	12N	101E
11867	594	2921 Jul 11	07:38:47	3851	11397	144	N	-t	-1.1700	0.7422	-0.3194	248.2	-	-	23S	96W
11868	594	2922 Jan 04	22:55:38	3854	11403	149	P	-a	0.7529	1.4748	0.4781	289.4	144.1	-	23N	33E
11869	594	2922 Jun 30	12:22:26	3857	11409	154	T	-p	-0.3673	2.1923	1.1757	339.9	210.2	63.1	23S	168W
11870	594	2922 Dec 25	08:59:41	3861	11415	159	T+	pp	0.0713	2.7532	1.7017	354.5	224.9	101.2	23N	119W
11871	594	2923 Jun 20	00:15:08	3864	11421	164	T	a-	0.4056	2.0963	1.1308	315.4	198.0	53.0	23S	13E
11872	594	2923 Dec 14	11:49:27	3868	11427	169	P	t-	-0.6429	1.7265	0.6309	343.4	177.6	-	22N	163W
11873	594	2924 May 10	09:12:12	3871	11432	136	N	-a	-1.4130	0.2427	-0.7126	129.5	-	-	19S	123W
11874	594	2924 Jun 08	16:42:16	3871	11433	174	N	a-	1.1163	0.7815	-0.1626	218.9	-	-	22S	126E
11875	594	2924 Dec 02	11:03:33	3875	11439	179	N	t-	-1.3217	0.4795	-0.6132	206.6	-	-	21N	153W
11876	594	2925 Apr 29	23:42:01	3877	11444	146	P	-a	-0.7423	1.4951	0.4967	296.6	149.4	-	15S	20E
11877	594	2925 Oct 23	01:29:54	3881	11450	151	P	-a	0.8370	1.3251	0.3194	286.4	123.5	-	12N	10W
11878	594	2926 Apr 19	07:46:46	3884	11456	156	T-	pp	-0.0147	2.8598	1.8026	362.7	229.9	104.2	11S	101W
11879	594	2926 Oct 12	15:28:29	3888	11462	161	T+	pp	0.0907	2.6714	1.7113	320.4	211.5	97.5	8N	141E
11880	594	2927 Apr 08	09:00:57	3891	11468	166	P	t-	0.7186	1.5853	0.4944	336.3	161.7	-	7S	118W

Cat Num	Canon Plate	Calendar Date	TD of Greatest Eclipse	ΔT s	Luna Num	Saros Num	Ecl. Type	QSE	Gamma	Pen. Mag.	Um. Mag.	Eclipse Phase Durations Pen. m	Par. m	Total m	Greatest in Zenith Lat.	Lng.
11881	595	2927 Oct 02	08:08:17	3895	11474	171	P	a-	-0.6192	1.6991	0.7441	292.9	168.9	-	3N	108W
11882	595	2928 Feb 26	17:21:42	3897	11479	138	N	-t	-1.5081	0.1181	-0.9361	101.0	-	-	7N	118E
11883	595	2928 Mar 27	09:14:32	3898	11480	176	N	t-	1.3992	0.3278	-0.7462	170.2	-	-	2S	121W
11884	595	2928 Aug 22	12:22:23	3901	11485	143	N	-h	1.2807	0.5150	-0.4992	199.2	-	-	10S	169W
11885	595	2928 Sep 20	22:48:32	3901	11486	181	N	h-	-1.3831	0.3159	-0.6757	153.7	-	-	2S	33E
11886	595	2929 Feb 15	02:55:11	3904	11491	148	P	-a	-0.7522	1.4773	0.4781	290.8	144.6	-	12N	25W
11887	595	2929 Aug 11	17:32:45	3908	11497	153	P	-t	0.5564	1.8676	0.8070	346.1	194.2	-	14S	115E
11888	595	2930 Feb 04	17:39:03	3911	11503	158	T-	-p	-0.0617	2.7297	1.7594	319.4	210.2	97.3	16N	115E
11889	595	2930 Jul 31	18:10:59	3915	11509	163	T-	pp	-0.2036	2.5181	1.4511	370.2	230.6	94.9	18S	106E
11890	595	2931 Jan 25	09:05:59	3918	11515	168	P	a-	0.6127	1.7271	0.7402	299.9	170.3	-	20N	118W
11891	595	2931 Jul 20	21:32:13	3921	11521	173	P	t-	-0.9536	1.1222	0.0944	281.6	72.0	-	21S	56E
11892	595	2931 Dec 16	05:16:05	3924	11526	140	N	-a	-1.5420	0.0619	-1.0044	74.6	-	-	22N	64W
11893	595	2932 Jan 14	20:38:00	3925	11527	178	N	a-	1.3492	0.4018	-0.6371	177.8	-	-	23N	69E
11894	595	2932 Jun 10	00:03:59	3928	11532	145	P	-a	0.9122	1.1620	0.2058	260.3	97.2	-	22S	16E
11895	595	2932 Dec 04	06:27:49	3931	11538	150	P	-t	-0.8966	1.2627	0.1636	311.8	97.8	-	21N	83W
11896	595	2933 May 30	16:56:40	3935	11544	155	T+	-p	0.1800	2.5011	1.5539	315.8	208.4	92.1	22S	122E
11897	595	2933 Nov 23	06:10:08	3938	11550	160	T-	pp	-0.1945	2.5417	1.4612	367.5	227.8	94.1	20N	80W
11898	595	2934 May 20	08:33:11	3942	11556	165	P	a-	-0.5637	1.8142	0.8327	309.2	181.1	-	21S	113W
11899	595	2934 Nov 12	11:32:00	3945	11562	170	P	a-	0.5241	1.9087	0.8842	325.9	189.9	-	18N	161W
11900	595	2935 Apr 10	03:47:50	3948	11567	137	N	-t	1.4869	0.1665	-0.9066	123.4	-	-	7S	40W
11901	596	2935 May 09	18:31:22	3948	11568	175	N	t-	-1.3637	0.3755	-0.6640	176.2	-	-	19S	97E
11902	596	2935 Oct 03	15:19:37	3951	11573	142	N	-a	-1.3883	0.2866	-0.6658	139.6	-	-	3N	145E
11903	596	2935 Nov 02	00:04:20	3952	11574	180	N	a-	1.1953	0.6506	-0.3215	206.5	-	-	16N	11E
11904	596	2936 Mar 29	03:51:11	3955	11579	147	P	-t	0.8056	1.4267	0.3334	325.0	136.1	-	3S	40W
11905	596	2936 Sep 22	07:59:30	3958	11585	152	P	-a	-0.6977	1.5577	0.5973	288.0	156.2	-	1S	105W
11906	596	2937 Mar 18	05:07:34	3962	11591	157	T+	pp	0.0821	2.7389	1.6762	361.6	228.2	101.9	1N	59W
11907	596	2937 Sep 11	21:23:18	3965	11597	162	T+	pp	0.0376	2.7916	1.7865	342.7	222.2	102.2	4S	55E
11908	596	2938 Mar 07	13:11:04	3969	11603	167	P	a-	-0.6255	1.7143	0.7062	308.7	171.5	-	5N	179W
11909	596	2938 Sep 01	04:00:13	3972	11609	172	P	t-	0.8291	1.3646	0.3093	311.4	129.3	-	7S	43W
11910	596	2939 Jan 26	16:50:38	3975	11614	139	N	-a	1.3889	0.2946	-0.6760	141.7	-	-	20N	127E
11911	596	2939 Feb 25	03:13:32	3975	11615	177	N	a-	-1.2696	0.5137	-0.4575	183.9	-	-	8N	30W
11912	596	2939 Jul 22	14:10:52	3978	11620	144	N	-t	-1.2522	0.5904	-0.4692	224.2	-	-	21S	166E
11913	596	2940 Jan 16	07:38:56	3982	11626	149	P	-a	0.7602	1.4627	0.4633	289.4	142.5	-	22N	96W
11914	596	2940 Jul 10	19:20:01	3985	11632	154	T	-t	-0.4453	2.0470	1.0346	333.6	202.0	29.2	22S	89E
11915	596	2941 Jan 04	17:20:48	3989	11638	159	T+	pp	0.0797	2.7398	1.6843	355.8	225.3	100.9	23N	117E
11916	596	2941 Jun 30	07:38:52	3992	11644	164	T	a-	0.3319	2.2297	1.2678	318.1	203.7	72.4	23S	97W
11917	596	2941 Dec 24	19:47:18	3996	11650	169	P	t-	-0.6294	1.7528	0.6542	345.6	180.4	-	23N	79E
11918	596	2942 May 21	17:01:36	3999	11655	136	N	-a	-1.4688	0.1397	-0.8146	99.3	-	-	22S	121E
11919	596	2942 Jun 20	00:18:29	3999	11656	174	N	a-	1.0497	0.9029	-0.0397	232.3	-	-	22S	13E
11920	596	2942 Dec 13	18:56:57	4003	11662	179	N	t-	-1.3016	0.5168	-0.5767	213.4	-	-	22N	91E
11921	597	2943 May 11	07:24:37	4006	11667	146	P	-a	-0.7949	1.3987	0.4000	291.1	136.5	-	19S	95W
11922	597	2943 Nov 03	09:27:52	4009	11673	151	P	-a	0.8712	1.2622	0.2566	280.7	111.4	-	16N	130W
11923	597	2944 Apr 29	15:12:15	4013	11679	156	T-	pp	-0.0632	2.7707	1.7139	363.4	230.1	103.5	15S	148E
11924	597	2944 Oct 22	23:33:47	4016	11685	161	T+	-p	0.1330	2.5945	1.6330	319.4	210.2	95.2	12N	19E
11925	597	2945 Apr 18	16:19:09	4019	11691	166	P	t-	0.6742	1.6652	0.5773	341.1	172.5	-	11S	132E
11926	597	2945 Oct 12	16:05:42	4023	11697	171	P	a-	-0.5716	1.7885	0.8295	297.7	176.0	-	7N	132E
11927	597	2946 Mar 09	01:31:32	4026	11702	138	N	-t	-1.5198	0.0938	-0.9552	89.9	-	-	3N	4W
11928	597	2946 Apr 07	16:52:13	4026	11703	176	N	t-	1.3626	0.3922	-0.6763	184.4	-	-	6S	125E
11929	597	2946 Sep 02	19:28:48	4029	11708	143	N	-h	1.3545	0.3824	-0.6370	174.6	-	-	6S	84E
11930	597	2946 Oct 02	06:23:37	4030	11709	181	N	h-	-1.3333	0.4101	-0.5871	174.4	-	-	2N	81W
11931	597	2947 Feb 26	11:30:31	4033	11714	148	P	-a	-0.7613	1.4586	0.4634	288.6	142.4	-	8N	153W
11932	597	2947 Aug 23	00:10:11	4036	11720	153	P	-t	0.6384	1.7187	0.6549	339.2	179.9	-	11S	15E
11933	597	2948 Feb 16	02:27:42	4040	11726	158	T-	-p	-0.0695	2.7148	1.7458	319.1	210.1	97.1	12N	17W
11934	597	2948 Aug 11	00:41:12	4043	11732	163	T-	pp	-0.1176	2.6757	1.6091	371.8	233.7	102.7	15S	8E
11935	597	2949 Feb 04	17:51:55	4047	11738	168	P	a-	0.6082	1.7360	0.7477	301.0	171.2	-	17N	112E
11936	597	2949 Jul 31	04:23:42	4050	11744	173	P	t-	-0.8712	1.2719	0.2471	293.7	113.7	-	19S	47W
11937	597	2949 Dec 26	13:26:38	4053	11749	140	N	-h	-1.5575	0.0357	-1.0349	56.9	-	-	22N	175E
11938	597	2950 Jan 25	05:06:25	4054	11750	178	N	h-	1.3472	0.4071	-0.6348	179.5	-	-	20N	57W
11939	597	2950 Jun 21	07:35:21	4057	11755	145	P	-a	0.9784	1.0389	0.0860	248.8	63.8	-	22S	96W
11940	597	2950 Dec 15	14:18:10	4060	11761	150	P	-t	-0.9159	1.2287	0.1269	309.1	86.6	-	22N	161E

Cat Num	Canon Plate	Calendar Date	TD of Greatest Eclipse	ΔT	Luna Num	Saros Num	Ecl. Type	QSE	Gamma	Pen. Mag.	Um. Mag.	Eclipse Phase Durations Pen.	Par.	Total	Greatest in Zenith Lat.	Lng.
				s								m	m	m		
11941	598	2951 Jun 11	00:37:15	4064	11767	155	T+	-p	0.2448	2.3816	1.4354	314.1	205.7	85.7	23S	8E
11942	598	2951 Dec 04	14:01:47	4067	11773	160	T-	pp	-0.2179	2.4989	1.4179	366.1	226.1	91.0	22N	164E
11943	598	2952 May 30	16:06:21	4071	11779	165	P	a-	-0.5036	1.9248	0.9427	314.9	189.5	-	22S	135E
11944	598	2952 Nov 22	19:40:25	4074	11785	170	P	a-	0.4967	1.9587	0.9345	327.0	192.9	-	21N	78E
11945	598	2953 Apr 20	11:12:26	4077	11790	137	N	-t	1.5341	0.0792	-0.9928	86.0	-	-	10S	151W
11946	598	2953 May 20	01:45:10	4078	11791	175	N	t-	-1.3075	0.4788	-0.5610	197.6	-	-	21S	10W
11947	598	2953 Oct 13	23:20:11	4081	11796	142	N	-a	-1.4331	0.2052	-0.7492	119.0	-	-	7N	24E
11948	598	2953 Nov 12	08:22:40	4081	11797	180	N	a-	1.1629	0.7106	-0.2625	214.1	-	-	19N	113W
11949	598	2954 Apr 09	11:16:53	4084	11802	147	P	-t	0.8444	1.3536	0.2641	318.7	122.3	-	7S	151W
11950	598	2954 Oct 03	15:48:15	4088	11808	152	P	-a	-0.7528	1.4589	0.4940	282.9	144.6	-	4N	138E
11951	598	2955 Mar 29	12:58:25	4091	11814	157	T+	pp	0.1108	2.6835	1.6265	359.6	227.1	100.5	3S	176W
11952	598	2955 Sep 23	04:45:05	4095	11820	162	T-	pp	-0.0255	2.8167	1.8058	344.5	222.9	102.4	0N	56W
11953	598	2956 Mar 17	21:33:02	4098	11826	167	P	a-	-0.6048	1.7496	0.7467	309.3	174.7	-	0N	55E
11954	598	2956 Sep 11	10:48:39	4102	11832	172	P	t-	0.7562	1.5006	0.4407	322.6	151.6	-	3S	146W
11955	598	2957 Feb 06	01:42:04	4105	11837	139	N	-a	1.3951	0.2832	-0.6872	139.1	-	-	17N	5W
11956	598	2957 Mar 07	11:54:01	4105	11838	177	N	a-	-1.2550	0.5392	-0.4292	187.7	-	-	4N	160W
11957	598	2957 Aug 01	20:43:50	4108	11843	144	N	-t	-1.3338	0.4397	-0.6182	195.8	-	-	19S	69E
11958	598	2957 Aug 31	11:32:50	4109	11844	182	N	t-	1.5086	0.1267	-0.9466	109.7	-	-	7S	156W
11959	598	2958 Jan 26	16:21:51	4112	11849	149	P	-a	0.7672	1.4511	0.4493	289.5	140.9	-	19N	135E
11960	598	2958 Jul 22	02:19:20	4115	11855	154	P	-t	-0.5226	1.9034	0.8946	326.5	192.0	-	20S	15W
11961	599	2959 Jan 16	01:42:19	4119	11861	159	T+	pp	0.0872	2.7278	1.6688	357.1	225.7	100.7	21N	7W
11962	599	2959 Jul 11	15:02:55	4122	11867	164	T+	p-	0.2581	2.3636	1.4048	320.0	208.0	84.6	22S	153E
11963	599	2960 Jan 05	03:46:50	4126	11873	169	P	t-	-0.6174	1.7759	0.6749	347.5	182.8	-	22N	39W
11964	599	2960 Jun 01	00:45:28	4129	11878	136	Ne	-a	-1.5287	0.0294	-0.9242	46.1	-	-	24S	6E
11965	599	2960 Jun 30	07:52:16	4129	11879	174	P	a-	0.9811	1.0283	0.0866	244.6	63.4	-	22S	100W
11966	599	2960 Dec 24	02:55:37	4133	11885	179	N	t-	-1.2853	0.5468	-0.5469	218.6	-	-	22N	27W
11967	599	2961 May 21	14:56:55	4136	11890	146	P	-a	-0.8551	1.2886	0.2893	283.9	118.4	-	21S	152E
11968	599	2961 Nov 13	17:35:24	4139	11896	151	P	-a	0.8982	1.2125	0.2071	275.8	100.5	-	19N	109E
11969	599	2962 May 10	22:26:26	4143	11902	156	T-	pp	-0.1203	2.6657	1.6092	363.5	229.5	101.1	18S	40E
11970	599	2962 Nov 03	07:47:44	4146	11908	161	T+	-p	0.1683	2.5305	1.5675	318.4	208.8	92.5	15N	104W
11971	599	2963 Apr 29	23:27:36	4150	11914	166	P	t-	0.6218	1.7598	0.6750	346.4	183.5	-	14S	25E
11972	599	2963 Oct 24	00:10:50	4153	11920	171	P	a-	-0.5306	1.8657	0.9027	301.8	181.5	-	11N	11E
11973	599	2964 Mar 19	09:35:34	4156	11925	138	N	-t	-1.5366	0.0602	-0.9832	71.9	-	-	1S	125W
11974	599	2964 Apr 18	00:21:54	4157	11926	176	N	t-	1.3188	0.4696	-0.5932	199.6	-	-	10S	12E
11975	599	2964 Sep 13	02:37:10	4160	11931	143	N	-h	1.4252	0.2553	-0.7696	145.0	-	-	2S	23W
11976	599	2964 Oct 12	14:03:43	4161	11932	181	N	h-	-1.2881	0.4959	-0.5073	191.1	-	-	7N	163E
11977	599	2965 Mar 08	20:02:09	4163	11937	148	P	-a	-0.7738	1.4335	0.4427	286.0	139.3	-	4N	79E
11978	599	2965 Sep 02	06:49:42	4167	11943	153	P	-t	0.7177	1.5750	0.5077	331.3	162.5	-	7S	85W
11979	599	2966 Feb 26	11:12:54	4171	11949	158	T-	-p	-0.0800	2.6947	1.7275	318.8	209.9	96.8	9N	148W
11980	599	2966 Aug 22	07:15:12	4174	11955	163	T-	pp	-0.0345	2.8282	1.7616	372.3	234.9	105.9	11S	90W
11981	600	2967 Feb 16	02:35:56	4178	11961	168	P	a-	0.6030	1.7461	0.7569	302.2	172.3	-	13N	18W
11982	600	2967 Aug 11	11:18:05	4181	11967	173	P	h-	-0.7900	1.4195	0.3973	304.0	140.5	-	16S	150W
11983	600	2968 Jan 06	21:40:03	4184	11972	140	N	-h	-1.5698	0.0149	-1.0594	36.9	-	-	21N	53E
11984	600	2968 Feb 05	13:34:06	4185	11973	178	N	h-	1.3453	0.4118	-0.6325	181.0	-	-	17N	177E
11985	600	2968 Jul 01	15:04:44	4188	11978	145	N	-a	1.0464	0.9127	-0.0374	235.8	-	-	22S	153E
11986	600	2968 Jul 30	22:36:40	4188	11979	183	Nb	a-	-1.4880	0.1114	-0.8567	91.1	-	-	19S	41E
11987	600	2968 Dec 25	22:13:36	4191	11984	150	P	-t	-0.9309	1.2023	0.0984	306.9	76.6	-	22N	44E
11988	600	2969 Jun 21	08:12:26	4195	11990	155	T	-p	0.3140	2.2544	1.3088	311.8	201.9	75.6	23S	105W
11989	600	2969 Dec 14	22:00:51	4198	11996	160	T-	pp	-0.2358	2.4661	1.3852	364.7	224.7	88.3	23N	46E
11990	600	2970 Jun 10	23:33:32	4202	12002	165	T	a-	-0.4392	2.0435	1.0603	320.3	197.3	37.2	23S	24E
11991	600	2970 Dec 04	03:55:41	4205	12008	170	P	a-	0.4745	1.9991	0.9756	327.5	195.0	-	23N	44W
11992	600	2971 May 31	08:52:28	4209	12014	175	N	t-	-1.2467	0.5908	-0.4496	217.8	-	-	23S	116W
11993	600	2971 Oct 25	07:28:21	4212	12019	142	N	-a	-1.4719	0.1352	-0.8214	97.2	-	-	11N	98W
11994	600	2971 Nov 23	16:48:05	4213	12020	180	N	a-	1.1363	0.7597	-0.2142	219.9	-	-	21N	122E
11995	600	2972 Apr 19	18:34:31	4216	12025	147	P	-t	0.8899	1.2681	0.1826	310.8	102.9	-	11S	99E
11996	600	2972 Oct 13	23:42:46	4219	12031	152	P	-a	-0.8027	1.3698	0.4002	277.8	132.2	-	8N	19E
11997	600	2973 Apr 08	20:41:47	4223	12037	157	T+	pp	0.1463	2.6152	1.5642	357.3	225.6	98.1	8S	68E
11998	600	2973 Oct 03	12:12:54	4226	12043	162	T-	pp	-0.0829	2.7144	1.6974	345.8	222.7	101.0	4N	168W
11999	600	2974 Mar 29	05:46:43	4230	12049	167	P	a-	-0.5766	1.7987	0.8012	310.4	178.8	-	4S	68W
12000	600	2974 Sep 22	17:44:13	4233	12055	172	P	t-	0.6893	1.6258	0.5610	332.0	168.2	-	1N	110E

Cat Num	Canon Plate	Calendar Date	TD of Greatest Eclipse	ΔT s	Luna Num	Saros Num	Ecl. Type	QSE	Gamma	Pen. Mag.	Um. Mag.	Eclipse Phase Durations Pen. m	Par. m	Total m	Greatest in Zenith Lat.	Lng.
12001	601	2975 Feb 17	10:30:03	4236	12060	139	N	-a	1.4045	0.2655	-0.7043	135.0	-	-	13N	136W
12002	601	2975 Mar 18	20:27:49	4237	12061	177	N	a-	-1.2347	0.5749	-0.3904	192.9	-	-	0S	72E
12003	601	2975 Aug 13	03:19:41	4240	12066	144	N	-t	-1.4131	0.2936	-0.7631	161.8	-	-	16S	30W
12004	601	2975 Sep 11	18:13:12	4241	12067	182	N	t-	1.4331	0.2660	-0.8088	157.0	-	-	3S	103E
12005	601	2976 Feb 07	01:03:02	4244	12072	149	P	-a	0.7754	1.4370	0.4333	289.5	139.1	-	16N	5E
12006	601	2976 Aug 01	09:19:40	4247	12078	154	P	-t	-0.6000	1.7599	0.7543	318.4	180.1	-	18S	120W
12007	601	2977 Jan 26	10:04:13	4251	12084	159	T+	pp	0.0936	2.7173	1.6556	358.3	226.1	100.6	19N	131W
12008	601	2977 Jul 21	22:24:47	4254	12090	164	T+	p-	0.1818	2.5024	1.5460	321.3	211.1	92.9	20S	44E
12009	601	2978 Jan 15	11:50:00	4258	12096	169	P	t-	-0.6086	1.7930	0.6904	348.9	184.5	-	20N	158W
12010	601	2978 Jul 11	15:23:03	4261	12102	174	P	a-	0.9099	1.1587	0.2173	256.0	98.6	-	21S	149E
12011	601	2979 Jan 04	10:59:07	4265	12108	179	N	t-	-1.2722	0.5705	-0.5227	222.4	-	-	21N	147W
12012	601	2979 Jun 01	22:22:35	4268	12113	146	P	-a	-0.9200	1.1699	0.1697	275.1	92.5	-	23S	42E
12013	601	2979 Nov 25	01:51:15	4272	12119	151	P	-a	0.9195	1.1734	0.1682	271.7	90.8	-	21N	14W
12014	601	2980 May 21	05:32:38	4275	12125	156	T-	pp	-0.1833	2.5500	1.4937	362.9	227.8	96.1	21S	66W
12015	601	2980 Nov 13	16:08:41	4279	12131	161	T+	-p	0.1980	2.4768	1.5122	317.5	207.4	89.8	18N	132E
12016	601	2981 May 10	06:28:37	4282	12137	166	P	t-	0.5630	1.8660	0.7845	351.6	194.1	-	17S	80W
12017	601	2981 Nov 03	08:22:42	4286	12143	171	P	a-	-0.4957	1.9318	0.9645	305.1	185.8	-	15N	112W
12018	601	2982 Mar 30	17:33:19	4289	12148	138	Ne	-t	-1.5589	0.0164	-1.0211	37.6	-	-	5S	115E
12019	601	2982 Apr 29	07:44:37	4290	12149	176	N	t-	1.2686	0.5588	-0.4982	215.1	-	-	13S	98W
12020	601	2982 Sep 24	09:51:36	4293	12154	143	N	-h	1.4899	0.1396	-0.8911	108.8	-	-	2N	132W
12021	602	2982 Oct 23	21:51:18	4293	12155	181	N	h-	-1.2497	0.5695	-0.4397	204.2	-	-	11N	46E
12022	602	2983 Mar 20	04:28:31	4296	12160	148	P	-a	-0.7910	1.3995	0.4134	282.7	134.9	-	1S	48W
12023	602	2983 Sep 13	13:34:03	4300	12166	153	P	-t	0.7923	1.4399	0.3688	322.6	141.7	-	3S	173E
12024	602	2984 Mar 08	19:52:55	4303	12172	158	T-	-p	-0.0948	2.6664	1.7012	318.4	209.7	96.3	4N	82E
12025	602	2984 Sep 01	13:54:23	4307	12178	163	T+	pp	0.0446	2.8098	1.7430	371.7	234.5	105.6	8S	170E
12026	602	2985 Feb 26	11:14:26	4311	12184	168	P	a-	0.5935	1.7637	0.7739	303.8	174.1	-	9N	148W
12027	602	2985 Aug 21	18:16:56	4314	12190	173	P	h-	-0.7114	1.5628	0.5427	312.4	160.0	-	12S	105E
12028	602	2986 Feb 15	21:57:16	4318	12196	178	N	h-	1.3404	0.4215	-0.6245	183.6	-	-	14N	52E
12029	602	2986 Jul 12	22:32:25	4321	12201	145	N	-a	1.1162	0.7836	-0.1643	221.0	-	-	20S	42E
12030	602	2986 Aug 11	05:58:03	4321	12202	183	N	a-	-1.4132	0.2477	-0.7184	133.7	-	-	16S	69W
12031	602	2987 Jan 06	06:11:45	4324	12207	150	P	-t	-0.9434	1.1800	0.0748	304.9	67.0	-	22N	74W
12032	602	2987 Jul 02	15:44:11	4328	12213	155	T	-a	0.3856	2.1229	1.1772	308.7	196.8	59.8	22S	144E
12033	602	2987 Dec 26	06:06:34	4332	12219	160	T-	pp	-0.2484	2.4426	1.3623	363.4	223.5	86.3	23N	74W
12034	602	2988 Jun 21	06:52:51	4335	12225	165	T	p-	-0.3691	2.1730	1.1882	325.6	204.4	63.4	24S	84W
12035	602	2988 Dec 14	12:19:11	4339	12231	170	T	a-	0.4584	2.0282	1.0056	327.6	196.3	11.6	24N	168W
12036	602	2989 Jun 10	15:50:11	4342	12237	175	N	t-	-1.1789	0.7155	-0.3258	237.6	-	-	24S	141E
12037	602	2989 Nov 04	15:44:27	4345	12242	142	N	-a	-1.5048	0.0761	-0.8830	73.3	-	-	14N	138E
12038	602	2989 Dec 04	01:21:19	4346	12243	180	N	a-	1.1157	0.7979	-0.1767	224.2	-	-	23N	5W
12039	602	2990 May 01	01:43:20	4349	12248	147	P	-t	0.9427	1.1692	0.0880	301.1	72.3	-	14S	8W
12040	602	2990 Oct 25	07:44:40	4353	12254	152	P	-a	-0.8460	1.2927	0.3182	273.2	119.6	-	11N	102W
12041	603	2991 Apr 20	04:19:45	4356	12260	157	T+	pp	0.1872	2.5370	1.4921	354.6	223.6	94.5	11S	47W
12042	603	2991 Oct 14	19:45:08	4360	12266	162	T-	pp	-0.1358	2.6204	1.5973	346.8	221.9	98.1	8N	78E
12043	603	2992 Apr 08	13:56:10	4364	12272	167	P	a-	-0.5442	1.8554	0.8635	311.7	183.2	-	8S	169E
12044	603	2992 Oct 03	00:44:12	4367	12278	172	P	t-	0.6264	1.7438	0.6740	340.1	181.3	-	5N	4E
12045	603	2993 Feb 27	19:14:41	4370	12283	139	N	-a	1.4166	0.2428	-0.7260	129.4	-	-	9N	93E
12046	603	2993 Mar 29	04:55:44	4371	12284	177	N	a-	-1.2097	0.6191	-0.3431	199.0	-	-	5S	55W
12047	603	2993 Aug 23	09:59:37	4374	12289	144	N	-t	-1.4893	0.1534	-0.9024	118.1	-	-	12S	130W
12048	603	2993 Sep 22	01:00:19	4374	12290	182	N	t-	1.3622	0.3969	-0.6796	189.4	-	-	1N	1E
12049	603	2994 Feb 17	09:41:01	4378	12295	149	P	-a	0.7857	1.4188	0.4137	289.1	136.7	-	13N	124W
12050	603	2994 Aug 12	16:24:46	4381	12301	154	P	-t	-0.6741	1.6226	0.6196	309.7	166.3	-	15S	134E
12051	603	2995 Feb 06	18:22:27	4385	12307	159	T+	pp	0.1025	2.7020	1.6382	359.4	226.4	100.3	16N	106E
12052	603	2995 Aug 02	05:50:01	4388	12313	164	T+	pp	0.1077	2.6375	1.6830	321.9	212.9	97.7	17S	67W
12053	603	2996 Jan 26	19:52:23	4392	12319	169	P	t-	-0.5990	1.8109	0.7075	350.2	186.4	-	18N	83E
12054	603	2996 Jul 21	22:52:28	4396	12325	174	P	a-	0.8376	1.2915	0.3499	266.4	122.0		10S	37E
12055	603	2997 Jan 14	19:05:35	4399	12331	179	N	t-	-1.2609	0.5907	-0.5013	225.4	-	-	20N	93E
12056	603	2997 Jun 12	05:40:06	4402	12336	146	P	-a	-0.9909	1.0405	0.0389	264.0	45.2	-	24S	66W
12057	603	2997 Dec 05	10:15:08	4406	12342	151	P	-a	0.9348	1.1451	0.1403	268.5	83.0	-	23N	138W
12058	603	2998 Jun 01	12:29:09	4410	12348	156	T-	-p	-0.2536	2.4210	1.3646	361.5	224.6	87.2	22S	169W
12059	603	2998 Nov 25	00:38:12	4413	12354	161	T+	-p	0.2210	2.4354	1.4693	316.7	206.2	87.3	21N	5E
12060	603	2999 May 21	13:21:13	4417	12360	166	P	t-	0.4971	1.9853	0.9069	356.7	204.2	-	20S	178E

Cat Num	Canon Plate	Calendar Date	TD of Greatest Eclipse	ΔT s	Luna Num	Saros Num	Ecl. Type	QSE	Gamma	Pen. Mag.	Um. Mag.	Eclipse Phase Durations Pen. m	Par. m	Total m	Greatest in Zenith Lat.	Lng.
12061	604	2999 Nov 14	16:41:25	4421	12366	171	T	a-	-0.4669	1.9868	1.0153	308.0	189.1	18.5	18N	124E
12062	604	3000 May 10	15:00:16	4424	12372	176	N	t-	1.2117	0.6604	-0.3907	230.6	-	-	17S	153E
12063	604	3000 Oct 05	17:10:38	4427	12377	143	Ne	-h	1.5497	0.0329	-1.0039	53.5	-	-	6N	118E
12064	604	3000 Nov 04	05:45:18	4428	12378	181	N	h-	-1.2170	0.6326	-0.3828	214.8	-	-	14N	72W

Astropixels Publications

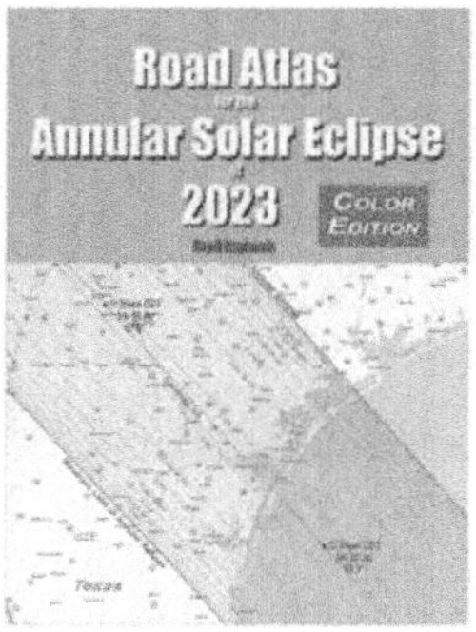

Below is a partial list of some of the books Fred Espenak has written through Astropixels Publishing:

21st Century Canon of Solar Eclipses

The complete guide to every solar eclipse occurring from 2001 tom 2100 (224 eclipses in all). It includes information and maps for all total, annular, hybrid, and partial eclipses. A special world atlas shows detailed full page maps of all central eclipse paths (total, annular and hybrid).

Road Atlas for the Annular Solar Eclipse of 2023

Detailed road maps of the entire eclipse path from the western USA, through Mexico, Central and South America. Information printed on the maps makes it easy to estimate the duration of annularity from any location in the eclipse path. This is the next annular solar eclipse visible from the USA.

Atlas of Central Solar Eclipses in the USA

When was the last total eclipse through the USA and when is the next? How often do they happen? What total eclipse tracks passed across the USA during the 17th, 18th, and 19th centuries, etc., and what states did they include? And how often is a total solar eclipse visible from each of the 50 states? The Atlas of Central Solar Eclipses in the USA answers all of these questions and more with hundreds of maps and tables.

Road Atlas for the Total Solar Eclipse of 2024

This book contains detailed road maps of the entire eclipse path from Mexico, through the USA and Canada. Information printed on the maps makes it easy to estimate the duration of totality from any location in the eclipse path. This is the next total solar eclipse visible from the USA.

21st Century Canon of Lunar Eclipses

The complete guide to every lunar eclipse occurring from 2001 tom 2100 (228 eclipses in all). It includes information and maps for all total, partial, and penumbral eclipses. The predictions use a new model for Earth's elliptical shadows.

For information on these books and more, visit *Astropixels Publishing*: *http://astropixels.com/pubs/index.html*

All Astropixels Publications are available from Amazon.com

Made in the USA
Monee, IL
05 November 2026

67811460R10155